Carbon Nanom
Hydrogen Syst

NATO Science for Peace and Security Series

This Series presents the results of scientific meetings supported under the NATO Programme: Science for Peace and Security (SPS).

The NATO SPS Programme supports meetings in the following Key Priority areas: (1) Defence Against Terrorism; (2) Countering other Threats to Security and (3) NATO, Partner and Mediterranean Dialogue Country Priorities. The types of meeting supported are generally "Advanced Study Institutes" and "Advanced Research Workshops". The NATO SPS Series collects together the results of these meetings. The meetings are co-organized by scientists from NATO countries and scientists from NATO's "Partner" or "Mediterranean Dialogue" countries. The observations and recommendations made at the meetings, as well as the contents of the volumes in the Series, reflect those of participants and contributors only; they should not necessarily be regarded as reflecting NATO views or policy.

Advanced Study Institutes (ASI) are high-level tutorial courses intended to convey the latest developments in a subject to an advanced-level audience

Advanced Research Workshops (ARW) are expert meetings where an intense but informal exchange of views at the frontiers of a subject aims at identifying directions for future action

Following a transformation of the programme in 2006 the Series has been re-named and re-organised. Recent volumes on topics not related to security, which result from meetings supported under the programme earlier, may be found in the NATO Science Series.

The Series is published by IOS Press, Amsterdam, and Springer, Dordrecht, in conjunction with the NATO Public Diplomacy Division.

Sub-Series

A.	Chemistry and Biology	Springer
B.	Physics and Biophysics	Springer
C.	Environmental Security	Springer
D.	Information and Communication Security	IOS Press
E.	Human and Societal Dynamics	IOS Press

http://www.nato.int/science
http://www.springer.com
http://www.iospress.nl

Series C: Environmental Security

Carbon Nanomaterials in Clean Energy Hydrogen Systems

Edited by

Bogdan Baranowski

Institute of Physical Chemistry of PAS,
Warsaw, Poland

Svetlana Yu. Zaginaichenko

Institute of Hydrogen and Solar Energy,
Kiev, Ukraine

Dmitry V. Schur

Institute for Problems of Materials Science of NAS,
Kiev, Ukraine

Valeriy V. Skorokhod

Institute for Problems of Materials Science of NAS,
Kiev, Ukraine

and

Ayfer Veziroglu

International Association for Hydrogen Energy,
University of Miami, Miami, U.S.A.

Published in cooperation with NATO Public Diplomacy Division

Proceedings of the NATO Advanced Research Workshop on
Using Carbon Nanomaterials in Clean-Energy Hydrogen Systems
Sudak, Crimea, Ukraine
22–28 September 2007

Library of Congress Control Number: 2008932434

ISBN 978-1-4020-8897-1 (PB)
ISBN 978-1-4020-8896-4 (HB)
ISBN 978-1-4020-8898-8 (e-book)

Published by Springer,
P.O. Box 17, 3300 AA Dordrecht, The Netherlands.

www.springer.com

Printed on acid-free paper

CONTENTS

PREFACE

The 2007 ARW "Using Carbon Nanomaterials in Clean-Energy Hydrogen Systems" (UCNCEHS'2007) was held in September 22–28, 2007 in the remarkable town Sudak (Crimea, Ukraine) known for its heroic and unusual fate. In the tradition of the earlier conferences, UCNCEHS'2007 meeting served as an multidisciplinary forum for the presentation and discussion of the most recent research on transition to hydrogen-based energy systems, technologies for hydrogen production, storage, utilization, carbon nanomaterials processing and chemical behavior, energy and environmental problems. The aim of UCNCEHS'2007 was to provide the wide overview of the latest scientific results on basic research and technological applications of hydrogen interactions with carbon materials. The active representatives from research/academic organizations and governmental agencies could meet, discuss and present the most recent advances in hydrogen concepts, processes and systems, to evaluate current progress and to exchange academic information, to identify research needs and future development in this important area. This ARW should help further the progress of hydrogen-based science and promote the role of hydrogen and carbon nanomaterials in the energy field.

The UCNCEHS'2007 was the ARW, where a related new important topic of considerable current interest on fullerene-related materials as hydrogen storage was included into the conference program. This meeting gave an opportunity for researchers to cover the entire range of basic and applied materials focusing on synthesis, structure, properties and applications of diverse carbon materials ranging from nanotubes and fullerenes to carbon fiber composites and sorbents. Papers on related topics and studies devoted to new methods, modelling, theory, computational simulation, design, experimentation and measurement were welcome. Thus, the UCNCEHS'2007 ARW was unique in bringing together hydrogen and carbon materials researchers, scientists, engineers and practitioners from developed countries of Europe and America, new independent states of FSU and other countries for discussions in advanced materials development and applications.

Many world-known scientists in the field of hydrogen energy and carbon nanomaterials made a presentations at the ARW. Scientists discussed the following problems:

- The global rise in a temperature and its effect on an environment
- The necessity of transition from fossil fuels to sun-hydrogen energy
- The use of hydrogen as a power medium and energy source supposes its storage and transportation
- The use of compressed gas, cryogenic storage and storage in a solid for hydrogen storage
- The hydrogen storage in a solid on the modern stage of development of human society can be divided into three methods of storage:

    1. As compounds
    2. In metals and alloys
    3. In nanostructural modifications of carbon

At the ARW the particular attention was given to the last moment.

The book with UCNCEHS'2007 Proceedings brings together the research papers of invited and contributed speakers. We hope that they will serve as both a useful reference and resource material for all the participants and for those whose interest in the subject matter may develop after the event.

Finally, this conference was generously supported by the Scientific and Environmental Affairs Division of NATO as an the NATO Science for Peace and Security Programme of the NATO Science Programme. Their contribution is gratefully acknowledged and the Organizing and all ARW participants want to overflow with effusive thanks to NATO Committee for the financial support of our UCNCEHS'2007 ARW and to **Mr. Jean Fournet**, Assistant Secretary General, Chairman of NATO Science Committee, and **Mr. Fausto Pedrazzini**, Programme Director, NATO Scientific Affairs Division, for the displayed mutual understanding and the comprehension of significance of problems under discussions at the UCNCEHS'2007 ARW.

Nejat T. Veziroglu
Svetlana Yu. Zaginaichenko
Dmitry V. Schur
Bogdan Baranowski
Anatoliy P. Shpak
Valeriy V. Skorokhod
Ayfer Veziroglu

80th ANNIVERSARY OF THE BIRTHDAY OF ACADEMICIAN B. BARANOWSKI

Professor Bogdan Baranowski was born in 27.10.1927 in Kępno, to the north-east of Wrocław.

His personality was formed by interwar conditions on territory of former Prussian conquest, and in the difficult war and occupation times. As from the youth years he was interested in chemistry, he undertook the chemical studies at the University and Technical University in Wrocław (1947–1951). His scientific career developed quickly. Yet before graduation, as outstandingly talented, he became an assistant at the new created Department of Physical Chemistry, directed by Professor K. Gumiński. In 1954 they both moved to the Jagiellonian University in Kraków, where he worked at the new created Department of Theoretical Chemistry. Two years later, invited by the Institute of Physical Chemistry B. Baranowski came to Warsaw to the Department of Physicochemistry of Electrode Processes, directed by Professor M. Śmiałowski. Soon after he became the candidate of sciences (1956) and in 1964 – at the age of 37 years, received professor title.

From 1956 his career, characterized by numerous stays abroad, has been strictly connected with the Institute of Physical Chemistry. Since 1960 he directed a department and 5 years later became Head of his own scientific group – the Department of Physical Chemistry of Solids.

The scientific papers of Professor B. Baranowski are characterized by a large variety. At the beginning they concerned mainly the thermo- and electrothermodiffusion in liquid solutions. B. Baranowski elaborated a thermodiffusion method for separation of electrolytes, and as the first applied the irreversible thermodynamics and statistical mechanics to surface systems. The next achievement was the phenomenological theory of thermodiffusion in liquid metallic alloys. In electrolyte solutions he extended Deby's salt effects theory and discovered the influence of chemical reactions on thermodiffusion. Among his first papers one finds also the semi-empirical theory of antyraoultic properties of liquid solutions.

Professor B. Baranowski was greatly interested in irreversible thermodynamics. In the middle 1960s, his most important theoretical papers arised on this discipline. One of the achievements was the extent of theory of dissipative structures, (defined by I. Prigogine), and the proposition of their classification. He explained, that the development of biological organisms are in agreement with the II law of thermodynamics.

Besides the adoptions mentioned above, he extended a contribution to the description of thermodiffusion in multiatomic gases; described transfer processes in multicomponent two-phase systems and in liquids under non-hydrostatic pressures, gave the phenomenological theory of electrotransport in solid metallic alloys. Later, the irreversible thermodynamics was adopted by him for membrane processes and for the description of diffusion in metals under pressure. As the fruit of numerous papers arised the monograph entitled: "Non-equilibrium Thermodynamics in Physical Chemistry" was published in Polish and German in 1974.

The crucial event in scientific work of Professor B. Baranowski was the discovery of the hydride phase in electrochemically saturated nickel (1958), what led to the construction of the first hydrogen high pressure vessel. This opened a new trend, causing, that his research from this time was connected mainly with metal hydrides under high pressures. There were investigations of thermodynamic properties of metal-hydrogen systems, the electrical and magnetic properties under hydrogen pressures, the absorption and desorption kinetics, and the penetration depths of hydride phases. B. Baranowski was the first, who synthesized by high pressure method hydrides of nickel (1966), chromium (1972), manganese (1975) and aluminium (1983) and gave their thermodynamic characteristics. After the hydrides synthesis followed deuterides of nickel and chromium. He also examined the properties of systems, such as: Ni-Mn-H, Ni-Co-H(D), Ni-Cu-H, Ni-Fe-H, Ni-Cr-H, Pd-Rh-H, Pd-Pt-H, Pd-Ag-H and Pd-Ni-H. Using the high pressure apparatus he obtained the stoichiometric palladium hydride and contributed to the discovery of its superconductivity. Besides transition metals, the subject of his investigations were also light and rare earth metals, solid organic compounds and at least – the phase transitions in crystals of acidic salts.

Among the multitude (340) of his papers, one finds also those devoted to the metallic hydrogen, the electrochemical analog on of the Benard problem, and strictly polemic papers. All of these present his extended scientific interests and large assiduity.

Due to his investigations on Me-H systems, a new field of research was created – the high pressure hydrogen apparatus. Devices constructed in his laboratory enables not only the synthesis, but also in-situ measurements of: X-ray diffraction, calorimetry, compressibility of liquids and solids, electrical transport properties (including low-temperature measurements), and the electrochemical measurements. Many of them are now used in other laboratories.

Professor is a member of editorial committees: Polish Journal of Chemistry, Journal of Non-equilibrium Thermodynamics, Journal of Alloys and Compounds and High Pressure Research. Especially he was active in Polish Journal of Chemistry, as editor in chief in the years 1991–2004. When he took on this function, this journal was in difficult conditions.

His talent and personal features caused, that he was many times invited abroad as well as performed different functions in scientific – Polish and International organizations. Thus, he was a member of executive committees of European High Pressure Research Group (1976–1980 and 1982–1986) and Federation of European Chemical Societies (1977–1984). Since 1975 he was a member of the "International Association for the Advancement of High Pressure in Science and Technology" – AIRAPT, where between 1981–1985 he was Vice President, and in the years 1989–1993 – he was its President. Independently he was a participant in the International advisory and organizing committees, last years as a chairperson of International conferences "Hydrogen Materials Science and Chemistry of Carbon Nanomaterials" (ICHMS) in Ukraine.

Since 1951 Professor B. Baranowski is a member of Polish Chemical Society, for many years being active in its Executive Committee, and between 1974–1979 playing the role of President. This organization gives him inspiration of every year congresses (first from cycle in 1964) as well as the creation of many new sections. In proof of his great merities, in 1987 he was granted the Honorary Member and in 1997 – the Honorary President of this society.

The outstanding scientific achievements brought to Professor B. Baranowski many awards and honours. The first, admitted already in 1957 by Polish Chemical Society, was the prize for young scientists; the first prestigious were: the Award of Maria Skłodowska-Curie (1973), admitted by Polish Academy of Sciences and the Burke medal of Faraday Society (1973).

In 1974 Professor B. Baranowski was granted by membership of Polish Academy of Sciences, where in years (1981–1984) he was the President of Committee of Chemical Sciences. Later he also became the membership of Deutsche Akademie der Naturforscher "Leopoldina" (1976), Polish Society for Advanced of Science and Arts – TPKN (1981), Warsaw Learned Society (1981), the International Academy of Sciences (1986), Deutsche Bunsen Geselschaft für Physikalische Chemie (1989) Academy of Sciences of GDR (1990), and the foreign member of Ukrainian National Academy of Sciences (1994).

Between his numerous awards the most prestigious are: doctor honoris causa of the Chalmers University in Göteborg (Sweden), admitted in 1983, and the medal of Bridgman, received from AIRAPT (1995). One should mention also gold decoration for merited for Warsaw (1974) J. Śniadecki medal (1984) by Polish Chemical Society, award of the Prime Minister of Republic of Poland (1994), and the medal admitted by Institute of Physical Chemistry PAS (1997).

Many times Professor B. Baranowski delivered invited lectures. Besides of those, connected with awards (1973 and 1983), there were: the lecture in Moscow (1975), the lecture of Świętosławski (1980), the lecture in Leopoldina (1983), the "August-Wilhelm-von-Hofmann-Lectureship" of German Chemical Society (1987), the lecture of Basiński (1997) and the Jost lecture of German Bunsen Society (1998).

Professor collaborated with scientists from many countries, especially intensive was his collaboration with institutes in Belgium, Netherlands, Germany, Northern Ireland, United States and Sweden. Among his honours, the separate place have positions of invited professor: at Mining Academy in Freiberg, at the University of Hanower, and at the Solid State Physics Institute of Max Planck in Stuttgart and the research grant of the Royal Society. In the group of his many foreign coworkers are such scientists as: the Nobel prizer – I. Prigogine, F.A. Lewis, T.B. Flanagan and E. Wicke.

Professor B. Baranowski has educated many theoretical and experimental scientists, who are now mostly laboring in other departments at the Institute and other cities. Six of them are Professors. He was a reviewer or superreviewer of several dissertations, qualifying for assistant-professor-ship, and promoted 20 doctoral theses. During the stay in Freiberg 1971–1972 he lectured on irreversible thermodynamics.

Professor B. Baranowski has already now come to the history, as one of the most outstanding physico-chemists of XX century, known from the fields of solid state chemistry, thermodynamics of irreversible processes, thermodiffusion and high hydrogen pressure research.

Professor. B. Baranowski is a man of sharply outlined personality and non-hesitative opinions. Gifted by the Nature by shrewd and critical brain; requiring for the other people, but helpful in the face of difficulties. Always he is full of creative energy and ideas. For those, who indeed love science, he is the best friend, like the best father.

L. Dębowska

ERWIN SCHRÖDINGER (12.08.1887, VIENNA–04.01.1961, VIENNA)

In August 12, 2007 120 years have passed from the birthday of Erwin Schrödinger, the Austrian physicist-theorist, one of the creators of quantum mechanics and, without exaggeration, one of the most outstanding physicists in XX century. The greatest E. Schrödinger's achievement was elaboration of the most adequate and efficient from the applied viewpoint variant of quantum mechanics which basis was the *wave equation* suggested by E. Schrödinger in 1926 and bearing his name. This differential equation for the wave function thought by E. Schrödinger defines entirely the change in the states of quantum systems in the course of time. For nascent quantum mechanics, E. Schrödinger equation was of the same importance as Newton equation for classical mechanics or Maxwell equation for electrodynamics. These basic equations reflecting the nature laws could not be deduced mathematically, they were *found* by genius of their creators. We thus have the right to put the name of E. Schrödinger in one row with these great names of creators of physics.

E. Schrödinger equation illustrated clearly the nature of corpuscular-wave dualism of material world postulated by de Broglie. E. Schrödinger is said to devise this equation in searching for the method of setting forth intelligibly his students the matter of the famous article written by Louis de Broglie "Investigation into theory of quanta" (1925) which was presented to him by Peter Debye. Although in 1925 V. Heisenberg, M. Born and P. Jorgan had already elaborated so-called "matrix mechanics", it was E. Schrödinger equation which was marching triumphally along many fields of physics and chemistry in XX century and became the basis for penetrating into them the concepts and methods of quantum mechanics. One of the first impressive results of the employment of E. Schrödinger equation was the complete calculation of energetic levels of a hydrogen atom. Hereafter this equation allowed the explanation and the prediction of a number of phenomena in atomic physics, and the *first principle* calculation of general energetic and spectral characteristics of atomic systems measured before experimentally. E. Schrödinger equation proved to be rather efficient for description of many phenomena in atomic physics, such as α - decomposition, γ - nuclear radiation, neutron scattering at nuclei etc.

However, E. Schrödinger quantum mechanics employed in theory of solids and quantum chemistry, which are of outstanding importance as a scientific basis for synthesis of new substances and creation of new materials, gave the most prominent and basic results. Based on the concepts of E. Schrödinger quantum mechanics, the quest for the understanding of general regularities of interrelation between structure and properties of substances and materials became a general paradigm of theoretical materials science in XX century. It was supported and developed actively by outstanding materials men, I.N. Francevich and G.V. Samsonov at the Institute for Problem of Materials Science of NAS of Ukraine.

E. Schrödinger's genius was not restricted to only physical problems, it took a fancy to searching for the answer a principal question of natural science, the question of the nature of living matter and a possibility to interpret physiological processes on a basis of physics. In 1944 E. Schrödinger wrote an excellent book about the relation between physics and biology "What is life?". It is conceivable that the active contacts with outstanding geneticists including N.V. Timofeev-Resovskiy instigated him to write this book. In this book E. Schrödinger showed convincingly the role of quantum mechanics in formation and stability of biologically active molecules considered as an aperiodic crystal, and substantiated systematically the materialistic approach to the explanation of the phenomenon of existence on a basis of laws of physics and physical chemistry.

The name of Erwin E. Schrödinger will always remain among the names of coryphaei in science, and his quantum mechanics will remain as in former times the basis for cognition of nature and quantitative description of phenomena and processes occurring in infinitely diverse micro-world.

Academician V.V. Skorokhod

150th ANNIVERSARY OF THE BIRTHDAY OF GERTS, GENRICH RUDOLF (22.02.1857, HAMBURG– 01.01.1894, BONN)

This year the mankind celebrate 150th anniversary of the birthday of the famous German physicist – Genrich Gerts. He had been graduated from the Berlin university. In 1885–1889 he worked as a professor of physics at the Technical University of Karlsruhe. From 1889 he became the professor of physics at the Bonn university.

The principle achievement of G. Gerts – experimental confirmation of electromagnetic theory of light by James Maxwell. G. Gerts discovered existence of electromagnetic waves in Karlsruhe. He investigated reflection, interference, diffraction and polarization of electromagnetic waves in details; proved their rate to equal the rate of light and that light represents by itself a kind of electromagnetic waves.

The main scientific papers of G. Gerts are: "About induction in a rotating sphere", "About rather prompt electrical oscillations", "About relationship between light and electricity", "About the principle equations of electrodynamics for moving bodies", "About influence of ultraviolet light on electrical discharge", "Principles of mechanics stated in new communication".

Genrich Gerts was born on 22nd of February 1857 in the family of lawyer. G. Gerts studied fine and was unsurpassed pupil by his ingenuity. He liked all the subjects, liked to compose verses and to work on turning machine tool. His weak health harmed to Gerts during all his life, to regret.

In 1875 after graduating gymnasium G. Gerts entered at first to Dresden, and then to Munhen high technical school. Everything was all right till the time when general subjects were studied. But when specialization began G. Gerts changed his decision. He did not want to be narrow specialist, he aspired to scientific work and he entered to Berlin University. Here famous Helmholtz became by his direct instructor. With great wish G. Gerts aspired to study in physical laboratory. Helmholtz suggested to G. Gerts the problem from electrodynamics: has an electric current the kinetic energy? Helmholtz wanted to direct efforts of his pupil into the field of electrodynamics, because he considered it to be the most confused.

G. Gerts begins to solve the task which was expected for 9 months. He made the devices himself and adjusted them. During this work characteristics of researcher were revealed: persistence, rare diligence and art of an experimenter. The task was solved in 3 months. The result coincided with Helmholtz's point of view although it was erroneous. But Helmholtz was not mistaken in ability of the young man. " I could see that I had matter with the pupil of unusual talent" – he noticed later. The work of G. Gerts was awarded by premium.

When he returned after summer vocations of 1879 G. Gerts had got permission to work with other theme "About induction in a rotating sphere" taken as doctor dissertation. It was a theoretical work. He assumed to finish it in 2–3 months, to defend it and get rank of doctor, although the university was not finished. He worked with large raising and enthusiasm and he finished his investigation quickly. The defence was successful, he got rank of doctor with "excellent insignia". Such phenomenon was very rare, especially for a student.

From 1883 to 1885 G. Gerts was a chief of theoretical physics chair in provincial Keel, where physical laboratory was absent at all. G. Gerts studied here the theoretical problems and tried to prove universality of electrodynamic theories of long-range interaction. As a result of the work G. Gerts wrote his own system of equations from which equations of Maxwell could get easily.

In 1885 G. Gerts took invitation of technical school in Karlsruhe, where his famous experiments associated with electric force spreading were fulfilled. He found two inductive coils in physical laboratory of the school and carried out demonstrations with them on lectures. G. Gerts discovered that with their help one can produce the fast electrical oscillations with period of 10^{-8}. In these experiments G. Gerts created not only high-frequency generator, but resonator too – receiver of these oscillations.

After numerous experiments G. Gerts made the conclusion about the existence of electromagnetic waves which spread with definite speed. He proved the complete analogy of electromagnetic waves and light ones after studying laws of reflection and diffraction, and establishing polarization phenomenon. All of that were given in the paper "About rays of electrical force", printed in December of 1888. This year is considered to be the year of discovering electromagnetic waves and experimental supporting of Maxwell theory.

This hard work of G. Gerts did not pass unpunishedly for his weak health. At first, his eyes refused, then ears, tooth and nose were ill. Soon the general blood infection began, due to it famous 37 years old scientist Genrich Gerts died.

G. Gerts finished large work which was begun by Faraday. If Maxwell transformed imaginations of Faraday into mathematical symbols, but G. Gerts converted them into visible and heard electromagnetic waves, which became the eternal monument for him. We remember G. Gerts when we listen to the radio, watch television, when we hear information about the launching of new spaceships and the stable radio communication existing with them. It was not occasionally that the first words of the Russian physicist A.S. Popov which he transferred with the help of wireless line were: "Genrich Gerts".

Assistant Professor, Ph.D. *A. Pomytkin*

140th ANNIVERSARY OF THE BIRTHDAY OF MARIA SKŁODOWSKA-CURIE (07.11.1867, WASSAW–04.07.1934, SAVOY)

Maria Skłodowska-Curie was born in 07.11.1867 in Warsaw as the fifth child of grammar-school teachers. It was a difficult time of Russian repressions after uprising in 1863 in occupied Poland. Her parents lived under the great influence of positivism, that meant the trend to society education, reconstruction of Polish intellectual elite and emancipation of women. The positivism of Comté, which rejected metaphysics and all theses, which are impossible to prove by logical, unambiguous evidences, substituted the religion in Maria's life and became the motto of her scientific work. Her youth interests to the equal degree contained exact sciences, as sociology, medicine and belles-lettres. She was writing poems and reading literature in Russian, French, German, and English languages. She completed of her education with a gold medal (1882). Soon after, the financial situation of the family forced her to give private lessons, to earn money for own livelihood and for her older sister medical studies. It was the time of self-education, when she got access to well equipped laboratory, directed by her cousin – J. Boguski, where she was learning experimental work. These probes consolidated her patience and interest for experimental work. Soon she became a participant of the "Flying University" – a secret high school for women, for those the legal high schools were closed. After sister's marriage, Skłodowska was invited by her to Paris (1891) and started studies at Sorbona. In spite of difficult financial and social conditions, in 1893 she achieved licentiate in physics, and 1 year later – in mathematics. In group of her teachers were professors, such as: P. Appel, G. Lippmann and H. Poincaré. In 1894 she met Pierre Curie – her later husband (1895) and main collaborator, who had already scientific achievements, such as formulation of Curie-law and the discovery of piezoelectricity. After licentiate Maria started her work on: "Magnetic properties of hardened steel", proposed by G. Lippmann. The decisive role in her carrier was the discovery of radiation by H. Becquerel. On husbands suggestion, in 1897 she reported to Becquerel for the doctoral work in the branch of radioactivity. Using Pierre's extremely sensitive quartz electrometer, she started the examinations of radiation, emitted by all contemporary known elements. Firstly stated, that the most intensive is the radiation emitted by uranium. Secondly, that the radiation of the black blende after extraction of uranium is much more intensive, than of pure uranium. The result suggested the existence of a new, unknown element. In 1898 she found in the black blende an extremely active substance, associated with bismuth. The consecutive laborious extractions and Pierre's measurements, with the use of electrometer and ionizing chamber, (made of tin after canned food) led to the discovery of polonium (1898). The similar way the couple Curie found the second element, accompanied by barium, active several million higher, in comparison to uranium (1898). The new element, called by Maria of "radium" was supported by Demarçay's spectrum analysis.

In the next years the couple was concentrated on the separation of the new elements from black blende and on prove, that they are indeed new. One has to emphasize, that the amount of polonium and radium in blenda is extremely low, and the discovery was possible only, due to a very strong radiation. The work was very hard, and performed in difficult local and social conditions. Later, with the help of professor Suesse from Vienna and industrialist E.A. de Lise, they organized industrial production of radioactive samples. Maria was purifying these products and condensing to extract pure radium chloride. Pierre's role relayed on the examinations of the physical properties of radiation. Efforts succeeded in 1902 by confirmation of the radium existence and determination of its properties. In 1903 she received doctor degree with annotation *très honorable* (*summa cum laude*); the same year, commonly with Becquerel and her husband, she received the Nobel Prize in physics for her research on radiation.

After Pierre's death (1906), Maria took over his department in Sorbona, and started lectures as assistant professor. In 1909 she became professor. In 1910 she extracted metallic radium; she could not extract polonium, but this was later done in her laboratory, whereas Maria's merit relayed on the exact determination of polonium decay constant. Her next important achievement was the elaboration of the method for the determination of the radium amount from measurements of radonium – gas, emitted from radium. Together with A. Debierne she constructed a vacuum apparatus for the extraction of radonium from aqueous solutions of radium chloride.

Besides that, she also investigated the properties of α-rays, determined the optical spectrum of actinium, and the electromagnetic spectrum of α-rays, emitted by the elements of actinide series. In 1910 Maria participated in the Conference of Radiologists and Engineers, where she proposed the radioactivity standard, known as "Curie". Simultaneously it was decided the improvement of the standard, containing a strictly defined amount of radium. This standard, (passed to Sevrés) in the form of the thin glass pipe, containing a radium salt was serving for the determination of radium atomic weight. In 1911 Maria participated in the first Solvay's Congress in Brussels, in the circle of famous scientists, such as: Goldschmidt, Planck, Sommerfeld, de Broglie, Rutherford, Einstein, Langevin, Nernst, Brillouin, Warburg, Lorentz, Perrin and Poincaré. Since then, she was a participant of all successive Solvay Conferences on Physics until 1933.

In 1911 Maria received the second time the Nobel Prize (in chemistry), for the discovery of new elements, isolation of radium and the study of the nature and compounds of this element. In 1914, in Paris was opened the Pierre Curie Radium Institute – the institution for research of radiation and its biological meaning. During the war, Maria was actively organizing and laboring in camp radiological stations, giving services for the army. The strong radium samples imported by her served for the preparation of pipes with radonium, used in therapy. After the war, Maria, who was already the director of the Radium Institute was traveling the_world, when by means of her foundation she helped in establishments of many medical institutes for cancer therapy. In 1932, with the help of the Polish President such an institute was opened in Warsaw.

The popularity of Maria's achievements in the word caused, that in 1920 she was invited to America, where the American women organized considerable means for her further research. In 1921 she was entertained in the White House by President Harding. During this stay (when received 1 g of radium, the expensive laboratory apparatus and money) she delivered many lectures at universities and scientific institutes. In 1922 she participated in activity of the International Commission of the Intellectual Cooperation in Geneva, where she was a member, and subsequently – vice chairman of this organization. In 1925 she participated in devotion of the foundation-stone under the Radium Institute in Warsaw, where she met President St. Wojciechowski and many Polish physicists and chemists. Four years later, she made the second trip to USA, where, as the guest of President H. Clark, she lived in the White House. During this visit she received the second gram of radium, visited the Saint-Lawrence University and participated in Edison's jubilee. The gift was transferred by her to Poland for the Radium Institute in built. In 1932 the Radium Institute was opened and Maria participated in the opening celebration as the honorary master. It was her last journey to Poland. She died on July 04, 1934 in Savoy, France, for the reason of leukemia, caused by long lasting contacts with radioactive elements. In 1995, the ashes of Maria and Pierre Curie were placed in Pantheon in Paris.

Maria left 80 scientific articles, 3 monographs: *La radioactivité et la guerre* (1921), *L'isotopié et les éléments isotopes* (1923), *Les rayons alpha, beta et gamma des corps radioactifs en rélation avec la structure nucléaire* (1933), and 2 extensive works on radioactivity: *Traité de Radioactivité* (1910) *and Radioactivité* (appeared after her death, in 1935). The outstanding scientific achievements brought her 10 scientific awards and many honours. So far she remained the first woman, twice awarded by the Nobel Prize. She was given 14 doctorates honoris causa from abroad universities and 16 medals, admitted by institutes and scientific societies. She was a member of 11 Academies of Sciences, and was the first woman, honored a full member of the French Medical Academy (1922). On the whole she was a honorary member of 72 scientific societies and honorary citizen of 3 cities: Warsaw, Glasgow and Riga.

The work of Maria Skłodowska-Curie became the base of new scientific disciplines, such as: nuclear physics, nuclear chemistry and radiochemistry. She created the base under oncology, as a new separate discipline in medical science. Under her personal supervision, the first in the word research on cancer therapy by radioactive preparations were performed.

Her greatest scientific achievements are: the introducing the concept of radioactivity and the elaboration of its theory, the discovery of polonium and radium, and the elaboration of radioactive isotopes separation technique.

Maria Skłodowska-Curie achieved great successes in spite of difficult circumstances. She had underwent discrimination as woman, as well as scientist and a Polish immigrant. In the first covering letter to Nobel Committee, only H. Becquerel and P. Curie were mentioned, as the first men who extracted radioactive elements. In 1910 -due to the sex, she was rejected to become a member of the French Academy of Sciences. At the beginning her life level was very modest, what forced her to earn money as a teacher and laboratory assistant. Her first laboratory in France was an old shed. Later she worked as the teacher at girls' school at Sévres, and her husband's assistant in laboratory. Only after the first Nobel Prize, Pierre received a laboratory at Sorbona and Maria became its director. Later she directed the laboratory in Pierre's Institute, which employed scientists from 25 countries. In 1923 the government admitted her – as the gift from the nation – the salary of ₣40,000 a year, inheriting by children.

Maria Skłodowska-Curie was mother of two daughters: Irena Joliot-Curie – the Nobel Prizer in chemistry (1935) and Ewa, talented in music, the author of Maria's biography.

L. Dębowska

ON CENTENARY OF VERA IVANOVNA MIKHEEVA (1906–1981)

May 7, 2006 was centenary since the birthday of Vera Ivanovna Mikheeva, the outstanding scientist in the field of physico-chemical analysis of inorganic systems, chemistry and technology of the light alloys, of hydrides of light and transition metals, Professor, Doctor of Chemistry, Honored Scientist of USSR.

V.I. Mikheeva was born in the town Gorbatov of the Nizhnii Novgorod's province in the family of teacher. The normal education she has obtained in Nizhnii Novgorod and then in 1924 she has entered the chemical department of Nizhnii Novgorod's University. From 4th year, she has been transferring to the chemical department of Leningrad's polytechnic. There she worked in the chair of general chemistry under the leadership of academician N.T. Kurnakov and, at the same time, she examined the theoretical electrochemistry under the leadership of professor V.A. Kistyakovskii.

From 1934 up to the end, Vera Ivanovna worked in the N.S. Kurnakov Institute of General and Inorganic Chemistry of RAS, where she become leader of the laboratory of the chemistry of boron and hydrides (now is the part of the laboratory of energy-capacious substances and materials). V.I. Micheeva has organized and effectuated the complex of works dealing with the technology of synthesis and investigation of magnesium and aluminium alloys for the air-space industry, with the investigation of chemical nature of the phases with variable composition, of highest oxygen compounds (e.g., of perchlorates) of the light metals, of different boron compounds (e.g., of borohydrides), of materials for nuclear technology. The was the very active and consistent propagandist of the method of physico-chemical analysis. Her monograph "Method of physico-chemical analysis in inorganic synthesis" (Moscow, 1975) was by right awarded by the N.S. Kurnakov's Premium of RAS.

After 1950, under the leadership of V.I. Mikheeva, the basic researches in the field of physico-chemistry of hydrides of metals and their alloys were started in USSR. The works dealing with the synthesis and investigation of binary hydrides of transition metals, of hydrides of alloys and intermetallic compounds of the systems Ce-Mg, Ce-Mg-Al; Ti-V, Nb, Ta, Mo; Zr-Ni, Co, Cr, V; La-Ni are widely known. During the working process, the number of unique installations was constructed, the very original techniques and methods of processing the chemically active hydrides were developed. Samples of hydrides, deuterides, and deuterohydrides synthesized in Mikheeva's laboratory have made the material basis for divers investigations (structure, lattice dynamics, catalysis, etc.) effectuated in a great number of scientific centers in Russia and abroad.

The concept of involving in hydriding of "passive" (in respect to hydrogen) metals was formulated by V.I. Mikheeva as well as the basic principles about character of the chemical bond and the limiting hydrogen capacity in metal-hydride systems. Such (obvious now) methods of data treating as the use of Gibbs's triangles, the comparison of composition and properties of hydrides with the corresponding phase diagrams of metallic systems, the graphical representation of the "hydriding regions" of complex alloys were also introduced by V.I. Mikheeva for the first time.

V.I. Miheeva was the organizer of many large-scale conferences on metal hydrides, the scientific editor of the Russian translation of K. Mackay's monograph "Hydrogen compounds of the metallic elements" (Moscow, 1968), and the author of the first (in our country) monograph "Hydrides of transition metals" (Moscow, 1960) which was held in century of the teacher of V.I. Mikheeva – N.S. Kurnakov.

Many pupils of V.I. Mikheeva work successfully in scientific centers of Russia and abroad (more than 40 Ph.D. and 9 Doctors of Sciences). All of them keep the grateful memory of this remarkable scientist and person.

N.T. Kuznetsov, L.N. Padurets, A.L. Shilov

C_{60} FULLERENE

TO 35TH ANNIVERSARY OF THE FIRST THEORETICAL CALCULATION OF C_{60} MOLECULE. E.G. GALPERN AND D.A. BOCHVAR "ON HYPOTHETICAL SYSTEMS: CARBODECAHEDRON, S-ICOSAHEDRON AND CARBO-S-ICOSAHEDRON". THE ARTICLE WAS ACCEPTED 26.06.1972 AT DAS OF USSR

CHERNOZATONSKIY L.A.
N.M. Emmanuel Institute of Biochemical Physics RAS,
Russia, Moscow, 119334, Kosigin Str.,4.

Number and fantasy, law and abundance are alive creative forces of nature. *K. Chapek.*

Fullerene C_{60} molecule consists of 60 carbon atoms located at the vertexes of a truncated icosahedron (Fig. 1) with 12 pentagons and 20 hexagons.

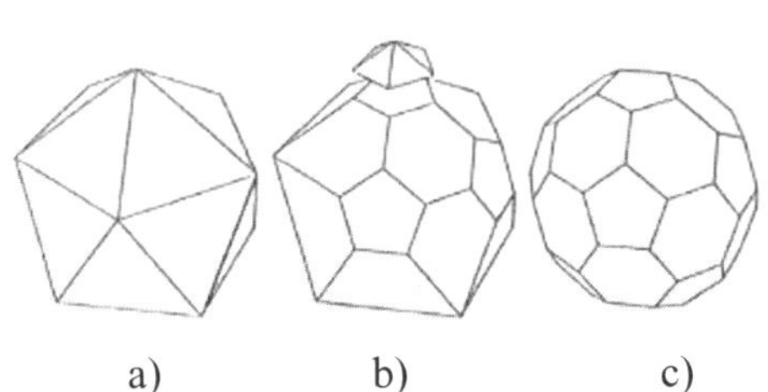

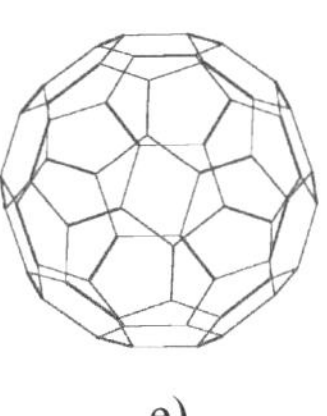

a) b) c) d) e)

Figure 1. (a) Icosahedron formed by 20 equilateral triangles, (b) diagram of "truncation" process, (c) truncated icosahedron, (d) fragment of the picture (1,480) of the famous Italian mathematician and painter, Pietro della Francheska, (e) the most stable structure of C_{60} with double bonds connecting pentagons from single bonds

The most surprising property is its high symmetry involving 120 operations of rotation about axes and reflections in the symmetry planes. This polyhedral shape was known from ancient times as one of Archimedes bodies. John Kepler (1571–1630) named this shape as *truncated isocahedron*.

In the early twentieth century in *Annales de Chemie* André-Marie Ampère (1775–1836) attempted to explain chemistry in terms of geometry. He made an assumption that the main molecules consisted of atoms located at the vertexes of regular polyhedrons, and the reactions between molecules are possible in the case that new molecules, polyhedrons with the certain symmetry, are formed.

In the last 40 years we have witnessed the embodiment of this outstanding hypothesis. Boranes and carboranes, which were the first predicted polyhedral molecules in 60th, marked the beginning of the chemistry of carcass nano-sized structures.

Late in the 1960s after the discovery of ferrocene, the investigations of its derivatives were started at Institute of Inorganic Compounds of Academy of Sciences of USSR. Academician A.N. Nesmeyanov, the head of this institute, set his collaborators the task of finding the ways of synthesis and modeling of new compounds in the form of polyhedrons from a carbon cage having one or several atoms of other elements inside. He named these compounds as "a bird in the cage". This problem was also discussed in the laboratory of quantum chemistry leaded by Professor D.A. Bochvar. This simingly easy problem of connecting several ferrocene molecules with hydrocarbon bridges turned out to be not so simple. This model had a flaw and gave some dangling bonds. D.A. Bochvar suggested to consider the C_{20} dodecahedron. However the calculations, performed basically by E.G. Galpern, showed that the resulting C_{20} in the ground state had an unclosed electron system and only C_{20}^{-2} dication was stable. In addition, very little space remained inside C_{20} for other atoms. In the course of discussion of these results in the laboratory, I.V. Stankevich suggested to consider the larger "cage", located it on a football and arranged carbon atoms at 60 vertices of this cage. "This molecule should be very firm because 22 players kick a ball for several hours and nothing happens to it!" – said Ivan Viacheslavovich, an inveterate football player. Everybody liked this idea, but it took 1 year to write a computing program and make calculations even within the simple Hukkel scheme because such polyatomic systems could not be calculated "by hand". In 1970 based on the obtained results, the paper was finally written and given to A.N. Nesmeyanov to review before publication. I.V. Stankevich refused to be among the authors of the paper and agreed only to be mentioned in acknowledgment although he participated in discussing the results and some details of calculations. However Alexandr N. Nesmeyanov was embarrassed by the revolutionary models and said that this was not his subject-matter, he did not know theory well and could not stake his life on that, everything had to be discussed and considered again. D.A. Bochvar was very upset, he had to speak about the unusual models of molecules several times and defend them at seminars and scientific councils. Unusualness of these molecules for chemists and their skepticism concerning the existence of these molecules continued after the publication of the famous paper written by Kroto H.W. et al. in 1985 [1] and until the appearance of the real material from C_{60} prepared by W. Kratschmer's group in 1989 [2]. Time passed. In February, 1972 prestigious Journal of American Chemical Society published information about peristilane, the crown from 6 carbon 5-membered cycles, incomplete C_{20} dodecahedron [3] what convinced A.N. Nesmeyanov in the importance of publishing the work performed by his collaborators. Hence, the paper "On hypothetical systems: carbodecahedron, s-icosahedron and carbo-s-icosahedron" was sent to Doklady Academii Nauk of USSR in June 26, 1972 and published in 1973 [4]. The most stable C_{60} structure was found to be the structure with double bonds among pentagons (Fig. 1(e)). Although this work represented the results of calculations performed by Hukkel method for plane unsaturated systems, the general conclusions were confirmed by further calculations by the more precise and broadened method of Hukkel [5]. The authors of these two papers became aware of Osawa's prediction of C_{60}, which was published in Japanese in 1970, only in 1987 from paper of H.W. Kroto et al. [7]. E. Osawa showed [8] that he have assumed the stability of truncated C_{60} icosahedron only on the basis of symmetry considerations without any calculations.

In 1996 the Nobel Prize in chemistry was awarded to H.W. Kroto, R.E. Smalley and R.F. Curl for the discovery of fullerenes. Over a period of several years these compounds were intensively studied in laboratories of different countries, the scientists attempted to establish the conditions of their formation, structure, properties, and possible applications. In particular, it was found that fullerenes are contained in great quantities in the soot produced by arc discharge at graphite electrodes evaporation and they had not been observed previously. At present the only method of fullerenes production is their artificial synthesis (for more details, see the excellent books [9, 10]).

Macro-amounts of C_{60} and C_{70} fullerenes are synthesized and sold by several enterprises in Russia, such as "Fullerene-Center" in Nijniy Novgorod and Russian Scientific Center "Kurchatov Institute".

Fullerenes are the basis for the new carbon nanomaterials (both C_{60} crystals as fullerites and their polymeric phases as fullerides ($C_{60}A_x$, x = 16, A are metal atoms, and others); for complexes of C_{60} with transition metal atoms; for heterofullerenes; for hydride carbon forms based on C_{60} fullerenes and nanotubes, and for C_{60}-based derivative molecules).

Solar batteries and accumulators can be noted among interesting new applications where in one way or another the fullerene additives are used. Also, fullerenes have a wide application in the creation of new medicines [11]. The processes of their formation, structure and transformation are under careful study of researchers [12].

The extremely encouraging results of scientists working in this field allow to hope that the designed by E.G. Galpern structure of C_{60} fullerene will become the basis of materials of XXI century.

Author is much obliged to E.G. Galpern for a large help in historical reconstruction of prediction of a remarkable C_{60} molecule.

1. Kroto H.W., Heath J.R., O'Brien S.C., Curl R.F. and Smalley R.E. C_{60}: Buckminsterfullerene // Nature, 1985, **318**: 162–163.
2. Kratschmer W., Lamb L.D., Fostiropoulos K., Huffman D.R. Solid C_{60}: a new form of carbon // Nature, 1990, **347**: 354–358.
3. Eaton P.E., Mueller R.H. J. Am. Chem. Soc., 1972, **94**: 1014.
4. Bochvar D.A. and Gal'pern E.G. Electronic structure of the molecules C_{20} and C_{60} // Proc. Acad. Sci. SSSR, 1973, **209**: 239–241.
5. Stankevich I.V., Nikerov M.V. and Bochvar D.A. Structural chemistry of crystalline carbon: geometry, stability, electronic spectrum // Russ. Chem. Rev., 1984, **53**(7): 640.
6. Osawa E. Kagaku (Kyoto), 1970, **25**: 854. In Japanese; Chem. Abstr.,1971, **74**.
7. Kroto H.W. The stability of the fullerenes C_n with n = 24, 28, 32, 36, 50, 60 and 70 // Nature, 1987, **329**: 529–531.
8. Osawa E. Philos. Trans. Royal Soc. London Ser. A – Phys. Sci. Eng. 1993, **343**: 1.
9. Trefilov V.I., Schur D.V., Tarasov B.P., Shul'ga Yu.M., Chernogorenko A.V., Pishuk V.K., Zaginaichenko S.Yu. Fullerenes - the basis of materials in future.
Kiev: ADEF-Ukraine, 2001, 148 p. (in Russian).
10. Sidorov L.N., Yurovskaya M.A., Borschevskiy A.Ya. et al. Fullerenes. Moscow: Examen, 2005, 668 p. (in Russian).
11. Piotrovsky L.B., Kiselev O.I. Fullerenes in biology. St. -Petersburg: Rostok, 2006, 336 p. (in Russian).
12. Schur D.V., Matysina Z.A., Zaginaichenko S.Yu. Carbon nanomaterials and phase transformations in these materials. Dnepropetrovsk: Nauka i obrazovanie, 2007, 680 p. (in Russian).

SAGA OF HYDROGEN CIVILIZATION

T.N. VEZIROGLU
President, International Association for Hydrogen Energy
Clean Energy Research Institute, University of Miami
Coral Gables, FL 33124, USA
E-mail: veziroglu@miami.edu

Abstract. Fossil fuels (i.e., petroleum, natural gas and coal), which meet most of the world's energy demand today, are being depleted fast. Also, their combustion products are causing the global problems, such as the greenhouse effect, ozone layer depletion, acid rains and pollution, which are posing great danger for our environment and eventually for the life in our planet. Many engineers and scientists agree that the solution to these global problems would be to replace the existing fossil fuel system by the Hydrogen Energy System. Hydrogen is a very efficient and clean fuel. Its combustion will produce no greenhouse gases, no ozone layer depleting chemicals, little or no acid rain ingredients and pollution. Hydrogen, produced from renewable energy sources, would result in a permanent energy system, which we would never have to change. This article presents the history of Hydrogen Energy from its conception through the dissemination of the idea and the beginning of the implementation.

Keywords: Hydrogen Energy, Hydrogen Energy System, Hydrogen Economy, Hydrogen Civilization

After having my elementary and middle school education in Izmir, Turkey, and high school education in Istanbul, I attended the Istanbul Technical University for one and a half years. Then I entered the Imperial College London to complete my university education, where I received the Associateship of City and Guilds Institute "A.C.G.I.", B.S. Degree in Mechanical Engineering, Diploma of the Imperial College "D.I.C.", and in 1951, Ph.D. Degree in Heat Transfer from the University of London. After my graduation, in Turkey I did my military service in the Ordnance Division, worked at the Office of Soil Products, first as a consultant and then as the Deputy Director of Steel Silos, and in the management of a major construction company.

In 1962, I joined the University of Miami Engineering Faculty as an Associate Professor of Mechanical Engineering. My first research project involved the Investigation of Two Phase Flow Instabilities in Nuclear Hydrogen Rocket Engines for Travel to Mars. Dr. Alan Stenning, Chairman of the Department of Mechanical Engineering, was the Project Director. We discovered three different instabilities in boiling two phase flows, and also the precautions to be taken to eliminate them. This study lasted five years – 1962 through 1967. During that time, I became quite familiar with the properties of liquid and gaseous hydrogen. It was the lightest fuel, and the best fuel for long distance journeys.

In the 1960s, air pollution in cities had become an important issue hurting the health of people. I became interested in environmental issues in general, and air pollution in

B. Baranowski et al. (eds.), *Carbon Nanomaterials in Clean Energy Hydrogen Systems.*

particular. I started studying possible vehicle fuels, with a view of determining the fuel which would cause little or no pollution. Particularly, I studied methanol, ethanol, ammonia and hydrogen, as well as the gasohols (i.e., mixtures of gasoline and methanol and/or ethanol). My investigation of fuels for transportation lasted five years, 1967 through 1972. The result was that hydrogen was the cleanest fuel, and it was also the most efficient fuel. It would not produce CO, CO_2, SO_x, hydrocarbons, soot and particulates. If hydrogen was burned in oxygen, it would not produce NO_x either. If it was burned in air, then there would be some NO_x produced.

Early in 1973, energy crisis started. Petroleum producing countries in the Middle East stopped exporting oil. The world economy almost came to a standstill. Some factories stopped and/or worked part time. Transportation was adversely affected. There were big lines at gas stations, few cars on the roads, and air traffic was curtailed.

As I always believed that engineers and scientists should strive to find solutions to the problems facing the humanity and the world, in 1973 I established the Clean Energy Research Institute (CERI) at the University of Miami. The mission of the institute was to find a solution or solutions to the energy problem, so that the world economy could properly function and provide all of the humankind with high living standards. It was also the mission of the institute to find clean forms of energy, so that they would not produce pollution and damage the health of flora, fauna and humans, and the environment of the planet Earth.

CERI started work in earnest. We quickly looked at all the possible primary energy sources, including solar, wind, currents, waves, tides, geothermal energy, nuclear breeders and thermonuclear energy. They were much cleaner than fossil fuels. They would last much longer than fossil fuels. We began using fossil fuels with the start of the Industrial Revolution in 1860s. First we used coal, and then we started using oil, and later natural gas. Petroleum companies expect that most of the fluid fossil fuels (i.e., petroleum and natural gas) will have been used by 2060s. Coal, tar sands and shale oil might last another 100 years, if a solution to CO_2 emission problem could be found. However, renewable energy sources will last as long as we have the sun – another five billion years!

Although, the new primary energy sources would last much longer than the fossil fuels, although they were much cleaner than fossil fuels, they were not practical to use. They were not storable or transportable by themselves, except nuclear. They could not be used as a fuel for transportation by themselves, except nuclear for marine transportation. In order, to solve the problem, it became clear that we had to manufacture a synthetic fuel using the new primary energy sources. As a result of my ten years of research and investigation (i.e., 1962 through 1972), I already knew that the best synthetic fuel was hydrogen. It was the lightest fuel, the most efficient fuel, the cleanest fuel, and the best fuel for transportation. I decided to call the resulting energy system "Hydrogen Energy System" or "Hydrogen Economy", since energy is the economy's locomotive. I was quite sure that this was the best solution to the depletion of fossil fuels and the global environmental problems they were causing, such as the global warming, climate change, ozone layer depletion, acid rains, air pollution, oil spills, etc. I was convinced that it was inevitable that the Hydrogen Energy System would replace the present fossil fuel system sooner or later.

In order to inform the scientific community about the proposed solution and get their reaction and input, I decided to organize an international conference, to be named The Hydrogen Economy Miami Energy (THEME) Conference. I went to Washington,

DC, and explained my idea at the National Science Foundation. They gave the Clean Energy Research Institute $70,000 to organize the conference. We sent letters and brochures to all the scientists around the world, who were interested and/or would be interested to solve the energy and environmental problems facing the humankind. We worked overtime to organize the conference in the shortest possible time, in about eight months.

The landmark THEME Conference opened at the Playboy Plaza Hotel in Miami Beach on 18 March 1974, with the participation of more than 700 scientists from some 80 countries. At my Opening Address, I described and proposed the Hydrogen Energy System as the permanent solution to the depletion of fossil fuels and the global environmental problems being caused by their utilization. After my opening address, we had a refreshment break. Some ten scientists from around the world came up to me. They said "Dr. Veziroglu, we agree with you, the Hydrogen Energy System is the best possible system to replace the fossil fuels". That evening eleven of us met at the Penthouse of the Hotel, which was my residence as the Chairman of the Conference.

The 11 scientists, who were later known as "Hydrogen Romantics", represented almost all regions of the World. They were **Hussein K. Abdel-Al** of Egypt and Saudi Arabia, **John O'M Bockris** of United Kingdom and Australia, **William J. D. Escher** of U.S.A., **Cesare Marchetti** of Italy and Austria, **Anibal R. Martinez** of Venezuela, **Tokio Ohta** of Japan, **Walter Seifritz** of Switzerland, **William D. Van Vorst** of U.S.A., **T. Nejat Veziroglu** of Turkey and U.S.A., **Kurt H. Weil** of Germany and U.S.A., and **Robert M. Zweig** of U.S.A. At the Penthouse meeting, there was a passionate, yet deliberate debate. It was agreed that the Hydrogen Energy System was an idea whose time had arrived. It was the permanent solution to the depletion of conventional fuels. It was the permanent solution to the global environmental problems. Then, the discussion turned to whether there was a need for a formal organization. It was Anibal Martinez of Venezuela (incidentally the one who took part in establishing the petroleum cartel OPEC), who urged the founding of a society dedicated to crusade for the establishment of what seemed to be to the gathering, and which later proved to be, the inevitable and universal energy system, the Hydrogen Energy System. It was ironic that he was proposing the establishment of an organization, which would make OPEC obsolete. As a result of the Penthouse meeting, the International Association for Hydrogen Energy (IAHE) was established by the end of the year 1974. I was elected President of IAHE, William J.D. Escher was elected Secretary, and most of the Hydrogen Romantics served on the Board of Directors.

IAHE started working enthusiastically. In 1975, we published two newsletters. I negotiated with Perganon Press to start a scientific journal on Hydrogen Energy, and CEO of Perganon Press, Bob Maxwell, agreed to be the publisher. The first issue of the International Journal of Hydrogen Energy – of which I became the Editor-in-Chief – published in January 1976. For the first three years, IJHE was quarterly. Three years later, in 1979 it became bi-monthly, and after another three years in 1982 it became monthly. This year IJHE will publish 18 issues. Starting in January 2008, IJHE will become forthnightly, in keeping up with the growing importance of Hydrogen Economy.

In order to provide a platform for Hydrogen Energy scientists from around the world, we decided to organize the biennial World Hydrogen Energy Conferences (WHECs). The first one was held in Miami Beach in 1976. Thereafter the WHEC conferences were held in Zurich, Tokyo, Pasadena, Toronto, Vienna, Moscow, Honolulu, Paris, Cocoa Beach, Stuttgart, Buenos Aires, Beijing, Montreal, Yokohama and Lyon, every other even year. The forthcoming WHEC conferences will be held in Brisbane, Australia, in 2008, Essen, Germany, in 2010, and Calgary, Canada, in 2012.

Of course, no technology can take roots without standards, and no universal technology can be established without international standards. Gustav Grob of Switzerland approached the International Standards Organization (ISO) based in Geneva, Switzerland, requesting ISO to consider preparing International Standards for Hydrogen Energy Technologies. A meeting of interested ISO Members was organized in 1990, which I attended as the Head of the U.S. Delegation. As a result of this meeting, ISO/TC-197 Committee was established to prepare the standards. For several years, Switzerland hosted the Secretariat of this committee. Some six years ago, the Secretariat has been moved to Canada. The Committee has already prepared nine International Standards for Hydrogen Energy Technologies, and are working on several others.

Because of the growing interest in Hydrogen Energy, IAHE Board of Directors decided that we should have technology oriented meetings during the odd years. First, World Hydrogen Technologies Convention (WHTC) was held in Singapore in June 2005. The second WHTC was held in MonteCatini, Italy, in November 2007. They will be followed by WHTC 2009 in Delhi, India, and WHTC 2011 in Glasgow, United Kingdom.

It was clear that the Hydrogen Energy System was inevitable. Through the research carried out around the world, through WHEC conferences and through the papers published in IJHE, the foundations of the Hydrogen Energy System were being established. The time had arrived to start implementation. I decided that a United Nations Centre would be the best organization to implement the establishment of the Hydrogen Energy System throughout the world. In 1988, I proposed such a center to the United Nations Industrial Development Organization (UNIDO), for which organization I was the consultant on Hydrogen Energy System.

The UNIDO Officials liked the idea. Now, it had to be approved by the UNIDO General Assembly. I together with UNIDO representatives visited many countries (including USA, Japan, China, India and EU) in order to explain the center proposal and get their agreement, which we did. The countries which we did not visit, including Russia, Ukraine, Brazil and Argentina, were invited to a meeting in Istanbul, organized by Engin Ture (who later became my deputy in the center) on behalf of the Turkish Scientific and Technological Research Organisation. At the conclusion of this meeting, the countries participating also agreed to the center proposal. In 1996, UNIDO General Assembly unanimously approved the establishment of the center in Istanbul, Turkey, because of its central location between the developing countries and the industrial countries. After that, negotiations started between UNIDO and the Turkish government for the final agreements. Yılmaz, Ecevit and Erdogan Governments, as well as the National Security Council, approved them. In October 2003, the Minister of Energy and Natural resources, Hilmi Guler, who strongly believes in the importance of Hydrogen Energy for Turkey and the World, signed the agreements with UNIDO in Vienna.

In April 2004, I took a three years' leave of absence from the University of Miami, and was appointed the Founding Director of the center, and started working vigorously in temporary facilities. In the ensuing three years, a truly international center, UNIDO International Centre for Hydrogen Energy Technologies (UNIDO-ICHET), has been established, consisting of scientists and staff from many countries of the world, including Turkey, USA, UK, France, Croatia, Greece, Bulgaria, Kirgizstan, India, Pakistan and Egypt. The center has been working to establish Hydrogen Energy pilot projects around the world. It is also organizing international conferences, training courses and workshops. As my leave of absence from the University of Miami has ended on 1 May 2007, I left the Associate Director Professor Engin Ture in charge of the Centre. He is doing admirably well to fulfill the Centre's mission to help convert the world to Hydrogen Economy.

It was not always smooth going. There were obstacles on the way – some of them formidable. When 11 of us established the IAHE, the petroleum interests became worried. They labeled 11 of us "Hydrogen Romantics". They said that we were dreaming, it could never happen. After the 1973 energy crisis, a bill was being discussed in the U.S. Congress to support electric cars, alcohol cars and hydrogen cars. Petroleum lobby made sure that "hydrogen cars" were taken off the bill. Then, petroleum companies established a consortium to fight Hydrogen Energy. The purpose of the consortium was to prove that fossil fuels were clean, and hydrogen was not feasible. They lavishly supported research, which agreed with them. They claimed that air pollution in the cities was not caused by emissions from the cars, buses and trucks, but it was caused by volcanoes erupting far away; the smoke and dust the volcanoes produced would travel through the atmosphere and fall on the cities to cause the air pollution. They claimed that the global warming was not caused by CO_2 produced from burning of the fossil fuel, but by regurgitating cows in meadows.

By the 1990s, most of the industry was looking at hydrogen as the fuel of the future, and they have started building prototypes to use hydrogen. There were prototypes of hydrogen cars, hydrogen buses, hydrogen air conditioning systems and hydrogen appliances. Aircraft companies were working on hydrogen fueled airliners. But there was one industry which was missing from the effort, and which seemed aloof to hydrogen, was the petroleum industries.

From time to time, I would think about the absence of interest in Hydrogen Energy amongst the petroleum industry. During an Organising Committee Meeting for the 12 WHEC Conferences, the Conference Chairman, Dr. Juan Carlos Bolcich, and I dwelt in this problem and planned a strategy! As a result, I sent letters to the presidents of major oil companies, saying that in June 1998 we were going to have the 12 World Hydrogen Energy Conference in Buenos Aires, Argentina; in the conference, we would have a session devoted to petroleum; and we would like them to come to this session or send their representatives to tell us what they would sell when the petroleum deposits were depleted! None of the oil company presidents answered. I later heard that one of the presidents asked his advisors, if it was true that the oil deposits would be depleted. Although none of the oil companies answered, Shell Oil Company sent a handful of their engineers to the Conference in Buenos Aires. Two months after the conference, in August 1988, Shell Oil Company left the Anti-Hydrogen Consortium of oil companies and as the fourth division in the company they established Hydrogen Division, in addition to the divisions on petroleum production, petroleum transportation and refineries. August 1998 was the date when the petroleum companies saw the "light", and started jumping on the Hydrogen caravan. Shell Oil Company was followed by the British petroleum, after that the anti-hydrogen consortium was dissolved, and all the petroleum companies – large and small – joined the Hydrogen caravan.

Of course, economy is important. Cost of the fuel to energize the economy must be as low as possible. Amongst others, cost depends on the utilization efficiency of the fuel and the environmental damage the fuel causes. Our investigations at the Clean Energy Research Institute have shown that Hydrogen is the most efficient fuel. When you consider all the energy applications in the world at the consumer end to produce thermal energy, mechanical energy and electrical energy, hydrogen is 26% more efficient than fossil fuels. The environmental and health damages the fossil fuels are causing worldwide through the global warming, climate changes, acid rains, ozone layer depletion, air pollution and oils spills add up to some six trillion US Dollars a year, and is growing. When one takes into account the higher efficiency of hydrogen, and the environmental and health damages caused by fossil fuels, hydrogen is the most cost effective fuel.

I was invited to a United States Congressional Committee in Washington, DC, to testify about the Hydrogen Energy System as the permanent answer to the depletion of fossil fuels and the environmental damage they are causing. At the end of my testimony, the Chairman of the Committee Congressman Brown of California told me "Dr. Veziroglu in the United States we have a free economic system, if hydrogen is such a good energy carrier, let it compete in the marketplace and win". My answer to the Congressman was "Mr. Brown, the United States may have a free economic system, but it does not have a fair economic system. If you will pass a law making all the goods responsible for their environmental and health damages, nobody will sell petroleum but hydrogen". In short, petroleum, natural gas and coal are relatively cheap, because they don't pay for their huge environmental and health damages. They are paid – directly and/or indirectly through taxes and insurance premiums, etc. – by the people.

As a result of research and development activities around the world, World Hydrogen Energy Conferences, and publication and dissemination of research and development results through the International Journal of Hydrogen Energy, foundations of the Hydrogen Energy System has been established during the quarter century of 1974 through 2000. Starting with the twenty first century implementation of the Hydrogen Energy System began. Some hydrogen fuel cells became commercially available. All major car companies came up with various models of experimental hydrogen fuelled cars. In several major cities of the world, hydrogen fuelled buses started being operated on a trial basis. Airbus and Boeing companies started programs for building hydrogen fuelled subsonic, supersonic and hypersonic passenger planes. Home appliances running on hydrogen have been built and tested. Hydrogen electric batteries have been commercialized.

At the Clean Energy Research Institute we carried out a model study in order to determine, when the conversion to hydrogen economy could be completed under the existing laws, regulations and rules. For this business as usual case, our model study indicated that the conversion to Hydrogen Economy would take another three quarters of a century, i.e., it would be completed by 2074.

Let us now view the planet Earth from a distance towards the end of the twenty first century, after the Hydrogen Energy System has replaced the present fossil fuel system. There will be no pollutants, no chemicals to damage the ozone layer, no acid rains, no oil spills. Earth's temperatures will return to pre-industrial levels. Global warming will end. Earth's climate will return to normal. It will be the most hospitable planet for flora, fauna and humans. Each country will be using the renewable energy sources available to it to manufacture the fuel "hydrogen" it needs for its economy. As a result, the countries' development will be hastened, the population increase will level off and cease, they will achieve higher quality of life, and illegal immigration from developing countries to industrial countries will end. Petroleum wars will be history. Planet earth will become a paradise. As Viktor Golstov of Ukraine has foreseen it early in 2000, the ultimate has now been reached for the planet Earth and for its inhabitants, i.e., **Hydrogen Civilization**!

A SIGHT AT THE FUTURE OF SCIENCE AND TECHNICS

F.P. SANIN
The Ukrainian Scientific Research and Development Institute of Mechanical Engineering Technique of National Space Agency of Ukraine, prosp. Kirova 46, Dnipropetrovsk, 49054 Ukraine;
A.M. Makarov Yuzhy Machine – Building Plant, Dnipropetrovsk, 49008 Ukraine

Abstract. In this article the author states his views at the future of science and technics, which based on articles and monographs of academician, Nobel Prize Laureate V.L. Ginzburg, other scientists, printed out within last 20 years, and more than 50 years operational experience of the author in rocket-space and nuclear power engineering, as well as his monographs and articles written and published for students of higher education institutions [1–10].

Keywords: thermonuclear fusion, superconductivity, fullerene, crystal, hydrogen, magnetic field, laser, nucleus, quark, neutrino, proton, the Universe, vacuum, quantum, planet, space, holography, rocket

1. Introduction

In the middle of 50th years we peered into the small 8 x 12 sm screen of the TV KVN. We thought up even a lens to increase the image. It was the convex flask under the form of a usual lens which filled with water. Then everyone only dreamed of a home phone, about mobile even to conversation was not. And in fact what is a mobile phone? It actually is a radio station which carried in special cars earlier. All radio equipment has been mounted on radio tubes.

Modern radio and the electronic equipment are constructed on semiconductors, transistors, but they have also undergone colossal changes aside miniaturizations. They can be made even less, but a problem is how they may be connected though also it is overcome in many cases.

The first computers with operation speed of about 50 ones per second occupied the huge areas equipped with systems of cooling and ventilation.

And in fact 50 years have passed only. Almost every schoolboy and student has a computer with operation speed more in millions time then those first computers have and which can execute many new operations in additional. It is both mail, and phone, and the newspaper, and, it is possible to tell, world library, etc. I have resulted only two examples from what humanity has achieved due to development of fundamental and applied science and technics. But there is a lot of unknown, undeveloped, not created still ahead. We shall make attempt to stop on some problem questions. First of all on fundamental ones, but here I shall be based on articles of the academician of Russian Academy of Science V.L. Ginzburg which have been printed in magazines «Successes of physical sciences» and «Science and life» on a threshold of 21 centuries, and articles of other scientists.

B. Baranowski et al. (eds.), *Carbon Nanomaterials in Clean Energy Hydrogen Systems.*

2. Global problems

2.1. CONTROLLED NUCLEAR SYNTHESIS

This problem all is well-known, and it was solved due to application of powerful magnetic fields for deduction of plasma in installations of tokamak and stellarator types. The tokamaks were developed in Russia more, the stellarators – in Ukraine (KhPhTI).

The interests to the stellarators come down during one time but today it has started to come back again. Some scientists solved a thermonuclear problem even by creation of so-called cold synthesis. These works were regarded a little bit skeptically, and suddenly this works become confidential in the USA.

In article published by academician R.I. Nigmatulin in magazine "Energy" (№3, 2003г.), there was underlined an opportunity of application of acoustic waves for reception of ultrahigh pressure in a liquid, formation and maintenance of plasma in liquid bubbles, true, during very short time. But it is already new way.

The variant of laser "arson" of deuterium and confinement of plasma in powerful magnetic fields which can be created in high-temperature superconductors is considered also. As it is possible to replace tritium with helium-3 in reaction of synthesis in the future it will has to be delivered from the Moon where lunar regolith contains it.

The matter is that tritium is not present in the nature and it needs to be received artificially, helium-3 is difficult for extracting on the Earth too therefore the cheaper way – to bring it from the Moon. Only 18 t of helium-3 in a year is enough to provide whole humanity with the electric power at the expense of controlled thermonuclear synthesis.

It is thought that researches of antisubstance will proceed from only scientific interests and in the far future, probably, and for accumulation of energy.

2.2. HIGH-TEMPERATURE AND ROOM-TEMPERATURE SUPERCONDUCTIVITY

Superconductivity was open in 1911 at temperature of liquid helium, i.e. less than 4.2 K. During 75 years the necessary critical temperature has been increased all up to 23.5 K, i.e. up to temperature which hardly higher than liquid hydrogen temperatures. Scientists were guided to superconductivity of metals and some alloys. And here, in 1986 messages about superconductivity of the some ceramics at temperatures above liquid nitrogen temperature have appeared. Such superconductivity has already considerably promoted aside its wide use as cost of liquid nitrogen approximately is less in 100 times than cost of liquid hydrogen, all the more helium. The main disadvantage of ceramics is its fragility. Now scientists work above the form for its application. Besides the new theory of superconductivity has appeared which allows, more truly, does not forbid reaching temperature of superconductivity almost up to room. It is the second fundamental problem.

Obviously, it is not necessary to stop in detail on application spheres for the phenomenon of superconductivity. It both new types of transport, and the electrical engineer, both electronics, and space communication on IR-waves, powerful magnetic fields, accelerators of the charged particles. For an outer space exploration – new electrodynamic rockets launches, power supplies of space vehicles, aerials, space communication, astrophysical researches, supersensitive receivers and super-power transmitting devices on frequencies up to hundreds of gigahertz, etc.

Except for mentioned, certainly, medicine, light industry, metallurgy, etc.

2.3. SOME QUESTIONS OF SOLIDPHYSICS

Physics of a solid body and science of materials as a whole have made very big step for last 50 years in the theoretical and practical views. Especially it is visible on elements of electronics. The majority of the transmitting and accepting devices are constructed on semi-conductor materials that, finally, have led to their essential miniaturization.

It is enough to tell, that computers which occupied the big rooms earlier and demanded special service, today fit into a desktop computer with much more operation speed. New control and telemetry systems, equipment for space research, microsatellites which can execute high volume research programs have appeared in space-rocket technics.

Many things were made for creation of the new materials which allow design rather easy rockets and satellites. But a lot of work is to do for reaching theoretical durability of many materials.

2.4. FULLERENES

Today many researches on fullerene structures of carbon have appeared but obviously fullerenes on the basis of other elements will be created. It is a new kind of molecular structures and today it is impossible to expect completely, what practical results they will cause.

We hope that fullerenes will take the place among semiconductors and super-conducting materials.

2.5. LIQUID CRYSTALS

It is known already today that liquid crystals are very sensitive to variables electric and to magnetic fields, temperature, acoustic fluctuations, chemical influence. New TV and computers screens practically relieved from soft x-ray unhealthy radiation are already created. The supersensitive devices reacting to signals of about 10^{-20} wt are created. Long-distance communication with the American space crafts “Pioneer”, “Voyager” etc. is carried out with their help. As in wildlife many cellular formations represent liquid crystals, for example cholesterol, it is possible to expect occurrence of the materialistic theory of biofields and their connection with other fields. The person really can operate biology of animals, plants, well and, certainly, itself.

2.6. FERROELECTRICS

Ferroelectrics were studied by as far back as I.V. Kurchatov who has established that their dielectric permeability depends on temperature, an electric field.

They are applied in radio electronics, optics. Recently it is revealed that they behave similarly to piezoelectric, i.e. react to variable acoustic fields, other unknown properties are marked also. So we can find a lot of new in ferroelectrics.

2.7. BEHAVIOUR OF SUBSTANCE IN MAGNETIC FIELDS

Due to occurrence of high-temperature superconductivity superstrong magnetic fields can be created and the new phenomena are open though, apparently, in this area already it is much made. Behaviour of alive organisms, including plants, in usual constants and variable electromagnetic fields is of high interest. It is connected to a nuclear magnetic resonance due to which the wildlife except for power perceives weak space radio emissions which bear the managing information. This question is not idle, insufficiently studied and also is of high interest for medicine and biology as a whole. About application of strong fields in power it is already told above.

2.8. METAL HYDROGEN

It is known, that there is firm hydrogen under very high pressure inside big planets such as Jove and Saturn. Preliminary researches of molecular hydrogen at very high pressures show that unusual properties can be found out here, up to liquid and firm hydrogen at temperatures about 3,000 K with metal conductivity. Besides now many scientists are studied a problem of hydrogen power, but even liquid hydrogen in a usual condition has density in 10 times less than density of water. There is a problem of explosion hazard of gaseous hydrogen, therefore scientists search for ways of its accumulation in the forms of hydrides or special sorbents, etc. So metal hydrogen could solve many problems, including a problem of fuel and protection at distant interplanetary travel.

Obviously, ultrahigh pressure can be received due to acoustic compression in liquid hydrogen or other means.

2.9. SUPER-POWER LASERS, X-RAY LASERS, GAMMA-RAY LASERS

Now many types of lasers are known.

But the question is super-power lasers and intensity (density) of laser radiation. Today the laser radiation density of 10^{20}–10^{21} W/sm^2 is achieved already. At such intensity electric fields up to 10^{12} V/sm arise and at that the magnetic field reaches 10^9–10^{10} A/m.

But for support of laser radiation interaction with atomic nucleus intensities on some orders above are required. It is known that with the more shortly the length of a wave of laser radiation the more capacity can be achieved in a pulse. Therefore lasers in x-ray and gamma-ray scale ranges i.e. x-ray lasers and gamma-ray lasers are required. Certainly powerful lasers will find application not only in nuclear physics but also in other areas of a science and technics.

2.10. SUPERHEAVY NUCLEUS

Theoretical developments indicate that transuranium elements with serial number **Z** more than 114 and mass number **A** about 300 can exist or can be received in the nature. It was informed that the element with **Z** = 114, A = 298 and lifetime about 30 s has been synthesized in Dubna. Even there was a message that group of physicists under the direction of academician N. Flerov, then not yet Nobel winner, has found out an element with a serial number 116 in waters of Caspian Sea about Cheleken Island. Certainly the reception problem of transuraniums with such numbers and life time even in 1,000 years is only scientific, but it is necessary to solve it.

2.11. QUARKS AND GLUONS

This problem concerns to physics of elementary particles. Theorists predict existence of quarks fractional particles from which such elementary particles as electrons, nucleons and mesons becoming habitual already consist.

Experimenters have not found out yet fractional particles at existing accelerating technics though have already for a long time found out antiparticles (a positron, an antiproton) and even have created antisubstance with shot lifetime.

About 30 years ago there was an attempt to find these particles in space probably they have remained there from the Universe birth as relic electromagnetic radiation.

For this purpose Proton satellite of 17-t weight where the data-acquisition equipment has been protected from usual space radiation has been launched. But it was not possible to find out fractional particles in space.

Scientists set the big hopes on the accelerator in which energy of $14 \cdot 10^{12}$ eV will be achieved building in CERNE. It should be put in operation in 2005–2007.

One more of many other particles – gluon is predicted. It ostensibly provides stability of atomic nucleus.

2.12. NEUTRINO WEIGHT, PROTON DISINTEGRATION

This problem interests scientists since time of opening of this particle, i.e. since 1934. The neutrino weight if it takes place would help to explain many phenomena in physics and astrophysics. It is supposed that the proton half-life is more than 10^{30} years. If it is true the theory of the Universe origin turns out falls as the proton is older it on many orders of years. Probably there are other problems connected to the disintegration of protons and neutrinos. Except for electron and muonic neutrinos there are relic ones in the Universe. They were formed at an early stage of elementary particles birth. Their detection will help to solve many questions connected with the birth of the Universe and substance which it contains.

2.13. FUNDAMENTAL LENGTH AND TIME

Theoretical calculations show that existing space-time conceptions are true up to distances of $\sim 10^{-17}$ sm and time $\sim 10^{-27}$ s. And what occurs in smaller scales? It is enough to recollect time of the Universe birth which amounts from 10^{-44} to 10^{-33} s i.e. there are unusual ideas about “granular” space-time, etc. We already spoke that the accelerator on energy of 14 TeV should be put into operation in 2005–2007. It corresponds to length $l = \frac{nc}{e_c} = 1,4 \cdot 10^{-18}$ sm. There are particles with maximal energy E = $3 \cdot 10^{20}$ eV in space beams though such particles exist in very small quantities and they have no practical importance for high power physics.

There is even a gravitational length $l_g = 10^{-33}$ sm but it can take place only inside “black holes”.

Certainly the fundamental length if it is precisely determined will play the certain role in physics. Interaction of particles at high and ultrahigh powers including on colliders, i.e. on installations with colliding beams where the above-stated length and time can be realized, attract the big interest. Probably fundamental physical sizes at which it is impossible to use the general relativity theory (GRT) will be determined.

2.14. VACUUM AND SUPERSTRONG MAGNETIC FIELDS

The vacuum and superstrong magnetic fields are very urgent. Probably in this case on the basis of GRT it is possible to endow space with certain properties of ether, zero fluctuations of electromagnetic and others still unknown fields. There is a item of creation of the quantum gravitation theory and its association with the theory of other interactions.

2.15. CREATION OF THE GENERAL THEORY OF ALL INTERACTION KINDS

Now weak, strong and gravitational interaction are known. On the basis of the general theory the space theory will be created, the new understanding and approximations of biology and electronics, creation of an artificial intellect, etc. will appear.

2.16. A STOP OF LIGHT

Now scientists have as though stopped light, i.e. have received the speeds of light measured in several kilometers a second. It is known that light speed in the environment is less, than in vacuum.

$C_{\text{env}} = \dfrac{C_{\text{vac}}}{n}$, where **n** is factor of refraction, for water it is equal approximately 1.5.

If **n** is increased to infinity light speed in environment $\mathbf{C_{env}}$ will aspire to zero. According to all available information due to strong increase **n** light speed $\mathbf{C_{env}}$ = 17 km/s is received for some materials at cryogenic temperatures.

What consequences expect us and what applied advantage we take from this? It is necessary to solve.

2.17. A PROBLEM OF AN EXTRATERRESTRIAL LIFE

Here it is necessary to define presence of any wildlife on other planets in abstracto and certainly presence of the advanced civilizations, i.e. extraterrestrial mind.

Besides completely new concept of alive substance is required then the contradiction between materialism and idealism can be taken off, other problems will be solved also.

2.18. PROBLEMS OF SPEED, TIME AND SPACE

The general relativity theory limits speeds of any particles, and therefore bodies to light speed or speed of any electromagnetic fluctuations distribution irrespective of frequency or wave length.

Soviet physics Saharov A.D. and Zeldovich Ja.B. as though bequeathed the resolve of two problems:

- The first is finding the “missing” weight in the Universe. Probably there are the gravitational strings containing huge masses.
- The second is a speed problem. If the space and time are quantum systems there should be in space antigravitational tunnels in which it is possible to move with any speed i.e. with the speed more than light speed.

Probably tunnel effects will be open in space and time where the relativity theory is inapplicable, superfast movings will appear, i.e. movement with speeds more than light speed.

2.19. QUANTUM GENERATORS

Today quantum generators are widely known as lasers.

But as they are also amplifiers of electromagnetic radiation with their help it will be possible to take (swing) energy from Space due to the relic radiation or neutrino. And may be other particles will appear also.

There is also a number of other unresolved problems in physics.

Here concern:

- Experimental check of the general relativity theory
- Gravitational waves and their detection
- Neutron stars and pulsars
- Black holes and space strings
- Quasars, formation of galaxies
- A problem of the latent matter and its detecting
- An origin of space beams with ultrahigh energy

3. Applied problems

3.1. SEMICONDUCTORS

Approximately for 50 years the science and practice of semiconductors have completely changed electronics. It is enough to tell that today miniaturization has reached such degree that 10 million transistors fit on the area of some square centimeters. The radio station which had to be carried in the special automobile now is located in clockwork and in a mobile phone. But following from professor's I. Alferov Nobel lecture and the literature on the topic question it is not limit yet. The matter is not only in miniaturization but in completely new materials and technology, processing speed, an opportunity of regulation by the "forbidden" zone, efficiency, radiation resistance, etc. also. Appearance of organic clusters and ferrimagnetics on their basis is possible. And that means high-current emitters for lasers and gyrocones (lasers in the microwave range), thin coverings for protection of biological objects against radiation, etc.

3.2. SUPERCONDUCTIVITY

We considered it in the first part as demanding theoretical development. But at the same time already today possible appendices of high-temperature superconductivity in practice are visible.

Such appendices are seen:

(a) Superfast transport.
(b) Electrodynamic launch of rockets and planes.
(c) Electrodynamic masse accelerators, for small masses they are available already today.

(d) Creation of strong magnetic fields for accelerators of elementary particles, plasma generators and thermonuclear installations.
(e) Small-sized electrogenerators and engines, magnetohydrodynamic cars. Electric motors with capacity from 100 W up to 2 KW have already been created.
(f) Inductive energy stores for space technics and industrial electronics. They already are created today.
(g) Application of high-speed elements of computer facilities, superconducting quantum generators and magnetometers, i.e. optical converters.
(h) Superfast transfer of the information along superconducting lines (up to 10^{13} bats per second). At that in case of application of optical fibre lines the transformation of electric signals into optical ones is excluded.
(i) Other applications: acoustic and a nuclear magnetic resonance – tomographs, long-distance communication, protection from ionizing radiations, magnetic shielding, etc.
(j) Difficult problems are input of energy in superconducting stores and its lost-free extraction, transportation, operation of superconducting installations, devices, devices, cooling in weightlessness condition, liquid-free cooling, epitaxial growth of crystals, etc.

3.3. APPLICATION OF AMORPHOUS, STRAINED METALS

It is known, that amorphous materials or glass materials have high strength properties but basically they are created on a basis oxide glasses. Amorphous metals are received also but manufacture and methods of their receiving are still very labour-consuming and not so accessible to wide introduction. Strained metals are metals with the increased strength properties and in some cases with "memory". They have not still been widely disseminated and practically applied, for example, there where explosions are applied, in space-rocket technics, etc.

3.4. HYDRIDE FUEL

Now the big attention is devoted hydrogen and its compounds. As hydrogen is pollution-free fuel its application is very attractive and perspective. But its application in the pure form in a liquid or gaseous state is unprofitable because of its very low density even in a liquid phase (100 kg/m^3) and explosibility. Therefore accumulation of hydrogen in its compounds or sorbents, for example fullerenes, zeolites are studied widely now.

3.5. THE MOON AND PLANETS

In the 21st century humanity will master the Moon, i.e. build moon bases and manufactures, solar power stations on the Moon and artificial stations. Probably first of all it will be reception of rocket fuel and further – metals for production of rockets and satellites. In fact there is a lot of titan, aluminium and other metals in the form of their oxides on the Moon. Regarding solar power stations on the Moon and at artificial stations it is possible to dedicate the following. As predicted scientists not less than 10 million tons of helium-3 is saved up in Moon regolith. Helium-3 can be used as thermonuclear fuel together with deuterium. If such controlled thermonuclear reaction $He_2^3 + D_1^2 \rightarrow He_2^4 + H_1^1$ be realized that 20 t of He-3 in a year which may be delivered from the Moon to the Earth will be enough for providing all humanity of the Earth with the electric power. On the Moon completely new materials can be received in conditions of low gravitation, ultrahigh vacuum and space radiation. Flights to other planets, their satellites, asteroids and comets which can be realized from the Moon are of the big scientific interest and probable practical one.

3.6. IONIC TECHNOLOGIES

Ionic technologies apply the accelerated ions of metals and nonmetals. Already now there were received superfirm films, sheetings, eximer chemical compounds, for example compounds of so-called neutral or noble elements such as NeCl, ArF and others having been used in laser technics. The firm films and sheetings are applied in tools, at work in excited environments. So about 20 years ago a semi-production version of installation for sputtering wearproof coverings on cutting tool has already been created in ChPhTI.

The ionic technologies application is not confined wit that. They will be applied in chemistry, at creation of new materials, semiconductors and superconducting films.

3.7. NEW NUCLEAR REACTORS

Chernobyl accident has caused an unfavorable altitude to atomic power stations and it proceeds from people far from nuclear physics and power industry.

In rocket technics sometimes we use elements as speak «designed for the fool» in order to prevent emergency rockets launch and satellites flights. Fortunately there are less units of risk in atomic engineering and today about 500 nuclear reactors works already in the world.

Everybody should understand that failure on Chernobyl atomic station has happened not on the most technical ideology, the human factor has played. Certainly after that reliability of work at atomic power stations has considerably increased. Scientists have developed new principles of trouble-free operation of nuclear reactors. First of all it is adjustable reactors criticality due to external "injection" of neutrons or protons from accelerators which can be stopped during second parts. There are also other principles of additional charging of an active zone with neutrons from an external source. By the way to tell Ukraine is included into number of the key countries in this direction. The big problem is storage or processing of used-up fuel elements and structure materials. There are also corresponding proposals for transmutation, new fuel elements, etc.

Reactors on fast neutrons have already worked today. Their future development will bring the big benefits since here all uranium is used completely but not only uranium-235 with concentration of 0.7% in ore. Now everybody speaks a lot about so-called renewed energy sources (a wind, the sun, the sea, etc.) but they can give no more than 15% of the necessary electric power. In the future nuclear rocket engines operating outside of the Earth atmosphere only will be applied. Certainly new proposals for example nuclear-chemical reactors and others will appear.

3.8. BURNABLE LAUNCH VEHICLES

Arthur Tsander and Jury Kondratyuk have already suggested ideas of combustible rockets after fuel consumption or during joint combustion of fuel and the structure. Now works in this direction have appeared including Dnepropetrovsk National University. Such works are carried out in Italy and in other countries. Now this problem becomes actual as the near-Earth space is littered with the rests of last rockets stages, the worked-out satellites and other subjects.

Besides the combustible design of fuel systems is an additional fuel.

3.9. MICROSATELLITES

Because of the big successes achieved in electronics due to semi-conductor microcells the opportunity for receiving from microsatellites the information which earlier received only from multiton spacecrafts has appeared. Certainly we don't consider here space vehicles with people or for delivery of special mechanisms to space stations, the Moon, planets.

Already now in Dnepropetrovsk microsatellites are created, they are the future for special researches and communications.

3.10. LASER TECHNICS

Today lasers are applied widely in technics, medicine, communication, etc. They have 1,000 specialities, can be used even as the weapon. But their opportunities are not confined by it. As the beam of the optical laser can be focused in very thin bunch it is possible to receive high density of energy, over 10^{20} J/sm^2.

At such density of energy the substance turns to radiation. It can be used in photon engines, thermonuclear reactors and even in many processes which even specified with difficulty today. And certainly laser technologies are still far from being confined in mechanical engineering and other branches.

3.11. BIOLOGY AND ELECTRONICS

Now electronic devices become so sensitive that react even to weak electromagnetic fields which are formed by the person. Moreover due to presence in an organism of a radioactive potassium isotope ${}^{40}_{19}K$ the person with weight of 70 kg radiates every second approximately 8,000 electrons and potassium turns into ${}^{40}_{20}Ca$.

Perhaps not all physicians know that ordering калиевые preparations to heart sufferer they so way stimulate heart work for the account of potassium-40. Electric fields forming in clothes and cellular membranes bias also considerably.

So in the future or already today it is necessary to create protection systems for computers, measuring devices and other electronics from operator radiation.

3.12. NEW TYPES OF TRANSPORT

Using electric motors on transport began probably approximately 100 years ago since the first tram. Then trolley buses, electric trains, sea nuclear ships, intrafactory electrotransport have appeared. But electric motors are not applied in motor industry practically for the lack of energy stores even on 500 km. Scientists work above this problem and the first electromobiles on fuel elements and solar converters have already appeared. Therefore the main task consists in power-intensive sources: stores or generators of the electric power.

It is necessary to do it to not break thermal balance of our planet which the big problems and cataclysms follow.

3.13. AN OZONE CLOUD AND A HOTBED EFFECT

The problem of an ozone cloud rescue is discussed at many conferences, the international meetings. The decisions are made but there are more conversations here than practical actions.

Soviet Union has already refused the superfast plane TU-144 which flied in the bottom layers of a stratosphere. In October 2003 last flight of similar French plane "Concord" has been accomplished.

But it is not enough also. It is necessary to forbid flights of jet planes on near distances. Why it is necessary to fly from Kiev to Moscow or Warsaw by the jet plane? Today it is already clear that the ozone cloud is destroyed with emissions of jet planes and first of all water and nitrogen oxides. Every year about 300 million tons water is thrown out in a stratosphere by planes.

We want it whether or not, but it is necessary to fly by fine turbo-propeller planes such as AN-70, AN-140.

It is necessary to forbid also using solid-propellant rocket engines for flights in space and basically it is time to plan space flights internationally. Except for above-stated the big contribution in ozone destruction is brought with mine emissions and outflow of methane and ammonia which is widely applied in the chemical industry and as fertilizer well and certainly CFCs. It was believed early that only CFCs responsible for destruction of an ozone cloud but as now it became clear their component does not exceed 20%.

There are assumptions about opportunity of ozone cloud rescue with the help of radiowaves of the certain frequency at which negative ions of chlorine, waters and other substances are formed. They can prevent ozone destruction. But it needs to be checked in order to not destroy connections, to not cause other unforeseen phenomena probably even earthquakes.

The second problem which related to emissions of gases and a dust is a hotbed effect. It is discussed and written so much therefore we shall not stop in detail here. Though it is necessary to tell that the radiophobia will gradually pass and people will address the needs of atomic engineering and further thermonuclear one which will rescue us from a gassed condition and a dust.

3.14. SPACE AND A LIFE

The majority of people today do not doubt any more that the life is more likely not the terrestrial phenomenon but the space one.

First, everyone knows that the sun, air and water are necessary for a life. The Earth is located to the Sun such way that the temperature of its atmosphere, ground and water allows existing wildlife in the big variety: from monocelled, plants, animals and birds and at last up to reasonable subjects – the person.

We shall not stop on a life origin on the Earth here. But the aforesaid it appears insufficiently. In fact our atmosphere and magnetosphere protect all alive from fatal space radiation.

In the first offer we have told that a life on the Earth is the space phenomenon. Our Earth represents the big magnet with average intensity of a magnetic field about 40 A/m (0.5 gausses in old system of units). Intensity of a magnetic field varies from equator, where it amounts 28 A/m approximately, to poles where it is equal 78 A/m. It is necessary to take into account at residence change on breadth.

For change processes in the lifeless nature energy is required only and here for wildlife is necessary the managing information also. Objects of wildlife receive this information from a radiating background of the Earth, radioactive isotopes which in necessary quantity are available in organisms. This is for example, potassium-40, sodium-23 which people and animals receive with food. This radioactive isotopes come from ground to plants which we basically eat.

But it appears this information insufficiently. The wildlife receives the main information from space as superlow-frequency radiowaves. Each planet, the Moon as well as a space objects arrangement represent space aerials which radiate these radiowaves.

It is known from physics that nucleus of various atoms radiate or accept so-called spin electromagnetic waves of very low frequency approximately from 50 up to 2,500 Hz. In a magnetic field of the Earth these frequencies amplify and resonate with the radio emissions which have come from space or from other sources.

This phenomenon has received the name of a nuclear magnetic resonance.

Now the nuclear magnetic resonance and just magnetic fields were applied in medicine, true, the nuclear magnetic resonance – for diagnostics, and the fields – for so-called magnetic therapy. But both methods are still applied without a due theoretical substantiation and experience.

There are conclusions from that:

1. It is necessary to study theoretically and experimentally an interaction of wildlife and space. Here the astrology, some phenomena in the lifeless nature, medicine can receive a scientific substantiation also.
2. It is necessary to connect physics to medicine for studying connections of Space and many, not only magnetic, fields.
3. A biofield and other phenomena, even a sight or ponderomotive movements, should receive a scientific materialistic explanation and a substantiation.

3.15. OXYGEN-FREE CERAMICS

Carbide, nitrides, silicides, borides, aluminides and others intermetallic compounds concern to this kind of materials.

All they are basically refractory heat resisting and firm materials which are applied as coverings for the tool, heat resisting heaters or refractory filling materials in composite materials. In space-rocket technics they are used as ablators of descent vehicles or heat-stressed units of engines.

Common fault of these materials is low resistance to often temperature changing (thermostability) and fragility.

Now the nitrides possessing self-lubricating properties have already begun to be applied in mechanical engineering as the details working without greasing.

It is supposed that scientists all the same will sometime create elastic above-named materials and they begin to be applied widely in motor industry, rocket technics as not cooled chambers for liquid-propellant rocket engines and certainly in solid-propellant engines. There is a hope that some of them will be applied as semiconductors and high-temperature superconductors. Bioceramics for example calcium hydroxyapatite and others will find application also though they contain oxygen.

3.16. BURIAL OF RADIOACTIVE WASTE

Constructional materials such as steel, zirconium do not cause special problems since their half-decay periods estimated in months, for example, for iron it is equal to 45 days, therefore they can be stored during certain time in special pools. The most serious problem is processing and burial of the spent fuel and structures. Dry salt mines are suitable for this purpose though temperature conditions should be supported strictly here since due to proceeding disintegration in fuel elements these waste products are heated up. They need to be transported and stored in special containers.

If reactors on fast neutrons and so-called breeders are introduce the waste products quantity will considerably decrease. In addition waste products transmutation methods are already developed but their introduction will take place probably not soon since that demands the big financial charges. At the transmutation it is possible to transform radioactive isotopes, as fission products, to quickly breaking up ones or on the contrary in long-living ones.

It is pleasant, that the Ukrainian scientists take leading place in the world in this areal. Except for the specified actions variants of radioactive waste products throwing out to far space are possible. But at that it is necessary to reduce their weight by the way of burning, dissolution, etc.

3.17. MATERIAL SCIENCE

We have stopped above on some materials which, at first sight, are less known and less mastered on a broad scale.

What should be priority in 21st century in the materials areal? First of all, it should be wide development of powder technologies; these are materials with “memory”, programmed and gradient materials.

Except for rocket technics as it was till now, powder materials and technologies will be widely introduced in mechanical engineering. Blast furnaces will sometime disappear, and more progressive metallurgical technologies, including powder will replace them.

The reinforced composite materials with using of moustaches, strings, refractory fillers will find application.

Polymeric semi-conductor materials, i.e. plastic electronic schemes will be very interesting. The huge telescreen can be rolled and replaced.

3.18. RESOURCES-ECONOMY TECHNOLOGIES

Today everybody speak much about them but do a little.

It is very simple decision: to replace the usual resistance furnaces working up to 600–700°C with aerodynamic ones or as they are named, with furnaces of aerodynamic losses which unfortunately not spread. They are at least twice more economic than usual ones with ohmic heating.

We have not learned yet to burn usual fuel (gas, coal, oil). Electromagnetic fields application will raise caloric content due to completeness of burning. The temperature rise at gas turbines work can give 15–20% fuel economy or efficiency. At last usual membrane technologies and ones with application of an electroosmosis can give appreciable economy of power resources. A lot of energy can be saved due to application of new lighting lamps as well as diodes, etc. Certainly nanotechnologies or as speak structural engineering will be widely used.

3.19. CENTENARY OF THE FIRST EARTH ARTIFICIAL SATELLITE

Obviously the centenary will be celebrated on the Earth, on the Moon, Mars, Europe and Ganimed (satellites of the Jove), the Titan (the satellite of Saturn) and at orbital stations. Certainly new technologies and materials can be created on many these objects.

3.20. HOLOGRAPHY

Optical and ultrasonic holography became widespread in technics and medicine.

The volumetric cinema and TV and even appointment on distance will appear, i.e. it will be possible to meet image as with the present person.

3.21. MEDICINE AND PHYSICS

Above we spoke that the life is the phenomenon more space than terrestrial, paid attention to a magnetic field of the Earth and in this connection to a magnetotherapy. Today the ultrasound and an ultrasonic tomography which simply name ultrasonic are widely applied in medicine. The nuclear magnetic resonance, an x-ray tomography and many other things are applied also.

But, basically, these methods are applied for diagnostics, except for a magneto-therapy, radiation, radiochemistry which are used for the medical purposes.

It is not enough of it. The special attention will be inverted on quantitative (dose) and the qualitative sides and, certainly, the new equipment and specialization of physicians and physicists will appear.

3.22. CONVERSION OF FUEL

Everybody knows that power generators are burdened non-uniformly within day. Especially it is inefficient and even harmful for nuclear and to usual power stations. There were many offers how to use or accumulate energy during the lowered burden. As far back as D.I. Mendeleev dreamed about gasification of underground coals, i.e. their transformation to окись carbon oxide CO or other gas which then burn.

Certainly for this purpose some energy has to be spent and it can be taken during the lower burdening of power stations, in the future – of solar ones also. The provisional scheme:

$$2C+O_2 \rightarrow 2CO;$$
$$CO+H_2O \rightarrow H_2+CO_2.$$

If it is necessary to receive hydrogen we use the second reaction if we need CO only for burning the first reaction will be used.

There is also a second kind of conversion of usual natural gas, methane. This method is already developed in laboratories, introduction is required. It is necessary to note that the increase in fuel caloric content turns out up to 25% minus all charges.

The hydrogen received as a result of conversion will be used in fuel elements, but there are schemes where hydrogenous compounds, for example, CH_3OH- methanol or methyl spirit are used simply instead of hydrogen. Scientists trend to application of hydrogenous compounds, i.e. hydrocarbonic fuels, to a greater extent. For example, it is supposed about an identical development level of atomic power stations and fuel elements in the future power of Russia.

3.23. MICROMINIATURE ELECTROMECHANICAL ENERGY CONVERTERS

Capacitor and induction converters play the basic role in a microcosm and in alive organisms. For example, a fiber of a bacterium is the most low-power natural capacitor "electromachine" of capacity 10^{-17} W. Capacitor electromechanical converters are the most low-power.

Among the most powerful converters induction ones, for example the turbogenerator, can have capacity more than 10^9 W (1 million of kilowatts).

Capacitor converters use induction – magnetic electric fields. By the way to tell the Earth is the induction converter.

Chemical energy of food is at first transformed in electric power and then it is spent for manufacture of adenosine-triphosphoric acid (ATP) which is a store and a carrier of energy.

Cellular membranes in organisms are capacitor generators with capacity about 10^{-17} W (approximately 1,000 protons in a second).

If physicians and physics study all that and learn to operate these processes humanity will get rid of many illnesses and medicines.

But the microminiature power will not be limited by it. The new control system of vegetative and fauna development, pest control, etc. will appear. And certainly new devices, micromotors, microgenerators will appear in comparison with which even today's will seem giants.

Many years ago E.K. Tsiolkovsky in a little bit vulgar form represented the future person eating electricity.

3.24. RECYCLING RUBBER

The basic problem here consists in non-polluting processing the worn out trunks of automobiles, planes, etc.

Now in some countries rubber of old trunks is burned simply on power stations, heating plants, applied at construction of roads, the bases also. One of problems is crushing rubber. Various methods are applied, including preliminary freezing, but all of them are very expensive.

In the future obviously pyrolysis or conversion methods will be developed for reception of new carbon-bearing products or combustible gases and probably even fullerenes.

3.25. SPACE "DUST"

Now there are about 50,000 of the subjects in a circumterraneous space which deserted there by the person. These are last stages of rockets, the accelerators, used satellites, fairings, fragments of heat-shielding coverings and other subjects having sufficient sizes for break-up of a new satellite or a rocket at collision.

Various ways of clearing are offered but any of them is realized yet.

The problem of withdrawal of possible large meteorites and asteroids adjoins to that also.

3.26. SCIENTIFIC SCHOOLS AND EDUCATION

Because of Soviet Union destruction and the arisen crisis situations in the young states, former republics, many scientific research institutes, laboratories of high schools have practically stopped their works and so the big scientific schools have ceased to exist.

Some of them have been created in Russia and have blossomed in Soviet Union and always differed from the West, especially from the USA where the local scientists are not present practically and immigrants work only.

There is even a joke. The scientist from the former USSR asks in America: "Well, how is your higher school? ». The answer: «Excellent, Russian professors read lectures to the Chinese ».

And in fact the school is not simply association, these are the traditions produced on decades and inherited by generations. Schools existed both in fundamental and applied science. Such school has been created in Yuzhnoye Design Office also.

The education system disturbs very much. We try to reconstruct too much per western sample with ease. It is pleasant that now the computer technical equipment is introduced widely, technologies are discussed everywhere. Here it is impossible to not go too far, in fact technologies are different. If the question is production of goods it would be better to name it «the automated and mechanized technologies with application of computer technical equipment». The main thing is to not break the best in the world system of training at schools and high schools! We well know what students and experts – trainees come to us from other countries.

Certainly, we do not apply for completeness in a statement of the problems taking a place in various branches of a science and technics.

To receive more panoramic sight at the future of humanity we invite all wishing to state the vision and understanding concerning mentioned and all not mentioned perspective directions.

4. Conclusions

1. The view of a scientist on future development of basic research in physics, rocket-space engineering, materials technology is stated in the article.
2. The author touches upon some practical problems including ecology, space "garbage", radioactive waste, connection of physics with medicine.
3. The issue of scientific schools and education is touched upon.
   The author doesn't consider social and political problems.

References

1. F.P. Sanin, A.F. Sanin, "Space and Technologies", 2005, Art-Press, Dnepropetrovsk, 456 p.
2. F.P. Sanin, Perspective directions in rocket-space materials science, IX International Conference ICHMS'2005 "Hydrogen Materials Science & Chemistry of Carbon Nanomaterials", September 4–10, 2005, AHEU, Kyev-2005 P.
3. Materials of XV International Conference "Nuclear Materials Science", September 2004, Alushta, Crimea.
4. Materials of XIV International Conference "Physics of Radioisotopic Phenomena and Radioactive Study of Materials", June 12–17, 2000, Alushta, Crimea.
5. F.P. Sanin, Atomic energetic problems and their influence to national safety, ICHMS'97 "Hydrogen Materials Science & Hydrides Chemistry", September 2–8, 1997, Kathiveli, Crimea.
6. Ukrainian Aerospace Herald, May 2005, pp. 32–37.
7. Yu. Krupnov, "Nuclear Doctrine of Russia", Special Issue, October 26, 2004.
8. Materials of scientific-technical and popular science editions: Rocket-Space engineering, Space Science and Technology, Heralds of the Science Academy of Ukraine and Russia, Reports of the Science Academy of USSR (Russia), Successes of Physical Sciences, In World of Science (USA), Power, Earth and Universe, Scientific World, Science and Life. Period – About 20 years.
9. V.L. Ginzburg, "About Physics and Astrophysics", 1980, Nauka, Moscow.
10. Operational experience in rocket-space and nuclear power engineering for 55 years.

SEARCH AND DEVELOPMENT OF NEW MATERIALS FOR HYDROGEN STORAGE IN THE CONNECTED STATE

B.P. TARASOV
Institute of Problems of Chemical Physics of RAS
Prospectus of the Academician Semenov, 1, Chernogolovka
Moscow Region, Russian Federation, 142432
Tel./Fax: +7(496) 5221743; E-mail: btarasov@icp.ac.ru

Abstract. Various materials for hydrogen storage in the connected condition are considered and their merits and demerits are discussed.

Keywords: hydrogen storage, connected state, intermetallic compound, hydrogen-accumulating materials, hydrogen accumulators

The problem of mobile storage and transportation of hydrogen is one of the basic problems, which interfere with its wide application in energy. None of the existing methods of hydrogen storage now satisfies showed operational requirements [1–4]. At the present the hydrogen storage in the connected condition is considered in quality of the most compact and safe way. This method uses the hydride-forming metals and alloys, hydrogen-sorbing materials having high specific surface area, metal hydride-carbon and amide-imides composites, metals and hydrides which are able to be subjected to hydrolysis, reversibly hydrogenated organic compounds (Fig. 1).

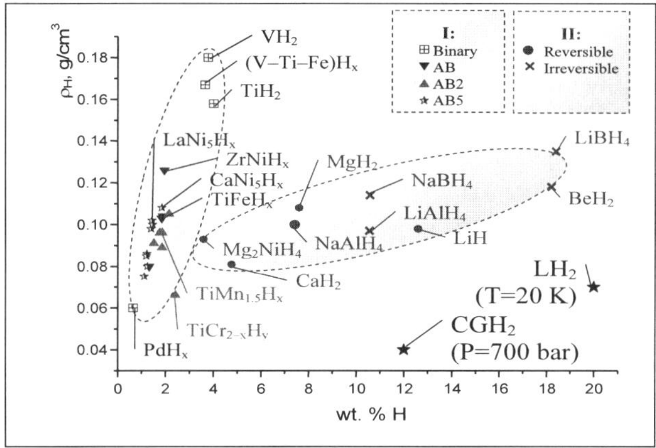

Figure 1. The weight contents of hydrogen in various environments

Unfortunately, till now there is no the universal hydrogen-accumulating material satisfying all operational requirements. So, the most convenient low-temperature (working temperature -20–100°C) hydrogen sorbents of repeated action – the intermetallic compounds of such as AB_5 (A – La, Mm, Ce; B – Ni, Co, Fe, Cu, Mn,

Al), AB_2 (A – Ti, Zr; B – Mn, Cr, Fe, V), AB (A – Ti, Zr; B – Fe, Co) and vanadium based composites – satisfy the requirements on working temperature and the volumetric contents of hydrogen, but have convertible hydrogen capacity less than 3 wt%. Besides kinetics of hydrogen absorption and the amount of hydrogen that is reserved convertibly at recycling "sorption/desorption H_2" strongly depend on impurity presence in hydrogen. Despite of these lacks, accumulators of hydrogen on the base of such metal-hydrides find wide application as sources of high-pure hydrogen. Apparently, the further efforts should be directed on search of the new hydrogen-accumulating materials absorbing hydrogen at an ambient temperature and pressure up to 35 MPa and isolating it at pressure not less 0.2 MPa at temperature of -30°C as last years the easy composite metal-polymeric balloons are actively used. Such hybrid system can come nearer on their operational characteristics to modern requirements to mobile systems of hydrogen storage.

Special interest can represent hydrogen-sorbing materials with high specific surfaces (zeolites, metal-organic skeletons, carbon nanostructures etc.), which adsorb a significant amount of hydrogen at temperature of liquid nitrogen. At increase of hydrogen-sorption capacities are higher than 10 wt% the cryoadsorption systems will be quite perspective, as modern heat-insulating materials allow to keep low temperature with high efficiency.

The similar situation is observed with boron hydrides and aluminium hydrides of alkaline and alkali-earth metals. Despite of the big expectations of that modified alanates can convertibly absorb more than 5 wt% of hydrogen, there is no any confirmation that they can meet the requirements on cyclic stability, safety, interaction kinetics, and cost. In addition extra high sensitivity toward water and oxygen, the interaction with which can be accompanied by explosion, is scarcely favour for wide employment of these materials.

In a case of boron hydrides of sodium and magnesium there are significant doubts whenever possible the organization of convertible process and the formation probability of highly toxic borane does not cause optimism.

Last years the system lithium amide – magnesium hydride is actively studied as environment for hydrogen storage. On reaction $2LiNH_2 + MgH \leftrightarrow Li_2Mg(NH)_2 + 2H_2$ it should be allocated more than 5 wt% of hydrogen. However there are no sufficient data, confirming high efficiency amide-imides systems. Apparently, wide use of such systems will be complicated because of problems of mechano-chemical activation, presence of impurity of ammonia and sensitivity to moisture.

To develop the middle-temperature (hydrogen release temperature is less than 300°C) hydrogen accumulators of repeated action one uses mainly magnesium and its alloys Mg–Ni and Mg–La(Mm)–Ni which are able to reversibly absorb 5.5–7.6 wt% of hydrogen. However such alloys have bad kinetics of hydrogenation, are sensitive to impurities, inclined to sintering. By creation of polymetallic and metal-carbon composites it is possible to remove a part of these lacks, but main lacks remain, i.e. high temperature of hydrogen isolation and the big heat of dehydrogenation which must be compensated by the heat comparable to heat of oxidation of 25% of stored hydrogen. Such accumulators of hydrogen are perspective only when there is a cheap source of heat, allowing organizing the fast heating of system.

"Organic hydrides" – reversibly hydrogenated organic compounds and polymers – yet have not found application because of technical difficulties on the organization of process of hydrogen isolation at moderate temperatures, and also the irreproducibility of hydrogen-accumulating characteristics at repeated recurrence of cycles "hydrogenation-dehydrogenation". Apparently, it is necessary to develop the techniques of forming of composites "metal hydride – unsaturated hydrocarbon" and "metal hydride – unsaturated polymer" which under the characteristics can satisfy showed requirements to materials for hydrogen storage.

The most suitable environment for hydrogen storage is the modified aluminum hydride, which is meeting the requirements under the hydrogen contents (7–10 wt%) and to conditions of its isolation (100–150°C) (Fig. 2). However till now it is not clear, whether it will be possible to overcome the main lack of this system – difficulty of direct synthesis of aluminum hydride. It is not excepting that by modification and creation of nanodimensional state it will be possible to develop new methods of direct synthesis of aluminum hydride. The further progress in this direction can be connected to creation of introduction phases, complex compounds of cluster type and pseudo-alloys on the basis aluminum, which could is convertible to cooperate with hydrogen.

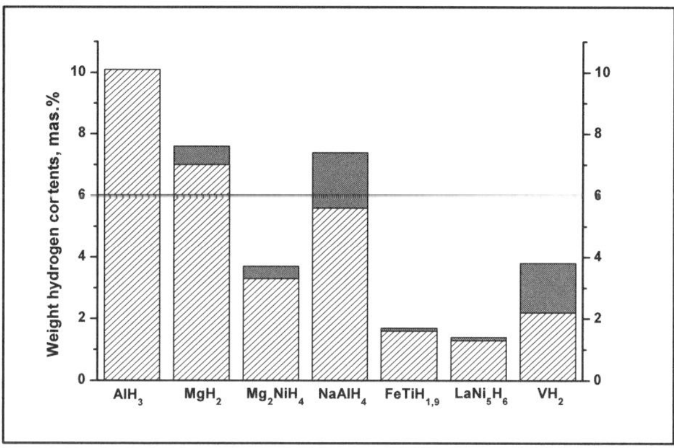

Figure 2. The weight contents of hydrogen in hydrides

As hydrogen source for portable fuel cells method of hydrogen reception by interaction of metals and hydrides with water is most perspective and close to practical realization. On potential ability to formation hydrogen at interaction with water and to its quantity on gram of compound, and also in view of density and availability, the most perspective materials are $NaBH_4$, Al, Mg and their hydrides. However here too there are the serious problems connected to difficulties of the organization of controllable process of hydrogen generation.

Acknowledgement

The work is executed at financial support of the Russian Basic Research Foundation (grant No 05-08-18130).

References

1. Sandrock G., Thomas G. (2001) The IEA/DOE/SNL *On-Line Hydride Databases. IEA Task 12: Metal Hydrides and Carbon for Hydrogen Storage*, pp. 36–38.
2. Schlapbach L. (September 2002) Hydrogen as a fuel and its storage for mobility and transport, *MRS Bulletin,* pp. 675–679.
3. Tarasov B.P., Lototskii M.V., Yartys' V.A. (2006) A problem of hydrogen storage and prospect of use of metal hydrides for hydrogen accumulation, *Rossiiskii Khimicheskii Zhurnal*, **50**: 34–48 (in Russian).
4. Tarasov B.P. (2006) Problems and prospects of creation of materials for hydrogen storage in the connected state, *Al'ternativnaya Energetika i Ekologiya*, **2**: 11–17 (in Russian).

HYDROGEN TECHNOLOGY FOR POROUS METALS (GASARS) PRODUCTION

V.I. SHAPOVALOV
Materials and Electrochemical Research Corporation
7960 S. Kolb Road Tucson, AZ 85706, USA

J.C. WITHERS
Materials and Electrochemical Research Corporation
7960 S. Kolb Road Tucson, AZ 85706, USA

Abstract. The paper presents a history and broad background of porous gas-eutectic materials (gasars) based on experimental data obtained by the authors as a result of the work conducted in the former USSR, Russia, Ukraine and United States from 1979 to date. Gasar synthesis methods and system designs for laboratory and industrial production of gasars are discussed. New processing technology recently developed in the U.S. that offers potential for low cost manufacturing of power gasar materials is presented.

Keywords: hydrogen, gasar, synthesis method, structure, commercial application, gasar-device

1. Introduction

The history of gasars began in 1980, when the first publication [1] appeared in the former Soviet Union indicating that a gas-eutectic reaction can be used to produce porous metals with controlled porosity with an example of a copper-based gasar. A gasar patent was obtained in the same year [2]. In the former USSR, the gasar technology developed quite rapidly, and within 5–6 years the technology was close to industrial production. However, the use of these materials in the USSR was strictly military-oriented and no information could be made available to the public. After «Perestroika», gasar research in Ukraine as well as in Russia was practically suspended due to lack of financing. On the other hand, it became possible to publish information about gasars in the open editions as well as abroad. As a result, a first publication about gasars in English appeared in 1989 [3]. This publication signified the beginning of a second period in the history of gasar development, which expanded to an international level. In 1993, the first U.S. and German patents were published [4, 5]. In 1995, Ukrainian specialists constructed a small gasar furnace for the Navy Research Laboratory. In 1996, Ukrainian scientists aided building a large-size facility (capable of producing 50 kg of material) at the Sandia National Labs, thus enabling commercial demonstration of gasars. In 1996, H. Nakajima (Osaka University) visited the Ukraine to become familiar with gasar technology. This lead to the first gasar furnace for Japan being designed in the Ukraine, and a number of Japanese colleagues were trained by Ukrainian specialists in the technology of gasar production. Gasar research was initiated and has been conducted at the Osaka University (Japan) since 1997. It is believed, gasar research is also being conducted in China, Korea and Brazil since 2000.

Based on the results from the many gasar investigations, a variety of very attractive properties were produced separating gasar porous materials from a number of known porous materials. However, no large scale commercial production has been reported anywhere in the world. Practical utilization of gasars includes tens to hundreds of articles intended for rocket and space technology, military aircraft, oil-processing industry and medicine. The 28 years since gasar inception, no rapid progress in this paradigm technology development has been reported. The question is: WHY? This paper attempts to summarize accumulated information and provide the answer to this uneasy question.

2. Scientific basis of gasar technology

As it is known [6–11], the main principle of producing gasars is a gas-eutectic reaction, which takes place during crystallization of metals and alloys saturated with hydrogen (or other active gas). Hydrogen is the most universal gas, which can be used with a majority of metals, except for strong hydride-forming metals, such as titanium, zirconium, niobium, vanadium, etc. It is also possible to use gaseous hydrocarbons, e.g. methane [12], although in this case the porous materials may become saturated with carbon. A combination of hydrogen and oxygen (water vapor) may also be used to produce gasars within acceptable range of pressures to facilitate mass production of gasars [13].

Nitrogen can also be used for producing gasars [14], however, it can form chemical compounds with a number of metals, which seriously limits its use in gasar technology. It is more advantageous to use nitrogen in combination with hydrogen in the form of ammonia [15] and similar compounds.

Oxygen was proposed for gasar production application in 1987 [16]. It is even more chemically active and can be used in pure form only for producing silver-based gasars or possibly to form oxide ceramics. However, due to interaction with carbon and formation of gaseous carbon oxides, it can be used indirectly. For example, since the 19th century a cellular structure of unkilled steel is known, which is formed as a result of gas-eutectic reaction: liquid iron, saturated with oxygen and hydrogen, breaks down into hard steel and CO gas when solidified.

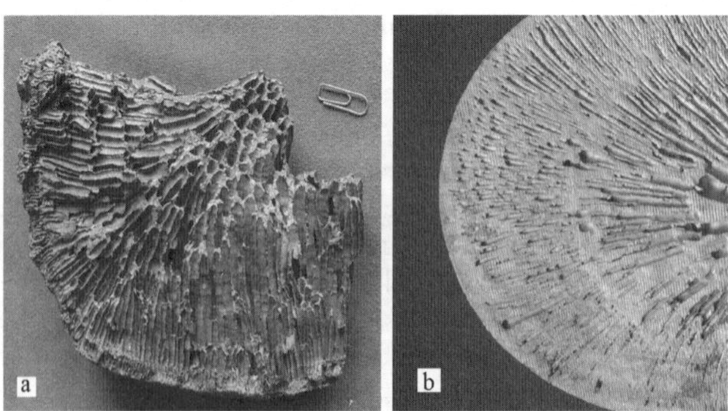

Figure 1. Oxygen-produced steel gasar (a) and hydrogen-produced magnesium gasar (b), average pore size ~ 2 mm

If crystallization proceeds directionally, an ordered cellular structure, typical for gasars, is formed (Fig. 1). This takes place naturally in millions of tons of unkilled steel without utilization of any special equipment or technologies. However, this phenomenon is considered to be rather harmful, and in order to produce dense steel, the latter is densified using the process of hot plastic deformation. Hence, millions of tons of steel gasars vanish without even revealing themselves. There is no doubt that this phenomenon will be used in the near future for producing steel gasars commercially.

Even such a superficial analysis reveals that an in-depth knowledge of interaction between gases (oxygen, hydrogen, nitrogen) and metals/alloys is required in order to scientifically substantiate gasar production as well as predict process conditions. The framework of such knowledge comprises metal/gas equilibrium diagrams. These diagrams provide a correct and quick answer to the primary question about possibility of producing gasars within given system. However, there is a significant shortage in such diagrams to date, since this involves a rather complicated experimental work, especially for the "metal-hydrogen" systems in the vicinity of the melting point. For example, equilibrium diagram for the "iron-hydrogen" system was only created in 1972 [17]. Later, a number of practically important equilibrium diagrams were created [18] for the following hydrogen systems: copper-, chromium-, aluminum-, magnesium-, manganese-, beryllium-, molybdenum-, titanium-, tungsten-, nickel-, cobalt-, vanadium- and zirconium-. It was the shape of these equilibrium diagrams, closely resembling ordinary eutectics, that made Ukrainian scientists think of a possibility of obtaining ordered "metal-hydrogen" structures in future gasars in 1980. It is also important to note that successful construction of the "metal-hydrogen" equilibrium diagrams was a result of inventing the original and reliable device for measuring hydrogen content in metals [19].

Due to these diagrams, practical methods were developed for producing gasars based on copper, tungsten, molybdenum, nickel, aluminum, iron, cast iron, magnesium, cobalt, beryllium, chromium, Ni-Ti (50%/50%) alloy, and high-temperature super alloys. All the engineering solutions (about 150 patents) were patented in USSR in 1981–1989.

In addition to the equilibrium diagrams, it is very important to know crystallization patterns of metals, and specifically, alloys. Producing gasars based on complex-doped alloys poses a serious problem. This results from the fact that in complex-doped alloys, dendrite crystallization is well established, while to produce gasars, a flat cellular crystallization front is required. How to make complex alloys crystallize in such a way? Such research studies are also quite tedious and expensive, but are a prerequisite to develop reliable commercial gasar technology.

Research studies in the field of gas phase formation in the liquid and along the "crystal-liquid" border are of particular importance. This is the process that determines the number of pores per unit area during the initial stage of structure formation as well as during production of gasars with alternating porous and monolithic layers. By controlling the process of gas bubble formation, gasar structure can be varied significantly. However, there are numerous problems in this field as well. Specifically, scientists are still debating whether homogeneous formation of gas in the liquid is possible near the melting point, or the presence of solid phase is required.

There is insufficient information concerning the growth mechanism of the gas phase in contact with both liquid and solid phases, especially in presence of impurities. This process determines the intensity of pore coagulation, growth halt and change in pore shape in the process of growth. Without this knowledge, it is practically impossible to reliably produce pores of a given shape in bulk of the material. Usually, a noticeable enlargement of pores is seen immediately after initiation of the solidification process, which continues until its completion. Significant effort is necessary to obtain pores with comparably identical diameters and a length of up to 500 mm.

Thus, at the present time, there is a serious lack of fundamental knowledge required to develop reliable, scientifically proven technology for producing gasar materials on the commercial scale. Of course, the equipment used to carry out gas-eutectic reaction is of great importance.

3. Equipment for producing gasars

The first apparatus for producing gasars was a hermetically sealed cylinder (Fig. 2) capable of rotating by 180 degrees end on end [20]. Installed at one end of this cylinder was a smelting furnace, at the other – a crystallizer, and in between – a funnel. Metal was melted and after its saturation with hydrogen the apparatus was tuned by 180 degrees allowing the melt to flow into the crystallizer, where the formation of the porous structure took place. This apparatus had one advantage – it had a very simple design and was easy to handle.

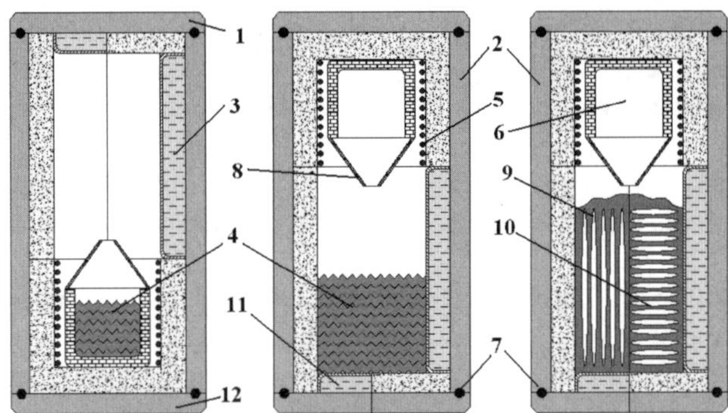

Figure 2. Schematic of gasar-furnace rotating by 180 degrees (left – view during melting of metal and saturation with hydrogen; center – after turning by 180 degrees; right – after solidification and obtaining radial porosity): 1 and 2 – lids; 2 – casing; 3 – radial crystallizer; 4 – molten metal or alloy; 5 – heater (resistance or induction furnace); 6 – melting crucible; 7 – seal between covers and casing; 8 – pouring funnel; 9 – gasar with radial porosity; 10 – gasar with parallel porosity; 11 – water-cooled crystallizer for producing parallel pores; 12 – bottom cover

It was found that the quality of gasars depends on the rate, at which metal was poured into the crystallizer. However, this design did not allow for controlling the pouring rate, nor did it allow for controlling the movement of the liquid inside the crystallizer in the process of pouring as well as upon its completion. Turbulent motion of the melt before the front of crystallization disturbed the ordered pore distribution and uniformity of pore sizes. In order to reduce turbulence, the melt was supplied to the crystallizer in portions using an intermediate cavity, or incident to a moving tilted plane made of refractory material [21]. These measures produced positive result, however were still insufficient to reliably control the pouring rate of the melt and its turbulence.

As a result, a stationary apparatus was designed [22], in which the smelting furnace was located in the upper portion, crystallizer – in the lower portion, and a funnel – in between the two (Fig. 3). The bottom of the melting crucible had an opening covered with a stopper attached to the end of ceramic rod. In the cold zone, this rod was connected to a steel rod extending outside through the top cover. After melting and saturation with hydrogen, the stopper was lifted and the melt was poured into the crystallizer. The pouring rate of the melt could be controlled by the height, to which the stopper rod was lifted. However, this design also revealed serious disadvantages. The main disadvantage resulted from the stopper developing such a strong connection with the crucible opening, that it was impossible to lift the stopper, or the stopper rod would lift, but not the stopper, which would get destroyed, thus blocking the opening. As a result, the metal would solidify inside the crucible causing the latter to be discarded.

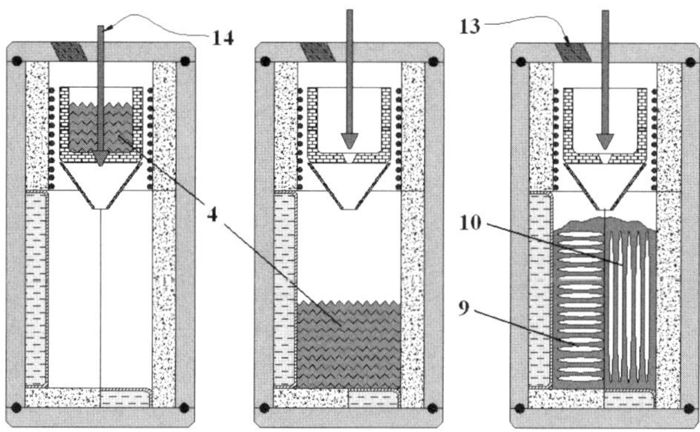

Figure 3. Schematic of stationary gasar-furnace with the stopper mechanism for pouring melt into the crystallizer (left – before pouring metal inside the crystallizer; center – after pouring; right – after solidifying): 13 – window for observing melting and saturation with hydrogen; 14 – stopper rod; remaining positions correspond to those shown in Fig. 2

In addition, such a design made it practically impossible to produce gasars based on steel and other higher-temperature refractory alloys, since the lifting rod would get destroyed during lifting, or chemically reacted with the melt, or melted.

Taking into consideration the shortcomings of the apparatus shown in Fig. 3 with a stopper and the apparatus capable of rotating by 180 degrees, a new apparatus was designed [23], in which the crucible was placed at an angle of 70–120 degrees with respect to the axis of the funnel with the crystallizer (Fig. 4). During melting and saturation with hydrogen, the crucible was positioned vertically, then the apparatus was rotated by about 90 degrees and the melt was pored into the crystallizer through the funnel. The pouring rate in this apparatus could be controlled either by speed of its rotation, or by the diameter of the funnel opening.

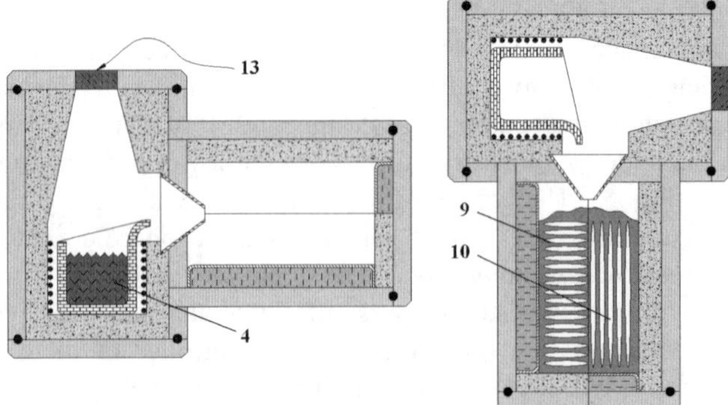

Figure 4. Schematic of gasar-furnace, rotating by 90 degrees, for pouring metal into a crystallizer (positions on the drawing correspond to those shown in Figs. 2 and 3): left – before pouring; right – after pouring and solidification

In addition, a dosed pouring could be performed in several steps. A big advantage of this apparatus was a possibility of observing melting and crystallization of the metal through windows located at a 90 degree angle with respect to each other. This design is the most suitable for the technological scheme of "melting – saturation with hydrogen – crystallization in the casting mould".

For articles in the form of tubes with radial porosity, a promising design was the one described in [24], in which filling of a crystallizer takes place from the bottom up through a ceramic tube inserted into a crucible with the melt (Fig. 5). This happens due to the suction effect, which occurs when the pressure inside the hermetic chiller drops. In this case, molten metal acts as a hydroseal. Experiments have shown that when using this method, filling occurs rather rapidly with a minimum turbulence of the melt.

Devices were also developed for producing gasars without the traditional casting mould (Figs. 6 and 7) – by "frosting" metal onto water-cooled chillers of different shapes immersed into the melt [25]. In case of such a design, the crystallizer is lowered into the melt after melting and saturation with hydrogen. Gasar is directionally crystallized on the chiller surface forming either parallel pores (in case of a flat chiller), or radial pores (in case of a cylindrical chiller). The flat chiller can be raised as gasar "frosting" takes place, which is similar to how crystals are grown according to the Chokhralskiy method.

In the 1980s, devices and methods were also developed (Fig. 8) for continuously gasar production in the form of a tape, sheets and cylinders [26]. Those devices are not described in details here, since they are very similar to the traditional equipment for continuous casting of steel or zone induction re-melting of metals with the only difference being that the entire device is enclosed within a hermetically sealed housing with controlled atmospheres.

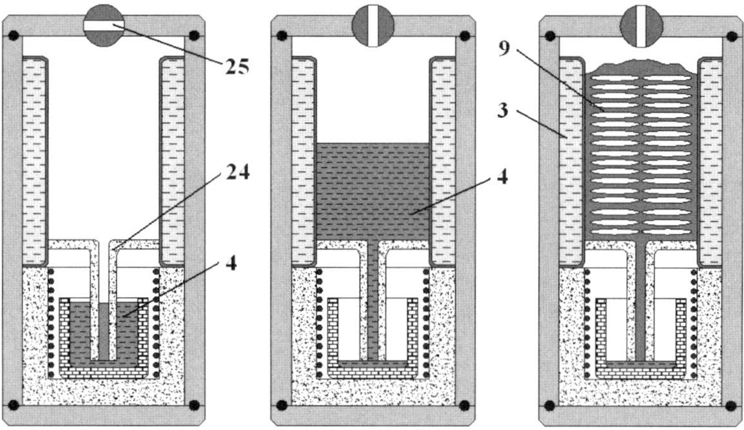

Figure 5. Schematic of stationary gasar-furnace with metal filling of a crystallizer using suction: left – before pouring; center – after pouring; right – after solidification: 24 – ceramic tubing; 25 – valve, which opens to suck in the melt into a crystallizer

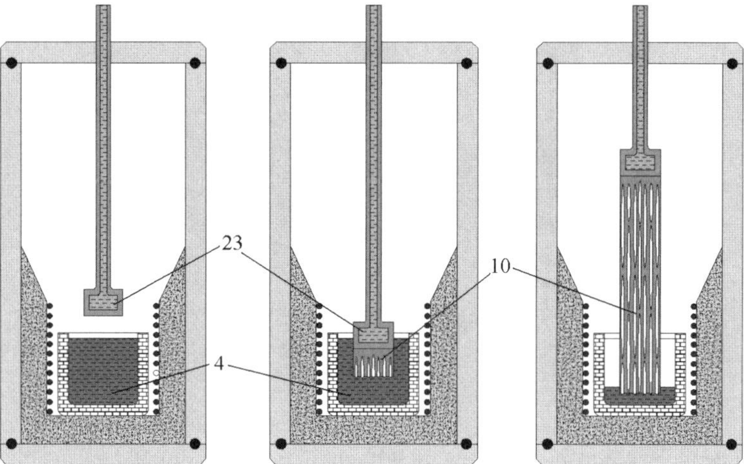

Figure 6. Schematic of gasar-furnace for producing gasars by drawing from the melt: left – before the beginning of solidification process; in the center, gasar formation begins at the end of the water-cooled chiller; right – end of solidification of gasar with parallel pores (23 – water-cooled copper disk-shaped crystallizer)

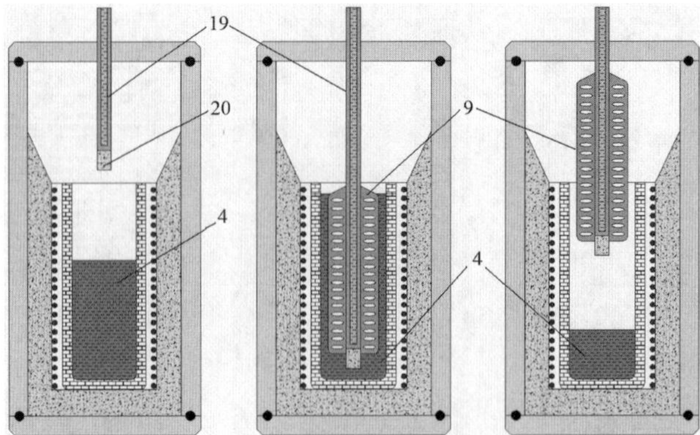

Figure 7. Schematic of gasar-furnace for producing gasars by “frosting” onto cylindrical chiller: left – before the beginning of solidification; in the center, gasar formation begins at the surface of the water-cooled chiller; right – end of solidification of gasar with radial pores and extraction the gasar piece from the melt

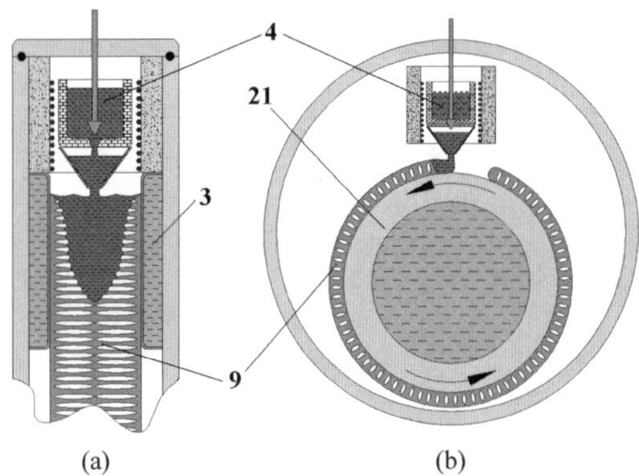

Figure 8. Schematics of gasar-devices for producing gasars using a semi-continuous method: (a) pouring inside a stationary crystallizer and drawing of an ingot downwards; (b) pouring onto the surface of a rotating water-cooled cylinder: 21 – water-cooled rotating crystallizer

It is also possible to produce gasars without a casting mold and a crucible – by re-melting a monolithic rod inside the annular induction furnace (Fig. 9) in the atmosphere of hydrogen. By varying the rod movement rate and position of the molten zone with respect to the inductor, it is possible to change pore direction in the gasar rod. The shape of the articles in this case depends on the shape of the cross-section of the monolithic rod and a shape of the corresponding inductor. For such type of products, this method can be considered the most applicable, since there is no contamination of the melt with foreign surfaces from a crucible and a casting mold. A disadvantage of this method is the difficulty of coordination between the movement rate of a monolithic rod and that of the gasar rod.

A high-temperature plasma torch (Fig. 10) can be used for producing gasars [26, 27]. Due to ionization of hydrogen, nitrogen or other active gas, a high concentration of the gas within a small molten portion of metal is achieved. Therefore, in order to produce gasars using such a method, a high-pressure active gas is no longer required. Utilization of this method allows producing gasars based on practically any metal or alloy, including such refractory metals as tungsten and molybdenum.

A modification of this method [28] can be used to produce porous coatings on monolithic products by melting just a surface layer of the material (Fig. 11). In this case hydrogen or other active gas is supplying in arc plasma (3,000–5,000°C). The gas is ionizing and saturates the small liquid pool quite rapidly. Then during cooling and solidification the gas-eutectic reaction takes place, and the gasar structure is formed in the surface layer. In that way closed and open porosity can be obtained (Figs. 12 and 13).

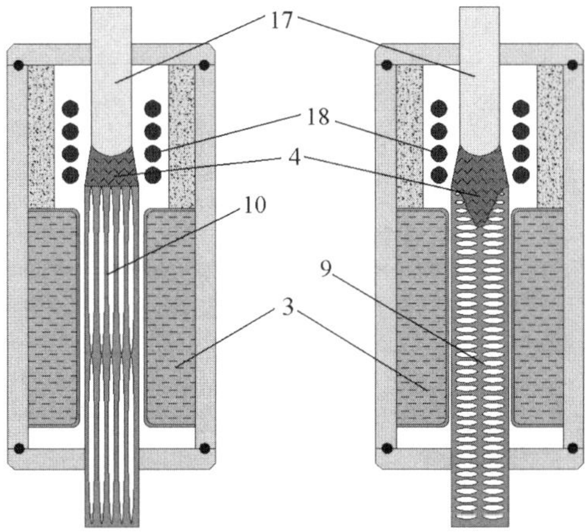

Figure 9. Left: re-melting of a monolithic rod inside the annular inductor resulting in parallel pore formation (rod and gasar move downwards); right: re-melting of a monolithic rod inside the annular inductor resulting in radial pore formation (rod and gasar move downwards); 17 – base material in the form of a monolithic rod or a pipe; 18 – annular inductor

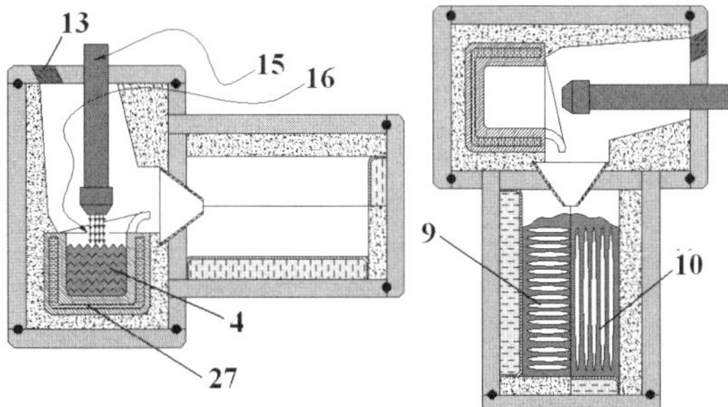

Figure 10. General schematic of producing gasars by high-temperature plasma torch melting: 15 – high-temperature plasma torch; 16 – plasma flow; 4 – liquid saturated with hydrogen or other active gas; 9 and 10 – gasar bodies; 27 – water cooling copper scull crucible

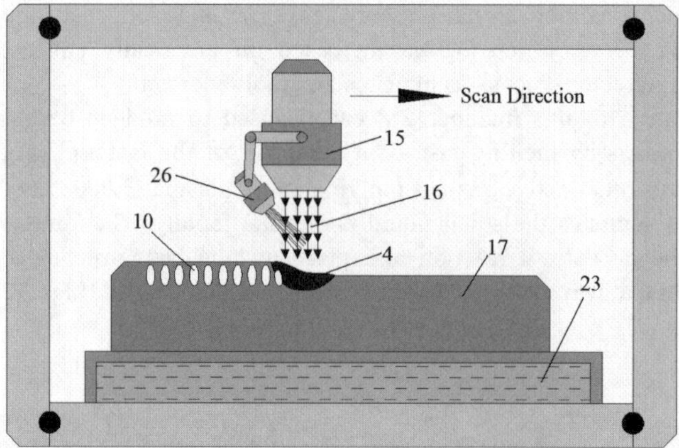

Figure 11. General schematic of producing gasar coatings by high-temperature plasma scanning system (PSS): 23 – water-cooled chiller (if it is necessary); 15 – high-temperature plasma torch; 16 – plasma flow; 4 –liquid saturated with hydrogen or other active gas; 10 – gasar coating; 26 – device for supplying hydrogen in form of a gas, liquid (for example, water) or solid particles (for example, hydrides) to the arc burning zone

Figure 12. Macrostructure of porous coatings having solid skin: upper – Ni-Fe bronze, pore diameter about 0.5 mm; bottom – stainless steel

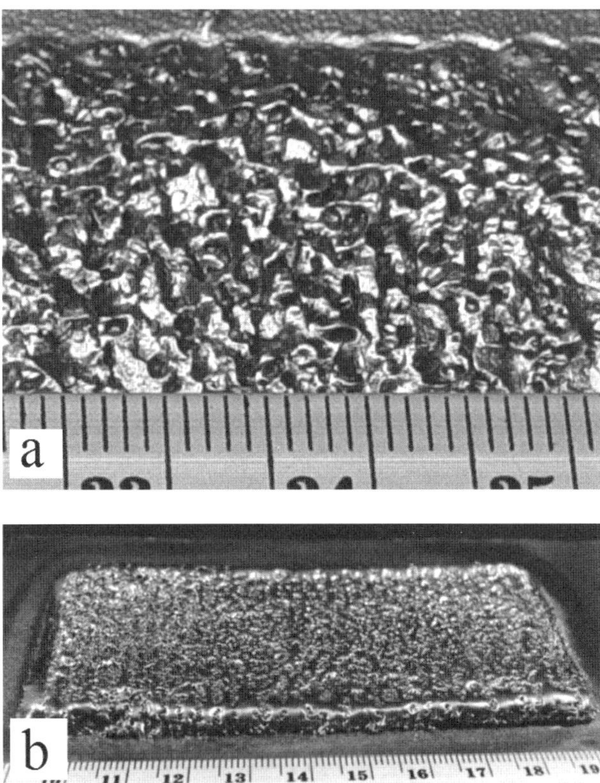

Figure 13. Macrostructure of porous coatings having pores open to surface: upper – stainless steel; bottom – titanium

In this case, base material is supplied in high temperature plasma zone (powder, wire, strip, and so on) which permits building the gasar structure layer by layer in any shape and any size [29]. The active gas (water, hydrides, hydrocarbons, oxides) is supplied in arc zone which decomposes at high temperatures producing the active gases. For the plasma scanning system (PSS) gasar method it is not necessary to utilize high gas pressure because the active gas is ionized which is dissolved in the liquid pool under the plasma torch. A fabrication chamber can be filled with an inert gas or any active gas at standard pressure. Another advantage of this PSS method is the capability of producing gasars of practically any size and shape with a uniform structure throughout the entire volume. In order to produce gasars using this method, no casting mould is required, since the product itself is formed in a layer-by-layer manner, while the movement of plasma torch is controlled by a computer (Figs. 14–17). The PSS method can produce gasars based on any metal or alloy includes titanium, tungsten, and molybdenum, etc. As can be seen in Fig. 15, another advantage of this method is relatively low cost of PSS and producing gasars.

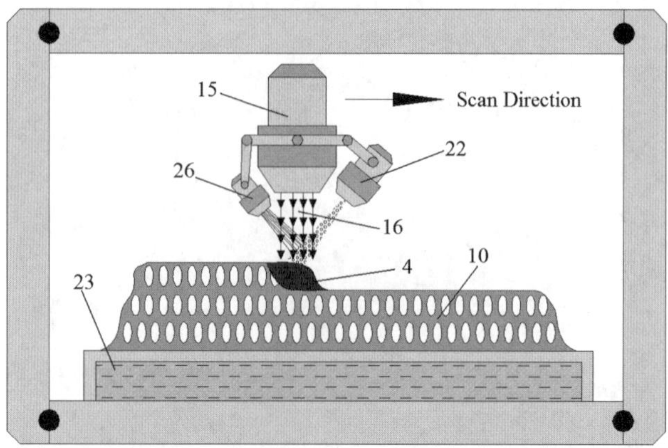

Figure 14. General schematic of producing gasars by high-temperature plasma scanning system (PSS): 23 – water-cooled chiller or graphite plate or ceramic plate or any metal substrate; 15 – high-temperature plasma torch; 16 – plasma flow; 4 – liquid pool saturated with hydrogen or other active gas; 10 – gasar article; 26 – device for supplying hydrogen in form of a gas, liquid (for example, water) or solid particles (for example, hydrides) to the arc burning zone; 22 – device for supplying base material in the form of a powder, wire or tape to the arc burning zone

Figure 15. General view of a plasma scanning system (PSS) device: 1 – computer control; 2 – wire feeder; 3 – low leaking work chamber (plasma torch inside); 4 – powder feeder

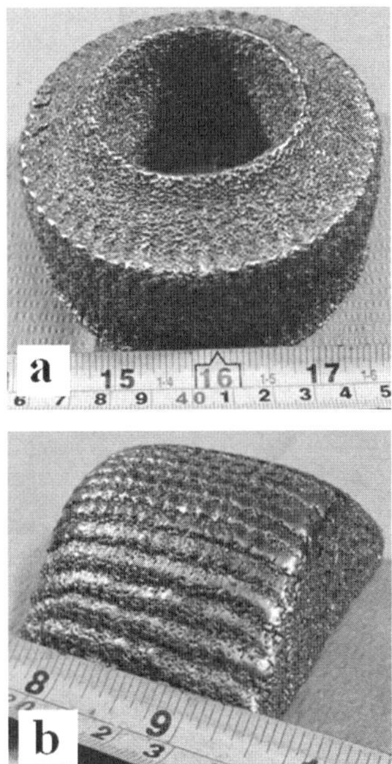

Figure 16. Net shape titanium gasar articles containing open porosity

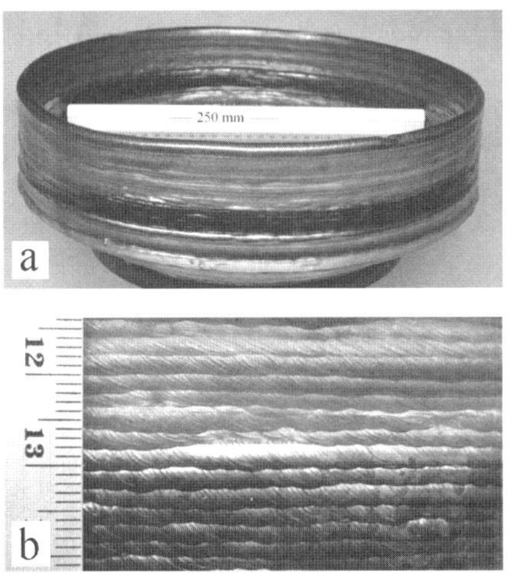

Figure 17. Helicopter titanium gasar part (a) made by PSS having closed porosity. The part wall at higher magnification for seeing layer structure is at bottom (b)

4. Technological stages

4.1. MELTING AND SATURATION WITH HYDROGEN OR OTHER ACTIVE GAS

Melting of a metal and its saturation with hydrogen usually takes place in the same melting crucible, although it is also possible to first melt the metal in a dedicated melting device at a normal pressure and then transfer the melt into a gasar apparatus for saturation with hydrogen and subsequent solidification. Different heating methods can be used to facilitate the melting process: resistance furnace [22]; induction furnace [30]; electric arc furnace [26]. The most desirable is the induction furnace, since it heats just the metal (not the crucible) and provides less contamination to the liquid metal with non-metallic inclusions and impurities.

In the case of small volumes of metal, saturation with hydrogen to an equilibrium state does not pose a serious problem, since the diffusion rate in a metal liquid is high. However, for a large amount of metal, achieving a gas-metal equilibrium may turn out to be a lengthy process. Besides, quite often an oxide layer is formed on the surface of the melt, thus retarding or even preventing hydrogen from penetrating into the metal. Therefore, for the commercial scale, it is necessary to first establish the kinetics of metal saturation with hydrogen and define the conditions to facilitate rapid saturation. A number of technical solutions have been proposed to accelerate the saturation of large amounts of melts with hydrogen [31] which provides a basis in gasar technology utilization.

Increased gas saturation in a liquid metal can be achieved with increased pressure. Therefore, an apparatus for producing gasars has to be not only hermetically sealed, but also be capable of withstanding pressures up to 50 atmospheres at very high temperature inside the furnace. This significantly complicates the entire technology and makes it expensive with low productivity. Attempts to solve this problem by using preliminary saturation with hydrogen or introducing hydrides (or other hydrogen-containing compounds) into the melt prior to crystallization [32] were not effective due to two reasons. First of all, no gas-metal equilibrium can be reached in this case, which is critical for a gas-eutectic reaction to occur. Secondly, in order to control porosity and pore sizes, high pressure is required anyway practically for all metals.

In this respect, PSS gasar production [29] is very promising because gas ionization sharply increases saturation speed and gas concentration in the small liquid pool up to the gas-eutectic composition. The PSS can build virtually any size gasar component.

4.2. POURING AND SOLIDIFICATION

When the melt is simply poured from the crucible into a crystallizer, liquid motion inside the crystallizer is very intensive and disorderly turbulent. Besides, this process is accompanied by a mass entrainment of the atmospheric gas inside the furnace into the melt. This negatively affects the gasar structure making it disordered on a macro- and micro-scale.

In order to produce high-quality structures, a dedicated pouring unit is required, which would provide a laminar and symmetric movement of the melt inside the crystallizer in the beginning and throughout the process of crystallization. There is no simple solution to this problem. Siphon (the most quiet) pouring method, when the metal enters the crystallizer from the bottom, is not applicable to gasars since the flow of liquid metal destroys gasar structure that is being formed.

Solidification of gasars is the final and most complicated step, which determines the final structure and properties of the material. As a result of the gas-eutectic reaction, two phases of a different physical nature are formed, which are governed by different laws. The solid phase is incompressible and the effect of the surrounding medium on it is very weak even under very high pressures. The gas phase, on the other hand, is a volatile material without shape, and even a slight change in pressure or flow rate of the liquid in contact with a growing bubble can very easily affect the size and shape of the pores. Ideally, a gas-eutectic reaction should be conducted in a completely steady liquid in absence of impurities and gravitational field. This is theoretically possible, but can not be achieved in practice and one can only try to minimize the effect of these factors. To overcome this, MER has developed several methods for influencing solidifying gasars. Each of these methods demonstrated positive results, however, no universal method was found, which would allow controlling pore size and shape throughout the gasar ingot with 100% reliability. Most likely, such method just does not exist.

However, a number of other factors were unexpectedly discovered in the process of these experiments, which significantly affects the structure of gasars. These factors enabled development of a series of methods for producing gasars with pores characterized by a vast variety of shapes, sizes, orientation and different distribution throughout the ingot volume [32].

The gasar structure is very sensitive to the rate and direction of heat removal from solidifying ingot. Typically, as the cooling rate goes up, porosity increases and the average pore size decreases. The direction of heat removal determines the pore orientation in gasars. Therefore, it is very important to reliably control these parameters.

All gasar crystallizers can be divided into two categories: water-cooled and massive (not water-cooled). Massive crystallizers are much simpler and less expensive (for example, copper or graphite molds and pipes), however gasar structures in them is characterized by significant non-uniformity in the direction of crystallization. Very strong pore coagulation is also observed, since the cooling rate decreases rapidly after a few seconds from the beginning of crystallization and continues to drop until it ends. Therefore, massive crystallizers can be recommended only for small gasar articles with the cross-section thickness of 40 mm or less.

Water-cooled crystallizers are free of such disadvantage. However, in the first such crystallizers it was difficult to control uniformity of surface heat removal from one point to another. After numerous experiments, a satisfactory solution to this problem was found [33] which translated into multi-channel design of the water-cooled surface of the chiller.

5. Gasar structure

5.1. GENERAL CONFIGURATION OF GASAR INGOT

As opposed to known materials, crystallization of gasars is accompanied by a two to three times increase in volume. Therefore, the shape and configuration of gasar ingots are not trivial. The typical gasar ingot is covered with a monolithic crust and does not have a traditional shrink cavity. Moreover, it has a residual shape in the upper part, formed by a liquid squeezed out in the process of gasar formation (Fig. 18). L.V. Boyko [34] successfully named it "a turban". A turban is formed as a result of periodic squeezing and subsequent solidification of the liquid metal on the ingot surface. Inside of the turban there are gas cavities, the number of which indicates how many times liquid metal was squeezed from underneath the hard crust onto the surface of the solidifying ingot.

Inside the ingot, different pore distributions can be seen. In the best case, porosity throughout the entire volume of the ingot is uniform (Fig. 19(a)). However, deviations are quite frequent. Monolithic areas or large cavities may be formed inside the ingot (Fig. 19(b, c)).

5.2. PORE STRUCTURE

The gasar structure is characterized by its diversity and may very well be compared with the structures of multi-phase alloys. This is somewhat amazing, since it is not expected to see a vast variety in the "gas bubble – liquid" or "gas bubble – crystal" systems. However, it turned out to be true.

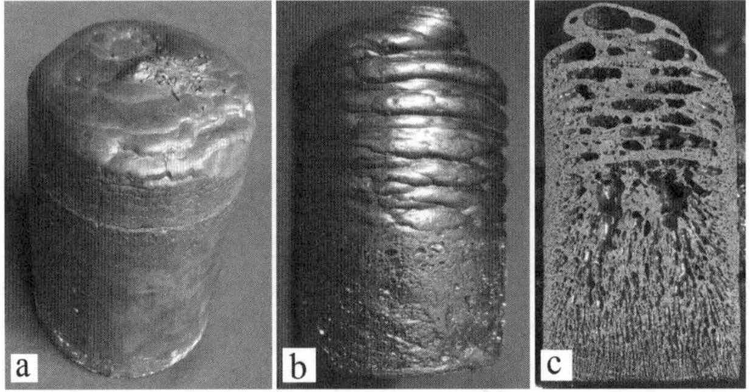

Figure 18. Examples of gasar ingot appearance (a, b) and longitudinal section of an ingot with a turban (c), ingot diameter ~100 mm

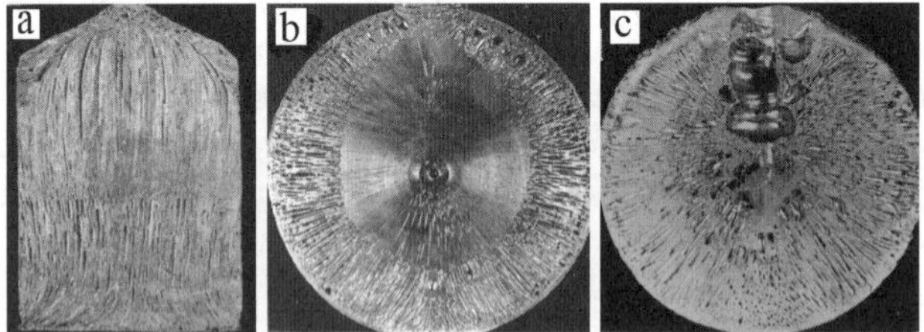

Figure 19. Basic types of gasar ingot macrostructure: (a) uniform porosity throughout the entire volume (cylindrical ingot, diameter ~100 mm); (b) monolithic area in the central part of the ingot (spherical ingot, diameter ~200 mm); (c) shapeless cavity in the central part of the ingot (spherical ingot, diameter ~200 mm)

The classical type of gasar structure, formed as a result of directional crystallization, contains ellipsoidal pores oriented in the direction of crystallization (Fig. 20(a)). The length-to-diameter ratio can be varied according to the above-mentioned variables. This would allow obtaining spheroidal pores (Fig. 20(b)) or practically cylindrical pores (Fig. 20(c)).

Typically, the pore shape in the cross-section is perfectly round. However, if the concentration of hydrogen deviates from the gas-eutectic point to the negative side, formation of pores along the boundaries of metallic dendrites is observed resembling typical features of their appearance (Fig. 21(a)). If the hydrogen pressure deviates considerably from the gas-eutectic point to the positive side, formation of coagulating pore structure is observed (Fig. 21(b)) resembling a foam structure. Structures of this type can also be formed in the case of a pressure decrease during crystallization. In case of intermediate conditions, oval shape pores are formed (Fig. 21(c)).

Variation in pressure during solidification usually causes formation of cone-shaped convergent or divergent pores. However, under different conditions (e.g., in case of a different hydrogen concentration in the melt) variation in pressure causes pore growth to stop, whether the pressure is increasing or decreasing. As a result, structures with alternating porous and monolithic layers can be obtained (Fig. 22).

In multi-component alloys, formation of faceted pores or pores of various shapes within one macro-volume can be seen (Fig. 23).

The pore surface in gasars is always free of foreign particles and is usually mirror-smooth (Fig. 24(a)). However, it is also possible to create a rough surface when faceted crystals protrude through the pore surface (Fig. 24(b)), or micro-rough (Fig. 24(c)).

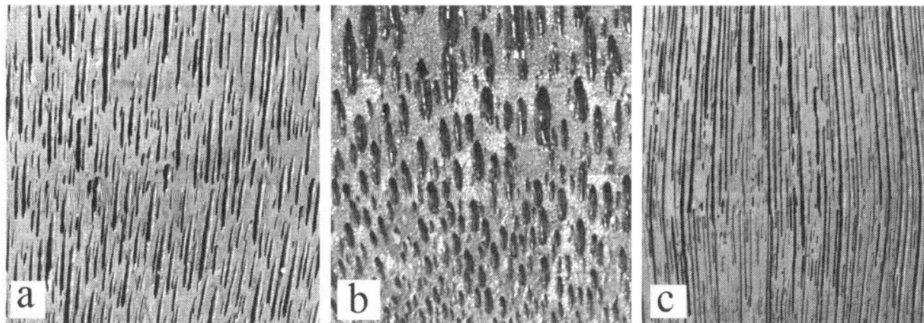

Figure 20. Ellipsoidal (a), spheroidal (b) and cylindrical (c) pores in Mg-based gasars (pore diameter ~1 mm)

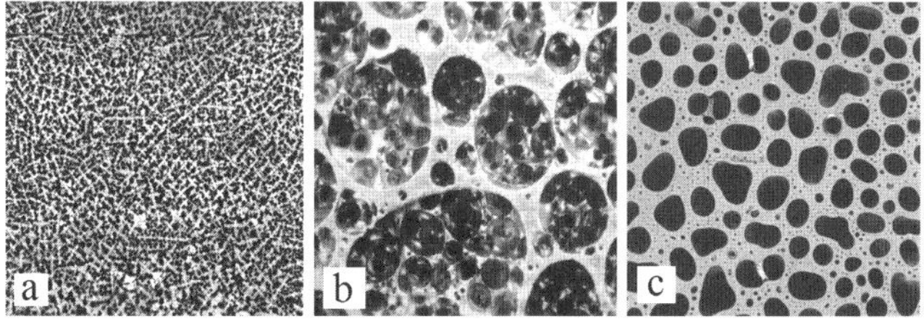

Figure 21. Dendrite porosity (a), quasi-foam structure (b), oval pore shape (c)

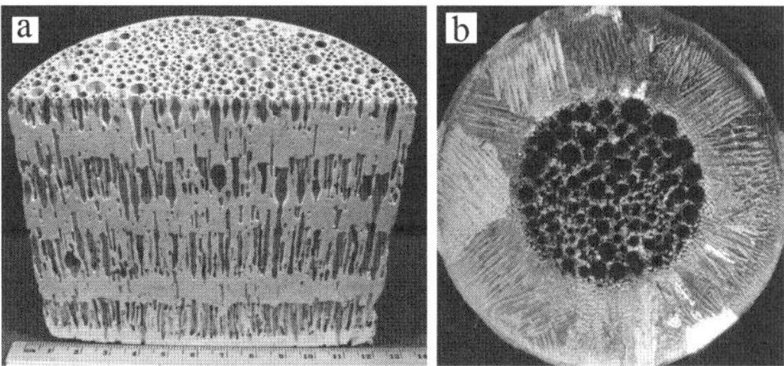

Figure 22. Examples of porous-monolithic structures in Mg-based (a) and Ni-based (b) gasars

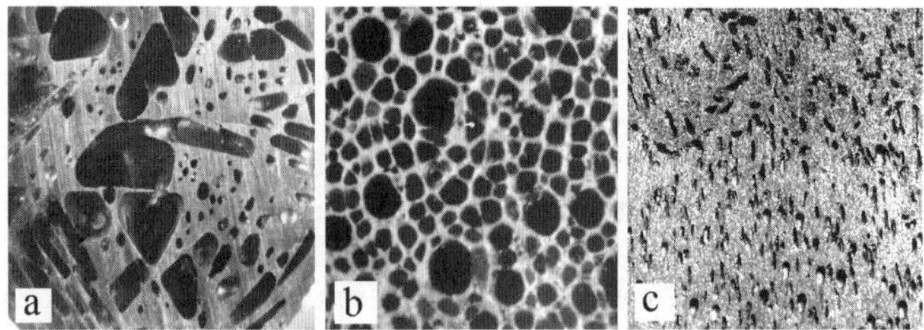

Figure 23. Faceted pores in gasars: (a) diamond faceting; (b) tetrahedral faceting; (c) combination of various pores shapes within one volume

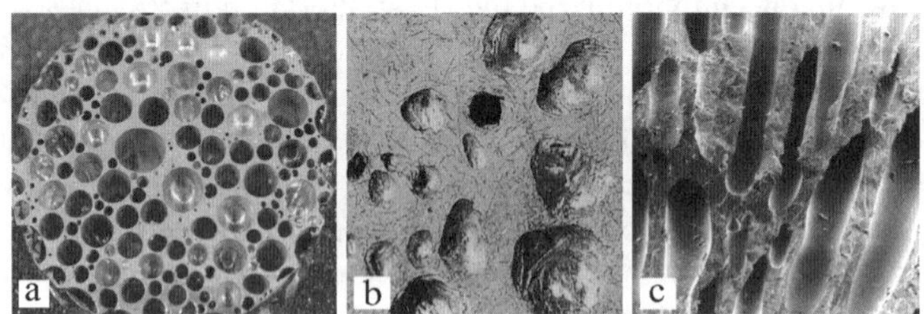

Figure 24. Pore surface in gasars: (a) mirror-smooth; (b) macro-rough; (c) micro-rough (mat)

Under certain conditions, formation of a rippled surface takes place (Fig. 25) with various general pore geometries. Ripples are caused by fluctuations in pressure of the liquid as well as hydrogen concentration at the crystallization front resulting in size variation of the growing bubble. The pore diameter changes accordingly. In principle, this process can be controlled allowing producing shapes with desired ripple effects.

Pore sizes in gasars can be varied in a wide range from ~7 μm to ~100 mm (Fig. 26). This allows producing gasars in the form of a pipe with a regular shape without special equipment required to produce pipe molding [35].

Plastic deformation of formed gasars using a variety of methods results in material densification which is accompanied by considerable change in structure and properties. After compacting deformation (cold or hot), a trace in the form of a zigzag line is left in place of each pore (Fig. 27). This trace is not exactly a grain boundary. This is a very stable formation, which serves as a barrier blocking any movement of vacancies and dislocations. As a result, the strength of material after deformation (especially cold) increases significantly and by far exceeds the strength of a monolithic material of the same chemical composition. For example, the yield strength of dense cast copper is 60–70 MPa. The ultimate strength of a Cu-based gasar with porosity of ~20% and pore diameter of ~30–50 μm is about 90 MPa. After cold compacting, the strength of the Cu-gasar is 300–350 MPa. This is explained by accumulation of dislocations as a result of plastic deformation as well as appearance of new formations – ex-pores, which pose a practically insuperable obstacle for dislocations to cross. The structure of ex-pores has not yet been studied using high-resolution electron microscopy.

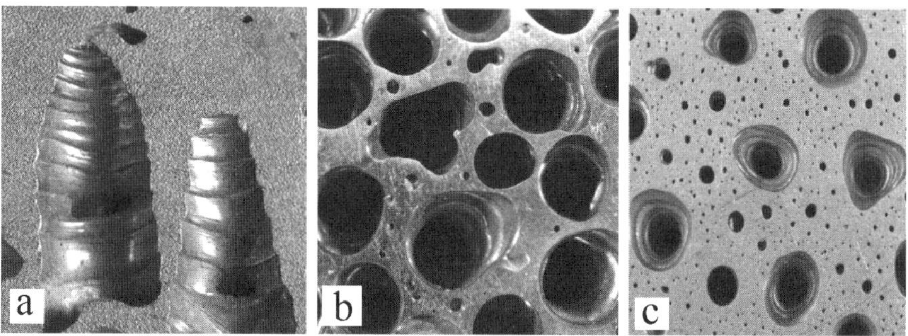

Figure 25. Ripple effect on the surface of converging (a), cylindrical (b), and diverging (c) pores (pore diameter ~3 mm)

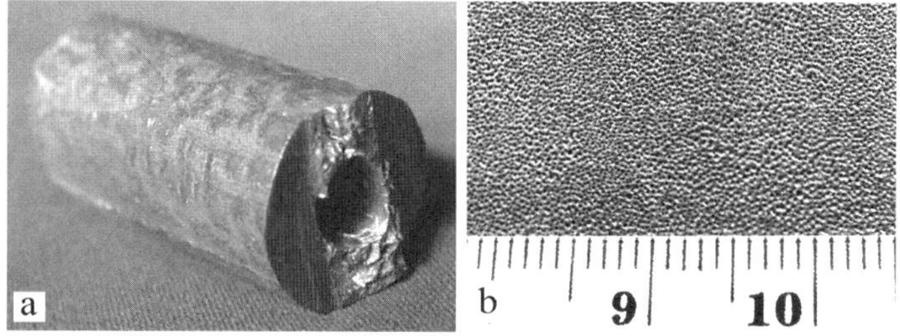

Figure 26. Pore size variation in gasars: (a) single pore in cylindrical molding (pore diameter ~53 mm); (b) gasar with pore diameter ~25 μm

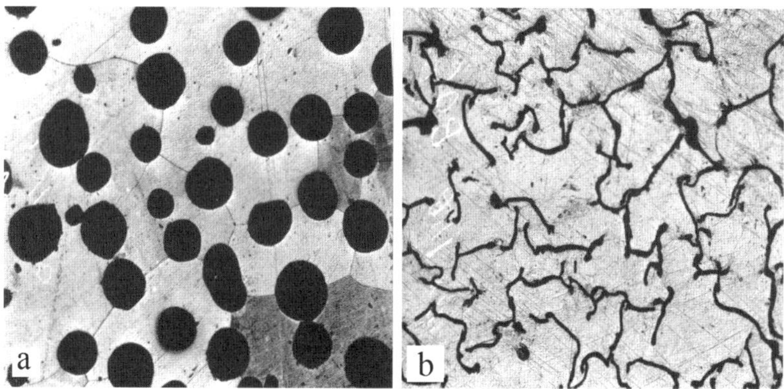

Figure 27. Left – structure of a Cu-based gasar before deformation (porosity ~35%, average pore diameter ~50 μm); right – same gasar after cold plastic deformation by drawing (porosity ~1%)

Ceramic gasars are possible with almost the same structures as metal gasars (Fig. 28(a)). However, in some case a lineage structure (Fig. 28(b)) was observed which had not been seen in metal gasars. At this time, there is no explanation for this structure geometry yet.

It is worth noting in Japan the gasar structure is being referred to as "lotus structure" which deviates from the original name proffered in 1986 [38] and the San Francisco Conference on "Porous Materials" in 1998 where there was a section devoted to "Gasar Technology" [36].

5.3. STRENGTH OF GASARS

Since the gasar structure is formed from a molten phase, it exhibits much higher strengths that porous structures formed by other techniques and especially sintering of powders. Figure 29 compares the strength of a steel gasar with the sintered structure. A comparison of the strengths of titanium formed using PSS technology is shown in Table 1.

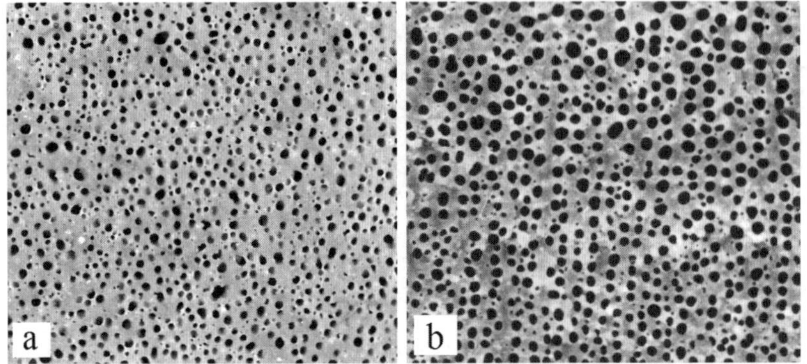

Figure 28. Typical structure of ceramic gasars (a) and lineage structure (b) (pore diameter about 1 mm)

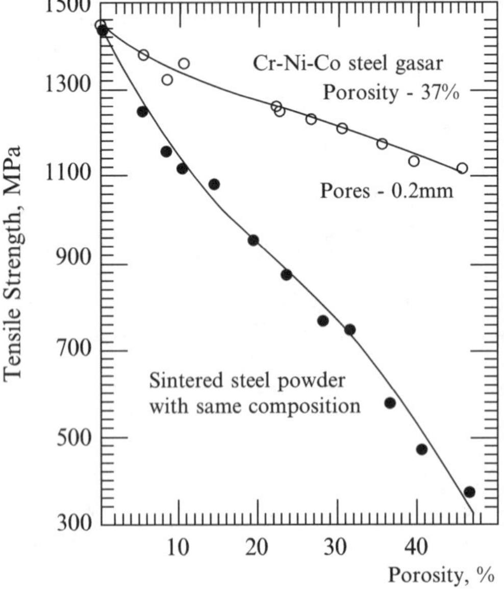

Figure 29. Strength of gasar produced steel compared to sintered steel

TABLE 1. Compressive strength of Ti-6-4 foams

Foam process	Density (%)	Initial diameter (in.)	Initial length (in.)	Final Diameter (in.)	Final length (in.)	Load max (lbf)	Stress (Ksi)
Sinter	39	0.4	0.4	0.43	0.37	6,300	43.5
PTA	43	0.5015	0.5535	0.5455	0.4480	19,001	96.2
PTA	43	0.5010	0.7350	0.5530	0.5340	19,050	96.6

6. Conclusions

Because of the diverse porous structures obtainable via the gasar process and the exceptional attractive properties that can be achieved in virtually any metal/alloy and some ceramics, gasars are expected to attract many diverse applications [37–39]. The high gas pressure requirement to produce bulk gasars as well as a lack of detailed hydrogen metal eutectic fundamental data has contributed to limited commercial utilization of the gasar technology. Also because hydrogen is the primary gas to produce the eutectic, hydride forming metals cannot readily be produced in a gasar structure. None-the-less, gasar produced porous metals offer attractive properties for commercial applications such as steel can be produced in the same density of titanium, and titanium in the same density as aluminum, with very attractive strength properties. The PSS technology to produce the metal foam including the hydride sensitive metals in a layer by layer processing that permits producing near net shapes, provides an approach to translate the technology into commercial applications.

Acknowledgements

Our thanks to Ukrainian "Splav" Laboratory Research Assistant: N. Serduk, L. Boyko, A. Chuprina, V. Trofimenko, L. Poltoratzkuy, N. Eremenko, A. Timchenko and many others.

Big thanks to Sandia National Laboratories colleagues (1995–2000) M. Maguire, B. Hammetter, F. Zanner, R. Decker for starting developing gasar technology in USA.

References

1. Shapovalov V., Serdyuk N. About Copper-Hydrogen Phase Diagram. Izvestiya Vuzov. Tsvetnaya Metallurgiya, 1980, (2): 90–93.
2. Shapovalov V.I., Serdyuk N. Method of Metal Foam Manufacturing. USSR Patent 1725485. March 11, 1980.
3. Shapovalov V.I., Eremenko N.D. The Structure and Properties of Composite Porous Materials with Monolithic Framework for Slide Bearing Units. 1989. Development and Application of Advanced ceramics and P/M Materials, International New business and High-Tech Research Conference, September 3–5, 1989, Jyvaskyla, Finland, European Advanced Material Institute (EAMI), P. 1–12.
4. Shapovalov V. Method for Manufacturing Porous Articles. U.S. Patent 5,181549. April 4, 1991.
5. Shapovalov V., Bichuya A., Eryomenko N. Exhaust Gas Catalyst for Combustion Engines and Process for Making Same. German Patent DE4134337. October 17, 1991.
6. Shapovalov V. Structure Formation Behavior of Alloys During Gas-Eutectic Transformation and Prospects of the Use of Hydrogen in Alloying. Microstructure Design by Solidification Processing. Proceedings of a symposium Materials Week, Chicago, Illinois, USA, November 1–5, 1992, P. 207–216.
7. Shapovalov V. Porous Metals. 1994. MRS Bulletin, April 4, P. 24–29.

8. Boyko L., Shapovalov V. Anisotropic Porous Metals Production by Melt Processing. 1997. Proceedings of the 1997 International Symposium on Liquid Metal Processing and Casting, Santa Fe, New Mexico, February 16–19, 1997, P. 417–426 (USA).
9. Shapovalov V. Formation of Ordered Gas-Solid Structures via Solidification in Metal-Hydrogen Systems. 1998. Porous and Cellular Materials for Structural Applications. Symposium held April 13–15, 1998, San Francisco, CA. MRS Symposium Proceedings, Vol. 521, P. 281–290.
10. Shapovalov V., Apprill J., Baldwin M., Maguire M., Miszkiel M. Production of Gas-Solid Structures in Aluminum and Nickel Alloys by Gasar Processing. 1999. Proceedings of the 1999 International Symposium on Liquid Metal Processing and Casting, Santa Fe, New Mexico, February 21–24, 1999, P. 322–329.
11. Shapovalov V.I. Forming of Ordered Gas-Solid Structures via Solidification in Metal-Hydrogen Systems. Materials Research Society, Vol. 521, 1998, P. 281–290.
12. Shapovalov V., Eremenko N.D. Method for Manufacturing Foamed Steel and Cast Iron. 1981. USSR Patent 3028859. Filed: November 2, 1981.
13. Shapovalov V., Dolzenkov V. Method for Gasar Manufacturing. 1985. USSR Patent 3109117. Filed: March 5, 1985.
14. Shapovalov V., Bulaniy P., Zuev V., Yakubovich P. Method for Gasar Manufacturing. 1987. USSR Patent 3166577. Filed: March 3, 1987.
15. Shapovalov V., Eremenko N.D. Method for Gasar Manufacturing. 1987. USSR Patent 3165964. Filed: March 31, 1987.
16. Shapovalov V., Karpov V. Method for Gasar Manufacturing. 1987. USSR Patent 3164289. Filed: January 30, 1987.
17. Bunin K., Shapovalov V., Trofimenko V. Investigation of Fe-H System at High Temperatures and Pressures. 1972. VICHP: 3 Molecular Interactions and Thermodynamics, P. 639–640.
18. Shapovalov V. Metal-Hydrogen Phase Diagrams in the Vicinity of Melting Temperatures. 1999. Proceedings of the 1999 International Symposium on Liquid Metal Processing and Casting, Santa Fe, New Mexico, February 21–24, 1999, P. 330–343.
19. Shapovalov V., Poltoratzkiy L. Apparatus for Determination Amount of Gases in Metals. 1979. USSR Patent 2809097. Filed: August 14, 1979.
20. Boyko L., Shapovalov V. Apparatus for Gasar Manufacturing. 1984. USSR Patent 3084013. Filed: January 3, 1984.
21. Shapovalov V., Serduk N., Trofimenko V., Chuprina A. Method for Gasar Manufacturing. 1981. USSR Patent 3032290. Filed: December 28, 1981.
22. Shapovalov V., Serduk N., Trofimenko V., Chuprina A., Semik A. Apparatus for Gasar Manufacturing. 1985. USSR Patent 3106071. Filed: January 18, 1985.
23. Boyko L., Shapovalov V., Shmatovskiy A. Apparatus for Gasar Manufacturing. 1988. USSR Patent. Filed: June 8, 1988.
24. Shapovalov V., Semik A., Timchenko A. Apparatus for Gasar Manufacturing. 1987. USSR Patent 3161044. Filed: January 19, 1987.
25. Serduk N., Shapovalov V., Titkov, A. Onufrienko V. Method for Gasar Filters Manufacturing. 1986. USSR Patent 3136536. Filed: February 12, 1986.
26. Pereloma V., Shapovalov V., Lenda Y. Naidek V. A method for the production of porous cast products. 1996. World Patent WO9811264. Filed: September 16, 1996.
27. Boyko L., Shapovalov V. Method for Light Ceramics Manufacturing. 1985. USSR Patent 3131008. Filed: April 4, 1985.
28. Shapovalov V., Bulaniy P., Yakubovich P., Zyev V. Method for Making Porous Layers on Solid Material Surfaces.1989. USSR Patent 4515188. Filed: May 29, 1989.
29. Shapovalov V., Withers J. Method and Apparatus for Producing Porous Articles. 2007. Patent Application Serial No. 60/956,374. Filed: August 16, 2007.
30. Shapovalov V., Krukov V., Tishin O. Method for Gasar Manufacturing. USSR Patent 3170492. May 13, 1987.

31. Shapovalov V., Serduk N., Boyko L. Method for Gasar Manufacturing. USSR Patent 3092029. June 26, 1984.
32. Shapovalov V. Method for Gasar Manufacturing. USSR Patent 3081685. January 10, 1984.
33. Shapovalov V., Chuprina A. Crystallizer for Direct Gasar Solidification. USSR Patent 4503933. December 13, 1988.
34. Boyko L., Shapovalov V. Gasar – A New Class of Porous Materials: Syntheses, Structure, Properties and Prospective Application. 2003. Cellular Metals: Manufacturing, Properties, Applications; Edited by J. Banhart, N. Flak, A. Mortensen; MetFoam 2003, Berlin, P. 203–209.
35. Danchenko V., Chus A., Semik A., Shapovalov V. Method for Continuous Tube Casting. USSR Patent 3067533. June 8, 1983.
36. Porous and Cellular Materials for Structural Applications. Symposium held April 13–15, 1998, San Francisco, CA. MRS Symposium Proceedings, Vol. 521.
37. Boyko L. Shapovalov V., Advantages of Gasar-Materials for Brake Shoes and Plates. 2002. 20th Annual Brake Colloquium & Exhibition, Phoenix, Arizona, October 6–9, 2002, Paper #72.
38. Shapovalov V.I. Prospective Applications of Gas-Eutectic Porous Materials (Gasars) in USA. 2007. Materials Science Forum, Vols. 539–543, P. 1183–1187, online at http://www.scientific.net, © 2007 Trans Tech Publications, Switzerland.
39. Shapovalov V., Loutfy R. Method and Apparatus for Storing Compressed Gas. US Patent 6520219. August 31, 2001.

THE FORMING PECULIARITIES OF C_{60} MOLECULE

D.V. SCHUR, S.Yu. ZAGINAICHENKO, E.A. LYSENKO,
T.N. GOLOVCHENKO
Institute for Problems of Materials Science of NAS of Ukraine,
3 Krzhizhanovsky Str., Kiev, 03142 Ukraine
Fax: 38-044-424-0381; E-mail: shurzag@materials.kiev.ua
N.F. JAVADOV
Experimental – Industrial Factory of Institute of Petrochemical
Processes after Yu.H. Mammadaliyev of NAS of Azerbaijan

Abstract. The structure, transformation and behaviour of C_{60} fullerene molecule is not sufficiently understood by the scientific community and was unresolved problem until the present day. This not provide a possibility of correct interpretation of its physical and chemical properties. In this work we have substantiated the advisability of consideration of pentatomic cycles in the C_{60} fullerene, as pentatomic molecules forming this frame. The mechanism of formation of the C_{60} fullerene frame from pentatomic molecules has been proposed. The levels of matter formation with description of fullerene production mechanism have been examined.

Keywords: C_{60} fullerene, pentatomic molecule, levels of matter formation, frame of C_{60} molecule

1. Introduction

The existing models of fullerene molecule formation are based on assumption of the key role of graphene sheets or their constituent elements [1–9]. With these lines of attack the researchers examine the possibility of folding of graphene sheets growing at the expense of attachment of C_2 molecules. Richard E. Smalley proposed the "pentagon rule" [4], which in fact is the variant of early model of graphene sheets folding. The distinctive feature of this model is the pentagons rise in necessary positions on annealing in order to close the system, to form the C_{60} molecule by natural way, to make the redistribution, to minimize defects, and so on. The stack of rings is one more model [10] and according to present views the acceptable number of rings of different size go into stack with formation of C_{60}.

Starting from the assumption of hexatomic origin of fullerene molecules, the authors of papers [11, 12] showed with the help of mass spectrometry that C_{60} and C_{70} molecules (probably fullerenes) can be produces in the gaseous phase from the cyclic carbon molecules of C_{18}, C_{24} and C_{30} in the ion-molecule reaction. In this area of study at the systems condensation from large rings the way of formation process is taken to be cationic. The research group, engaged also in such studies [2], noted by the use of ionic gaseous chromatograph the following progression: linear element → monocycle → bicycle, and so on. It was suggested that the small fullerenes C_{30} are formed at the carbon atoms concentration in the cluster from rings containing more than 30 atoms. Then, the C_{60} molecules grow from C_{30} due to the attachment of small fragments.

As experimental investigations of all authors of performed works have shown in the course of discussion of scientific results, the lack of graphite building units of small size, i.e. hexagons or hexagonal bi- and tricycles, cause the embarrassment.

In this work we have demonstrated the possibility of production of fullerene molecule frame from pentatomic cyclic molecules.

2. Pentatomic cyclic molecules of carbon in the cluster of C_{60}

The energetic non-equivalence of carbon atoms surroundings in aromatic hydrocarbons (in distinction to graphene) is understandable and caused mainly by the different electronic load of added elements and compounds to each carbon atom. So, if in the benzol of ring-like structure it is reasonable to consider the activity of each atom, then it is difficult to identify the individuality of each carbon atom at the construction of graphene layer in connection with featureless property of hexagonal rings and their unification in one big flat molecule.

The presence of strained and conjugated double bonds in carbon nanotubes (CNT) does not make them more chemically active in comparison with graphene. Even the availability of half-fullerene caps at the ends of closed CNT does not provide the possibility to them to dissolve in the organic solvents. The reason is that the percent content of pentatomic rings in the frame of closed CNT accounts for the millionth fractions of percent. So, their contribution is not felt especially.

With the advent of pentatomic rings in graphene the energetic equivalence of carbon atoms disappears. As a matter of fact the pentatomic ring is a defect in the graphene structure. This defect strains energetically the surrounding carbon atoms, this is manifested in the form of convexity or convacity of graphene sheet.

The structural distinction of fullerene molecule from the graphite-like formations is the presence of pentatomic cycles in its frame. This provides evidently an explanation for its high chemical activity and solubility.

Since the activity is conditioned by the presence of pentatomic cycle, it is necessary to examine of peculiarities of influence of this component on the overall frame of fullerene molecule for the comprehension of this structural nature as a whole.

The exerted attempts at consideration of fullerene frame by analysis of the state of its graphite-like component (hexagons) were too cumbersome and have not been successful. The examination of action of broken double bond on energetic state of carbon hexagons (the breaking has been occurred at their boundaries) has been hampered by the fact that both carbon atoms belong simultaneously to two neighbouring hexagons and the account of influence of each carbon atom on the state of both hexahedrons makes the complicated problem. In addition, they are not individual hexagonal structures, but present the completed circuit of hexagons, in which the adjacent cycles have the common sides (in distinction to pentagons). This makes them the single band chain of graphene hexagons or molecule, or simply the circuit of hollows, filled by electronic gas. The pentatomic rings constitute the isolated structures standing separately. They respond easily to description.

Thus, on examination of the C_{60} fullerene frame it is more advisable to speak about the frame formation from pentatomic molecules, resulting in forming of hexatomic emptinesses. These emptinesses form round each pentatomic molecule the intermolecular cavity, filled by electronic gas.

3. The composition of carbon plasma

If pentagon has been formed in the net of graphene sheet instead of hexagon, so the graphene will be transformed into the cone, being the centre of nanofunnel or nanoburner. The presence of one pentagon in the graphene net corresponds to the cone with an angle of 112.9°, two pentagons – to an angle of 83.6°, three pentagons – to an angle of 60.0°, four pentagons – to an angle of 38.9°, five pentagons – to an angle 19.2°. At the introduction of six pentagons the graphene roll up into tube or hemisphere. By fulfilling the so-called "rule of isolated pentagons" the more tolerant structures come into existence.

In such a manner due to the structural reconstruction the "insoluble" form of carbon can transform into soluble (spherical modification), littlesoluble (at low ratio of CNT length to its diameter of 3 ÷ 10) and conditionally soluble modification of carbon (when these values for CNT are in ratio of >10).

The polygons with odd amount of carbon atoms in the graphene sheet can be produced by the removal or introduction of carbon atoms into hexagonal carbon molecules. Such way is possible already at the formed graphene sheet. The more simple variant of formation of carbon structures, containing polygons with the odd number of carbon atoms, can be the path-way of their formation at the stage of nucleation of future modification, i.e. in the carbon plasma.

The experiments have demonstrated that by graphite evaporation in electric arc the arc temperature in the reaction zone may be as much as 1.2×10^4 K, that is responsible for the full atomization of carbon compounds and generation of excited carbon atoms (Fig. 1) [13].

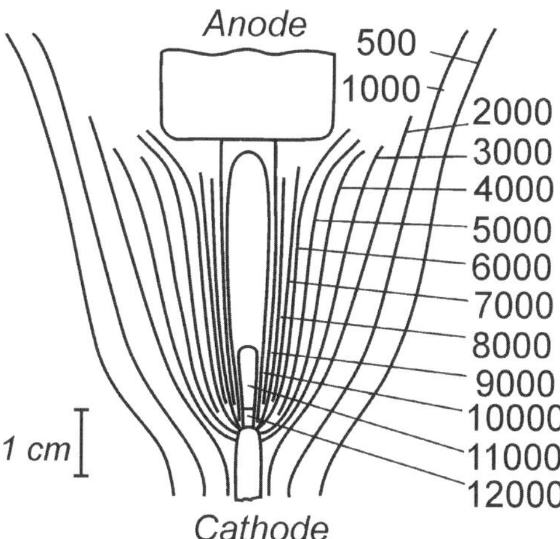

Figure 1. Temperature distribution (in K) in different regions of the electric arc between the carbon electrodes at strength of current equal to 200 A

The mass-spectrometric analysis of carbon vapour in plasma conditions is illustrated in Fig. 2 [1]. The authors have noted two "magic numbers" (as they named them) 60 and 70 in this spectrum.

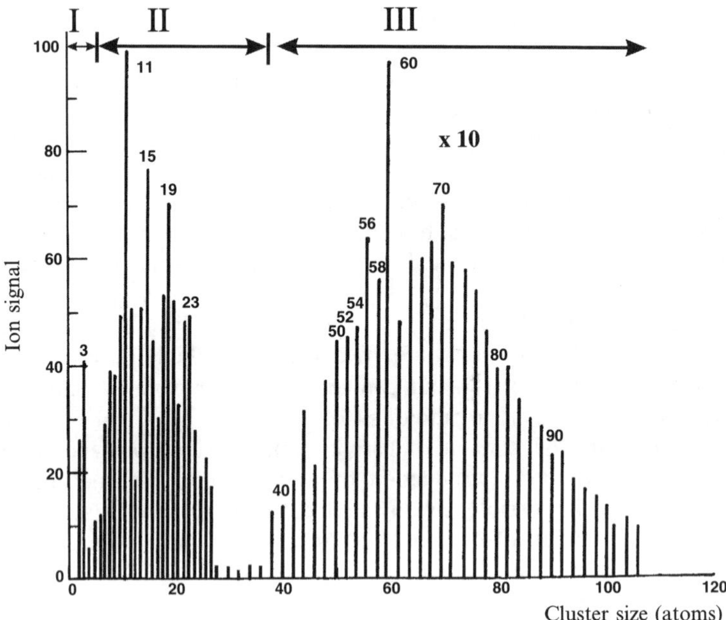

Figure 2. The mass spectrum for carbon clusters obtained with doubled Nd: YAG vaporizing laser energy and with unfocused ArF ionizing laser energy [1].

This spectrum is a combination of different spectra:

I – Atomic and molecular level

II – Cyclic and chain-like structures

III – Spheroidal clusters and molecules

However, in our opinion, the existence of three groups of particles in this spectrum is no less magic. Their presence reflects the three-stage reactions of formation of spherical clusters or molecules. We have numbered them as I, II, III. These stages correspond to the three basic groups of structural formations, formed at the atomic-molecular level during the construction of fullerene frame (I – for molecules, II – for cycles and III – for spheres).

In the initial stage I of plasma-chemical interactions (Fig. 2, I) the carbon atoms form the di- and triatomic molecules, that in combination with atoms participate in cycloformation at the second stage (Fig. 2, II). The interval of molecule masses, appearing on this stage, is considerably wider and includes in itself the structures with masses from 3 to 30 and 40 atoms (Fig. 3). As this takes place, the structures, containing from three to six carbon atoms, are intermediate or transitional between I and II stages, since they can have both the chain and cyclic structure of molecules.

The spheric molecules start to shape only at the III stage. They can contain 30 atoms and more. But only molecules with atoms in amounts of 50 and over are in fact stable spherical molecules thermodynamically.

In spite of the fact that spectrum of Fig. 2 is made of two different spectrums, it is undifficult to notice that at the I stage–molecular dimensions reach a maximum of di- and triatomic, at the II stage – molecules containing 11, 15, 16 and 19 atoms, and at the III stage – molecules with 60 and 70 atoms.

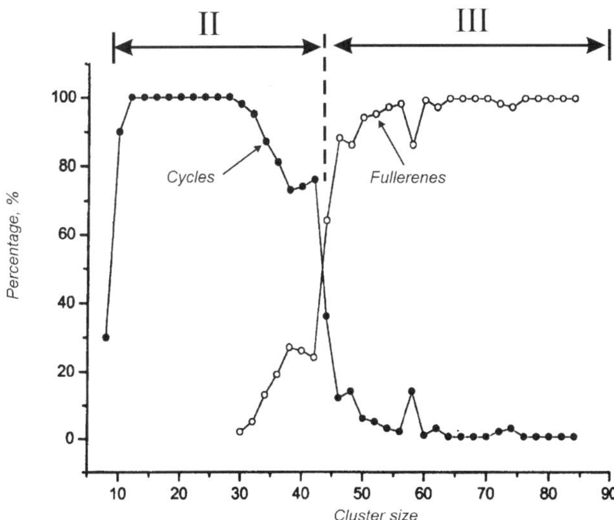

Figure 3. The confirmative experimental evidence for the cyclic nature of fullerene molecules [2], obtained from investigation of positive ions

At the binding together the maximums of all three stages in logical succession it may be deduced that for synthesis of C_{60} molecules it is necessary to create such thermodynamic conditions at which in atomic carbon vapour the di- and triatomic molecules can be formed and they make up the pentatomic cycles. These cycles are combined into bi- and trimers, that allow them in their turn to form clusters and spherical molecules with atoms in amounts of 60 and more.

In addition, it should be noted that the absence of clusters with the odd amount of atoms at the III stage testifies to the pairing process of clusters appearing at the second stage.

4. The structure of the frame of C_{60} molecule

The frame of C_{60} molecule is made up of 60 carbon atoms connected together by the single and double bonds.

Theoretically 12,500 variants of arrangement of double and single bonds is possible in the C_{60} molecule [4], but their disposition must follow the certain logic of struc ture and thermodynamics of existence of the stable spherical system. The most stable isomer (Fig. 4) has a structure of truncated icosahedron in which the double bonds are missing from pentagons. It is agreed to assume that this isomer is the C_{60} fullerene molecule (Fig. 4(b)).

At the discussion of its chemical activity let take a look at the energy state of carbon atoms in each hexagon and pseudoaromaticity of the last. But this is a sufficiently hard task by virtue of the fact that in distinction to the benzol molecule each hexagon holds three π-electrons, but not six. For this reason the system can not follow the Hückel rule and interpretation of its state is hampered. Under examination of carbon atoms only not in adjacent hexagons, but in two nearby pentagons, connected by the double bond, it is possible to identify sharply the impact of each atom on energy state of atoms, surrounding this atom, and fullerene molecule as a whole.

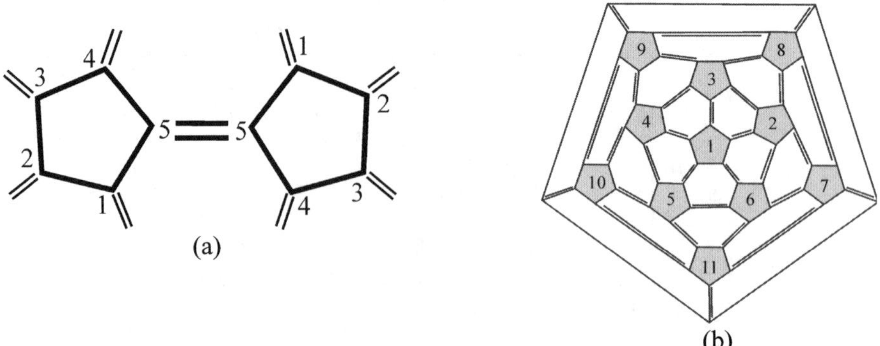

Figure 4. The combination of pentatomic molecules in the frame of C_{60} molecule: (a) pentatomic carbon molecules; (b) C_{60} fullerene molecule

From the aforesaid, it may be assumed that in the fullerene frame of C60 molecule free of defects there can be carbon atoms only in the five energy states 1, 2, 3, 4, 5 (Fig. 4(a)). As the energy state of one atom of pentagon (i.e. the energy state of pentagon as a whole) changes, the overall electronic system will tend to introduce the similar alterations in analogical position of each remaining eleven pentagons or to return the carbon atom in the initial state in the first pentagon in order for the system to be again stable thermodynamically (balanced energy-wise).

Such method of attack shows the possibility of fullerene molecule formation from 12 pentatomic molecules connected by double bonds. Therefore in the frames of the stablest C60 fullerene molecules the structural units are multiple of 5 and 12 except of 2.

5. The mechanism of carbon structures formation

The region of thermodynamical stability of gaseous carbon is in the interval of near-plasma temperature ~8,000 K. That is why the transition of carbon atom to the excited state is possible only in the temperature condition exceeding this interval. According to Fig. 1, in the interelectrode arc there are temperature zones allowing to transfer the carbon atoms, making up the graphite electrodes, in the free (atomic) state.

According to our preliminary experimental and theoretical investigations the prepared scheme, presented in Fig. 5, is one of variants of such process.

However the process of cooling of carbon plasma and product forming will be more complicated. The levels of matter organization, considered in physical chemistry (atom, molecule, substance), provide for the different types of interaction (interatomic, intramolecular and intermolecular and so on), but their nature remains general. At formation of substance from great quantity of particles any system tends to the maximal number of chemical bonds, to the most number of interactions that eventually reduce the general energy of the system.

In plasma the carbon atoms and cations are long-lived particles recombined mainly with formation of diatomic molecules. However the collision of C_2 molecules and C* excited atoms can give the appearance of C_5^+, C_5^- and C_5^* (in excited state) molecules which at lowering temperature tend to pass into the more stable thermodynamical state. As this takes place, the clusters from 2, 3, 4 and more similar molecules can appear and these clusters are more stable thermodynamically than each of components (molecules) individually.

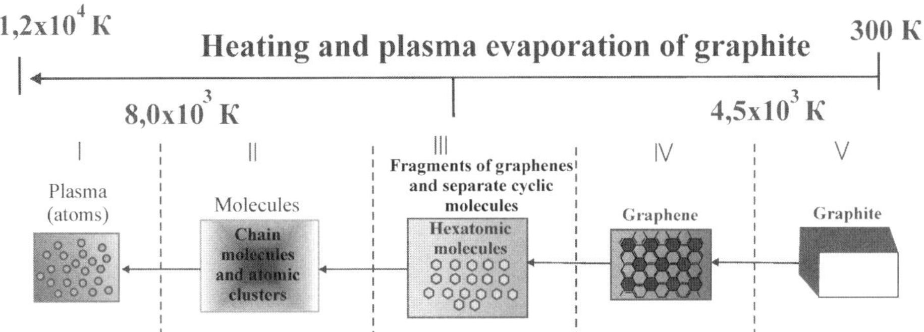

Figure 5. Levels of matter organization which allow the consideration of mechanism of carbon vapour formation with increasing temperature from the near-room to about 1.2×10^4 K

The tricycle formation occurs by the addition of three pentatomic molecules to one carbon atom with production of unstable C_{16} molecule, but experimentally their quantity is half as many as the number of C_{15} (Fig. 2), formed by intramolecular bonds between pentatomic molecules.

Figure 6 demonstrates (by the use of Schlegel diagram) one of great number of variants of formation mechanism of C_{30} molecule, representing hemisphere of C_{60} molecule, from bi- and tricycles of pentatomic molecules. According to the quantum-chemical calculation of C_{30} structures (Fig. 6), formed during the course of interaction of bi- and tricyclic molecules, are stable [5, 6].

As experimental investigations showed [1], the C_{60} molecules at times are formed with separation of carbon atoms and di- and triatomic molecules of carbon (Fig. 6(b)).

The authors of work [7] through a series of experiments achieved that with increasing time of clusters location in the collisions zone the spectrum turned into a single peak of C_{60} and in so doing they confirmed the C_{60} formation from products of the foregoing stages.

However by the use of mass spectrometry and another methods of physical and chemical analysis the possibility exists of identification of only stable relatively long-living intermediate carbon structures, formed in the course of carbon vapour transformation into macro-product. If complementary energy is imparted to these structures, over the long run the necessary molecules of C_{60} can be produced. In the paper [7] this process was said to be "annealing".

The process rate of fullerene molecules formation in the course of arc synthesis of these molecules is close to the velocity of light and because of this there is no way to trace this process by any devices. For this reason all arguments "pro and con" of one or other mechanism of the fullerene molecules production are the circumstantial evidence.

Uniquely it can be said with assurance that with fast decreasing temperature of the system the fullerene molecule is formed from thermodynamicly unstable structures having very small "life time".

On examination of the further transformation of particles, participating in the process, it should be pointed out that an increase of cluster mass brings into existence the spherical molecules. One of mechanisms of their formation is given in Fig. 7. The data on fragmentation of free C_{60} molecules at the bombardment by particles of high energy testify to the fact that clusters with 25–40 atoms have the least "life time", interacting with each other and forming more long-living particles [4].

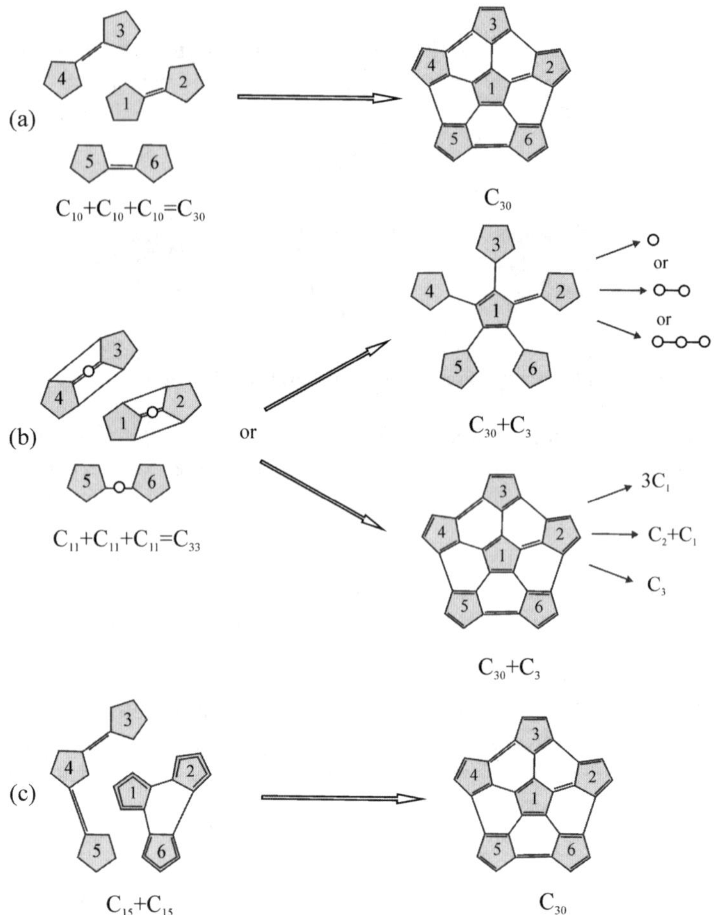

Figure 6. Formation of C_{30} molecule from bi- (a, b) and tricycles (c) of pentatomic molecules

As indicated above, the similar reactions proceed quickly enough. For production of product in large quantities it is necessary to transfer the reacting system into the certain state of "annealing" when concentration of cycles with the necessary number of atoms can be considerable.

An active searches for mechanism of the fullerene molecules formation by ionic gas chromatography have been carried out principally by two groups of scientists [8, 9]. It was observed experimentally [8] that three-dimensional fullerene clusters appear after joining up of more than 40 carbon atoms. Thirty atomic clusters are mono-, bi- and tricyclic planar structures. The combining of pentatomic cycles into molecules ($c_5 + c_5 = c_{10}$ and $c_5 + c_5 + c_5 = c_{15}$) (Fig. 6) can be interpreted as a single complex. Two or three such complexes can form tricycle ($c_{10} + c_{10} + c_{10} = c_{30}$).

The results of the experiments [9] also provide support for this view that formation of the C_{60} molecule occurs from polycyclic particles. Their concentration changes in antiphase as to the concentration of C_{60} molecules. The papers [8, 9] are experimental verification of our suggestion about formation of spherical molecules from mono-, bi-

and tricycles of carbon atoms, followed from analysis of Fig. 2. However pent- and hexatomic cycles were not found by scientists of papers [8, 9].

Impossibility to determine the presence of pentatomic and hexatomic clusters by the ionic gas chromatography talks not about the absence of the last or impossibility of their formation in the carbon vapor, but about their high chemical activity and consequently about the small "life time".

As our experimental investigations have shown, at the arc evaporation of graphite the percentage of spherical structures in a product (under favourable thermodynamic conditions) varies in the range from 3% to 12%. This testifies that in these energetic conditions molecules with a content of five carbon atoms appear no more than 20–30%

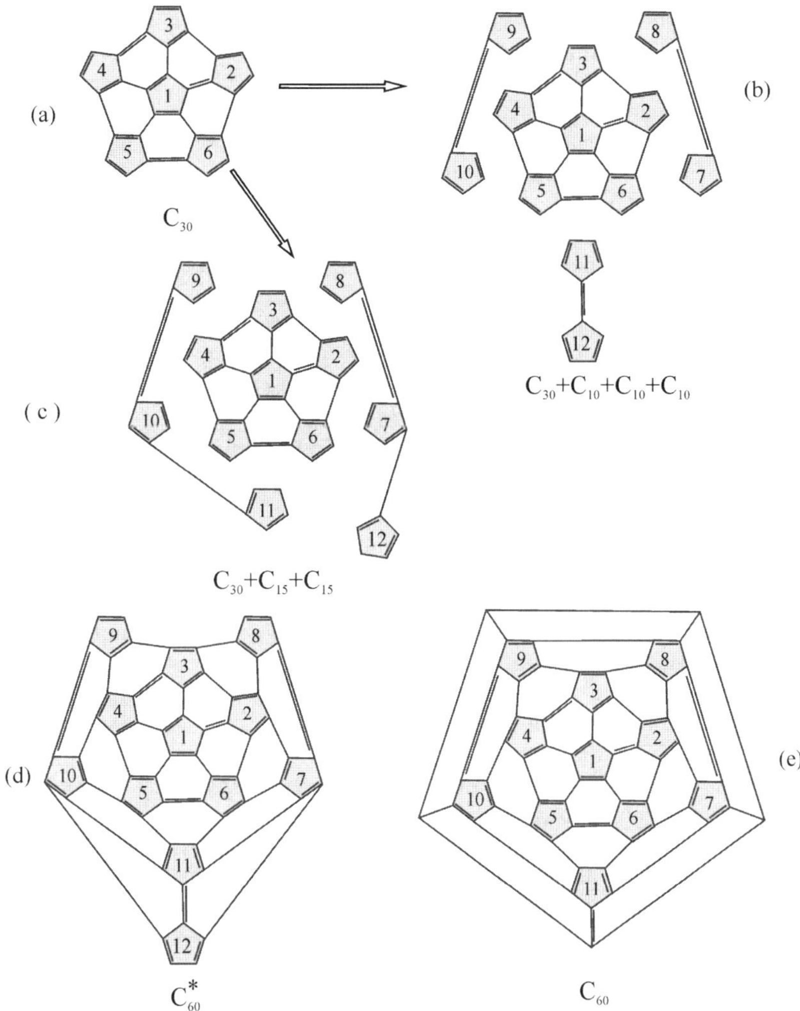

Figure 7. The sequence of formation of molecular frame of the C_{60} fullerene from pentatomic cyclic carbon molecules and C_{30} molecules (Schlegel diagram): (a) C_{30} hemisphere; (b) attachment of C_{10} bicycles to C_{30}; (c) attachment of C_{15} tricycles to C_{30}; (d) the last stage of C_{60} formation; (e) C_{60} molecule

(if all possible losses of these cycles at the formation of carbon nanostructures are taken into account). The rest 70% of product are cycles, chains, cones, tubes of different modification and other hybrid structures, combining in itself the above-listed modifications. All of them contain the hexatomic molecules to a greater or lesser extent (Fig. 2).

Our experimental data and results of experiments of another researchers allow us to draw the conclusion that the spherical carbon C_{60} molecules in the initial stage of their formation can be organized from pentatomic molecules in the course of clustering. The pentatomic molecule is building block of the C_{60} molecule.

On this basis, transforming the rule of isolated pentagons on the language of pentatomic carbon molecules, it is possible to conclude the following. At the clusters formation from pentatomic cyclic molecules the last unite between themselves by covalent bonds. Each pentatomic molecule so can add to itself five other (Fig. 6) and each of these five molecules also form five covalent bonds between themselves and subsequent pentatomic molecules. At the formation of double bonds between all atoms of pentatomic molecules the illusion is created that all pentatomic molecules are separated from each other by aromatic hexagons or by graphene-similar strip of polygons (Fig. 7(e)).

This illusion assisted to formulate the rule of “isolated pentagons”, which correctly take into account the noticed feature, but not properly interpret the crux of the problem.

Hence, in the C_{60} frame the pentatomic cycles are carbon molecules forming between themselves the hexagonal intermolecular emptinesses and interstices.

6. The sequence of reactions, proceeding at the formation of C_{60} molecules

As mentioned previously [14], the processes, in the course of which the carbon vapor and nanostructures are formed, have the nanosecond persistence and proceed in highly non-equilibrium conditions.

The explanation for the fact of carbon nanostructures formation by a system of kinetic equations creates no principle difficulties. These equations provide a way of decisions resulting in the self-organization of examined systems in the form of new spatial and temporal patterns within the framework of “strongly nonequilibrium thermodynamics” of I.R. Prigozhin [15].

At the same time the creation of well-composed chain of conversion processes, going on at matter transformation at all levels of its organization at the carbon nanostructure formation from the carbon vapor, leads to some problems up till now.

The molecules formation and their aggregation in a matter at reduced temperatures of the system is possible in principle by the change of enthalpy and entropy or of one and another simultaneously. The mechanism of C_{60} fullerite formation from carbon vapor, simulated by the temperature change from plasma up to room, is shown in Fig. 8. This is not in contradiction with the data of paper [1].

The overall process can be separated into six main stages: I – plasma interaction between atoms, II – formation of di- and triatomic molecules due to formation of chemical bond, III – formation of cyclic and chain molecules, IV – ordering of pentatomic molecules in clusters with the formation of C_{60} molecule by covalent bond, V – clustering of C_{60} molecules. The stage VI is conducted after extraction of C_{60} molecules from soot and their crystallization from solutions. The molecular crystal with f.c.c. lattice of fullerite appears as the result of ordering.

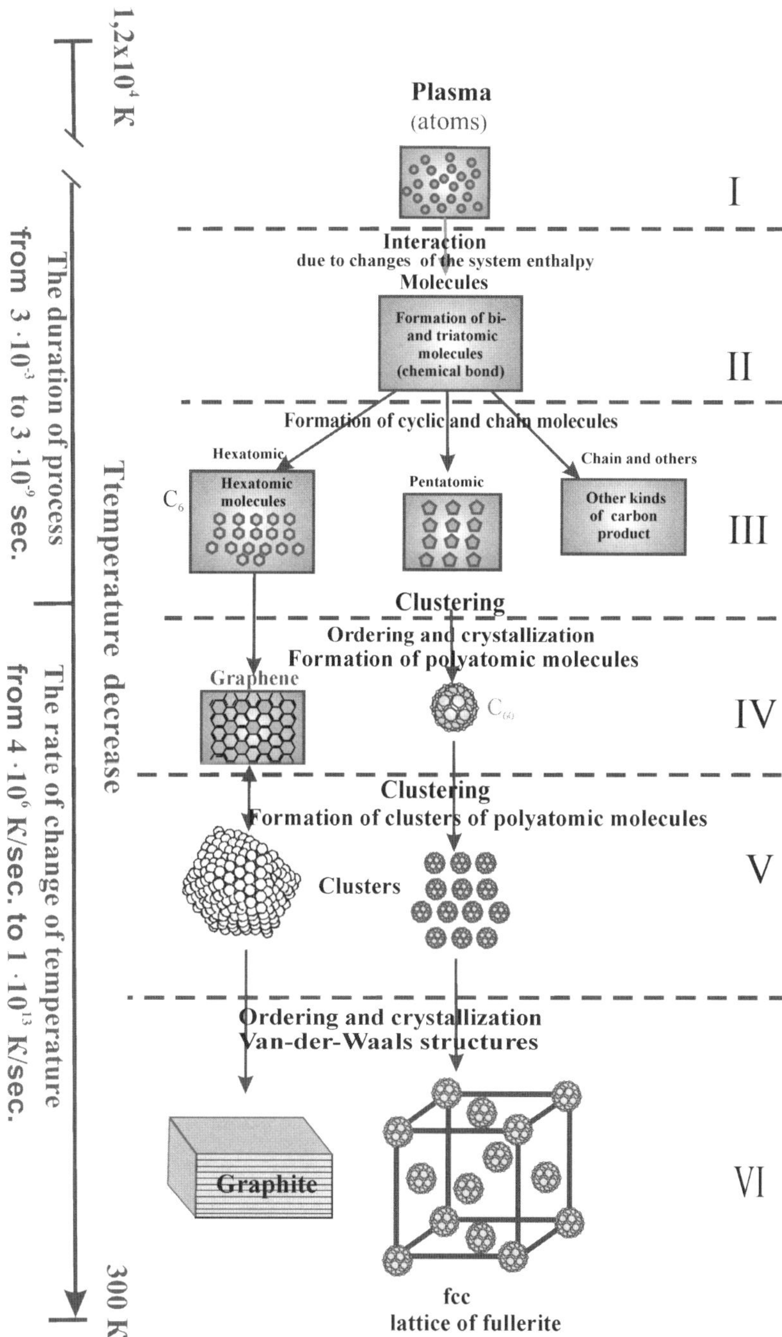

Figure 8. Levels of matter organization, reflecting the mechanism of C_{60} fullerite formation from carbon vapour under changes of temperature from 1.2×10^4 K to the room

The obtained information about this process and presented in Fig. 8 may be combined with data, collected in the Table 2.1 of paper [14]. Figure 9 is scheme with such integration of data about conventional levels of matter organization, processes types at each level at the structure formation and classification of carbon structures by dimensional levels.

The proposed scheme allows one to trace the transformations undergone by the system from carbon atoms, susceptible to structural and energetic conversions at the define level of matter organization.

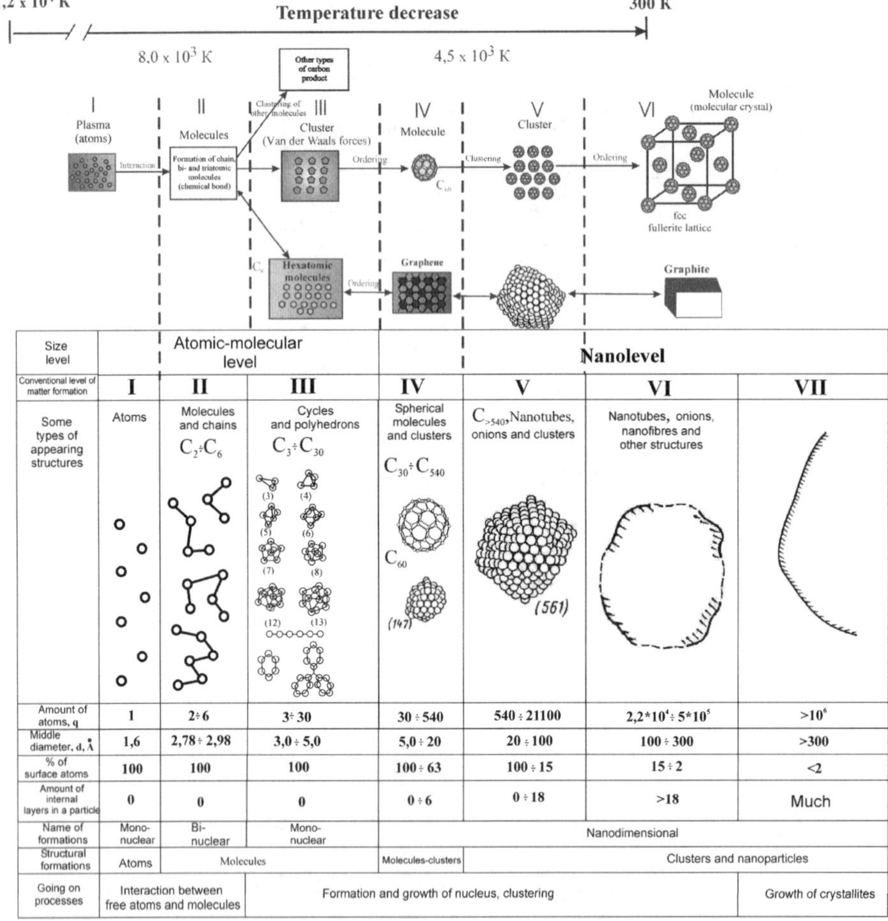

Size level	Atomic-molecular level				Nanolevel		
Conventional level of matter formation	I	II	III	IV	V	VI	VII
Some types of appearing structures	Atoms	Molecules and chains $C_2 \div C_6$	Cycles and polyhedrons $C_3 \div C_{30}$	Spherical molecules and clusters $C_{30} \div C_{540}$	$C_{>540}$, Nanotubes, onions and clusters	Nanotubes, onions, nanofibres and other structures	
Amount of atoms, q	1	2÷6	3÷30	30÷540	540÷21100	$2{,}2 \cdot 10^4 \div 5 \cdot 10^5$	$>10^6$
Middle diameter, d, Å	1,6	2,78÷2,98	3,0÷5,0	5,0÷20	20÷100	100÷300	>300
% of surface atoms	100	100	100	100÷63	100÷15	15÷2	<2
Amount of internal layers in a particle	0	0	0	0÷6	0÷18	>18	Much
Name of formations	Mono-nuclear	Bi-nuclear	Mono-nuclear	Nanodimensional			
Structural formations	Atoms	Molecules		Molecules-clusters	Clusters and nanoparticles		
Going on processes	Interaction between free atoms and molecules		Formation and growth of nucleus, clustering				Growth of crystallites

Figure 9. The conventional levels of matter organization, types of processes, going on each of them at the structure formation and classification of carbon structures by the size levels

7. Conclusions

The advisability of pursuing consideration of pentatomic cycles in the frame of C_{60} fullerene as pentatomic its constituent molecules has been justified.

It has been suggested the formation mechanism of the C_{60} fullerene frame from pentatomic molecules.

The levels of matter organization, represented the mechanism of fullerene formation, have been considered in close detail.

References

1. Rohlfing E.A., Cox D.M., Kaldor A.J. Production and characterization of supersonic carbon cluster beams. *J. Chem. Phys.* 1984. **81**: 3322.
2. von Helden G., Gotts N.G., Bowers M.T. Experimental evidence for the formation of fullerenes by collisional heating of carbon rings in the gas phase. *Nature.* 1993. **363**: 60–63.
3. Kim A.M. Organic chemistry – Novosibirsk: Siberian University Publishing House, 2002, 970 p. *(in Russian).*
4. Sidorov L.N., Yurovskaya M.A., Borshchevskii F.Ya., Trushkov I.V., Ioffe I.N. Fullerenes – Moscow, Ekzamen, 2004, 688 p. *(in Russian).*
5. von Helden G., Hsu M.-T., Kemper P.R., Bowers M.T. Structures of carbon cluster ions from 3 to 60 atoms: linears to rings to fullerenes. *J. Chem. Phys.* 1991. **95**(5): 3835–3837.
6. Brabec C.J., Anderson E.B., Davidson B.N., Kajihara S.A. et. al. Precursors to C_{60} fullerene formation. *J. Phys. Rev.* 1992. **B46**(11): 7326–7328.
7. Kroto H.W., Health J.R., O'Brien S.C., Curl R.F., Smally R.E. C_{60}: Buckminsterfullerene. *Nature.* 1985. **318**: 162–163.
8. Bowers M.T., Kemper P.R., von Helden G., Van Koppen P.A. Gas – phase ion chromatography: transition metal state selection and carbon cluster formation. *Science.* 1993. **260**: 1446–1451.
9. Hunter J.M., Fye J.L., Roskamp E.J., Jarrold M.F. Annealing carbon cluster ions: a mechanism for fullerene synthesis. *J. Phys. Chem.* 1994. **98**: 1810–1818.
10. Wakabayashi T.S., Achiba Y. A model for the C_{60} and C_{70} growth mechanism. *Chem. Phys. Lett.* 1992. **190**(5): 465–468.
11. Rubin Y., Kahr M., Knobler C.B., Diederich F., Wilkins C.L. The higher oxides of carbon $C_{8n}O_{2n}$ (n = 3 – 5): synthesis, characterization and X-ray crystal structure. Formation of cyclo [n] carbon ions C_n^+ (n = 18, 24), C_n^- (n = 18, 24, 30) and higher carbon ions including C_{60}^+ in laser desorption Fourier transform mass spectrometric experiments. *J. Am. Chem. Soc.* 1991. **113**: 495–500.
12. McElvany S.W., Ross M.M., Goroff N.S., Diederich F. Cyclocarbon coalescence – mechanisms for Tailor-made fullerene formation. *Science.* 1993. **259**: 1594–1596.
13. Schur D.V., Dubovoy A.G., Zaginaichenko S.Yu., Adejv V.M., Kotko A.V., Bogolepov V.A., Savenko A.F., Zolotarenko A.D. Production of carbon nanostructures by arc synthesis in the liquid phase. *Carbon.* 2007. **45**(6): 1322–1329.
14. Schur D.V., Matysina Z.A., Zaginaichenko S.Yu. Carbon nanomaterials and phase transformations in these materials. Dnepropetrovsk: Nauka i obrazovanie, 2007, 678 p. *(in Russian).*
15. Prigozhin I.R., Kondeludi D.I. Modern thermodynamics from thermal engine to dissipative structures – Moscow: Mir., 2002, 461 p. *(in Russian).*

THE SPECIAL FEATURES OF FORMATION OF CARBON NANOSTRUCTURES, THEIR CLASSIFICATION AND SITE ON THE STATE DIAGRAM OF CARBON

D.V. SCHUR, S.YU. ZAGINAICHENKO
Institute for Problems of Materials Science of NAS of Ukraine, 3 Krzhizhanovsky Str, Kiev, 03142 Ukraine
Fax: 38-044-424-0381; E-mail: shurzag@materials.kiev.ua
Z.A. MATYSINA
Dnepropetrovsk National University, Dnepropetrovsk, st. Nauchnaya,13 49050, Ukraine

Abstract. The peculiarities of possible formation of low-temperature carbon structures (fullerite F, crystalline carbyne CC, nanotubes crystals NTC) in the area of graphite G realization in the state diagram P-T have been described in the present paper. In addition, the regions of existence of high-temperature structures (molecular carbyne C, graphene Gr, molecules of nanotubes NT, fullerene molecules Fn, pent- and sixatomic cyclic molecules) in the area of gaseous carbon have been marked.

In response to changing state of carbon atoms the hypothetic areas of these structures formation have been indicated on the P-T diagram. The temperature levels I ... IV of existence of different carbon structures have been marked off. Discussion has been presented concerning the possibility of formation not only of single-layered or single-wall nanotubes (SWNT) and fullerenes Fn, but of multi-layered or multi-wall nanotubes (MWNT) and fullerenes (so-called onion) as well. Their location has been shown on the P-T diagram.

The results of statistical theory of phase transformations in carbon crystals as: diamond $\rightleftarrows$ fullerite (D $\rightleftarrows$ F), fullerite $\rightleftarrows$ graphite (F $\rightleftarrows$ G), graphite $\rightleftarrows$ carbyne (G $\rightleftarrows$ C) have been presented. The overall formula for Gibbs thermodynamic potential has been given and it defines the dependence on temperature, pressure, order parameters, energetic constants. The equilibrium values of order parameters have been evaluated as a function of temperature and external all-round pressure. The plots of temperature dependence of thermodynamic potential φ_i(T) have been constructed for different pressures for the examined phases i = D,F,G,C, the state diagram of carbon has been plotted by the intersection points of the φ_i(T) curves. The qualitative correspondence between experimental and design diagram of carbon state has been received.

Keywords: state diagram, carbon nanostructure, fullerene, fullerite, nanotube, graphene

1. Introduction

Variety in properties of carbon materials is caused by the electron structure of a carbon atom. The mechanism of reconstruction in the crystal lattices of carbon, which

B. Baranowski et al. (eds.), *Carbon Nanomaterials in Clean Energy Hydrogen Systems.*

determine the formation of different carbon modifications, consists in changing configurations of external valence electron clouds. This mechanism is controlled by the energetic state of carbon atoms and their different valence [1–5]. Carbon atoms can be tetra-, tri- and bivalent. The type of the bond between carbon atoms can be different. It can be single- (realized by the pair of electrons) as well as double or triple. Difference in the electron density of carbon atoms can cause the presence of carbon atoms of different rating C', C'' in the crystal what, in its turn, can determine the ordering process of these atoms.

Figure 1 illustrates the experimental phase diagram for carbon with the temperature and pressure ranges in which the phases of diamond D, graphite G, melt M and the probable metal phase MP are realized [3–9]. The diagram does not show other known crystalline carbon forms: carbyne C and fullerite F. The diversity of carbon nano-structures is not illustrated also.

In the present paper the attempt to compensate the presence of such gap has been made.

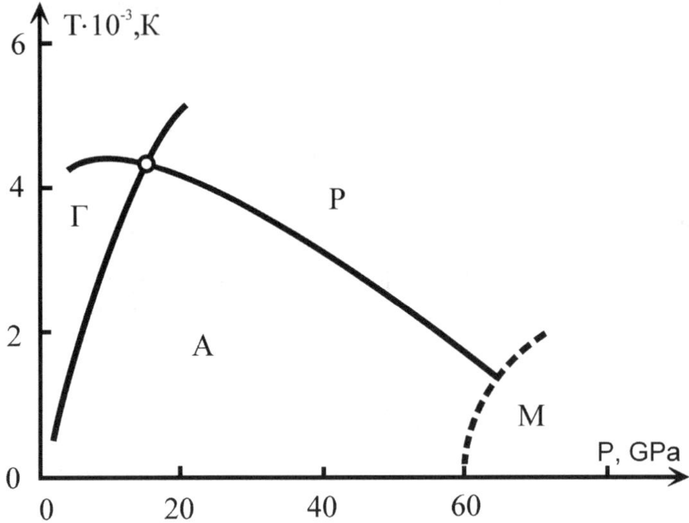

Figure 1. Experimental state diagram for carbon phases. D – diamond, G – graphite, M – melt, MP – probable metal phase. The triple site is marked with the circle

2. Van der Waals structures

Sometimes in publications authors don't make any distinctions between the crystalline and molecular state (two levels of matter organization) of carbon structures. All crystalline structures, as crystallites of nanotubes, fullerite, crystallites of carbyne and graphite, are formed from molecules, presenting nanotubes, fullerene and carbyne molecules, graphene, which are bound together in crystals by Van der Waals forces.

The graphite, crystalline carbyne, crystallites of nanotubes and fullerite can be related to the low-temperature modification. These modifications exist at normal conditions and humanity take them to a number of uses as materials (i.e. in macro-quantities). They differ from one another by the value of Van der Waals interaction between molecules forming them and consequently by the thermodynamics of systems.

By virtue of the fact that forces of Van der Waals interaction is the main parameter defined the physicochemical properties of low-temperature modifications of carbon, one can give the name of Van der Waals structures or Van der Waals carbon to these structures.

The known low-temperature carbon structures (considering the hybridization degree of electron orbitals of carbon atoms in them) can be entered on the experimental state diagram (Fig. 2(a)) setting aside from it the state diagram of carbon Van der Waals structures (Fig. 2(b)).

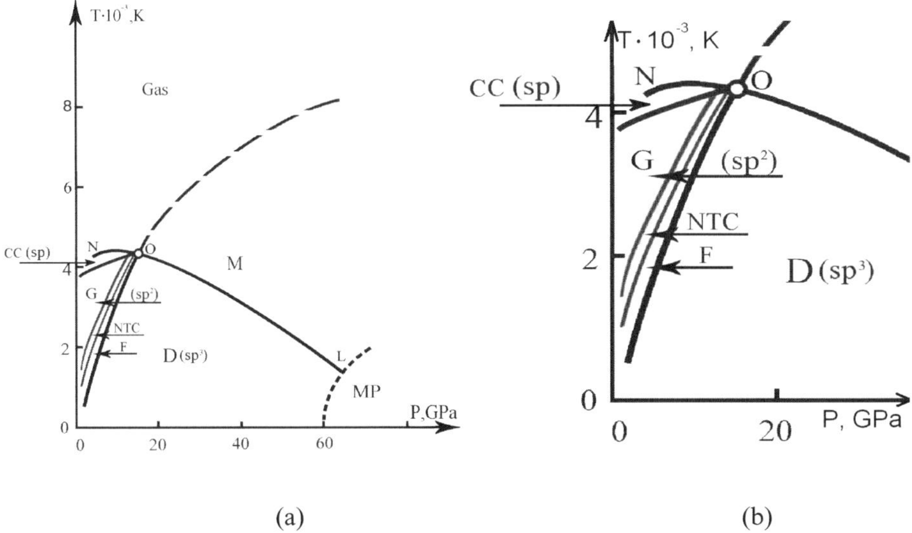

Figure 2. The state diagram of carbon: (a) the complete state diagram, (b) the diagram with low-temperature Van der Waals structures (NTC, F, G). D – diamond, G – graphite, M – melt, MP – supposed metallic phase; F – fullerite, CC – crystalline carbyne, NTC – crystals of nanotubes. "Gas" is carbon atoms in the excited state. The triple point is marked with the circle

Carbyne with *sp*-hybridization and its phase change into graphite at about 3×10^3 K and melting at 4×10^3 K can have the area of realization in the region, belonging a graphite at these temperatures (Fig. 2, CC area).

All crystalline structures with sp^2-hybridization are bound to be arranged also in the region, prescribed previously to the graphite. They can be arranged hypothetically on the basis of their thermal resistance in oxygen. So, at the temperature of T = 820 K fullerite interact with oxygen, at T = 970–1,000 K the interaction of nanotubes crystals with oxygen occurs and only after T = 1,100 K graphite starts its intensive interaction [5].

Thus, the fullerite with sp^2-hybridization of orbitals will occupy the regions of low temperatures at the moderate pressures. It has to be located between diamond and graphite, as at high pressures the fullerite turns into diamond and at high temperatures it transforms into graphite [5].

In virtue of the fact that crystallites of carbon nanotubes are combined by Van-der-Waals interaction, change into diamond at the high pressures and their heat stability and physical nature is between the properties of fullerite and graphite, the carbon nanotubes should be placed in the intermediate field between three phases as diamond, graphite and fullerite.

3. Covalent modifications of carbon

The analysis of experimental data, obtained from investigation the peculiarities in processes going on at the plasma temperatures, has demonstrated that the resulting product nature in the course of synthesis depends on the structure of molecules, forming at the high temperatures, and on their mutual location in the case of construction from them of more compound nanosystems, existing above the triple point.

Taking into consideration the fact that the stable atoms of carbon are in existence at the temperatures above 8×10^3 K and graphite crystallizes at the temperature of 4.5×10^3 K, it is reasonable to suppose that the transition areas of existence of stable fullerene molecules and other molecular high-temperature nanostructures should be in the area of different pressures and in the temperature range of $(4.5 \div 8) \times 10^3$ K.

Since it turns out that the soluble carbon nanostructures differ from insoluble by the molecules structure of their formation, so it is advantageous to note the field of existence of pentatomic cyclic molecules “5”, hexatomic cyclic molecules “6” and their mixture (“5” + “6”), where nanotubes are under formation. This make the interpretation of physical effects, associated with transformation of different modifications of carbon, much easier.

The state diagram (Fig. 3) presents between the gaseous and solid states the transition temperature areas (I–IV), in which the structures, completing the building of their frame at the molecular level, are realized. The known today structures, as molecules of carbyne, fullerene, nanotube and graphene (area IV) fall into these zones. The pent- and hexatomic cyclic molecules “5”, “6” can be related to the area III. The di- or triatomic carbon molecules can be formed in the area II and the atomic carbon gas exists in the area I.

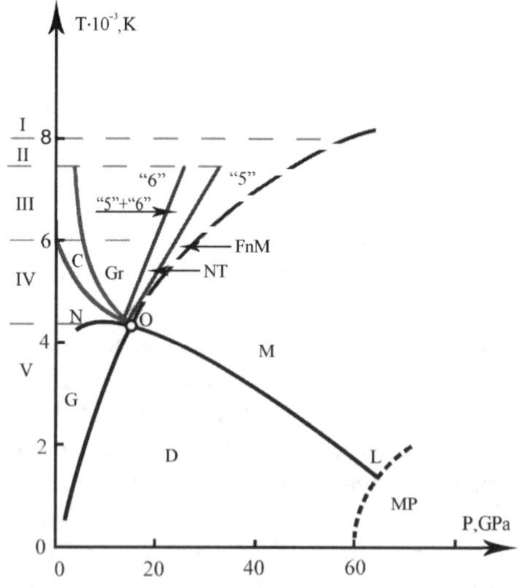

Figure 3. The state diagram of carbon with indication of the hypothetical areas of formation of covalent high-temperature nanostructures. I, II, III, IV, V – the temperature levels of organization of carbon structures; D – diamond, M – melt, G – graphite, NOL – the curve of crystallization, C – molecular carbyne, Gr – graphene, NT – molecules of nanotubes, FnM – fullerene molecules, “5” – pentatomic cyclic molecules, “6” – hexatomic cyclic molecules, “5” + “6” – area of co-existence of pent- and hexatomic molecules, MP – the hypothetical metallic phase

As the temperature decreases from 8×10^3 K, the carbon gas shall be turned to the di- and triatomic molecules, forming the pent- and hexatomic cycles, which in their turn produce spherical molecules, graphene sheets or single-wall nanotubes (SWNT) by clustering.

In all these structures their constituent carbon atoms are combined only by covalent bonds. For this reason we can name these structures as covalent structures or covalent carbon.

The electronic atomic orbit of atoms, constituent these molecules, are nonhybridizated, i.e. carbon atoms are in the excited state.

The carbyne chains are clustered at the lower pressures. The appearance of carbyne chains between the graphene sheets brings into existence the diamond-like layer-chain carbon LC2 and another modifications.

The further clustering of molecules of covalent carbon, as the temperature decreases, brings the formation of crystalline carbene, fullerite, crystal, consisting of nanotubes, and graphite, including graphene molecules. All of them are combined into clusters by Van-der-Waals interaction and, as mentioned above, make up Van der Waals structures.

4. Joint state diagram of carbon

On the basis of the state diagram of carbon (Fig. 1), it is believed that only diamond in crystal from all carbon allotropies forms one large molecule, representing simultaneously both molecule and crystal. Diamond is the only allotropic modification of carbon, in which Van der Waals interaction between layers is transformed into covalent bond. For this reason the intermolecular bonds in crystals and intermolecular interaction are equivalent. As evident from the state diagram diamond does not have two modifications differing by the type of bond between its constituent particles. The equivalence of bonds allows diamond the transformation into the liquid-fluid state. All other allotropies have high-temperature and low-temperature modifications.

At the combination of covalent and Van der Waals carbon structures into one state diagram (Fig. 4) the question arises as to the existence of triple point for these structures; that is, the graphite, fullerite, nanotube crystallite undergo transformation in the liquid-fluid state (with the loss of long range order) or not during the transition of Van-der-Waals structures to covalent.

As far as is well known, the transformation of structures to the melt (i.e. the disruption of long range order) is bound to imply the chaotic arrangement of particles in the crystallite.

By virtue of the fact that particles, forming the crystallite of Van der Waals structure, present the molecules, containing in their composition not a few, but tens (their number is equal to 60 in the fullerite F_{60}), thousands and more atoms of carbon (as in graphene) and bound to each other by the weak bonds, the crystallite, on being heated, changes into cluster with the chaotic disposition of these molecules. As this takes place, the internal structure of these molecules, forming the crystal, is conserved.

The extreme adverse kinetic factor prevents them from the transition to the liquid-fluid state under these thermodynamic conditions. This factor is concerned with the weak intermolecular interaction and with very high activation energy of destruction reaction of intramolecular covalent bonds C-C in these molecules. At the disordering the Van der Waals bonds only disrupt, but the covalent bonds, forming these molecules, persist. In this case the usual transformation into the liquid-fluid state shall not be observed on heating and the molecules shall sublimate.

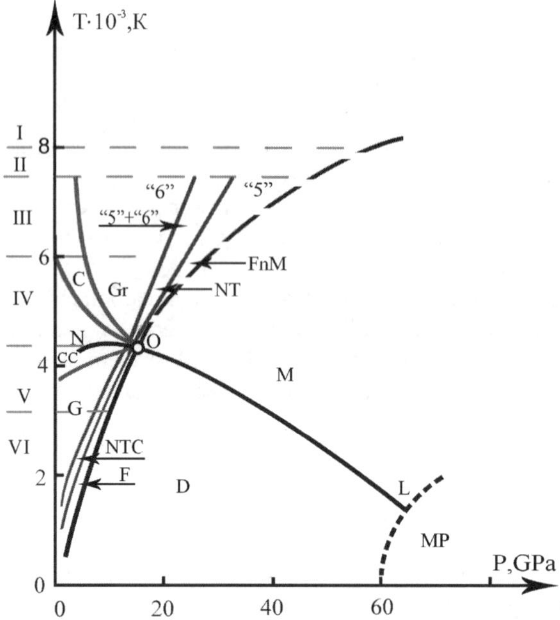

Figure 4. The state diagram of carbon with the hypothetical areas of existence of carbon nanostructures. I, II, III, IV, V – the temperature levels of organization of carbon structures; D – diamond, M – melt, F – fullerite, NTC – nanotubes crystallite, G – graphite, CC – crystalline carbyne, NOL – the curve of crystallization, C – carbyne molecular, Gr – graphene, NT – molecules of nanotubes, FnM – fullerene molecules, "5" – pentatomic cyclic molecules, "6" – hexatomic cyclic molecules, MP – the hypothetical metallic phase, "5" + "6" – area of co-existence of pent- and hexatomic molecules

Seemingly only the fullerite among all existing carbon nanostructures has the feasibility to turn into the liquid state. This is due to the fact that in contradistinction to the rest the fullerene molecules make polyatomic formations, have the completed geometry with the rigid frame and are of the same geometric size, defined by the number of carbon atoms. The fullerene molecules pass into solution and can not be confined in the liquid-fluid state in view of the specificity of the Van der Waals interaction between the molecules. The polymers of fullerenes are capable to this. But in the case of polymerization at the high pressure the covalent bonds are formed between the carbon atoms and the structure converts into the diamond at the sufficient degree of polymerization.

For these reasons the carbon structures with sp^2–hybridization have no the triple point, in spite of the fact that at the temperature of phase transformation with increasing pressure the carbon atoms in each structure can be carried to the sp^3–hybridizated state of diamond and then to the liquid state. This liquid state shall be due to the disordering of carbon atoms, but not of the covalent structures, forming the Van der Waals carbon.

On the basis of the geometrical structure of polyatomic molecules of fullerene and graphene, the fullerene molecules would be expected to be five times loosely bound in crystal than graphene forming the equivalent bonds in all directions. All other carbon nanostructures are anisotropic, have molecules of various size, differing from one another a hundred and thousands times. The graphene molecules are free to leak only in the basal plane!

Since different number of carbon atoms of each molecule (each atom in graphite, every fifth atom in fullerite, the average number between two first in the nanotube crystallite) participate in the formation of Van der Waals bonds in the each type of crystals, destruction thermodynamics of each crystal and the temperature of its realization will be different. So, at all other equal conditions this temperature is equal to 1.1×10^3 K for fullerite and to 4.5×10^3 K for graphite at the normal pressure.

Hence, at the constant temperature the disordering of Var der Waals crystalline structures shall occur at the different pressures, i.e. line of phase change (NO) exists at the state diagram for these structures (Fig. 4) and phase transformations of these structures are distributed along this line. The NO line originates in the triple point.

With increased temperature the further destruction of formed molecules (fullerene molecule, nanotube molecule and graphene) seemingly shall proceed in nearly equal thermodynamic conditions, because at this level of destruction the less strong (from the rest) covalent bonds, formed by 2s*-electrons in the excited state, will be disrupted in all structures.

The existent theories of atom structure and formation of hybrid electron clouds fit and adapt all physical effects to the hybridization theory in view of the Pauli-Fermi principle. It is believed that at the sp^2-hybridization 2s*-electron orbitals (or 2s*-electrons) are equivalent energy-wise to 2p-orbitals. This conclusion emerge from the equality of valence angles and isotropy of some mechanical properties.

However in practice this is not at all true. At the formation of carbon cyclic molecules each carbon atom uses two 2p-electron at a time for the production of molecular frame (of cyclic molecule). The excited 2s*-electron is used for the formation of intermolecular bond between these molecules in the course of formation of molecules frames of fullerene, nanotube or graphene. The intramolecular bond is always more strong than intermolecular. For this reason the destruction thermodynamics of fullerene frame is bound to differ from the destruction thermodynamics of its constituent pentatomic molecule.

By virtue of the fact that the geometry of frames of forming molecules is different (this is a planar layers of hexatomic molecules – for graphene, the same planar layers, rolled up into tube, with hemispheres from pentatomic molecules at the ends – for nanotube, the sphere collected from pentatomic molecules – for fullerene molecule), the thermodynamics of their destruction is to be different also.

As a consequence these structures can be realized at different pressures (at T = const) or at different temperatures (at P = const) and each of them has the area of its own on the state diagram in the conversion region.

Furthermore, the formation thermodynamics of pent- and hexatomic cycles is also different.

At the creation of the combined state diagram we propose to your notice the additional hypothetical areas of existence of carbon nanostructures at all levels of their formation and they have been plotted on the state diagram of carbon (Fig. 4). Within the limits of this diagram six temperature levels of matter formation have been suggested for consideration. At the first level I at a temperature higher 8×10^3 K the carbon atomic gas exists and it transforms into vapour as the pressure increases.

In the course of temperature decrease the di- and triatomic molecules are formed (level II). The further combination of these molecules into chains and cycles take place at the level III. At this level the chains and both pent- and hexatomic cycles can appear depending on pressure. At the boundary between these regions there exist the area, in which the mixtures of pent- and hexatomic cycles can be in existence thermodynamically.

At the level IV (below 6×10^3 K) all molecules, formed at the level III, make up clusters, are arranged in an order (with decreasing temperature) and form the more complex polyatomic molecules designated as graphene, molecule of fullerene and molecule of nanotube. The carbyne molecules also can form clusters. At the formation of all these structures three valence electrons of each carbon atom are used for the covalent bonds production. The bonds of other type are absent in the structures, the atomic orbitals are nonhybridizated. As pointed out above, the structures are said to be covalent.

The levels V and VI of organization of carbon structures are located below the temperature of graphite crystallization (4.5×10^3 K). The processes of clustering and ordering of molecules occur at these levels.

At the intersection of the curve of phase transformation (NO) the clustering of covalent structures brings about the hybridization of electronic orbitals of carbon atoms and the further formation of crystals at the expense of Van der Waals interaction between the covalent structures.

In this situation the Van der Waals structures (graphite, fullerite, crystallites of nanotubes and others) are formed and these structures are realized at normal conditions.

5. Multi-wall nanotubes and onions

Among all known carbon nanostructures at present only fullerite can dissolve in the organic solvents owing to the limitedness of atoms number in its molecule and to the completeness of molecule frame.

Even short single-wall carbon nanotubes (SWNT) do not share such property. The reason is that these nanotubes possess the properties, which do not allow them to form the three-dimensional crystal lattices. All interactions between molecules of nanotubes in crystals (cord, wire and cable) are two-dimensional. At the formation of multi-wall nanotubes (MWNT) they possess many properties of nanotube and graphite. In this case the interlayer interactions of π-electron shells exert the additional influence on the properties of nanotubes.

The same situation takes place also with fullerene onions, in which π-electron cloud of fullerene molecule grows as the number of shells increases. So one of the smallest molecules of such type (three-layered onion $C_{60}@C_{240}@C_{560}$) contains 860 atoms of carbon, the layers of which are bound together by the complicated intermolecular bonds.

Seemingly for this reason the fullerene onions and multi-layer nanotubes are bound to have more tolerant structure than graphite because a closed frames, formed by covalent bonds, participate in the structure construction in addition to the Van der Waals interaction between the layers and these frames give the mechanical strength to the construction.

However it turns out that bonds between the shells of MWNT and onions are of less power than in graphite, since π-electron density of each layer is not the same above and under a layer. This is because the different number of carbon atoms is contained in the neighbouring layers, resulting in more weak interlayer bond than in graphite. The mechanical strength of closed shell of onion can fall off by the chemical activity of its external shell and by processes that are responsible for this activity.

Hence, as experiments show, the character of interlayer interaction in MWNT and onions, contingency and stress level of chemical bonds, forming their shells, make them more chemically active. This is why they remain more thermostable than the molecules of fullerene and nanotubes, but of less thermostability than graphite.

From the aforesaid, it might be well to point out that the coexistence structures of MWNT and onions, possessing the properties both of covalent and Van der Waals structures, begin the formation in the covalent area IV (Fig. 5) and crystallization already at the level V. After that they form clusters and crystallites from MWNT and onions by Van der Waals interactions and in so doing they form the Van der Waals structures.

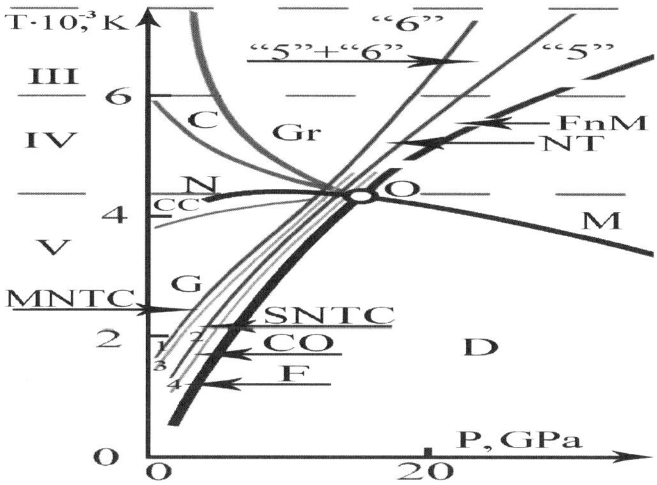

Figure 5. The fragment of state diagram of carbon with the hypothetical areas of existence of carbon nanostructures. D – diamond, M – melt, F – fullerite, SNTC – crystallite of single-wall nanotubes, G – graphite, CC – crystalline carbyne, NO – curve of crystallization, C – carbyne molecular, Gr – graphene, NT – molecules of nanotubes, FnM – fullerene molecules, MNTC – crystallite of multi-wall nanotubes, CO – clusters of onions, "5" – pentatomic cyclic molecules, "6" – hexatomic cyclic molecules, "5" + "6" – area of co-existence of pent- and hexatomic molecules, curve 1 – boundary between graphite and HNTC, curve 2 – boundary between crystallites of fullerene-like structures and nanotubes, curve 3 – boundary between MNTC and SNTC, curve 4 – boundary between regions of fullerite and cluster of onions

Figure 5 demonstrates the fragment of state diagram of carbon with indication of the hypothetical areas of existence of carbon nanostructures. This figure differs from Fig. 4 in that the areas of existence of clusters, crystallites of SWNT, MWNT and onions are entered on it. As illustrated in Fig. 5, between the curves "1" and "2" the area of nanotubes existence stand out and this area is divided into two section (area SWNT and area MWNT) by the curve "3".

The area of existence of spherical molecules is positioned between the boundary line with diamond and curve "2" and this area breaks down into two section by the curve "4". The area, adjacent to the triple point, corresponds to the realization of fullerite the only one of its kind soluble Van der Waals structure. The area between the curves "2" and "4" is appropriate to the realization of onions.

From the above discussion it should be mentioned that for the carbon SWNT production the process of their formation should be interrupted on the process-dedicated stages of SWNT molecules forming, exclusive of their interaction with reagent.

6. Hybridization of electron orbitals of carbon atom

On the basis of hybridization conditions of electron orbitals all known allotropic modifications of carbon can be arranged in the vertexes of spatial diagram, presented the triangular bipyramid (Fig. 6).

The proposed scheme has been constructed on the principle that takes into account the state of four valence electrons of carbon atom in carbon allotropies. We have combined five energy states of electrons in carbon atom into the unified conventional diagram (Fig. 6). Modifications with incomplete hybridization of orbitals are arranged at the corners of the pyramid base, but atomic carbon (not exposed to hybridization) and diamond (with the top degree of hybridization of atomic sp^3-orbitals) are bound to be located in the opposite corners of bipyramid.

The quantity of electrons, being in each allotropic modification in the starting energy state of free atom (i.e. before the processes of hybridization or excitation), are indicated by numerals at the pyramid vertexes. So, as an example, diamond with four sp^3-hybridizated electron orbitals is to be found in the vertex "0", i.e. neither of four valency electrons is in the initial energy state. In the vertex "1" (graphite) the electron orbitals of carbon atom are exposed to sp^2-hybridization, i.e. one of four electrons is in the initial state. Similarly, in the vertex "2" we have the structure of carbine with sp-hybridization, i.e. already two electrons are in the starting energy state.

The volume of spatial diagram, that is enclosed in the pyramid "1" – "2" – "3" – "4" and appropriate to the molecular state of carbon modifications, can be named as the zone of existence of covalent structures.

In this zone molecules of all modifications are in gaseous state. The electron orbitals of carbon atoms are not hybridizated. The straight line "4" – "K" – "0" indicates the direction of hybridization and the process of formation of Van der Waals structures. The plane "1" – "2" – "3" is the boundary of hybridization or the demarcation of phase transition between covalent and Van der Waals structures. The point "K" on this plane is the triple point, in which the carbon vapour, transforming into the liquid-fluid state, turns into diamond. The change in position of point "K" in the plane "1" – "2" – "3" disrupts the energetic equivalence of electrons in carbon atom and brings into existence the different modifications of carbon.

The volume of pyramid "0" – "1" – "2" – "3", corresponding to the crystalline state, can be designated as the zone of existence of Van der Waals structures. These modifications during the process of their transformation may be turned to diamond, reaching the utmost level of hybridization of electron orbitals of carbon atom.

7. The statistical calculation of state diagram for carbon crystals

The results of statistical theory of state diagram for carbon with consideration for the formation of widely known structural modifications, as diamond D, fullerite Ft (designate as F), graphite G and carbyne C, are given below.

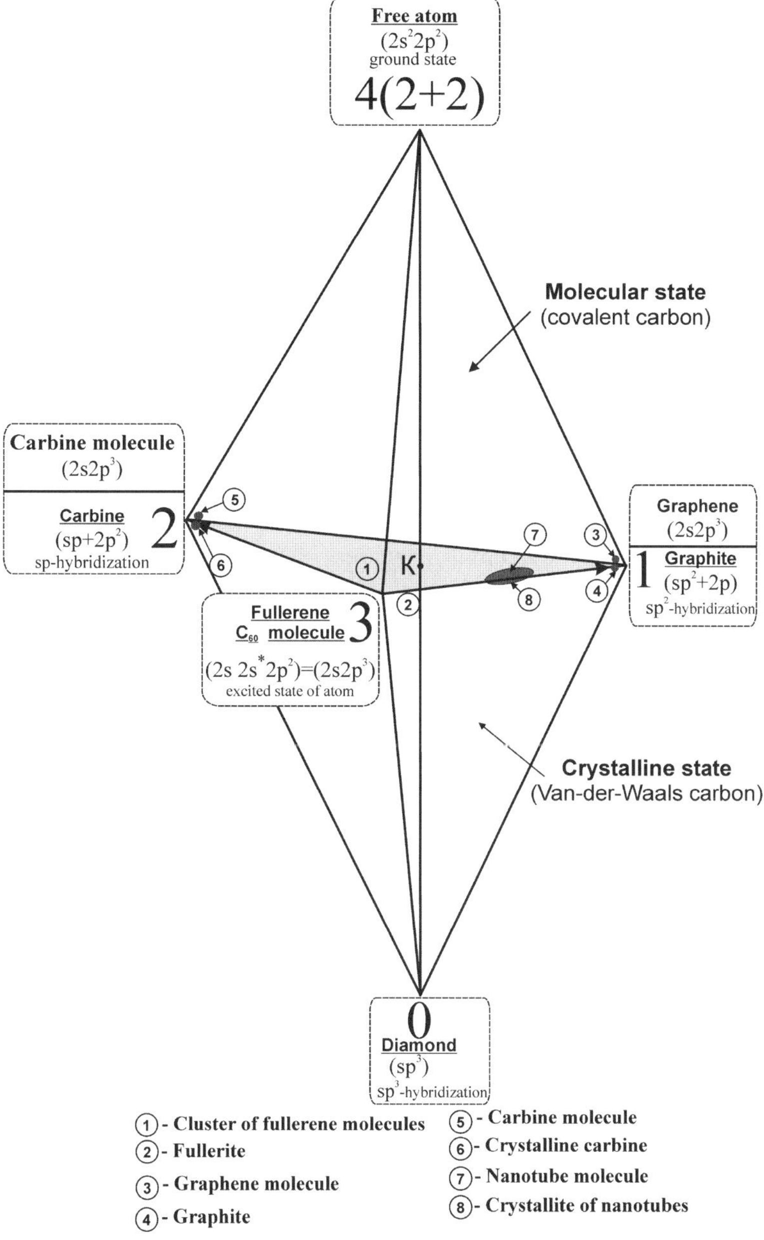

Figure 6. The spatial diagram of carbon allotropic modifications on two levels (molecular and crystalline) of matter organization. "4"–"0"– direction of hybridization, "1"–"2"–"3" – plane of hybridization

The functional dependences of Gibbs thermodynamic potential of all phases D, F, G, C have been calculated on the basis of the molecular-kinetic concept in accordance with the formula from our paper [10]

$$\Phi = E - kT\ln W + PV . \quad (1)$$

These derived dependences dictate the values of Φ_i (i = D, F, G, C) as a function of the absolute temperature T, pressure P, composition of each two-component phase (of c_1, c_2 concentrations), order parameter η and energetic constants. In formula (1) E is the configuration internal energy, defined from the sum of energies of interaction between all nearest pairs of atoms (molecules in fullerite), W is thermodynamic probability, equal to the number of distinguishable distributions corresponding to different energies of atoms (molecules) of any kind, which is determined by the combinatorial rules, k is Boltzmann's constant, T is the absolute temperature, P is the external all-round pressure, V is volume of crystal.

The calculation of thermodynamic potential for one lattice site in the case of stoichiometric composition ($c_1 = c_2$) gives the overall formula for all phases [4, 5, 10]

$$\varphi_i = (e_i - w_i\eta_i^2)[1 + \frac{P}{P_i'} + n_i\left(\frac{P}{P_i'}\right)^2] + \frac{1}{2}kT\Delta_i + \omega_i(1 - æ_i P)P, \qquad i = \text{D, F, G, C}, \quad (2)$$

where e_i, w_i are energetic parametres for i phases, η_i is the parameter of long range order in distribution of atoms (molecules) over the sites of crystal lattice, P_i' and η_i are the empirical constants which can be both positive and negative [11, 12], ω_i is the atomic volume of the crystal, $æ_i$ is the compressibility of the crystal.

In approximation of equilibrium condition $\partial\varphi_i/\partial\eta_i = 0$ we get the equilibrium equation as follows:

$$kT\ \ln\frac{1+\eta_i}{1-\eta_i} = 4w_i\eta_i\left[1 + \frac{P}{P_i'} + n_i\left(\frac{P}{P_i'}\right)^2\right], \quad (3)$$

which defines the equilibrium value of the degree of long range order depending on temperature and pressure.

The curve plots for the equilibrium values of order parameter in dependence on temperature and pressure are presented in Figs. 7 and 8. These curves have been plotted according to for formula (3).

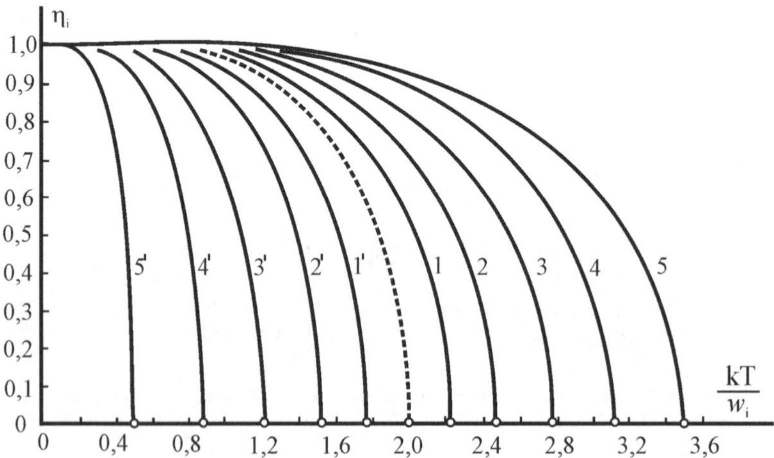

Figure 7. The curve plots of the temperature dependence of equilibrium order parameters constructed by formula (3) for different values of $P/P_i' = 0$; ±0.1; ±0.2; ±0.3; ±0.4; ±0.5 (curves 1–5 for $P_i' > 0$, $n_i = 1$ and curves 1′ – 5′ for $P_i' < 0$, $n_i = -1$). The dotted line corresponds to $P/P_i' = 0$. The ordering temperatures for different pressures are marked with circles on the abscissas axis

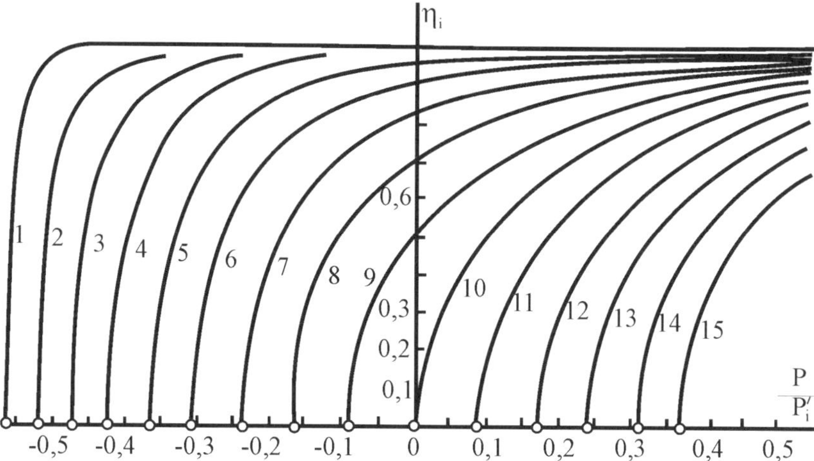

Figure 8. The curve plots of the dependence of equilibrium order parameters on pressure constructed by formula (3) for $P_i' > 0$, $n_i = 1$ and $P_i' < 0$, $n_i = -1$, and for different temperatures equal to $kT/w_i = 0.2$; 0.4; 0.6; 0.8; 1; 1.2; 1.4; 1.6; 1.8; 2; 2.2; 2.4; 2.6; 2.8; 3 (curves 1–15). The ordering pressures at different temperatures are marked with circles on the abscissas axis

The construction of curve plots for thermodynamic potentials of the phases has been performed with the use of the equation (3) and the plots of Figs. 7 and 8.

To evaluate temperature and pressure for possible phase transitions by formula (2), the curves for temperature dependences of thermodynamic potential have been plotted for different values of external pressure for the purpose of determination of their intersection points, which correspond to the points of phase transformations.

The energetic parameters e_i and w_i have been chosen according to the relationships between atomic volumes and average values of atomic radii of carbon phases (see Table 1) and also taking into account the experimental fact that fullerite transforms into diamond with increased pressure and into graphite with increased temperature.

TABLE 1. Structural parameters of crystals for different allotropic phases of carbon

Crystal	**Crystal lattice**	**Parameters (nm)**	**Atomic (molecular for fullerite) volume (nm^3)**	**Atomic radius (nm)**
Diamond	fcc, three-dimensional	$a = 0.357$	$a^3/8 = 0.006$	0.179
Fullerite	fcc, three-dimensional	$a = 1.417$	$a^3/4 = 0.711$	0.893
Graphite	hcp, quasi-two-dimensional	$a = 0.246$, $c = 0.671$	$a^2c\sqrt{3}/8 = 0.009$	0.206
Carbyne	hcp, quasi-one-dimensional	K_1: $a = 0.295$ $c = 0.508$	$\sim a^2c = 0.044$	0.35

The following values for energetic parameters e_i and ordering energies have been chosen (in eV):

$$\begin{array}{llll} e_D = 0.2, & e_F = 0.168, & e_G = 0.16, & e_C = 0.09, \\ w_D = 0.2, & w_F = 0.21, & w_G = 0.16, & w_C = 0.17. \end{array} \quad (4)$$

According to the chosen values for energetic parameters and the revealed regularities in the variation of order parameters with temperature and pressure, the numerical values for order parameters have been evaluated for every of four phases, to be taken into consideration later on at the construction of curve plots for temperature dependence of thermodynamic potentials for different pressures.

The case of $P_i' > 0$ and $n_i = 1$ has been studied.

The curves plots for temperature dependence of thermodynamical potentials φ_i ($i = D, F, G, C$), plotted by formula (2) for different pressures are shown in Fig. 9. As indicated by Fig. 9, for all the phases $i = D, F, G, C$ the thermodynamical potential decreases with increasing temperature. On the other hand, the growth of pressure increases the thermodynamical potential: curves $\varphi_i(T,P)$ fall with increasing temperature and rise with increasing pressure. This means that the elevation of temperature stabilizes each phase: the temperature range, in which the phase exists, can be increased. The rise of pressure tends to decrease the phase stability: the pressure range, in which the phase exists, can be decreased. Moreover, as evident from Fig. 9, at the temperatures above the points of φ_i curves intersection, the values of thermodynamical potentials φ_i, in transforming from the first phase A to the second one F and further to the third one G and the fourth one C, show a decrease what indicates the increase of stability of the fourth phase in comparison with to the third one, of the third phase in comparison with the second one, of the second phase in comparison with the first one.

The phase diagram given in Fig. 10 has been plotted by the intersection points of the $\varphi_i(T)$ curves for different pressures. Figure 10 shows that beginning from the values $P/P_i' = 0{,}15$ the temperatures of phase transitions rise with increasing pressure what corresponds to experimental data. At low pressures, the temperature of phase transition can slightly drop with increasing pressure as it has been found out for the equilibrium curve between phases F and G. The temperature range, in which every of the phases exists, increases with a rise in pressure (because from the first phase D to the fourth one C the tangent of canting angle of equilibrium curves increases). The exception is the second phase F at low pressures, for which with increasing P/P_i' up to 0.1 the temperature range of its realization shows a decrease, but further at $P/P_i' > 0.1$ this range also undergoes a rise. It is quite possible that this change is stipulated by the formation of fullerene clusters.

We can suppose that at the design diagram phase D corresponds to the diamond phase. Like in the experimental diagram (Fig. 1), the growth of pressure tends to increase the temperature, at which the diamond phase is realized. In this case the pressure range of this phase existence is diminished with increasing temperature, but the temperature range increases with pressure rise. On the contrary, for the graphite phase with increasing temperature the pressure range undergoes a widening.

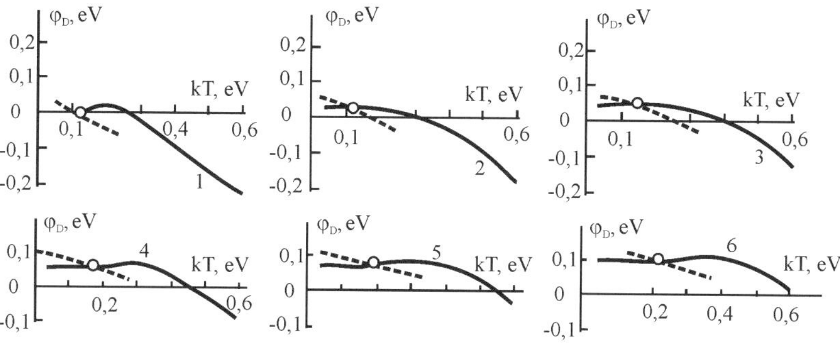

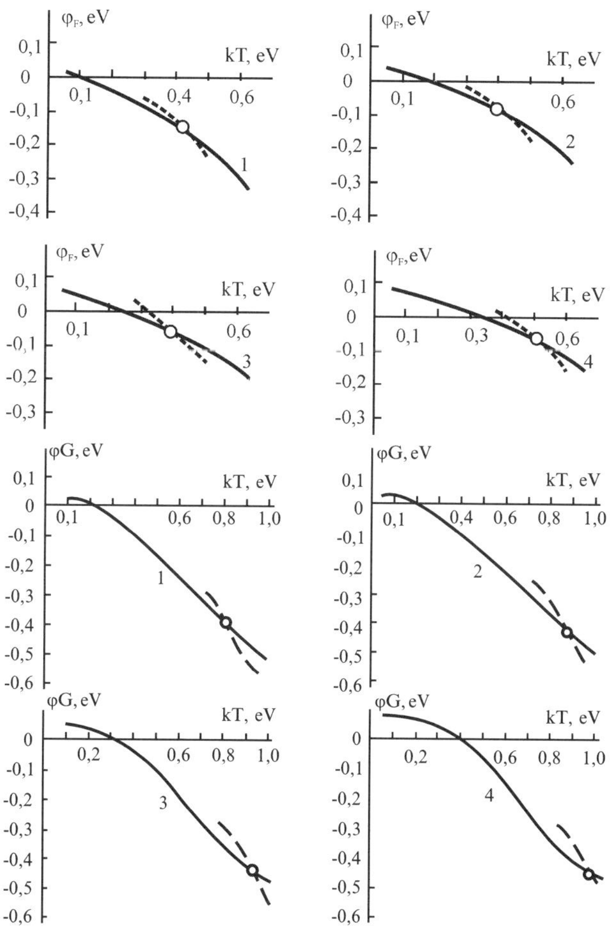

Figure 9. The curve plots of the temperature dependence of thermodynamic potentials φ_i for carbon phases constructed by formula (2) for energetic constants (4) and different pressures $P/P_i' = 0; 0.1; 0.2; 0.3; 0.4; 0.5$ (curves 1–6). The intersection points of functions φ_A (T) and φ_F (T), φ_F (T) and φ_G (T) and φ_G (T) and φ_C (T) are marked with circles. The dotted lines around the intersection sites show the sections in the plots for thermodynamic potentials with which the intersection occurs, respectively with φ_F, φ_G, φ_C

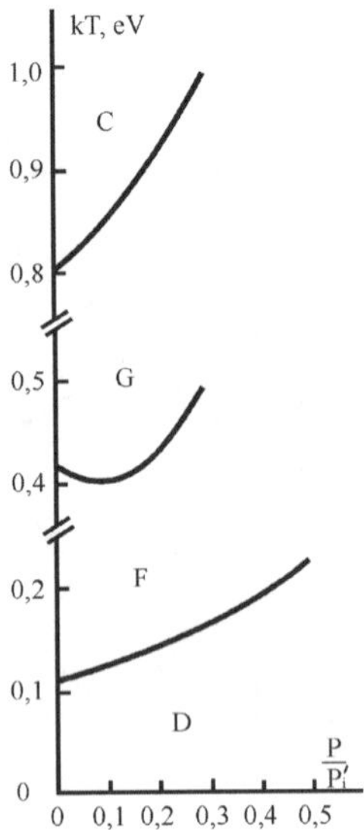

Figure 10. The phase diagram calculated for carbon

8. Conclusions

Hence the molecular-kinetic theory for phase transitions between allotropic carbon modifications (D, F, G, C) has been developed. The possible ordering of atoms and fullerenes in these phases has been considered. The statistical calculation of thermodynamic potentials for carbon phases: diamond, fullerite, graphite, carbyne has been carried out, their dependences on the phase composition, temperature, pressure, order parameter and energetic constants have been established.

The curves for temperature dependence of thermodynamic potentials at different values of external pressure have been plotted using the chosen energetic parameters. The possible values of temperature and pressure for phase transitions have been determined by the intersection points of these curves.

The equations for thermodynamic equilibrium, defining the dependence of the order parameter on temperature and pressure, have been derived. The curves of these dependences have been plotted. As illustrated, the increase of temperature at the certain constant pressure tends to decrease the order parameter. The degree of ordering can both undergo a rise and a decrease due to pressure depending on the values of energy constants of the phase.

The phase diagram has been constructed with estimation of the temperature and pressure ranges, in which different carbon phases can be realized. As found out, the pressure interval of the diamond phase existence narrows with temperature increase. For other phases this range tends to the widening. For the diamond and graphite phases this result is in agreement with experimental data. The region of the fullerite phase is sandwiched between the regions of diamond and graphite and this is also consistent with the data of experimental investigations. And finally, the region of carbyne realization has practicability at high temperatures (up to 2×10^3 K according to experimental data). However it should be mentioned that correspondence between the results of calculation and experiment is only qualitative.

References

1. Koulson Ch. Valency. - Moscow: Mir, 1965. 426 p. (*in Russian*).
2. Levin A.A. Introduction to the quantum chemistry of solid. - Moscow: Khimiya, 1974. 240 p. (*in Russian*).
3. Trefilov V.I., Schur D.V., Tarasov B.P., Shul'ga Yu.M., Chernogorenko A.V., Pishuk V.K., Zaginaichenko S.Yu. Fullerenes as a basis of materials of the future. - Kiev: ADEF, 2001. 146 p. (*in Russian*).
4. Matysina Z.A., Zaginaichenko S.Yu., Schur D.V. Orders of different type in crystals and phase transformations in carbon materials. - Dnepropetrovsk: Nauka i obrazovanie, 2005. 524 p. (*in Russian*).
5. Schur D.V., Matysina Z.A., Zaginaichenko S.Yu. Carbon nanomaterials and phase transformations in these materials. - Dnepropetrovsk: Nauka i obrazovanie, 2007. 680 p. (*in Russian*).
6. Bundy F.P., Bovenkerk H.P., Strong H.M., Wentorg P.H. Diamond-graphite equilibrium line from growth and graphitization of diamond. *J. Chem. Phys.* 1961. **35**(2): 383–391.
7. Strong H.M., Hanneman R.E. Crystallization of diamond and graphite. *J. Chem. Phys.* 1967. **49**(9): 3668–3676.
8. Kennedy C.S., Kennedy G.C. The equilibrium boundary between graphite and diamond. *J. Geophys. Res.* 1976. **81**(14): 2467–2470.
9. Novikov N.V. Physical properties of diamond. Spravochnik. - K.: Naukova dumka, 1987. 192 p. (*in Russian*).
10. Matysina Z.A., Molodkin V.B., Zaginaichenko S.Yu., Schur D.V. Phase transformations in carbon materials. *J. Nanosyst. Nanomater. Nanotechnol.* - K.: RVV IMF NANU, 2005. **3**(1): 11–37 (*in Russian*).
11. Smirnov A.A. Molecular-kinetic theory of metals. - M.: Nauka, 1966. 488 p. (*in Russian*).
12. Smirnov A.A. Theory of phase transformations and distribution of atoms in interstitial alloys. - K.: Naukova dumka, 1992. 280 p. *(in Russian)*.

SOLUBILITY AND TRANSFORMATION OF FULLERENE C_{60} MOLECULE

D.V. SCHUR, S.YU. ZAGINAICHENKO, A.D. ZOLOTARENKO
Institute for Problems of Materials Science of NAS of Ukraine,
3 Krzhizhanovsky Str., Kiev, 03142 Ukraine
Fax: 38-044-424-0381; E-mail: shurzag@materials.kiev.ua
T.N. VEZIROGLU
Clean Energy Research Institute, University of Miami,
P.O. Box 248294, Coral Gables, Miami, FL, 33124-0622, USA

Abstract. This paper deals with the research of solubility of fullerene C_{60} molecule and from our investigation we have concluded that the degree of solubility of fullerene molecules depends on their chemical activity and their chemical activity is defined by the energy state of fullerene molecule. Among the set of states three ground resonance structures (A, B, C) are in existence in fullerene molecule.

In our opinion the extremality of solubility of molecule C_{60} near the room temperature is explained by the structure change of double bonds in fullerene frame. These bonds status modify the chemical activity of molecule C_{60} and consequently the capacity to add the different amount of solvent molecules. The status of double bonds depends on the state and interaction of pentatomic molecules, forming the frame of molecule C_{60}, with each other. Here we show that on cooling the five atomic molecules in the C_{60} frame transform from the state of radialene into pentaene and then into pentadiene. In the course of this process the double bonds in the C_{60} frame are turned from extramolecular to intramolecular for five atomic molecules.

This conclusion follows from the physical and chemical features of effects observed both in a solid and in solution for the definite interval of temperatures.

Keywords: fullerene C_{60} molecule, solubility, pentatomic cyclic molecule, transformation of fullerene frame, pyracyclene unit

1. Introduction

The paragraph with the title, corresponding to the paper title of this presentation, made its appearance in 1993 year in the section "Letters to Nature" of journal "Nature" [1]. It presented experimental results of research in the field of solubility of molecules C_{60}. The authors were of opinion to believe anomalous the fact that the solubility of C_{60} in the organic solvents, given in this paper, has a maximum at near-room temperature (280 K). The established dependence didn't fit well within the existing limits and usual concept about solubility of organic bodies in organic solvents. An increase of solubility with increased temperature was believed to be regular. The assumption that the dissolution process is endothermal was suggested.

Neither this paper nor the works [2–11] and others, published during the ensuing 15 years, did not provide an answer on a raised question and this question is still an open question.

B. Baranowski et al. (eds.), *Carbon Nanomaterials in Clean Energy Hydrogen Systems.*

2. Results and discussion

In the paper [12] (well before the experimental finding of fullerenes) the structure was calculated and the possibility of existence of stable fullerene C_{20} and C_{60} molecules was predicted.

In addition, in this work according to the calculations of electronic structure of fullerene molecule (by the method of M.O. Huckel) and orders of bonds between carbon atoms one of important conclusions was made. It resides in the fact that three basic resonance structures (Fig. 1(a–c)) exist among the large family of fullerene isomers "C_{60}".

The structure (a) (Fig. 1) is more tolerant from them. As a matter of fact these structures take account of the stepwise graphitization (annealing) of fullerene molecule in the reverse sequence (c → b → a) [13–14]. During the course of graphitization the rearrangement of double intramolecular bonds into intermolecular takes place for pentatomic molecules of carbon, forming the fullerene molecule. The cyclopentadiene carbon molecule become cyclopentaene and then pass into radialene (Fig. 2).

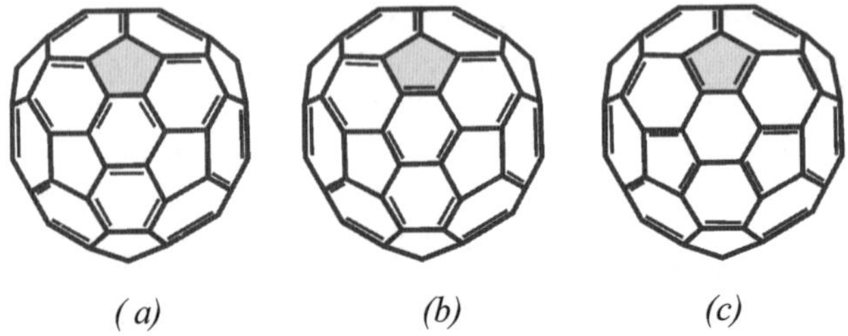

Figure 1. The covalent resonance structures of carbo-s-icosahedron: (*a*) with the axis of fifth order (one structure), (*b*) and (*c*) with the axis of the third order (10 equivalent structures in each). All of structures are with 30 double bonds [12]

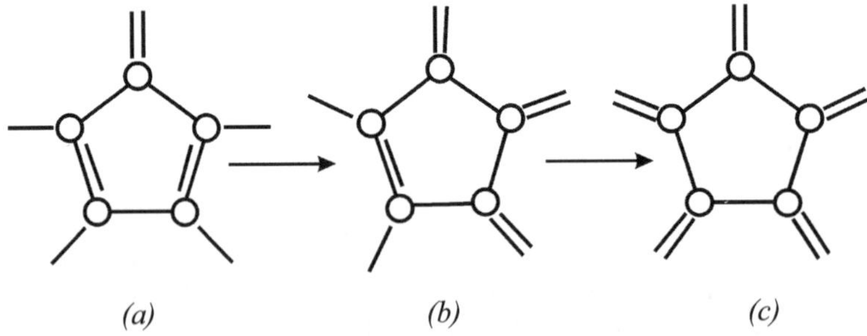

Figure 2. The transformation of pentadiene molecule to radialene: (*a*) pentadiene, (*b*) pentaene, (*c*) radialene

These molecules owing to the availability of unpaired electrons, combined in six pairs in the C_{60} frame, can form the stable structures in the form of pyracyclene units. The existence of such pairs or so-called pyracyclene units distinguishes the fullerene molecule from all other allotropies of carbon and exerts influence on its chemical activity. The state of cyclopentadiene molecule is shown in Fig. 3 at pairing or formation of pyracyclene units.

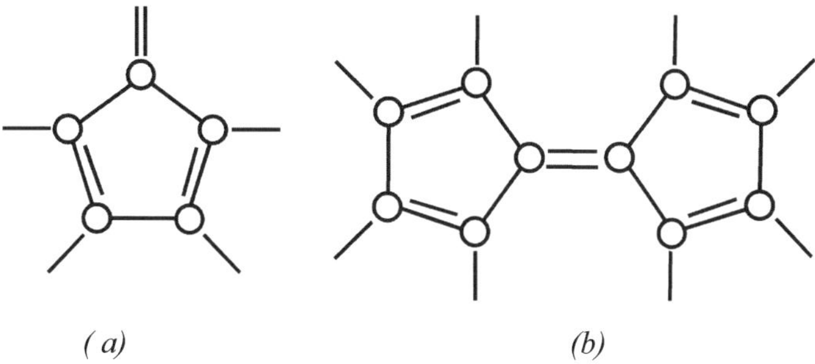

Figure 3. The state of pentatomic molecule at the formation of cyclopentadiene pairs: (*a*) single pentatomic molecule with one double (external) intermolecular bond; (*b*) pyracyclene-similar unit in the frame of fullerene C_{60} molecule

The remaining four electrons in each pentatomic cyclic molecule in the frame of fullerene molecule are used for intra- and extramolecular bonds [15] (depending on thermodynamic conditions). A total of 48 bonds ($4 \times 12 = 48$) is in each molecule C_{60}. They determine physical and chemical properties of fullerene molecule [15–16].

2.1. THE STRUCTURAL TRANSFORMATIONS IN FULLERITE

At the phase transition f.c.c. → s.c. in fullerite the process finds its way through the intermediate stage of the b.c.c. lattice formation. In this case in the f.c.c. lattice the pentatomic cyclic molecule in the frame of fullerene molecule exists at high temperatures in the form of radialene (Fig. 4(a)).

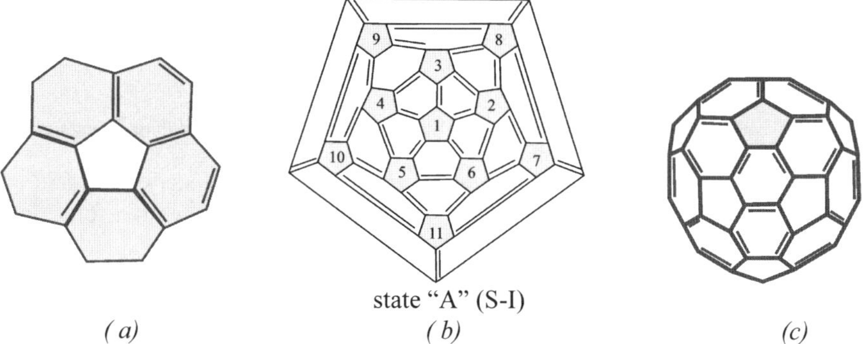

Figure 4. The covalent resonance structure of carbo-s-icosahedron with the axis of fifth order, forming the f.c.c. lattice of fullerite: (*a*) pentatomic molecule, (*b*) Shlegel diagram, (*c*) frame of molecule C_{60}

Twelve radialenes in the frame of fullerene molecule interact with radialenes of 12 fullerene molecules, forming the b.c.c. structure. The radialene is turned to the pentaene molecule (Fig. 5) as the temperature decreases below 280 K.

These molecules, interacting with each other, form eight cyclic hexatomic structures in the frame that follow the rule of Huckel. These structures due to the magnetic interaction, excited by ring currents, attract to itself eight neighbouring fullerene molecules, forming the b.c.c. structure (transitional state "B"). The ring currents flow in the direction to each other.

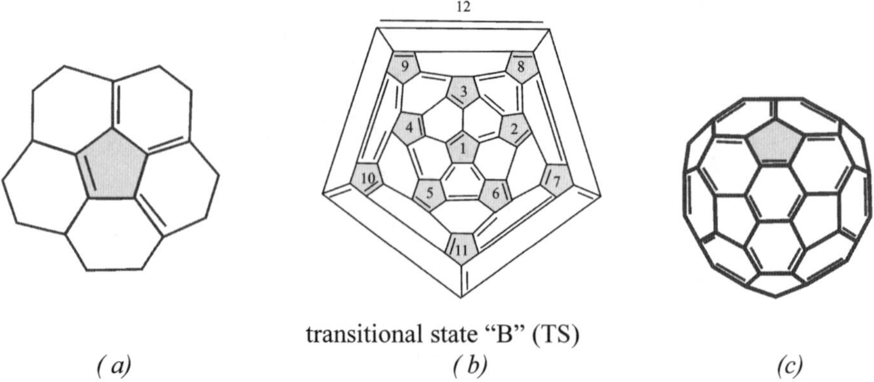

Figure 5. The covalent resonance structure of carbo-s-icosahedron with the axis of third order, forming the b.c.c. lattice of fullerite

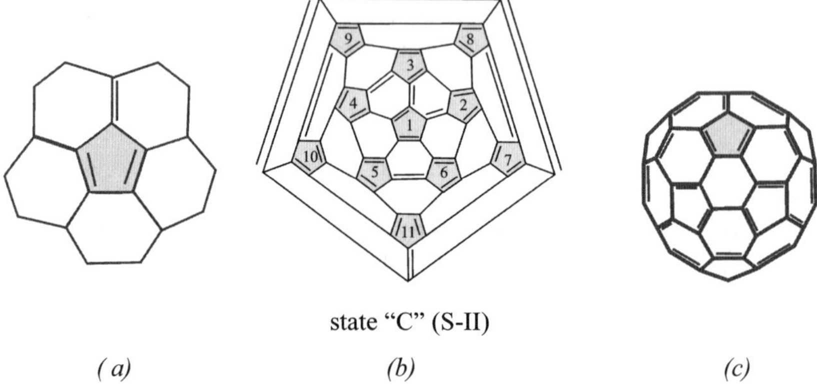

Figure 6. The covalent resonance structure of carbo-s-icosahedron with the axis of third order, forming the s.c. lattice of fullerite

At the temperature below 260 K the phase transition b.c.c. → s.c. occurs and all extramolecular bonds (except of pairing) of pentatomic molecule transform into intramolecular. The molecule itself is turned to pentadiene (Fig. 6(a)).

In these conditions the fullerene molecule with the structure presented in Fig. 6(b, c) is formed. As is obvious from the Shlegel diagram of Fig. 6(b), this structure has 30 double bonds, 24 of which are intramolecule for pentadiene molecules and 6 another intermolecular combine molecules into 6 pyracyclene-similar pairs.

In this case the pairing bonds between pentatomic molecules become the chemical active centers. Six pyracyclene units represent aromatic structures and, interacting with the similar structures in the six neighbouring fullerene molecules, form the s.c. structure.

Hence modifications A, B and C of the fullerene molecule exist correspondingly at the temperatures of $T_A \geq 280$ K, $T_B = 273$ K and $T_C \leq 260$ K (Fig. 7).

But these temperatures correspond to the moments of existence of molecules of one kind. The mixtures of fullerene molecules of two types are found over intervals between these temperatures. The relationship among the concentrations of fullerene molecules defines the type of crystal lattice in fullerite (Fig. 7). At the temperature of orientational phase transition (260 K) the concentration of one or another molecule kind is over 50%. This dictates the fullerite structure in the present thermodynamic conditions. In this manner the solid phase transition in fullerite is possible at the existence in the system of more than 50% of any modification of fullerene molecule (Fig. 7).

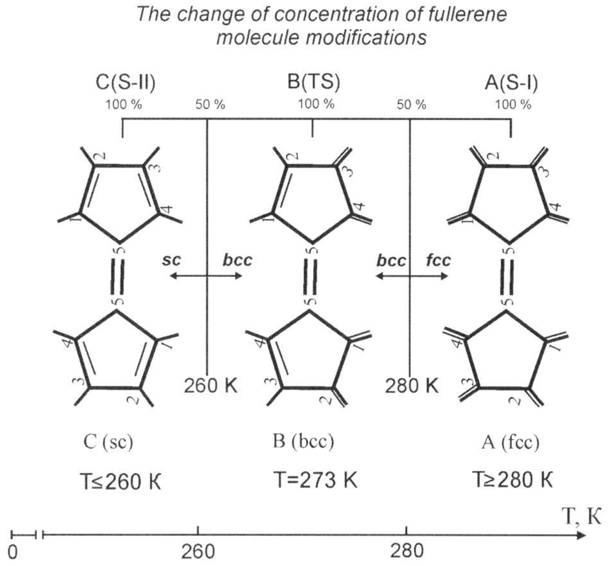

Figure 7. The structural transformations in the cyclopentadiene molecules of fullerene molecule at the phase transition s.c. → b.c.c. → f.c.c. in the molecular crystal of fullerite

As the temperature increases the s.c. → f.c.c. phase transition in the fullerite lattice also occurs as on cooling by the way of formation of intermediate phase with the b.c.c. lattice. On heating of fullerite to the temperature of 260 K and above the bonds reconstruction takes place in the frame of fullerene molecule. In consequence of this, the fullerene molecules rebuild their frame, producing the change of lattice structure. The fullerene molecule transforms from the “C” state into the “B” intermediate state, which with further rise in temperature give place to the more steady state “A” with the resulting formation of f.c.c. lattice of fullerite (b.c.c. → f.c.c. phase transition).

In the f.c.c. lattice of fullerite in the pentatomic molecule all double bonds are directed outward. The formation of five equivalent intermolecular double bonds in each pentatomic molecule of tetracycline unit transforms the pentatomic molecule into independent structural radialene unit.

The combined effect of A ↔ B ↔ C transition is well defined experimentally and notes the step change of physical properties. However for the present the attempts to note each semi-stage individually have not met with success.

2.2. FULLERENE MOLECULES IN SOLUTION

The effects, different from solid phase, are observed experimentally in the solutions of fullerenes in the absence of short and long range ordering between molecules.

The structural transformations in fullerene molecule are easily observable in solutions. Anomalies (Fig. 8) in their solubility make it possible to fix the formation temperatures of one or another modification of fullerene molecule A (T ≥ 300 K), B (T = 280 K) and C (T ≤ 260 K).

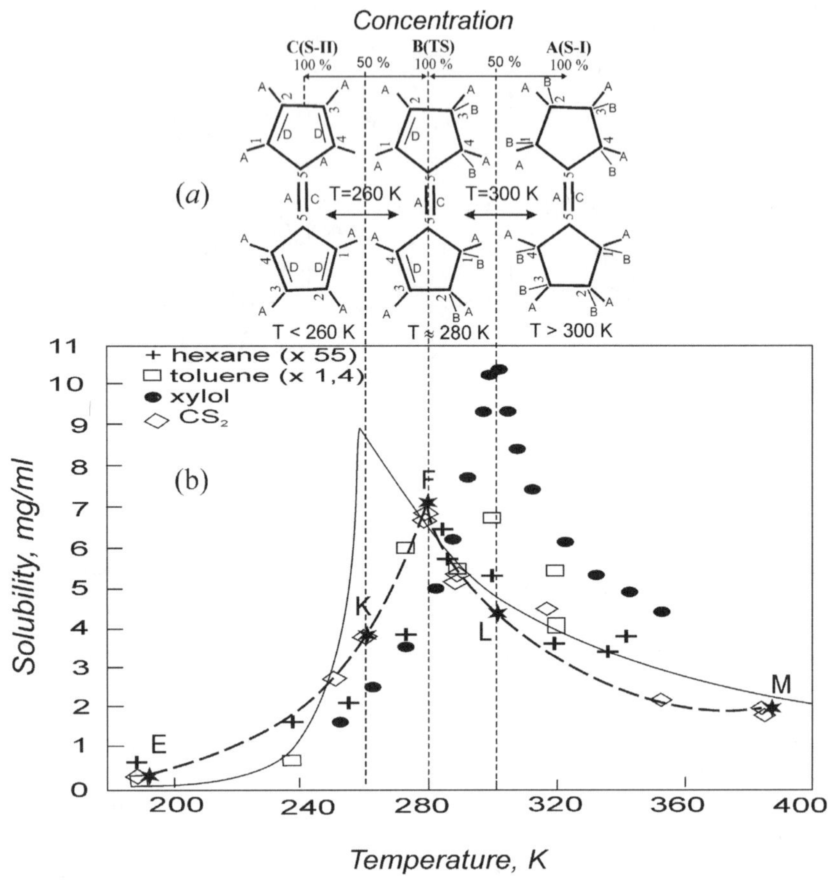

Figure 8. The structural modifications in fullerene C_{60} molecule with the change of solution temperature: (*a*) structural changes in pentatomic molecule at thermal change (A – bond with neighbouring carbon atom in fullerene molecule; B – bond with solvent molecules; C – bonds, forming pyracyclene molecules; D – intramolecular bonds of pentatomic molecules); S-I, S-II, TS – states (I, II and transient) of fullerene molecule; (*b*) temperature dependence of fullerene C_{60} solubility. Solid line is calculated curve of clusterization processes within the framework of drop model [7], EKFLH – idealized curve of solubility of fullerene C_{60} molecule in CS_2

In the paper [7] on the basis of present knowledge of R.S. Ruoff [1] the authors undertook the attempt to provide an explanation for the observed effect by the processes of clasterization (Fig. 8(b)). We amplify this figure by Fig. 8(a), where the temperature dependent structural transformations in the pentatomic molecule are shown. As illustrated in Figs. 7 and 8, the processes in solutions are analogous to those going on in a solid. Radialene in fullerite not fully corresponds to modification “A” in solution. However the pentatomic molecule in both cases (in fullerite and solution) don’t have the double intramolecular bonds. As this takes place, its extramolecular bonds can be intramolecular or extramolecular for the frame of fullerene molecule. This makes them nonequivalent energy-wise.

At high temperatures each pentatomic molecule forms four "B"-bonds with a solvent molecules, forming the donor-acceptor bonds [17, 18] or bonds with charge transfer [19]. In transition from the state "A" to the state "B" two from four bonds of pentatomic molecules, forming with solvent, are changing from extramolecular to intramolecular.

In this situation the half of solvent molecules, participating in interaction, become free and the released solvent is in a position to dissolve the same amount of fullerene molecules. For this reason at the temperature of 280 K the solubility is increased in two times.

At the temperature below 260 K (the phase transition B ↔ C) the concentration of pentadiene molecules is dominant. The paired pentadiene molecules in the frame of fullerene make up six stable structures thermodynamically (under these thermodynamic conditions) conforming to the Huckel rule. In that case the fullerene molecules are insoluble.

We concluded, on the basis of the above line of reasoning, that the degree of conversion of fullerene molecules depends on their chemical activity and the chemical activity is determined by the energy state of fullerene molecule. In each energy state the fullerene molecule is definite in structure. Among the sent of states three ground resonance structures are in existence, the conventional symbols of these states are A, B and C (Figs. 7 and 8).

In such a way the degree of solubility of fullerene molecules is related directly to the concentration of one or another kind of fullerene molecules in the solution. In that case on examination of Fig. 8 it is pertinent to note that at the phase transition C ↔ B ($T = 260$ K) the concentrations of fullerene molecules of C and B kind are bound to be equal to $(50 \pm 1)\%$. As concentration of "B"-kind molecules approaches zero, the fullerene solubility also approaches zero.

On consideration of the phase transition B ↔ A it should be pointed out that in this situation when concentration of "B"-kind fullerene molecules tends to zero, the solubility of fullerene molecules tends to the definite limit value, at which the concentration of "A"-kind molecules corresponds to 100% ($T \geq 300$ K). The phase transition B ↔ A comes at the temperature of $T = 300$ K, corresponding to the concentrations of both modifications of fullerene molecules equal to $(50 \pm 1)\%$.

The change of a value of solubility maximum (i.e. transference of "F" point along the ordinate axis) for different dissolvents is defined by the solvents nature, that is by their chemical activity with respect to the fullerene molecules.

The moving of "F" point parallel to the abscissa is determined by the influence of surroundings of fullerene molecule on the temperature dependent processes of transformations in fullerene frame. This manifests itself in the fact that the behaviour of fullerene molecule is different in the fullerite and solution.

2.3. THE STRUCTURAL TRANSFORMATIONS IN FULLERENE MOLECULE AT FULLERITE FORMATION

The fullerite crystallization from gaseous phase occurs at the temperature about 1,000 K and graphite formation happens at the test temperature of around 4,000 K. On cooling of carbon vapour the forming fullerene molecules are adsorbed at the surface of just forming soot particles, most of which represent the graphene molecules or their clusters. The carbon atoms in these molecules of fullerene are free of hybridizated orbitals.

In practice desorption (release) of fullerene molecules is performed by sublimation in gaseous phase or by extraction (Fig. 9).

At the sublimation fullerene molecule (Fig. 10(a)) goes through the graphitization process which result in the hybridization of electron orbitals of carbon atoms (Fig. 10(b)).

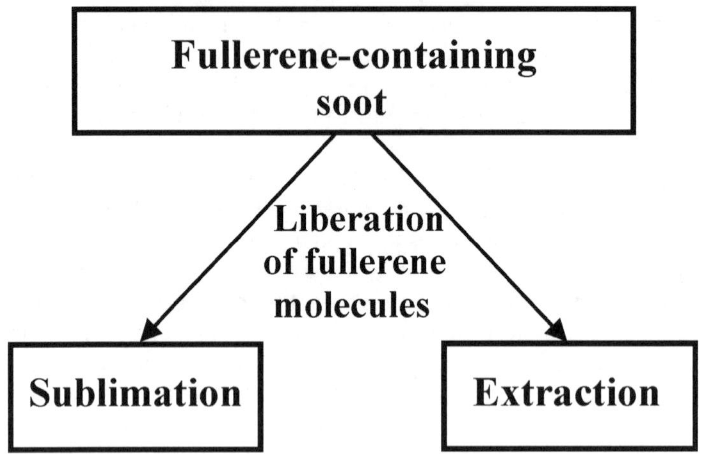

Figure 9. The methods of fullerene molecules release from fullerene-containing soot

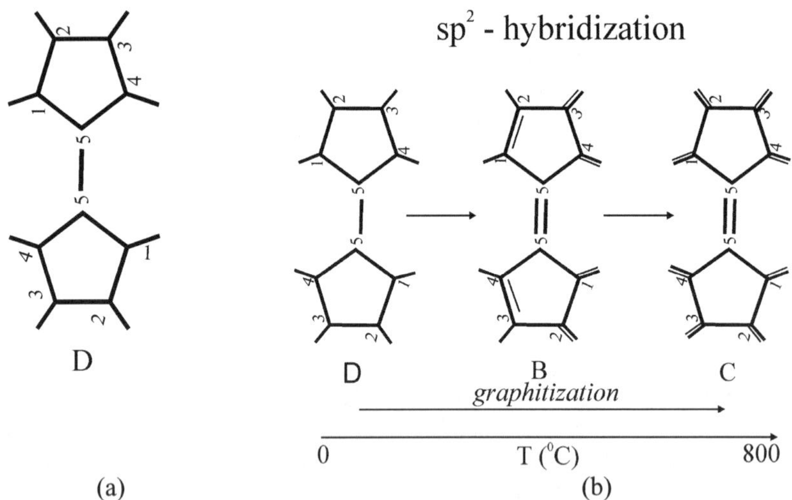

Figure 10. The structural transformation in cyclopentadiene fullerene frame at the sublimation of fullerene molecules from soot: (a) pentatomic molecules, atoms of which have no hybridizated orbitals ("D"); (b) the sequence of stages of fullerene molecule graphitization prior to sublimation

At the extraction fullerene molecules form more strong bonds with solvent molecules, than those they are bonded with grapheme molecules, and pass into solution (Fig. 11).

In solution the chemical activity of fullerene molecule vary with the temperature of solution, determining the structure of fullerene molecule (Fig. 8). The crystallization of fullerene molecules from solution is possible by two methods: salting-out of fullerenes or evaporation of solvent (Fig. 12).

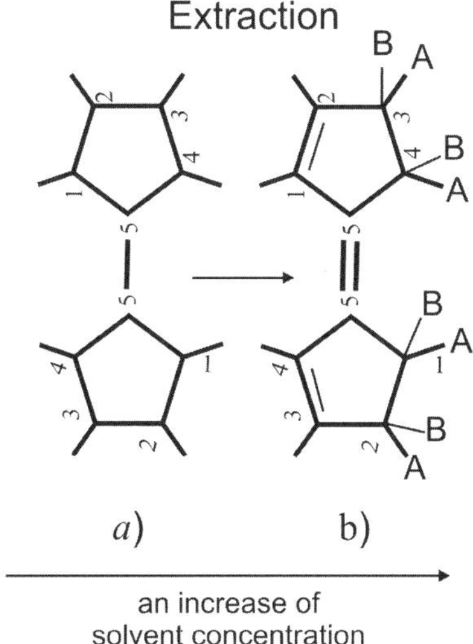

Figure 11. The extraction of fullerene molecules from fullerene-containing soot by solvents: (*a*) molecule, adsorbed on soot particles; (b) molecule in solution; A – bond between carbon atoms in fullerene frame; B – bond between fullerene molecule and solvent molecule

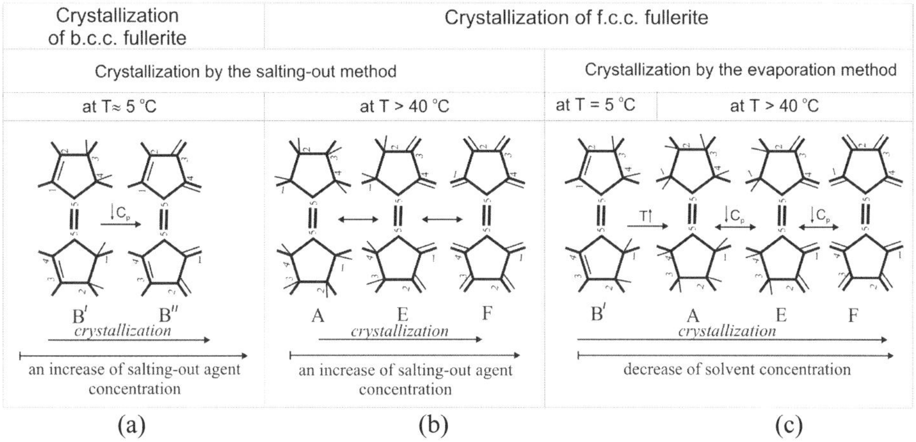

Crystallization of b.c.c. fullerite	Crystallization of f.c.c. fullerite		
Crystallization by the salting-out method		Crystallization by the evaporation method	
at T≈ 5 °C	at T > 40 °C	at T = 5 °C	at T > 40 °C
C_p B′ B″ *crystallization* an increase of salting-out agent concentration	A E F *crystallization* an increase of salting-out agent concentration	T↑ C_p C_p B′	A E F *crystallization* decrease of solvent concentration
(a)	(b)	(c)	

Figure 12. The structural transformations in pentatomic molecules of fullerene frame at crystallization of fullerite from solution (C_p – concentration of solvent): (a) crystallization of b.c.c. fullerite, B′ – fullerene molecule in solvent, B″ – fullerene molecule in fullerite; (b) crystallization of f.c.c. fullerite by the method of salting-out, "E" – fullerene molecule forming the bonds with molecules of solvent a half as many as molecule "A", "F" – fullerene molecule in f.c.c. fullerite; (c) crystallization of f.c.c. fullerite by the method of evaporation

At the salting-out more strong bonds between molecules of solvent and salting-out agent allows the last to break bonds between molecules of fullerene and solvent. In consequence of this, fullerene molecules interact with each other, forming clusters and crystallize with precipitation.

The evaporation method is capable of matching the same result. Depletion of solution by molecules of solvent results from vaporization of solvent molecules.

3. Conclusions

- Thus, the structural transformations in crystalline lattice of fullerite is the consequence of reconstruction of fullerene molecule. Owing to this reason both in fullerite and in fullerene solution the mentioned process is accompanied by physical effects as crystalline lattice rearrangement or solubility change.
- As it follows from the above, each molecule of fullerene C_{60} is surrounded by three layers of solvent molecules. First of them has 48 molecules at the temperature above the room. The number of interactions reduces as the temperature decreases. So, the molecule C_{60} interacts with 24 molecules of solvent at the room temperature and does not have interaction at the temperature below 260 K.
- The special feature of examined process is the fact that the temperatures of two transformations in the molecule C_{60}, staying in solution, are near to the temperature of medium of habitation of living and vegetable world. The first exists at the temperature near to the temperature of human body (310 K), and the second – near the temperature of 0ºC (273 K), which is the threshold of existence of the most representatives of living and vegetable world.
- A similar physical effect can be used in pharmaceutics for the preparation of different medicine solutions, having applied the change of chemical activity of molecule C_{60} with the temperature change within the range of room temperatures.

References

1. Ruoff R.S., Malhotra R., Huestis D.L., Tse D.S., Lorents D.C. Anomalous solubility behaviour of C_{60}. *Nature* 1993. **362**: 140–141.
2. Ying Q., Marecek J., Chu B. *Chem. Phys. Lett.* 1994. **219**: 214.
3. Suzuki E. et al. Molecular Nanostructures. - Singapore: World Scientific, 1998.
4. Sivaraman N., Dhamodaran R., Kaliappan I., Srinivasan T.G., Rao P.R.V. Solubility of C_{60} in organic solvents. *J. Org. Chem.* 1992. **57**: 6077–6079.
5. Doome R.J., Fonseca A., Nagu J.B. ^{13}C–MAS–NMR investigationn of solutions saturated with fullerenes: study of the anomalous solubility behaviour. *Colloids Surf. A Physicochem. Eng. Asp.* 1999. **158**(1–2): 137–143.
6. Beck M.T., Mandi G. *Fullerenes Science Technol.* 1996. **3**: 32.
7. Bezmel'nitsyn V.N., Eletskii A.V., Okun' M.V. Fullerenes in solutions. *Uspekhi Fiz. Nauk.* 1998. **168**(11): 1195–1220 (*in Russian*).
8. Ruof R.S., Tse D.S., Malhotra, R., Lorents D.C. Solubility of fullerene (C_{60}) in a variety of solvents. *J. Phys. Chem.* 1993. **97**: 3379–3383.
9. Scrivens W.A., Tour J.M. *J. Chem. Soc., Chem. Commun.* 1993. **15**: 1207–1209.
10. Beck M.T., Mandi G. *Fullerenes Science Technol.* 1996. **5**(2): 291–310.
11. Letcher T.M. et al. *S.-Afr. J. Chem.* 1993. **46**: 41.
12. Bochvar D.A., Gal'pern E.G. About hypothetical system: carbododecahedron, s-icosahedron and carbo-s-icosahedron. *Dokl. Akad. Nauk SSSR.* 1973. **209**(3): 610–612 (*in Russian*).
13. Schur D.V., Zaginaichenko S.Yu., Veziroglu T.N. The peculiarities of hydrogenation of pentatomic carbon molecules in the frame of fullerene molecule C_{60}. *Int. J. Hydrogen Energy* 2008. **33**(13): 3330–3345.
14. Schur D.V., Zaginaichenko S.Yu., Lysenko E.A., Golovchenko T.N., Vlasenko A.Yu. The peculiatities of molecule C_{60} formation. Proceedings of 10th International Conference

"Hydrogen Materials Science and Chemistry of Carbon Nanomaterials", Sudak, Crimea, Ukraine, 2007, P. 716–721.
15. Schur D.V., Zaginaichenko S.Yu., Matysina Z.A. Effects of stimulation of hydrogen solubility in fullerite C_{60}. *Nanosyst. Nanomater. Nanotechnol.* 2007. **5**(2): 385–409.
16. Schur D.V., Matysina Z.A., Zaginaichenko S.Yu. Carbon nanomaterials and phase transformations in them. - Dnepropetrovsk: Nauka i obrazovanie, 2007. 678 p. (*in Russian*).
17. Zaginaichenko S.Yu., Anikina N.S., Zolotarenko A.D., Krivushenko O.J., Schur D.V. Regularity of C_{60} fullerene dissolving in methyl derivatives of benzene. Proceedings of 10th International Conference "Hydrogen Materials Science and Chemistry of Carbon Nanomaterials", Sudak, Crimea, Ukraine, 2007, P. 668–671.
18. Anikina N.S., Schur D.V., Zaginaichenko S.Yu., Zolotarenko A.D., Krivushenko O.Ya. On the donor-acceptor mechanism of C_{60} fullerene dissolving in aromatic hydrocarbons. *Ibid.* P. 676–679.
19. Anikina N.S., Schur D.V., Zaginaichenko S.Yu., Zolotarenko A.D., Krivushenko O.Ya. The role of chemical and physical properties of C_{60} fullerene molecules and benzene derivatives in processes of C_{60} dissolving. *Ibid.* P. 680–683.

THERMALLY CONTROLLED HYDROGEN STORAGE SYSTEM USING NOVEL CARBON MATERIALS

L.L. VASILIEV, L.E. KANONCHIK, V.A. BABENKO
Laboratory of Porous Media, Luikov Heat & Mass Transfer Institute, National Academy of Sciences, P. Brovka, 15, 220072 Minsk, Belarus
Tel./Fax: (375-17) 284-21-33; E-mail: kanon@htmi.ac.by

Abstract. Successful development of sorption storage technologies for hydrogen assumes an active temperature control and special properties of the materials, capable to adsorb hydrogen in a reversible way.

The goal of this paper is the development of the sectional vessel with heat pipe (HP) for hydrogen sorption storage at average pressures 3.5–6 MPa, every separate section of which has the case made from an aluminium (or reinforced plastics) and filled with briquettes of the sorbent material where hydrogen is situated in adsorbed and compressed states. A two-dimensional transient model was developed to analyse discharge/charge process of a flat rectangular vessel with sorbent and hydrogen. Such vessels for hydrogen storage and transportation are interesting to be applied in fuel cells vehicle, or dual-fuel engine car (hydrogen/gasoline, hydrogen/methane. The suggested sorbent bed for the storage system is based on compressed carbon fiber as a host material which includes the micro- and nanoparticles of different metal hydrides disposed on the surface of the active carbon and inside macropores.

Keywords: carbon material, metal hydride, hydrogen, heat pipe, sorption storage

<u>*Nomenclature:*</u> a_v – volumetric capacity of hydrogen storage using physicsorption, C_a – specific heat capacity of adsorbed gas, g_i, – gas output from the elementary cell, used for computer modeling, K_{s0} – pre-exponent constant in the equation of kinetic of sorption, *M*– mass of the gas in the cylinder, M_i – mass of the gas in the calculated cell, *m* – dynamic coefficient of filling the cylinder; *N* – number of calculated cells in the cylinder; *Q* – heat flow, *r* and *z* – cylindrical coordinates, *R* – external radius of the cylinder shell, R_0 – internal radius of the heat pipe, r_0 and r_1, internal and external radii of the annular layer of the sorbent, *2S* – fins step, W_0 – maximum microporous specific volume, *v* – component of the velocity vector, STP – standard of temperature (273 K) and pressure (0.1 MPa), β – thermal coefficient of expansion, 2δ – fin thickness, η – dynamic viscosity, ρ_s – density of the solid sorbent, ρ – total density of the free and adsorbed gases in the cylinder.

<u>*Subscript:*</u> a, adsorbate; ch, charge; cr, critical state; e, finite value; env, environment; eq, equilibrium conditions; ex, extractive; hp, heat pipe; 0, initial value; s, sorbent; f, fin; t, transfer.

1. Introduction

Hydrogen has the potential to be a very attractive alternative energy carrier. It is clean, efficient, and derived from diverse domestic resources, such as fossil resources.

B. Baranowski et al. (eds.), *Carbon Nanomaterials in Clean Energy Hydrogen Systems.*

Success in development of a new hydrogen technologies, such as fuel cells, transport systems on hydrogen, metal hydride based heat pumps has shown, that the use of hydrogen results in qualitatively new solutions of ecological and power problems.

One of the ways of gradual introduction of hydrogen on motor transport consists in application of dual-fuel internal combustion engines (hydrogen – gasoline, hydrogen – methane) with a gas sorption storage system. Such approach provides minimal weight increase due to storage system and does not demand application of electromotors, current converters, powerful accumulators and the advanced infrastructure of hydrogen refuellings. Sorption storage is free of necessity for bulky and metal-consuming tanks, for large energy expenses on compression up to high pressures (20–60 MPa) or liquefaction of gas, so it cuts down expenses for the compressor and cryogenic equipment. In service of the dual-fuel motor transport a change is possible from hydrogen to hydrocarbon fuel or to their mixture, depending on accessibility of this or that fuel. Realization of any scheme of hydrogen usage in transport depends on making a cost-effective system of gas storage. Successful development of a getter system of hydrogen storage presumes a presence of an active thermal regulation and special properties of materials capable to store up hydrogen efficiently. The inexpensive microporous carbon obtained by thermal treatment of raw material (wood, sawdust, cellulose, straw, paper for recycling, peat, etc.) after impregnation is especially attractive [1]. In this way a number of perspective sorbents were created in Belarus: the activated carbon fiber "Busofit-M8" (a product of pyrolysis of the impregnated cellulose) and the activated carbon made of waste products of wood "WAC 3-00" [2, 3]. A heat pipe, a thermosyphon, a single-phase heat exchanger or an electric element can be used as the heat exchange element. The heat pipe has advantages to other devices. The heat pipe allows us to use different sources of energy (a gas flame, electricity, vapor and/or liquid circuits). The automobile air conditioning system, exhaust gases or an automobile radiator can also serve for sorbent preheating. Moreover, the heat pipe being many factors of ten better "superconductor" of heat, than silver or copper ensures homogeneity of temperature field at a heating surface.

This work is devoted to development and research of the thermally regulated onboard system of hydrogen storage applying the novel carbon sorbents. Modeling of thermodynamic process of discharge/charge of hydrogen from the flat rectangular vessel with heat pipes (HP) was carried out to choose the best performance. A new sorbent bed (composite sorbent – "metal-hydride particles on the fibre") for hydrogen storage and transportation vessels was suggested. Experimental verification of the efficiency of the composite sorbent was validated.

2. Heat and mass transfer in the sorbent bed

Modeling of thermodynamic process of gas discharge from sorption storage system was carried out to choose the best performance. The efficient system to perform a sorbent bed thermal control during its sorption/desorption is heat pipe heat exchanger.

Figure 1 shows the sectional vessel with heat pipes [4] for hydrogen sorption storage at average pressure 3.5–6 MPa. Every separate section has the case made from an aluminum (or reinforced plastics) and filled with briquette sorbent where hydrogen is situated in adsorbed and compressed states. Heat pipes can easily be implemented inside sorption storage vessels due to its flexibility, high heat transfer efficiency, cost-effectiveness, reliability, long operating life, and simple manufacturing technology. Heat pipe metal fins serve for intensification of heat transfer in sorbent bed. A flat rectangular, rather than cylindrical form is convenient for gas vessel location in the automobile. The collector channels for gas supply and extraction were manufactured by extrusion inside of aluminum case. Longitudinal grooves in the case can be replaced with cylindrical channels in a sorbent body at the center of each section in parallel to the axis of heat pipe (HP). The thin perforated pipes which are connected then to the common gas collector can be entered into them.

Figure 2 illustrates a calculation domain of cylinder with breakdown to the two calculation areas: S1 (the cylinder case, the body of HP, fins/partitions), and S2 (sorbent) with indication of boundaries between them. The calculation domain is bounded by external surface of the case 2, the internal wall of HP (boundary 5), the symmetry planes at the middle of fins (boundaries 1 and 4) and at the middle of sorbent layer (boundary 3).

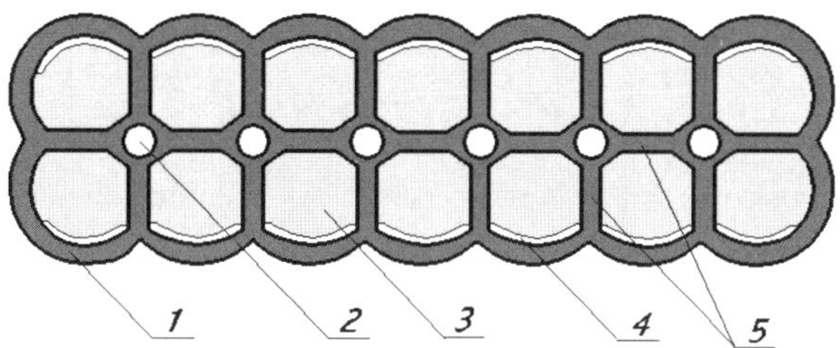

Figure 1. The flat sectional vessel for hydrogen sorption storage: 1 – vessel case, 2 – HP, 3 – sorbent, 4 –channel for gas removal, 5 – longitudinal fins/partitions

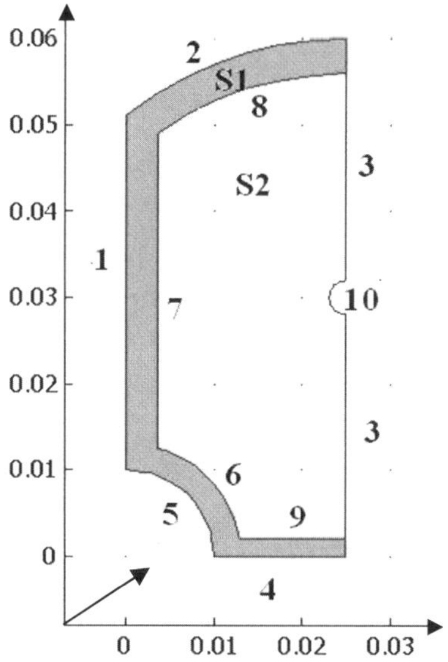

Figure 2. Scheme of calculation domain inside flat sectional cylinder for adsorbed hydrogen storage with indication of the calculation areas S1 and S2 and boundaries: S1 – metal, S2 – sorbent; 1, 3, 4 – symmetry planes; 2, 8 – external and internal surfaces of cylinder case, 5, 6 – internal and external surfaces of HP body; 7, 9 – boundaries between sorbent and metal fins/partitions; 10 – channel for gas supply/extraction

The mathematical model of heat transfer and gas sorption processes in the reactor is based on the following assumptions: (1) the gas in the cylinder is ideal; (2) the temperature of the solid phase is equal to the temperature of the gas phase at each point, because of the high coefficient of the volumetric heat transfer between them; (3) heating and cooling of the sorbent is carried out by heat pipe (HP) with inner heat transfer coefficient $\alpha_{HP} = 10^3 - 10^4$ *W*/(m^{2}K). This coefficient is uniform along the surface and large in comparison with the thermal resistance of interface HP-sorbent bed.

The dynamic model of the sorbent bed has the components [5–7]:
(1) Dubinin – Radushkevich equation of the state of gas

$$a_{eq} = \frac{W_0}{v_a}\exp\left\{-\left[\frac{R_\mu T\ln\left(P_{sat}/P\right)}{E}\right]^2\right\}, \tag{1}$$

where R_μ – gas constatnt, E – activation energy, v_a – specific volume of adsorbed medium

$$P_{sat} = P_{cr}(T/T_{cr})^2; \tag{2}$$

(2) The equation of sorption kinetic

$$\frac{\partial a}{\partial \tau} = K_{s0}\exp\left(-\frac{E}{R_\mu T}\right)\left(a_{eq} - a\right), \tag{3}$$

where a – adsorption capacity, $K_{s0} = 15D_{s0}/R_p^{\ 2}$, D_{s0} – phenomenological constant, R_p – the average radius of particle;
(3) Energy conservation equation

$$(\varepsilon\rho_g C_g + \rho C + \rho a C_a)\frac{\partial T}{\partial \tau} + \nabla\cdot\left(-\lambda\nabla T + \rho_g C_g T\mathbf{u}\right) = q_{st}\rho\frac{\partial a}{\partial \tau} - \beta_g T\varepsilon\frac{\partial P}{\partial \tau}, \tag{4}$$

ε – porosity determined as a part of the volume occupied by the free gas, ρ_g – density of free gas, C – solid sorbent specific heat capacity, C_g – specific heat capacity of free gas, λ – effective thermal conductivity of the sorbent layer, τ – time, q_{st} – latent isosteric heat of sorption,
(4) Mass balance

$$\frac{\partial(\varepsilon\rho_g)}{\partial \tau} + \nabla\left(\rho_g\mathbf{u}\right) + \nabla\left(-D\nabla a\right) = -\frac{\partial}{\partial \tau}\left(\rho a\right); \tag{5}$$

(5) Momentum balance

$$\rho_g\frac{\partial \mathbf{u}}{\partial \tau} + \frac{\eta}{K}\mathbf{u} = \nabla\left[-P\mathbf{I} + \eta\left(\nabla\mathbf{u} + \left(\nabla\mathbf{u}\right)^T\right)\right]. \tag{6}$$

K – permeability

The heat of the phase transition of sorption can be calculated using Clausius-Clapeyron equation with some corrections. In accordance with [8], the isosteric sorption heat of a gas mole is determined by the formula

$$q_{st} = \left(1 - \frac{V_a}{V_g}\right) T \ R_\mu z_g \left[\frac{\partial \ \ln P}{\partial \ \ln T}\right]\Bigg|_{a=const}. \tag{7}$$

z_g – coefficient of gas compressibility, V_a – partial molar adsorption volume, V_g – molar gas-phase volume.

For the ideal gas ($z_g = 1$, $V_a << V_g$), the isosteric heat is calculated from the relation

$$q_{st} = R_\mu T \left[\frac{\partial \ \ln P}{\partial \ \ln T} \right] \Bigg|_{a=constt} \tag{8}$$

The limits of applicability of Eq. (8) are depend on the degree to which the gas can be approximated as ideal, i.e., on the ratio between the partial molar volumes of the adsorbate and gas and on the compressibility coefficient. In the case of a nonideal gas, the estimation obtained with (8) should be considered as being approximate.

Equation system (1–8) defines four variables $T, P, a, \mathbf{u}$, dependent from space coordinates and time.

Initial conditions for them are set as:

$$T(\tau_0) = T_0, \quad P(\tau_0) = P_0, \quad \mathbf{u}(\tau_0) = 0, \quad a(\tau_0) = a_{eq}(T_0, P_0). \tag{9}$$

Boundary conditions are those. At the external side of the case (boundary 2) the conditions of convective heat exchange with environment are considered which are characterized by the resulted heat transfer coefficient:

$$\mathbf{n} \cdot \mathbf{q} = \alpha_{env}(T - T_{env}), \quad \mathbf{q} = -\lambda \nabla T, \tag{10}$$

where $\mathbf{n}$ is the unit vector of exterior surface normal.

Depending on the conditions at the internal surface of the HP body (boundary 5) the temperature T_{HP} or density of heat acting from fins and sorbent is set:

$$\mathbf{n} \cdot \mathbf{q} = q_{HP}, \quad \mathbf{q} = -\lambda \nabla T, q_{HP} = \alpha_{HP}(T_{HP} - T), \tag{11}$$

where α_{HP} is the heat transfer coefficient from HP working fluid (steam) to the internal surface of HP body.

The heat-flux density is equal to zero at symmetry planes (boundaries 1, 3, 4)

$$\mathbf{n} \cdot \mathbf{q} = 0, \quad \mathbf{q} = -\lambda \nabla T = 0. \tag{12}$$

On internal boundaries of calculation area S_1 and on external boundaries of calculation area S_2 (boundaries 6–9) the coupling conditions are accepted (an equality of heat fluxes and temperatures). For boundary 10 where the cylindrical aperture for gas supply is located, the heat flux is ordered which is caused by the receipt of gas with the temperature T_i

$$-\mathbf{nq} = q_0, \ \mathbf{q} = \lambda \nabla T - \rho_g C_p \mathbf{u}, \ q_0 = -\rho_g C_p T_i (\mathbf{u} \cdot \mathbf{n}). \tag{13}$$

The boundary conditions of mass flow absence at external surfaces of calculation domain S_2 (boundaries 6–9) are formulated as zero scalar product of the unit vector of exterior surface normal and diffusive mass flux

$$\mathbf{n} \cdot \mathbf{N} = 0; \quad \mathbf{N} = -D \nabla a, \tag{14}$$

where $\mathbf{N}$ is the vector of diffusive mass flux (the walls are impermeable), D – diffusivity. At surface 10, where gas is extracted, we neglect diffusive mass flux ($-D\nabla a$).

Normal component of flow velocity is set to zero at vertical symmetry axis (boundary 3)

$$\mathbf{n} \cdot \mathbf{u} = 0; \tag{15}$$

The velocity is zero at the walls (boundaries 6–9)

$$\mathbf{u} = 0. \tag{16}$$

In the case of the presence of axial channel for gas supply at the center of calculation cell (boundary 10), the supply pressure is set, which is equal to hydrogen storage pressure for charge process

$$P = P_i = P_{ch}. \tag{17}$$

Discharge process usually occurs at a constant mass rate outward the cylinder at the opening

$$\mathbf{u} = \mathbf{u}_0 = \frac{\mathbf{n} \cdot G}{NL_b \int \rho_c dl}, \tag{18}$$

G – gas output from the cylinder.

To solve the set of equations (1–8) the method of finite elements for fixed mesh was used [9]. Convergence precision was equal to 10^{-6}. The system of equations with a sorption source is stiff and allows integration only with a small time step. For achievement of required accuracy and numerical stability the integration step did not exceed 2 s.

3. Experimental results and discussion

The experimental set-up is shown on Fig. 3. The cylindrical vessel of 0.023 m inner diameter and 0.3 m length was used to simulate full-scale conditions of the experiment. This experimental rig was made from the stainless steel and served as a simulator of the real vessel in the ratio 1:100. This vessel is filled with some porous cylindrical samples – sorbent bed, fabricated from the composite sorbent. The vessel in disposed on the electronic balance and is connected through the valve with the hydrogen storage container by the flexible coil made from the thin stainless steel capillary. The vessel has an electric heater and a set of thermocouples for temperature registration of the sample during gas adsorption/desorption. Thermocouples were attached to the sample through the sorbent bed (0.012 m thick). Pressure sensor was used for pressure measurements in the experimental chamber.

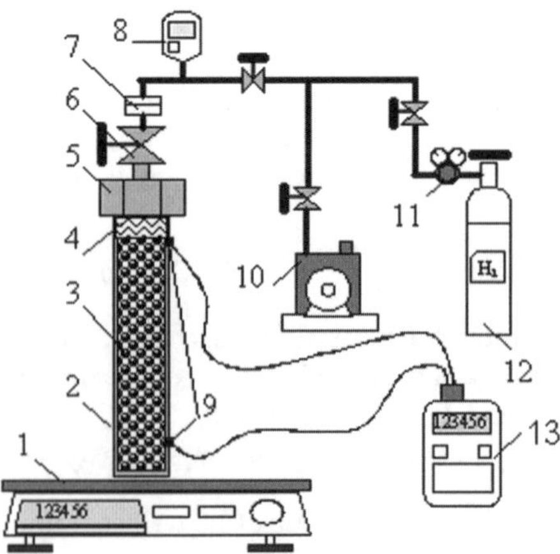

Figure 3. Experimental apparatus: 1 – high precision electronic balances, 2 – test reactor, 3 – sorbent bed, 4, 5 – reactor cap screw, 6 – valve, 7 – quick-release coupling, 8 – pressure sensor, 9 – thermocouples, 10 – vacuum pump, 11 – reducer, 12 – cylinders with hydrogen, 13 – temperature indicator

There are two independent modes of the sorption capacity registration – by gas volume control and by gas mass control. The hydrogen sorption capacity of the sorbent material was controlled as the vessel mass increase/decrease in time. The final sorption capacity of the sorbent bed was controlled volume by gas pressure change inside the vessel at the room temperature. The advantage of such experimental set-up is a possibility to perform the control of the gas vessel mass before and after gas adsorption. During the tests the pressure inside the vessel (up to 6 MPa) was controlled by the precisely calibrated pressure sensor with the accuracy near 1 kPa. Such combined system of hydrogen sorption capacity measurements (direct mass and volume control) was encompassed on the test performance of the samples with the mass 50–400 g. Before the start the measurements the samples were several times "filled" with hydrogen. Then hydrogen was removed from the vessel by its heating up to 500 K. The system was connected to the vacuum pump to reduce the pressure down to 0.7 Pa for some hours. After this procedure was accomplished the experimental set-up was ready for the experiments.

The results of experiments are shown in Table 1 and Fig. 3. The experiments were performed with four different samples: metal hydride $La_{0.5}Ni_5Ce_{0.5}$, active carbon fibre "Busofit-M8" and two composite sorbents COM 10-80 (10% "Busofit-M8" + 80% $La_{0.5}Ni_5Ce_{0.5}$ + 10% of binder) and COM 40-50 (40% "Busofit-M8" + 50% $La_{0.5}Ni_5Ce_{0.5}$ + 10% of binder). The cylindrical samples of the sorbent bed were made from the compressed porous carbon fibre with its further activation to ensure a high surface for gas adsorption.

TABLE 1. Results of tests of sorbents

Sample	Sorbent composition	a_{max} (wt%)	τ (min)
MH	$La_{0.5}Ni_5Ce_{0.5}$	1.32	35
AC	"Busofit-M8"	0.58	10
COM 10-80	10%"Busofit-M8" + 80%$La_{0.5}Ni_5Ce_{0.5}$ + 10% of binder	1.3	34
COM 40-50	40%"Busofit-M8" + 50%$La_{0.5}Ni_5Ce_{0.5}$ + 10% of binder	1.1	15

As an example, Fig. 4 shows the experimental isotherms of the adsorption and desorption of hydrogen on "Busofit-M8" and "WAC 3-00" (active carbon material made from waste wood) at the liquid nitrogen temperature. The carbon fiber "Busofit" can be considered as a universal microporous adsorbent with pore diameter of 1–2 nm and at the same time as a material with high gas permeability. Impregnated cellulose – containing raw for manufacture special activated carbon materials for methane and hydrogen storage systems with high micro porosity, surface area and narrow micropore size distribution is the attractive host material for adsorption of different gases. To increase the adsorption capacity and the bulk density of material we compressed active carbon fiber together with a binder. Briqueted "Busofit" disks have a high effective thermal conductivity and a large surface area. Figure 2 shows the texture of "Busofit-M8", which can be performed as a loose fibers bed or felt or as monolithic blocks with binder to have a good thermal conductivity along the filament. Wood-based carbons were produced by controlled pyrolysis of waste wood and special activation. Some samples from "Busofit" have been prepared by selective thermal processing at high temperature near 850°C. In this way, some of the carbon atoms are removed by gasification, which yields a very porous structure. Numerous pores, cracks were formed in the carbon material increasing a specific surface area due to the growth of micropore

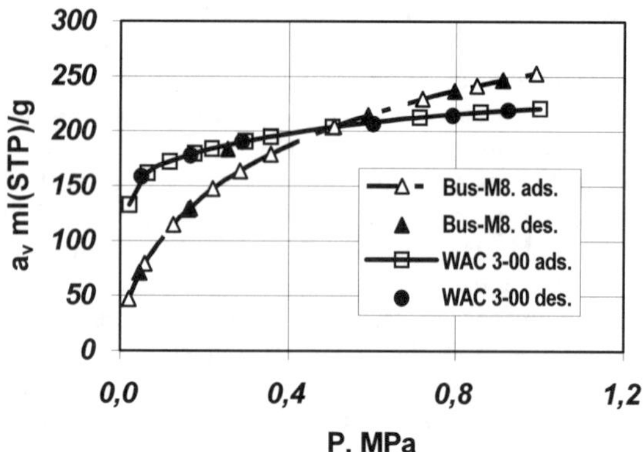

Figure 4. Isotherms of hydrogen adsorption and desorption on carbon materials at temperature 77 K, measured by the High Speed Gas Sorption Analyser NOVA 1200

volume. Additional activation of the samples was carried out at the presence of carbonic gas. The atmosphere of carbonic gas appeared more preferably oxygen for growth of a specific surface and sorption capacity.

The hydrogen sorption isotherms were measured with the High Speed Gas Sorption Analyser NOVA 1200 at 77 K in the pressure range 0–0.1 MPa. Absence of an appreciable hysteresis confirms that reversible physisorption exclusively takes place with all investigated materials. The best sorbent in nitrogen temperature is found the activated fiber "Busofit-M8", having major effective porosity (0.78), bulk density of 500 kg/m^3 and an advanced surface area (S_{BET} = 1,935 m^2/g). At pressure 6 MPa its capacity of hydrogen storage due to a physical sorption achieves 482 ml/g.

If we combine the carbon material with metal hydride microparticles disposed in the same volume, we can solve the problem of efficient gas storage and transportation. The total volume V, associated with an active carbon adsorbent may be split up into its components:

$$V = V_c + V_v + V_{void} + V_{\mu}, \tag{19}$$

where V_c – the volume of the carbon atoms of which the adsorbent is composed; V_{μ} – micropores volume; V_v – meso- and macropores volume; V_{void} – the space inside the vessel free from adsorbent bed. This latter V_{void} can be eliminated by making the solid block of adsorbent.

Composite sorbent bed was performed as a set of microparticles of metal hydrides attached to the filament surface to increase the total sorption capacity of the sorbent bed. Due to the high density of metal hydrides the systems is very compact. Metal hydride applications are best suited for systems, where power and volume are more critical than weight. The use of additives (active carbon fibre) to metal hydride has two functions: the increase of the total sorption capacity (adsorption + absorption) and the maintenance of high porosity of the medium during the solid-gas reaction. The addition of the active carbon fibre to metal hydride enhances the performances of the gas storage vessel. It is particularly convenient to use carbon fabrics because they possess the following main qualities: 2D high thermal conduction (if the precursor fibbers have good thermal conductivity); gas permeability into space of fibber, no discontinuities of the thermal conductivity between the reactor core and the heat exchanger wall.

The fast adsorption of hydrogen was checked for "Busofit-M8" – 5 min (Fig. 4). The slow rate of adsorption was determined for pure metal hydride $La_{0.5}Ni_5Ce_{0.5}$ – 35 min. The mean time of adsorption was noticed for the composite sorbent COM 40-50 – 15 min. It means, that the permeability and kinetic of adsorption of "Busofit-M8" and composite sorbent is higher to compare with $La_{0.5}Ni_5Ce_{0.5}$. The maximum temperature rise of the sorbent bed was determined for pure metal hydride $La_{0.5}Ni_5Ce_{0.5}$ – 370 K. It means the isosteric heat of hydrogen sorption for "Busofit-M8" or combination of "Busofit-M8" and $La_{0.5}Ni_5Ce_{0.5}$ is lower to compare with metal hydride. So, we obtained fast sorbent bed (compound 40% "Busofit-M8" + 50% $La_{0.5}Ni_5Ce_{0.5}$ + 10% binder), which sorption capacity of hydrogen is near the same as pure metal hydride, but the cycle of adsorption/desorption is three times less.

4. Results of the numerical investigation

The feature of developed storage system is combination of heat transfer and sorption processes in one volume. An effective use of internal volume can be achieved with the stuff made of carbon microporous sorbent, having advantage before metallic powder structure because of developed system of micropores and super micropores (6–7 Å < r < 15–16 Å), the volume of which can be 40–90% of the total pores volume. The main problem arising at developing of SSSG is the organization of effective temperature control in sorbent layer. On the one hand, the sorbent is cooled at the expense of sorption of phase transformation heat at hydrogen release. The part of energy is removed at this process with the flow of extracted gas. On the other hand, the heat is supplied to the sorbent layer due to external heat transfer of the case, and also inside by HP. Consequently, the sorption rate is determined by the ratio of heat fluxes directed inside and outside the sorbent layer. Sorbent briquetting promotes to increasing of packed density (up to 500 kg/m^3) and effective heat conductivity, inclusion of metal fins intensifies heat transfer inside sorbent. Low potential heat transferred by HP from warmed massive details or exhaust gases is redistributed through fins to sorbent and is expended to heat capacity of construction materials and caring reaction of desorption. In this connection the influence of power of heat pipe to the SSSG performance was investigated.

As the results of calculations 2D fields of temperature, hydrogen concentration, pressure and velocity during discharge where obtained. They are a basis for the determination of volumetric storage density (the ratio of gas volume at normal conditions to the geometrical vessel volume), discharge time and the rest of gas.

Figure 5 shows comparison between the total volume storage for the activated carbon fiber "Busofit-M8" (a product of the impregnated cellulose) and the activated carbon made of waste products of wood "WAC 3-00". The initial hydrogen mass is determined from a condition of equilibrium. The variation of the parameter is determined by the dependence of physical sorption on temperature and pressure for the sorbent, and also by the vessel design. At temperature below 150 K the adsorbed volumetric storage density of hydrogen nearly does not vary. The increase of stored hydrogen occurs basically due to filling of meso – and macropores with the compressed gas. Suggested design provides a hydrogen-supply of 124–145 nm^3/m^3 at average pressure 6 MPa and temperature 193 K, admitting cheap thermal isolation of the vessel case made of foamed polyurethane. The best parameters of 313 nm^3/m^3 correspond to "Busofit-M8" and liquid nitrogen temperature demanding heavy expenses for cylinder cooling and maintenance at a cryogenic level. The reduction of volumetric storage density down to 44–102 nm^3/m^3 is observed at 273 K.

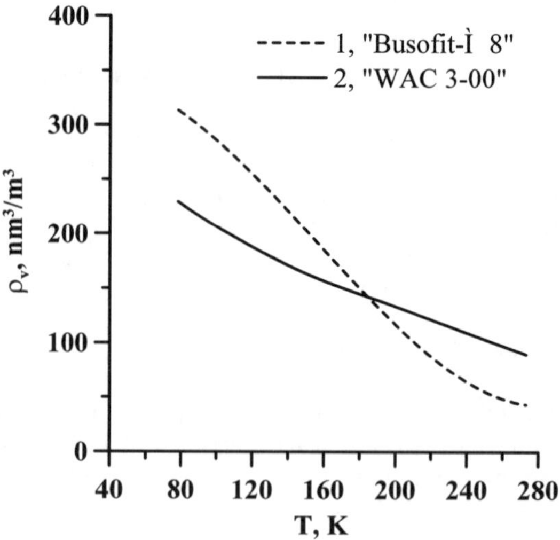

Figure 5. Volumetric storage density of hydrogen vs. temperature for the flat sectional vessel with sorbent: 1 – “Busofit-M8”, 2 – “WAC 3-00”

The basic filling performances of a temperature-controlled vessel are shown in Fig. 6. The data of base calculation variant concerned to the activated carbon fiber “Busofit-M8”. A peculiarity of the designed storage system is the combination of heat transfer processes and accumulation of hydrogen in one volume. The rate of charge by hydrogen (time of filling) is determined by the reaction rate of gas adsorption and heat transfer processes in a sorbent bed. Initially hydrogen and sorbent were in an equilibrium state

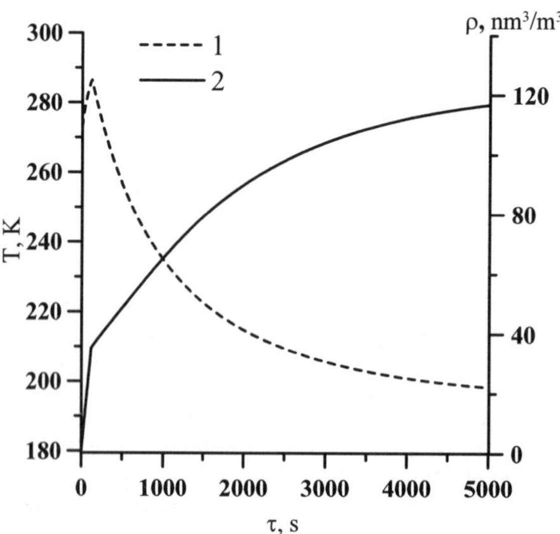

Figure 6. The mean temperature of the sorbent (1) and the volumetric storage density of hydrogen (2) vs. the time of filling in the process of charge (filling) of the SSSG (T_{HP} = 193 K: α_{env} = 0 W/m^2/K)

which was characterized by pressure $P_0 = 0.1$ MPa, temperature $T_0 = 273$ K and sorption capacity $a_0 = f(P_0, T_0)$. The gate was unclosed, gas moved in a cylinder throughout the channels. Charge pressure with the help of a reducer is supported to stationary values $P_i = P_{ch} = 6$ MPa. The temperature inside sorbent could be controlled by heat pipes ($T_{HP} = 193$ K) and convective heat transfer with an environment. During the gas filling there is a rise of the volumetric storage density of hydrogen, its variation being depending on the charge rate, pressure and sorbent temperature.

Numerical experiments have shown, that heating factor defined by HP power and condition of heat exchange with an environment, affects on duration of vessel discharge, helps to extend run of a vehicle at the expense of increasing average sorbent temperature due to reduction of gas rest. Note, at statement of the problem it was considered that the reducer holds constant the discharge rate of gas from vessel. During the hydrogen consumption by the car engine the pressure in the vessel decreases from 6 MPa down to 0.1 MPa. In the technical gas storage application, only a part of the stored mass of hydrogen can be extracted from the vessel. This fraction of hydrogen is the gas mass reversibly stored/consumed during a charge-discharge cycle between the considered pressure levels. An alternative to this is the gas heating procedure inside the sorbent bed with heat pipe heat exchanger. Figure 7 shows the pressure change, the mean temperature of the sorbent and the volumetric storage density of hydrogen extraction ($\rho_v^{ex} = [\rho_v(T_0, P_{ch}, 0) - \rho_v(T, P, \tau)]$) during the vessel discharge for three variants of sorbent bed heating conditions. In the variant where the heat pipe was switched off ($\alpha_{env} = 0$ W/m^2/K, $Q_{HP} = 0$ W), sorbent temperature has decreased by 35 K as a consequence of the endothermic reaction of desorption. Turning on of the heat pipe (Q_{HP} – 500 W) results to practically isothermal conditions of desorption. The heat flow Q_{HP} increases the gas pressure and temperature inside the vessel. Taking away of gas reduces pressure in the vessel but the heating raises the pressure, on the contrary because of reduction of the adsorption capacity and renewal output of gas molecules, which was absorbed earlier by the sorbent. The resulting character of the pressure change in the vessel depends on these two opposite factors. As temperature increases, the pressure in the cylinder during the gas desorption decreases more slowly at the same gas flow rate. Heat flux input (Q_{HP} = 1,000 W) to the sorbent bed has the noticeable influence on the degree of gas extraction (109–121 nm^3/m^3) and on the gas discharging time (2,900–32,000 s), as compared with the variant where the heat pipe is switched off.

The counteraction to cooling of the sorbent bed during desorption can be provided as a result of the sorption storage system of gas (SSSG) heating by the natural convection from an environment. For this analysis the no-load of a heat pipe ($Q_{HP} = 0$ W) and the six different heat transfer coefficient transfer between the cylinder and an environment [0, 3 (natural convection), 10, 20, 30, 100 W/m^2/K] were examined. The α_{env} = 0 W/m^2/K situation correspond to a thermally insulated cylinder. Figure 8 shows changes of average values of pressure and temperatures in a sorbent bed during discharge of the SSSG for an ambient temperature 273 K and gas flow rate G = 0.2 g/s. The results for this analysis are summarized in the Fig. 9. We observed the latent heat of sorption influences of discharge of the gas from the SSSG. For the heat-insulated storage gas system ($\alpha_{env} = 0$ W/m^2/K, $Q_{HP} = 0$ W) sorbent temperature is equal to 160 K. It is interesting to note, that for higher heat transfer coefficient, for example, equal 100

W/m²/K, the factor of gas extraction ($m_{ex} = 1 - m$, where m represents the ratio of gas mass in vessel to its initial value) is above 1. This is because the heat transfer intensification due to the higher convection coefficient increases the sorbent bed temperature above the isothermal limit. The bed temperature increase causes an increase in the cylinder pressure maintaining constant the mass flow rate at the cylinder opening. This pressure increase can extend the discharge duration in up to 3,200 s. It can be seen that the outside convection heat transfer and additional heat flux applied by heat pipes play an important role in the discharge duration.

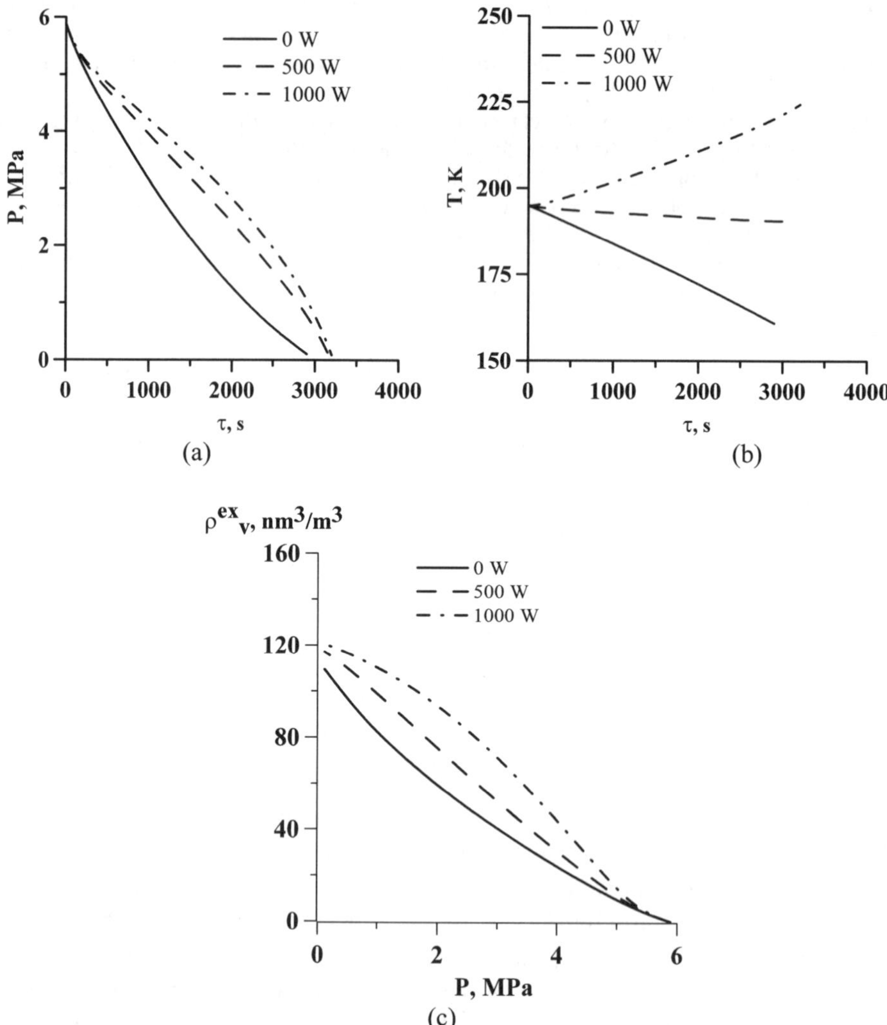

Figure 7. Changes of the pressure (a), sorbent average temperature (b) and volumetric storage density of hydrogen extraction (c) in the discharge process of SSSG for different additional heat flow applied by heat pipes: 1 – Q_{HP} = 0 W; 2 – 500; 3 – 1,000 ($\alpha_{env} = 0$ W/m²/K)

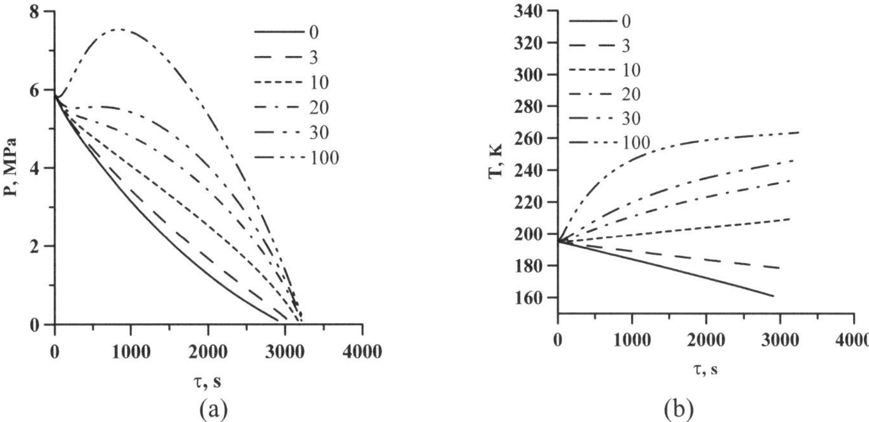

Figure 8. Changes of pressure (a) and temperature (b) during discharge of the flat section cylinder filled "Busofit-M8", for different values of an outside heat transfer coefficients

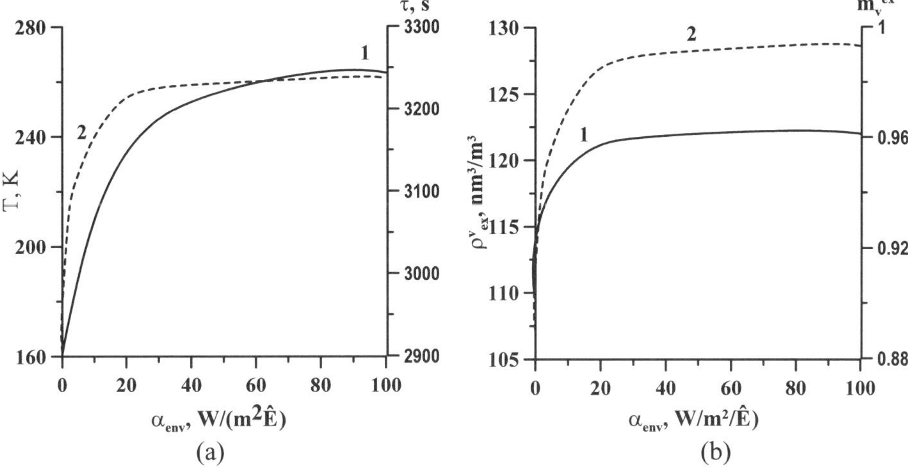

Figure 9. Sorbent bed temperature (a, 1), time of discharge (a, 2), volumetric density of gas extraction (b, 1) and the factor of hydrogen extraction (b, 2) vs. an outside heat transfer coefficient

5. Conclusions

The flat sectional vessel with heat pipes and composite sorbent ("metalhydride particles on the fiber") are promising for vehicles with internal combustion engine (especially for the dual-fuel automobile) and for the vehicles with fuel cells. Application of heat pipes as heat exchanger in the SSSG enables one to control the temperature distribution in the sorbent bed, influence of the degree and time of gas extraction.

The theoretical 2-D non-equilibrium model of heat and mass transfer processes in the flat sectional vessel with sorbent and hydrogen was developed and used to analyze the influence of power of heat pipe and outside heat transfer coefficient on the discharge/charge capacity.

Acknowledgements

The work was done in frameworks of Scientific Program "Hydrogen" in NAN of Belarus, the task N 13/12.

References

1. Nijkamp M. G., Raaymakers J. E. M. J., van Dillen A. J., de Jong K. P., Hydrogen storage using physisorption – materials demands, *Appl. Phys.*, 2001, **A72**: 619–623.
2. Vasiliev L. L. et al., Activated carbon fiber composites for ammonia, methane and hydrogen adsorption, *Int. J. Low Carbon Technol.*, Manchester University Press, Manchester, 2/1, 2006, pp. 95–111.
3. Vasiliev L. L., Kanonchik L. E., Storage and Transportation of Hydrogen in the Adsorbed State Preprint № 2, Luikov Heat & Mass Transfer Institute, National Academy of Sciences, Byelorussia, Minsk, 2006, 36 p.
4. Vasiliev L. L., Kanonchik L. E., Mishkinis D. A., Rabetsky M. I., A new method of methane storage and transportation, *Int. J. Environ. Conscious Design Manufact.*, 2000, **9**: 35–62.
5. Barbosa Mota J. P., Saatdjian E., Tondeur D., Rodrigues A. E., A simulation model of a high-capacity methane adsorptive storage system, *Adsorption*, 1995, **1**: 17–27.
6. Zhang L. Z., Wang L., Performance estimation of an adsorption cooling system for automobile waste heat recovery, *Appl. Therm. Eng.*, 1997, **17**: 1127–1139.
7. Babenko V. A., Kanonchik L. E., Mathematical modeling of a cylinder with a sorbent and natural gas, *J. Eng. Phys. Thermophys.*, 2000, **73**(3): 529–541.
8. Young D. M., Growell A. D., Physical Adsorption of Gases, Butterworths, London, 1962.
9. Zenkevich O. S., Method of Finite Elements in Engineering Moscow: Mir, 1975, 536 p. (in Russian).

COMPOSITE SORBENTS BASED ON CARBON FIBRE "BUSOFIT" FOR HYDROGEN STORAGE AND TRANSPORTATION

L.L. VASILIEV, L.E. KANONCHIK, A.A. ANTUH
Laboratory of Porous Media, Luikov Heat & Mass Transfer Institute, National Academy of Sciences, P. Brovka, 15, 220072 Minsk, Belarus
Tel./Fax: (375-17) 284-21-33; E-mail: kanon@htmi.ac.by

Abstract. A new hydrogen storage and transportation system based on composite sorbent application (metal-hydride particles on the fibre) was suggested and tested. This new design of the sorbent bed with heat pipe thermal control is capable to enhance thermodynamic efficiency of the gas storage vessels and increase its sorption capacity. The inexpensive microporous carbon obtained by thermal treatment of raw material (wood, sawdust, cellulose, straw, paper for recycling, peat etc.) after impregnation is especially attractive. In this way a number of perspective sorbents were suggested in Belarus: the activated carbon fiber "Busofi-M8" (a product of pyrolysis of the impregnated cellulose) and the activated carbon obtained from waste products of wood "WAC 3-00". The goal of this presentation is the development of the sectional vessel with heat pipe (HP) for hydrogen sorption storage at average pressures 3.5–6 MPa, every separate section of which has the case made from an aluminium (or reinforced plastics) and filled with briquettes of the sorbent material where hydrogen is situated in adsorbed and compressed states. Such vessels for hydrogen storage and transportation are interesting to be applied in fuel cells vehicle, or dual-fuel engine car (hydrogen/ gasoline, hydrogen/methane).

Keywords: carbon fibre, hydrogen, adsorption, carbon fibre, metal-hydride

Nomenclature: G – gas output from the cylinder, g_i – gas output from the elementary cell, used for computer modeling, M – mass of the gas in the cylinder, M_i – mass of the gas in the calculated cell, m – dynamic coefficient of filling the cylinder, N – number of calculated cells in the cylinder, q_{st} – latent isosteric heat of sorption, Q – heat flow, r and z – cylindrical coordinates, R – external radius of the cylinder shell, R_0 – internal radius of the heat pipe, r_0 and r_1 – internal and external radii of the annular layer of the sorbent, v – component of the velocity vector, V_a – partial molar adsorption volume, V_g – molar gas-phase volume, z_g – coefficient of gas compressibility.
α – heat transfer coefficient, β – thermal coefficient of expansion, 2δ – fin thickness, η – dynamic viscosity, ρ_γ – density of the free gas, ρ_s – density of the solid sorbent.
Subscript: a, adsorbate; cr, critical state; e, finite value; env, environment; eq, equilibrium conditions; hp, heat pipe; 0, initial value; s, sorbent; f, fin; t, transfer.

1. Introduction

Today none of the existing methods of hydrogen storage is efficient in terms of energy density, of neither a volume nor a mass and gas release rate at the same time. From the

B. Baranowski et al. (eds.), *Carbon Nanomaterials in Clean Energy Hydrogen Systems.*

application point of view it is vital to find the most effective way to store hydrogen and then, to replace the current fuel systems. Hydrogen storage by solid materials is the most recent system proposed [1]. Initially the research was based on cryogenic systems [2], which are unprofitable from an economic point of view. Recently the studies have focused on the search of the high capacity adsorbent to be used at room temperature. Due to the high density of the adsorbed phase (it corresponds to density of pressurized gas at 60–80 MPa) the gas storage capacity in the vessel with sorbent material can be increased significantly. Activated carbon fibre with metal-hydride particles on its surface can be considered as a promising material for hydrogen storage, Fig. 1. The objective of this work is to analyze hydrogen storage in several porous carbon-based materials with different porous structures to propose perspective activated carbons (carbon fibers) and metal hydrides compositions for high performance hydrogen storage system. Another interrelated work objective is development of thermally regulated adsorption storage system for dual-fuel (hydrogen and natural gas) automobile. Activated carbon is well known as one of the best adsorbents for gases [3]. In contrast to the chemisorption in metal hydrides [4], the phenomenon of physical adsorption is essentially accumulation of the undissociated hydrogen molecules on a surface of microporous carbon materials. This property is due to ability of carbon to be prepared in a very fine powdered or fiber form with highly porous structure and due to specific interactions between carbon atoms and gas molecules. The total amount of adsorbed hydrogen on the activated carbon fibre surface strongly depends on the pore geometry and pore size distribution as well as the storage pressure and temperature. Recently many improvements have been made to obtain microporous carbonaceous materials with extremely high adsorbing properties [5].

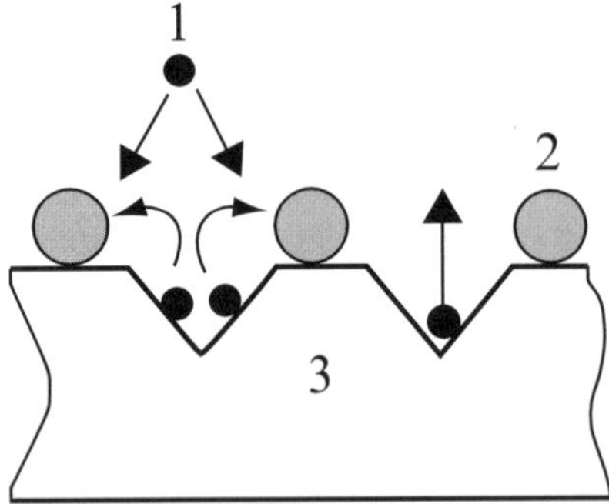

Figure 1. The molecules of the hydrogen (1) adsorbed by metal-hydride particles (2) and micropores of surface of the activated carbon fibre (3)

Adsorption of methane and hydrogen usually takes place in micropores. For conventional activated carbon, the hydrogen uptake is proportional to its surface area and pore volume, while, unfortunately, a high hydrogen adsorption capacity (4 ~ 6 wt%) can be only obtained at cryogenic temperatures. The cheap activated carbon fabricated by special thermal treatment of impregnated raw (wood, sawdust, cellulose, straw, paper for recycling, peat etc.) is attractive for modern sorption technologies. The use of specific organic and non-organic compounds as raw impregnates offers production of activated carbons with controlled porous structure and high yield (up to 50 wt%). Advanced technology allows to produce the homogeneous carbon adsorbents with benzene pore volume 0.3–0.6 cm^3/g (70–80% volume of micropores), nitrogen surface area up to 1,500 m^2/g, iodine adsorption capacity 40–70 wt% and methane adsorption capacity up to 160 mg/g (3.5 MPa, 293 K).

The activated carbon fibre "Busofit" and activated wood-based carbon are perspective materials for gas storage systems. Active carbon fibre "Busofit" is a universal adsorbent, which is efficient to adsorb different gases (H_2, N_2, O_2, CH_4, and NH_3). Figure 2 shows the texture of "Busofit-M8", which can be performed as a loose fibers bed or felt or as monolithic blocks with binder to have a good thermal conductivity along the filament.

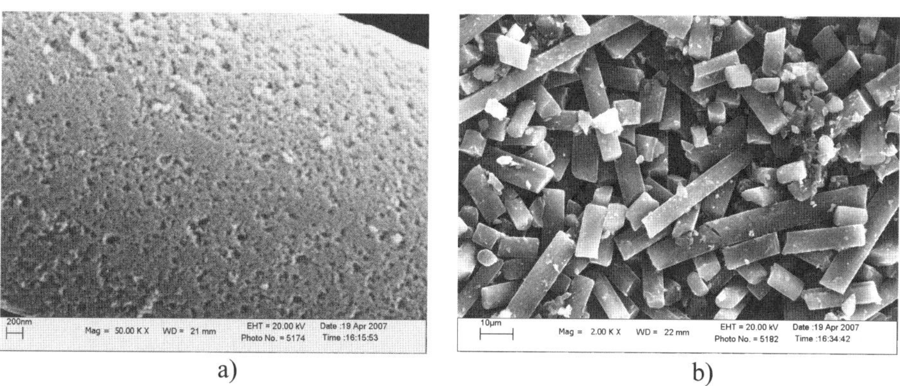

Figure 2. Photo of the active carbon fibre "Busofit-M8": (a) micro- and mesoporous surface structure of additionally activated fibre, multiplied by 50,000; (b) briquett "Busofit-M8", multiplied by 200

2. Hydrogen physisorption on the active carbon

In our experiments some samples of activated carbon "Busofit" obtained by the new technology were investigated. The surface area of the commercially available "Busofit" was measured with "Micromeritics AccuSorb 2100" and BET Sorbtometer NOVA and varied from 1,140 up to 1,570 m^2/g. It can be considered as a typical microporous adsorbent with pore diameter near 1–2 nm and at the same time as material with high gas permeability. The micropore distribution is performed mostly on the carbon filament surface. Modified "Busofit" has such advantages as high rate of adsorption and desorption; uniform surface pore distribution (0.6–1.6 nm); small number of macropores (100–200 nm) with its specific surface area 0.5 m^2/g; small number of mesopores with its specific surface area 50 m^2/g.

The total volume V, associated with an active carbon adsorbent may be split up into its components:

$$V = V_c + V_v + V_{void} + V_{\mu}, \tag{1}$$

where V_c – the volume of the carbon atoms of which the adsorbent is composed; V_μ – micropores volume; V_v – meso- and macropores volume; V_{void} – the space inside the vessel free from adsorbent bed. This latter V_{void} can be eliminated by making the solid block of adsorbent. Porous texture of different materials was characterized using nitrogen (N_2) physisorption at 77 K and up to a pressure of 0.1 MPa. From the nitrogen physisorption data, obtained with NOVA 1200, the BET-surface area, total pore volume, microporous volume were derived. The hydrogen sorption isotherms were measured with NOVA 1200 at 77 K in the pressure range 0–0.1 MPa.

Table 1 summarizes the results of the N_2 and H_2 physisorption measurements of the materials analyzed. All samples are highly micro-and mesoporous carbon materials. In our experiments four samples of carbon "Busofit-AYTM" and three samples of wood-based activated carbon obtained by new technology were investigated. The activated carbon 207C is made in the Great Britain from coconut shell. Samples 9 and 10 – granular activated carbons, specially developed for effective storage of gas. According to the offered technology some samples from "Busofit–AYTM" have been prepared by selective thermal processing at high temperature 850°C. In this way some of the carbon atoms are removed by gasification, which yields a very porous structure.

TABLE 1. Textural characteristics and hydrogen-sorption capacities at 77 K and 0.1 MPa for the investigated active carbon materials

No	Sorbent	a_v, ml (STP)/g	a_H, g/g	S_H, m^2/g	S_{BET}, m^2/g	S_{DR}, m^2/g	V_{DR}, ml/g	$R_{DR}\cdot 10^8$, m
1	Busofit 191-5	199.9	0.0176	462	1,691	2,496	0.887	49.9
2	Busofit-M2	203.9	0.0179	465	1,702	2,507	0.89	41.5
3	Busofit-M4	225.1	0.0198	536	1,715	2,547	0.9	42
4	**Busofit-M8**	**252.9**	**0.0223**	**571**	**1,939**	**2,985**	**1.04**	**51**
5	WAC 97-03	115	0.0101	271	715	1,050	0.37	33.4
6	WAC 19-99	172.1	0.0151	393	1,005	1,486	0.53	41.7
7	**WAC 3-00**	**221.1**	**0.0195**	**575**	**1,383**	**2,142**	**0.74**	**50**
8	207C	209.2	0.0184	502	1,300	1,944	0.69	41
9	Norit sorbonorit-3	193.8	0.0171	458	1,361	2,044	0.73	50
10	Sutcliff	236.6	0.0208	527	1,925	2,864	1.02	53.6

Note: WAC – wood-based active carbon; a_v – volume capacity of hydrogen storage using physisorption; a – capacity of hydrogen storage using physisorption; S_H – BET surface area determined on hydrogen; S_{BET} – BET surface area determined on nitrogen; S_{DR} – surface area, determined on Dubinin-Radushkevich method; V_{DR} – micropore volume, determined on Dubinin-Radushkevich method; R_{DR} – size of pore, determined on Dubinin-Radushkevich method

Numerous pores, cracks were formed in the carbon material increasing a specific surface area due to the growth of micropore volume. As follows from Table 1 the increase of time of activation from 2 until 8 h in an atmosphere of carbonic gas promotes increase to sorption capacity almost in 1.5 times (samples 2 and 4). To increase the adsorbent capacity and the bulk density of material active carbon fiber was compressed together with a binder. Briquetted "Busofit-M8" disks have a high effective thermal conductivity and a large surface area. Wood-based carbons were produced by controlled pyrolysis of waste wood and special activation. As seen in Table 1, the high surface area and micropore volume among carbon fibrous materials has "Busofit-M8"; from wood-based activated carbon it is "WAC 3-00", from granular carbon – "Sutcliff".

For carbon samples a linear relationship between BET surface area and volume sorption capacity of hydrogen a_v at 77 K and 0.1 MPa was found as:

$$a_v = 0.0783\, S_{BET} + 84.02 \tag{2}$$

Influence of micropore volume on sorption capacity of hydrogen for various carbon materials under the chosen conditions (77 K, 0.1 MPa) is described by the following linear correlation:

$$a_v = 119.12\, V_{DR} + 115.41 \tag{3}$$

As follows from Table 1, sorbents "Busofit-M8" and "Sutcliff" have micropore volume more than 1 ml/g. They are the best storage systems for hydrogen adsorption (accordingly 253 ml (STP)/g and 237 ml (STP)/g), STP – standard of temperature (273 K) and pressure (0.1 MPa). Our results demonstrate that a high capacity of adsorbed hydrogen by physisorption under chosen conditions was obtained with sorbents containing a large volume of micropores and a high BET surface area. Optimization of sorbent material and sorption conditions is expected to lead to storage capacity up to 500–600 ml (STP)/g, close to targets set for mobile applications. The analysis of isotherms of gas at the temperature interval 233–293 K and pressure interval 0.1–6 MPa was realized by the gravimetric control of the sample (200–400 g) during adsorption/desorption cycle.

3. Composite sorbent bed

Composite sorbent bed was performed as a set of metal-hydride particles attached to the filament surface to increase the total sorption capacity of the sorbent. Due to the high density of metal hydrides the systems is very compact. Metal hydride applications are best suited for systems, where power and volume are more critical than weight. For example a hydrogen compressor, or reservoir for hydrogen storage based on a reversible solid sorption is attractive if the process ensures high gas density storage. For that, the reactor must contain a large quantity of reactive mixture and the kinetics of the gas-solid reaction need to be fast. Heat and mass transfer ought to be high, which implies good porosity and high thermal conductivity of the sorbent bed. For this reason the metal hydride reaction beds inside the storage vessel are temperature controlled by heat pipe heat transfer devices, or pressure controlled (isobaric). To keep the temperature of an absorbing bed constant, the pressure has to raise due to the increasing absorbed gas concentration that result in a higher pressure at a given temperature To keep the pressure of an absorbing bed constant, the temperature of the hydride must be decreased. During absorption/adsorption the reactor operates as absorber in which hydrogen and metal hydride reacts with an exothermic reaction:

$$S_{(sol)} + n \cdot G_{(gas)} \rightarrow S \cdot G_{n(sol)}; \quad \Delta H_{react} < 0\,. \tag{4}$$

The enthalpy of reaction $\Delta H_{react.}$ in Eq. (4) is about 50 kJ/NH_3 mol, and is a reason for sorbent bed heating. The low thermal conductivity of the metal hydride and the very high expansion factor *S* during its reaction with the gas *G* are two important coefficients being able to reduce the sorption capacity. Heat flow has to be evacuated outside the vessel during the chemical reaction of absorption, and gas diffusion has not to be slackened to reach the solid. Conversely, in desorption phase, the reactor operates as regenerator in an endothermic reaction.

The use of additives (activated carbon fiber) to metal hydride has two functions: the increase of the total sorption capacity (adsorption + absorption) and the maintenance of high porosity of the medium during the solid-gas reaction. The addition of the active carbon fibre to metal hydride enhances the performances of the gas storage vessel. It is particularly convenient to use carbon fabrics because they possess the following main qualities: 2D high thermal conduction (if the precursor fibers have good thermal conductivity); gas permeability into the inter-fibers space, no discontinuities of the thermal conductivity between the reactor core and the heat exchanger wall. Table 2 contains some of the low temperature metal hydrides parameters.

TABLE 2. Metal hydride materials for the hydrogen storage and transportation

Metal hydride	Storage capacity, wt%	Temperature, K
$LaNi_{4.5}Al_{0.29}Mn_{0.21}$	0.8–1.0	373–433
$LaNi_{4.85}Sn_{0.15}$	1.0	279–303
$Ti_{0.9}Zr_{0.10}V_{0.43}Fe_{0.09}Cr_{0.09}Mn_{1.5}$	1.0	283–323
$Ti_{0.99}Zr_{0.01}V_{0.43}Fe_{0.09}Cr_{0.05}Mn_{1.5}$	1.0	268–303

4. Experimental results and discussion

The experimental set-up is shown on Fig. 3. The cylindrical vessel of 0.025 m inner diameter and 0.4 m length was used to simulate full-scale conditions of the experiment. This experimental rig was made from the stainless steel and served as a simulator of the real vessel in the ratio 1:100 (Fig. 4). Thermocouples were attached to the sample through the sorbent bed (0.012 m thick). Pressure sensor was used for pressure measurements in the experimental chamber. As an example, Fig. 5 shows the experimental isotherms of the adsorption and desorption of hydrogen on "Busofit-M8" and "WAC 3-00" at the liquid nitrogen temperature. Absence of an appreciable hysteresis confirms that reversible physisorption exclusively takes place with all investigated materials. The best sorbent in nitrogen temperature is found the activated fiber "Busofit-M8", having major effective porosity (0.78), bulk density of 500 kg/m^3 and an advanced surface area S_{BET}= 1,939 m^2/g . At pressure 6 MPa its capacity of hydrogen storage due to a physical sorption achieves 482 ml/g. During the tests the pressure inside the vessel (up to 6 MPa) was controlled by the precisely calibrated pressure sensor with the accuracy near 1 kPa. Such combined system of hydrogen sorption capacity measurements (direct mass and volume control) was encompassed on the test performance of the samples with the mass 50–400 g. Before the start the measurements the samples were several times "washed" with hydrogen. Then hydrogen was removed from the vessel by its heating up to 500 K. The system was connected to the vacuum pump to reduce the pressure down to 0.7 Pa for some hours. After this procedure was accomplished the experimental set-up was ready for the experiments. The experiments were performed with four different samples (Table 3, Fig. 6): metal hydride $La_{0.5}Ni_5Ce_{0.5}$, active carbon fibre "Busofit-M8", compound (10% "Busofit-M8" + 80% $La_{0.5}Ni_5Ce_{0.5}$ + 10% binder), and compound (40% "Busofit-M8" + 50% $La_{0.5}Ni_5Ce_{0.5}$ + 10% binder). The fast adsorption of hydrogen was checked for "Busofit-M8" 5 min. The slow rate of adsorption was determined for pure metal hydride $La_{0.5}Ni_5Ce_{0.5}$ 35 min. The mean time of adsorption was noticed for the combination (40% "Busofit-M8" + 50% $La_{0.5}Ni_5Ce_{0.5}$ + 10% binder) 15 min. It means, that the permeability and kinetic of adsorption of "Busofit-M8" and composite sorbent is higher to compare with $La_{0.5}Ni_5Ce_{0.5}$. The maximum temperature rise of the sorbent bed was determined for pure metal hydride $La_{0.5}Ni_5Ce_{0.5}$ – 370 K. It means the isosteric heat of hydrogen sorption for "Busofit-M8" or combination of "Busofit-M8" and $La_{0.5}Ni_5Ce_{0.5}$ is lower to compare with metal hydride. So, we obtained fast sorbent bed compound (40% "Busofit-M8" + 50% $La_{0.5}Ni_5Ce_{0.5}$ + 10% binder), which sorption capacity of hydrogen is near the same as pure metal hydride, but the cycle of adsorption/desorption is three times less.

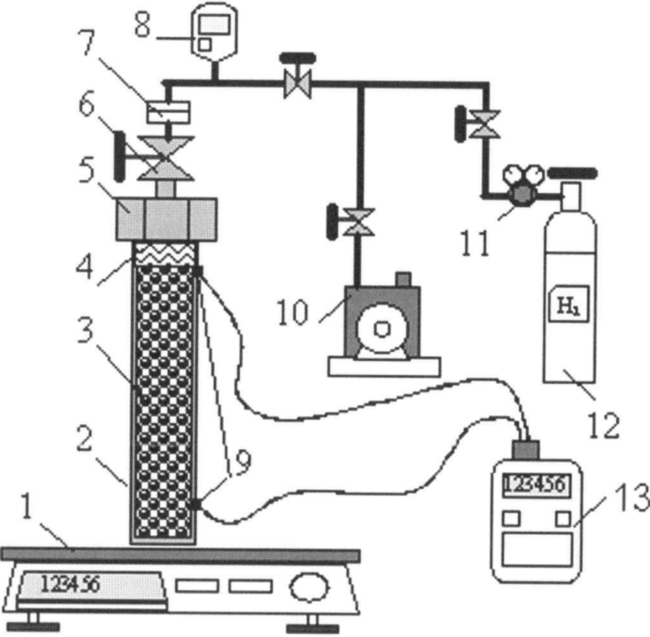

Figure 3. Experimental apparatus: 1 – high precision electronic balances, 2 – test reactor, 3 – sorbent bed, 4, 5 – reactor cap screw, 6 – valve, 7 – quick-release coupling, 8 – pressure sensor, 9 –thermocouples, 10 – vacuum pump, 11 – reducer, 12 – cylinders with hydrogen, 13 –temperature indicator

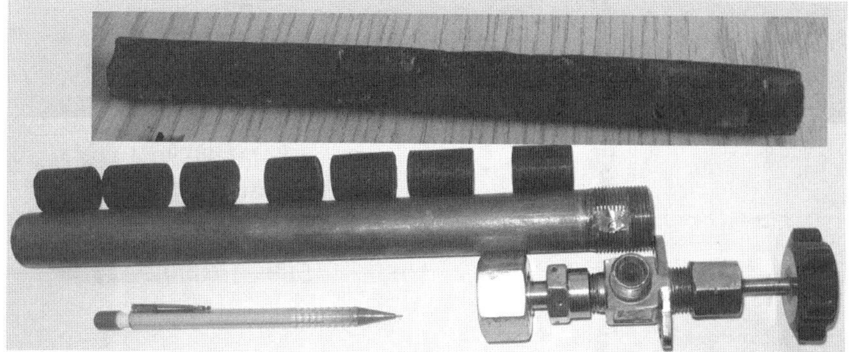

Figure 4. The samples of sorbents after pressing and a test reactor with the valve

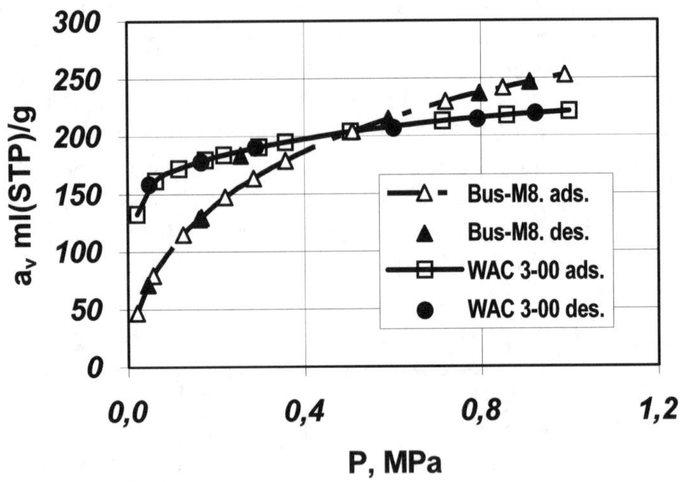

Figure 5. Isotherms of hydrogen adsorption and desorption on carbon materials at temperature 77 K, measured by the High Speed Gas Sorption Analyser NOVA 1200

TABLE 3. Results of tests of composite sorbents

Sample	Sorbent composition	a_{max}, wt%	τ, min
MH	$La_{0.5}Ni_5Ce_{0.5}$	1.32	35
AC	"Busofit-M8"	0.58	10
COM 10-80	10% "Busofit-M8" + 80% $La_{0.5}Ni_5Ce_{0.5}$ + 10% of binder	1.3	34
COM 40-50	40% "Busofit-M8" + 50% $La_{0.5}Ni_5Ce_{0.5}$ + 10% of binder	1.1	15

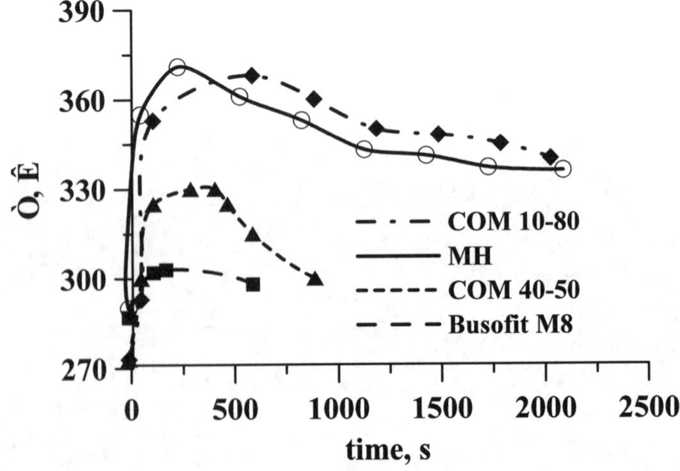

Figure 6. Average temperature of the sorbent bed during hydrogen adsorption for separate components and composites

5. Heat and mass transfer in the sorbent bed

Composite sorbent ("Busofit-M8" + metal hydride) can be used as a compact sandwich with cylindrical or flat heat pipes, applied as thermal control system. The dynamic model of the sorbent bed [6], has the components: Dubinin – Radushkevich equation of the state of gas the equation of sorption kinetic, conservation of energy, mass and momentum balance:

$$a_{eq} = \frac{W_0}{v_a} \exp\left\{ -\left\{ \frac{R_\mu T}{E} \ln\left[\left(\frac{T}{T_{cr}} \right)^2 \left(\frac{P_{cr}}{P} \right) \right] \right\}^2 \right\}, \quad (5)$$

where W_0 – maximum microporous specific volume, v_a – specific volume of adsorbed medium, E – activation energy, R_μ – gas constant, P – pressure, T – temperature,

$$\frac{\partial a}{\partial \tau} = K_{s0} \exp\left(-\frac{E}{R_\mu T} \right) \left(a_{eq} - a \right), \quad (6)$$

a – adsorption capacity, τ – time

$$(\varepsilon \rho_g C_g + \rho C + \rho a C_a) \frac{\partial T}{\partial \tau} + \nabla \cdot \left(-\lambda \nabla T + \rho_g C_g T \mathbf{u} \right) = q_{st} \rho \frac{\partial a}{\partial \tau} - \beta_g T \varepsilon \frac{\partial P}{\partial \tau} \quad (7)$$

ε porosity determined as a part of the volume occupied by the free gas, ρ, total density of the free and adsorbed gases in the cylinder, C – solid sorbent specific heat capacity, C_g – specific heat capacity of free gas, C_a – specific heat capacity of adsorbed gas, ; λ – effective thermal conductivity of the sorbent layer,

$$\frac{\partial (\varepsilon \rho_g)}{\partial \tau} + \nabla \left(\rho_g \mathbf{u} \right) + \nabla \left(-D \nabla a \right) = -\frac{\partial}{\partial \tau} (\rho a), \quad (8)$$

$$\rho_g \frac{\partial \mathbf{u}}{\partial \tau} + \frac{\eta}{K} \mathbf{u} = \nabla \left[-P \mathbf{I} + \eta \left(\nabla \mathbf{u} + (\nabla \mathbf{u})^T \right) \right], \quad (9)$$

where K – permeability, K_{s0}, – pre-exponent constant in the equation of kinetic of sorption, $K_{s0} = 15 D_{s0} / R_p^2$, R_p, – mean radius of the particles, D – diffusivity.

The suggested model gives us a possibility to obtain the temperature field and gas concentrations during the charge-discharge procedure of the vessel.

The efficient system to perform a sorbent bed thermal control during its charging/ discharging is heat pipe heat exchanger. Figure 7 shows the sectional vessel with heat pipes (HP) for hydrogen sorption storage at average pressure P = 3.5–6 MPa and comparison between the total volume storage density and the density of adsorbed phase of hydrogen in the vessel for "Busofit-M8". The variation of these parameters is determined by the dependence of physical sorption on temperature and pressure for the sorbent, and also by the vessel design. As indicated in the graphs (Fig. 8), it is

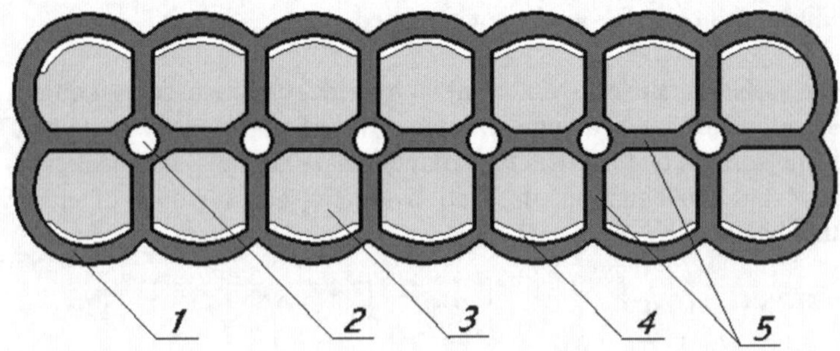

Figure 7. The flat sectional vessel for hydrogen sorption storage: 1 – vessel case, 2 – HP, 3 – sorbent, 4 – channel for gas removal, 5 – longitudinal fins/partitions

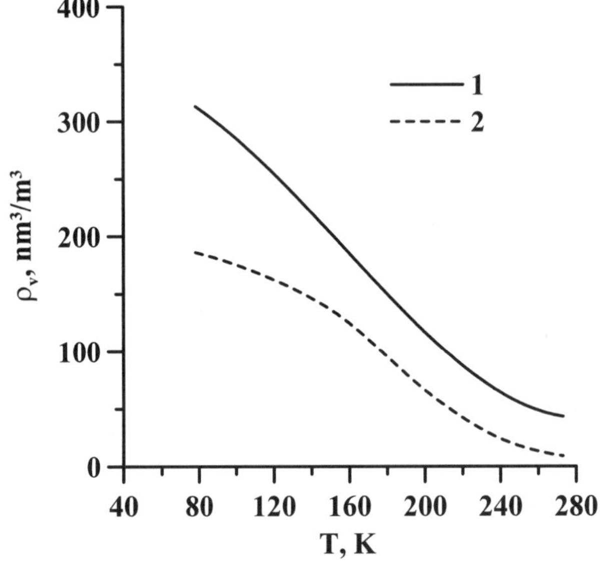

Figure 8. Volume storage density of hydrogen on "Busofit-M8" vs. temperature for the flat sectional vessel (1 – the adsorbed and compressed gases; 2 – the adsorbed gas)

necessary to take into account the presence of compressed gas in the vessel, which fraction can be measured up to 30–50% at pressures 6 MPa. At temperature below 150 K the adsorbed volume storage density ρ_v of hydrogen nearly does not vary. The increase of stored hydrogen occurs basically due to filling of meso – and macropores with the compressed gas. Suggested design provides a hydrogen-supply of 124 nm^3/m^3 at average pressure 6 MPa and temperature 195 K, admitting cheap thermal isolation of the vessel case made of foamed polyurethane. The best parameters of 313 nm^3/m^3 correspond to liquid nitrogen temperature demanding heavy expenses for cylinder cooling and maintenance at a cryogenic level. The reduction of volume storage density down to 44 nm^3/m^3 is observed at 273 K.

6. Conclusions

A new sorbent bed (composite sorbent "metal-hydride particles on the fiber") for hydrogen storage and transportation vessels was suggested. Experimental verification of the efficiency of the composite sorbent was validated. Application of HP based thermal control and high thermal conductive sorption materials promotes to reduce the time of adsorption/desorption of the gas and increase the efficiency of the vessel for hydrogen storage and transportation. Some sorption capacity data for new materials were obtained.

Acknowledgments

The work was done in frameworks of Scientific Program "Hydrogen" in NAS of Belarus, the task N 13/12.

References

1. Hynek S., Fuller W., Bentley J. Hydrogen storage by carbon sorption. *Int. J. Hydrogen Energ.*, 1997, **22**: 601–610.
2. Carpetis C., Peschka W. A study on hydrogen storage by use of cryoadsorbents. *Int. J. Hydrogen Energ.*, 1980, **5**: 539–554.
3. Cheng H.M., Yang Q.H., Liu C. Hydrogen storage in carbon nanotubes. *Int J Hydrogen Energ.*, 2002, **27**: 193–202.
4. Kuznetsov A.V., Vafai K. Analytical comparison and criteria for heat and mass transfer models in metal hydrides packed beds. *Int. J. Hydrogen Energ.*, 1995, **38**: 2873–2884.
5. Vasiliev L.L., Kulakov A.G., Mishkinis D.A., Safonova A.M., Luneva N.K. Activated Carbon for Gas Adsorption. Proceedings of III International Symposium "Fullerene and Semifullerene Structures in the Condensed Media", Minsk, Belarus, 2004, P. 110–115.
6. Vasiliev L.L., Kanonchik L.E., Mishkinis D.A., Rabetsky M.I. Adsorbed natural gas storage and transportation vessels. *Int. J. Therm. Sci.*, 2000, **39**: 1047–1055.

PHYSICAL AND CHEMICAL BASES OF METAL HYDRIDES' SYNTHESIS

U. MIRSAIDOV
Nuclear and Radiation Safety Agency (Regulatory Authority) of the Academy of Sciences of the Republic of Tajikistan, 33, Rudaki Ave., 734025, Dushanbe, Tajikistan
Tel.: (992 372) 27 77 91; E-mail: ulmas@tajik.net

A. BADALOV
M.S .Osimi Tajik Technical University, Dushanbe, Tajikistan

B.A. GAFUROV
Tajik Kurgonteppa University, Kurgonteppa, Tajikistan

Abstract. We have done systematic investigation of synthesis and properties of borohydrides and metals' alumohydride of I–III metals' groups and aluminum hydride with involvement of theory and methodology of physical and chemical analysis and preparative methods. We have determined the main directions and interaction mechanism in such key systems for development of these compounds as alkaline borohydride – metal's halogenide and binary hydride – aluminum halogenide. We have developed the effective methods of synthesis and crystallization of alumohydride and borohydride and compound complex on their basis and have been identified new hydrogen compound of boron and aluminum.

Keywords: metal hydrides, tetrahydroaluminates of metals, synthesis, properties

1. Introduction

Chemistry of metal hydrides has its benchmark – beginning of 20th century when A. Shtock [1] had created a special equipment and working methodology for super air-active substances. Beginning since 50th of last century the chemistry of complex metal hydrides developed widely [2].

Beginning of researches of hydride metal compounds in Tajikistan is connected with development of needs of new technologies and relates to the end of 70th of 20th century. By the decision of the Presidium of the Academy of Sciences of Tajik SSR the laboratory of inorganic hydrides within the structure of the V.I. Nikitin Institute of Chemistry had been created and the first works on AlH_3 and metals' tetrahydroborates had been started. During that period our scientists developed and mastered methods of work with metal hydrides.

The laboratory of inorganic hydrides during 70th–90th years worked hardly on the development of chemistry of tetrahydroborate and tetraaluminate of metals. During investigation of metal hydrides they widely used methods of physical and chemical analysis. Physical and chemical analysis gives the possibility of:

1. Synthesis of new compounds as it is shown in the works of V.I. Mikheeva [3]
2. Choosing optimal criteria for carrying out of reactions and optimal conditions for them
3. On the basis of analysis of fusion diagram and solubility isotherm to find conditions for purification of metal hydrides

A number of works done on 70th–90th of 20th century belong to the chemistry of binary and complex aluminum hydrides – potential components of solid rocket fuels, hydrogen sources, catalysts and selective reducing agents, as well as effective materials for military equipment [4].

For the first time in the V.I. Nikitin Institute of Chemistry of the Academy of Sciences of Tajikistan the systematic investigation of phase equilibriums in triple systems "borohydride of intransitive metal – borohydride of transitive element-solvent" have been carried out. On the basis of these investigations it have been created the synthesis method for individual borohydrides of rare-earth metals (RAM), which is based on use of most cheap and available sodium borohydride. With the use of the developed method for the first time the whole row of RAM, their complexes with borohydrides of alkaline metals and tetraalkylammonium have been synthesized. The reliability of received results have been confirmed by the detailed investigation of physical and chemical properties of extracted compounds [5, 6].

2. Results and discussion

The physical and chemical analysis of a number of systems on the basis of hydride compounds of aluminum gave us a possibility to prove the existence of complex binary hydride complexes, carry out a simple synthesis method of unknown and known but not extracted in individual state hepta- , hexa- and tetrahydroaluminates of alkaline and alkaline-earth metals with the use of binary hydrides of the most cheap and available sources of hydride hydrogen [7].

On the basis of systematic interaction investigation of binary hydrides and alumohydrides with electron-seeking reagents of different accepting power (chlorine-containing reagents, aryl (alkyl) halogenides, borohydride and alumohydride of lithium and etc.) the original and effective synthesis methods have been proposed for different hydride complexes [4].

The chemistry of metals borohydrides had particular development in the V.I. Nikitin Institute of Chemistry. These works have been carried out in following directions: synthesis of new compounds and development of existing synthesis methods, investigation of physical and chemical properties and composition using modern investigation methods (NMR, IR-spectroscopy, X-ray structure analysis and etc.).

First of all we want to mention the works [5, 6] on synthesis of borohydrides of RAM. On the basis of investigation of RAM chlorine reactions with sodium borohydride in the medium of pyridine, tetrahydrofuran, benzol and toluol the synthesis methods for RAM borohydrides have been developed [5]. We succeeded to synthesize delocalized tetrahydroborates of cerium group [5] having solved RAM borohydrides' solvates extracted from reaction solutions with tetrahydrofuran (THF) in the medium of benzol and toluol with further extraction of solvent and drying at 110–120ºC at degasifying treatment. Taking into account some solubility of RAM borohydrides in aromatic hydrocarbons it is possible to carry out the synthesis in planetary-type mill at 40–100º C in the medium of benzol and toluol:

$$LnCl_3 + 3MBH_4 = Ln(BH_4)_3 + 3MCl.$$

We did not succeed in desolvation of borohydrodes' tetrahydrofuranates of yttrium subgroup metals which shows less strong tie of oxygen atom of solvating solvent with molecules of cerium subgroup borohydrides.

RAM borohydrides are well soluble in THF. According to the data solubility polytherms of the systems RAM borohydrides-THF show on the diversity of solvate forms of RAM tetrahydroborates. The tetrahydroborates of scandium, neodymium and europium form the solvates $M(BH_4)_3 \cdot nTHF$, where n = 2, 4, 8 for M = Sc and 4, 6, 8 for M=Nd, Eu. The borohydrides of gadolinium, terbium, ytterbium and lutetium form octa- and trisolvates. For other lanthanoids along with octa- and trisolvates we have found the tetrasolvates. With increase of temperature we observed the solubility increase with different degree for RAM of cerium and yttrium subgroups, that can be used for separation of RAM. The solubility of RAM borohydrides changes in the presence of borohydrides of metals IA and IIA groups of periodic system. The solubility isotherms of the systems $M(BH_4)_3$ –$LiBH_4$-THF, where M=Gd, Dy, Ho, Er, Tm, Y, have crystallization branches of solvates and complexes $LiLn(BH_4)_4 \cdot nTHF$. For elements of yttrium subgroup the complexes of the mentioned composition are not extracted [6].

The works on synthesis of metals alumohydrides [7-9] are of considerable interest for chemical engineers.

In works [8, 9] the synthesis methods for metals tetrahydroaluminates were proposed:

$$MAlH_4 + MCl \rightarrow MAlH_4 + MCl,$$

$$MAlH_3 + M^nH_n \rightarrow Mn(AlH_4)_m.$$

These reactions goes with high speed. $MAlH_4$ have been extracted by organic solvents from extracted mixture of alumohydrides and metal chlorides.

The employees of the inorganic hydrides laboratory found new sources of aluminum hydride – conversion of alumohydrides under influence of haloid alkyls. It have been proposed chlorine-benzol method of AlH_3 synthesis, which excludes adhesion and ensures high quality of product by purity, thermal stability, crystalls' habitus and grain-size classification. The potential of clorous benzyl have been detected to fix AlH_4 groups by complex-formation, it allows to increase the effective concentration of AlH_3 solutions and productivity.

For the first time the ability of highly concentrated aluminum hydride 20–60 g/l for conversion in the medium of Et_2O of binary hydrides of sodium and alkaline-earth metals into alumohydrides was established. The scientists revealed the existence, synthesized and characterized the physically and chemically solvable in ether complexes of alumohydrides of alkaline-earth metals with aluminum hydride – polyhydroaluminates $M(AlH_4)_2 \cdot nAlH_3$ (M=Ca, Sr, n = 2…8), as well as the composite binary complexes with alumo- and lithium borohydrides.

3. Conclusions

The problem of utilization of cheap hydrides of sodium and calcium in the synthesis of hydroaluminates and aluminum hydride has been solved.

The direct crystallization of aluminum hydride during one stage at interaction of binary metal hydride with aluminum halogenides in the medium of ether and toluol at 60–100° C and distillation of solvent has been carried out.

In the reaction of LiH with $AlCl_3$ the output of pure crystallic AlH_3 of hexagonal modification close to quantitative has been reached.

References

1. Stock A. Hydrides of boron and silicon. – NY: Cornell University Press, 1933, 322 p.
2. Schlesinger H.J., Brow H.C., *J. Am. Chem. Soc.*, 1953, **75**(1): 192.
3. Mikheeva V.I., Method of physical and chemical analysis in inorganic synthesis. Moscow, Publishing House "Nauka", 1975, 272.
4. Mirsaidov U.M., Synthesis and properties of aluminum hydride. Dushanbe, Publishing House "Donish", 2004, 105.
5. Mirsaidov U.M., Dymova T.N. Borohydrides of transitive metals. Dushanbe, Publishing House "Donish", 1985, 123.
6. Mirsaidov U.M., Borohydrides of metals. Dushanbe, Publishing House "Donish", 2004, 138.
7. Mirsaidov U.M., Alumohydrides of metals. Dushanbe, Publishing House "Donish", 2004, 86.
8. Bakum S.I., Synthesis and properties of hydroaluminates of alkaline metals. Candidate thesis, Moscow, 1970, 168.
9. Mirsaidov U.M., Synthesis, some physical and chemical properties and reactions of tetrahydroaluminates of alkaline metals. Candidate thesis, Moscow, 1973.

COMPOSITE ELECTRODES BASED ON PLATINUM NANOPARTICLES, CARBON NANOTUBES AND NICKEL MICRO-WIRE ARRAYS PRODUCED BY TEMPLATE SYNTHESIS TECHNIQUE

P. NDUNGU, W. DAVIDS, R. FAKIR, M. WILLIAMS,
G. VAIVARS, A. NECHAEV, V. LINKOV
South African Institute for Advanced Materials Chemistry, Department of Chemistry, University of the Western Cape, Private Bag X17, Bellville, 7535, South Africa

Abstract. Ion track poly-ethylene-terephthalate membranes (pore diameter 200-, 400-, 1,000-nm), were employed as sacrificial templates for the preparation of micro-wire arrays by the electrodeposition of nickel. After etching of the template, the Ni micro-wire arrays were used as a scaffold for the growth of carbon nanotubes by chemical vapour deposition. The growth of carbon nanotubes on the nickel micro-wire arrays was found to enhance the electrochemically active surface area, redox properties in basic solution, and catalytic activity towards the hydrogen evolution reaction. The subsequent deposition of Pt onto the consolidated composite carbon material by electroless deposition enhanced the catalytic activity of the composite materials further. The significance towards applications as high surface area electrodes, especially for the hydrogen evolution reaction in water electrolysis, is also presented.

Keywords: micro-wire array, carbon nanotubes, chemical vapour deposition, platinum

1. Introduction

Materials with dimensions in the nanometre or micrometer domain have potential technological applications in areas such as nano-electronics [1–4]; drug delivery; biological applications [1, 4]; and energy conversion devices [2, 3]. This is a result of their abilities to exhibit significantly improved physical and chemical properties compared to their corresponding bulk counterparts, as well as to offer opportunities for the development of new materials with improved or unique characteristics and applications.

Metal micro-wires, in particular, are technologically attractive due to their enhanced thermal, electrical, optical, mechanical properties, and ability to increase the available surface area. Micro-wires consist of metallic crystals that are infinite in only one direction, and are hence classified as one-dimensional materials [5]. In addition, micro-wires can easily be contacted to macroscopic structures in the preparation of novel micro-devices. These micro-devices include electrochemical sensors in biomedicine and analytical sciences for the detection of various chemical species. Micro-wires assembled in an array like configuration can be used as high surface area electrodes [6], with applications in various energy conversion devices, such as fuel cells, lithium ion batteries, and electrolysers.

Metal nano- and micro-wires with very homogeneous diameter and length are easily attainable through the use of "hard-template" synthesis, which involves deposition of the metal into the pores of templates with very narrow pore size distributions [1, 7, 8]. The method entails the growth of the wires, with aspect ratios (length-to-diameter ratio) as high as 1,000, using membranes with pores of the desired length and diameter [2, 4]. The membrane template may include inorganic-based or ion track polymeric membranes [9]. The technique allows for easy control of the structural properties (i.e. length, diameter) through careful selection of templates with defined thickness and pore diameter [1–4, 8]. Importantly, the technique allows for the preparation of ordered micro- and nano-wire arrays which could be further used as secondary templates for the preparation of other micro- or nano-scale structures. The preparation of these arrays is conducted by using a sacrificial template method in which the template is etched in aggressive media to produce the residual micro-array. Use of polymeric templates should result in the formation of nickel micro-wires which are well-ordered, uniform with regards to length and diameter, and perpendicularly aligned with the plane of the template [7]. In particular, ion track polymeric membranes are attractive for the preparation of these types of micro-wire arrays as their preparation, by ionizing radiation of polymeric membranes and subsequent chemical etching, results in templates with very narrow pore size distributions, controlled porosity, and well-aligned channels.

Micro-wire hard-template synthesis in the pores of a polymeric membrane can be accomplished using either electroless deposition or electroplating [1–4, 8]. With electroplating, the pores are continuously filled and control over the length of the resulting wire can be easily achieved by varying the time of the applied potential [1–3, 8].

Growth of carbon nanotubes (CNTs) onto high-surface area, catalytically-active, nickel micro-wire arrays can be used to fabricate high-surface area nickel-carbon composite electrodes with improved electrical, mechanical, and thermal properties or unique characteristics and applications. The growth can be conducted using chemical vapour deposition (CVD) which increases adhesion between CNTs and the nickel secondary structure [7, 10]. To prepare these composite materials as electrodes for use in electrochemical devices, especially those used in water electrolysis, the addition of highly-active catalytic materials such as platinum group metals can be undertaken. This is due to the high surface area provided by CNTs facilitating the deposition of metal nanoparticles. The resultant high-surface area composite materials may have the capacity to be incorporated as electrode materials for lithium ion batteries or various electrochemical sensors.

The following article presents the synthesis of nickel micro-wires in the pores of ion track-etched membrane. The nickel wires were modified by the CVD growth of CNTs, the subsequent functionalization of the CNTs by deposition of a platinum catalyst, and extensively characterized.

2. Experimental

The preparation of the template and the nickel micro-wires was adapted from methods by Nkosi [11]. Ion track-etched poly-ethylene-terephthalate membranes (PET), with a thickness of 20 μm and pore diameter of 200 nm, were activated on one side by sputter coating a 50 nm thick layer of platinum. The second step was the electrodeposition of a Ni backing layer on one side of the PET membrane. Before electrodeposition, the side of the membrane without any sputter deposited platinum was covered with a layer of Parafilm®. Electrodeposition was carried out by placing the activated membrane in a nickel hard bath consisting of nickel sulphate (180 g/l), ammonium chloride (25.0 g/l), and boric acid (30.0 g/l). The pH was between 5.6–5.9, the bath temperature was maintained between 40–55°C, the anode was a 5.0 × 6.0 cm nickel plate (Aldrich technical grade), the distance between the electrodes was 5.0 cm, the total volume of the deposition cell was 0.50 l, the current density was maintained at 75 mA/cm^2, and the deposition was carried out for 15 minutes.

To grow nickel micro-wires, the membrane was supported on pyrolytic graphite and then used as the working electrode in a two electrode system. Nickel micro-wires were grown from a nickel hard bath using the same parameters described earlier. After growth of the nickel micro-wires the template was removed in a 25% KOH solution for 30 minutes; followed by thorough rinsing in de-ionized water.

To grow CNTs on nickel micro-wires, the sample was secured in a stainless steel frame, which was then loaded into a quartz tube located inside a horizontal tube furnace. Once appropriately sealed, the system was flushed with nitrogen and then heated to 800.0°C under nitrogen. The nitrogen flow was then terminated and liquid petroleum gas was introduced into the system for 5.0 minutes. Samples were cooled to ambient under nitrogen.

Pt was deposited onto the Ni-CNT composite material by substrate-enhanced electroless deposition (SEED) [17]. The electrochemical bath consisted of 1 mM of H_2PtCl_6 and 0.1 M HCl made up to 1 l. 1.2 cm^2 clippings of the consolidated composite electrode were attached to a 2 cm^2 stainless steel mesh using a steel paper clip. The deposition of Pt was performed at room temperature by placing the consolidated composite material in 50 ml of the Pt bath solution and depositing for 30 minutes.

Samples were characterized by scanning electron microscopy (SEM), X-ray diffraction, and cyclic voltammetry. For SEM characterization (Hitachi 650 SEM) the accelerating voltage used was 25.0 kV. XRD (Bruker D8 Advance) was done to evaluate structural and crystallographic properties of the materials (Cu$K_{\alpha 1}$ – 1.5406 Å; step size of $2\theta = 0.04°$; 1 second/step measurement time; voltage – 40 kV; tube current – 40 mA). Cyclic voltammetry was done on a scanning electrochemical microscopy in alkaline solution (0.1 M KOH), the counter electrode was platinum, the reference electrode was Ag/AgCl, and the working electrode consisted of the sample material supported on a highly oriented pyrolytic graphite current collector. For all CV experiments the sweep rate was 50 mV/s and the potential was swept from –1.0 to 0.6 V. Electrochemical impedance measurements were adapted from a method by Shervedani et al. [10]. Electrochemical Impedance spectra were recorded using an A.C. signal of 1 mV, over a 10 mHz–100 kHz frequency range. These measurements were carried out using an AUTOLAB PGSTAT30 (FRA) instrument, and a 6.0 M KOH solution. Catalytic activity of composite materials after the surface-enhanced electroless deposition of platinum was elucidated using a chonoamperometric technique which facilitated electrochemical analysis of the kinetics and mechanism of the surface reactions. Chronoamperometric analyses for the hydrogen evolution reaction were conducted by the application of a series of potential steps of 100 mV amplitude for 10 second intervals between –1.4 and –2.0 V. 40% KOH solution at 70°C was used as an electrolyte with Pt mesh as counter electrode.

3. Results and discussion

Initial characterization of the Ni micro-wires was done using SEM (Fig. 1(A)). The examination revealed straight cylindrical structures with diameters larger than the pore sizes of the template, a distribution of the structures similar to that of the original template, and lengths shorter than the membrane thickness. The template used was a polymer and can therefore be stretched by the growth of the metal wires, thus the increase in micro-wire diameter is expected. The wires were in general free standing structures and no significant bends or breaks were observed within the arrays. EDS analysis on the structures showed no significant amount of oxygen, indicative of nickel formation as opposed to nickel oxide.

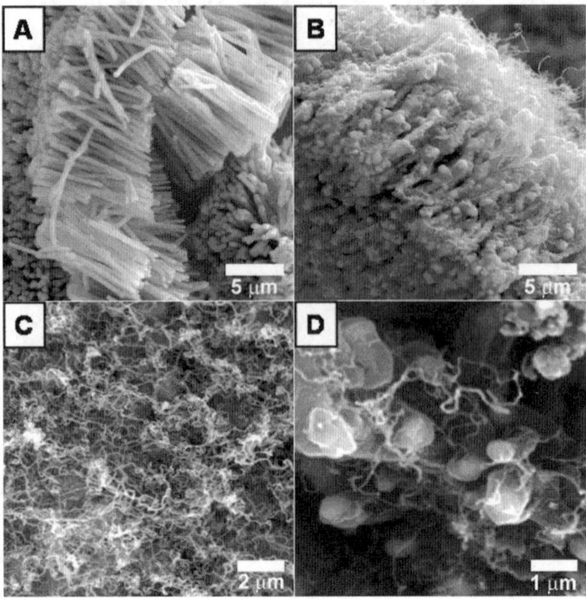

Figure 1. (A) Ni micro – wires grown in a 400 nm PET template, before CVD growth of CNT; (B) Ni micro-wire + CNT composite; (C and D) CNT mat on Ni micro-wires produced after CVD synthesis. Micro-wires were electrodeposited inside 400 nm pores and then exposed to LPG at 800°C for 5 minutes

XRD analysis corroborated the initial EDS assessment by concluding that the cylindrical structures were crystalline Ni with no nickel oxide peaks (Fig. 2). Common to all Ni micro-wire structures were Ni (111) at 2θ = 44.50°, Ni (200) at 2θ = 51.88°, Ni (220) at 2θ = 76.50°, Ni (311) at 2θ = 92.96°, and Ni (222) at 2θ = 98.64°. Flat Ni substrates exhibited identical peaks; however, there was a notable difference in the peak intensities of the Ni (111) peak of the micro-wires compared to that of the flat Ni substrate. With the micro-wires the dominant peak is the Ni (111), whereas on the flat Ni the dominant peak is the Ni (200). This difference is attributed to the change from a bulk nickel structure to a micro-crystalline Ni structure. XRD analysis on the samples after CVD revealed a peak at 2θ = 24.6°, identified as the C (002) which is characteristic for ordered or crystalline graphite in a curved configuration such as nanotubes or nanofibers. It is interesting to note that the crystalline structure of the Ni micro-wires is preserved through the CVD process.

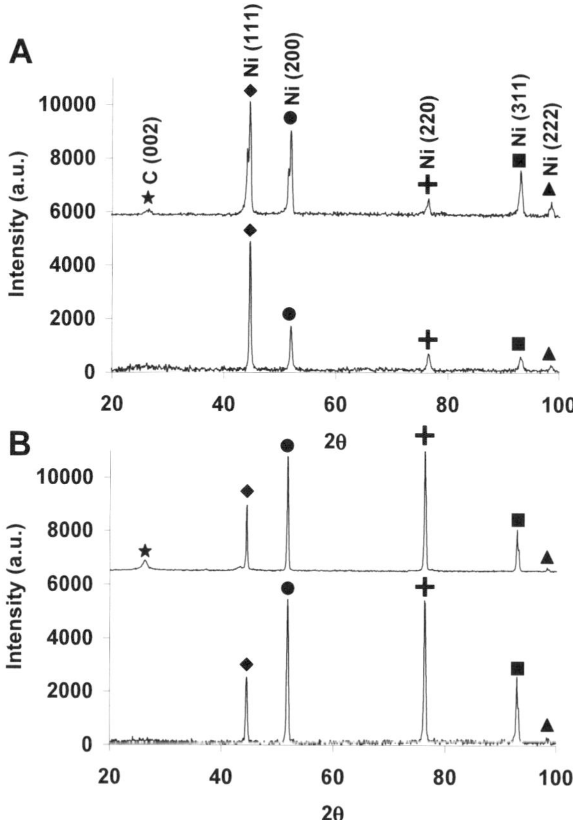

Figure 2. XRD spectrums of composite materials: (A) 200 nm Ni micro-wires before CVD growth of the CNT (lower spectrum) and Ni micro-wires with CNT after CVD (upper spectrum) (identical for 400 and 1,000 nm); (B) Ni foil (lower spectrum) and after CVD growth of CNT (upper spectrum)

SEM micrographs of the samples after CVD (Fig. 1(C, D)), found the presence of an extensive mat of CNT on the nickel micro-wires. The CNTs grew preferably from the nickel micro-wires and not the nickel backing layer, the latter showing no CNT arrays but mainly amorphous carbon structures. This result highlights the importance of the nickel micro-wires in the distribution and growth of CNTs. In current models for CNT growth in a CVD setup, the hydrocarbon source decomposes to carbon. This is followed by the carbon dissolving into the metal catalyst, saturating the metal catalyst, and finally precipitating to form the CNT structures [10, 12]. The rate of the process is a function of the crystallinity of the metal catalyst where CNT growth occurs rapidly on structures possessing micro-crystalline structure (i.e. Ni micro-wires). The effect of nickel crystal structure on the CVD growth of CNTs has been clearly demonstrated [10] and confirmed in the current results.

The nickel backing layer provides a very good electrical contact to the nickel wires and the nickel wire CNT composite. Thus the nickel wires or the composite can be easily addressed in any electrochemical system of choice. Initial experiments in this direction were done by using cyclic voltammetry (CV) and electrochemical impedance spectroscopy (EIS).

For the CV experiments there were seven separate electrodes that could be classified into three categories: flat Ni substrate electrode, Ni micro-wire (1.0, 0.4, 0.2 μm micro-wires) based electrodes, and CNT + Ni micro-wire composite (CNT grown on the 1.0, 0.4, or 0.2 μm micro-wires) electrodes. A general trend was seen between the flat Ni substrate, the Ni micro-wires, and the CNT + Ni micro-wire composite electrodes (Fig. 3). The anodic peak current density (j_{PA}) increased in the following order: Flat Ni substrate electrode ≤ Ni micro-wires electrode ≤ CNT + Ni micro-wires electrode. This increase in the j_{PA} may be indicative of the higher electrocatalytic activities of the Ni micro-wire electrodes towards the anodic surface reaction compared to the Flat Ni electrode. In addition the anodic peak potential (E_{PA}) shifts to more cathodic potentials (-0.75 V vs. Ag/AgCl) when using the Ni micro-wire electrodes, compared to the more anodic E_{PA} of the Flat Ni electrode (–0.63 V vs. Ag/AgCl). The higher activity in the cathodic peak (E_{PC} = –0.26 V vs. Ag/AgCl) and the more elaborate current tailing (higher peak current density at the maximum cathodic potential of 1.10 V vs. Ag/AgCl) of the Ni micro-wire based electrodes and the CNT + Ni micro-wire composite electrodes are also indicative of higher electrocatalytic activity of these electrodes towards the hydrogen evolution reaction, when compared to that of the flat Ni electrode [13–15].

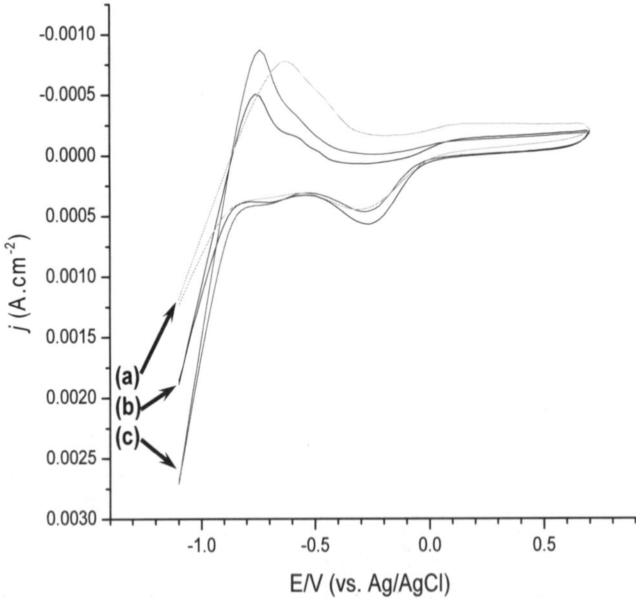

Figure 3. CV on: a – flat Ni foil, b – Ni micro-wires, and c – Ni micro-wires + CNT composite

Increasing the electrochemically accessible surface area of an electrode will increase its electrocatalytic activity [16], and any such increase can be readily ascertained through the use of electrochemical impedance spectroscopy (EIS) measurements. The double-layer capacitance C_{dl}, obtained from the impedance measurements, was calculated using Eq. 1, and typical Nyquist plots along with the capacitance is presented in Fig. 4(A) and (B).

$$C = \frac{-1}{2\pi f Z''} \qquad (1)$$

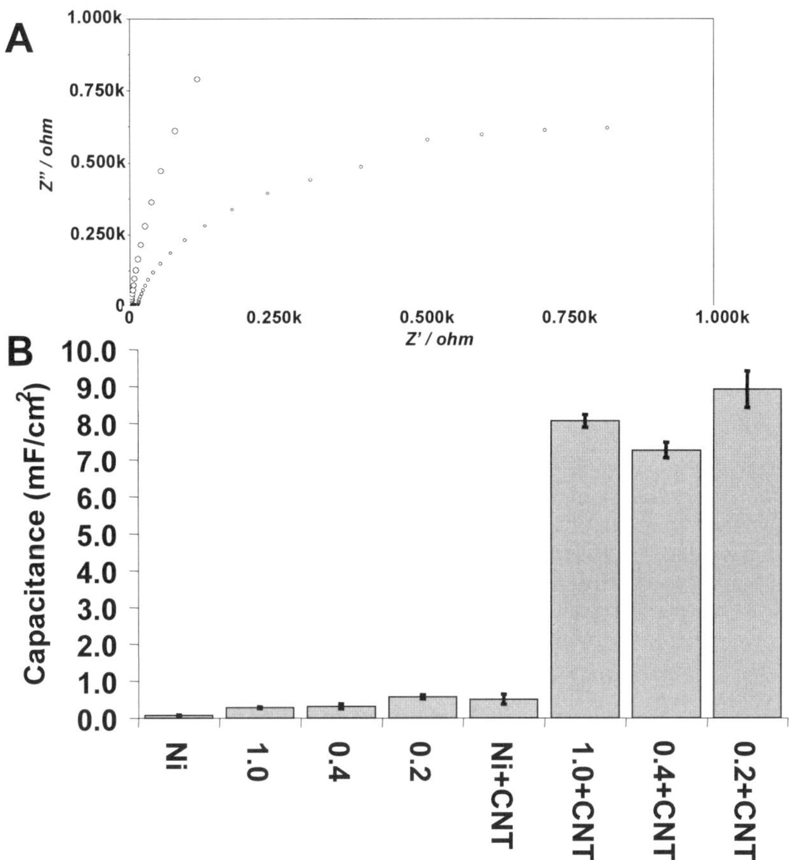

Figure 4. Typical Nyquist plots on Ni micro-wires (large circles in (A)), and the Ni micro-wires + CNT composite (small circles in image A); (B) difference in capacitance between flat Ni substrate ("Ni"), 1,000 nm Ni micro-wires ("1.0"), 400 nm Ni micro-wires ("0.4"), 200 nm Ni micro-wires ("0.2"), flat Ni substrate with CVD deposited CNT (labelled "Ni + CNT"), 1,000 nm Ni micro-wires with CVD deposited CNT ("1.0 + CNT"), 400 nm Ni micro-wires with CVD deposited CNT ("0.4 + CNT"), and 200 nm Ni micro-wires with CVD deposited CNT ("0.2 + CNT")

From Fig. 4, the C_{dl} increases from 0.078 on the Ni plate to 0.282, 0.318, and 0.576 mF on the 1.0, 0.4, and 0.2 μm Ni micro-wire arrays. This slight increase in the C_{dl} can be attributed to the increase in surface area from the micro-wires, and correlates with the simple treatment presented above. The greatest increase in the C_{dl} is observed after the CVD growth of CNTs on the Ni micro-wires, were the C_{dl} is 8.062, 7.267, and 8.916 mF on the 1.0, 0.4, and 0.2 μm Ni micro-wire + CNT composite electrodes. It is interesting to note that the growth of CNT on the Ni plate (with no micro-wires) resulted in an increase in the C_{dl} comparable to that seen on the Ni micro-wires (0.505 mF). This result highlights the importance of the synergistic effect of the Ni micro-wires and the CNT in the composite. The higher activity in the hydrogen evolution reaction is only seen in the composite arrays of Ni micro-wires + CNT and not in the Ni plate + CNT electrode, thus the higher activity in the hydrogen evolution reaction can be attributed to the increase in the C_{dl}.

Energy-dispersive spectroscopic analysis, following SEED of Pt onto the composite CNT-Ni micro-wire material, confirmed the presence of appreciable quantities of Pt (22.0 wt%) on the surface of the composite exposing the CNTs (Fig. 5). The Ni and carbon constituents of the composite electrode were maintained after the deposition process was completed. This observation confirmed the effectiveness of the SEED of Pt onto the composite CNT-Ni micro-wire material, in that the Pt ions were reduced onto the CNT surface without the addition of a reducing agent.

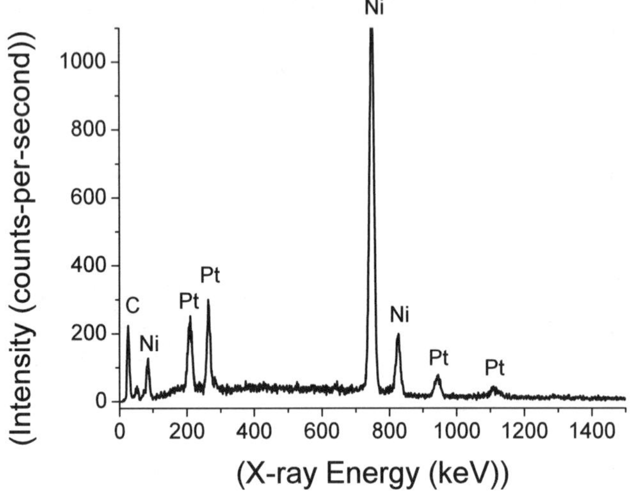

Figure 5. EDS spectrum SEED of platinum onto the composite CNT-Ni micro-wire material; analysis time – 100 live seconds, voltage – 25 kV, Amp time – 100 μs

Results from the chronoamperometric analysis of the composite Pt-CNT-Ni micro-wire material following the substrate-enhanced electroless deposition of platinum are presented in the form of potential-current density curves (Fig. 6). Nickel micro-structures modified with CNT and CNT + Pt each showed enhanced catalytic activity for hydrogen evolution compared to Ni wires, under identical experimental conditions. The catalytic activity obtained from structures modified with CNT + Pt exhibited the highest measured current densities (160 mA/cm^2 at 2.0 V) compared to nickel microstructures modified with CNT (100 mA/cm^2 at 2.0 V) and unmodified nickel microstructures (75 mA/cm^2 at 2.0 V). Thus, the catalytic activity for hydrogen evolution can be enhanced by modifying the nickel structures with CNTs and Pt nanoparticles. This observation demonstrates the potential effectiveness of this type of electrode in electrochemical devices such as that used in water electrolysis.

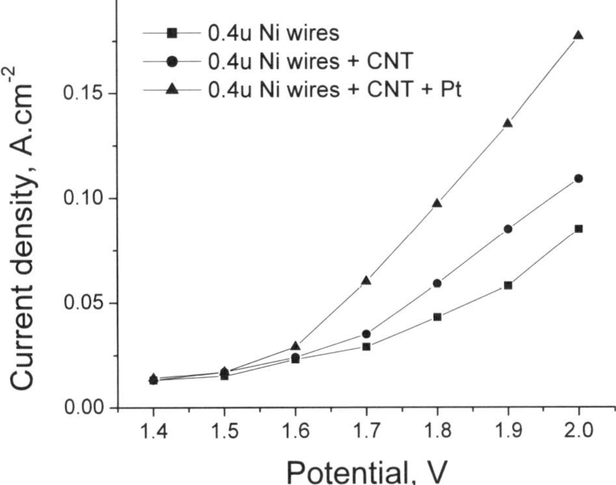

Figure 6. Chronoamperometric analysis following the SEED of platinum onto the composite CNT-Ni micro-wire material; electrode surface area – 1.2 cm^2; step – 100 mV, step time – 10 seconds, range – 1.4 to –2.0 V, electrolyte – 40% KOH, temperature – 70°C, counter – platinum mesh

4. Conclusions

Secondary structures based on nickel micro-wire arrays were successfully prepared using the galvanostatic electrodeposition of nickel into pore channels of ion track polymcric tcmplatcs. Morphological studies revealed the presence of well-aligned cylindrical micro-wires with a surface distribution similar to that of the ion track polymeric template. Slightly smaller aspect ratios were observed for the nickel micro-wires as a result of the elasticity of the polymeric template material.

It was also found that the CNTs grew mainly on the top of the array, with some CNT along the sides of the micro-wires. Nickel micro-wire arrays and CNT/micro-wire composites exhibited higher electrocatalytic activity towards the hydrogen evolution reaction, compared to bulk Ni foil. It was also determined that the increase in the electrochemical active surface area of the composite arrays by a factor of 10 was due to the synergistic effect of the CNT and Ni micro-wires.

The CNT are easily addressable on the array, which does suggest the arrays can be modified by the electrodeposition of metal nanoparticles, polymers or altered using electro-synthesis. The unique arrangement of the carbon atoms, allows for the modification of the outer wall of the nanotube with specific chemical functionalities that may or may not alter the general properties of the nanotube.

Using substrate-enhanced electroless deposition appreciable quantities of platinum were deposited onto the surface of the CNTs on the composite CNT-Ni micro-wire material Chronoamperometric analysis demonstrated that nickel microstructures modified with CNT and CNT + Pt showed enhanced catalytic activity for hydrogen evolution compared CNT + Ni micro-wires and unmodified Ni wires.

The nano-composites could thus be used as sensors, bio-analytical probes, nano-composite reactors, or as electrodes in any electrochemical energy conversion method. The array and micro-wire composite can be easily connected to an electrical power source, and as a result provides an alternative method for the fabrication of nano-composite based devices for electrochemical applications.

References

1. Hernandez-Velez M. Micro-Wires and 1D Arrays Fabrication: An Overview, *Thin Solid Films*, 2006, **495**: 51–63.
2. Kline T.R., Tian M., Wang J., Sen A., Chan M.W.H., Mallouk T.E. Template-Grown Metal Micro-Wires, *Inorganic Chemistry*, 2006, **45**: 7555–7565.
3. Huczko A. Template-Based Synthesis of Nanomaterials, *Applied Physics*, 2000, **A 70**: 365–376.
4. Wanekaya A.K., Chen W., Myung N.V., Mulchandani A. Nanowire-Based Electrochemical Biosensors, *Electroanalysis*, 2006, **18** (6): 533–550.
5. Allen P.B. Nanocrystalline Nanowires: III. Electrons, *Nano Letters*, 2007, **7** (5): 1220–1223.
6. Metz K.M., Tse K.-Y., Baker S.E., Landis E.C., Hamers R.J. Ultrahigh-Surface-Area Metallic Electrodes by Templated Electroless Deposition on Functionalized Carbon Nanofiber Scaffolds, *Chemistry of Materials*, 2006, **18**, 5398–5400.
7. Lee J.S., Gu G.H., Kim H., Suh J.S., Han I., Lee N.S., Kim J.M., Park G.-S. Well-Ordered Co Nanowire Arrays for Aligned Carbon Nanotube Arrays, *Synthetic Metals*, 2001, **124**: 307–310.
8. Shantha S.K., Raychaudhuri A.K. Fabrication of Micro-Wires of Multicomponent Oxides: Review of Recent Advances, *Materials Science and Engineering*, 2005, **C 25**: 738–751.
9. Bercu B., Enculescu I., Spohr R. Copper Tubes Prepared by Electroless Deposition in Ion Track Templates, *Nuclear Instruments and Methods in Physics Research*, 2004, **B 225**: 497–502.
10. Kukovitsky E.F., L'vov S.G., Sainov N.A., Shustov V.A. CVD Growth of Carbon Nanotube Films on Nickel Substrates, *Applied Surface Science*, 2003, **215**: 201–208.
11. Nkosi M.M. Preparation and Physico-Chemical Properties of Nickel Nanostructured Materials Deposited in Etched Ion-Track Membrane. Ph.D. thesis. Department of Chemistry, University of the Western Cape, South Africa, 2005.
12. Shervedani R.K., Lasia A. Evaluation of the Surface Roughness of Microporous Ni-Zn-P Electrodes by In Situ Methods, *Journal of Applied Electrochemistry*, 1999, **29**: 979–986.
13. Dupuis Anne-Claire. The Catalyst in the CCVD of Carbon Nanotubes - A Review, *Progress in Materials Science*, 2005, **50**: 929–961.
14. Bard A.J., Faulkner L.R. Electrochemical Methods: Fundamentals and Applications, 2nd Edition, Vol. 4, New York: Wiley, 2001, 137.
15. Zhou J., Zu Y., Bard A.J. Scanning Electrochemical Microscopy. Part 39. The Proton/ Hydrogen Mediator System and Its Application to the Study of the Electrocatalysts of Hydrogen Oxidation, *Journal of Electroanalytical Chemistry*, 2000, **491**: 22–29.
16. Mirkin M.V., Horrocks B.R. Electroanalytical Measurements Using the Scanning Electrochemical Microscope, *Analytica Chimica Acta*, 2000, **406**: 119–146.
17. Qu L., Dai L. Substrate-Enhanced Electroless Deposition of Metal Nanoparticles on Carbon Nanotubes, *Journal of the American Chemical Society*, 2005, **127**: 10806–10807.

SYNTHESIS OF CARBONIC-NICKEL NANOSTRUCTURES AND THEIR APPLICATION FOR PROTON PUMPS

V.A. BORTYSHEVSKYY, D.S. KAMENSKYH,
V.A. YEVDOKYMENKO, R.V. KORZH, T.V. TKACHENKO,
S.L. MELNYKOVA, V.G. MOTORNIY
Institute of Bioorganic Chemistry and Petrochemistry of NAS of Ukraine Kharkivs'ke shose, 50, Kiyv, 02121 Ukraine
Fax: 38(044)559-04-95; E-mail: bort2001@ukr.net

Abstract. High porous carbon material was prepared by template synthesis. Ni-carbon materials were formed by deposition of Ni^{2+} salts into the carbon supporter pores and by following reducing of adsorbed Ni^{2+} to Ni. The structure peculiarities of nickel–carbon materials were studied by small-angle X-ray scattering methods. Obtaining Ni-carbon materials used as electrodes in proton pump were enabled to raise its power and to reduce the price. The Ni-nanocarbon electrodes in proton pump have shown the high catalytic activity in electrode processes of hydrogen ionization. It was demonstrated the influence of proton current on the rate of some catalytic reactions such as propene hydration, isopropyl alcohol dehydration, synthetic gas to dimethyl ether conversion and isopropylbenzene hydrocracking. The use of Ni-carbon nanostructured materials as electrodes and proton current increases essentially the hydrogenating function of industrial catalysts in isopropylbenzene hydrocracking process.

Keywords: fuel cell, proton pump, high-temperature proton conductive materials, nanocarbon, electrodes, hydrocracking, olefin hydration, alcohol dehydration, DME obtaining

1. Introduction

Solid materials with high specific surface are widely used as catalysts for acidic catalytic processes such as hydration, dehydration, cracking and alkylation. These materials are getting acidic properties thanks to surface hydroxyl groups with mobile proton. The proton mobility in catalytic reactions is determined with chemical structure of catalyst matrix, reactants and reaction products; it also depended on the permittivity of reaction medium directly related with the presence of electric and magnetic fields.

The carrying out of the catalytic reactions with hydrated proton current could be considered as the combination of the influence of the reaction medium and electric field on acidic active sites [1]. The hydrated protons are generated by proton pump being typical hydrogen fuel cell under voltage. This cell doesn't product but consumes electric current with increasing hydrogen concentration in cathode chamber [2]. For reaction realization the catalyst layer is disposed between the cathode of such cell and proton conductive membrane. After raw materials feeding inside cell-reactor at condition of proton pump operation the catalytic reaction will carry out in hydrated proton current.

B. Baranowski et al. (eds.), *Carbon Nanomaterials in Clean Energy Hydrogen Systems.*

Proton generation ease and proton current intensity are firstly determined by the used electrodes' material. The best one is noble metals such as Pt and Pd. Cheaper metals (Ni, Fe, Cu) are low effective for the hydrogen decomposition of molecular form to atomic one and its ionization under normal conditions. Presented paper is devoted to the attempt to increase the activity of cheap metals by its deposition on nanostructured carbon base and by increasing of proton generation temperature. The best last one with nickel surface is 240–360°C. The nickel electrodes application in proton pump under this temperature is faced to the absence of the standard set of middle temperature proton conductive membranes [3]. So, for our researchers we have made such solid conductive materials (with thermal stability and with protonconductivity in the above mentioned temperature range) on the basis of polyacetylene and polycyanamide. The membranes obtained on the basis of synthesized materials and of deposited Ni-nanocarbon electrodes were the basic elements of proposed construction for effective and relatively cheap proton pump able to generate the proton current up to 1.0 A. This pump was used for the investigation of some catalytic reactions such as olefin hydration, alcohol dehydration for ether obtaining, dimethyl ether one-step production from syngas and also alkylaromatic hydrocarbons hydrocracking.

2. Experimental

2.1. NANOCARBON PREPARATION

A nanostructured high porous carbon material was prepared by template synthesis had been presented elsewhere [4, 5]. Apparent density of the "parent" silica gel (Fluka, cat. N 60752) was 0.52 g/sm^3, specific surface was 325 m^2/g. Furfuryl alcohol (99%, Fluka) was used as a chemical that had considerable carbon yield upon pyrolysis. Saturation of the silica gel with furfuryl alcohol was lasted up to full cessation of gas bubbling out of alcohol in vacuum. The polymerization of furfuryl alcohol was performed at 80°C. The cross-linking of polymer chains in silica gel pores was carried out at 150°C. For carbonization the sample was placed into a quartz flask in a furnace under argon (99.9%, Iceblick) flowing at temperature 700°C. The obtained silica carbon composite was washed by HF (40%, Merck; ratio used of about 20 cm^3 acid per g of composite) and after that by water to neutral pH. As a result, a template synthesized carbon material with apparent density 0.2 g/sm^3 and specific surface about 1,120 m^2/g was prepared. The content of acid groups on template synthesized carbon surface was about 0.9 mmol/g (0.6 mcmol/m^2).

2.2. NI-NANOCARBON ELECTRODES MAKING

Nickel was deposited on carbon support by method [6] from aqueous solutions of nickel nitrate of different concentrations with subsequent reduction of nickel ions to metal state in flowing hydrogen gas at temperature 390–400°C. Three samples of catalyst with content (by weight) of deposited metal nickel on nanostructured carbon inner surface in a quantity of 1.25%, 2.5% and 5.0% were prepared.

2.3. CHARACTERIZATION METHODS

Small-angle X-ray scattering (SAXS) experiments were performed using a Kratky camera. The FFSAXS package [7] was used to smooth out the SAXS curves and to correct them for parasitic scattering. The global unified equation of Beaucage [8] that describes the scattering by multi-level systems has been used to fit the measured SAXS intensity curves.

The nitrogen adsorption-desorption isotherms were measured by Sorptometer "KELVIN-1042" (Costech Microanalytical). Samples were degassed at 105°C during 2 h in helium flow before nitrogen adsorption. The pore size distributions were calculated using BJH method.

The content of functional groups on a carbon surface was measured with use of Boehm method [9].

2.4. HIGH-TEMPERATURE PROTON CONDUCTIVE MEMBRANES OBTAINING

Proton conductive materials were obtained on the basis of polyacetylene by [10] and polycyanamide by [11]. The synthesized materials were entered into the pores of macroporous ceramic tube (Al_2O_3/ZrO_2 = 7/3) by the filtration. The tube length was 210 mm, the external diameter was 12 mm and the wall's thickness was 1.0 mm. The input proton conductive material weight was 40–60 mg.

2.5. MEASUREMENT OF MEMBRANES PROTON CONDUCTIVITY AT DIFFERENT TEMPERATURES AND PRESSURES

The principle of measuring cell operation is based on the generation of protons from molecular hydrogen on the surface of the deposited platinum anode (0.5% Pt/Al_2O_3), on its transport through proton conductive membrane to cathode chamber and on its reduction to molecular hydrogen with electrons of external dc source (Fig. 1). The electron current discharge on the proton reduction is determined with responsive electrometer.

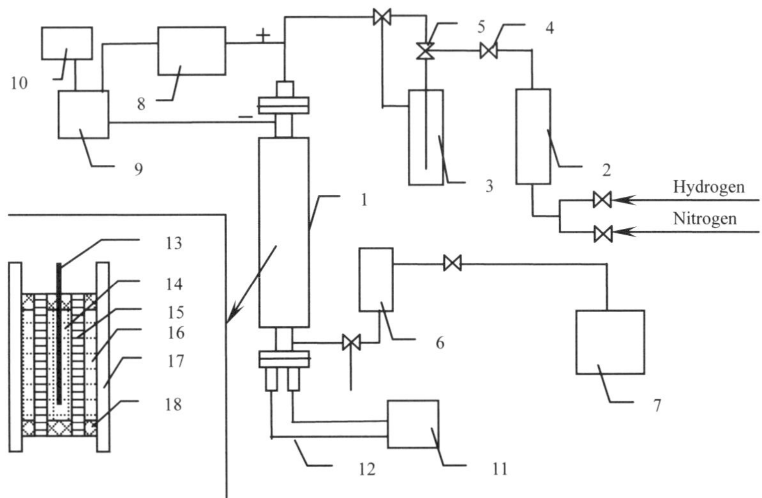

Figure 1. Laboratory installation for the measurement of proton conductivity: 1 – measuring cell; 2 – dehydrator; 3 – saturator; 4 – stop-cock; 5 – three-way cock; 6 – absorber; 7 – vacuum pump; 8 – dc source; 9 – electrometer; 10, 11 – signal registration; 12 – thermocouples; 13 – anode current contact jaw; 14 – anode poured (0.5% Pt/Al_2O_3); 15 – ceramic membrane with proton conductive material; 16 – cathode poured (0.5% Pt/Al_2O_3); 17 – cathode current contact jaw; 18 – quartz nozzle

The cell temperature and pressure are controlled with respective apparatus.

Conditions for measuring cell are:

(a) Temperature (°C) – 20 ÷ 600
(b) Pressure (MPa) – 0.1 ÷ 10.0
(c) Proton current (A) – 0.0001 ÷ 1.0
(d) Medium – oxidative, neutral, deoxidizing

2.6. CATALYTIC INVESTIGATION WITHIN PROTON CURRENT

Catalytic equipment for reactions within proton current consists of catalyst, solid proton conductive membrane (sometimes it is as catalyst) and two electrodes. It also has some canals for raw materials feeding and reaction products removal. In the catalytic processes proton pump is an integral part of the catalytic reactor unit (Fig. 2). It consists of the anode and cathode zones disengaged with the proton conductive membrane. The raw stock is fed on the catalyst layer sited in the cathode zone. The

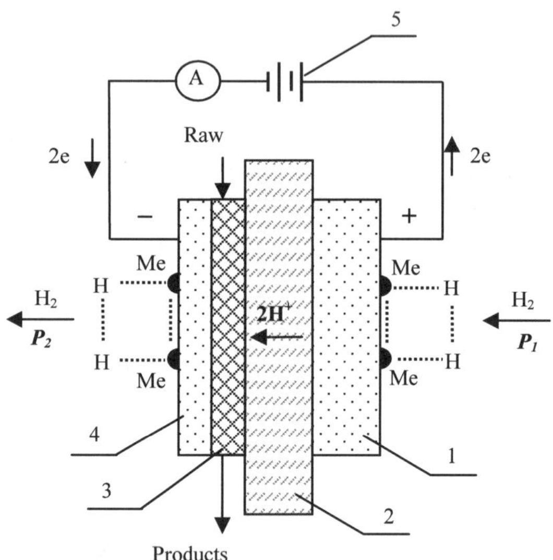

Figure 2. Proton pump use for the investigation of acidic catalytic processes: 1 – complex anode; 2 – membrane; 3 – catalyst; 4 – complex cathode; 5 – dc source

anode zone is transmitted with molecular hydrogen. The last one is decomposed to atomic form on the metal containing anode surface. Under external electrical potential for electrodes the formed hydrogen atoms are ionized to protons with following electrons' transition into external electric circuit. The generated protons from the anode zone are transported through the proton conductive membrane and "catalyst–raw stock" system into the cathode zone. There they are reduced on the surface of the metal entering into the composition of complex cathode.

2.6.1. Olefin hydration and alcohol dehydration

The films based on cross-linked chain polymer sulfocationic (MSC-H, Dow, USA) or linear perfluorsulfonic (Nafion) cationite exchanges being both the proton conductive membrane and the acidic catalyst were investigated in the reaction of propene hydration and propanol-2 dehydration.

2.6.2. Dimethyl ether one-step obtaining from synthetic gas

The industrial catalyst for methyl alcohol production (CJSC "Severodonetsk AZOT", Ukraine) with adding γ-Al_2O_3 and thus consisting of $2CuO{:}1ZnO{:}0.16Al_2O_3/\gamma\text{-}Al_2O_3$ = 1/1 and plating on ceramic tube containing inside its pores the proton conductive material from polycyanamide was investigated in the process of dimethyl ether obtaining from synthetic gas.

2.6.3. Alkylaromatic hydrocarbons hydrocracking

The catalyst of isopropylbenzene (≥98%, Fluka) hydrocracking consisting of $NiO{:}MoO_3{:}\gamma\text{-}Al_2O_3$ = 1:2:7 (is equal to industrial) was prepared by mixing water solutions corresponding salts with following calcination in hydrogen atmosphere at 450°C during 4–5 h [12]. The proton conductive material was made from polyacetylene.

2.6.4. Finely porous catalyst plating on ceramic membrane surface

The catalysts for dimethyl ether production were crumbled up, selected with fraction 0.2–0.4 mm and plated evenly on the external surface of ceramic tube with proton conductive material inside tube pores. The plated catalyst weight was 140–200 mg and ones thickness was 0.2 ÷ 0.35 mm.

2.7. ANALYSIS OF REACTION PRODUCTS

Analysis of reactions products was realized by gas-liquid chromatography ("Chrom-5") with flame-ionization detector. The glass columns with length 3.5 m and internal diameter 3 mm were filled with 15% KW-1500 on Chromaton N-AW-HMDS (0.2–0.25 mm) for alcohols and ethers analysis and 3% OV-225 on Chromaton N-super (0.2–0.25 mm) for hydrocarbons analysis. The analysis temperature was 80°C and 100°C respectively.

3. Results

3.1. NI-CARBON MATERIALS NANOSTRUCTURE

Pore size distribution of templated carbon with deposited Ni nanoclusters of different content by nitrogen adsorption-desorption measuring is showed in Fig. 3. Parameters of their pore structure are presented in Table 1 [13].

Figure 4 shows that effective pore diameter of initial templated carbon decreases on about 1.6 nm after nickel deposition [13]. So, an effective size of Ni nanoclusters is about

0.8 nm. As the content of deposited Ni increases more, a pore diameter of carbon weakly decreases and a width of pores size distribution some spreads (Table 1).

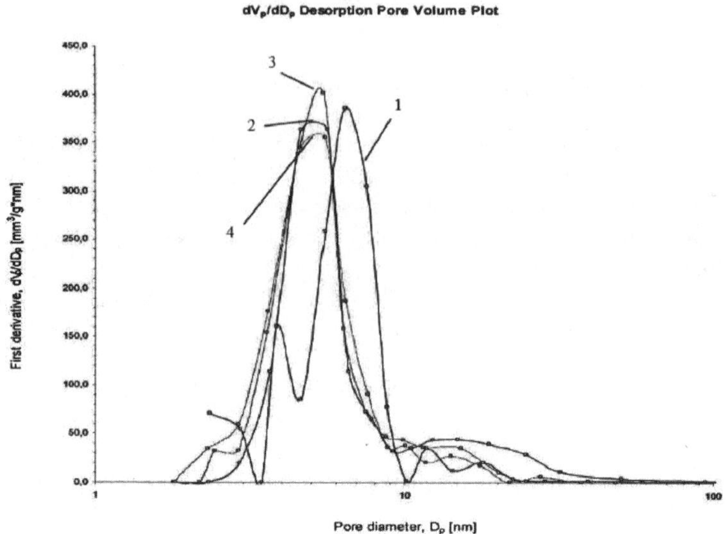

Figure 3. Pore size distribution of templated carbon with deposited Ni nanoclusters: 1 – initial templated carbon; 2 – 1.25% Ni on carbon; 3 – 2.5% Ni on carbon; 4 – 5.0% Ni on carbon

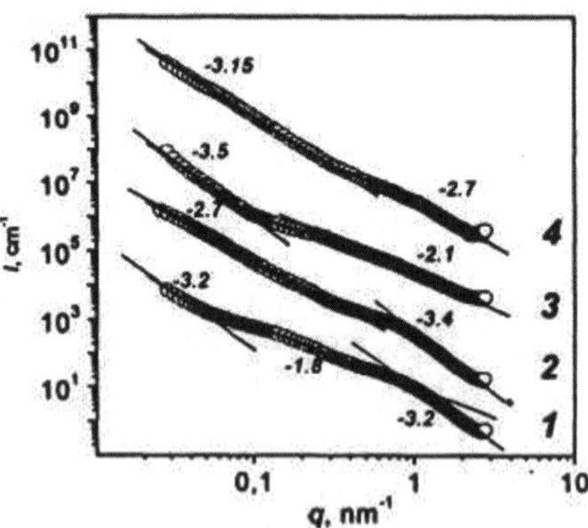

Figure 4. Normalized to absolute units SAXS curves in log(I) – log(q) scale: 1 – initial templated carbon; 2 – 1.25% Ni on carbon; 3 – 2.5% Ni on carbon; 4 – 5.0% Ni on carbon [13]

SAXS curve of initial templated carbon (curve 1, Fig. 4) has three rectilinear regions that correspond to three levels of fractal aggregation in material's structure. The first rectilinear region corresponds to the primary pores with mean diameter about 6 nm that have wavy surface. The second rectilinear region corresponds to mass fractal aggregates of pores with mean diameter about 30 nm (Table 1). The size of third level aggregates diameter is not identifiable because it is enlarged and has no well-defined limit within the region of method registration (more than 520 nm).

TABLE 1. Pore structure of templated carbon with deposited Ni [13]

Sample	Fitting SAXS curves parameters						N_2 adsorption data	
	Fractal aggregation level	Fractal aggregation type	S	D	R_g (nm)	d_s (nm)	D_p	S_{BET} (m^2/g)
Carbon	1	S	3.2	2.8	2.2	5.7	6.7	1,120
	2	M	1.8	1.8	11.0	28.4		
	3	S	3.2	2.8	>200	>520		
1.25% Ni + Carbon	1	S	3.4	2.6	2.3	5.9	5.1	990
	2	M	2.7	2.7	>200	>520		
2.5% Ni + Carbon	1	M	2.1	2.1	5.0	12.9	5.1	1,060
	2	S	3.5	2.5	>200	>520		
5% Ni + Carbon	1	M	2.7	2.7	2.7	7.0	4.8	900
	2	S	3.2	2.8	>200	>520		

Notes: Fractal aggregation type: **M** – mass fractal, **S** – surface fractal; **s** – slope of rectilinear regions in SAXS log-log plots; **D** – fractal dimension; $\mathbf{R_g}$ – gyration radius; $\mathbf{d_s} = \mathbf{R_g}2.58$, mean diameter of aggregates of correspondent level; S_{BET} – BET surface area; $\mathbf{D_p}$ – pore diameter.

The rate of SAXS curve of supported 2.5% Ni on carbon is quite different (curve 3, Fig. 4) nevertheless that it has also two rectilinear regions. The size of primary particles (first level, Table 1) enlarges and the slope of correspondent segment in the curve (curve 3) fits to mass fractal aggregation. Possibly, dendritic structure of Ni clusters appears on the inner surface of carbon pores. As mentioned above, X-ray scattering intensity by Ni nanoclusters much more than by templated carbon support, so it cannot identify scattering contribution of carbon support. One can view the scattering only by layer of Ni nanoclusters and the size of "primary particles" of dendritic shape Ni nanoclusters some greater than mean size of carbon material's pores. The size of surface fractal aggregates of second level enlarges and has no well-defined limit within the region of method registration (Table 1).

The rate of SAXS curve of supported 5% Ni on carbon (curve 4, Fig. 4) is similar to curve 3. But, because of compacting of Ni nanoclusters layer with increasing of metal content the size of primary particles decreases (first level, Table 1), possibly, directed to the size of initial primary pores in carbon support. The branchy dendritic shape aggregates of Ni nanoclusters at high concentrations are converted into condense metal layer that forms pores of catalyst apparently in borders of initial carbon support pores [13].

3.2. TESTING THE NI-NANOCARBON MATERIALS AS ELECTRODES

The samples of Ni-nanopore carbon with different Ni-content (1.25%; 2.5%; 5.0%) was tested as electrodes of electrochemical cell for the laboratory installation of the proton conductivity measurement (Fig. 5). The rate changes of proton generation from molecular hydrogen at Ni-nanocarbon electrodes surface at 45 ÷ 92°C (Fig. 5, curve 2–4) are shown in comparing with palladium-black (curve 6) and Raney nickel (curve 5). The MK-40-L sulfocationic polymer membrane (CJSC "Shchekinoazot", Russia) being the analogue to MSC-H (Dow, USA) was the solid electrolyte in this experiment.

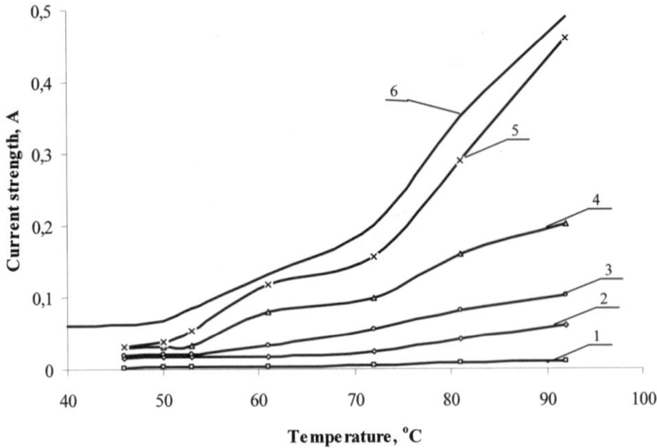

Figure 5. Proton current intensity versus temperature (U = 12 V): 1 – initial templated carbon; 2 – 5% Ni on carbon; 3 – 1.25% Ni on carbon; 4 – 2.5% Ni on carbon; 5 – Raney nickel; 6 – Palladium black

The best catalyst for the hydrogen ionization among the examined samples was the material with 2.5% Ni inside the pores of carbon matrix (Fig. 5, curve 4). Its activity is approaching to the corresponding characteristics of Raney nickel (curve 5) and palladium-black (curve 6). Low Ni content and high rate of proton formation for this sample indicate that the nanoporous structure of carbon matrix apparently has an influence upon nickel atoms distribution at internal surface of pores. This impact is to reach the optimum of interatomic distances metal – metal in the nanoclusters of 2.5% Ni-carbon sample:

$$\begin{array}{ccc} H & \cdots & H \\ | & & | \\ Me & \cdots & Me \end{array}$$

The most researchers have pointed the necessity of this optimum because it is the limiting factor for the proton generation process over some metals such as Ni, Fe, Cu [14, 15].

3.3. PROTON CONDUCTIVITY OF DIFFERENT MATERIALS

The measurements of the proton conductivity of some standard and laboratory-made materials used for the next catalytic processes are presented in Fig. 6. The logarithms of top proton conductivity for the cation exchange resins (in the likeness of MK-40-L or MSC-H) are higher than ones for ceramic membranes with input proton conductive materials based on polyacetylene and polycyanamide in three or four orders. Despite this fact the advantage of last two ones is their thermostability and proton conductivity up to 500°C. The development of such materials can fill up the deficiency of proton conductive materials for medium temperatures of temperature (200 ÷ 500°C) [3].

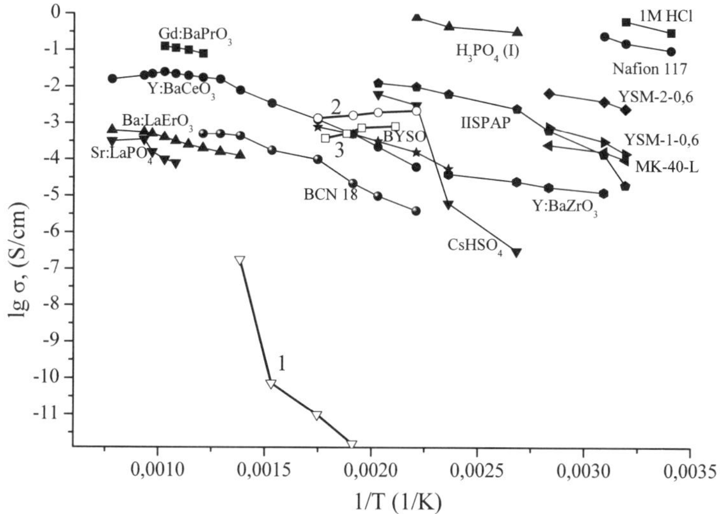

Figure 6. Proton conductivity of some materials

It was shown that the polyacetylene derivative material had been in order effective than the polycyanamide one. But the last one preparation technique is simpler and cheaper. Moreover, a side benefit of polycyanamide derivative materials is the absence of additional costs for their storage and operation whereas the polyacetylene based membranes are needed especial atmosphere and the severe maintenance of explosion precautions.

3.4. CATALYTIC INVESTIGATIONS

The making of high temperature proton conductive materials and cheap effective Ni-nanocarbon electrodes allows approaching to reactionary apparatus creation for the investigation of catalytic acidic-basic reactions within proton current. These researches results are presented in Figs. 7–9.

Hydration reaction was carried out over MF-4-SK perfluorsulfonic membrane (SPA "PlastPolymer", Russia) (0.89 g) as was both the catalyst and proton conductive membrane. The propene hydration was run under conditions: t = 160°C, P = 2.0 MPa, propene weight space velocity was 2 h^{-1}, molar ratio water/propene was (3.5 ÷ 4)/1 and U = 12 V. The propanol-2 formation rate in propene hydration steadily increases according to proton current intensity rising (Fig. 7, curve 1). The rate maximum is reached at proton current of 0.25 A. At the same time the end product yield is in 2.0–2.5 times higher than the results obtained without proton current.

The propanol-2 dehydration for isopropyl ether obtaining was passed over the catalyst based on cross-linked chain sulfocationic resin MSC-H (1.085 g) at t = 140°C, P = 1.6 MPa, propanol-2 weight space velocity was 10 h^{-1}, U = 12 V and proton current intensity up to 0.25 A. The effectiveness of dehydration within top proton current was increased in 1.5 times against the reaction without any current (Fig. 7, curve 2).

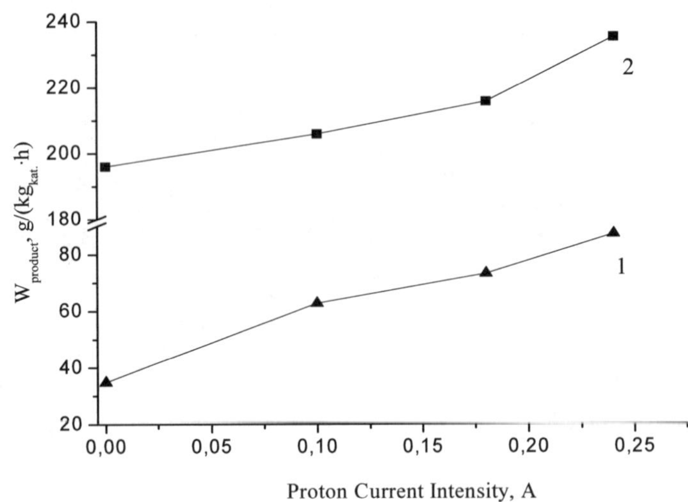

Figure 7. Products formation rate versus proton current intensity: 1 – propanol-2 in propene hydration; *2* – isopropyl ether in propanol-2 dehydration

The made high temperature proton conductive materials based on polyacetylene and polycyanamide in addition to Ni-nanocarbon electrodes were used for the analysis of proton current influence on the behavior of catalytic processes with temperature of operation higher than 200°C. The catalysts for such process were corresponded to industrial ones.

The dimethyl ether obtaining from synthetic gas was carried out over $1(2CuO{:}1ZnO{:}0.16Al_2O_3)/1\ \gamma\text{-}Al_2O_3$ (0.271 g) at t = 260°C, P = 4.0 MPa, CO weight space velocity was 1.0 h^{-1}. The CO conversion rate was increased even at low level of proton current equal to 0.03–0.05 A (Fig. 8, curve 1). The main contribution for total CO conversion increment with proton current is brought in the reaction of CO hydrogenation to methane (curve 2). The rate of dimethyl ether obtaining increases in 2.5 times at voltage up to 40 V and proton current 0.15 A (curve 3).

The proton current has the most potent influence on the nickel-alumina-molybdenum catalyst ($1Ni/2.5Mo/9Al_2O_3$, 0.465 g) efficiency for isopropylbenzene hydrocracking. The reaction was run at conditions: t = 230–260°C, P = 4.0 MPa and isopropylbenzene weight space velocity was 2.5 h^{-1}. Under these conditions the activity of industrial hydrocracking catalyst is negligible. At proton current increasing the total rate of hydrocracking process sharply augments achieving the level exceeded in 8–12 times initial one at 0.16 A (Fig. 9, curves 1 and 8). The absence of propene, the low level of benzene and the high quantity of toluene, xylene and methylisopropylbenzene among the hydrocracking products testifies the digression of hydrocracking mechanism from classical scheme [16].

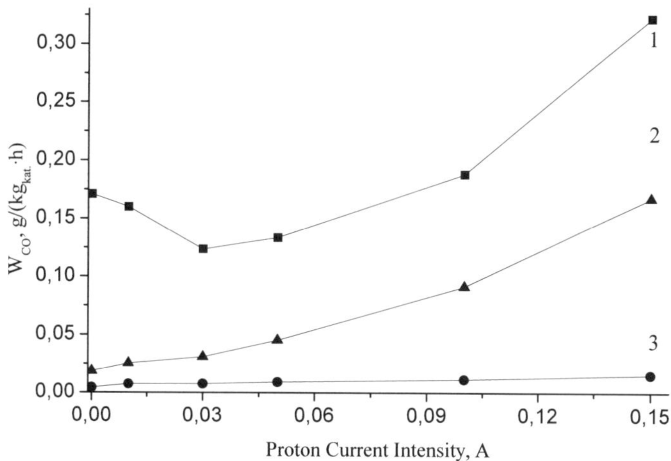

Figure 8. CO conversion rate versus proton current intensity: 1 – total; 2 – to methane; 3 – to dimethyl ether

If the proton current increase up to 0.05 A the primary reactions of isopropylbenzene cracking to benzene, toluene and xylene are beginning (Fig. 9, curves 5, 4, 3 respectively). Concurrently the rate of secondary reactions of alkylation or disproportiontion to methylisopropylbenzene is increasing (curve 2). The further proton current increase causes the rate's intensity of hydrogenation reaction – so the cyclohexane and the cyclohexene formation in the reaction products is increases (curvcs 6 and 7).

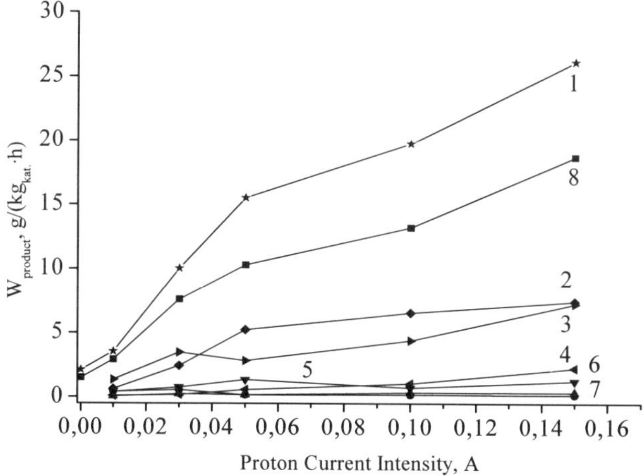

Figure 9. The formation rate of hydrocracking products versus proton current intensity: 1 – total; 2 – methylisopropylbenzene; 3 – xylenes; 4 – toluene; 5 – benzene; 6 – cyclohexene; 7 – cyclohexane; 8 – total without 2.5% Ni on carbon

The presented results of catalytic investigations are evidence of the proton current influence on the initial reactants conversion rate. It was shown by special experiments that transported protons don't affect directly to the catalyst activity. The essence of proton current influence concludes in possibility to reduce and keep up the stability of the hydration surroundings of acidic active site owing to the reciprocity and leveling of water molecules content between the hydration sphere of acidic centre proton and the hydration covering of proton transported through catalyst. Particularly, the high activity of perfluorsulfonic catalyst in propene hydration was shown to be reached at presence around the $-SO_3H$ acidic centre by the hydration sphere of 14–18 water molecules under proton current condition of 0.05–0.08 A [17]. The high activity of cross-linked chain sulfocationic catalyst in propanol-2 dehydration in the proton current was reached at the presence around the $-SO_3H$ the hydration sphere of 10–14 water molecules corresponding to proton current of 0.01–0.03 A [18]. The decrease of proton current intensity in the last case is caused by braking of proton transport with the hydration covering of the catalytic proton.

Both the used nickel-alumina-molybdenum and cuprum-zinc-alumina catalysts for hydrocracking and synthetic gas conversion are the bifunctional catalysts containing acidic (γ-Al_2O_3) as well as oxidation-reduction (Ni, Cu) centres. The acidic component is activated by proton current by mechanism described above for hydration and dehydration.

Transported through the catalyst protons do not take immediate part in hydrogenation reactions occurring at the hydrogenate centers of bifunctional catalysts. The intensive increase of hydrogenate function of analyzed catalysts is connected with the impact of transported proton on active hydrogenate centers. Mainly, the limiting stage in hydrogenation mechanism is considered to be the molecular hydrogen disintegration. This stage is depended on Me–Me bond length. The observed increase of hydrogenation intensity is possible to be related to the increase of the activity of Ni from Ni-nanocarbon electrodes because of the contact of cathode material with reactants is not excluded. To clarify the possible localization places of hydrogenate component the Ni-nanocarbon electrodes was substituted for the electrodes made from industrial hydrocracking catalyst. The investigations of the isopropylbenzene hydrocracking with the proton pump containing the new electrodes have shown the course of activity change being previous but for all that the absolute meanings were lower (Fig. 9, curves 8). These results are indicated that the hydrogenation is proceeding on both the catalyst and cathode centers. So the activity increase is the result of increase of hydrogenate centers number. But the investigations of hydrocracking in the flow reactor without proton current under mentioned above conditions have shown the absence any hydrogenation products. This fact is the evidence of essential interdependence of the proton current with hydrogenate centers. The proton current increase is caused by the rising of hydrogen atoms thanks to change of cathode reduction mechanism. The detected heightened activity of Ni atoms of made Ni-nanocarbon materials shows the wide perspectives for their use as effective electrode materials and catalysts of hydrogenation processes.

4. Conclusions

High porous carbon material was prepared by template synthesis. Ni-carbon materials were formed by deposition of Ni^{2+} salts into the carbon supporter pores and following reducing of adsorbed Ni^{2+} to Ni. Use of obtained Ni-carbon material as electrodes in proton pump enable to raise its power and to reduce price. Application of Ni-carbon nanostructured electrodes in proton pump have shown high catalytic activity in electrode processes of hydrogenization. It was demonstrated influence of proton current on the rate of the catalytic reactions of propene hydration, isopropyl alcohol dehydration, synthetic gas to dimethyl ether conversion and isopropylbenzene hydrocracking. Use of Ni-carbon nanostructured materials as electrodes in the presence of proton current increases essentially hydrogenating function of industrial catalysts in isopropylbenzene hydrocracking process.

References

1. Yevdokymenko V.O., Bortyshevskyy V.A., Melnykova S.L. Membrane catalyst of olefins hydration. Hydrated proton influence into productivity. *Chemical Industry of Ukraine* 2005; **1**: 44–47.
2. Tkachenko T.V., Bortyshevskyy V.A., Motorniy V.G. New method for investigation of proton transport in organic ion-exchanging membranes. *Catalysis and Petrochemistry* 2003; **12**: 84–88.
3. Smotkin E.S. Hydrogen permeable membrane for use in fuel cells and partial reformate fuel cell system having reforming catalysts in the anode fuel cell compartment. Pat. US 2002/0031695A1. Publ. 2002, Mar 14.
4. Barata-Rodrigues P.M., Mays T.J., Moggridge G.D. Structured carbon adsorbents from clay, zeolite, and mesoporous aluminosilicate templates. *Carbon* 2003; **41**(12): 2231–2246.
5. Vasiliev A.N., Golovko L.V., Povazhny V.A., Zlotnikov E., Chen J., Khinast J.G. Functionalized nanoporous carbon as a catalyst for Suzuki coupling reactions. *Microporous and Mesoporous Materials* 2007; **101**(3): 342–347.
6. Stiles A.B. Catalyst supports and supported catalysts: Theoretical and applied concepts. Stoneham, MA: Butterworths, 1987.
7. Vonk C.G. FFSAXS's Program for the processing of small-angle Xxray scattering data. Geleen: DSM, 1974.
8. Beaucage G. Small-angle scattering from polymeric mass fractals of arbitrary mass-fractal dimension. *Journal of Applied Crystallography* 1996; **29**(2): 134–146.
9. Boehm H.P. Some aspects of the surface chemistry of carbon blacks and other carbons. *Carbon* 1994; **32**(5): 759–769.
10. Korshak V.V., Sladkov A.M., Kudryavtsev Yu.P. Synthesis of acetylene polymers. *High Molecular Compounds* 1960; **2**(12): 1824–1827.
11. Lunin A.F., Paushkin Ya.M. Homopolycondensation of carbamide to polycyanamide - nitrogen-containing polymer with conjugated bonds. *High Molecular Compounds* 1964; **6** (8): 1467–1469.
12. Radchenko E.D., Nefedov P.K., Alyiev R.R. Industrial catalysts of hydrogenation processes refining. Moscow, RU: Chemistry, 1987.
13. Golovko L., Gomza Yu., Bortyshevskyy V., Nesin S., Povazhny V., Kamenskyh D. Structure peculiarities and catalytic activity of nickel nanoclusters in carbon nanopores. Thesis of the 2nd International Conference on Surfaces, Coatings and Nanostructured Materials. Alvor, Algarve, Portugal, 9–11 July 2007, P. 306.
14. Frumkin A.N. Selected works. Electrode processes. Moscow, RU: Science, 1987.
15. Koryta J., Dvorak J., Bogachkova V. Electrochemistry. Moscow: Mir, 1977, 472 p. (in Russian).
16. Faro Jr. A.C., dos Santos A.C.B. Cumene hydrocracking and thiophene HDS on niobia-supported Ni, Mo and Ni-Mo catalysts. *Catalysis Today* 2006; **118**: 402–409.
17. Yevdokymenko V.O. Isopropanol obtaining over proton conductive catalytic membranes. Abstract of the candidate's dissertation. Kiyv, UA, 2005.
18. Tkachenko T.V., Bortyshevskyy V.A. Proton pump use for alcohol dehydration. Thesis of the V Ukrainian Science Conference. "Modern Problems of Chemistry". Kiyv, 2004. P. 123.

THE ROLE OF INTERFACE IN THE MECHANISM OF HYDROGEN ABSORPTION BY METAL-POLYMER COMPOSITES

G. CAROTENUTO
Institute of Composite and Biomedical Materials, National Research Council, Piazzale Tecchio 80, I-80125 Napoli, Italy
Fax: (+39) 081 7682104; E-mail: giancaro@unina.it

R. CHECCHETTO, N. BAZZANELLA, A. MIOTELLO
Department of Physics, University of Trento, Via Sommarive 14, I-38050 Povo (TN), Italy

Abstract. We have studied the hydrogen storage capacity and sorption kinetics of composite materials made of $LaNi_5$ particles dispersed into hydrogen permeable polymers. Samples consisted of metal particles well separated and completely embedded in the polymeric matrix. Experimental results by Sievert analysis show that the composite material made with polysiloxane presents negligible H_2 storage capacity while in that made with polyethylene the $LaNi_5$ particles result completely hydrogenated. The hydrogenation of the metallic phase depends on the type of chemical interaction at metal-polymer interface. Strong interfacial interactions occur in the composite made with polysiloxane due to the formation of dative bonds between polymer side-groups and the metal surface atoms. Weak Van der Waals interfacial interactions occur on the contrary in the composite material made with polyethylene. We suggest that strong metal-polymer interfacial interactions impede the H_2 molecular dissociation at the metal surface and the H transfer to the metal lattice.

Keywords: hydrogen-storage, $LaNi_5$, polyethylene, polysiloxane, composite materials, interface

1. Introduction

The development of hydrogen storage systems based on metal hydrides represents a very important technological challenge because metal hydrides have high volumetric efficiency and present low cost of H_2 storing operation. Upon repeated H_2 absorption and desorption cycles metal hydride particles generally transform in very fine powders (decrepitation, this is the well known problem with the $LaNi_5$ hydride): the handling of these fine powders is difficult and their presence can damage the storage system. Therefore a consolidation treatments is required by, for example, embedding the metal particles into a hydrogen-permeable matrix such as polymers or porous material [1, 2]. Thermoplastic polymers show high permeability to small molecules like hydrogen [3–5] and therefore their use as binder for the metal hydride grains has been suggested [6]. It is worth remembering that the metal to hydride phase transition is an exothermic reaction giving rise to a significant expansion of the metal lattice (up to 30%): the embedding matrix must thus also have good thermal stability and, in particular, rubbery nature thus suggesting the use of elastomeric polymers.

B. Baranowski et al. (eds.), *Carbon Nanomaterials in Clean Energy Hydrogen Systems.*

In a previous study we analyzed the hydrogen storage capacity of metal-polymer composite material made of $LaNi_5$ embedded in polysiloxane with two different metal filling ratios [6]: the sample with lower filling ratio consisted of metallic particles embedded in the continuos polymeric matrix (PL: polymer-like sample) while in the sample with larger filling ratio the metal particles touched each other forming a percolative network (ML: metal-like sample). Although polysiloxane is highly permeable to hydrogen molecules, we observed that only the ML sample shows high H_2 storage capacity, that was anyway lower than as expected on the basis of the metal content. In the PL composite hydrogen reaches the metal particles in molecular form: the negligible H_2 storage capacity was thus attributed to the chemistry of the metal-polymer interface that impedes the H_2 dissociation and thus the H absorption by the metallic particles. The not complete hydrogenation of the ML composite was explained by assuming that a fraction of the $LaNi_5$ particles embedded in the composite were excluded from the percolative network formed by the connected metallic particles where at the metal-metal interfaces hydrogen atomically migrates after H_2 dissociation at the surface of the composite sample.

In this paper we present a study to clarify the role of the chemical interactions at the interface between the polymer chains and the surface of the metal particles: we have analyzed the hydrogen storage capacity and sorption kinetics of two composite materials made with the same metallic phase ($LaNi_5$) embedded in two different polymeric matrix: polysiloxane (PS-$LaNi_5$ sample) and polyethylene (PE-$LaNi_5$ sample). We have prepared composite samples with metal content lower than the percolation threshold: samples thus consist of metal particles completely embedded in the polymeric matrix by only forming a few number of metal-metal interfaces. Polyethylene and polysiloxane have been selected as polymers since they present different metal-matrix interface states: polyethylene exhibits weak Van der Waals interactions with metal surfaces while polysiloxane may coordinate the oxygen atoms in the polymer back-bone to the metal surface. These two types of polymers were combined with fine $LaNi_5$ powders which is a standard hydride-forming metal with well-known hydrogenation properties.

2. Experimental

Powdered $LaNi_5$ (SAES-Getter, average size of 100 μm) was fractioned by decantation from a viscous solution of polystyrene/chloroform and then it was accurately washed by pure chloroform. The composite samples were prepared using the metal fraction having average particle size lower than 30 μm. The PE-$LaNi_5$ composites were prepared by dissolving pellets of low-density polyethylene (LD-PE) with low molecular weight in boiling heptane and then adding the fine $LaNi_5$ powder to this solution. Heptane was removed from the system by evaporation and the obtained composite material was homogenized by melting and then shaped in form of disks by hot-pressing. In the preparation of PS-$LaNi_5$ composites, the metallic powder was accurately dispersed into a self-curing acetic silicon polysiloxane formulation (Saratoga) by using a steel spatula, then the obtained homogenous solid paste was cured in air at room temperature. In this paper we present the analysis carried out on samples with metal content lower than the percolation threshold: it was 50 and 77 wt% in the PS-$LaNi_5$ and PE-$LaNi_5$ composite, respectively.

Scanning Electron Microscope (SEM) analysis have been carried out by using JEOL JSM-7001F Field Emission Scanning Electron Microscope after the deposition of a thin graphite film on the sample surface. Micrographs were obtained using 8.0 or 10.0 keV electron energy for PS-$LaNi_5$ and PE-$LaNi_5$ composites, respectively, by acquiring the signal of the backscattered electrons.

The H_2 storage capacity and desorption kinetics were studied by measuring the hydrogen desorption flux at room temperature (295 K) from composite samples in equilibrium with H_2 gas at pressures in the 1 ÷ 10 atm range as function of time using calibrated mass flow meters. The PS-$LaNi_5$ composites were in form of nearly spherical fragments with size ~1 mm while the PE-$LaNi_5$ composites were in form of thin foils (thickness ~0.5 mm) with area ~10 mm^2.

The experimental apparatus and the measurement procedures as well as details on the sample preparation are presented in a previous paper [6].

3. Results and discussion

SEM micrographs of the surface of as prepared PE-$LaNi_5$ and PS-$LaNi_5$ samples are presented in Figs. 1 and 2, respectively. The micrographs show that the samples consist of a continuous polymeric phase, dark background, embedding separated metal particles that we can see as white spots and island. The small spots in the micrographs are buried $LaNi_5$ particles with only a small fraction of the surface emerging from the polymer. Some metal particles emerge from the polymeric matrix and can be well distinguished. The hydrogen transport to the $LaNi_5$ particles in the present polymer-like samples thus occurs by molecular diffusion through the polymer: the two different composites allow us to investigate the role of the metal-polymer interface in the H_2 sorption properties of $LaNi_5$.

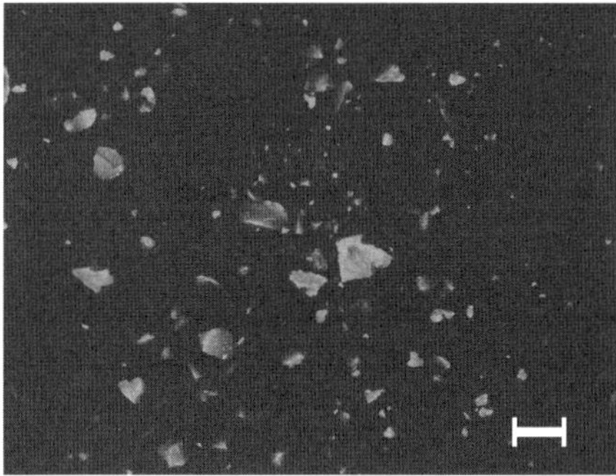

Figure 1. SEM micrographs of the surface of as prepared PE-$LaNi_5$ composite material. The white scale bas denote 10 μm

Figure 3 shows the H_2 absorption properties of both pure $LaNi_5$ powders and composite samples. The horizontal axis of the figure reports the hydrogen content in wt% of the metal fraction of the samples and allows a comparison between the hydrogenation level of the metallic phase in the pure powders and that in the composite materials. We can observe that the hydrogenation of the metallic phase of the PE-$LaNi_5$ composite (open squares) resembles that of the pure metallic powders (solid triangles) and is much larger than that of the PS-$LaNi_5$ composite (open circles).

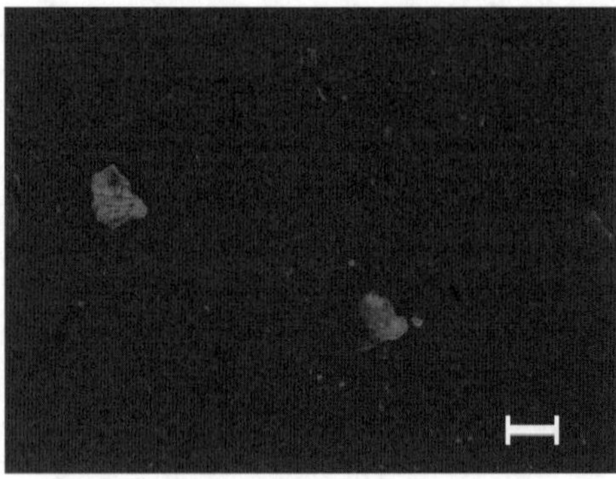

Figure 2. SEM micrographs of the surface of as prepared PS-$LaNi_5$ composite material. The white scale bas denote 10 μm

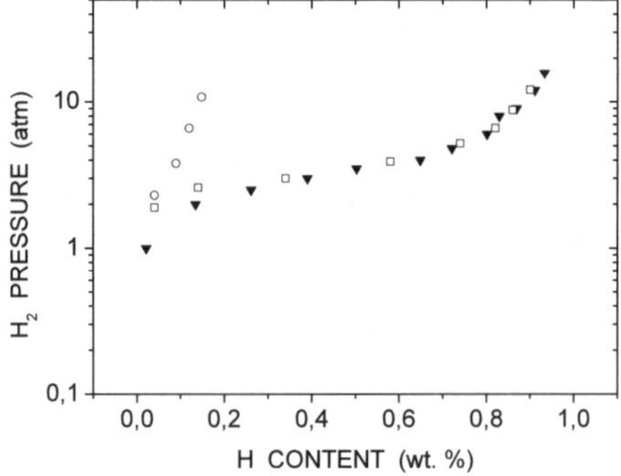

Figure 3. H_2 Pressure-Composition (P-C) isotherm at 295 K of pure $LaNi_5$ powder (solid triangles), PE-$LaNi_5$ (open squares) and PS-$LaNi_5$ (open circles) composite material

We can explain this evidence considering the chemical structure of the two different metal-polymer interfaces.

Because the nucleophilicity of a σ H-H bond is very low, a strong competition occurs in presence of other nucleophilic species like polar groups in the polymer. Most techno-polymers have an non-polar nature, which makes them stable to moisture, however (with exceptions of polyolefins) they have side-groups containing hetero-atoms with lone-pare electrons (e.g., oxygen, nitrogen), such as the polysiloxane matrix used in the present experiment. The nucleophilicity of these functional groups is high enough to generate stable coordinative complexes with surface metal atoms [8]. When metal atoms are involved in such dative bonds with polymer, H_2 molecules compete with these groups in the chemisorption process and consequently the H_2 dissociation results effectively impeded [9]. On the contrary, when the polymer embedding the metal phase is not able to chemically interact with the metal surface, as in the case of a polyethylene matrix, the capability of the metal surface to dissociate the H_2 molecule is weakly affected and the H_2

storage capacity of the metallic phase in the composite does not change. The low hydrogenation of the metallic phase in the PS-$LaNi_5$ composite can be explained assuming that a limited fraction of the metal particles are able to form hydride phase, reasonably those at the composite surface because part of their surface is exposed to the H_2 gas.

In Fig. 4 we present curves showing the hydrogen desorption kinetics of the $LaNi_5$ powders line (a), of the PS-$LaNi_5$ composite, line (b), and of the PE-$LaNi_5$ composite, line (c). Measurements have been carried out at 295 K with the samples in equilibrium with H_2 gas at 6.0 ± 0.1 atm (pure $LaNi_5$ powders) and at 6.6 ± 0.1 atm (composite materials). The vertical axis in the figure reports the H_2 desorbed amounts in wt% of the metallic fraction of the composite material. Figure 4 shows that the typical desorption time of the pure $LaNi_5$ powders is much shorter than that observed with both the composite samples. The slower H_2 desorption kinetics of the PS-$LaNi_5$ composite, as compared with that of the $LaNi_5$ powders, can be tentatively explained by the fact that the hydrogenated metal particles have only part of their surface not coated by the PS matrix and thus reliable for the H_2 recombinative desorption process.

The much longer H_2 desorption time of the PE-$LaNi_5$ composite, see line (b), has a different explanation. H atoms dissociated from the hydride phase in the $LaNi_5$ particles recombine at the metal-polymer interface and then migrate to the composite surface where desorption occurs: the slower kinetics can be thus explained by the slower diffusion process of hydrogen in molecular form into the PE-$LaNi_5$ composite. Given a value $\sim 5 \times 10^{-7}$ cm^2/s [10] for the room temperature diffusion constant of H_2 in PE and assuming as representative diffusion length the PE-$LaNi_5$ foil thickness, ~0.5 mm, we can evaluate a typical desorption time well compatible with that indicated by line (b) of Fig. 4.

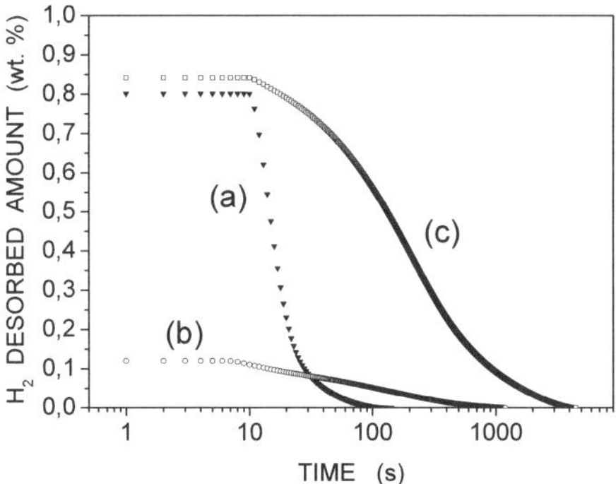

Figure 4. H_2 desorption curves at 295 K. Line (a): $LaNi_5$ powders. Line (b): PS-$LaNi_5$ composite. Line (c): PE-$LaNi_5$ composite. The vertical axis reports the H_2 desorbed amounts in wt% of the metallic fraction of the composite. Measurements have been carried out with samples in equilibrium with H_2 gas at 6.0 ± 0.1 atm (pure $LaNi_5$ powders) and at 6.6 ± 0.1 atm (composite materials)

To conclude this discussion we remark the important point emerging from this work and indicate perspectives for future studies. In order to completely hydrogenate the metallic phase of a metal-polymer composite, weak chemical interaction of the polymeric chain with the metal surface is required. When this requirement is satisfied, the H_2 storage capacity of the composite depends only on the metal filling ratio. We plan to study the hydrogenation kinetics of composite samples where: (i) the H_2 diffusivity in not lower than the H diffusivity in the metal or metal hydride phase, and (ii) minimizing the size of the composite foils being this size the critical diffusion length for H_2 molecules in the sorption process.

4. Conclusions

Polymeric composites based on hydride-forming metals like $LaNi_5$ can be conveniently used as hydrogen storage systems. The basic requirements for an adequate polymeric matrix are: (i) diffusivity of the H_2 molecules not lower than in the metal phase, (ii) thermal stability and elasticity and, according to the results presented in this paper, (iii) weak chemical interactions of the polymeric chains with the metal phase ($LaNi_5$). Using low density polyethylene only weak Van der Waals forces are involved at interface between the polymer matrix and the metal particles and thus all the metallic fraction of the composite material results hydrogenated because the metal surface sites are active for H_2 molecular dissociation. The poisoning of metal surface sites, by dative bond formation between metal atoms and oxygen atoms of polymer back-bone (Si-O-Si), impedes in PS-$LaNi_5$ composites the H_2 dissociative absorption.

Acknowledgments

This research activity was financially supported by the Hydrogen-FISR Italian project.

References

1. Mainzer R.A. US Patent 5, 360: 461.
2. Heung L.K. and Wicks G.G. *Journal of Alloys and Compounds* 1999, **293–195**: 46.
3. Park J.K. and Paul D.R. *Journal of Membrane Science* 1997, **125**: 23.
4. Merkel T.C., Gupta R.P., Turk B.S. and Freeman B.D. *Journal of Membrane Science* 2001, **191**: 85.
5. Pinnau I. and Toy L.G. *Journal of Materials Science* 1996, **116**: 199.
6. Checchetto R., Carotenuto G., Bazzanella N. and Miotello A. *Journal of Physics D: Applied Physics* 2007, **40**: 4043.
7. Redhead P.A., Hobson J.P. and Kornelsen E.V. in "*The Physical Basis of Ultrahigh Vacuum*", American Institute of Physics, New York, 1995, 37 p.
8. Bartholomew C.H. *Applied Catalysis A: General* 2001, **212**: 17–60.
9. Jewell L.L. and Davis B.H. *Applied Catalysis A: General* 2006, **310**: 1–15.
10. Miyake H., Matsuyama M., Ashida K. and Watanabe K. *Journal of Vacuum Science & Technology* 1983, **A1**: 1447.

SOME PHYSICAL ASPECTS OF BCC HYDROGEN STORAGE ALLOYS

G. MAZZOLAI
University of Perugia, Department of Physics, Via A. Pascoli 5, 06123 Perugia, Italy
Tel.: +39 3381305695; Fax: +39-075-44666; E-mail: giovanni.mazzolai@fisica.unipg.it

Abstract. Hydrogen absorption/desorption has been investigated in the three series of solid solution bcc alloys $Ti_{35}V_xCr_{65-x}$ (x = 18.22), $Ti_{40}V_xMn_{50-x}Cr_{10}$ (x = 32.36) and $Ti_xCr_{97.5-x}Mo_{2.5}$ (x = 43.46). It has been found that H absorption at pressures smaller than one bar can only occur after elimination of the oxide films by heating the alloys to temperatures higher than 600 K. Hydrogen desorption from pre-loaded materials (n_H = H/Me $\leq$ 0.27) takes place on heating at much lower temperatures in the $Ti_{40}V_xMn_{50-x}Cr_{10}$ and $Ti_{35}V_xCr_{65-x}$ than in the $Ti_xCr_{97.5-x}Mo_{2.5}$ alloys. The H diffusion parameters W and D_o deduced from high temperature (>450 K) absorption experiments are: W = 0.318 ± 0.005 eV, D_o= (4 ± 1) × 10^{-7} m^2/s for $Ti_{40}V_xMn_{50-x}Cr_{10}$; W = 0.32 ± 0.02 eV, D_o= (3 ± 2) × 10^{-7} m^2/s for $Ti_{35}V_xCr_{65-x}$; W = 0.79 ± 0.06 eV, D_o = (4 ± 2) × 10^{-8} m^2/s for $Ti_xCr_{97.5-x}Mo_{2.5}$. The higher value of the activation energy for H diffusion in Mo containing alloys is most likely due to H trapping by Mo atoms.

Keywords: H absorption, H diffusion, solid solution bcc alloys, PCT isotherms, internal friction

1. Introduction

Solid solution bcc alloys have been found to be able to store large amounts of H, thus attracting a lot of attention in recent years [1–27]. As a result of the studies carried out so far several properties of these materials are now reasonably well known such as lattice structures of the hydrides [6, 9, 11, 18], PCT isotherms [3, 6, 8, 18, 20] including their dependence on alloy composition [2, 7, 11, 12, 14, 15, 18, 20], thermal history of the materials [8, 13], absorption-desorption cycling [9, 10, 15] and microstructure [5, 8]; also the sequence of phase transformations taking place on progressive H charging at constant temperature is rather well established [11, 18]. Not much seems to be known about H diffusivities, H localisation and also about the mechanisms of phase transitions occurring within the hydrides. Therefore, a research program has been undertaken to measure the diffusion coefficient of H at high and low temperatures in the alloy systems TiVCrMn, TiVCr and TiCrMo. Some results primarily dealing with calorimetric, thermodynamic and low temperature anelastic (internal friction) measurements have already been reported [26, 27] and some others will be the subject of a forthcoming paper [17]. In the present work attention will be focussed mainly on H diffusion at high temperature and on thermal decomposition spectroscopy (TDS).

B. Baranowski et al. (eds.), *Carbon Nanomaterials in Clean Energy Hydrogen Systems.*

2. Experimental

Three series of alloys $Ti_{40}V_xMn_{50-x}Cr_{10}$ (x = 32.36), $Ti_{35}V_xCr_{65-x}$ (x = 18.22) and $Ti_xCr_{97.5-x}Mo_{2.5}$ (x = 43.46) were prepared by arc melting at IENI-CNR. The compositions were chosen in such a way to attain homogeneous bcc solid solution alloys of high H mass capacity according the indications given by data available in literature. The buttons were re-melted several times to improve the alloy homogeneity. The samples cut from the ingot by spark erosion were first chemically etched with a water solution of hydrofluoric and nitric acid (10 and 40 vol%, respectively) and then exposed to H_2 atmospheres in a reaction chamber of known volume for absorption/desorption experiments. During the measurements the temperature and pressure were monitored as a function of time by a Pt-Pt10%Rh thermocouple and a precision Datametrics capacitance manometer, respectively. The temperature was stable typically within ±2 K. The pumping system consisted of mechanical and turbo-molecular pumps and the chamber sealing was made by copper gaskets. The final H content n_H = H/Me was determined from pressure changes during absorption and turned out to be included between 0.003 and 0.27 depending on the H-loading conditions. Differential scanning calorimetry measurements (DSC) have been carried out at a heating rate of 20 K/min by a Pyris apparatus of Perkin Elmer Co.

3. Results and discussion

3.1. SURFACE ACTIVATION PROCESS

Under an H_2 gas pressure of about 0.1 MPa (1 bar) the etched samples did not absorb hydrogen at room temperature. Thus, a surface activation treatment was given to the materials which consisted in heating them to high temperatures (T) in an H_2 gas atmosphere; the used pressures varied from 0.014 to 0.08 MPa. A single heating/cooling cycle was enough to activate the surface. Typical activation processes for the $Ti_{43}Cr_{54.5}Mo_{2.5}$ and $Ti_{40}V_{36}Mn_{14}Cr_{10}$ alloys are illustrated by Figs. 1 and 2, where p(T) curves are reported together with their derivatives with respect to temperature. As seen, the H pressure first increases with increasing T, as expected from the gas law, then decreases, to finally increase again. The higher slope of the curves in the third stage with respect to those in first one is due to the combined effect of the gas law and of the progressive dehydrogenation of the sample at the highest temperatures caused by a progressive diminution with increasing T of the H solubility. The pressure decrease starts at around 720 K for the $Ti_{43}Cr_{54.5}Mo_{2.5}$ alloy and 780 K for the $Ti_{40}V_{36}Mn_{14}Cr_{10}$ one and is associated with the dissolution of surface oxides, which at lower temperatures prevent H uptake by the samples. The observations made clearly show that unless appropriate precautions are taken, H diffusion data below these temperatures may not be fully reliable since the rate of absorption may be controlled by surface barriers rather than by bulk diffusion.

3.2. THERMAL DECOMPOSITION SPECTROSCOPY (TDS) DATA

A desorption curve for the $Ti_{43}Cr_{54.5}Mo_{2.5}$ alloy pre-charged to an H content n_H = 0.08 is plotted in Fig. 3 together with its derivative with respect to temperature. As seen

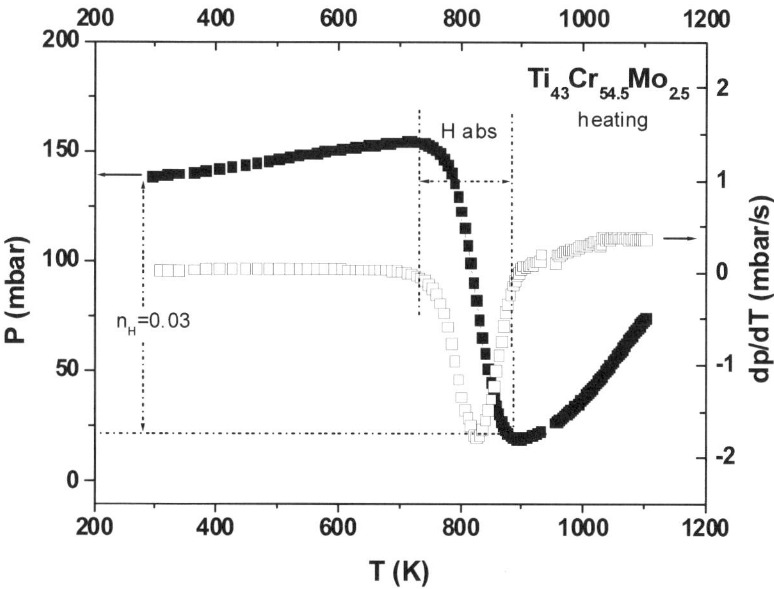

Figure 1. First H absorption process of the $Ti_{43}Cr_{54.5}Mo_{2.5}$ alloy on heating

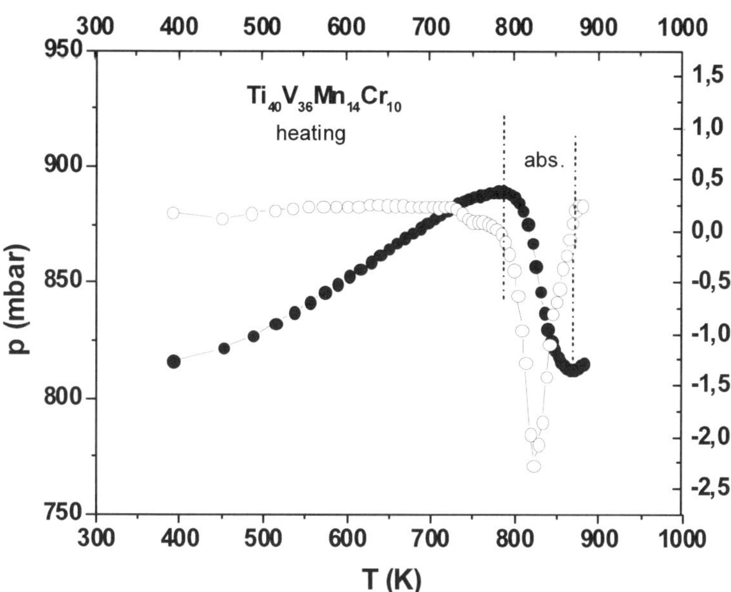

Figure 2. First H absorption process of the $Ti_{40}V_{36}Mn_{14}Cr_{10}$ alloy, on heating

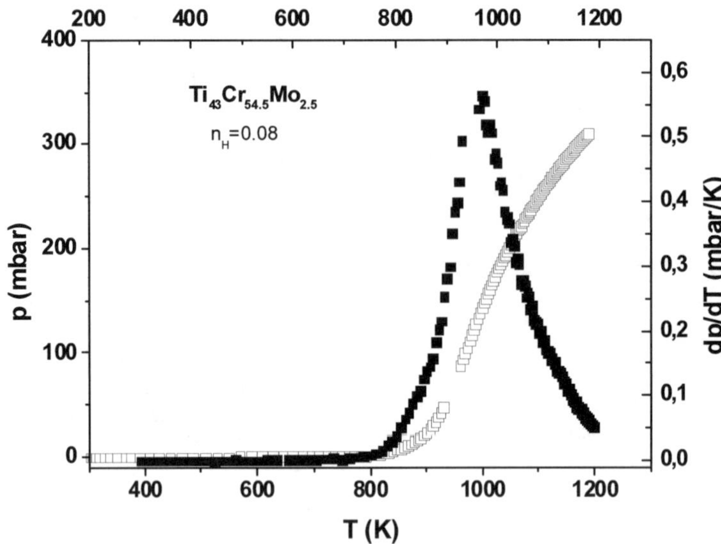

Figure 3. Thermal desorption spectroscopy of the alloy $Ti_{43}Cr_{54.5}Mo_{2.5}$ pre-charged with H to a value of n_H = H/Me = 0.08

from the temperature dependence of $\frac{dp}{dT}$ desorption takes place in a single stage which begins at around 800 K and goes through a maximum at 960 K. This desorption stage is most likely associated with the dissolution of the H solid solution (α-phase), which is the only phase expected to be present within the material for n_H = 0.08 [3, 7]. A similar behavior is exhibited, on heating, by the $Ti_{40}V_{36}Mn_{14}Cr_{10}$ alloy charged to a value of n_H equal to 0.27 (Fig. 4). In this case the H desorption rate starts increasing at 690 K and gets its maximum value at 750 K. The single-peak structure of the TDS spectrum of this quaternary alloy is again attributable to the dissolution of the H solid solution. Accordingly, the data indicate that the homogeneity field of the α-phase in this alloy extends at least up to values of n_H = 0.27.

For n_H = 0.26, H desorption in the $Ti_{35}Cr_{47}V_{18}$ alloy occurs in two stages, respectively taking place over the temperature ranges 650–720 K and 720–800 K (Fig. 5). A double peak is also seen in the differential scanning calorimetry (DSC) curve of the same alloy for n_H = 0.08.

On equilibrating the alloys of the three investigated series with H_2 gas at increasingly higher pressures and fixed temperature T, the H solid solution phase is the first to form while it is the last to disappear on decreasing p. The same behavior is expected on increasing and decreasing T during a thermal cycle. This means that the solid solution is the most stable phase. Thus, the desorption peak occurring at the higher temperature in Fig. 5 is attributable to the decomposition of the solid solution, the one occurring at lower temperature to the decomposition of the bcc β hydride. The appearance of two peaks indicates the coexistence of two phases and, consequently, that the limit solubility of H in the solid solution is lower than 0.08 for the $Ti_{35}Cr_{47}V_{18}$ alloy.

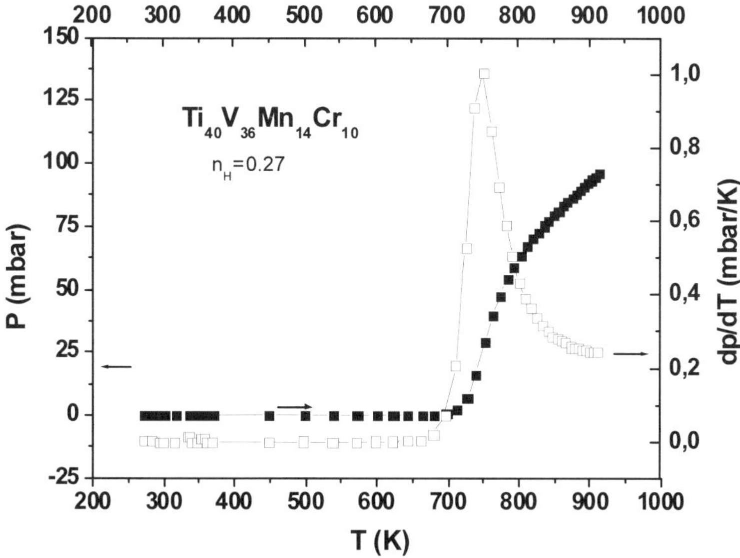

Figure 4. Thermal desorption spectroscopy of the alloy $Ti_{40}V_{36}Mn_{14}Cr_{10}$ pre-charged with H to a value of $n_H = 0.27$

β

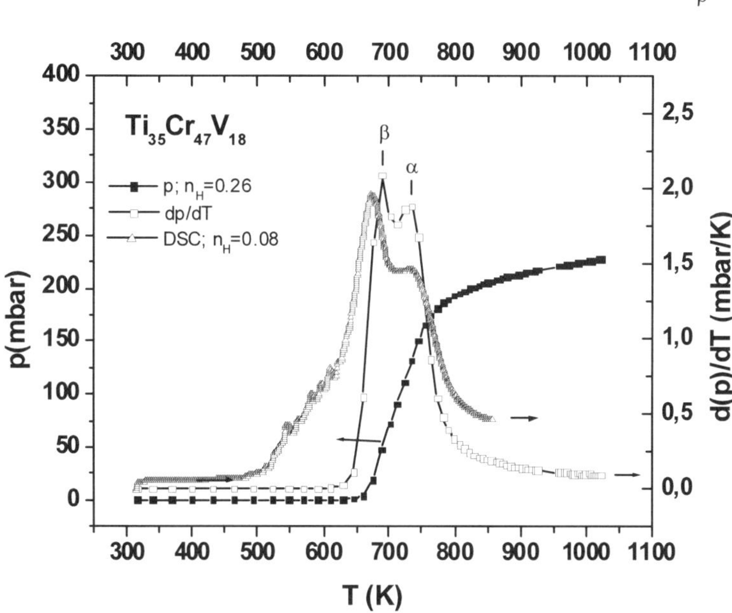

Figure 5. Thermal desorption spectroscopy and differential scanning calorimetry (DSC) of the alloy $Ti_{36}Cr_{47}V_{18}$ pre-charged with H to the indicated values of n_H

The TDS data show that the H solid solutions of the $Ti_{40}V_{36}Mn_{14}Cr_{10}$ and $Ti_{36}Cr_{47}V_{18}$ alloys dissolve at about the same temperature (see Figs. 4 and 5), thus indicating similar stabilities of this phase for the two materials. With respect to the previous ones the H solid solution of the $Ti_{46}Cr_{51.5}Mo_{2.5}$ alloy dissolves at a higher temperature thus exhibiting a higher stability, which is most likely due to trapping of H by Mo atoms. This explanation is also supported by H diffusion data as will be discussed in a moment.

3.3. H DIFFUSION

Isothermal absorption of hydrogen has been investigated at temperatures between about 450 and 1,200 K by following pressure changes within the calibrated volume as a function of time. A typical set of normalised curves for the $Ti_{35}Cr_{43}V_{22}$ alloy is plotted in Fig. 6, where the lines are fits to the following relationship:

$$P(t) = P_f + (P_0 - P_f)\frac{8}{\pi^2}\sum_{j=0}^{\infty}\frac{1}{(2j+1)^2}\exp\left\{-[\frac{(2j+1)\pi}{h}]^2 D_c t\right\}, \quad (1)$$

which has been deduced from the second Fick's equation solved for a thin bar shaped sample, as required by the present experimental conditions.

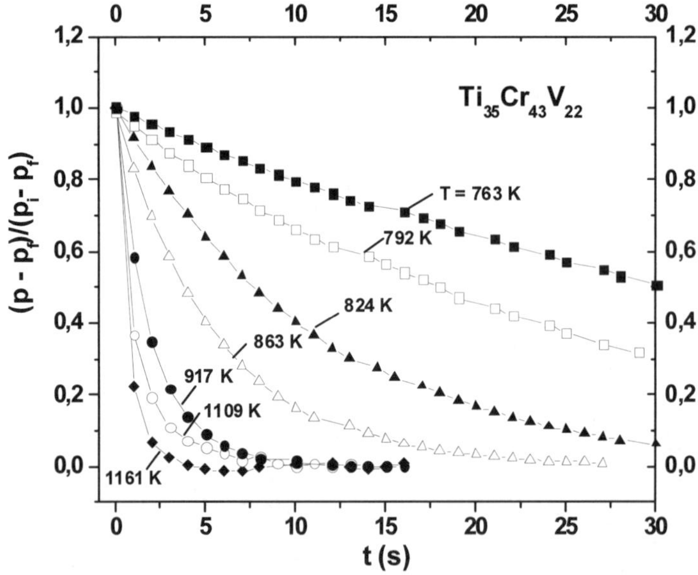

Figure 6. Normalized isothermal absorption curves for the $Ti_{35}Cr_{43}V_{22}$ alloy at the indicated temperatures and their fits (full line) to relation (2)

In this relation D_c is the chemical diffusion coefficient, which has been used as the fitting parameter, h the sample thickness, P_0 and P_f the initial and final pressures, and t the time. As seen the fits are very good proving that the rate controlling factor for absorption is bulk diffusion; they provide for D_c the values reported in the Arrhenius plot of Fig. 7, which refers to both the investigated alloys of the series $Ti_{35}Cr_{65-x}V_x$. Within the limit of the experimental accuracy no change with the alloy composition is

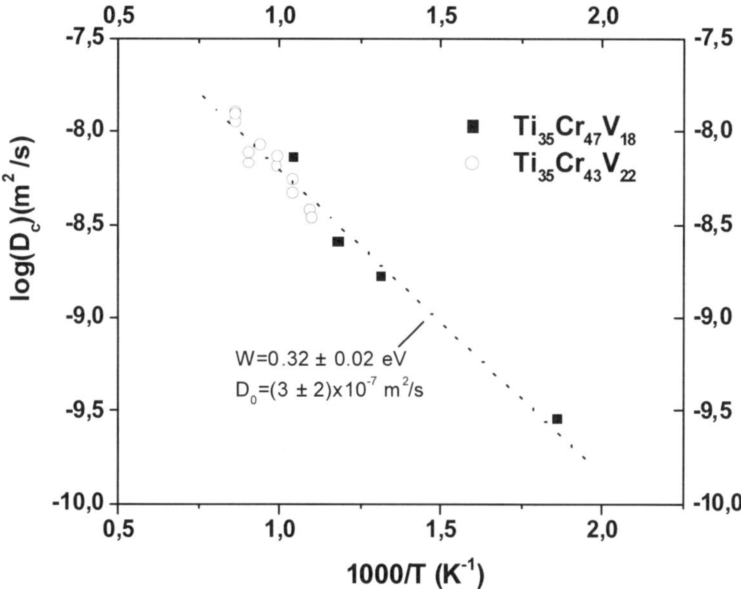

Figure 7. Chemical diffusion coefficient of H in the alloys of the series $Ti_{35}Cr_{65-x}V_x$ as deduced from fits of the absorption curves to relation (2)

observed for D_c. The values of the diffusion coefficient are high and the activation energy, W, derived from the linear regression of the overall experimental points is rather small (W = 0.32 ± 0.02 eV) as happens with bcc metal alloys, usually exhibiting occupancy of tetrahedral interstitial sites by H. This is in agreement with the fact that the $Ti_{35}Cr_{65-x}V_x$ alloys have a homogeneous bcc structure as proved by X-ray diffraction patterns [16]. It is to be noted that the experimental data points are aligned along a straight line including the one taken at the lowest temperature of 550 K. This means that the oxide film did not form on the surface of the sample, which during the sequence of the measurement runs at the selected temperatures were kept in ultrahigh vacuum or in contact with the H_2 gas atmosphere.

The results of a series of measurements carried out with the $Ti_{40}V_{32}Mn_{18}Cr_{10}$ and $Ti_{40}V_{36}Mn_{14}Cr_{10}$ alloys are shown in Fig. 8, where are also indicated the diffusion parameters D_0 and W obtained by a linear fit to the overall data points. Also in this case no appreciable dependence of D_c on the alloy composition is observed at the temperatures of the present measurements. The H diffusion parameters are quite similar to the ones of the ternary TiCrV alloys indicating that substitution of Cr for Mn does not play a remarkable role on the long-range H mobility.

Data of D_c for the two alloys of the series $Ti_xCr_{97.5-x}Mo_{2.5}$ are shown in Fig. 9. For this alloy system the activation energy W for H diffusion is surprisingly high (W = 0.79 ± 0.06 eV). Judging from data in Fig. 1 surface oxide films should not be present for values of T > 950 K ($T^{-1} < 1.1 \times 10^{-3}$ K^{-1}). Hence, surface energy barriers should not have affected H uptake by the sample. Therefore, it can be inferred that the high value for the activation energy W is due to the fact that Mo atoms act as trapping centers for H. Consequently, the measured value of W will be approximately the sum of the diffusion and the binding energies. This view is also supported by the observed higher stability of the H solid solution of $Ti_xCr_{97.5-x}Mo_{2.5}$ alloys as suggested by the TDS measurements.

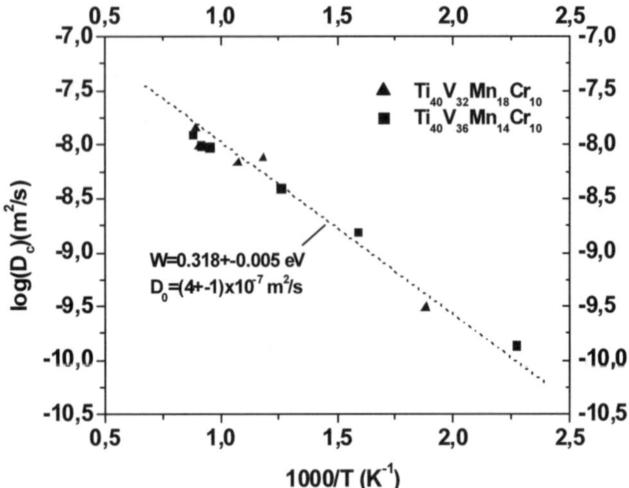

Figure 8. Chemical diffusion coefficient of H in the $Ti_{40}V_{32}Mn_{18}Cr_{10}$ and $Ti_{40}V_{36}Mn_{14}Cr_{10}$ alloys

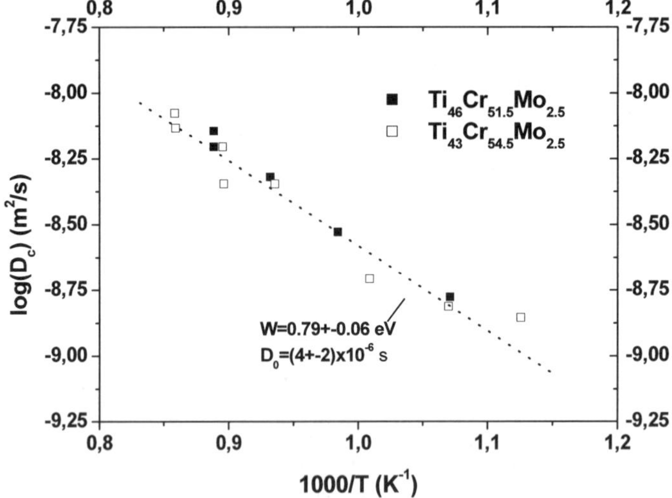

Figure 9. Chemical diffusion coefficient of H in the alloys of the series $Ti_xCr_{97.5-x}Mo_{2.5}$

The room temperature values of the H diffusion coefficient as deduced from an extrapolation of the present data are: 1.3×10^{-12} m^2/s, 0.9×10^{-12} m^2/s and 1×10^{-21} m^2/s for the $Ti_{40}V_xMn_{50-x}Cr_{10}$, $Ti_{35}V_xCr_{65-x}$ and $Ti_xCr_{97.5-x}Mo_{2.5}$ alloys, respectively. Therefore, assuming that absorption times of the order of 5 min are needed for refueling a H storage tank on board of a vehicle, the powders particles of the $Ti_{40}V_xMn_{50-x}Cr_{10}$ and $Ti_{35}V_xCr_{65-x}$ alloys should have dimensions of the order of the diffusion length $x = \sqrt{2Dt}$, that is about 60 μm. Powders of this particle size can easily be obtained. On the other hand, the alloys containing molybdenum would need nano-particles to get comparable absorption times; this would be unpractical. However, at the high H concentrations of interest in H storage applications, trapping effects at Mo content as low as 2.5 at.% should actually have a minor effect on the absorption kinetics.

The prediction of fast kinetics for H absorption as derived from the high-temperature diffusion data has been confirmed by absorption measurements carried out at 273 K with a powder of the average particle size of 40 μm of the quaternary alloy $Ti_{40}V_{32}Mn_{18}Cr_{10}$. As seen in Fig. 10, where normalized pressure vs time plots are reported for various initial pressures and H contents. It is to be noted that the 80% of the entire H absorption takes places in less than 150 s.

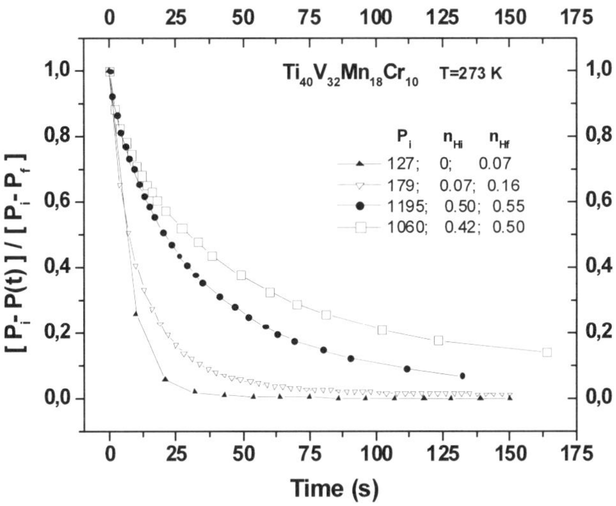

Figure 10. Kinetics of H absorption at 273 K by a powder of the $Ti_{40}V_{32}Mn_{18}Cr_{10}$ alloy

4. Conclusions

The main results of the present investigation can be summarised as follows:

(a) The applied procedure for surface activation has led to the identification of the temperatures needed for oxide films dissolution; thermal decomposition spectroscopy has allowed the determination of the temperatures for the decomposition of the α and β hydrides. It turns out that the stability of these two phases is too high for H extraction from the alloys at ambient pressures and temperatures. This feature reduces the usable mass capacity of the investigated alloys and constitutes a drawback for their use in H storage applications.

(b) The chemical diffusion coefficient of H, which has been measured at high temperature for all the three series of investigated alloys, exhibits an Arrhenius-type temperature dependence, as expected. Surprisingly, however, the activation energy for H diffusion in the $Ti_xCr_{97.5-x}Mo_{2.5}$ alloys turns out to be high, indicating that Mo atoms act as strong trapping centres for H. An extrapolation of the obtained diffusivity data towards room temperature suggested that rapid kinetics of H absorption/desorption could be attained at room temperatures with $Ti_{40}V_xMn_{50-x}Cr_{10}$ and $Ti_{35}V_xCr_{65-x}$ alloys. This expectation has been confirmed by absorption experiments carried out at 273 K with the $Ti_{40}V_{32}Mn_{18}Cr_{10}$ alloy.

Acknowledgements

The author gratefully acknowledges financial support from the Ministry of Education and Research in the framework of the 2004 PRIN project entitled "Sintesi e caratterizzazione di materiali per l'immagazzinamento di idrogeno allo stato solido" and Dr. A. Tuissi for providing him with the alloy.

References

1. Akiba E., Iba H., *Intermetallics*, 1998. **6**: 461–470.
2. Cho S.W., Han C.S., Park C.N., Akiba E., *J. Alloy. Compd.*, 1999. **288**: 294–298.
3. Tamura T., Kamegawa A., Takamura H., Okada M., *Mater. Trans.*, 2001. **42**: 1862–1865, ibid. 2002. **43**: 2753–2756.
4. Okada M., Kuriiwa T., Kamegawa A., Takamura H., *Mater. Sci. Eng.*, 2002. **A329-331**: 305–312.
5. Itoh H., Arashima H., Kubo K., Kabutomori T., *J. Alloys Compds.*, 2002. **330-332**: 287–291.
6. Tamura T., Kamegawa A., Takamura H., Okada M., *Mater. Trans.*, 2002. **43**: 410–413.
7. Tamura T., Kazumi T., Kamegawa A., Takamura H., Okada M., *Mater. Trans.*, 2002. **43**: 2753–2756.
8. Okada M., Kuriiwa T., Tamura T., Takamura H., Kamegawa A., *J. Alloys Compds.*, 2002. **330-332**: 511–516.
9. Kazumi T., Tamura T., Kamegawa A., Takamura H., Okada M., *Mater. Trans.*, 2002. **43**: 2748–2752.
10. Shirasaki K., Tamura T., Kuriiwa T., Goto T., Kamegawa A., Takamura H., Okada M., *Mater. Trans.*, 2002. **43**: 1115–1119.
11. Kamegawa A., Shirasaki K., Tamura T., Kuriiwa T., Takamura H., Okada M., *Mater. Trans.*, 2002. **43**: 470–473.
12. Okada M., Chou T., Kamegawa A., Tamura T., Takamura H., Matsukawa A., Yamashita S., *J. Alloy. Compd.* 2003. **356-357**: 480–485.
13. Arashima H., Takahashi F., Ebisawa T., Itoh H., Kabutomori T., *J. Alloy. Compd.*, 2003. **356-357**: 405–408.
14. Cho J.S., Park C.N., Yoo J.H., Choi J., Park J.S., Suh C.Y., Shim G., *J. Alloy. Compd.*, 2005. **403**: 262–266.
15. Itoh H., Arashima H., Kubo K., Kabutomori T., Ohnishi K., *J. Alloy. Compd.*, 2005. **404-406**: 417–420.
16. Mazzolai G., Coluzzi B., Biscarini A., Mazzolai F.M., Tuissi A., Palade P., Principi G., Lo Russo S., *J. Alloy. Compd.*, In preparation.
17. Yu X.B., Wu Z., Xia B.J., Huang T., Chen J.Z., Wang Z., Xu N., *J. Mater. Res.*, 2003. **18**: 2533–2536.
18. Yu X.B., Xia B.J., Xu N.X., *J. Alloy. Compd.*, 2004. **372**: 272–277.
19. Yu X.B., Chen J.Z., Wu Z., Xia B.J., Xu N., *Int. J. Hydrogen Energ.*, 2004. **29**: 1377–1381.
20. Yu X.B., Xia B.J., Wu Z., Xu N., *Mater. Sci. Eng.*, 2004. **A, 373**: 303–308.
21. Yu X.B., Wu Z., Xia B.J., Xu N.X., *J. Alloy. Compd.*, 2004. **373**: 134–136.
22. Yu X.B., Wu Z., Xia B.J., Xu N.X., *J. Alloy. Compd.*, 2005. **386**: 258–260.
23. Yu X.B., Feng S.L., Wu Z., Xia B.J., Xu N.X., *J. Alloy. Compd.*, 2005. **393**: 129–134.
24. Yu X.B., Dou T., Wu Z., Xia B.J., Xu N.X., *Nanotechnology*, 2006. **17**: 268–271.
25. Mazzolai G., Biscarini A., Coluzzi B., F.M. Mazzolai, Tuissi A., *Chem. Eng. Trans.*, 2005. **8**: 165–170.
26. Mazzolai G., Coluzzi B., Biscarini A., Mazzolai F.M., Tuissi A., *Proc. Mater. Res. Soc. Symp.*, 2006. **885E**: 215–220.
27. Kamegawa A., Tamura T., Takamura H., Okada M., *J. Alloy. Compd.*, 2003. **356–357**: 447–451.

VERY LARGE C_n – EXPECTATIONS AND REALITY

N. KOPRINAROV, M. KONSTANTINOVA
Central Laboratory for Solar Energy and New Energy Sources – Bulgarian Academy of Science, Boul. Tzarigradsko shausee 72, 1784 Sofia, Bulgaria
E-mail: koprin@phys.bas.bg

Abstract. All known methods for fullerene synthesis produce a variety of C_n molecules, indicating that fullerene with other sizes than C_{60} exist and can be stable. However, the range and variety of those molecules is largely unexplored. It is also unknown if there is an upper limit for *n*. The size and the shape of those molecules become crucial when they are surrounded by a gas or a liquid since the molecules are exposed to elevation force and pressure. The resistance to pressure and the elevation force are the highest if the shape is spherical. Spherical or almost spherical shape can be achieved if the carbon network includes a certain amount of defects (pentagon-heptagon rings). Those spherical structures can be expected since computer simulation of the C_n growth predicts that the number of the defects is significant during the synthesis. In this paper we show that spherical C_n structures, which are large enough to be lighter than air, can potentially withstand atmospheric pressure.

Keywords: fullerene, carbon nanoparticles, new materials, hollow carbon molecules

1. Introduction

With the discovery of C_{60} [1] began the exploration of a new kind of carbon nanoparticles. Their unique chemical, physical and electrical properties raised great interest and expectations. The demand for those new materials and their high price encouraged the search for effective methods for their synthesis [2, 3] All of those methods, in addition to the C_{60}, form a variety of carbon molecules with *n* bigger than 60. Despite the fact that there are also other molecules, C_{60} is considered as representative to the whole C_n family and almost all of the studies are based on it. However, the production methods show that the probability obtaining different C_n molecules is not drastically different. In mass spectra, the number of the C_n molecules decrease with increasing *n*, however, no limit can be observed. The obtained C_n molecules include values of *n* up to several hundred but the theoretical calculations of A. Siber [4] show, that C_n up to $n = 10^6$ must be still stable. That minces that C_n up to this n value will not collapse due to van der Waals forces in difference to the carbon nanotubes [5].

The dimensions and the shape of the fullerenes were never considered as an important characteristic. However, this becomes crucial when C_n is surrounded by a medium since the C_n molecules are exposed to elevation force and pressure [6]. In this situation, the fullerene molecule can be modeled by a balloon that has a carbon shell consisting of five- and six-atom rings. According to the rule for building of a spherical structure from faces with 5 and 6 edges, the number of the faces with 5 edges must be at least 12. The number of five-atom rings in C_n becomes negligible in relation to the number of six-atom rings when the value of carbon atoms, *n*, becomes large enough. Therefore, it can be assumed that, after a certain size, the balloon surface is a graphite plane bent as a sphere.

2. Results and discussion

The elevation force depends on the weight difference between C_n molecule and the displaced medium. This difference is a function of the size and the density of the both materials. Figure 1 shows this relationship between C_n (marked as WC_n) and spheres made out of H_2O, air, and H_2. The curve WC_n crosses the H_2O, air, and H_2 curves in different points and determines two zones. All spheres below WC_n are lighter than C_n. All spheres above are heavier. The values of *n* and *R* at the cross points represent the situation, when the weight of the C_n equals the sphere from the corresponding substance.

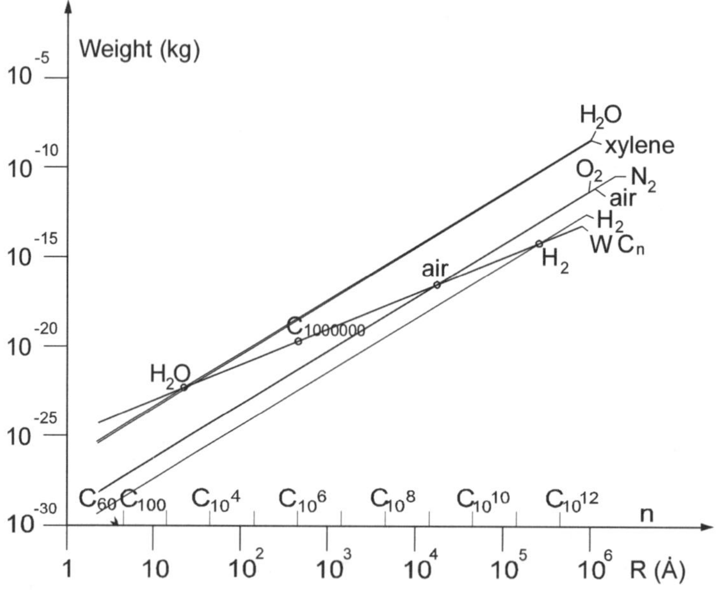

Figure 1. The C_n weight (marked as WC_n) and the weight of spheres made out of different materials as function of their radiuses

The cross points of WC_n with H_2O, air, and H_2 are $n = 2{,}497$, $n = 1.45 \times 10^9$, and $n = 3.1 \times 10^{11}$ respectively. Those correspond to the radiuses $R = 22.82$ Å, $R = 1.74 \times 10^4$ Å, and $R = 2.56 \times 10^5$ Å. Due to the difference in density between gases and liquids, *n* for the corresponding cross differs quite substantially. The parameters of the spheres where WC_n crosses the H_2O, air, and H_2 curves are shown in Fig. 2 for better comparison. It becomes obvious that all values of *n* are much bigger than those of the fullerene molecules studied so far.

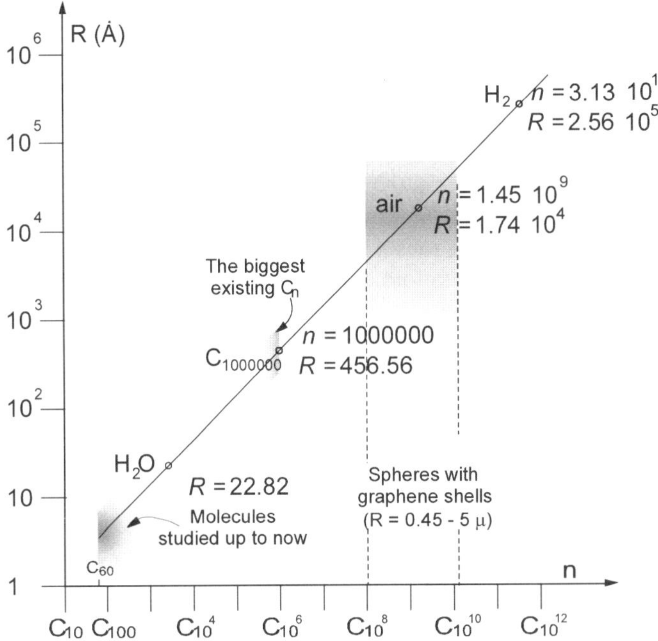

Figure 2. Relationship between the radius of a hollow spherical molecule and the number *n* of its carbons atoms. The C_n sizes commonly studied, the hollow multilayer structure, $C_{1,000,000}$, and the cross-point values discussed in the text are marked on the graph

If the shape of the C_n molecules is not spherical, the cross points will diverge from the values in Fig. 1. Therefore, the shape of the C_n molecules needs to be studied in more detail.

According to the continuum elasticity theory, the shape is a function of parameter γ

$$\gamma = 4\pi R^2 Y/k,$$

where Y is the two-dimensional Young's modulus of the sheet, k is the Helfrich bending constant and R the curvature radius [7]. The same approach was used earlier to calculate the relation between the shape of a virus and the parameter γ [8–10] as well as to solve the Thomson problem [11, 12]. The proteins of the viral coating are organized in units that consist of five (pentamers) or six proteins (hexamers), which protect the enclosed RNA or DNA genome molecules against enzymatic digestion. The Thomson's problem is to find the lowest energy of N Coulomb charges distributed on the surface of a unit sphere. The initial shape in the both cases is an icosahedron. The similarity between the problems allows the exchange of results. For example, the shape deformation during the viral maturation was calculated for the cases when the virus keeps its inside pressure and when the volume reduction is fixed. Those results could be of interest for the C_n molecules when they are filled with some medium. The results obtained by solving of the Thomson problem indicate that the icosahedral surface becomes more spherical by including of the pentagonal – heptagonal face segments and that their optimal number and place depend on the dimensions of the icosahedron.

Similar results were derived for fullerenes [13], which show that the icosahedral C_n with five to seven ring defects will result in more spherical structures.

The icosahedral shape is simple and symmetrical. However, it is not the only possibility. Euler's theorem describes spatially closed structures consisting of faces with different number of edges according to the equation

$$3N_3 + 2N_4 + 1N_5 + 0N_6 - 1N_7 - 2N_8 - 3N_9 - 4N_{10} = 12, \quad (1)$$

where N_i is the number of polygons with i sides. In nature, carbon forms ring with different number of atoms. Therefore, such rings need to be considered as well as building blocks of the C_n structure.

The biggest deformation of the icosahedral fullerene shape is in the areas around the vertices. In [14, 15], those regions were removed and surrounding atoms ware rearranged via energy minimization to get a quasi-spherical arrangement with holes. Subsequently, the holes were filled with heptagonal and pentagonal rings in the right proportion to preserve the topology. Such modifications of the icosahedral shape allow nearly spherical structure to be obtained. For example, M. I. Heggie at al. [16] demonstrated a decrease in radius variations of C_{240} from 0.44 to 0.12 Å and in the case of C_{540} from 1.21 to 0.63 Å. Moreover, H. Terrones et al. [17] calculated that the quasi-spherical fullerenes are more stable than C_{60}.

If we consider that the perfect icosahedral shape has only 12 pentagons and that the incorporation of additional rings with carbon atoms different than 6 will result in a spherical shape, it is necessary to evaluated how many defects needs to be present in the network during the synthesis of C_n and how fast the rearrangement needs to occur to obtain one of both shapes. The model of C_n formation by means of molecular dynamics [18, 19] comes to conclusion that the fullerene growth follows the steps: nucleation, ring-condensation, chain-growth, opening-closure. The energetic difference between pentagons and hexagons does not matter very much and both five- and six-membered rings are readily created with a ratio of about 1:1. Over time, this ratio approaches approximately 1:2 at 2,000 K, favoring hexagons as a consequence of continued annealing. Seven atom rings are present in acceptable amount too. The authors show that the stability of the molecules with $n > 60$ is even higher than that of C_{60}. Those conclusions indicate that the spherical C_n molecules have a good chance of arising and surviving.

The synthesis and the existence of hollow C_n molecules depend on the resistance of the structure to pressure. The tangential compressive stress in air of a spherical shell can be calculated from the equation, given by Li et al. in [20]

$$\sigma = pR/2h, \quad (2)$$

where σ is the circumferential stress, h is the thickness of the graphene, p is the atmospheric pressure, and R is the radius of C_n molecule. However, the thickness of graphene found in the literature is controversial. The values published are 0.075 [21] 0.07 [22] and 0.34 nm [23] where the last one corresponds to the spatial extend of the π orbitals.

By applying the above formula to C_n with a radius of 1.74×10^4, we obtain that σ in air equals 1.2590 GPa for $h = 0.07$ nm or 0.2592 GPa for $h = 0.34$ nm. We can compare those values to different permissible values for the compressive strength available in the literature/Table 1/: 100–150 GPa for single MWNT in [24] 100–170 GPa for NT in [25]

153 GPa for SWNT in [26] and 20 GPa for graphite whiskers in [23]. It becomes clear that the compressive stress in C_n with $R = 1.74 \times 10^4$ Å, when exposed to air, would be much lower compared to that permissible for carbon sheets or for other similar structures.

We believe that a more realistic approach would be to include high-pressure studies of carbon nanostructures. For example, Patterson et al. [27] studied the C_{70} conductivity as function of pressure. They reported that the conductivity is reversible up to 10 GPa and the structure collapses between 20–35 GPa. Tang et al. [28] measured the linear elasticity of SWNT under hydrostatic pressures of up to 1.5 GPa. Li et al. [20] have determined the hydrostatic pressure for buckling of SWNT (diameters 0.6–2 nm) to be 1.5–11 GPa. If we calculate the compressive stress for the cited carbon structures by the same way as by C_n with $R = 1.74 \times 10^4$ Å (see Table 2), it becomes clear that the studied nanostructures have been successfully exposed to much grater pressures than the effect of the atmosphere would be on C_n.

TABLE 1. Calculated σ for C_n with $R = 1.74 \times 10^4$ Å at 1 atm, compared to theoretical σ limit for carbon structures

Theoretical σ limit	Calculated σ for C_n with $R = 1.74 \times 10^4$ Å at 1 atm	According to
–	1.2590 GPa	$pR/2h$; $h = 0.07$ nm
–	0.2592 GPa	$pR/2h$; $h = 0.34$
20 GPa for graphite		J.-P. Salvetat et al. [23]
100–150 Gpa for SWNT		O. Lourie et al. [24]
100–170 GPa for SWNT		P. Zhang et al. [25]
153 GPa for SWNT		D. Srivastava et al. [26]

TABLE 2. Calculated σ for C_n with $R = 1.74 \times 10^4$ Å at 1 atm, compared to σ from high-pressure studies of carbon nanostructures

R	P, [GPa]	σ, [GPa] (for $h = 0.34$)	σ, [GPa] (for $h = 0.07$)	According to
1 nm	11	16.18	78.57	Chunyu Li [20]
0.7 nm	1.5	1.54	7.5	Tang et al. [28]
6.83 nm	10	100.44	487.86	Patterson et al. [27]
1.74 μm	1×10^{-4}	0.26	1.26	$pR/2h$

To achieve maximum elevating force by using hollow carbon molecules immersed in some liquid, they should be empty. In the upper consideration C_n it was assumed that in the hollow part of the C_n molecules is vacuum. In reality that will not be the case, if they have been synthesized in some atmosphere, resulting in gas molecules trapped inside.

The molecule $C_{1,000,000}$ with n = 1,000,000, obtained by B. I. Yacobson and R. E. Smalley [29] is the largest molecule we know from the literature. We assume that it is the biggest one, although, we do not know if it is an ideal hollow (vacuum) sphere and if its shell consists of only one layer. On the curve in Fig. 2, $C_{1,000,000}$ is higher than the cross points of water.

Whereas the synthesis of molecules lighter than liquids could be considered as possible (being smaller than $C_{1,000,000}$), molecules with very large radiuses, lighter than gases, seem to be problematic or even unrealistic. This can be concluded if we judge on the bases of the C_n production methods developed up to now. However, carbon layers that are fully or partially bent with large radiuses are commonly observed. P. J. F. Harris [30] proposes that pyrographite consists from such structures. Recent studies also gave new insights into the large C_n molecules. Z. Lou et al. [31] used CO_2 as a carbon source and Li as an oxidant in their experiments. The team obtained near perfect carbon spheres with diameters ranging from 360 to 600 nm in reactions lasting 10 h at 650^0C and 1,000 atm. A few larger spheres with sizes around 900 nm have also been found. The typical spheres observed consisted of an amorphous core surrounded by a shell, which was clearly crystalline. K. Niwase et al. [32] started from C_{60} powder and, after shock-compression up to 57 GPa, obtained hollow spheres with diameters up to 10 μm. Studies by Raman have shown that the spherical particles are in a highly graphitized state. V. D. Blank et al. [33] synthesized carbon nanotubes by applying isostatic pressures up to 1,300 atm in an environment of pure Ar or an Ar-N_2 mixture. The obtained nanotubes are composed of closed structures resembling deformed spheres (with diameters up to 1–2 μm), which are stuck to each other, surrounded by the nanotube walls. It must be mentioned that the spherical structures are hollow with very thin shells of only 5–10 layers. If the nanotube walls are removed (e.g. by the destruction technique described in Collins et al. [34], the big hollow structures could be released and studied. One can even go further and strip down the spheres shell by shell till only one shell is left to gain insight into the properties of such structures.

The hollow spherical structures described in [31–33] have diameters near or larger than the one, which is necessary for C_n to be lighter than air (see Fig. 1). They might be in terms of size and stability what we are looking for but, so far, they cannot be considered as the ultimate proof. Their stability could solely be due to their heavy, multilayered shells. Nevertheless, such hollow spheres, even if heavier than air, would be very useful as a material which is lighter than some liquids. For example, the spheres obtained by V. D. Blank [33] with 10 layers and radius of 200 nm would be about 9 times lighter than water.

3. Conclusions

It can be concluded that the theoretical results support the existence of very light spherical C_n molecules. However, the methods developed so far for synthesis of C_{60} do not seem to be suitable for their production. One can only benefit from the new low-weighted materials when their synthesis becomes possible, which still remains a challenge.

References

1. Kroto H.W., Heath J.R., O'Brien S.C., Curl R.F., Smalley R.E. C_{60}: Buckminsterfullerene. *Nature* 1985, **318**: 162–163.
2. Curl R.F., Smalley R.E., Probing C_{60}. *Science* 1988, **242**: 1017–1021.
3. Parker D.H., Wurz P., Chatterjee K., Lykke K.R., Hunt J.E., Pellin M.J., et al. High-yield synthesis, separation, and mass-spectrometric characterization of fullerenes C_{60} and C_{266}. *J. Am. Chem. Soc.* 1991, **113**: 7499–7503.
4. Siber A., Energies of sp^2 carbon shapes with pentagonal disclination and elasticity theory. *Nanotechnology*. 2006, **17**: 3598–3606.
5. Hertel T., Walkup R.E., Avouris P. Deformation of carbon nanotubes by surface van der Waals forces. *Phys. Rev. B* 1998, **58**: 13870–13873.
6. Koprinarov N., Konstantinova M. Can solid carbon materials be lighter than water and air? *J. Nanopart. Res.* 2007 (DOI 10.1007/s11051-007-9229-3).
7. Siber A. Shapes and energies of giant icosahedral fullerenes. Onest of ridge sharpening transition. *Eur. Phys. J. B* 2006, **53**: 395–400.
8. Lidmar J., Miny L., Nelson D.R. Virus shapes and buckling transitions in spherical shells. *Phys. Rev. E* 2003, **68**(051910): 1–10.
9. Nguyen T.T., Bruinsma R.F., Gelbart W.M. Elasticity theory and shape transitions of viral shells. *Phys. Rev. E* 2005, **72**(051923): 1–19.
10. Siber A. Bucking transition in icosahedral shells subjected to volume conservation constraint and pressure: relations to virus naturation. *Phys. Rev. E* 2006, **73**(061915): 1–10.
11. Perez-Carido A., Dodgson M.J.W., Moore M.A. Influence of discussions in Tomson's problem. *Phys. Rev. B* 1997 **56**(7): 3640–3643.
12. Perez-Carrido A., Moore M.A. Symmetric patterns of dislocations in Tomson`s problem. *Phys. Rev. B* 1999, **60**(23): 15628–15631.
13. Perez-Carrido A. Giant multilayer fullerene structures with symmetrically arranged defects. *Phys. Rev. B* 2000, **62**(11): 6979–6981.
14. Terrones H., Terrones M., Moran-Lopes J.L. Curved nanomaterials. *Curr. Sci.* 2001, **81**(8): 1011–1020.
15. Terrones M., Terrones H. The Role of Defects in Graphitic Structures http://www.informaworld.com/smpp/title(content=t713597253)
16. Heggie M.I., Terrones M., Eggen B.R., Jungnikel G., Jones R., Lathan C.D., et. al., Quantitative density-functional study of nested fullerenes. *Phys. Rev. B Third Series* 1998, **57**(21): 13339–133342.
17. Terrones H., Terrones M. The transformation of polyhedral particles into graphitic onions. *J. Chem. Solids* 1997, **58**(11): 1784–1796.
18. Irle S., Zheng G., Wang Z., Mokoruma K. The C60 formation puzzle "solved": QM/MD simulations reveal the shrinking hot giant road of the dynamic fullerene self-assembly mechanism. *J. Phys. Chem. B* 2006, **110**: 14531–14545.
19. Zheng G., Irle S., Mokoruma K. Towards formation of buckminsterfullerene C_{60} in quantum chemical molecular dynamics. *J. Chem. Phys.* 2005, **122**(014708): 1–7.
20. Li C., Chou T.-W. Elastic properties of single-walled carbon nanotubes in transverse directions. *Phys. Rev. B* 2004, **69**(073401): 1–3.
21. Qi H.J., Teo K.B.K., Lau K.K.S., Boyce M.C., Milne W.I., Robertson J., et. al. Determination of mechanical properties of carbon nanotubes and vertically aligned carbon nanotube forests using nanoindentation. *J. Mech. Phys. Sol.* 2003, **51**: 2213–2237.
22. Xin Z., Jianjun Z., Ou.-Yang Zhong-Can. Strain energy and Young's modulus of single-wall carbon nanotubes calculated from electronic energy-band theory. *Phys. Rev. B* 2000, **62**: 13692–13696.
23. Salvetat J.-P., Bonard J.-M., Thomson N.H., Kulik A.J., Forró L., Benoit W., Zuppiroli L. Mechanical properties of carbon nanotubes l, *Appl. Phys. A* 1999, **69**(3): 255–260.

24. Lourie O., Cox D.M., Wagner H.D. Buckling and collapse of embedded carbon nanotubes. *Phys. Rev. Lett.* 1998, **81**(8): 1638–1641.
25. Zhang P., Lammert P.E., Crespi V.H. *Phys. Rev. Lett.* 1998, **81**(24): 5346–5349.
26. Srivastava D., Menon M., Cho K. Nanoplasticity of single-wall carbon nanotubes under uniaxial compression. *Phys. Rev. Lett.* 1999, **83**(15): 2973–2976.
27. Patterson J.R., Catledge Sh.A., Vohra Y.K. Nanodentation and x-ray diffraction studies of pressure-induced amorphization in C-70 fullerene. *Appl. Phys. Lett.* 2000, **77**(6): 851–853.
28. Tang J., Qin Lu.-Ch., Sasaki T., Yudasaka M., Matsushita A., Iijima S. Compressibility and polygonization of single-walled carbon nanotubes under hydrostatic pressure. *Phys. Rev. Lett.* 2000, **85**(9): 1887–1889.
29. Yacobson B.I., Smalley R.E. Fullerene Nanotubes: C1,000,000. American Scientist Online—The Magazine of Sigma XI, The Scientific Research Society, 1997.
30. Harris P.J.F. Fullerene-related structure of commercial glassy carbons. *Philos. Mag.* 2004, **84**(29): 3159–3167.
31. Lou Z., Chen Q., Gao J., Zhang Y. Preparation of carbon spheres consisting of amorphous carbon cores and graphene shells. *Carbon* 2004, **42**: 219–238.
32. Niwase K., Homae T., Nakamura K.G., Kondo K. Generation of giant carbon hollow spheres from C60 fullerene by shock-compression. *Chem. Phys. Lett.* 2002, **362**: 47–50.
33. Blank V.D., Gorlova I.G., Hutchison J.L., Kiselev N., Ormont A.B., Polyakov E.V., Sloan J., Zakharov D.N., Zybtsev S.G. The structure of nanotubes fabricated by carbon evaporation at high gas pressure. *Carbon* 2000, **38**: 217–1239.
34. Collins P.G., Arnold M.S., Avouris P. Engineering carbon nanotubes and nanotube circuits using electrical breakdown. *Science* 2001, **292**: 706–709.

ENERGETICS OF METAL HYDRIDES FORMATION IN ELECTROCHEMICAL SYSTEMS

A.V. ZVYAGINTSEVA, Y.N. SHALIMOV
Voronezh State Technical University, 394026, Voronezh, Russia

Abstract. We have determined that in low-temperature annealing conditions for Cr-samples ($t_{annealing} < 300°C$) the secondary hydrogenation is possible which is proved by the fixation of damping peak on the graphic chart $Q^{-1} = f(T)$. The amount of the absorbed hydrogen accurate within the experimental error proves that all hydrogen is fixed on the borders of structure's defects. The character of distribution confirms with the result received by the experiment. Current density distribution is rather homogeneous (difference between the center and angular point does not exceed 12%) and is characterized by smooth variation along electrode surface.

Keywords: electrochemical boron influence, Ni-film, hydrogen saturation

1. Introduction

In previous works [1, 2] it was specified that atomic hydrogen and metals interaction is realized across the places of potential wells (structure defects). The amount of such areas is determined by the existence of voltage gradients in metals' crystals and in the areas of their borders.

Thus, we should expect that binding metal-hydrogen energy can be defined by the value of potential barrier. This suggestion is proved by the fact that samples' annealing temperatures exceed the temperature of metals' recrystallization thus the inclusion of hydrogen into electrolytic precipitates is practically impossible. We have determined that in low-temperature annealing conditions for Cr-samples ($t_{annealing} < 300°C$) the secondary hydrogenation is possible which is proved by the fixation of damping peak on the graphic chart $Q–^{1} = f(T)$. The amount of the absorbed hydrogen accurate within the experimental error proves that all hydrogen is fixed on the borders of structure's defects [2].

2. Experimental procedures

One of the most important evidence of the suggested process of metal hydrides formation is that with the increase of the thickness of electrolytic precipitate the amount of hydrogen falling at metal unit weight is decreasing with time.

One more fact of the suggested process of metal hydrides formation is that with the increase of the thickness of electrolytic precipitate the amount of hydrogen falling at metal unit weight is decreasing with time. We have examined the hydrogenation of

B. Baranowski et al. (eds.), *Carbon Nanomaterials in Clean Energy Hydrogen Systems.*

Ni-coatings received by electrocrystallization from sulfuric ($NiSO_4$ $7H_2O$ – 220 g/l; NaCl – 10 g/l; H_3BO_3 – 30 g/l) and sulphamate electrolytes ($Ni(NH_2SO_3)_2 \cdot 4H_2O$ – 400 g/l; $NiCl_2$ $6H_2O$ – 18 g/l; H_3BO_3 – 30 g/l). Coatings were depositing on grade M-1copper. The content of hydrogen was determined by the method of vacuum extraction [3]. Figure 1 and Table 1 show the kinetic curve, indicating the decrease of the amount of absorbed hydrogen with time of electrolysis.

3. Results and discussion

Regularities which were given in Fig. 1 and Table 1 can be explained by the fact that in initial moment of electrolysis the structure of substrate has maximum amount of defects through which metal-hydrogen interaction takes place. During further electrolytic precipitate's structure normalization the amount of defects reduces and correspondingly the total amount of H_2 falling into the mass of a sample will be lower. The attempt to explain thc dccrcasc of the amount of absorbed hydrogen by the idea that hydrogen is entrapped in clusters from which it can't be removed during vacuum extraction is slightly justified. If any crack network exists, vacuum deepness equals to 10^{-4} tori and the fluidity of hydrogen is rather high, the process of hydrogen extraction approaches the level of 100%.

TABLE 1. The dependence of hydrogen content in Ni-coating, received from sulphamate electrolytes by the duration electrolysis. Electrolysis regimes: $i_k = 2$ A/dm^2; t = 40°C; pH = 4.0

τ, min	d, micrometer	V_{H2}, cm/100G
10	4	104
38	15	80

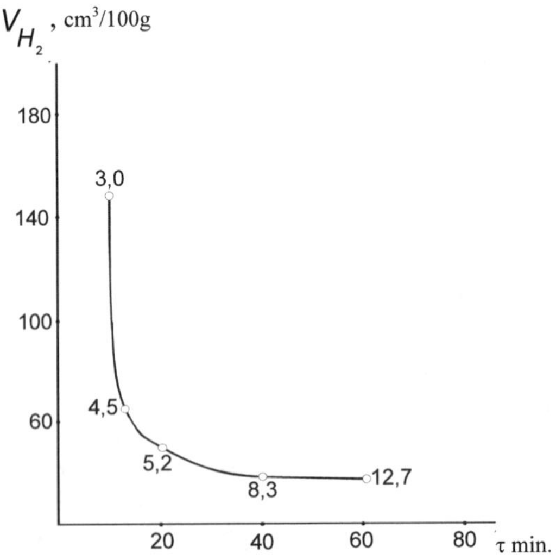

Figure 1. The influence of the duration electrolysis on the content of hydrogen in Ni-coatings received from sulfuric electrolytes. Electrolysis regimes: $i_k = 2$ A/dm^2; $t_{electrolytes} = 25$°C; pH = 3.0. Numerals on the curve show the thickness of the coating

The degree of the surface preparation before cathodic reduction influences greatly the content of hydrogen in precipitate. The evidence of this is the results of studying the processes of hydrogenation of Ni plates subjected to mechanical polishing. We can arrive to such conclusion by conducting experimental hydrogenation of Ni samples subjected to mechanical polishing, as it was defined [4]. For example, during the ideal degree of surfaces' preparation (nine surface finish class) samples don't practically absorb hydrogen (the definition is lying within the limits of measurement error). During the preparation of polished samples in the solution of dimethyl formamid (4% solution) the amount of the absorbed hydrogen dramatically increases, as the structure of Ni becomes defective due to the interaction of Ni surface atoms with dimethyl formamid molecules with the formation of complex compounds.

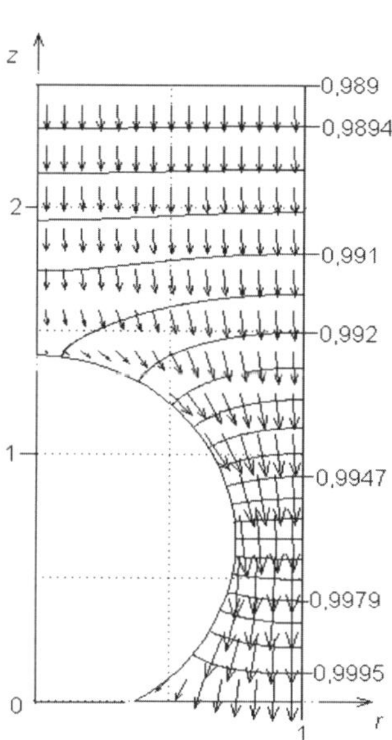

Figure 2. Initial distribution of reference potential and current near hydrogenous bubbles during $i_k = 5$ A/dm^2

Thus, the conducted tests one more time prove the fact that the highest opportunity of hydrogen-metal interaction is observed by the structure's defects.

The impact of environmental acidity is one of the characteristic features of metal hydrides formation by electrochemical method. For example, the plate made of Cr-Ni steel containing as a basic component Ni and subjected to electrochemical polishing in universal electrolyte of nonferrous and ferrous metals (composition, g/l: H_3PO_4 – 60–80, H_2SO_4 – 10–20, CrO_3 – 5–6, t = 60°C) practically is not hydrogenated during electrochemical cathodic processing, and a large aggregation of hydrogenous bubbles generated on the surface does not form a big amount of hydrogen nucleation. The situation dramatically changes in case with alkaline solutions during electrochemical cathodic procession in electrolyte (composition NaOH – 18–22%). Hydrogenous bubbles generate solid layer and their sizes get smaller. According to our studies the density of nucleating centers is 10^6–10^7 units on the surface [5].

In order to confirm our hypotheses we have conducted mathematical simulation of the process of gas evolving on Ni-electrodes (grade HA) in different electrolysis conditions (i_k; $t_{electrolyte}$; $pH_{electrolyte}$; chemical composition; existence of gas evolving membranes). While examining the process of indifferent electrolyte degrading consisting of 18–20% solution of NaOH the following results have been established (results are presented on Figs. 2 and 3). On Figs. 2 and 3 we can see an explicit heterogeneity of primary current-density distribution vanishing in the point of bubble and electrode contact and reaching its maximum between the bubbles (Fig. 3).

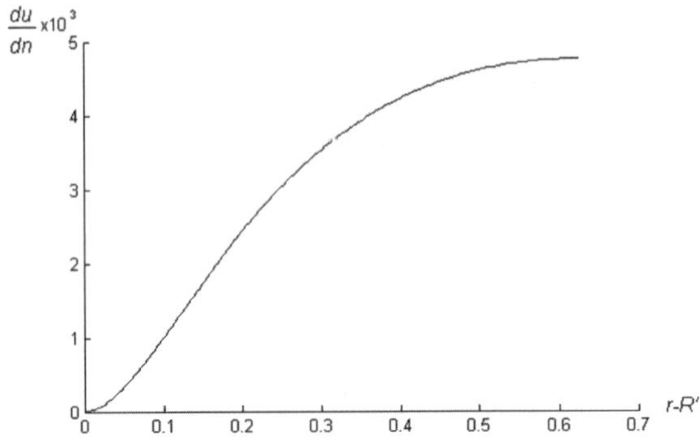

Figure 3. Initial distribution of normal derivative of electropolarization potential on the electrode with hydrogenous bubble (r-R` – distance between bubbles' nucleating centers)

Changing of the potential according to the height of calculated cells along the axis between bubbles is shown on Fig. 4.

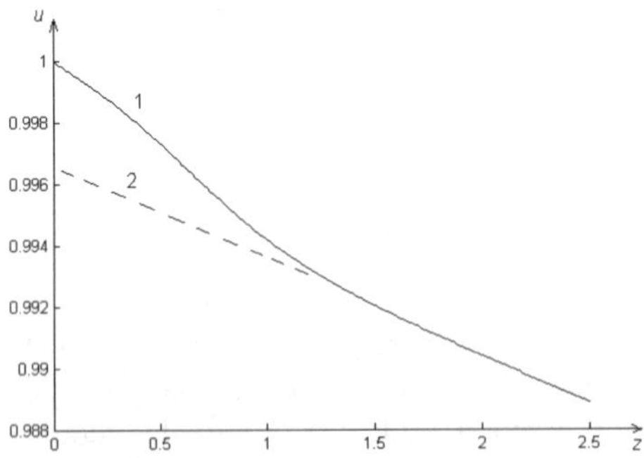

Figure 4. Changing of the potential along the axis between bubbles: 1 – exact solution for the bubble; 2 – linear extrapolation from the area of homogeneous potential changing

Outside the bubble region the potential is changing linearly and near the bubble there is an additional increase of the potential relative to the level of linear interpolation (lines 1 and 2 on Fig. 4). This increase represents the ohimic overvoltage induced by the decrease of the area of the conduction region at the expense of its blocking by the bubble [6]. During the initial distribution there is a zone near the bubble in which current density diminishes up to 0 (Fig. 3) and during the secondary distribution we can't directly use Taifel approximation. We can surmount this difficulty by using generalized Taifel approximation. Thus we can simultaneously define the initial potential distribution and current density:

$$\nu_0 = 1\text{-}1/\beta \; \mathrm{Arsh}(\lambda \; \partial u/\partial n), \tag{1}$$

where ν_0 – initial-approximation function; β – parameter of surface overvoltage «stiffness»; λ – characteristic test length of current density near electrode surface, A/cm^2; $\partial u/\partial n$ – mean value of dimensionless potential normal derivative.

$$\beta = \alpha \cdot F \cdot V/R \cdot T\,;$$
$$\lambda = k \cdot V/2a \cdot i_o\,;$$

where k – interectrode distance in electrochemical cell, equal to 135 cm; i_o – exchange current for the electrode depending on metal nature, taken for Ni 0.113 A/dm^2; V – standardized potential drop, B; a – distance between the holes of the membrane separating cathode and anode chambers in electrochemical cell has been taken as 1 cm in the present calculations; α – parameter of electrochemical cell equal to 0.5.

In order to decrease the amount of independent parameters of the task taking into consideration similarity properties we have introduced the following dimensionless derivatives: $u = \varphi/V$ – dimensionless potential; φ – electric field potential, B; V – standardized potential drop, B.

Equal current distribution through electrode surface is defined as current magnitude and existence of additional factors influencing the initial and secondary current distribution. The adequacy of current distribution is mainly defined by current line-of-force density normally directed to the surface of the sample. The objective of the definition of thermal sources density was reduced to the calculation of the distribution of electric fielding electrochemical cell. On the bases of different invariants of models with the usage of well-known distribution methods such as Laplace and Landau-Lifshitz methods [7] we have received the following picture (Figs. 5 and 6).

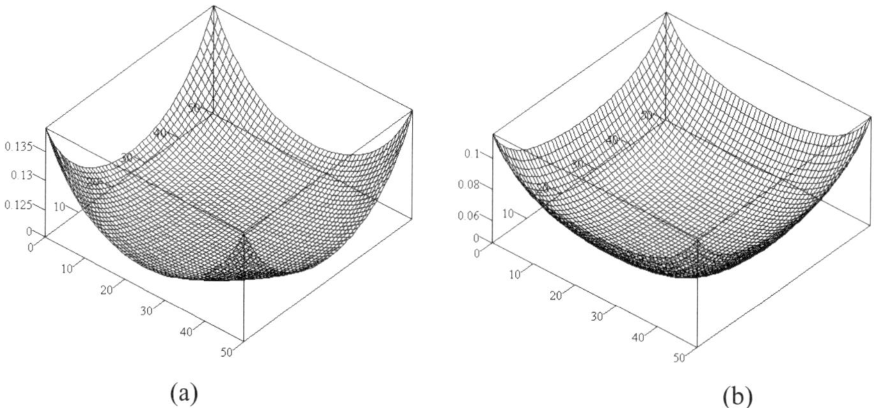

Figure 5. Secondary current density distribution on front (a) and back (b) electrode surfaces for $i_k = 5$ A/dm^2

While developing the model we have taken into consideration the influence of current distribution along electrode surface depending on several different factors (electrolyte concentration, the existence of surface active substances in the electrode, nature of electrode). During the increasing of current density realized in the model by changing the electrode V potential parameters, β and λ are increasing in boundary condition:

$$U=1-1/\beta*\text{Arsh}[\lambda/(1-\theta')*\partial u/\partial n](\theta)/\sqrt{\rho}*\partial u/\partial n. \quad (2)$$

The second summand of the right part in this formula represents the so-called surface overvoltage conditioned by phase transformation kinetics corrected taken into account the coating θ' of the surface. The third summand describes ohmic overvoltage connected with the coating of the surface with gas beads.

Designations in the formula: θ' – grade of metal electrode surface closure by beads calculated through the area of cell's surfaces by the expression:

$$\theta'=(S_{r2}-S_{r1})/S_{r2},$$

α – coefficient of proportionality depending on the grade of electrode surface closure by beads θ,

$$\theta=\pi*R^2/S,$$

where R – bead radius, S = 1/ ρ – sectional area of full cell, ρ – surface density of beads.

Now we will consider the part of parameters β and λ in forming picture of current density distribution.

In previous calculation we have received minimal value of dimensionless current density j = 0.046, which identifies minimal value of dimensionless complex $\lambda/(1-\theta')*\partial U/\partial n = 11.6$. With these values it is possible to use asymptotic formula $\text{Arsh}(x) \approx \ln(2x)$ and boundary condition (2) is simplifying:

$$u \approx 1-1/\beta*\ln(2\lambda/(1-\theta'))-1/\beta*\ln(\partial U/\partial n). \quad (3)$$

Expression (3) corresponds to Taifel approximation in boundary condition on the electrode. The second summand of the right part (3) is a constant for the examined problem and therefore it doesn't have any influence on the character of distribution and is just expressed in electrode V potential value transfer. Thus, in the examined regime parameter λ doesn't influence the distribution of currency density which is defined only by parameterβ. With the increasing of the last contribution of the third summand of the right part (3) the distribution of current density approaches the initial value.

In order to check this situation we have conducted the calculation of current density distribution with average density i_k = 50 A/dm^2, which according to Table 2 corresponds value of β ≈ 24. The received picture of distribution is presented on Fig. 6.

As may be inferred from Fig. 6 there is a dramatically expressed heterogeneity which is typical for initial distribution (approximately 150% on front side and 1,200% on back side of the electrode). At the same time there is practically no current on the back side (lesser in several times nine than on the front side).

The character of distribution confirms with the result received by the experiment [5]. Current density distribution is rather homogeneous (difference between the center and angular point does not exceed 12%) and is characterized by smooth variation along electrode surface. In secondary distribution in contrast to initial distribution there is an equalizing of current density on the back side of the plate which composes approximately 45% from current density on the front side. However, its distribution is characterized by more inhomogeneity (approximately 80%).

TABLE 2. Table of dimensionless parameters transfer

V, V	0.32	0.45	0.81	1.35
i_k (A/dm^2)	5	10	25	50
β	5.63	7.92	14.23	23.75
λ	191	269	483	806
k*V/a (A/dm^2)	43.2	60.7	109.3	182.2
k*V^2/a (W/dm^2)	13.82	27.32	88.53	245.97
k*V^2/a^2 (W/dm^3)	138.20	273.15	885.30	2,459.70

Designations in Table 2:

V – standardized potential fall, V; i_k – cathodic current density (A/dm^2)
k*V/a – current density taking into consideration linear electrodes sizes (A/dm^2)
k*V^2/a – power density of heat current leaking per unit volume (W/dm^2)
k*V^2/a^2 – power density of heat current leaking per unit volume (W/dm^3)
k – interelectrode spacing in electrochemical cell, equal to 135 cm
a – characteristic linear size of the problem (number of sections of electrodes = 25)

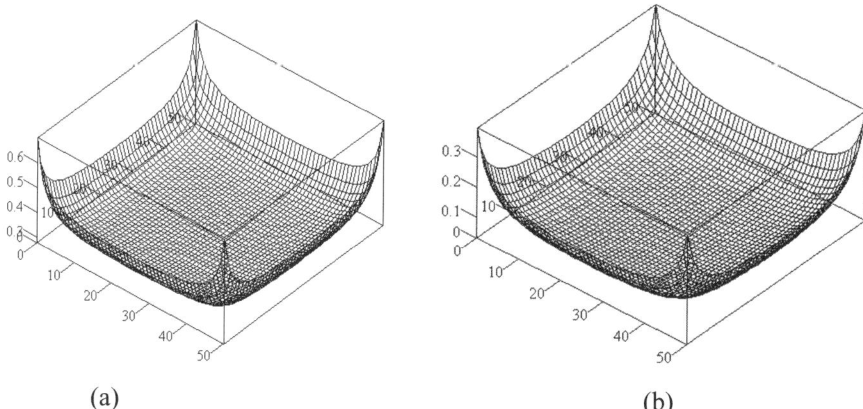

(a) (b)

Figure 6. Secondary current density distribution on the front (a) and back (b) electrode surfaces for i_k = 50 A/dm^2

As numerical analysis showed the received current density distribution can be accurately described by the equation:

$$j(x,y)=j_2+(j_1-j_2)*(f(x)+f(y))+(j_0+j_2-2*j_1)*f(x)*f(y), \quad (4)$$

where j_0, j_1, j_2 – values of current density in angular point, in the center of the edge and in the center of the plate, j – standardized dimensionless current density, $j = \partial u/\partial \xi$, where $\xi = x/a$ – dimensionless coordinate; i = k V/a; j – real current density, A/dm^2; f(t) – function of the coordinate for which the current is defined is set by the expression:

$$f(t)=(d*(2*l+d)/l^2)*((l+d)^2/(t+d)*(2*l-t+d)-1. \quad (5)$$

Here the origin of coordinate is matched with the angular point, 2l – the width of the plate, sm, d – characteristic length of inhomogeneity distribution of current density, sm. For example, during the initial distribution we have received d ≈ 0.1 sm which corresponds to the thickness of the plate and agrees with the solution of Dirichlet's problem. During the secondary current density distribution d = 2 sm on the front side d = 1.5 sm on the back side of the electrode, that is the area of inhomogeneity during the secondary current density distribution increases in 20 times.

The character of distribution which is very close to the initial is proved by the value d ≈ 0.2 sm, (for the initial distribution it was defined as d ≈ 0.1 sm). At the same time during the decreasing of β we should expect the smoothing the character of current distribution.

Checking calculation for $i_k = 1A/dm^2$ ($\beta \approx 2$) has shown that in such case within the limits of calculation accuracy the distribution of current density is homogenous through the surface of the plate.

Thus we can establish that the character of current density distribution and surface thermal sources is defined by the criterial parameter $\beta = (\alpha*F*V)/(R*T)$ (α – parameter of electrochemical cell, equal to 0.5). During $\beta \approx 1$ the distribution is homogenous, during $1 < \beta < 10$ there is a slight inhomogeneity with characteristic scale of changing $d/h \approx 10$, where h – characteristic linear size, for flat electrode – plate thickness. Finally, during $\beta > 10$ a dramatic inhomogeneity of distribution is forming. Such inhomogeneity is typical for Dirichlet's problem, for which $d/h \approx 1$.

4. Conclusions

Conducted mathematical simulation of gas evolving processes on Ni-electrodes has shown the following results:

1. We have developed a model which enables to predict hydrogen absorption be every electrode element depending on cathodic current density.
2. The value of cathodic current density is defined by the existence of defects of electrochemical surfaces structure and uniformity of H_2 distribution in electrode elements.
3. During high cathodic density on the electrode a significant concentration of current density and thermal sources density are forming due to edge effects. Besides the back side of the electrode is eliminated from the process.

References

1. Makkey K. Hydrogen metals' compounds. Moscow: Mir Press: 1968, 244 p.
2. Povetkin V.V., Kovenski I.M. The structure of electrolitic coatings. Moscow: Metallurgiya, 1989, 136 p. (in Russian).
3. Sadakov G.A. Galvanoplastics. Moscow: Energoatomizdat, 1987, 288 p. (in Russian).
4. Zvyagintseva A.V., Falicheva A.I. Physico-mechanical properties of Ni coatings alloyed by boron/electroplating and surfaces processing. *Galvanotekhnika i obrabotka poverkhnosti*, 1997, **5** (2): 24–31. (in Russian).
5. Soltsev Y.P., Pryakhin E.M. Materials science: Text book for higher educational institutions. Saint-Petersburg: Himizdat Press: 2004, 736 p.
6. Dukovic I., Tobias Ch.W. The influence of attached bubbes on potential drop and current distribution at gasevolving electrodes, *Journal of Electrochemical Society*, 1987, **134** (2): 331–342.
7. Zvyagintseva A.V. Boron influence on Ni film hydrogen saturation, *Alternative Power Engineering and Ecology*, 2006 (5): 85–87.

THEORY OF CHEMISORPTION OF A SINGLE HYDROGEN ATOM ON GRAPHENE

V.A. MARGULIS
Department of Physics, N.P. Ogarev Mordovian State University, Saransk 430000, Russian Federation
E.E. MURYUMIN
Department of Chemistry, N.P. Ogarev Mordovian State University, Saransk 430000, Russian Federation

Abstract. A theory of chemisorption relevant to a single hydrogen atom on a monolayer graphene is presented. The electronic structure of graphene is described within an effective-mass approximation. The chemisorption problem is investigated by using the Anderson-Newns model, which takes account of the intra-atomic electron-electron interaction on the adatom. Three different adsorption positions of the adatom are considered: directly above a C atom (position I), over a C–C bond (position II), and above the center of a hexagon of carbon atoms (position III). The adatom Green's function, the charge transfer between the chemisorbed H atom and the substrate graphene, and the adsorption energy of a single H atom are calculated. It is found that at reasonable values of the parameters entering the model, position I is the most stable one for the H atom binding to graphene.

Keywords: graphene, chemisorption, Green's function methods, semi-empirical models and model calculations

1. Introduction

The recent discovery of graphene [1] – a thermodynamically stable two-dimensional (2D) crystal, which represents a flat monolayer of carbon atoms tightly packed into a 2D honeycomb lattice – has opened up a new area of fundamental research with great potential for many important practical applications (the state of art in the field has been quite recently reviewed by Geim and Novoselov [2]). One of these potential applications is hydrogen storage with chemisorption as a possible mechanism.

In this paper, we present a theory of chemisorption of a single hydrogen atom on a graphene sheet. The chemisorption problem is treated within the Anderson-Newns model [3, 4], which takes account of the intra-atomic electron-electron interaction on the adatom. The main objective of this study is to identify the energetically most favorable adsorption sites for an individual hydrogen atom and to assess the intrinsic chemical reactivity of graphene with respect to atomic hydrogen. Recently, a similar approach based on the Anderson-Newns model has been successfully used to study the chemisorption of different single atoms on the surface of single-walled carbon nanotubes [5–8]. Up to now, however, such an approach has not been applied to the calculation of the chemisorption energy ΔE_{ads} of a single H atom on a graphene plane.

Also, to the best of our knowledge, no conclusive experimental values of the ΔE_{ads} in this case are known at presented.

The paper is organized as follows. In Sect. 2, we give a brief description of the model used and present a calculation of the adatom Green's function. The formulas and procedure for a calculation of the adsorption energy and charge transfer are also presented in this section. The numerical results are given and discussed in Sect. 3. We conclude in Sect. 4 by summarizing our findings.

2. Basic theory

The chemisorption system we consider consists of a single H atom interacting with an infinite graphene plane. When the adatom interacts with the substrate graphene, charge transfer may occur between them, resulting in the need to take account of the intra-atomic electron-electron interaction U on the adatom. It is then necessary to consider the occupancy $\langle n_{a\sigma}\rangle$ of the adatom orbital $|a\sigma\rangle$ of spin σ and energy ε_a, which is expressed in terms of the imaginary part of the adatom Green's function $G^{\sigma}_{aa}(\varepsilon)$ (see further below). To find $G^{\sigma}_{aa}(\varepsilon)$ we start, in line with the Anderson-Newns approach [3, 4], from the Hamiltonian of the system, writing it down in second-quantized form as

$$H_{\sigma} = E_{a\sigma}n_{a\sigma} + \sum_{\mathbf{k}}\varepsilon_{\mathbf{k}}n_{\mathbf{k}\sigma} + \sum_{\mathbf{k}}\left(V_{a\mathbf{k}}c^{\dagger}_{a\sigma}c_{\mathbf{k}\sigma} + \text{H.c.}\right), \tag{1}$$

where the effective adatom level of spin σ is shifted to

$$E_{a\sigma} = \varepsilon_a + V_{\text{im}} + (U - 2V_{\text{im}})\langle n_{a-\sigma}\rangle \tag{2}$$

due to the image potential V_{im} which renormalizes ε_a and U. The other notations used in Eq. (1) are as follows: $n_{a\sigma}(n_{\mathbf{k}\sigma})$ is the occupation number operator for the adatom (graphene) electronic states, $c^{\dagger}(c)$ is the Fermi creation (annihilation) operator, and $V_{a\mathbf{k}}$ is the hopping matrix element connecting the adatom state $|a\sigma\rangle \equiv \Psi_{a\sigma}(\mathbf{r})$ with the electronic state $|\mathbf{k}\rangle \equiv \Psi_{\mathbf{k}\sigma}(\mathbf{r})$ of energy $\varepsilon_{\mathbf{k}}$ of graphene, that is

$$V_{a\mathbf{k}}(\mathbf{R}_a) = \int \Psi^{*}_{a\sigma}(\mathbf{r})V(\mathbf{r}-\mathbf{R}_a)\Psi_{\mathbf{k}\sigma}(\mathbf{r})\,\mathrm{d}^3r, \tag{3}$$

where $V(\mathbf{r}-\mathbf{R}_a)$ is the effective self-consistent Hartree-Fock potential of the entire (graphene plus adatom) system and $\mathbf{R}_a(0,0,z_a)$ denotes the position vector of the adatom. Note that the presence of $V(\mathbf{r}-\mathbf{R}_a)$ in Eq. (3) implies that all the electrons of the system are generally involved in the formation of a chemisorption bond between the adatom and the graphene.

Treating the Hamiltonian (1) in the manner prescribed by Anderson [3] and Newns [4], we obtain a spin-dependent adatom Green's function

$$G^{\sigma}_{aa}(\varepsilon) = \left[\varepsilon - E_{a\sigma} - \Sigma_a(\varepsilon)\right]^{-1}, \tag{4}$$

where $\Sigma_a(\varepsilon) = \Lambda_a(\varepsilon) - i\Delta_a(\varepsilon)$ is a self-energy. The so-called chemisorption functions $\Delta_a(\varepsilon)$ and $\Lambda_a(\varepsilon)$ are defined as

$$\Delta_a(\varepsilon) = \pi \sum_{\mathbf{k}\sigma} \left|V_{a\mathbf{k}}\right|^2 \delta(\varepsilon - \varepsilon_{\mathbf{k}}), \tag{5}$$

$$\Lambda_a(\varepsilon) = \frac{1}{\pi} \text{V.P.} \int_{-\infty}^{\infty} \frac{\Delta_a(\varepsilon')}{\varepsilon - \varepsilon'} d\varepsilon', \tag{6}$$

and have a clear physical meaning – namely, $\Lambda_a(\varepsilon)$ is the shift of the adatom level due to the real part of the self-energy, whereas $\Delta_a(\varepsilon)$ is the half-width of the adatom level. The density of states associated with this level is then equal to

$$\rho_{aa}^{\sigma}(\varepsilon) = -\frac{1}{\pi} \operatorname{Im} G_{aa}^{\sigma}(\varepsilon) = \frac{1}{\pi} \frac{\Delta_a(\varepsilon)}{\left[\varepsilon - E_{a\sigma} - \Lambda_a(\varepsilon)\right]^2 + \Delta_a^2(\varepsilon)}. \tag{7}$$

In order to calculate the chemisorption functions, we need to know the electronic states of graphene in an explicit form. Graphene has a hexagonal structure with two carbon atoms in the unit cell and very strong sp^2 bonds, causing a threefold coordinated planar structure. The remaining p_z orbitals are perpendicular to the plane, forming π (bonding) and π^* (antibonding) states. The overlap of π orbitals with the intraplane σ orbitals is small and π electronic states dominate the physical properties of graphene at low energies around the Fermi level, which is located at K and K' points of the first Brillouin zone. The wave vectors associated with those points are given by

$$\mathbf{K} = \frac{2\pi}{a_0}\left(\frac{1}{3}, \frac{1}{\sqrt{3}}\right), \quad \mathbf{K}' = \frac{2\pi}{a_0}\left(\frac{2}{3}, 0\right) \tag{8}$$

with $a_0 = 2.46$ Å being the lattice constant.

In an effective-mass approximation, π electronic states in the vicinity of the K point are described by the equation, which is similar to the Weyl equation for a massless neutrino:

$$\mathcal{H}_0 \mathbf{F}(\mathbf{r}) = \varepsilon \mathbf{F}(\mathbf{r}) \tag{9}$$

with

$$\mathcal{H}_0 = \gamma(\boldsymbol{\sigma}\hat{\mathbf{k}}) = \gamma \begin{pmatrix} 0 & \hat{k}_x - i\hat{k}_y \\ \hat{k}_x + i\hat{k}_y & 0 \end{pmatrix}, \tag{10}$$

where $\gamma = \sqrt{3}\left|V_{pp\pi}\right| a_0/2$ is the $\mathbf{k}\cdot\mathbf{p}$ interaction parameter ($V_{pp\pi} \simeq -3.03$ eV is the π-electron transfer integral), $\boldsymbol{\sigma} = (\sigma_x, \sigma_y)$ is the Pauli spin operator, and $\hat{\mathbf{k}} = (\hat{k}_x, i\hat{k}_y) = (-i\partial_x, -i\partial_y)$ is a wave-vector operator.

Equation (9) can be readily solved. The corresponding eigenvectors and energy eigenvalues are given by

$$\mathbf{F}_{sk}(\mathbf{r}) = \frac{1}{\sqrt{2S}} \begin{pmatrix} s \\ \exp(i\phi_{\mathbf{k}}) \end{pmatrix} \exp(i\mathbf{kr}), \tag{11}$$

$$\varepsilon_{s\mathbf{k}} = s\gamma\sqrt{k_x^2 + k_y^2}, \tag{12}$$

where S is the normalized area of the system, $s = +1(-1)$ refers to the conduction (valence) band, $\mathbf{k}(k_x, k_y)$ is the 2D electron wave-vector, and $\phi_{\mathbf{k}} = \tan^{-1}(k_y/k_x)$.

Further we shall consider three possible stable adsorption positions of the adatom – namely, directly above a C atom (position I), over a C–C bond (position II), and above the center of a hexagon of carbon atoms (position III). Carrying out the calculation of the matrix element $V_{a\mathbf{k}}$ of Eq. (3) in a manner similar to that used in our previous papers [5–8], we obtain the following results:

(a) For position I

$$|V_{a\mathbf{k}}|^2 = V_0^2/N, \tag{13}$$

(b) For position II

$$|V_{a\mathbf{k}}|^2 = \frac{4|V_0|^2}{N}\cos^2\left[\frac{(\mathbf{K}+\mathbf{k})\mathbf{d}}{2}\right], \tag{14}$$

(c) For position III

$$\begin{aligned} |V_{a\mathbf{k}}|^2 = \frac{4|V_0|^2}{N}\Bigg\{ & 3 + 4\cos\left[\frac{(\mathbf{K}+\mathbf{k})\mathbf{d}_1}{2}\right]\cos\left[\frac{(\mathbf{K}+\mathbf{k})\mathbf{d}_2}{2}\right] \\ & + 4\cos\left[\frac{(\mathbf{K}+\mathbf{k})\mathbf{d}_1}{2}\right]\cos\left[\frac{3(\mathbf{K}+\mathbf{k})\mathbf{d}_2}{2}\right] \\ & + 2\cos\left[(\mathbf{K}+\mathbf{k})\mathbf{d}_1\right]\cos\left[(\mathbf{K}+\mathbf{k})\mathbf{d}_2\right] + 2\cos\left[(\mathbf{K}+\mathbf{k})\mathbf{d}_1\right] \\ & + 2\cos\left[(\mathbf{K}+\mathbf{k})\mathbf{d}_2\right] + \cos\left[2(\mathbf{K}+\mathbf{k})\mathbf{d}_2\right]\Bigg\}, \end{aligned} \tag{15}$$

where N is the number of the lattice sites contained in the area S, the interaction parameter

$$V_0(\mathbf{R}_a) = \int \Psi_{a\sigma}^*(\mathbf{r}) V(\mathbf{r} - \mathbf{R}_a) \Phi(\mathbf{r}) d^3r, \tag{16}$$

describes an effective coupling between the adatom and a monolayer graphene ($\Phi(\mathbf{r})$ is the atomic p_z function), $\mathbf{d}$ is the vector connecting the two surface nearest-neighbors of the adatom in a bridge configuration, $\mathbf{d}_1$ and $\mathbf{d}_2$ are the vectors defined as

$$\mathbf{d}_1 = \mathbf{e}_x a_0, \quad \mathbf{d}_1 = \mathbf{e}_y a_0/\sqrt{3}, \tag{17}$$

$\mathbf{e}_x$ and $\mathbf{e}_y$ being the unit vectors of the x and y axes, respectively.

Since neither an effective self-consistent potential $V(\mathbf{r}-\mathbf{R}_a)$ nor an equilibrium distance $\mathbf{R}_a$ are known, we shall further treat V_0 as a phenomenological parameter. This is in agreement with the general philosophy of the Anderson-Newns approach [3, 4].

Using Eqs. (13–15) in Eq. (5), we obtain an explicit form of the chemisorption function $\Delta_a(\varepsilon)$:

$$\frac{\Delta_a(E)}{\eta}=\frac{2\sqrt{3}}{3}B^2\begin{cases}|E| & \text{for position I,}\\ 2|E|\left[1-\frac{1}{2}J_0(E)\right] & \text{for position II,}\\ 6|E|\left[1-J_0(E)\right] & \text{for position III,}\end{cases} \tag{18}$$

where $B=|V_0|/|V_{pp\pi}|$, $E=\varepsilon/\eta$ with $\eta=(3/2)|V_{pp\pi}|$, and $J_0(E)$ is the Bessel function of zero order.

Inserting Eq. (18) into Eq. (6) and performing the integration over ε', we obtain with the logarithmic accuracy

$$\frac{\Lambda_a(E)}{\eta}=\frac{4\sqrt{3}}{3\pi}B^2E\begin{cases}g_c(E)\ln|E/E_c| & \text{for position I,}\\ 2g_c(E)\ln|E/E_c|-(\pi/2)N_0(|E|) & \text{for position II,}\\ 6g_c(E)\ln|E/E_c|-3N_0\left(\sqrt{3}|E|\right) & \text{for position III,}\end{cases} \tag{19}$$

where $N_0(|E|)$ is the Neumann function of zero order, $g_c(E)$ is the cut-off function defined as

$$g_c(E)=\frac{E_c^2}{E^2+E_c^2}. \tag{20}$$

Here the parameter E_c represents a cut-off energy, which is chosen to be equal to $\gamma k_c/\eta$ with the cut-off wave number $k_c=\delta_c(2\pi/a_0)$, where δ_c is about 0.1–0.2.

The knowledge of the chemisorption functions $\Delta_a(\varepsilon)$ and $\Lambda_a(\varepsilon)$ allows one to calculate the average electron occupation number of the adatom orbital, which is expressed through $\rho_{aa}^{\sigma}(\varepsilon)$ of Eq. (7) as follows:

$$\langle n_{a\sigma}\rangle=f\langle n_{a-\sigma}\rangle\equiv\int_{-\infty}^{0}\rho_{aa}^{\sigma}(\varepsilon)g_c(\varepsilon)\,\mathrm{d}\varepsilon. \tag{21}$$

The chemisorption energy ΔE_{ads} and the charge transfer $\Delta q/e$ may then be found by using the following expressions:

$$\Delta E_{\text{ads}}=\sum_{\sigma}\Delta E_{1\sigma}-U\langle n_{a\sigma}\rangle\langle n_{a-\sigma}\rangle-\varepsilon_a-V_{\text{im}}, \tag{22}$$

$$\Delta q/e = 2\langle n_{a\sigma}\rangle - 1, \tag{23}$$

where e is the electron charge and

$$\Delta E_{1\sigma} = \frac{1}{\pi}\int_{-\infty}^{0} g_c(\varepsilon)\xi(\varepsilon)\,\mathrm{d}\varepsilon \tag{24}$$

with

$$\xi(\varepsilon) = \tan^{-1}\left[\frac{\Delta_a(\varepsilon)}{\varepsilon - E_{a\sigma} - \Lambda_a(\varepsilon)}\right]. \tag{25}$$

Thereby $\langle n_{a\sigma}\rangle$ of Eq. (21) should be determined self-consistently from the equation

$$f\left[f\left(\langle n_{a\sigma}\rangle\right)\right] - \langle n_{a\sigma}\rangle = 0. \tag{26}$$

3. Numerical results and discussion

To carry out the numerical calculations of the chemisorption energy ΔE_{ads} and the charge transfer $\Delta q/e$, we first of all need to specify the values of the essential parameters entering our model. The latter is characterized for the adatom by the electronic level ε_a and the intra-atomic Coulomb potential U, and for the substrate graphene by the band parameter γ and the Fermi energy E_{F}. In the following, we take E_{F} to be the zero-energy reference. Using the above-mentioned values of $V_{pp\pi}$ and a_0, we get γ = 7.45 eV · Å. The appropriate values of ε_a and U can be obtained if we take into account the relations [9]

$$\varepsilon_a = W - I, \qquad U = I - A, \tag{27}$$

where W is the graphene work function, I and A are the first ionization potential and the electron affinity for an isolated atom, respectively. In the case of a hydrogen atom I = 13.6 eV and A = 0.7 eV, so that we take $\varepsilon_a = -8.8$ eV and U = 12.9 eV, assuming that W = 4.8 eV – the value that is typical for bulk graphite.

The accurate estimation of the two remaining parameters of the model, V_{im} and V_0, describing, respectively, the ionic and the electronic components of the interaction between the adatom and the substrate graphene, goes beyond the Anderson-Newns model itself. Consequently, it is reasonable to treat them as adjustable parameters, at least at the present stage when no experimental information relevant to hydrogen adsorption on graphene is available. Choosing a reasonable value for the parameter V_{im} of 6.5 eV, we shall calculate the chemisorption energy ΔE_{ads} as a function of the dimensionless parameter $B^2 = \left(V_0/V_{pp\pi}\right)^2$, varying it in the range 0.5–1 that also seems to be reasonable.

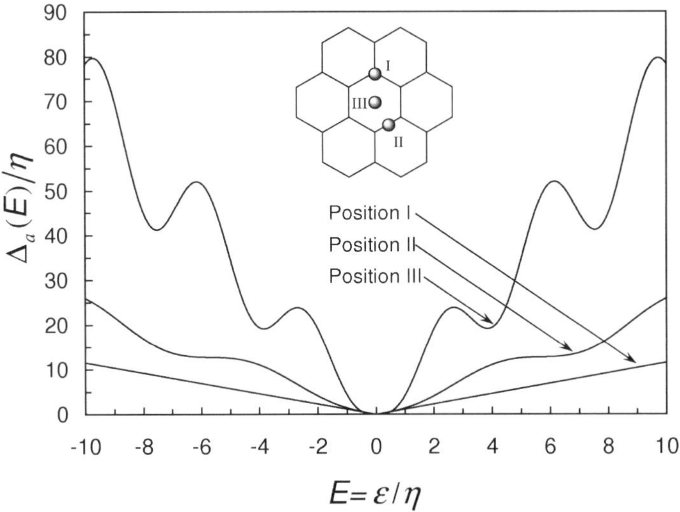

Figure 1. Plot of the chemisorption function $\Delta_a(E)$ defined in Eq. (18) for three possible adsorption position of a H adatom on a graphene plane. The dimensionless hoping parameter B is 1

Figure 1 shows the chemisorption function $\Delta_a(E)/\eta$ versus the dimensionless energy $E = \varepsilon/\eta$ for the three above-mentioned positions of the adatom. As seen from this figure, the behavior of $\Delta_a(E)/\eta$ for position I is the same as that of the electronic density of state in graphene (see, e.g., [10]). For both other positions (II and III), the width of the adatom level is larger than that in position I, exhibiting pronounced oscillations with E. This is a direct consequence of the fact that in positions II and III the width of the adatom level is augmented by a factor 2 and a factor 6, respectively (see Eq. (18)). This in turn reflects the fact that in position II(III) an adatom electron can hop to any of the two (six) nearest-neighboring C atoms of the graphene plane. The above-mentioned oscillations of $\Delta_a(E)/\eta$ in positions II and III are due to the presence of the Bessel functions in Eq. (18) for those positions. It is evident that such oscillations, along with the oscillations of the function $\Lambda_a(E)/\eta$ (see further below) might, in principle, lead to the oscillations of an adatom charge. This is, however, not the case, and the oscillations are smeared out after averaging over ε in calculating $\langle n_{a\sigma} \rangle$ (Eq. (21)).

Figure 2 shows the adatom-level shift $\Lambda_a(E)/\eta$ against the dimensionless energy $E = \varepsilon/\eta$ for the three considered positions of the adatom. A global N-like shape of the curves in that figure, as well as the oscillations of $\Lambda_a(E)/\eta$ in positions II and III, are a direct consequence of the fact that, according to Eq. (6), the chemisorption function $\Lambda_a(\varepsilon)$ is the Hilbert transform of the function $\Delta_a(\varepsilon)$.

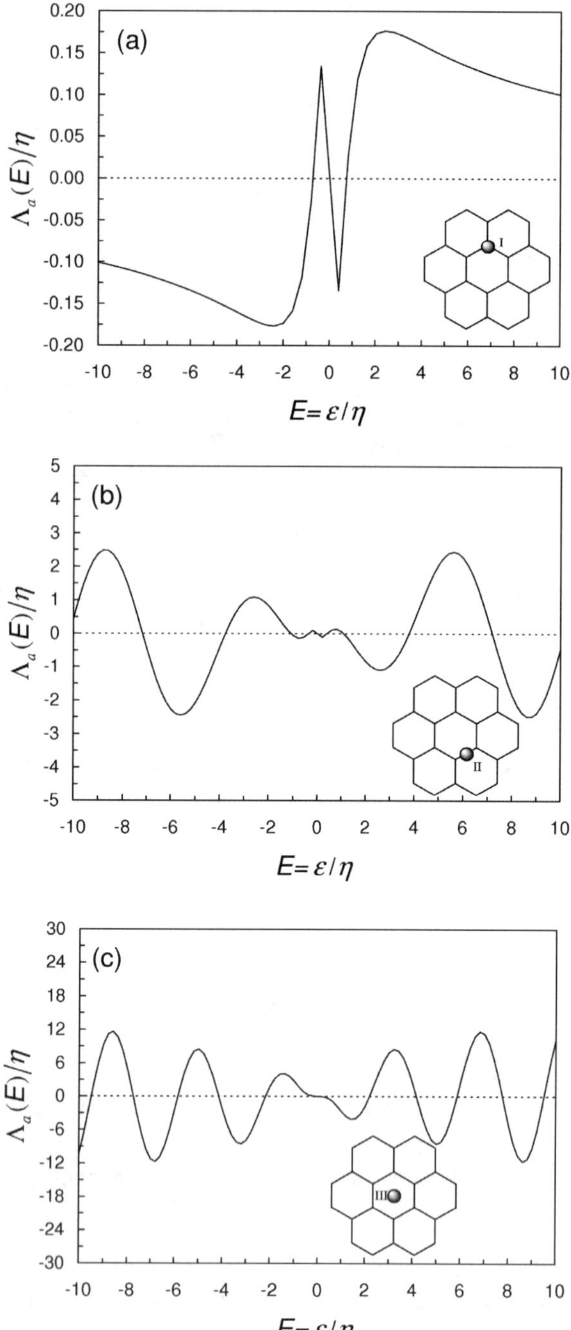

Figure 2. Plot of the chemisorption function $\Lambda_a(E)$ defined in Eq. (19) for three possible adsorption position of a H adatom on a graphene plane: (a) – position I, (b) – position II, and (c) – position III. Parameter B is the same as Fig. 1

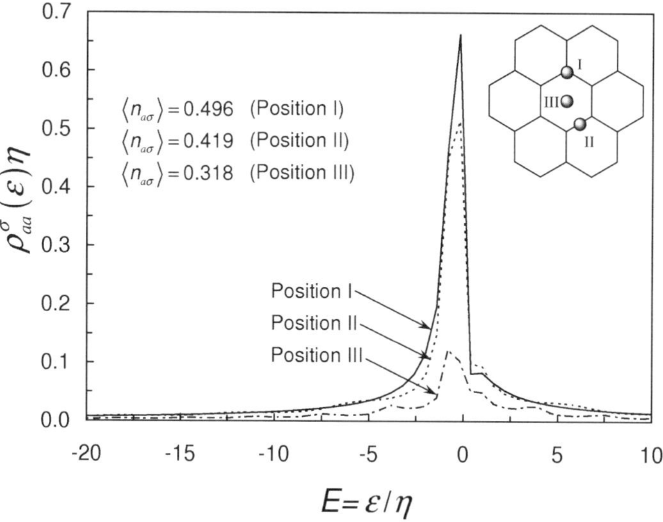

Figure 3. Adatom density of states $\rho^{\sigma}_{aa}(\varepsilon)$ defined in Eq. (7) for three possible adsorption positions of a H adatom on a graphene plane. The solid, dotted and dot-dashed curves correspond to position I, II and III, respectively. The dimensionless hoping parameter B is 0.78

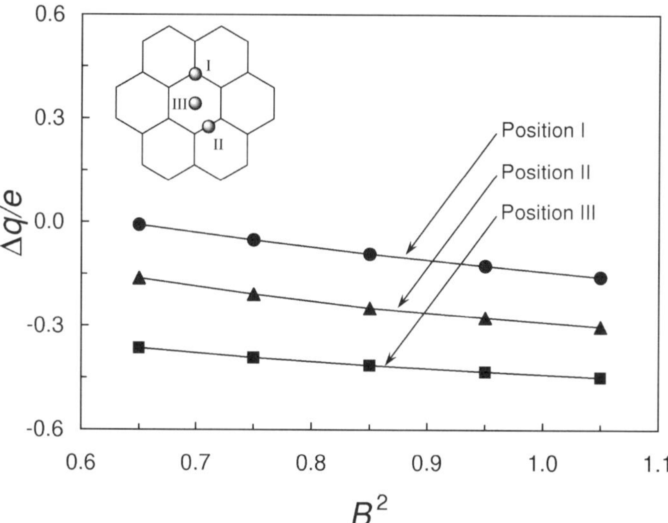

Figure 4. Charge transfer $\Delta q/e$ from the H adatom to the substrate graphene versus the dimensionless interaction parameter B^2. The solid circles, triangles and squares refer to positions I, II and III, respectively. The solid lines are intended as a guide to the eye

The curves of the adsorbate spectral density $\rho^{\sigma}_{aa}(\varepsilon)$ for positions I, II and III are plotted in Fig. 3. The curves are calculated for the values of $\langle n_{a\sigma}\rangle$ indicated in that figure. For all the three positions, one can clearly see the main resonant peak $\rho^{\sigma}_{aa}(\varepsilon)$, which is located just below the Fermi level. This peak arises from the interaction between the discrete atomic state of the adatom and the continuum of the valence-band electronic states of the substrate graphene. As seen from Fig. 3, the height of the peak is the largest for position I of the adatom.

Figure 4 depicts the calculated charge transfer $\Delta q/e$ between the adatom and the substrate graphene. The negative sign of $\Delta q/e$ indicates that the charge transfer is directed from the adatom to the substrate. As seen from Fig. 4, the charge transfer for all the three considered positions of the adatom is not large and does not change significantly with increasing the interaction parameter B.

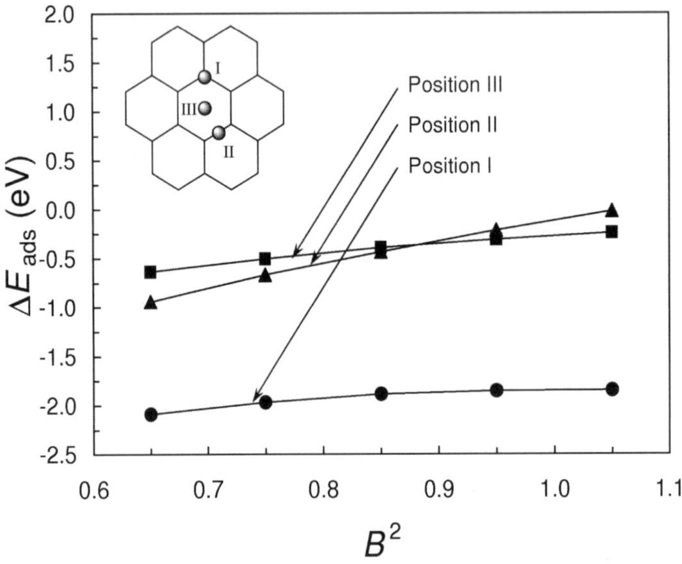

Figure 5. Chemisorption energy ΔE_{ads} of a single hydrogen atom an a graphene surface versus the dimensionless interaction parameter B^2. The solid circles, triangles and squares refer to positions I, II and III, respectively. The solid lines are intended as a guide to the eye

Figure 5 shows the calculated values of the chemisorption energy ΔE_{ads} against the dimensionless interaction parameter B^2. Note that, by definition, negative ΔE_{ads} values correspond to an energetically favorable adsorption, while the positive ones indicate that adsorption is not possible. As seen from Fig. 5, position I is characterized by larger $|\Delta E_{\text{ads}}|$ values, and, hence, it is this position that is the most stable one for the H atom binding to graphene.

4. Conclusions

In this paper, we have studied the atomic hydrogen chemisorption on graphene using the Anderson-Newns model. The electronic structure of graphene has been treated within an effective-mass scheme. In a number of numerical examples we have demonstrated systematic trends in the charge transfer and the chemisorption energy as a function of a dimensionless parameter describing the interaction between a single hydrogen atom and a substrate graphene. We have considered three possible stable adsorption positions of the adatom – namely, on the top of a C atom, over a C–C bond, and alove the center of a carbon hexagon. Our results show that the first of them is energetically the most favorable position for atomic hydrogen adsorption.

References

1. Novoselov K.S., Geim A.K., Morozov S.V., Jiang D., Zhang Y., Dubonos S.V., Grigorieva I.V. and Firsov A.A. (2004) Electric field effect in atomically thin carbon films, *Science* **306**, P. 666–669.
2. Geim A.K. and Novoselov K.S. (2007) The rise of graphene, *Nat. Mater.* **6**(3), P. 183–191.
3. Anderson P.W. (1961) Localized magnetic states in metals, *Phys. Rev.* **124**(1), P. 41–53.
4. Newns D.M. (1969) Self-consistent model of hydrogen chemisorption, *Phys. Rev.* **178**(3), P. 1123–1135.
5. Margulis Vl.A., Muryumin E.E. and Tomilin O.B. (2004) Theoretical study of atomic chemisorption on single-walled carbon nanotubes. Application of Anderson-Newns model, *Physica B* **353**(3–4), P. 314–323.
6. Margulis Vl.A. and Muryumin E.E. (2007) Atomic oxygen chemisorption on the sidewall of zigzag single-walled carbon nanotubes. *Phys. Rev. B* **75**(3), 035429/1-035429/12.
7. Margulis Vl.A. and Muryumin E.E. (2007) Chemisorption of single fluorine atoms on the surface of zigzag single-walled carbon nanotubes: A model calculation. *Physica B* **390**(1–2), P. 34–142.
8. Margulis Vl.A., Muryumin E.E. and Tomilin O.B. (2007) Binding of a fluorine atom to the sidewall of single-walled carbon nanotubes. *In: Hydrogen Materials Science and Chemistry of Carbon Nanomaterials*, edited by T.N. Veziroglu et al., Springer, Berlin, P. 789–793.
9. Gadzuk J.W., Hartman J.K. and Rhodin T.N. (1971) Approach to alkali-metal chemisorption within the Anderson model, *Phys. Rev. B* **4**(2), P. 241–255.
10. Ando T. (2006) Anomaly of optical phonon in monolayer graphene, *J. Phys. Soc. Jpn.* **75**(12), 124701/1-124701/5.

CARBON-CARBON COMPOSITES – CARRIERS FOR HETEROGENEOUS CATALYSTS FOR DEHYDROGENATION OF HYDROCARBONS

M.N. EFIMOV, L.M. ZEMTSOV, G.P. KARPACHEVA, M.M. ERMILOVA, N.V. OREKHOVA
Topchiev Institute of Petrochemical Synthesis, Russian Academy of Sciences, 29 Leninsky prospect, Moscow, 119991 Russia
E.L. DZIDZIGURI, E.N. SIDOROVA
Moscow State Institute of Steel and Alloys, 4 Leninsky prospect, Moscow, 119049 Russia
O.N. EFIMOV
Institute of Problems of Chemical Physics, Russian Academy of Sciences, 1 prospect. Semenova, Chernogolovka, 142042 Russia

Abstract. Highly-dispersed metals supported on oxide carriers are active in the most of catalytic reactions of hydrocarbons dehydrogenation. Coking is the serious problem of such catalysts. It occurs due to undesirable further dehydrogenation of transition surface forms of hydrocarbons. To solve this problem, additives (Pb, Sn, Re, etc.) are introduced into the catalyst. They reduce the ability of a metal to form multiple bonds of adsorbed forms with the active phase of the catalyst. Recently, carbon materials were proposed as carriers for the task in question [1].

Among new forms of carbon carriers allowing high specific surface of metal and stable dispersion of its particles, materials obtained by carbonization of polymers hold high interest. Introduction of catalytically active metals into the carbon matrix can be performed as early as in the stage of carbonization, simultaneously increasing their activity and stability.

In this work carbon-carbon composites containing carbons of different origin are studied as carriers for platinum group metals with the purpose of efficient dehydrogenation of hydrocarbons.

Keywords: hydrocarbon, composite, carbonization, dehydrogenation, stability

1. Experimental

Attachment of catalytically active metals to the carbon carrier was performed under the conditions of IR-pyrolysis of the precursor based on polyacrylonitrile (PAN) and salts of Pt, Ru, Re, Rh. Tri-component precursors were prepared by joint dissolution of PAN and salts of two catalytically active metals ($PtCl_4$ and $RuCl_3$) in DMFA. Total concentration of metals was 1.0, 2.5, 5.0, 10.0, and 30.0 wt%. The ratio Pt:Ru was 9:1. In preparation of precursors containing carbon multiwalled nanotubes (MWNT) or coal SKT-6A, an additional stage was taken: ultrasonic dispersion of carbon powders in solution of the mixture of PAN and metal salts for 1 h. The metal content was calculated in relation to the total carbon content in the system. Prepared solutions were placed in stainless steel cups and subjected to IR-pyrolysis in a laboratory set-up [2].

IR-calcination was performed in two steps: preliminary calcination in air at 150°C and 200°C (for 15 min each) during which evaporation of the solvent and preliminary structuring of PAN take place, and the main calcination in atmosphere of an inert gas at required intensity of IR radiation. Intensity of the main calcination was varied from 400°C to 1,000°C, and duration – from 10 to 120 s.

Microphotographs were obtained with the electronic microscope EM-301 (Philips), accelerating voltage being 60–80 kV. Hystograms were built on the basis of measurements of particle size on microphotographs.

Phase and structural studies were made at room temperature on an X-ray diffractometer (Rigaku) with focusing by Bragg-Brentano using FeK_{α}-radiation in a non-interrupted mode. Results were processed by approximation method and Fourier analysis [3]. We also used the method of calculation of diameter distribution of spherical crystallites from XRD data [4].

2. Results and discussion

In the present work, IR-pyrolyzed PAN with graphite-like structure and carbon-carbon composites based on IR-pyrolyzed PAN and MWNT or purified finely-dispersed coal SKT-6A were used as carbon carriers for preparation of a Pt-, Ru-based bimetallic catalyst. In the course of IR-pyrolysis, carbonization processes occur leading to the formation of a graphite-like structure. At low intensities of IR-pyrolysis, the structure is amorphous due to the irregular shift of graphene planes between one another and small size of regions where scattering of crystallites is coherent [5]. With an increase in IR-pyrolysis intensity, the degree of orderliness of the graphite-like structure grows. Interplanar distances become smaller, from $d_{002} = 3.56$ Å at T = 500°C to $d_{002} = 3.42$ Å at T = 1,200°C. Simultaneously, sizes of regions with coherent scattering of crystallites grow. At high intensities of IR-pyrolysis, graphite-like structure transforms into a nanocrystalline one. At T = 1,000°C–1,200°C L_c = 37–45 Å.

Under the conditions of IR-pyrolysis of precursors, aside from the formation of orderly carbon structures, reduction of metals occurs with participation of hydrogen being released during dehydrogenation of the main polymeric chain of PAN.

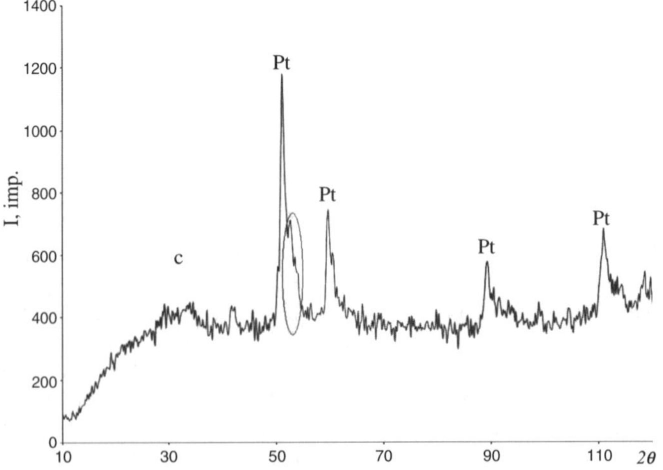

Figure 1. A diffractogram of catalyst after IR-annealing at 700°C. Asymmetric reflection peaks of Pt indicates formation the solid solution of substitution

By X-ray structural and phase analysis it is established that IR-pyrolysis with intensity corresponding to sample heating above 700°C gives bimetallic Pt-Ru nanoparticles which exist as an alloy or intermetallic compound $Pt_{13}Ru_{27}$. Asymmetry of peaks corresponding to the Pt phase (Fig. 1) and also increased noise level in the whole range of homogeneity of Pt-based solid solutions testify to this.

Diffractograms of samples prepared at T < 700°C show that peaks of Pt reflection are symmetrical, and those of the intermetalloid cannot be registered even at low rates of filming at angles ~50° (Fig. 2).

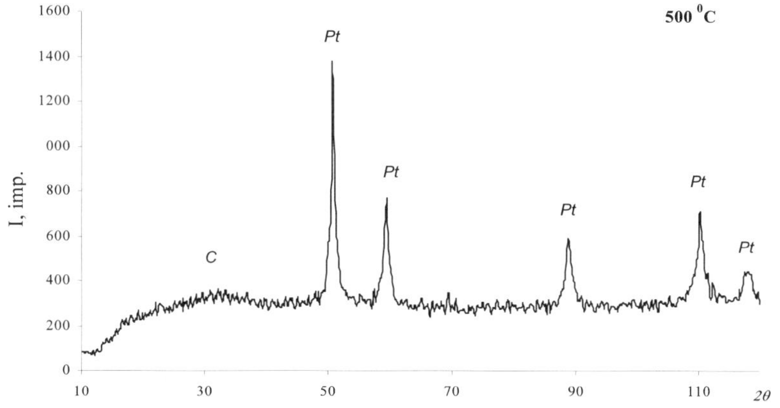

Figure 2. A diffractogram of the sample obtained at T = 500°C

For quantitative estimation of the composition of metallic phase forming in the carbon matrix, the plot for the composition dependence of lattice period in Pt-Ru system in coarse crystalline state is given (Fig. 3, continuous line). ASTM data were used for the border points, lattice periods of pure Pt and the intermetalloid $Pt_{13}Ru_{27}$. To determine lattice periods of Pt phase in nanostate, a sample was prepared at T = 700°C with Pt nanoparticles in the carbon matrix. The obtained value «a» is taken for the initial point in the composition dependence of lattice period of Ru solid solution in Pt in nanomaterials. The value of the intermetalloid lattice period a^* at this intensity of

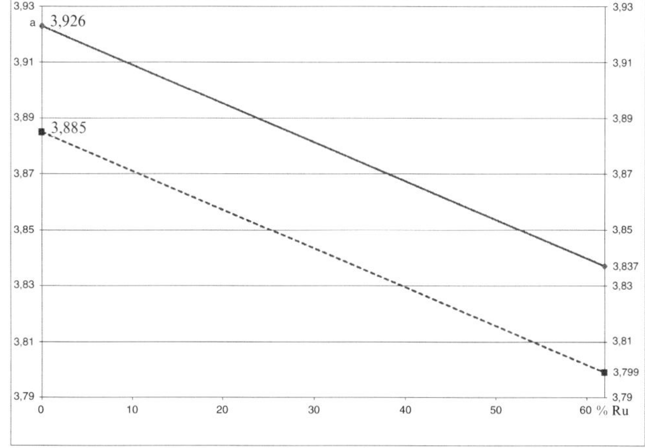

Figure 3. An estimation of lattice periods in the system $Pt-Pt_{13}Ru_{27}$. Solid line – massive pattern, dotted line – ultradispersed pattern

IR-pyrolysis is 3.799 Å (Fig. 3, dotted line). As seen from the figure, plots for dependences of lattice periods on the composition of Ru solid solution in Pt in large crystalline and nanostructured states are parallel to each other.

Size of metal nanoparticles was determined from studies of obtained nanocomposites by TEM. Size distribution plots for metal nanoparticles were created by applying the procedure of random dissection to microphotographs (Fig. 4).

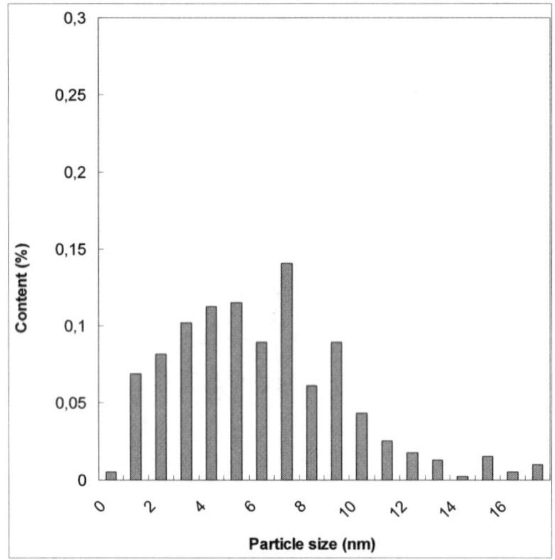

Figure 4. Size distribution plot for metal nanoparticles in the IR-PAN/Pt-Ru sample

Within the whole temperature range of pyrolysis, nanoparticles sizes are in the range 1 < d < 16 nm, the average size being d = 6–8 nm. TEM studies showed that metal nanoparticles are located as inside the nanotubes in the IR-PAN-MWNT/Pt-Ru sample, so on their surface (Fig. 5).

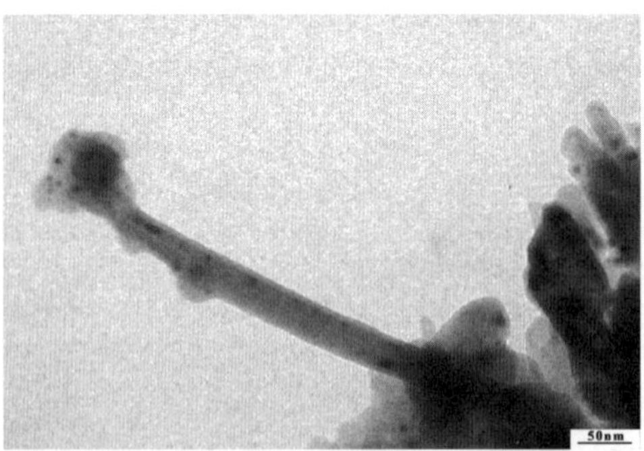

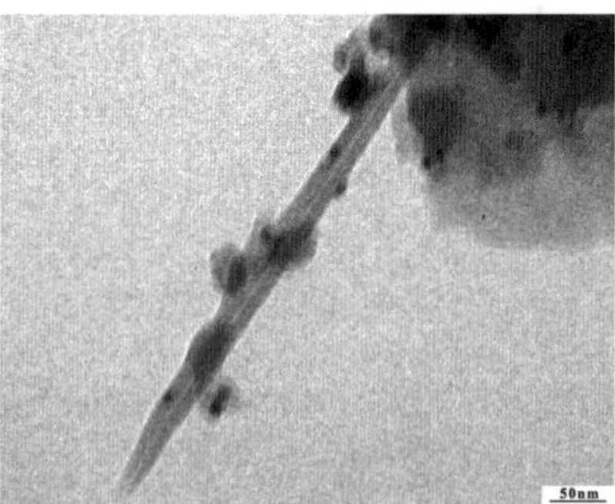

Figure 5. Microphotographs of the IR-PAN-MWNT/Pt-Ru sample

Size distributions of regions of coherent scattering for nanocomposite samples obtained at different temperatures were calculated from XRD data (Fig. 6).

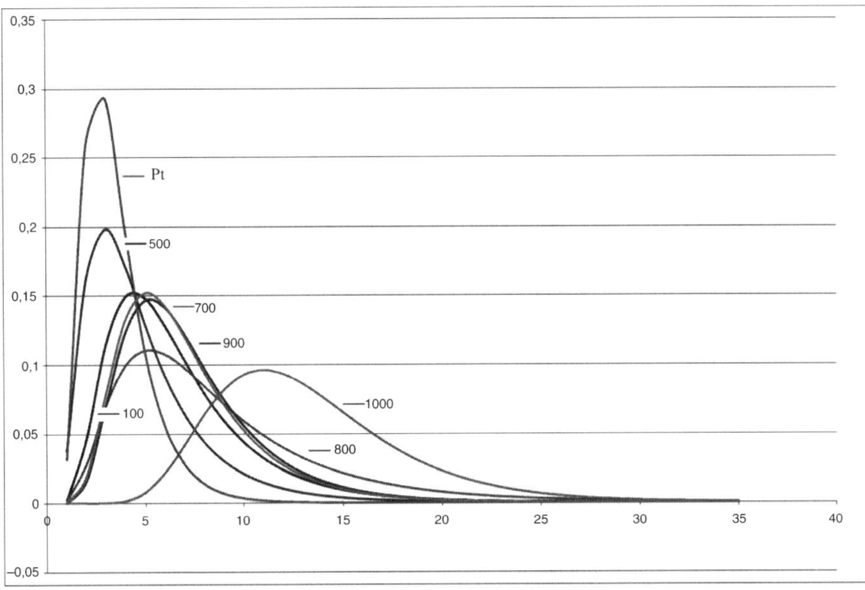

Figure 6. Size distribution of crystallites in nanocomposite IR-PAN/Pt-Ru with temperature marks

The narrowest distribution and smallest size of region of coherent scattering was found for the sample which contained Pt only. Modification by Ru gives wider size distribution. Temperature makes similar effect.

From TEM and XRD data it is found that one particle corresponds to one region of coherent scattering.

Thus, the method developed allows introduction of catalytically active metals into the carbon matrix directly in the process of its formation providing sufficient concentration and finely disperse distribution of active centers (alloys of platinum group metals) of the catalyst.

Metal-carbon nanocomposites prepared as described above were tested as heterogeneous catalysts of hydrocarbons dehydrogenation. Nanoparticles of catalytically active metals were supported on nanocrystalline IR-PAN, on nanocomposite on the basis of IR-PAN and MWNT, on nanocomposite on the basis of IR-PAN and finely-dispersed carbon SKT-6A.

Specific surface area values for studied catalysts form the sequence IR-PAN/SKT-6A/Pt-Ru > IR-PAN-MWNT/Pt-Ru > IR-PAN/Pt-Ru (639, 40, 15 m^2/g correspondingly).

Studied catalysts were found to be active in the reaction of cyclohexane dehydrogenation. The most active was IR-PAN/SKT-6A/Pt-Ru. Selectivity into benzene was 100% in the range 220–520°C.

Full conversion of cyclohexane was achieved over the catalyst IR-PAN/SKT-6A/Pt-Ru(2.8%) at T > 450°C (Fig. 7).

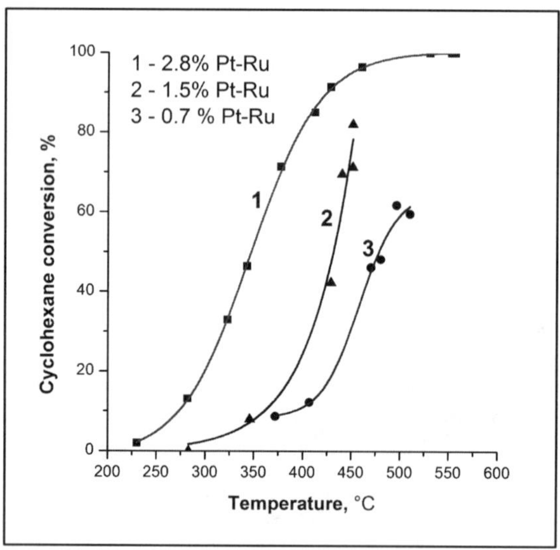

Figure 7. Cyclohexane conversion

3. Conclusions

The method is developed for preparation of metal-carbon nanocomposites which consist of carbon-carbon composites on the basis of IR-pyrolyzed PAN with nanoparticles of platinum group metals introduced into the carbon matrix. On their basis efficient catalysts of hydrocarbons dehydrogenation are prepared. Activity of these catalytic systems is determined as by their specific surface area so by the nature of carbon components of the carrier.

Acknowledgements

This work is done with the financial support of Russian Foundation for Basic Research (Grant No. 07-03-00309) and DCSM RAS no. 8.

References

1. Radovic L.R., Rodriguez-Reinoso F. Carbon materials in catalysis. In: Thrower PA, editor. Chemistry and Physics of Carbon. 25. New York: Marcel Dekker. 1997. 243–358.
2. Zemtsov L.M., Karpacheva G.P., Efimov M.N., Muratov D.G., Bagdasarova K.A. Carbon nanostructures based on IR-pyrolyzed polyacrylonitrile. *Polymer Science A* 2006, **48**(6): 633–637.
3. Gorelik S.S., Rastorguev L.N., Skakov U.A. Radiographic and Electro-optical Analysis. The Practical book. Moscow: Metallurgy. 1970: 366 p.
4. Selivanov V.N., Smyslov V.F. Shortcut methods of x-ray analysis of crystallites distribution of deformated polycrystallites dislocation structure. *Materialovedenie.* 1998, **4–5**: 10–14. (in Russian).
5. Kozlov V.V., Korolyov Y.M., Karpacheva G.P. Structural changes of composite based on polyacrylonitrile and fullerene C_{60} after IR-radiation. *Vysokomolekulyarnye Soedineniya A* 1999, **41**(5): 836 (in Russian).

STRUCTURE AND PHYSICAL PROPERTIES OF NANOCARBON CONTAINING POLYMER SYSTEMS

J. RASHIDOV, U. SHOIMOV, J. SALOMOV, SH. TUICHIEV, S. TABAROV
Tajik State National University, Dushanbe, Tajikistan
E-mail: jab.s04@mail.ru
B.M. GINZBURG
Research Institute of Mechanical Engineering Problems, RAS, St. Petersburg, Russia
E. OSAWA
NanoCarbon Research Institute, Ltd., Kashiwa, Chiba, Japan

Abstract. Structure, mechanical, thermophysical and dielectrical properties of set nanocarbon-containing polymers were investigated. It is shown in dependence the sizes of aggregates of nanocarbons, character of their distribution in volume of a matrix, activity and concentration and an applied technology of receiving of nanocarbon containing polymers a change of structure and physical properties of these systems are observed.

Keywords: nanocarbon, polymer, dielectric, structure, physical properties

1. Introduction

As is known in the filled polymers in dependence of the sizes of particles of fillers, character of their distribution, activity and concentration nonmonotonic changes of physical properties of materials are observed. In this aspect nanocarboncontaining polymers, on visible, are not an exception to the rules [1–4].

In the present work influence of nanocarbons on structure and physical properties of amorphous-crystalline polymers are investigated.

2. Objects and experimental methods

The polymer films were formed from solutions, melts and by application of a method of the limited swelling. As objects of research were served low density polyethylene (LDPE), polymethylmethacrilate (PMMA), synthetic isoprene rubber (SIR-3), the block-copolymer styrene-buthadiene-styrene (SBS) marks of Cariflex, etc. As the filler used nanocarbon (NC) C60 and nanocarbon soot (NCS). The polymer films were formed from a mixture of solution of polymers and nanocarbons, from mets mixture of powders of polymers both NC C60 and NCS, by application of method of the limited swelling of polymers in solutions of nanocarbons. The concentration of C60 and NCS in polymers were varied at the wide range C = 0–10% mass.

Mechanical characteristics of samples: explosive strength, lengthening, the initial elastic moduli E were defined from deformation curves $\sigma(\varepsilon)$ by the standard explosive machine at speed of a stretching of samples at 12 mm/min. Thermal properties of samples have been studied on DSC Netzsch at speed of heating and cooling of 5 grade/min. Pictures of SAXRD and WAXRD have received by diffractometers KRM-1 and DRON-2, correspondingly. The copper radiation (λ = 0.1542 nm) filtered by nickel were used. Electrophysical measurements: tgδ-dielectric losses and ε – dielectric permeability at frequency 1–10 kHz at the temperature interval 10–160°C spent on the bridge of alternating current P5079.

3. Results and discussion

The films of LDPE + C60 have received from a solution in xylole (the samples 1), from melt of to LDPE + C60 (the samples 2) and from melt of LDPE + NCS (the samples 3).

The concentration C60 and NCS in samples 1–3 were varied at the intervals of C = 0–10%. The observe of a surface of samples in microscope have shown that at samples 1 and 3 there are aggregations of C60 and NCS with the sizes ~1–2 a micron, in samples 2 sizes particles are not less ~25–30 a micron. Pictures of WAXRD and SAXRD of samples 1 and 3 and their changes with growth C of additives are almost identical. In interval of C = 0–5% on WAXRD in a range of scattering angles 2θ = 8–40° intensive reflexes PE (110) and (200) are observed only, traces of aggregations of C60 and NCS are absent. However at C = 10% on WAXRD on background of reflexes of PE weak enough and indistinct reflexes of C60 (2θ ≈ 10.5°, 23–33°) and graphite (2θ ≈ 26°) start to be shown. The sizes of crystallites of PE estimated on radial width of reflexes (110) and (200) in samples 1 and 3 are made with 0.5–0.6 nm. In pictures of WAXRD of the samples 2 for C = 1–10% along with reflexes (110) and (200) of PE set of intensive reflexes from aggregations of C60 are observed. The size of crystallites of PE and C60 make ~0.5–0.6 and ~20–25 nm, accordingly. The received results testify that, the first, molecules C60 do not enter into crystalline lattice PE; in second, they form individual substructures of a different degree of orderliness and perfection, etc.

In the pictures of SAXRD of samples 1 and 3 (Fig. 1(a) and (c)) with increase of concentration C60 and NCS in interval of C = 0–10% almost same changes are observed. On SAXRD of the samples 1 for C = 0–3% there is an increase in intensity of discrete reflex I_m growth with C and then for C = 3–10% the tendency of decrease $I_m(C)$ is appreciable. On SAXRD of the samples 3 value of I_m with growth of C in the beginning (C = 0–3%) increases and then it does not change. But for all this beginning of C = 1% on a background of discrete scattering is shown a diffuse scattering. The increasing of C is accompanied by strong growth of intensity of a diffuse scattering (Fig. 1(c)). On SAXRD patterns of the sample 2 (Fig. 1(b)) with increase of concentration C60 the tendency of reduction I_m is appreciable. The analysis of pictures SAXRD of samples 1–3 on the basis of use two phase model of structure of polymers shows that in the first, only a small part of molecules C60 (~10%) get into amorphous regions of fibrils and increase their densities; in second, a big part of molecules C60 (~90%) penetrate into interfibrillar amorphous space, they form aggregations and promote moving apart of fibrill of matrix, to occurrence of internal pressure due to which there is a deformation of amorphous domain of fibrills, i.e. there are partial of reduction of densities, etc. Some increase in density of amorphous region of fibrills in the first case, and reduction of them in the second case, are cause reduction and increase at intensity of discrete scattering I_m

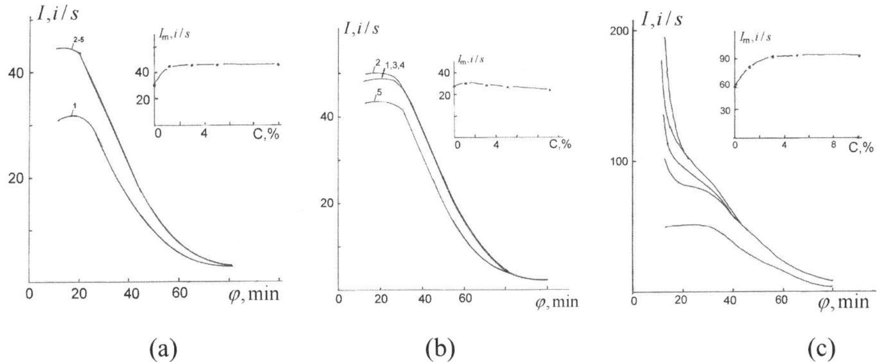

Figure 1. SAXRD of samples 1 (a), samples 2 (b), samples 3 (c): 1 – C = 0; 2 – C = 1; 3 – C = 3; 4 – C = 5; 5 – C = 10% C60

on SAXRD, accordingly. Penetration of molecules C60 into amorphous regions is accompanied by formation of microdeffects of types of pores, cracks, etc. also. The scattering from them and nanocarbons is caused with occurrence and increase in intensity of the diffuse scattering on SAXRD.

DSC researches of samples 1–3 have shown that growth of concentration of additives in interval of C = 0–10% do not influence on temperature of melting and crystallization of a matrix of samples 1 and 2. In case of the samples 3 with growth of concentration NCS reduction of temperature of melting of matrix from 112° up to 108°C at an invariance of temperature of crystallization (90°) is observed. Results of mechanical tests of samples 1–3 are lead in the Table 1.

TABLE 1. Mechanical characteristics nanocarboncontaining samples of LDPE

	Samples 1			Samples 2			Samples 3		
C (%)	σ (Mpa)	ε (%)	E (Mpa)	σ (Mpa)	ε (%)	E (Mpa)	σ (Mpa)	ε (%)	E (Mpa)
0	3.2	185	42	16	330	72	11	350	16
1	13	80	175	10	170	100	11	300	46
3	7.5	57	220	10	170	153	12,8	38	67
5	7.0	30	300	8,4	105	153	9.4	10	160
10	6.8	21	420	7.8	105	160	6.1	2.5	240

For them the general feature is the monotonous increase in value of E and nonmonotonic changes of σ and ε with growth of concentration C60 and NCS. For samples 1 and 3 value of σ increases for C = 0–3% a little, and then for C = 5–10% it decreases; the value of ε in interval of C = 0–10% decreases sharply. For the samples 2 growth of concentration of the additive is accompanied monotonous and not sharp decreases of σ and ε. Apparently, such behaviour of mechanical properties are caused by influence of set of factors: by structure of samples, the sizes of dispersed particles, their distribution, concentration, development of processes of sailing and destruction of chain molecules of matrix under influence of nanocarbons, etc. The shielding role of nanocarbons are shown in changes of thermal properties of crystalline lattices of matrix (PE). As have shown WAXRD researches of samples 1–3 in a combination with thermal influence in interval of temperatures 10–100° with increase of concentration of additives (C = 0–10%) the value of thermal coefficient of linear expansion of crystalline lattice of PE $\alpha_b = 0.8\cdot10^{-4}$ K^{-1} = const, but the value α_a gradually decreases from $6.5\cdot10^{-4}$ K^{-1} up to $2.5\cdot10^{-4}$ K^{-1}.

In thin samples (thickness less than 20 μm) PMMA + C60 with growth of concentration C60 (C = 0–10%) the tendency of increase σ, ε and E in comparison with initial samples are observed [1]. But in massive samples (thickness more than 40 μm) a situation is another. The increase of concentration C60 (C = 0–5%) is accompanied by monotonous decrease of σ and ε on 45% in comparison with the initial sample. Researches kinetics of photoageing of initial and nanocarboncontaining samples under irradiation of UV-light with λ = 254 nm have shown that at identical and final times of irradiation light resistance of initial samples decreases on 60%, for the sample with C = 1% on 50%, for the sample with C = 3% light resistance at first raises on 30% and then it decreases, for the sample with C = 5% light resistance of samples increases on 35% and then it decreases. For samples of PMMA + C60 with C = 3% and 5% maxima of light resistance are displaced at aside greater times of irradiation. From these dates follows that on the one hand, nanocarbons stimulate processes of structurization and destruction of macromolecules, and on the another hand, nanocarbons play a role of a defended screen and protecting chain molecules from destruction. From the received results it is possible to draw a conclusion and to expect that the lightstabilized properties of nanocarbons will even more strongly be shown in partially destructed and sailed polymers as they on specificity of chemical structure of a chain are lightresistanced and introduction of nanocarbons in them can be accompanied by improvement not only light resistance, but other physical properties of polymers also.

The films of samples from SIR-3 + C60 have received by method of swelling in saturated solution of C60 in toluene. Maximal concentration of C60 in the sample has made C = $2.5\cdot10^{-2}$%. Within the limits of an error of the lead measurements (10–15%) the deformation curves σ(ε) of initial and nanocarboncontaining samples practically do not differ. However crystallization of nanocarboncontaining samples a few more brightly is shown in comparison with initial sample. For example, the crystallization of the initial sample begins at value of deformation of ε = 400% and for nanocarbon-containing sample it begins at ε = 500% and is shown more brightly.

The films of block-copolymer SBS + C60 have received from solutions in toluene. The concentration of C60 varied in interval of C = 0–0.5%. Deformation curves σ(ε) of samples are lead on Fig. 2.

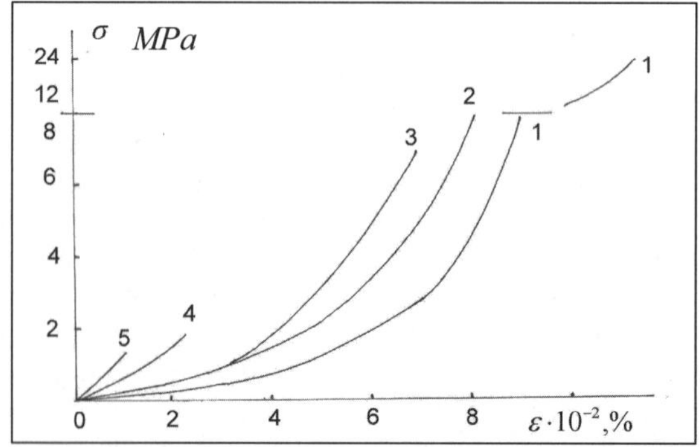

Figure 2. Deformation curves of nanocarboncontaining SBS marks «Cariflex». 1 – initial, C = 0%; 2 –10^{-2}%; 3 – $5\cdot10^{-2}$%; 4 – 10^{-1}%; 5 – 0.5% C60

As seen from figure with growth of maintenance of C60 sharp decrease of σ and ε and increase of E of samples is observed. In the pictures of WAXRD of all samples traces of C60 are not shown, small changes in intensity amorphous halo and its radial width are observed only. In the picture of SAXRD (Fig. 3) the initial sample the discrete scattering with the long period of ~300 nm is observed. The increase of concentration of C60 in samples is accompanied first all growth the intensity of discrete scattering and then with its output on the established value. The changes of SAXRD pictures of samples SBS + C60 are similar to changes SAXRD of LDPE + C60 (the samples 1).

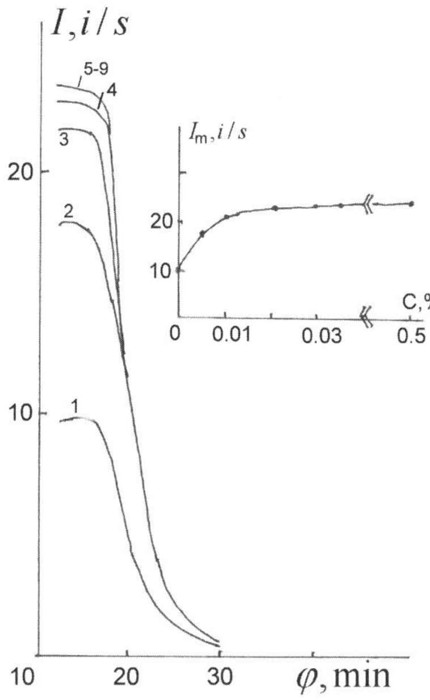

Figure 3. SAXRD of SBS + C60: 1 – C = 0; 2 – C = $5{\cdot}10^{-3}$; 3 – C = 10^{-2}; 4 – C = $2.5{\cdot}10^{-2}$; 5 –C = $5{\cdot}10^{-2}$; 6 – C = 10^{-1}; 7 – C = $2{\cdot}10^{-1}$; 8 – C = $3{\cdot}10^{-1}$; 9 – C = $5{\cdot}10^{-1}$ % C60

4. Conclusions

Changes of electrical properties of all nanocarboncontaining systems have the general features and distinctions. For example, at first, with growth of concentration of C60 or NCS the conductivity of systems increases and has ionic character; secondly, in the certain temperature areas on dependence tgδ(T) at frequencies 1 and 10 kHz maxima of dielectric losses tgδ are observed. Their values sharply decrease and disappear with growth of concentration of nanocarbons; thirdly, the dependences of ε(T) have strongly extended S-form and with growth of maintenance C-60 differ, the levels of values of ε(T) decrease. In PMMA on dependences tgδ(T) one maximum of dielectric (dipole-segmental) losses in the field of 80–90° is observed. The increase of concentration C60 (C = 0–5%) is accompanied by gradual reduction of tgδ in a maximum and for the sample with C = 10% there is its splitting on two maxima at 90° and 110°. The analysis of available literary and structural dates allows to assume that in polymer can exist nanocarbons in the free and connected states. Then, the first maximum at 90° can be carried to pure PMMA and the second maximum at 110° should be carried to connected system PMMA with C60. Thus, from the lead researches follows that in dependence of technology of reception of samples, the sizes and concentration of nanocarbon additives, character of their distribution in volume of a matrix, etc. the structure and physical properties of polymers can strongly change. The work was executed by project ISTC T - 1145 and the science plan of TSNU.

References

1. S.Tuichiev, etc. *Doklady AS RT*, 2005. **48**(7): 92–96.
2. S.Tuichiev, etc. *Doklady AS RT*, 2005. **48**(7): 110–115.
3. S.Tuichiev, etc. *Doklady AS RT*, 2005. **48**(7): 16–120.
4. S.Tujchiev, B.M.Ginzburg, D.Salomov. The program and the collection of reports of the international conference «Nanocarbon and Nanodiamond - 2006», P. 25. St. Petersburg, Russia.

SIMULATION OF HYDROGEN POLYCONDENSATION BY SINGLE-WALLED CARBON NANOTUBES

S.A. BEZNOSYUK, O.A. MASLOVA, S.V. VAZHENIN
Altai State University, 6556049 Barnaul, Russia
M.S. ZHUKOVSKY, T.M. ZHUKOVSKY
Altai State Technical University, 656038 Barnaul, Russia

Abstract. The prediction of self-organizing of hydrogen on the carbon nanotubular surface was suggested. The analysis of stability quantum-sized systems of nanotubular carbon and hydrogen adsorbate was given. Theoretical results of computer simulation reveal the mechanism of high hydrogen storage of carbon nanotubular accumulators. It is shown that effectiveness of atomic and molecular hydrogen adsorptions on the nanotubular surface is a result of contact exchange supramolecular adhesion forces acting in polycondensate. These bonds provide a stability of hydrogen polycondensate to its decay to hydrogen gas molecules at overflow of critical abundance of hydrogen over weight 11% in the adsorbate monolayer.

Keywords: hydrogen polycondensation, nanotubular carbon accumulator, computer nanotechnology

1. Introduction

Recently the nanoparticles attract heightened interest of physics, chemistry, biology experts. It is bounds up with particular physicochemical properties of nano-sized particles. These properties are distinct from properties of macroparticles and microparticles [1]. Especially one can stand out carbon nanotubes among such particles Their discovery, receiving and study of their properties arouse appearance of many works to detect their application fields. One of the most perspective directions of their use is making the effective accumulative mediums for hydrogen storage on the base of nanotubular carbon. Economical profitability demands hydrogen capacity of accumulators to be not less than 6.5 mass%. It is possible to achieve it with help of light materials. The use of these materials provides both a thermodynamic stability of compounds with high hydrogen concentration and an easy reversibility of hydrogen sorption and desorption.

Attempts of making of hydrogen storages on the basis of carbon nanotubes are realized in many experimental researches. Obtained by different authors test data about an amount of hydrogen capacity are greatly differ each from other.

Impossibility of getting hydrogen capacity over more than 6.5 mass% is underlined in most cases. So, for example, in experiment [2] amount of adsorbed hydrogen has been not exceeded 0.92 mass% at temperature 77 K and pressure 10^7 Pa. According to [3] hydrogen capacity of carbon nanotubes is 1.84 mass% at temperature 298 K and

B. Baranowski et al. (eds.), *Carbon Nanomaterials in Clean Energy Hydrogen Systems*.

standard atmosphere pressure. In the report [4] hydrogen capacity of nanotubular carbon exceeds 6 mass% at pressure in 10^5 Pa and temperature 77 K. It is possible under more high pressure to achieve 7.7 mass% that corresponds of receiving of the surface compounds C–H with stoichiometry 1:1. It has been shown that amount of adsorbed hydrogen is more than 7 mass% is realized at high temperatures and pressure 9 GPa [5]. In work [6] it has been informed about a possibility of 8.25 mass% of hydrogen content in carbon nanotubes at temperature 80 K and pressure 10 MPa. Opinion of experimenters and supporters of impossibility of receiving of high hydrogen concentration is confirmed by a standard quantum-chemical arguments about that atoms and molecules of hydrogen have only two probable paths of bonding with nanocarbon's surface. First, physical adsorption of hydrogen molecules appears at the expense of intermolecular Van-der-Waals interactions. In this case hydrogen capacity of carbon nanotubes should not be more than 2.7 mass% under normal conditions. Second, there is chemical adsorption like as for molecules C – H. And here hydrogen content cannot exceed 7.7 mass% under normal conditions.

At the same time, there are works which shows possibility of getting hydrogen capacity of nanotubular carbon over limit of 10 mass%. According to [7] concentration of adsorbed hydrogen raises from 7 to 13 mass% at ambient temperature and pressure in 10^6 Pa. In carbon nanotubes with admixture of alkali metals, as Chen informs, [8] the hydrogen capacity is in interval from 10 to 20 mass% at temperature 473–673 K and pressure in 0.1 MPa. In the review [9] such experimental researches are considered from the point of view of concepts about existence of abnormal «super»- Van-der-Waals or «weak» chemical interactions between hydrogen and carbon nanomaterials in hydrogen «polycondensates». Thus, it has been marked, that interaction energy between hydrogen and carbon nanostructure (20–40 kJ/mol) exceed interaction (distraction) energy of Van-der-Waals bonds in ten times, which is characteristic for physical hydrogen adsorption by carbon materials. This energy is ten times lower than distracting energies of covalent C-H bonding, which are characteristic for chemisorptions. It is marked by authors of review [9], that quantum mechanisms of such «weak» chemical or «super» physical interaction were not considered in no one works.

In the given paper quantum-sized mechanisms of the abnormal interaction in hydrogen polycondensates on nanocarbon's surface are considered by us for the first time. Computer modeling of process of formation high-concentration aggregates of hydrogen atoms and molecules on a carbon nanotubes' surface is carried out.

2. The theoretical footing behind this polycondensation of hydrogen on carbon nanotubular surfaces

For the first time mechanism of appearance of abnormal molecules interaction intermediate between physical intermolecular and chemical intramolecular bonds has been justified with help of the concept of quantum-field chemistry. These abnormal molecules interactions appear as a result of quantum-size effects of strong interelectronic correlation in condensed systems [10–12]. The electronic mechanism of these abnormal bonding shows appearance of quantum-sized compressing of electronic density at boundary atoms and molecules as a result of effect of the strong interelectronic pairing. As a consequence of this effect overlapping of such compact molecules becomes the intermediate one between degrees of overlapping in case of usual chemical bonds and usual intermolecular bonding. The contact exchange interactions (β–bond) between compact hydrogen molecules differs from well-known covalent bonds (α–bond) between hydrogen atoms in a molecule and physical Van der Waals forces (γ–bond) binding hydrogen molecules in a gas. Such exchange interaction is appeared without forming of common supramolecular electron orbitals.

Now these interactions are also known as a supramolecular interaction of the type «key – lock». They describe specific intermediate bonding between the supermolecules, considered in biochemistry, polymer chemistry, nanochemistry and fermentation catalysis. Detailed arguing of a contact exchange interaction in nanosystems is given in works [1, 13–14].

In this paper we consider simulation of hydrogen accumulators on the basis of hydrogen quantum-sized condensation process on nanotubular carbon under exposure of contact exchange interactions between hydrogen and carbon. Simulation of hydrogen adsorption process by carbon nanotubes's surface is realized using gradient density method.

The surface of potential energy was represented as a superposition of pair inter-atomic interactions:

$$E = \frac{1}{2}\sum_{i=1}^{M}\sum_{j=1}^{M} S_{ij}\varepsilon_{ij}(r_{ij}) \,, \tag{1}$$

where M – total number of atoms in a cluster, S_{ij} – an element of an adjacent matrix of its bonding graph, r_{ij} – interatomic distances, E_{ij} – interaction energy of i- and j-atoms as approximation of bonding by Morse potential. Used in our papers diatomic Morse potentials were obtained by analytical approximating of diatomic binding energies calculated by a nonlocal density functional method [10–12]. Figure 1 shows example of approximating of electronic term of β-bond CH by Morse potential. All received results of approximating of electronic term in binding energy, the distance between atoms and cyclic frequency of fluctuation of dimers CC, CH, HH are in Table 1.

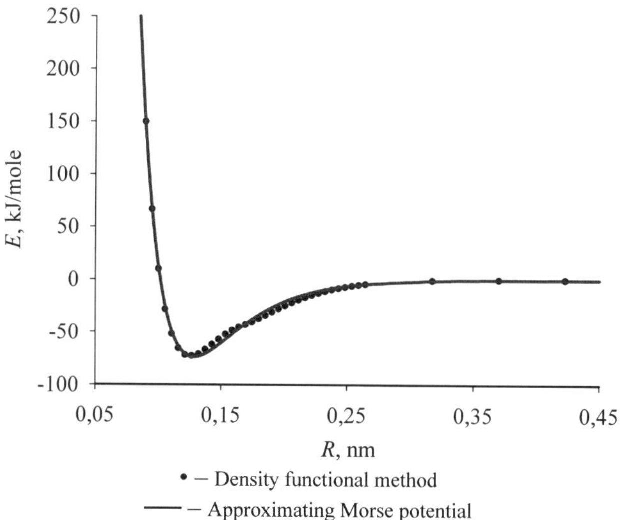

Figure 1. Analytical approximation of numerically calculated electronic term of C—H supramolecular dimer via Morse potential

TABLE 1. Equilibrium parameters of α – и β – bonds of dimmers CC, CH, HH

Bond of atoms	α–bond		
	Uo (kJ/mol)	Ro (nm)	ω_0 (1/cm)
C–C	-316	0.142	1,246
Bond of atoms	β–bond		
	Uo (kJ/mol)	Ro (nm)	ω_0 (1/cm)
C–H	-72	0.127	1,744
C–H[a]	-36	0.165	933
H–H	-51	0.101	2,110
H–H[a]	-40	0.124	1,567

[a]Different electronic terms of supramolecular bond, calculated taking into account molecular state of hydrogen

We considered five models of hydrogen accumulation by carbon nanotubular with a diameter of 1.2 nm. Atomic hydrogen takes part in two of them and molecular hydrogen takes part in other three. Each of the studied systems is characterized by special position of hydrogen on nanotube's surface. Fact of regular lattice forming by hydrogen atoms has been taken as a hypothesis. During a relaxation the lattice can be distorted by contact exchanged interactions. Lattices of hydrogen atoms and molecules and carbon atoms, forming a nanotube's surface are commensurate. So hydrogen atoms and molecules lattice should be also hexagonal lattice, which is lied on nanotube's surface.

3. Analysis of results of computer experiment

Polycondensation of atoms and polycondensation of molecules are competiting processes. Calculated values of C-H and H-H contact exchange bonds are differ in depending on embodying atomic or molecular model of the adsorption. In case of atomic adsorption energy of C-H bond (72 kJ/mol) is higher then energy of H-H bond (51 kJ/mol). On the contrary, for the molecular process, C-H bond (36 kJ/mol) is more weak then energy of H-H bond (40 kJ/mol). It is necessary to mark proximity of calculated parameters of contact exchange bonds to guessed phenomenological parameters of the abnormal bonding (20–40 kJ/mol) in case of molecular and atomic hydrogen condensation on a surface of nanocarbon tubes.

Among all considered models of hydrogen sorption two of them corresponds endothermic processes. These models are the atomic sorption with hydrogen capacity 7.7 mass% (Fig. 2) and the molecular sorption with hydrogen capacity 7.7 mass% (Fig. 4).

Figures 2–6 show an initial state (a) and final state (b) of the polycondensate nanostructure. Interaction energies of these states are calculated for one mole of hydrogen molecules. The other three models corresponds exothermic processes. So, the first start structure of atomic model consists of 390 carbon atoms and 570 hydrogen atoms (Fig. 3(a)). In this case the first part of hydrogen atoms is disposed above C-C binding and the second part of hydrogen atom is situated between two neighbored carbon atoms. After the reorganization nanotube is deformed. Local hydrogen accumulation is appeared (Fig. 3(b)).

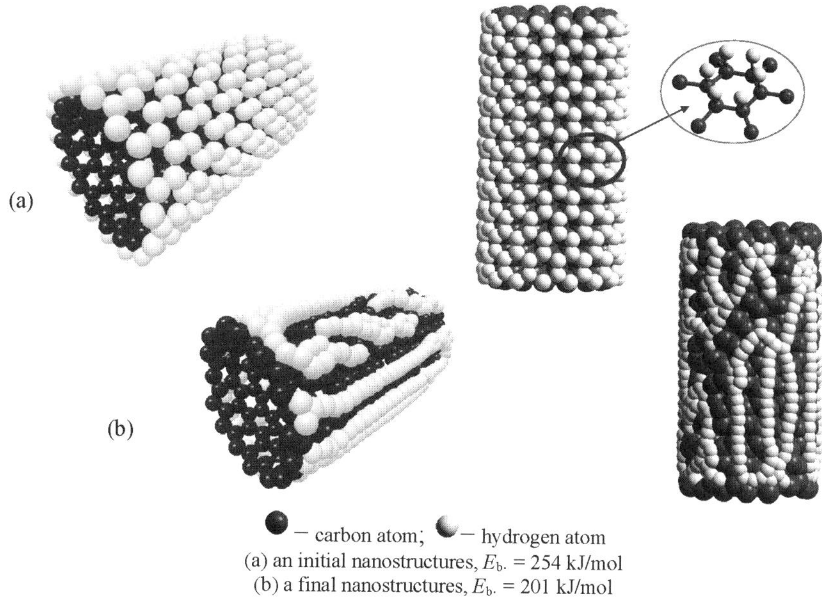

Figure 2. Carbon nanotubular accumulators of hydrogen atoms (hydrogen content ω = 7.7 mass%)

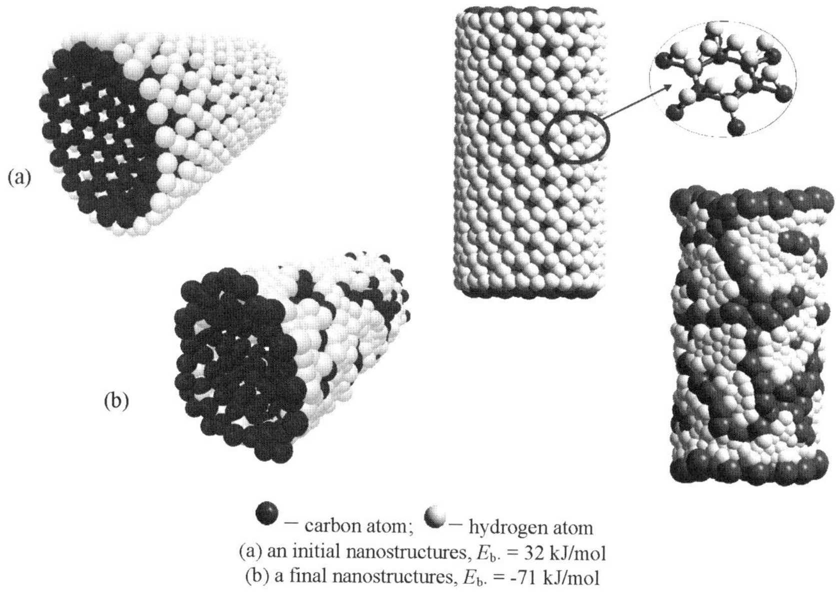

Figure 3. Carbon nanotubular accumulators of hydrogen atoms (hydrogen content ω = 10.9 mass%)

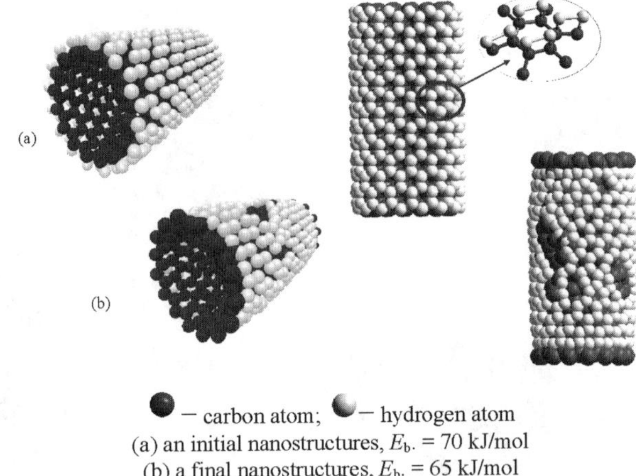

Figure 4. Carbon nanotubular accumulators of hydrogen molecules (hydrogen content ω = 7.7 mass%)

In the second start structure the model consists of 390 carbon atoms and 300 hydrogen molecules (Fig. 5(a)). In this case some of hydrogen molecules are disposed above carbon atom and others are situated above middle of C-C bonds. After processes of optimization carbon nanotube is slightly strained. Hydrogen molecules are condensed on the external surface of nanotube. Part of nanotubular carbon atoms becomes vacant and they become accessible for other hydrogen molecules (Fig. 5(b)).

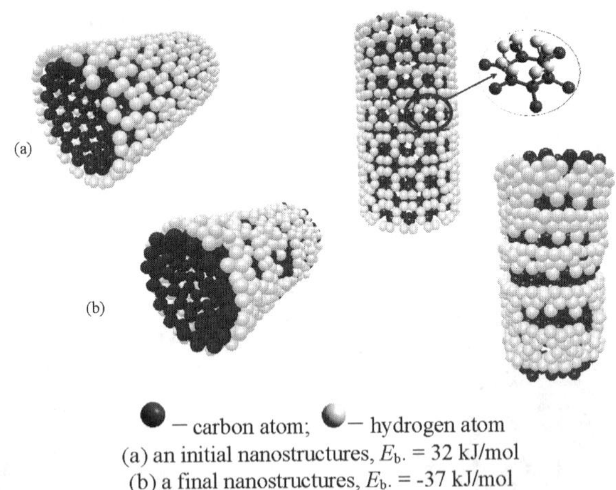

Figure 5. Carbon nanotubular accumulators of hydrogen molecules (hydrogen content ω = 11.4 mass%)

In the third start structure the model consists of 390 carbon atoms and 390 hydrogen molecules (Fig. 6(a)). In the initial geometry of condensed model molecules of hydrogen monolayer have been disposed exact above nanotubular carbon atoms. During a computer experiment hydrogen molecules are being dismissed from their initial positions. As a result of relaxation a character of hydrogen distribution on carbon nanotubular surface are sufficiently exchanged. In this case hydrogen monolayer has unstable form (Fig. 6(b)).

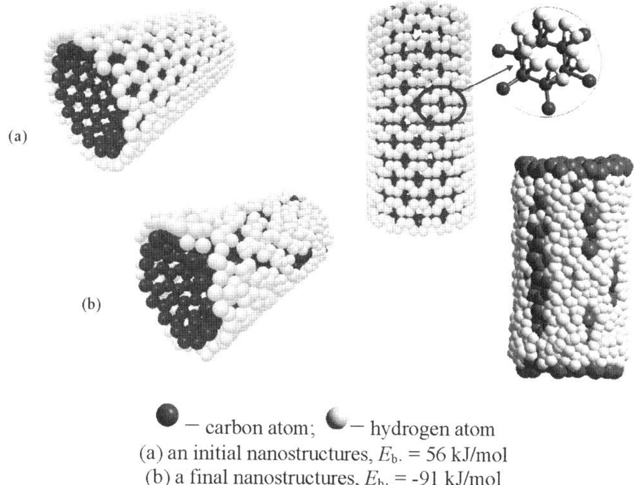

— carbon atom; — hydrogen atom
(a) an initial nanostructures, E_{b}. = 56 kJ/mol
(b) a final nanostructures, E_{b}. = -91 kJ/mol

Figure 6. Carbon nanotubular accumulators of hydrogen molecules (hydrogen content ω = 14.3 mass%)

In above-mentioned models all three processes of adsorption are exothermic. It is important to remark that in these models density of hydrogen is more then density one in case when a relation carbon – hydrogen is equal to 1:1. The density of hydrogen atoms exceeds 11%, and for hydrogen molecules one exceeds 14%. It is known, that binding energy of hydrogen molecule in a gas is equal to -456 kJ/mol. After optimization in above-mentioned nanotubular polycondensates binding energy: -527, -493, and -547 kJ/mol falls on each of hydrogen molecules, correspondingly. For each of three models the process of reconstruction is accompanied by releasing energy: 71, 37, and 91 kJ/mol relatively.

4. Conclusions

We considered quantum-sized mechanisms of abnormal interactions in hydrogen polycondensates on the nanotube's surface. We present the computer simulation of atomic and molecular hydrogen aggregate adsorption on the carbon nanotube's surface with receiving of high hydrogen content. Receiving results show that:

- By computer simulation it is shown that effectiveness of atomic and molecular hydrogen adsorptions on the nanotubular surface is a result of contact exchange supramolecular adhesion forces. These bonds resulted from quantum-sized effect.
- Computer optimization of polycondensate structures showed the existence of changes in a hydrogen distribution and nanotube's form. Polycondensate hydrogen nanostructures, which provide its stability at molecular hydrogen desorption in a gas has been found.
- Conducted investigations explain receiving of hydrogen content which is more then theoretical limit of 7.7 mass% in some experiments. It's shown that existence of high hydrogen capacity in the case of its polycondensation on the carbon nanotude's surface needs reaching of initial critical hydrogen concentration in monolayer which is not less then 11 mass%.

Acknowledgements

This work was supported in part by Program “Grants of the President of the Russian Federation”. Number of grant MK-2077.2007.3

References

1. Beznosyuk S.A., Potekaev A.I., Zhukovsky M.S., Zhukovsky T.M., Fomina L.V. Multilevel Structure, Physicochemical and Information Properties of Substance. NTL: Tomsk, 2005, 264 p.
2. Churilov G.N., Fedorov A.S., Sorokin P.B., Novikov P.V., Bulina N.V., Marchenko S.A., Martinez Yu.S., Gedanken A. In: Veziroglu T.N., Zaginaichenko S.Yu., Schur D.V., Baranowski B., Shpak A.P., Skorokhod V.V., Kale A. editors. Hydrogen Material Science and Chemistry of Carbon Nanomaterials. Dordrecht: Springer, 2005. P. 127–137.
3. Dai G.P., Liu C., Liu M., Wang M.Z., Cheng H.M. Electrochemical hydrogen storage behavior of ropes of aligned single-walled carbon nanotubes. *Nano Lett.*, 2002; **5**(2): 503–506.
4. Pradhan B.K., Sumanasekera G.U., Adu K.W., Romero H.E., Williams1 K.A., Eklund P.C. Experimental probes of the molecular hydrogen–carbon nanotube interaction. *Physica B.*, 2007; **323**: 115–121.
5. Bashkin I.O., Antonov V.E., Bazhenov A.V., Bdikin I.K., Borisenko D.N., Krinichnaya E.P., Moravsky A.P., Ossipyan Yu.A., Ponyatovsky E.G., Fursova T.N., Harkunov A.I., Shulga Yu.M. In: Schur D.V., Zaginaichenko S.Yu., Veziroglu T.N. editors. Hydrogen Material Science and Chemistry of Carbon Nanomaterials. IHSE: Kiev, 2003. P. 796–797.
6. Ye Y., Ahn C., Witham C., Fultz B., Lin J., Rinzler A., Colbert D., Smith K., Smalley R. Hydrogen adsorption and cohersive energy of single-walled carbon nanotubes. *Appl. Phys. Lett.*, 1999; **74**: 2307–2309.
7. Chen Y., Shaw D.T., Bai X.D., Wang E.G., Lund C., Lu W.M., Chung D.D.L. Hydrogen storage in aligned carbon nanotubes. *Appl. Phys. Lett.*, 2001; **78**(9): 2128–2130.
8. Chen P., Wu X., Lin J., Tan K.L. High H_2 Uptake by alkali-doped carbon nanotubes under ambient pressure and moderate temperatures. *Science*, 1999; **285**: 91–93.
9. Nechaev Yu.S., Alexeeva O.K., Gusev A.L., Filippov G.A. «Unsolved» problems of nature and characteristic of hydrogen sorption by carbon nanomaterials and path of solving. *Int. Sci. J. Alternative Energy and Ecology*, 2006; **36**(4): 15–18.
10. Muldachmetov Z.M., Minaev B.F., Beznosjuk S.A. Theory of Molecular Electronic Structure (New Aspects). Nauka: Alma-Ata, 1988.
11. Beznosjuk S.A., Minaev B.F., Dajanov R.D., Muldachmetov Z.M. Approximating quasiparticle density functional calculations of small active clusters: strong electron correlation effects. *Int. J. Quant. Chem.*, 1990; **38**(6): 779–797.
12. Beznosjuk S.A., Minaev B.F., Muldachmetov Z.M. Informative energetic structure and electronic multistability of condensed state. *J. Mol. Struct.* (Theochem) 1991; **224**: 125–127.
13. Beznosyuk S.A. Modern quantum theory and computer simulation in nanotechnologies: Quantum topology approaches to kinematical and dynamical structures of self-assembling processes. *Mat. Sci. Eng. C.*, 2002; **19**(1): 91–94.
14. Beznosyuk S.A., Stobbe I.A., Zhukovsky M.S. Quantum-sized mechanism of hydrogen polycondensation on carbon nanotubular surfaces *Mat. Sci. Eng. C.*, 2007; **27**(5–8): 1277–1279.

TRANSFORMATION EVOLUTION OF GRAPHENE AND NICKEL NANOPARTICLES

S.A. BEZNOSYUK, S.V. VAZHENIN, O.A. MASLOVA
Altai State University, 6556049 Barnaul, Russia
M.S. ZHUKOVSKY, T.M. ZHUKOVSKY
Altai State Technical University, 656038 Barnaul, Russia

Abstract. A new description of ultra-fast femtosecond quantum processes with energy, entropy and information exchange in open substantial nanostructures is considered. The ultimate aim of the approach is to obtain the new nanotechnologies based on the properties of electron-field multistable transformers (MT) which are the peculiar nanolevel objects. These nanoparticles have an ability of discrete exchange energy, entropy and information in time intervals of the range of $0.001 \div 10$ fs, which are very small in comparison with atoms' vibration period in substance. The results of the research of MTs' features causing unusual nanosystems' traits and their adaptive and cybernetic behavior properties are reported. MT systems give limitless resource for devices created within the same nanolevel to make measurements in femtosecond scale and to manipulate with substantial nanosystems. Computer experiments to research of femtosecond multikinetic processes with graphene and magnetic nickel clusters as the perspective new-generation nanodevices materials are carried out.

Keywords: nanoparticles' evolution, femtosecond kinetics, multistable transformers, graphene, nickel

1. Introduction

The laws of microlevel quantum processes as well as the laws of macrolevel classical processes are the two simplified asymptotics of the complicated laws of processes which are took place in fundamental nanoworld level [1, 2]. In latest twenty years the "first-generation" nanotechnologies were developed. Progress at this stage was made as a result of probe and laser-spectroscopic tool improvement as well as in consequence of increasing of computers efficiency in standard calculation schemes of molecules and condensed matter theory. Today, these conventional methods have reached their own limits in urgent nanotechnology problems-solving, when measuring, control and operation processes have to be made within the same nanolevel just as in vivo. The present problems have the next level of complexity e.g. assembling nanosystems in situ using adaptive nanorobots in real-time.

At the new level of developmental of second-generation nanotechnologies knowledge of complicated mechanisms and laws of quantum-sized nanoparticles is required. One can never reduce them to the "2 in 1"-like "sum" of classical and quantum properties. According to quantum-field chemistry [3, 4] compound nanoparticles appear as various finite discrete-field objects, namely, multistable transformers. Their typical sizes lie in the

B. Baranowski et al. (eds.), *Carbon Nanomaterials in Clean Energy Hydrogen Systems.*

range from 0.1 to 10 of nanometers. MTs specify peculiar physicochemical and informational properties of open substantial nanosystems [1].

The unity of multilevel processes with quantum microparticles and classical macroparticles participation is provided by quantum-sized multiparticles which are nanoworld's transformers. The basic feature of multiparticle transformers is topological discrete-distinctive mode of their structure and motion. Coherence in random micro- and macroscopic dynamic processes is provided on substantial nanolevel for algorithmic of nanotransformers' evolution kinetics. As an example it can be mentioned compound ("amorph" – "crystal"), polymorphous ("crystal" – "crystal") and polyamorphous ("amorph" – "amorph′") transitions occurring in time in compact fragment of material [5, 6] as a result of bifurcation of transformers' evolution algorithms. One can explain the topological distinguishability of particle's final states in terms that they are outcomes of distinct algorithms of its processing evolution. Every algorithmic evolution process ends with reaching of one of the energy degenerated and topologically distinguishable ground states.

In the paper the problems of theoretical modeling and the results of computer experiment research of multikinetics of discrete-algorithmic nanosystems' evolution at ultra-fast femtosecond processes of energy, entropy and information exchange in MT environment are considered. Nanoparticles' transformation evolution kinetics is studied by the example of graphene [7] and nickel atoms metal clusters [8]. It is shown that its shape are determined not by only the gradients of thermodynamic potentials but also by nanotransformers' algorithmic evolution femtosecond kinetics. The kinetics factor specifies information content of quantum-sized processes which is differ from dynamic unitary evolution of quantum and classical mechanics.

2. Basic approaches

It is well known that chemical reactions in molecules and materials as the reactions of breaking and creating of chemical bonds are the result of nuclear shifts. Speed of nuclei motion is defined by their vibration periods. There is no reaction which is faster than the nuclei vibration period and therefore it is a typical time quantum of chemical reactions in substance. Typical total elementary reaction time for any material lies in the range of $10 \div 10^4$ fs. It lasts over the time from one to hundreds of nuclei vibration periods.

Since the end of 1980th for studying fast reaction acts it is used femtosecond laser spectroscopy which is based on time compression of laser light pulse. The lowest time limit of laser pulse compression was obtained today is 4 fs. It is supposed that there is no sense to keep on shortening pulses since even the most rapid elementary chemical reactions are no longer than 10 fs. It is considered that there is no any other reason except of chemical reactions for quantization of processes' kinetics in substance. Since in the absence of chemical reactions atomic systems' and bonds' motion are described by continuous evolution, then theoretical computer modeling according to continuous molecular dynamics is applicable. However in the end of 1990th within the framework of quantum-field concepts of condensed state (thermo-field dynamics, quantum-field chemistry) we forecasted the field effects of electron plasma's evolution quantization in nanostructures even in the absence of chemical reactions [3, 4].

The effects of supplementary discreteness of nanostructural processes' kinetics are caused by the presence in the field plasma systems of temporary-lived compound quantum particles – compactons (swarming electron pairs, electrons together with antielectrons, ternaries, tetrads of them and so on) along with elementary quantum particles. Compactons theory was considered in one of the first paper about basics of quantum-field chemistry at the beginning of 1990th [3].

MT's evolution kinetics has three variants of discrete-by-time effects of energy exchange. The acts occurring at compound particles decay and formation are: first, the acts of transformers' unitary transformations with energy emission, second, the acts of transformers' nonunitary transformations with energy emission and entropy production, third, the acts of nonequivalent transformations with energy emission and information production. These three discrete acts types of nanosystems' motion multi-kinetics are named as emission, dissipation and MT reaction respectively. These acts supplement the discreteness of quantum kinetics of chemical nanomaterial conversion on the atomic level. Typical quanta of multi-kinetics duration have boundaries which lie below the time of elementary chemical reactions. Three quantum levels referred above bound within the range from thousandths to tens of femtosecond. In other words they lie beyond the femtosecond spectroscopy and probe microscopy reach.

One can view the complex discrete-algorithmic character of processes' multi-kinetics in its falling into sequence of discrete acts of ultra-fast emissions, dissipations, MT reactions and rapid elementary chemical reactions. This is an uniqueness of nanoobjects' functional properties, including their potential to adaptability and cybernetic intellectuality.

In the absence of catastrophic nonequivalent transformations in quantum systems one can consider two principal evolution ways which are differ one from another in dominating energy fluctuations channel, namely, unitary and dissipative ways. Using a general description of algorithmic evolution motion stated below in well-known matrix form of affine transformation from theory of optimization

$$\mathbf{X}^{i+1} = \mathbf{X}^{i} - \Delta S_i \frac{\nabla E(\mathbf{X}^i)}{\| \nabla E(\mathbf{X}^i) \|}, \tag{1}$$

one has to say that differences in the ways first of all explicitly affects the system's image point translation ΔS_i which is more or less changed parameter. Classical mechanics interpretation of image point's motion in configuration space gives simple expression for displacement at uniformly accelerated motion without initial velocity

$$\Delta S_i = (\Delta t_i)^2 \sum_p \frac{F_{i,p}}{2M_p}. \tag{2}$$

Notice that such an intermittent manner of evolution motion is in agreement with discrete character of energy fluctuation fluxes utilization by the boson component which is a swarming base of multistructural particle. Using these and some other common considerations, specifically the quantum-field concept of the least upper bound of time of coherence motion, it is not difficult to obtain the expression for time quantum in unitary evolution way:

$$\Delta t_i = \left(\hbar \Big/ \sum_p \frac{(F_{i,p})^2}{M_p} \right)^{1/3}. \tag{3}$$

Such a strong enough limitation allows to separate unitary evolution mode from another nonunitary ones which are characterized by a new thermostatistic mechanism of energy dissipation. One can deduce a formula for translation required to keep the evolution unitarity by correlating the expression (3) with the general law (2):

$$\Delta S_i = \left(\hbar^2 \sum_p \frac{(F_{i,p})^2}{M_p} \right)^{1/3} \sum_p \frac{F_{i,p}}{M_p} \Big/ \sum_p \frac{(F_{i,p})^2}{M_p} . \tag{4}$$

In case of homonuclear N-atomic cluster with mass M formula (4) is simplified to:

$$\Delta S_i = \left(\hbar^2 N \sum_p (F_{i,p})^2 \Big/ M \right)^{1/3} \sum_p F_{i,p} \Big/ \sum_p (F_{i,p})^2 . \tag{4'}$$

In dissipative mode equations of time and spatial displacement quanta include temperature dependence and one can deduce them under certain assumptions of balance between an internal quantum contribution and an external thermal contribution to energy of multiparticle's swarming component fluctuations.

$$\Delta t_i = \left(3NkT \Big/ \sum_p \frac{(F_{i,p})^2}{M_p} \right)^{1/2} ; \tag{5}$$

$$\Delta S_i = \frac{3}{2} NkT \sum_p \frac{F_{i,p}}{M_p} \Big/ \sum_p \frac{(F_{i,p})^2}{M_p} . \tag{6}$$

In case of homonuclear cluster with mass M one can simplify the expressions (5) and (6) as below:

$$\Delta t_i = \left(3MkT \Big/ \sum_p (F_{i,p})^2 \right)^{1/2} ; \tag{5'}$$

$$\Delta S_i = \frac{3}{2} NkT \sum_{p=1}^{N} F_{i,p} \Big/ \sum_{p=1}^{N} (F_{i,p})^2 . \tag{6'}$$

3. Computer experiment results and discussion

The demonstrative instance in theoretical sense is transformation of planar graphene sheet in a free state. Starting from an initial planar structure (Fig. 1(a)) one will always fall into thermodynamically and structurally more stable quasi-two-dimensional configuration (Fig. 1(b)) as a result of unitary or dissipative evolution. The effect is in complete agreement with natural experiment [4]. Wave generation over the entire graphene surface exemplifies the peculiarity of evolution motion of covalent two-dimensional crystal.

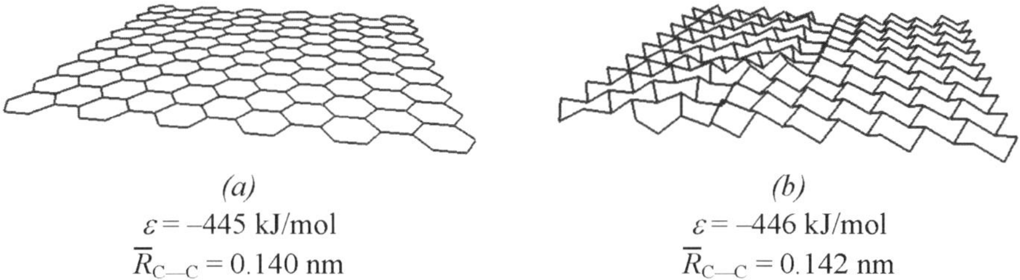

Figure 1. Unitary evolution of graphene C_{228}: (*a*) initial configuration; (*b*) optimized configuration; ε – binding energy per atom; $\overline{R}_{C—C}$ – average C–C bond length

One of the most vivid demonstration of two-dimensional instability is revealed in algorithmic evolution of planar metal monolayers (films) in a free state. Simulating the metallicity effects by specifying the complete binding graph of an atomic system allows one to point the essential differences between evolution of metal two-dimensional nanosystems and covalent ones (e.g., graphene). So, whereas although graphene is not stable in an "ideal" planar state it however keeps its quasi two-dimensional structure, metal monolayer surfaces on the contrary never keep quasi two-dimensional form. This is shown on the Fig. 2 by the example of dissipative evolution of nickel monolayer which is isomorphic with crystallographic plane (111) of nickel FCC lattice. Notice that reconstruction of two-dimensional metal nanoparticles with other atomic packing results in the same spheroid attractor.

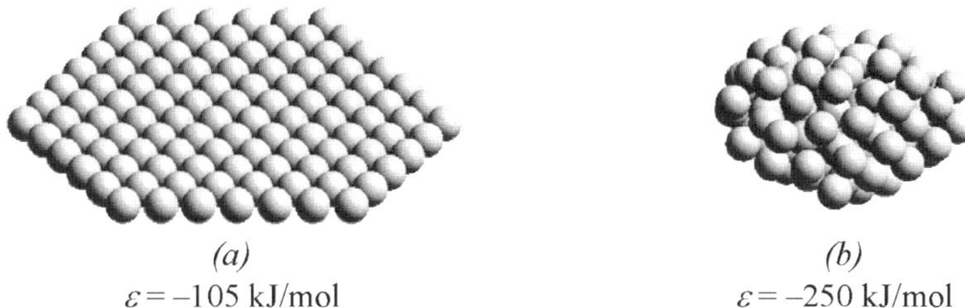

Figure 2. Reconstruction of crystal plane (111) Ni_{108}: initial configuration (*a*); optimized configuration (*b*); ε – binding energy per atom

The most significant feature of quantum-sized multiparticles is their discrete-distinctive variability i.e. transformability. The distinguishing character of transformer nanoparticles is degeneration of their ground state leading to non-dynamical but algorithmic way of transitions between their topologically distinguishable forms. This determines essentially bifurcational character of materials' evolution in nanometer range and is the key point in control of transformers' (e.g., nanorobots') "intellectual" movement to aim.

Figure 3 represents cubic nickel nanoclusters Ni_{108} and Ni_{500} (a) as well as topologically discrete-distinctive products (b, c) of their dissipative evolution under various temperature conditions. For the clusters with particles number of about 100 one can see almost full degeneration of final states and therefore they are dynamically equivalent at the absence of topological and informational identity. Particles number range where practical degeneracy keeps is bounded both above and below which allows one to specify the region with the largest informational resource.

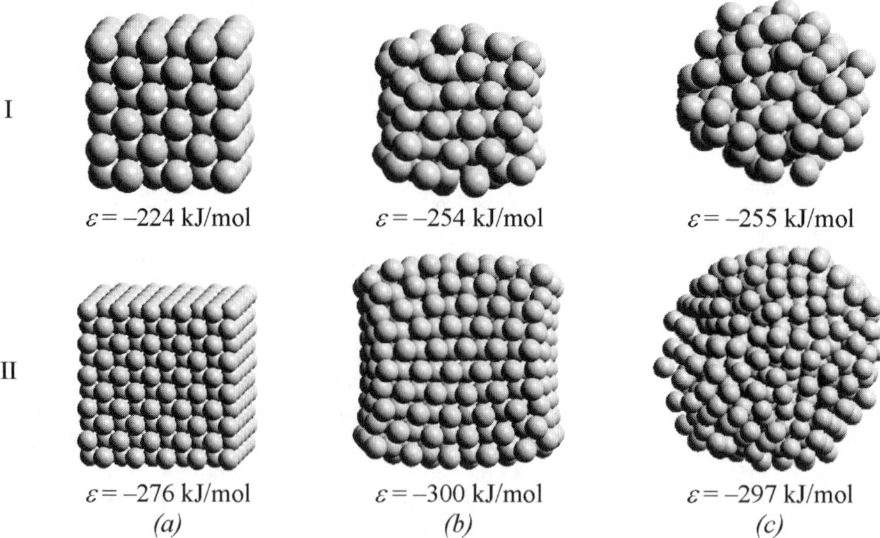

Figure 3. Cubic nanoclusters Ni_{108} (I) and Ni_{500} (II) (*a*) and their evolution products: quasi-crystals (*b*) and metglasses (*c*)

View of evolution may be obtained in form of algorithmic sweeps representing internal characteristic parameter (e.g. of dynamical nature) variation depending on path of image point in configuration space. Figure 4 shows energy dependence on system's path in case of dissipative mode resulting in quasi-crystal attractor. For not large clusters it is clearly visible typical slope even under large-scale oscillations conditions. This region corresponds to cluster's surface reconstruction with the formation of resulting typical quasi-crystal shape. Sweep gradation is the result of algorithmic-controlled temperature changing (forced cooling after energy stabilization). As the temperature decreases oscillations range reduces as a result of translations step decreasing ($\Delta S \propto T$).

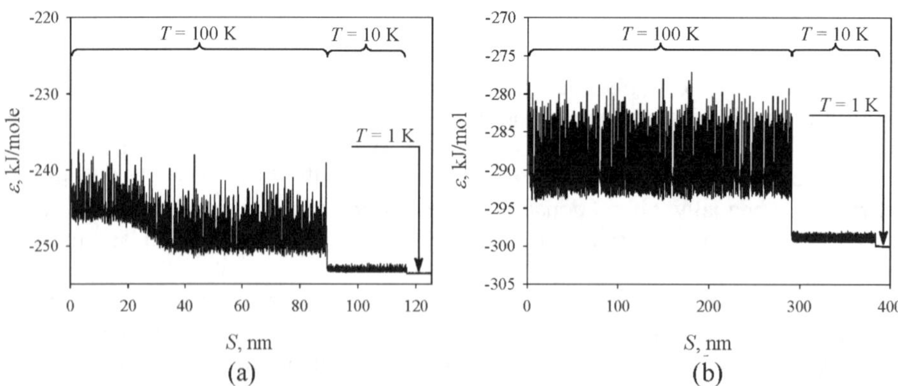

Figure 4. Temperature-energy sweep of dissipative evolution towards the quasi-crystalline attractor Ni_{108} (*a*) and Ni_{500} (*b*)

High-temperature mode results in complete loss of structure regularity and ultimately in metglass (spheroid attractor) formation. Through the high temperatures step magnitude increases causing energy oscillations intensification (see Fig. 5).

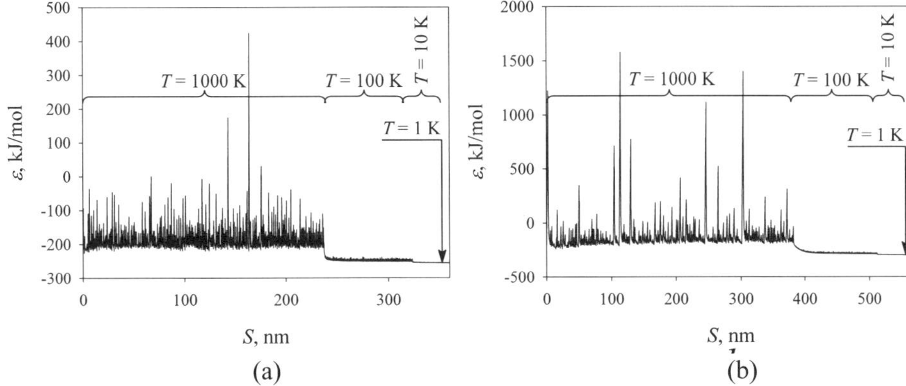

Figure 5. Temperature-energy sweep of dissipative evolution towards the spheroid attractor Ni_{108} (*a*) and Ni_{500} (*b*)

By estimation of time quantum for the considered above cases it is possible to reveal a regularity of processes accelerating through increasing of particles number (see Figs. 6 and 7). One has to refer this to competitive growth of gradient (there is an inverse proportionality from the norm of gradient in (5')). It is also clear the oscillation range reducing as particles number increases. On the contrary, at fixed number of particles one can note a small enough dependence of average time quantum on the temperature mode variations which only changes oscillations range. All the considered processes lie roughly speaking in the time range from a few to tens of femtosecond i.e. beyond any scanning microscope reach where time resolution not exceeds a gauge's atoms vibration period which is more than 10 fs.

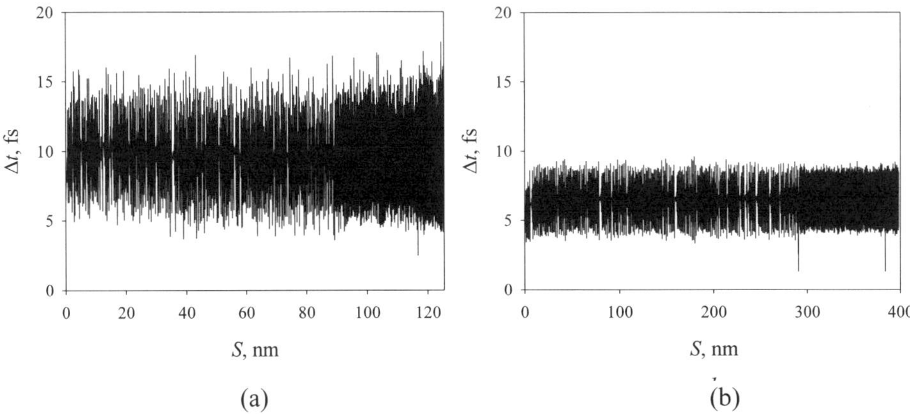

Figure 6. Time-base sweep of dissipative evolution towards the quasi-crystalline attractor Ni_{108} (*a*) and Ni_{500} (*b*)

Pattern of change in atomic packing regularity can be viewed via pair correlation function. At the initial state (Fig. 8(a)) the function has well-separated peaks corresponding to the maximum probability of particles detecting at the given distance from the coordinate origin. The quasi-crystal state's function (Fig. 8(b)) indicates the keeping of correlations in particles' positions, though peaks broadening and light overlapping can be seen. A significant loss of correlations in case of spheroidal state (Fig. 8(c)) manifests in the evident spreading and overlapping of peaks starting with the second coordination sphere.

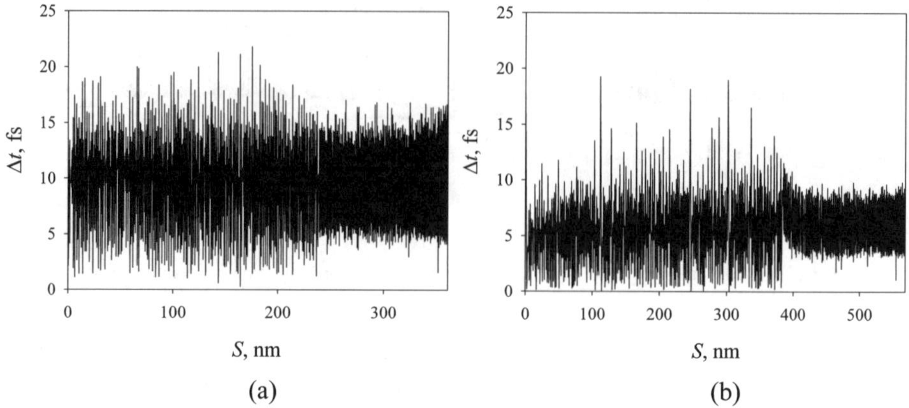

Figure 7. Time-base sweep of dissipative evolution towards the spheroid attractor Ni_{108} (*a*) and Ni_{500} (*b*)

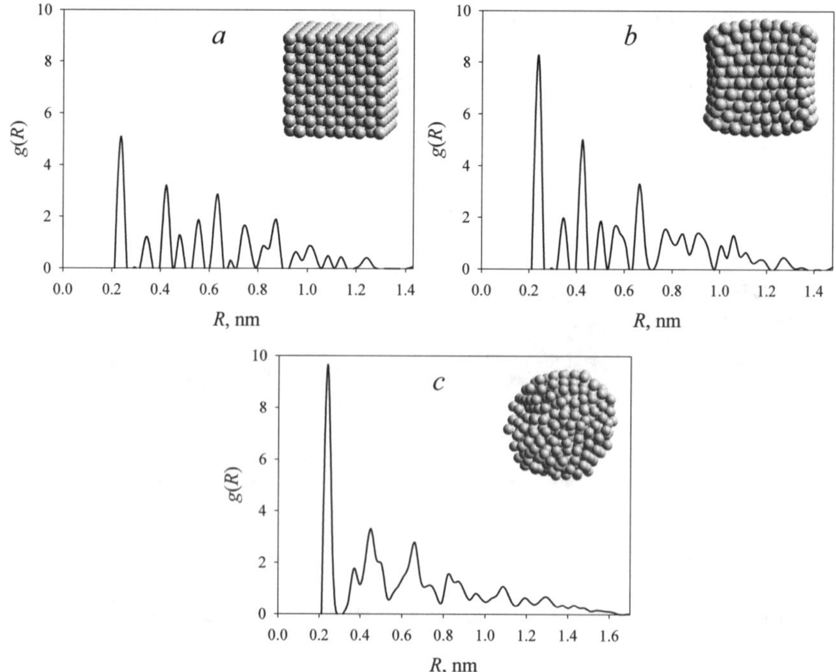

Figure 8. Pair correlation function of nanocluster's Ni_{500} multi-stable forms

4. Conclusions

The paper is headed for solving of one of the most central and actual problem for engineering nanosciences' and technologies' grounds advancement, to be exact, providing a new basis for the design of the next-generation nanodevices. The basic resources for the problem solving may be found in researching of mechanisms and laws of ultra-fast quantum processes of energy, entropy and information exchange in open substantial nanosystems. The problem solving makes possible creating the closed processing lines within the same nanolevel and constructing nanofactories on self-reproducing of nanodevices of various functional destinations. This is of special importance for the design of self-organized nanodevices which will provide on the one hand effectiveness enhancement in energy, entropy and information production and utilization and on the other hand the new fast-adaptive nanomaterials synthesis and super-high-speed intellect resources nanodevices producing. Just the long-term objectives of adaptive cybernetic nanodevices framework producing are the most foreground at a given line of nanotechnology development.

Getting an understanding of the physicochemical timing behavior mechanisms and laws of discrete-algorithmic quantum kinetics of energy, entropy and information exchange in nanosystem MT is one of the central problems within the common high nanotech problem. Solving of the former gives good grounds for the new nanotechnologies' development aimed at production of the new high-speed nanosystem devices with cybernetic adaptability to femtosecond changes in service conditions. This is the way to produce the new-generation nanomaterials and nanodevices efficient even under aggressive and difficult environmental conditions.

According to quantum-field theory of condensed states the exact description of open nanosystems' cooperative motion requires taking into account the existence of short-lived compound quantum-field quasi-particles (multistable transformers – MT) within "cold" electronic plasma. MTs' specific feature is discrete-distinctive variability of their motion mode as a result of discrete interchangement of energy, entropy and information with environmental "cold" electronic plasma. Thus one can view the complex discrete-algorithmic character of processes' multi-kinetics in its falling into sequence of discrete acts of ultra-fast emissions, dissipations, MT reactions and conventional elementary chemical reactions. Evolution acts supplement the discreteness of quantum kinetics of chemical reactions with bonds breaking or making within nanostructural matter what defines unusual multi-kinetic properties of the latter in comparison with a crystal or a molecule.

One should mention the difficulties at this fundamental problem solving both in the view of state-of-the-art experiment and in the view of state-of-the-art theory. First, MT transformations take up from thousandths to tens femtosecond and so lie below the times of any elementary chemical reactions, duration of which exceeds 10 fs. Spectrum of ultra-fast pulsating discrete motion of nanostructures lies beyond any scanning microscope reach where time resolution not exceeds a gauge's atoms vibration period which is more than 10 fs. Moreover it lies out of the super-high-speed femtosecond laser spectroscopy time resolution (all-time record value of 4 fs was obtained by the method's author the Nobel Prize laureate of 1999th A. Zewail). That is why the state-of-the-art nanotachnological tooling is incapable of reliable detection of MTs' discrete motion acts within nanosystem materials all the more it is incapable of algorithmic manipulation with such super-high-speed processes within nanosystems.

Second, all the known computer simulation methods treating continuous evolution of quantum-mechanical systems (e.g. quantum chemical methods, molecular dynamics methods etc.) are in principle inapplicable at such problem solving as description of discrete-algorithmic MT evolution. Owing to the fact that these methods reduce the discrete plasma-field dynamics of MT to continuous quantum mechanics of molecules or microcrystals with the same atomic composition, the discrete peculiarities of this problem are out of these methods' sight. As a result whereas atomic composition of the molecule or the microcrystal is the same, structure, evolution nature, physicochemical and informational properties are diverse. This difference looks like DNA's properties in vivo versus similar macromolecule's properties in biochemist's flask or inside of a crystallographer fluoroscopic camera.

By the referred above reasons the problems of theoretical modeling and computer researching of nanosystems' discrete-algorithmic evolution kinetics in terms of super-high-speed processes with energy, entropy and information exchange over the range from 0.001 to 10 fs are considered in the paper in details.

Within this approach's framework femtosecond multi-kinetics problem solving allows one to develop the new simulation, computation and design methods headed for open MT nanosystems and underlain the new breakthrough nanotechnologies of producing the cybernetic nanodevices framework with the resources of super-high-speed information processing and advanced adaptability to the dynamic environment.

Acknowledgements

This work was supported in part by Program "Grants of the President of the Russian Federation". Number of grant MK-2077.2007.3.

References

1. Beznosyuk S.A., Potekaev A.I., Zhukovsky M.S., Zhukovsky T.M., Fomina L.V. Multilevel Structure, Physicochemical and Information Properties of Substance, NTL, Tomsk, 2005, 264 p.
2. Fomin A.S., Zhukovsky M.S., Beznosyuk S.A. Modeling of nanomaterial structure based on quantum-sized mesoparticles. *Russ. Phys. J.* 2006, **49**(7): 754–757 (in Russian).
3. Beznosjuk S.A., Minaev B.F., Dajanov R.D., Muldachmetov Z.M. Approximating quasiparticle density functional calculations of small active clusters: strong electron correlation effects. *Int. J. Quant. Chem.* 1990, **38**(6): 779–797.
4. Beznosjuk S.A., Minaev B.F., Muldachmetov Z.M. Informative energetic structure and electronic multistability of condensed state. *J. Mol. Struct. (Theochem)* 1991, **224**: 125–127.
5. Beznosyuk S.A., Stobbe I.A, Zhukovsky M.S. Quantum-sized mechanism of hydrogen polycondensation on carbon nanotubular surfaces. *Mat. Sci. Eng. C.* 2007, **27**(5–8): 1277–1279.
6. Beznosyuk S.A., Fomina L.V., Perezhogin A.A., Zhukovsky M.S. Computer modeling of hydrogen and methane transport in cellular nanostructures of amorphous ice. *Mat. Sci. Eng. C.* 2007, **27**(5–8): 1390–1392.
7. Novoselov K.S. et al. Two dimensional atomic crystals. *Proc. Natl Acad. Sci. USA* 2005, **102**: 10451–10453.
8. Beznosjuk S.A., Dajanov R.D., Kuljanov A.T. Density functional calculation of transition metal cluster energy surfaces. *Int. J. Quant. Chem.* 1990, **38**(5): 691–698.

CARBON NANOTUBES FROM GRAPHENES IN THE PORES OF INORGANIC MEMBRANES

A.P. SOLDATOV, O.P. PARENAGO
A.V. Topchiev Institute of Petrochemical Synthesis of Russian Academy of Science, 119991 Moscow, Russia
E-mail: soldatov@ips.ac.ru, parenago@ips.ac.ru

Abstract. The new technique of synthesis of new carbon nanostructures is proposed: the oriented carbon nanotubes with graphene walls (OCNTG). To synthesize these structures the consistent covering of surface of pores of ultrafiltration inorganic membranes with monolayers of graphenes was held. These monolayers were formed during pyrolysis of the definite quantity of methane. This resulted in formation of OCNTG inside the nano-size pores (D_{av}.= 50 nm). Formation of the monolayer of graphenes was identified with the help of X-ray photoelectron spectroscopy (XPS). The depth of covering with monolayer was controlled with scanning electron microscope investigation (SEM). It was shown that OCNTG which were formed inside the pores of membrane were able to sorb and desorb hydrogen in the same quantities as traditional carbon nanotubes do.

Keywords: monolayer, carbon nanotubes with graphene walls, storage of hydrogen, ultrafiltration membranes

1. Introduction

Nowadays it is paid much attention to investigation of nano-size effects. Researches of carbon nanostructures (nano- crystallites, nanotubes, nanofibers, graphenes etc.) are among the most important works in this field [1–3]. Such attention to carbon nanostructures is connected with their unique surface-adsorptive and electron-emissive properties [4–6].

The main factors which determine these properties of nanomaterials are multiplicity of interfaces, existence of dislocations, vacancies and so forth [7–8] which are determined by particular qualities of their structure and construction. So it is interesting to synthesize and investigate of hybrid carbon nanostructures which might perform new promising properties. That's why investigations of synthesis of oriented carbon nanotubes with walls which were constructed with graphenes (OCNTG) were started to continue our previous researches concerned modification of pore structure of inorganic membranes with nanocrystallites of pyrocarbon (NCP) [9–10]. Synthesis of monolayer and multilayer nanotubes was possible at that.

Technique of synthesis of OCNTG was in step be step preparation of one or several carbon layers which consisted of the set of graphene fragments. Their precipitation on the surface of pores of membranes with oxide composition (Al_2O_3, SiO_2, TiO_2, MgO etc.) was performed by means of topochemical reaction of dehydrogenation of methane and its homologues (Knudsen diffusion procedure in the pores) with subsequent separation of nanotubes in alkali or hydrofluoric acid [11]. In the present paper the first results of the investigation are shown. These results include technique of synthesis and identification of monolayer and multilayer carbon nanotubes in pores of membranes and investigation of their adsorptive and transport properties. So at the first step of the research system of nanotubes which were "solded" into the nano-porous medium was investigated. Such approach lets one to get new information about nature of nano-size effects and about regularities of the interference of oriented nanotubes and nanoporous solid material.

2. Experimental procedures

Ultrafiltration composite membranes "TRUMEM" ($TiO_2 + Cr_2O_3$ on the porous steel) were used in the investigation. Thickness of selective layer of membranes was ~20 μm. Investigation of their porous structure by dynamic desorption porosimetry (DDP) [12] showed that $D_{av.}$= 50 nm, porosity ~25–30%, area of surface of pores ~1.7 m^2.

Preparation of OCNTG was performed at 800°C using device with quartz reactor ($D_{in.}$ = 54; L = 990 mm) and electrical oven using domestic methane with methane percentage ~99% like pyrolyzed reagent. For programming of temperature on the different stages and for determination of heating and cooling down rates the microprocessor-based controller was used which had 12 ramps.

The process thermocouple which was located inside the oven was calibrated by means of control thermocouple which was placed in the reaction area with the help of the devoted reactor tip. Its regulus was fixed at the distance $\leq$1.0 mm from the surface of membrane. Membrane was settled in the fixative frame [9] and was placed in the process zone of the oven. The set of calibration experiments which were performed in the conditions of the real process (process vacuum and delivery of methane) made it possible to find the exact correlation between process and control thermocouples.

Formation of OCNTG was performed by means of step be step plating surface of pores with monolayers of graphenes. Experiments were performed according to the scheme: heating up to 800°C → isothermal soaking (plating with monolayer) → cooling down to 500°C or to the room temperature → heating up to 800°C etc.

Scanning electron microscope investigation of chips of membranes was performed with the field emissive raster electron microscope JSM-6700F with add- on for energy-dispersive spectrometry (EDS) JED-2300F (JEOL, Japan). The apparatus was equipped with so called "cold" cathode and new "semi-in-lens" objective lenses with electron filtration. Analysis was performed with accelerating voltage of 15 kV, emission current of 10 μA, primary electron bunch current of 5 nA and process distance of 8 mm.

Add-on JED-2300F was used for qualitative and quantitative analysis using EDS technique. In the investigation quantitative analysis with reference spectra of elements was used (function "differential filter + least-squares method + ZAF method").

X-ray photoelectron spectroscopy investigation was performed on the spectrometer PHI 5500 ESCA (Perkin Elmer) using Mg Kα radiation (hv = 1,253.6 eV) with 300 W/14 kV power. Pressure of the residual gases in the measuring camera was 8–9 x 10^{-10} Torr. Fine spectrums were got with analyzer transmission energy of 11.75 eV and frequency of data acquisition of 0.1 eV/step. Data processing was performed with approximation with non-linear least-squares method using Gauss-Laurence function.

Correction of charging effects was performed by means of calibrating of binding energies scale relative to carbon $C1_s$ – 285.0 eV.

To change the depth of analysis in the surface zone spectrums were got at different angles of the sample to the axis of analyzer [13]. For purification of surface the ion gun was used in the blow off mode with energy of ions 150 eV, raster 2 x 2 mm^2, partial pressure for argon 12 MPa.

Quantity of hydrogen absorbed with OCNTG was determined gravimetric technique up to hundredths of mg. Such approach is rather correct as two membranes were used for saturation with hydrogen: the initial one and that one with OCNTG. The initial pressure of hydrogen at saturation was 10–13 MPa.

Electro-surface properties of membranes were studied using current potential technique with chlorine-silvery (Ag-AgCl) electrodes [14]. In the present work dependence between potential difference ΔE and pressure difference ΔP on the membrane were studied while passing 0.01 M solution of KCl through the membrane. Current potential ($\Delta E/\Delta P$) was used for determination of ξ-potential according to Smoluchowski equation. The quantity of the latter was used in Gui equation for determination of charge density on the surface of pores. The sign of the surface charge was determined according to the slope ratio of ($\Delta E/\Delta P$).

Study of transport characteristics of membranes was performed using standard filtration set up PVF-47 at temperature about 20–22°C and process pressure difference ΔP = 100 kPa.

3. Results

It is evident that it is necessary to form the layer of carbon on the whole depth of pores during forming of OCNTG in pores of membranes. For this purpose diffusion of molecules of methane inside the pores should be in line with Knudsen diffusion procedure when free path of molecules (λ) is much larger that diameter of pores. Evaluation of quantity of λ [15] at process pressure of 4.9 kPa showed that for membranes under consideration ratio λ/D is λ/D = 94.

(a)

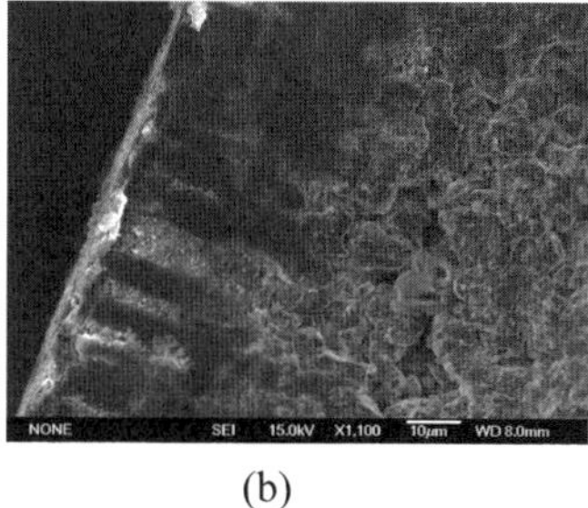

(b)

Figure 1. Photos of chips of ultrafiltration membranes «TRUMEM»: (a) initial membrane; (b) membrane plated with NCP at P = 4.9 kPa

Figure 1(a) shows the photo of chip of membrane where selective layer (the left side) and substrate are well defined. Their morphologies are pretty different – the selective layer has fine-grained structure while the substrate is formed with large massive fragments. It is worth saying that during investigation of the depth of plating of pores with carbon their surface was plated not with graphene layers but with nanocrystallites of pyrocarbon (NCP). L_c quantity of the latter was 2–6 nm.

Plating of surface of pores with NCP resulted in appearance of dark fragments of carbon on the photos of chips of membranes (Fig. 1(b)). One can see that depth of plating with NCP corresponds to the thickness of selective layer. Such morphology is quiet explicable as during breaking of membrane after immersion of the latter into liquid nitrogen line of fracture passes through pores and interpore partition walls, and chips of interpore partition walls have structure of the initial membrane.

Moreover energy-dispersive spectrum of the chip of membrane with NCP layer which is shown in Fig. 2(a) lets one to evaluate the elemental composition of its surface (the sondage region is at the depth of ~18–20 μm from the membrane surface). One can see that together with elements forming the oxide selective layer (Ti, Cr, O) the surface of the chip contents 16.27% of carbon while there is no carbon in the similar spectrum of the initial membrane (Fig. 2(b)). Presence of aluminium and some quantity of potassium is the result of composition of the raw material used fir preparation of membranes.

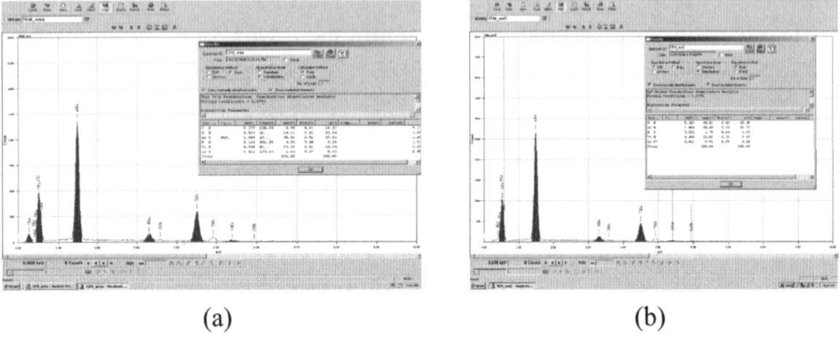

(a) (b)

Figure 2. Energy-dispersive spectrums of the chips of membranes «TRUMEM»: (a) membrane plated with NCP at P = 4.9 kPa; (b) initial membrane. Scanning at the depth ~18–20 μm

It is evident that plating with NCP at 4.9 kPa takes place for the whole depth of pores (Knudsen diffusion procedure). Hence we can form OCNTG in the pores of membranes if we can work through the technique of plating with monolayers of NCP and identification of the latter which were precipitated on the surface of pores.

In [16] authors showed that NCP covering membrane have the ideal crystalline structure of graphite while lattice of pyrocarbon is characterized with hexagonal symmetry with the following parameters: d_{002} = 3.368 A, L_c = 40.0 nm, L_a = 80.0 nm. Hence we get the set of carbon nanotubes which are oriented transversely to the surface of membrane during plating of the surface with monolayer of pyrocarbon. The walls of nanotubes are formed from graphene fragments (OCNTG) at that. It is worth mentioning that at the initial stage of precipitation of NCP their size L_a is significantly smaller [11]. So in the following equation (1) we suppose that average unit graphene forming on the surface of pore consists of 12 condensed hexamerous rings.

Using approaches described in [17] and proved that their application in calculations resulted in rather satisfactory compliance with experimental data one can derive the equation for determination of mass of pyrocarbon which is necessary to form the monolayer of carbon (m_{mono}, g) depending of the area of surface of pores (S_{ps}, m^2):

$$m_{mono} = S_{ps} M n / S_c N, \quad (1)$$

where N – Avogadro constant, S_c – area of 12 condensed hexamerous carbon rings (m^2), M – carbon atomic mass, n – number of carbon atoms in 12 rings.

Quantity m_{mono} for the membranes under consideration according to the equation (1) was 0.0026 g. The process parameters of plating with OCNTG were determined using kinetic data for topochemical reaction of dehydrogenation of methane which are listed in [18].

To fix the monolayer of graphenes the comparative analysis of spectrums of the initial membrane (sample No. 0), of membrane which was heated in vacuum at the temperature of plating with pyrocarcon (sample No. 2) and of membrane which was plated with the layer of graphenes (0.0028 g) was performed with the help of XPS.

XPS investigation was performed taking into consideration the positive charge of the surface of samples due to emission of electrons. For this purpose spectrums were shifted on ΔU during determination of binding energies. ΔU quantity was determined using the adsorbed carbon (AC) in the sample No. 0 as carbon state on the surface in the other samples is unknown. For this sample position of maximum of oxygen peak corresponding to oxygen atoms in TiO_2 (O1s – 529.75 eV) (Fig. 3(a)) was determined by its shift relatively to AC. The obtained value was used in samples No. 1 and No. 2 for determination of ΔU.

According to the results of analysis spectrums of titanium were the same for all samples (Fig. 3(b)) and they correspond to titanium dioxide. Spectrums of carbon are the same for samples No. 0 and No. 2 and they correspond to AC while spectrum of sample No. 1 is shifted in relation to the first ones to the direction of lower energies – 284.3 eV (Fig. 3(c)). This fact tells that there is the layer of pure carbon on the surface of sample No. 1.

So the technique of formation of monolayer and multilayer OCNTG in the pores of inorganic membranes is developed and tested.

One of the unique and very important from the applied point of view properties of carbon nanotubes is their ability to absorb and to store hydrogen [19–20]. That's why existence of the similar property of the OCNTG which were placed to the surface of pores is very interesting problem.

Table 1 shows the experimental results received during saturation of OCNTG with hydrogen. Experiments were held on four samples of membranes with different plating conditions including the initial membrane which allows evaluating of adsorption of hydrogen with gravimetric method. It is worth saying that after saturation with hydrogen which lasted for 23 hours for all experiments the initial mass of the membrane increased on 0.01 mg. However this variation corresponds to the uncertainty of weighting and is not taken into consideration in Table 1.

Table 1 shows that there is significant different in adsorption of hydrogen from sample to sample and membrane with maximum mass of OCNTG is characterized with minimal adsorption. The consequent plating with definite quantity of carbon is necessary during synthesis of OCNTG to form the monolayer of graphenes (0.0026 g for membranes under consideration). The most ideal is sample No. $1H_2$ which was plated with 0.0031 g of carbon during formation of the first layer of graphenes while the average quantity of precipitated carbon forming the following 8 layers was ~0.0021 g per layer. For the samples $4H_2$ (16 layers) and $5H_2$ (8 layers) plating with carbon during formation of 2–3 initial monolayers was significantly larger than necessary quantity while it was 40–60% lesser during formation of the following layer. This might be the reason of the found differences in the quantity of adsorbed hydrogen.

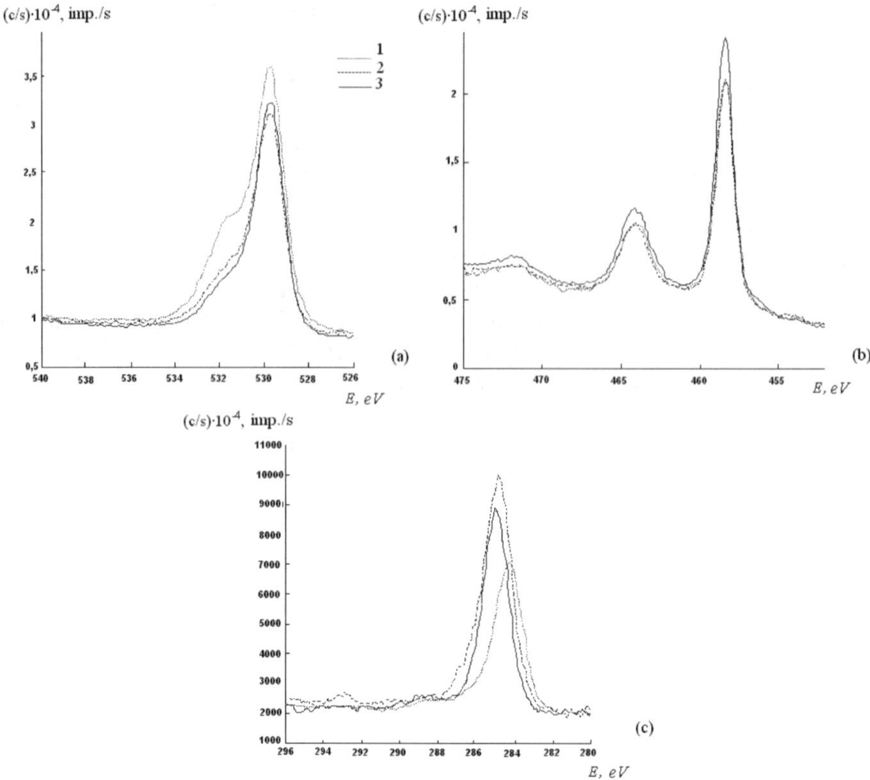

Figure 3. Fine spectrum of oxygen O1s (a), titanium Ti 2p (b) and carbon C1s (c) at the angle equal to 45°; samples 0 (1), 2 (2), 1 (3)

As one can see from the Table 1 desorption of hydrogen from OCNTG occurs rather slowly. Its quantity was only ~6% after storage during 40 days at temperature 20–22°C and atmospheric pressure. At the same time heating of the sample No. $1H_2$ in the atmosphere of helium up to 450°C and keeping for 10 minutes at this temperature resulted in total desorption of hydrogen.

The preliminary evaluation of influence of OCNTG on transport properties of membranes was performed. Performance of membrane on ethanol increased from 36.0 to 341.2 ml/s cm^2 cm Hg after plating of surface of pores with monolayer of graphenes (0.0026 g). Electro-physical characteristics of this surface are significantly changed as well at that: the electrostatic charge quantity reduces from 49.0 to 6.6 x 10^{-4} C/m^2 and ξ-potential decreases from 20.6 to 2.8 x 10^{-3} V.

TABLE 1. Adsorption and desorption of hydrogen with OCNTG

Sample	OCNTG mass (mg)	H_2 press. (MPa)	Adsorbed H_2 (% from OCNTG)	Adsorbed H_2 (% from OCNTG after 24 hours)	Adsorbed H_2 (% from OCNTG after 120 hours)	Adsorbed H_2 (% from OCNTG after 40 days)
Initial	–	10.7	–			
$1H_2$	18.61	10.7	5.37	5.37	5.37	5.05
$5H_2$	19.57	11.8	2.30	2.30	2.30	
$4H_2$	24.82	12.9	0.93	0.93	0.93	

4. Conclusions

The technique of synthesis of new carbon nanostructures – mono- and multilayer oriented carbon nanotubes with walls which are formed with graphenes is developed. Conditions developed for plating of inorganic ultrafiltration membranes with graphenes on the whole depth of pores. Formation of monolayer of graphenes on the surface of pores is detected. Synthesis of monolayer and multilayer OCNTG in pores of membranes «TRUMEM» is performed. It is shown that they able to absorb and to store hydrogen and their adsorption capacity is in line or larger than similar value of traditional carbon nanotubes. The first result received which illustrate the significant increase of performance of membranes after plating them with multilayer OCNTG. Also changes in electrosurface properties are studied.

And at least its worth saying that it seems to be very interesting to utilize the adsorption ability of OCNTG synthesized in the pores of membranes to hydrogen to create the membrane microreactors in which one of the reagents (hydrogen) is stored in OCNTG and is desorbed to participate in the reaction as required.

References

1. Sun X., Li R., Stansfield B., Dodelet J-P., Menard G., Desilets S. Controlled synthesis of pointed carbon nanotubes. *Carbon*, 2007. **45**: 732–737.
2. Harris P.J. Solid state growth mechanisms for carbon nanotubes. *Carbon*, 2007. **45**: 229–239.
3. Gao L., Zhou X., Ding Y. Effective thermal and electrical conductivity of carbon nanotube composites. *Chem. Phys. Lett.*, 2007. **434**: 297–300.
4. Zuttel A., Sudan P., Mauron P., Wang O.X., Wu Y.L., Yu H. *Appl. Phys. A.*, 2004. **78**: 941–946.
5. Hagen A., Moos G., Talalaev V., Hertel T. Electronic structure and dynamics of optically excited single-wall carbon nanotubes. *Appl. Phys. A.*, 2004. **78**: 1137–1145.
6. Zhang M., Atkinson K.R., Baughman R.H. Multifunctional carbon nanotubes yarns by downsizing an ancient technology. *Science*, 2004. **306**: 1358–1361.
7. Wang W.H., Hong T.H., Kuo C.T. Super growth of vertically aligned SWCNTs using self-assembled nanoparticles from $CoCrPtO_x$ ultra-thin film. *Carbon*, 2007. **45**: 97–102.
8. Bachtold A., Hadley P., Nakanishi T., Nakanishi T., Dekker C. Logic circuits based on carbon nanotubes. *Physica E*, 2003. **16**: 42–46.
9. Soldatov A.P., Shkolnikov E.I., Rogailin M.I., Rodionova I.A., Parenago O.P., Volkov V.V. Technique of modification of porous stricture of inorganic membranes with pyrocarbon. *RF Patent 2179064*, 2002 (in Russian).

10. Soldatov A.P., Parenago O.P. Modification of inorganic membranes with nanocrystallites of pyrocarbon. *Russ. Chem. J.* (J. of D.I. Mendeleev Russian Chem. Soc.), 2006. **L**: 60–63.
11. Gorodetskii A.E., Tesner P.A., Lukyanovich V.M., Poliakova M.M., Zaharov A.P., Iagmin A.G. Structure of thin layers of pyrocarbon synthesized from methane. *Proc. Acad. Sci.*, 1972. **203**: 1336–1338.
12. Shkolnikov E.I., Elkina I.B., Volkov V.V. Technique of analysis of pore structure. *RF Patent 2141642*, 1999 (in Russian).
13. Moulder J.F., Stickle W.F., Sobol P.E., Bomben K.D. In: Chastain J. (Ed.), Handbook of X-ray photoelectron spectroscopy. Physical Electronics, Eden Prairie, MN, 1992. 439.
14. Berezkin V.V., Volkov V.I., Kiseleva O.A., Mitrofanova N.V., Sobolev V.D. Charge of pores of track membranes from polyethylene terephthalate. *Colloid J.*, 2003. **65**: 129–131 (in Russian).
15. Mulder M. Introduction in membrane technology. Moscow, Mir, 1999. **212**: 266 (in Russian).
16. Soldatov A.P., Rodionova I.A., Parenago O.P. Influence of pyrocarbon modification on physico-chemical characteristics of surface of pores and on transport properties of inorganic membranes. *Russ. J. Phys. Chem.*, 2006. **80**: 500–506 (in Russian).
17. Soldatov A.P., Rodionova I.A., Shkolnikov E.I., Parenago O.P., Volkov V.V. Pyrocarbon modification of composite inorganic membranes. *Russ. J. Phys. Chem.*, 2004. **78**: 1458–1462 (in Russian).
18. Soldatov A.P., Berezkin V.V., Gontar I.V., Evtugina G.N. Parenago O.P. Knudsen diffusion procedure of methane in the pores of inorganic membranes: kinetics and depth of plating with pyrocarbon, its influence on transport properties. *Russ. J. Phys. Chem.*, 2008. **82**(6): 1124–1130. (in Russian).
19. Ning G.Q., Wei F., Luo G.H., Wang Q.X., Wu Y.L., Yu H. Hydrogen storage in multi-wall carbon nanotubes using samples up to 85 g. *Appl. Phys. A.*, 2004. **78**: 955–959.
20. Shiraishi M., Takenobu T., Kataura H., Ata M. Hydrogen adsorption and desorption in carbon nanotubes systems and its mechanisms. *Appl. Phys. A.*, 2004. **78**: 947–954.

EFFECT OF PORE STRUCTURE OF ACTIVATED CARBON FIBERS ON HYDROGEN ADSORPTION

R.G. DING, X.D. YAO, Z.H. ZHU, G.Q. LU
ARC Centre of Excellence for Functional Nanomaterials, The University of Queensland, St Lucia, QLD 4072, Australia
Z.F. YAN
State Key Laboratory of Heavy Oil Processing, University of Petroleum, Dongying, Shandong 257061, China
E-mail: r.ding@uq.edu.au

Abstract. The activated carbon fibers were air-oxidized and their textural properties were well characterized by N_2 and CO_2 adsorption. The correlation of pore size distribution of raw and treated activated carbon fibers with hydrogen adsorption performance indicated that ultramicropores dominate the hydrogen adsorption at low pressure. However, hydrogen molecules are not only trapped in the ultramicropores, but also adsorbed in the supermicropores at 77 K and high pressure. The narrow ultramicropores represent the higher energy adsorption sites that can adsorb gas molecules at high temperature and very low pressure. CO_2 adsorption is proven to be necessary to provide complementary characterization of narrowest pores.

Keywords: activated carbon fibers, hydrogen adsorption, pore structure, ultramicropore, supermicropore

1. Introduction

Hydrogen adsorption on various carbon materials has been extensively investigated for many years and the discovery of carbon nanotubes attracted a great deal of attention in the fundamental mechanism and potential high adsorption capacity for the novel carbon nanomaterials. Nevertheless, carbon nanotubes have failed to fulfill their initial promise. Generally, supercritical adsorption of hydrogen is limited by the low interaction energy between hydrogen molecules and carbon surfaces, which is only enhanced in porosity where the pore dimensions are similar to those of hydrogen. Some theoretical calculations show that the effective pore diameter for hydrogen physisorption is less than 0.56 nm [1], while further experimental investigation found that 0.66 nm is the optimum pore size in activated carbon and activated carbon fibers for hydrogen adsorption under 10 MPa and ambient temperature [2].

In the previous studies, various carbon adsorbents under different treatments were used so that the effect of functional groups, pore shape and carbon type on the determination of optimum pore size for hydrogen adsorption can not be excluded. Zhao et al. found that functional groups have a small detrimental effect on hydrogen adsorption, and this is related to decreased adsorbate-adsorbent and increased adsorbate-adsorbate interactions [3]. On the other hand, for a given pore size, cylindrical and spherical pores may result in a different total sorbate-sorbent interaction compared to slit-shaped pores. In this article, only one typical microporous carbon, activated carbon fiber (ACF), was selected for air oxidation treatment and hydrogen adsorption investigation in order to avoid the influence of pore shape, functional groups and carbon type while determining correlation of pore size with hydrogen adsorption.

2. Experimental procedures

Commercially available activated carbon fiber (ACF, A-20) made by Osaka Gas Co. Ltd, Japan was used in this research. The raw activated carbon fiber was treated in an air flow of 100 ml/min under different temperatures (400°C, 500°C, and 600°C) for 30 minutes. The heat treated samples are denoted as from ACF400 to ACF600, with numbers indicating the treatment temperature. The textural property of raw and heat treated samples were characterized by N_2 adsorption at 77 K and CO_2 adsorption at 273 K on an Autosorb-1 analyzer from Quantachrome Instruments. The samples were degassed for 12 hours at 473 K before each isotherm measurement. The micropore volumes were calculated from the application of the Dubinin-Radushkevitch equation to the N_2 adsorption at $P/P_0 < 0.1$ and the CO_2 adsorption at low relative pressure ($P/P_0 < 0.03$). This choice enables us to estimate the micropore volume (<2 nm) using N_2 adsorption and ultramicropore volume (<0.7 nm) using CO_2 adsorption.

Hydrogen adsorption isotherms under low pressure range up to 1 bar were also obtained on the Autosorb-1 analyzer. The measurement temperature ranges from 77 to 298 K. The samples were usually degassed for 10 hours at 473 K prior to the adsorption analysis. Hydrogen adsorption measurements at low pressure ranges provide good accuracy of capacities due to the high resolution of measurements of pressure and temperature. Hydrogen adsorption at high pressures was conducted on a PCT rig made by Suzuki-Shokan Co. Ltd, Japan. The sample was in-situ degassed at 473 K for 4 hours using the associated vacuum system and furnace. The blank test was conducted under the same conditions as the actual measurements to minimize the measurement error caused by thermal expansion effects and room temperature fluctuations.

3. Results and discussion

Figure 1 shows the N_2 adsorption isotherms of raw and air-treated ACFs. Table 1 lists textural characteristics of the raw and air-treated ACF samples characterized by N_2 and CO_2 adsorption.

Specific surface areas of heat treated samples are reduced after the air oxidation under different temperatures. The pore volumes of the heat treated samples remain nearly unchanged except ACF500. However, the micropore volumes decrease and mesopore volumes increase with rising oxidation temperature. It is evident that there is a pore size distribution shift with the heat treatment. Adjacent walls of the micropores may collapse during air oxidation and form wider pores, such as supermicropores and small mesopores. Together with the reduction of small micropores, there is also a widening of the existing micropores. It is seen that high oxidation temperatures have a detrimental effect on the enhancement of the surface area and micropore volume of the oxidized samples probably due to the destruction of pores under high temperatures.

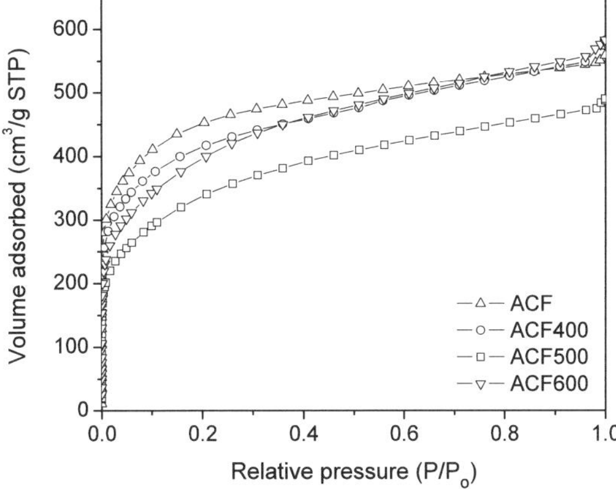

Figure 1. N_2 adsorption isotherms of raw and air-treated ACFs

TABLE 1. Textural properties of raw and air-treated ACFs

Sample	S_{BET} (m^2/g)	V_t (cm^3/g)	V_{mes} (cm^3/g)	$V_{N2\text{-}DR}$ (cm^3/g)	$V_{CO2\text{-}DR}$ (cm^3/g)
ACF	1,649	0.889	0.116	0.773	0.319
ACF400	1,506	0.903	0.255	0.648	0.220
ACF500	1,218	0.760	0.098	0.662	0.209
ACF600	1,418	0.905	0.302	0.603	0.224

It is well known that diffusion of nitrogen molecules into carbon micropores is very slow at 77 K. Moreover, it was pointed out that diffusional limitations at this temperature might influence adsorption in ultramicropores (pores smaller than 0.7 nm). It has been long time recognized that problems of this type can be eliminated by using CO_2 adsorption analysis at 273 K. The Dubinin-Radushkevich (DR) micropore volume from the CO_2 adsorption data at 273 K has been proven to be useful for the assessment of narrow microporosity [4] (width below 0.7 nm). The measured ultramicropore volumes of the air-treated samples undergo a significant reduction as shown by CO_2 DR micropore volume in Table 1.

Figure 2 shows that hydrogen adsorption isotherms at 77 K and high pressure are clearly of Type I isotherms. However, hydrogen adsorption at room temperature demonstrate nearly linear isotherm as shown in Fig. 3. The isotherm of the sample at 77 K and high pressure exhibits a maximum point. The presence in the isotherms of a maximum at high pressure and low temperature is a typical feature of excess adsorption isotherms under supercritical conditions.

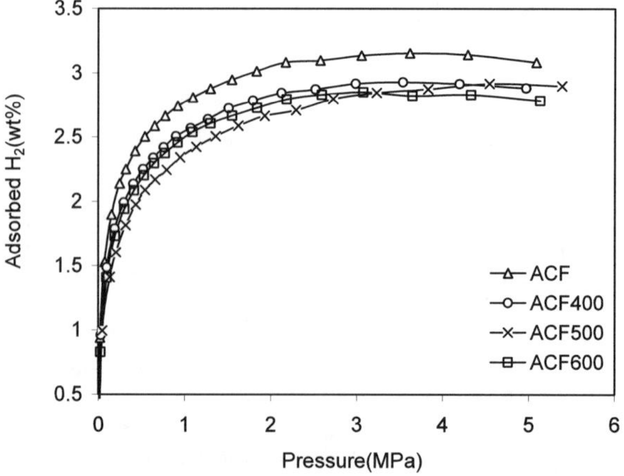

Figure 2. Hydrogen adsorption isotherms at 77 K on raw and air-treated ACFs

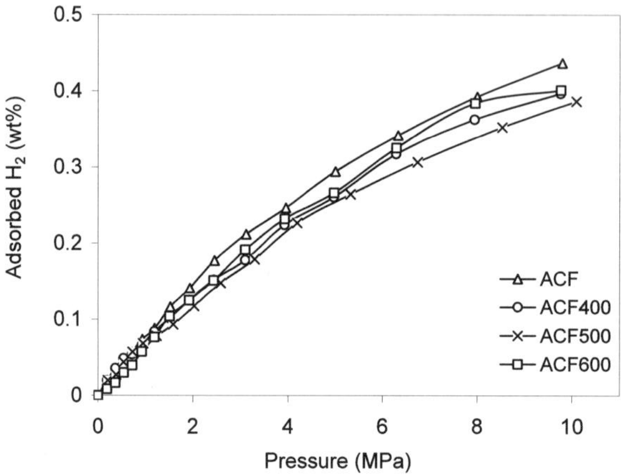

Figure 3. Hydrogen adsorption isotherms at 295 K on raw and air-treated ACFs

The appearance of maximum can be explained by the surface excess adsorption equation. Adsorption itself, according to Gibbs's description, is an excess quantity. The relationship between the measured excess amount of adsorption and the so-called absolute adsorption can be described by the Gibbs definition of adsorption

$$n = n_{abs} - \rho_g V_a = V_a(\rho_a - \rho_g), \tag{1}$$

where ρ_a is the density of the adsorbed phase, ρ_g is the density of the gas phase, and V_a is the volume of the adsorbed phase. Equation (1) clearly shows that n is a density excess amount, while n_{abs} is the total mass confined in the adsorbed phase and named as absolute adsorption.

Density of adsorbed hydrogen (ρ_a) increases with the gas pressure, so does the density of the compressed hydrogen gas (ρ_g). Density of adsorbed phase is a function of gas pressure and adsorption energy distribution of solid surface, while density of compressed gas is a function of only gas pressure. Increase of ρ_a is limited by the adsorption energy distribution of solid surface and up to a limit, but increase of ρ_g is continuous with the increase of the pressure. Therefore, the difference between the two phase density (ρ_a - ρ_g), becomes smaller and smaller with increasing adsorption pressure, until the appearance of isotherm maximum. At this maximum point, ρ_g equals to ρ_a. The excess adsorption capacity will decrease with the continuously increased pressure beyond the maximum point.

At 295 K, the H_2 adsorption capacity is approximately a linear function of the pressure which can be explained with Henry's law. The ACFs show the hydrogen adsorption capacities range from 0.386 to 0.437 wt% at 10 MPa. No saturation occurs in the investigated pressure below 10 MPa. At the room temperature, the interaction based on van der Waals force between H_2 and carbon is the same order as the thermal motion energy of H_2 molecule on the surface. In order to increase the H_2 storage capacity, one should operate at much lower temperature or under higher pressure.

The hydrogen adsorption capacities of ACFs at 77 and 298 K and at low and high pressures are listed in Table 2. Considering the raw and air-treated ACF samples exhibit the similar surface chemistry characteristics, it is reasonable to compare the effect of the pore structure changes on the hydrogen adsorption performance. The hydrogen adsorption capacity at ambient conditions is plotted as a function of the ultramicropore volume measured by CO_2 adsorption and micropore volume determined by N_2 adsorption and the linear dependence of hydrogen uptake on ultramicropore volume is observed for the ACF samples (Fig. 4(a)). However, the micropore volume does not correlate well with the hydrogen uptake of the tested samples. The narrow ultramicropores represent the higher energy adsorption sites that can adsorb gas molecules at high temperature and very low pressure.

TABLE 2. Hydrogen adsorption capacities on ACFs at low and high pressures

Sample	Adsorbed H_2 at low P (wt%)		Adsorbed H_2 at high P (wt%)	
	77 K	298 K	77 K	295 K
ACF	1.782	0.045	3.154	0.437
ACF400	1.659	0.037	2.928	0.397
ACF500	1.615	0.032	2.916	0.386
ACF600	1.677	0.035	2.850	0.401

The highest hydrogen capacity values for high pressure are taken from the isotherm maximum points.

Similarly, the ultramicropores still dominate the hydrogen adsorption at 77 K and atmospheric pressure (Fig. 4(b)).

Anson et al. investigated the hydrogen adsorption of CNTs and found that pores of about 0.45 nm adsorb H_2 at a minimum pressure of about 125 mmHg and 0.5 nm of about 250 mmHg at 298 K [5]. They suggested that the best material to adsorb hydrogen at 298 K and atmospheric pressure would have micropores that should be a bit larger than 0.5 nm of maximum width and as dense as possible. The supermicropores are less

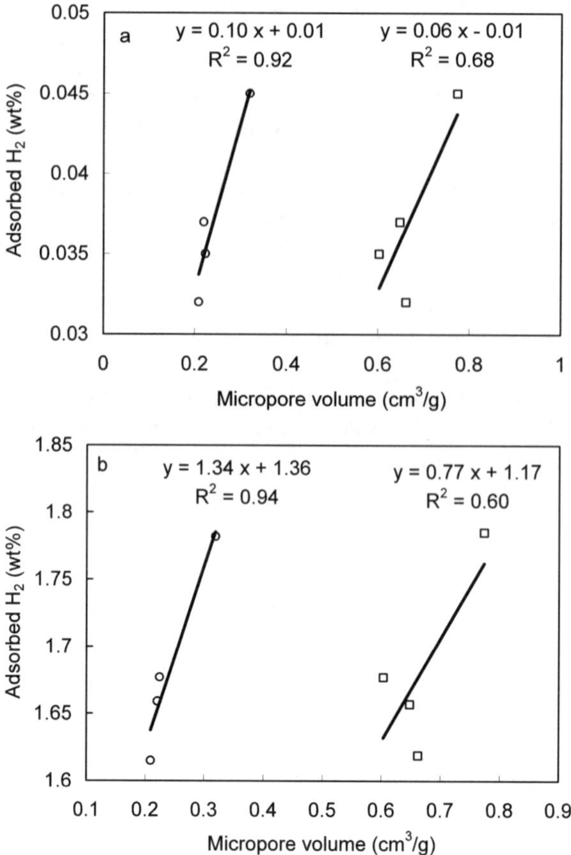

Figure 4. Correlations between $V_{N2\text{-}DR}$ and $V_{CO2\text{-}DR}$ with hydrogen uptakes at 298 K (a) and 77 K (b) at low pressure (○: $V_{CO2\text{-}DR}$; □: $V_{N2\text{-}DR}$)

energetic sites for adsorbing H_2 at 298 K and contribute less to the hydrogen adsorption. Nevertheless, the presence of wide micropores can compensate the lack of more energetic sites if the conditions are adequate. The increase of pressure will take advantage of the supermicropores and small mesopores thus help enhance the hydrogen adsorption.

Figure 5 shows H_2 adsorption capacities versus N_2 and CO_2 micropore volumes by DR method at 77 and 295 K, respectively. It is very interesting to note the difference of the relationship at these two temperatures. A good linear correlation between micropore volumes determined by CO_2 and hydrogen adsorption capacities at 295 K has been achieved. However, N_2 micropore volume does not correlate well with hydrogen uptake at 295 K. In the contrary, the micropore volumes measured by N_2 adsorption exhibits better correlation with hydrogen adsorption capacities at 77 K than the CO_2 micropore volumes although the latter also exhibit a reasonably good linear relationship with hydrogen uptake. Considering that V(DR-CO_2) measures ultramicropore volume

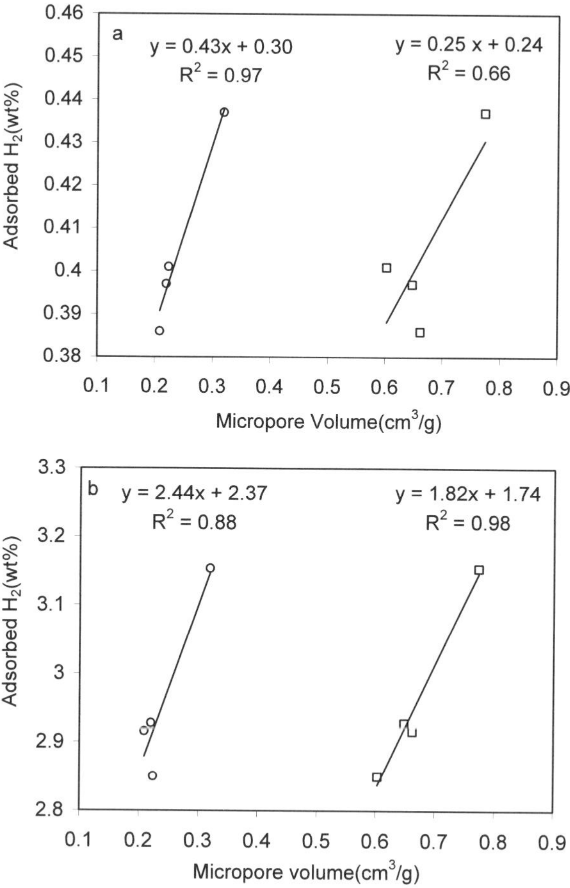

Figure 5. Correlations between $V_{N2\text{-}DR}$ and $V_{CO2\text{-}DR}$ with hydrogen uptake at 295 K (a) and 77 K (b) at high pressure (○: $V_{CO2\text{-}DR}$; □: $V_{N2\text{-}DR}$)

(<0.7 nm) and V(DR-N_2) micropore volume (<2 nm), it is clear that ultramicropores with the diameter lower than 0.7 nm dominate the hydrogen adsorption at room temperature. However, hydrogen molecules are not only trapped in the ultramicropores, but also adsorbed in the supermicropores at 77 K (0.7–2 nm). That is why hydrogen uptake shows a stronger dependence on micropore volume than the ultramicropore.

The ACF sample is a kind of microporous carbonaceous adsorbent exhibiting slit-shaped pores with centralized pore size distribution. Smaller pores are clearly more efficient in H_2 sorption due to a stronger interaction with the H_2 molecules. The total interaction between the adsorbate molecule and a solid is greater if the molecule can interact with a larger number of surface atoms, as happens in small curved pores or narrow slit pores. When a molecule is located in a slit pore, the sorbate-sorbent interaction potentials from both sides of the pore wall overlap to a degree determined by the pore width. The adsorption energy in smaller pores is much larger compared to larger mesopores or to the nonporous surface because of the overlapping of adsorption forces from the opposite walls of the micropores.

When the low and high pressure hydrogen adsorption at 77 and 295–298 K are compared (Figs. 4 and 5), it is found that at low pressure the ultramicropores contribute the most to the hydrogen uptake at both cryogenic and room temperatures. Nevertheless, at high pressure and 77 K, supermicropores also contribute to the hydrogen adsorption in spite of their low efficiency. The chemical potential of hydrogen will change with the varied temperature and pressure, thus optimum pore size which provides sufficient adsorption potential of adsorbents will vary accordingly. As a result, the dependence of hydrogen uptake on the micropore volume also changes accordingly.

4. Conclusions

Ultramicropores with the diameter lower than 0.7 nm dominate the hydrogen adsorption at low pressure. However, hydrogen molecules are not only trapped in the ultramicropores, but also adsorbed in the supermicropores at 77 K and high pressure. The narrow ultramicropores represent the higher energy adsorption sites that can adsorb gas molecules at high temperature and very low pressure. CO_2 adsorption is proven to be necessary to provide complementary characterization of narrowest pores.

References

1. Rzepka M., Lamp P., De la Casa-Lillo M.A., *J. Phys. Chem. B*, 1998. **102**: 10894–10898.
2. de la Casa-Lillo M.A., Darkrim F.L., Cazorla-Amoros D., Linares-Solano A., *J. Phys. Chem. B*, 2002. **106**: 10930–10934.
3. Zhao X.B., Xiao B.A., Fletcher J., Thomas K.M., *J. Phys. Chem. B*, 2005. **109**: 8880–8888.
4. Lozano-Castello D., Cazorla-Amoros D., Linares-Solano A., *Carbon*, 2004. **42**: 1233–1242.
5. Anson A., Callejas M.A., Benito A.M., Maser W.K., Izquierdo M.T., Rubio B., Jagiello J., Thommes M., Parra J.B., Martinez M.T., *Carbon*, 2004. **42**: 1243–1248.

ESR STUDY OF HYDROGEN SORPTION/DESORPTION KINETICS IN POLY(ORTHO-ANISIDINE) AND POA/SWNTs COMPOSITE FILMS

A.A. KONCHITS, S.P. KOLESNIK, V.S. YEFANOV, F.V. MOTSNYI
Institute of Semiconductor Physics, NAS of Ukraine, 45 Pr. Nauki, 03028 Kyiv, Ukraine

E. TAMBURRI, M.-L. TERRANOVA
Dipartimento di Scienze e Technologie Chimiche and INFM Universita di Roma "Tor Vergata", Via del politecnico 1, 00133 Roma, Italy

Abstract. The cyclic hydrogen storage properties of poly(ortho-anisidine) (POA) and POA/SWNTs composite films have been investigated using the ESR spectroscopy. The ESR signals with g ≈ 2.0033 were observed in both types of the initial films for the first time. The temperature dependencies of ESR parameters suggest the polaron origin of the paramagnetic states in these films. A substantial effect due to oxygen, stored within the films, has been observed in the ESR response of composite films as well.

A strong increase of the ESR signal has been detected for composite samples kept in hydrogen atmosphere at pressure of ~1.1 bar at room temperature. The less pronounced effect of hydrogen has been detected for the pure polymer films. The kinetics of the hydrogen adsorption – desorption processes in the samples was investigated and the results was interpreted in the frame of the hydrogen physisorption processes. The discussion of the H-induced effects in polymer and composite was carried out taking into account the morphology of the samples and the presence of paramagnetic defects.

Keywords: poly(ortho-anisidine), carbon nanotubes, electron spin resonance, polaron, hydrogen adsorption

1. Introduction

Hydrogen is one of the most promising alternative fuels because of its cleanness and high energy density. However, storage is the most important problem in its application. Therefore, hydrogen storage materials have attracted intense research interest last decade.

Hydrogen storage capacity of various carbon nanostructures, including single-walled nanotubes (SWNTs), multi-walled nanotubes (MWNTs) and other nano-carbon materials, has been widely investigated experimentally and theoretically [1]. Hydrogen storage properties of the number of conducting polymers have been investigated as well [2]. Recently suitability of composite materials obtained by embedding CNTs into conducting polymeric matrices for hydrogen storage applications has been demonstrated also [3–4].

Since the electronic properties of CNTs and CNTs-based composites depend on presence of both the delocalized (free carriers) and localized (defect states) paramagnetic centers (PC) [5–8], the electron spin resonance (ESR) spectroscopy is widely used techniques for characterization of such systems. The localized states originate usually from the breaking of some carbon bonds (paramagnetic defects with the factor of spectroscopic splitting in the range of g = 2.00–2.01) that influence the electronic properties directly or by means of the subsequent adsorption of gases (oxygen, hydrogen etc.) [6, 9–10]. ESR spectroscopy enables to directly "see" paramagnetic charge carrier species, the so-called polarons in conducting polymers [11–14] as well.

In this paper the ESR spectroscopy was used to study the origin of paramagnetism and hydrogen storage ability in conducting polymer poly(ortho-anisidine) (POA) and POA/SWNTs nanocomposite films. Effect of hydrogen sorption on the electron properties, the adsorption/desorption kinetics and role of defects were studied in these materials.

2. Experimental

POA and POA/SWNTs composite films (thickness d ~ 100 nm) were deposited on a platinum foil by means an oxidative potentiostatic electropolymerization of the monomer (ortho-anisidine) and adding of the purified SWNTs. To have POA and POA/SWNTs samples with the polymer in the emeraldine oxidation state, both the polymer and the composite films have been subjected to reduction treatments. The morphology of the films was examined ex situ by atomic force microscopy.

Magnetic resonance measurements were carried out in the temperature range of 10–300 K using X-band (microwave frequency $\nu \cong 9.4$ GHz) ESR spectrometer Radiopan X-2544 with 100 kHz modulation of magnetic field. In the hydrogenation experiments the samples were placed in a previously evacuated quartz ampoule connected to a chamber containing dry hydrogen gas (purity > 99.9%) at pressure of ~1.1 bar. In some experiments oxygen and helium gases (purity > 99.9%) were used also. The kinetics of adsorption/desorption processes was studied by performing in-situ ESR measurements.

3. Results

The typical ESR spectra (the derivative of imaginary part of the magnetic susceptibility vs. magnetic field) of POA film are shown in Fig. 1. At room temperature (RT) the ESR signal is characterized by g = 2.0032 ± 0.0002, the peak-to-peak linewidth $\Delta H_{pp} = 11 \pm 0.8$ G. and lineshape is very close to the Lorentz type. The concentration of PC, determined from the comparison with the reference sample, is $N_s \cong (2.1 \pm 0.3) \times 10^{19}$ cm^{-3}.

As temperature decreases the parameters of ESR spectrum changes significantly. The ESR line broadening is observed so that at T = 17 K the ESR linewidth $\Delta H_{pp} = 16 \pm 0.8$ G and lineshape is close to Gauss type.

The RT ESR spectrum obtained for the POA/SWNTs composite film in ambient air presented in Fig. 2 (curve 1) is characterized by g = 2.0033 ± 0.0002, linewidth $\Delta H_{pp} = 15 \pm 1$ G and almost Lorentz type lineshape (dotted line near curve 1).

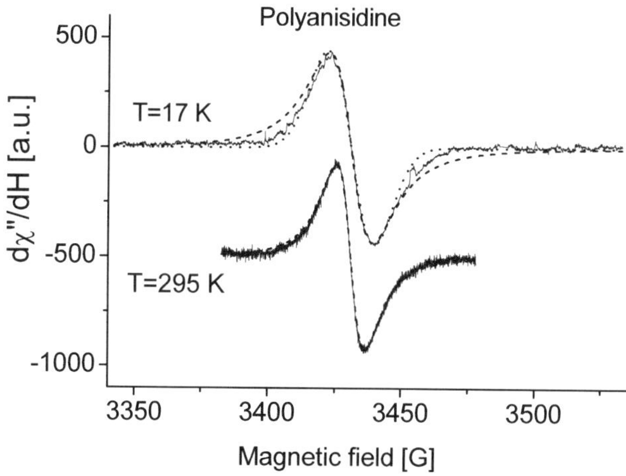

Figure 1. ESR spectra of the POA films. Note at T =17 K the ESR lineshape is intermediate between Gaussian and Lorentzian line shapes (upper dotted line and lower dashed one correspondingly), while at T = 295 K the lineshape is Lorentzian strictly

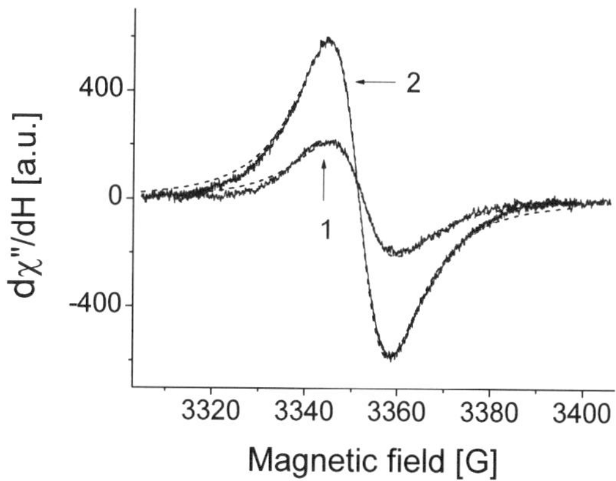

Figure 2. ESR spectra ($\nu \cong 9.4$ GHz, T = 295 K) for POA/SWNTs composite film in ambient air (curve 1) and in hydrogen atmosphere at a pressure of ~1.1 bar (curve 2). The dotted lines correspond to a Lorentzian line shape

The PC concentration in composite samples is $N_s \cong (1.8 \pm 0.3) \times 10^{19}$ cm^{-3}, that is only slightly less than in the polymer films.

The temperature dependences of the ESR parameters (integral intensity, g-factor, ESR linewidth) for both POA and composite POA/SWNT films are shown in Fig. 3.

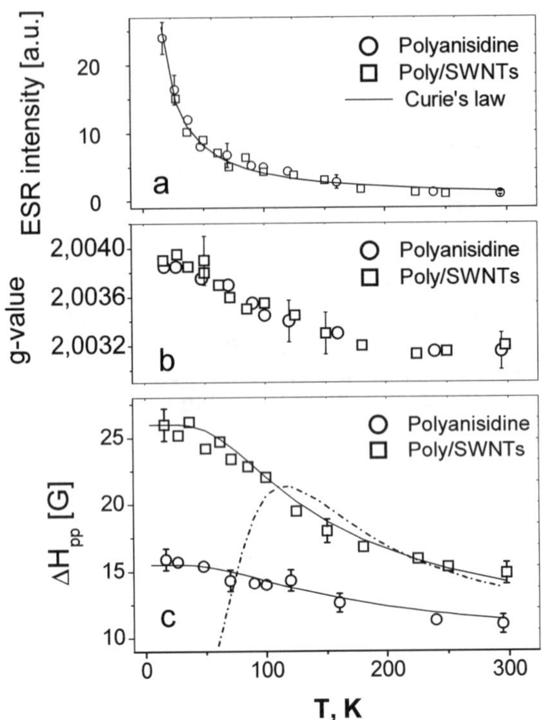

Figure 3. Temperature dependences of : (a) integral intensity I = $A(\Delta H_{pp})^2$, (b) g-value and (c) ESR linewidth ΔH_{pp} for the POA films (circles) and POA/SWNTs composites (squares) The solid line in Fig. 3(a) corresponds to the Curie's law: I $\propto$ C/T. The solid lines in Fig. 3(c) are fits to the experimental points obtained using Arrhenius type equations (see Eq. 2 in the text). The dashed line corresponds to the "exchange" mechanism [11] (see Eq. 1 in the text)

The ESR integral intensity (Fig. 3(a)) follows the Curie's law (I(T) = C/T where C is the Curie constant) for both type of the films within the experimental errors. At the same time, the ESR linewidth in the composite sample is appreciably greater than that in the polymer sample over all temperature range (Fig. 3(c)). Note also the slight gradual increase of the g-factor (Fig. 3(b)) in both types of films with temperature decreasing.

Curve 2 in Fig. 2 presents the room temperature ESR spectrum for the composite sample after placing it in a H_2 gaseous atmosphere at the pressure of ~1.1 bar for 72 h. It is clearly seen the ESR intensity increases considerably after hydrogenation. In addition, the value of g-factor at T = 300 K gradually increases up to g = 2.0038 ± 0.0002. At the same time it should be emphasized that the ESR spectra remain in air and in vacuum or in helium atmosphere are almost the same (curve 1 of Fig. 2).

Figure 4 shows the ESR measurements of hydrogen absorption-desorption kinetics in both polymer and composite samples at room temperature. The samples were preliminary degassed by heating at T ≅ 60°C in vacuum during 30 h. At the moment t = 0 the hydrogen gas was introduced (sign "on" in Fig. 4), after 45 h the chamber was opened on air (sign "off" in Fig. 4). It is clearly seen that the hydrogen effect in the composite samples is much more evident than in the polymer ones.

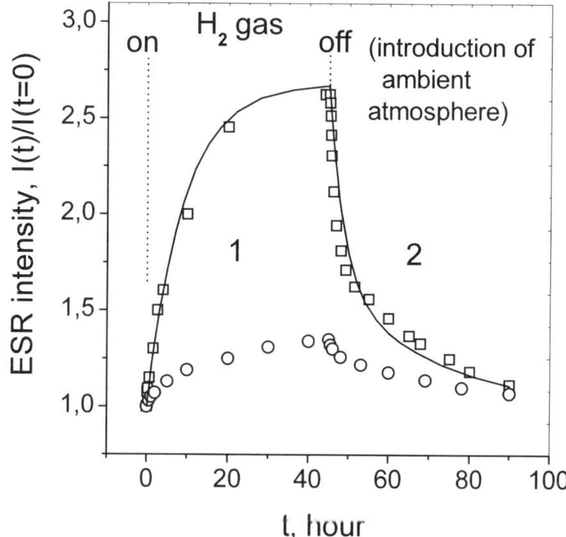

Figure 4. ESR intensity kinetics curves of POA films (circles) and POA/SWNTs composite films (squares) registered at T = 295 K. At t = 0 s hydrogen (~1.1 bar) has been introduced (sign "on"). Sign "off" corresponds to air inlet. Curves 1 and 2 correspond to Eqs. (3a) and (3b), respectively

4. Discussion

Let us discuss the origin of the paramagnetic states and factors which determine the observed ESR linewidth and lineshape and their temperature behavior.

Paramagnetic centers with g ≅ 2.002–2.004 which are typical for conducting polymers are commonly ascribed to polaron states, i.e. the paramagnetic charge carrier species [11–16]. We suppose that the paramagnetic centers in our POA films have the same origin. The similarity of the ESR parameters for the POA and the POA/SWNTs films (Fig. 3(a, b)) suggests that PC in composite films are mainly due to polaron states as well.

It is well known that in conducting polymers ESR linewidth and shape strongly vary depending on various factors, such as synthesis methods, doping and morphology of the polymer, PC concentrations, polaron mobility in individual polymer chains, chain aggregation, presence of oxygen and humidity [11–20]. In high conductive polymers with effective inter-chain interaction, the contribution of the two subsystems, delocalized and localized charge carriers, usually determines the presence of two overlapping ESR lines, a broad and a narrow one, respectively [14, 18, 21]. In our case the inter-chain interaction is not effective probably, because in both POA and POA/SWNTs films only the single ESR line with well definite shape is observed (Fig. 1) and the Pauli contribution to spin susceptibility from disordered metallic phase is negligible (Fig. 3(a)).

Another factor which may explain the ESR line broadening is related to the physical adsorption of molecular oxygen on the film surface or inside the film. In this case we have two interactive paramagnetic subsystems, namely A and B, with $S_A = 1/2$ (polaron spins) and $S_B = 1$ (oxygen molecule spins), respectively, where oxygen molecules can be considered as a fast-relaxing subsystem [11, 17, 19, 22]. According to [22] the interaction of polarons with oxygen paramagnetic molecules should lead to the ESR line broadening:

$$\delta(\Delta H_{pp}) = P_{ex}\delta(\Delta H_{pp,ex}) + P_d\delta(\Delta H_{pp,d}), \quad (1)$$

where P_{ex} and P_d represent the weight of exchange and dipole contributions, respectively; $(P_{ex} + P_d) = 1$.

The term $\delta(\Delta H_{pp,ex})$ describes the line broadening due to exchange interaction between polaron and oxygen molecules [11, 19]. The term $(\Delta H_{pp,d})$ in Eq. (1) origins from dipole-dipole interactions between spins S_A and S_B [22] and gives the contribution to the motional narrowing of the ESR line due to polaron hopping. If we take into account the super-hyperfine interactions (SHFI), the temperature dependence of the ESR linewidth ΔH_{pp} (T) can be described by an Arrhenius type equation [23–24]:

$$\Delta H_{pp}\,(T) = \Delta H_{pp}(0)/((1+b\cdot\exp(-E_A/kT)), \quad (2)$$

where $\Delta H_{pp}(0)$ is the inhomogeneous of the ESR linewidth at T = 0 K due to both dipole-dipole interactions and SHFI between PC and nuclear spins of hydrogen and nitrogen; E_A is the activation energy of the polaron hopping and b is a fitting parameter.

The solid lines of Fig. 3(c) have been obtained according to Eq. (2) by assuming the following set of values: $\Delta H_{pp}(0) = 15$ G, b = 0.8, $E_A = 20$ meV for the pure polymer film, and $\Delta H_{pp}(0) = 26$ G, b = 1.8, $E_A = 20$ meV for the composite film. The dashed line in the Fig. 3(c) represents the contribution of the first term only in Eq. (1) at the same value of E_A. A confirmation to the fact that oxygen stored in the films influences the signal broadening comes from the observed sharp decrease of the ESR linewidth (e.g. from 13 to 2.9 G at T = 300 K) in some of our composite samples after few cycles of pumping. On the other hand, the gradual modification of the ESR spectra from a Gaussian to a Lorentzian line-shape with temperature increasing (Figs. 1 and 2) agrees with the activation of a motional narrowing mechanism.

As a result, the temperature dependence of the ESR linewidth may be explained by the polaron hopping which average effectively both dipole-dipole and SHF interactions.

It should be noted that carbon dangling bond defects with similar values of g-factor (2.002–2.004) and ESR linewidth ($\Delta H_{pp} = 2$–30 G) were found recently in the series of chemically modified CNTs [5, 25]. The similarity among the g-values observed for the polaron paramagnetic subsystem in both the POA and POA/SWNTs films and for the carbon dangling bond defects, together with the small shift of g with T founded in our samples (see Fig. 2(b)) makes it difficult to discriminate the signals due to defect-related subsystem from the total ESR signal. Thus we are not allowed to exclude the presence of some carbon dangling bonds in our composite films.

The most interesting effect observed in this study is the influence of hydrogen on the paramagnetic properties of both POA and POA/SWNTs films. The strong increase of ESR intensity in the composites exposed to hydrogen (Fig. 2) evidences the effective hydrogen adsorption in the samples. The crucial question to consider is how the non-paramagnetic H_2 molecules can enhance the paramagnetic properties of the systems under consideration? This point may be rationalized by taking into account a hydrogen spillover phenomenon [26–28]. This effect consists in the dissociation of hydrogen molecules on a metal supporting catalyst (Pt in our case) and the subsequent spillover as atomic hydrogen towards carbon surfaces. Such hydrogen atoms furthermore may be stored in the host material by physisorption and chemisorption processes.

Considering our adsorption – desorption measurements, one can see from Fig. 4 that the processes are approximately exponential and may be described as follows:

$$I(t)/I(t{=}0) = A - B\exp(-t/\tau), \qquad \text{(adsorption)} \tag{3a}$$

$$I(t)/I(t{=}0)=0.6\,(B\cdot\exp(-t/\tau_1)+1)+0.4\,(B\cdot\exp(-t/\tau_2)+1), \quad \text{(desorption)} \tag{3b}$$

where $A = 2.68$, $B = 1.68$, $\tau = 9$, $\tau_1 = 3.5$ h and $\tau_2 = 25$ h are fitting parameters.

It is seen that the time constant of the signal rise, τ, is larger by a factor of ~2.5 with respect to the small time constant τ_1 of the signal droop. One can therefore suppose that the detected discrepancy could be due to hydrogen spillover processes, which are not significant in the desorption process. On the other hand, the large time constant τ_2 in desorption process may be caused by differences in the initial conditions (vacuum and air for adsorption and desorption processes, respectively). Since the chemisorption processes are thermally stable up to 670 K [29], we can conclude that the physisorption processes dominate in our POA/SWNTs films.

Finally, let us discuss the origin of the extra paramagnetic centers observed in the samples exposed to hydrogen. In a number of experiments [6, 30–31] the influence of hydrogen on the ESR response for SWNTs and multi-walled CNTs samples has been studied. With the exception of Ref. [31], for SWNTs samples after hydrogen treatment of in all cases a decreasing of ESR signals has been reported. Subject to this, the authors [6, 30–31] ascribed the PC to the defects, and argued that the role of hydrogen was to passivate of such defects. In our experiments we detected an increase of the ESR signals for both POA and POA/SWNTs samples (see Figs. 2, 4). The results obtained indicate that hydrogen atoms are the source of extra PC, while the role of defects is incidental.

Hydrogen in POA samples may be embedded either in polymeric chains or in the interchains space. In the case of POA/SWNTs samples hydrogen may be stored onto nanotube surfaces and inside the intratube space also. If one takes into account the similarity of the ESR parameters of the untreated and hydrogen-treated composite film one could suppose that the origin of the H-induced PC is may be due to the polaron states in polymeric chains also. Such interpretation would indicate that in the presence of hydrogen more polaron states are formed in the polymer chain, in other words the occurrence of oxidation reactions (from leucoemeraldine to emeraldine state) or reduction reactions (from pernigraniline to emeraldine state) of the polymer. Anyway these are redox chemical reactions not compatible with a physisorption process.

The other possibility for the presence of the extra PC would be hydrogen physisorption within polymer nanopores as well as intratube space and on nanotube surfaces. In this case the stronger effect produced by hydrogen adsorption in composite films (see Fig. 4) could be explained by the different microstructure of these films compared with that of the only polymer films. In fact the strong effects of the material porosity on the efficiency of the hydrogen physisorption process have been recently established [32–34]. According with this explanation, the role of defects, acting for pinning of hydrogen atoms, may therefore turn out determinant [35]. However, the small differences between the PC values do not allow at this moment to draw more definitive conclusions regarding the modification of PC structure under hydrogen treatments.

5. Conclusions

The ESR signals for POA and POA/SWNTs composite films prepared by electro polymerization on Pt electrodes and with the polymer in a controlled oxidation state (emeraldine base) were registered for the first time. The basic ESR line with $g \approx 2.0032$ originates from polaron states in polymeric chains. Linewidth and lineshape depend on polaron mobility, presence of oxygen and SHF interaction of electron spins with nuclear moments of hydrogen and nitrogen. As a result the inhomogeneous broadening of ESR line takes place at low temperatures. As temperature increases the linewidth decreases gradually and the lineshape changes from Gauss to Lorenz one due to the motional narrowing effect.

The strong increase of the ESR intensity was found in composite samples exposed to hydrogen gas, indicating on the effective hydrogen adsorption in the films. It is found that H-induced extra paramagnetic centers have characteristic very close to the signals detected in the unhydrogenated samples. The kinetics of adsorption – desorption processes is explained taking into account the domination of hydrogen physisorption processes. The effect of hydrogen is less evident in pure polymer samples. Even it is difficult to conclude more precisely about the PC structure modification under hydrogen treatment, the hypothesis is that the reduced effect in pure POA layers is due to the more compact structure of such samples with respect to the POA/SWNTs nanocomposites.

The results of the present experiments indicate that the ESR technique may be the useful tool for study of adsorption–desorption processes of hydrogen in various materials and for the setting up of innovative strategies for hydrogen storage in nanocomposite systems.

Acknowledgment

This research has been partially supported by the Italian MUR through the PON Project No13569.

References

1. Han S.S., Kim H.S., Han K.S., Lee J.Y., Lee H.M., Kang J.K., Woo S.I., van Duin A.C.T., Goddard W.A. III, *Appl. Phys. Lett.*, 2005, **87**: 213113-1–213113-3.
2. Jurczyka M.U., Kumara A., Srinivasanb S., Stefanakos E., *Int. J. Hydrogen Energ.*, 2007, **32**: 1010–1015.
3. Banerjee S., Murad S., Puri I.K., *P. IEEE*, 2006, **94** (10): 1806–1814.
4. MacDiarmid A.G., Heben M.J., Venancio E.C., Manohar S.K., Dillon A.C., Gilbert K.E.H., Conducting Polymers as Potential New Materials for Hydrogen Storage (*IPHE International Hydrogen Storage Technology Conference: June 2005: Lucca, Italy*).
5. Wang C., Guo Z.X., Fu S., Wu W., Zhu D., *Prog. Polym. Sci.*, 2004, **29**: 1079.
6. Shen K., Tierney D.L., Pietraß T., *Phys. Rev.*, 2003, **B 68**: 165418.
7. Goze-Bac C., Latil S., Lauginie P., Jourdain V., Conard J., Duclaux L., Rubio A., Bernier P., *Carbon*, 2002, **40**: 1825.
8. Huang H., *JEOL News*, 2003, **38** (2): 70–72.
9. Dag S., Gülseren O., Yildirim T., Ciraci S., *Phys. Rev.*, 2003, **B 67**: 165424.
10. Lim S.C., Jo C.S., Jeong H.J., Shin Y.M., Lee Y.H., Samayoa I.A., Choi J., *J. Appl. Phys.*, 2002, **41** (9): 5635.
11. Houze E., Nechtschein M., *Phys. Rev.*, 1996, **B 53** (21): 14309.
12. Kahol P.K., Pinto N.J., *Synth. Met.*, 2004, **140**: 269–272.
13. Graeff C.F.O., Brunello C.A., *Synth. Met.*, 2001, **119**: 327–328.
14. Sitaram V., Sharma A., Bhat S.V., Mizoguchi K., Menon R., *Phys. Rev.*, 2005, **B 72** (3): 035209.1–035209.7.
15. van Haare J.A.E., Havinga E.E., van Dongen J.L.J., Janssen R.A.J., *J. Cornil, J. Bredas, Chem. Eur. J.*, 1998, **4** (8): 1509–1522.
16. Paasch G., Scheinert S., Petr A., Dunsch L., *Russ. J. Electrochem.*, 2006, **42** (11): 1161–1168.
17. Krinichnyi I., Tokarev S.V., Roth H.K., Schrodner M., Wessling B., *Synth. Met.*, 2005, **152** (1–3): 165–168.
18. Kon'kin A.L., Shtyrlin V.G., Garipov R.R., Aganov A.V., Zakharov A.V., Krinichnyi V.I., Adams P.N., Monkman A.P., *Phys. Rev.*, 2002, **B 66**: 075203–0752011.
19. Houzé E., Nechtschein M., Pron A., *Phys. Rev.,* 1997, **B 56**: 12263–12267.
20. Han M.G., Byun S.W., Im S.S., *Polym. Adv. Technol.*, 2002, **13**: 320–328.
21. Kulikov A.V., Komissarova A.S., Ryabenko A.G., Fokeeva L.S., Shunina I.G., Belonogova O.V., *Russ. Chem. B.*, 2005, **54** (12): 2794–2804.
22. Kanemoto K., Yamauchi J., *J. Phys. Chem.*, 2001, **B 105**: 2117–2121.
23. Chipara M., Zaleski J.M., Hui D., Du C., Pan N., *J. Polym. Sci. Pol. Phys.*, 2005, **43**: 3406–3412.
24. Druz B., Zaritskiy I., Yevtukhov Y., Konchits A., Valakh M., Shanina B., Kolesnik S., Yanchuk I., Gromovoy Yu., *Diamond Relat. Mater.*, 2004, **13**: 1592.
25. Niyogi S., Hamon M.A., Hu H., Zhao B., Bhowmik P., Sen R., Itkis M.E., Haddon R.C., *Acc. Chem. Res.*, 2002, **35** (12): 1105.
26. Lueking A.D., Yang R.T., *Appl. Catal. A-Gen.*, 2004, **265**: 259–268.
27. Zacharia R., Kim K.Y., Fazle Kibria A., Nahm K.S., *Chem. Phys. Lett.*, 2005, **412** (4–6): 369–375.
28. Li Y., Yang R.T., *Am. J., Chem. Soc.*, 2006, **128** (3): 726–727.
29. Pekker S., Salvetat J.P., Jakab E., Bonard J.M., Forro´L., *J. Phys. Chem.*, 2001, **B 105**: 7938–7943.
30. Clewett C.F.M., Kombarakkaran J., Pietraß T., *Phys. Stat. Sol.*, 2006 **(b) 243** (13): 3242–3245.
31. Musso S., Porro S., Rovere M., Tagliaferro A., Laurenti E., Mann M., Teo K.B.K., Milne W.I., *Diamond Relat. Mater.*, 2006, **15**: 1085–1089.

32. Han S.S., Kim H.S., Han K.S., Lee J.Y., Lee H.M., Kang J.K., Woo S.I., van Duin A.C.T., Goddard III. W.A., *Appl. Phys. Lett.*, 2005, **87**: 213113-1–213113-3.
33. Panella B., Hirscher M., Roth S. Hydrogen adsorption in different carbon nanostructures, *Carbon*, 2005, **43** (10): 2209–2214.
34. Terrés E., Panella B., Hayashi T., Kim Y.A., Endo M., Dominguez J.M., Hirscher M., Terrones H., Terrones M. Hydrogen storage in spherical nanoporous carbons, *Chem. Phys. Lett.*, 2005, **403**: 363–366.
35. Yoo E., Habe T., Nakamura J. Possibilities of atomic hydrogen storage by carbon nanotubes or graphite materials, *Science & Technol. Adv. Materials*, 2005, **6** (6): 615–619.

PRODUCTION OF NICKEL-CARBON FIBRES WITH THE INCREASED CONTENT OF HIGHLY DISPERSIVE METAL AND NANOSIZED STRUCTURALLY-ORDERED CARBON

A.M. SAFONOVA, O.V. BEZHOK, L.E. SHPILEVSKAYA
Institute of general and inorganic chemistry of National Academy of Sciences of Belarus, 220072 Minsk, Belarus

Abstract. Nickel-carbon fibres with the high content of nanosized metal and structurally-ordered carbon are obtained and their physic-chemical properties are studied. It is shown, that the increased content of metal in carbon fibres improves their milling and wettability by polymer binders. The properties of composite materials on the basis of these fibres are studied and the efficiency of their application as antifriction materials and as conductive elements of different purpose are shown.

Keywords: metal-carbon fibers, carbonisation, structurally-ordered carbon, highly dispersive metal, X-ray diffraction, composites

1. Introduction

The method of obtaining of earlier developed metal-carbon fibres (Me-CF) which include the thermal treatment in an inert atmosphere of various metal-containing polymeric precursors [1–3], has allowed to produce the wide assortment Me-CF, containing highly dispersive (3–20 nm) particles of metals or their alloys.

It has been established that metals of iron subgroup at temperature of heat treatment (THT) 800–900°C catalyze the formation of the phase of nanosized structurally-ordered carbon. It exerts the essential influence on the physical and chemical properties of obtained Me-CF and opens up the new areas of their use. Occurrence of structurally-ordered carbon phases is accompanied by decrease of Me-CF durability. The occurrence of structurally-ordered carbon phases is accompanied by decrease of Me-CF strength. This factor is negative in the case of production of Me-CF, intended for use as independent products (fabrics, strings, etc.) since does not allow to introduce into the Me-CF a significant amount of metals without reduction of their strength. At the same time when Me-CF are applied as the fillers for obtaining of functional composites, the decrease of strength are even the positive factor because that facilitates the process of milling. Consequently, Me-CF with the high content of metals and of structurally-ordered carbon may be successfully used as the fillers of polymeric matrixes. In this connection the process of Ni-CF production with the high content of metal has been investigated at the given work.

2. Experimental

Ni-CF samples were obtained as follows: hydrated cellulose (HC) in the form of viscose silk was impregnated in a water solution of nickel chloride, then wrung out for removal of surplus of the solution, dried on air and subjected to the carbonization to 900°C in an inert atmosphere on the certain temperature-time mode. The conditions of carrying out of chemical and X-ray diffraction analyses and also the methods of determination of breaking strength and adsorption pores volume are resulted in work [2].

The chosen conditions of Ni-CF production allowed to increase considerably (in 1.7 times) the contents of highly dispersive nickel in Ni-CF structure in comparison with Ni-CF* samples which have been obtained earlier [2] (see the Table 1).

On the data of X-raying it is established that the reducing of nickel chloride up to free metal at the process of carbonization of system HC-$NiCl_2$ occurs at enough low temperature (360°C). On X-rayograms of these samples there are small broad reflexes of metal nickel ($2\theta \cong 44°30'$, 52°, 76°40′, 93°20′), that testify to the high dispersity of the phase of reduced nickel (the size of particles is 3–4 nm). With THT rise the size of nickel particles increases and makes 20 nm at 900°C. On X-rayograms of Ni-CF samples with THT 600°C the broad reflex at $2\theta \cong 24°$ with $d_{002} = 0.37$ nm is fixed, that is typical for turbostrate carbon. The reflex intensity rises with temperature growth. At temperature 760°C on the background of this reflex there is a reflex of structurally-ordered carbon at $2\theta \cong 26°10'$ with $d_{002} = 0.337$ nm. Its intensity grows with THT increase. The formation of a new phase of structurally-ordered carbon is connected with catalytic influence of nickel on the process of structurization of carbon. The increase of the content of highly dispersive nickel results in increase of the content of structurally-ordered carbon in structure of the Ni-CF (see the table). It is necessary to note, that the process of catalytic structurization of carbon is accompanied by the formation of enough developed porosity that causes the essential decrease of Ni-CF strength (see the table).

Before introduction into the polymeric matrixes, the researched Ni-CF samples (and also CF without nickel) were ground in vibrational mill of mark DDR-GM9458. The length of pieces of milled fibres was measured with the help of a microscope MIN-8. The distribution of the obtained pieces of fibres depenging on their length is resulted in Fig. 1 From the resulted data one can see that the dispersion of milled Ni-CF with the increased content of nickel has narrow distribution on length of the pieces of fibres and basically consists of particles with length 40–60 μm. At the same time the dispersion of milled CF without nickel is characterized by the wide distribution on length of pieces of fibres. Thus a maximum of distribution is shift in the side of more long fibres.

The estimation of a contact angle of Me-CF wettability by different polymer binders has shown that the increase of the metal content in Me-CF structure results to essential (more than in 1.5 times) improvement of Me-CF wettability compared to initial CF. It is connected with the increase of Me-CF specific surface and with the change of it functional composition. The improvement of Me-CF wettability promotes the uniform distribution of fibre dispersions in binder. It allows at identical filling with unmodified CF and with Me-CF to use Me-CF filler for more effective decrease of the electrical resistance, shrinkage and the coefficient of linear thermal expansion of composites.

On the basis of various polymeric matrixes and researched Ni-CF a number of solid and liquid compositions of various purpose has been obtained. So, on the basis of aromatic polyamide phenilon C and Ni-CF the carbonic plastic with good tribological characteristics at high loadings and temperatures [3] has been obtained. It is shown that the friction coefficient of such plastic is reduced in 3 times, and wear-resistance increases more than 12 times in comparison with the initial polymer. It is caused by the presence of highly dispersive metal and of nanosized ordered carbon at Me-CF structure. The combination of these elements, as it is known, gives the high antiwear effect for composite materials which have been obtained on their basis. At friction of such composites the microdefects on the surfaces of friction "are healed" and the steady film is formed. It lowers a level of contact strains and prevents the phenomena of the "sticking up" and microsetting of the friction surfaces.

The introduction of Me-CF into various polymeric film-forming systems (alkyd, acrylate, nitrocellulose, etc.) has allowed to form liquid compositions for obtaining current-conductive paint and varnish coverings on the surface of metal, plastic, concrete, wood, etc. Such coverings may be used as effective screens for protection against electromagnetic radiation (EMR) with coefficient of EMR weakness 20–40 dB in a range of frequencies of 0.15 MHz–2 GHz, and also as current-conductive elements of various purpose, including the materials for removal of the static electricity.

It is necessary to note that the great volume of sorption pores (0.11–0.23 $cm^3 \cdot g^{-1}$), the high content of nanosized nickel and structurally-ordered carbon in obtained materials cause the potential possibilities of their application in various adsorption-catalytic processes of hydrogen adsorption.

TABLE 1. Physic-chemical properties of Ni-CF

Sample	Ni content (mass%)	Strength (MPa)	Volume of sorbtion pores (cm^3/g)	I (Arb. Unit)
Ni-CF*	22.0	200	0.16	7
Ni-CF	37.6	30	0.23	12

I – X-ray intensity of (002) reflex of structurally-ordered carbon phase in arbitrary units.

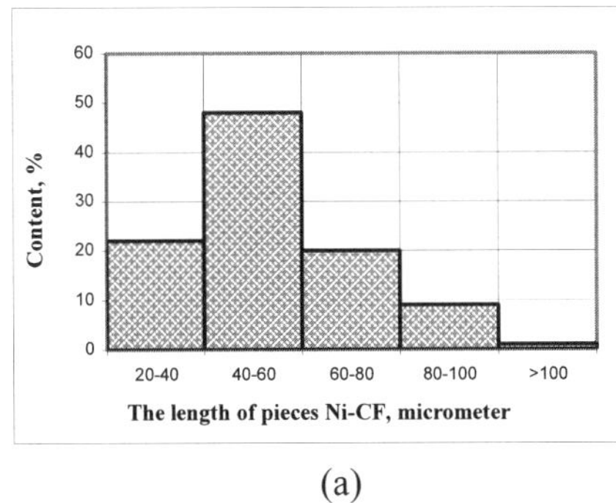

(a)

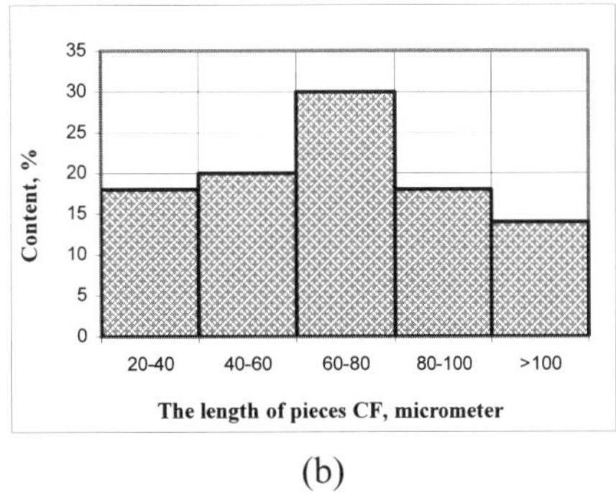

(b)

Figure 1. The distribution of the milled pieces of Ni-CF (a) and CF (b) depending on their length

3. Conclusions

Nickel-carbon fibrous materials with the high content of nanosized metal and structurally-ordered carbon have been received and investigated at this work. It is shown the perspectivity of their use as active fillers of polymeric matrixes for obtaining of composite materials with special properties.

References

1. Safonova, A.M., Lukomskaja, O.A., Ermolenko, I.N., et al. The investigation of thermal conversions of the salt forms of copolymer of cellulose, polyacrylonitrile and polyacrylic acid, *Khimia drevesiny*, 1989 (5): 53–57 (in Russian).
2. Safonova, A.M., Ermolenko, I.N., Apanasenok, V.I., et al. The formation of the structure and properties of nickel-carbon fibres, *J. Pricl. Khim.*, 1991 (11): 2447–2451. (in Russian)
3. Safonova, A.M., Shpilevskaja, L.E. Byrja, A.I. The formation of highly dispersive metals and structurally-ordered carbon in carbon fibres, Minsk, Belarus, *The collection of proceedings of carbon nanostructures*, 2006, P. 342–348.

CREATION OF NEW COMPOSITE MATERIALS FOR HYDROGEN ENERGY PURPOSES. I. NEW LINES OF MEMBRANE PRODUCTION TECHNOLOGY

O.K. ALEXEEVA, M.M. CHELYAK, A.A. KOTENKO, B.L. SHAPIR
Russian Research Centre "Kurchatov Institute", Hydrogen Energy & Plasma Technology Institute, Kurchatov sq. 1, Moscow, 123182 Russia
YU.S. NECHAEV
Kurdjumov Institute of Metals Science and Physics, Bardin Institute for Ferrous Metallurgy, Vtoraya Baumanskaya Str., 9/23, Moscow 105005, Russia

Abstract. One of the main problems of hydrogen energy is separation and purification of hydrogen produced by various conversion methods from raw hydrocarbons. Carbon membranes can become real alternative both to inorganic and polymeric ones and enlarge possibilities of membrane methods of gas separation. Investigations aimed at the creation of composite carbon membranes suitable for operation at high temperatures in the presence of organic vapors, in acid or alkaline (non-oxidizing) medium are presented. Our approach is based on the carbonization of polymeric layers on the porous inorganic high temperature supports with definite structure which opens up new possibilities for carbon nanostructure production. Main trends of future developments of composite carbon gas separation and filtration membranes are also discussed.

Keywords: hydrogen energy, carbon, nanostructures, membranes, composite materials

1. Introduction

Investigations and development of new functional materials for hydrogen energy goal have been carried out at Hydrogen Energy and Plasma Technology Institute RRC "KI" for many years. Lately the most active researches are those for creation multilayer composite materials (materials with functional coatings) necessary for hydrogen production, evolving, separation and purification. Universal approach to production of functional composite materials has been developed. Materials consisting of chemo- and thermostable supports coated with one or several catalytic and/or selective surface layers have been synthesized. Special attention here is focused on the possibility to produce surface functional layers with certain micro (nano) structure. Depending on the specific targets the developed technology involves different methods of coating formation, both physical and chemical ones. Methods of vacuum deposition (ion magnetron sputtering and electron beam evaporation) are used in combination with fluorination in the gas phase. The latter has been developed partly to enhance catalytic activity of the materials [1]. Development of these works resulted in the creation of tubular inorganic catalytic membranes and catalytic membrane reactors for hydrogen production from hydrocarbons and hydrogen sulfide [2]. Oxide, metallic, intermetallic surface layers can be obtained including various combinations. Thus, nickel was used as intermediate layer in multilayer composite membranes with surface Pd layer (these membranes are alternative to dense Pd membranes) [3]. Method of carbon functional layer preparation by carbonization of polymers on chemo- and thermostable supports has been developed recently [4].

Here we present investigations and main trends of future developments of composite carbon gas separation and filtration membranes.

2. Results and discussion

Rapid development of hydrogen energy in recent years resulted in the marked progress in the field of membrane technology as one of the main problems is separation and purification of hydrogen produced by various conversion methods from raw hydrocarbons. However existing polymeric and inorganic membranes have limitations: e.g. polymeric membranes can not be used at high temperatures, high permeability of these membranes corresponds to low selectivity and vice versa; permselectivity of porous ceramic membranes is insufficient etc., thus creation of novel membrane materials is necessary. Membranes based on these materials should provide better selectivity in combination with high permeability, thermal and chemical stability.

Carbon membranes for gas separation can become real alternative both to inorganic and polymeric membranes and enlarge possibilities of membrane methods of gas separation. In some processes selectivity of such membranes is close to that of polymeric ones, or even better in a number of cases, and permeability, thermal and chemical stability are comparable with corresponding characteristics of the best inorganic membranes.

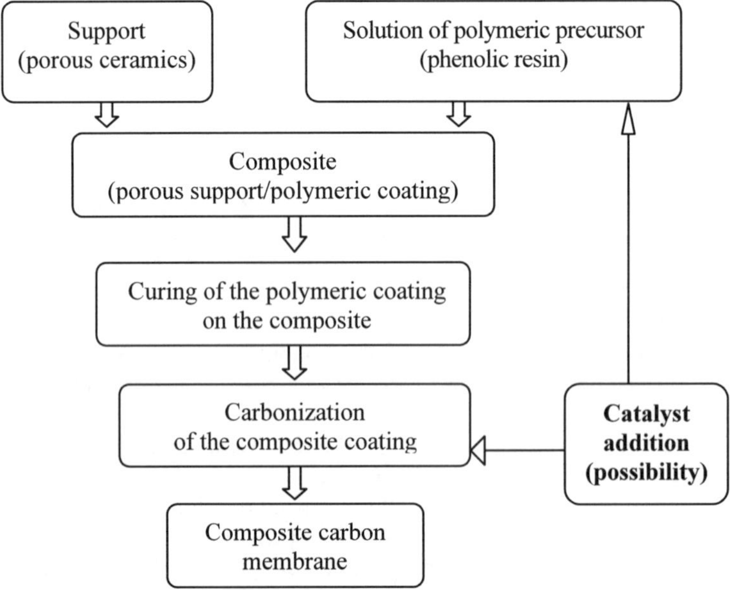

Figure1. The scheme of composite membrane fabrication

Depending on the type of precursors (polymers) and the carbonization conditions selective high-temperature membranes characterized by the different gas separation mechanisms can be produced: molecular sieve membranes; membranes with surface selective flow; combined ones (membranes with shape-selected transport) [5–10]. However until now there are no reliable methods of production of membranes with specified mechanical strength and high functional performance. Moreover, attention is not focused on carbon structural peculiarities though they may be of significant importance as discussed below.

Our investigations are aimed at the creation of composite carbon membranes suitable for operation at high temperatures in the presence of organic vapors, in acid or alkaline (non-oxidizing) medium. Composite membranes with carbon selective layer are more durable than brittle and fragile carbon membranes. Our previous results concerning the influence of type and concentration of precursor solutions, pretreatment and carbonization conditions on the characteristics of the selective layer are presented in [4, 11–14]. The gas phase method of porous alumina membranes modification has been developed in our early experiments. This method consists of the preliminary treatment by volatile fluorides (to add catalytic activity) followed by hydrocarbon pyrolysis. Such treatment allows to obtain disperse distribution of carbon in ceramic structure and impart electric conductivity. Later we began investigations aimed at creation of thermo- and chemical stable carbon selective membranes by means of carbonization of polymers coated on porous ceramic supports. The following polymers have been used: membrane polymers seragel (polysulfone – polybutadiene block-copolymer), ipplone (rodlike chain heteroarylene polynaphtoelenimidobenzimidasol), polysulfone, polyestersulfon and also ED-20 epoxy-based compounds and/or fluoroelastomers with functional groups cured by amines and later polyamidoimides, and some types of phenol-formaldehyde resins.

Based on these data porous alumina ceramic tubes with the outer diameter of 8–10 and 14–16 mm and the length of 30–60 mm have been selected as the most appropriate ones due to their stability, strength, commercial availability and some structure characteristics. Supports with additional intermediate layers have been also investigated. Tubular membrane configuration is preferable for various industrial applications. Phenol-formaldehyde resins PFR (resol and novolac types) in our opinion are the optimal polymeric precursors (especially because they are promising for carbon nanostructure production) and we use them as main precursors for membrane production in our works.

Carbon membranes were produced according to the following scheme (Fig. 1). Ceramic tubes were coated with polymers from solutions in organic solvents via multiple dipping (with and without pressure differential) with intermediate drying at room temperature. After preliminary curing carbonization was conducted in wide temperatures range in the presence of flowing nitrogen (for more details see [15]). The possibility to control carbon structure (controllable carbonization) results from use of certain supports, necessary surface modification of these supports, choice of polymeric precursors and carbonization conditions.

Structure and morphology of the initial supports and deposited coatings have been investigated by means of scanning electron microscopy (S-570 Hitachi), X-ray diffraction measurements using Cu K_α radiation. Gas permeation properties of membranes were tested by nitrogen, helium, oxygen by volumetric method using a soap-film rotameter or GSK-4 gasometer (Russia).

Method of polymer carbonization can be applied to flat and tubular supports of various shapes and sizes. Some of the tubular supports of different diameters used in our experiments are presented in the Fig. 2(a). Ceramic alumina tubes (mean pore size of 0.2 μm) with the outer diameter of 8 mm (initial and with PFR – layer after carbonization) and ceramic macroporous (mean pore size of 5 μm) alumina tubes with the outer diameter of 16 mm with cured phenol resin layer and the same after carbonization are shown in the Fig. 2(b).

(a)

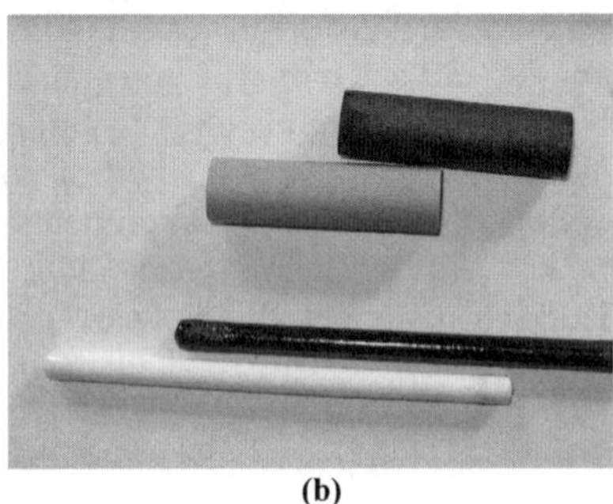

(b)

Figure 2. (a) Some of ceramic alumina supports investigated in our experiments; (b) Supports mainly used in our works: α-Al_2O_3-based ceramic tubes with the outer diameter of 8 mm (initial and with PFR carbonized coating) and macroporous α-Al_2O_3-based ceramic tubes with the outer diameter of 14 mm (with cured PFR coating and part of the same one after carbonization)

Analysis of the XRD data have shown that carbonization of the same precursor under the same conditions depends on the support. It can be seen even at the first stage of carbonization (heating up to 923 K with soaking during 1 h). X-ray patterns of several samples are presented in the Fig. 3 (along with the pattern of PFR cured at 423 K during 1 h). XRD data in details will be presented elsewhere.

In general the influence of the support on the membrane which is formed as a result of carbonization is complicated and should be investigated in each specific case. As we shall discuss below the support structure can play important role in carbon nanostructure production thus improving membranes properties.

Lately special attention was given to the investigation of behavior of fluids inside carbon nanotubes (CNT) and other nanostructures. Though available data are insufficient and sometimes contradictory, it is clear that CNT–based membranes (composite, hybride et al.) would differ markedly from polymeric and zeolite membranes. First it was shown theoretically that the transport of gases inside single-walled CNT is orders of magnitude faster than in any known nanoporous materials. Then first membranes have been fabricated from nanotubes embedded in impermeable (polymer or silicon nitride) film, experimental data have been obtained, supported by further simulations [16, 17].

Experimental data show that carbon nanostructures (including CNT and nanofibers) can be formed from definite polymeric precursors. Their synthesis can be promoted by use of catalysts. For example, carbonization of a mixture containing poly(ethylene glycol) as the carbon source and $NiCl_2$ as the catalyst precursor leaded to the synthesis of platelet graphite nanofibers [18]. Carbon nanotubes were obtained from polyethylene precursors in the presence of Ni-catalysts [19]; catalytic graphitization of polyvinyl alcohol produced multiwalled graphite nanotubes [20]. Authors of [21] also focused on formation of CNT by thermal conversion of polymeric precursors containing metal catalysts (e.g. Fe). Recently multiwalled carbon nanotubes were synthesized from novolac type phenol-formaldehyde resin mixed with ferrocene (Fe precursor) [22]. It is especially interesting as carbonization of these polymers without iron gives only amorphous carbon. Some other examples of nanocarbon materials production from polymers are listed in [12].

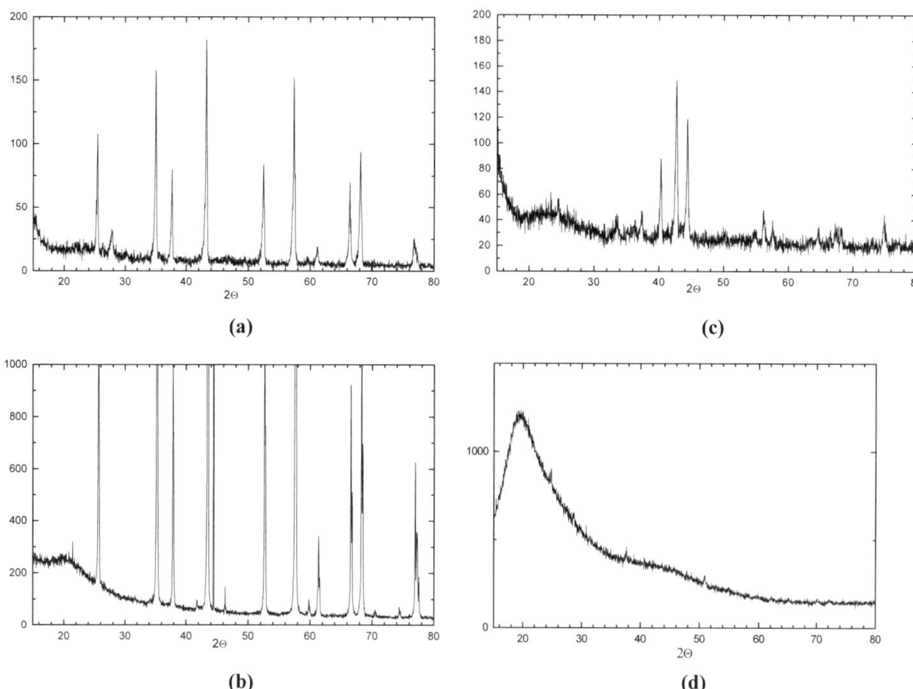

Figure 3. XRD patterns of PFR (resol type) coatings after carbonization at heating up to 923 K with soaking during 1 h. (a) α-Al_2O_3-based ceramic tubes with the mean pore size of 0.2 μm; (b) α-Al_2O_3-based ceramic tubes with the mean pore size of 40 nm; (c) α-Al_2O_3-based ceramic tubes with the mean pore size of 0.2 μm with additional surface layer (Cr-C); (d) PFR (resol type) cured at 423 K during 1 h

Our approach to membrane production is based on the carbonization of thin continuous polymeric layers on the porous inorganic high temperature supports with definite structure which opens up new possibilities for carbon nanostructure production [12]. However few works study substrate effects. Authors of [23] has shown that unique carbon nanostructures can be formed at temperatures less than 1,000°C without any catalyst, for example, by carbonization of phenolic resin on ceramic substrates. What is important, the carbon synthesized from the same precursor under the same conditions but not coated on ceramic, does not show such structure. Carbons formed on different phases of alumina from different precursors revealed strong effect of ceramic substrate on porous carbon structure [24]. Our results also showed that ordered carbon structure can be formed at carbonization of polymer coatings on alumina ceramics in contrast with the same polymers carbonization without substrate.

Our technology allows promoting formation of necessary nanostructures by introduction of catalysts in precursor or at the later stage of carbonization. Methods of introduction of fine Ni powder and Co from salts into polymer have been recently developed. The details will be published elsewhere. There is also an opportunity to promote formation of nanostructures by producing additional catalytic sites on the surface of the support before polymer coating (gas phase fluorination, deposition of additional metal-containing layers by magnetron sputtering etc.)

However it is insufficient to carry out only experimental researches. Another important condition of successful forecasting of membrane efficiency and purposeful synthesis of carbon nanostructured membranes with necessary selectivity/permeability is study of gas transport mechanisms.

Though many researches are devoted to carbon nanoporous membranes, few of them consider mechanisms of gas permeation. This process combines adsorption and diffusive transport effects and both are important. In general for porous membranes the following kinds of flow take place (according to permselectivity increase): viscous (Poiseuille) flow, Knudsen flow, surface diffusion/selective adsorption, multilayer diffusion, capillary condensation, molecular sieving. These mechanisms can also work together. Conclusions concerning molecular transport in carbon nanotubes and other nanostructures are still contradictory. For example authors [25] indicated that nitrogen permeance was consistent with Knudsen diffusion, but according to [16] gas (air) flow exceeded Knudsen model values by more than an order of magnitude. However single-component selectivity for hydrogen, nitrogen, oxygen and other gases corresponded to Knudsen model with the exception of hydrocarbons whose selectivities were higher. Authors attributed it to surface diffusion or possibly solubility/diffusion mechanism. Recently Skoulidas et al. have used atomic simulations to study the adsorption and diffusion of CO_2 and N_2 in SWNT and concluded that the observed diffusion is not Knudsen type [17]. Our analysis shows that significant issues concerning nature, kinetics (diffusion) and limiting sorption of hydrogen by carbon nanostructures [11, 26, 27] which are closely related to the creation of new generation membrane, should be taken into consideration.

3. Conclusions

The developed technology of production of composite membrane with carbon selective layer has principal distinction, though conventional polymeric precursors are used. Our approach to membrane production is based on the application of thin continuous polymeric layers transforming into necessary carbon structure (including certain nanostructures) at carbonization on the specified support. This technology can be also used for production of polymeric membranes with introduced nanotubes. However it is extremely important to clarify transport mechanisms which would allow purposeful synthesis of gas separation and filtration membranes.

Acknowledgements

The work was supported by RFBR, grant number 06-08-00614.

References

1. Amirkhanov D.M., Alexeeva O.K., Alexeev S.Yu., et. al. Gaseous fluorine modification of mesoporous ceramic membranes. *Extended Abstracts Book of VII ICHMS'2001*, Alushta, Ukraine, September 16–22, 2001, P. 736–737.
2. Alexeeva O.K., Alexeev S.Yu., Shapir B.L., Tulskii M.N. Modified tubular catalytic membrane reactor for hydrogen production from hydrocarbons. In: Eds. M.D. Hampton et al. *Hydrogen Materials Science and Chemistry of Metal Hydrides*, Kluwer, NATO Science Series, 2002, **71**: P. 339–347.
3. Alexeeva O.K., Amirkhanov D.M., Kotenko A.A., Chelyak M.M. Creation of hydrogen - selective tubular composite membranes based on Pd-alloys: I. Improvement of ceramic support with Ni layer deposition. In: *Hydrogen Materials Science and Chemistry of Carbon Nanomaterials Science*, Springer, 2007, P. 95–103.
4. Alexeeva O.K., Amirkhanov D.M., Kotenko A.A., et. al. Carbon membranes produced from polymeric compositions deposited on porous ceramic supports. In: *Extended Abstracts Book of IX International Conference ICHMS'2005 Sevastopol*, Ukraine, September 5–11, 2005, P. 588–591.
5. Fuertes A.B., Centeno T.A. Carbon molecular sieve membranes derived from a phenolic resin supported on porous ceramic tubes. In: *Book of Abstracts 6 International Conference on Inorganic Membranes*, Montpellier, France, June 26–30, 2000, P. 89.
6. Sedigh M.G., Xu L., Tsotsis T., Sahimi M. Transport and morphological characteristics of polyetherimide-based carbon molecular sieve membranes. *Ind. Eng. Chem. Res*, 1999, **38**: 3367–3380.
7. Sedigh M.G., Onstot W.J., Xu L., et. al. Experiments and simulation of transport and separation of gas mixtures in carbon molecular sieve membranes. *J. Phys. Chem.*, 1998, **102**(44): 8580–8589.
8. Rao M.B., Sircar S. Performance and pore characterization of nanoporous carbon membranes for gas separation. *J. Membr. Sci.*, 1996, **110**: 109–118.
9. Acharya M., Foley H.C. Spray-coating of nanoporous carbon membranes for air separation. *J. Membr. Sci.*, 1999, **161**: 1–5.
10. Acharya M., Foley H.C. Transport in nanoporous carbon membranes: experiments and analysis. *AIChE J.*, 2000, **46**(5): 911–922.
11. Alexeeva O.K., Amirkhanov D.M., Kotenko A.A., et. al. Prospects of creation of new composite materials based on carbon nanostructures for gas separation and filtration membranes. In: *Materials of IV Conference of Russian Material Science Societies*, November 21–24, 2006, Ershovo, Russia, P. 64–65.
12. Alexeeva O.K., Amirkhanov D.M. Progress and prospects in the field of creation inorganic gas separating membranes with carbon selective layer. *Russ. Chem. J.*, 2004, **XLVIII**(5): 82–89 (in Russian).
13. Amirkhanov D.M., Alexeeva O.K., Kotenko A.A., et al. Review on current production of carbon composite gas selective membranes using controllable polymers carbonization. *Membranes*, 2006, **4**(32): 19–39 (in Russian).
14. Shapir B.L., Alexeeva O.K., Alexeev S.Yu., Amirkhanov D.M., Tulskii M.N. Possibility of carbon coating formation on mesoporous ceramic membranes. In: *Extended Abstracts Book of VII ICHMS'01*, Alushta, Ukraine, September 16–22, 2001, P. 744–745.
15. Alexeeva O.K., Kotenko A.A., Chelyak M.M. High temperature filters and gas separation membranes produced under polymer controllable carbonization. *Membranes*, 2007, **4**(36): 3–16. (in Russian).
16. Holt J.K., Park H.G., Wang Y., et. al. Fast mass transport through sub-2-nanometer carbon nanotubes. *Science*, 2006, **12**: 1034.
17. Skoulidas A.I., Sholl D.S., Johnson J.K. Adsorption and diffusion of carbon dioxide and nitrogen through single-walled nanotube membrane. *J. Chem. Phys.*, 2006, **124**: 054708.

18. Huang C.-W., Li Y.-Y. In situ synthesis of platelet graphite nanofibers from thermal decomposition of poly(ethyleneglycol). *J. Phys. Chem. B*, 2006, **110**: 23242–23246.
19. Kiselev A., Sloan J., Zakharov D.N., et. al. Carbon nanotubes from polyethylene precursors: structure and structural changes caused by thermal and chemical treatment revealed by HREM. *Carbon*, 1998, **36**: 1149–1157.
20. Krivoruchko O.P., Maksimova N.I., Zaikovskii V.I., et. al. Study of multiwalled graphite nanotubes and filaments formation from carbonized products of polyvinyl alcohol via catalytic graphitization at 600-800^{0}C in nitrogen atmosphere. *Carbon*, 2000, **38**: 1075–1082.
21. Keller T.M., Laskoski M., Quadri S.B. Ferrocene catalyzed carbon nanotubes formation in carbonaceous solid. *J. Phys. Chem. C*, 2007, **111**: 2514–2519.
22. Stamatin I., Morozan A., Dimitru A., Ciupina V., Prodan G., Niewolski J., Figiel H. The synthesis of multi-walled nanotubes (MWNTs) by catalytic pyrolysis of the phenol-formaldehyde resins. *Physica E*, 2007, **37**: 44–48.
23. Gadkaree K.P. Growth of unique carbon nanostructures on ceramic substrates. *J. Mater. Sci. Lett.*, 2002, **21**: 1081–1084.
24. Inagaki M., Kobayashi S., Kojin F., Tanaka N., Morishita T., Tryba B. Pore structure of carbons coated on ceramic particles. *Carbon*, 2004, **42**: 3153–3158.
25. Hinds B.J., Chopra N., Rantell T., Andrews R., Gavalas V., Bachas L.G. Aligned multiwalled carbon nanotube membranes. *Science*, 2004, **303**: 62–65.
26. Nechaev Yu.S. The nature, kinetics and ultimate storage capacity of hydrogen sorption by carbon nanostructures. *Physics-Uspekhi*, 2006, **49**(6): 563–591.
27. Nechaev Yu.S., Alexeeva O.K. Methodological, applied and thermodynamic aspects of hydrogen sorption by graphite and related carbon nanostructures. *Russ. Chem. Rev.*, 2004, **73**(12): 1211–1238.

STRUCTURAL INVESTIGATION OF CARBONMETALLIC NANOCOMPOSITIONS ON THE BASIS OF LaMnNi

E.A. SAMEDOV
Azerbaijan National Aerospace Agency
AZ1106 Baku, av. Azadlig 159, Azerbaijan
Fax: +99412 4479616; E-mail: ea_samedov@rambler.ru

Abstract. In this paper the structural characteristics nanocarbonmetallic compositions on the basis of LaMnNi, received by arc evaporation, have been investigated by the application of the methods of X-ray phase differential thermal and gravimetric analysis and IR-spectroscopy. The LaMnNi compositions free of carbon and containing the carbon nanotubes in its structure have been studied. In diffraction patterns of nanocarbonmetallic composition the lines corresponding to $LaNi_5$ intermetallic phase and to nanostructural carbon have been observed. In hydrogenated samples, the X-ray phase and spectral researches have shown also the presence of MnNiH hydride phases breaking up at 350°C. On the basis of advanced researches the conclusion about biphase structure of compositions has been made, thus carbon nanoparticles a part of compositions raise thermal stability of a metal phase, increasing sorption capacity of hydrogen-containing compositions.

Keywords: nanocarbonmetallic nanocomposition, structural characteristic, X-ray phase, IR-spectroscopy, nanostructural carbon, metalhydride, hydrogensorption

1. Introduction

In recent years the ways of hydrogen accumulation on the basis of its reversible sorption are most widely investigated by heat-sink matrixes. Nanocarbon materials, hydrideforming metals and intermetallic connections are considered now as the most actual sorbents of hydrogen [1].

2. Results and discussion

The structural characteristics of metal-carbon compositions received by a method of arc evaporation are investigated in the present work. Owing to the high vaporization temperature (4,000–5,000°C), the high speed of cooling (10^9–10^{14} degrees/sec), the high degree of dispersion (1–100 nm) of material and the big degree of germforming this method allows the synthesis of materials with unique properties.

At the evaporation of metal together with carbon it is possible the formation of metal-carbon nanostructural compositions in the form of the metallic particles containing carbon nanotubes. Such material incorporates the nanostructural carbon having the high capacity on hydrogen and hydrideforming metals and alloys which can carry out a role both of hydrogen sorbents and of catalysts of dissociation of hydrogen molecules.

For research of structural characteristics the X-ray phase, differential gravimetric (DTG) and differential thermal (DTA) analyses have been used. The X-ray phase

B. Baranowski et al. (eds.), *Carbon Nanomaterials in Clean Energy Hydrogen Systems.*

analysis has been performed with the use of Co-radiation on the spectrometer ДРОН-3, the DTG-, DTA-analyses have been made on derivatograph of the model Q-1500D in conditions of dynamic heating on air in an interval of temperatures 20–1,000°C. IR spectra have been investigated on the spectrophotometer Specord 71 IR in the field of 4,600–650 cm^{-1}.

We have undertook an investigation of the LaMnNi compositions received by arc evaporation and free from carbon and of the LaMnNi compositions containing the carbon nanotubes in its structure. The X-ray spectra of dispersion of LaMnNi compositions free from the carbon (1), the metals received by joint evaporation with carbon (2) and nanocarbonmetallic composition LaMnNi after crushing (3) are shown on the Fig. 1.

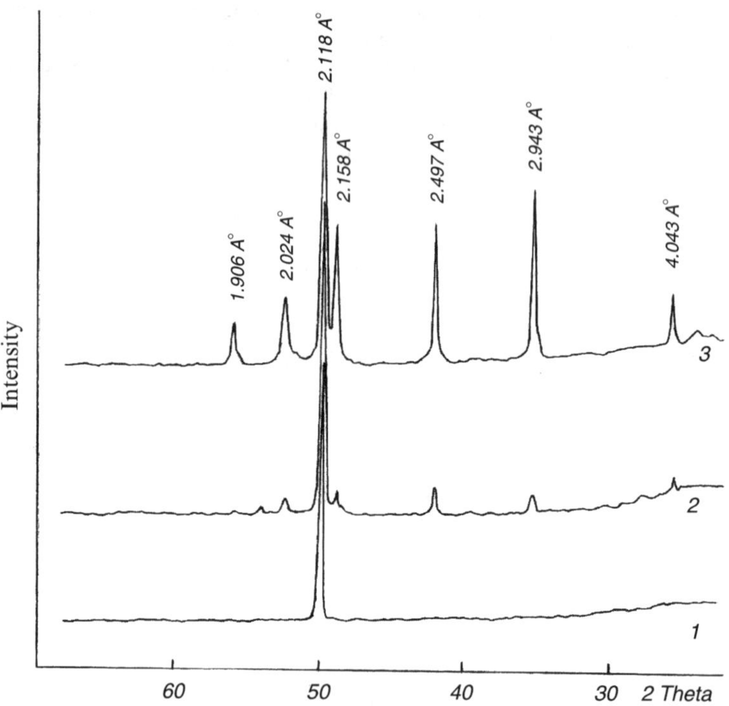

Figure 1. Diffractograms of nanocompositions LaMnNi: 1 – free from carbon; 2 – received by joint evaporation with carbon; 3 – after crushing

Apparently, in diffractograms the observed intensive lines corresponds to the interplane distance d = 2.118 Å and carried Ni. In diffractograms of nanocarbonmetallic compositions (Fig. 1 – 2) there is a weak line with d = 2.024 Å carried $LaNi_5$ and lines with d = 1.906, 2.158, 2.497, 2.943, 4.043 Å which have been carried to nanostructural carbon.

At samples crushing (Fig. 1 – 3) the change of a parity intensities of observable lines is observed: intensity of lines corresponding to the nanostructural carbon and Ni carbide increases approximately for the order and intensity of a line corresponding to Ni distinctly decreases.

Such change of diffractograms after the samples crushing can be explained by biphase structure of the received compositions which consist of intermetallic phase and the phase of nanostructural carbon.

Figure 2 presents the derivatograms of compositions LaMnNi and compositions of the metals received by joint evaporation with carbon. As illustrated, in both types of compositions the exothermal oxidation beginning from 450°C and accompanied with an increment on curves TG is observed.

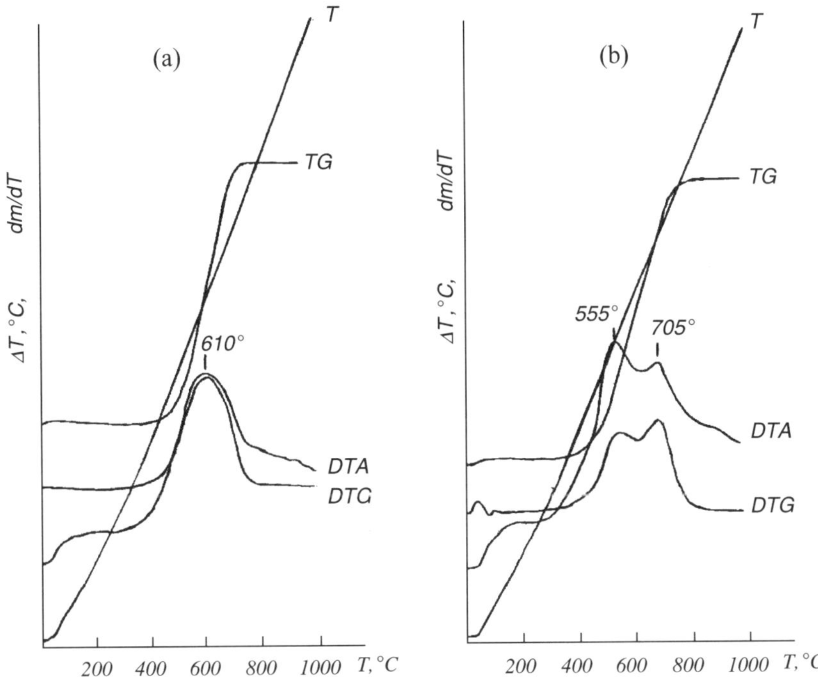

Figure 2. Derivatograms for oxidation of the nanocompositions LaMnNi: (a) free from carbon; (b) received by joint evaporation with carbon

In the case of LaMnNi composition (Fig. 2(a)) a precise maximum on the curves DTG (a) and DTA (b) takes place at temperature 610°C, in the case of nanocarbon-metallic compositions (Fig. 2(b)) on curves DTG and DTA two precisely resolved maxima are observed accordingly at T = 555° and T = 705°C, suggesting that biphase structure of compositions exists.

The maximum at T = 555°C should be carried to the oxidation of carbon particles of nanometric sizes. It is possible to explain the displacement of oxidation temperature of metallic components to high-temperature area (705°C) by the presence of various carboncontaining compounds of these metals in them or carbon nanoparticles of the various form.

The X-ray phase diffractograms and infra-red spectra compositions have been investigated also after process desorption and hydrogen containing nanocarbonmetallic compositions.

Figure 3 shows the diffractograms of carbonmetallic nanocompositions LaMnNi. In diffractogram of hydrogen containing compositions the line corresponding to the interplane distance d = 4.075 Å, related to MnNiH hydride phase, is observed. The lines with d = 2.158, 2.434, 2.943 Å have been carried to the nanostructural carbon. After hydrogen desorption and thermal influence in diffractograms there are peaks with d = 2.086 and 2.408 Å related to the oxide nickel NiO. The lines corresponding to nanostructural carbon disappear and there are lines with d = 2.68 and 3.099 Å corresponding to unstructured soot.

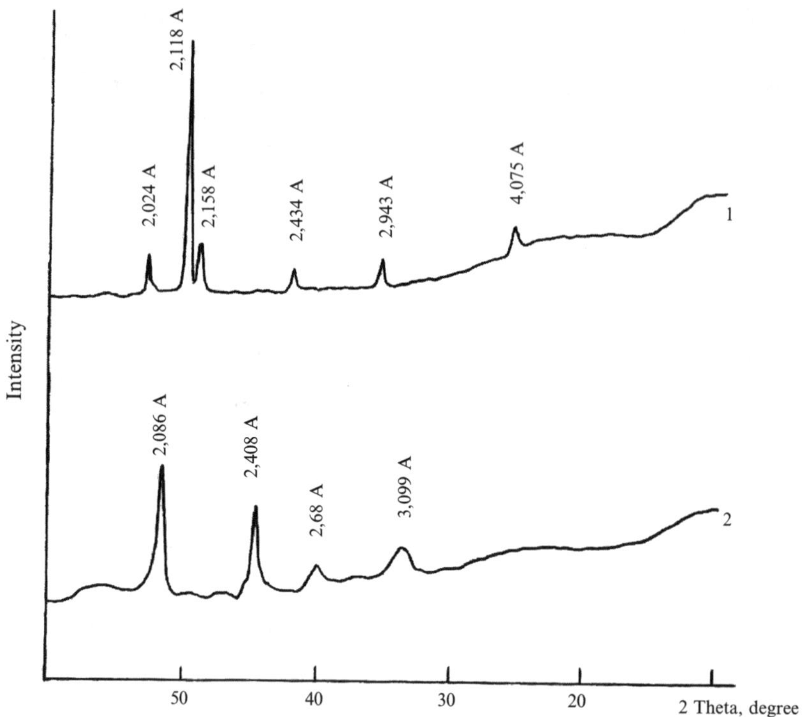

Figure 3. Diffractograms of nanocompositions LaMnNi: 1 – after hydrogen sorption, 2 – after desorption and thermal influence

There are bands of 1,620, 1,450 and 1,100 cm^{-1} carried to fullerenes in the IR-spectrum of an initial composition. The bands of 1,450 and 1,100 cm^{-1} have been carried to fluctuations of surface hydrocarbonic groups $C{=}CH_2$ and C=CH, accordingly, which disappear at the oxidation. The band of 1,620 cm^{-1} is carried C=C to nonlimiting bonds in fullerensimilar structures.

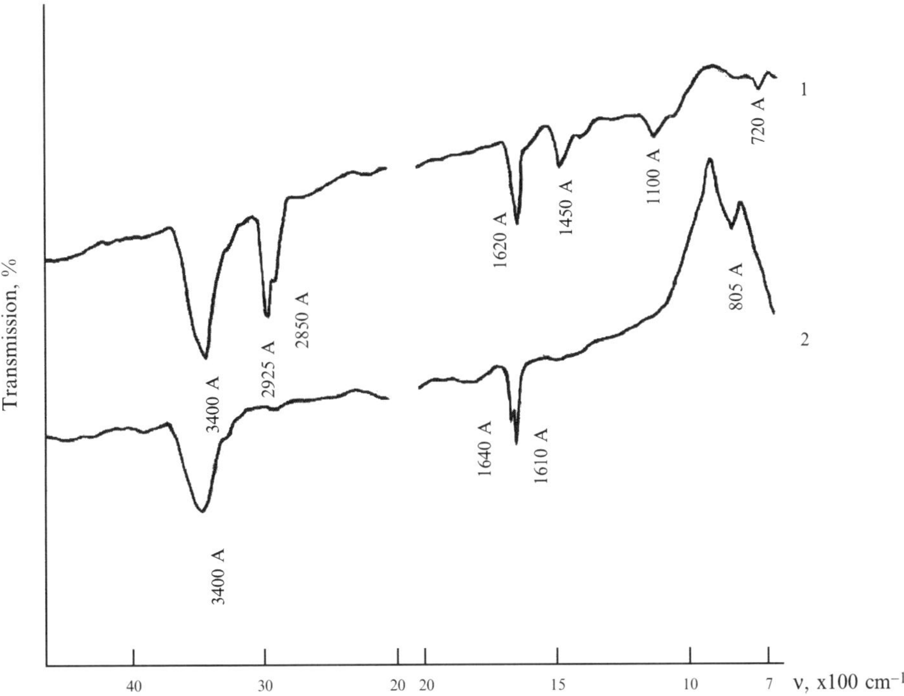

Figure 4. IR-spectra nanocompositions LaMnNi: 1 – after sorption hydrogen, 2 – after desorption and thermal influence

In the oxidized samples there is a band of 805 cm^{-1} related to fluctuation of Ni-O bond, the band of 1,620 cm^{-1} is split on 1,640 and 1,610 cm^{-1}. It is possible to explain such splitting by decomposition of fullerensimilar structures on linear nonlimiting carbon components.

Investigation of hydrogensorption characteristics of obtained LaMnNi compositions has shown the increase of sorption capacity in the case of compositions containing nanostructural carbon that can be explained by additional sorption of hydrogen by carbon nanotubes.

The temperature dependences of electroresistance of initial and hydrogen containing metalhydride compositions have been investigated. The temperature dependences of electroresistance of cast alloy LaMnNi and after saturation by hydrogen are characterized by linear dependence on which the maximum of electroresistance is observed at T ~ 350°C. It is possible to explain disintegration of hydrogen containing phases, it is probable MnNiH. At T ~ 550°C the sharp increase of electric resistance, which is possibly connected with disintegration of spending phase MnNi, is observed.

3. Conclusions

Thus, matrixes on the basis of the LaMnNi compositions containing carbon nanotubes, represent biphase system for hydrogen sorption. Thus, carbon nanoparticles, a part of compositions, raise thermal stability of metal compositions increasing it sorption capacity on hydrogen. The X-ray phase and spectral researches have shown also the presence of MnNiH hydrogen containing phases in the hydrogenated samples, breaking up at 350°C.

References

1. Trefilov V.I., Schur D.V., Tarasov B.P. Fullerenes as a basis of materials of the future. – Kiev: ADEF, 2001. - 148 p. (in Russian).

MODIFICATION OF ULTRATHIN α-C:H FILMS BY SHORT VOLTAGE PULSES IN STM

A.S. BATURIN, A.A. CHOUPRIK, E.P. SHESHIN
Moscow Institute of Physics and Technology, Institutskiy per. 9, Dolgoprudniy 141700, Russia
Fax: +7 495 4099543; E-mail: chouprik@mail.ru

V.D. FROLOV, E.V. ZAVEDEEV
A.M. Prokhorov General Physics Institute, Vavilov Str. 38, Moscow 119991, Russia

Abstract. This article discusses mechanisms of scanning-tunneling-microscopy-based nanolithography of ultrathin diamond-like α-C:H films. Results of films modification by short voltage pulses in scanning tunneling microscope (STM) was verified in constant current STM and conductive atomic force microscope (AFM). Critical parameters for STM-nanolithography by this method were determined. Results of modification in ambient conditions and in low vacuum are compared.

Keywords: α-C:H films, scanning-tunneling-microscopy-based nanolithography, graphitization

1. Introduction

Now the most effective way of scanning probe nanolithography of inorganic carbon materials is a local impact by an electric field applied between a probe of scanning probe microscope (SPM) and a surface of samples. It is so called electric-stimulated scanning probe lithography (ESPL). ESPL was carried out on hydrogenated layers of diamond, graphite and various types of graphite- and diamond-like films [1–6]. Fulfilled researches have shown that electric SPM-impacts on carbon films lead to nanoscale transformation of their initial surface (formation of nanohills or nanoholes), and in some cases also lead to change of electroconductivity in a vicinity of electric impact. All these achievements cause the great scientific and practical interest to the further studying of mechanisms of carbon nanostructuring by ESPL method.

α-C:H films are the most convenient "modelling" material for studying all known modification mechanisms of carbon materials by ESPL method. In the present work we study this prominent aspect of nanostructuring of ultrathin α-C:H films.

2. Experiment and discussion

2.1. SAMPLES

Diamond-like ultrathin α-C:H films have been deposited on silicon substrates by PE CVD method in C_2H_4 atmosphere. In this process, a bias voltage (200–300 V) was applied to the substrate. The deposition rate of α-C:H films was 8–12 nm/min. According to ellipsometry the thickness of the films were 6 and 13 nm for two set of samples.

2.2. THE EQUIPMENT AND TECHNIQUES

Research was made on a multimode scanning probe microscope NTegra (NT-MDT, Russia). STM tips were prepared by stretching of Ag wire. Instability of tip shape causes some differences in modification results at the same conditions.

Scanning of surface and its modification were performed with grounded (through a current meter) STM probe and a positive bias voltage V_0 on a sample. For all samples a current I_0 was held equal to 0.1 nA during scanning. A bias voltage V_0 was chosen for each sample individually.

We adopted a technique for local STM-modification of diamond-like ultrathin α-C:H films, deposited on a silicon substrate. Modification of a surface was carried out by application of series of identical positive voltage pulses on a sample with the variable parameters: amplitude V_{max}, duration t and on-off time ratio. This procedure was carried out in nodes of a rectangular matrix on a sample surface. Measurements were performed in ambient conditions (temperature ~23°C, humidity ~40%, pressure 10^5 Pa) and in low vacuum (pressure of residual gases ~$2.5 \cdot 10^{-2}$ Torr). Results of modification were investigated by STM in a constant current method. "Holes" observed on a sample surface can be interpreted as area with lower conductivity or/and as real deepenings of a surface. Also "hills" can be interpreted as area with larger conductivity or/and as real relief of a surface. Influence on film by negative voltage pulses up to -6 V on a sample at ground tip is not resulting in any modification.

2.3. EXPERIMENT

Experiments were carried out with 6 and 13 nm thick α-C:H films. In ambient conditions it has been shown that "deepenings" were formed during modification procedure. Typical threshold values of a bias voltage amplitude and duration of bias voltage pulses for surface modification were $V_{max} = 3.8$ V, $t = 70$ ms or $V_{max} = 4.0$ V, $t = 40$ ms or $V_{max} = 4.2$ V, $t = 20$ ms for number of pulses $N = 10$ and on-off time ratio was equal 1. At those conditions a minimal lateral size of the modified areas was about 100 nm and depth was less than 1 nm. Threshold amplitude of bias voltage pulses leading to the film modification was lower in the case of preliminary for annealed films.

The signal of the z-scanner in a constant current STM method contains the combined information on topography of the sample and on its electrical properties. "Holes" on STM image could be treated as areas with lower conductivity. To check up this assumption we investigated the modified areas by means of atomic force microscopy. Due to the difficulties of precise positioning of an AFM probe into the area previously modified in STM the large objects were chosen for AFM investigation. The lateral size was 1–1.5 μm. Modified areas was positioned in nodes of a rectangular matrix with step of 3.8 μm. "Effective depth" of the obtained structures ("holes") was 20 nm by results of constant current STM method. While the AFM analysis shows that actually they are hills with height less than 1 nm (Fig. 1). This modified areas have lower conductivity (Fig. 1(c)), than the original diamond-like α-C:H film.

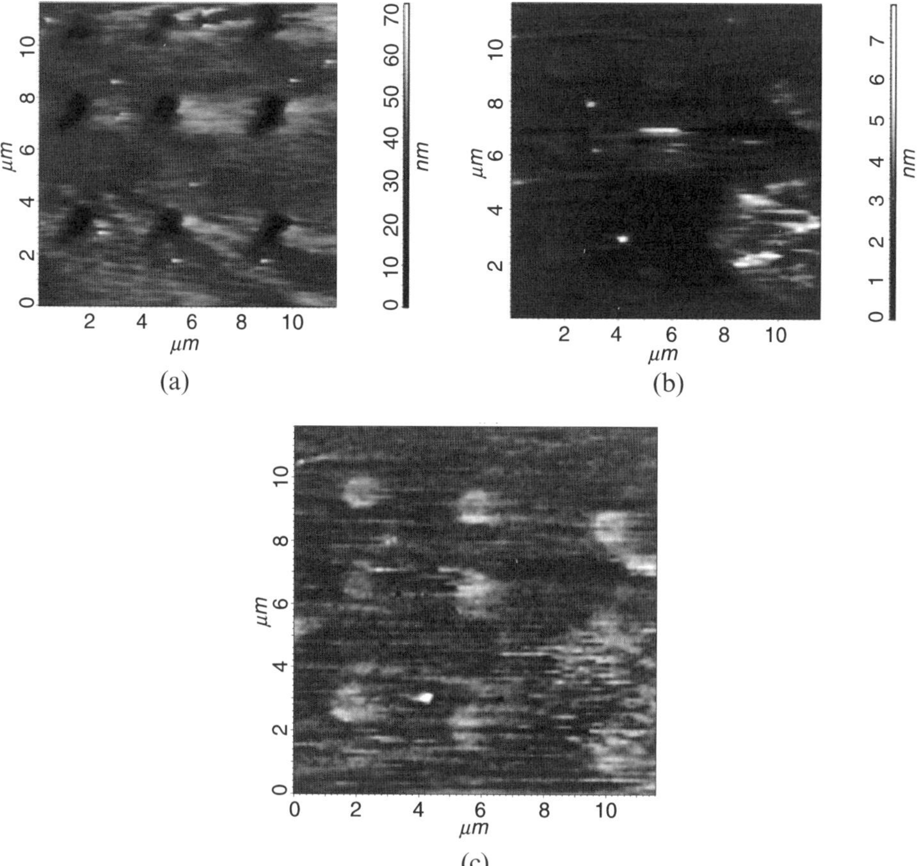

Figure 1. Comparison of the modified area of 6 nm thick a C:H film in STM and AFM: (a) the surface topography measured in STM by a constant current method, I_0 = 0.1 nA, V_0 = 1.7 V; (b) the surface topography measured in AFM (probe with 10 nm radius) in a contact method at normal force about 1.5 nN; (c) distribution of spreading resistance on area corresponding to (b), V = -8 V

The dependence of geometrical parameters of modified areas on the exposition $T = Nt$ were measured. Here N is a number of pulses in a series and t is a pulse duration. Figure 2 shows dependence of area and depth of individual modified region.

It is known [7–9] that in ambient conditions between a tip and a sample, separated on a tunnel distance, a water bridge is condensed. Moreover, the bias voltage applied to this gap intensifies and accelerates this process [10, 11]. So, it is certainly that the modification process takes places in the liquid meniscus.

To avoid a liquid meniscus formation second part of experiments were performed. In contrast to the ambient conditions modification of α-C:H films in vacuum results in the formation of “hills”. Typical threshold values of a bias voltage amplitude and duration of bias voltage pulses for surface modification were $V_{max} = 3.0$ V, $t = 20$ ms for number of pulses $N = 10$ and on-off time ratio was equal 1.

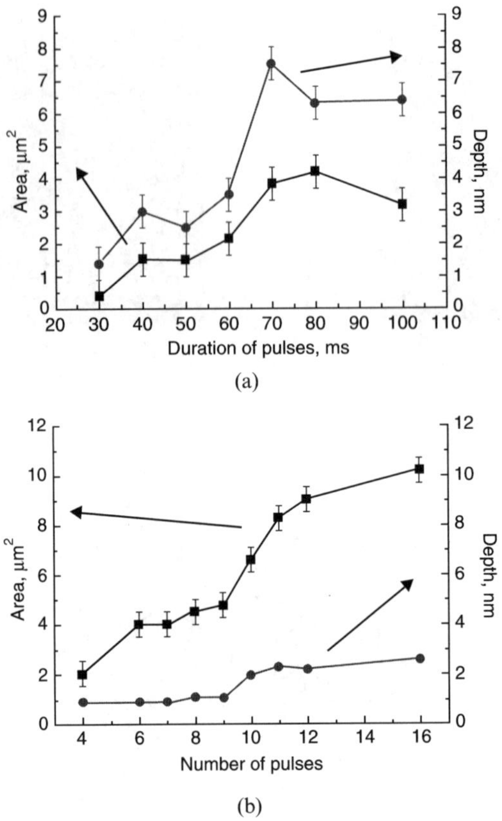

Figure 2. Dependence of area (squares) and depth (circles) of modified regions on the number of pulses for $t = 50$ ms (a) and on the pulse duration for $N = 10$ (b)

Figure 3 demonstrates the example of such critical experiments. Conductive AFM analysis confirmed that the modified areas have higher conductivity then untreated film. Similar effect was observed early [5], that experiment was carried out in ultrahigh vacuum (~$5 \cdot 10^{-10}$ mbar) with tetrahedral amorphous carbon tα-C films. The modified areas were 10 nm in diameter and up to 5 nm in height. Conductive AFM demonstrated ohmic conductance of those areas. The experimental results [5] indicate that field-emitted free electrons enter the tα-C film and are absorbed within the film. That is strong electric fields applied to metal-contacted DLC films lead to irreversible changes in the *I–V* characteristics due to the creation of sp^2 islands embedded in the sp^3 matrix. We suppose that in our case STM impact induced local graphitization at α-C:H film.

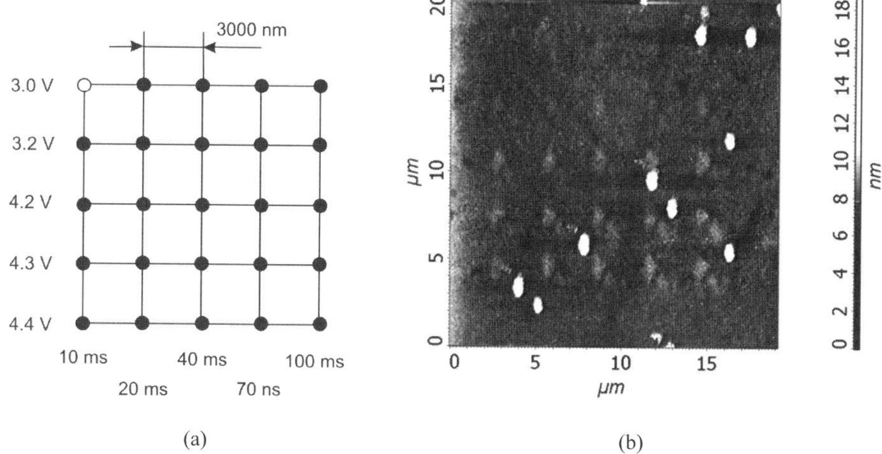

Figure 3. Determination of characteristic threshold values of a bias voltage amplitude and duration of bias voltage pulses for surface modification. (a) Scheme of distribution of modification points; (b) STM image after modification

The second threshold of bias voltage amplitude was observed in low vacuum. Modification with parameters $V_{max} = 4.5$ V, $t = 10$ ms, $N = 10$ provides a trench arounded "hills" formation on STM image (Fig. 4). We explain this fact as a condensation of liquid meniscus under bias voltage higher then second threshold. In ambient conditions the characteristic time of field-induced water bridge condensation is over ~20 ms [10, 11]. Apparently in low vacuum the time of water bridge condensation is larger. Thus at low bias voltage graphitization happened before liquid meniscus condensation and trench formation processes are similar to the surface modification in ambient conditions.

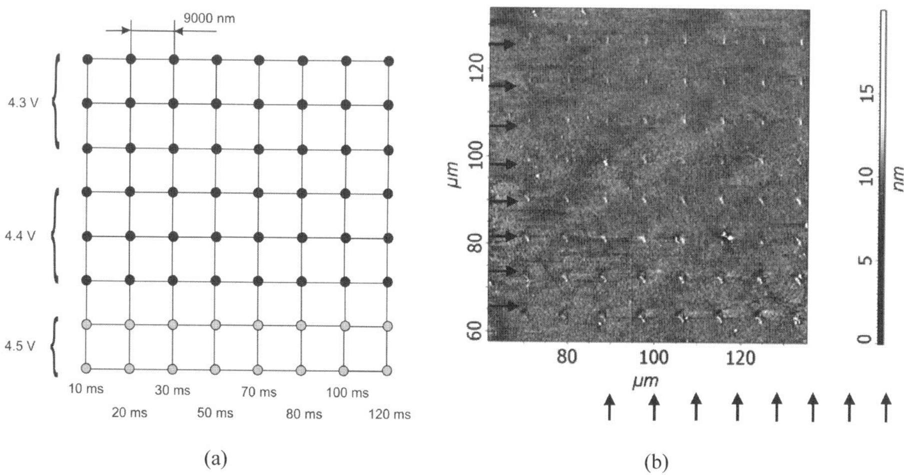

Figure 4. Determination of second characteristic threshold value of a bias voltage amplitude for trench arounded "hills". (a) Scheme of distribution of modification points; (b) STM image after modification

3. Conclusions

Short positive pulses of a bias voltage applied between STM tip and α-C:H film (at the grounded probe) leads to the formation of poorly conducting channels in the α-C:H film under ambient conditions. The lateral size of such channels is in the range from 100 nm up to 2 μm. High conducting channels could be formed during modification in low vacuum. However application of bias voltage over than 4.5 V provides formation of structures with complex topography.

Acknowledgments

The work was supported by Russian Foundation for Basic Research under grant 05-02-16317 and Russian Federal Agency of Education under grant NRP.2.1.2.9353.

References

1. Tachiki M, Fukuda T, Sugata K, Seo H, Umezawa H, Kawarada H. Nanofabrication on hydrogen-terminated diamond surfaces by atomic force microscope probe-induced oxidation. Jpn. J. Appl. Phys. 2000, **39**: 4631–4632.
2. Kondo T, Yanagisawa M, Jiang Lei, Tryk DA, Fujishima A. Nanolithogrphic modification of diamond. Diamond Relat. Mater. 2002, **11**: 1788–1796.
3. Albrecht TR, Dovek MM, Kirk MD, Lang CA, Quate CF, Smith DPE. Nanometer-scale hole formation on graphite using a scanning tunneling microscope. Appl. Phys. Lett. 1989, **55**: 1727–1729.
4. Weeks BL, Villmer A, Welland ME, Rayment T. High-pressure nanolithography using low-energy electrons from a scanning tunnelling microscope. Nanotechnology 2002, **13**: 38–42.
5. Mühl T. Scanning-tunneling-microscopy-based nanolithography of diamond-like carbon films. Appl. Phys. Lett. 2004, **85**: 5727–5729.
6. Frolov VD, Konov VI, Pimenov SA, Zavedeev EV. Field-induced modifications of hydrogenated diamond-like carbon films using a scanning probe microscope. Diamond Relat. Mater. 2004, **13**: 2160–2165.
7. Israelachvili J. Intermolecular and Surface Forces. Academic, San Diego, CA. 1997.
8. Restagno F, Bocquet L, Biben T. Metastability and nucleation in capillary condensation. Phys. Rev. Lett. 2000, **84**: 2433–2436.
9. Bhushan B. Adhesion and stiction: Mechanisms, measurement techniques, and methods for reduction. J. Vac. Sci. Technol. B 2004, **21**(6): 2262–2296.
10. Gomez-Monivas S, Saenz JJ. Field-induced formation of nanometer-sized water bridges. Phys. Rev. Lett. 2003, **91**: 056101-1–056101-4.
11. Calleja M., Tello M., Garcia R. Size determination of field-induced water menisci in noncontact atomic force microscopy. J. Appl. Phys. 2002, **92**: 5539–5542.

INFLUENCE OF THE MECHANICAL ACTIVATION ON THE THERMAL STABILITY OF THE AlH_3 IN Al-C-H SYSTEM

R.V. LUKASHEV
Institute of Problems of Chemical Physics of RAS, Prospectus of the Academician Semenov, 1, Chernogolovka, Moscow Region, 142432 Russia
Tel./Fax: +7(496) 5221743; E-mail: btarasov@icp.ac.ru
S.N. KLYAMKIN, B.M. BULYCHEV
Department of Chemistry, M.V. Lomonosov Moscow State University, Moscow
B.P. TARASOV
Institute of Problems of Chemical Physics of RAS, Prospectus of the Academician Semenov, 1, Chernogolovka, Moscow Region, 142432 Russia
Tel./Fax: +7(496) 5221743; E-mail:btarasov@icp.ac.ru

Abstract. It is shown, that mechanical activation of aluminum hydride reduces its thermal stability. The additive of carbon at milling process prevents decomposition AlH_3.

Keywords: aluminum hydride, carbon, mechanical activation, thermal stability

1. Introduction

Nowadays, the important problem is the creation of the portable hydrogen sources for different technical application. One of the perspective materials for hydrogen storage application is the aluminum hydride AlH_3. Hydrogen contention in aluminum hydride is 10 wt%. Synthesis of the aluminum hydride is a multistage process at ether environment. As initial components the solutions of the $LiAlH_4$ and $AlCl_3$ is used [1, 2]. Thermodynamically AlH_3 is unstable ($\Delta G > 0$ at 298 K) even at room temperature but because of the kinetics difficulties the temperature range of the decomposition is 150–180°C (for the most stable α-modification). Thus the main aim of this investigation is the development of the different methods of modification of the aluminum hydride AlH_3, which can allow to decrease the temperature of the dehydrogenation while minimal loosing in hydrogen capacity.

One of the famous methods of increasing of the reactive ability is its mechanical activation (m/a) [3], which results to defect formation and surface increasing. However while mechanical activation in different planetary ball mills local heating takes place, what can lead to the AlH_3 decomposition with hydrogen evolution. Thus the critical goal is searching of the optimal activation conditions and composition of mixtures for obtaining of the mechanical activated aluminum hydride with high hydrogen capacity and minimal temperature of the dehydrogenation.

B. Baranowski et al. (eds.), *Carbon Nanomaterials in Clean Energy Hydrogen Systems.*

2. Experimental

Aluminum hydride and its mixtures with 10 wt% of carbon (graphite of mark СЭУ) were subjected to mechanical treatment at the planetary ball mill Fritsch "Pulverissete 6" in argon atmosphere.

3. Results and discussions

While mechanical activation (from 5 till 30 min) of the aluminum hydride and its mixture with carbon the formation of new compounds was not observed. Great widening of the α-AlH_3 peak at difractorgam can be seemed, what is evidence of the defects and micro stress increasing and decreasing of the coherence scattering area (CSA) of the samples (Table 1).

TABLE 1. Influence of the mechanical activation on the microstructure parameters

Activation time, min	AlH_3		AlH_3 – carbon (10 wt%)	
	CSA data of the α-AlH_3 phase (nm)	Micro stresses (%)	CSA data of the α-AlH_3 phase (nm)	Micro stresses (%)
0	>200	0.01 (*a*) 0.02 (*c*)	–	–
5	50 (*a*) 53 (*c*)	0.01 (*a*) 0.07 (*c*)	100 (*a*) 110 (*c*)	0.02 (*a*) 0.04 (*c*)
30	19 (*a*) 17 (*c*)	0.12 (*a*) 0.17 (*c*)	35 (*a*) 43 (*c*)	0.02 (*a*) 0.18 (*c*)

The 5 min mechanical activation of the aluminum hydride leads to greatly decreasing of the decomposition temperature of the AlH_3 (Fig. 1).

However at the mechanical treatment on planetary ball mill partial decomposition took place. Thus if initial powder contained 95% AlH_3, then after 5 min milling, contention of the AlH_3 was 74%. Further time increasing of the milling results to decreasing of the dehydrogenation temperature (DSC curves – 140°C) and hydrogen contention of the AlH_3 phase up to 68%. The carbon addition prevents to decomposition of the aluminum hydride while mechanical activation.

The temperature of the decomposition of the AlH_3 at such composites doesn't differ from desorption temperature of the activated hydride with analogies milling times. Thus for AlH_3–carbon sample (m/a 5 min) peak of the desorption corresponds to 147°C, and phase contention of the AlH_3 after treatment is 85 wt%. Prolonged milling (30 min) doesn't lead to the desorption temperature (desorption peak at 142°C), and visibly hydrogen contention at composite (83 wt%) decreasing. Only the longest activation (more than 45 min) leads to slump of the hydride phase contention for both samples (4 and 4.3 wt% correspondingly).

On the Fig. 2 curves of the hydrogen desorption from initial AlH_3 and mechanical activated samples of the AlH_3 and AlH_3–carbon at constant temperature can be observed. As it shown from Fig. 2 prior mechanical activation promotes to the increasing of the desorption rate.

For the mathematical treatment of the experimental data equation of Avrami-Erofeeva was used:

$$\alpha = 1 - \exp(-k{\cdot}\tau n), \tag{1}$$

where α – fractional conversion, τ – time, k – constant of the rate, n – parameter, which depends from the reaction mechanism. The parameter n can determine as a experimental dependence of the lg(lg (1/(1-α))) from the lg(τ). Thus, the reaction of the hydrogen desorption from untreated AlH_3 controlled by diffusion without formation of new nucleus (n = 1.7). Hydrogen desorption from mechanically activated samples of the aluminum hydride with and without carbon has the same mechanism (n is equal to 3.3 and 3.2 correspondingly) and limited by process on the phase interface.

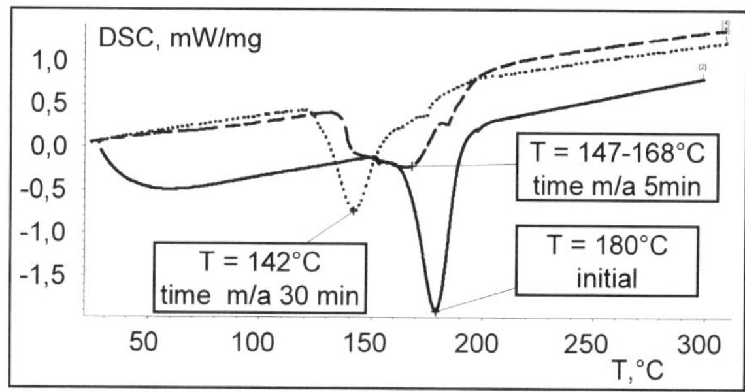

Figure 1. Thermogram of the decomposition of the initial and mechanically activated samples of the AlH_3 (The heating rate is 5 grad/min)

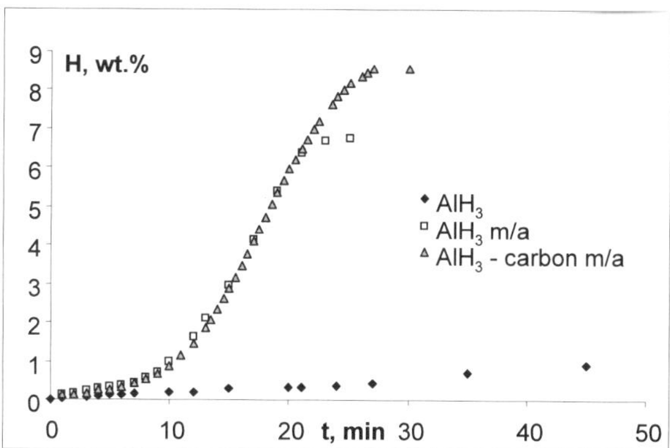

Figure 2. Hydrogen desorption curves of initial and mechanically activated samples (5 min) at temperature 140°C

4. Conclusions

Mechanical activation of the aluminum hydride leads to the decreasing of the thermal stability of the AlH_3, however partial decomposition takes place. Addition of the carbon at milling process prevent from AlH_3 decomposition.

The kinetic of the hydrogen desorption from the mechanical activated samples at constant temperature has been studied. Mechanical activation changes the mechanism of the hydrogen desorption from aluminum hydride cardinally, but carbon addition does not influence on the process mechanism.

Acknowledgement

The work is supported by INTAS (Grant № 05-1000005-7665).

References

1. Sandrock G., Reilly J., Graetz J., Zhou W.-M., Johnson J., Wegrzyn J. Alkali metal hydride doping of α-AlH_3 for enhanced H_2 desorption kinetics, *J. Alloy. Compd.*, 2006, **421**: 185–189.
2. Konovalov S.K., Bulychev B.M. The P,T-state diagram and solid phase synthesis of aluminium hydride, *Inorg Chem.*, 1995, **34**: 172–175.
3. Lukashev R.V., Klyamkin S.N., Tarasov B.P. Preparation and properties of hydrogen-storage composites in the MgH_2–C system, *Neorganicheskie Materialy*, 2006, **42**: 726–732 (in Russian).

ELECTRODES FOR FUEL CELLS BASED ON CARBON NANOTUBES AND CATALYSTS

M.O. DANILOV, G.YA. KOLBASOV
Institute of General and Inorganic Chemistry of the Ukrainian National Academy of Sciences, Prospekt Palladina 32/34, Kyiv 142, Ukraine
Fax: (044) 4243070; E-mail: danilovmickle@rambler.ru
A.V. MELEZHYK
TM Spetsmash Ltd., Viskoznaya Str. 5, Build. 23, 02660 Kyiv, Ukraine
Fax: (044) 5010620; E-mail: nanocarbon@rambler.ru

Abstract. It has been shown that for oxygen electrode based on multiwalled carbon nanotubes lead and manganese dioxides, which have high oxygen evolution overpotential, are good catalysts in alkaline electrolyte. For hydrogen electrode based on carbon nanotubes palladium, aluminium-nickel and magnesium-nickel alloys, which have high hydrogen evolution overpotential, are good catalysts in alkaline electrolyte. These investigations demonstrate that multiwalled carbon nanotubes are good support for catalysts for fuel cells. Hypo thesis about localization of electrochemical reactions is suggested, which allows to predict catalytic properties of different materials in the electrochemical reactions considered.

Keywords: carbon nanotubes, catalysts, electrocatalysis, fuel cells

1. Introduction

Electrochemical power sources and fuel cells, in which electrocatalytic processes take place, belong to alternative or minor energetics. At the present time the power generated by all current sources in the world is equal to the power generated by all nuclear, thermal and water power stations. Therefore, research in the area of electric current generation is very topical at present. Further we shall consider electrocatalytic processes which occur on multiwalled carbon nanotubes, containing deposited catalysts, in alkaline fuel cells.

All electrocatalytic processes are based on the compromise potential of the electrochemical reactions proceeding at electrode consisting of a catalyst and its support. The value of this compromise potential depends on the potentials of the electrochemical reactions which proceed simultaneously at this electrode. If one assumes that reactant adsorption and electron attachment are localized and occur simultaneously either on the catalyst or on its support, and that reactant particles diffuse through the catalyst/support interface, then by selecting appropriate catalyst–support pair one can influence the electrocatalytic properties of such an electrode. We shall consider the possibility of this influence on the catalytic properties of electrodes consisting of a catalyst and its support for the following reactions:

1. Oxygen reduction reaction, which proceeds at the oxygen electrode of fuel cell
2. Hydrogen oxidation reaction, which proceeds at the hydrogen electrode of fuel cells

The aim of our study was to examine the possibility of evaluating the electrocatalytic properties of nanostructured compositions for their use as electrode materials for alkaline fuel cells.

2. Experimental

Multiwalled carbon nanotubes (MWNT) were used as the catalyst support. They were prepared by CVD method from ethylene with using Fe-Mo/Al_2O_3-SiO_2 catalyst [1]. The product was obtained as black powder with bulk density of 25–30 g dm^{-3}. The outer diameter of nanotubes was about 10–30 nm, specific surface was 230 m^2 g^{-1}. Content of minerals (catalyst) in unpurified product was 15–20 wt%. The catalyst impurities were removed from nanotubes by treatment with hydrofluoric acid solution.

The double–layer oxygen electrodes were prepared by pressing. The hydrophobic layer contained 0.07 g cm^{-2} acetylene black with 25% polytetrafluoroethylene, and the active layer contained 0.02 g cm^{-2} MWNTs, modified with various catalysts, with 5% polytetrafluoroethylene. Method of deposition of manganese dioxide on carbon nanotubes was described in [2]. The hydrogen electrodes were prepared by pressing of various compositions based on MWNTs and catalysts with addition of 25% polytetrafluoroehylene as binder. The investigations were carried out in a fuel cell mockup. The mockup for testing gas-diffusion electrodes was described in Ref [3]. Six molar solution of KOH (analytical grade) was used as electrolyte. For the oxygen electrode zinc was used as auxiliary electrode. For the hydrogen electrode nickel oxide was used as auxiliary electrode. All potentials were measured and are given relatively to silver-chloride reference electrode connected through a salt bridge. The electrochemical plots were recorded under galvanostatic conditions. The hydrogen and oxygen source was a U-shaped electrolyzer with alkaline electrolyte. Hydrogen and oxygen were delivered to the gas-diffusion electrodes under an excess pressure of 0.01 MPa. Before operation, the gas-diffusion electrodes were blown through with an appropriate gas for an hour. In electrochemical tests the electrolyzer current exceeded by a factor of three the current generated by the gas-diffusion electrode.

Electron images were recorded with use of transmission electron microscope JEM-100 CXII.

3. Results and discussion

We have studied the oxygen reduction reaction, on which the operation of oxygen electrodes of fuel cells and air–metal current sources is based. Manganese dioxide deposited on MWNT possesses high catalytic activity in this reaction.

Electron images of initial MWNT (a) and MWNT modified with manganese dioxide (50% MnO_2) are shown in Fig. 1(a–c). For the sample of MWNT modified with MnO_2 prepared by reduction of potassium permanganate with carbon of MWNT fine nanoparticles of MnO_2 are observed, which are pasted and uniformly distributed on the surface of nanotubes (Fig. 1(b)). When potassium permanganate was reduced with formic acid, the surface of nanotubes is clean, and MnO_2 is present in form of much larger particles as mechanical mixture with nanotubes (Fig. 1(c)). It should be noted also that modified MWNT samples prepared by reduction of permanganate by carbon of MWNT at boiling are black at all mass contents of MnO_2, while reduction of permanganate with formic acid gives brown hydrated manganese dioxide. The cheapest

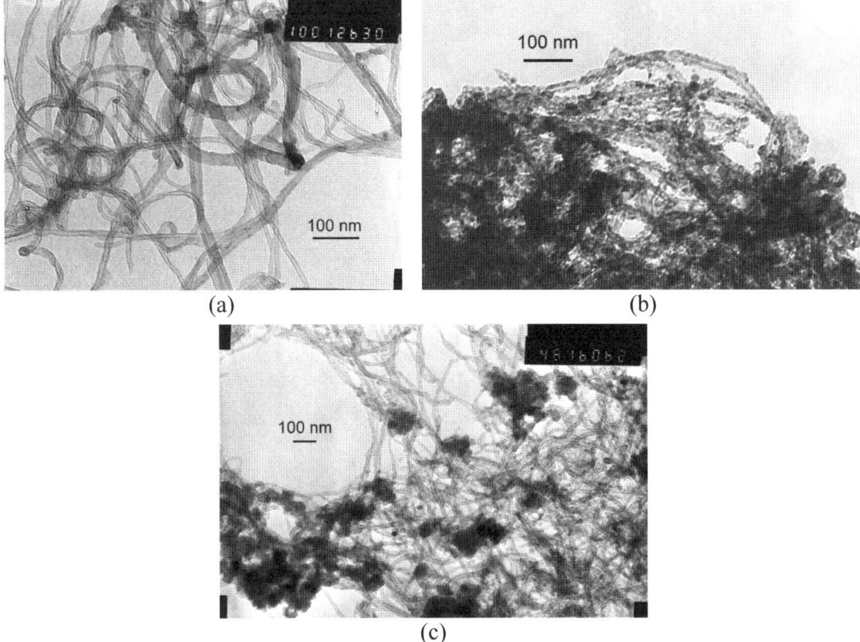

Figure 1. Electronic images: a – initial carbon nanotubes; b – carbon nanotubes modified with 50% manganese dioxide obtained by reduction of permanganate with carbon of nanotubes without addition of formic acid; c – carbon nanotubes modified with 50% manganese dioxide obtained by reduction of permanganate with formic acid

material (lead), which has a high value of molecular oxygen evolution overpotential, was chosen for comparison. A nanocomposition of carbon nanotubes with 20% lead content was prepared by electrodeposition. Figure 2 shows current–potential curves for oxygen electrodes based on a composition of MWNTs with 20% manganese dioxide (Fig. 2, curve 2). It turned out that the electrodes containing 20% lead-MWNT nanocomposition operate at current densities of 150 mA cm^{-2} better than the composition based on MnO_2/MWNTs. Thus, catalytic activity of lead, which has high molecular oxygen evolution overpotential, in the oxygen reduction reaction has been revealed. The electrical characteristics of electrodes based on a composition of MWNTs and galvanically deposited nickel have been studied. Such oxygen electrodes had much worse characteristics than the electrodes based on compositions with lead and manganese dioxide (Fig. 2, curve 3). The experimental data obtained may be explained as follows. During the electrocatalytic oxygen reduction reaction at the electrode, the reactant delivery stage, the electrochemical act of electron transfer, and reactant withdrawal take place. According to our assumption of electrochemical reaction localization, by using a heterogeneous catalyst/support composition with different energy barriers for different electrochemical reaction stages one can change the catalytic activity of this composition. This is supported by the fact of higher catalytic activity at composite electrodes in comparison with pure materials [4]. There is no bond rupture in oxygen molecule on activated carbon, pyrographite and hence on carbon nanotubes. Therefore, a catalyst with high adsorption energy, where bond rupture in oxygen molecule occurs more readily, must be used for carbon materials. Bond rupture in oxygen molecule will most likely occur on materials with high oxygen evolution overpotential, i.e. on materials with high value of the coefficient a in the Tafel equation:

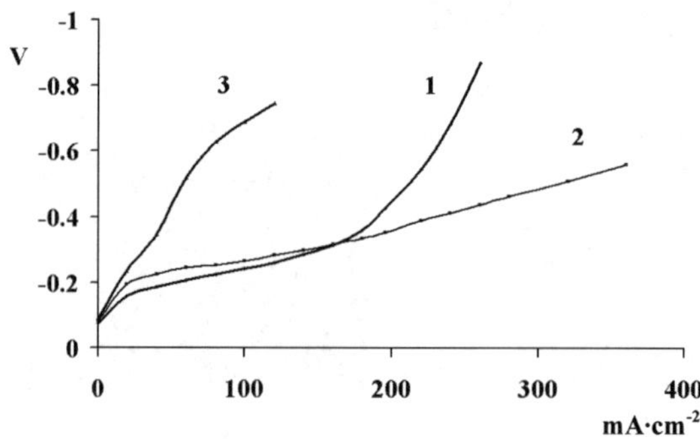

Figure 2. Current-potential plots for oxygen electrodes for fuel cells with different compositions: 1 – MWNTs containing 20% manganese dioxide; 2 – MWNTs containing 20% lead; 3 – MWNTs containing 20% nickel

$$\eta = a + \mathrm{b}\ \lg \mathrm{I}\ . \tag{1}$$

The series of metals in order of increasing coefficient *a* for oxygen evolution in an alkaline medium at low current magnitude is following [5, 6]:

$$\mathrm{Co}<\mathrm{Fe}<\mathrm{Ni}<\mathrm{Cd}<\mathrm{Pb}<\mathrm{Au}<\mathrm{Pt}\ . \tag{2}$$

The value of the coefficient *a* depends on the nature of material, specific surface area, orientation of the crystallite faces, solution composition, degree of surface oxidation [5, 6]. It follows that in the series (2) the catalytic activity of the catalysts deposited on the carbon support must increase from Co to Pt in the oxygen reduction reaction in alkaline medium. It is known from literature that platinum possesses a higher catalytic activity towards oxygen reduction reaction in alkaline media than other metals when they are used as catalysts deposited on a carbon support [7]. The fact, discovered by us, of catalytic activity of electrodes based on lead/MWNTs composition, which has high molecular oxygen evolution overpotential in alkaline solutions, also proves the assumption made above. Manganese dioxide, which possesses high catalytic activity in oxygen reduction reaction, also has high values of oxygen evolution overpotential in alkaline media. The overpotential value for oxygen evolution reaction is much lower for metallic nickel compared to lead and manganese dioxide. This corresponds to decrease of the catalytic activity of these electrodes in the oxygen reduction reaction. The assumption of localization of electrochemical reactions is corroborated also by the fact that the oxide compounds of Ni, which have also a high oxygen evolution overpotential in alkaline solutions, exhibit catalytic properties. Iron, which is known to be a catalytic poison for manganese dioxide, has, correspondingly to the above series, a lower oxygen evolution overpotential in the oxygen reduction reaction; this also corroborates the assumption of reaction localization.

Let's consider catalysts for the hydrogen electrode of a fuel cell with alkaline electrolyte. According to our assumption of reaction localization, catalysts with low hydrogen evolution overpotential in alkaline medium must be used here. In this case, adsorbed hydrogen atoms diffuse across the catalyst/support interface onto carbon surface, where electron detachment with simultaneous desorption of hydrogen ions into the electrolyte takes place. The series of metals in order of decreasing coefficient *a* for hydrogen evolution in an alkaline medium is following [5, 6]:

$$\begin{array}{llllllllllllll} \text{Metal} & Pb > & Sn > & Zn > & Cd > & Cu > & Ti > & Fe > & Ag > & Ni > & Al > & Co > & Pd > & Pt \\ \boldsymbol{a} & 1.36 & 1.28 & 1.2 & 1.05 & 0.96 & 0.83 & 0.76 & 0.73 & 0.65 & 0.64 & 0.6 & 0.53 & 0.31 \end{array} \quad (3)$$

As early as 1924, Bonhoffer found that the higher the catalytic activity of metal towards the atomic hydrogen recombination reaction, the lower hydrogen overpotential they possess, i.e. these parameters correlate in the series of metals in an opposite manner [6]. Materials consisting of MWNTs with 10% palladium content and mixtures of powders of the Ni_2Al_3 and Mg_2Ni alloys containing MWNT were chosen for investigations. From these materials, hydrogen electrodes for fuel cell were made. Figure 3 shows current-potential plots for these electrodes, which were recorded in 6 M KOH electrolyte. As follows from the electrochemical characteristics of the electrodes, these materials can be arranged according to activity for hydrogen electrode in the following order: Pd (curve 4, Fig. 3), Ni_2Al_3 (curve 3, Fig. 3), Mg_2Ni (curve 2, Fig. 3), pure carbon nanotubes (curve 1, Fig. 3). This series fully coincides with the arrangement of these metals in the order of decreasing hydrogen evolution overpotential in alkaline medium.

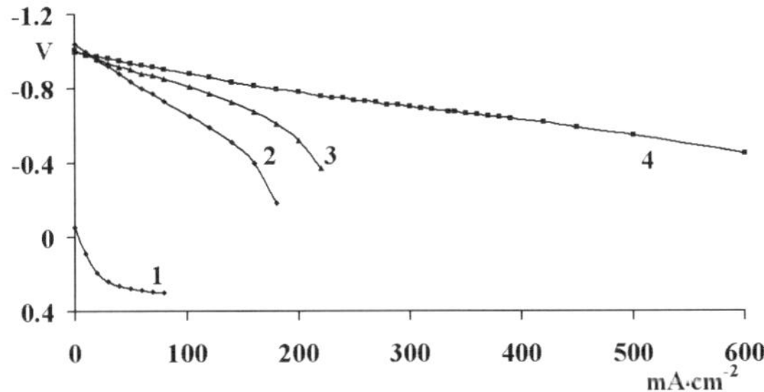

Figure 3. Current-potential plots for hydrogen electrodes for fuel cells with different active-layer compositions: 1 – 0.029 g cm^{-2} MWNT; 2 – 0.029 g cm^{-2} Mg_2Ni with 0.029 g cm^{-2} MWNT; 3 – 0.029 g cm^{-2} Ni_2Al_3 with 0.029 g cm^{-2} MWNT; 4 – 0.029 g cm^{-2} MWNT with 10% palladium

Let's now see how the catalytic properties of the MWNT-based composite hydrogen fuel cell electrode are affected by introduction of a catalyst with high hydrogen evolution overpotential in alkaline medium. Following the principle of electrochemical reaction localization, lead should be such catalyst. Using the above method for the deposition of metals on carbon nanotubes, we have obtained compositions containing: MWNTs with 5% Pd (Fig. 4, curve 2) and MWNTs with 5% Pd and 10% Pb (Fig. 4, curve 1). As follows from Fig. 4 (curve 1), introduction of lead into Pd/MWNT composition deteriorates the characteristics of the fuel cell hydrogen electrode. Thus, the assumption of reaction localization is also corroborated by the investigation of catalysts for fuel cell hydrogen electrode.

Analyzing the research results, one can formulate the following assumption. For any electrocatalytic reaction proceeding in a particular electrolyte at an electrode consisting of an inhomogeneous catalyst–support system, materials for the catalyst and support with diametrically opposite properties in this electrolyte with respect to the catalyzed reaction must be selected. For example, for oxygen reduction reaction in an alkaline electrolyte, if carbon material with relatively low oxygen evolution overpotential in alkaline medium is chosen as a catalyst support, then a material with high oxygen evolution overpotential in alkaline medium must be taken as a catalyst in this medium. For hydrogen ionization, which takes place at the fuel cell hydrogen electrode in alkaline electrolyte, good catalysts are materials having low hydrogen evolution overpotential.

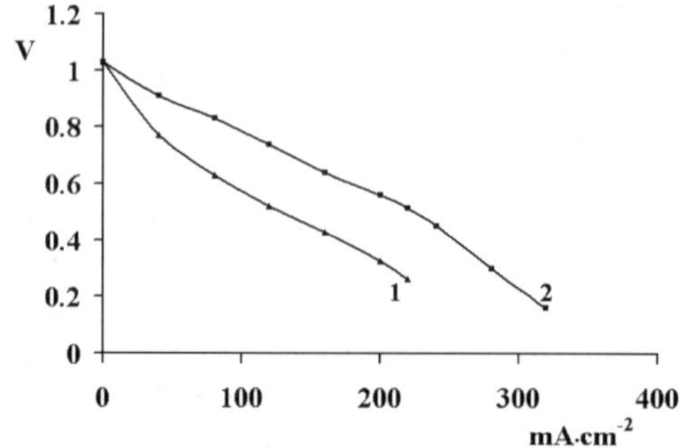

Figure 4. Current-potential plots for hydrogen electrodes for fuel cells with different active-layer compositions: 1 – 0.029 g cm^{-2} MWNT with 5% Pd and 10% Pb; 2 – 0.029 g cm^{-2} MWNT with 5% Pd

This assumption allows us to explain the electrocatalytic activity of different materials. The electrocatalytic activity of any inhomogeneous composition can be predicted from the difference of the electrochemical evolution overpotential between the catalyst and support. The larger this difference, the higher the catalytic activity of the composition obtained in the reactant ionization reaction. On the basis of this principle one can select efficient catalytic materials for electrochemical power sources and other electrochemical systems and reactions. A limitation in the selection of materials for different electrochemical reaction is the corrosion stability of the chosen materials in particular electrolyte.

4. Conclusions

1. Efficient catalysts for the fuel cell oxygen electrode based on MWNT in alkaline medium are MWNT supported lead and manganese dioxide, which have high molecular oxygen evolution overpotential.
2. Efficient catalysts for the fuel cell hydrogen electrode based on MWNT in alkaline medium are MWNT supported palladium, aluminium–nickel and magnesium–nickel alloys, which have low hydrogen evolution overpotential in alkaline medium.
3. An assumption of the localization of stages of electrochemical reactions at the catalyst and support has been made, which makes it possible to estimate the catalytic activity of the nanocomposite electrodes in reactant ionization reactions from the values of reactant evolution overpotential of the electrode components.

References

1. A.V. Melezhyk, Yu.I. Sementsov, V.V. Yanchenko, *J. Appl. Chem.* (Russian), 2005, **78**: 938–944.
2. M.O. Danilov, A.V. Melezhyk, *Power Sources*, 2006, **163**(1): 376–381.
3. M.O. Danilov, A.V. Melezhyk, *J. Appl. Chem.* (Russian), 2004, **77**(12): 1980–1984.
4. S. Cattaria, M. Musiani, *Electrochim. Acta*, 2006, **52**(3): 1339–1348.
5. A.N. Frumkin, Selected Works: Electrode Processes, Nauka, Moscow, 1987, 336 p. (in Russian).
6. L.I. Antropov, Theoretical Electrochemistry, Vysshaya Shkola, Moscow, 1975, 568 p. (in Russian).
7. X. Wang, M. Waje, Y. Yan, *Electrochem. Solid-State Lett.*, 2005, **8**: A42–A44.

THE PROBLEM OF CARBON NANOTUBES USING IN LITHIUM-ION BATTERIES

A.M. SKUNDIN
A.N. Frumkin Institute of Physical Chemistry and Electrochemistry of the RAS, 31, Leninskii prospekt, 119991, Moscow, Russia
Fax: 07 (495) 952 0846; E-mail: askundin@mail.ru

Abstract. Short review. The application of carbon nanotubes in lithium-ion batteries as matrix for lithium intercalation, and as conductive additive in negative and positive electrodes are considered

Keywords: single-walled and multi-walled carbon nanotubes, lithium-ion batteries, lithium intercalation, conductive additive

1. Introduction

The only thing which is common to carbon nanotubes (CNT) and lithium-ion batteries (LIB) is the fact that the both appeared simultaneously. Usually the origin of CNT is referenced to 1991 [1] and the first mention of modern LIB could be dated by the same year [2]. At the same time, such remarkable inventions as CNT and LIB were doomed to meet each other. The first idea of CNT application in LIB consisted in attempts to use CNT instead of graphite material in negative electrodes. Further it was proposed to use CNT as a conductive additive in active mass of positive and negative electrodes.

The modern LIB contains negative electrode made from carbon material, positive electrode made from proper oxide material (strong oxidizer), and solution of lithium salt in mixed aprotic solvents as an electrolyte. The active compounds of both electrodes are able to insert lithium ions. In the charged state negative electrode is enriched, and positive electrode is depleted by lithium. During the discharge lithium ions are de-intercalate from the carbon material on the negative electrode and are inserted into oxide on the positive electrode. The charge involves the reverse processes

2. CNT as intercalation matrices

The change of carbon materials (including graphite) for lithium metal was a key point in creation of LIB [3]. In the course of LIB charge, i.e. upon cathodic polarization of its negative electrode lithium ions are intercalated into crystal structure of graphite. Lithium activity in such intercalation compounds is more or less close to that of plain lithium, so the potential of intercalation electrode is close to the potential of lithium metal. Carbon was shown to be a very convenient matrix for lithium intercalation. The point is that specific volumes of many graphitized materials changed by no more than 10−12% upon intercalation of sufficiently large amounts of lithium, whereas other matrices (e.g. aluminum or tin) demonstrate about threefold volume increase. The most important index of carbon as the host material for lithium intercalation is so-called intercalation capacity. As a rule, intercalation of lithium into well pronounced graphite structures results in the formation of a thermodynamically stable compound, namely LiC_6. In this case the intercalation – de-intercalation could be presented by the following equation:

$$x\,Li^+ + x\,e + 6\,C \leftrightarrow Li_xC_6 \qquad (0<x<1)$$

According to Faraday's laws intercalation of 1 mol of lithium per 6 mol of carbon corresponds to specific capacity 372 mAh/g. Of course, real figures are less; they are dependent upon various factors and in modern LIB come nearer to the theoretical value.

The dependence of Li_xC_6 potential E on the intercalation degree x (this dependence is charge curve of an electrode) contains several almost horizontal steps corresponding to the following compounds: LiC_6, LiC_{12}, LiC_{18} etc. Lithium intercalation into graphite occurs at rather negative potentials, the main share of lithium is intercalated at potentials more negative than 0.5 V *vs.* Li/Li^+. On the one hand this is an advantage because the cell voltage is high; on the other hand this is a drawback because reduction of electrolytes proceeds more actively. Lithium intercalation into non-graphitized (amorphous) structures occurs at more positive potentials, in the range from 0.9 to 0.0 V. In this case E,x-dependence represents smooth curve typical to homogeneous systems.

The first short mention about CNT evaluation for LIB application appeared in the review [4], and the first papers described at length electrochemical lithium intercalation into CNT were published in 1999 [5–10]. Authors of [5, 6] used CNT obtained by pyrolysis of acetylene or ethylene catalyzed by iron or iron oxide nanoparticles, whereas group of F. Beguin [7, 8] prepared CNT by acetylene catalytic decomposition over silica supported cobalt. In both cases the objects under study were multiwalled CNT (MWNT). These first reports on lithium intercalation into CNT seemed to be hopeful. Authors of [5, 6] noted notable dependence of intercalation behavior of CNT on their graphitization degree: strongly graphitized samples demonstrated more stable cycling behavior. In all cases discharge curves were smooth notwithstanding graphitization. In all above cited works rather high values (more than 400 mAh/g) of reversible capacity were declared, however these high values referred to potential range from 0 to 3 V. The capacity released in practically important potential range, e.g. 0–1 V was much modesty.

Authors [5–8] noted also that electrodes made from MWNT possess very large irreversible capacity of the first cycle. The cathodic charge corresponding the first cycle is much more that following anodic charge and more than cathodic charge of the second cycle. This phenomenon is very well known for common carbon materials. Upon the first contact of an aprotic electrolyte with a carbon material the latter acquires a potential in the range from 2.5 to 3.5 V, the value being determined by the presence of compounds with certain oxygen-containing groups in the electrolyte. During the first cathodic polarization the potential shifts in negative direction, and slow lithium intercalation begins. Simultaneously, the electrolyte is reduced with formation insoluble, soluble, and gaseous products. Insoluble products of the reduction form a film on the carbon surface. This film is known as *solid electrolyte interface* (SEI). SEI possesses properties of electrolyte with conductance by lithium ions. This SEI prevents direct contact between the electrolyte and carbon, thus virtually eliminating further reduction of the electrolyte. At the same time, being ionic conductor SEI does not hinder lithium intercalation and de-intercalation. It is the charge consuming for electrolyte reduction during the first cycle that presents irreversible capacity.

The investigations fulfilled after the pioneered works [5−10] confirmed by and large the above conclusions. It is important that both single walled CNT (SWNT) [10−15] and MWNT [16–26] were studied, the latter having some advantages. As a rule, lithium intercalation capacity of CNT amounts to 150–250 mAh/g. Very large irreversible capacity was documented in [8−10, 12, 18, 27−29].

One can find in the literature some data about very high intercalation capacity of CNT (see, e.g., [9, 30]), but these data were not confirmed.

In principle, poor intercalation ability of CNT, and especially, SWNT is not surprising. Lithium intercalation in carbon consists in placing lithium ions in between graphene planes [31]. Entering lithium ions through cylindrical surface of nanotube is low probable [32]. Entering such ions from tube's end is connected with very long diffusion distance. The most probable places of lithium ions are vacancies, nanoscopic cavities and voids [6, 26]. At the same time, side surface of CNT is ideal place to irreversible reactions occurring [33, 34].

3. Composite electrodes based on CNT: negative electrodes

There are various host materials able to insert much higher amounts of lithium than carbon. However, as it was noted in the previous section all these materials suffer from large volume changes upon lithium insertion. For instance, aluminum, tin, and silicon are able to form intermetallic compounds Li_9Al_4, $Li_{17}Sn_4$, and $Li_{21}Si_5$, which corresponds to limiting insertion capacity 2,235, 959, and 4,010 mAh/g, respectively. Volume change in these cases amounts to 238%, 257%, and 297% (!).

With due account for the excellent mechanical properties of CNT it was proposed to use they as some matrix for compensating volume change of above mentioned materials. Surely, composite materials with CNT must contain counterparts in nano-structured forms as well. Very interesting approach was presented in [35]. CNT synthesized by the catalytic decomposition were purified, oxidatively opened, and filled with a tin salt by capillary action. Then tin ions were reduced with such strong reducer as $NaBH_4$ or by hydrothermal process. Material thus obtained had specific capacity about 800 mAh/g at low discharge rate and about 600 mAh/g at rate 0.3 C. It was the unique work describing filling inner cavity by a metal. Unfortunately, this work had no continuation. The work [36] describes nanocomposites of CNT with Sb and $SnSb_{0.5}$ particles, which were prepared by chemical reduction of $SnCl_2$ and $SbCl_3$ precursors in the presence of CNT. In this case Sb and $SnSb_{0.5}$ particles are uniformly deposited on the CNT exterior and in the CNT web. The composite electrodes showed improved cyclability compared to unsupported Sb and Sn–Sb particles and higher reversible specific capacities than plain CNT. The reversible capacities were as high as 462 mAh/g for CNT–36 wt% Sb and 518 mAh/g for CNT–56 wt% $SnSb_{0.5}$. After 30 cycles, the capacity was 62.1% of the initial capacity for the former and 67.2% of the initial capacity for the later. The improvement in cyclability may be attributed to the nanoscale dimension of the metal particles and CNT role as a buffer in relieving the mechanical stress induced by specific volume changes in electrochemical lithium insertion and extraction reactions.

In the years gone attention of many researchers is compelled to silicon−carbon composites as very promising materials for negative electrodes of LIB. No wonder that CNT were proposed as carbon component of the composites [37−43].

Such composites can be manufactured by various methods. For instance, MWNT/Si composites were produced from purified MWNT and Si-powder by high-energy ball-milling [39–42]. It was shown in [40] that electrochemical behavior of the composites is very sensitive to the weight ratio of CNT to Si. The highest reversible capacity (1,770 mAh/g) and the lowest irreversible capacity (470 mAh/g) were found in fifty-fifty composites. Most of lithium is inserted in silicon and role of CNT consists in composite stabilization and preventing silicon crumbling.

In [37, 38] CNT were catalytically grown by a chemical vapor deposition at silicon particles covered by nickel. Thus obtained core/shell structure provides very good electrochemical performance of the composite. This beneficial feature was ascribed to the conductive buffering role of $CNT_{surface}$ layer. The void space and the flexible characteristics in this surface layer allow volume expansion of the Si core without severe electrode swelling. Similar technique for CNT/Si composite preparing was used in [43], where CNT were deposited onto silicon doped with chromium. Such doping is known to increase silicon conductivity. Composites with doped silicon possess enhanced cycling stability.

Negative electrodes based on composites CNT with CuO are described in [5]. Specific capacity of these electrodes was too low.

4. Composite electrodes based on CNT: positive electrodes

When CNT are used in the composites for positive electrodes the main importance is their high conductivity. Matrices from CNT ensure even distribution of electrochemical process intensity along electrode's depth and more effective use of active material. CNT were offered for positive electrodes with $LiCoO_2$ [44, 45], doped lithium-manganese spinel [46], V_2O_5 [47], polyaniline (PANI) [48] and sulfur [49]. The positive electrodes in [44–46] were prepared by traditional technology, specifically application of active mass, consisting from active oxide substance, CNT as conductive additive, and proper binder. The electrodes in [47] were prepared by sol-gel method with using SWNT and vanadium oxide aerogel. The similarity in morphology as well as dimensional scale for SWNT and V_2O_5 ribbons enabled the excellent electrical contact to be made between both components of the composite without seriously affecting the aerogel nanostructure. In this case intimate contact was established between the both phases at the nanodimensional level. The V_2O_5/SWNT composite electrodes exhibited specific capacity more than 400 mAh/g at high discharge rates and good cyclability.

The principle of CNT action in $LiCoO_2$/MWNT composites was suggested in [44] to be the same as in Si/MWNT composites [38]. CNT are a much better conductive additive than carbon black. The improved performance of such composites appears to be largely due to resilience of the MWNT aggregates that form conductive bridges between particles of $LiCoO_2$.

PANI/CNT nanocomposite [48] was synthesized by *in situ* chemical polymerization of aniline in well-dispersed MWNT suspension. Such a composite electrode was found to have the better performance in terms of cyclability and discharge capacity than mechanical mixture of PANI and MWNT with the same composition. The composite electrode delivered discharge capacity of 86 mAh/g with good capacity retention.

Elemental sulfur is known to be not a very popular, although promising active material for LIB. Authors of [49] proposed a novel material named as MWNT–sulfur nanocomposite (MSN). This material was synthesized by heating the mixture of sublimited sulfur and MWNT in certain conditions. The positive electrodes made from MSN demonstrated rather high discharge capacity (670 mAh/g at the second cycle and more than 500 mAh/g at the 60th cycle). Authors believed that MWNT play a vital role on adsorbing sublimited sulfur and polysulfides within the electrode body, and furthermore provide an excellent electrical contact.

5. Conclusions

The hopes to utilize CNT as insertion material for negative electrodes of LIB were not justified. However, CNT have good prospects for using in the composite materials for both negative and positive electrodes.

References

1. S. Iijima, *Nature*, 1991, **354**: 56.
2. T. Nagaura, in Materials of the 5th International Seminar on Lithium Battery Technology and Application, Deerfield Beach, FL, 1991, p. 255.
3. A.M. Skundin, O.N. Efimov, O.V. Yarmolenko, *Russ. Chem. Rev.*, 2002, **71**: 329.
4. J.C. Withers, R.O. Loutfy, T.P. Lowe, *Fullerene Sci. Technol.*, 1997, **5**: 1.
5. G.T. Wu, C.S. Wang, X.B. Zhang, H.S. Yang, Z.F. Qi, W.Z. Li. *J. Power Sources*, 1998, **75**: 175.
6. G.T. Wu, C.S. Wang, X.B. Zhang, H.S. Yang, Z.F. Qi, P.M. He, W.Z. Li, *J. Electrochem. Soc.*, 1999, **146**: 1696.
7. E. Frackowiak, S. Gautier, H. Gaucher, S. Bonnamy, F. Beguin, *Carbon*, 1999, **37**: 61.
8. F. Leroux, K. Metenier, S. Gautier, E. Frackowiak, S. Bonnamy, F. Beguin, *J. Power Sources*, 1999, **81–82**: 317.
9. G. Maurin, Ch. Bousquet, F. Henn, P. Bernier, R. Almairac, B. Simon, *Chem. Phys. Lett.*, 1999, **312**: 14.
10. B. Gao, A. Kleinhammes, X.P. Tang, C. Bower, L. Fleming, Y. Wu, O. Zhou, *Chem. Phys. Lett.*, 1999, **307**: 153.
11. B. Gao, C. Bower, J.D. Lorentzen, L. Fleming, A. Kleinhammes, X.P. Tang, L.E. McNeil, Y. Wu, O. Zhou. *Chem. Phys. Lett.*, 2000, **327**: 69.
12. A.S. Claye, J.E. Fischer, C.B. Huffman, A.G. Rinzler, R.E. Smalley, *J. Electrochem. Soc.*, 2000, **147**: 2845.
13. D. Aurbach, J.S. Gnanaraj, M.D. Levi, E.A. Levi, J.E. Fischer, A. Claye. *J. Power Sources*, 2001, **97–98**: 92.
14. R. Scott Morris, B.G. Dixon, T. Gennett, R. Raffaelle, M.J. Heben. *J. Power Sources*, 2004, **138**: 277.
15. S.H. Ng, J. Wang, Z.P. Guo, J. Chen, G.X. Wang, H.K. Liu, *Electrochim. Acta*, 2005, **51**: 23.
16. Zhan-hong Yang, Hao-qing Wu. *Solid State Ionics*, 2001, **143**: 173.
17. T. Ishihara, A. Kawahara, H. Nishiguchi, M. Yoshio, Yu. Takita, *J. Power Sources*, 2001, **97–98**: 129.
18. Heon-Cheol Shin, Meilin Liu, B. Sadanadan, A.M. Rao, *J. Power Sources*, 2002, **112**: 216.
19. I. Mukhopadhyay, N. Hoshino, S. Kawasaki, F. Okino, W.K. Hsu, H. Touhara, *J. Electrochem. Soc.*, 2002, **149**: A39.
20. G.X. Wang, Jung-ho Ahn, Jane Yao, M. Lindsay, H.K. Liu, S.X. Dou, *J. Power Sources*, 2003, **119–121**: 16.
21. T. Ishihara, A. Kawahara, H. Nishiguchi, M. Yoshio, Yu. Takita, *J. Power Sources*, 2003, **119–121**: 24.
22. A. Odani, A. Nimberger, B. Markovsky, E. Sominski, E. Levi, V.G. Kumar, M. Motiei, A. Gedanken, P. Dan, D. Aurbach, *J. Power Sources*, 2003, **119–121**: 517.
23. Ji Yong Eom, DongYung Kim, Hyuk Sang Kwon, *J. Power Sources*, 2006, **157**: 507.
24. M. Zabloudi, J. Makovička, M. Sedlaříková, J. Vondrák, In Proceedings of 8th International Conference Advanced Batteries and Accumulators, Brno, June 3–7, 2007, p. 40.
25. Shubin Yang, Huaihe Song, Xiaohong Chen, A.V. Orotrub, L.G. Bulusheva, *Electrochim. Acta*, 2007, **52**: 5286.
26. M. Wang, Z.H. Li, G.T. Wu, *Russ. J. Electrochem.*, 2005, **41**: 946.

27. T. Prem Kumar, A. Manuel Stephan, P. Thayanananth, V. Subramanian, S. Gopukumar, N.G. Renganathan, M. Raghavan, N. Muniyandi. *J. Power Sources*, 2001, **97–98**: 118.
28. M. Sharon, W.K. Hsu, H.W. Kroto, D.R.M. Walton, A. Kawahara, T. Ishihara, Y. Takita, *J. Power Sources*, 2002, **104**: 148.
29. S. Yoon, C. Park, H. Yang, Y. Korai, I. Mochida, R.T.K. Baker, N.M. Rodrigues, *Carbon*, 2004, **42**: 21.
30. H. Shimoda, B. Gao, X.P. Tang, A. Kleinhammes, L. Fleming, Y. Wu, O. Zhou, *Phys. Rev. Lett.*, 2002, **88**: 0155021.
31. G. Maurin, Ch. Bousquet, F. Henn, P. Bernier, R. Almairac, B. Simon, *Solid State Ionics*, 2000, **136–137**: 1295.
32. Z.H. Yang, H.-Q. Wu, *Chem. Phys. Lett.*, 2001, **343**: 235.
33. Z.H. Yang, H.-Q. Wu, *Chem. Phys. Lett.*, 2001, **343**: 225.
34. Z.H. Yang, H.-Q. Wu, *Mater. Chem. Phys.*, 2001, **71**: 7.
35. T. Prem Kumar, R. Ramesh, Y.Y. Lin, G.T.-K. Fey, *Electrochem. Comm.*, 2004, **6**: 520.
36. Wei Xiang Chen, Jim Yang Lee, Zhaolin Liu, *Carbon*, 2003, **41**: 959.
37. T. Kim, Y.H. Mo, K.S. Nahm, S.M.S. Oh, In Proceedings of 207th ECS Meeting, Quebec City, Canada, May 15–20, 2005, Abstract 163.
38. T. Kim, Y.H. Mo, K.S. Nahm, S.M. Oh, *J. Power Sources*, 2006, **162**: 1275.
39. J. Shu, H. Li, R. Yang, Y. Shi, X. Huang, *Electrochem. Comm.*, 2005, **8**: 51.
40. J.Y. Eom, J.W. Park, H.S. Kwon, S. Rajendran, *J. Electrochem. Soc.*, 2006, **153**: A1678.
41. Y. Zhang, X.G. Zhang, H.L. Zhang, Z.G. Zhao, F. Li, C. Liu, H.M. Cheng, *Electrochim. Acta*, 2006, **51**: 4994.
42. W. Wang, P.N. Kumta, In Proceedings of 212th ECS Meeting, Washington, DC, October 7–12, 2007, Abstract 706.
43. T. Ishihara, M. Nakasu, M. Yoshio, H. Nishiguchi, Yu. Takita, *J. Power Sources*, 2005, **146**: 161.
44. K. Sheem, Young Hee Lee, H.S. Lim, *J. Power Sources*, 2006, **158**: 1425.
45. M. Sysel, J. Makovička, M. Sedlaříková, J. Vondrák, In Proceedings of 8th International Conference Advanced Batteries and Accumulators, Brno, June 3–7, 2007, p. 106.
46. Qian Lin, J.N. Harb, *J. Electrochem. Soc.*, 2004, **151**: A1115.
47. J.S. Sakamoto, B. Dunn, *J. Electrochem. Soc.*, 2002, **149**: A26.
48. S.R. Sivakkumar, Dong-Won Kim, *J. Electrochem. Soc.*, 2007, **154**: A134.
49. W. Zheng, Y.W. Liu, X.G. Hu, C.F. Zhang, *Electrochim. Acta*, 2006, **51**: 1330.

INFLUENCE OF CARBON NANOFIBERS ON CYCLIZATION AND CARBONIZATION PROCESSES OF POLYACRYLONITRYLE

G.N. GUBANOVA, T.E. SUKHANOVA, A.I. GRIGORIEV, YU.N. SAZANOV, A.V. NOVOSELOVA, A.V. GRIBANOV, G.N. FIODOROVA, A.A. KUTIN
Institute of Macromolecular Compounds Russian Academy of Sciences, Bolshoy pr.31, Saint-Petersburg, 199004, Russia

Abstract. Polyacrylonitryle synthesized by anion polymerization mechanism (a-PAN), and their compositions with carbon nanofibers (CNF) were investigated by X-ray diffraction analysis, electron microscopy, differential scanning calorimetry and thermo-gravimetric analysis. The influence of carbon nanofibers on cyclization and carbonization processes was detected, as well as on transformation of PAN's and their compositions structure under the heat treatments.

Keywords: carbon nanofiber, polyacrylonitryle, cyclization, carbonization, structure transformation

1. Introduction

Investigation of polyacrylonitryle (PAN) cyclization and carbonization processes with addition of different fillers is very important problem at present. Compositions based on PAN and their carbonized products have broad applications in industry [1–2].

It is well known that carbonized PAN is the leader among polymers precursors for carbon materials owing to his high cox residue. Development of optimal carbon structure after PAN's and their compositions temperature's treatments are determined by forming pre-structure at temperature up to 600°C, where cyclization is the main chemical reaction. Cyclization is accompanied by high quantity heat release, which can lead to destruction of polymers and decreasing of cox residue. Decrease of exo-effect and the temperature of PAN's cyclization – is very important practical and scientific problem. For this purpose polymer's modification by fillers are often used [3, 4].

For investigation of PAN's cyclization and initial carbonization process polymer, synthesized by radical as well anionic polymerization were chosen [5, 6]. It was shown by DTA, IR- and mass spectroscopy [7, 8] that PAN synthesized on anionic initiators may be used as convenient model. The disadvantages of radical methods are: chemical additives, oxides, catalysts. Investigations of a-PAN show absence of volatile products at temperature 20–300°C [8].

Last years carbon nanofilers often used for polymer's properties modification [9]. It may be supposed that introduction carbon nanofillers may increase thermostability and cox residue of polymer.

The goal of this work is the study cyclization's and carbonization's features of a-PAN and their compositions, modified by carbon nanofibers (CNF).

2. Experimental

For investigations the a-PAN powder, synthesized by means of anion catalyst were used as described [7]. Thermochemical study were carried out using the samples with molecular weigh $M_w \sim 20{,}000$. Combinations of two methods – thermogravimetric and differential thermal analysis were used for estimation mass loss and temperature's transitions during heating up to 1,000°C. For thermogravimetric analysis (TGA) derivatograph 'MOM'-Hungary was used. The scanning rate during experiment was 10°/min.

Modifications of polyacrylonitryle was performed using carbon nanofibers (CNF), obtained in Tambov State Technical University according with [10]. A-PAN compositions with CNF were crushed to powder by mechanical method.

The transmission electron microscopy data of CNFs under different magnifications are presented in Fig. 1. Figure 1(a) presents non-fractionated fragments of nanofibers. Carbon nanofibers are curved. The diameters of CNFs varied from 30 to 70 nm, and length attained 1,000 nm. It is clearly seen so called 'kinks' on the carbon fiber's surfaces due to stresses originated under bending (Fig. 1(b)).

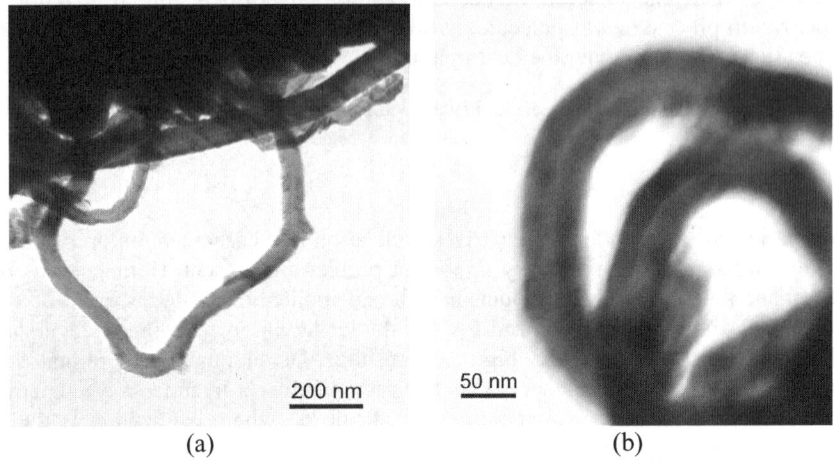

Figure 1. Transmission electron microscopy data for CNF's at different magnifications

Wide-angle X-ray diffraction (WAXD) analysis was carried out using a DRON-3 (Russia) diffractometer with Ni-filtered CuK_α radiation (λ = 1.54 Å). Measurements were performed at scattering angle 2θ = 4–36°. Cyclization and carbonization of a-PAN and their compositions with CNF were carried out on differential scanning calorimeter DSM-2M (Pushino, Russia) at scanning rates 8° and 16°/min. Indium was used for temperature's and enthalpy calibration.

3. Results and discussion

Figure 2 presents TGA (a) and DTA (b) data for a-PAN (curve 1), composition's a-PAN: CNF in wt% ratio 80:20 (curves 2), 50:50 (curves 3) and CNF (curves 4). Mass losses were recoded for all samples at 160°C, then multistep processes, connected with evaporation of low molecular weigh products, were observed up to 350°. During heating up to 370°C a-PAN (curve 1) lost 4% of mass related to evaporation of adsorbed water. In temperature region 270–308°C drastic mass loss (about 12%) connected with release of low molecular weigh products of polymerization is observed. During heating from 310°C till 800°C TGA curve reflects typical carbonization process, connected with dehydration of heterochain polymer. The coke residue value after heating a-PAN up to 800°C is equal to 44%.

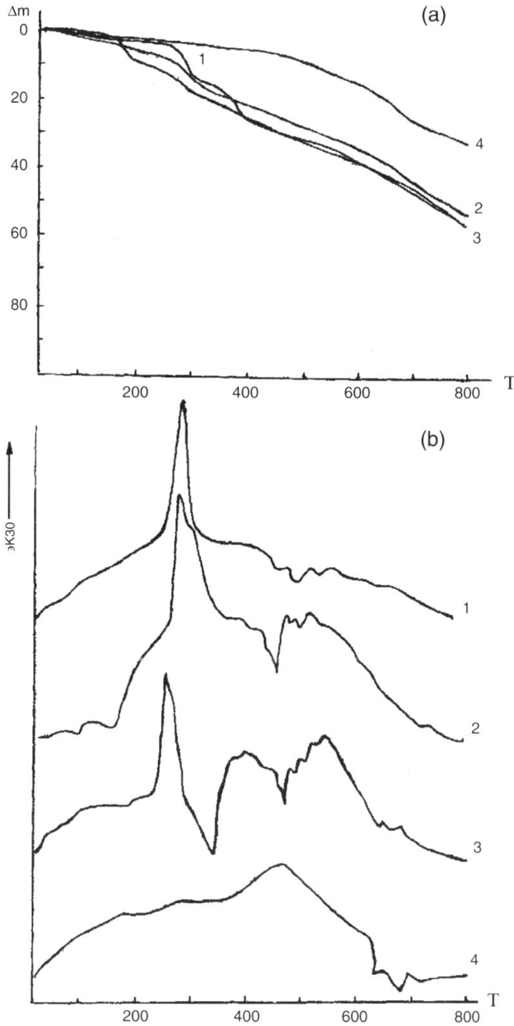

Figure 2. TGA (a) and DTA (b) curves for a-PAN (1) and compositions PAN:CNF (80:20) – (2); (50:50) – (3); CNF- (4)

The coke residue value after heating up to 800°C is equal to 43–45% for composition 80:20 and 50:50 respectively.

For all DTA curves in region 200–350°C one can observed exo-effects, connected with cyclization of a-PAN and their compositions. The exo-effect's maximums change from 250° till 300°C. Notable shift of cyclization's temperature, observed for a-PAN:CNF (50:50), may be connected with influence of CNF on a-PAN cyclization. After cyclization process (when temperature of heating attain 400°C) carbonization of a-PAN starts and followed by slow release of volatile products. Rebuilding of virgin cycling structure of carbonization product in region 400–550°C has more complex character.

On the DTA curve for pure a-PAN two weak and non symmetric exo-peaks are observed at 456°C, 470°C and 537°C (Fig. 1(b), curve 1). On the thermograms for a-PAN compositions, modified CNF (Fig. 1(b), curves 2, 3) in temperature region 400–550°C endo effects observed, that can be explained as melting of ordered regions of polymer's chain in a-PAN and their compositions with CNF. It is known that cyclization of a-PAN accompanied by forming of oriented PAN's macromolecules with crystalline structure fragments. These fragments were synthesized and characterized in [11].

DSC study of pure a-PAN and compositions a-PAN with CNF (50:50) was carried out for estimating of nano filler's influences on physical transition. Exo effect due to cyclization of pure a-PAN characterizes by semi-width about 50°C with maximum at 310°C. The beginning of cyclization exotherm coincides with 250°C. After mixing a-PAN with CNF, exotherm of cyclization became bimodal and at high scan rate (16°/min) both peak became separate. Maximum of exo-peaks are 280°C and 330°C at scan rate 8°/min. It was shown that energy of a-PAN cyclization at presence of CNF in order lower than for pure PAN and equal to 600 and 2,200 J/g, respectively.

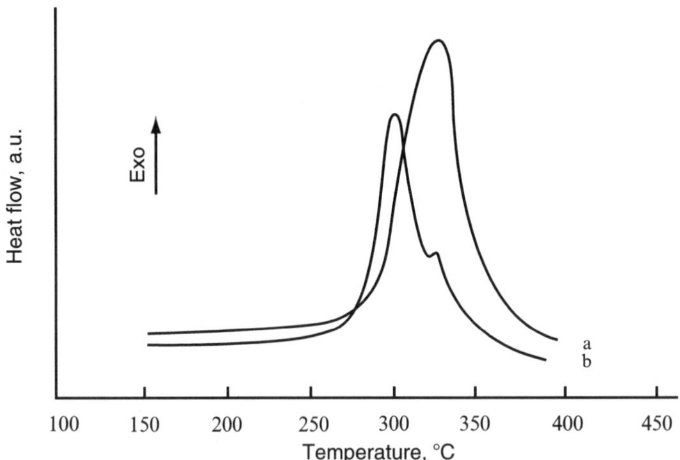

Figure 3. DSC data for pure a-PAN (a) and composition a-PAN with CNF (50:50 wt%); scan rate 8°/min

Figure 4 presents the X-ray diffraction data for a-PAN (a), CNF (b), and composition (c) a-PAN:CNF (50:50 wt%).

X-ray investigation of initial a-PAN (a) show, that very sharp intensive reflex is observed at the diffraction angle 2θ = 17°, which correspond d = 5.2 Å, and two broad reflexes 2θ = 25°30' (d = 3.5 Å) and 2θ = 29° (d = 3.1 Å).

Diffractogramm from CNF contains narrow reflex at 2θ=26°15'(d = 3.4Å), corresponding graphite structure (b). On the diffractogramm from composition a-PAN:CNF in initial state (c) one sharp reflex with d = 5.2 Å, corresponding to initial a-PAN, and reflex with d = 3.4 Å, corresponding CNF are observed, which confirmed that we dealt with mechanical mixture of a-PAN and CNF.

After heating of pure a-PAN at 400°C during 3 h reflexes for initial a-PAN escaped (Fig. 5(a)), only weak halo at $2\theta = 15°$ (d = 5.9 Å) is registered and new intensive reflex $2\theta = 25°$ (d = 3.55 Å) appeared. This fact proved that structure of initial a-PAN is destroyed and carbon's turbostratic structure arose at the beginning of carbonization.

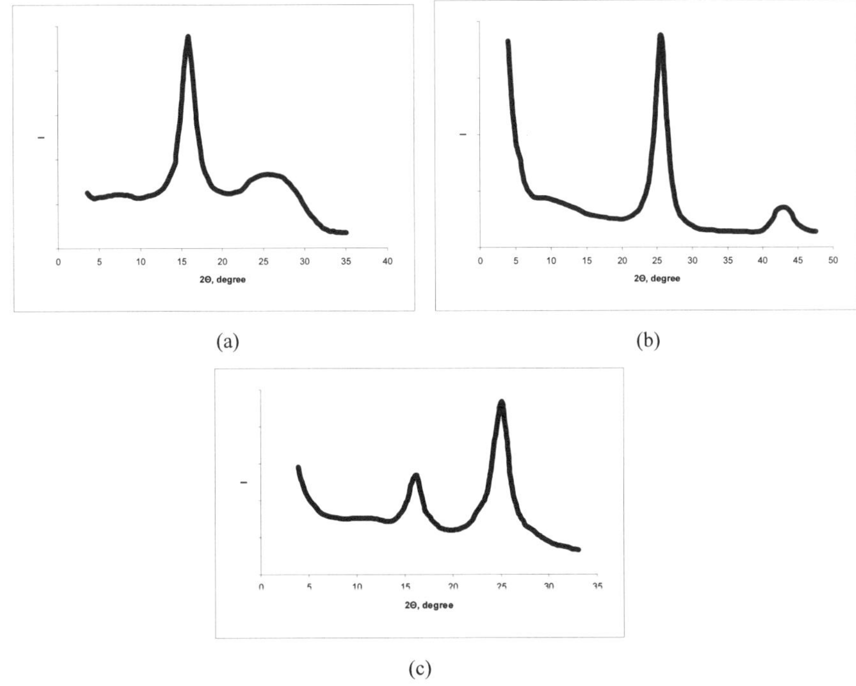

Figure 4. Wide-angle X-ray scattering pattern for a-PAN (a), CNF (b), composition a-PAN:CNF (50:50 wt%) – (c)

After heating of pure a-PAN at 400°C during 3 h reflexes for initial a-PAN escaped (Fig. 5(a)), only weak halo at $2\theta = 15°$ (d = 5.9 Å) is registered and new intensive reflex $2\theta = 25°$ (d = 3.55 Å) appeared. This fact proved that structure of initial a-PAN is destroyed and carbon's turbostratic structure arose at the beginning of carbonization.

Temperature treatment of a-PAN:CNF (50:50 wt%) composition at 400°C during 3 h leads to fracture of initial ordered structure a-PAN. On diffractogramm (Fig. 5(b)) remains only one reflex at $2\theta = 26°15'$ (d = 3.4 Å), with more broad width at the foundation, then for the same reflex for unheated composition with CNF. Hence, this reflex is originates from superposition of CNF' reflex and reflex from heated a-PAN (with carbon' turbostratic structure). All this data allowed to conclude that after cyclization of PAN and compositions ordered regions with turbostratic structure of carbon are formed, and then they undergo phase transitions after heating at 400–550°C.

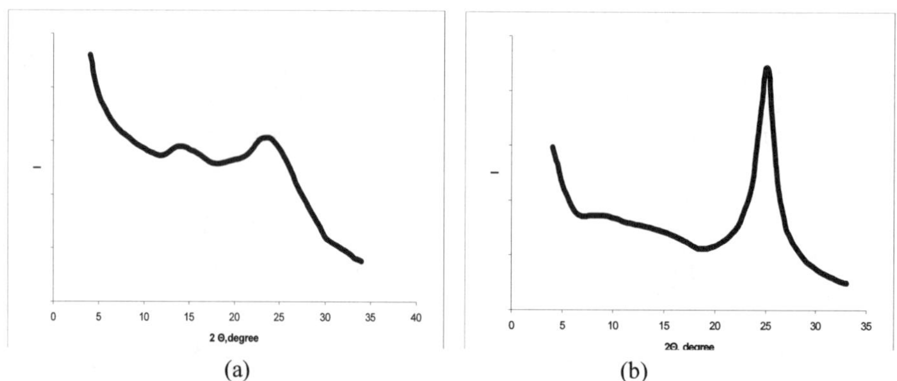

(a) (b)

Figure 5. Wide-angle X-ray scattering pattern for a-PAN (a), composition a-PAN: CNF (50:50 wt%) – (b), after heating at 400°C during 3 h

4.Conclusions

1. The influence CNF on cyclization of a-PAN were investigated.
2. It was shown that a-PAN's cyclization energy in CNF presence in order low than for pure a-PAN. Adding of CNF into the a-PAN leads to decreasing cyclization temperature also.
3. By Wide-angle X-ray scattering method carbon's turbostratic structure formation was found in a-PAN and their compositions at initial stage of carbonization (till 400°C).

References

1. Kasatochkin V.I., Kargin V.A.//Dokl Akad Nauk SSSR, 1970, **19** (5):1004–1007 (in Russian).
2. Fialkov A.S. *Carbon, interlaminar compounds and composites on their base* (Moscow: Aspect Press: 1997), 717 p. (in Russian).
3. Konkin A.A. *Carbon and other thermostable fiber's materials* (Moscow: Chemistry: 1974), 544 p. (in Russian)
4. Varshavskii V.Ya. *Polymer Sci.*, 1983, **25** (ser. A (4)):823–830 (Russia).
5. Erusalinsky B.L., Novoselova A.V. *Faserforsch.und Textiltech.*, 1975, **B. 26**: 81–87.
6. Fedorova L.A., Mazurek V.V., Shirokov N.A., Turkova L.D. *React. Kinet. Catal. Lett.*, 1983, **23** (3/4):343–350.
7. Novoselova A.V., Zgonnik V.N., Cazanov Yu.N. et al. *Polym. Sci.*, 1992, **34** (ser. A (3):59–63 (Russia).
8. Spirina T.N., Sazanov Yu.N., Shibaev L.A., et al. *J. Thermal Anal.*, 1990, **36**:2329–2338.
9. Jeffrey J., Karl J., et al. Experimental trends in polymer nanocomposites – a review. *Mat. Sci. Eng.*, 2005, **A 393**:1–11.
10. Zolotuchin I.B. et al., *Lett. J. Tech. Phys.*, 2006, **32**: 5.
11. Takata T., Hiroi I., *J. Polym. Sci. A.*, 1964, **2**: 1367–1385.

GROWTH FEATURES OF CARBON NANOSTRUCTURES ON Co PARTICLES DEPOSITED ON SiO_2 AND Al_2O_3

A.A. VOLODIN, P.V. FURSIKOV
Institute of Problems of Chemical Physics of RAS, Prospectus of the Academician Semenov, 1, Chernogolovka, Moscow Region, 142432 Russia
Tel./Fax: +7(496) 5221743; E-mail: btarasov@icp.ac.ru
YU.A. KASUMOV
Institute of Microelectronics Technology and High Purity Materials of RAS, Institute Str. 6, Chernogolovka, Moscow Region, 142432 Russia
B.P. TARASOV
Institute of Problems of Chemical Physics of RAS, Prospectus of the Academician Semenov, 1, Chernogolovka, Moscow Region, 142432 Russia

Abstract. The synthesis of carbon nanostructures by Co catalyst based on SiO_2 and Al_2O_3 was investigated. It was established that the catalyst Co/Al_2O_3 is more effective for the synthesis of carbon nanostructures than Co/SiO_2. Increasing the pyrolysis temperature favors the formation of the more uniform carbon nanostructures with smaller diameters.

Keywords: carbon nanostructures, cobalt, catalyst, pyrolysis, temperature

1. Introduction

For the pyrolytic synthesis of carbon nanostructures (CNS) one needs to avoid aggregation of metal particles of a catalyst during the process [1]. For this purpose clusters of a catalyst are usually deposited on different supports using the methods of impregnation, co-deposition, photolithography, laser ablation, etc. [2]. Oxides (SiO_2, Al_2O_3, MgO, and La_2O_3), nitrides, and soot are of wide use as a support [3, 4].

One of factors having effect on the growth on CNS is the interaction of the catalyst with the surface of the support. This is of utmost importance when metal catalysts are deposited on Al_2O_3 and SiO_2 [4]. A different effect of the support may be also due to a distinction in the surface area or density of defects on the substrate [5].

The aim of the present work is to study the influence of the chemical nature of support of catalyst particles on the structure and diameter of carbon nanofibers and nanotubes formed during the pyrolysis.

2. Experimental

In the work we used a Co catalyst deposited on the SiO_2- and Al_2O_3-support, which were granules having the diameter of 1–3 mm. The deposition was performed by impregnation with a water solution of $Co(NO_3)_2$ with subsequent drying and heating up to the decomposition of the starting salt. Additional reduction to Co^0 was performed by heating of the samples in a hydrogen flow at 500ºC for 30 min. After the reduction the Co contents in the samples was 3 wt %. The surface area of the catalysts was 120 m^2/g in all cases.

Carbon nanostructures were synthesized in a gas flow reactor at 0.1 MPa. The catalyst was placed in the middle of the reactor as a thin bed with the thickness about of 0.3 mm. At the pyrolysis temperature of 500ºC a gas mixture C_2H_4:H_2 = 5:1 was used, at 750°C – mixtures: C_2H_4:H_2 = 5:1 and C_2H_4:H_2 = 1:2. The duration of the process was 30 min in all experiments. Pyrolysis products were examined by TEM and SEM.

3. Results and discussion

When the catalyst Co/Al_2O_3 was used the ethylene pyrolysis at 500°C resulted in a formation of two kinds of fibers: thick ones being of the diameter 40–50 nm, thin ones – 8–10 nm (Fig. 1).

Figure 1. Image of the pyrolysis products of the gas mixture C_2H_4:H_2 = 5:1 at 500°C over Co/Al_2O_3

In the case of the catalyst Co/SiO_2 at the same conditions fibers were not formed. According to TEM data all the surface of the catalyst were covered by a layer of amorphous carbon.

At 750°C and the gas ratio C_2H_4:H_2 = 5:1 over Co/Al_2O_3 single-type fibers were formed with the average diameter of 20–30 nm (Fig. 2).

Over Co/SiO_2 at the same conditions fibers were not formed, while the catalyst particles covering by carbon shells (Fig. 3). The diameter of these capsules ranged within 50–200 nm.

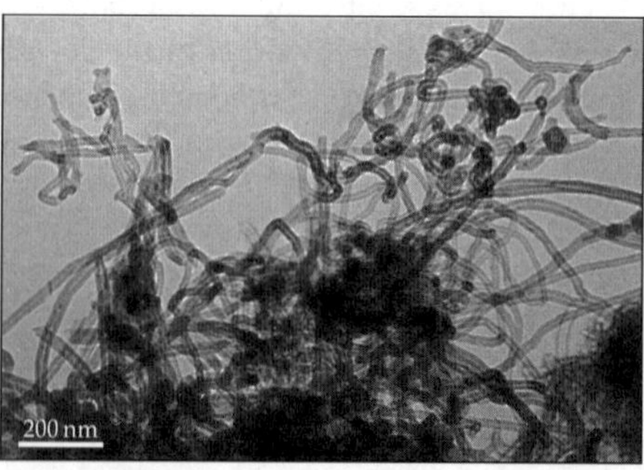

Figure 2. Image of the pyrolysis products of the gas mixture C_2H_4:H_2 = 5:1 at 750°C over Co/Al_2O_3

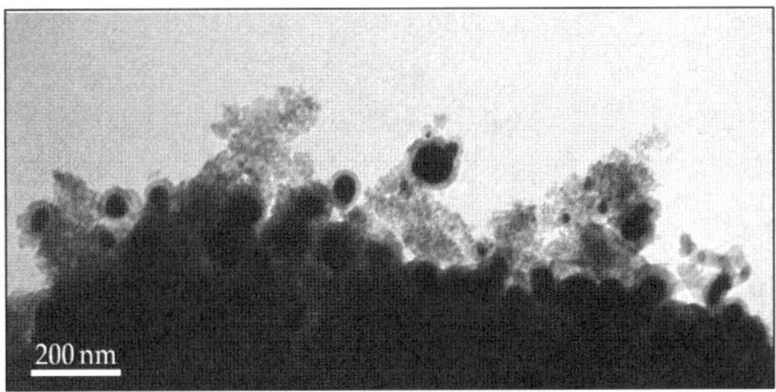

Figure 3. Image of the pyrolysis products of the gas mixture C_2H_4:H_2 = 5:1 at 750°C over Co/SiO_2

An addition of hydrogen in the gas mixture prevents coking of the catalyst particles. Increasing the hydrogen content (C_2H_4:H_2 = 1:2) promotes a growth on Co/Al_2O_3 of thinner and uniform fibers with diameter of 10–20 nm (Fig. 4). Such influence of the temperature and hydrogen content on the products composition could be explained within the framework of the model "decomposition-diffusion-precipitation" [6].

Increasing of the hydrogen content favors the growth of carbon nanofibers in the case of utilizing the catalyst Co/SiO_2 as well. At the gas ratio C_2H_4:H_2 = 1:2 on the surface of Co/SiO_2 nanofibers with the diameter of 20–30 nm are formed along with encapsulated particles and amorphous carbon (Fig. 5).

Analysis of the obtained data shows that the chemical nature of the support plays an essential role in the growth of carbon nanostructure. Al_2O_3 is known to have more pronounced Lewis acidity than SiO_2, therefore the interaction of the Co particles with the surface of Al_2O_3 is stronger than that of SiO_2 [7]. This discrepancy would result in realization of whether the base or tip model of the fibers growth [2].

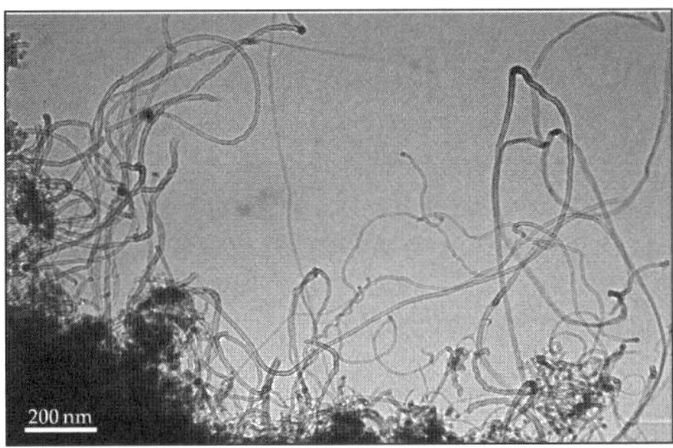

Figure 4. Image of the pyrolysis products of the gas mixture C_2H_4:H_2 = 1:2 at 750°C over Co/Al_2O_3

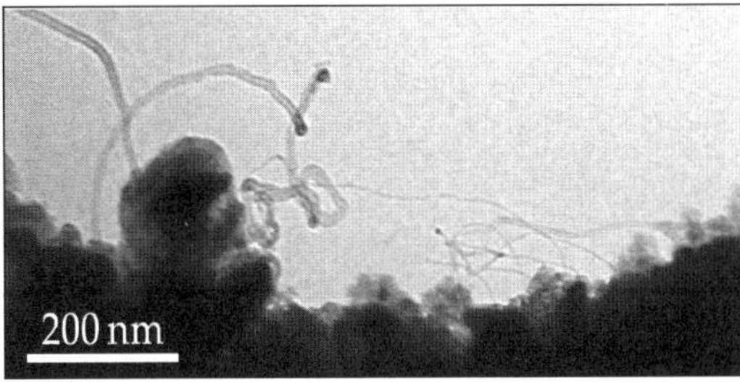

Figure 5. Image of the pyrolysis products of the gas mixture C_2H_4:H_2 = 1:2 at 750°C over Co/SiO_2

4. Conclusion

The catalyst Co/Al_2O_3 is more effective for the synthesis of CNS than Co/SiO_2. Increasing the pyrolysis temperature favors the formation of the more uniform carbon nanostructures with smaller diameters.

Acknowledgement

The work is supported by the Program of the Presidium of the RAS "Encouragement of young scientists".

References

1. Rakov E.G. Preparation of thin carbon nanotubes by catalytic pyrolysis on a support, *Russ. Chem. Rev.*, 2007, **76**: 1–22.
2. Fursikov P.V., Tarasov B.P. Catalytic synthesis and properties of carbon nanofibers and nanotubes, *Altern. Energy Ecol.*, 2004 **(10)**: 5–21.
3. Liu B.C., Tang S.H., Yu Z.L., Zhang B.L., Chen T., Zhang S.Y. Catalytic growth of single-walled carbon nanotubes with a narrow distribution of diameters over Fe nanoparticles prepared in situ by the reduction of $LaFeO_3$, *Chem. Phys. Lett.*, 2002, **357**: 297–300.
4. Su M., Zheng B., Liu J. A scalable CVD method for the synthesis of single-walled carbon nanotubes with high catalyst productivity, *Chem. Phys. Lett.*, 2000, **322**: 321–326.
5. Ward J.W., Wei B.Q., Ajayan P.M. Substrate effects on the growth of carbon nanotubes by thermal decomposition of methane, *Chem. Phys. Lett.*, 2003, **376**: 717–725.
6. Volodin A.A., Fursikov P.V., Kasumov Yu.A., Khodos I.I., Tarasov B.P. Synthesis of carbon nanofibers by catalytic pyrolysis of ethylene and methane on hydrides of intermetallic compounds of lanthanum with nickel, *Russ. Chem. Bull.*, 2005 **54**(10): 2281–2285.
7. Cassell A.M., Raymakers J.A., Kong J., Dai H. Large scale cvd synthesis of single-walled carbon nanotubes, *J. Phys. Chem. B*, 1999, **103**: 6484–6492.

EQUIPMENT AND TECHNOLOGY OF NANOSTRUCTURED CARBON MATERIALS PRODUCTION

A.G. TKACHEV
Tambov Innovative–Technological Center of Mechanical Engineering Ltd., 392000 Tambov, Russia

Abstract. The materials on experimental production of carbon nanomaterials (CNM) are presented. The choice of gaseous-phase sedimentation of carbon in the process of catalytic pyrolysis of hydrocarbons is proved as an optimum method of nanotube production in industrial volumes. The descriptions of the technological circuit of production, the design of the reactor for CNM synthesis and the characteristics of the produced nanotubes are given.

Keywords: carbon nanomaterials, production in industrial volumes, reactor for CNM synthesis

1. Introduction

The analysis of the conditions and directions of development of potential nano-industrial objects allows to draw a conclusion, that one of the most perspective nanoindustrial components is manufacture of carbon, fullerenlike structures (carbon nanotubes and nanofibres).

It is necessary to mention, that by the present time a significant mental potential in the field of carbon nanomaterials (CNM) synthesis has been collected, which allows to proceed from the laboratory to the industrial ways of nanomaterial production.

An industrial technology of CNM production has been developed jointly by "Tambov Innovative Technological Centre of Mechanical Engineering" (TSTU) and joint stock company "Tambov Plant Komsomolets" based on research and development results.

2. Results and discussion

In order to find out the rational technological parameters of the process of CNM synthesis the experimental researches were carried out with the usage of several constructive variants of reactors – tubular, capacitor with flat substrate and reactors with fluidized bed (made by vibration) of the catalyst [1–4].

As a result of this work the values of key parameters of the process – time, temperature, charge of hydrocarbon, thickness of the catalyst layer were established (Figs. 1–3).

B. Baranowski et al. (eds.), *Carbon Nanomaterials in Clean Energy Hydrogen Systems.*

In parallel with carrying out the experiments morphological and physico-mechanical CNM properties were investigated as well. The results of diagnostics received in cooperation with our partners have shown that synthesized CNM – are nanoscaled threadlike formations of the polycrystalline graphite, representing multilayered packaged nanotubes with mainly conic form of graphenic layers as agglomerates with the micrometric sizes (Fig. 4(a, b)).

Numerical values of some parameters are resulted in Table 1.

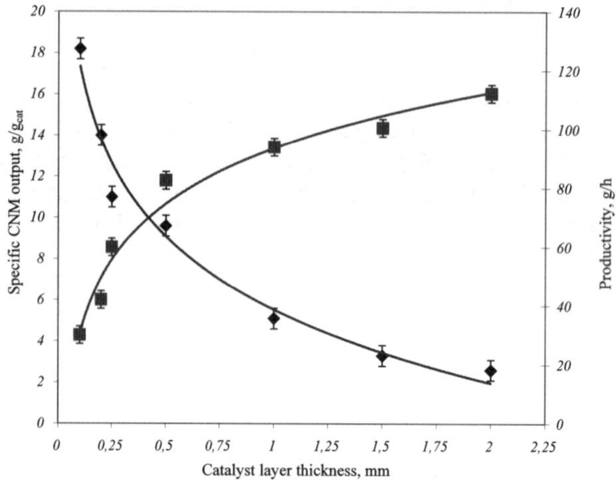

Figure 1. The dependence of productivity ■ and specific CNM output ◆ of thickness of the catalyst layer

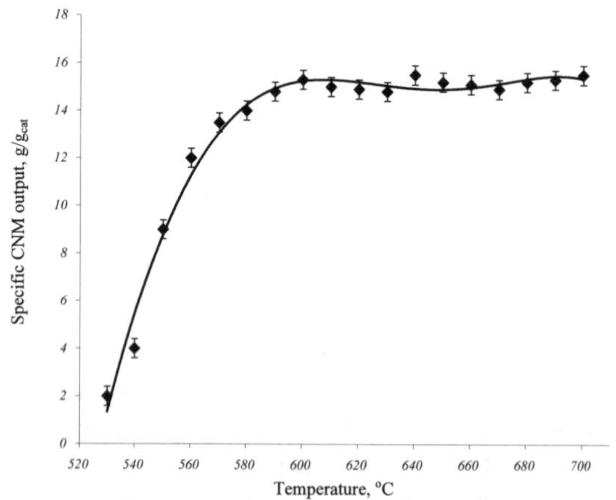

Figure 2. The dependence of the specific CNM output of the pyrolysis temperature

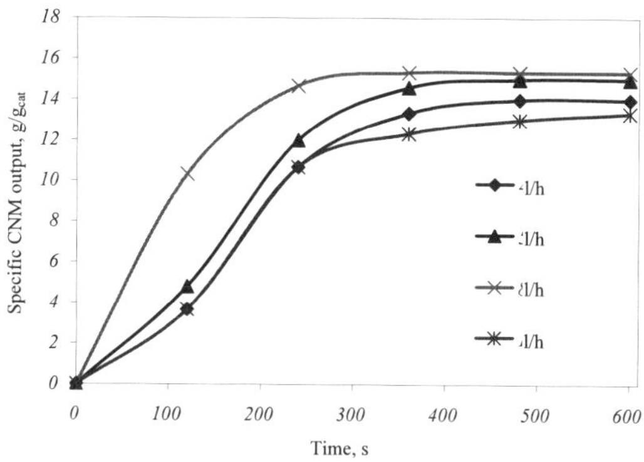

Figure 3. The dependence of specific CNM output of pyrolysis time at various values of the charge of carbon-containing gas

We managed to solve rather challenging task to maintain periodic loading of the catalyst and unloading of the product without depressurization of reactionary space, and also to stabilize the temperature mode in the CNM synthesis zone.

TABLE 1. CNM characteristics

Characteristic	Value
External diameter (nm)	10 ÷ 60
Internal diameter (nm)	10 ÷ 20
Length (μm)	2 and more
Purity (%)	Up to 1.5
including amorphous carbon	0.3 ÷ 0.5
Bulk density (g/cm³)	0.4 ÷ 0.5
Specific geometrical surface (m²/g)	120 and more
Thermal stability (°C)	Up to 700
Average bulk of pores (cm³/g)	0.22
Average size of pores (A)	70

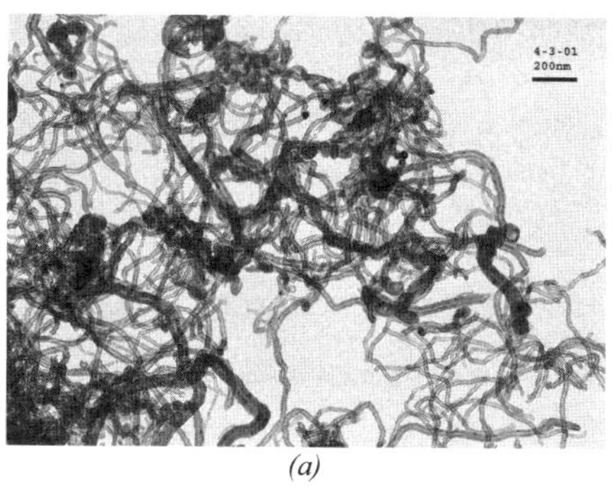

(a)

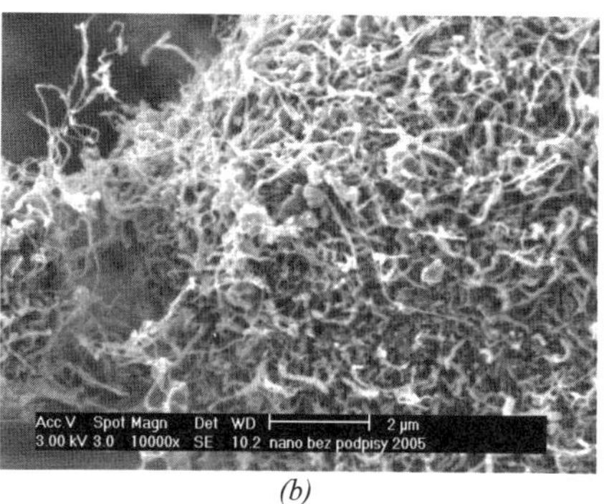

(b)

Figure 4. Microstructure of the synthesized CNM: (a) TEM; (b) SEM

The CNM synthesis is carried out in the basic device of the technological circuit – a capacitor type reactor of semicontinuous action (Fig. 5).

The key parameters of the reactor are submitted in Table 2.

TABLE 2. The reactor key parameters

Capacity (kg per year)	Up to 2,000
Temperature (°C)	600–650
Pressure	Atm
Raw materials	C_xH_y
Power (kW)	35
Dimension (m)	3.7 × 2.8 × 2.0

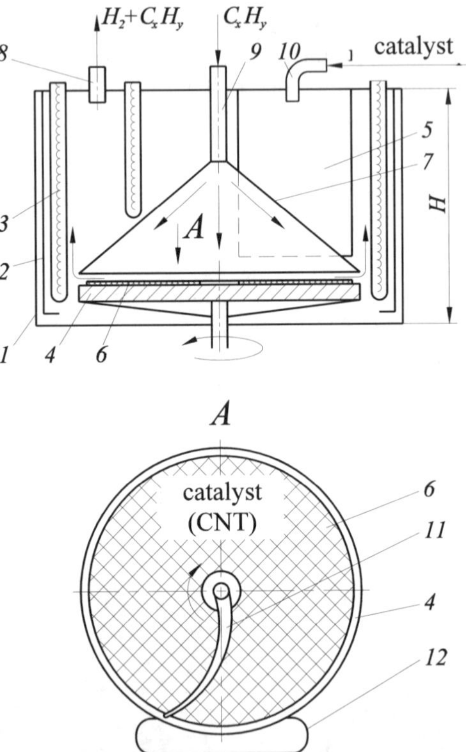

Figure 5. Circuit of a reactor: *1* – the case, *2* – the screen, *3* – a heating element, *4* – a disk-substrate, *5* – a case of the sedimentation chamber, *6* – a layer of the catalyst, *7* – a gas distributer, *8* – a union of the pyrolysis products removal, *9* – a union of the carbon-containing gases supply, *10* – a spray of the catalyst, *11* – a scraper, *12* – a product collector

The technology of production of the carbon nanostructured formations of polycrystalline carbon used in the reactor requires catalytic pyrolysis of the carbon containing mixture [5,6]. Diffusing through a layer of a catalyst with small dispersivity, hydrocarbon adsorbs on a surface of nanosized particles of the catalyst metal, where a large number of subsequent chemical reactions take place, the final products of which are carbon and hydrogen. At achievement of the necessary concentration carbon graphitizes in the form of threadlike formations (carbon nanotubes).

In the reactor the following processes are conducted: heterocatalytic, endothermic, mass exchanging processes in inert gases, at a motionless layer of catalyst periodically loaded in the reactor and internal heating of a reactionary zone (Fig. 6).

Management of the technological process – automatic control of the parameters (temperature, time of separate operations, pressure) at manual regulation by gas streams (CxHy, Ar).

Figure 6. Experimental industrial reactor

The work of the reactor includes the following stages:

1. Purge with inert gas (Ar)
2. Making a catalyst layer *6*, accompanied by turning of a disk – substrate *4* on a corner 315°
3. Heating of a reactionary zone to the temperature 630–650°C with the heaters
4. Carbon containing gas mixture supply through the union *9*
5. The hydrocarbon pyrolysis accompanied by the CNM synthesis (30 min)
6. Unloading of the final material with the help of a scraper *11* at rotation of a disk – substrate *4* avoiding the depressurization of the device

Further recurrence of a cycle at performance of the actions 2, 4, 5, 6.

Thus, in the reactor a semi continuous operating mode is used. The number of consistently executed cycles of synthesis is limited by the bunker volume only (if its periodic unloading is not stipulated), and is limited by the necessity of periodic preventive maintenance of all systems of the device (clearing of a reactionary zone, purge of highways and locking armatures, etc.) which is stipulated in production schedules of the reactor operation.

The experience of laboratory researches of experimental installations (chap. II) and an experimental reactor have shown, that the specified technology allows to synthesize CNM with stable qualitative characteristics.

By present time the technologies jointly developed with partners from the leading research organizations and industrial enterprises, are at various stages of introduction into industry where the manufactured CNM is being realized under the trade mark "Taunit". First of all there are various polymeric composites, nanomodified building materials, filters, additives to lubricants, carriers of catalysts, membranes of fuel elements, dyes with absorbing microwave properties, etc.

3. Conclusion

The technology and the experimental industrial equipment for CNM manufacture have been developed, intended to be used in various industries.

References

1. The patent of the Russian Federation № 2108287. The way of manufacture of carbon material and hydrogen. Kuvshinov G.G., Mogilnyh J.I., Avdeeva L.B., Zavaruhin S.G., Korotkih V.N., Kuvshinov D.G., Liholobov V.A.
2. The patent of the Russian Federation № 59557. Device for manufacture of carbon material. Tkachev A.G., Baranov A.A., Memetov N.R., Negrov V.L., Pasko A.A., Shubin I.N.
3. The patent of the Russian Federation № 59936. Device for manufacture of carbon material. Tkachev A.G., Baranov A.A., Memetov N.R., Negrov В.Л., Pasko A.A., Shubin I.N.
4. Memetov N.R., Tkachev A.G., Zelenin A.D. The design procedure of reactors for manufacture of carbon nanostructured materials in fluidized bed (made by vibration) // Questions of modern science and practice. University in honour of V.I. Vernadsky, 2006, **3** (5): 124–131.
5. Tkachev A.G., Mishchenko S.V., Konovalov V.I., Catalytic synthesis of carbon nanotubes from gas-phase products of hydrocarbons pyrolysis. *Russian Nanotechnologies*, 2007, **2**: 100–108.
6. Tkachev A.G., Experimental industrial reactor for synthesis of carbon nanostructured materials by gas-phase sedimentation on the catalyst. *Chemical and Oil-and-Gas Mechanical Engineering*, 2007 (6): 3–5.

ON POSSIBLE SYNTHESIS OF SOME ORGANIC MICROSTRUCTURES IN A PROCESS OF FULLERENE FABRICATION BY MEANS OF AN ARGON ARC DISCHARGE

V.I. PODGORNYI, V.P. KISELEV, A.N. YAKOVLEV
Petrozavodsk State University, Lenin Avenue, 33, Petrozavodsk, 185910, Russia
Fax: (8142)711000, E-mail: podgvlad@psu.karelia.ru
B.Z. BELASHEV, A.N. TERNOVOI
Geology Institute of Karelia Scientific Center, Pushkin Str., 11, Petrozavodsk, 185910 Russia

Abstract. Some colourless long formations similar to fibres are detected in the discharge chamber. The results of investigation of its morphology and element composition are presented. An infrared spectrum of these objects to be measured coincides practically with the spectrum of cellulose. The possible causes and mechanisms of fabrication of these objects in the arc discharge are discussed.

Keywords: graphite, arc discharge, electron microscopy, infrared spectroscopy, microstructure

1. Introduction

It is known various methods of a carbon nanostructure manufacture. The most simple and widely spread one is an arc discharge in inert gas. Our works [1–3] have been described a cheap and attractive method of fullerene preparation by means of argon arc discharge. This report informs about the further investigation in this direction.

2. Experimental

An argon arc discharge was burned in the discharge chamber, which was placed into the vacuum bell jar [1–3]. A plate and rods manufactured from the high purity graphite were used as electrodes. The arc discharge burning time was varied from 20 to 40 minutes. The sputtered products which appeared on the walls of the discharge camber and the vacuum bell jar, on the electrode supports, on other details of the discharge camber were the objects of our investigation.

Morphology of these substances was studied with some electron and optical microscope. Element analysis was performed with an electron X-ray Vega-Teskan spectroscopy. The infrared spectra of an optical transmission of a pressed sample tablet in KBr were measured on a Specord M80 spectrometer.

B. Baranowski et al. (eds.), *Carbon Nanomaterials in Clean Energy Hydrogen Systems.*

3. Results and discussion

In our investigation the interelectrode gap l was mostly retained at the value $l \approx 0.1$ mm because this operating mode of the arc was optimal for the fullerene production. Increasing of a height of the cathode deposit (or the ledge on the cathode [3]), which was observed visually through the peepholes in the bell jar, was an adding indicator of the operating mode.

We established that the morphology and inner structure of the cathode ledge-deposit strongly depended upon the argon gas pressure p. At the values $p \geq 3\cdot10^4$ Pa the inner part of the formations was rather a friable structure mostly consisting of thin (~0.04 mm diameter) carbon fibers. The latter easily could be separated from one to another. If the gas pressure p was in the range ($8\cdot10^3...2\cdot10^4$) Pa, the inner structure of a ledge-deposit was more compact, and the adjacent carbon fibers were very close one to another without any gap.

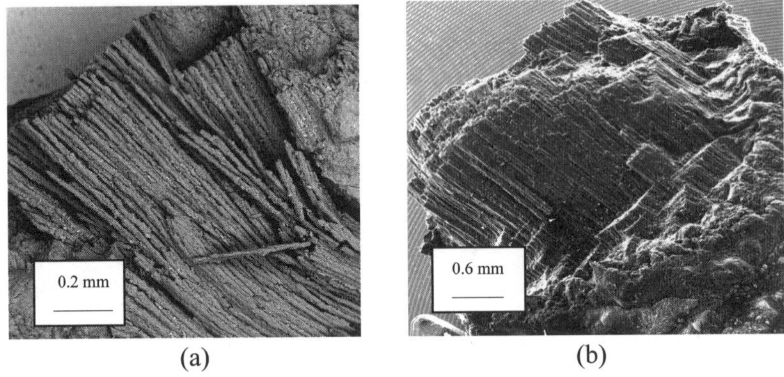

Figure 1. Electron micrograph of a ledge-deposit fracture: (a) $P = 3\cdot10^4$ Pa, (b) $P = 1\cdot10^4$ Pa

Figure 1 illustrates the general view and internal structure of two ledges. The first specimen was performed in the arc discharge at the argon pressure $P = 3\cdot10^4$ Pa (Fig. 1(a)) and the second was grown at the pressure $P = 1\cdot10^4$ Pa (Fig. 1(b)). It is seen that these formations consist of outer envelope and a beam of carbon fibers. The latter are mostly directed along the symmetry axis of a ledge.

In the sputtered products we often noticed some long objects with a length about several millimeters and a diameter about tens microns. These objects were observed in the soot appearing on the water-cooled walls of the discharge chamber, on the cathode surface, on the walls of the vacuum bell jar. With an optical microscope, these formations looked as the bright and colorless fibers bended by some manner. We established that they possessed some mechanical elasticity of its form. Sometimes, the fibers of red-brown colour and the objects similar to rods were observed as well.

Figure 2 shows the electron micrographs of the most typical specimens of these objects. The fibers with thickness of about 20 μm are observed in the most quantity in the sputtered products. Their morphology has great variety of forms (Fig. 2(a)). The objects similar to a rod with a thickness of about 150 μm (Fig. 2(b)) are detected there more rarely. The 50 mkm fibers (Fig. 2(c)) and the fiber with a thickness of less 10 mkm (Fig. 2(d)) are observed ever more seldom. In the micrographs some of these formations are looked as a solid but other objects consist of more thin fibers screwed on a common axis. However, Fig. 3 shows that the ends of these objects have a complicated structure.

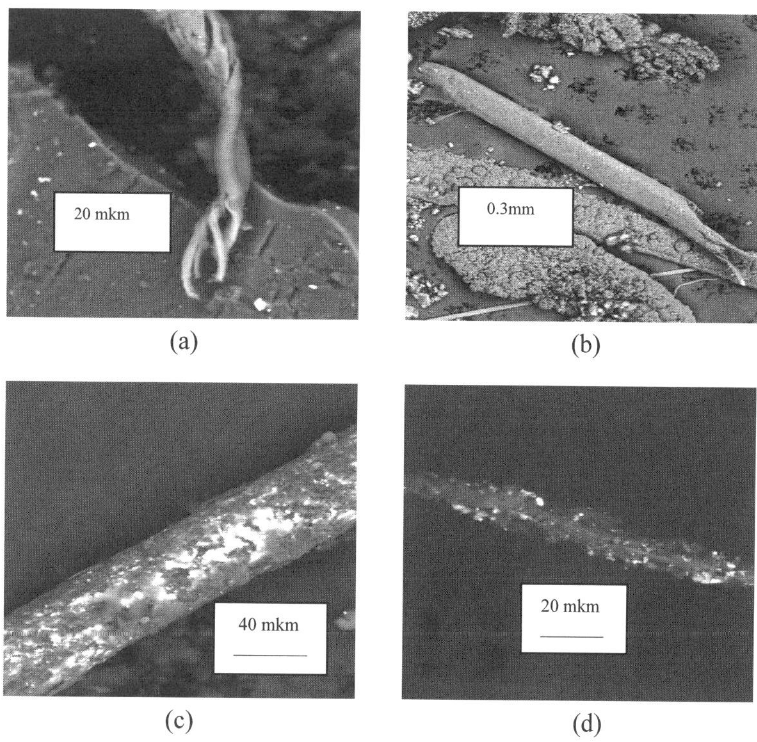

Figure 2. Electron micrographs of some objects to be observed in the sputtered products

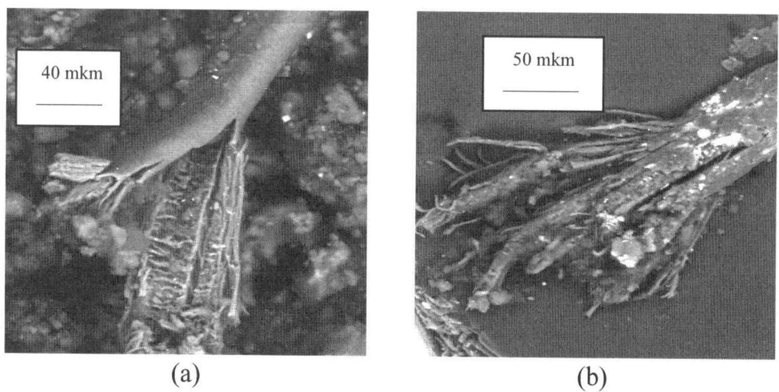

Figure 3. Electron micrographs of the ends of some fibers

Our preliminary investigation of some mechanical properties of these fibers points out that they are not strong structures. All the fibers are easily became flat at a small compression and also easily divided into more short fragments. Their contact with a water drop leaded to a whole loss of elasticity. The fibers with diameter 20 μm were wholly dissolved by the strength sulphuric acid during several tens seconds. The more thick fibers were dissolved by this acid with a less rate.

A chemical composition of the objects to be observed in the Figs. 2, 3 was investigated in more than 40 points by an electron X-ray spectroscopy. Carbon and oxygen are the main components of the most thin fibers (see Fig. 2(a)). However, the relative content of these elements in the fibers is varied in rather big limits. So a carbon concentration is in the range from 65 to 85 at.% and an oxygen concentration from 35 to 15 at.%. Besides, the X-ray lines of silicon, chlorine, sodium, molybdenum, calcium and zinc are detected in these fibers. Their relative content is also varied from one point to another in the range 0…1, 5 at.%. Seemingly, all these atoms are random admixtures; another words, they are not the structural elements. At the same time, it is registered about to 25 at.% nitrogen in some fibers.

In our investigation it was noticed that the intense electron bombarding led to grow dark of the part of a fiber in the electron picture (or to decreasing of its electron emission) and to increasing of its profile dimension. Moreover, after this process the oxygen content has always decreased to about 5 at.% in comparison with its primary value

In the short and fit rods (see Fig. 2(b)) the carbon content is (83…92) at.%, the oxygen content (7… 15) at.%. In addition to these elements there are S (1…2 at.%) and Na (0, 1 at.%).

In the 50 μm fibers (Fig. 2(c)) the carbon content is (60…80) at.%, the oxygen content (6…10) at.%, the nitrogen content 20 at.%, the sulphur content (0, 2…1) at.%, the chlorine content (0, 1…0, 9) at.%. Also here it was registered Cu (0, 2…0, 9 at.%), Zn (up to 8 at.%) and X-ray line Sn (0, 04 at.%) in the brightest parts of fiber surface.

In the sputtered products we noticed some particles of the plate form which consisted from a conglomerate of more fine particles of several microns dimension (see Fig. 2(b)). There the average carbon content is 95 at.% and the oxygen content 5 at.%. Other elements are not found in these formations.

In our opinion the clue moment to make clear the data is a presence of surplus oxygen in the fibers in comparison with its content in the carbon plate conglomerate to be obtained in the same conditions. Oxygen, carbon and, possible, hydrogen, which is not registered by an electron X-ray spectroscopy, are capable of forming of an organic fibrous structure. But it is not understood a nature of manufacturing of this structure in the plasma conditions. Perhaps, the observing fibers and rods appear due to its releasing from the cathode or anode materials in the process of argon arc burning. The presence of these structures in the graphite materials may be a consequence of known tendency to reserve the morphology of an organic carbon-hydrogen molecule after graphitization [4]. Other words, a carbon-oxygen frame of these molecules may be ever reserved in the conditions of thermal processing.

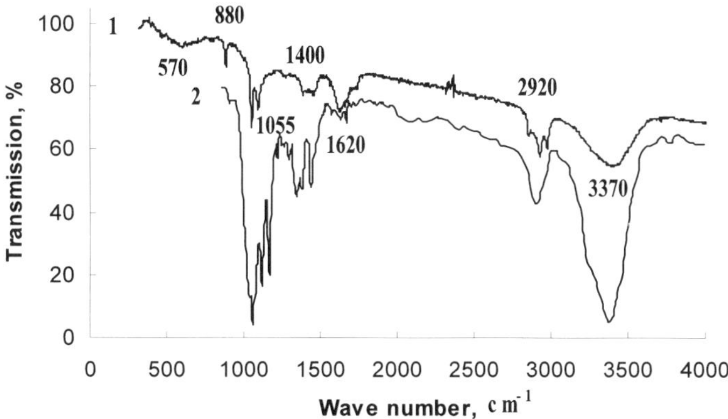

Figure 4. Infrared spectra of thin fibers (1) and cellulose (2) [6]

In Fig. 4 it is shown the infrared spectra of the 20 μm fibers (curve 1) and cellulose $(C_6H_{10}O_5)_n$ (curve 2) to be published in the work [5]. One may see a good coincidence the maximums of wave number at 3,370, 2,920, 1,400 and 1,055 cm^{-1}. Moreover, the hydrophilate properties of cellulose and its dissolving in the strength sulphuric acid is good known [6]. Thus, the observing thin fibers containing no nitrogen may be macromolecules of cellulose.

However, the forming mechanism of these macromolecules is not understood. These molecules have to destroy into components due to its low thermal stability (~500 K) [6] in a process of graphitization of a wood-coke. The alternative mechanism of forming of these fibers may be a plasma synthesis in the conditions of an arc discharge.

The following facts testify in favour of this proposal. At first, it is the increasing of a quantity of fibers in the soot nearly in ten times if a source of oxygen is put in the discharge camber, and, on the contrary, it is the decreasing of their quantity if the residual vacuum in the chamber is improved to the power two (from 10^{-2} to 10^{-4} Torr). Second, it is an existence of fibers in the soot if the cathode ledge-deposit is utilized as a cathode. Third, it is the absence of effect of fiber accumulation if a mass of sputtered graphite is increased in three times, or, another words, if the discharge time is increased from 20 to 60 minutes. In our opinion, oxygen atoms take the lead to the fiber formation in the arc discharge. If these atoms are absent, the synthesis of the fiber formations is not occurred.

More fit fibers and “rods” are practically characterized by the same quantity of atoms of oxygen and carbon as the thin fibers. Hence, they consist of the same structural elements and have some common nature. However, the objects containing nitrogen, which not is in the composition of cellulose, may have another origin.

The observing objects could be accepted for a bundle of carbon nanotubes or carbon nanofibers. However, a chemical composition of these objects and its morphology do not agree with this supposition. Moreover such the objects are usually detected only in the cathode deposit.

Note that we could not detect any fibers in initial electrode materials.

4. Conclusions

1. It is detected the long structures in a form of fibers in the sputtered products.
2. Infrared spectrum points to an organic nature of these formations.

References

1. Aleshina L.A., Podgorny V.I., Stefanovich G.B., Fofanov A.D. Arc discharge sputtering of schungites, *Technical Physics* 2004; **74**(9): 1146–1149.
2. Podgorny V.I., Yakovlev A.N., Belashev B.Z. About some special features of fullerite manufacture by means of an argon arc discharge. Extended abstracts, IX International Conference on Hydrogen Materials Science and Chemistry of Carbon Nanomaterials. Sevastopol (Ukraine), Kiev, 2005: 394–395.
3. Podgorny V.I., Yakovlev A.N. Some special features of fullerene production by means of argon arc discharge, *Materialovedenie* 2005; **9**: 42–47. (In Russian).
4. Ubbelohde A.R. Graphite and its crystal compounds. Moscow: Mir. 1965: 31–36. (In Russian).
5. Nakanishi K. Infrared absorption spectroscopy. Moscow: Mir. 1965: 152. (In Russian).
6. Cellulose and cellulose derivatives. Edited by Bikales N.M. and Segal L. Moscow: Mir. 1974: 1, 2. (In Russian).

FERROMAGNETIC NANOPARTICLES PRODUCED BY ARC-DISCHARGE EVAPORATION OF Fe-Ni-GRAPHITE ELECTRODES

V.E. MURADYAN, YU.M. SHULGA
Institute of Problems of Chemical Physics of RAS
Prospectus of the Academician Semenov, 1, Chernogolovka, Moscow Region, 142432 Russia
Tel./Fax: +7(496) 5221743; E-mail: muradyan@icp.ac.ru

M. LEONOWICZ
Faculty of Materials Science and Engineering, Warsaw University of Technology, 141 Woloska, 02-507 Warsaw, Poland

Abstract. Arc-discharge method of carbon-shells nanoparticles encapsulated with a various ratio of iron and nickel have been synthesized. Samples by a thermogravimetry, transmission electron microscopy, X-ray diffractometry and magnetic measurements have been studied.

Keywords: metal-carbon nanoparticles, arc-discharge evaporation, magnetic properties

1. Introduction

Recently nanoparticles (NP) attract steadfast attention of the explorers with the unique properties and potential applications, including new types of very active and selective catalysts, chemical and biological sensors, components of medicinal preparations, magneto- and optoelectronic devices etc. This work presents the results of the investigation of NP encapsulated in the carbon shell, which can be the technologically important magnetic materials. NP were produced by arc-discharge evaporation of Fe-Ni-graphite electrodes by the procedure, which was described earlier [1–3].

2. Experimental

Eight millimeter graphite rods with the drilled 4 mm channel filled by the mixture of nickel and iron powders were used as the anode. The products of arc-discharge evaporation of electrodes (soot from walls of the chamber (W) and around of the cathode (C)) at the iron and nickel ratios of 1:3, 2:2 and 3:1 have been studied (Fig. 1, Table 1).

S31WA and S31WH samples have been obtained by heating of S31W in the muffle at 250°C during 30 min and subsequent refluxing. S31WA sample was refluxed in conc. HCl, and S31WH was refluxed in solution of 30% H_2O_2–HCl (10:1). After refluxing, the samples were washed with water and dried.

The thermograms were obtained by the Netzsch STA 409 PC Luxx thermo-analyzer in airflow, in the temperature range of 25–1,000°C, at the heating rate of 10°/min.

B. Baranowski et al. (eds.), *Carbon Nanomaterials in Clean Energy Hydrogen Systems.*

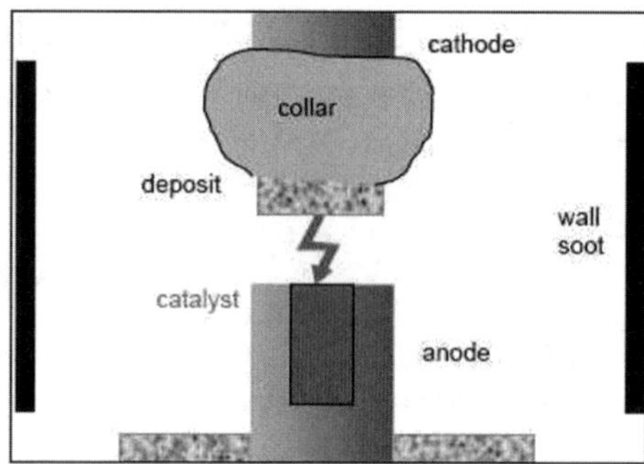

Figure 1. The schema of arc-discharge reactor

TABLE 1. Conditions of arc-discharge evaporation

Ratio Fe:Ni in electrodes	The content Fe and Ni (wt%)	Erosion rate of electrode (g/min)	Samples
1:3	2.9	0.32	S13
	8.7		
2:2	5.4	0.32	S22
	5.4		
3:1	8.6	0.16	S31
	2.9		

The investigation of magnetization curves has been carried out on PARC M4500 vibrating magnetometer. Morphology of products was observed by Jeol 3010 transmission electron microscope. The X-ray diffractograms have been recorded by DRON-2 diffractometer (monochrome CrK_α-radiation).

3. Results and discussion

The results of magnetic measurings are presented in the Table 2. The quantities of the fire-resistant residual obtained from thermogravimetric investigations are given ibid. It is visible, that there is a quite good correlation between σ and amount of ashes.

The TEM image of S13C sample is presented in Fig. 2. Besides the carbon shells, there have been observed the metal nanoparticles of round shape of 30–60 nm size. The carbon shells covering the nanoparticles differ in the structure. After acid treatment, there are only those nanoparticles, which had a continuous graphite-like shell.

In Fig. 3 the fragments of powder X-ray patterns of soots with different ratio of metals in the area of the most intensive peaks are presented. The analysis has shown, that these peaks are stipulated by reflections of 111 and 200 fcc lattice of the solid solution between Ni and γ-Fe. The absence of the peaks appropriated to α-Fe bcc lattice and/or to Fe_3Ni and FeNi compounds means that the solid solution formation is accompanied by enough prompt hardening.

TABLE 2. The magnetization of saturation (σ), coercivity (ΔH) and mass remaining after heating of samples up to 1,000°C (ashes)

Samples	σ (emu/g)	ΔH (mT)	Ashes (wt%)
S13C	24.1	11.9	34.7
S13W	14.4	10.5	23.0
S22C	27.1	12.1	34.8
S22W	19.3	11.2	27.0
S31C	10.8	9.0	19.4
S31W	10.4	8.9	20.1
S31WA	2.54	12.5	7.3
S31WH	1.68	13.1	5.7

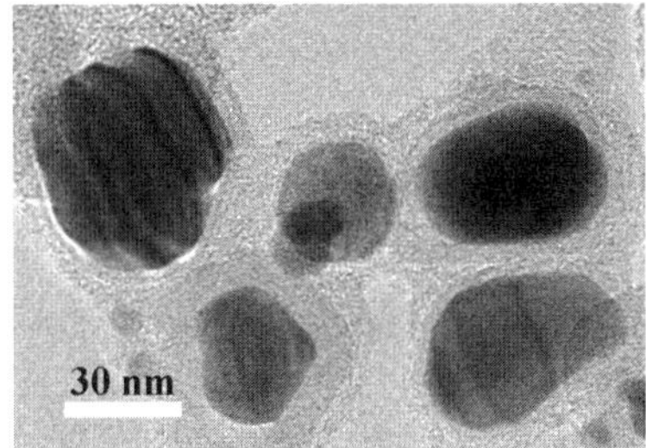

Figure 2. TEM image of soot S13C

Λccording to thc Vcgard's rulc for fcc constant latticc of thc solid solution ($\mathbf{a}_S$) it can bc possible to write the following equation: $\mathbf{a}_S = \mathbf{a}_{Ni} + 0.036 \cdot C_{Fe}$, where $\mathbf{a}$ = 3.524 Å, C_{Fe} – iron content in atomic fractions. The calculation of $\mathbf{a}_S$ for S31C sample yields the value of 3.551 Å, and the measured value is equal to 3.59 Å, that essentially exceeds the value of **a** for clean γ-Fe (3.56 Å). Therefore, obtained by us NP of the solid solution contain a large quantity of carbon atoms.

The estimation of the average size of the nanoparticles in the direction perpendicular to the plane (111), has been carried out according to the formula $D_{111} = \lambda/\beta\cos\Theta$. The obtained in such way D_{111} values are presented in the Table 3. It is apparent, that the nanoparticles sizes in the wall-soot are less, than in the collar.

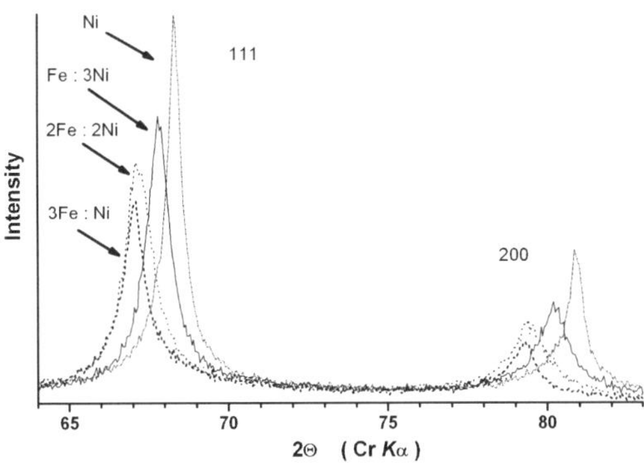

Figure 3. X-ray patterns of collar soots

TABLE 3. D_{111} value

Samples	D_{111} (nm)
S13C	21.6
S13W	12.6
S22C	16.7
S22W	12.5
S31C	17.7
S31W	12.2
S31WA	15.0
S31WH	14.7

4. Conclusions

It is shown for the first time, that the evaporation of the mixture of Fe and Ni powders pressed in graphite electrode, is accompanies by formation of Ni and γ-Fe solid solution. The solid solution is probably stabilised by carbon atoms and is kept safe at the room temperature.

It is determined that the nanoparticles size in the wall-soot is less than those in the collar. The treatment in acid dissolves unclosed (not covered with the dense carbon shell) nanoparticles. The large NP turn out to be unprotected.

The magnetic properties of NP, after acid treatment, are sufficient to carry out their trial as a potential component of magnetic pompes of aggressive fluids.

Acknowledgement

The work is supported by INTAS (grant 05-1000008-7834).

References

1. Leonowicz M., Woźniak M., Shulga Y.M., Muradyan V.E., Liu Z., Davies H.A., Kaszuwara W., Grabski J. Proccessing and properties of magnetic nanoparticles encapsulated in carbon shells, *Mater. Lett.*, 2006, **60**: 442–446.
2. Fnidiki A., Lemarchand D., Talbot E., Pascard H. Characterization of carbon and iron nanostructures synthesized by the DC arc discharge method: influence of the location in the reactor and of the pressure, *Eur. Phys. J.: Appl. Phys.*, 2005, **32**: 177–185.
3. Wozniak M.J., Wozniak P., Bystrzejewski M., Cudzilo S., Huczko A., Jelen P., Kaszuwara W., Kozubowski J.A., Lange H., Leonowicz M., Lewandowska-Szumiel M. Magnetic nanoparticles of Fe and Nd-Fe-B alloy encapsulated in carbon shells for drug delivery systems: Study of the structure and interaction with the living cells, *J. Alloy. Compd.*, 2006, **423**: 87–91.

OPTIMIZATION OF SYNTHESIS METHODS OF SINGLE-WALLED CARBON NANOTUBES USING 3*d* TRANSITION METALS

V.E. MURADYAN, B.P. TARASOV, N.S. KUYUNKO
Institute of Problems of Chemical Physics of RAS, Prospectus of the Academician Semenov, 1, Chernogolovka, Moscow Region, 142432 Russia
Tel./Fax: +7(496) 5221743; E-mail: muradyan@icp.ac.ru

Abstract. Arc-discharge synthesis of single-walled carbon nanotubes was carried out using as catalyst only mixtures of 3*d* transition metals – Ni, Co, Fe, Mn, Cr and V. For two- and three-component catalytic systems the best yields of SWCNT was observed in the following row – 3Ni-Co-Mn > 4Ni-Mn > 3Ni-Co-Cr > 4Ni-Cr. At the highest SWCNT yields the average size of nanoparticles achieved 14 nm by results of X-ray diffraction.

Keywords: carbon nanotubes, arc-discharge synthesis, transition metals

1. Introduction

The metals of iron group are known to be catalysts of the arc-discharge synthesis of single-walled carbon nanotubes (SWCNT). At this the yield of SWNT decreases in the row Ni > Co > Fe. Such dependence of the SWCNT yield from metal is related to their activity in the reaction: C + M → M_3C → C(SWCNT) + M. At this the mentioned metals produce metal-like carbides [1]. The carbides of manganese, chromium and vanadium belong to the same group. Usually for the SWCNT synthesis one employs bimetallic systems: 3*d*-metal with Y, Ca or rare-earth metals (Re) [2]. We believe that at the use of the 2Ni-Re system a significant SWNT yield, could be related to the high thermal stability of ReC_2 carbides (1,600–2,300°C), and the synergetic influence of binary system expressing by that one metal catalyzes SWCNT nucleation and the other promotes the growth of tubes.

At using 3Co-Ni catalytic system the carbides of both metals – Co_3C and Ni_3C – with a near thermal stability (300–500°C) can be formed, therefore to separate their role in the synergetic influence on the SWNT growth is inconvenient. It is known, that the SWCNT maximal yield is observed for synthesis parameters resulting in the formation of metal nanoparticles with the size of 5–20 nm.

The purpose of the work was a detection of bimetallic synergism for Ni–X mixtures, where X stands for Co, Fe, Mn, Cr and V, definition of dependence of the diameter of synthesized SWCNT from the chemical nature of the second metal, detection of dependence of the SWCNT yield from the size of catalytic particles.

2. Experimental

The experiments on arc-discharge synthesis were carried out at the helium pressure of 500 Torr and current intensity ~90 A. The content of Ni–X metals in the anode achieved 4 and 1 at.%, accordingly. For filling the anode with manganese or vanadium were used MnO oxide and VH_2 hydride, accordingly. Powders of Ni, Co, Fe and Cr with a size 10–20 μm and purity 99.5–99.8 wt% have been used. In all experiments the electrodes were warmed up in dynamic vacuum in the regime of electrical short at the temperature ~1,000°C up to achievement of 0.01 Torr vacuum.

The thermal oxidation have been studied with a Netzsch STA 409 PC Luxx thermoanalyzer through linear heating (10 C/min) of a weighed portion of the sample in airflow 100 ml/min. UV-viz-NIR spectra of soot suspensions by a Shimadzu UV-3101PC spectrophotometer were recorded. X-ray diffraction patterns of soots were recorded by a DRON UM-2 diffractometer (CuK_α-radiation).

3. Results and discussion

During the synthesis around the cathode a "collar" which is richest with SWCNT is formed. SWCNT content in the "collar" was estimated by the methods of thermogravimetry and UV-vis-NIR spectroscopy [3,4].

SWCNT contents in the formed "collars" and their weight gradually decreased in the row Mn > Cr > Co > Fe ~ V. Using Mn and Cr has shown quite high yields in the "collars" which weight achieved 6–7%. To increase the total quantity of SWCNT an increase of the arc discharge stability with the simultaneous increasing the inetrelectrode gap and the weight of chamber soot is required.

At using Co and Fe the SWCNT yields were much lower, but they stabilized an arc discharge better. At this the cobalt activity was higher, than the one of iron. Therefore in systems Ni–Mn and Ni–Cr a part of nickel was replaced with cobalt: the metal content was 3, 1 and 1 at.%, accordingly. Application of such mixtures has led to a more uniform arc "burning" and a better control over the parameters of the process.

To estimate the SWCNT amount UV-viz-NIR spectra of soot suspensions (0.1 mg/ml) in a dimethylformamide were recorded. The SWCNT content was determined by the ratio of integrated intensity of the ABC area to the total integrated intensity ABCDE (Fig. 1(a)) using the following formula [5]:

$$RP = \frac{S_{ABC}}{0.141 \cdot S_{ABCDE}} \times 100\%.$$

Fig. 1(b) shows absorption bands of the soots after subtraction of the basic line. The intensity of the absorption band in the area 8,000–14,000 cm^{-1} related to the S_{22} second interband transition of semiconductor SWCNTs, evidently shows the increase in the nanotubes yield at a partial replacement of nickel with cobalt in ternary catalysts.

The SWCNT content in the "collar" (see Fig. 1(b)) achieves 25% for 3Ni-Co-Mn (1), 14% for 4Ni-Mn (2), 10% for 3Ni-Co-Cr (3) and 9% for 4Ni-Cr (4). For 3Ni-Co-Mn the weight of the "collar" has achieved 10–12%, and the weight of the chamber soot – 25–30% from the total amount of the condensed products.

The obtained results assume that catalytic activity of 3Ni-Co-Mn, 4Ni-Mn, 3Ni-Co-Cr and 4Ni-Cr mixtures in arc-discharge synthesis of SWCNT is connected to a formation of more homogeneous metal-carbon vapor in arc region and formation of catalytic particles of the optimum size.

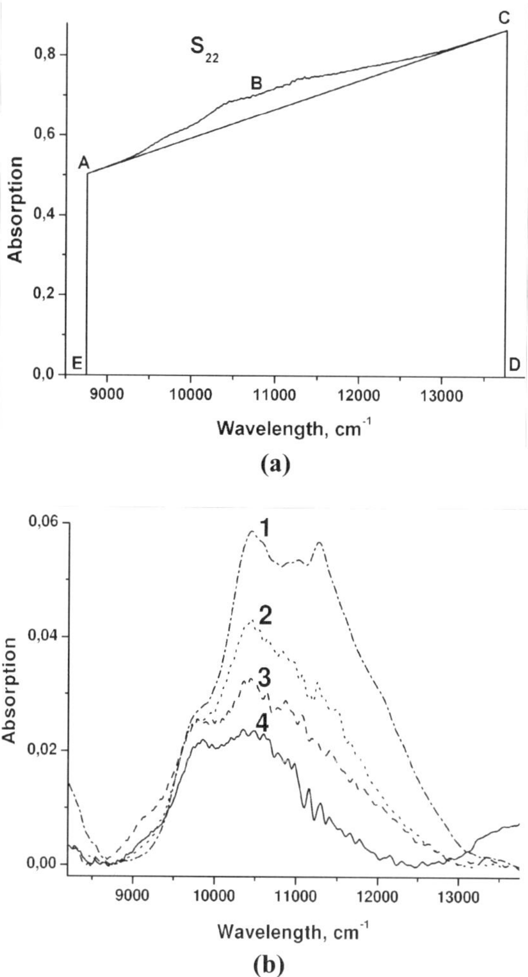

Figure 1. Near-IR spectra of soot suspensions in DMF

On X-ray patterns of soots the peaks related to the 111 and 200 reflections of nickel particles with fcc lattice and manganese carbides are observed. The medium-sized estimation of nickel particles on 111 peak for two-componental mixtures has shown values D_{111} equal – 14, 18, 24 and 18 nm for mixtures with Cr, Co, Fe and V, accordingly. For ternary mixtures D_{111} values are about 14 nm.

At the highest SWCNT yield the average size of particles achieved 14 nm that was in a good agreement with the Gavillet's paper [4], and the obtained value can be considered as one of the optimum parameters of the SWCNT synthesis.

From Raman spectra of the "collars" obtained with ternary mixtures the average diameters of the formed SWCNTs are determined as 1.45–1.15 nm.

Optical spectra (Fig. 1(b)) show an increase of the SWCNT yield with the simultaneous increase in a share of nanotubes with smaller diameters. For 4Ni-Cr the SWCNT diameter ranges within 1.38–1.22 nm, and for 3Ni-Co-Mn – 1.38–1.13 nm [6].

Use of manganese and chromium for SWCNT synthesis will demand a selection of conditions for fuller removal of their carbides during purification. Purification results will allow to estimate availability of the studied ternary catalysts for obtaining products with a high SWCNT content (more than 70 wt%).

4. Conclusions

Using two- and three-component catalytic systems single-walled carbon nanotubes were synthesized on the basis of nickel, cobalt, iron, manganese, chromium and vanadium. At the use of 3Ni-Co-Mn, 4Ni-Mn, 3Ni-Co-Cr and 4Ni-Cr mixtures high SWCNT yields are shown.

Acknowledgements

The work is supported by Russian Fund Basic Research (project 06-03-33089).

References

1. Kosolapova T.Ya. *Carbides*, Metallurgy, Moscow, 1968, 300 p. (in Russian).
2. Yao M., Liu B., Zou Y., Wang L., Cui T., Zou G. Effect of rare-earth component of the Re/Ni catalyst on the formation and nanostructure of single-walled carbon nanotubes, *J. Phys. Chem. B*, 2006, **110**: 15284–15290.
3. Tarasov B.P., Muradyan V.E., Shul'ga Yu.M., Krinichnaya E.P., Kuyunko N.S., Efimov O.N., Obraztsova E.D., Schur D.V., Maehlen J.P., Yartys V.A., Lai H.J. Synthesis of carbon nanostructures by arc evaporation of graphite rods with Co-Ni and YNi2 catalysts, *Carbon*, 2003, **41**: 1357–1364.
4. Gavillet J., Loiseau A., Journet C., Willaime F., Ducastelle F., Charlier J.-C. Root-growth mechanism for single-wall carbon nanotubes, *Phys. Rev. Lett.*, 2001, **87**: 275504.
5. Itkis M.E., Perea D.E., Niyogi S., Rickard S.M., Hamon M.A., Hu H., Zhao B., Haddon R.C. Purity evaluation of as-prepared single-walled carbon nanotube soot by use of solution-phase near-IR spectroscopy, *Nano Lett.*, 2003, **3**: 309–314.
6. Jost O., Gorbunov A.A., Pompe W., Pichler T., Friedlein R., Knupfer M., Reibold M., Bauer H.-D., Dunsch L., Golden M.S., Fink J. Diameter grouping in bulk samples of single-walled carbon nanotubes from optical absorption spectroscopy, *Appl. Phys. Lett.*, 1999, **75**: 2217–2219.

DEPENDENCE OF ELECTRONIC STRUCTURE PECULIARITIES OF CARBON NANOTUBES ON DIAMETER AND PRESENCE OF DEFECTS

YA.V. ZAULYCHNYY, YU.M. SOLONIN, E.V. PRYLUCKIJ, S.S. ZVEZDA

Frantsevich Institute for Problems of Materials Science of National Academy of Science of Ukraine, 03680, Kiev, Ukraine

Abstract. Investigations of the electronic structure in multi-wall carbon nanotubes obtained by different methods with different diameters were performed by means of the ultra-soft X-ray emission spectroscopy. It has been revealed that the narrowing of the CK_{α}-spectra of nanotubes was due to decreasing the degree of the πp_z-overlapping above the nanotubes surface. It has been established that the narrowing of the emission bands induced by the great number of broken bonds, which appeared after purifying from catalysts, was significantly greater than that due to decreasing nanotubes diameters. It has been found that the overlapping of the πp_z-orbitals became insensitive to changes of nanotubes diameters at thickness exceeded 70 nm.

Keywords: nanotubes, X-ray emission spectroscopy, emission spectrum, valence electron

1. Introduction

New structural modifications of carbon such as nanotubes are matter of steady study. A great deal of attention has been devoted to investigations of structural characteristics, properties and peculiarities of the electronic structure, which determine these properties in carbon nanotubes. Structural characteristics of these objects were widely studied [1–5] while for study of the electronic structure of these carbon nanosize materials was not given enough attention.

Quantum-chemical calculations by the LCAO method [6] and some other methods [7, 8] showed that the electronic structure essentially depends on diameters of nanotubes. At the same time, distinctions between shapes of emission bands obtained using different methods were revealed in [9]. In papers [10, 11] devoted to investigations of X-ray emission bands of nanotubes mainly theoretical spectra were studied. Experimental X-ray spectra in [9, 11] were obtained in the energy region 271-285 eB with low resolving capacity (0.5 eV) that did not allow obtaining complete information about the electronic structure of objects under investigation. At the same time, narrowing and change of shapes of the X-ray emission bands were revealed in some crystal materials [12–14] when going from large to nanosize objects. It was concerned with the energy localization and redistribution of the electronic states belonging to the broken bonds. Furthermore, the nanosize effect was revealed in some carbon nanomaterials [15], which have practically all σ-bonds and most of the π-bonds closed. The energy redistribution mainly of the π-states was conditioned on differences in their π-overlapping in nanomaterials with different sizes.

Therefore, it is important to elucidate the influence of sizes and the presence of defects in nanotubes on the energy redistribution of the Cp-valence electrons.

B. Baranowski et al. (eds.), *Carbon Nanomaterials in Clean Energy Hydrogen Systems.*

2. Experimental procedures

Since *p*-electron states make the main contribution into the interatomic interaction in modifications of carbon it is necessary to investigate the energy distribution of valence electrons in carbon nanotubes with different diameters using the ultrasoft X-ray emission spectroscopy method. CK_α-emission bands investigated in present paper were obtained by means of the spectrometer PCM-500 with the energy resolution of about 0.2 eV. In order to prevent additional underlying of the CK_α-bands from hydrocarbons vapors the trap to freeze these vapors out was used.

Multi-wall nanotubes (MWNT) obtained in arc discharge and in the process of low-temperature conversion of carbon monoxide over a catalyst were chosen for present investigations. Catalytic nanotubes contain a great deal of defects formed when moving off metal-catalysts whereas MWNT obtained in arc discharge were low-imperfect.

Catalytic nanotubes with 5–10 walls were obtained by workgroup of E.V. Priluckij in the process of low-temperature conversion of carbon monoxide due to the reaction:

$$2CO = CO_2 + C$$

Reaction was carried out over catalysts – Fe (33%) and Co (11%). Diameters of MWNT obtained over the Co-catalyst were 10–30 nm and diameters of nanotubes synthesized over the Fe-catalyst were 60–80 nm.

Double-wall nanotubes with diameter 4 nm and multi-wall nanotubes with diameter 140 nm were synthesized in arc discharge by 'MER-corporation'.

3. Results and discussion

Comparison of the CK_α-emission bands of MWNT obtained using Co and Fe-catalysts and purified from them with spectra of catalytic nanotubes synthesized over 2.5 wt% Co and 2.5 wt% Fe catalysts from [9] (Fig. 1) showed that the last is most closely resemble to the spectrum of MWNT with Fe obtained by us, but in the spectra [9] intensity of "*a*" and "*e*" peak is much higher. Since in [9] the CK_α-emission band was shown just in region 275 ÷ 285 eV, and CK_α-bands from carbon contaminants could underlay it is not possible to compare widths of spectra and analyze their low-energy parts. At the same time spectra of MWNT purified from Co, differ even greater from results obtained by authors [12].

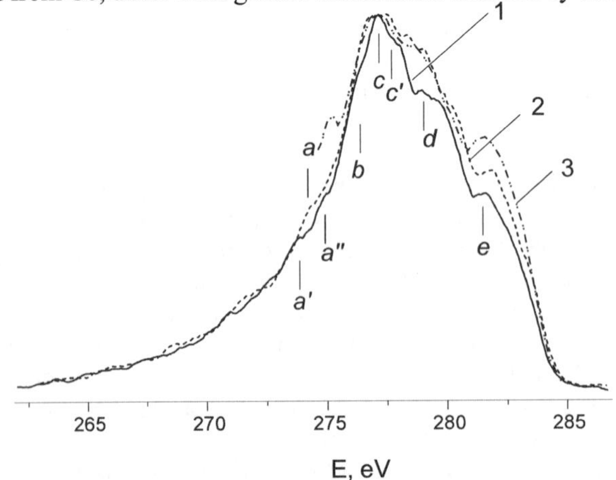

Figure 1. Comparison of emission bands of nanotubes synthesized over the Co-catalyst (1), Fe-catalyst (2) obtained in present paper with results [9] (3)

From comparison of spectra of MWNT obtained in present paper (Fig. 1, curves 1 and 2) it is clear that the low-energy part of MWNT obtained over the Co catalyst noticeably differ from the CK_α of MWNT obtained over the Fe catalyst due to the splitting of "*a*" peculiarity into features "*a*" and "*a*" at 275.5 eV. Peaks of the density of electronic states obtained from calculations of the electronic structure [6] correspond to the fine structure of the CK_α of MWNT obtained over Co. $sp\sigma$ ("*a*") and $sp\sigma + pp\sigma$-гібридні states ("*a*") concentrate in these states. Then near-peak features "*b*" appear in spectra of all MWNT due to the formation of *sp*-hybrid bonds between atoms of carbon and metals when overlapping high-energy $3d$ + $4s$-states with sp^2-orbitals. Such isolation of these peaks is connected with differences in energies of states involved in σ-bonds directed along the nanotube axis and those directed on the angle. However such a split was observed just in MWNT with diameter $d \approx 20$ nm (purified from Co), whereas in thicker nanotubes with $d \approx 70$ nm (purified from Fe) this split disappear since curvature of nanotubes surfaces decrease and angles between σ-bonds approach to those in graphite and differences between energies of when increasing diameter of

MWNT. In the near-peak region of the high-energy part of the CK_α-bands features "*c*" and "*d*" appear in addition to "*c*". According to calculations [6] top of the CK_α-band of MWNT reflects $pp\sigma$-hybrid states, which mixed with $pp\pi$-states when rolling graphene layer into nanotube. $pp\pi$-states concentrate in two divided maximums since π-overlapping along the axis of MWNT and in radial direction are different.

In radial plane inside nanotube π-overlapping is greater, therefore $pp\pi$-states have smaller energy and add to the $pp\sigma$-states concentrated in peak reflected by "*c*" peculiarity of the CK_α-band of MWNT. Degrees of the $pp\pi$-overlapping along the axis of MWNT inside and over the surface of nanotubes are equal but smaller than overlapping inside MWNT in radial direction. Therefore states involved into the overlapping inside MWNT in radial direction concentrate in maximum corresponding "*d*" peculiarity of the CK_α of MWNT. Width of bands of MWNT purified from Co decrease in high-energy region corresponding to $pp\pi$-states in comparison with nanotubes purified from Fe.

In high-energy part of spectra of all nanotubes at 282.5 eV sharp divided peculiarity "*e*" was revealed. Its intensity increases in the spectrum of nanotubes purified from Fe. This peculiarity reflects πp_z-states weakly overlapped in radial plane over the surface of nanotubes. Intensity of "*e*" peak increases due to increase of the population of the π-sub-band as a result of additional electronic density introduced by residual atoms of metal after purifying.

In order to determine influence of metals used in synthesis of nanotubes on the change of width of the bands when changing diameters of MWNT it is necessary to investigate spectra of nanotubes before purifying from catalysts.

Comparison of emission bands of MWNT unrefined from metals (with 11% Co and 33% Fe) (Fig. 2) showed that low-energy parts of MWNT spectra are similar in the region 267–275 eV.

It is the evidence of the similar energy redistribution of their valence states involved into sp^2-hybrid bonds of the lower half of valence band. High-energy parts of bands significantly differ. Increase of intensity of "*e*" feature due to the additional electron density introduced by Fe atoms with three times higher concentration than in Co atoms was revealed. From comparison of widths on the half-height of spectra obtained from unrefined nanotubes it is clear that the width of the CK_α-bands of nanotubes decrease when decreasing their diameters so as in purified nanotubes but much smaller.

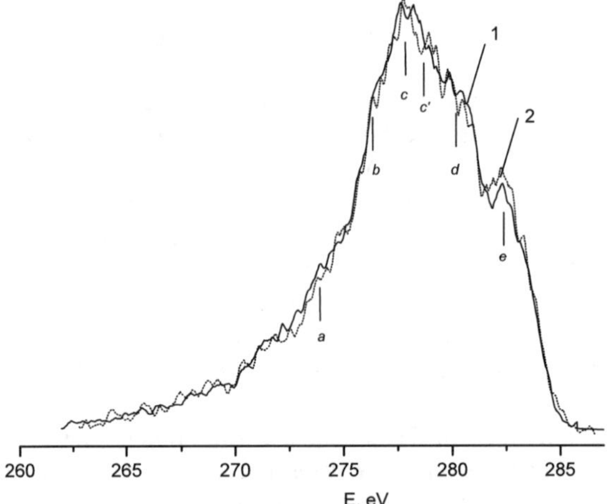

Figure 2. Emission bands of unrefined nanotubes synthesized over the Co-catalyst (1), Fe-catalyst (2)

In order to avoid influence of the interaction of carbon atoms with atoms of metal-catalysts on width of the spectrum the CK_{α}-bands of nanotubes synthesized without catalysts were obtained.

Comparison of spectra of multi-wall carbon nanotubes with different diameters (double-wall (DWNT) with diameter 4 nm and multi-wall with diameter 140 nm), obtained in arc discharge (Fig. 3) revealed that low-energy parts of spectra are practically similar.

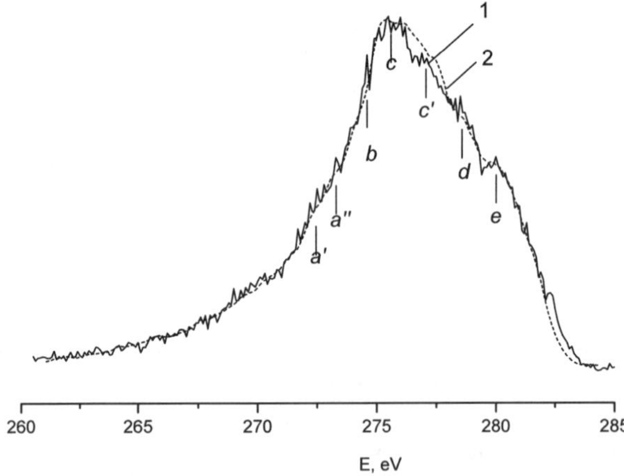

Figure 3. Emission bands of non-catalytic nanotubes: 1 – double-wall, 2 – multi-wall with d = 140 nm

Noticeable differences are observed in short-wave parts of spectra. Main peculiarities of the CK_{α}-spectra of non-catalytic MWNT are less divided than those of DWNT but their intensities are equal. Widths of bands are different in the energy region 276–278 eV. In the DWNT spectrum "*c*", "*c*", "*d*" and "*e*" elements of fine structure are much better divided than those in MWNT with $d \approx 140$ nm. High-energy contours of the CK_{α}, corresponding to "*c*" and "*c*" peculiarities shift towards the low-energy part. As a result the width of the top of the CK_{α} in DWNT in the energy region

276–277.5 eV is by 0.2–0.6 eV smaller than in MWNT with $d \approx 140$ nm. It is a result of the great contribution of overlapping of $pp\pi + pp\sigma$-states in the greater number of walls in MWNT in comparison with DWNT. Due to the latest the greater number of splitted levels overlap the wider energy region.

From comparison of non-catalytic MWNT with $d \approx 140$ nm with spectra of catalytic nanotubes with $d \approx 20$ nm and $d \approx 70$ nm significant broadening of spectrum was revealed when going from the thinnest nanotube to MWNT with $d \approx 70$ nm but widths of spectra did not change during the further increase of diameter (Fig. 4).

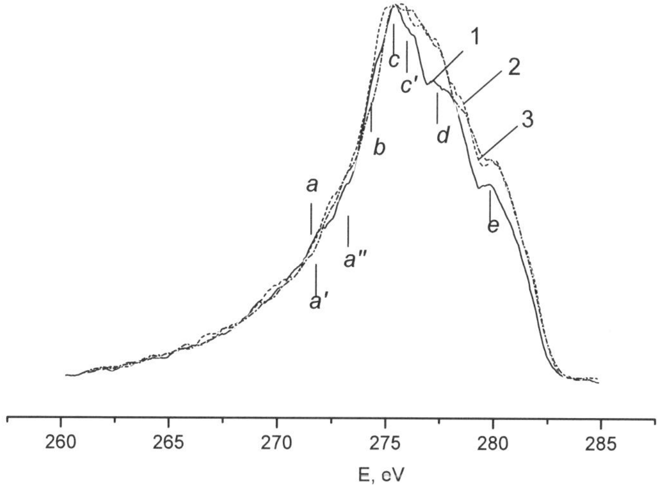

Figure 4. Emission bands of nanotubes with diameters: 1 – 20 nm, 2 – 70 nm, 3 – 140 nm

So at diameters larger then 70 nm curvature of surface decrease when increasing diameter so that overlapping of the πp_z-orbitals will not change during the further increasing of diameter of nanotubes (Fig. 5).

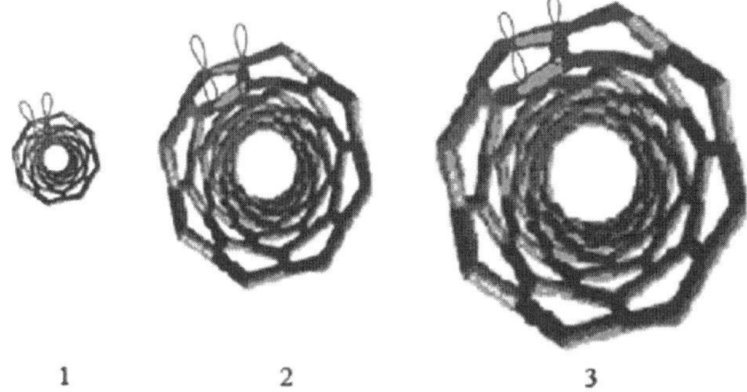

Figure 5. Scheme of the overlapping of the Cp_z-orbitals in nanotubes with diameters: 1 – 20 nm, 2 – 70 nm, 3 – 140 nm

Greater narrowing of the CK$_\alpha$-emission band of MWNT with $d \approx 20$ nm in addition to smaller degree of the π-overlapping of p_z-orbitals in radial planes over nanotubes surfaces is conditioned by contribution of the great number of broken σ-bonds that formed when removing metal-catalysts from nanotubes.

4. Conclusions

It was revealed that narrowing of the CK_{α}-bands of multi-wall carbon nanotubes due to the decreasing of the degree of the π-overlapping of Cp_z-orbitals over the nanotubes surfaces when decreasing their diameter is much smaller than narrowing of the emission bands due to the great number of broken bands that appeared after purifying of nanotubes from catalysts.

It was established that overlapping of the πp_z-orbitals became insensitive to changes of diameters of MWNT at diameters larger than 70 nm.

References

1. Qin L.C. Measuring the true helicity of carbon nanotubes. // Chem. Phys. Lett., 1998, V. 297, P. 23–28.
2. Qin L.C., Ichihashi T., Iijima S. On the measurement of helicity of carbon nanotubes. // Ultramicroscopy, 1997, V. 67, P. 181–189.
3. Qin L.C., Iijima S., Kitaura H., Maniwa Y., Suzuki S., Achiba Y. Helicity and packing of single-wall carbon nanotubes studied by electron nanodiffraction. // Chem. Phys. Lett., 1997, V. 268, P. 101–106.
4. Kane C.L., Mele E.J. Size, shape and low energy electronic structure of carbon nanotubes. // Phys. Rev. Lett., 1997, V. 78, P. 1932–1936.
5. Lambin Ph., Loiseau A., Culot C., Biro L.P. Structure of carbon nanotubes probed by local and global probes. // Carbon, 2002, V. 40, P. 1635–1648.
6. Östling D., Tománek D., Rosén A. Electronic structure of single-wall, multi-wall, and filled carbon nanotubes. // Phys. Rev. B, 1997, V. 55, P. 3980–13988.
7. Reich S., Thomsen C. Electronic band structure of isolated and bundled carbon nanotubes. // Phys. Rev. B, 2002, V. 65, P. 1554411-1–1554411-11.
8. Bulusheva L.G., Okotrub A.V., Romanov D.A., Tománek D. Electronic structure of (n,0) zigzag carbon nanotubes: Cluster and crystal approach. // J. Phys. Chem. A, 1998, V. 102, P. 975–981.
9. Okotrub A.V., Gusel'nikov A.V., Bulusheva L.G. X-ray spectroscopy characterization of carbon nanotubes and related structures. // In: Eds. S. Guceri et al. Nanoengineered nanofibrous materials, Kluwer, NATO Science Series, 2004, P. 349–362.
10. Brzhezinskaja M.M., Bajtinger E.M., Kormilec V.I. Ozhe- i emissionnye spektry uglerodnyh nanotrubok. // Fizika tverdogo tela, 1999, V. 41, P. 1511–1514 (in Russian).
11. Okotrub A.V., Bulusheva L.G., Tomanek D. X-ray spectroscopic and quantum–chemical study of carbon tubes produced in arc-discharge. // Chem. Phys. Lett., 1998, V. 289, P. 341–349.
12. Zhurakovskij E.A., Zaulychnyy Ya.V., Neshpor V.S. Osobennosti elektronnogo strojenija ultradispersnyh poroshkov kubicheskogo nitrida bora. // Poroshkovaja metallurgija, 1991, V. 1, P. 72–76 (in Russian).
13. Zaulychnyy Ya.V. Rentgenospekrtalnoe issledovanie elektronnoj struktury i himicheskoj sviazi v ulrtadispersnyh poroshkah I poluchennyh iz nih melkokistallicheskih materialah. I. Nitrid titana // Poroshkovaja metallurgija, 1999, V. 7/8, P. 75–85 (in Russian).
14. Zaulychnyy Ya.V. Osoblyvosti elektronnoji struktury metastabilnyh ultradyspersnyh system ta oderzhanyh z nyh materialiv // Fizyka i himija tverdogo tela, 2002, V. 3, P. 312–314 (in Russian).
15. Zaulychnyy Ya.V., Solonin Yu.M., Zviezda S.S., Pryluckij E.V. Osoblyvosti elektronnoji struktury ta jiji zalezhnist vid romiriv v vuglecevyh nanomaterialah u riznyh atomnostrukturnyh stanah. // Nanostructurnoje materialovedenie (to be published) (in Russian).

BIOCOMPATIBILITY OF MULTIWALL CNT AND NANOCOMPOSITES ON THE BASE OF POLYMERS

YU. I. SEMENTSOV, N.A. GAVRILUK, G.P. PRIKHOD'KO
Institute of Surface Chemistry NAS Ukraine, 17 Naumova Str., Kyiv, 03164, Ukraine
T.A. ALEKSYEYEVA
Institute for Metal Physics by G. Kurdumov NAS Ukraine 36 Boul., Kyiv-142, 03680, Ukraine
O.N. LAZARENKO
National Medical Academy of Postgraduate Education by P. Shupika, 9 Dorogogitskaya Str., Kyiv, 04112, Ukraine
V.V. YANCHENKO
TM Specmash Ltd., 5 Viskoznaya, Build № 23, Kyiv, 02660, Ukraine

Abstract. The behavior of multiwall carbon nanotubes (MWCNT) conglomerates in water solution depending on concentration electrolytes and/or biomolecules was studied by methods of evaluation of electroconductivity, atome-force microscopy (AFM) and force microscopy (FM). Comparative analysis of AFM and FM determination of polymer surfaces: polytetrafluorineethylene (PTFE), polypropylene (PP) and nanocomposites – PTFE –MWCNT, PP-MWCNT in vitro and reaction of living body to implanted samples. There were found that: FM study could be the method of preliminary testing new materials for biocompatibility; MWCNT in studied polymer matrixes changes the nanocomposites surface chemistry and influence on their biocompatible properties.

Keywords: biocompatibility, carbon nanotubes, polymer nanocomposites, atome-force microscopy, force microscopy, tissue histological analyze

1. Introduction

Conception "biocompatibility" has no certain determination. The most important mandatory is the level of inflammation reaction of recipient's body to exogenous material. In our study it is system "artificial implant – surrounding tissue". The inflammation reaction and its activity could be determined by the thickness of fibrous-connective capsule (pocket) that formed around foreign body [1]. On the base of this knowledge there was suggested to determine rejection reaction by:

- Measurement the thickness of fibrous-connective capsule (pocket) that formed around implant in 4 weeks after installation
- Determination of biochemical parameters of recipients' blood plasma before and at the end-point of the study (may be take in account not the absolute value but their rate before and after)
- Histological study of surrounding tissue

Suggested method of implants' biocompatibility evaluation is reasonable for in vivo experiments. In the case of correlation between histological data, biochemical parameters and physical data this method could be the effective instrument for in lifetime diagnostic of implants biocompatibility with recipient's body. So, in our study we use common and cheep clinical biochemical methods, we could suppose that this method of biocompatibility evaluation will use in medical practice. More over, it would be a good method for new implants' biocompatibility correction. In FM study the forces interaction between tip and samples' surface are depended on properties of tip, sample and medium, where study performed, and in general are complex. In the case of neutral surfaces study in atmosphere (in air) the main contribution in forces interaction between tip and sample are: repulsive forces, that induced by mechanic contact of edge atoms of tip and the sample, van der Waals forces, and also capillary forces which connected with the presence of adsorbate (water) on the sample's surface. Based on interaction characteristics between tip and surface that have similar origin, evidently, we could speculate about their "activity". So, the surface evaluation by FM (force interaction of tip and surface) could be compared with biochemical data that were received in "in vivo" study for compatibility. In the case of establish of correlation bonds between FM and histological data this method could be used as express analyze new materials for biocompatibility.

At last years the carbon nanotubes are attracted for biological and medical purpose – from diagnostic to pharmacology [2]. The use of carbon nanoparticles, nonmaterial and systems that they include are named "nanomedicine". The main goal for it is to determine toxicity and pharmacological properties of carbon nanotubes in ever mode. The possible dangerous of CNT is directly connected with their size. It means that CNT has high specific surface that cause high chemical activity and high ability for penetration into the living body. Cancerogenity of disperse materials commonly connected with the presence of free radicals, in other words, with presence of surface's atoms and molecules that could oxidize living cell. Inhalation with carbon nanofibers guinea pig and rats shows that nanotubes resulted in lung granuloma formation. Subcutaneous implantation in mice nanocarbon materials for 3 months leads to granuloma formation and more over to the blood cells changes. But none of animals died, and fibrosis was lower than in the case of asbestos fibers implantation. At the same time in vivo experiments [2,3] show that CNT does not arise complications and even could be used as carrier for vaccination. Modification of MWCNT [4] leads to their desegregation and promote their uniform distribution in medium. Polymer materials filled with MWCNT develop unique physical-mechanical properties (strongly transcend the characteristics of initial materials) and electron properties [5] and could be used as prospective materials for endoprosthesis manufacturing.

At present work we study in vivo the organism reaction to implantation of purified MWCNT conglomerates and nanocomposites – PP and PTFE filled with MWCNT in different concentration. FM studied the changes of the nanocomposites' surface in comparison with initial matrixes. So as the state of the surface of implanted in living body material is affected by the chemical reactions at interface of solid/liquid phases, in other words the body reaction and the level of implant's biocompatibility.

Beside this in our work we investigated the influence of concentration of electrolytes and biomolecular in solution on the level of MWCNT desegregation.

2. Experiment

Method of MWCNT obtaining (Fig. 1) and characterization are described [5]. The average diameter of MWCNT was 10–20 nm, specific surface determined by argon desorption was – 200–400 m^2/g, and bulk density was at the range 20–40 g/dm^3. Ash residue of unpurified MWCNT was 6–20%, in purified ones – lower than 1%.

According to the data of x-Ray analyze the MWCNT d_{002} parameter of graphitelike crystal lattice is in the 0.3431–0.3451 nm. The low size of coherent-scattering region and microstress lead to significant enlargement of X-ray reflection. We must mark that analyze TEM images, X-ray diffraction, Raman spectroscopy data and temperature value of the start intensive mass loss (580–610°C) does not show the presence of noticeable amount of amorphous phase.

XPS data show about 0.6 % atom oxygen presence on the MWCNT surface that according analysis C 1s spectra represented in different oxygen containing groups at such mode: phenol, alcohol – 49.1%, carbonyl, kinone – 17.2%, carboxyl, ether – 17.2%, carbonate and/or absorbed CO, CO_2 – 16.5%.

Nanocomposites [5] PP-MWCNT with MWCNT 0.05; 1.0; 3.0; 5.0 mass % concentration, PTFE-MWCNT with concentration 15 wt % and initial polymer matrixes were chosen for in vivo experiments.

Suspension MWCNT was prepared in twice-distillate and deionized water by treating with ultrasonic dispergator УЗДН-2 (Tochelectropribor, Kyiv, Ukraine). The value of working chamber was 2 ml. The goal of the study was to investigate interaction of MWCNT and living body and all measurements were carried out at 37.5°C by the use thermostat. The same reason was taken in account during study of electrical resistance of the systems 0.1 mass % MWCNT – water solution of albumin (up to 2%), heparin (1%). In all cases the measurements were provided in the presence of 0.9% NaCl. Electrical resistance was measured with apparatus P-5068 (Tochelectropribor, Kiev, Ukraine) at frequency f=1 kHz, in the chamber with two vertical platinum electrodes with surface 1 cm^2. The electrical resistance data are shown in Table 1.

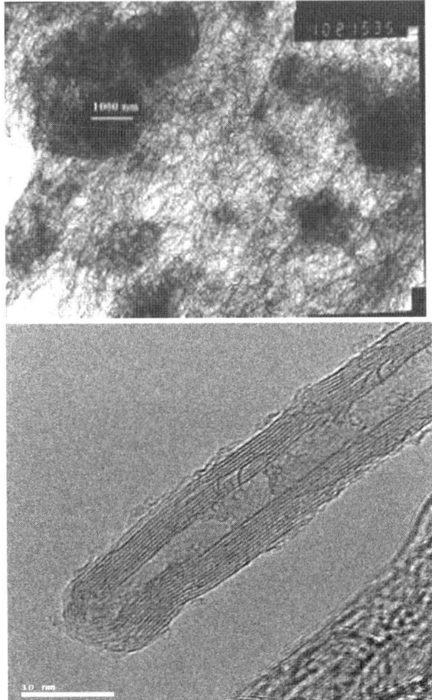

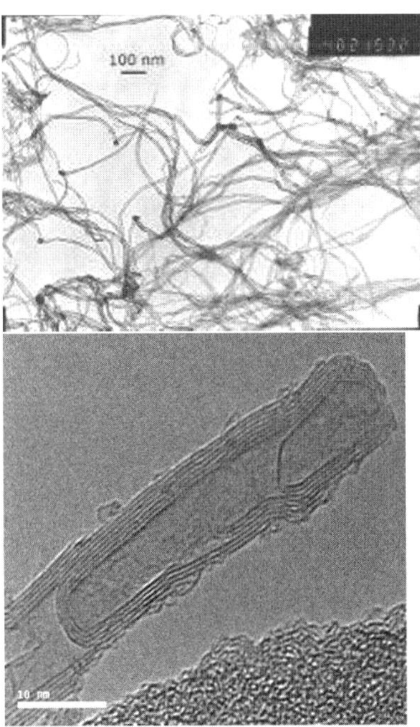

Figure 1. TEM of MWCNT

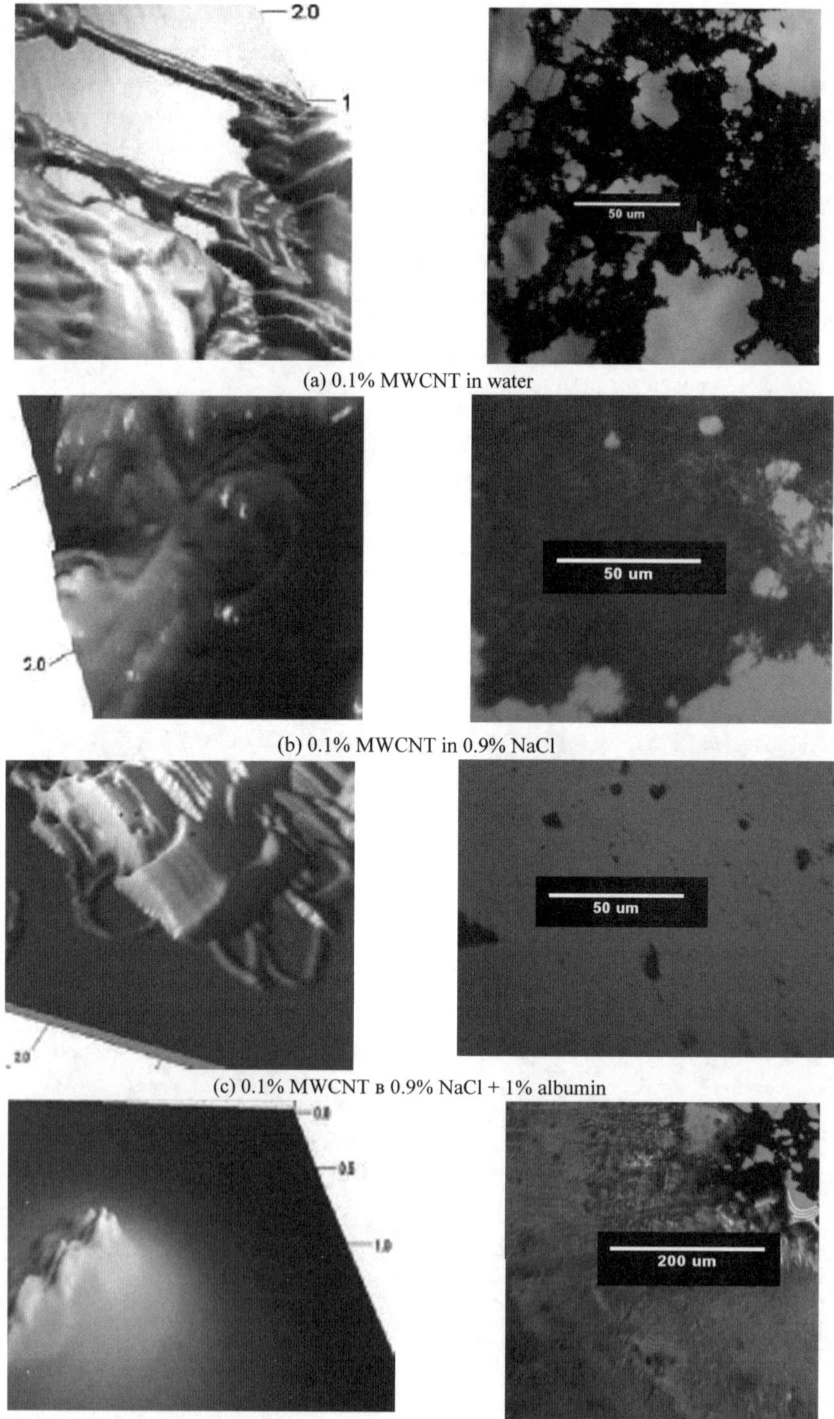

(a) 0.1% MWCNT in water

(b) 0.1% MWCNT in 0.9% NaCl

(c) 0.1% MWCNT в 0.9% NaCl + 1% albumin

(d) 0.1% CNT в 0.9% NaCl + 1% heparin

Figure 2. AFM and optical microscope images of MWCNT in systems

The level of MWCNT desegregation (the mode and size of conglomerates) was determined by AFM (NANOSCANE II, Veeco corp) and optical microscopy Carl Zeiss, NY/2E (Germany) investigation. MWCNT modification by solution of different compounds (from ion to biomolecules) resulted in changing the mode of initial MWCNT conglomerates (Fig. 2).

At Fig. 2(a) you can see that in water solution MWCNT are arranged in rather big compact aggregates. At the conglomerates edge there are observed single elongated clusters, obviously formed of a few MWCNT. Adding of 0.1% of MWCNT leads to reduction of system electrical resistance in 25 times. Dissolving of 0.9% NaCl in water resulted in drop of electrical resistance in ~10^3 times. Adding to the system 0.1% of MWCNT reduces electrical resistance by 0.4%. This effect accompanies with conglomerates size and compactness decreasing in comparison with system MWCNT-water. Data from Table 1 show that uniform reduction of NaCl concentration leads to the corresponded increasing of electrical resistance.

TABLE 1. Electrical resistance of the system water-NaCl-MWCNT-albumin

Sample	Electrical resistance (Ω)
H_2O	42.13×10^3
H_2O +1% MWCNT	1.641×10^3
H_2O + 0.9% NaCl	41.4
H_2O + 0.9% NaCl + 0.1% MWCNT	41.25
H_2O + 0.675% NaCl + 0.1% MWCNT	49.03
H_2O + 0.45% NaCl + 0.1% MWCNT	67.42
H_2O + 0.225% NaCl + 0.1% MWCNT	111.8
H_2O+ 0.9% NaCl + 2% albumin	44.61
H_2O + 0.9% NaCl + 2% albumin + 0.1% MWCNT	43.85
H_2O + 0.9% NaCl + 1.5% albumin + 0.1% MWCNT	43.25
H_2O + 0.9% NaCl + 1.0% albumin + 0.1% MWCNT	42.55
H_2O + 0.9% NaCl + 0.5% albumin + 0.1% MWCNT	42.12

Adding 1% of albumin (60 kD) to the system water – NaCl resulted in increasing electrical resistance by ~7%, but MWCNT reduce it slightly. At Fig. 2(b) one can see that parameters of MWCNT aggregates in albumin the presence becomes significant smaller. The sedimentation process inhibits in the albumin presence. In 5 minutes later in supernatant one could observe MWCNT aggregates. Albumin replacement in the system for 1% heparin (14 kD) leads to greater MWCNT aggregates decomposition. More over at the presence of heparin, which molecular is smaller than MWCNT, the ones exfoliation is observed. This could be explained that heparin molecule in contrast to albumin, is rather small and can fill the space between aggregates that resulted in dispersion phenomenon. The same situation was registered during the surfactant use – nonionic detergent like Triton X-305 (1.528 D) [6].

Thus AFM study of MWCNT in water system indicates that in correspondence with [2, 6] for destruction (dispersion) MWCNT aggregates and stabilizing the system MWCNT-water it is necessary the presence of one more component with the size more than 1.500 D, but lower than 13.000 D.

The interaction FM tip with samples surface (Fig. 3) was described means of jump-to and pull-off forces that are represented at Table 2. The interaction forces between tip and samples' surface were measured for PTFE initial, PTFE-MWCNT initial and for samples after 4 weeks stay in living body. The measurement was provide in air and in liquid medium – 0.9% NaCl – in order to approximate situation to in vivo experiments.

In study were used as clear tip and modified with protein IgG one. The tip modification was carried out according [7] from IgG solution concentration 70 mcg/ml. In liquid the value of interaction forces were significant lower in comparison with ones in air. This phenomenon could be explained by surface masking with nearby water lays. The jump-to forces are connected with application (imposition) to the tip definite potential that changes depend on the type of surface. Therefore more the value of pull-off forces would be more informative and significant. Forces with which composites surface hold modified with IgG practically do not detect. May be it could be explained that IgG molecule is rather large (150 kD) and has unpliable space view like letter Y.

TABLE 2. Force interaction between tip and surface PTFE and nanocomposite PTFE-MWCNT (15 mass %)

Sample	Air		Water	
	Jump to	Pull off	Jump to	Pull off
	«Clear» Si_3N_4 tip/force interaction (nN)			
PTFE	0.79 ± 0.26	10.64 ± 0.33	0.28 ± 0.06	3.72 ± 0.22
PTFE-MWCNT	0.13 ± 0.01	8.32 ± 0.38	0.08 ± 0.01	0.92 ± 0.06
PTFE initial	0.22 ± 0.02	10.32 ± 0.28	Doesn't detected	2.43 ± 0.07
	Si_3N_4 +IgG tip/ force interaction (nN)			
PTFE	2.37 ± 0.22	26.28 ± 0.28	Doesn't detected	1.82 ± 0.06
PTFE-MWCNT	1.29 ± 0.03	9.63 ± 0.21	Doesn't detected	0.48 ± 0.06
PTFE initial	3.86 ± 0.11	23.83 ± 0.70	Doesn't detected	4.72 ± 0.15
PTFE-MWCNT initial	1.93 ± 0.14	13.10 ± 0.22	Doesn't detected	Doesn't detected

Data from Table 2 show that forces with which PTFE-MWCNT surface hold clear tip are smaller than initial PTFE (8.32 vs 10.64 nN respectively). Significant difference between forces with which initial PTFE and PTFE samples in 4 weeks after implantation surfaces are not detected (10.32 and 10.64 nN, 0.22 and 0.79 nN respectively).

Interaction forces between modified tip and samples' surfaces are rather differed from ones in the case of clear tip. For example, the force with which PTFE surface hold modified tip near two times higher than ones for PTFE-MWCNT surface (23.83 vs.13.10 nN respectively). The state of surface in 4 weeks after implantation brightly demonstrates the effect of living body on the foreign body. The PTFE surface after implantation holds modified tip strongly in comparison with ones before implantation 26.28 vs. 23.83 nN, respectively, but in the case of nanocomposites we receive the certain force decreasing (9.63 vs. 13.10 nN respectively).

Thus, in all cases the MWCNT incorporation into polymer matrix resulted in decreasing of forces interaction between samples' surfaces and tip. Modification tip with IgG demonstrated more general difference. This result is very important because IgG is the first biomolecules that trigger mechanism of living body identification the foreign material. The more the value of interaction forces between modified tip and investigated surface the lager bodies replay would call implantation.

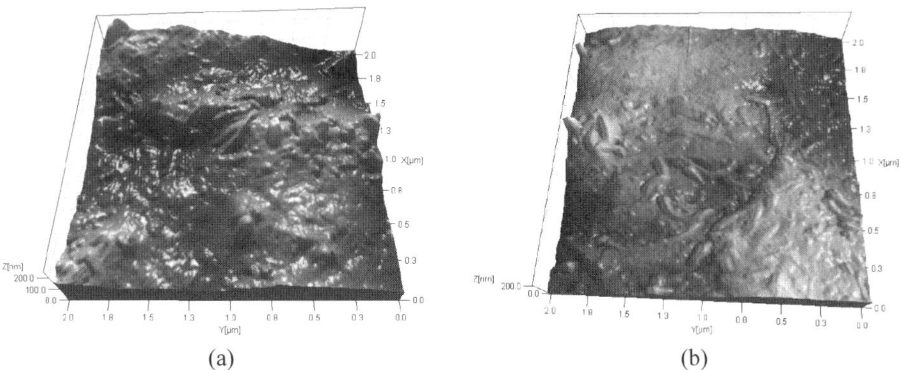

(a) (b)

Figure 3. AFM data of samples PTFE-MWCNT: (a) before implantation; (b) after implantation

From obtained data we can expected that the lowest "reaction ability" to the organism of experimental animals would show the nanocomposites in comparison with initial matrixes.

Samples were implanted into animals' back muscles. In order to reduce the amount of animals but receive statistical reliable data we implanted up to four samples of polymer with different MWCNT concentration per animal. In 4 weeks samples were excised with surrounding tissue for the follow-up histological study and investigation of samples' surfaces. Level of experimental animal organism reaction to sample was evaluated by the thickness of fibrous-connective capsule that formed around it. The histological study results show that recipient tissue reaction to implanted samples significant differs. In the case of initial matrix PTFE there was formed thick and compact capsule (362.6 ± 24.7 mcm), that was specified with plentiful artificial inclusion. In the case of composite with MWCNT the thickness of fibrous-connective capsule was 58.4 ± 8.2 mcm. In surrounding tissue also were found artificial particles, but rather rare in comparison with initial polymer matrix. MWCNT concentration in matrix does not uniformly influence on its biocompatible properties. Data from Table 3 argue that the lowest thickness of fibrous-connective capsule (as marker of the level of organism reaction on foreign body) was in the case of 1% MWCNT in the sample.

TABLE 3. Thickness of fibrous-connective capsule around nanocomposites PP-MWCNT

№	Concentration of MWCNT in PP matrix (%)	Thickness (mcm)
1	0.05	58.9 ± 12.6
2	1.00	32.1 ± 5.2
3	3.00	72.4 ± 14.8
4	5.00	152.6 ± 24.8

The level of body reaction also could be determined by the changes of blood biochemical parameters: Glucose, C-reactive protein and Cholesterol. The data of biochemical parameters in animals' blood serum in Table 4 show that at 1 day after operation concentration of Glucose and C-reactive protein markedly increase. Such elevation is typically for postoperative period and indicate that organism actively react on intervention. In 4 weeks after all three parameters do not differ from initial value. It's say that tissue reaction is local and aseptic. The difference in fibrous-connective capsule thickness that formed around sample depends on its surface chemistry.

TABLE 4. Changes of biochemical animals' blood parameters after implantation of nanocomposites samples

Parameters	Before operation	1 day after operation	In 4 weeks after operation
Glucose	5.3 ± 0.52 g/dl	17.6 ± 2.8 g/dl	6.1 ± 0.37 g/dl
C-reactive protein	2.4 ± 0.09 mg/l	16.2 ± 4.1 mg/l	3.8 ± 0.12 mg/l
Cholesterol	36 ± 4.8 mM/l	Not detected	42 ± 12.1 mM/l

3. Conclusions

The behaviors of MWCNT' conglomerates in water solution depending on electrolytes and/or biomolecules concentration were studied by the conductometry, AFM and FM methods. There was shown that dispersion (desegregation) of MWCNT conglomerates and stabilization of system MWCNT – water, require one more component. Namely, molecule which has stronger ability to form hydrogen bond with MWCNT oxygen containing centers than at the system MWCNT-water.

Comparative analysis of AFM, FM data of in vitro polymers' surface investigation – PTFE, PP and nanocomposites based on these matrixes – and the results of living body reaction to implanted ones, show that: (a) FM would be useful method for primary new material testing for biocompatibility; (b) MWCNT in studied polymer matrixes change nanocomposites surface chemistry and influence on their biocompatible properties.

References

1. M. Sumer, M. Muglali, E. Bodrumlu et al. Reactions of connective tissue to amalgam, intermediate restorative material, mineral trioxide aggregate, and mineral trioxide aggregate mixed with chlorhexidine // *J. Endodont.* 2006; **32**: 1094–1096.
2. L. Lacerda, A. Bianco, M. Prato, K. Kostarelos Carbon nanotubes as nanomedicines: From toxicology to pharmacology // *Adv. Drug Deliv. Rev.* 2006; **58**:1460–1470.
3. D. Pantarotto, C.D. Partidos, J. Hoebeke, et al. Immunization with peptide-functionalized carbon nanotubes enhances virus-specific neutralizing antibody responses // *Chem. Biol.* 2003; **10**: 961–966.
4. G. Pastorin, Wei Wu, S. Wieckowski et al. Double functionalisation of carbon nanotubes for multimodal drug delivery // *Chem. Commun.* 2006; 1182–1184.
5. Yu. Sementsov, O. Melezhek, G. Prihod'ko, et al. in book Physical chemistry on nanomaterials and supramolecular structures // vol. 2 edited by A. Shpak, P. Gorbyk – Kiev: Naukova Dumka, 1007; pp 116–158.
6. M. Lisunova, N. Lebovka, O. Melezhyk, et al. Stability of the aqueous suspensions of nanotubes in the presence of nonionic surfactant // *J. Colloid Interf. Sci.* 2006; **299**: 740–746.
7. J. Gu, Z. Xiao, C.-M. Yam, et. al. Attaching single biomolecules selectively to the apex of AFM tips for measuring specific interactions // *Biophys. J.* 2005; **89**(5): L31–L33.

STUDIES OF ELECTROCHEMICAL KINETICS OF HYDROGEN OXIDATION ON THE NANODISPERSED DIAMOND

G.P. BOGATYREVA, M.A. MARINICH, G.A. BAZALIY
V.N. Bakul Institute of Superhard Materials of National Academy of Sciences of Ukraine, 2, Avtozavodskaya Str., Kiev, Ukraine
Tel./Fax: +38(044) 468-85-99; E-mail: bogatyreva@ism.kiev.ua

Abstract. The object of this work was investigate of electrochemical kinetics of hydrogen oxidation on the nanodispersed diamond in solutions of sulphuric acid, sodium chloride and sodium hydroxide in the temperature range 20–60°C.

The initial nanodispersed diamond powders of ASUDCh have been selected as subject of investigations. It has been ascertained that values of exchange current density of hydrogen oxidation reach maximum on heat-treated diamond in sulfuric solutions, whereas these values reach minimum in solutions of sodium chloride. Value i_0 depends on concentration of solution. It has been shown that optimal concentration is 1.0 N. It has been ascertained that carrying out of two-stage thermal treating results in decreasing of activation energy nearly by two times.

The water quantity, which desorbed by diamond after heat-treated presipication in hydrogen, increases in three times. The increase of quantity waters confirms about significant increase adsorbing activity of a diamond surface. It has been shown that heat-treated presipication in hydrogen forms on diamond surface the amorphous-like carbon phase promotional to chemical dissociation of molecular hydrogen.

The special heat-treated processing in hydrogen allowed us create effective catalysts for hydrogen electrodes.

Keywords: nanodispersed diamond, electrooxidation of hydrogen, exchange current density, heat-treated treatment

1. Introduction

Carbon materials are very promised in development of various electrochemical current sources. High adsorbility and possibility to form various types of adsorptive compounds as well as wide spectrum of electro-physical parameters, developed porous structure and good chemical stability determine catalytic activity of carbon materials in very varied reactions. A number of researches are dedicated to study of process of electrooxidation of hydrogen on carbon materials [1–3].

B. Baranowski et al. (eds.), *Carbon Nanomaterials in Clean Energy Hydrogen Systems.*

Tungsten and vanadium carbides are well-known electrode materials for electro-oxidation of hydrogen. Well promise of use of nanodispersed diamond as hydrogen electrode has been shown by researches of ISM [4–6].

The presence of electronegative impurities (O, Cl etc.) on surface of diamond powders results in worsening of kinetics of hydrogen adsorption kinetics and reduction of its capacitance.

The object of this research is study of hydrogen oxidation processes on surface of initial and heat-treated nanodispersed diamond powders.

2. Experimental

Studies were carried out on potentiostate P-5848 in solutions of sulphuric acid, sodium chloride and sodium hydroxide in the temperature range 20–60°C. The following diamond powders have been selected as subject of investigations:

1. Initial nanodispersed diamond powders of ASUDCh series (firm ALIT production) synthesized by detonation of explosives with negative oxygen balance
2. Powders subjected to thermal treatment in argon-saturated atmosphere
3. Powders subjected to special two-stage treatment in hydrogen saturated atmosphere [7]

The experiments were carried out film electrodes applied to a substrate of isotropic graphite [8].

Value of exchange current density (i_0) is measure of electrocatalytic activity of electrooxidation of hydrogen. Value of exchange current density (i_0) is determined by extrapolation of hydrogen overvoltage curve (η) from current density (i) value $\eta = 0$.

The parameters of electrooxidation of hydrogen were compared using the materials like tungsten and vanadium carbides.

Surface condition of initial and heat-treated nanodispersed diamond powders was studied by mass-spectrometry method [9]. Mass-spectra and thermodesorption spectra of monatomic and molecular oxygen, mono oxide of carbon (CO), two oxide carbon (CO_2), water were drawn using mass spectrometer MI-1201 in the temperature range 20–600°C.

3. Results and discussion

The dependence of electrocatalytic activity of hydrogen oxidation process taken place on nanodispersed diamond surface (exchange current density, i_0) working electrolytes was studied. Figure 1 illustrates dependences of value i_0 on initial and heat-treated nanodispersed diamonds in 1н solution of a sulfuric acid, sodium hydrooxide and sodium chloride. It has been ascertained that values of exchange current density of hydrogen oxidation process reach maximum on heat-treated diamond in sulfuric solutions, whereas these values reach minimum in solutions of sodium chloride.

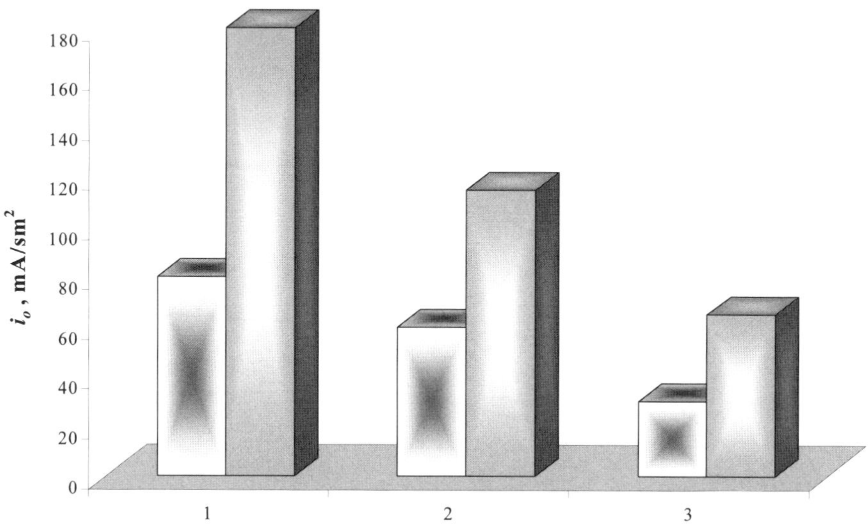

Figure 1. Influence of thermal processing nanodispersed diamond powders on density of a current of an exchange of process of oxidation of hydrogen in solutions 1 N sulfuric acid (1), 1 N hydrooxide sodium (2), 1 N chloride sodium (3)

The value i_0 of nanodispersed diamond after two-phasic heat treatment in atmosphere of hydrogen has increased in 2–2.5 times. The first stage of heat treatment was a clearing of diamonds of organic materials impurity, water, and also activation of their surface. Results of hydrogen process of electrooxidation at two-phasic processing in a hydrogen atmosphere were compared with results processing in argon under the same conditions.

Value i_0 depends on concentration of solution. Figure 2 illustrates dependences of values i_0 in 1H solutions from a sulfuric acid and hydrooxide sodium from their concentration on initial powders (1) and heat-treated in atmospheres of argon (2) and hydrogen (3). In the field of concentration from 0.1 up to 1.0 N rectilinear growth of value density of a current of an exchange is observed. It has been shown that optimal concentration is 1.0 N. From the resulted data follows, that carrying out of two-phasic thermal processing of powders nanodispersed diamonds in an atmosphere of hydrogen allows to increase i_0 in a solution of a sulfuric acid almost by two times.

It has been established, that with growth of temperature of solution from 20°C up to 60°C the value i_0 increases by 1.5–1.7 times.

Values of activation energy of hydrogen oxidation process (ε) were calculated from $\lg i_0$ – 1/T (K) dependences. It has been ascertained that carrying out of two-stage thermal treating results in decreasing of ε nearly by two times. Figure 3 illustrates dependences $\lg i_0$ from 1/T for process of electrooxidation of hydrogen on initial and nanodispersed diamonds after heat treatment in hydrogen. For process of electrooxidation on initial powders ε equally 15 kJ/mol, and at nanodispersed diamonds after heat treatment at in hydrogen equally 7.5 kJ/mol are resulted. Thus, It has been ascertained that carrying out of two-stage thermal treating results in decreasing of ε nearly by two times. It has been shown that this treatment forms on diamond surface the amorphous-like carbon phase promotional to chemical dissociation of molecular hydrogen. Just combination of amorphous-like monolayer and diamond core results in significant increasing of catalytic hydrogen oxidation rate.

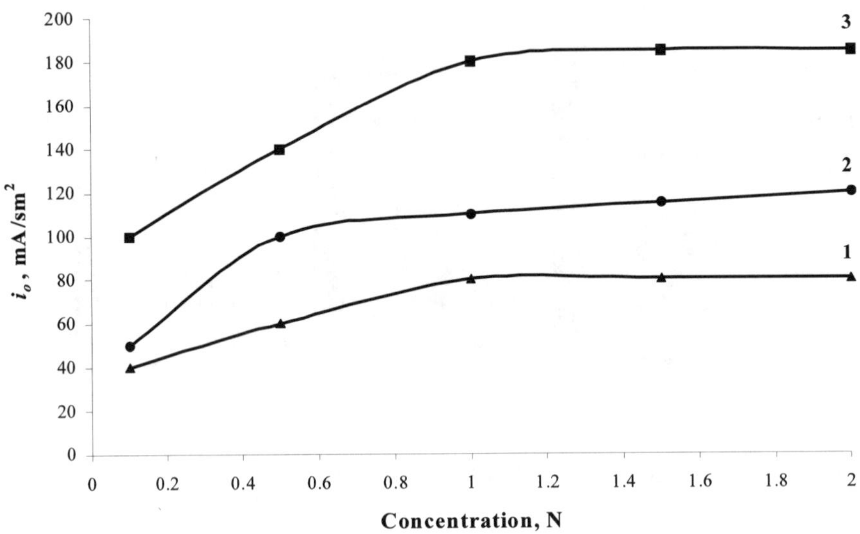

Figure 2. The exchange current of hydrogen electrooxidation from of initial nanodispersed diamond (1), nanodispersed diamond after argon heat-treated (2), nanodispersed diamond after hydrogen heat-treated (3) from concentration of sulfuric acid

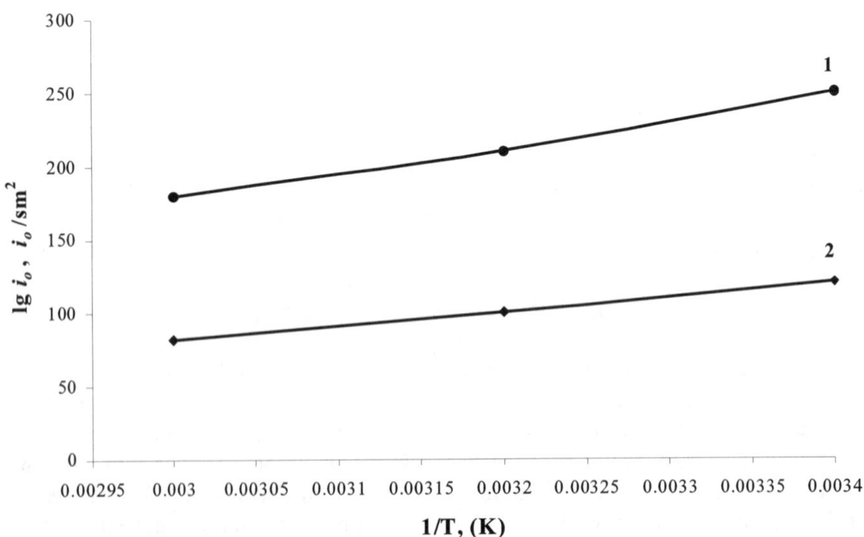

Figure 3. Dependences lgi_0 from 1/T (K) for process of electrooxidation of hydrogen on initial (1) and nanodispersed diamonds after heat treatment in hydrogen (2) in sulfuric acid

Thermodesorption spectra allow to obtain extensive informative material. The thermodesorption spectra have been considered: water, atomic and molecular oxygen mono oxide carbon and two oxide carbon from the surface of initial and heat-treated diamonds taken in 20–600°C.

It is confirmed maximal intensity of water in the temperature range 25–100°C. Figure 4 shows the thermodesorption spectra of water from the surface the initial diamond (1), and diamond after heat-treated presipication in argon (2) and diamond after heat-treated presipication in hydrogen (3). It is seen from figures that there is a great difference in quantity of water on the initial and diamond after heat-treated presipication in hydrogen. The water quantity, which desorbed with diamond after heat-treated presipication in hydrogen increased in 3. The increase quantity waters testified about significant increase adsorbed activity of a surface.

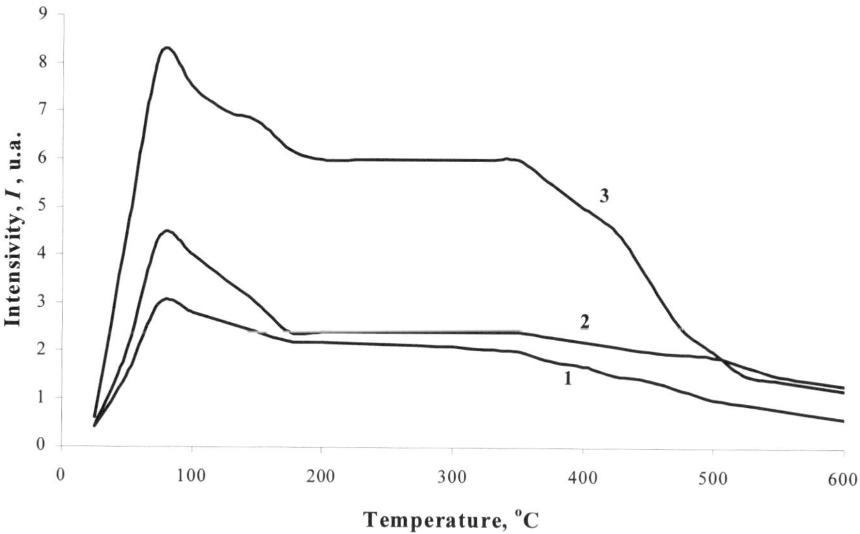

Figure 4. Thermodesorption spectra of water from the surface: the initial nanodispersed diamond (1), nanodispersed diamond after heat-treated in argon (2) and nanodispersed diamond after heat-treated in hydrogen in sulfuric acid (3)

Figure 5 shows the thermodesorption spectra of mono oxide carbon from the surface the initial diamond (1), and diamond after heat-treated presipication in hydrogen (2). The maximum of CO, which desorbed with diamond after heat-treated presipication in hydrogen, has been set at temperature of 100°C lower.

Maximal value of exchange current density i_0 = 180–200 mA/cm^2. Values of i_0 obtained on surfaces of nanodispersed diamond, tungsten and vanadium carbides have been compared. All catalysts have close values of specific surface. It has been shown that catalytic activity of diamond catalyst exceeds activity of carbide catalysts nearly by two times. It has been shown that carrying out of two-stage thermal treating in hydrogen atmosphere forms on diamond surface the amorphous-like carbon phase promotional to chemical dissociation of molecular hydrogen. Just combination of amorphous-like monolayer and diamond core results in significant increasing of catalytic hydrogen oxidation rate.

Active carbon enters into oxidizing process by means of hydrogen dissociation and transition of H_3O^+ as proton.

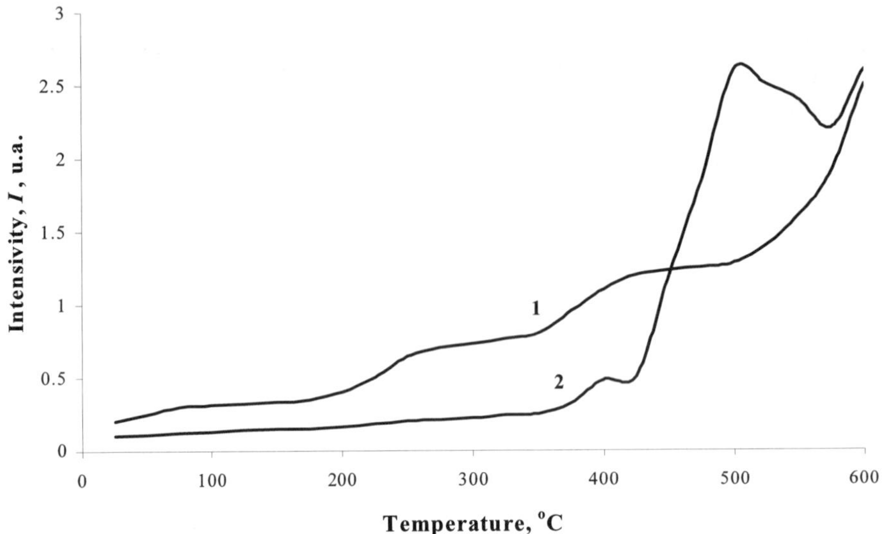

Figure 5. Thermodesorption spectra of mono oxide carbon from the surface: the initial nanodispersed diamond (1), nanodispersed diamond after heat-treated in hydrogen in sulfuric acid (2)

4. Conclusions

1. Kinetics of electrochemical oxidation of hydrogen on surface of initial and heat-treated nanodispersed diamond powders in solutions of sulphuric acid, sodium chloride and sodium hydroxide in the temperature range 20–60°C were studied. It has been ascertained that values of exchange current density of hydrogen oxidation process reach maximum on heat-treated diamond in sulfuric solutions, whereas these values reach minimum in solutions of sodium chloride.
2. Dependence of value of density of a current of an exchange of process of electrooxidation of hydrogen on temperature of electrolite is studied. The values of activation energy of hydrogen oxidation process have been calculated. It has been ascertained that as-two-stage-thermal treating activation energy value is decreased nearly by two times.
3. The water quantity, which desorbed by diamond after heat-treated presipication in hydrogen increased in three times. The maximum of CO, which desorbed by diamond after heat-treated presipication in hydrogen has been set at temperature 100°C lower.

4. It has been shown that heat-treated presipication in hydrogen forms on diamond surface the amorphous-like carbon phase promotional to chemical dissociation of molecular hydrogen. Just combination of amorphous-like monolayer and diamond nucleus results in significant increasing of catalytic hydrogen oxidation rate. This special heat-treated presipication in hydrogen has allowed us effective for hydrogen electrodes.

References

1. Tarasevich M.R. Electrochemistry of carbon materials. Moscow: Science, 1984.- 253 p.
2. Koros Y., Schwartz S. Long term hydrogen oxidation catalysts in alkaline fuel cells // *J. Power Sources*, 2000, **87** (1–2): 101–105.
3. Volfkovich Yu.M., Bagotzky V.S. Research of porous structure density of adsorbed hydrogen in carbon nanomaterials and electrochemical generation adsorption-desorption and electrooxidation in them of hydrogen // *J. Power Sources*, 1994, **48**: 327–339.
4. Bogatyreva G.P., Marinich M.A., Bazaliy G.A. et al. Studies of catalytic reactions proceeding on the diamond surface in liquid and gaseous media // *J. Electrochem. Soc.*, 2002, **5**: 23–28.
5. Bogatyreva G.P., Marinich M.A., Ishchenko E.V., Gvyazdovskaya V.L., Bazaliy G.A. Adsorption and catalytic processes on the surface of nanodispersed diamonds diamond // *J. Superhard Materials*, 2002, **6**: 10–15.
6. Zhutaeva G.V., Marinich M.A., Bogatyreva G.P., et al. Electrocatalist on electrodes on the basis of promotive ultradisperse diamond // *J. Superhard Materials*, 2001, **6**: 55–60.
7. Novikov N.V., Bogatyreva G.P., Marinich M.A., Bazaliy G.A., Olejnic N.A. Catalysts of electrochemical hydrogen evolution oxidation and method of their application. Patent of Ukraine PAT. 50370 A, 2002, Bul. 10.
8. Bogatyreva G.P., Marinich M.A., Gvyazdovskaya V.L. The use of cathode potentiodynamic pulses for a quantitative assessment of efficient conducting centres at the surface of diamond // 7th European Conference on Applications of Surface and Interface Analysis, Goteborg, 16–20 June 1997, P. 263–266.
9. Bogatyreva G.P., Ishchenko E.V., Marinich M.A., et al. Catalytic properties of Pd, Pd–Ag, and Ag deposited onto diamond submicron powders // *J. Superhard Materials*, 2000, **2**: 72–77.

DIELECTRIC PROPERTIES OF EPOXYAMINE COMPOSITES MODIFIED BY CARBON NANOFIBERS IN MICROWAVE RANGE

V.E. MURADYAN
Institute of Problems of Chemical Physics of RAS, Prospectus of the Academician Semenov, 1, Chernogolovka, Moscow Region, 142432 Russia
Tel./Fax: +7(496) 5221743; E-mail: muradyan@icp.ac.ru

E.A. SOKOLOV, S.D. BABENKO
Institute of Energy Problems of Chemical Physics (branch) of RAS, Prospectus of the Academician Semenov, 1, Chernogolovka, Moscow Region, 142432 Russia

A.P. MORAVSKY
MER Corporation, Tucson, AZ, USA

Abstract. Dielectric resonator-based measurements for determination of the dielectric properties of epoxy-composites containing carbon nanofibers with various diameters and aspect ratio in microwave range are described.

Keywords: carbon nanofibers, epoxy composites, dielectric properties

1. Introduction

It is known, that the additives of carbon nanofibers (CNF) up to 2 wt% in polymers considerably improve their strength and thermal characteristics. In works [1–3] was reported about using composites with carbon nanomaterials as radio adsorbing materials. The materials were used without additional enrichment, and the content of nanofibers in them was contained 50–70 wt%. The contribution of alone nanofibers in dielectric properties of composites remains to vague.

2. Experimental

The complex permittivity of polymeric composites has been studied on the basis of epoxy resin with the additive up to 2 wt% of filler with nanofibers concentration in carbon nanomaterial more than 80 wt%. The materials produced by CVD [4] had the following characteristics:

9CL7T (86 wt% CNF, diameter 12 ± 3 nm, aspect ratio 3,300)
MRCMW (92 wt% CNF, diameter 35 ± 15 nm, aspect ratio 2,000)
CT1H (82 wt% CNF, diameter 18 ± 3 nm, aspect ratio 170)
CT2H (80 wt% CNF, diameter 22 ± 3 nm, aspect ratio 140)
VGCF162 (85 wt% CNF, diameter 162 ± 95 nm, aspect ratio 65)
The Fig. 1 shows SEM images on which corresponding bundles of CNF are visible.

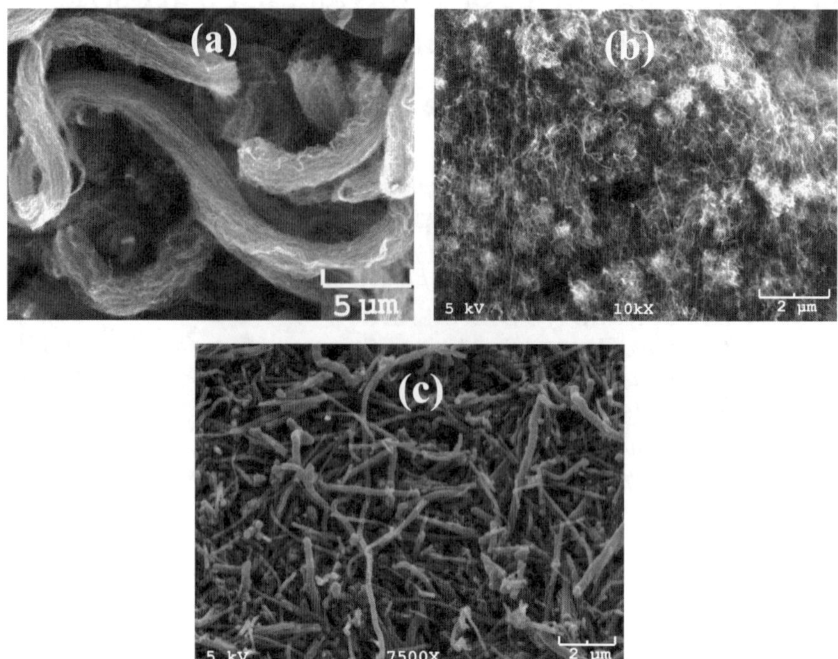

Figure 1. SEM images of samples 9CL7T (a), CT1H (b) and VGCF162 (c)

Samples CT1H and CT2H were heated on air up to 450°C for removal of the chemisorbed impurity and amorphous carbon. Uniform loading of filler in epoxyamine composites was obtained by ultrasonic dispergation of CNF in epoxy resin ED-20 and subsequent addition of hardener PEPA (6:1).

Curing of composites were carried out in teflon profile board step by step increasing temperature up to 50°C, 100°C and 150°C. The permittivity measuring (ε' and ε'') of samples were carried out on frequency 2.73 GHz by resonator method.

3. Results and discussion

In value of permittivity two features are observed at filler concentrations ~0.4–0.5 and 0.8–1.0 wt% for composites with MRCMW (Fig. 2(a)). For 9CL7T of feature are observed at filler concentrations ~0.4–0.5 and 0.7–0.9 wt%.

For CT1H the second feature is in interval of CNF concentrations 0.95–1.15 wt%. That is at smaller aspect ratio of CNF the second feature is shifted in the major side of concentrations.

At the same time at similar observation of single-walled nanotubes at a filler concentration of ~0.73 wt% the deep dip in quantity of a complex permittivity is observed [5]. The causes of appearance of detected features yet do not set. It is possible that this are bound to reorganization of formed reticular structure of nanofibers in composites and depend on CNF aspect ratio. Let‘s mark, that these features become noticeable at nanofibers concentration in filler not less than 70 wt%.

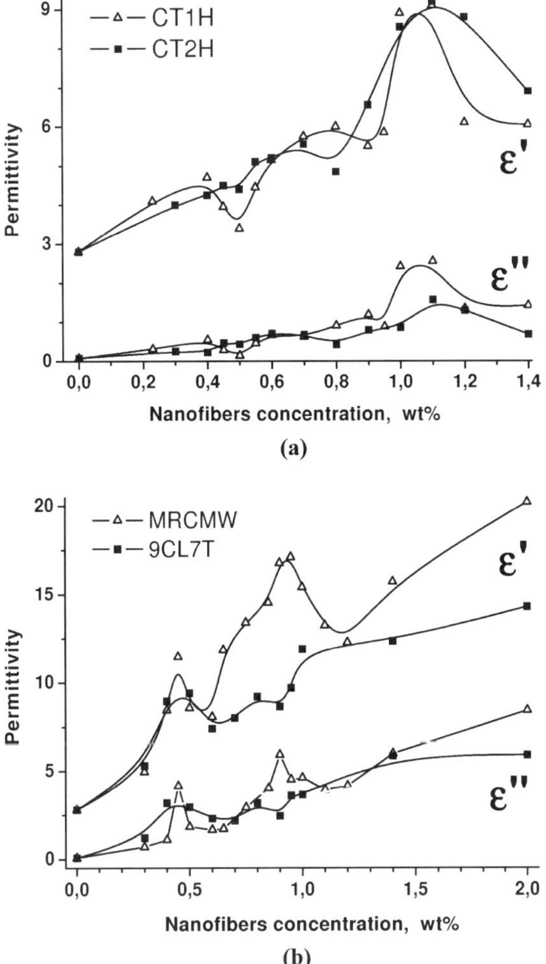

Figure 2. Dependence of composite permittivity on CNF concentration

The composites with VGCF162 behave similarly to soot at average value ε' and ε" 3.5 and 0.1, accordingly. Probably, it is bound to fragility of these nanofibers, which are shattered with ultrasonic treatment of CNF in epoxy resin.

As contrasted with soot (ε' = 3.0 and ε" = 0.08–0.14) nanofibers already at 0.2 wt% of filler in composites show noticeable increase a complex permittivity. At 2 wt% of filler ε' and ε" are increase: for MRCMW – in 7 and 50 times, for 9CL7T – in 5 and 20 times, accordingly.

4. Conclusions

It is shown availability of applications of the studied fillers at small concentrations for complex improvement of mechanical and dielectric properties of radio adsorbing materials.

Acknowledgements

The work is supported by Russian Basic Research Foundation (project 06-03-33089).

References

1. Grimes C.A., Mungle C., Kouzoudis D., Fang S., Eklund P.C. The 500 MHz to 5.50 GGz complex permittivity spectra of single-wall carbon nanotube-loaded polymer composites, *Chem. Phys. Lett.*, 2000, **319**: 460–464.
2. Xiang Ch., Pan Y., Liu X., Sun X., Shi X., Guo J. Microwave attenuation of multiwalled carbon nanotube-fused silica composites, *Appl. Phys. Lett.*, 2005, **87**: 123103.
3. Saib A., Bednarz L., Daussin R., Bailly C., Lou X., Thomassin J.-M., Pagnoulle C., Detrembleur C., Jérôme R., Huynen I. Carbon nanotube composites for broadband microwave absorbing materials, *Proceedings of 35 European Microwave Conference (Paris, October 4–6, 2005)*. P. 285–288.
4. www.mercorp.com
5. Babenko S.D., Kislov M.B., Krestinin A.V., Raevsky A.V., Sokolov E.A., Cherepanov V.V., *Proceedings of 3 All-Russia Conference "Irreversible Processes in Nature and Technique" (Moscow, January 24–26, 2005)*. P. 51.

ELECTROLYTIC Ni-BASED COMPOSITE COATINGS CONTAINING MOLYBDENUM AND SILICON FOR HYDROGEN EVOLUTION REACTION

J. KUBISZTAL, J. PANEK, A. BUDNIOK
Institute of Materials Science, University of Silesia, 12 Bankowa, 40-007 Katowice, Poland
Tel.: 48-32-359-1954; E-mail: jpanek@us.edu.pl

Abstract. Ni + Mo + Si_{Ni} composite coatings were prepared by codeposition of Ni with powders of molybdenum and silicon (covered with the electroless plated nickel – Si_{Ni}) on a steel substrate from the nickel bath in which Mo and Si_{Ni} particles were suspended by stirring. Deposition was conducted under galvanostatic conditions. Deposits were characterized by the presence of Mo and Si phases embedded into the nickel matrix. For comparison Ni + Mo composite coatings without silicon were obtained under comparable conditions. Incorporation of Mo and Si_{Ni} powders into electrolytic nickel matrix causes an increase in the real surface area of the deposits.

The obtained composite coatings were tested as electrode materials for hydrogen evolution reaction (HER) in an alkaline environment. Electrochemical characterization of the composites was carried out by steady-state polarization method and EIS measurements. It was ascertained, that Ni + Mo + Si_{Ni} coatings are characterized by enhanced electrochemical activity for this process which was confirmed by considerable decrease in the hydrogen evolution overpotential compared to nickel electrode as well as to the Ni + Mo composite coatings. Basing on the results of EIS measurements the rate constants of the HER were determined. The calculated values of surface roughness factor allowed to study the apparent and intrinsic activity of investigated coatings for the HER. It was stated that the improvement of the electrocatalytic performance of Ni + Mo and Ni + Mo + Si_{Ni} composite coatings could be attributed both to the increase in their real surface area as well as to the synergetic effect of molybdenum with the nickel matrix.

Keywords: nickel, molybdenum, silicon, electrolytic composite coatings, hydrogen evolution reaction, electrochemical impedance spectroscopy

1. Introduction

The course of electrochemical processes occurring on the composite coatings depend markedly on their surface morphology as well as on the chemical and phase composition. Composite coatings could serve as electrode materials and additional components in the coatings may improve their electrochemical characteristics. Molybdenum as a component of Ni-based composite coatings seems to be particularly interesting for cathode materials because it could play a role of an activator of hydrogen evolution reaction (HER) [1–2]. In order to improve the utilization of these materials various methods of their modifications could be applied. Such modifications are often carried out to obtain electrodes with very developed, rough or porous surface [1–10]. One of these methods is embedding of solid particles into metallic matrix.

Electrolytic deposition seems to be an interesting method of the production of composite coatings containing powders of molybdenum and silicon. Such a hybrid composite coating consisting of transition metal and metalloid powders embedded in the nickel matrix could be very interesting as electrode materials for the HER. The present study was undertaken in order to investigate electrochemical properties of the Ni + Mo + Si_{Ni} composite coatings for the HER in an alkaline environment related to the electrodeposited Ni electrode and electrolytic Ni + Mo composite coatings. The presence of small amount of Si_{Ni} particles could increase the surface roughness of the electrode and in this way affect its activity towards the HER.

2. Experimental

Ni + Mo + Si_{Ni} composite coatings were obtained from a nickel bath having the following composition (concentrations in g dm^{-3}): $NiSO_4$ $7H_2O$ – 84, H_3BO_3 – 8, CH_3COONa – 10, $C_6H_5O_7Na_3$ $2H_2O$ – 30, NH_4Cl – 10. Reagents of analytical purity were used for bath preparation. In this bath a mixture of 10 g of Mo metallic powder (Aldrich, 99.9+%, 100 mesh) and 20 g of Si_{Ni} powder per 1 dm^3 was dispersed by stirring. The silicon powder (Aldrich, 99.9%, 325 mesh) was subjected to preliminary nickel electroless plating, according to generally accepted procedure [11]. Such a procedure should protect Si particles in aggressive alkaline environment during the investigation of the HER. All coatings were deposited on a carbon steel substrate (St3S, 0.3% C). The platinum grid (1 dm^2) served as an anode. The process of deposition was carried out at the temperature of 293 K. pH of the suspension was kept in the range from 5.4 to 5.8 and the rate of stirring was 150 rpm.

Electrodeposition was performed under galvanostatic conditions at the current density j = 100 mA cm^{-2} and the electric charge Q = 360 C cm^{-2}.

For comparison Ni coating was deposited from the pure nickel bath under the same current conditions as well as Ni+Mo composite coatings from the bath containing 10 g dm^{-3} of dispersed Mo powder at the current density of 225 mA cm^{-2} [12].

Surface morphology of deposited coatings was investigated by Hitachi S-4200(4) scanning electron microscope (SEM). Qualitative and quantitative phase composition of Ni + Mo + Si_{Ni} and Ni + Mo coatings was determined by X-ray diffraction (XRD) method using Philips diffractometer and $Cu_{K\alpha}$ radiation (U = 40 kV, I = 20 mA). Registration of diffraction patterns was carried out by step-scanning method. Qualitative phase identification was carried out on the base of ICDD cards. For quantitative phase analysis, both external standard method and matrix method were used. The strongest, nearest lying and not overlapping reflexes coming from three phases: Si (111), Mo (110), Ni (111) were chosen to the analysis [13–15].

The investigated Ni + Mo + Si_{Ni} composites were subjected to electrochemical characterization in an alkaline environment and their activity towards hydrogen evolution reaction in comparison to Ni electrode and Ni + Mo composites was analyzed. The study of the HER was done by means of steady-state polarization measurements and electrochemical impedance spectroscopy (EIS).

For data registration electrochemical system Autolab®/PGSTAT30 (Eco Chemie B.V. the Netherlands) was used with Frequency Response Analyser (FRA) and

Differential Electrometer Amplifier. Prior to the electrochemical experiments the tested electrodes were activated at a constant cathodic current of 320 mA cm^{-2} for 24 h. Electrochemical tests were performed in 5 M KOH solution using three-electrode cell with platinum grid as a counter electrode and the Hg/HgO/6M KOH as a reference electrode. Geometric surface area of working electrode was 1 cm^2.

Steady-state polarization curves were registered galvanostatically. For this purpose a constant current density $-j$ ranging between 320 mA cm^{-2} and 0.1 μA cm^{-2} was applied to the electrode and after 60 s the corresponding steady-state overpotential value $-\eta$ was registered. Basing on recorded $\eta = f(\log j)$ dependences the kinetic parameters for this process (i.e. Tafel slope b, exchange current density j_0 and overpotential at the current density of 100 mA cm^{-2} – η_{100}) were determined.

Impedance measurements were performed under the potentiostatic control in a frequency range from 10 kHz to 0.1 Hz (10 points per decade). The amplitude of *ac* signal was 5 mV. Complex nonlinear least squares method (CNLS) was used to approximation of the obtained impedance results [16–18].

Comparative tests were carried out on the Ni coating as well as on the Ni + Mo composites containing the comparable amount of incorporated Mo.

3. Results and discussion

The results of chemical composition analysis of the Ni + Mo + Si_{Ni} coatings revealed that under applied electrolytic conditions the content of Mo is about 40 wt% and the content of Si is about 10 wt%. Ni + Mo composite coatings contain about 39 wt% Mo.

The presence of embedded particles of Mo and Si_{Ni} in the nickel matrix distinctly enlarges the surface roughness of composite coatings in comparison with Ni coating (Fig. 1). On the SEM micrographs of Ni + Mo + Si_{Ni} and Ni + Mo composite coatings it is possible to recognize the characteristic Mo/Si particles covered with Ni electrodeposit, which confirms the composite, inhomogenous character of obtained coatings (Fig. 1(b, c)).

Figure 1. Surface morphology of: (a) Ni, (b) Ni + 39% Mo, (c) Ni + 40% Mo + 10% Si_{Ni} coatings

X-ray diffraction patterns show the presence of reflexes corresponding to the crystalline Ni matrix and embedded Mo and Si crystalline phases (Fig. 2). This fact confirms the multiphase nature of these coatings.

In order to determine the kinetic parameters of the HER steady-state polarization curves were recorded. From a linear part of Tafel plot the characteristic parameters of the electrode system providing information about the mechanism and the rate of the process i.e. Tafel slope – b and exchange current density – j_0 could be determined. Moreover, the apparent activity of different kinds of coatings could be evaluated by comparison of the overpotential values at selected current density [7, 19, 20]. In this study the HER overpotential at j = 100 mA cm^{-2} – η_{100} was chosen. Steady–state polarization curves after correction for the ohmic drop were presented in Fig. 3 and the obtained kinetic parameters are listed in Table 1.

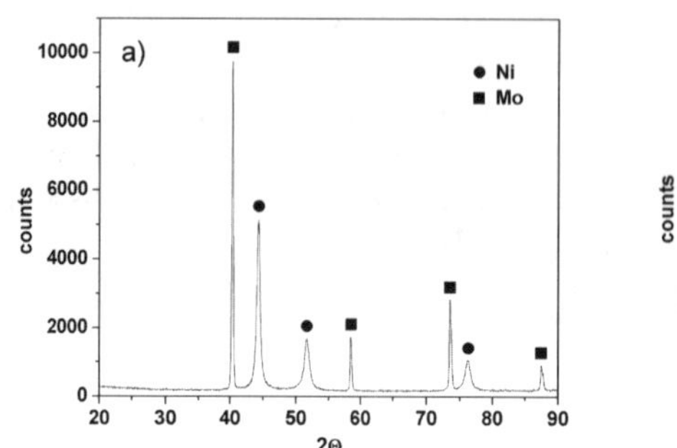

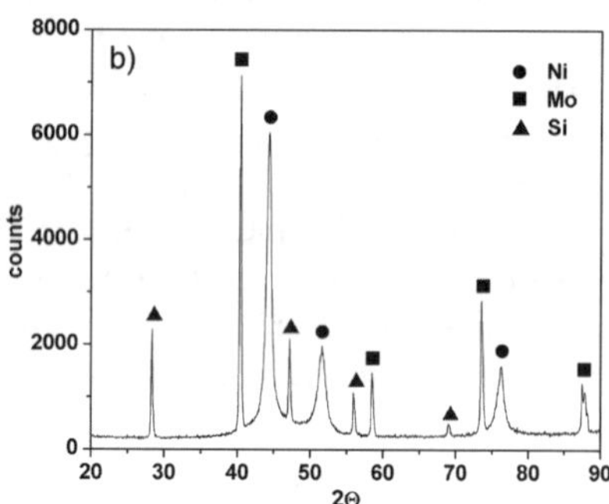

Figure 2. X-ray diffraction patterns of: (a) Ni + 39% Mo, (b) Ni + 40% Mo + 10% Si_{Ni} composite coatings

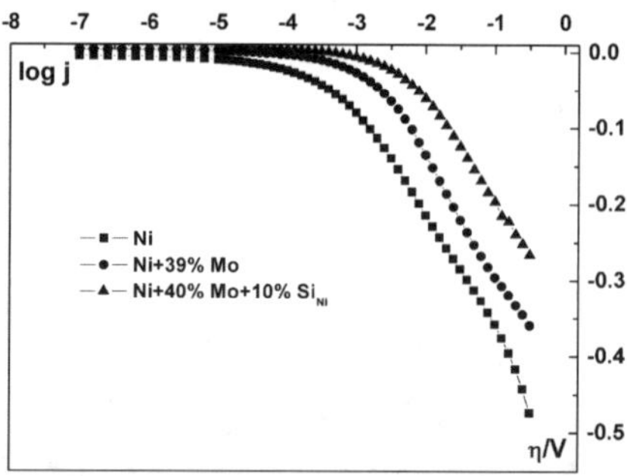

Figure 3. Tafel polarization curves for the HER on the investigated electrodes in 5 M KOH

TABLE 1. Tafel slope b, exchange current density j_0 and hydrogen evolution overpotential at 100 mA cm^{-2} η_{100} on Ni, Ni + 39% Mo and Ni + 40% Mo + 10% Si_{Ni} coatings

Electrode	*b/V dec^{-1}*	*j_0/mA cm^{-2}*	*η_{100}/mV*
Ni	-0.121	0.18	-358
Ni + Mo	-0.153	2.15	-255
Ni + Mo + Si_{Ni}	-0.136	3.36	-196

Assuming Volmer-Heyrovsky mechanism for the HER [21–24] according to equations (1)–(2) the linear part of Tafel curve should yield a slope of 116.3 mV dec^{-1}

when Volmer reaction is the rate determining step (*rds*) or 38.8 mV dec^{-1} for Heyrovsky reaction as the *rds*. But, it has been shown that formally it is impossible to distinguish which reaction is the *rds* when the slope is equal 116.3 mV dec^{-1} [24–26]. However, as the surface coverage by adsorbed hydrogen should increase with the increase in a negative overpotential, the rate limiting step should be that of Heyrovsky. When Tafel slope is 29.1 mV dec^{-1} the rate determining step is the Tafel reaction (3)

$$H_2O + M + e^- \leftrightarrow MH + OH^-, \tag{1}$$

$$H_2O + MH + e^- \leftrightarrow H_2 + M + OH^-, \tag{2}$$

$$2MH \leftrightarrow H_2 + 2M\,. \tag{3}$$

Tafel slopes obtained for all coatings show that the HER proceeds via the same Volmer-Heyrovsky mechanism and its rate is controlled by Heyrovsky step (Table 1). It should be noticed that obtained Tafel slopes are slightly higher in comparison with theoretical value, probably due to the fact that electrode surface is covered with thin film of oxides, which causes additional potential drop.

The values of the exchange current density – j_0 for the investigated composite coatings are significantly higher compared to Ni electrode suggesting their higher activity for the HER (Table 1). It could be related to the increase in the real surface area of composite electrodes arising from the incorporation of Mo and Si_{Ni} particles into Ni matrix. The considerable decrease in the η_{100} value on Ni + Mo and Ni + Mo + Si_{Ni} electrodes by a nearly 100–160 mV compared to the nickel electrode is an additional proof for higher apparent activity of these materials (Table 1).

In order to further characterization of electrode-electrolyte processes the electrochemical impedance spectroscopy technique was applied. The recorded impedance spectra in the complex plane $Z''(\omega) = f(Z'(\omega))$ are presented in Fig. 4. It should be noticed that nickel electrode is characterized by one semicircle on the Nyquist plot in whole frequency range what indicates on process with one time constant τ (Fig. 4(a)). Complex impedance plots obtained for Ni + Mo and Ni + Mo + Si_{Ni} composite coatings reveal two semicircles, first at high-frequencies (HF, τ_1) which diameter does not change with the overpotential and the second one at low frequencies (LF, τ_2), what refers to process with two time constants. The zoomed inset of the high-frequencies part of Nyquist plot more clearly demonstrate the presence of two semicircles (Fig. 4(b)).

The impedance data obtained for the nickel electrode were fitted using one-CPE model (1CPE). In this model the double-layer capacitance C_{dl} is substituted by a constant phase element (CPE), which impedance is given by $Z_{CPE}(\omega) = 1/T(j\omega)^{\phi}$, where ω is the angular frequency of the *ac* voltage, T is a capacity parameter which may depend on the electrode potential and ϕ is a dimensionless parameter related to the constant phase angle, $\alpha = 90^{\circ}(1-\phi)$. The CPE is in parallel with the faradaic impedance Z_f, usually represented as a simple charge-transfer resistance $Z_f = R_{ct}$. The solution resistance R_s is in series with the parallel connection of CPE and R_{ct}. As a result of approximation of experimental data by this model four parameters could be obtained: R_s, R_{ct}, T, ϕ [19, 24].

For Ni + Mo and Ni + Mo + Si_{Ni} composite coatings, two-CPE serial model was applied (2CPE). This equivalent circuit predicts two time constants and consists of the resistance of solution in series with two parallel Z_{CPE}-R elements [19, 21, 27, 28]. It has been postulated that potential independent time constant τ_1, which is described here by R_1, T_1, and ϕ_1 is attributed to the porosity of the electrode material, whereas potential dependent time constant τ_2 is related to kinetic of the HER process (R_{ct}, T, ϕ) [19, 21, 27, 28].

For composite coatings the diameter of low-frequency semicircle diminishes with the increase in the overpotential which points at the decrease in the charge transfer resistance – R_{ct}. The values of R_{ct} obtained for Ni + Mo + Si_{Ni} composite coatings were from 4 to 14 times lower then that obtained for Ni electrode (Fig. 4).

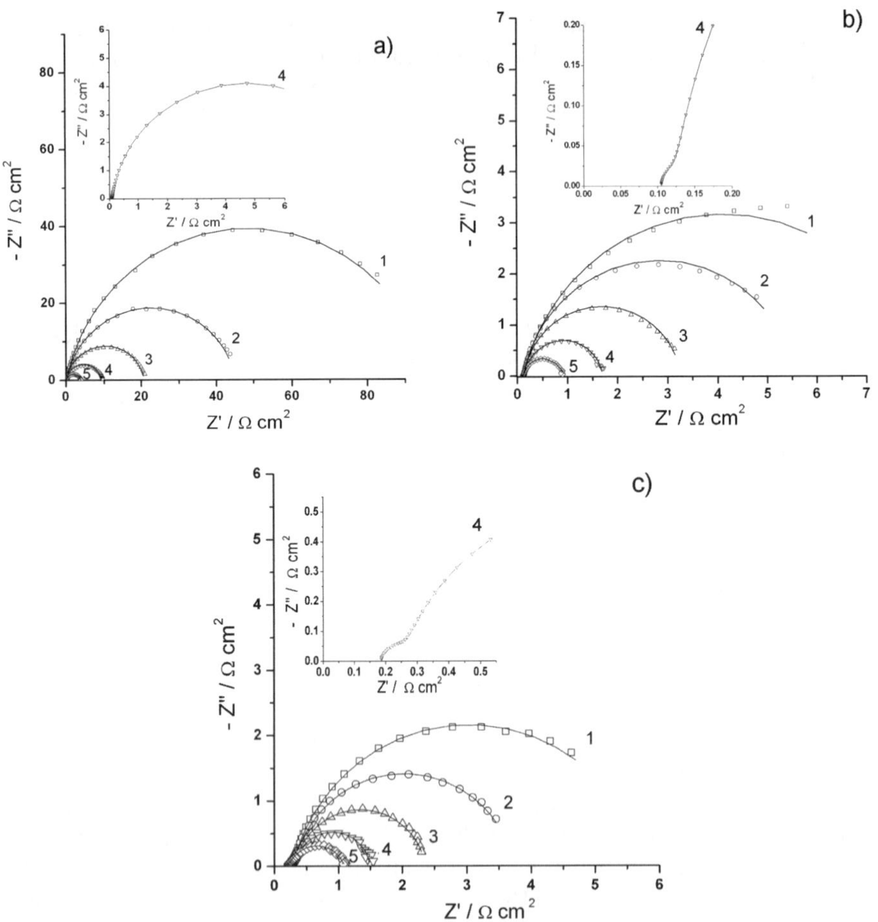

Figure 4. Experimental complex plane plots obtained on the (a) Ni, (b) Ni + 39% Mo, (c) Ni + 40% Mo + 10% Si_{Ni} electrodes in 5 M KOH at the overpotentials 50 mV (1), 100 mV (2), 150 mV (3), 200 mV (4), 250 mV (5). Symbols are the experimental data, continuous lines were fitted using CNLS method

In the case of 1CPE model capacitance parameter T is associated with the double layer capacitance C_{dl} by relation: $C_{dl} = [T/(R_s^{-1} + R_{ct}^{-1})^{(1-\phi)}]^{1/\phi}$ [7, 19, 21, 29]. In the case of serial 2CPE model C_{dl} could be determined using relation: $C_{dl} = [T/((R_s+R_1)^{-1} + R_{ct}^{-1})^{(1-\phi)}]^{1/\phi}$. Figure 5 shows the relation between the double-layer capacitance and the overpotential for all obtained coatings. Double-layer capacitance of Ni + Mo + Si_{Ni} coatings insignificantly decreases with the increase in the overpotential. Electrochemically accessible surface area is proportional to C_{dl} values, what indicates that the presence of Mo and Si_{Ni} in Ni + Mo + Si_{Ni} coatings causes an distinct increase in electrochemically accessible area in comparison with Ni coating (from 4 to 14 times).

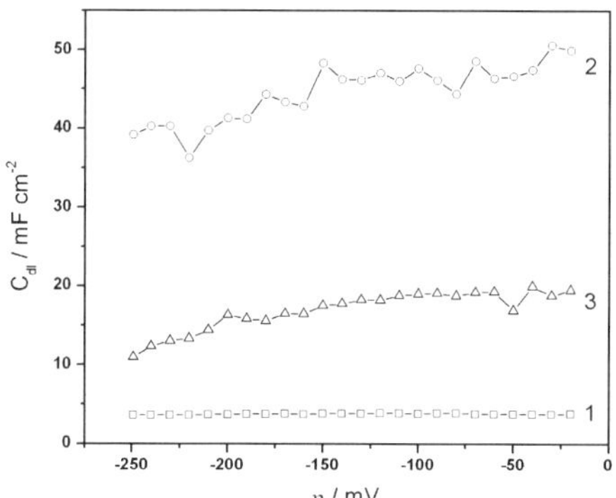

Figure 5. Double-layer capacitance C_{dl} in dependence on the overpotential of the HER in 5 M KOH at 25°C on the Ni (1), Ni + 40% Mo + 10% Si_{Ni} (2) and Ni + 39% Mo (3) coatings

The rates of reactions (1)–(3) are given by the following expressions:

$$v_1 = k_1 (1-\Theta) \exp(-\beta_1 f\eta) - k_{-1}\, \Theta \exp[(1-\beta_1) f\eta]\,, \tag{4}$$

$$v_2 = k_2\, \Theta \exp(-\beta_2 f\eta) - k_{-2} (1-\Theta) \exp[(1-\beta_2) f\eta]\,, \tag{5}$$

$$v_3 = k_3\, \Theta^2 - k_{-3}(1-\Theta)^2\,. \tag{6}$$

where β_1 and β_2 are the transfer coefficients, η is overpotential, Θ is the surface coverage by adsorbed hydrogen, and f = F/RT. Assuming Volmer-Heyrovsky mechanism the rate constants k_i and transfer coefficients β_i of the HER could be determined. The rate constants were obtained assuming that the transfer coefficients β_i of both reactions (4), (5) are the same. The rate constants k_1, k_{-1} and k_2 were evaluated using the NLS method [7]. For exemplary Ni + Mo + Si_{Ni} coating approximations of both current density j (j = -F(v_1 + v_2)) and reciprocal of charge-transfer resistance R_{ct} ($1/R_{ct} = -F(\partial r_0/\partial\eta)_\Theta$; $r_0 = v_1 + v_2$) as functions of η are shown in Fig. 6. Approximations were carried out for large overpotentials (η > 50 mV) in a linear part of the curves. The k_i and β values are gathered in Table 2.

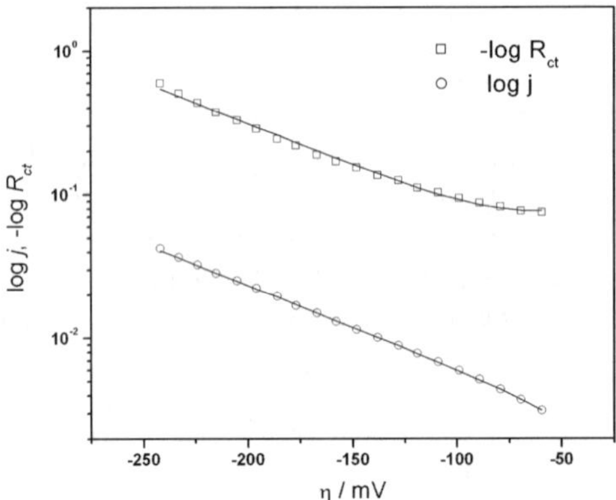

Figure 6. log j and –log R_{ct} in dependence on the overpotential of the HER in 5 M KOH at 25°C on Ni + 40% Mo + 10% Si_{Ni} coating. Symbols are the experimental data, continuous lines were fitted using NLS method

TABLE 2. The kinetic parameters of the HER obtained for the investigated electrodes k_i/mol s^{-1} cm^{-2}

Electrode	k_1	k_{-1}	k_2	$\beta_1 = \beta_2$	R_f	k_{rds}/R_f
Ni	1.16×10^{-9}	6.49×10^{-8}	7.48×10^{-10}	0.55	190	3.94×10^{-12}
Ni + Mo	2.33×10^{-8}	8.04×10^{-7}	1.64×10^{-8}	0.42	840	1.95×10^{-11}
Ni + Mo + Si_{Ni}	3.23×10^{-8}	2.75×10^{-7}	2.72×10^{-8}	0.47	2,200	1.24×10^{-11}

Making comparison of the k_1 and k_2 values it should be noticed that the Heyrovsky reaction is the rate-determining step for the HER at all investigated electrodes. For Ni + Mo and Ni + Mo + Si_{Ni} composite electrodes the values of the k_1 and k_2 are higher than for Ni electrode, which point at their higher apparent activity toward the HER. It could be attributed to the increase in the real surface area of these materials.

The intrinsic activity may be estimated as a ratio of the rate constant of *rds* – k_{rds} and the roughness factor $R_f = C_{dl}/20$ μF cm^{-2} [7, 22]. Roughness factors and kinetic parameters of the HER on the investigated materials per unit of a real surface area are displayed in Table 2.

The results show that the addition of the Mo powder increases the intrinsic activity as the values of k_{rds}/R_f for Ni + Mo and Ni + Mo + Si_{Ni} are higher than for Ni electrode. The increase in the intrinsic activity could be related to the catalytic interaction of Mo with the Ni matrix, as Ni/Mo materials are considered, according to the literature data [1, 2, 19, 30], typical electrocatalysts for the HER. The incorporation of Si_{Ni} particles significantly increases the surface roughness of the electrode but it does not influence the intrinsic activity.

4. Conclusions

It was stated that all Ni + Mo and Ni + Mo + Si_{Ni} composite coatings exhibit composite character. Incorporation of Mo and Si_{Ni} particles results in the increase in development of surface topography in comparison with Ni coating obtained at the same current conditions.

Hydrogen evolution reaction on all obtained coatings can be successfully explained through Volmer-Heyrovsky mechanism with Heyrovsky reaction as the rate determining step (*rds*).

The apparent activity of the composite coatings towards the HER is much greater then that obtained for Ni electrode. The reason of this is mainly connected with the increase in the electrochemically active surface area resulting from the incorporation of Si_{Ni} and Mo particles into the electrolytic Ni matrix. The higher intrinsic activity of the Ni + Mo and Ni + Mo + Si_{Ni} composite coatings could be related to the synergetic effect of molybdenum dispersed in the Ni matrix.

Acknowledgements

This work was financially supported by KBN Project Nr 3 T08C 009 28.

References

1. R. Karimi-Shervedani, A. Lasia, *J. Electrochem. Soc.*, 1998, **145**(7): 2219–2225.
2. J. M. Jaksic, M. V. Vojnovic, N. V. Krstajic, *Electrochim. Acta*, 2000, **45**: 4151.
3. J. Niedbała, A. Budniok, D. Gierlotka, J. Surówka, *Thin Solid Films*, 1995, **226**: 113.
4. B. Łosiewicz, A. Stępień, D. Gierlotka, A. Budniok, *Thin Solid Films*, 1999, **349**: 43–50.
5. M. Popczyk, A. Budniok, A. Lasia, *Int. J. Hydrogen Energy*, 2005, **30**: 265.
6. I. Paseka, J. Velicka, *Electrochem. Acta*, 1997, **42**: 237.
7. M. Popczyk, A. Serek, A. Budniok, *Nanotechnology*, 2003, **14**: 342.
8. R. Karimi-Shervedani, A. Lasia, *J. Electrochem. Soc.*, 1997, **144**: 511.
9. E. Bełtowska-Lehman, E. Chassaing, *J. Appl. Electrochem.*, 1997, **5**: 568–572.
10. C. Fan, D.L. Piron, A. Sleb, *Surf. Coat. Technol.*, 1995, **73**: 91.
11. S. Bagdach, T. Biestek, T. Biestekowa, K. Czajka, S. Gębalski, A. Krokosz, T. Paszek, J. Rudzki, A. Samołyk, J. Socha, J. Syrek, K. Szmidt, J. Weber, T. Żak, *Poradnik galwanotechnika*, PWT, Warszawa, 1961.
12. J. Kubisztal, A. Budniok, A. Lasia, *Appl. Surf. Sci.*, 2006, **252**: 8605–8610.
13. C. R. Hubbard, E. H. Evans, D. K. Smith, *J. Appl. Cryst.*, 1976, **9**: 169–174.
14. T. Knudsen, *X-Ray Spectrom.*, 1981, **10**(2): 54–56.
15. L. S. Zevin, G. Kimmel, Quantitative X-Ray Diffractometry, Springer, New York, 1995, ISBN-0-387-94541-5.
16. J. R. Macdonald Impedance Spectroscopy, Wiley, New York, 1987.
17. J. R. Macdonald, J. Schoonman, A. P. Lehner *J. Electroanal. Chem.*, 1982, **131**: 77.
18. B. A. Boukamp, *Solid State Ionics*, 1986, **20**: 31–44.
19. L. Birry, A. Lasia., *J. Appl. Electrochem.*, 2004, **34**(7): 735–749.
20. J. Panek, A. Serek, A. Budniok, E. Rówiński, E. Łągiewka, *Int. J. Hydrogen Energy*, 2003, **28**: 169–175.
21. E. Navarro-Flores, Z. Chong, S. Omanovic, *J. Mol. Catal. A-Chem.*, 2005, **226**: 179.
22. B. Borresen, G. Hagen, R. Tunold, *Electrochim. Acta*, 2002, **47**: 1819.
23. E. Ndzebet, O. Savadogo, *Int. J. Hydrogen Energy*, 1995, **20**: 635.
24. A. Lasia, A. Rami, *J. Electroanal. Chem.*, 1990, **294**: 123.
25. F. Berthier, J. P. Diard, C. Montella, L. Pronzato, E. Walter, *J. Chim. Phys.*, 1993, **90**: 2069.
26. F. Berthier, J. P. Diard, L. Pronzato, E. Walter, *Automatica*, 1996, **32**: 973.
27. L. Chen, A. Lasia, *J. Electrochem. Soc.*, 1992, **139**: 3214.
28. B. Łosiewicz, A. Budniok, E. Rówiński, E. Łągiewka, A. Lasia, *J. Appl. Electrochem.*, 2004, **34**: 507–516.

29. G. J. Brug, A. L. G. Van Der Eeden, M. Sluyters-Rehbach, J. H. Sluyters *J. Electroanal. Chem.*, 1984, **176**: 275.
30. J. Kubisztal, A. Budniok, A. Lasia, *Int. J. Hydrogen Energy*, 2007, **32**: 1211–1218.

FORMATION OF CARBON NANOSTRUCTURES DURING IR-PYROLYSIS OF POLYACRYLONITRILE IN PRESENCE OF Fe AND Co

L.M. ZEMTSOV, G.P. KARPACHEVA, K.A. BAGDASAROVA
Topchiev Institute of Petrochemical Synthesis, Russian Academy of Sciences, 29 Leninsky prospect, Moscow, 119991 Russia
Fax: 7-495-633-85-20; E-mail: lm@ips.ac.ru

E.L. DZIDZIGURI, E.N. SIDOROVA
Moscow State Institute of Steel and Alloys, 4 Leninsky prospect, Moscow, 119049 Russia

Abstract. It is shown that IR-pyrolysis of compositions based on polyacrylonitrile and $Co(C_5H_7O_2)_2$, $Fe(C_5H_5)_2$ results in formation of a metal-carbon nanocomposite in which nanosized particles of metal are dispersed finely in the structure of the carbon matrix. Carbonaceous phase of this composite is a carbon-carbon nanocomposite of a structure where carbon nanoparticles (carbon nanospheres, or nanoswirls) are embedded into the main graphite-like material.

Keywords: nanocomposite, carbon, IR-pyrolysis, Fe-, Co-compounds

1. Introduction

Scientific and engineering interest in nanostructured carbon materials which increased steadily during this decade is rooted in a wide range of their potential use (for example, nanoelectronics, fuel cells, hydrogen accumulation, membrane and strengthening materials, thermoinsulators, carriers for heterogeneous and electrocatalysts, antifriction additives, etc. [1–7]). To this moment, various carbon nanomaterials are synthesized (single- and multiwalled nanotubes, nanohorns, nanospheres, nanoonions, etc.). Different catalytic and non-catalytic methods of synthesis of carbon nanoparticles are developed. The most often used methods are: laser and arc evaporation (also in liquid phase), vacuum (including plasma-assisted) precipitation, catalytic pyrolysis of gaseous hydrocarbons. Nanostructured carbon particles can be obtained also by pyrolysis of ferrocene [2, 8] and also of polymers like polyacrylonitrile (PAN), PAN-methylmetacrylate copolymer, polytetrafluoroethylene (PTFE) [9, 10].

This work is dedicated to synthesis of carbon nanostructures by pyrolysis of PAN/$Co(C_5H_7O_2)_2$ or $Fe(C_5H_5)_2$–based compositions under non-coherent IR radiation. Specific effect of IR radiation on the vibrational energy of single bonds of a PAN macromolecule allows to increase rapidly the rate of chemical transformations, thus increasing the efficiency of pyrolysis.

It has been shown by our research that under IR radiation, PAN samples undergo chemical and structural transformations which lead to the formation of a graphite-like layered structure [11–13]. It is amorphous due to an irregular shift between graphene planes and small size of regions where scattering of crystallites is coherent.

2. Experimental

PAN was synthesized in presence of a redox catalytic system [14]. Composition of PAN and cobalt acetylacetonate $Co(C_5H_7O_2)_2$ or ferrocene $Fe(C_5H_5)_2$ was prepared by joint dissolution in DMFA ("Fluka", 99.5%) of the polymer and $Co(C_5H_7O_2)_2$ or $Fe(C_5H_5)_2$ (chemically pure) in mass ratio PAN:(Co, Fe) = 5:1, total concentration $c = 5$ wt%.

Exposure of samples to IR radiation was achieved in a special IR chamber of a laboratory set-up for pulse photonic calcination. Halogen lamps KG-220, KG-2000-2 were used as source of radiation. Their maximum intensity is in the range 0.9–1.2 mcm. The sample (as a film or a powder) in a graphite cassette was placed into a cylindrical quartz reactor. In order to ensure uniform heating of the sample, the inner wall of the chamber was made from polished stainless steel. Intensity of IR radiation was controlled by the sample temperature measured with a chromel-alumel thermocouple placed directly under the sample. A controller unit allowed programming of IR radiation intensity. Temperature was maintained with the precision 0.25°C.

Calcination was done in two stages: preliminary calcination in air at 150°C for 15 min and at 200°C for 15 min, then the main procedure in argon at 600–1,200°C preceded by heating in Ar under its increased pressure at 150°C and 200°C for 15 min.

Microphotographs of nanocomposites were obtained with the electronic microscope EM-301 (Philips) with accelerating voltage 60–80 kV. X-ray studies were done with the diffractometer "Geiger Flex" (Rigaku), the radiation source being a lamp with Fe anode (current – 25 mA, voltage – 25 kV), producing radiation at the wavelength 0.193728 nm. A Mn filter was used. Focusing was done by the Bragg-Brentano method with two Soller slits.

3. Results and discussion

It has been shown by us [11] that addition to the initial PAN solution of transition metal compounds, such as Fe or Co chlorides, changes substantially the dynamics of formation of the initial carbon material. In this case IR-pyrolysis of PAN proceeds at a higher rate and at lower temperatures when compared to IR-pyrolysis of pure PAN.

In the course of IR-pyrolysis of precursor compositions based on PAN and $Co(C_5H_7O_2)_2$ or $Fe(C_5H_5)_2$ at IR radiation intensity providing heating of the sample up to 600–1,200°C, two simultaneous processes occur which lead to the formation of metal-carbon nanocomposites. One is the formation of orderly carbon structures, another is metal reduction in presence of atomic hydrogen released during the dehydrogenation of the main polymeric chain of PAN. It is shown that metal (Co, Fe) particles are dispersed finely in the rigid carbon matrix, as confirmed by TEM data. Figure 1(a, b) depict microphotographs of Co-, Fe-nanocomposites obtained at 900°C ($t = 2$ min).

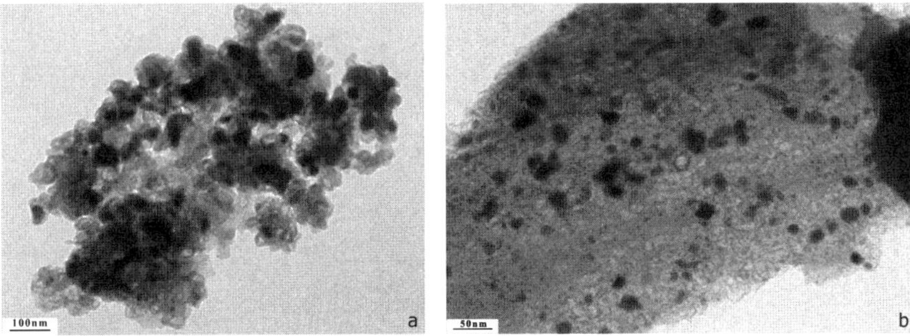

Figure 1. Microphotograph of a Fe/IR-PAN (a) nanocomposite and Co/IR-PAN (b) obtained by IR-pyrolysis at 900ºC

Figure 2 shows the size distribution of Co-, Fe-nanoparticles.

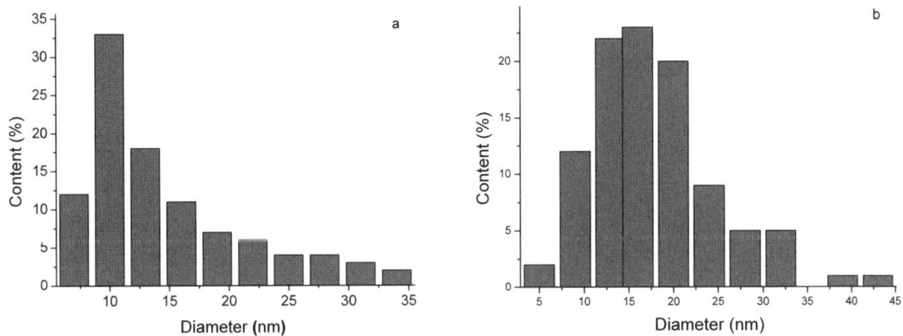

Figure 2. Fe/IR-PAN and Co/IR-PAN nanocomposites (a and b, correspondingly) obtained by IR-pyrolysis at 900ºC. (a) About 80% of nanoparticles are sized in the range 7–16 nm, (b) more than 85% of nanoparticles are sized in the range 9–24 nm

With addition of Co-, Fe-compounds into the system, carbonaceous phase of the resulting composite contains nanostructured carbon particles. If the IR radiation intensity allows heating of the samples up to 900–1,200°C, TEM studies show the presence of carbon nanoobjects in the pyrolyzed PAN structure which may contain metal particles. The morphology type of carbon nanostructures depends on metal. Thus, a Fe/IR-PAN nanocomposite obtained by IR-pyrolysis was found to contain nanospheres of diameter 22–290 nm (Fig. 3).

Formation of such nanostructured carbon particles in the structure of bulk carbonaceous material can be explained as follows. In the course of IR-pyrolysis various gaseous products of destruction of PAN polymeric chain including gaseous hydrocarbons such as ethylene and propylene are confirmed to exist in the reaction zone by mass-spectroscopy and chromatography data [15, 16]. At high temperatures they are pyrolyzed in presence of metals with formation of carbon nanostructures depicted in Figs. 3–5.

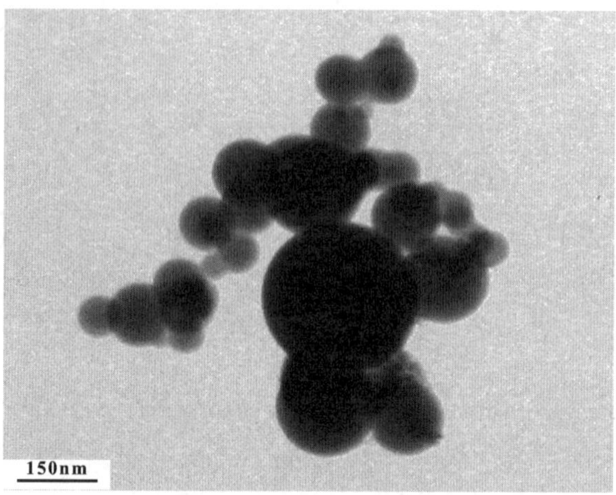

Figure 3. A Fe/IR-PAN nanocomposite obtained by IR-pyrolysis at 1,000°C

With an increase in the IR radiation intensity, nanospheres form chains of length up to 17 mcm (Fig. 4).

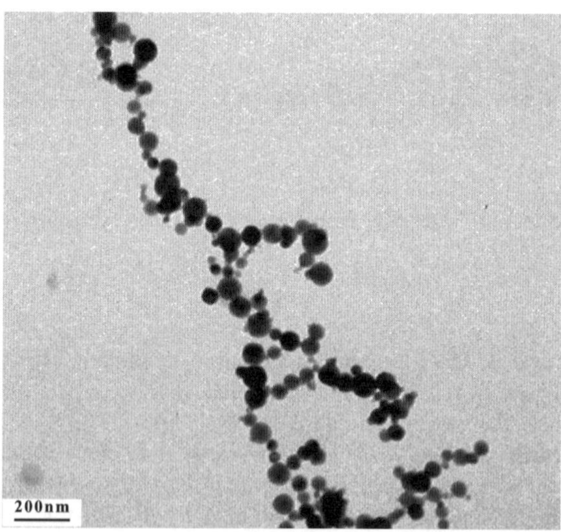

Figure 4. A Fe/IR-PAN nanocomposite obtained by IR-pyrolysis at 1,200ºC

Co/IR-PAN composite was found to contain cone-shaped objects – nanoswirls of diameter 50–100 nm which consist of carbon and may include Co nanoparticles (Fig. 5).

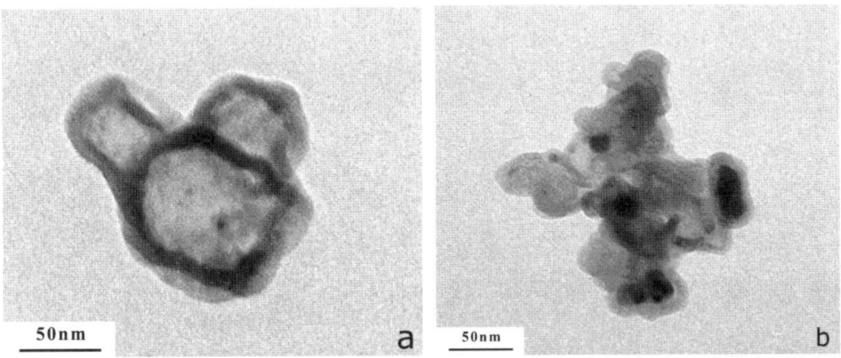

Figure 5. (a) A Co/IR-PAN nanocomposite obtained by IR-pyrolysis at 900ºC; (b) at 1,100ºC

It is known that for the synthesis of carbon nanostructures (nanotubes, nanofibers) Fe, Co, Ni are used most often as metal catalysts of high-temperature pyrolysis [17, 18]. Most recently, it has been shown that bimetallic compounds are more efficient catalysts [19, 20], especially those containing rare earth metals (Y, Ce, Tb, La, and Ho) [21–23]. Authors [14] have described the catalytic effect of Gd on the formation of carbon nanostructures. Depending on the IR-pyrolysis intensity, the precursor based on PAN and $GdCl_3$ yields bamboo-shaped nanotubes, nanospheres, and octahedral carbon particles [14].

According to XRD data, the structure of a Fe-PAN nanocomposite contains, aside from the amorphous carbon phase, Fe carbides. When intensity of IR-pyrolysis is increased up to 1,200ºC, the carbon peak in Fig. 6 becomes more narrow wide, which corresponds to an increase in crystallinity of the carbon phase.

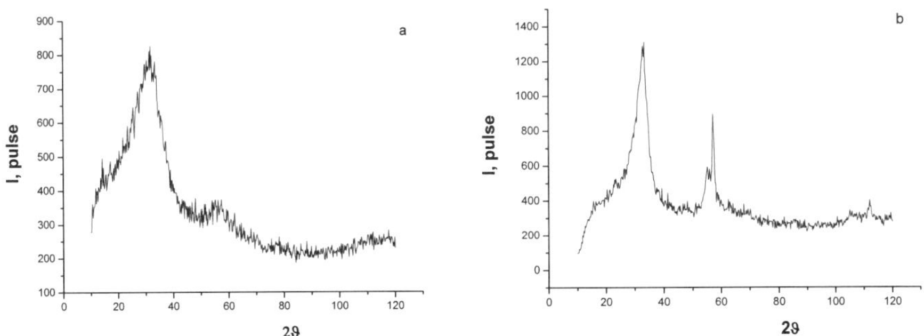

Figure 6. A diffractogram of the Fe/IR-PAN sample obtained at (a) 700°C, (b) 1,200°C

Co/IR-PAN nanocomposite has at least three identifiable phases: amorphous and crystalline carbon and metallic cobalt with face-centered cubic lattice. The higher the IR-pyrolysis intensity, the less amorphous is the composite. Simultaneously, structural changes occur in the crystalline phases, identified by narrowing of peaks in diffractograms (Fig. 7). At pyrolysis intensity equaling 1,200ºC, carbonaceous phase consists of crystalline graphite with interplanar distance d = 3.38 Å.

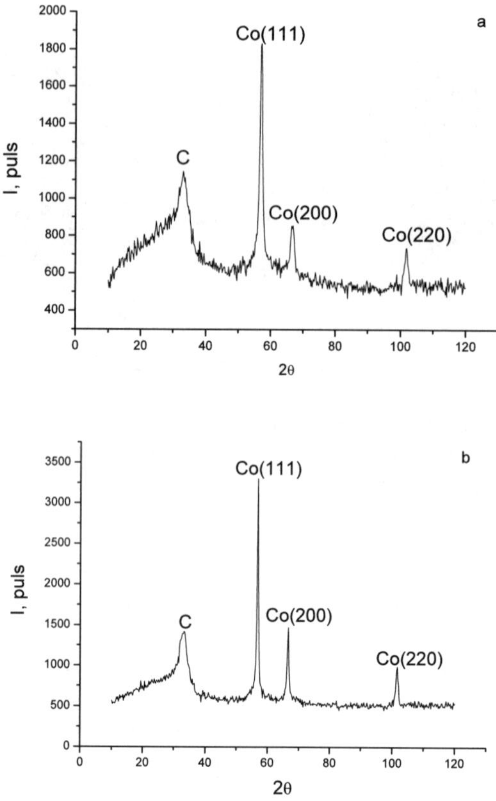

Figure 7. A diffractogram of the Co/IR-PAN sample obtained at (a) T = 700°C, (b) T = 1,200°C

It is shown for Co/IR-PAN composites that the lattice period of Co nanoparticles dispersed in the carbon matrix is smaller for all samples obtained at 700–1,200ºC than for metallic Co (a = 0.35441 nm); depending on the temperature of calcination, it lies in the range from 0.3540 to 0.3517 nm.

4. Conclusions

Thus, in the course of IR-pyrolysis of the precursor based on PAN and Fe-, Co-compounds, metal-carbon nanocomposites are formed in which the carbonaceous phase consists in its turn of a carbon-carbon nanocomposite structured in the way where carbon nanoparticles (bamboo-shaped carbon nanotubes, nanospheres, nanoswirls) are inserted into the main graphite-like carbon material. It is shown that the morphology type of carbon nanoparticles depends on the nature of metal in the composite.

Thus, by varying the conditions of IR-pyrolysis and the nature of metal in the PAN-metal composite, it is possible to obtain nanostructured carbon materials of different orderliness.

References

1. Popov V.N. Carbon nanotubes: properties and application, *Mater. Sci. Eng. R: Reports*, 2004. **43**(3): 61.
2. Xu L., Zhang W., Yang Q., Ding Y., Yu W., Qian Y. A novel route to hollow and solid carbon spheres, *Carbon*, 2005. **43**(5): 1090.
3. Perspectives of Fullerene Nanotechnology. Ed. by Osawa E. Dordrecht; Boston; London: Kluwer Acad. Publ. 2002.
4. Senderova O.A., Zhirnov V.V., Brenner D.W. Carbon nanostructures, *Crit. Rev. Solid State Mater. Sci.*, 2002. **27**(3–4): 227.
5. Inagaki M., Kaneko K., Nishizawa T. Nanocarbons—recent research in Japan, *Carbon*, 2004. **42**(8–9): 1401.
6. Iijima S. Carbon Nanotubes: past, present, and future, *Phys. B-Condens. Mat.*, 2002. **323** (1–4): 1.
7. Harris P. Carbon Nanotubes and Related Structures. New Materials for the Twenty-First Century. London: Cambridge University Press, 1999.
8. Urones-Garrote E., Avila-Brande D., Ayape-Katcho N., Gomez-Herrero A., Landa-Canovas A.R., Otero-Diaz L.C. Amorphous carbon nanostructures from chlorination of ferrocene, *Carbon*, 2005. **43**(5): 978.
9. Hulikova D., Hosoi K., Kuroda S., Oya A. Carbon nanotubes prepared from three-layered copolymer microspheres of acrylonitrile and methylmethacrylate, *Carbon*, 2005. **43**(6): 1246.
10. Cudzilo S., Bystrzejewski M., Lange H., Huzcko A. Spontaneous formation of carbon-based nanostructures by thermolysis-induced carbonization of halocarbons, *Carbon*, 2005. **43**(8): 1778.
11. Zemtsov L.M., Karpacheva G.P. Polyacrylonitrile chemical formations under incoherent IR-radiation, *Vysokomolekulyarnye Soedineniya A*, 1994. **36**(6): 919 (in Russian).
12. Shul'ga Y.M., Rubtsov V.I., Efimov O.N., Karpacheva G.P., Zemtsov L.M., Kozlov V.V. Investigation of polyacrylonitrile pyrolized films by X-ray photoelectronic spectroscopy, electronic AUGR-spectroscopy and electron loss of energy spectroscopy, *Vysokomolekulyarnye Soedineniya A*, 1996. **38**(6): 989 (in Russian).
13. Kozlov V.V., Korolyov Y.M., Karpacheva G.P. Structural changes of composite based on polyacrylonitrile and fullerene C_{60} after IR-radiation, *Vysokomolekulyarnye Soedineniya A*, 1999. **41**(5): 836 (in Russian).
14. Zemtsov L.M., Karpacheva G.P., Efimov M.N., Muratov D.G., Bagdasarova K.A. Carbon nanostructures based on IR-pyrolyzed polyacrylonitrile, *Polym. Sci. A*, 2006. **48**(6): 633–637.
15. Heiderich M.A. Research of polyacrylonitrile thermal formation. Moscow, Topchiev Institute AS USSR, Ph.D. thesis, 1965.
16. Kozlov V.V., Karpacheva G.P., Petrov V.S., Lazovskaya E.V. Features of polyacrylonitrile polyconjugated bonds formation in vacuum during thermal treatment, *Vysokomolekulyarnye Soedineniya A*, 2001. **43**(1): 20 (in Russian).
17. Iigima S., Ichihashi T. Single-shell carbon nanotubes of 1-nm diameter, *Nature*, 1993. **363**(6430): 603.
18. Betune D.S., Kiang S.H., de Vries M.S., Gorman D., Savoy R., Beyers R. Cobalt-catalysed growth of carbon nanotubes with single-atomic-layer walls, *Nature*, 1993. **363**(6430): 605.
19. Tarasov B.P., Muradyan V.E., Shul'ga Y.M., Krinichnaya E.P., Kuyunko N.S., Efimov O.N., et al. Synthesis of carbon nanostructures by arc evaporation of graphite rods with Co–Ni and YNi_2 catalysts, *Carbon*, 2003. **41**(7): 1357.
20. Chen B., Wu P. Aligned carbon nanotubes by catalytic decomposition of C_2H_2 over Ni–Cr alloy, *Carbon*, 2005. **43**(15): 3172.
21. Journet C., Maser W.K., Bernier A., Loiseau M., de la Chapelle M.L., Lefrant S., et al., *Nature*, 1997. **388**(6644): 756.

22. Liu B.B., Wagberg T., Olsson E., Yang R., Zou G.T., Sundqvist B. Synthesis and characterization of single-walled nanotubes produced with Ce/Ni as catalysts, *Chem. Phys. Lett.*, 2000. **320**(3–4): 365.
23. Yao M., Liu B., Zou Y., Wang L., Li D., Cui T., et al., Synthesis of single-wall carbon nanotubes and long nanotube ribbons with Ho/Ni as catalyst by arc discharge, *Carbon*, 2005. **43**(14): 2894.

INFLUENCE OF IMPURITIES AND DEFECTS ON ELECTRONIC STRUCTURE OF CARBON NANOTUBES

A.P. POPOV, I.V. BAZHIN
Pedagogical Institute of South Federal University Bolshaya Sadovaya Str. 33, Rostov-on-Don, 344082 Russia
E-mail: nanosys@mail.ru

Abstract. The influence of vacancies and defects on equilibrium configuration and electronic structure of single-wall carbon nanotubes is investigated. In the framework of semi-empirical PM3-method the IR and UV absorption spectra of fragment (12, 0)-tube C_{192} and of the same fragment with the vacancies of simplest kinds and with Stone-Wales defect are calculated.

Keywords: nanotubes, defects and vacancies, quantum-chemical calculations, IR and UV spectra

1. Introduction

Here we continue the theoretical investigations of influence of impurities and defects on equilibrium geometry and electronic structure of carbon nanotubes in framework of semi empirical methods of calculations.

The problem of classification of defects types for nanotubes is essentially more complicated that the same one for graphite monolayer. Removal of only one carbon atom can lead to appearance of two different kinds of defects, and removal of two neighbouring atoms it can create two different types of defects also. As well known appearance of defects is accompanied by local and sometimes global changes of nanotubes geometry (see, for example [1–8]).

2. Results and discussions

The computations of equilibrium configuration, total energy, heat of formation, IR and UF absorption spectra for fragment of (12, 0) tube C_{192} and the same fragment with the different kinds of local vacancies and Stone-Wales defects [9] are performed in the framework of semi empirical PM3 method [10–13]. Here we use the notifications for local vacancies and defects proposed in our previous paper [14]. All possible kinds of local vacancies and defects to appear in nanotubes after removing not more than two neighbouring carbon atoms are showing in Fig. 1.

The results of calculations of total energy and heat of formation are presented in the Table 1.

B. Baranowski et al. (eds.), *Carbon Nanomaterials in Clean Energy Hydrogen Systems.*

TABLE 1. Physical and chemical characteristics for (12,0) tube with different local defects

	C_{192}	C_{191} $V_{1,1}D_1\alpha$	C_{191} $V_{1,1}D_1\beta$	C_{190} $V_{2,2}D_2\alpha$	C_{190} $V_{2,2}D_2\beta$	C_{192} $D_0D_2\alpha$
E_{tot} (eV)	-22,655	-22,529	-22,528	-22,413	-22,412	-22,651
ΔH (kcal/mol)	2,847.3	3,025.4	3,047.9	2,972.6	2,990.1	2,928.5

To get a possibility of direct comparison with experimental data there are calculated not only DOS but IR and UF absorption spectra of all the objects (see Figs. 2, 3 below).

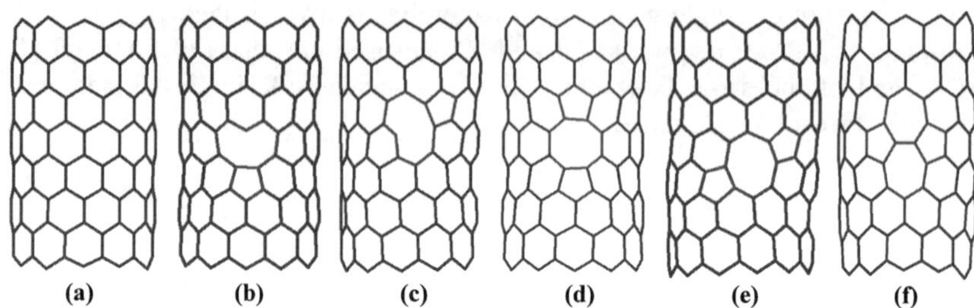

Figure 1. Fragment of (12, 0) tube C_{192} (a) and the same fragment with vacancies $V_{1,1}D_1\alpha$ (b), $V_{1,1}D_1\beta$ (c), $V_{2,2}D_2\alpha$ (d), $V_{2,2}D_2\beta$ (e) and Stone-Wales defect $D_0D_2\alpha$ (f).

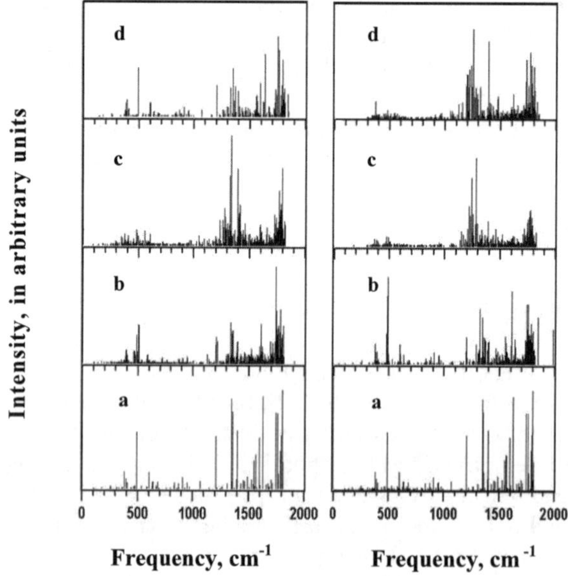

Figure 2. On the left hand there are shown IR spectra of the fragment C_{192} of ideal tube (a), and of the same fragment with vacancies (defects) $V_{1,1}D_1\alpha$ (b); $V_{2,2}D_2\alpha$ (c); $D_0D_2\alpha$ (d). On the right hand there are shown IR spectra of the fragment C_{192} of ideal tube (a), and of the same fragment with vacancies (defects) $V_{1,1}D_1\beta$ (b); $V_{2,2}D_2\beta$ (c); $D_0D_2\beta$ (d)

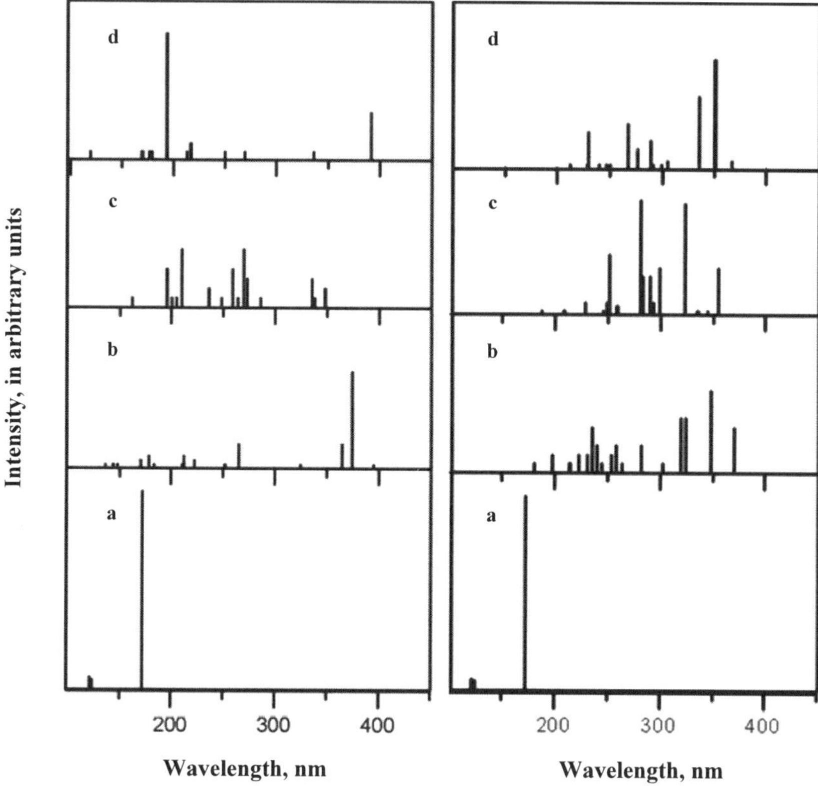

Figure 3. On the left hand there are shown UV spectra of the fragment C_{192} of ideal tube (a), and of the same fragment with vacancies (defects) $V_{1,1}D_1\alpha$ (b); $V_{2,2}D_2\alpha$ (c); $D_0D_2\alpha$ (d). On the right hand there are shown IR spectra of the fragment C_{192} of ideal tube (a), and of the same fragment with vacancies (defects) $V_{1,1}D_1\beta$ (b); $V_{2,2}D_2\beta$ (c); $D_0D_2\beta$ (d)

Only the modifications of IR spectra caused by the appearance of local vacancies and defects in single wall carbon nanotubes allow relatively simple explanation. As analysis shows these modifications are connected at the first with the appearance of oscillations modes localized in the nearest neighbourhood of vacancies or defects. The collective oscillations modes are modified also and here the accordance of local and global symmetries is very important. Unfortunately the low symmetry of investigated objects does not allow getting any complete classification of types of oscillations modes. The problem of interpretation of UV spectra of nanotubes with vacancies and defects is more complicated. The creation of vacancies or defects is accompanied by the appearance of large number of new lines in UV spectra of nanotubes. As is clear from the general point of view these modifications are connected with the shifts and splitting of energetic levels caused by falling down of nanotubes symmetry.

3. Conclusions

As follows from the results of computations the peculiarities (lines positions and their intensities), observed in IR and UV absorption spectra, can be used for identification of local vacancies and defects in single-wall carbon nanotubes.

References

1. A.H. Nevidomskyy, G. Csanyi, M.C. Payne. *Phys. Rev. Lett.* 2003, **91**: 105502.
2. K.N. Nicolski, A.S. Baturin, V.S. Bormashov, et al. Field emission investigation of carbon nanotubes doped by different metals. Proceedings of ICHMS'2003. NATO Science Series, ed. by T.N. Veziroglu et al., 2004, **172**: 123–130.
3. P. Ajayan, V. Ravikumar, J. Charlier. *Phys. Rev. Lett.* 1998, **81**: 1437.
4. D. Orlikowski, M. Nardelli, J. Bernholc, C. Roland. *Phys. Rev.* 2000, B**61**: 14194.
5. T. Belytschko, S. Xiao, G. Schatz, R. Ruoff. *Phys. Rev.* 2002, B**65**: 235430.
6. S. Mielke, D. Troya, S. Zhang, J. Li, S. Xiao, et al. *Chem. Phys. Lett.* 2004, **390**: 413.
7. C. Wei, K. Cho, D. Srivastava. *Phys. Rev.* 2003, B**67**: 15407.
8. A. Andriotis, M. Menon, D. Srivastava. *J. Chem. Phys.* 2002, **117**: 10245.
9. A. Stone, D. Wales. *Chem. Phys. Lett.* 1986, **128**: 501.
10. M.J.S. Dewar, W. Thiel. *J. Am. Chem. Soc.* 1977, **99**: 4899.
11. J. J. P. Stewart. *J. Comput. Chem.* 1989, **10**: 209.
12. J. J. P. Stewart. *J. Comput. Chem.* 1989, **10**: 221.
13. T. Clark, A. Breindl, G. Rauhut. *J. Mol. Model.* 1995, **1**: 22.
14. A.P. Popov, I.V. Bazhin. Influence of impurities and defects on electronic structure of carbon nanotubes. NATO Science Series, ICHMS'2005, ed. by T.N. Veziroglu et Al., Springer, 2007, 795–799.

ABOUT QUASI-LIQUID SPONTANEOUS DEFORMATION IN IRON-HYDROGEN SYSTEM

V.I. SHAPOVALOV
Materials and Electrochemical Research Corporation, Tucson, AZ, 85706, USA

Abstract. Features of gas-eutectoid reaction in iron-hydrogen system are described. The pore formation and spontaneous plastic deformation are controllable by changing the pressure of hydrogen and cooling-heating conditions (thermocycling range, heating speed and direction, cooling speed and direction). This phenomenon can be a basis for entirely new methods of porous metal production, having not only closed but open porosity.

Keywords: deformation, hydrogen, iron, concentration, microstructure, phase diagram

1. Introduction

In 1980 a new scientific discovery was made in former USSR [1, 2]: "Phenomenon of Occurrence of Mobile Hydrogen-Saturated Metastable Zones during Polymorphic Transformations in Metals."

This new phenomenon pertains to physical metallurgy and physics of metals: specifically, the study of structural and phase transformations (gas-eutectoid transformation). It appeared that in the presence of dissolved hydrogen in the metal crystal lattice the mechanism of polymorphous transformation is changed radically. Mobile, highly hydrogen-saturated unstable zones with very unusual properties emerge at sites between growing and disappearing crystals [3–9].

2. Gas-eutectoid reaction description

One of the unusual effects of the phenomenon is complete loss of shape stability during rearrangement of the lattice – solid samples flow under internal forces and deform like a viscous liquid (Figs. 1, 2). The second important effect is the formation of closed porosity inside solid metal samples up to gas-eutectoid reaction (Fig. 3).

The pore formation and deformation are controllable by changing the pressure of hydrogen and cooling-heating conditions (thermo cycling range, heating speed and direction, cooling speed and direction). This phenomenon can be a basis for entirely new methods of porous metal production, having not only closed but open porosity (Fig. 4).

B. Baranowski et al. (eds.), *Carbon Nanomaterials in Clean Energy Hydrogen Systems.*

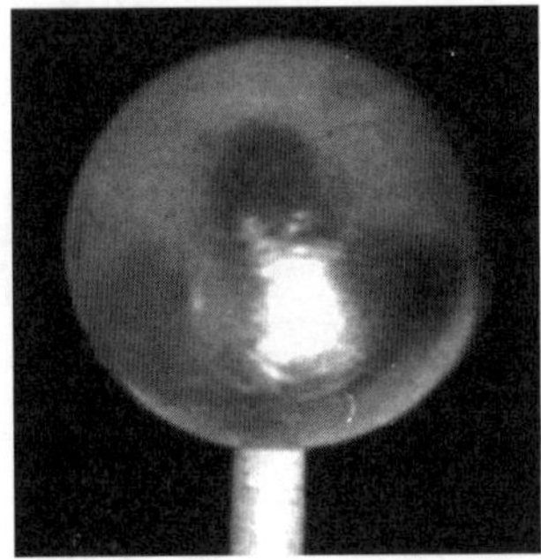

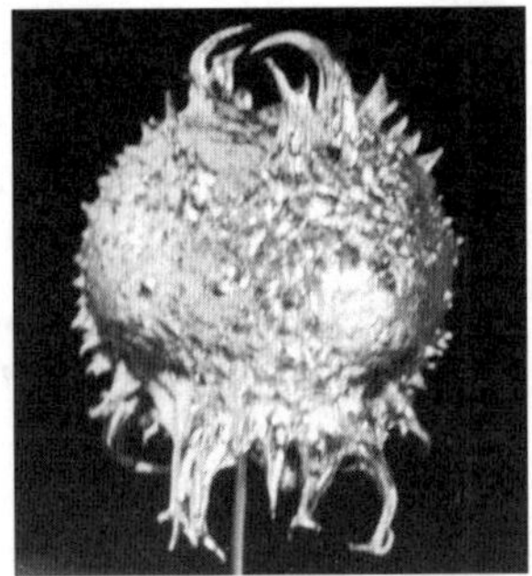

Figure 1. Quasi-liquid deformation of iron balls (initial diameter is 25 mm) after multiple gas-eutectoid reaction

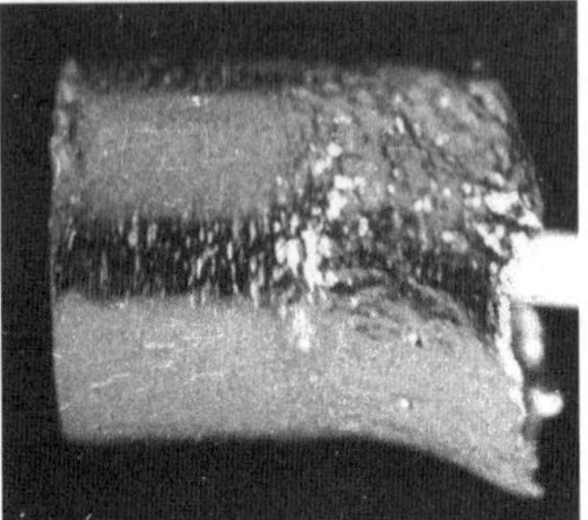

Figure 2. Different kind of quasi-liquid deformation of iron and steel cylinders (initial diameter and length about 25 mm) after multiple gas-eutectoid reaction – left; First stage of quasi-liquid deformation of steel cylinder (left side of the cylinder did not have gas-eutectoid reaction) – right

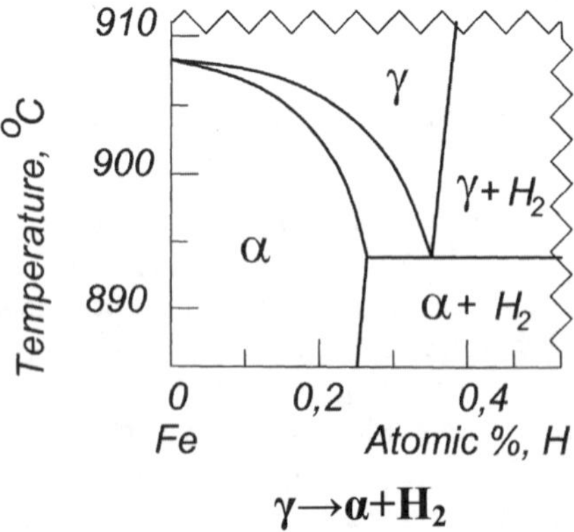

Figure 3. A part of Fe-H phase diagram including gas-eutectoid equilibrium and gas-eutectoid transformation scheme

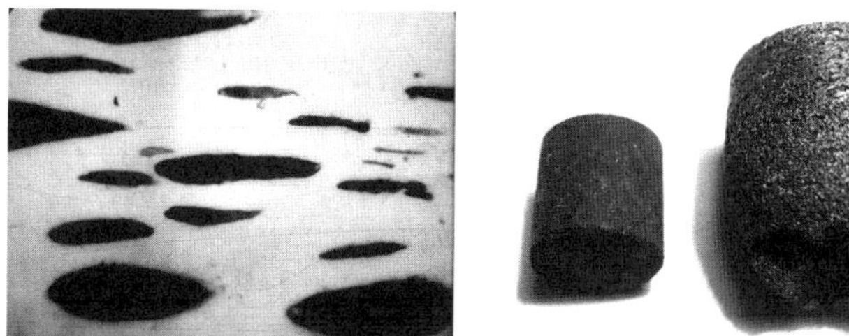

Figure 4. Microstructure of steel sample after multiple gas-eutectoid reaction (closed porosity 35%, pore diameter about 25 μm) – left; Steel sample before (left) and after (right) multiple gas-eutectoid reaction (open porosity 65%) – right

This new phenomenon pertains to physical metallurgy and physics of metals studying the features of structural and phase transformations. As is known, rearrangement of the crystal lattice in metals is one of the most important processes determining their structure and properties; this provides the basis for many kinds of heat treatment of metals and alloys.

Iron, being the most widely used material for engineering purposes, has heretofore seemed to be extensively studied. The new phenomenon, however, was discovered in iron.

It appeared that in the presence of hydrogen solved in the crystal lattice of the metal, the mechanism of polymorphous transformation is changed radically. Movable, highly hydrogen-saturated unstable zones with properties very similar to those of a liquid emerge at sites between growing and disappearing crystals. This provides a good basis for ascertaining that the new phenomenon is a breakthrough in metal science and in the study of the effect of impurities on the mechanism of rearrangement of the crystal lattice of metals.

The phenomenon is manifest in unusual external effects, e.g., in a complete loss of the shape stability by metal crystals during rearrangement of the lattice – they flow under internal forces and deform like a viscous liquid.

This spontaneous flow of the metal has, however, appeared controllable by changing the pressure of hydrogen and by certain temperature variations. This phenomenon is the basis for entirely new methods of changing the shape of metallic materials, connecting metals with ceramics, obtaining metallic capillaries and monocrystals, high-velocity diffusion welding, detection of inner flaws in articles; 15 inventor's certificates have been granted to the author as the patent protection.

Ideally, the technique can be represented as follows. An article of a simple shape is inserted into a gas-tight, hydrogen-filled chamber; by inputting a pre-developed computer program for built-in heaters, a finished piece of an entirely new shape is obtained without any mechanical external effect on the metal.

Another accompanying outcome is the impulse evolution of hydrogen from the metal during its lattice rearrangement. This contributed to development of efficient methods of degassing of steel articles for heavy-duty uses.

A transition from cutting metals to their deforming resulted in a huge economy of metals in machine building. Today, plants use high-power presses, rolling mills, big forging machines and explosion techniques for metal shaping.

The results of recent investigations modify existing technologies considerably. As found out by researchers, metals can change their shape spontaneously without applying any external forces.

For more than a quarter of a century physicists of many countries have been trying to master the most promising source of energy, nuclear fusion. We have, however, a long way to cover before thermonuclear power plants are erected and used. A number of problems in science and engineering have to be solved. In particular, the reactors of such thermonuclear power stations will operate under exceptionally heavy conditions. A great deal of knowledge has to be accumulated: the effects of high-energy particles, super high temperatures, etc. on construction materials.

We investigated characteristics of metals during their heat cycling in a hydrogen atmosphere. The outcomes had been predictable in experiments with nickel, cobalt and tungsten. With iron, surprising transformations began. On processing in a heater, an iron cylinder transformed into a tube, a ball into a "hedgehog" and a cube into a branched dendrite. Was that a scientifically queer thing or an unknown phenomenon?

Theoretically, such spontaneous transformations had not been predicted. The results of these investigations provided a foundation for a new course in physical metallurgy. Here we mean investigations into the effect of hydrogen on structural transformations in metals and alloys for imparting desirable properties to them.

It appeared that hydrogen, on penetrating the metal, alters the mechanism of rearrangement of the metal crystal lattice. Highly gas-saturated layers form between growing and disappearing crystals. These active zones are responsible for stress growth and inducing intensive plastic deformation of the metal without any external force applied.

As shown by further research, the effect was observed only in a certain temperature range and only in the presence of hydrogen. Which internal forces cause the plastic deformation of iron and make it flow "spontaneously" like a viscous liquid?

3. Conclusions

The gas-eutectoid reaction is of great value for industry. By varying the pressure of hydrogen and by producing required temperature differences, we can control the flow of the metal. This effect may provide the basis for entirely new methods of metal deformation, high-velocity diffusion welding, pore metal formation, connecting metals with ceramics and detection of inner flaws in articles. In particular, some laboratories are already making metallic capillaries and thread-like monocrystals without affecting the metal mechanically, but by altering the value and orientation of the temperature gradient.

Another important aspect of the discovery consists in it is used as the basis for an efficient method of hydrogen removal from crucial machine parts made of steel. As is known, higher hydrogen concentration results in steel embrittlement and dangerous inner cracking by flake formation [10, 11].

Recently gas-eutectoid reaction was discovered in Ti-H alloys [12–14]. So, it extends the range of application of the phenomena in metal industry.

References

1. Shapovalov V.I. Karpov V.Y. Discovery of the USSR under No. 313 on 26 December 1985 with the priority dates of 10 February 1980 in regard to the experimental detection of the phenomenon and 19 May 1982 in regard to the theoretical explanation of the phenomenon according to scientific discovery application No. OT-10562, filed 9 April 1982.
2. Shapovalov V.I. Structure formation behavior of alloys during gas-eutectic and gas-eutectoid transformation and prospects of the use of hydrogen in alloying. Proceedings of a symposium Materials Week'92 in Chicago, IL, November 1–5, 1992, 207–216.
3. Shapovalov V.I., Karpov V.Yu. The Effect of Anomalous Plastic Self-Deformation of Fe-H Alloys. Doklady AN Ukr.SSR, 1981, **7** (in Russian).
4. Shapovalov V.I., Karpov V.Yu. Anomalous Superplasticity of Iron. Izvestiya Vuzov. Chernaya Metallurgiya, 1981, **8** (in Russian).
5. Shapovalov V.I., Karpov V.Yu. Self-Deformation of Iron in the Presence of Hydrogen. Fizika metallov i Metallovedenie, 1981, **52**(6) (in Russian).
6. Shapovalov V.I., Karpov V.Yu. Possibility of Iron Change to a Quasiliquid State on Polymorphic Transition in Hydrogen Atmosphere. Izvestiya AN SSSR. Metally, 1982, **1** (in Russian).
7. Shapovalov V.I., Karpov V.Yu. Anomalous Dynamic Superplasticity and Quasiliquid State of Iron in the Presence of Hydrogen. Fiziko-Khimicheskaya Mekhanika Materialov, 1982, **3** (in Russian).
8. Shapovalov V.I., Grigorovich V.K. On the Nature of Quasiliquid State of Iron. Doklady AN SSSR, 1982, **267**(4) (in Russian).
9. Shapovalov V.I., Karpov V.Yu. The Nature of Anomalous Spontaneous Deformation of Iron in the Presence of Hydrogen. Fizika metallov i metallovedenie, 1983, **55**(4) (in Russian).
10. Shapovalov V.I., Trofimenko V.V. On Treatment of Wrought Steel Wheels to Prevent Flakes. Izvestiya Vuzov. Chernaya Metallurgiya, 1981, **8** (in Russian).
11. Shapovalov V.I., Staroseletskii M.I., Trofimenko V.V., Antipova N.Y. Hydrogen Behaviour in Steel During Manufacture of Wrought Railway Wheels. Stal, 1982, **6** (in Russian).
12. Schuh C., Dunald D.C. Internal Stress Plastisity Due to Chemical Stresses, Acta Materialia, 2001, **49**: 3387–3400.
13. Frary M., Schuh C., Dunald D.C. Strain Ratchetting of Titanium upon Reversible Alloying with Hydrogen, Philosophical Magazine A, 2001, **81**(1): 197–202.
14. Murraya N.G.D., Schuh C.A.B., Dunand D.C. Solid-State Foaming of Titanium by Hydrogen-Induced Internal-Stress Superplasticity, Scripta Materialia, 2003, **49**: 879–883.

EQUILIBRIUM CONFIGURATION AND ELECTRONIC STRUCTURE OF THE $C_{32}H_8$ MOLECULE ISOMERS

A.P. POPOV, I.V. BAZHIN
Pedagogical Institute of the South Federal University
Bolshaya Sadovaya Str., 33, Rostov-on-Don, 344082, Russia
E-mail: nanosys@mail.ru

Abstract. In the framework of two methods – electron density functional and semi-empirical PM3-metod – for different isomers of molecule $C_{32}H_8$ there are calculated geometrical parameters of equilibrium configuration, full energy and heat of formation, and absorption spectra in infrared and ultraviolet region also.

Keywords: fullerenes, hydrofullerenes, quantum-chemical calculations, IR and UV spectra

1. Introduction

At the ICHMS'2005 conference, the attention was attracted to the report [1] that has introduced new nanostructures, synthesized by means of modified electro-chemical method. Molecular ions with the masses from 720 to 1,157 were detected in the structure of carbon fibers, which accord to formulae of the C_{60} and C_{70} fullerenes.

Polycyclic hydrocarbon ions were detected in mass-specters of the benzene solution reaction products, and the compound with the molecular mass of 392 and formulae $C_{32}H_8$ prevails.

Authors [1] suggested a model of $C_{32}H_8$ molecule in a form of a strength structure, in the middle of which the belt of four six-member cycles composes a deformed cylinder pipe (Fig. 1(a)). The structure in general has the C_{2v} symmetry.

However, the chemical formula of the compound allows the existence of isomers, and in our opinion, it is not possible to make any unequivocal conclusions about the compound structure at this stage of research.

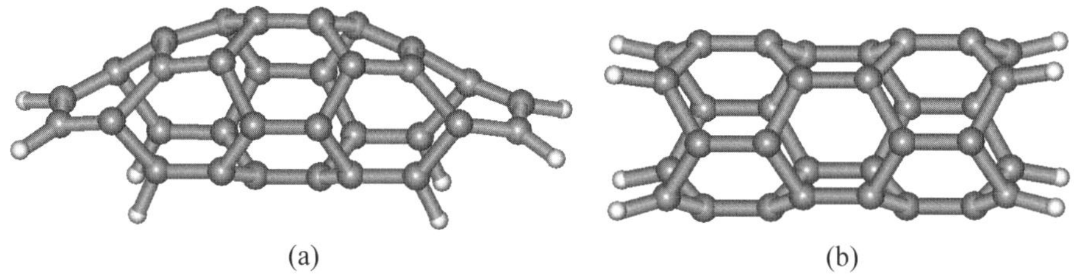

(a) (b)

Figure 1. Isomers of molecule $C_{32}H_8$ with the symmetry: (a) C_{2v} и (b) D_{4h}/C_{2h}

2. Results and discussions

Calculations, performed by means of the DFT electronic density functionality method [2–4] and semiempirical PM3 method [5–8] revealed the fact that even a structure in a form of short open (4,0) nanotube with the hydrogen atoms joined to tube's edges (Fig. 1(b)) has lower full energy than the structure [1]. It must be denoted that during the process of finding an optimal geometry, the initial symmetry of D_{4h} structure is lowed to C_{2h} due to nanotube flattening.

Further calculations have showed that hydrofullerene composed by joining the hydrogen atoms to isomers of small fullerene C_{32} that have D_3 and D_{3h} symmetry have the lowest full energy among the isomers of the $C_{32}H_8$ molecule (Fig. 2).

To make the direct comparison with the experiment possible, the absorption IR and UV spectra of all described isomers $C_{32}H_8$ have been calculated (Fig. 3). The positions and intensities of absorption lines allow confident identify of every described above isomers of $C_{32}H_8$ molecule.

The results of computation of some physical and chemical characteristics of isomers of molecule $C_{32}H_8$ are presented in Table 1. In the left part of this table the results of the calculation of total energy (in Hartree units), average energy of a chemical binding (per one atom) and the potential of ionization of all isomers of $C_{32}H_8$ molecule, produced using DFT electronic density functionality method (DMol3 software) are displayed. In the right part of this table the values of total energy, heat of formation and the ionization potential calculated using PM3 method (from the renewed version of Mopac software) are displayed.

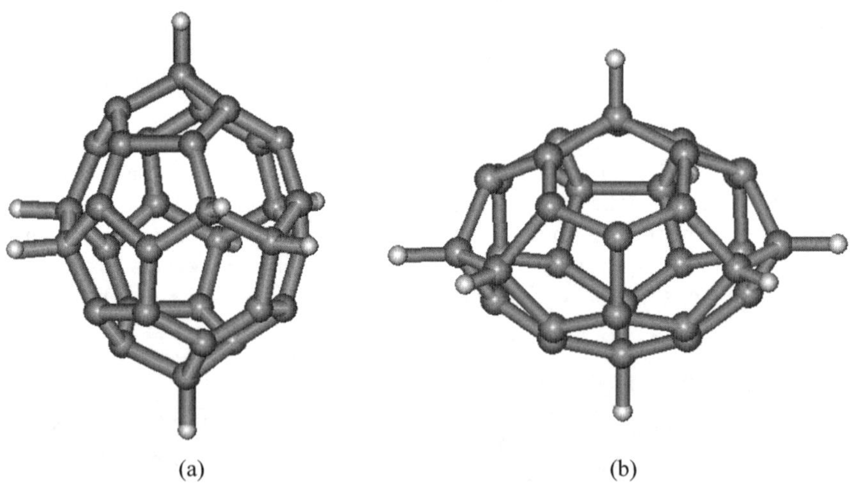

Figure 2. Molecule of hydrofullerene $C_{32}H_8$ with symmetry D_3 (a) и D_{3h} (b)

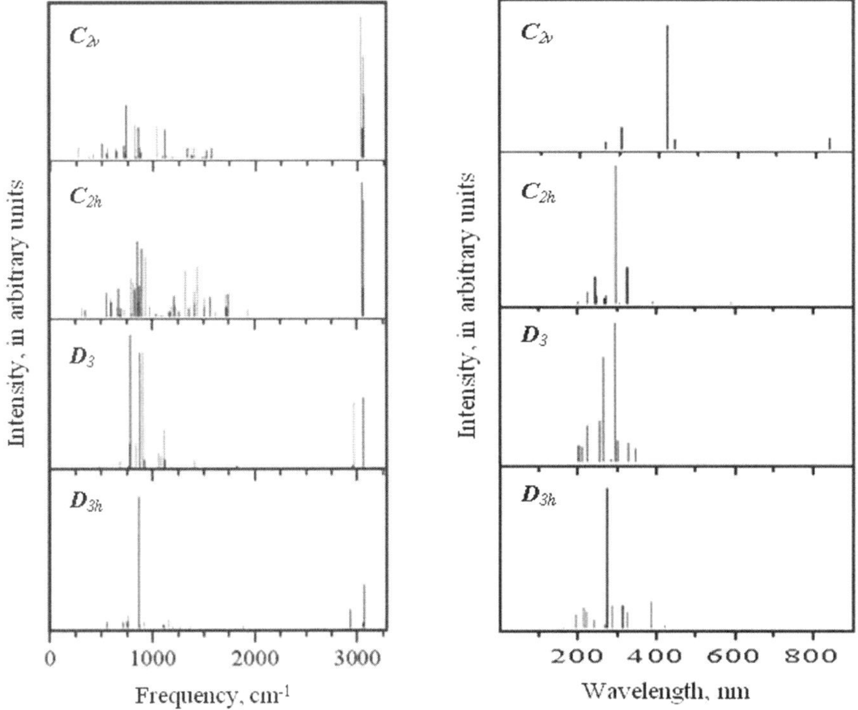

Figure 3. The calculated IR and UV spectra of isomers of molecule $C_{32}H_8$

TABLE 1. Physical and chemical characteristics for isomers of molecule $C_{32}H_8$

	DFT			PM3		
	E_{tot} (Ha)	ΔE_{bind} (Ev)	IP (eV)	E_{tot} (Ha)	ΔH (kcal/mol)	IP (eV)
D_3	-1,213.263	285.32	5.44	-3,894.676	564.50	8.10
D_{3h}	-1,213.278	285.74	6.48	-3,893.557	603.87	7.77
D_4/C_{2h}	-1,213.033	279.08	5.64	-3,885.911	766.62	8.12
C_{2v}	-1,213.003	278.26	5.37	-3,888.593	704.77	8.25

3. Conclusions

The results of computer modeling allow us to make a confident conclusion that the compound $C_{32}H_8$ synthesized by authors [1] using the electro-chemical method is, most likely, a hydrofullerene based on a small fullerene C_{32} with D_{3h} symmetry.

References

1. V.M. Ogenko, L.S. Lysyuk, S.V. Volkov, A.P. Shpak, Novel carbon nanostructures produced by electrochemical method. Book of Abstracts of IX International Conference ICHMS'2005, *edited by D.V. Shur et al.*, Kiev: AHEU, 2005. P. 364.
2. B. Delley, *J.Chem. Phys.* 1990, **92**: 508.
3. B. Delley, *J.Chem. Phys.* 1991, **94**: 7245.
4. B. Delley, *J.Chem. Phys.* 2000, **113**: 7756.
5. M.J.S. Dewar, W. Thiel, *J. Am. Chem. Soc.* 1977, **99**: 4899.
6. J.J.P. Stewart, *J. Comput. Chem.* 1989, **10**: 209.
7. J.J.P. Stewart, *J. Comput. Chem.* 1989, **10**: 221.
8. T. Clark, A. Breindl, G. Rauhut, *J. Mol. Model.* 1995, **1**: 22.

PROTON-CONDUCTING MEMBRANES BASED ON COPOLYMER COMPOSITES

G.F. MYACHINA, T.G. ERMAKOVA, N.P. KUZNETSOVA, L.I. VOLKOVA, B.A. TROFIMOV
A.E. Favorsky Irkutsk Institute of Chemistry, Siberian Branch, Russian Academy of Sciences, 664033, Irkutsk, Russia
D.M. MOGNONOV
Baikal Institute of Nature Management, Siberian Branch, Russian Academy of Sciences, 670047, Ulan-Ude, Russia

Abstract. Copolymers of 1-vinyl-1,2,4-triazole with methyl methacrylate and fluoroalkylmethacrylates have been modified with phosphorus- and silicon-molybdenum heteropolyacids as well as copolymer of ethylene glycol vinyl glycidyl ether with N-phenylmaleimide to furnish new composite materials. The elastic films obtained show proton conductivity of about 10^{-3}–10^{-1} S/cm (after doping with orthophosphoric acid). They are promising candidates for the design of proton-conductive membranes.

Keywords: 1-vinyl-1,2,4-triazole, fluoroalkylmethacrylates, copolymers, heteropolyacids, proton-conductive polymer membranes

1. Introduction

The synthesis and investigation of new polymer materials to develop proton-conductive membranes of fuel cells is an urgent scientific task. Along with the known perfluorinated electrolyte membranes such as «Nafion», «Flemion», «Neosepta-F», the membranes based on other fluorinated polymers, as well as on polybenzimidazoles are under extensive investigations [1–4].

The present work is aimed at the synthesis of functionalized copolymers of 1-vinyl-1,2,4-triazole with methyl- and fluoromethylmethacrylates to develop new composite materials, promising candidates for proton-conductive membranes of low- and middle temperature fuel (hydrogen and methanol) cells.

2. Experimental procedures

The starting monomers, 1-vinyl-1,2,4-triazole [5], methylmethacrylate, 1,1-dihydrotrifluoroethylmethacrylate, 1,1,3 – trihydrotetrafluoropropylmethacrylate and 1,1,5-trihydrooctafluoroamylmethacrylate [6], were preliminarily distilled in vacuum.

Radical copolymerization of 1-vinyl-1,2,4-triazole with the copolymers aforementioned was performed in sealed glass ampoules under inert atmosphere in DMFA solution in the presence of dinitrile of azoisobutyric acid at 60°C for 6 h. The copolymers were isolated and purified by reprecipitation from DMFA solutions, and dried in vacuum at 50–70°C until the constant weight. Structure and composition of the copolymers were proved using elemental analysis, IR and ^{13}C NMR techniques.

The copolymers of 1-vinyl-1,2,4-triazole were modified with heteropolyacid hydrates: $H_8[Si(Mo_2O_7)_6]\cdot nH_2O$, $H_7[P(Mo_2O_7)_6]$, as well as copolymer of ethylene glycol vinyl glycidyl ether with N-phenylmaleimide, synthesized by radical copolymerization of the corresponding co-monomers in methyl ethyl ketone at 75°C.

The modified copolymers were used to prepare polymer films, which further were doped with orthophosphoric acid of various concentrations (7–13 mol/l).

IR spectra were recorded on a «Bruker» instrument in acetonitrile solution and in KBr pellets. ^{13}C NMR spectra were run on VXR-500S spectrometer, "Varian" (DMSO-d_6). Molecular weight was determined using GPC technique («Water», USA).

3. Results

Radical copolymerization of 1-vinyl-1,2,4-triazole with methylmethacrylate, 1,1,3-trihydrotetrafluoropropylmethacrylate, 1,1,5-trihydrooctafluoroamylmethacrylate and 1,1-dihydrotrifluoroethylmethacrylate affords new functionalized copolymers I, II, III, IV, respectively [7]:

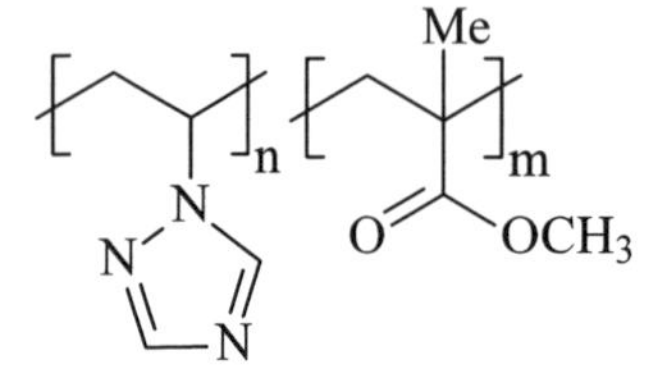

I (m: OCH_3)

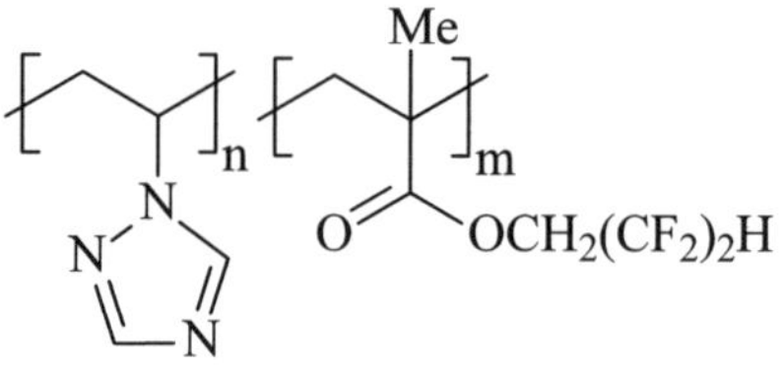

II (m: $OCH_2(CF_2)_2H$)

III (m: $OCH_2(CF_2)_4H$)

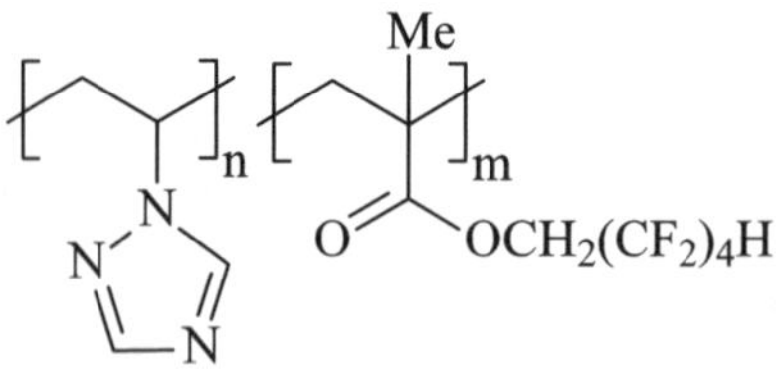

IV (m: OCH_2CF_3)

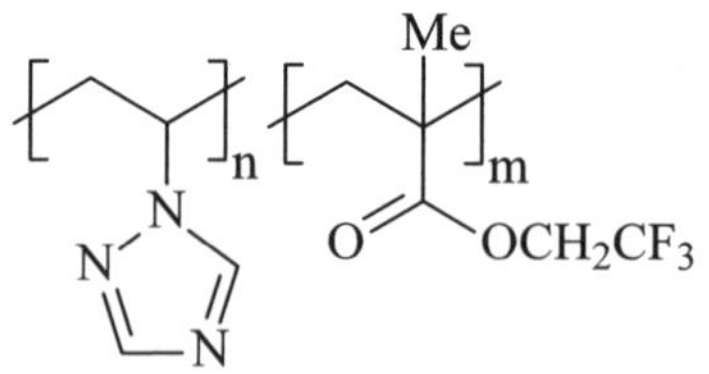

The copolymers obtained are white powders, well soluble in acetone, dimethylformamide, dimethylsulfoxide and N-methylpyrrolidone. The copolymers composition was proved using data of elemental analysis (N, F) and ^{13}C NMR technique (Table 1). The copolymers obtained are characterized by molecular weight of 26,300–94,300 and melting point of 180–250°C (Table 1).

TABLE 1. Characteristics of 1-vinyl-1,2,4-triazole-based copolymers (I–IV)

Copolymer	Content of 1-vinyl-1,2,4-triazole, mass%	Molecular weight	Melting point (°C)
I	18	83,700	180–200
II	36	94,300	205–216
III	49	26,300	200–250
IV	27	80,600	185–202

The copolymers I–IV synthesized are modified with phosphorus- and silicon-molybdenum heteropolyacids to furnish new composite materials. The latter have been employed for the preparation of elastic films, which were doped with orthophosphoric acid [8]. The dependence of the films doping degree on orthophosphoric acid concentration has been studied (Fig. 1).

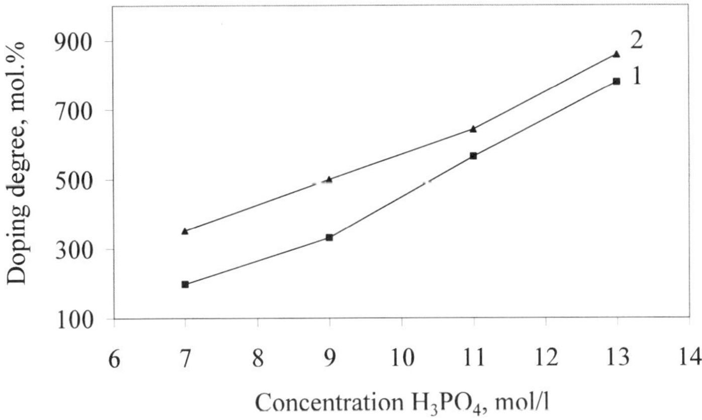

Figure 1. Dependence of copolymer (1) and the copolymer-based composite (2) doping degree on orthophosphoric acid concentration

The effect of doping degree of the composite films on specific conductivity has also been investigated (Fig. 2).

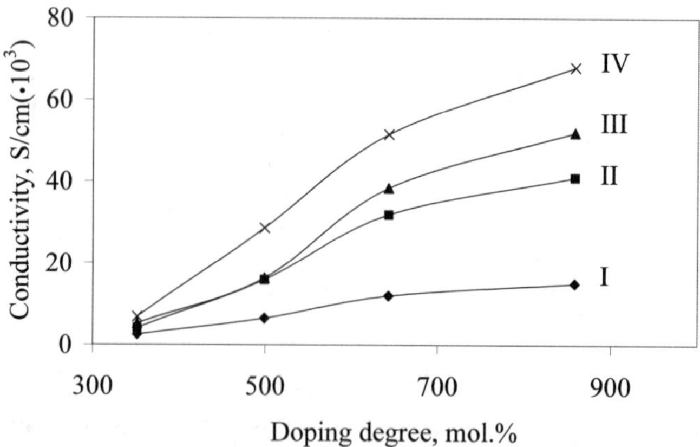

Figure 2. Dependence of specific conductivity of composite films based on copolymers I–IV on orthophosphoric acid doping degree

It has been found that the composite materials studied show the growth of proton conductivity with an increase of temperature to 130°C followed by its stabilization (Fig. 3).

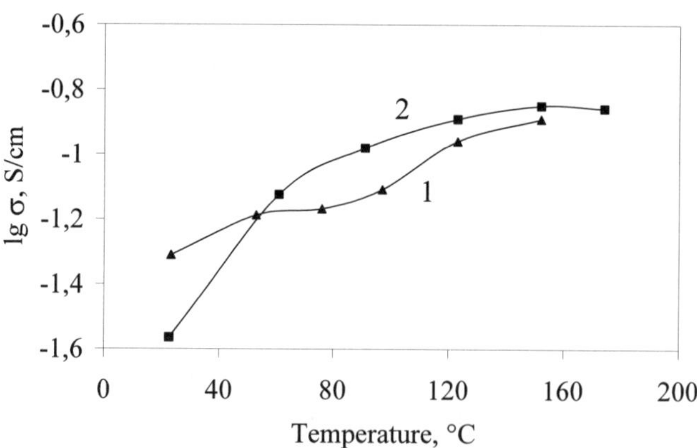

Figure 3. Temperature dependence of specific conductivity of composite films of the copolymer II with silicon (1) and phosphorusmolybdenum (2) heteropolyacids

Furthermore, the possibility of the copolymers modification by their cross-linking with thermosets followed by thermal hardening at 100–230°C has been investigated [9].

For example, copolymer of ethylene glycol vinylglycidyl ether with N-phenylmaleimide (V) is used as thermoset:

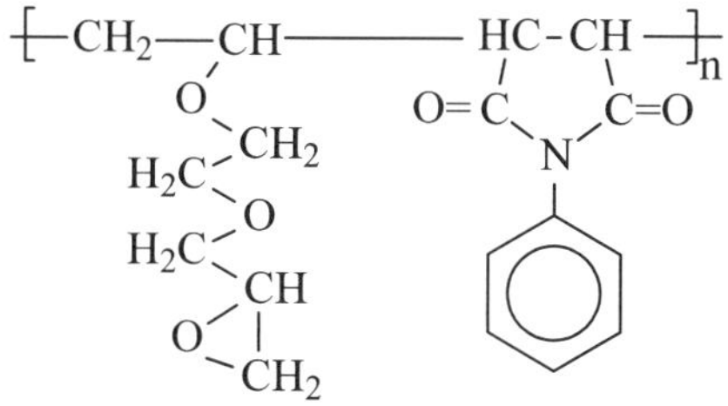

V

4. Discussion

IR spectra of the copolymers I–IV synthesized show absorption bands at 1,100, 1,450, 1,510 cm^{-1}, assigned to the 1-vinyl-1,2,4-triazole heteroring stretching vibrations as well as absorption bands assigned to fluoromethylmethacrylates at 1,130, 1,200, 1,240, 1,270 cm^{-1} (ν CF_2 moieties), at 420–1,000 cm^{-1} (δ CF_2 moieties) and at 1,740 cm^{-1} (ν C=O groups). At the same time the absorption bands attributable to the C=C bond stretching vibrations disappear in the spectra. This fact is indicative of the comonomers double bond cleavage during the copolymerization and formation of the copolymers I–IV.

Elemental analysis and ^{13}C NMR data were employed to prove the structure of the copolymers I–IV. It has been shown that methyl and fluoromethylmetacrylate moieties prevail in the copolymer macromolecules (Table 1).

Modification of the copolymers I–IV with phosphorus- and silicon-molybdenum heteropolyacids has enabled to obtain the composite materials capable of the formation of elastic strong films [8]. Doping the films with orthophosphoric acid involves the growth of doping degree with the rise of acid concentration from 7 to 13 mol/l (Fig. 1). Further increases of the acid concentration leads to partial decomposition of the films.

The composite films doped are characterized by specific conductivity of about 10^{-3}–10^{-2} S/cm (room temperature), which drastically increases with the growth of the doping degree (Fig. 2). Notably that the composite materials studied show the growth of ionic conductivity up to 10^{-1} S/cm under the increase of temperature to 130°C followed by stabilization at 150–160°C (Fig. 3). This fact is of prime importance for proton-conductive membranes of fuel cells. The composite membranes prepared from the copolymers I–IV cross-linked with the thermoset (copolymer of ethylene glycol vinylglycidyl ether with N-phenylmaleimide (V) possess also high proton conductivity after the doping with orthophosphoric acid: 10^{-3}–10^{-1} S/cm [9]. The increase of orthophosphoric acid concentration involves the rise of the doping degree as well as the values of specific conductivity of the films. The latter value increases with the rise of temperature to 120–130°C.

The copolymers I–IV-based composite membranes (both modified with heteropolyacids and cross-linked with thermosets) are characterized by chemical stability to aggressive media (acids, alkali, hydrogen peroxide), thermal stability (up to 160–350°C), high moisture absorption as well as mechanical strength: breaking strength is $(2–7)\cdot 10^7$ N/m^2, elongation – 5–7%.

The valuable properties of the composite materials mentioned are due to the presence of both triazole cycle and fluorine moieties in the copolymers macromolecule.

5. Conclusions

1. New copolymers on the basis of 1-vinyl-1,2,4-triazole with methyl methacrylate and fluoroalkylmethacrylates have been synthesized. The copolymers have been modified with phosphorus- and silicon-molybdenum heteropolyacids and cross-linked with thermoset, copolymer of ethylene glycol vinyl glycidyl ether with N-phenylmaleimide, to deliver new composite materials capable of the formation of elastic strong membranes.
2. High proton conductivity of the composite films based on the modified copolymers, increasing up to 10^{-2}–10^{-1} S/cm after doping with orthophosphoric acid, has been revealed.
3. It has been found that proton conductivity of the composite films increases to 10^{-1} S/cm with the growth of their temperature to 130°C.
4. The results obtained point to the fact that new copolymers based on 1-vinyl-1,2,4-triazole with methyl- and fluoroalkylmethacrylates are promising building blocks to design proton-conductive membranes of fuel cells.

References

1. Hickner M.A., Ghassemi H., Kim Y.S., Einsla B.R., McGrath J.E. Alternative Polymer Systems for Proton Exchange Membranes (PEMs), *Chem. Rev.*, 2004, **104**: 4587–4612.
2. Joachim K., Oemer U., Euiwong C., Michael S., Brian B., Gordon C. *Proton – conducting polymer membrane comprising sulfonic acid – containing polyazoles, and use there of in fuel cells*. US Patent WO2004034499, 2004.
3. Dobrovolskii Yu.A., Volkov E.V., Pisareva A.V., Fedotov Yu.A., Likhachev D.Yu., Rusanov A.L. Proton Exchange Membranes for Hydrogen Air Fuel Cells, *Russ. Chem. J.*, 2006, **6**: 95–104.
4. Tarasevich M.P., Karichev Z.R., Bogdanovskaya V.A., Kuznetzova L.N., Efremov B.N., Kapustin A.V. Electroconductivity and Permeability of Polybenzimidazole Films in Alkaline Solutions, *Elektrokhimia*, 2004, **40** (6): 745–749.
5. Ermakova T.G., Tatarova L.A., Kuznetsova N.P. Vinylation of 1,2,4-triazole., *J. Obsch. Chem.*, 1997, **67** (5): 859–861.
6. Goldin G.S., Averbakh K.O., Nekrasova L.A., Lavygin I.A., Leitan O.V., Chalbysheva N.V. Synthesis and Physical-Chemical Properties of Organofluorine Ethers of Acrylic, Methacrylic and Maleinic Acids, *Zh. Prikl. Khim.*, 1985, **58** (6): 1349–1358.
7. Kuznetsova N.P, Kanitskaya L.V., Fedorov S.V., Ermakova T.G. The radical copolymerization of 1-vinyl-1,2,4-triazole with fluoro-containing alkylmethacrylates, *Vysokomolek. soed. B*, 2001, **43** (9): 1565–1569.
8. Trofimov B.A., Mognonov D.M., Ermakova T.G., Kuznetsova N.P., Myachina G.F. et al. Method for preparation of proton-conducting polymer membranes. RF Patent No 2279906. Bull. Izobr., 2006, 20.
9. Trofimov B.A., Mognonov D.M., Ermakova T.G., Kuznetsova N.P., Myachina G.F., Volkova L.I., et al. Proton-conducting compositional polymer membranes and method of their preparation. RF Patent No 2284214. Bull. Izobr., 2006, 27.

SELECTIVE SYNTHESIS OF CARBON NANOTUBES BY METHANE PYROLYSIS AND PREPARATION Pt/C ELECTROCATALYSTS ON THEIR BASIS

E.V. GERASIMOVA, A.A. VOLODIN, B.P. TARASOV
Institute of Problems of Chemical Physics of RAS, Prospectus of the Academician Semenov, 1, Chernogolovka, Moscow Region, 142432 Russia
Tel./Fax: +7(496) 5221743; E-mail: btarasov@icp.ac.ru

Abstract. Carbon nanotubes with a diameter of 10–50 nm was obtained by methane pyrolysis over a Fe/MgO catalyst and investigated as supports for Pt clusters for elaboration of electrocatalysts for fuel cells.

Keywords: carbon nanotubes, methane pyrolysis, electrocatalyst, fuel cell

1. Introduction

Catalytic pyrolysis of carbon compounds is known to allow one to produce carbon nanotubes (CNT) in large-scale quantities and to perform their selective synthesis by variation of reaction parameters, catalysts and deposition on various supports [1–3]. The particle size of the metal catalyst defines the nanotubes diameter, while the support preventing sintering of small catalyst particles [4]. Thus, varying the synthesis parameters one could obtain carbon nanotubes with high selectivity and narrow diameter distribution.

One of the most perspective areas of application of carbon nanotubes is their usage as supports for many kinds of catalysts. For example, Pt clusters deposited on carbon nanotubes are effective electrocatalysts in the processes proceeding in air-hydrogen fuel cells (FC) [5].

The aim of the present work was selective synthesis of carbon nanotubes and their investigations as supports for Pt clusters for elaboration of electrocatalysts for fuel cells.

2. Experimental

For the growth of carbon nanotubes a catalyst Fe/MgO, containing 3 wt% of Fe was taken. The catalyst was prepared by impregnation of a highly dispersed MgO powder with a solution of the iron(III) nitrate with subsequent heating at the temperature 500°C in a hydrogen flow for 1 hour for the complete reduction of the iron. CNT were synthesized in a gas flow reactor by feeding the gas mixture CH_4:H_2 = 5:1 at 900°C and the pressure of 0.1 MPa during 1 hour. The pyrolysis products were purified by a treatment in concentrated hydrochloric acid and oxidized by nitric acid. Platinum was deposited by the chemical reduction of H_2PtCl_6 with formate-ion.

The obtained materials were studied by TEM, XRD, TG, DSC. The prepared catalyst Pt/CNT was examined in the content of a hydrogen-air fuel cell in a measuring cell ElectroChem with the active area of 1 cm^2 using a potentiostate PS-9.

3. Results and discussion

TEM data show that during the methane pyrolysis thin carbon nanotubes are formed on the surface of the catalyst. These nanotubes are bounded in bundles with diameters ranging within 10–50 nm. In addition to the bundles sharply shaped cubic crystals of magnesia are present in solid pyrolysis products. The magnesia crystals have the average size of about 50 nm. Assessment of the CNT content in the soot samples gives the value of ~10%.

By the treatment of the pyrolysis products with hydrochloric acid the particulates of magnesia are practically removed. Besides the CNT bundles there remain encapsulated catalyst particles with the size of 10–20 nm, a little amount of amorphous carbon, and compact particulates with sizes ranging within 5–20 nm (Fig. 1). The latters appeared to be remains of MgO encapsulated in a matrix of amorphous carbon.

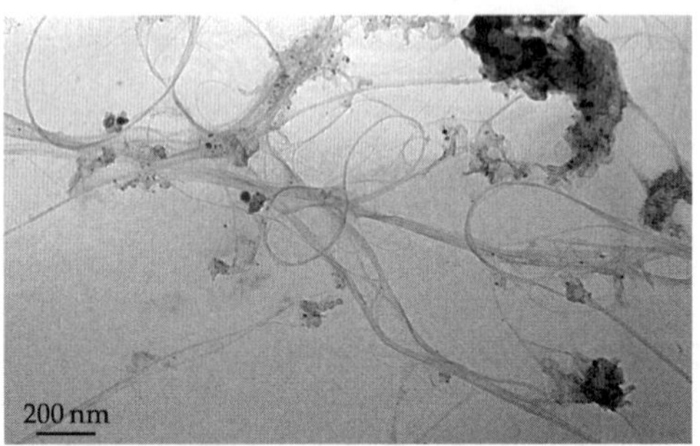

Figure 1. TEM image of the purified CNT

According to data of the oxidative thermo-gravimetry the sample is oxidized within the temperature range of 550–650°C (Fig. 2).

The maximal oxidation rate was observed at 580°C. This temperature range is reported to be peculiar to bundles of single-walled carbon nanotubes. A solid residue after the oxidation had the mass of ~3%.

The TG and TEM analysis of the pyrolysis products both pristine and purified allowed us to evaluate the content of the carbon nanotubes in the products to be 70%. The surface area of the purified sample measured by the BET method was 280 m^2/g.

After the Pt deposition on the CNT the Pt content in the sample was about 20 wt% as measured by the oxidative thermogravimetry data. The size of Pt crystallites calculated by making use of the Sherrer's equation is as big as 5 nm (Fig. 3).

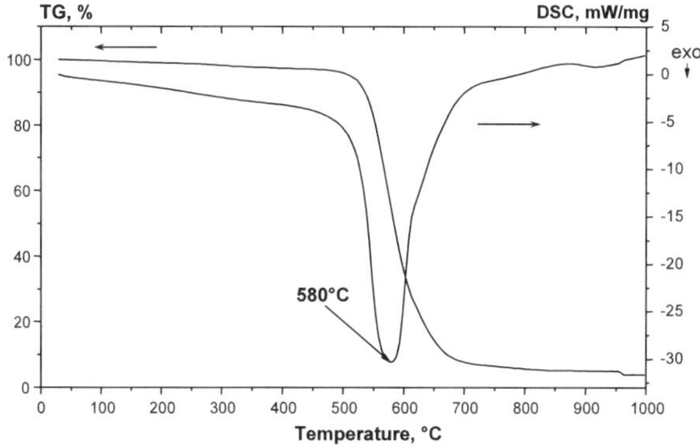

Figure 2. Thermal analysis data of the CNT sample obtained by methane pyrolysis at 900°C

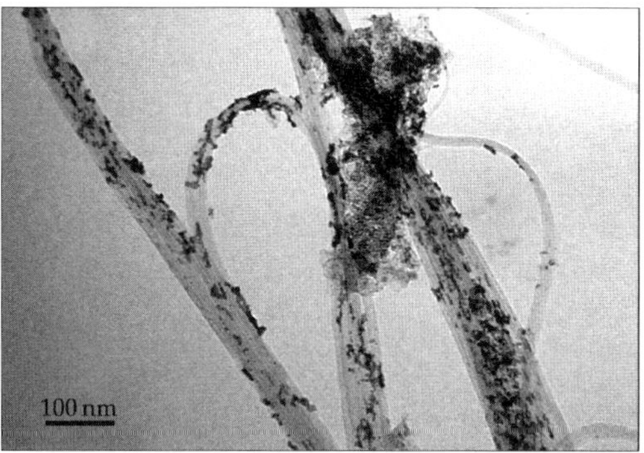

Figure 3. TEM image of the sample Pt/CNT

The obtained Pt/CNTs were employed to make membrane-electrode assemblies (MEA). An Investigation of the volt-ampere characteristics of the made MEA showed that their maximal power was 50 mW/cm^2 (Fig. 4).

4. Conclusions

Selective synthesis of CNT as bundles with a diameter of 10–50 nm was performed by methane pyrolysis over a Fe/MgO catalyst. The CNT content in the pyrolysis products was about 70 wt%. The surface area of the purified material was 280 m^2/g.

MEA made on the basis of Pt/CNT catalysts containing 20% of Pt have the maximal power 50 mW/cm^2.

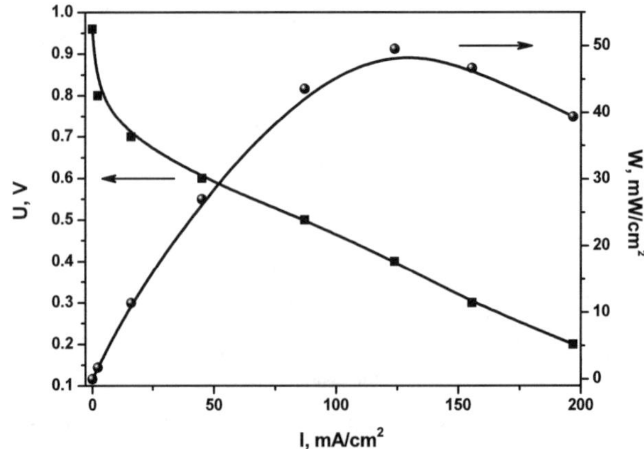

Figure 4. Volt-ampere and power characteristics of the MEA with the catalyst Pt/CNT

Acknowledgement

The work is supported by the grant "UMNIK" and the Program of the Presidium of RAS "Encouragement of young scientists".

References

1. Harutyunyan A.R., Pradhan B.K., Kirn U.J., Chen G., Eklund P.C. CVD synthesis of single wall carbon nanotubes under soft conditions, *Nano Lett.*, 2002, **2**: 525–530.
2. Volodin A.A., Fursikov P.V., Kasumov Yu.A., Khodos I.I., Tarasov B.P. Synthesis of carbon nanofibers by catalytic pyrolysis of ethylene and methane on hydrides of intermetallic compounds of lanthanum with nickel, *Russ. Chem. Bull.*, 2005 **54**(10): 2281–2285 (in Russian).
3. Rakov E.G. Preparation of thin carbon nanotubes by catalytic pyrolysis on a support, *Russ. Chem. Rev.*, 2007, **76**: 1–22 (in Russian).
4. Chesnokov V.V., Buyanov R.A. The formation of carbon filaments upon decomposition of hydrocarbons catalysed by iron subgroup metals and their alloys, *Russ. Chem. Rev.*, 2000, **69**: 623–638 (in Russian).
5. Li X., Hsing I.M. The effect of the Pt deposition method and the support on Pt dispersion on carbon nanotubes, *Electrochim. Acta*, 2006, **51**: 5250–5258.

Pt NANOCLUSTERS ON CARBON NANOMATERIALS AS HYDROGENATION CATALYSTS

S.D. KUSHCH, N.S. KUYUNKO, B.P. TARASOV
Institute of Problems of Chemical Physics of RAS
Prospectus of the Academician Semenov, 1, Chernogolovka, Moscow Region, 142432 Russian
Tel./Fax: +7(496) 5221743; E-mail: ksd@icp.ac.ru

Abstract. Platinum fixation via functional groups of carbon support is necessary criterion for effective catalyst creation. Fullerene black inclusive unconjugated double bonds (one it per three to five carbon atoms) is preferable support than carbon material with graphene structure wherein about 4 wt% of oxygen as carboxyl groups may be introduced by oxidative treatment. Platinum on fullerene black fixing as the clusters with particles size about 3.5–4 nm whereas particles sizes for Pt clusters on carbon material with graphene structures is more and equal of 6–7 nm. Pt nanoclusters supported on carbon nanomaterials catalyze of hydrogenation of both multiple bonds as decene-1 and nitroaromatics as nitrobenzene. The catalysts based on fullerene black exceed catalysts based on carbon material with graphene structure and conventional Pt/C catalyst with respect to catalytic activity. Pt clusters sizes have more influence with respect to catalytic activity than electroconductivity and specific surface of carbon support.

Keywords: platinum, nanocluster, carbon nanomaterial, hydrogenation, catalysis, decene-1, nitrobenzene

1. Introduction

In recent time an interest to catalysts including noble metals supported on carbon nanomaterials (CNM) has grown considerably. One allows that the sizes of metal clusters coating nano-sized carbon clusters lie within the nano-scale range that results in a high catalytic activity of the metals. CNM as a support for catalysts of electron transfer are also perspective due to their electroconductivity determined by an extensive conjugation system of the graphene structure in the network.

2. Experimental

The following carbon materials were studied as a Pt support [1]: single-walled (SWNT) and multi-walled (MNT) carbon nanotubes produced by the arc-discharge method, nanofibers (CNF) obtained by catalytic pyrolysis of both propane-butane mixture (CNF1) and ethene (CNF2), and fullerene black (FB) that is a soot after fullerenes extraction.

Pt was supported on CNM from an aqueous solution. An H_2PtCl_6 solution was added to an ultrasonically treated CNM aqueous suspension during 2–3 h. The reduction was performed by formate-ion.

3. Results and discussion

A strong fixing of the metal on the support is a necessary criterion for the effective catalyst preparation. To realize the criterion one need both the presence of functional groups on the surface of the support and the use of metal precursor suitable for a formation of covalent bonds with these groups.

SWNT, MNT, and CNF do not contain functional groups suitable for Pt fixing. Otherwise, they include plain (CNF1 and CNF2) or curved (SWNT and MNT) graphene sheets with the aromatic conjugation system. These groups may be introduced by oxidative treatment. The presences of carboxyl groups after the treatment follow from moderate temperatures of decarboxilation (Fig. 1, for example).

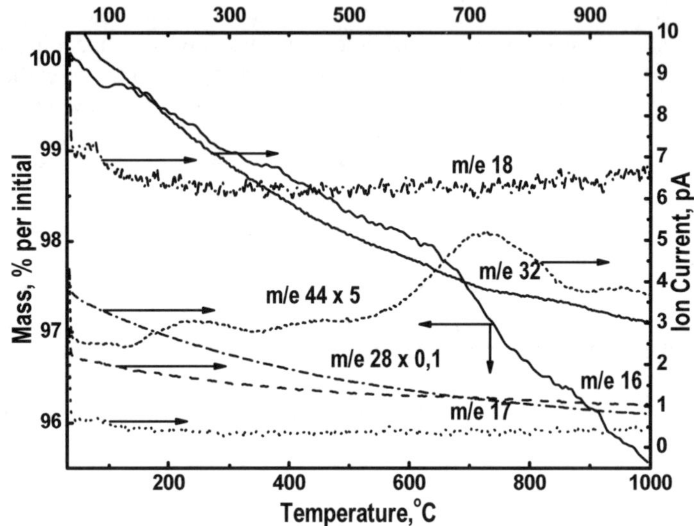

Figure 1. Thermogramm combined with mass spectrum analysis of SWNT oxidated in Ar flow

In the presence of a base the carboxyl groups suitable for the interaction with H_2PtCl_6 may be introduced in SWNT, MNT and CNF by oxidizing acids (HNO_3, HNO_3 + H_2SO_4) treatment. The fixing of the metal reduced by formate ion may be realized by donor-acceptor interaction of the metal with an unshared electron pair of an oxygen atom in the functional groups. However, the amount of these groups introduced by treatment is low and metal can be supported only when the Pt content is as low as 18–20%.

The network of FB composed of fullerene fragments doesn't contain the aromatic conjugation system but it has unconjugated double bonds (one bond per three to five carbon atoms). These bonds are a cause of oxygen selective adsorption (Fig. 2).

Therefore FB as an alkene can react with agents of the HX type and halogens [2]. Platinum may be added to the FB double bonds by an action of H_2PtCl_6. The other method includes bromination (one bromine atom per six to seven C atoms) and following alkaline saponification to form reactive –OH groups [2, 3]. After this technique a double bonds are absent and oxygen does not adsorbed (Fig. 3).

The content of Pt which could be supported on FB amounts to 60 wt%.

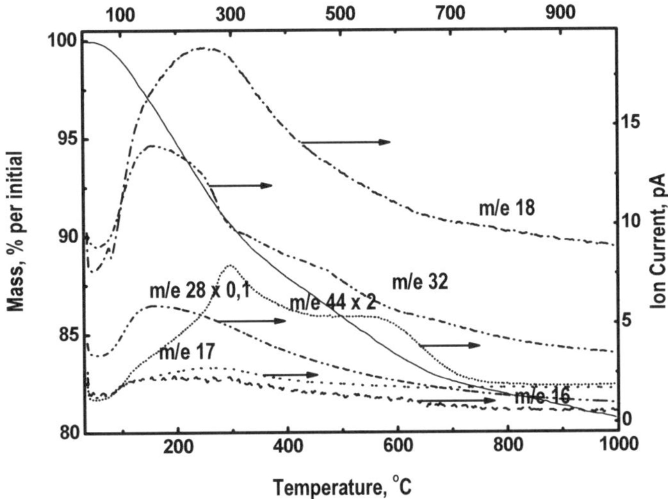

Figure 2. Thermogramm combined with mass spectrum analysis of FB in Ar flow

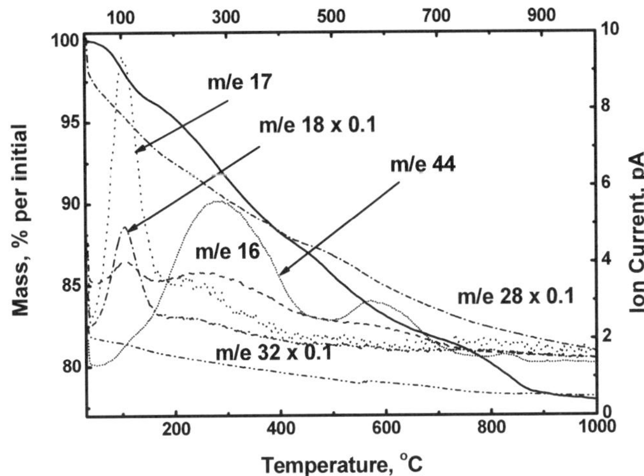

Figure 3. Thermogramm combined with mass spectrum analysis of FB-OH in Ar flow

Pt supported on CNM catalyzes hydrogenation of -C=C- bonds and nitroaromatics. The hydrogenation was carried out at the temperature of 45°C and hydrogen atmospheric pressure in a temperature-controlled shaker (600 min^{-1}) with H_2 uptake registration. Decene-1 was hydrogenated in bulk or as isopropanol solution. Nitrobenzene was used as a 10% isopropanol solution.

In all experiments the catalyst mass were 0.03 g and the substrate/catalyst molar ratios were more than 100 in all cases.

Figure 4 shows that during the hydrogenation of a terminal alkene such as decene-1 the activity of all catalysts based on CNM exceeds that of the conventional Pt/C catalyst based on the active charcoal AG-3, which has the specific surface of three to five times larger than the one of the CNM.

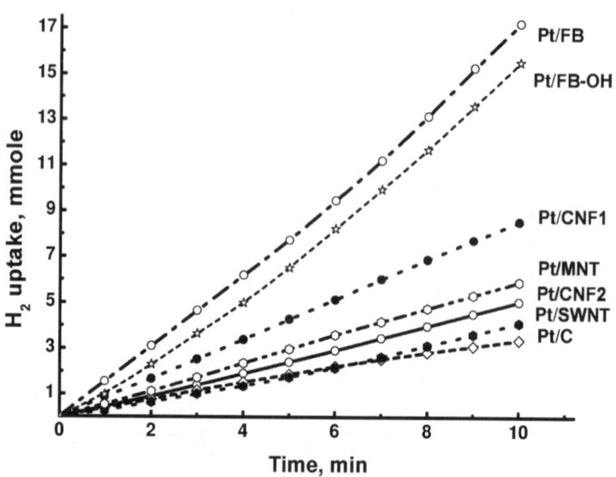

Figure 4. Hydrogen uptake during the hydrogenation of decene-1 at 45°C over catalysts based on CNM coated by 10 wt% of Pt

During the hydrogenation of nitrobenzene the activity of all catalysts based on CNM exceeds that of conventional Pt/C catalyst (Fig. 5). Alkene was hydrogenated more readily than nitrobenzene. The catalysts based on FB have high activity though the electroconductivity of this carbon support is low.

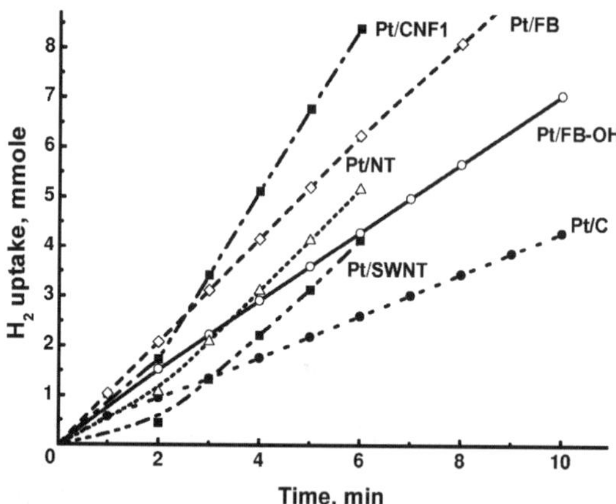

Figure 5. Hydrogen uptake during the hydrogenation of nitrobenzene at 45°C over catalysts based on CNM coated by 10 wt% of Pt

The rate of the decene-1 hydrogenation in bulk is lower than that of the reaction in solution whereas the initial concentration of decene-1 in bulk (~5.07 M) is surely higher. Thus, during the hydrogenation in bulk the specific activities of the catalysts 5.25% Pt/FB, 12.82% Pt/FB, and 10% Pt/C were (in mol/Pt mol min) 94.9, 77.7, and 37.4 correspondingly. At the same time, during the hydrogenation of decene-1 in isopropanol at the initial concentration of 1.014 M the specific activities of both

12.82% Pt/FB and 12.8% Pt/FB-OH are equal to 83.4 mol/Pt mol min. In addition, during the hydrogenation of decene-1 in bulk the specific activity of the catalyst decreases dramatically from 77.7 to 24.2 mol/Pt mol·min after the H_2 uptake of about 20% of its rated amount. This fact was not observing during the hydrogenation of decene-1 in an isopropanol solution.

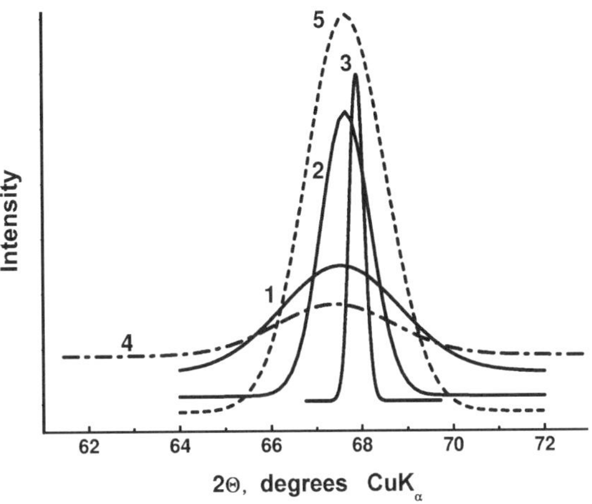

Figure 6. Fitted profiles of the diffraction line Pt 220 for 10% Pt/FB (1), Pt/FB-OH (2), Pt/C (3), 5% Pt/FB (4) и 10 % Pt/CNF1 (5) catalysts

The minimal size of Pt clusters reached 3.5–4 nm on the Pt/FB catalyst as calculated using the width of the diffraction line 220 of the Pt phase (Fig. 6). On the other CNM the minimal sizes of Pt clusters were 6–7 nm. For comparison, the size of Pt clusters on the conventional Pt/C catalyst is more and equal to 60–70 nm. This fact affects the catalytic activity though both reactions are insensitive toward the structure. Our earlier results evidence that the size of Pt clusters affects the catalytic activity greater than both the electroconductivity of support and the surface area of catalyst [3].

Hydrogenation in solvents is carried out easier than in bulk (Fig. 7).

On CNM with graphene structures having low oxygen-containing groups high Pt content can not be supported. So, H_2 uptake rate is not proportional to Pt content after of 19 wt% (Fig. 8).

To summarize, Pt nanoclusters supported on carbon nanomaterials are active, effective and stable hydrogenation catalysts. Fullerene black is preferable support for deposition of Pt clusters though its electroconductivity is lower then for CNM with graphene structures. The particle size of Pt clusters on FB is lowest among of CNM studied. For CNM with graphene structures the rate of H_2 uptake is not proportional to Pt content when Pt content is more then 19 wt%. The particle size of Pt clusters influences on catalytic activity in hydrogenation larger than electroconductivity and specific surface of supports.

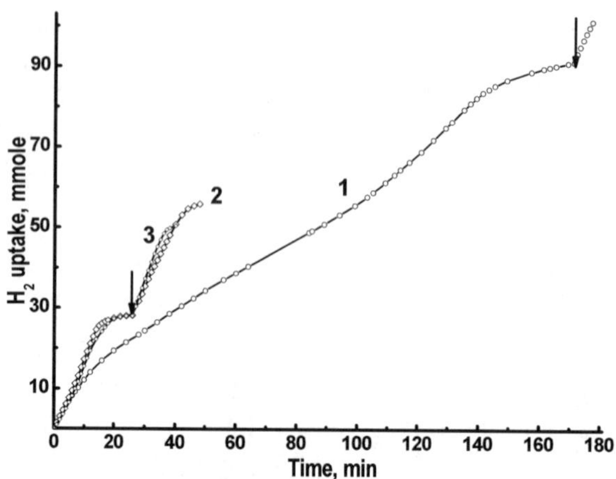

Figure 7. Hydrogen uptake during the decene-1 hydrogenation in bulk (1) and in propanol-2 media (2, 3) at 45°C over catalyst 12.81% Pt/FB. Arrows indicates time of decene-1 introducing

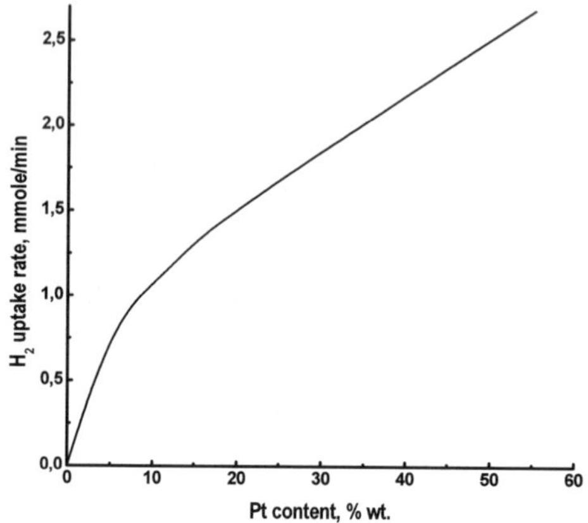

Figure 8. Relation H_2 uptake rate vs Pt content on NF 2 in decene-1 hydrogenation in the bulk at 45°C

References

1. Kuyunko N.S., Kushch S.D., Muradyan V.E., Volodin A.A., Torbov V.I., Tarasov B.P. Pt nanoclusters on carbon nanomaterials for hydrogen fuel cells. T.N. Veziroglu et al. (eds.). *Hydrogen Materials Science and Chemistry of Carbon Nanomaterials*, NATO Security Through Science. Series A. Chemistry and Biology. Springer Science + Business Media BV, Dordrecht, 2006, P. 205–211.
2. Kushch S.D., Fursikov P.V., Kuyunko N.S., Kulikov A.V., Savchenko V.I. Fullerene Black: Relationship between catalytic activity in n-alkanes dehydrocyclization and reactivity in oxidation, bromination and hydrogenolysis, *Eurasian Chem.-Techn. J.*, Almaty, Kazakhstan, 2001 **3**(2): 131–139.
3. Kuyunko N.S., Kushch S.D., Burlakova T.V., Tarasov B.P. Pt nanoclusters on X-ray amorphous carbon as effective hydrogenation catalysts, *Carbon Nanostructures*. Institute of Heat and Mass Exchange BAS, Minsk, 2006, P. 306–311 (in Russian).

ELECTROCHEMICAL CHARACTERIZATION OF Ni-Mo AND Ni-Mo + PE_{Ni} COATINGS AS MATERIALS FOR HYDROGEN EVOLUTION

JOLANTA NIEDBAŁA
Institute of Materials Science, University of Silesia
40-007 Katowice, Bankowa 12, Poland
E-mail: jniedbal@us.edu.pl

Abstract. The Ni-Mo + PE_{Ni} coatings were prepared from a galvanic nickel-molybdenum bath containing a suspension of polyethylene subjected to preliminary nickel plating (PE_{Ni}). Surface morphology was examined using scanning electron microscope. Electrocatalytic properties for the hydrogen evolution reaction (HER) were investigated on polyethylene-modified Ni-Mo coatings in 5 M KOH, using steady-state polarization and electrochemical impedance spectroscopy methods. Based on the obtained results parameters of hydrogen evolution process have been determined. The value of a rate constant is almost five times higher for the coating with polymer than for the Ni-Mo alloy. Therefore, it was found that the process proceeds better on the coatings containing an embedded PE_{Ni} than on the Ni-Mo coatings. Thus obtained coatings may be useful in application as electrode materials for the hydrogen evolution reaction.

Keywords: electrodeposition, composite coatings, polyethylene, Ni-Mo matrix

1. Introduction

The production of hydrogen, one of the most promising candidates for "the fuel of the future" [1, 2], by the electrolysis of water represents an ecologically clean and sustainable technology whose essential problem is an effective, durable and economical electrocatalyst. Nickel alloys with an addition of molybdenum are used as electrode materials and are characterised by catalytic properties during various electrochemical processes, including hydrogen evolution process [3–8]. In order to improve their catalytic properties suitable modifiers are being introduced or heat treatment is applied [9–12]. The most frequently used modifiers of the alloy coatings are metal powders or oxides [13, 14]. It is interesting, therefore, to investigate an effect of co-deposited polymer on modification of the properties of the alloy coating. The composite coatings containing a polymer embedded into metallic matrix were prepared by electrodeposition in a galvanic bath and used for different purpose [15–20]. It was found that this addition improved some properties of the obtained coatings, e.g. the nickel coatings exhibited higher microhardness compared to those prepared using pure nickel [17].

In this paper was aimed to investigate the role of polyethylene as a modifier embedded into the Ni-Mo matrix in the process of hydrogen evolution in an alkaline environment.

2. Experimental

The Ni-Mo + PE composite coatings were prepared from electroplating bath with the following composition (mol/dm^3) [6, 21]: Na_2MoO_4 0.035; $NiSO_4$ 0.75; $Na_3C_6H_5O_7$

B. Baranowski et al. (eds.), *Carbon Nanomaterials in Clean Energy Hydrogen Systems.*

0.45, in which 10 g/dm^3 of polyethylene powder was dispergated. The powder was preliminarily chemically nickel plated (PE_{Ni}).The Ni-Mo + PE_{Ni} coatings were deposited under galvanostatic conditions j = 50, 100, 150, 200 and 300 mA/cm^2. Electrochemical tests of hydrogen evolution in an alkaline environment (5 M KOH) were performed on the obtained Ni-Mo + PE_{Ni} and on the Ni-Mo coatings.

Investigations of electrolytic hydrogen evolution on the prepared coatings were conducted in a three-electrode cell, using steady-state polarization and electrochemical impedance spectroscopy (EIS) methods. These measurements were carried out in 5 M KOH solution, at the temperature of 293 K, using Autolab® electrochemical system. The auxiliary electrode was a platinum mesh and the reference electrode was of the type Hg/HgO/6 M KOH. The exposed geometric surface area of the working electrode was 1 cm^2. Steady-sate polarization measurements as well as EIS experiments were conducted using Autolab®/PGSTAT20 with Frequency Response Analyser (FRA) and Differential Electrometer Amplifier (Eco Chemie B.V. the Netherlands), combined with one of the software packages and controlled by an IBM compatible PC. Prior to the electrochemical experiments the tested electrodes were activated at a constant cathodic current of 320 mA cm^{-2} for 24 h.

Steady-state polarization curves were registered galvanostatically. For this purpose a constant current density ranging between 320 mA cm^{-2} and 0.1 μA cm^{-2} was applied to the electrode and after 45 s the corresponding steady-state overpotential value was registered. Basing on recorded $\Delta E = f(\log j)$ dependences the kinetic parameters for this process (i.e. Tafel slope b, exchange current density j_0 and overpotential at the current density of 100 mA cm^{-2} – ΔE_{100}) were determined.

The EIS measurements were performed under the potentiostatic control in a frequency range from 10 kHz to 0.1 Hz and was covered with 10 points per decade. At the steady-state, the electrode potentials were corrected for iR_s drop determined by EIS. In the EIS measurements the amplitude of *ac* signal was equal 0.005 V. A The real and imaginary components of the complex plane plots were analysed using a modified version of a complex non-linear least-squares fitting program (CNLS) [22] from which the experimental parameters of an electrical equivalent circuit were determined. Kinetics of hydrogen evolution process was determined on basis of the dependences of both the inverse the charge-transfer resistance, R_{ct}, and the total current density, j, registered in steady-state polarization measurements, on overpotential ΔE. The average rate constants ($k_{av.} = 1/(1/k_1 + 1/k_2)$) were evaluated using the NLS method [5].

Surface morphology of the coatings was examined using JSM-648 electron microscope. The Micro-Scan computer program and a system for morphometric measurements composed of the NIKKON SMZ-2T stereoscopic microscope and Panasonic GT-KR22E type colour camera were used for the assessment of polyethylene content in a coating. The contents of Ni and Mo were determined by the X-ray fluorescent spectrometry using TUR-M-62 diffractometer with suitable attachment.

3. Discussion of results

The obtained Ni-Mo + PE_{Ni} coatings adhered well to the substrate, and were grey, matt and velvety, with visible particles of an embedded polymer covered with nickel coating. Surface morphology of the Ni-Mo + PE_{Ni} and Ni-Mo coatings is shown in Fig. 1. It was found that structure of the Ni-Mo coatings (Fig. 1(a)) is relatively regular of a island character. When polymer is embedded into the Ni-Mo matrix, its structure changes and is no more island.

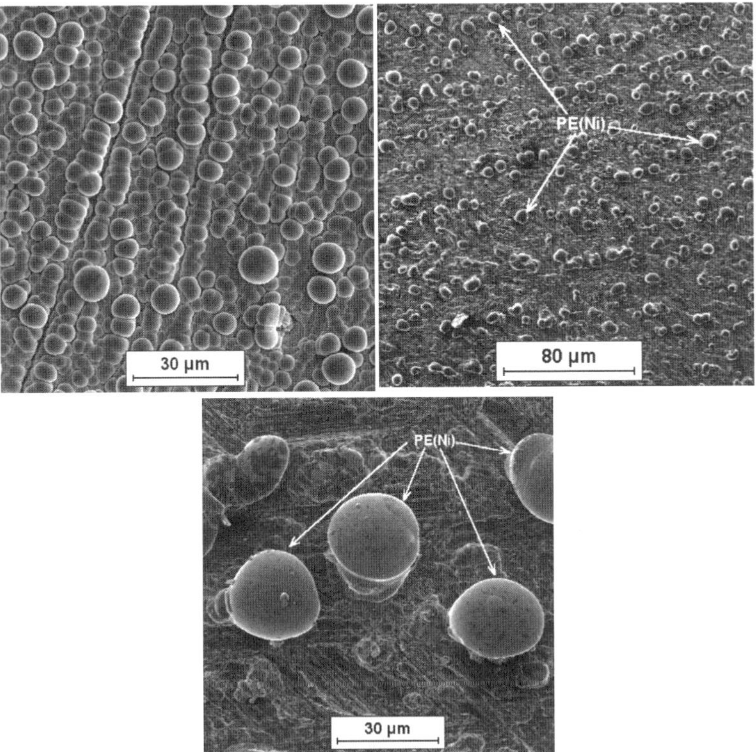

Figure 1. Surface morphology of the alloy matrix and of composite coatings with an embedded polymer: (a) Ni-Mo matrix (j = 150 mA/cm^2), (b), (c) Ni-Mo + 27.2 vol% PE_{Ni} (j = 50 mA/cm^2)

Morphometric measurements were made in order to determine polyethylene content in the Ni-Mo + PE_{Ni} coatings. They showed that it ranged from about 5–27% (Table 1). It was also found that an increase of deposition current density resulted in the decrease of polyethylene content in the coating, which was particularly high at the densities of 200 and 300 mA/cm^2. This effect can result from intensive co-evolution of hydrogen, which prevents polymer from accessing an electrode surface and its embedment.

TABLE 1. Chemical composition of the Ni-Mo + PE_{Ni} coatings, ant the exchange current values determined for the hydrogen evolution process

Current density during coating deposition (j/[mA/cm^2])	Ni-Mo + PE_{Ni} coatings		
	α phase		PE_{Ni}, (vol%)
	Ni, (%)	Mo, (%)	
50	74.7 ± 2.2	25.3 ± 0.8	27.2 ± 1.4
100	74.2 ± 2.2	25.8 ± 0.6	24.2 ± 1.2
150	72.2 ± 2.1	27.8 ± 0.8	17.2 ± 0.9
200	75.4 ± 2.3	24.5 ± 0.8	9.9 ± 0.5
300	83.2 ± 2.4	16.8 ± 0.5	4.7 ± 0.3

The behaviour of thus obtained coatings was studied in the process of hydrogen evolution from 5 M KOH solution. The obtained results were used to estimate the ability of a given materials to facilitate hydrogen evolution in an alkaline solution. The research were conducting on the coating with maximum PE_{Ni} – Ni-Mo + 27.2% PE and by the comparison on the Ni-29.7% Mo (the higher kinetic parameters of hydrogen evolution process for Ni-Mo coatings [8]). The steady-state polarization curves obtained on different electrodes are presented in Fig. 2, and the kinetic parameters are presented in Table 2.

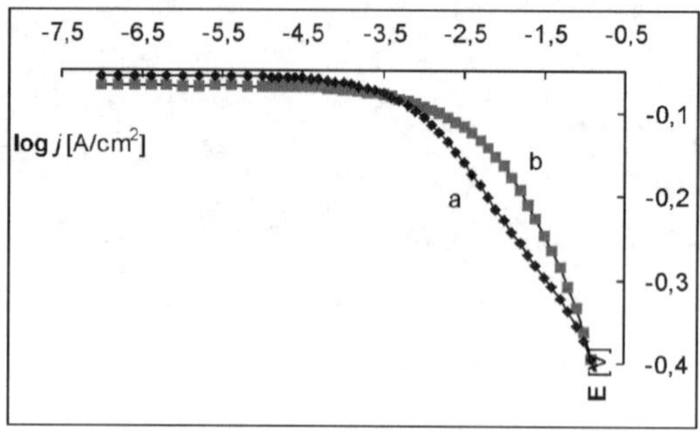

Figure 2. The steady-state polarization curves obtained for HER in 5 M KOH on obtained coatings: (a) Ni-Mo matrix, (b) Ni-Mo + 27.2 vol% PE_{Ni}

TABLE 2. Kinetic parameters of the HER in dependence from the type of coating

Type of coating	B (mV dec^{-1})	j_0 (A cm^{-2})	ΔE_{100} (V)
Ni-Mo	0.116	$3.4 \cdot 10^{-4}$	-0.283
Ni-Mo + 27.2% PE	0.126	$11.5 \cdot 10^{-4}$	-0.242

On the base of steady-state of the hydrogen evolution, the rate of this process was estimated depending on the type of coatings. It was found, that in the Ni-Mo + PE_{Ni} coatings the increase in the rate of hydrogen evolution was observed. The value of the Tafel slope for the tested electrodes change between -116 and -126 mV dec^{-1} and are almost the same like the theoretical value -120 mV dec^{-1}. It suggest a Volmer-Heyrovsky mechanism for HER (Table 2). The value of hydrogen evolution overpotential at the current density of 100 mA cm^{-2} – ΔE_{100} is lower for the Ni-Mo + PE_{Ni} coatings than for the Ni-Mo coatings and value of exchange current density – j_0 for the Ni-Mo + PE_{Ni} coatings is higher than for the NiMo coatings. It suggests an influence of the PE_{Ni} presence in the Ni-Mo matrix on the rate of hydrogen evolution.

The complex impedance was analysed using a modified version CNLS fitting program [22]. Character of dependence of imaginary impedance component on electrode overpotential and real impedance component is show in the three-dimensional diagrams (3D): $-Z_{im}$ = f (Z_{re}, ΔE) (Fig. 3). These diagrams show surface topography of mutual influence of real impedance component and electrode overpotential on values of imaginary impedance component, which are lower for the Ni-Mo + PE_{Ni} coatings than for the Ni-Mo coatings.

With the increase of overpotential of hydrogen evolution, the impedance spectra for the Ni-Mo coating assume a form of semi-circles, the radius of which diminishes with overpotential decrease. Existence of a single semi-circle has been found over the whole frequency range, which proves that the process of hydrogen evolution is controlled by the rate of charge transfer. It has been found, that the *ac* impedance behaviour of obtained coatings may be well describes using the one-CPE electrode model. This is typical model for the HER on solid, rough or porous electrodes. Such equivalent circuit is characteristic of electrode materials composed of cylindrical pores of radius *r* and length *l*. As was shown in a paper [30] for short and wide pores l^2/r is very small and only one semicircle on the complex plane plot was observed (model CPE1).

In the case of Ni-Mo coatings with embedded PE particles, at the potentials below 100 mV a semi-circle is observed at the high frequency range confirming that the process is controlled by the charge transfer rate, similarly as in the Ni-Mo alloy coatings. At low frequencies the dependence is linear, indicating that the charge transfer resistance is negligibly low and the process is affected mainly by the diffusion of depolarizer to the electrode surface. With the increase of overpotential, this effect of diffusion becomes increasingly smaller, and over 100 mV fully disappears. At the overpotentials higher than 100 mV the character of impedance spectra are the same as those for the Ni-Mo coatings. Perhaps at the small value of overpotential occur the mask of the electrode surface connected with embedded PE_{Ni} in Ni-Mo matrix.

In case of the Ni-Mo + PE coatings, however, reduction of a radius of recorded semi-circles proceeds faster than in the coatings from the Ni-Mo alloy (Fig. 3), which indicates that the values of R_{ct}. resistance are lower (by about four times).

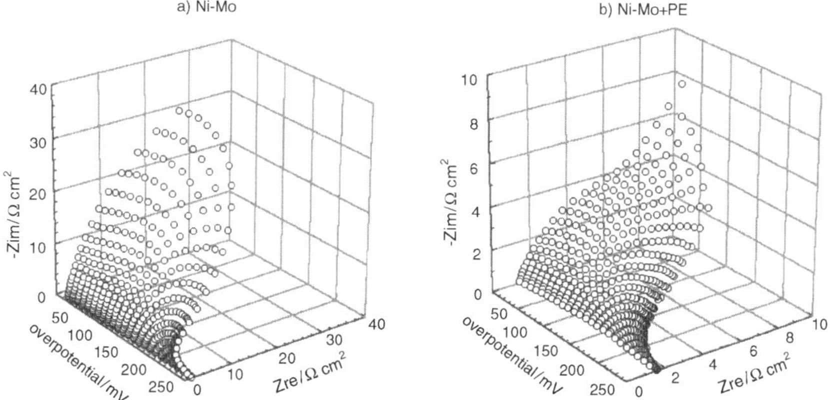

Figure 3. Impedance diagrams illustrating the changes of $-Z_{im} = f(Z_{re})$ in dependence on the potential change, obtained for the coatings: (a) Ni-23.2% Mo (j = 150 mA/cm²), (b) Ni-25.3% Mo + 27.2 vol% PE_{Ni} (j = 50 mA/cm²)

The impedance of CPE element is given as $Z_{CPE} = 1/(Tj\omega)^{\phi}$, where ω is the angular frequency of *ac* voltage. As a result of approximation of experimental data in the case of CPE1 model the following parameters could be obtained: R_{ct}, R_s, T, ϕ. The double-layer capacitance – C_{dl} was calculated according to [22] $T = C_{dl}^{\phi}\ (1/R_s + 1/R_{ct})^{1-\phi}$, and in consequence roughness factor R_f, in order to characterize the influence of the surface of composite coatings on the hydrogen evolution reaction. Assuming that the average double-layer capacitance of a smooth metal surface is 20 μF cm^{-2} [22], the surface roughness factor R_f was determined. The parameters of the CPE1 model after approximation of experimental data for the overpotential ΔE = 100 mV and for the Ni-Mo and Ni-Mo + PE coatings are presented in Table 2. These parameters have been determined with an error of 3–7%.

At overpotential equal ΔE = 100 mV the value of charge transfer resistance R_{ct} for the Ni-Mo-PE coating is about 5 times smaller than that for the Ni-Mo alloy (Table 3), and is over 50 times smaller than for pure Ni, for which R_{ct} = 497 Ω cm^2 [11]. Comparison of the parameters shows that the polymer present in a Ni-Mo matrix results in a decrease of double layer capacitance and the value of R_f. It can be expected, therefore, that the embedded polymer shields locally the electrode surface.

TABLE 3. Parameters of the CPE1 model after approximation of experimental data for the overpotential ΔE = 100 mV for the Ni-Mo + PE coatings

Coating type	R_{ct} (Ω cm^2)	C_{dl} (F cm^{-2})	R_f	$k_{av.} \cdot 10^9$ (mol cm^{-2} s^{-1})
Ni-Mo	41.1	0.0097	526	1.56 ± 0.07
Ni-Mo + PE_{Ni}	8.5	0.0080	423	7.64 ± 0.38

In an alkaline electrolyte solution, hydrogen atoms are generated by the reduction of water molecules (discharge of hydrogen from water molecule): $2H_2O + 2e^- \leftrightarrow H_2 + 2OH^-$. It is generally accepted that the HER mechanisms is controlled by the following reactions [22–31]:

$$H_2O + M + e^- \leftrightarrow MH + OH^- , \tag{1}$$
$$H_2O + MH + e^- \leftrightarrow H_2 + M + OH^-, \tag{2}$$
$$2MH \leftrightarrow H_2 + 2M , \tag{3}$$

where M and MH denote the metal and the metal-hydrogen adsorbed on the electrode surface, respectively. The reaction (1) involves the formation of adsorbed hydrogen (Volmer reaction) followed by the electrochemical desorption (Heyrovsky reaction, according to Eq. (2)) and/or by the chemical desorption (Tafel reaction, according to Eq. (3)). The promote this reaction step, electron transfer must occur between the conducting alloy (metal) substrate and the reacting solution spices. On the surface Ni-Mo - PE_{Ni} coatings are presence micro galvanic cell. One electrode – Ni-Mo solid solution on a high activity and the second electrode – Ni (on a PE particles) on a low activity. The presence of this cells on the all surface of the coating results presumably in directing electrochemical process towards more activity areas on the electrode surface thus activating them and improving kinetics of hydrogen evolution on the Ni-Mo + PE coatings. In the absence these micro cells (when process coming on the surface Ni-Mo coating) the mechanism HER is other and activity this coating could be lower.

Based on the analysis of $\log(1/R_{ct})$ dependence on overpotential for the Ni-Mo and Ni-Mo + PE_{Ni} coatings, an effect of charge transfer resistance on the hydrogen evolution process has been characterised (Fig. 4). In both cases, overpotential increase results in the increase of the $\log(1/R_{ct})$ values. However, over the whole range of overpotentials studied the values of are higher for the Ni-Mo + PE coatings than for the Ni-Mo alloy. It proves, that the value of charge transfer resistance R_{ct} for Ni-Mo + PE is smaller. The R_{ct} resistance is closely related with the rate of processes taking place on the electrode. Small values of R_{ct} indicate that the kinetic processes on an electrode proceed rapidly.

Based on the linear dependence $\Delta E = f(\log(1/R_{ct}))$ in the range of medium overpotentials (Fig. 4), the values of the k_{av} were calculated similarly as in other works [5, 23–26, 28]. An approximation of the values of $1/R_{ct}$ in function of ΔE and j registered from the steady-state polarization in function of ΔE, was carried out by adjusting the average rate constant, $k_{av.}$, and are presented in Table 2. This approximation was carried out at more negative potentials where the reverse reactions could be neglected (k_{-1} and k_{-2} much lower than k_1 and k_2) and only the average rate

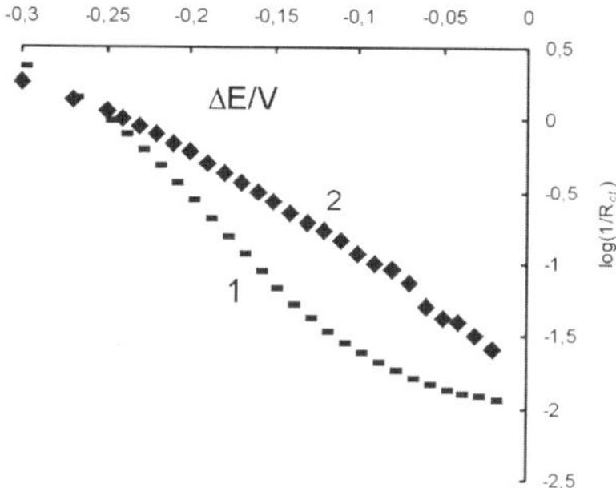

Figure 4. Dependence of the log(1/R$_{ct}$) on overpotential for the Ni-Mo coatings (1) and Ni-Mo + PE coatings (2)

constant ($k_{av.} = 1/(1/k_1 + 1/k_2)$) could be determined [24]. In this papers it was found, that the examined HER process is controlled by the Volmer-Heyrovsky mechanism for the reciprocal of charge transfer resistance one can write:

$$\frac{1}{R_{ct}} = -F[\frac{\partial(v_1 + v_2)}{\partial\eta}]_\theta = \frac{2\beta F^2}{RT} k_{av} \exp(-\beta f\eta) \; , \qquad (4)$$

where $k_{av} = k_1k_2/(k_1 + k_2)$ has a sense of the mean average reaction rate constant. Using this equation one can determine the rate constant k_{av} of HER process (Table 3).

This value of a rate constant for the Ni-Mo + PE coating is almost five times higher than for the Ni-Mo coatings Therefore, it was found that in spite of lower capacitance of a double layer and smaller value of R_f, shielding of the electrode surface by an embedded polymer does not affect kinetics of the process. It is clear that Ni-Mo + PE coatings have the higher apparent towards hydrogen evolution reaction. The presence of the polymer particles with Ni on the surface, embedded into the coating, results presumably in directing electrochemical process towards more electro-activity areas on the electrode surface thus activating them and improving kinetics of hydrogen evolution on the Ni-Mo + PE coatings.

4. Conclusions

Ni-Mo + PE coatings have the higher apparent towards hydrogen evolution reaction in comparison with Ni-Mo. This is evidenced by higher values of an exchange current j_0

and k_{av}. Lover overpotential value (ΔE_{100}) at the current density of -100 mA cm^{-2} and lower value of the charge transfer resistance R_{ct} for the coatings with PE pointed on influence of polymer at hydrogen evolution process. It was concluded that embedding the polymer into the coating results in local shielding of the electrode surface and directing the electrochemical process to the free areas thus improving kinetics of hydrogen evolution process.

Acknowledgements

This work was supported by the State Committee for Scientific Research (KBN) under the project No PKZ/KBN/013/T08/10.

References

1. S. Dunn, Int. *J. Hydrogen Energy*, 2002, **27**: 235.
2. I. Dincer, Int. *J. Hydrogen Energy*, 2002, **27**: 265.
3. E. Bełtowska-Lehman, E. Chassaing, *J. Appl. Electrochem.*, 1997, **5**(27): 568–572.
4. D. Landolt, E.J. Podlaha, N. Zech, *Zeitschrift fur Physikalische Chemie* 1999, **208**(1–2): 167–182.
5. R. Karimi-Shervedani, A. Lasia, *J. Electrochem. Soc.*, 1998, **145**: 2219.
6. J. Niedbała, *Mater. Sci. Forum*, 2006, **514–516**: 465.
7. J. Niedbała, *Acta Metall. Slov.*, 2005, **2**(11): 166.
8. J. Niedbała, M. Popczyk, J. Panek, A. Budniok, E. Łągiewka, *Inżynieria Materiałowa*, 2005, **1**: 31.
9. D.M. Lenz, M. Delamar, C.A. Ferreira, *J. Electroanal. Chem.*, 2003, **540**: 35–44.
10. J.O. Iroh, K. Levine, *Eur. Polym. J.*, 2002, **38**: 1547–1550.
11. S. Sotiropoulos, I.J. Brown, G. Akay, E. Lester, *Mat. Lett.*, 1998, **35**: 383–391.
12. P.J. Kulesza, K. Miecznikowski, M.A. Malik, M. Galkowski, M. Chojak, K. Caban, A. Wieckowski, *Electrochim. Acta*, 2001, **46**: 4065–4073.
13. J. Niedbała, J. Panek, A. Budniok, E. Łągiewka, *Kompozyty*(Composites), 2004, **4**(9): 83–87.
14. J. Kubisztal, A. Budniok, J. Niedbała, *Kompozyty* (Composites), 2004, **4**(11): 293–296.
15. B. Bozzini, I. Sgura, *J. Appl. Electrochem.*, 2006, **36**: 983–989.
16. B. Bozzini, C. Mele, L. D'Urzo, B. Bozzini, C. Mele, L. D'Urzo, *J. Appl. Electrochem.*, 2006, **36**: 87.
17. Z.A. Hamid, J.M. Ghayad, *Mat. Lett.*, 2002, **53**: 238–243.
18. F. Wang, S. Arai, M. Endo, *Mat. Trans.*, 2004, **45**(4): 1311–1316.
19. Y. Liu, Q. Zhao, *Plat. Surf. Finish.*, 2004, **91**(4): 48–51.
20. J. Niedbała, A. Budniok, E. Łągiewka, *Mater. Sci. Forum*, 2006, **514–516**: 1176.
21. J. Niedbała, K. Wykpis, A. Budniok, E. Łągiewka, *Archiwum Nauki o Materiałach,* 2002, **2**: 123.
22. JR. Macdonald, Impedance Spectroscopy. Wiley, New York, 1987.
23. R. Karimi-Shervedani, A. Lasia, *J. Appl. Electrochem.*, 1999, **29**: 982.
24. C. Hitz, A. Lasia, *J. Electroanal. Chem.*, 2001, **500**: 213.
25. L. Birry, A. Lasia, *J. Appl. Electrochem.*, 2004, **34**: 735.
26. S. Rodrigues, N. Munichandraiah, A.K. Shukla, *Bull. Mater. Sci.*, 2000, **23**: 383.
27. B. Mahdavi, D. Miousse, J. Fournier, H. Ménard, J. Lessard, *Can. J. Chem.*, 1996, **74**: 380–388.
28. R. Bocutti, M.J. Saeki, A.O. Florentino, C.L.F. Oliveira, A.C.D. Ângelo, *Int. J. Hydrogen Energy*, 2000, **25**: 1051–1058.
29. E. Navarro-Flores, Z. Chong, S. Omanovic, *J. Mol. Catal. A-Chem.*, 2005, **226**: 179–197.
30. A. Lasia, A. Rami, *J. Electroanal. Chem.*, 1990, **294**: 123.
31. A. Lasia, *J. Electroanal Chem.*, 2001, **500**: 30.

ALGORITMS OF PARAMETERS ESTIMATION OF HYDROGEN PERMEABILITY MODEL

YU.V. ZAIKA, E.P. BORMATOVA
Institute of Applied Mathematical Research of Karelian Research Centre of Russian Academy of Sciences,
185910 Petrozavodsk, Karelia, Russia

Abstract. The inverse problem of the parametric identification in the nonlinear model of hydrogen permeability is considered. A few algorithms for estimation of the transport parameters (sorption, desorption and diffusion) for different amount of experimental data are presented.

Keywords: hydrogen permeability, mathematical modeling, parameters estimation

1. Introduction

The experimental method of permeability is a classical method and it is described in detail, for example, in [1, 2]. The corresponding boundary-value problems are mathematically justified and investigated. At the same time there is a certain dispersion in estimations of the parameter values of the models. One of the basic causes is: the inverse problems of mathematical physics are characterized by high sensitivity to experimental and computation errors. Besides, the traditional 'fitting method' usually does not guarantee the uniqueness of the set of parameters that acceptably approximates the experimental permeability curves. Therefore different (even wrong) qualitative conclusions and practice recommendations can follow from modelling out of the experimental area. So, firstly, it is necessary to define the amount of data sufficient for unique parametric identification of the model. Secondary, the algorithm for the parameters estimation must be well-defined (in particular, it must not involve differentiation of noisy measurements).

We'll keep in mind the following version of the experiment. A membrane thickness l from investigated material (the metal or alloy) is a vacuum vessel barrier. The temperature of the sample is fixed. At the inlet side the pressure of molecular hydrogen $\bar{p}_0 = const$ is stepwisely created. At the outlet side vacuum is maintained. The outlet desorption flux $J = J(t)$ (t is for the time) is measured by mass-spectrometer. We mean the fluxes per unit of area, i. e. we talk about densities. To be exact at the outlet side the pressure is measured:

$$p_l(t) = \theta_1 \int_0^t \exp\{(\tau - t)\,\theta_0^{-1}\} J(\tau)\, d\tau, \qquad J(t) = \left[\dot{p}_l(t) + p_l(t)\theta_0^{-1}\right]\theta_1^{-1},$$

where θ_i are the characteristics of vacuum apparatus. To determine J using p_l is an inverse problem, which we omit (see, for example, [3]). When stationary state

B. Baranowski et al. (eds.), *Carbon Nanomaterials in Clean Energy Hydrogen Systems.*

$J(t) \approx \bar{J} = const$, $t \geq t_*$, is arrived, the pressure at inlet jumps to value $\bar{p}_0^+ > \bar{p}_0$ and we wait the next stationary state $t \geq t^* > t_*$. Such experiment is preferable than two classical experiments, because there is no need in repeated degassing of the sample and a second phase starts not from the null initial atomic hydrogen distribution, but from the preceding steady state (the diversity gives more information). Note that the steady state has an asymptotical nature. But t_* and t^* should not be too large so that transient phenomena 'not lost' against the stationary processes.

The parameters dependence on the temperature is assumed to be Arrhenius type. For example, for the diffusion coefficient $D = D_0 \exp\{-E_D/[RT]\}$. Theoretically, two values at the different temperatures are sufficient for determination the pre-exponential factor D_0 and activation energy E_D. But it's desirable to have more due to statistical considerations. If it's necessary, other dependences on T can be considered; this is not important in the sequel. The pressures and concentrations $c(t,x)$ of dissolved hydrogen are assumed to be relatively small so that D does not depend on c.

As far as we talk about inverse problems and experimental errors are estimated at the best as 10–20%, we keep the number of the coefficients as low as possible. The correct identification of simple models is the initial approximation, if the theoretical analysis of complex systems is difficult and we are restricted by numerical experiments.

2. The mathematical model

Let us consider the following model with bulk desorption [1]:

$$\frac{\partial c}{\partial t} = D\frac{\partial^2 c}{\partial x^2}, \quad (t,x) \in (0,t_*) \times (0,l), \quad c(0,x) = 0, \quad x \in [0,l], \quad P \equiv \mu s \bar{p}_0, \tag{1}$$

$$P - bc_0^2(t) = -D\frac{\partial c}{\partial x}\bigg|_{x=0}, \quad bc_l^2(t) = -D\frac{\partial c}{\partial x}\bigg|_{x=l}, \quad c_0(t) \equiv c(t,0), \quad c_l(t) \equiv c(t,l). \tag{2}$$

Here μ is kinetic constant, s – adhesion coefficient, b – desorption coefficient. The vacuum system is sufficiently strong to neglect resorption: $\mu s p_l \approx 0$. The function $J \equiv J_l(t) = bc_l^2(t)$ is known from the measurements. For the second stage of experiment ($t \in [t_*, t^*]$) the initial distribution $c(t_*, x)$ is stationary and $\bar{p}_0$ is substituted for $\bar{p}_0^+$.

The defects effect. Instead of diffusion equation we have the system:

$$\frac{\partial c}{\partial t} = D\frac{\partial^2 c}{\partial x^2} - a_1 c + a_2 w, \quad \frac{\partial w}{\partial t} = a_1 c - a_2 w.$$

Here $w(t,x)$ is the concentration of hydrogen atoms in traps (in particular, in microvoids), a_i – reversible capture coefficients. If necessary, it is possible to restrict the traps capacity: $a_1 c \to a_1 c[1 - w/w_{\max}]$. Such a rough model can only answer to the question how much hydrogen is inhibited. For the practice it is often enough. The permeability curves have the standard S-shaped form of the saturation curve. The values a_i do not influence on the stationary desorption flux value $\bar{J}$. In case $a_i > 0$ the plot $J(t)$ firstly sags a little (because of accumulation in the defects), and later becomes more convex upwards. So if there is success in coincidence of the model and the experimental levels of $\bar{J}$, then, in case the influence of defects can be seen, it is not difficult to match a_i numerically. Taking $a_i > 0$ into account considerably complicates the mathematical calculations.

Solving the direct problem. Let us briefly discuss the numerical solving of the boundary-value problem. The initial condition and boundary conditions do not agree in the classical sense. Therefore formally the solution is in the weak sence. At the beginning the inlet side is subjected to impact load due to the pressure jump. The plot of the inlet flux $F_0(t) \equiv -Dc_x(t,0)$ looks like a splash with subsequent attenuation till the stationary value $\bar{J}$. This fact and nonlinearity of the problem are the reasons why the computational scheme is implicit and iterative. In numerical modelling the stiffness of the problem was reduced: we modelled fast (linear and within a few seconds) establishment $p_0(t)$ till level $\bar{p}_0$. Firstly, it corresponds to experiment reality. Secondary, it permits to smooth the δ-function effect: a theoretical initial splash F_0 can produce the big integral component, which is not characteristic for the real material with limited capacity. Therefore it's necessary to 'truncate' the peak of F_0 by physically permissible maximum ($\leq P$), this also accelerates the establishment $c_0(t) \to \bar{c}_0$. We emphasize that information about initial process is not analyzed in identification algorithms, only integral type relations are used.

The difference scheme (Crank-Nicolson method) has six-point template. In standard notations $c_i^k \approx c(t_k, x_i)$, $t_0 = 0$, $x_0 = 0$, $x_n = l$, $\Delta x = h$. At each time level we approximate $c_x(t_k, 0) \approx [-3c_0^k + 4c_1^k - c_2^k]/2h$. Substituting it to boundary condition (2) with $x = 0$, we obtain the expression $c_0^k = f_0(c_1^k, c_2^k)$. In the same way we get $c_n^k = f_n(c_{n-2}^k, c_{n-1}^k)$. At the preliminary time level all $c_i^k = 0$, $k = -1$. When $k = 0$ at first we set $c_i^0 = 0$, $1 \leq i \leq n-1$, $c_0^0 = f_0(c_1^0, c_2^0)$, $c_n^0 = f_n(c_{n-2}^0, c_{n-1}^0)$. Then the tridiagonal linear equation system is solved by sweep method. We get new values c_i^0, $1 \leq i \leq n-1$. Again $c_0^0 = f_0$, $c_n^0 = f_n$ are computed and the system is solved. We proceed iterations till the boundary values established. At the next time levels values c_0^{k-1} and c_n^{k-1} are used as initial approximates of c_0^k and c_n^k. Input perturbation reaches the outlet side during finite time. Discrete approximation 'smoothes down' the defects of model (1)–(2) and is adapted to real experiment conditions. A computation check was done by reaching the linear stationary state (4) of the distribution $c(t,x)$ together with keeping balance:

$$\int_0^l c(t,x)\,dx = -\int_0^t Dc_x(t,0)\,dt + \int_0^t Dc_x(t,l)\,dt. \tag{3}$$

For the second stage constructions are similar. Numerical experiments were performed in wide range: D from 10^{-9} to 10^{-3} cm^2/s; b from 10^{-21} to 10^{-5} cm^4/s; s from 10^{-12} to 10^{-2}; $\bar{p}_0$ from 0.1 to 10 $Torr$; $\mu = 1.46\cdot 10^{21}$ $1/cm^2 sTorr$. Scaling was made because of spread orders of the values. Parameters, generating permeability curves, were 'forgotten' and then were regenerated by the described below method.

3. Equilibrium and stationary state

The equilibrium pair ($\bar{p}, \bar{c}$) is defined by equating to zero the derivatives:

$$\mu s\bar{p} = b\bar{c}^2 \Rightarrow \bar{c} = \gamma\sqrt{\bar{p}}\,,\, \gamma \equiv \sqrt{\mu s b^{-1}}.$$

By that, the experiment is carried out in region $\bar{c} \sim \sqrt{\bar{p}}$. If the coefficient γ is known, we can take away the parameter $s = b\gamma^2/\mu$ from further considerations.

The stationary distribution of hydrogen atoms is linear:

$$c_t = 0 \Rightarrow c(t_*,x) = -D^{-1}\bar{J}\,x + \bar{c}_0,\; \bar{c}_0 = \sqrt{b^{-1}(P-\bar{J})}. \tag{4}$$

From the relation $\bar{J} = b\bar{c}_l^2$ we derive the stationary value $\bar{c}_l$ and equate it to $c(t_*,l)$:

$$\sqrt{\bar{J}} - \sqrt{P-\bar{J}} + l\sqrt{b}D^{-1}\bar{J} = 0 \quad (P \equiv \mu s\bar{p}_0). \tag{5}$$

From two such equations for pressures $\bar{p}_0$ and $\bar{p}_0^+$ we eliminate μs:

$$\left(\frac{\bar{J}_+^2}{\bar{p}_0^+} - \frac{\bar{J}^2}{\bar{p}_0}\right)z^2 + 2\left(\frac{\bar{J}_+^{3/2}}{\bar{p}_0^+} - \frac{\bar{J}^{3/2}}{\bar{p}_0}\right)z + 2\left(\frac{\bar{J}_+}{\bar{p}_0^+} - \frac{\bar{J}}{\bar{p}_0}\right) = 0, \quad z \equiv \frac{l\sqrt{b}}{D}. \tag{6}$$

Roots have different signs, so the complex $z = l\sqrt{b}/D$ is defined uniquely. Then from equation (5) we define s, calculate $b = \mu s/\gamma^2$ and $D = l\sqrt{b}/z$. It is rational to divide equation (6) to $\bar{J}$ and use the dimensionless variable $z\sqrt{\bar{J}}$.

For this amount of information the parametric identification problem is solved by elementary manipulations. Let us consider the next phase: let us try to economize on saturation-degassing experiments. Furthermore, other experiments are characterized by their own errors. Is it possible to uniquely determine the parameters without preliminary specifying the coefficient of equilibrium dissolubility γ?

4. Delay time

The linear problem. Instead of boundary conditions (2) let us consider the linear boundary conditions:

$$c_0(t) \equiv c(t,0) = \tilde{c}_0 = const, \qquad c_l(t) \equiv c(t,l) = 0. \tag{7}$$

Denote the boundary problem (1), (2) by I, and (1), (7) by II. The plot of the function (it is the amount of atoms H, desorbed from unit area at $x = l$)

$$S_{II}(t) = \int_0^t J_{II}(\tau)\,d\tau\,,\quad J_{II}(t) = F_l(t) \equiv -D\frac{\partial c}{\partial x}\bigg|_l\,,$$

has an asymptote. The cross point of this asymptote with axis t is named the delay time: $t_0 = l^2/6D$ [1]. Equation of the tangent to plot of the function $S(t)$ in the point (t_*, S_*) (we omit an index II at t_*, J, S): $S = S(t_*) + \bar{J}\cdot(t - t_*)$, $\dot{S} = J$. Hence, the value of t_0 is calculated sufficiently accurately as $t_* - S_*/\bar{J}$, $S_* = S(t_*)$. It is important that t_0 does not depend on $\tilde{c}_0$. Usually accepted $c_0(t) = \bar{c}$, where $\bar{c} \sim \sqrt{\bar{p}_0}$ is the equilibrium with pressure $\bar{p}_0$ concentration . The other variant: $\tilde{c}_0$ is equal to the steady state concentration $\bar{c}_0$ in problem I. As far as $P = b\bar{c}^2$, $P - b\bar{c}_0^2 = \bar{J}$, then $\bar{c} > \bar{c}_0$, $\bar{J} = b\bar{c}_l^2 \Rightarrow \bar{c}^2 = \bar{c}_0^2 + \bar{c}_l^2$. Formally the problem II is of the first kind. But if $\tilde{c}_0 = \bar{c}_0 = \sqrt{b^{-1}(P - \bar{J})}$ it is connected indirectly with nonlinear problem I that is of the third kind. When flux of atoms to inlet side exceeds significantly the membrane capacity ($P >> \bar{J}$), we have $\bar{c}_0 \approx \bar{c}$. Finally, it follows from the formula (4) that $\bar{c}_0 = \bar{c}_l + \bar{J}lD^{-1}$, and if $\bar{c}_l << \bar{c}_0$, then $\bar{c}_0 \approx \bar{J}lD^{-1}$ (in the problem II $\tilde{c}_0 = \bar{J}lD^{-1}$).

<u>Note 1</u>. Numerical experiments in wide region of parameters show that the value of diffusion coefficient weakly affects to accuracy of approximation $\bar{c}_0 \approx \bar{J}lD^{-1}$. The parameter bP is determinant – the less it is, the less the models I and II have in common ($\tilde{c}_0 = \bar{c}_0$), surface processes become limiting factor too. Really, if we considerably decrease the incident flux to inlet side $P = \mu s\bar{p}_0$, then permeating flux $\bar{J}$ will be very small and $\bar{c}_l \approx \bar{c}_0$. If we decrease b, then at the outlet side the desorption will be very weak, which promotes the growth of $\bar{c}_l$ ($c_l(t)$ monotonically increases until level $\bar{c}_l$). Inversely, if b is relatively big, then at the outlet side an active desorption decreases the concentration of $\bar{c}_l$, and as result we get the problem II with $c_l(t) = 0$. The growth of bP implies the decreasing of the stabilization time for stationary concentration: $c_0(0) = 0 \to \bar{c}_0$. When the models overlap, (the I degenerates to II), big (but not too big) variations of coefficients s, b practically don't affect the delay time $t_0 \approx l^2/6D$. Here the ill-posedness of the inverse problem appears. In qualitative reasoning (strongly, weakly, practically) authors oriented conditionally to the error up to 10% in the determination of the penetrate flux of hydrogen.

For the second stage ($t \in [t_*, t^*]$) we do the substitution $c(t,x) = c(t,x) - c(t_*,x)$ and take t_* as the zero time. In linear approach II ($\widetilde{c}_0 = \overline{c}_0$) we get the same boundary problem for c, only with $\Delta\overline{c} = \overline{c}_0^+ - \overline{c}_0$ and $\Delta J_{II} = J_{II} - \overline{J}_{II}$ instead of $\overline{c}_0, J_{II}(t)$. The new delay time is calculated by the formula

$$\Delta t_0 = \Delta t_* - \frac{\Delta S_*}{\Delta \overline{J}_{II}}, \quad \Delta S_* \equiv \int_0^{\Delta t_*} \Delta J_{II}(\tau)\, d\tau, \quad \Delta t_* \equiv t^* - t_*, \quad \Delta \overline{J}_{II} \equiv \overline{J}_{II}^+ - \overline{J}_{II}.$$

In the origin time the integration is performed on $\tau \in [t_*, t^*]$.

Let us describe the identification scheme. Remember that s and $z = l\sqrt{b}/D$ are known from analysis of the stationary states. Using the desorption density known from the experiment we calculate $t_1 = t_* - S_*/\overline{J}$ and $t_2 = \Delta t_* - \Delta S_*/\Delta\overline{J}$, where instead of $\Delta J_{II}(t)$ we use $\Delta J(t) = J(t) - \overline{J}$ and $\Delta\overline{J} = \overline{J}^+ - \overline{J}$. If t_i coincide with high accuracy, we reasonably suppose that the experiment is in the area where the model I degenerates into II. Though theoretically t_i can coincide in the model I for special $\overline{p}_0$, $\overline{p}_0^+$. From $t_i = l^2/6D$ we determine the estimation of the diffusion coefficient D. Then we calculate $\overline{c}_0 = \overline{J}lD^{-1}$, $\overline{c}_0^+ = \overline{J}^+lD^{-1}$. Comparing with $\overline{c}_0 = \sqrt{b^{-1}(P - \overline{J})}$, $\overline{c}_0^+ = \sqrt{b^{-1}(P^+ - \overline{J}^+)}$, we evaluate the estimation of desorption coefficient b. More reliable value is $\sqrt{b} = Dz/l$. Finally, we solve numerically the problem I with obtained s, b, D and compare the model curve $J(t) = bc_l^2(t)$ with experimental one. If the result is not satisfactory, then we accept obtained estimations of s, b, D as initial approximation for basic model I.

The nonlinear model. The more the models I and II differ, the more well-posed is the inverse problem. Rates of diffusion and (de)sorption processes must have comparable orders (note, then $[\sqrt{bP}] = cm/s = [D/l]$). Therefore, we talk about rather low temperatures because chemical processes are activated faster as temperature growth compared to diffusion. In applications, the model I is aimed, for example, for investigation thin-films or walls, when concentrations are very small, and it's necessary to estimate slow (but on significant interval time) accumulation of hydrogen (or it is isotopes) at the outside.

Note 2. It is impossible to do unique conclusion about commensurability of diffusion and (de)sorbption from a classic permeability curve. Just so we offer the experiment with two pressures and with starting not from zero, but from the achieved stationary state. If we take $\overline{J}$ and $t_0 = t_* - S_*/\overline{J}$ from the problem I and accept in II $D = l^2/6t_0$, $\widetilde{c}_0 = \overline{J}l/D$, then we get the same steady state value $\overline{J}$ in II. Transitional processes are also hardly indistinguishable. For example, at $bP \sim 10^{-7}$, $D = 10^{-6}$ the concentration $\overline{c}_0$ is nearly twice greater than $l\overline{J}/D$, t_0 is more than three times greater than $l^2/6D$. Though the plots $J(t)$ in models I and II (D and $\widetilde{c}_0$ are calculated as mentioned above) coincide up to 5% error.

Let us proceed to estimating the delay time in model I, $t \in [0, t_*]$. Denote

$$R(t) = \int_0^t F_0(\tau)\, d\tau, \quad F_0(t) = -D\frac{\partial c}{\partial x}\bigg|_0, \quad R_* = R(t_*).$$

We define the function $\eta(t) \equiv F_0(t) - \bar{J}$ and represent $R(t)$ in the form of

$$R(t) = M(t) + \bar{J}t, \quad M(t) \equiv \int_0^t \eta(\tau)\, d\tau,$$

where the integral is asymptotically finite ($\approx M_* = M(t_*)$) due to big velocity of decreasing the function $|\eta(t)|$, $t >> 1$ ($\dot{R} = F_0 \to \bar{J}$).

At $t = t_*$ we derive from the material balance (3) and presentation (4)

$$M_* + \bar{J}t_* - S_* = \int_0^l c(t_*, x)\, dx = \bar{c}_0 l - \frac{\bar{J}l^2}{2D}.$$

Substituting $\bar{c}_0 = \bar{c}_l + \bar{J}l/D$, $\bar{c}_l = \bar{I}/\sqrt{b}$, we have for the delay time $t_0 = t_* - S_*/\bar{J}$ the expression

$$t_0 - \frac{l^2}{2D} + \frac{l}{\bar{I}\sqrt{b}} - \frac{M_*}{\bar{J}}, \quad I = \sqrt{J}. \tag{8}$$

Taking into account the equation for the steady state (5):

$$\frac{l\sqrt{b}\bar{I}}{D} = \chi - 1, \quad \chi^2 \equiv \frac{P}{\bar{J}} - 1 \;\Rightarrow\; t_0 = \frac{l^2}{2D}\cdot\frac{\chi+1}{\chi-1} - \frac{M_*}{\bar{J}}. \tag{9}$$

In considered range of model parameters $M_* > 0$, so we have the estimation $t_0 < (\chi+1)(\chi-1)^{-1} l^2/2D$. The less relative is the splash $(F_0 - \bar{J})/\bar{J}$ and the faster is the fixing $F_0 \to \bar{J}$, the more accurate is the estimation. When in the experiment $P >> \bar{J}$, then $t_0 \le l^2/2D$ is valid with high accuracy. If the inequality $(M_* - \bar{c}_l l)/\bar{J} << l^2/2D$ holds, then $t_0 \approx l^2/2D$ follows from representation (8).

Note 3. When the model I degenerates into II ($\widetilde{c}_0 = \bar{c}_0$), from $\bar{c}_0 = \bar{J}l/D$ and $t_0 = l^2/6D$ it follows $M_* = \bar{J}l^2/3D$. At numerical modeling in the referred region $M_* > \bar{c}_l l$, i. e. the estimation $D \in (l^2/6t_0, l^2/2t_0)$ is valid. The increasing of the delay time with the growth of (de)sorption processes takes place through increasing of t_*, the stabilization time of $c_0(t) \to \bar{c}_0$ and decreasing of $\bar{J}$. For example, in the model I at $l = 2\cdot 10^{-2}$, $D = 10^{-6}$, $b = 10^{-20}$, $s = 10^{-8}$, $\bar{p}_0 = 1$ (i. e. $\sqrt{bP} \approx 4\cdot 10^{-4} \sim D/l$) we have $\bar{J} = 1.2\cdot 10^{12}$, $\bar{c}_0 = 3.6\cdot 10^{16}$,

$\bar{c}_l = 1.1 \cdot 10^{16}$, $t_* \sim 500$, $t_0 = 130$, while in II (with steady state value of concentration $\bar{c}_0$ in model I) $\bar{J} = 1.8 \cdot 10^{12}$, $t_* \sim 300$, $t_0 = 67$.

Let us formulate the question: when can we assume that $F_0 \approx \bar{J}$ (this means that $M_* \approx 0$) under the integral? Let us write conservation law (3) for the degenerative model II after stabilization:

$$\frac{\tilde{c}_0 l}{2} = \int_0^{t_*} F_0(\tau)\, d\tau - S(t_*) = R_* - S_*, \; J = J_{II}.$$

Using the expressions $t_0 = t_* - S_* / \bar{J}$, $\tilde{c}_0 = \bar{J}l / D$, $t_0 = l^2 / 6D$, we get $R_* = \bar{J}t_* + 2\bar{J}t_0$. Thus the answer is: if $t_0 << t_*$. The similar effect was found numerically in the model I.

The scheme of calculations to separate b, D is following. As initial approximation we take, for instance, $D = l^2 / 3t_0$ and calculate $\sqrt{b} = Dz / l$, M_* by solving the problem I. Then from equation (9) we find the next approximation for D and so on until stabilization. While M_* is less than actual value and grows, the approximations of D will be upper decreasing estimations, when M_* is greater – contrary. The algorithm has the iterative nature because there is no information about the concentration $c_0(t)$ and the flux $F_0(t)$ at the input.

In conclusion of the section let us consider the possibility to improve significantly well-posedness of the inverse problem, though it is connected with additional costs. Suppose that total hydrogen expense is known. Either it is measured at the inlet side, either is calculated as the integral from $J(t)$ plus the quantity H in the membrane (per unit area). The last supposes fast cooling of the sample with following degassing in vacuum. Or we quickly remove the hydrogen from the input camera (for example, by instantly opening the preliminary evacuated cylinder of high capacity). The experiment time can be reduced by connecting the input and output cameras, because after t^* we are only interested in the total amount of H atoms in the membrane.

Let us consider the second stage of the experiment: $\bar{p}_0 \to \bar{p}_0^+$, $t \in [t_*, t^*]$.

By the change of variables

$$c(t,x) = c(t,x) - c(t_*,x),\; c(t_*,x) = \bar{c}_0 - D^{-1}\bar{J}x,\; \frac{\partial \tilde{n}}{\partial t} = D\frac{\partial^2 c}{\partial x^2},\; t > t_*,\; c(t_*,x) = 0,$$

$$-D\frac{\partial c}{\partial x}\bigg|_{x=0} = F_0(t) - \bar{J} \equiv \Delta F_0(t),\quad -D\frac{\partial c}{\partial x}\bigg|_{x=l} = J(t) - \bar{J} \equiv \Delta J(t).$$

We obtain the following expressions for the new delay time:

$$\Delta t_0 = \frac{l\Delta\bar{c}_0}{\Delta\bar{J}} - \frac{l^2}{2D} - \frac{\Delta M_*}{\Delta\bar{J}},\quad \Delta\bar{c}_0 = \bar{c}_0^+ - \bar{c}_0,\; \Delta\bar{J} = \bar{J}^+ - \bar{J},\; \Delta M_* = \int_{t_*}^{t^*} [F_0(t) - \bar{J}^+]\, dt,$$

$$\bar{J}t_0 + \Delta\bar{J}\Delta t_0 = \bar{c}_0^+ l - \frac{l^2 \bar{J}^+}{2D} - (M_* + \Delta M_*). \tag{10}$$

The value of Δt_0 as $\Delta t_* - \Delta S_* / \Delta\bar{J}$; $M_* + \Delta M_*$ is calculated as the difference between the total hydrogen expense at the inlet side and $\bar{J}t_* + \bar{J}^+ \Delta t_*$. From equation (10) after substitution $\bar{c}_0^+ = \sqrt{b^{-1}(P^+ - \bar{J}^+)} = l\sqrt{P^+ - \bar{J}^+}/(zD)$ the diffusion coefficient D is obtained.

5. Conjugate equations

Let us describe the application of the conjugate equations in problems of identification hydrogen permeability models [4]. Define the conjugate to (1) equation $\psi_t = -D\psi_{xx}$. For any solution (conjugate function) by means integration by parts

$$0 = \int_0^{t_*}\int_0^l \psi[c_t - Dc_{xx}]\, dx\, dt = \int_0^l (\psi c)\Big|_{t=0}^{t_*} dx - D\int_0^{t_*} (c_x \psi)\Big|_{x=0}^{l} dt + D\int_0^{t_*} (\psi_x c)\Big|_{x=0}^{l} dt. \tag{11}$$

Sense of such choice $\psi(t,x)$ is that after transforms the double integral does not appear. There remained only quantities connected with boundary conditions and with measurements. In trivial case ($\psi = 1$) we get the material balance equation.

One can submit expressions $c(t_*,x)$ according to (4) to this identity, $c(t,l) = \sqrt{J/b}$, $c(0,x) = 0$. There is no information about $c_0(t)$ and $F_0(t) \equiv -Dc_x(t,0)$. One term in (11) can be removed by adding a restriction $\psi(t,0) = 0$ or $\psi_x(t,0) = 0$. The density of the input flux F_0 has a splash, which is often rather significant in integral measurement. The concentration $c_0(t)$ in a wide region of parameters is stabilized rather fast. So it is better to remove the density of input flux F_0 from the identity (11) ($\psi(t,0) = 0$).

For the simplest variant $\psi = x/l$ (denominator is for the normalization) we get

$$f(b,D) \equiv \frac{\bar{c}_0 l}{2} - \frac{\bar{J}l^2}{3D} + S(t_*) - \frac{DS_{1/2}(t_*)}{l\sqrt{b}} - \frac{D}{l}\int_0^{t_*} c_0(t)\, dt = 0, \tag{12}$$

$$S(t_*) = S_* = \int_0^{t_*} J(\tau)\, d\tau,\ S_{1/2}(t_*) = S_{1/2}^* = \int_0^{t_*} I(\tau)\, d\tau,\ I(t) = \sqrt{J(t)},\ \ \bar{c}_0 = \sqrt{b^{-1}(P - \bar{J})}.$$

Due to wide range of orders of the values let us introduce the dimensionless variables

$$z_1 \equiv \frac{l}{\bar{I}t_*\sqrt{b}},\quad z_2 \equiv \frac{Dt_*}{l^2},\quad Z \equiv z_1 z_2 = \frac{1}{\bar{I}z} = \frac{D}{\bar{I}l\sqrt{b}},\quad \chi^2 \equiv \frac{P}{\bar{J}} - 1,\ \chi > 1.$$

The value of Z is known – it is calculated using the values of $\bar{J}$, $\bar{J}^+$. In our notations

$$z_1\left[\frac{1}{3Z}-\frac{\chi}{2}\right]+\frac{z_2}{t_*}\int_0^{t_*}\frac{lc_0(t)}{\bar{J}t_*}\,dt=\frac{S_*}{\bar{J}t_*}+\frac{Z}{t_*}\cdot\frac{S^*_{1/2}}{\bar{I}},\quad z_1z_2=Z;\ \ c_0(t)=c_0(t;z_1,z_2).\quad (13)$$

Let us emphasize that the concentration $c_0(t)$ is unknown and is determined by the parameters b, D. We understand $S_*/\bar{J}$, $S^*_{1/2}/\bar{I}$ as following: at first we calculate $J/\bar{J}$, $I/\bar{I}$, and then the integrals.

Suppose initially that $c_0(t)\approx\bar{c}_0$ when the integral is calculated. The experiment conditions are such that the inlet size is saturated relatively fast. This can be easily achieved by increasing the pressure $\bar{p}_0$ and (or) thickness l. The truth is: the less stabilization time $c_0(t)\approx\bar{c}_0$, the nearer are the problems I, II and more sensitive is the value of b to errors (degree of ill-conditionality is growing). Numerically was founded that in logarithmical coordinates the dependence stabilization time for $c_0(t)$ on the parameter bP is linear decreasing. For example, for $D=10^{-6}$ even at $bP>10$ it is less than a few seconds, while t_* is a few hundreds.

By submitting the expression $\bar{c}_0=\chi\sqrt{\bar{J}/b}$ instead of $c_0(t)$ into equation (13), we get that the second (integral) term at left-hand side is equal to χZ, and z_1, i. e. desorption coefficient b, is calculated uniquely. If due to equation (5) for the steady state we change χZ to $1+Z$, and change $c_0(t)$ to $\bar{c}_0$, then the relation (13) appears in the form

$$z_1\left[\frac{\chi}{2}-\frac{1}{3Z}\right]=\frac{t_0+Zt^0_{1/2}}{t_*},\qquad t^0_{1/2}\equiv t_*-\frac{S^*_{1/2}}{\bar{I}},\tag{14}$$

where $t^0_{1/2}$ is the delay time for $I(t)=\sqrt{J(t)}$. In a general case the presented above is the first approximation. In the left-hand side the additional term appears:

$$z_1\left[\frac{\chi}{2}-\frac{1}{3Z}\right]+\frac{z_2}{t_*}\int_0^{t_*}\frac{l[\bar{c}_0-c_0(t)]}{\bar{J}t_*}\,dt=\frac{t_0+Zt^0_{1/2}}{t_*}.\tag{15}$$

Therefore, equation (14) gives a lower-bounded estimation for the coefficient b and an upper-bounded one for D because $Z=z_1z_2=D/(\bar{I}l\sqrt{b})$. With the current b, D we compute the function $c_0(t)$ from the boundary-value problem and the integral in left side of the equation (15). In considered range $[\ldots]<0$, so $z_{1,2}$ can either grow or decrease simultaneously. But it can not be because of $z_1z_2=Z$ is constant. Substituting $z_2=Z/z_1$, we get the quadratic equation with respect z_1; its roots have different signs. Using z_1, Z we find next values of b, D. The iterations are continued until stabilization. It is possible to start the iterations from $D \approx l^2/6t_0$ or $D \approx l^2/3t_0$ also, keeping in mind the a priori estimation $t_0 \in (l^2/6D, l^2/3D)$.

The conjugate equation for the second stage of experiment $t \in [t_*, t^*]$ is constructed in a similar way. One must only consider that instead of null initial data in identity of type (11) there will be stationary distribution $c(t_*, x)$. If we have only one permeability curve ($t \in [0, t_*]$), one must change Z to $z_1 z_2$ and solve (iterative) the system of equations (9), (15) with respect to the variables b, D.

The simplified computational scheme. For practice the next approach may be sufficient. The function $f(t) = \bar{c}_0 - c_0(t)$ decreases rapidly: $f(0) = \bar{c}_0$, $f(t) \approx 0$, $t < t_*$. Let us approximate, for example, $f(t) = \bar{c}_0 \exp\{-at\}$. The constant $a > 0$ is selected from $f(t) \approx 0$ for the a priori estimation of stabilization time t for the input concentration $c_0(t)$ ($at \approx 5-7$). Then the integral term in equation (15) will equal the known number $\approx \chi Z/(at_*)$. By integrating the boundary condition $P - bc_0^2(t) = F_0(t)$ upon $t \in [0, t_*]$ with substitution $c_0(t) \approx \bar{c}_0[1 - \exp\{-at\}]$ we get in equations (8), (9) $M_* \approx (P - \bar{J})3/2a$. In quality sense $f(t) = \bar{c}_0(1 - \exp\{-a/t\})$ ($a/t \approx 5-7$) seems more preferable.

The other possibility: by integrating upon $t \in [0, t_*]$ the boundary condition (2) at $x = 0$ we have

$$(P - \bar{J})t_* - M_* = \int_0^{t_*} bc_0^2(t)\,dt = b\alpha\bar{c}_0 \int_0^{t_*} c_0(t)\,dt.$$

The parameter α appeared from the theorem of mean: $\alpha\bar{c}_0 = c_0(\xi)$, $\xi \in (0, t_*)$ ($c_0(t) \approx \bar{c}_0 \Rightarrow \alpha \approx 1$). Substituting e integral of $c_0(t)$ from equation (13), we have the expression for M_*, which is be submitted into (8), (9). Using $z = l\sqrt{b}/D$, $z\bar{I} = \chi - 1$ we get the explicit expression for D. It remains only to variate α in a neighborhood of unit.

6. Conclusions

Parameters of surface and diffusion processes are determined in one experiment. It seams to be more correct than using 'table data' from different sources for each coefficient separately. It is only sufficient to have information about desorption flux density at the outlet size of the membrane, it is possible not to perform measurements at the entry of 'black box'. But, naturally, then the identification algorithm loses well-posedness, so ill-posedness of inverse problem increases. The considered range of orders of the values is too wide to give unique recommendations. In concrete case, when the sufficiently narrow range is known from physical considerations, one must introduce the dimensionless variables x_i: $D = \widetilde{D}x_1$, $b = \widetilde{b}x_2$, $s = \widetilde{s}x_3$ (typical orders are with the tilde). For example, the equation of stationary state (5) will be rewritten in the form

$$\sqrt{\alpha x_3 - 1} - \beta X = 1, \quad \alpha \equiv \frac{\mu \widetilde{s}\, \bar{p}_0}{\bar{J}}, \quad \beta \equiv \frac{\bar{I} l \sqrt{\widetilde{b}}}{\widetilde{D}}, \quad X \equiv \frac{\sqrt{x_2}}{x_1}.$$

In particularity, it permits to reveal small parameters at qualitative investigations. Errors in solution of inverse problem are inevitable. So at the final stage, when the solution in a neighborhood of 'true value' is obtained, it is reasonable to apply the locally convergent least-squares method.

References

1. Kunin L.L., Golovin A.I., Surovoi U.I., Hohrin V.M. (1972) Problems of metals degassing. Nauka, Moskow (in Russian).
2. Bekman I.N., Gabis I.E., Kompaneets T.I., Kurdyumov A.A., Lajsnikov B.N. (1985) The investigation of hydrogen permeability in production technology of electronic technics. The electronic technics reviews, series 7, is. 1 (1084). Moscow (in Russian).
3. Tikhonov A.N., Arsenin V.Y. (1977) Solutions of ill-possed problems. New York: *J. Wiley & Sons*, 258 p.
4. Zaika Yu.V. (1998) Determination of model parameters for the hydrogen permeability of metals, *J. Tech. Phys.*, **43**(11), 1304–1308.

MAGNETIC PROPERTIES OF $Gd_2Fe_{14}BH_x$ HYDRIDES

R. BEZDUSHNYI, R. DAMIANOVA
Department of Solid State Physics, Faculty of Physics, Sofia University, Sofia 1126, Bulgaria

S.A. NIKITIN, E.A. TERESHINA
Department of Physics, Moscow State University, Leninskie Gory, Moscow, 119992, Russia
E-mail: nikitin@psys.msu.ru

G.S. BURKHANOV, I.S. TERESHINA, O.D. CHISTYAKOV
Baikov Institute of Metallurgy and Material Science RAS Leninski prospekt 49, Moscow, 119991, Russia

Abstract. Magnetic properties of intermetallic compound $Gd_2Fe_{14}B$ and its hydrides $Gd_2Fe_{14}BH_x$ have been investigated. Measurements of the temperature dependencies of magnetization have been performed with the permanent control of the hydrogen content in the investigated samples. The dependencies of Curie (T_C) and spin-reorientation transition (T_{SR}) temperatures on the hydrogen pressure and concentration of the inserted hydrogen atoms in the crystal lattice have been determined. On the base of experimental data for T_C the exchange parameters and their change after hydrogenation have been calculated.

Keywords: rare-earth alloys and compounds, hydrogen absorption materials, magnetic measurements

1. Introduction

The intrinsic magnetic properties of $Gd_2Fe_{14}B$ intermetallic compound have been previously investigated [1, 2]. The crystal structure of this compound was described in the tetragonal space group *P4/mnm* [2, 3]. According to the literature data [2], the magnetic ordering temperature of this compound is $T_C = 647$ K. $Gd_2Fe_{14}B$ has an easy-axis type of magnetic anisotropy, i.e. magnetization is aligned along the c-axis in the whole range of magnetic ordering.

$Gd_2Fe_{14}B$ readily absorbs a copious amount of hydrogen and forms stable hydrides at room temperature [4–6]. As it was shown in Ref. [7–9], in $R_2Fe_{14}B$ with heavy rare-earth elements, hydrogenation leads to a filling of the pseudotetrahedral interstitial sites. Changes of magnetic characteristics of these compounds occurring due to the hydrogen absorption are of great fundamental and technological interest.

The permanent control of the hydrogen content in the investigated compounds during the measurements plays an important role when the influence of hydrogen on magnetic properties (especially at high temperatures) of intermetallic compounds is under study. The magnetization measurements under fixed hydrogen pressure in the temperature interval including T_C for the $RFe_{11}Ti$ [10–12] and $Er_2Fe_{14}B$ [13] compounds have been performed earlier. Similar investigations performed for the other types of intermetallic compounds, particularly $R_2Fe_{14}B$, are of interest.

The aim of this work is to study the effect of hydrogenation on the magnetic phase transitions (Curie and spin-reorientation transition temperatures) in the $Gd_2Fe_{14}B$ compound using the magnetization measurements with the continuous control of the hydrogen content in the sample.

2. Experimental details

The rare-earth metal Gd (purified by double vacuum distillation), Armco iron, and Fe-B alloying composition were used as starting materials. An original furnace has been designed and purification regimes were developed at Baikov Institute of Metallurgy and Materials Science, Russian Academy of Sciences [14, 15]. The purification process was realized in a resistor furnace at residual pressure of $1.33\cdot(10^{-8}–10^{-9})$ atm using a graphite heater. The impurity composition of metals has been determined using a laser mass spectrometry. The purified metal (of 99.956–99.983 wt% purity) is characterized by low contents of metallic and interstitial elements, in particular, oxygen, which concentration as a result of the purification has been decreased by two orders of magnitude.

The $Gd_2Fe_{14}B$ compounds have been prepared by arc melting in a high-purity helium atmosphere using a water-cooled copper bottom and a non-consumable tungsten electrode. Specific regime of rapid cooling has been used (that allows to exclude the primary crystallization of α-Fe). At the same time, the solidification should not be very rapid to allow the occurrence of a peritectic reaction for the formation of principal magnetic $Nd_2Fe_{14}B$-type phase in the arc furnace. As a result, we have obtained practically single-phase compounds with well-defined directional structure. After melting the as-cast samples were wrapped in a tantalum foil, sealed in quartz tubes and annealed in argon atmosphere at 700°C for 200 h. The phase composition of the compounds has been determined by the X-ray diffraction analysis, using a DRON-3 M diffractometer. The quantity of impurity phase (α-Fe) has been estimated by both X-ray and thermomagnetic analysis (TMA) to be about 3%.

The magnetization measurements have been carried out using an apparatus for magnetic measurements, constructed on the principle of a vibrating-sample magnetometer in the temperature range of 100–700 K in permanent magnetic fields up to 12 kOe under gas pressure up to 12 atm.

During the measurements of the temperature dependence of magnetization (σ(T)), the sample has been placed in a chamber with a constant hydrogen pressure. In order to determine the hydrogen concentration x in the sample at any temperature (x(T)), a volumetric type experimental set-up [16, 17] has been used to study the hydrogen absorption-desorption properties at hydrogen pressures up to 4 atm in the temperature range from 100 to 750 K.

3. Results and discussion

Previous investigations of magnetic properties of $Gd_2Fe_{14}BH_x$ hydrides, didn't take into account the hydrogen loss in the sample when the temperature increased. We used an apparatus designed for measuring magnetic properties at high temperatures in a hydrogen atmosphere with the pressure control.

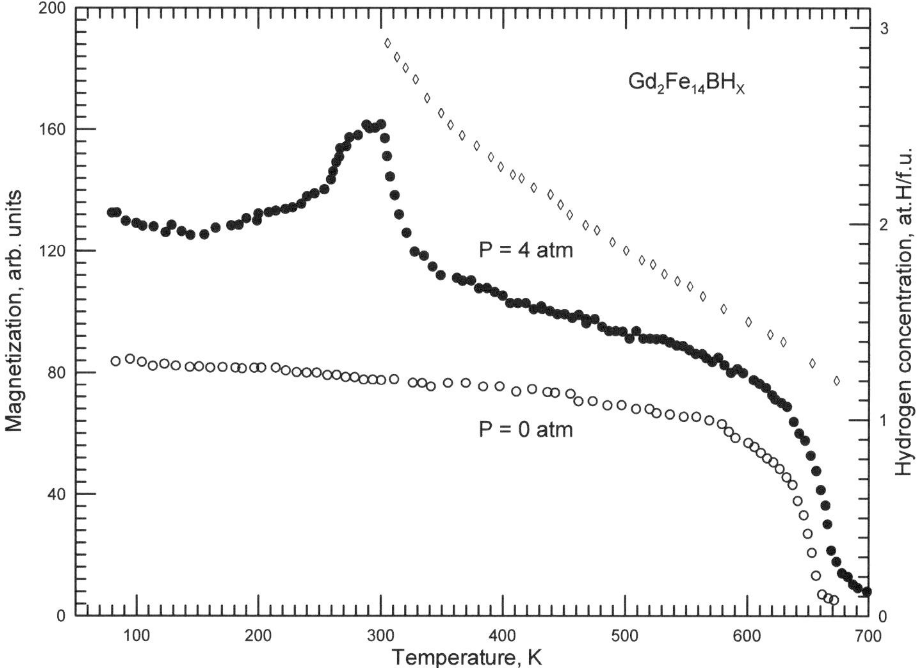

Figure 1. Representative curves σ(T, H = 1.6 kOe) at P = 0 atm and P = 4 atm and x(T) at P = 4 atm

Figure 1 shows the representative curves σ(T, H = 1.6 kOe) at P = 0 atm and P = 4 atm and x(T) at P = 4 atm, which allows to determine the hydrogen concentration in the temperature range 100–700 K. The hydrogen concentration in the chamber with the $Gd_2Fe_{14}BH_x$ sample was kept constant at lower temperatures.

As it seen from Fig. 1, the curve σ(T) at P = 0 atm has no peculiarities up to the Curie point at T = 652 K, which agrees well with the previous literature data. The hydrogen absorption leads to an appearance of a spin reorientation transition, which takes place at T_{SR} = 295 K at hydrogen pressure of 4 atm (corresponds to the hydrogen concentration x = 3.2). The introduction of hydrogen leads to an increase of both the Curie temperature and the saturation magnetization of the sample.

TABLE 1. Crystallographic parameters of $R_2Fe_{14}H_x$ (R = Y, Gd, x = 0; 2.5) compounds

Compounds	a (Å)	C (Å)	c/a	V (Å^3)	ΔV/V (%)
$Gd_2Fe_{14}B$	8.76	12.00	1.370	920.85	–
$Gd_2Fe_{14}BH_{2.5}$	8.84	12.09	1.367	944.78	2.6
$Y_2Fe_{14}B$	8.75	12.04	1.376	921.81	–
$Y_2Fe_{14}BH_{2.5}$	8.85	12.10	1.367	947.7	2.8

The experimentally determined dependence of Curie temperatures of $Gd_2Fe_{14}BH_x$ on the hydrogen pressure in the chamber containing the investigated sample is presented in Fig. 2. As it is shown in Fig. 2, the Curie temperature strongly depends on the hydrogen pressure: in the interval $0 < P < 0.5$ atm, the increase of hydrogen pressure leads to a fast increase of T_C from $T_C = 652$ K to $T_C = 665$ K. Later on, an increase of T_C slows down, at $P > 2$ atm, the $T_C(P)$ dependence is almost linear and at $P = 10$ atm, the Curie temperature reaches the value of 680 K.

As it was mentioned above, hydrogen insertion leads to an appearance of a spin-reorientation transition, which is not observed in the pure $Gd_2Fe_{14}B$ compound. Figure 3 shows the experimentally determined dependence of the spin-reorientation transition temperature on the hydrogen pressure. The spin-reorientation transition temperature rises quickly at a hydrogen pressure of about 1 atm. At pressures that exceed 2 atm up to hydrogen pressure 10 atm the T_{SR} rises slowly to 310 K atm.

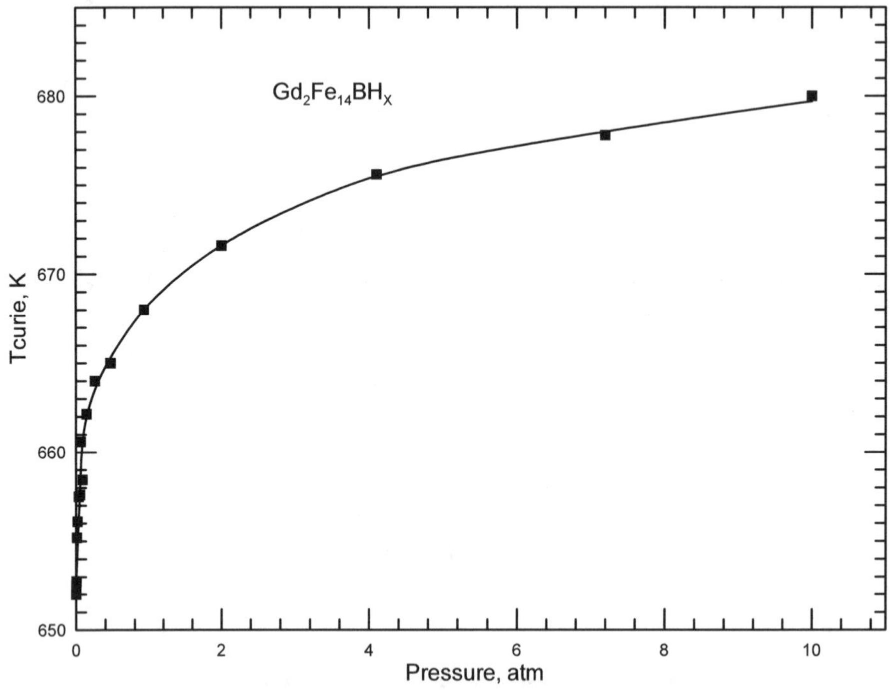

Figure 2. Dependence of the Curie temperatures T_C on the hydrogen pressure

T_C and T_{SR} dependencies on the hydrogen concentration x, derived from the isobars data are presented at Figs. 4 and 5, respectively. We emphasize the linear growth of $T_C(x)$ to the concentrations x ~ 1 at.H/f.u. This behavior follows from the linear dependence of the integral of exchange interactions on the unit cell volume. The similar situation was observed $Er_2Fe_{14}BH_x$ hydrides [13].

As it seen from Fig. 5, the spin-reorientation transition does not take place up to $x = 2$ (according to our experimental data extrapolation to $T = 0$). As it was mentioned before, hydrogenation of $R_2Fe_{14}B$ with heavy rare-earths was found to begin with the filling of the pseudotetrahedral interstitial sites H(1) in 8f (Wyckoff notation) sites [7] up to its infill. According to our magnetic measurements, this does not lead to an appearance of a spin reorientation transition, which starts with the filling of the pseudotetrahedral interstitial sites H(2) in 16k sites. Further hydrogenation is accompanied by filling of other interstitial sites as well and slows down the increase of T_{SR} at x > 3. So, the $T_{SR}(x)$ dependence differs from the $T_C(x)$ dependence. It is well known, that magnetic anisotropy constant is almost independent of the unit cell volume but strongly depends on the axial relation c/a [18] ($K \sim M_S^3[1-(c/a)^2]$ within the single-ion model for localized magnetic moments) which changes upon an insertion of hydrogen atoms in the crystal lattice of $Gd_2Fe_{14}BH_x$ compounds (see, for example, data in Table 1 for x = 2.5).

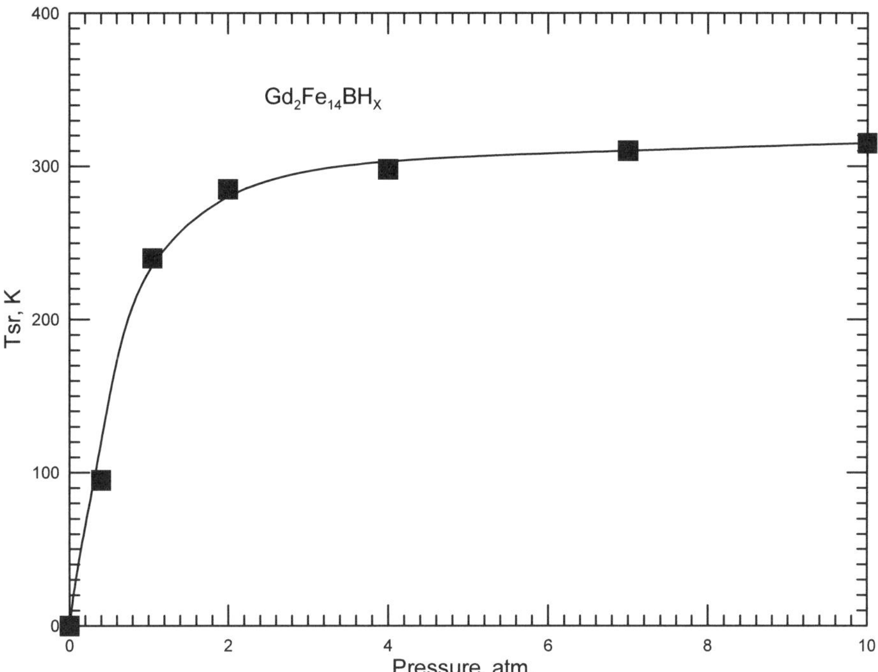

Figure 3. Dependence of the spin-reorientation transition temperatures T_{SR} on the hydrogen pressure

TABLE 2. Magnetic parameters of $R_2Fe_{14}H_x$ (R = Y, Gd, x = 0; 2.5) compounds

Compounds	T_C (K)	h_{11} (10^6 Oe)	H_{21} (10^6 Oe)	T_{SR} (K)
$Gd_2Fe_{14}B$	652	–	1.14	–
$Gd_2Fe_{14}BH_{2.5}$	737	–	1.12	230
$Y_2Fe_{14}B$	565	6.26	–	–
$Y_2Fe_{14}BH_{2.5}$	650	6.81	–	–

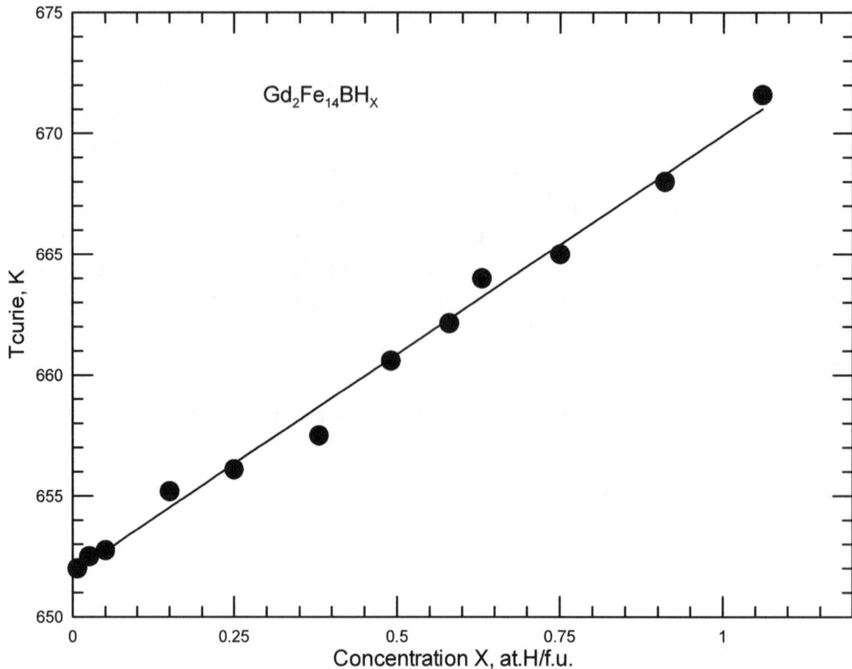

Figure 4. Dependence of the Curie temperatures T_C on the hydrogen concentration x

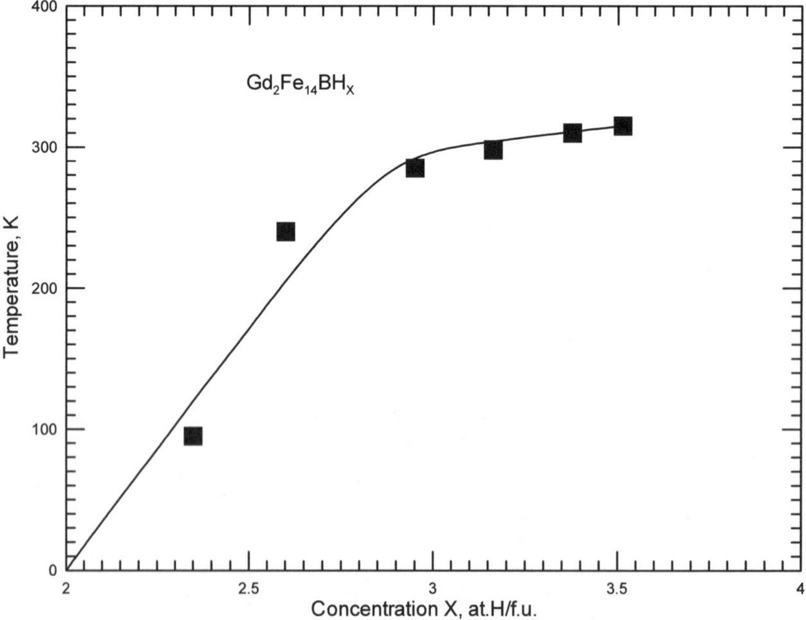

Figure 5. Dependence of the spin-reorientation transition temperatures T_{SR} on the hydrogen concentration x

The analysis of the T_C variations of $R_2Fe_{14}BH_x$ (R = Gd and Y, x = 0 and 2.5) performed in terms of a molecular field theory for two-sublattices ferrimagnets shows that the hydrogenation leads (see Table 2) to increase of the effective exchange field h_{11} (within the Fe-sublattice for $R_2Fe_{14}BH_x$ with the non-magnetic rare-earth atom, i.e. Y). Effective exchange field h_{21} (affecting the R-sublattice by the Fe-sublattice) is practically unchanged.

4. Conclusions

$Gd_2Fe_{14}B$ intermetallic compound and its hydrides have been synthesized and studied. The dependencies of the Curie temperatures and spin-reorientation temperatures on the hydrogen pressure and concentration have been investigated. It has been found that T_C and T_{SR} increase monotonously with the hydrogen pressure increase. The concentration dependencies of T_C and T_{SR} on the hydrogen atoms incorporated have a different character that can be explained taking into account the dependence of exchange integrals on the unit cell volume and dependence of the magnetic anisotropy constant on the axial relation c/a. The obtained experimental results are useful for a better understanding of the change in exchange and magnetocrystalline interactions upon the hydrogen insertion in the crystal lattice.

The work has been supported by the Federal Program on Support of Leading School Grant NSH -8701.2006.2, RFBR Grant № N 05-02-16361 and Science Research Foundation of Sofia University – Grant 57/2006.

References

1. Buschow K.H.J. In Handbook of Magnetic Materials, edited by E.P. Wohlfarth, North-Holland, Amsterdam, 1988.
2. Coey J.M.D., Li H.S., Gavigan J.P., Cadogan J.M., Hu B.P. Intrinsic magnetic properties of iron-rich compounds with the $Nd_2Fe_{14}B$ or $ThMn_{12}$ structure. Concerned European Action on Magnets [CEAM]. Elsevier Applied Science, London/New York, 1989, 76–97.
3. Givord, Li H.S., Moreau J.M. Magnetic properties and crystal structure of $Nd_2Fe_{14}B$, *Solid State Commun.*, 1984. **50**: 497–499.
4. Herbst J.F. $R_2Fe_{14}B$ materials: Intrinsic properties and technological aspects, *Rev. Mod. Phys.*, 1991. **63**: 819–898.
5. Zhang L.Y., Pourarian F., Wallace W.E. Magnetic behavior of $R_2Fe_{14}B$ hydrides (R = Gd, Tb, Dy, Ho and Er), *J. Magn. Magn. Mater.*, 1988. **71**: 203–211.
6. Fruchart D., Miraglia S. Hydrogenated hard magnetic alloys from fundamental to applications (invited), *J. Appl. Phys.*, 1991. **69**(8): 5578–5583.
7. Isnard O., Yelon W.B., Miraglia S., Fruchart D. Neutron-diffraction study of the insertion scheme of hydrogen in $Nd_2Fe_{14}B$, *J. Appl. Phys.*, 1995. **78**(3): 1892–1898.
8. Obbade S., Miraglia S., Wolfers P., Soubeyroux J.L., Fruchart D., Lera F., Rillo C., Malaman B., Le Caer G. Structural and magnetic study of $Ho_2Fe_{14}BH_x$ (x = 0-3.1), *J. Less-Common Met.*, 1991. **171**(1): 71–82.
9. Dalmas de Reotier P., Fruchart D., Pontonnier L., Vaillant F., Wolfers P., Yaouanc A., Coey J.M.D., Fruchart R., L'Heritier P. Structural and magnetic properties of $RE_2Fe_{14}BH(D)_x$, *J. Less-Common Met.*, 1987. **129**(1–2): 133–144.
10. Apostolov A., Bezdushnyi R., Stanev N., Damianova R., Fruchart D., Soubeyroux J.L., Isnard O. Magnetic properties and magnetic structure of $HoFe_{11}Ti$ and its hydrides, *J. Alloys Comp.*, 1998. **265**: 1–5.
11. Apostolov A., Bezdushnyi R., Damianova R., Stanev N., Naumova I. The effect of absorbed hydrogen on the magnetic properties of $DyFe_{11}Ti$, *J. Magn. Magn. Mater.*, 1995. **150**: 393–398.

12. Apostolov A., Bezdushnyi R., Damianova R., Stanev N., Naumova I., Gamari-Seale. Magnetic properties of some $TbFe_{11}TiH_X$ hydrides, *Phys. Stat. Sol.(a)*, 1994. **143**: 385–390.
13. Bezdushnyi R., Tereshina I.S., Damianova R., Nikitin S.A., Tereshina E.A., Burkhanov G.S., Chistyakov O.D. Magnetic Properties of Some $Er_2Fe_{14}BH_x$ Hydrides. ICHMS'2005, Sevastopol, Ukraine, 5–11 September 2005, P. 160, P. 162–163.
14. Burkhanov G.S., Chistyakov O.D., Kolchugina N.B. High-purity rare-earth metals for research and technology, *Elements*, 1998. **7**(1): 6–9.
15. Chistyakov O.D., Burkhanov G.S., Kol'chugina N.B., Panov N.N., Refining rare-earth metals by solidification from vapor, *Visokochistye veshchestva*, 1994. **3**: 57–65 (in Russian).
16. Satnev N., Damianov I. In: Proceedings of the XI National Conference DEFECTOSCOPY-96, Sozopol, Bulgaria, 1996, P. 279.
17. Satnev N., Damianov I., Annuaire de I'Universite de Sofia "St. Kliment Ohridski", 1995. **86**: 121.
18. Soubeyroux J.L., Fruchart D., Isnard O., Miraglia S., Tomey E. Role of the (H, C, N) interstitial elements on the magnetic properties of iron-rare earth permanent magnet alloys, *J. Alloys Comp.*, 1995. **219**: 16–24.

THERMODYNAMICS OF THERMOLYSIS OF ALKALI METALS BOROHYDRIDES

S.YU. ZAGINAICHENKO, D.V. SCHUR, T.A. TRIFONOVA
Frantsevich Institute for Problems of Materials Science of NASU
3, Krzhyzhanovsky Str., Kyiv, 03142 Ukraine
Fax: 38 (044) 424-0381; E-mail: shurzag@ipms.kiev.ua

Z.A. MATYSINA
Dnepropetrovsk National University, Str. Nauchnaya, 13, Dnepropetrovsk, 49050 Ukraine

Abstract. Theoretical investigation of phase transformation of borohydrides of alkali metals has been performed on the basis of molecular-kinetic theory in the approach of the pair interaction of nearest atoms. The elaborated statistical theory of phase transformations at the realization of chemical reactions between phases of borohydrides of alkali metals gives the explanation and justification for the course of these reactions and provides the investigation of phase diagram with full details for the components of studied reactions. The knowledge of phase parameters for each concrete reaction from independent experiments predetermines the conditions of such reactions performance and the conditions of extraction of hydrogen as power source. The comparison of theoretical results with the experimental data for phase transformations of borohydrides of alkali metals shows the good agreement.

Keywords: borohydride, hydrogen concentration, statistical theory, temperature, phase transformation, phase diagram

1. Introduction

The latest experimental investigations have established the possibility to use borohydrides of alkali metals, among other hydrides and hydroborides of another metals [1–15] as hydrogen accumulator and storage, what gives the promising use of energy. The $LiBH_4$, $NaBH_4$, KBH_4, $Be(BH_4)_2$, $Mg(BH_4)_2$, $Al(BH_4)_2$, LiH, NaH, NaD systems and another's are under extensive investigation. The hydrogen atoms in the named borohydrides form the tetrahedral configuration around the boron atoms. The weight content of hydrogen in these phases is correspondingly equal to 0,183; 0,106; 0,074; 0,207; 0,148; 0,168; 0,126; 0,041; 0,045 [9].

The phase $LiBH_4$ realized at the temperature of 20°C was under investigation in the paper [9] and it was proposed the following scheme of its dehydrogenation with increase in temperature up to 450°C

$$LiBH_4 \rightarrow LiBH_{4-\varepsilon} + \frac{1}{2}\varepsilon H_2 \rightarrow LiBH_{2+\varepsilon} + (1-\frac{\varepsilon}{2})H_2 \rightarrow LiH{+}B{+}\frac{3}{2}H_2. \quad (1)$$

The first transformation occurs from 108^{0}C, the phase $LiBH_2$ is realized at the 200°C and the realization of phase mixture LiH + B takes place at the 450°C. In this case both in hydroborides and in hydrides of alkali metals the deviation from stoichiometry is observed with increased temperature [9, 16, 17], i.e. the hydrogen is manifested in deficit as this is partially mentioned in the presented chemical reactions (1).

The hydrides of alkali and other metals were investigated in the papers [1–15, 18–25].

Our interest is in theoretical investigation of phase transformations in the course of thermolysis which is patterned after the mentioned chemical reactions, the elucidation of conditions of such reactions carrying out, the evaluation of temperatures of transitions between phases and the estimation of the temperature and concentration intervals of each phase realization as well as the construction of constitution diagram of the systems of the chemical reactions (1).

Taking into consideration the different degree of nonstoichiometry of phases we have studied the chemical reactions of the following type

$$MBH_{x_1} \rightarrow MBH_{x_2} + \frac{x_1 - x_2}{2} \cdot H_2 \rightarrow MH_{x_3} + B + \frac{x_1 - x_3}{2} \cdot H_2, \quad x_1 > x_2 > x_3, \qquad (2)$$

where M is the atom of alkali metal (Li, Na and others), x_1, x_2, x_3 are concentrations of hydrogen atoms in the ratio to the number of boron atoms in borohydrides and to the number of metal M atoms in hydrides, their values lie in the ranges

$$0 \le x_1 \le 4, \quad 0 \le x_2 \le 2, \quad 0 \le x_3 \le 1. \qquad (3)$$

For solution of the raised problem the calculation of free energies of three phases $\alpha = MBH_{x_1}$, $\beta = MBH_{x_2}$, $\gamma = MH_{x_3} + B$ (the gas component H_2 of the third phase γ has the insignificant energy, which is neglected) has been carried out, their dependences on phase composition, temperature and energetic parameters have been defined. The method of average energies has been used in the calculation having regard to the pair interatomic interaction between the nearest atoms BH and MH without considering the correlation in substitution of positions of different types by atoms of different kind as well as assuming that some positions of hydrogen atoms are vacant by virtue of nonstoichiometry of phases.

The lattice elementary cells of the phases α, β, γ are shown in the Figs. 1–3, [9, 16, 17, 26]. The phase α has a base-centered orthorhombic lattice (bco) with parameters a_1, $b_1 \approx a_1$, c_1 (Fig. 1). The phase β has a face-centered cubic lattice of LI_0 structure (fcc) of CuAu type with the lattice constant a_2 (Fig. 2). The first component of the phase γ has a simple cubic lattice of B1 structure of NaCl type with the lattice parameter a_3 (Fig. 3(a)) and the second component B of the phase γ has a base-centered tetragonal lattice (bct) with the lattice constants a_4, c_4 (Fig. 3(b)).

The calculation of free energies F_i ($i = \alpha, \beta, \gamma$) of these phases has been performed on the basis of molecular-kinetic theory in accordance with the following formula [27–32]

$$F_i = E_i^0 + E_i - kT \ln G_i, \quad i = \alpha, \beta, \gamma, \qquad (4)$$

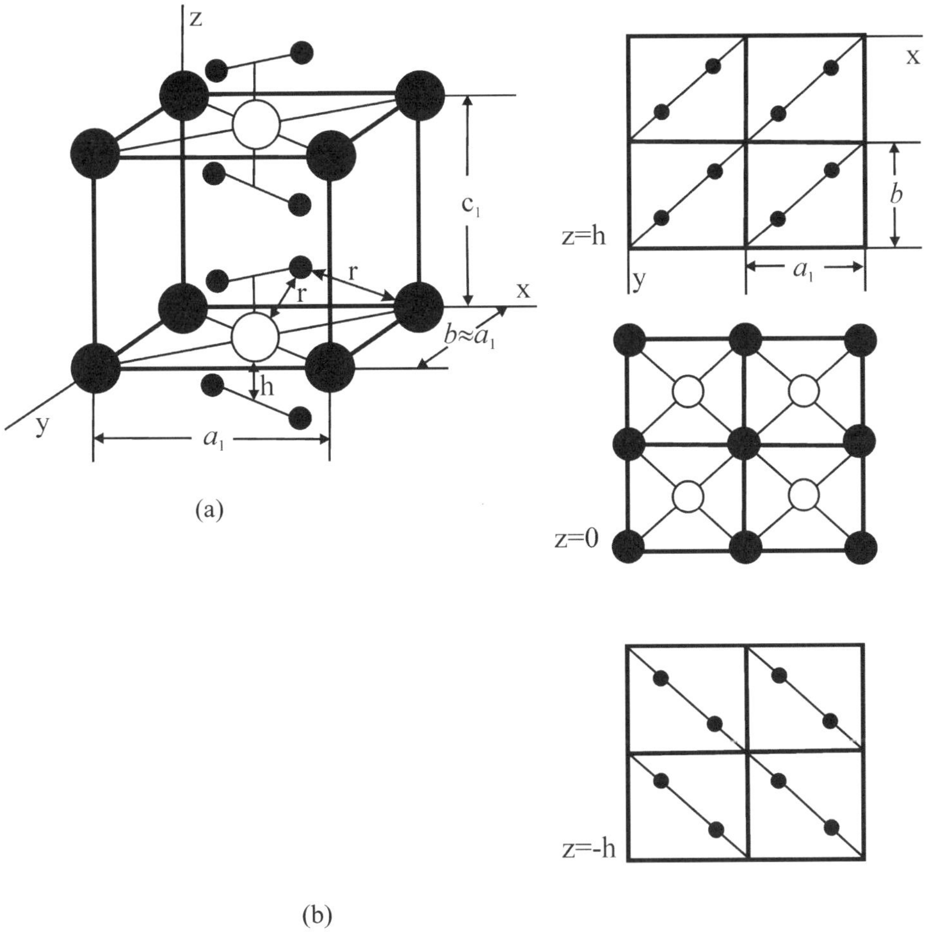

(a)

(b)

Figure 1. The elementary cell of base-centered orthorhombic lattice of the crystal $LiBH_4$ (a) and its projection at a reduced scale (b) on the plane xy for the coordinate values z = h, 0, –h. h is the distance between the basic planes, a_1, b_1, c_1 are the lattice parameters, r, h are interatomic distances between the pairs BH, LiH and distances between the planes of B, Li and H atoms arrangement.

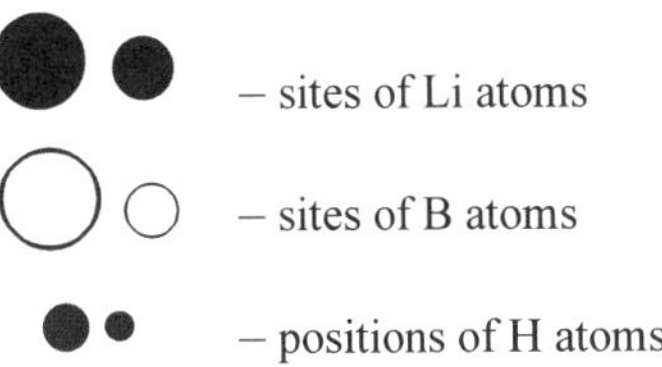

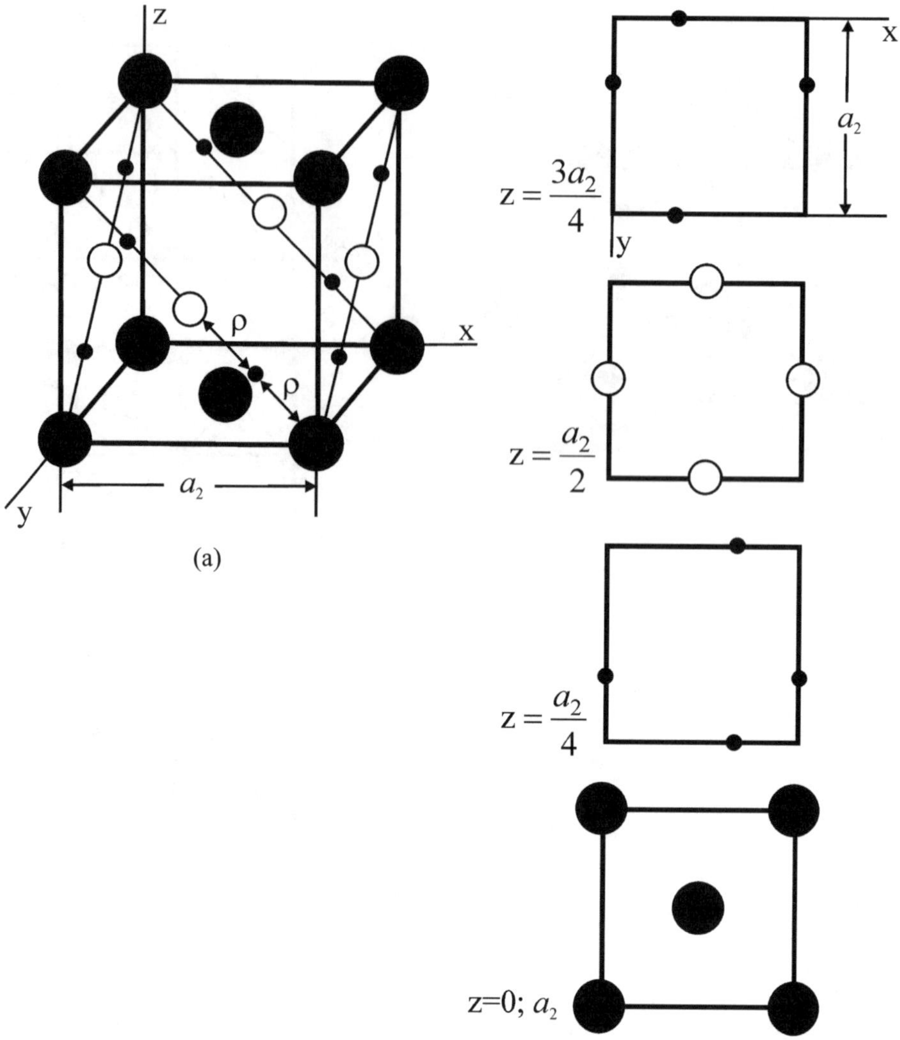

(a)

(b)

Figure 2. The elementary cell of face-centered cubic lattice of the crystal $LiBH_2$ (a) and its projection on the plane xy (b) for the coordinate values $z = 0, \frac{a_2}{4}, \frac{a_2}{2}, \frac{3a_2}{4}, a_2$.

$a_2$is the lattice parameter, ρ is the nearest interatomic distance between the pairs BH and LiH.

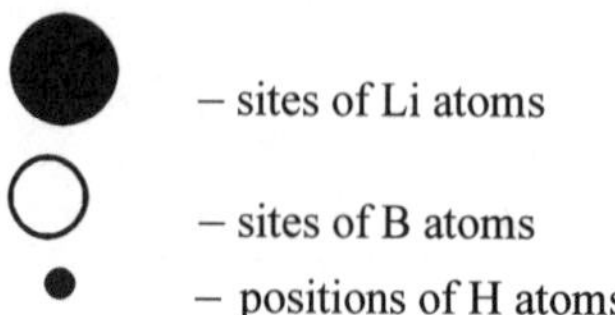

– sites of Li atoms

– sites of B atoms

– positions of H atoms

where E_i^0 is the internal configuration part of the free energy, determined by the sum of interaction energies of nearest atoms MM, BB, MB and is independent of hydrogen content, E_i is the same energy with consideration for the pair interaction of MH and BH atoms; G_i is the thermodynamic probability of the hydrogen atoms distribution at all their positions in each phase that is determined by the combinatorial rule (thermodynamic probability of distribution of atoms M and B in crystals is equal to one in the approximation being considered, but for hydrogen atoms in the gaseous component of the phase it is a negligibly small quantity), k is Boltzmann's constant and T is absolute temperature.

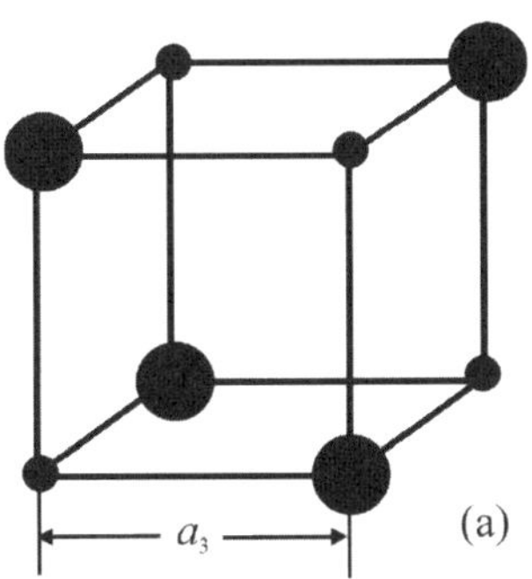

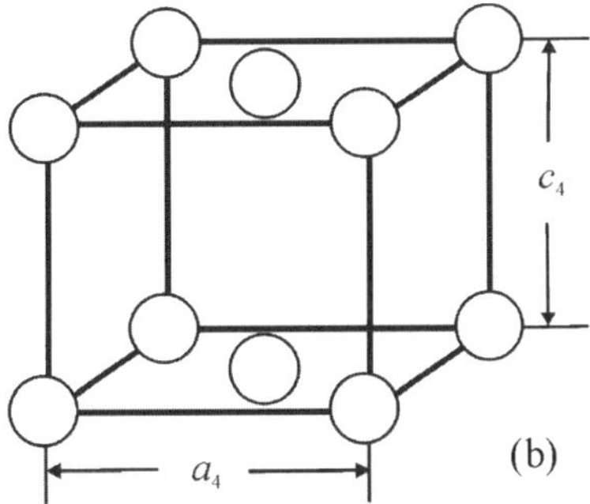

Figure 3. The elementary cells of the crystals LiH and B, the ultimate products of the chemical reaction (1): (a) simple cubic lattice B1 of NaCl type, (b) base-centered tetragonal lattice of boron.

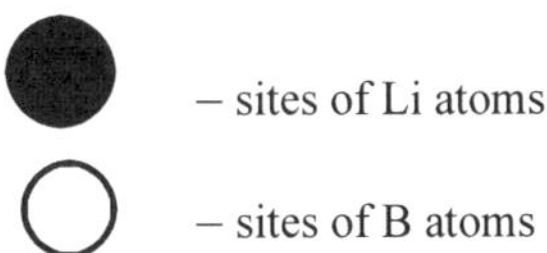

2. Free energy of the phase α

The hydrogen atoms in the α-phase are located in the vertex of a tetrahedron with the boron atom at the centre (Fig. 1), r is the interatomic distance for the pairs BH and MH (at $b \approx a_1$).

The following nomenclature is used: N is the number of the bco lattice sites equal to the number of B and M atoms, N/2 is the number of B and M atoms respectively, 2N is the number of positions of hydrogen atoms, N_H, N_V are the numbers of hydrogen H atoms and vacant positions V

$$2N = N_H + N, \tag{5}$$

c, c_V are concentrations of hydrogen atoms and vacancies in the ratio to the number of their positions

$$c = N_H / 2N, \quad c_V = N_V / 2N, \quad c + c_V = 1. \tag{6}$$

The relation

$$x_\alpha = 4c \tag{7}$$

is valid for the phase α and

$$0 \le c \le 1, 0 \le x_\alpha \le 4. \tag{8}$$

The numbers of the pairs BH and MH are equal to

$$N_{BH} = 2Nc, \quad N_{MH} = 2Nc. \tag{9}$$

The configuration internal energy can be written as

$$E_\alpha = E_\alpha^0 - N_{BH}\upsilon_{BH} - N_{MH}\upsilon_{MH}, \tag{10}$$

where υ_{BH}, υ_{MH} are the interaction energies for the indicated atomic pairs at the distance r, taken with the opposite sign.

The energy E_α can be written, in view of equation (9), as

$$E_\alpha = E_\alpha^0 - N\upsilon_\alpha c, \quad \upsilon_\alpha = 2(\upsilon_{BH} + \upsilon_{MH}). \tag{11}$$

The thermodynamic probability can be written as follows:

$$G_\alpha = (2N)!/\,N_H!N_V!. \tag{12}$$

Using the Stirling formula lnX! = X (lnX–1), which is valid for large numbers X, and formula (6) we find the natural logarithm of the thermodynamic probability

$$\ln G_\alpha = -2N[c\ln c + (1-c)\ln(1-c)]. \tag{13}$$

Substituting formulae (11) and (13) into (4), we find the free energy of the phase α

$$F_\alpha = E_\alpha^0 - N\upsilon_\alpha c + 2NkT[c\ln c + (1-c)\ln(1-c)], \tag{14}$$

which determines the dependence of the free energy F_α on the temperature T, concentration of hydrogen and energy parameters υ_α and E_α^0.

The study of functional dependence $F_\alpha = F_\alpha(c)$ for different temperatures and comparison of this dependence with the same for the phase β and γ makes it possible to define both the concentration and the temperature ranges of each phase existence and also the temperature of phase transformations realized in the chemical reactions (2).

The condition of thermodynamic equilibrium

$$\partial F_\alpha / \partial c = 0 \tag{15}$$

makes it possible to find the temperature dependence of equilibrium concentration of hydrogen atoms in the phase α

$$2kT\ln\frac{c}{1-c} = \upsilon_\alpha \tag{16}$$

or

$$c = \frac{1}{4}x_\alpha = [1 + \exp\frac{-\upsilon_\alpha}{2kT}]^{-1}. \tag{17}$$

The plots of temperature dependence of hydrogen concentration in the phase α are shown in Fig. 4, constructed by the formula (16) or (17) for different values of energetic parameter υ_α. As can be seen from Fig. 4 the hydrogen concentration decreases to the value c = 0.5 with the increase in temperature and this leads to the formation of the phase β, for which we have $x_\beta = 2c$. The curve of hydrogen concentration falls off more steeply with the increasing temperature for the smaller energy υ_α. For the high values of the energy υ_α the decrease in hydrogen concentration is found to be more smooth as the temperature increases.

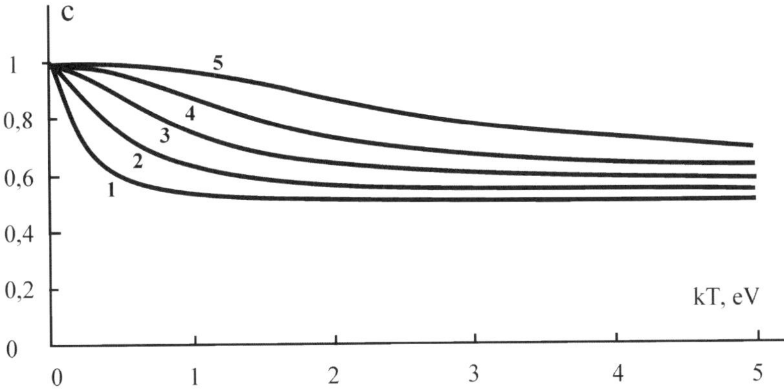

Figure 4. The plots of temperature dependence of equilibrium concentration *c* of hydrogen constructed by formula (16) for energetic parameter υ_α values: 0.2; 0.5; 1; 1.8; 3.2 eV (curves 1, 2, 3, 4, 5, correspondingly)

3. The free energy of the phase β

The number of positions of hydrogen atoms in this phase is less by half in comparison with the identical for the phase α. This number is symbolized by N

$$N = N_H + N_V \tag{18}$$

and the concentrations c, c_V of hydrogen atoms and vacancies are equal to

$$c = N_H / N, \qquad c_V = N_V / N, \qquad c + c_V = 1. \tag{19}$$

In this case these concentrations vary within the ranges

$$0 \le c \le 1, \quad 0 \le x_\beta \le 2, \tag{20}$$

and for the phase β concentration x_β is equal to

$$x_\beta = 2c. \tag{21}$$

In the phase β hydrogen atoms are arranged in pairs on the diagonals of cube faces (Fig. 2) near the boron atoms. Because of this, the each B and M atom have two nearest neighbouring atoms of hydrogen at a distance ρ. The numbers of such pairs are equal to

$$N_{BH} = Nc, \qquad N_{MH} = Nc. \tag{22}$$

The configuration internal energy E_β can be written as

$$E_\beta = E_\beta^0 - N\upsilon_\beta c, \qquad \upsilon_\beta = \upsilon_{BH} + \upsilon_{MH} \tag{23}$$

and the energies υ_{BH}, υ_{MH} are taken here at the distance ρ.

The calculation of the thermodynamic probability

$$G_{\beta} = N!/N_{H}!N! \tag{24}$$

gives the formula for the natural logarithm of it as follows

$$\ln G_{\beta} = -N[c\ln c + (1-c)\ln(1-c)]. \tag{25}$$

As a result the free energy of the phase β is found as

$$F_{\beta} = E_{\beta}^{0} - N\upsilon_{\beta}c + NkT[c\ln c + (1-c)\ln(1-c)], \tag{26}$$

by minimizing this expression

$$\partial F/\partial c = 0 \tag{27}$$

we have obtained the equilibrium concentration of hydrogen in the phase β and this formula is similar to equation (16). By this is meant that in the phase β the hydrogen concentration decreases with increasing temperature and this is a stimulator to the phase transformation into the phase γ.

4. The free energy of the phase γ

The phase γ is made up of two components: crystals of metal hydride MH_{X_3} and boron B (Fig. 3). In line with the scheme of chemical reactions (1) of the paper [9] the boron crystals don't contain the hydrogen atoms. We have been interested to find the dependence of free energy F_{γ} of the phase γ on the hydrogen content. Thus we can include the energy of boron crystals into the component E_{γ}^{o} and shall restrict our consideration to the free energy of crystals MH_{X_3}.

The numbers of atomic pairs MH according to the structure of Fig. 3(a) are equal to

$$N_{MH} = 3Nc, \tag{28}$$

where the number of all sites of crystal, occupied by the M and H atoms, is indicated by the symbol N and concentration of hydrogen atoms in the ratio to the number of their positions is

$$c = N_{H}\Big/\frac{1}{2}N. \tag{29}$$

In this case we have

$$x_{\gamma} = c \tag{30}$$

and

$$0 \le c \le 1, \qquad 0 \le x_{\gamma} \le 1. \tag{31}$$

The configuration internal energy E_{γ} is equal to

$$E_{\gamma} = E_{\gamma}^{0} - 3N\upsilon_{MH}c, \tag{32}$$

where υ_{MH} is the energy of pair interaction MH taken with the opposite sign at the distance of lattice parameter a_3 of the crystal MH_{x_3}.

The thermodynamic probability G_γ can be written as

$$G_{\gamma} = (\frac{1}{2}N)! \Big/ N_H ! N_V !, \tag{33}$$

from which we define its natural logarithm in view of formula (29) as

$$ln G_{\gamma} = -\frac{1}{2}N[c \ln c + (1-c) ln(1-c)]. \tag{34}$$

With allowance for formulae (32), (34) we have found the free energy F_γ of the phase γ

$$F_{\gamma} = E_{\gamma}^{o} - N\upsilon_{\gamma}c + \frac{1}{2}kTN[c \ln c + (1-c) ln(1-c)], \;\; \upsilon_{\gamma} = 3\upsilon_{MH}. \tag{35}$$

Substitution of free energy F_γ into the condition of thermodynamic equilibrium

$$\partial F_{\gamma} / \partial c = 0 \tag{36}$$

gives the equation of equilibrium, from which we have defined the hydrogen concentration in dependence on temperature. This concentration is determined by the expression of (16) type and shows the decrease of hydrogen concentration in the crystal MH_x with a rise in temperature (at $\upsilon_\gamma > 0$).

5. The discussion of calculation results. The interpretation of experimental data. The construction of constitution diagram

The analysis of formulae (14), (26), (35) for the free energies $F_i (i = \alpha, \beta, \gamma)$ shall be carried out and we shall rewrite these formulae calculating on one site of crystal lattice

$$f_i = F_i / N, \quad i = \alpha, \beta, \gamma \tag{37}$$

and in dependence on the concentration x_i of hydrogen atoms in the phases

$$\begin{aligned} f_{\alpha} &= e_{\alpha} - \frac{1}{4}\upsilon_{\alpha}x_{\alpha} + 2kT[\frac{x_{\alpha}}{4}\ln\frac{x_{\alpha}}{4} + (1-\frac{x_{\alpha}}{4})\ln(1-\frac{x_{\alpha}}{4})], \;\; 0 \le x_{\alpha} \le 4, \\ f_{\beta} &= e_{\beta} - \frac{1}{2}\upsilon_{\beta}x_{\beta} + kT[\frac{x_{\beta}}{2}\ln\frac{x_{\beta}}{2} + (1-\frac{x_{\beta}}{2})\ln(1-\frac{x_{\beta}}{2})], \;\; 0 \le x_{\beta} \le 2, \\ f_{\gamma} &= e_{\gamma} - \upsilon_{\gamma}x_{\gamma} + \frac{1}{2}kT[x_{\gamma}\ln x_{\gamma} + (1-x_{\gamma})\ln(1-x_{\gamma})], \;\; 0 \le x_{\gamma} \le 1, \end{aligned} \tag{38}$$

where $e_i = E_i / N$.

Using the condition of thermodynamic equilibrium $\partial f_i / \partial x_i = 0$, we find the temperature dependences of equilibrium concentrations $x_i = x_{\alpha}, x_{\beta}, x_{\gamma}$ of hydrogen in the phases α, β, γ as

$$x_\alpha = 4[1+\exp\frac{-\upsilon_\alpha}{2kT}]^{-1}, \; x_\beta = 2[1+\exp\frac{-\upsilon_\beta}{kT}]^{-1}, \; x_\gamma = [1+\exp\frac{-\upsilon_\gamma}{kT}]^{-1}. \quad (39)$$

The plots of temperature dependences of concentrations $x_\alpha = x_\alpha(T)$, $x_\beta = x_\beta(T)$, $x_\gamma = x_\gamma(T)$ are shown in Fig. 5, constructed by formulae (39) for the energies $\upsilon_i = \upsilon_\alpha$, υ_β or υ_γ equal to 0.5 eV. As can be seen from Fig. 5 in the phase α for this value of energy $\upsilon_i = 0.5$ eV the hydrogen concentration initially falls sharply as the temperature rises from the zero. However at the kT ≈ 1 eV the value x_α decreases smoothly, reaches the value $x_\alpha = 2$ at the temperature $T \to \infty$. In the phase β the picture of the $x_\beta(T)$ dependence is similar, but the value x_β has less drastic decrease at the low temperatures and it reaches $x_\beta \to 1$ at the temperature $T \to \infty$. In the case of the phase γ the plot $x_\gamma(T)$ is flat, i.e. the concentration x_γ falls smoothly with a rise of temperature and attains the value $x_\gamma = 0.5$ at the temperature $T \to \infty$.

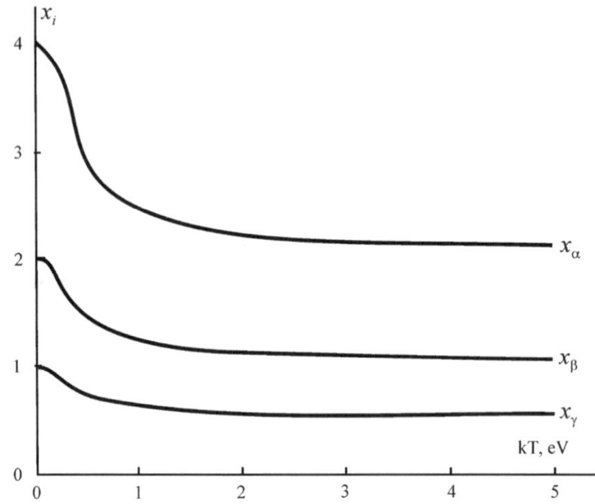

Figure 5. The plots of temperature dependence of equilibrium concentration of hydrogen $x_i = x_\alpha$, x_β, x_γ in the ratio to the number of atoms B in the phases α and β and of atoms Li in the phase γ, constructed by formula (39) for the energy $\upsilon_i = 0.5$ eV

The plots of concentration dependences of free energies $f_i = f_i(x_i)$ can be constructed for different temperatures using the formulae (38). For this purpose we shall make an estimate of the values of energetic parameters e_i, $\upsilon_i (i = \alpha, \beta, \gamma)$.

We shall consider the chemical reactions (2) for M = Li based on the corresponding values of hydrogen concentrations and temperatures of phase transformations [9]. Each phase transformation occurs at the equal concentrations of hydrogen in the phases and the definite temperature of these phases. In this case the free energies of these phases are equal in value.

According to the experimental data [9] the first phase transformation α↔β (PT1) takes place at the temperature $T_1 \approx 200°C$ and the concentration $x = x_\alpha = x_\beta \lesssim 2$. In this case we have $f_\alpha = f_\beta$. However at the temperature $T < T_1$ and the hydrogen content $x = 2$ the phase α is stable and conversely at the $T > T_1$ and $x = 2$ the phase β is to be in the equilibrium state ($f_\alpha > f_\beta$). The second phase transformation $\beta \leftrightarrow \gamma$ (PT 2) is realized at the temperature $T_2 \approx 450°C$ and the hydrogen concentration $x = x_\beta = x_\gamma \lesssim 1$ and for this case we have $f_\beta = f_\gamma$. At the temperature $T < T_2$ and concentration $x = 1$ the phase β is to be equilibrium ($f_\beta < f_\gamma$), but at the $T > T_2$ and $x = 1$ the phase γ is stable ($f_\beta > f_\gamma$).

The consideration of the intimated equalities and inequalities of the free energies f_α , f_β, f_γ

$$f_\alpha = f_\beta \quad \text{at } T = T_1, \quad x \lesssim 2, \quad f_\beta = f_\gamma \quad \text{at } T = T_2, \quad x \lesssim 1,$$

$$f_\alpha < f_\beta \quad \text{at } T < T_1, \quad x = 2, \quad f_\beta < f_\gamma \quad \text{at } T < T_2, \quad x = 1, \tag{40}$$

$$f_\alpha > f_\beta \quad \text{at } T > T_1, \quad x = 2, \quad f_\beta > f_\gamma \quad \text{at } T > T_2, \quad x = 1,$$

makes it possible to evaluate the values of energetic parameters e_i, υ_i of the phases α, β, γ. The calculation results give the following values for energies (in eV) equal to

$$e_\alpha = -0,2; \ \upsilon_\alpha = 3,796; \ e_\beta = -1,3; \ \upsilon_\beta = 4,434; \ e_\gamma = -2,1; \ \upsilon_\gamma = 2,424. \tag{41}$$

The plots of the concentration dependences of free energies for various values of temperature have been constructed by the formulae (38) with the use of the calculated values of energies e_i, υ_i. These curves are plotted in Fig. 6. The phase with the lesser value of free energy is bound to be realized at the each specific temperature. The intersection points of these curves determine the concentration of the each phase α, β, γ for the each temperature. The phase diagram defining the concentration dependence of phase transformation temperatures $kT(x_i)$ can be constructed by these points of intersection of the curves $f_i = f_i\,(x_i)$.

The constitution diagram of the phases α, β, γ for the chemical reactions (2) is presented in Fig. 7. As can be seen from this Fig. 7 the regions of realization of the phases β and γ get narrow and of the phase α is expanded with increase in temperature. The transitions $\alpha \leftrightarrow \beta$ and $\beta \leftrightarrow \gamma$ are realized with decreasing concentrations x_i and increasing temperature. At the low temperatures each phase α, β, γ has the greatest possible value of the concentration x_i (7), (20), (31).

6. Conclusions

The elaborated statistical theory of the phase transformations of the chemical reactions (2) provides the explanation and substantiation of realization of such phase transitions, allows the establishment of temperature dependence of hydrogen concentrations in each of phases α, β, γ and of the character of this dependence for each phase, permits the construction of the plots of the concentration dependences of free energies of investigated phases for different temperatures and the determination of phase diagram of the components states of the studied chemical reaction (2) in the case of lithium metal M = Li. In this case the energetic parameters of the phases have been estimated with the use of experimental data for the values of temperatures and hydrogen concentration at the realized phase transformations.

It has been established that all phases α, β, γ are to be realized at the low temperatures in the hydrogen concentrations ranges respectively

$$2 \le x_\alpha \le 4, \ 1 \le x_\beta \le 2, \ 0 \le x_\gamma \le 1. \tag{42}$$

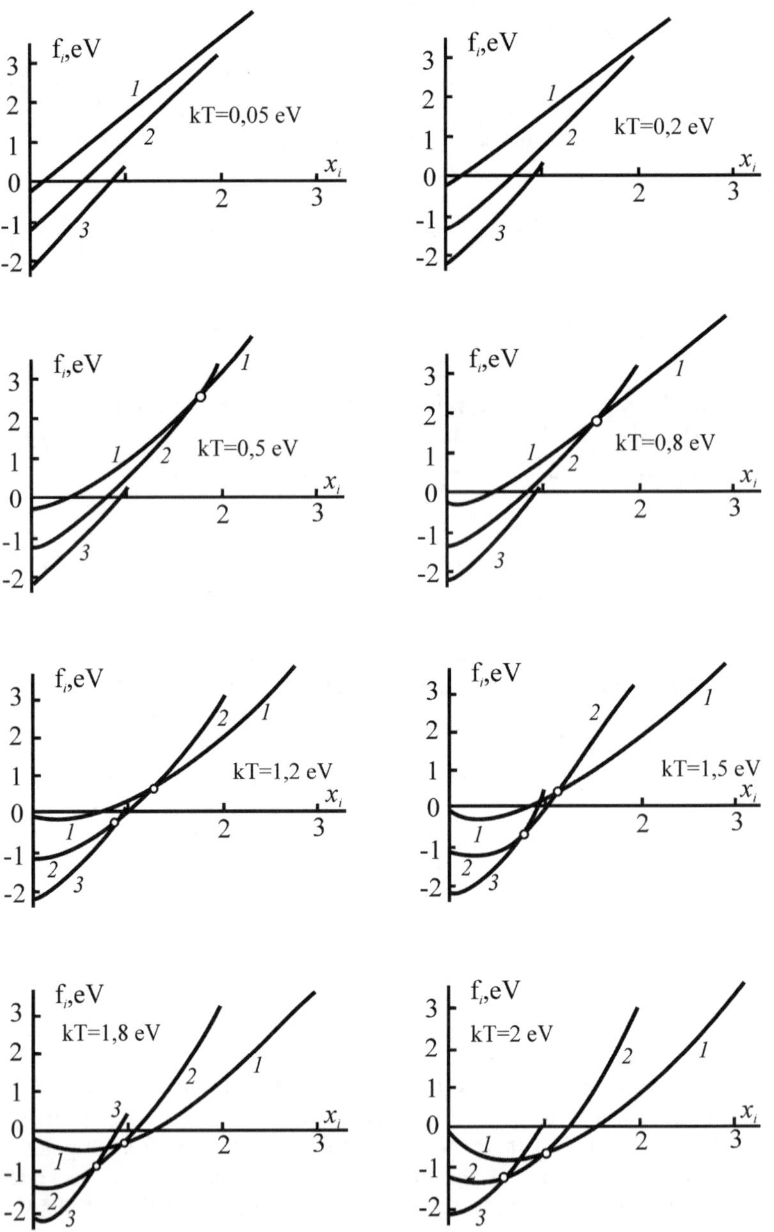

Figure 6. The plots of concentration dependence of free energies $f_i = f_\alpha(x_\alpha)$, $f_\beta(x_\beta)$, $f_\gamma(x_\gamma)$ (curves *1, 2, 3*, correspondingly) for the phases i = α, β, γ constructed for energetic parameters (41) and different temperatures. The intersection points of the curves f_α, f_β, f_γ, corresponding to the phase transformations $\alpha \rightarrow \beta \rightarrow \gamma$ are marked off by circles

The concentration range of formation of the phase α is increased and of the phases β and γ is decreased with a rise in temperature. This result is in agreement with experimental data because experiments established the decrease of hydrogen concentration in the each phase with the increased temperature and this stimulates the reconstruction of the crystal structures and the phase transformations by the chemical reactions (2) at the end.

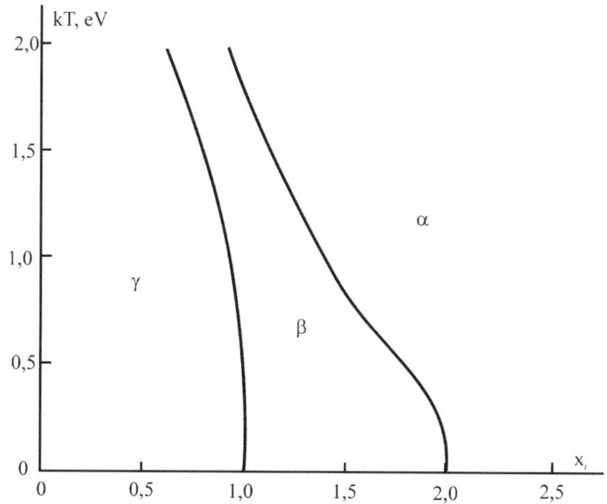

Figure 7. The diagram of the phase transformations $\alpha \rightarrow \beta \rightarrow \gamma$

The knowledge of phase parameters for each concrete system of borohydrides and hydrides of alkali and other metals from independent experiments can permit to predetermine the necessary temperature conditions of dehydrogation by the derived formulae for the purpose of production of hydrogen, an ecologically clean power source, in the required quantities.

It is interesting to note also that in recent years a start has been made on the investigations for solving the technical moments of problems of the use of borohydrides and hydrides of metals in transport as the source of hydrogen, energy carrier.

References

1. Amendola S.C., Sharp-Goldman S.L., Januja M.S., Spencer N.C., Kelly M.T., Petillo P.J., Binder M. Hydrogen storage by $NaBH_4$ // *Int. J. Hydrogen Energy*, 2004, **29**: 263.
2. Lodziana Z., Vegge T. Structural stability of complex hydrides: $LiBH_4$ revisited // *Phys. Rev. Lett.*, 2004, **93** (14): 145501.
3. Vajo J.J., Skeith S.L., Mertens F. Reversible storage of hydrogen in destabilized $LiBH_4$ // *J. Phys. Chem. B Condens. Matter. Mater. Surf. Interface. Biophys.*, 2005, **109** (9): 3719–3722.
4. Chandhuri S., Muckerman J.T. A critical first step for reversible hydrogen storage in $NaAlH_4$ // *J. Phys. Chem. B Condens. Matter. Mater. Surf. Interface. Biophys.*, 2005, **109** (15): 6952–6957.
5. Meisner G.P., Scullin M.L., Balogh M.P., Pinkerton F.E., Meyer M.S. Hydrogen release from mixtures of lithium borohydride and lithium amide: a phase diagram study // *J. Phys. Chem. B Condens. Matter. Mater. Surf. Interface. Biophys.*, 2006, **110** (9): 4186–4192.

6. Au.M. Jurgensen A., Zeigler K. Modified lithium borohydrides for reversible hydrogen storage // *J. Phys. Chem. B Condens. Matter. Mater. Surf. Interface. Biophys.*, **110** (13): 7062–7067.
7. Au.M. Jurgensen A., Zeigler K. Modified lithium borohydrides for reversible hydrogen storage // *J. Phys. Chem. B Condens. Matter. Mater. Surf. Interface. Biophys.*, 2006, **110** (51): 26482–26487.
8. Talyzin A.V., Anderson O., Sundqvist B., Kurnosov A., Dubrovinsky L. High-pressure phase transition in $LiBH_4$ // *J. Solid State Chem.*, 2007, **180** (2): 510–517.
9. Laversenne L., Bonnetot B. Hydrogen storage using borohydrides // *Ann. Chim. Sci. Mat.*, 2005, **30** (5): 495–503.
10. Luo W., Ronnebro E. Towards a viable hydrogen storage system for transportation application // *J. Alloy. Compd.*, 2005, **404–406**: 392–395.
11. Ranjard J.F., Glipa X. Procédé pour maintenir liquide a la température ambiante une solition aqueuse de borate de sodium. - French Pat. 2 859648-A1. (2003/9/03).
12. Liu J., Fang Z.Z., Sohn H.Y. A new Li-Al-N-H system for reversible hydrogen storage // *J. Phys. Chem. B Condens. Matter. Mater. Surf. Interface. Biophys.*, 2006, **110** (29): 14236–14239.
13. Gilbert M., Brown P.J., Joachim H., Schaeibel J.H. Complex hydrides for hydrogen storage studies of the $Al(BH_4)$ system. - Project ID # STP24. - Metal hydride Center of Excellence. - 2007.
14. Soloveichik G.I. Metal borohydrides as hydrogen storage materials // *Material Matters.*, 2007, **2** (2): 11–15.
15. Balema V.P. Mechanical processing in hydrogen storage research and development // *Material Matters.*, 2007, **2** (2): 16–18.
16. Goldshmidt H.J. Splavy vnedreniya. - Moscow: Mir, 1971. - V. 1 and 2. - 424 and 464 p.
17. Alefeld G., Felkl H. Vodorod v metallakh. - Moscow: Mir, 1981. - V. 1 and 2. - 480 and 432 p.
18. Stock A., Sfitterlin W., Kurzen F. Boron hydrides. XX. Potassium diborane $K_2(B_2H_6)$ // *Z. Anorg. Allgem. Chem.*, 1935, **225**: 225–242.
19. Schlesinger H.I., Sanderson R.T., Burg A.B. Metallo borohydrides I. Aluminium borohydride // *J. Am. Chem. Soc.*, 1940, **62**: 3421–3425.
20. Burg A.B., Schlesinger H.I. Metallo borohydrides II. Beryllium borohydride // *J. Am. Chem. Soc.*,1940, **62**: 3425–3429.
21. Schlesinger H.I., Brown H.C. Metallo borohydrides III. Lithium borohydride // *J. Am. Chem. Soc.*, 1940, **62**: 3429–3435.
22. Schlesinger H.I., Herbert C., Brown H.C., Abraham B., Bond A.C., Davidson N., Finholt A.E., Gilbreath J.R., et. al. New developments in the chemistry of diborane and the borohydrides // *J. Am. Chem. Soc.*, 1953, **75** (1): 186–190.
23. James B.D., Wallbridge M.G.H. Metal tetrahydroborates // *Prog. Inorg. Chem.*, 1970, **11**: 97–258.
24. Greenwood N.N. The chemistry of boron // *Comprehensive Inorg. Chem.*, 1973, **8**: 745–762.
25. Maybury P.C., Mitchell R.W., Hawthorne M.F. Hydrogen adducts of cobalt and nickel boride // *J. Chem. Soc., Chem. Commun.*, 1974, 534–535.
26. Smitls K. Metally. Spravochnik. - Moscow: Metallurgiya, 1980. - 448 p.
27. Smirnov A.A. Molekulyarno-kineticheskaya teoriya metallov.- Moscow: Nauka, 1966. - 488 p.
28. Matysina Z.A. Atomniy poryadok i svoystva splavov. - Dnepropetrovsk: DGU, 1981. - 112 p.
29. Matysina Z.A., Milyan M.I. Teoriya rastvorimosti primesey v uporyadochennyh fazah. - Dnepropetrovsk: DGU, 1991. - 180 p.
30. Matysina Z.A., Schur D.V. Vodorod i tverdofaznie prevrascheniya v metallakh, splavakh, fulleritakh. - Dnepropetrovsk: Nauka i Obrazovanie, 2002. - 420 p.
31. Matysina Z.A., Zaginaichenko S.Yu. Defecty structury kristallov. -Dnepropetrovsk: Nauka i Obrazovanie, 2003. -284 p.
32. Matysina Z.A., Zaginaichenko S.Yu., Schur D.V. Rastvorimost primesey v metallah, splavah, intermetallidah, fulleritah. - Dnepropetrovsk: Nauka i Obrazovanie, 2006. - 514 p.

INTERACTION PECULIARITIES OF HYDROGEN AND Ni-B GALVANIC ALLOYS

A.V. ZVYAGINTSEVA
Voronezh State Technical University, 394026, Voronezh, Russia

Abstract. Thus the introduction of boron into Ni fcc lattice leads to the change of its parameters, to the increase of internal stresses in the received Ni-B films. With the increase of boron concentration in the deposits the share of large blocks which are character for pure Ni decreases and the share of small block increases; the texture perfection decreases, Ni lattice microdeformations increase and lattice spacing reduces.

Keywords: electrochemical, Ni-B, electrocrystallization, hydrogen saturation

1. Introduction

Nickel alloys have become the subject of detailed research long time ago due to the existence of a number of mechanical, chemical and magnetic properties. Ni-B alloys are of great interest for the researchers. On the one hand, the introduction of boron into nickel coatings, received by electrocrystallization technique, improves physico-mechanical properties, such as wear resistance, heat resistance, solders ability, weldability, improves corrosion resistance and several other parameters. Thus, nickel coatings alloyed by boron are becoming very important for the quality of engineering materials for radioelectronic industry, instrument and mechanical engineering. On the other hand, it was found out that boron introduction into Ni coatings increases internal stress, fragility and can activate additional hydrogen saturation of nickel basis. In connection with this fact it is necessary to select experimentally the range of boron concentrations and modes of electrocrystallization which contributes to less fragility and hydrogen saturation.

2. Experimental procedures

The aim of the present work is to study the influence of boron on electrochemical hydrogen saturation, properties and structure of Ni-B coatings precipitated from sulphamate electrolytes.

The research was conducted in nickel-plating sulphamate electrolyte with the usage of boron compounds of the class of highest polyhedral borates [1]. Hydrogen saturation ($V_{H2,}$ sm^3/100 g) of the films was detected by vacuum extraction method. Polarization measurements were conducted by potentiodynemic method (4 MV/sec) on potentiostats П-5827M using recording potentiometer КСП. Partial curves of Ni disposition were plotted by the expansion of accumulation curves of nickel and hydrogen ions discharge.

3. Results and discussion

Hydrogen relative content in Ni – films is 104 sm^3/100 g (i_k = 2 A/dm^2; pH = 4.0; $t_{electrolyte}$ = 40°C; d = 4 μm). On Fig. 1 we can see data describing the dependence of the boron addition concentration on Ni-B coatings' hydrogen saturation.

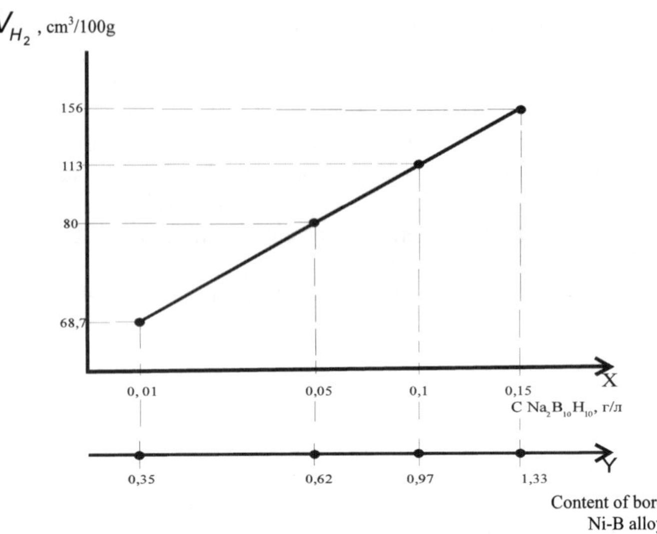

Figure 1. Dependence of Ni-B coatings hydrogen saturation from boron addition concentration in electrolyte. Electrolysis regomes: i_k = 2 A/dm^2; $t_{electrolyte}$ = 40°C; pH = 4.0; d = 4 μm; basis – copper grade M-1

The degree of Ni-B coatings hydrogen saturation increases with the increase of boron concentration from 0.1% to 1.5%. The introduction of polyhedral borate class boron addition agent into nickel-plating sulphamate electrolyte leads to considerable hydrogen discharge on the cathode. Hydrogen appearance is probably connected with the boron containing adding decay to hydrogen and elementary boron. Elementary boron penetrates to nickel fcc lattice and forms boron solid solution in nickel.

The increasing of H_2 discharge during Ni-B^0 precipitation in presence of boron containing addition is proved by experimental data. In Table 1 we can see partial densities of Ni-B ^{0} and H_2 discharge currents which characterize electroreduction speed.

TABLE 1. Partial densities of Ni-B^0 and H_2 current. Electrolysis regimes: i_k = 2 A/dm^2; $t_{electrolyte}$ = 40 °C; pH = 4.0; d = 4 μm

Boron containing adding concentration in electrolyte (g/l (C_B))	Containing of boron in the film Ni-B (%)	Currents' partial densities (A/dm^2)		BT_{H2} (%)
		i_{Ni-B}^{0}	i_{H2}	
0.01	0.35	1.961	0.039	1.96
0.05	0.62	1.911	0.089	4.35
0.10	0.97	1.857	0.143	7.05

It is clear from the Table 1 that during the increasing of boron containing adding concentration in electrolyte from 0.01 to 0.10 g/l partial share of the current falling on $Ni\text{-}B^0$ formation decreases from $i_{Ni\text{-}B}{}^0 = 1.961$ to $i_{Ni\text{-}B} = 1.857$ A/dm^2 and increases for H_2 discharge from $i_{H2} = 0.039$ to $i_{H2} = 0.143$ A/dm^2 ($i_k = 2$ A/dm^2; pH = 4.0; $t_{electrolyte} = 40°C$).

For H_2 emission it is necessary to consume the amount of current from 1.96% (C_B = 0.01 g/l) to 7.05% (C_B = 0.10 г/л) with $i_k = 2$ А/дм^2 which increases the opportunity to include hydrogen into the coating. Discharged hydrogen is adsorbed during metal electrocrystallization and after that it is included into metal coating activating its complementary hydrogen saturation. The degree of Ni-B films' hydrogen saturation increases from 68 to 113 sm^3/100 g ($i_k = 2$ A/dm^2; pH = 4.0; $t_{electrolyte} = 40°C$) with the increase of boron containing from 0.1% to 1% .

The analysis of the experimental data shows that hydrogen content in Ni-B alloy is lesser than in Ni coatings for identical electrolyses regimes. For example: $V_{Ni} = 104$ sm^3/100 g and $V_{Ni\text{-}B} = 80$ sm^3/100 g ($i_k = 2$ A/dm^2; $pH_{electrolyte} = 4.0$; $t_{electrolyte} = 40°C$) during the concentration of boron containing adding in electrolyte up to 0.05 g/l. Further increasing of the concentration of boron containing adding in electrolyte up to 0.15–0.20 g/l and more leads to the increasing of boron and hydrogen containing in Ni-B film and the amount of V_{H2} in Ni-B coatings increases in comparison with Ni-film. So, during the concentration of boron containing adding in electrolyte $C_B = 0.15$ g/l; $V_{H2,\,Ni\text{-}B} = 156$ sm^3/100 g ($i_k = 2$ A/dm^2; pH = 4.0; $t_{electrolyte} = 40°C$; d = 4 μm).

Comparing the influence of boron on hydrogen saturation process of Ni-B films in comparison with Ni^o (Fig. 1, Table 2) we can conclude that during the formation of electrochemical alloys of boron with d-elements the systems with rather ambiguous ability of hydrogen absorption are forming [2]. In one case, when the concentration of boron in the alloy is rather low (less than 1%) the ability of hydrogen absorption decreases and during the increasing of boron concentration in the alloy the ability also increases. When boron concentration in alloy is lower than 1% hydrogen absorption is lower than in pure Ni^o and when the concentration is higher than 1% hydrogen absorption increases in comparison with Ni^o [3]. The above-mentioned phenomenon is connected with structural changes. In Table 2 there are data about the influence of boron concentration in Ni-B coatings on the internal stress (σ), phase composition, structure, texture and hydrogen saturation. Internal macrostresses of Ni-B coatings increase during the increasing of boron content. Thus during the increasing of boron up to 1% the internal stresses increase approximately by a factor of 4 and lattice parameter decreases by 0.0003 A^o for 0.5% boron and by 0.0009 A^o for 1% boron [5].

According to the X-ray diffraction and electron microscopical analysis, diffraction pattern received for galvanic Ni-B films with boron content up to 1% are characterized by clear diffraction peaks and the same set of lines corresponding to Ni. Ni-B films received from sulphamate electrolyte are polycrystalline and consist of pure Ni with a thin layer of NiO. Identification shows that the continued proportion:

$$\theta = \frac{H_i^2 + K_i^2 + L_i^2}{H_k^2 + K_k^2 + L_k^2},$$

where θ – angle of reflection and H, K, L – indices of reflecting planes for the specific line in X-ray spectrum has values close to the series θ for Ni fcc lattice; 1; 1.33; 2.66; 3.67; 4, that indicate the phase composition of Ni-B film which is similar to

Ni phase composition. During the alloying of Ni coatings with boron up to 1% the solid solution of boron and hydrogen in crystallizing Ni with fcc lattice which is character for α-Ni forms. The peculiarities of microstructure and the formation of single-phase alloy are mainly determined by the type of Ni lattice on the basis of which the solid solutions form as the difference between Ni and B atomic radiuses according to Hume-Rothery's rule does not exceed 15% (r_{Ni} = 0.124 nm; r_B = 0.091 nm; r_H = 0.046) [4]. Inclusion of the alloying component – boron into α-Ni lattice increases the dispersion of the received coatings (Figs. 2 and 3).

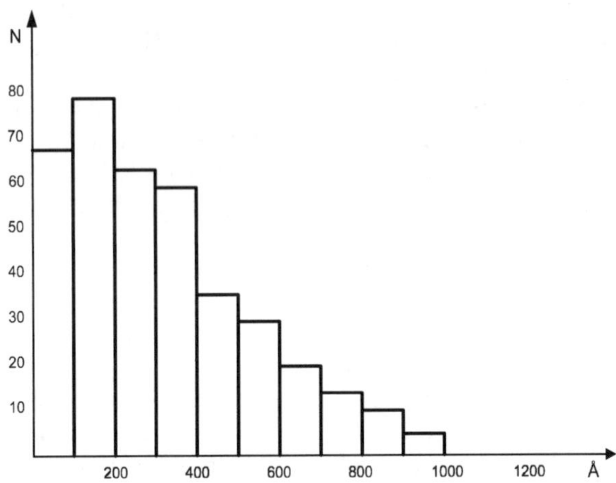

Figure 2. The histogram of the grains distribution with size for nickel film Electrolysis regimes: i_{cath} = 2 A/dm^2; pH = 4.0; t = 40°C

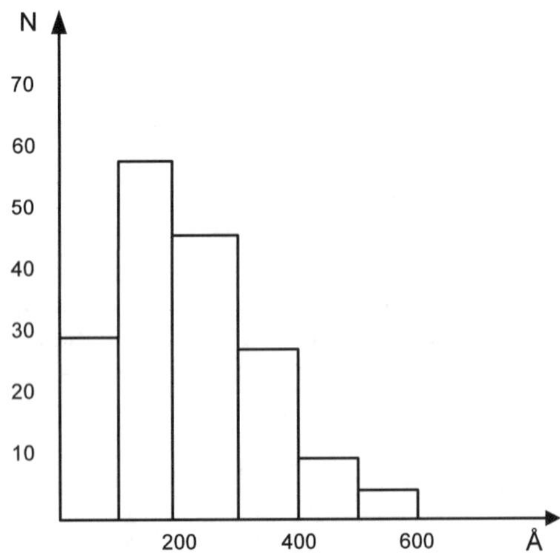

Figure 3. The histogram of the grains distribution with size for nickel-bor film Electrolysis regimes: i_{cath} = 2 A/dm^2; pH = 4.0; t = 40°C

TABLE 2. Dependence of the internal stresses (σ), texture, phase composition and hydrogen saturation from boron content in electrolytic Ni-B alloy Ni-B alloy hardness, internal stress, texture, phase composition and hydrogen amount in the Ni-B deposits

B content in the alloy (%	Phase composition	d_{311} (A$^{o)}$	A_{av} (A^{o})	Grain size (μm)	HKL Texture axis	Pa	V_{H2} (cm^3/100 g)
0	fcc-ni	1.0624	3.5236	0.06–0.20	<111>	5.7	80
0.1	One phase: solid solution	–	–	–	<111>	0	43
0.5	of B in fcc-Ni	1.0623	3.5233	0.02–0.12	<111> weak	35	60
1.0		1.0621	3.5227	–	none	80	88

Electrolysis regimes: $i_к = 2$ A/dm^2; pH = 4.0; $t_{electrolyte} = 40$°C; d = 15 μm

Notations: d_{311} – interplanar distance; a_{av} – lattice constant average value; H; K; L – indices of reflecting planes, 3;1;1 – correspondingly (index311 equals to 114°).

On Fig. 2 there is the histogram of the grains distribution according to the size of their quantity for Ni film received during $i_к = 2$ A/dm^2; pH = 4.0; $t_{electrolyte} = 40$°C. We can conclude that in this regime the major amount of grains (68%) is located within the limits of sizes from 0 to 400 Å and the received structure can be referred to fine-grained structure and a form of grains to equiaxed grains.

On Fig. 3 there is the hystogram of the grains distribution according to the size of their quantity for Ni-B film with 0.5%, B content received during $i_к = 2$ A/dm^2; pH = 4.0; $t_{electrolyte} = 40$°C. In this regime the major amount of grains (93%) is located within the limits of sizes from 0 to 400 Å and the received structure has equigenets form.

Thus the introduction of boron into Ni fcc lattice leads to the change of its parameters, to the increase of internal stresses in the received Ni-B films. With the increase of boron concentration in the deposits the share of large blocks which are character for pure Ni decreases and the share of small block increases; the texture perfection decreases, Ni lattice microdeformations increase and lattice spacing reduces.

The ambiguous boron influence on hydrogen saturation probably can be connected with the following phenomenon. During relatively small concentration (up to 1%), boron embeds into fcc α-Ni lattice and concentrates on grains boundaries and Ni lattice defects. And block them. This effect stops hydrogen inclusion. During the increasing of boron concentration (more than 1%) in Ni lattice new defects appear which are the basis of hydrogen inclusion and this increases the possibility of hydride phase formation in Ni-B electrochemical alloy. The increase of hydrogen content in electrochemical Ni during higher boron concentration in comparison with pure Ni is connected with hydrogen collection into boron surrounding and their better interaction than interaction of Ni and B. Hydrogen atoms are distributed in lattice defect places and they form compound with boron and nickel.

4. Conclusions

The research of boron influence on hydrogen saturation, Ni-film alloyed by boron (up to 1.5%) structure has shown the following results:

1. Hydrogen content in Ni-B coating is lesser 1% than in electrolytic nickel and make up ~ from 60 to 100 sm^3/100 g.
2. Ni coatings alloyed by boron up to 1.5% increases its hydrogen saturation up to 156 sm^3/100 g. Introduction of the specific concentration of boron into nickel coating activate the increase of the amount of the absorbed hydrogen as a number of defects in Ni-B alloy exceeds its amount than in pure Ni.
3. Boron and hydrogen introduction into electrochemical nickel occurs though α-Ni fcc lattice defects.

References

1. Patent № 2124072 Russia. Electrolyte for electrochemical Ni-B functional coatings precipitation. A.V. Zvyagintseva, A.I. Falicheva. Inquiry № 93036355/02. Published 27.12.98. Bulletin № 36.
2. Makkey K. Hydrogen metals' compounds, Moscow: Mir Press: 1968, 244 p.
3. Zvyagintseva A.V. Boron influence on Ni film hydrogen saturation, *Alternative Power Engineering and Ecology* 2006 (5): 85–87.
4. Soltsev Y.P., Pryakhin E.M. Materials science: Text book for higher educational institutions, Saint-Petersburg: Himizdat Press: 2004, 736 p.
5. Zvyagintseva A.V., Falicheva A.I. Physico-mechanical properties of Ni coatings alloyed by boron, *Electroplating and Surfaces Processing*, Moscow: 1997, **5** (2): 24–31.

MECHANISMS OF METAL HYDRIDES FORMATION IN THE PRESENCE OF BORON COMPOUNDS

A.V. ZVYAGINTSEVA, Y.N. SHALIMOV
Voronezh State Technical University, 394026, Voronezh, Russia

Abstract. We have made an attempt to receive aluminum hydride by electrochemical method during cathodic hydrogen saturation of aluminum electrode. It is very interesting to consider alloys of light metals and boron from the point of their hydrogen accumulating capacities.

Keywords: electrochemical, metal hydrides, hydrogen formation, metals borides, boron influence

1. Introduction

In the literature [1] there is an idea that for forth period transition-row metals (from Ti to Ni) there is the so-called “hydrogen gap” that is the metals of this row should not form the compounds with hydrogen. However, in the 1960s of the Однако в 60-х of the twentieth century new information has appeared [2]. It was studied that Cr can form hydrides on the cathode during its electrochemical reduction. These results were confirmed by scientific researches [3]. The reason of such different approaches to metal hydrides formation on the cathode is that physicochemical properties of electrochemical coatings differ greatly from metals received by other methods. According to classical scheme of metal structures formation on the cathode we can present the following mechanism of hydrogen formation:

$$2H_2O + 2\bar{e} \rightarrow 2H^o + 2OH^-$$
$$H^o + Me^o \rightarrow Me^+ H^-$$
$$H^o + H^o \rightarrow H_2$$

2. Experimental procedures

We have made an attempt to receive aluminum hydride by electrochemical method during cathodic hydrogen saturation of aluminum electrode.

The formation of hydrides in aluminum structure was defined by the method of internal friction. On the graphic charts $Q^{-1} = f(T)$ we have defined two characteristic “hydrogen” peaks under the temperature -35 ÷ -40°C. The analysis of relationships $Q^{-1} = f(T)$ shows (Fig. 1(a, b)), that during the increasing of current density the background of the internal friction on the cathode increases. This can be explained by the formation of hydroxide $Al(OH)_3$ adsorbed film on the electrode, which significantly damps oscillatory system.

It is very interesting to consider alloys of light metals and boron from the point of their hydrogen accumulating capacities.

3. Results and discussion

From the point of view of metal structures forming it is very interesting to add boron compounds into electrolytes for receiving alloys based on transition metals such as Ni, Cr. Initiation of hydrogen saturation process in presence of boron compounds is mentioned in several researches [4]. We consider that the causes of increased surfaces' hydrogen saturation are peculiarities of boron electronic structure (all there electrons of the outer shell are in excited state (sp^2-hybridization) on the higher energy level). Besides the peculiarity of the pattern geometry of electron clouds allows to suppose the increased activity for hydrogen bond formation. A lot of researchers mention that even metal borides of low periods are inclined to hydrogen absorption. Such systems are attracting more and more researchers studying the problems of the creation of hydrogen storage systems. Not high values of H-Me bond energy in these systems require additional experimental studies for their description. The objective of the present research is to study boron influence on the process of hydrogen saturation of electrochemical coatings containing boron as one of its components. The idea of hydrogen storage in metals as hydride compounds is very interesting for the users of hydrogen fuel because it has several advantages. First of all the usage of such method excludes high power inputs as in comparison with cryostatic method (hydrogen storage in liquid phase).

Secondly, it is safer as in comparison with the usage of high-pressure cylinder for transportation. Hydrogen storage is possible in terms of several energy effective compounds, for example, ammonia, hydrazine, methyl alcohol. However these substances are considered to be harmful and chemically dangerous. On the other hand the processes of fuel hydrogenation are very effective from the point of view of carbohydrates storage safety and increasing of their specific calorific power. In accordance to the data of academician V.A. Legasov [5] fuel calorific power is increasing proportionally to the ratio H/C (atoms content) in hydrocarbon (Table 1).

TABLE 1. Thermodynamic characteristics of several hydrides: enthalpy and entropy formation

Initial and final compositions	ΔH_f^0 (kcal/mol H_2)	ΔS_f^0 (cal/deg·mol H_2)
Li-LiH	-432	-32.8
Na-NaH	-270	–
Mg-Mg H_2	-17.8	-32.3
$Mg_2NiH_{0,3}$-	-15.4	-29.0
$Mg_2NiH_{4,2}$	-41.7	-30.4
Ca-CaH_2	-2.7	-30.9
Al-AlH_3	-49.6	–
La-LaH_2	-7.6	-26.0
$LaNi_5$-$LaNi_5H_6$	-29.9	-30.0
Ti-$TiH_{1,97}$	-6.7	-25.0
$TiFeH_{0,1}$-$TiFeH_{1,0}$	-39.0	–
Zr-ZrH_2	-15.3	-18.8
V-$VH_{0,2}$	-9.6	-34.0
$VH_{0,9}$-$VH_{2,0}$	-20.2	-28.9
U-UH_3		

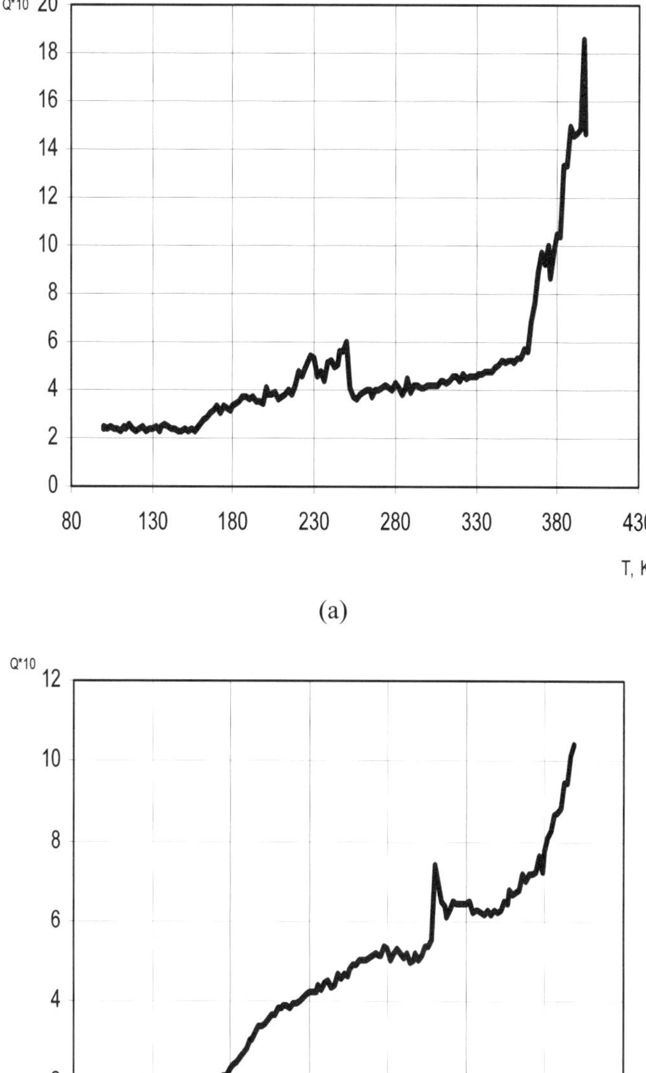

Figure 1. Temperature dependence of aluminum internal friction subjected to hydrogen saturation by electrochemical methods during 450 minutes with current cathodic density 5 A/dm^2 (a) and during 510 minutes with current cathodic density 80 A/dm^2 (b). Electrolyte: 0.5 M solution $Al_2(SO_4)_3$

Metal hydrides could be a secure source of hydrogen received by classical chemical or thermal methods. Recently the electrochemical method ahs become widespread and it has several advantages for separate families of metals and alloys. One of the most important properties of gases stored in hydride forms is the return of the major part of accumulated gas under the constant pressure. This fact proves validity of the suggestion that gas cannot be dissolved in metal structure. Though for some metals hydrogen dissolution is possible in terms of interstitial phases without changing interatomic spacing in lattice.

Defining the opportunity of the material usage as hydrogen accumulator it is necessary to take into consideration its thermodynamic characteristics (enthalpy, entropy, Gibbs energy). For example, data given for aluminum hydride in Table 1.

For the reaction $Al \rightarrow AlH_3$, $\Delta H_f = -2.7$ kcal/mol H_2; $\Delta S = -30.9$ cal/deg mol H_2.

However, the receiving of free aluminum hydride [1] is not a simple task. Chemical receiving of aluminum hydride is possible under the scheme mentioned below.

$$\begin{array}{l} 4\,Li\,AlH_4 + ZnJ_2 \rightarrow Li_2[Zn(AlH_4)_4)] \\ \quad\quad\quad\quad\downarrow \quad\quad\quad + Zn\,J_2 \\ \quad\quad 2Zn(AlH_4)_2 + 2LiJ \\ \quad\quad\quad\quad\quad\downarrow \\ ZnH_2 + AlH_3(\text{into solution.}) \end{array}$$

As for the purity of aluminum hydride in the solution, some researchers think that there can be some foreign substances in electrolyte which are formed by type of alanates [1].

Earlier we have established the increase of hydrogen content in electrochemical alloys Ni-B with the increase of boron concentration [6].

The capacity of the accumulated hydrogen for alloys based on boron received by metallurgical method will probably depend on the technology of the receiving of the accumulating system and not only on phase composition of Me-B alloy. Table 2 represents data of the composition and properties of metal borides received by metallurgical method.

According to the data indicated in Table 2 we can draw the following conclusion. First of all, the increasing of boron percentage in the series AlB_2 $^-AlB_{10}$ $^-AlB_{12}$ stimulates the rise of microdisorptions in the alloy, therefore the volume of the absorbed hydrogen by the alloy will increase with the increasing of the number of structure's defects. Secondly, the increase of these alloys accumulating capacities is connected with anomalous boron capacity to form bonds with hydrogen. Compounds of such type will have bigger capacity to occlude hydrogen if during the receiving of these alloys a branched structure is formed. This is connected with the increase of the diffusion speed of the released hydrogen atoms and further their disintegration in the process of gaseous H2 formation during thermal decomposition of complex hydrides. Alloys of such type have high chemical resistance so the number of hydrogen accumulation cycles will increase. During cathodic hydrogen saturation of such alloys in water electrolyte solutions there is no necessity in existence of hydrogen ions' transporting channels in the structure of electrode.

TABLE 2. Composition and properties of metals' borides

Metals borides	Composition	Density (g/sm^3)	Chemical properties
Beryllium	Be_5B	2.10	Decomposed by acids with the formation of boranes
	α-Be_2B	2.20	
	β- Be_2B	2.35	Chemically resistant, are nc decomposed in acids
	BeB_6	2.33	
Magnesium	MgB_2	2.48-2.67	Decomposed by water with the extraction of boranes
	MgB_6	2.45-2.47	Are not decomposed by water and acids
	MgB_{12}	2.44	
Calcium	CaB_6	2.45	Resistant to acidic solutions (except HNO_3).
Strontium			
Barium	SrB_6	3.42	Decomposed during alloyage with alkali and alkaline carbonates
	BaB_6	4.25	
Aluminum	AlB_2	2.84	Has high chemical resistance
	AlB_{10}	2.54	
	AlB_{12}	2.49	
		2.59	
Titanium	TiB_2	4.52	Resistant to acids in the cold except HNO_3
Zirconium	ZrB_2	6.17	Resistant in the cold to HCl and H_3PO_4
Vanadium	VB_2	5.10	Dissolving in the cold in HNO_3 and $HClO_4$
Niobium	NbB_2	6.60	Dissolving in boiling H_2SO_4, H_3PO_4 and $H_2C_2O_4$
Tantalum	TaB_2	11.70	Dissolving during long-term boiling in H_2SO_4 and H_3PO_4
Chrome	CrB_2	5.22–5.60	Dissolving in HCl, delude by H_2SO_4 and $H_2C_2O_4$ in the cold
Molybdenum	Mo_2B_5	7.01	Dissolving in the cold in HNO_3 and $HClO_4$
Tungsten	W_2B_5	13.0	Dissolving during long-term boiling in HNO_3, H_2SO_4 and $HClO_4$
Manganese	MnB	–	Dissloving during heating in the solution of KJO_4 in 5–10 minutes

The supposed mechanism of interaction can be presented as the following scheme: 1 stage – H_3O^+- $H_2O \rightarrow H^+$ represents dehydration of hydroxonium cation in electric double layer. 2 stage: H^+ free cation migration can be realized in the deep layer under the action of electric field of the alloy crystal. Migrating in the alloy cation H^+ fixes on the potential pit (structure defects) and forms element-hydrogen hydride. Saturation of deep layers can be realized at the expense of token passing of fixed hydrogen ions or free transportation to the depth of the alloy structure in connection with infinitely small size of dehydrated H^+ ion ($r_H^+ \sim 10^{-5}$ A°). Comparing the influence of boron on the process of hydrogen saturation of Ni-B alloy with Al-B alloys we can conclude that during the formation of electrochemical boron alloys with d-elements some systems appear and their hydrogen absorbing capacity is ambiguous. In one case when boron concentration in the alloy is rather small (less than 1%) the absorbing capacity decreases and with the rise of boron concentration in the alloy the capacity increases. If boron concentration in the alloy is lesser than 1% hydrogen absorption is lower than in pure Ni^{o}. If boron concentration in the alloy is more than 1% hydrogen absorption is higher than in pure Ni^{o}. This anomaly can be explained by the peculiarities of boron and nickel interaction depending on their concentration. It is obvious that in small concentrations of boron admixture in electrolyte the process of hydrogen disintegration (molization) is facilitated which can be indirectly seen visually during the increasing of the concentration of the evolved gaseous hydrogen during electrolysis. Interaction of hydrogen and metals is more possible only in atomic state. During the increasing of boron concentration in the alloy Ni-B the distortions in Ni lattice structure are growing and the number of defects is growing also. So the possibility of the formation of hybrid metal phase in electrochemical alloy Ni-B is increasing.

4. Conclusion

Anomalous alloys' behavior from the point of hydrogen absorption can be explained by the environmental change of the number of structural defects created during alloys electrocrystallization and by anomalous behavior of protonated hydrogen in these systems.

References

1. Makkey K. Hydrogen metals' compounds, Moscow: Mir Press, 1968: 244 p.
2. Soltsev Y.P., Pryakhin E.M. Materials science: Text book for higher educational institutions, Saint-Petersburg: Himizdat Press, 2004: 736 p. (in Russian).
3. Povetkin V.V., Kovenskiy I.M. The structure of electrolitic coatings. Moscow: 1989, 136 p. (in Russian).
4. Zvyagintseva A.V., Falicheva A.I. Physico-mechanical properties of Ni coatings alloyed by boron, *Electroplating and Surfaces Processing*, 1997, **5** (2): 24–31.
5. Sadacov G.A. Galvanoplastics. Moscow: 1987, 288 p. (in Russian).
6. Zvyagintseva A.V. Boron influence on Ni film hydrogen saturation, *Alternative Power Engineering and Ecology*, 2006 (5): 85–87.

SOME ASPECTS OF PRODUCING AND USING BORON COMPOUNDS

S.A. MELNIKOV, A.P. PARSHIN
Federal State Unitary Enterprise "All-Russian Research Institute of Chemical Technology", 115409, Moscow, Russia
E-mail: manu@hotbox.ru

V.A. POPOV
Moscow Institute of Steel and Alloys, 119049, Moscow, Russia

K.Yu. ZHIZHIN, V.M. RETIVOV, N.T. KUZNETSOV
Kurnakov Institute of General and Inorganic Chemistry, Russian Academy of Sciences, 119991, Moscow, Russia

Abstract. The methods of synthesis of salts of clozoboron anions $B_{10}H_{10}^{2-}$ and $B_{12}H_{12}^{2-}$ are studied. These salts can be used as strengthening particles of a metal matrix from copper, aluminium, magnesium. For reception of products with strengthening particles used mechanical alloy in the planetary activator. The products of thermal decomposition of boron salts products, surface and structure of particles of a metal matrix with strengthening particles are studied.

Keywords: boron–hydrogen, boron compounds, closo-borate, copper, aluminium and nickel matrix, mechanical alloying

1. Introduction

A reduction of weight of structural materials is a topical problem. This work is devoted to development of a new composite material using boron–hydrogen or boron–halogen compounds [1, 2] of 1 up to 2.5 g/cm^3 density as reinforcing particles. A boron compound is anion of the general formula $Cat_2B_nX_n$, where Cat is a metal cation or onium cation, B is boron, X is hydrogen or chlorine, n = 10 or 12. The compounds whose decomposition temperature exceeds 500°C were chosen for the researches.

2. Production, preparation and researches of the properties of the initial materials

The following materials were taken. Copper, aluminium and nickel were used for the matrix. The boron compounds were as follows: $Cs_2B_{10}H_{10}$, $[Me_2NH_2]_2B_{10}H_{10}$, $[Ph_4P]_2B_{10}H_{10}$, $[Et_3NH]_2B_{10}Cl_{10}$, $Cs_2B_{12}H_{12}$, $[Et_3NH]_2B_{12}H_{12}$, $[Bu_3NH]_2B_{12}H_{12}$, $[Bu_4N]_2B_{12}H_{12}$, $[Ph_4P]_2B_{12}H_{12}$, $((C_2H_5)_3NH)[CuB_{10}H_{10}]$.

Figures 1 and 2 show the general view of particles of some boron–hydrogen compounds used. We can see from the figure that the shape and size of boron compounds can be most diverse, but they have one feature in common: all large crystals are not monolithic strong material but consist of finer particles. These fine particles were studied on a Nanoscan scanning probe microscope-nanoindenter to determine the properties of a particle on the nanolevel. The scanning window had size

B. Baranowski et al. (eds.), *Carbon Nanomaterials in Clean Energy Hydrogen Systems.*

of 4.9 × 4.9 × 0.3 μm. Figure 3 presents the testing results. It is seen that even fine particles are not homogeneous, but are rather similar to agglomerates of even finer particles. This was confirmed by transmission microscope studies of their structure. After an ultrasonic treatment in hexane, boron–hydrogen compounds were broken to nanosize particles.

Figure 1. General view of boron compounds used in the study: (a) $Cs_2B_{10}H_{10}$; (b) $[Me_2NH_2]_2B_{10}H_{10}$; (c) $[Ph_4P]_2B_{10}H_{10}$; (d) $[Et_3NH]_2B_{10}Cl_{10}$; (e) $Cs_2B_{12}H_{12}$; (f) $[Et_3NH]_2B_{12}H_{12}$

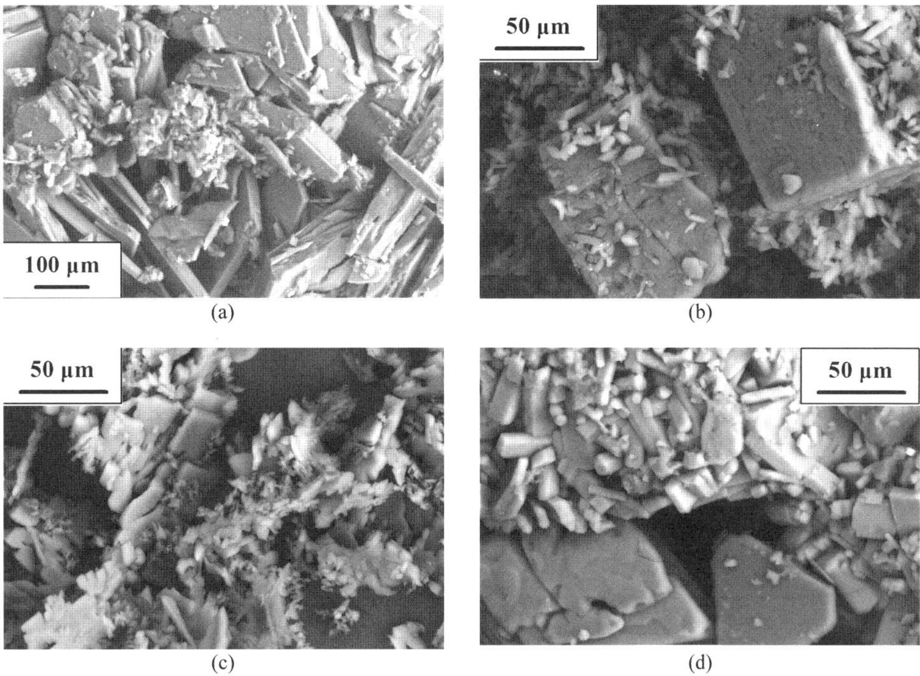

Figure 2. General view of boron compounds used in the researches: (a) $[Bu_3NH]_2B_{12}H_{12}$; (b) $[Bu_4N]_2B_{12}H_{12}$; (c) $[Ph_4P]_2B_{12}H_{12}$; (d) $((C_2H_5)_3NH)[CuB_{10}H_{10}]$

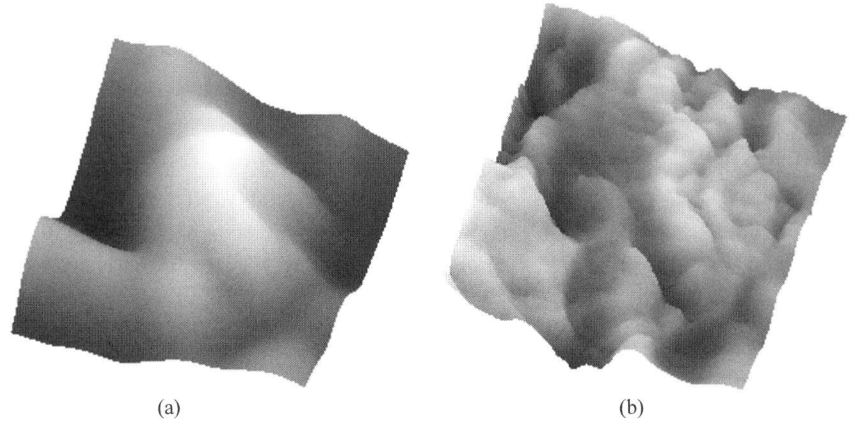

Figure 3. Results of particle of a boron compound scanning on a Nanoscan unit: (a) surface relief, (b) elasticity modulus map

Polyhedral boron–hydrogen compounds $B_{10}H_{10}^{2-}$ and $B_{12}H_{12}^{2}$ are bounded and convex polyhedra with triangular faces from boron atoms. The triangular shape is the most preferable fragment of constructing the skeleton of polyhedral boron–hydrogen anions. It ensures the maximum overlapping of B–B–B three-centre bonds and the maximum delocalization of electrons in the polyhedron, thus stabilizing the polyhedral particle on the whole. Experimentally was found that the closer the structure of the boron backbone of the polyhedral hydroborate ion to the icosahedron (B_{12}), the more stable the ion was. The most typical properties of the anions $B_{10}H_{10}^{2-}$ and $B_{12}H_{12}^{2-}$ is their kinetic stability and heat resistance, as well as the diversity of the substitution reactions, which enable using these anions in directed synthesis of substances with preset properties (Fig. 4). Researches of the pyrolysis products of borate compounds revealed formations similar to nanotubes (Fig. 5).

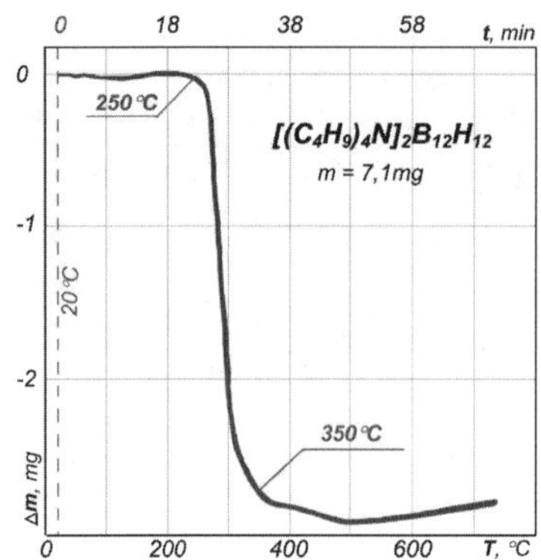

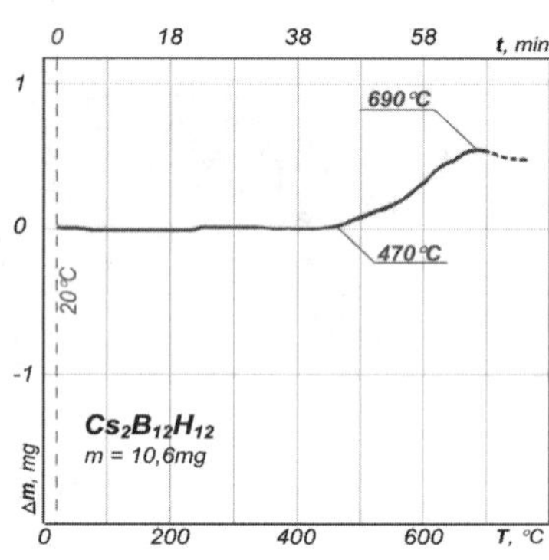

Figure 4. An example of changing the weight of closo-borate anion $[B_{12}H_{12}]^{2-}$ salts by heating

The closo-borate anions $B_{10}H_{10}^{2-}$ and $B_{12}H_{12}^{2-}$ were synthesized by various methods.

Anion $[B_{10}H_{10}]^{2-}$ was synthesized via the stage of synthesis of 1,6-*bis*-(triethylamine) decaborane according to the scheme:

$$B_{10}H_{14} + 2(C_2H_5)_3N \text{ xylol} \xrightarrow{\text{xylol, } 50\,^\circ C} 1{,}6\text{–}B_{10}H_{12}((C_2H_5)_3N)_2$$

$$1{,}6\text{–}B_{10}H_{12}((C_2H_5)_3N)_2 \text{ xylol} \xrightarrow{\text{xylol, } 100\,^\circ C} (C_2H_5)_3NH)_2[B_{10}H_{10}]$$

The residue $(Et_3NH)_2[B_{10}H_{10}]$ precipitated as the result of the synthesis was filtered, washed and recrystallized from hot water. The salt $(Et_3NH)_2[B_{10}H_{10}]$ was obtained with the yield of 87% in terms of boron.

The dodecahydro-closo-dodecaborate anion $[B_{12}H_{12}]^{2-}$ was produced by pyrolysis of a solution of decaborane in triethylamine borane according to the scheme:

$$B_{10}H_{14} + 2(C_2H_5)_3N{*}BH_3 \text{ xylol} \xrightarrow{\text{xylol, } 100\,^\circ C} ((C_2H_5)_3NH)_2[B_{10}H_{10}] + H_2.$$

After recrystallization from water, the yield of the end product was 85% in terms of $B_{10}H_{14}$.

Alkylammonium, tetraphenyl phosphonium and cesium salts of closo-borate anions $B_{10}H_{10}^{2-}$ and $B_{12}H_{12}^{2-}$ were obtained from the aqueous solution of respective sodium salts by reprecipitation by a respective halogenide. The salts obtained were dried up in a vacuum drying chamber (p = 0.05 Pa) at 50°C till complete removal of moisture. The yield of the products was quantitative. The composition of the salts $Cs_2B_{10}H_{10}$, $[(CH_3)_4N]_2B_{10}H_{10}$, $[(C_4H_9)_4N]_2B_{10}H_{10}$, $[(C_6H_5)_4P]_2B_{10}H_{10}$, $[(C_2H_5)_3NH]_2B_{10}H_{10}$,

$Cs_2B_{12}H_{12}$, $[(C_2H_5)_3NH]_2B_{12}H_{12}$, $[(C_4H_9)_4N]_2B_{12}H_{12}$, $[(C_6H_5)_4P]_2B_{12}H_{12}$ was confirmed by element analysis and IR spectroscopy data.

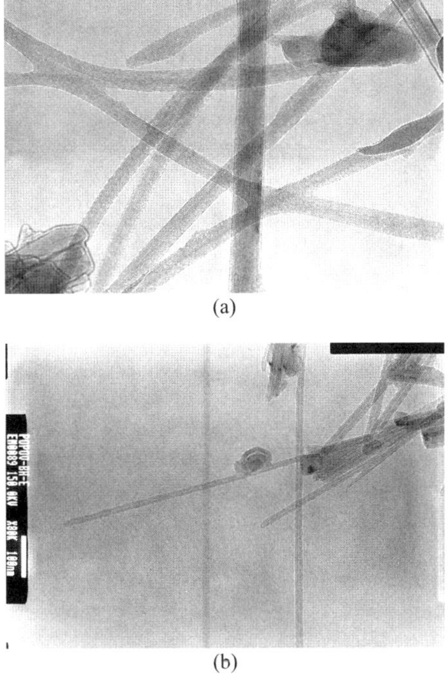

Figure 5. Nanotubes obtained from boron–hydrogen compounds

3. Methods

Composite materials with copper, aluminium and nickel matrix were obtained by mechanical alloying in a Gefest-11-3 planetary activator equipped with quasicylindrical milling bodies (Fig. 6). The power density of this unit reached 10 kW/dm^3, which exceeds that of gravitation ball mills by two or three levels of magnitude. The impact of large inertial forces on the milling bodies and treated material reduces the milling time considerably.

As in this unit we apply intensive cooling and lubrication of the drums with flow water, the extent of activation of treated substances is increased considerably due to the slowdown of the recovery processes and suppression of phase transformations.

Initial material for the matrix was prepared as shavings, which were preliminarily comminuted for 2 min. The size of powder particles of the matrix alloys after preliminary comminution was 0.5–3 mm.

After preliminary comminution, we prepared samples to produce dispersion-strengthened composite material in terms of 160 g per one drum, which were annealed in a vacuum drying cabinet at a temperature of 110°C to remove moisture. The fraction of reinforcing particles varied from 10 up to 20 vol%.

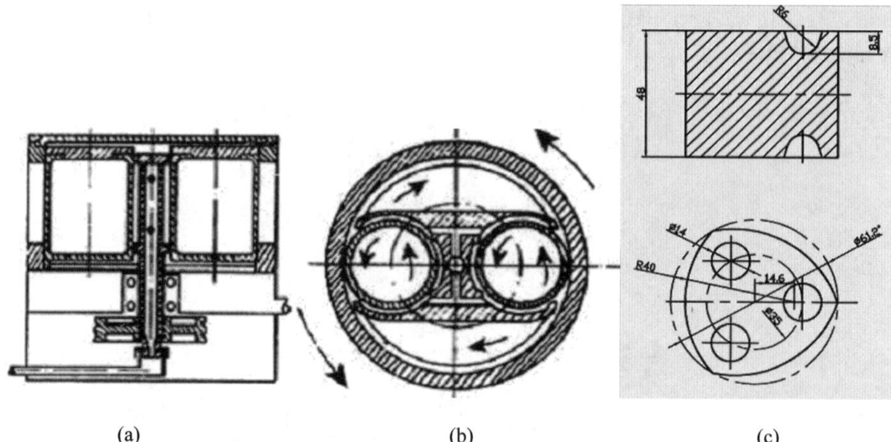

(a) (b) (c)

Figure 6. Longitudinal (a) and transversal (b) cross-sections of a Gefest-11-3 planetary activator and a quasicylindrical milling body (c)

The material was treated in a planetary activator in an argon medium. For the treatment in an argon medium, the drums were pre-evacuated for 30 min to rule out the oxidation of particles, and were filled with argon at an installation specially designed and fabricated for these works. The duration of the treatment of the powder mixtures in a planetary activator was changed from 30 up to 120 min.

The granules obtained by mechanical alloying were compacted on a 40-t press, which was followed by heat treatment in a vacuum heating furnace.

4. Results and their discussion

The material was studied both in the initial state and at all stages of treatment.

It is necessary to know the hardness indices of the initial materials to choose the regimes of mechanical alloying. The hardness of the initial materials was measured on a Nanoscan scanning probe microscope-nanoindenter. The hardness of some boron compounds is lower than that of metal components, and of some others approximately equal to metal components or slightly exceeds it. It should be noted that in the printed sources there is information on mechanical alloying of materials with a significant difference in the hardness of treated components [3].

Mechanical alloying was performed according to the above method. The produced granules were studied using a focused-beam scanning microscope. After treatment in a planetary mill, the granules can have two types of shape. In one case, the boron compound is totally distributed along the surface of metal particles as a uniform layer. In the other case, the boron compound is split into minute particles and evenly covers the surface of metal particles. In both cases, incorporation of boron compounds into metal is hindered, as the boron compounds are less hard as the metal used. For better mixing, we used metal particles shaped as shavings, not as spherical particles. During the treatment in a mill, plane particles of shavings are comminuted and kinked or bent. Herewith, particles of boron compounds get into the metal. However, during the mill treatment of up to 120 min, a considerable amount of boron compounds still remains on the surface of metal particles. This complicated the compaction process. Attempts to compact granules by chill pressing did not get a positive result. Therefore, the further studies were held in two directions: (1) improvement of the compaction process, (2) improvement of the process of mechanical alloying in planetary mills.

This work studied the ways to improve the process of mechanical alloying. A silica nanopowder or a nanodiamond powder in the amount from 1 up to 5 vol%. was added to the components. This made possible to mix the components much more efficiently. The size of the initial matrix particles was reduced considerably. As the result, the granules were successfully compacted. Figure 7 shows the general view of the surface of copper granules after mechanical alloying. Figure 8 presents the cross sections of granules from copper-based composites with various reinforcing particles.

Figure 9 is a photograph of a workpiece from copper powder produced by mechanical alloying with boron hydride.

5. Conclusion

We have shown a principle possibility of producing a composite material with reinforcing particles from boron compounds of a density of 1.3 up to 2.5 g/cm^3 in the case of mechanical alloying with addition of 1–2.5% silica nanopowder or nanodiamond powder.

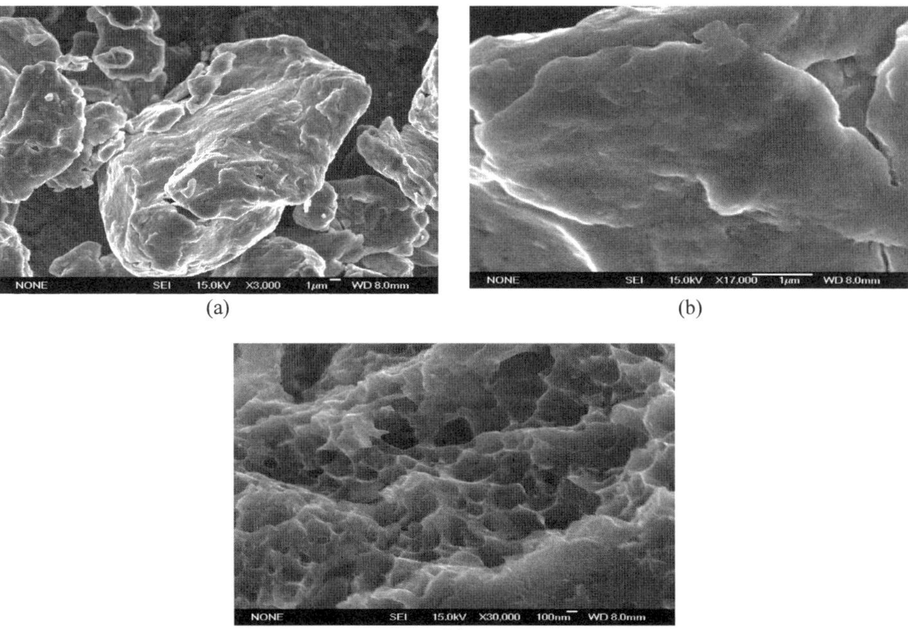

Figure 7. General view of the surface of copper granules after mechanical alloying with boron hydrides (a, b) and with addition of silica nanopowder (c)

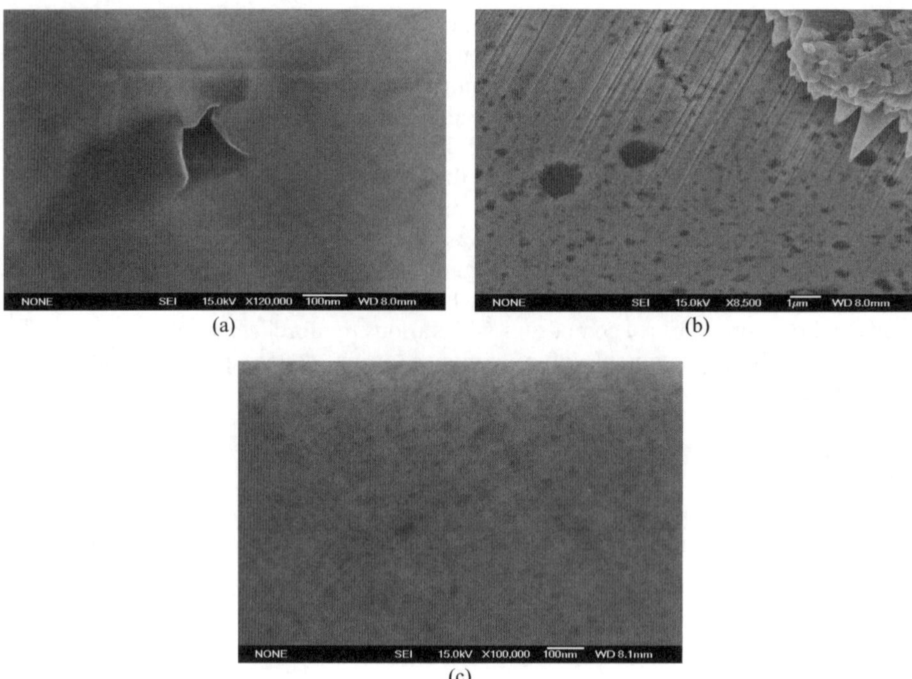

Figure 8. Cross-sections (obtained by an ion beam) of granules from copper-based composites with various reinforcing particles: (a) boron hydride, (b) diatomite (silica nanopowder), (c) nanodiamond

Figure 9. A workpiece from copper powder produced by mechanical alloying with boron hydride

References

1. E. Malinina, K. Zhizhin, L. Goeva, et al., Closo-decaborate anion as a lidand in complex of copper(I). EUROBORON II. France. Durham. 2001. P. 24.
2. E. A. Malinina, K. Yu. Zhizhin, et. al., Doklady Chemistry, 2001. **378**(1–3): 139–142.
3. J. Kaneko, M. Sugamata, L. Blaz, R. Kamei, Materials Science Forum, 2002. **396–402**: 161–166.

METALLOGRAPHIC INVESTIGATIONS AND HYDROGENATION PECULIARITIES OF THE ALLOY Mg-La-Ni

P.V. FURSIKOV, D.N. BORISOV
Institute of Problems of Chemical Physics of RAS, Prospectus of the Academician Semenov, 1, Chernogolovka, Moscow Region, 142432 Russia
Tel./Fax: +7(496) 5221743; E-mail: btarasov@icp.ac.ru
V.A. YARTYS
Institute for Energy Technology, Instituttveien 18, Kjeller, N-2027 Norway
B.P. TARASOV
Institute of Problems of Chemical Physics of RAS, Prospectus of the Academician Semenov, 1, Chernogolovka, Moscow Region, 142432 Russia

Abstract. At use of modern microscopic methods of the analysis structural features and the hydrogenation mechanism of ternary eutectic alloy of composition 72 wt% Mg – 8 wt% La – 20 wt% Ni are investigated.

Keywords: ternary eutectic alloy, hydrogen, structural peculiarities, hydrogenation mechanism

1. Introduction

Mg based alloys such as the ternary eutectic alloy Mg-La-Ni are perspective hydrogen sorbing materials [1]. Owing to the disperse structure, a reduced size of grains of the Mg phase, and catalytic effect of the rare-earth metal [2] this eutectic alloy has an improved hydrogen sorption kinetics. Detailed studies of hydrogen sorption properties of the alloy require metallographic investigations of structural changes proceeding during hydrogenation and dehydrogenation processes.

Thus the present work aimed at elucidation of structural peculiarities and the hydrogenation mechanism of the ternary eutectic alloy Mg-La-Ni.

2. Experimental

An alloy having composition of 72 wt% Mg – 8 wt% La – 20 wt% Ni (later referred to as Mg-La-Ni) was prepared by induction melting in vacuum. The purity of magnesium, nickel and lanthanum used were of 99.95%, 99.99% and 99.79% respectively. The melting point of the alloy was 500°C.

Metallographic sections of the alloy were investigated by means of the scanning electron microscopy (SEM) and energy dispersive X-ray spectroscopy (EDX) on a scanning electron microscope equipped with energy dispersive X-ray spectrometer Oxford Instruments.

Stepwise hydrogenations of two metallographic sections were carried out in a high-pressure chamber at 553 K and 3 MPa of hydrogen as well as at 623 K and 2 MPa of hydrogen. After every step of the hydrogenation the specimens were removed from the chamber and examined by means of SEM and X-ray diffraction (XRD) analysis with the use of a diffractometer STOE $\theta - \theta$, CuK_α radiation. The relative integral intensities (RII) of diffraction peaks were calculated by the normalization on the most intense peak (101) of the Mg phase.

3. Results and discussion

In a micrometer scale the alloy has a uniform structure which is peculiar to the eutectic alloys Mg-RE-Ni. However in the scale of about 100 μm the alloy is not uniform that is seen on the Fig. 1. Thus there are clearly distinct areas. An area of the first type is dominating and has a variegated-like fine structure. Areas of the second type look like large islands included within the dominating area.

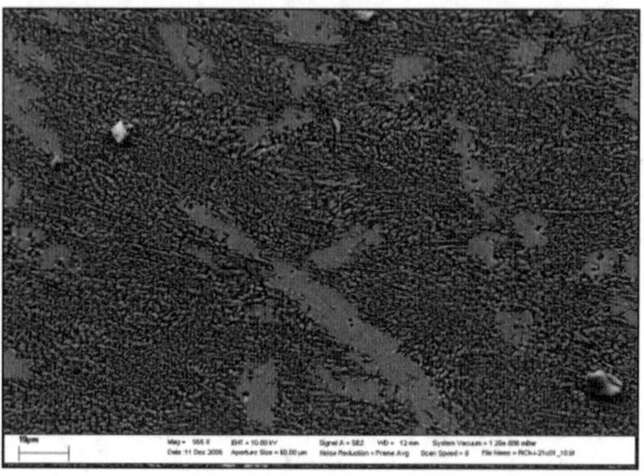

Figure 1. Overall SEM image of the alloy surface. Scalebar is 10 μm

Plotting EDX spatially resolved chemical images aided by so-called hydrogen etching allowed one to study in detail the spatial distribution of the constituent phases (Fig. 2). The first type areas comprise Mg, Mg_2Ni, and $Mg_{17}La_2$ phases. The second type areas consist of $Mg_{17}La_2$ and Mg phases, and the size of the Mg phase in these areas is substantially smaller than in the first ones.

Stepwise progressive hydrogenation of the Mg-La-Ni alloy investigated on every stage by means of the surface XRD allowed to study in detail the initial stage of the alloy interaction with hydrogen. Figures 3–5 show that the phase La_2Mg_{17} is decomposed at first. Then the phase LaH_3 is formed, while the formation of the MgH_2 phase proceeds only after the formation of the phase of lanthanum hydride. Analysis of the XRD patterns also evidenced the formation of the Mg_2NiH_4 phase during the hydrogenation. However, the method used in the present work doesn't allow one to make any reliable conclusions on the relative rate of this process.

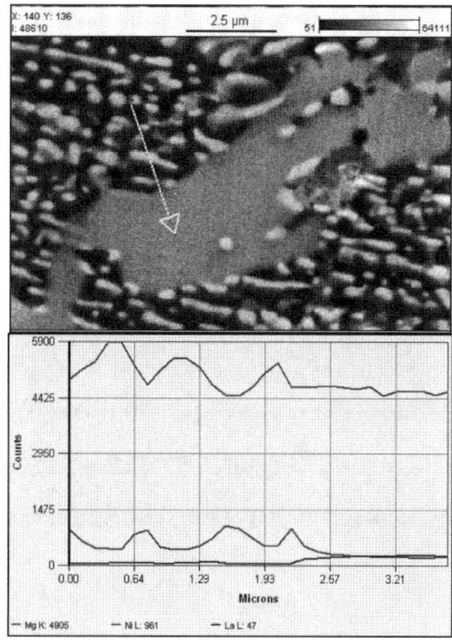

Figure 2. Elemental line-scan EDX of the alloy. Lines from up to down: Mg(K), Ni(K), La(L)

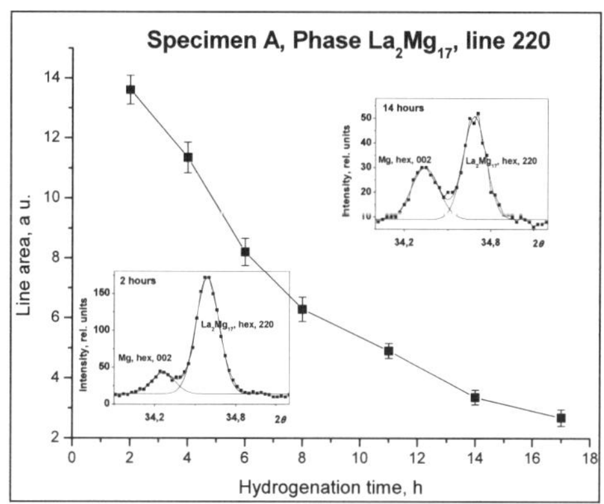

Figure 3. Hydrogenation time dependence of the RII of the 220 XRD peak of the phase La_2Mg_{17}; hydrogenation temperature 553 K

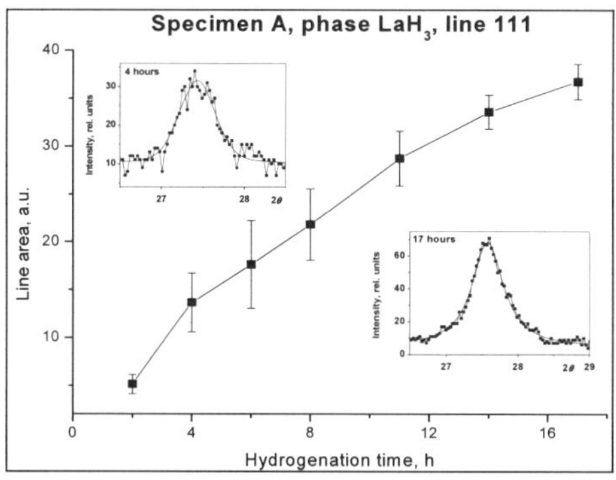

Figure 4. Hydrogenation time dependence of the RII of the 111 XRD peak of the LaH_3 phase; hydrogenation temperature 553 K

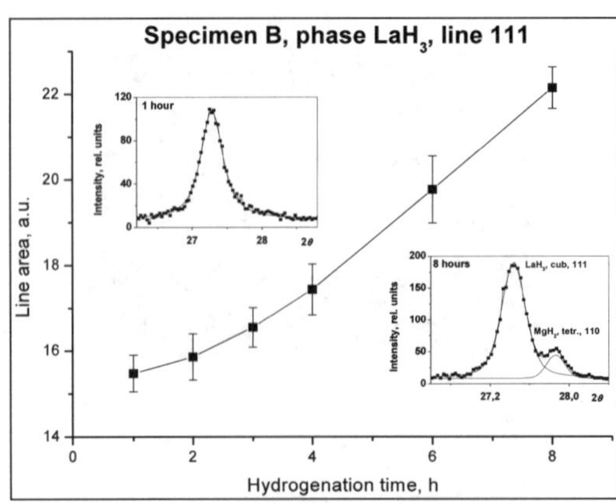

Figure 5. Hydrogenation time dependence of the RII of the 111 XRD peak of the LaH_3 phase; hydrogenation temperature 623 K

4. Conclusions

During hydrogenation of the alloy Mg-La-Ni the phase La_2Mg_{17} decomposes first, then the phase LaH_3 forms, while the formation of the phase MgH_2 proceeding only after that.

Acknowledgment

The work is supported by the funds Nordisk Energiforskning (NORSTORE 46-02) and RFBR (Project №05-08-18130).

References

1. Tarasov B.P., Fokin V.N., Borisov D.N., et. al. Hydrogen accumulation by alloys of magnesium and rare-earth metals with nickel, *Alternative Energy and Ecology*, 2004, **1**: 47–52.
2. Pezat M., Darriet B., Hagenmuller P. A comparative study of magnesium-rich rare-earth-based alloys for hydrogen storage, *Journal of Less-Common Metals*, 1980, **74**: 427–434.

INVESTIGATION OF HYDROGEN-THERMAL TREATMENT INFLUENCE ON THE PROPERTIES OF BORON NITRIDE BASED MATERIALS

I.A. MOROZOV, R.A. MOROZOVA, T.V. DUBOVIK, A.I. ITSENKO, V.M. PANASHENKO, A.A. ROGOZINSKAYA, V.S. TSYGANENKO, V.V. LYCHKO
I. Frantsevich Institute for Problems Materials Science of Ukrainian National Academy of Sciences, 3 Krzhyzhanovsky Str., 03142, Ukraine
E-mail: imorozov@i.ua

Abstract. The boron nitride BN and based on BN materials have high dielectric characteristics, high thermal and chemical stability. The products manufactured from these materials are used in high-temperature techniques. The problem of density, strength and chemical purity increasing of developed materials is very urgent now. With this aim the influence of preliminary hydrogen-thermal treatment of boron nitride precursor powders and the properties of BN-based composites has studied. The initial BN powder has hydrogen-thermal treatment at temperatures of 900°C and 1,100°C during 1–3 h at every temperature selected. Powder properties before and after processing were investigated. Different temperature regimes of hot pressing and cold-pressed samples sintering, made of precursor and processed powders, were also studied. The physical, structural and mechanical properties of manufactured samples were investigated.

Keywords: boron nitride, powder, hydrogen-thermal treatment, density, resistance, compression strength, thermal expansion coefficient

1. Introduction

Boron nitride (BN) is nonmetal refractory compound which crystallizes in ordinary conditions like a graphite having the hexagonal lattice with periods of a = 2,504 A, b = 6,671 A, c = 2,662 A. Such structure is characterized by heterodesmic of "boron-nitrogen" chemical bonds due to strong covalent interaction within one monolayer and weak van-der-vaaltz one between layers [1].

The features of boron nitride structure determine on the one hand is high properties: stability upon thermal shock, high electrical resistance, dielectric stability in wide frequency range, chemical inert relative to metals and alloys melts, and on another hand hinder the possibility of compact isotropic material sintering.

Low concentration and low mobility of lattice defects as well as hampering of diffusion processes is characteristic of both boron nitride and compounds with predominantly covalent bonding. In connection with this it is advisable to use the sintering methods that enable to control the structure and to produce the porousless materials with necessary set of properties.

In sintering of powder materials not all powder surface takes part. The great part of powder surface is passivated by the atoms of adsorbed impurities that also hamper the sintering. The efficient means of removing the adsorbed impurities from powder surface and effecting their structure is hydrogen-thermal treatment (HTT) of powder in the presence of catalysts at temperatures in the range of 900–1,200°C. Such treatment allows getting the purified powders with improved structure that in turn makes it possible to design the materials with definite physics-chemical properties [2].

2. Experimental details

At present, effect of HTT of nonmetal nitrides powders on the structure and properties of compact materials is little investigated. The goal of this work is to study the influence of HTT on the structure of BN powder and on the properties of BN-based compact materials.

As a precursor the boron nitride powder of technical grade was used. This powder was manufactured by the method of reduction-nitriding of B_2O_3 and carbon black at temperature 1,800°C. The average powder particles size was of 10–40 μ. HTT of this powder was carried out in the hydrogen ambient at temperatures of 900°C and 1,100°C during 1, 2 and 3 h at each temperature. Chemical composition of BN powders before and after treatment is listed in Table 1.

TABLE 1. Variation of BN powder chemical composition with treatment conditions in hydrogen environment

№	Temperature of treatment (°C)	Duration of treatment (h)	Content of elements (mass %)			
			B	N	B_2O_3	C
1	Initial	–	42.1	54.4	2.0	1.4
2	900	1	42.9	55.2	0.2	–
3	900	2	43.0	55.7	0.1	–
4	900	3	43.0	55.5	0.1	–
5	1,100	1	43.3	56.0	–	–
6	1,100	2	43.5	56.3	–	–
7	1,100	3	43.5	56.2	–	–

3. The results and discussion

It is evident from Table 1 that the quantity of B_2O_3 and C impurities decreases and of nitrogen increases in treated BN powders. At the same time the increase in the temperature and duration of treatment results in more efficiency of treatment. The BN powder treated at temperature 1,100°C for 2 h was used later on. It was established by electron microscopy that, in the case of BN, HTT favors the disintegration of large porous aggregates, the intensification of processes of mass transfer, and the formation of aggregates of round shape (Figs. 1 and 2).

The samples for laboratory examination of precursor powder and of hydrogen treated one were manufactured by hot pressing at temperatures 1,600–2,000°C (in each

100°C), pressure 30 MPa for 5–15 min exposure at maximum temperature (Table 2). During hot pressing the shrinkage of powders was registered and through the sample final height and the sample initial density the density at given instance was calculated.

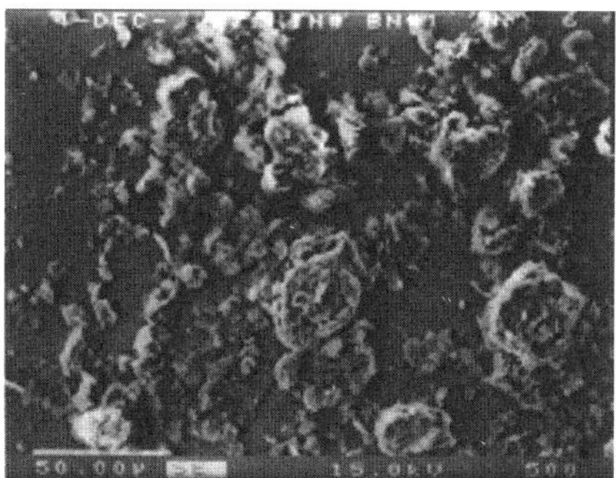

Figure 1. General view of structure components of precursor BN powder

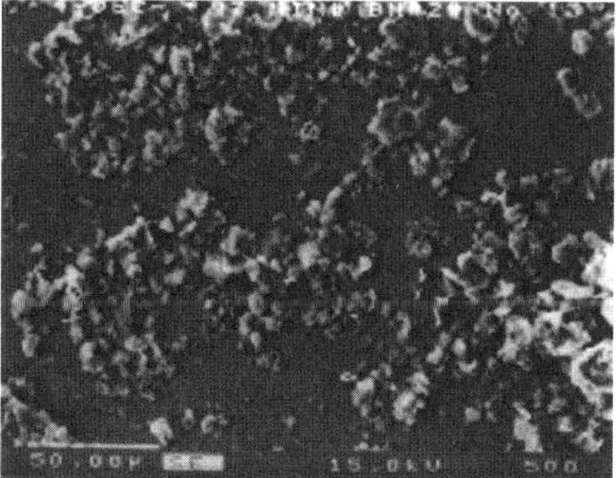

Figure 2. Structure components of BN powder after HTT at 1,100°C

From the data of Table 2 it follows that the character of density variation of initial and treated BN powders in hot pressing is mostly the same: the samples density quickly increases for 5 min then asymptotically approaches its maximum value at given temperature. The lover temperature of hot pressing the more difference between initial and final density of sample and the more time is necessary to reach the maximum density. The treatment time interval, above which the density does not increase, corresponds to each pressing temperature. This behaviour could be explained from the point of view of the theory of bulky-tough yielding of porous material created by M. Kovaltchenko [3].

The data listed in Table 2 show also that only samples of hydrogen treated BN powders have density close to theoretical one. This is because of considerable increase in reactive surface of BN due to refining of powder from surface impurities. Being the very active reagent the hydrogen removes oxides and other impurities from the powder surface and forms the volatile compounds with these impurities.

TABLE 2. Densification of BN powders in hot pressing (density value of 1.0 corresponds to 2,200 kg/m^3)

BN	Exposure time (min)	Hot pressing temperature (°C)				
		1,600	1,700	1,800	1,900	2,000
Initial	5	0.42	0.52	0.65	0.74	0.79
	10	0.53	0.56	0.68	0.77	0.76
	15	0.58	0.60	0.71	0.77	0.71
After HTT at 1,100°C (2 h)	5	0.71	0.79	0.83	0.85	0.85
	10	0.73	0.82	0.82	0.89	0.86
	15	0.74	0.86	0.85	0.90	0.90

The strength characteristics of hot pressed BN samples before and after HTT have been studied. The data are listed in Table 3. Data of Table 3 show that the preliminary hydrogen-thermal treatment results in an increase in the mechanical strength of material more than by two times.

TABLE 3. Comparative strength characteristics of hot pressed BN samples at different temperatures (exposure time, 10 min)

Hot pressing temperature (°C)	Compression strength (MPa)	
	Initial	Treated
1,600	14.5	27.4
1,800	18.8	41.6
2,000	24.3	49.5

The pressing of samples of initial and treated BN powders followed by sintering has been studied. The samples were pressed in steel press-forms at pressure 2 t/cm^2 and sintered in nitrogen ambient on the bed of B_2O_3 with carbon black in Tamman's furnace at temperature of 1,900°C (1 h). The results are listed in Table 4.

TABLE 4. Properties of sintered samples of boron nitride (1,900°C, 1 h, N_2)

Material	Properties		
	Relative density	Compression strength (Mpa)	Resistance (Ohm·cm)
BN, initial	0.6	10.8	10^{11}
BN, treated (H_2, 1,100°C, 2 h)	0.8	23.2	10^{13}

Data of Table 4 show the hydrogen-thermal treatment to increase the density, strength and electroinsulating properties of boron nitride. On the whole, HTT provides the cleaning of the powder surface and boundaries of cracks, which increases the activity of surface atoms before sintering. Sintered BN specimens made of powder subjected to HTT have a higher density and specific structural features. These are: the smaller grain size, the coalescence of particles proceeds to a higher degree along the basal planes, the less noticeable damage of the texture (Figs. 3 and 4).

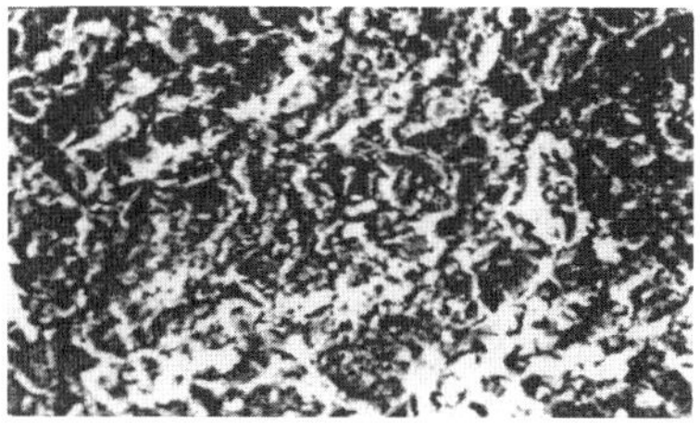

Figure 3. Structure of sintered initial BN at 1,900°C, 1 h

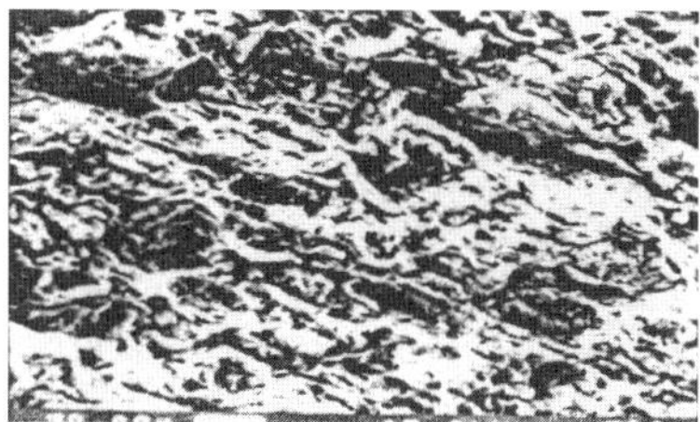

Figure 4. Structure of sintered BN (1,900°C, 1 h) after HTT (1,100°C, 2 h)

The effect of hydrogen treated BN powder on the properties of base on it the composite material – boron carbonitride – has been investigated [4]. The samples were manufactured by the abovementioned pressing and sintering regime for boron nitride. The data are listed in Table 5.

TABLE 5. Properties of the sintered boron carbonitride (BNC)

Property	Temperature (°C)	BNC based on: Initial BN	Treated BN
Resistance (Ohm·cm)	20	$>10^{13}$	$>10^{13}$
	1,000	$1.5 \cdot 10^{7}$	$1.6 \cdot 10^{9}$
	1,800	$2 \cdot 10^{4}$	$1 \cdot 10^{6}$
Coefficient of thermal expansion (degree^{-1})	200–1,800	$4.6 \cdot 10^{-6}$	$3.0 \cdot 10^{-6}$
Compression strength (MPa)	20	25.9	53.3

It is evident from Table 5 that HTT of initial BN increases physical and mechanical properties of boron carbonitride. On the base of such boron carbonitride it was developed the device for temperature measuring of metal melts [5]. Refractory entrance bushing made of boron carbonitride permits to execute continuous temperature measuring during entire melting process of steel #45 in crucible furnace (2 h).

4. Conclusions

The HTT of BN powders at 1,100°C results in their cleaning from oxygen and carbon impurities, which is accompanied by structural changes of powders effecting favorably their sintering. The investigations revealed the perspective of using the method of hydrogen-thermal treatment of initial powders for activation of densification processes of BN and the materials on it base. Further investigations in this field have to be carried out.

References

1. Kosolapova T.Ja., Andreyeva T.V., Bartnitskaya T.S., et al. Non-metal refractory compounds. Moscow: «Metallurgiya» Publishing House. 1985. 224 p. (in Russian).
2. Trefilov V.I., Morozov I.A., Morozova R.A., et al. Effect of Hydrogen on improving purity of WC and AlN Powders, *Int. J. Hydrogen Energy*, 1996. **21**(11–12): 1097–1099.
3. Kovaltchenko M.S. Theoretical bases of hot treatment of the porous materials with pressure. Kiev: «Naukova dumka» Publishing House. 1980. 240 p. (in Russian).
4. Grigoriev O.N., Dubovik T.V., Bega N.D., Kotenko V.A., Panashenko V.M., Lyashenko V.I., Rogozinskaya A.A., Chernenko L.I. Sintered ceramics based on boron nitride and carbide, *Powder Metall. Met. Ceram.*, 2007. **46**(1–2): 46–50.
5. Pat. 33418 A Ukraine, MPI[6] G 01 K 7/02. Device for melts temperature measuring / Itsenko A.I., Dubovik T.V., Morozov I.A., Morozova R.A. N 99020989 15.02.2001, Bull. N 1 (in Russian).

THERMAL STABILITY AND H-DESORPTION PROPERTIES OF $Mg_3MnNi_2H_x$ PREPARED BY REACTIVE MECHANICAL ALLOYING OF POWDERS Mg, Mn, Ni IN H_2-ATMOSPHERE

V.D. DOBROVOLSKY, O.G. ERSHOVA, YU.M. SOLONIN
Institute for Problems of Materials Science, National Academy of Sciences of Ukraine, Kiev, 03142, Ukraine
E-mail: dobersh@ipms.kiev.ua
R.V. DENYS, I.YU. ZAVALIY
Physico-Mechanical Institute of NASU, 79601, Lviv, 5 Naukova Str., Ukraine

Abstracts. In the present paper, $Mg_3MnNi_2H_x$ hydride has been synthesized using the reactive mechanical alloying (RMA) process, i.e. employing a simple one-stage technology instead of the complicated four-stage route used in [1] for synthesis of the same hydride (the milling powder of pure Mg, Ni and Mn metals and of Mg_2Ni alloy in argon atmosphere followed by their pressing, sintering and hydriding from the gaseous phase). The $Mg_3MnNi_2H_x$ hydride derived by the RMA method was studied by thermodesorption spectroscopy. Similarly to $Mg_3MnNi_2H_x$ hydride derived in [1], the hydride obtained in the present paper reveals decreased thermal stability and close hydrogenation capacity (1.35 wt%).

Keywords: magnesium alloys, hydrogenation, mechanical alloying

1. Introduction

As it has been shown in Ref. [1], a hydride of the Mg_3MnNi_2 alloy obtained by milling powders of pure Mg, Ni and Mn metals and of the Mg_2Ni alloy in argon atmosphere with following their pressing, sintering and hydriding from a gaseous phase, possesses decreased thermal stability: an intensive hydrogen release in vacuum begins already at 100°C with the maximum at 200°C. In addition, the Mg_3MnNi_2 alloy reveals better kinetics of hydrogen sorption-desorption. The alloy starts to absorb hydrogen already at room temperature and at rather low pressure (1–2 MPa). The similar phases were observed during the studies of the Mg-M-Ni alloys (M = Ti, Al) [2]. Mg_3MNi_2 (M = Mn, Ti, Al) are characterized by Nb_3Ni_2Si structure type (super structure to Ti_2Ni-type) [3]. The present work was aimed at obtaining the $Mg_3MnNi_2H_x$ hydride possessing the parameters similar to those in [1] (the same hydrogen content and decreased thermal stability) using a simplified technology of preparation involving only one stage (milling a mixture of Mg, Ni and Mn powders in hydrogen atmosphere at pressure of 1.2 MPa) instead of the four-stage technique applied in [1].

2. Experiment details

Pure elemental powders of Mg, Mn and Ni with their purity greater than 99.9% and with initial particle sizes of 3.2, 4, and 2 μm, respectively, were used in the present reactive milling alloying (RMA) experiment. The RMA experiment for the powder mixture Mg (29wt%) + Mn (22.44wt%) + Ni (47.82wt%) was carried out in hydrogen atmosphere under pressure of 1.2 MPa in a ball mill for 5 h with rotation speed 1,200 rpm. The weight ratio of steal balls to powder mixture was equal to 33:1. The tank filled by $TiFeH_x$ metal hydride was used as a source of high-pure hydrogen. Almost every 30 min of the milling, hydrogen was injected from the source into a reaction chamber in order to keep pressure in the chamber at about 1.2 MPa.

An X-ray diffraction (XRD) analysis of the mechanical alloys (MA's) was carried out with a DRON-3M diffractometer employing X-ray Cu Kα radiation. The present XRD measurements were made with a step of 0.1° and exposition time was 10 sec in every point.

Thermal decomposition of the hydride phase, a product of the Mg + Mn + Ni MA, was studied using Sievert's method in hydrogen and argon atmospheres at normal pressure. The device allows monitoring a dependency of quantities of hydrogen releasing from a hydride during its heating (TDS method). A sample (m = 0.2 g) was loaded to the volumetric setup equipped by a piston mechanism with a displacement sensor preliminary calibrated on changes of volume of gas in the system. The sample holder was placed into a quartz reactor equipped with a K-type thermocouple and heated with an electric furnace connected to a controller providing linear heating (ramp rate 5 deg/min). The output of the thermocouple and the displacement sensor were connected to a data acquisition system. After loading the sample, the system was evacuated by a rotary pump followed by filling with argon or hydrogen and heating. After the maximum temperature of the TDS experiment was reached, the furnace was switched off, and the reactor was cooled to room temperature (in so doing, the logging of gas volume change and temperature was continued).

3. Results and discussion

A possibility of synthesis of the $Mg_3MnNi_2H_x$ phase has been tested employing a new method, different from that used in [1] (millimg – sintering – hydriding Mg_3MnNi_2). The influence of the preparation conditions of the mentioned phase on the hydrogen-sorption properties and thermal stability has been determined. The $Mg_3MnNi_2H_x$ phase has been obtained in the ball mill as a result of high-energy milling of the powder mixture consisting of Mg, Mn and Ni under 1.2 MPa hydrogen pressure.

XRD analysis of the alloy obtained by reactive mechanical method (Fig. 1) shows the presence of the $Mg_3MnNi_2H_x$ phase which is analogous to that reported in Ref. [1]. In addition, the presence of Mg, Mn, Ni and MgO was also observed.

Curves of hydrogen desorption from the $Mg_3MnNi_2H_x$ hydride phase of the mechanical alloy heated in hydrogen and argon atmosphere with rate of 5 K/min are presented in Fig. 2. From an analysis of the desorption curve presented in Fig. 2(a) (heating in hydrogen atmosphere) it is obvious the $Mg_3MnNi_2H_x$ hydride phase obtained by our technique begins to decompose at very low temperature (240°C) and possesses decreased thermal stability (in this condition the MgH_2 hydride starts to decompose only at 450°C and at 350°C after 20 min milling [4]).

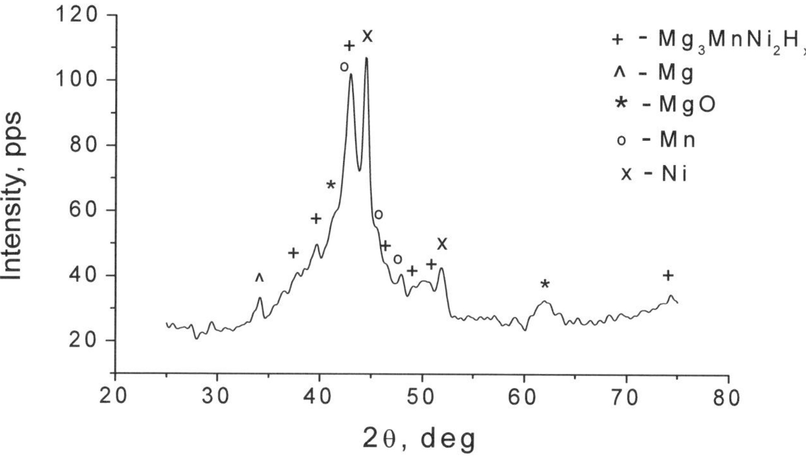

Figure 1. X-ray diffraction pattern of the mechanical alloy ((Mg (29wt%), Mn (22.44wt%), Ni (47.82wt%)) prepared by high-energy ball-milling in 1.2 MPa hydrogen for 5 h

Comparison of the hydrogen desorption curves presented in Fig. 2(c) shows differences in a dynamic of decomposition of the hydride phase during its heating in hydrogen and argon atmosphere. In the case of heating the hydride in hydrogen atmosphere, an intensive hydrogen release occurs at 240°C, above this temperature the hydrogen desorption rate increases rapidly with its maximum at 320°C. In the case of heating the hydride in argon atmosphere, the hydrogen release begins already at 50°C and its rate increases slowly up to its maximum at 320°C. This temperature was found to be higher by 120 K as compared with that observed for this case in Ref. [1]. The observed difference in the temperatures, corresponding to maximum rate of hydride decomposition, can be explained by different desorption conditions: dynamic vacuum in [1] and hydrogen (argon) atmosphere in our case.

In the both cases of desorbed hydrogen (in argon and hydrogen atmospheres) was about 140 cm^3/g. Taking into account the XRD data (see Fig. 1), a fraction of the hydride phase in the mechanical alloy does not exceed 85%. Therefore, hydrogen sorption capacity of the $Mg_3MnNi_2H_x$ hydride phase is about 1.35%, the value which is close to that obtained in [1].

After studies of the desorption curves, we investigated phase compositions of the decomposition products of the mechanical alloy during its heating in hydrogen atmosphere. Respective XRD pattern is presented in Fig. 3. The XRD analysis reveals the presence of the ternary Mg_3MnNi_2 compound, with face-centered cubic structure (Fd3m) belonging to the Nb_3Ni_2Si-type [1, 3], as well as Mg, MgO and Ni in the specimen after its dehydriding.

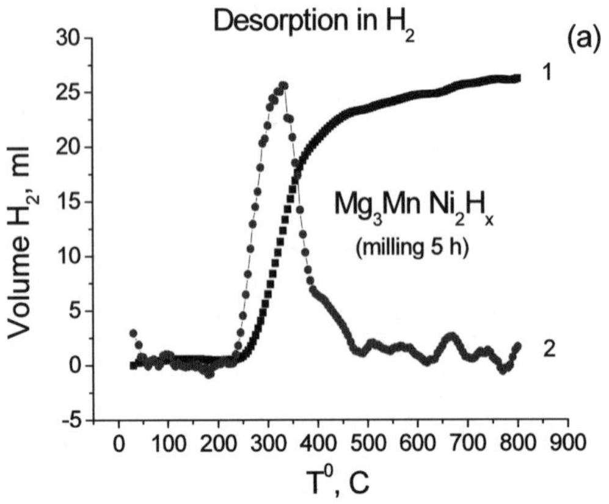

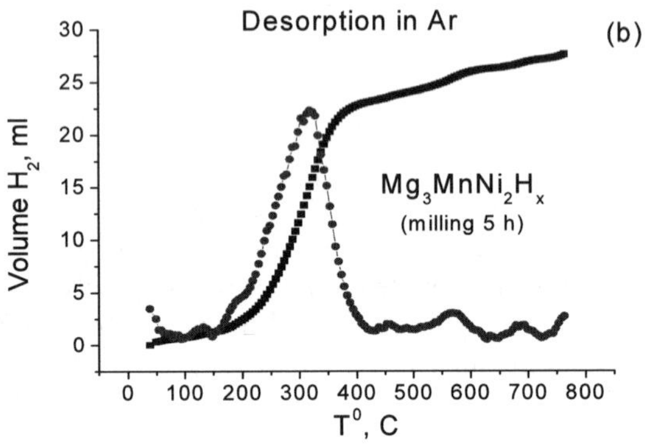

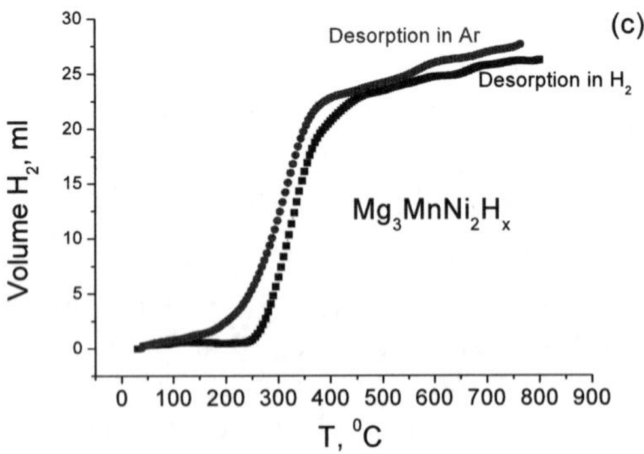

Figure 2. Curves of hydrogen release (1) and the corresponding differential plots (2) for the $Mg_3MnNi_2H_x$ hydride heated in hydrogen (a) and argon (b) atmosphere and their matching (c) in the same temperature range (rate 5 K/min)

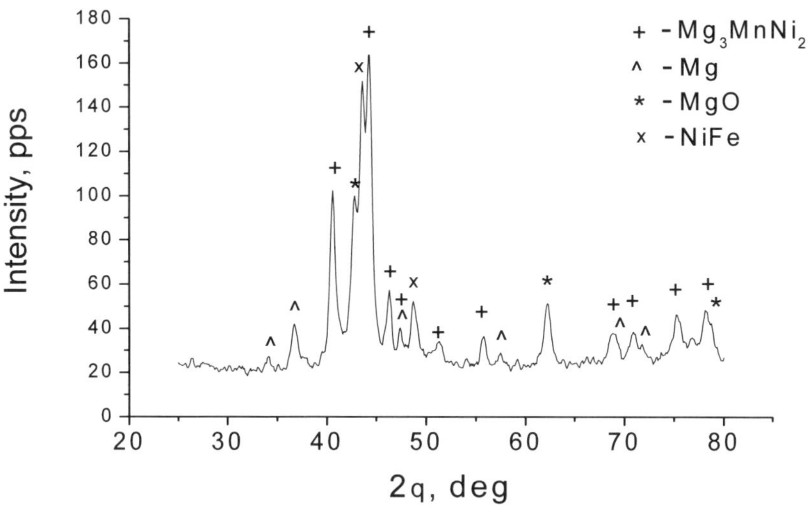

Figure 3. X-ray diffraction pattern of the mechanical alloy after its heating in hydrogen atmosphere

The presence of the Mg_3MnNi_2 intermetallic phase detected in the products of MA decomposition shows that during reactive alloying of Mg, Mn and Ni powder in hydrogen atmosphere simultaneous synthesis and hydriding of this phase takes place. This phase irrespective of the preparation method is able to sorb/desorb hydrogen without decomposition, i.e. "to work" in a wide temperature range starting from room temperature up to 800°C.

4. Conclusions

A possibility to prepare the $Mg_3MnNi_2H_x$ hydride using the simplified technique (high-energy milling a mixture of Mg, Mn and Ni powders under hydrogen pressure; i.e. employing the RMA method) was confirmed experimentally.

Acknowledgements

Financial support of INTAS (project # 05-1000005-7671) is gratefully acknowledged.

References

1. Denys R.V., Saldan I.V., Delaplane R.G., Berezovets V.V., Zavaliy I.Y. Structure and hydrogenation properties of new Mg-based alloys. Proceedings of IX International Conference on Hydrogen Materials Science & Chemistry of Carbon Nanomaterials (ICHM'S 2005), Sevastopol, Ukraine, P. 154–156.
2. Guanglie L., Linshen C., Lianbang W., Huantang Y. Study on the phase composition of Mg M Ni (M=Al, Ti) alloys. *J. Alloy. Compd.* 2001, **321**: L1–L4.

3. Gladyshevsky E.I. et. al. Crystal structure of the Mn_3Ni_2Si, V_3Ni_2Si, Mb_3Ni_2Si compounds. *J. Struct. Chem.* 1963, **4**(3): 372–379.
4. Ershova O.G., Dobrovolsky V.D., Solonin Y.M., Morozova R.A. About manner and mechanisms of reduction of thermal firmness of Mg-, Ti-, Y-based mechanical alloys. In: Veziroglu T.N. et al., editors. Hydrogen Materials Science and Chemistry of Carbon Nanomaterials. Dordrecht: Springer, 2007, P. 429–436.

THERMAL STABILITY AND HYDROGEN SORPTION PROPERTIES OF THE MgH_2 HYDRIDE DERIVED BY THE REACTIVE MILLING OF THE Mg + 10 wt% Ti MIXTURE

O.G. ERSHOVA, V.D. DOBROVOLSKY, YU.M. SOLONIN, O.YU. KHYZHUN
Institute for Problems of Materials Science, National Academy of Sciences of Ukraine, Kiev, 03142, Ukraine
E-mail: dobersh@ipms.kiev.ua

Abstracts. Thermal stability and hydrogen sorption properties of the MgH_2 phase derived by the reactive milling of the Mg + 10 wt% Ti mixture was studied by the thermodesorption spectroscopy method during the first five cycles of hydrogen sorption-desorption. The addition of Ti decreases decomposition temperature of MgH_2 by 115 K, if we conclude about decomposition temperature taking into account peaks of speed of a hydrogen release on the desorption spectra derived during the first cycle of heating-cooling in hydrogen atmosphere of the Mg + 10 wt% Ti mechanical alloy. During the second cycle of dehydriding-hydriding, the decomposition temperature of the MgH_2 phase of the Mg + 10 wt% Ti composite decreases additionally by 35 K, and the temperature does not change for three following sorption-desorption cycles. It has been established the influence of the method of synthesis of the hydride phase MgH_2 upon location sites of hydrogen in the phase, and, consequently, upon its decomposition temperature. The addition of titanium to magnesium increases the effect of the above influence and promotes increasing the process of dispersion of magnesium and the MgH_2 hydride synthesized due to the reactive milling. Additionally, the addition of titanium promotes decreasing the quantity of oxygen-containing structures, catalytic poisons preventing dissociative hydrogen chemosorption, adsorbed on the surface of MgH_2 particles.

Keywords: metal hydrides, thermal stability, mechanical milling, thermodesorption spectroscopy, X-ray photoelectron spectroscopy, scanning electron microscopy

1. Introduction

Reactive mechanical alloying (RMA) in H_2 of magnesium powder allows to obtain partial conversion of Mg to MgH_2 [1, 2]. The ratio MgH_2/Mg and the hydrogen storage properties can be increased particularly by adding 3d elements (e.g., Co, Fe, Ni) [3]. A decrease of thermal stability and decomposition temperature of the MgH_2 hydride of a mechanical alloy (MA) obtained by the reactive milling of the Mg + 10 wt% Fe mixture in hydrogen atmosphere at pressure 1.2 MPa was observed in Refs. [4, 5] employing the thermodesorption spectroscopy (TDS) method.

Influence of two elements, mainly Ni, a transition metal, and Al, a non-transition metal, adding to Mg upon the thermodynamic stability of the hydride phase of alloys obtained using the RMA method was investigated in Ref. [6]. The above results allows to assume that the reduction of decomposition temperature may be caused by the catalytic effect of Ni in the mixture and possibly Al dissolved in the Mg matrix. The influence of addition of two transition metals (Ni, Mn; 3 at.%) and one non-transition metal (Al; 7 at.%) to Mg upon the thermal stability of β-MgH_2 obtained by the RMA method was studied in Ref. [7]. A decrease by more than 15 K of dissociation (dehydriding) temperature with increasing the hydride content was detected. The effect has to be associated with the different thermodynamic stability of the hydride formed at the different stages of the RMA process. A significantly decreased temperature of hydride decomposition for the $Mg_{87}Ni_3Al_3M_7$ (M = Ti, Mn, Ce, La) alloys synthesized employing the RMA method [8] was assumed to be due to the combined influence of alloying, the nanocrystallinity and the reduced particle size of the alloys.

The purpose of the present work was to study the influence of small addition of titanium to magnesium on the hydrogen storage and on the decomposition temperature of the MgH_2 hydride, a component of the composite synthesized by the reactive milling of the Mg + 10 wt% Ti mixture in hydrogen atmosphere at pressure 1.2 MPa. The aim of the work was also to elucidate whether the effect of decreasing decomposition temperature of the MgH_2 phase observed in the first hydriding-dehydriding cycle will preserve in the second and a few following hydriding-dehydriding cycles.

2. Experiment technique

Pure powders of Mg and Ti (purity greater than 99.9%) with initial particle size of 3.2 μm and 4.3 μm, respectively, were used in the present RMA treatment. The treatment of the Mg + 10 wt% Ti powder mixture was carried out in hydrogen atmosphere at pressure 1.2 MPa in a ball mill for 5 h with rotation speed 1,200 rot./min. The ratio of steal balls weigh to initial burden's weight was equal to 25:1. The $TiFeH_x$ metal hydride was used as a source of high-purity hydrogen. Almost every 30 min of the milling, hydrogen was injected from the source into a reaction chamber in order to keep pressure in the chamber at about 1.2 MPa. Changes of powder sizes after 5 h RMA treatment were evaluated using BET measurements of powder specific surfaces. Uncertainty of the measurements did not exceed 10%.

Microstructures of initial magnesium powder and of the Mg + 10 wt% Ti powder mixture before and after its 5 h RMA treatment were studied using a Super-Probe 733 scanning electron microscope. The X-ray photoelectron spectroscopy (XPS) method was employed in the present work to study chemical states of the Mg + 10 wt% Ti mixture and the composite obtained due to 5 h RMA of the mixture. For the above specimens, the XPS O 1s, Mg 2p and Mg 2s core-level spectra were recorded using an ES-2401 spectrometer equipped with an ion-pumped chamber having a base pressure less than 5×10^{-8} Pa. The spectra were excited using Mg Kα radiation (E = 1,253.6 eV). The energy scale of the spectrometer was calibrated by setting the measured Au $4f_{7/2}$ binding energy of pure gold to 84.0 eV with respect to the Fermi energy of a spectrometer energy analyzer. Energy drift due to charging effects was calibrated by taking as a reference the C 1s line of hydrocarbons (285.0 eV).

An X-ray diffraction (XRD) analysis of the composite obtained due to 5 h RMA treatment of the Mg + 10 wt% Ti mixture in hydrogen atmosphere at pressure 1.2 MPa was carried out with a DRON-3M programming diffractometer employing X-ray Cu Kα radiation. The present XRD analysis was made with a step of 0.1° and accumulative time was 10 sec in every point.

Thermal stability of decomposition of the MgH_2 hydride, a component of the Mg + 10 wt% Ti alloy, was investigated using a digital device allowing to evaluate a volume of hydrogen desorbed from the specimen heated in hydrogen atmosphere at normal pressure. In the present experiments, the device operated with uniform heating the samples with speed 5 K/min.

3. Results and discussion

Figure 1 represents the XRD pattern of the composite obtained by 5 h reactive milling the Mg + 10 wt% Ti mixture in hydrogen atmosphere at pressure 1.2 MPa. The XRD analysis reveals in the composite under study the presence of the MgH_2 hydride as well as metallic magnesium, $TiFe_2$ and TiFeH. The XRD pattern presented in Fig. 1 shows the absence of reflexes of pure Ti and magnesium oxides. Content of the hydride phase in the composite obtained was estimated as ranging from 40 to 60 wt%.

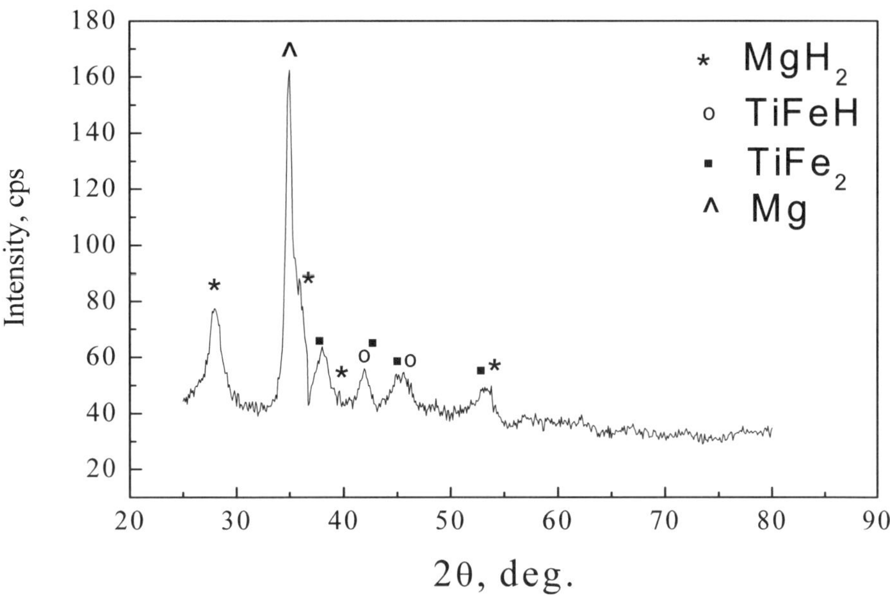

Figure 1. XRD pattern of the mechanical alloy derived after 5 h RMA treatment of the Mg + 10 wt% Ti mixture in hydrogen atmosphere at pressure 1.2 MPa

Figure 2 shows SEM images of magnesium particles used as a precursor for preparing the Mg + 10 wt% Ti powder mixture, particles of the composite obtained after 5 h reactive milling the mixture and of the MgH_2 hydride obtained by direct hydriding magnesium in hydrogen atmosphere at pressure 1.2 MPa. It is apparent that particles of the initial magnesium powder (containing some quantities of horseshoe-like scales) transform into layer particles (after treatment at 1.2 MPa in hydrogen atmosphere) with a special facet as a result of the milling of plastic magnesium (Fig. 2(b)). Data of specific surface measurements and the SEM image presented in Fig. 2 (c) reveal that after 5 h RMA treatment of the Mg + 10 wt% Ti powder mixture a significant dispersion occurs for particles of the mechanical alloy under consideration. The BET data indicate that the average size of the Mg + 10 wt% Ti alloy particles after the mentioned treatment equals to 0.27 μm. The mechanical alloy powder obtained possesses big quantities of small spherical particles (grains) with their sizes ranging from 0.1 to 0.3 μm and some quantities of big cakes with their sizes ranging from 2 to 6 μm.

Figure 2. SEM images of (a) the initial magnesium powder and (b) the MA2 (x 62) and (c) MA1 specimens investigated (×4,800)

Thermal stability of the MgH_2 phase in the Mg + 10 wt% Ti mechanical alloy under consideration (this alloy we shall refer to as the MA1 alloy for the convenience) was studied using the TDS method employing the software allowing to evaluate quantities of hydrogen desorbing from a specimen heated in hydrogen atmosphere at normal pressure. For comparison, the method was employed to investigate the thermal stability of MgH_2 treated at the same conditions and without alloying with the transition metal (we shall refer to as the MA2 alloy). Results of the TDS measurements are presented in Fig. 3 as desorption curves (i.e. dependencies of the volume of hydrogen releasing from the specimen upon its temperature; curves 1) and as desorption spectra (dV/dT dependencies; curves 2). The above TDS spectra were recorded when heating the samples with the rate 5 K/min in hydrogen atmosphere.

Figure 3(c) shows also the curve and the spectrum of hydrogen thermodesorption (HTD) of the MgH_2 hydride obtained by direct hydriding the magnesium powder in gaseous medium.

From the desorption spectra presented in Fig. 3, it is obvious that an intensive release of hydrogen begins in the mechanical alloys MA1 and MA2 (without addition of titanium) at temperatures 330°C and 350°C, respectively. Peaks of speed of hydrogen releasing in the mentioned sequence of mechanical alloys are observed at 350°C and 465°C, respectively. The analysis of the above mentioned temperatures allows to conclude that addition of Ti to magnesium leads to decreasing thermal stability and decomposition temperature of the MgH_2 phase of the MA1 specimen. Addition of Ti decreases decomposition temperature of MgH_2 by 115 K, if we conclude about decomposition temperature of MgH_2 taking into account peaks of speed of a hydrogen release on the desorption spectra. When decomposition temperature of MgH_2 is determined taking into account beginnings of an intensive release of hydrogen on the desorption spectra, the temperature decrease, ΔT, equals to 20 K.

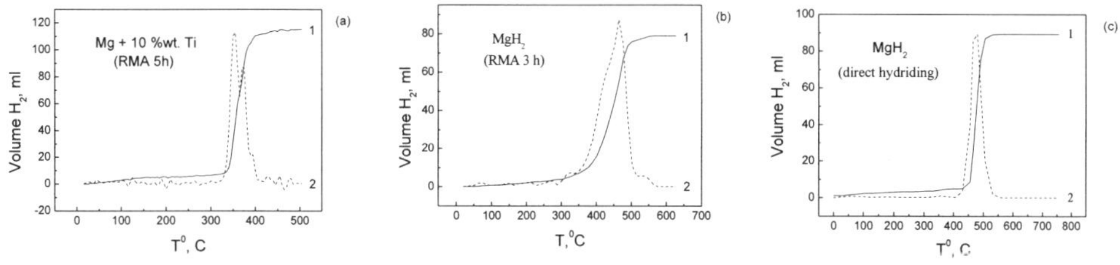

Figure 3. Curves (1) and spectra (2) of thermodesorption of hydrogen from the MgH_2 hydride of the (a) MA1 and (b) MA2 specimens obtained by the RMA treatment and from (c) the MgH_2 hydride phase derived by direct hydriding in gaseous medium

It is worth to mention that the HTD spectra presented in Fig. 3 differ from each other by shapes and fine-structure peculiarities, i.e. by quantities of lines (peaks) composing the desorption spectra, their intensities and positions in the temperature scale. The background width of the spectrum of MA1 (Fig. 3(a)) equals to 70 K, but that of MA2 150 K (Fig. 3(b)). The background width of the spectrum of MgH_2, obtained by direct hydriding in gaseous medium (Fig. 3(c)) equals to 100 K.

The spectrum of MA1 consists of almost two intense desorption lines (Fig. 3(a)). On the HTD spectrum of MgH_2 derived by the reactive milling for 3 h, in addition to a distinct peak at 465°C, another peak at 430°C can be also distinguished as a shoulder (Fig. 3(b)). There is only one rather narrow and symmetrical peak at 480°C on the HTD spectrum of MgH_2 obtained by hydriding in gaseous medium as one can see from Fig. 3(c).

As it has been established in Ref. [5], the existence of several lines (peaks) in the thermodesorption spectrum of a Mg-based mechanical alloy reveals the presence of several possibilities (positions) for a hydrogen localization in the alloy for the MgH_2 phase, synthesized by the reactive milling, with a characteristic depth of the potential well for every position. The presence of only one peak on the HTD spectrum of MgH_2 obtained by direct hydriding the magnesium powder in gaseous medium (Fig. 3(c)) indicates the existence of one type of crystal interstitial positions occupied by hydrogen atoms in the hydride lattice. Several peaks on the TDS spectrum of MgH_2 synthesized by the RMA method, in accordance with the results of Ref. [5], indicate the presence of several groups of hydrogen atoms with their energies of interaction depending significantly on the surrounding nearest matrix atoms (magnesium), bonding partners. For a complete removal of hydrogen atoms belonging to the different groups of the MgH_2 hydride, different temperatures are necessary for a thermal release of hydrogen atoms when heating the specimen. It is apparent that, for a removal of weak-bonded hydrogen from the MgH_2 hydride a lower temperature is necessary as compared with that when strong-bonded hydrogen is removed.

The both peaks (at 430°C and at 465°C) on the HTD spectrum of MgH_2 derived by the reactive milling are positioned at somewhat lower temperatures as compared with the peak at 480°C on the HTD spectrum of MgH_2 obtained by hydriding in gaseous medium. The above results indicate that, lower thermal stability is characteristic of the MgH_2 hydride synthesized by the reactive milling as compared with the hydride obtained by hydriding the magnesium powder in gaseous medium. From comparison of the above four-peak HTD spectrum and the two-peak HTD spectrum of MgH_2 synthesized by reactive milling magnesium powder for 3 h with the one-peak HTD spectrum (with its maximum at 480°C) of MgH_2 obtained by hydriding the magnesium powder in gaseous medium, the conclusion can be made about the influence of the method employed for a synthesis of the MgH_2 hydride on hydrogen locations in the hydride. The addition of titanium to magnesium increases significantly the above influence as our hydrogen desorption spectra reveal (see Fig. 3).

The mechanism of a decrease of thermal stability and decomposition temperature of the MgH_2 hydride obtained employing the RMA method as compared with those of the hydride derived by direct hydriding magnesium powder in gaseous medium can be attributed mainly to another location of hydrogen in MgH_2 synthesized during the reactive milling of magnesium powder (i.e., synthesized by a non-traditional method). Under the term 'another location of hydrogen' we assume an appearance of new sites of stable positions of hydrogen atoms in the regions of accumulation of high quantities of defects, grains and sub-grain boundaries and other imperfections of crystal lattice that are formed permanently during the reactive milling and their quantity increases with increasing time of the milling.

It has been observed in Ref. [5] that with decreasing sizes of powder particles of the mechanical alloys their HDT spectra reveal both increasing intensities of low-temperature peaks and decreasing intensities of high-temperature peaks representing decreasing stability of the MgH_2 phase of the alloy. The transition metal, titanium, added to magnesium is a dispergator [10] preventing agglomeration of particles of plastic magnesium. The use of the transition metals allows (within limits of time of mechanical treatments) to achieve smaller sizes of particles of mechanical alloys (including the MgH_2 hydride, a component of the alloys) synthesized using the RMA method [9, 11] and to reach more significant decreasing both thermal stability and decomposition temperature of MgH_2 [12–17]. Titanium is a dispergator mainly due to its catalytic activity with respect to the $H_2 \leftrightarrows H + H$ reaction. Synthesis of the MgH_2 hydride on surfaces of powder magnesium particles undergone to mechanical treatment in hydrogen medium (under pressure) becomes significantly faster in the case of the presence of a mixture of magnesium and titanium particles due to a formation of atomic hydrogen on the particles (catalytic active centers). A thin layer of MgH_2

formed on surfaces of magnesium particles is destroyed and it is removed from the surfaces under shock influence of milling balls because of high brittleness of the layer. The layer will not grow to the thickness limiting further formation of the MgH_2 hydride inside a magnesium particle in the case of the usual method of hydriding magnesium powder and it will be removed from the surface of magnesium particles. The smaller size of titanium particles and the bigger quantity of titanium ultra-fine particles covering every magnesium particle, the better function of this transition metal as a dispergator preventing agglomeration of magnesium particles and allowing to obtain faster a disperse powder of the mechanical alloy. In our case, titanium in the very beginning stages of reactive milling, due to formation of titanium hydride, becomes very brittle and thins rapidly its particle sizes. Due to the XRD pattern presented in Fig. 1, part of adding titanium interacts with milling iron and hydrogen and forms the $TiFeH_x$ hydride that disperses very fast.

It has been established [18–22] that in hydride-forming alloys based on transition and rare-earth metals the XPS O 1s core-level spectra can be decomposed on two features (lines) positioned at about 530 and 532 eV. The feature at 530 eV is formed by oxygen atoms belonging to a surface oxide layer while that at 532 eV is due to contributions of oxygen atoms belonging to oxygen-containing structures, catalytic toxic substances. Such structures adsorbed on alloy surfaces can block catalytic active centers and can prevent the dissociation of molecular hydrogen [18–20]. Therefore, the XPS O 1s core-level spectra presented in Fig. 4(a) reveal that the addition of 10 wt% Ti to magnesium (the MA1 sample) leads to significant decreasing the quantity of oxygen-containing structures, catalytic poisons preventing dissociative hydrogen chemosorption, adsorbed on the particle surfaces. As one can see from Fig. 4(b, c), the addition of 10 wt% Ti to magnesium does not alter the charge state of magnesium atoms on the surface of the mechanical alloy because values of the XPS Mg 2p and Mg 2s core-level binding energies do not change within accuracy of the measurement when going from the MA2 sample (the mechanical alloy without the addition of titanium) to

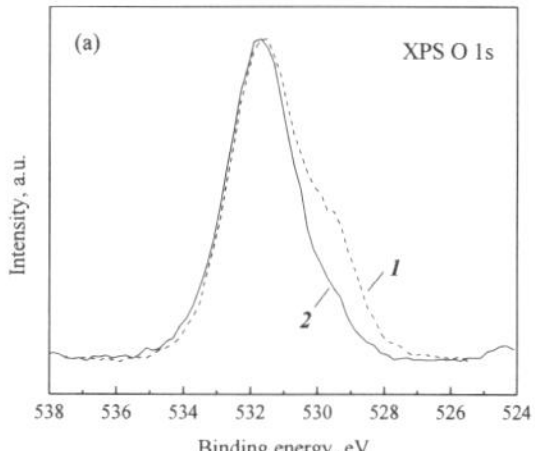

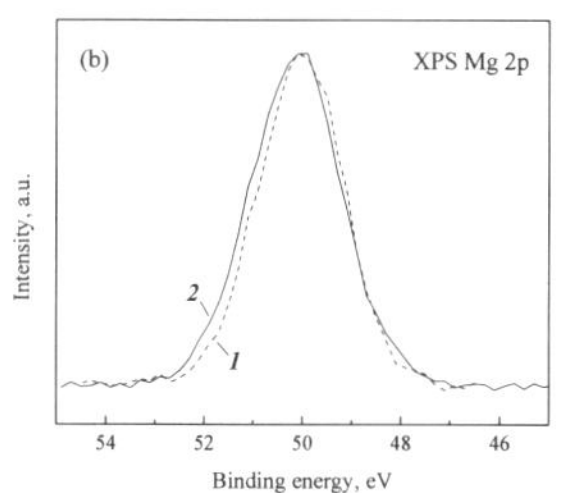

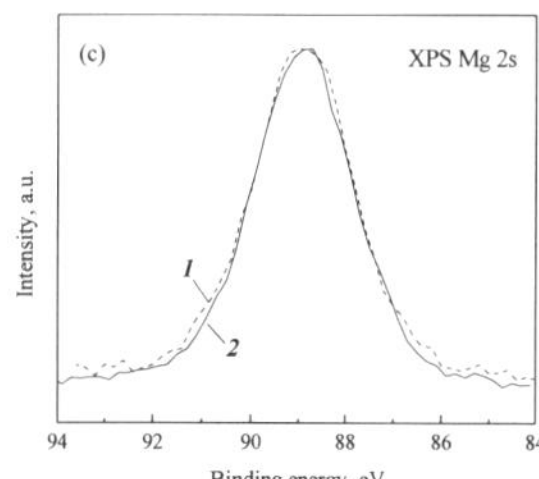

Figure 4. XPS (a) O 1s, (b) Mg 2p and (c) Mg 2s core level spectra of the MA1 (dashed curves 1) and MA2 (solid curves 2) specimens obtained by the RMA treatment

the MA1 sample. All the mentioned peculiarities of titanium behavior during the RMA processes probably are the main reason of the fact that the powder of the Mg + 10 wt% Ti alloy reveals a rather high specific surface (11.2 m^2/g) and big decreasing the decomposition temperature of the hydride phase of this alloy as well as nice kinetics of its hydriding-dehydriding.

In order to determine changes of decomposition temperature of the MgH_2 phase, a component of the Mg + 10 wt% Ti composite, when measuring the second and following cycles of hydrogen sorption-desorption (i.e., during the second, third etc. heating-cooling of the composite) the sample was not taken out of the reactor after its first heating up to 500°C (during registration of the desorption curve in the first cycle). After cooling the sample (together with the furnace) to room temperature (in such a case the Mg + 10 wt% Ti composite sorped almost the whole quantity of hydrogen desorped from it when measuring the first cycle), the sample was heated up to 420°C with the speed 5 K/min with the following registration of the desorption curve in the second cycle. When cooling the sample and the furnace, we was measuring again the sorption curve in the second cycle. We have made five cycles of such hydrogen sorption-desorption.

Figure 5 shows results of hydrogen desorption measurements for the MgH_2 phase of the Mg + 10 wt% Ti composite during the first, second and fifth heating-cooling cycles (curves, 1, 2 and 3, respectively). As one can see from Fig. 5, the beginning of sharp increase of hydrogen desorption on the curve 2 is shifted by 45 K towards lower temperatures as compared with that of the curve 1, i.e., during the second heating the composite its MgH_2 phase begins to decompose at lower temperature as compared with that during the first heating. The curves (and spectra) of hydrogen desorption measured for the third, fourth and fifth cycles were found to be similar to those measured for the second cycle. Additionally, the curves (and spectra) of the second, third, fourth and fifth cycles of hydrogen sorption registered during the sample cooling (after its respective heating) reveal their almost complete coincidence.

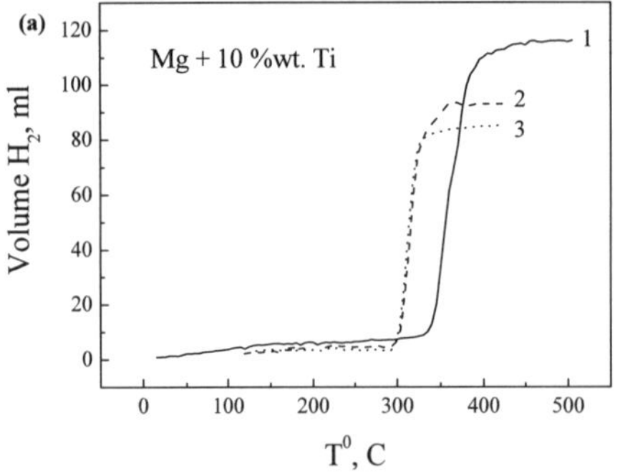

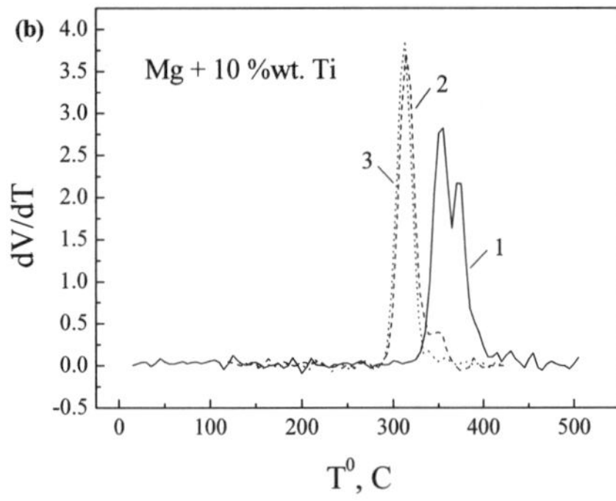

Figure 5. Curves (upper panel a) and spectra (bottom panel b) of thermodesorption of hydrogen from the MgH_2 hydride measured during the 1st (curve 1), 2nd (curve 2) and 5th (curve 3) cycles

As an example, Fig. 6 shows the curves and spectra of hydrogen sorption-desorption measured for MgH_2 of the Mg + 10 wt% Ti composite during the second and fifth hydriding-dehydriding cycles. From Fig. 6 it is obvious that when cooling the sample up to 314°C, an intense hydrogen absorption, caused by hydride forming, takes place.

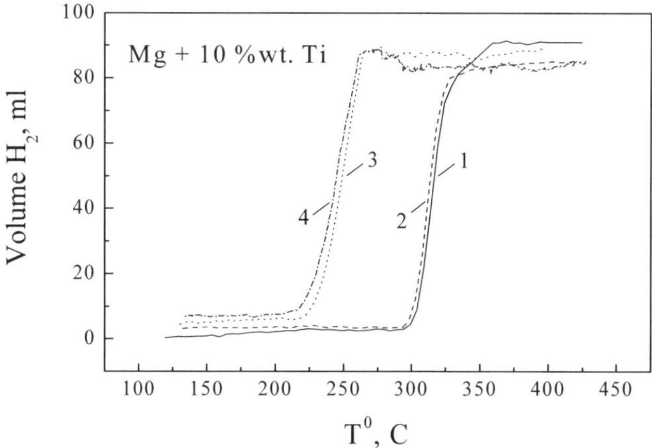

Figure 6. Curves of desorption (1 and 2) and sorption (3 and 4) of hydrogen from the MgH_2 hydride measured during the second (1 and 3) and fifth (2 and 4) cycles

From the above analysis of the desorption spectra derived for the first five cycles of heating-cooling the composite we can make the following conclusion: the effect of a decrease of decomposition temperature of the MgH_2 hydride derived due to the reactive milling of magnesium with addition of titanium preserves in the following cycles (after the first cycle) of hydrogen sorption-desorption. It is worth to mention also that during five cycles of H sorption-desorption under study, the Mg + 10 wt% Ti composite shows both good kinetics and reversibility of hydrogen sorption-desorption in spite of the low pressure in the reactor (normal pressure; 0.1 MPa).

4. Conclusions

Influence of addition of titanium to magnesium upon thermal stability of the hydride phase MgH_2 synthesized due to RMA in hydrogen atmosphere as pressure 1.2 MPa was studied employing the TDS method. The spectrum of hydrogen thermodesorption obtained during the first cycle of dehydriding reveals that the addition of 10 wt% of titanium to magnesium leads to additional decreasing by 115 K of decomposition temperature of the hydride phase MgH_2 derived employing the RMA method. During the second cycle of dehydriding, the decomposition temperature of the hydride phase of the Mg + 10 wt% Ti composite decreases additionally by 35 K, and the temperature does not change for three following sorption-desorption cycles. Therefore, the effect of decreasing the decomposition temperature of the MgH_2 phase obtained as a result of the reactive milling and titanium addition preserves after the first dehydriding-hydriding cycle.

Our experimental data reveal that the addition of titanium to magnesium promotes, due to the catalytic effect, good kinetics and reversibility of hydrogen sorption-desorption at low pressure of hydrogen in the reactor (0.1 MPa). The addition of titanium leads mainly to increasing the process of dispersion of magnesium and its hydride, MgH_2, synthesized by RMA. Additionally, as XPS data reveal, the addition of titanium promotes decreasing the quantity of oxygen-containing structures, catalytic poisons preventing dissociative hydrogen chemosorption, adsorbed on the surface of MgH_2 particles.

Acknowledgements

Financial support of INTAS (project # 05-1000005-7671) is gratefully acknowledged.

References

1. Chen H., Williams J.S. Formation of metal hydrides by mechanical alloying. *J. Alloy. Compd.*, 1995, **217**:181–184.
2. Huot J., Akiba E., Takada T. Mechanical alloying of Mg–Ni compounds under hydrogen and inert atmosphere. *J. Alloy. Compd.*, 1995, **231**:815–819.
3. Bobet J.L., Akiba E., Darriet B. Study of Mg-M (M = Co, Ni and Fe) mixture elaborated by reactive mechanical alloying: hydrogen sorption properties. *Int. J. Hydrogen Energy*, 2001, **26**:493–501.
4. Ershova O.G., Dobrovolsky V.D., Solonin Y.M., Morozova R.A. Proceedings of IX International Conference on Hydrogen Materials Science & Chemistry of Carbon Nanomaterials (ICHM'S 2005), 5–11 September 2005, Sevastopol, Ukraine, P. 178–179.
5. Dobrovolsky V.D., Ershova O.G., Solonin Y.M., Gorbachuk N.P. Thermal stability of the hydride phase of the mechanical alloy derived by treatment in hydrogen medium under pressure of the Mg + 10 %wt. Fe. *Powder Metall. Met. Ceram.*, 2006, **45**:381–386.
6. Doppiu S., Solsona P., Spassov T., Barkhordarian G., Dornheim M., Klassen T., Surinach S., Baro MD. Thermodynamic properties and absorption-desorption kinetics of $Mg_{87}Ni_{10}Al_3$ alloy synthesized by reactive ball milling under H_2 atmosphere. *J. Alloy. Compd.*, 2005, **404–406**:27–30.
7. Delchev P., Solsona P., Drenchev B., Drenchev N., Spassov T., Baro M.D. Direct hydriding of $Mg_{87}Al_7Ni_3Mn_3$ by reactive mechanical milling in hydrogen atmosphere and influence of particle size on the dehydriding reaction. *J. Alloy. Compd.*, 2005, **388**:98–103.
8. Spassov T., Rangelova V., Solsona P., Baro M.D., Zander D., Köster U. Hydriding/ dehydriding properties of nanocrystalline $Mg_{87}Ni_3Al_3M_7$ (M = Ti, Mn, Ce, La) alloys prepared by ball milling. *J. Alloy. Compd.*, 2005, **398**:139–144.
9. Ershova O.G., Dobrovolsky V.D., Solonin Y.M., Morozova R.A. About manner and mechanisms of reduction of thermal firmness of Mg-, Ti-, Y-based mechanical alloys. In: Veziroglu T.N. et al., editors. Hydrogen Materials Science and Chemistry of Carbon Nanomaterials. Dordrecht, Springer, 2007, P. 429–436.
10. Konstanchuk I.G., Ivanov E.Y., Boldyrev V.V. Interaction with hydrogen of alloys and intermetallics obtained by mechanochemical methods. *Uspekhi Khimii*, 1998, **67**: 75–86 (in Russian).
11. Bobet J.L., Even C., Nakamura Y., Akiba E., Darriet B. Synthesis of magnesium and titanium hydride via reactive mechanical alloying: Influence of 3d-metal addition on MgH_2 synthesize. *J. Alloy. Compd.*, 2000, **298**:279–284.
12. Rebinder P.A. Selection Publications: Surface Phenomena in Dispersion Media. Fuks G.I., editor. Moscow, Nauka, 1978 (in Russian).
13. Morokhov I.D., Petinov V.I., Trusov L.I., Petrunin V.F. Structure and properties of small metallic particles. *Usp. Fiz. Nauk*, 1981, **133**(4):653–692 (in Russian).

14. Zubov V.I. Some nano-size effects and properties of ultradispersive systems. *Zhurn. Mendeleev Vses Khim. Obshch.*, 1991, **36**:133–136.
15. Zaluska A., Zaluski L., Strom-Olsen J.O. Nanocrystalline magnesium for hydrogen storage. *J. Alloy. Compd.*, 1999, **288**:217–225.
16. Liang G., Huot J., Boily S., Van Neste A., Schulz R. Catalytic effect of transition metals on hydrogen sorption in nanocrystalline ball milled MgH_2-Tm (Tm=Ti, V,Mn, Fe and Ni) systems. *J. Alloy. Compd.*, 1999, **292**:247–252.
17. Huot J., Liang G., Schulz R. Mechanically alloyed metal hydride systems. *Appl. Phys.*, 2001, **72**:187–195.
18. Dobrovolsky V.D., Yendrzheevskaya S.N., Sinelnichenko A.K., Skorokhod V.V., Khyzhun O.Y. Analysis of the surface condition of $Ti_4Fe_2O_x$. *Int. J. Hydrogen Energy*, 1996, **21**(11/12): 1061–1064.
19. Dobrovolsky V.D., Solonin Y.M., Skorokhod V.V., Khyzhun O.Y. XPS and transmission electron microscopy of the multicomponent hydride-forming alloys. *Int. J. Hydrogen Energy*, 1999, **24**:195–201.
21. Solonin Y.M., Dobrovolsky V.D., Khyzhun O.Y., Skorokhod V.V., Galii O.Z., Correlation between surface chemical states and electrochemical activities of alloys AB_5 and AB_2. In: Hampton M.D. et al., editors. Hydrogen Materials Science and Chemistry of Metal Hydrides. Dordrecht, Kluwer, 2002, P. 415–422.
22. Dobrovolsky V.D., Solonin Y.M., Skorokhod V.V., Khyzhun O.Y. XPS and transmission electron microscopy of the multicomponent hydride-forming alloys for electrochemical applications. *J. Alloy. Compd.*, 1997, **253–254**: 488–491.

EFFECT OF BALL MILLING AND ADDITIONS OF METAL HYDRIDES AND AMIDES ON THE THERMAL STABILITY OF AlH_3

M.S. DULYA
Institute of Problems of Chemical Physics of RAS, Prospectus of the Academician Semenov, 1, Chernogolovka, Moscow Region, 142432 Russia
Tel./Fax: +7(496) 5221743; E-mail: btarasov @icp.ac.ru
B.M. BULYCHEV
Department of Chemistry, M.V. Lomonosov Moscow State University, Moscow
B.P. TARASOV
Institute of Problems of Chemical Physics of RAS, Prospectus of the Academician Semenov, 1, Chernogolovka, Moscow Region, 142432 Russia

Abstract. The influence of mechanical treatment on decomposition temperature of aluminum hydride in the compositions with LiH, MgH_2, TiH_2, VH_x, and $LiNH_2$ was investigated. It was shown, that the additives of LiH, MgH_2, TiH_2, and $LiNH_2$ reduce the thermal stability of AlH_3, and presence in a composite of VH_x does not render influence on decomposition temperature of aluminum hydride. The investigations were accomplished by methods of the thermal analysis, the electronic microscopy and the X-ray phase analysis.

Keywords: aluminum hydride, composite, additive, mechanical activation, thermal stability

1. Introduction

Aluminum hydride is one of the higher mass (10%) and volume (1.47 g/ml) hydrogen content material, but his thermodynamic stability depends from the methods of preparation and refine. Thermodynamically AlH_3 is unstable ($\Delta G > 0$ at 298 K) even at room temperature but because of the kinetics difficulties the temperature range of the decomposition is 150–180°C (for the most stable α-modification).

Aluminum hydride can be modified by a mechanical treatment and doping of the hydrides and amides of light metals by a mechanical activation in a planetary ball mill. Using these techniques the decomposition of the aluminum hydride phase with LiH, NaH and KH additions displace at lower temperature area (125–135°C) [1].

Catalytic influence of addition of the Ti and V compounds on hydrogen sorption properties, reversibility and hydrogenation/dehydrogenation kinetic of alanates and [2] MgH_2 [3] is also known.

The method of the mechanical alloying of the metals hydride by planetary ball mill Fritsch «Pulverisette 6» under hydrogen or argon at different milling energies had been suggested and approved. By means of the different methods of investigation (X-Ray diffraction, DSC, TG, mass spectroscopy) the influence of the LiH, TiH_x, VH_x and MgH_2 additions in different quantities on the thermal stability AlH_3 has been investigated in details.

2. Experimental

Aluminum hydride obtained by Schlesinger methods [4]. The lattice characteristic of obtained phase α-AlH_3 (a = 4.457 Å, c = 11.786 Å) corresponds with literatura data [4]. Aluminum hydride contains 0.01 wt% Cl, according to results of the chemical analysis.

The mechanical treatment was carried out in planetary spherical mill Fritsch "Pulverisette 6" in an atmosphere of hydrogen and argon. Acceleration of grinding bodies, time of a grinding, the relation of weight of a sample to weight of spheres, and quantity of doping component were varied. All operations with obtained powders were carried out without contact to air in inert boxing MBraun (contents of O_2 and H_2O are less than 0.1 ppm). Received materials were investigated with the help of X-ray phase analysis on diffractometer ARL X'TRA TERMO, scanning electronic microscopy on microscope JEOL JEM-100 CX and the thermal analysis on installation STA 409 Luxx companies Netzsch. Registration as DSC curves, and curve of weight losses (TG) was carried out at speed of heating of 10 degrees/min in a argon stream.

Hydrides LiH, TiH_x, VH_x, MgH_2, and $LiNH_2$ have been characterized by method of X-ray diffraction, element and thermal analyses.

3. Results and discussion

Under mechanical activation at different conditions of the treatment (vial volume, atmosphere Ar and H_2, energy of melting) AlH_3 partly decomposes and residuary hydride becomes less stable. The quantity of the AlH_3 phase reduces to 25 wt% at the time of the treatment in the range of 5–90 min (250 min^{-1}, Ar/H_2, 250 ml vial, 125:1) and temperature of decomposition is 134°C, whereas initial AlH_3 with 9.9 wt% H (99% of AlH_3) contents has a maximum decomposition temperature equal to 178°C (Fig. 1).

By means of the treatment in the planetary ball mill at different conditions the mixtures of the aluminum and lithium hydrides in the ratio of AlH_3:LiH = 1:1, 5:1 and 9:1 (mol) have been prepared. Figure 2 demonstrates the thermograms of such compositions. In the case of AlH_3:LiH = 1:1 ratio we clearly see the exothermal effect of the alanate formation (T = 130ºC) and two stages of its decomposition. The greatest effect on the thermal stability of the aluminum hydride shows the addition of the lithium hydride in 10–20 mol% quantity. Temperature of the decomposition decreases to 112ºC. It can be explained by formation of the $LiAlH_4$ phase and island mechanism of the atomic hydrogen transportation through alanate and aluminum oxide phases [1].

After mechanical activation of the hydrides mixture of AlH_3 and MgH_2 (250 min^{-1}, 1 h) we can observe AlH_3, Al, Mg, and intermetallic Mg_2Al_3 phases. The effects of the decomposition of hydrides: AlH_3 (145°C), γ-MgH_2 (280°C), α-MgH_2 (330°C) and melting of Mg_2Al_3 (444°C) can be observed on the thermorgams of DSC and TG of $2AlH_3 + MgH_2$ and $3AlH_3 + 2MgH_2$ compositions.

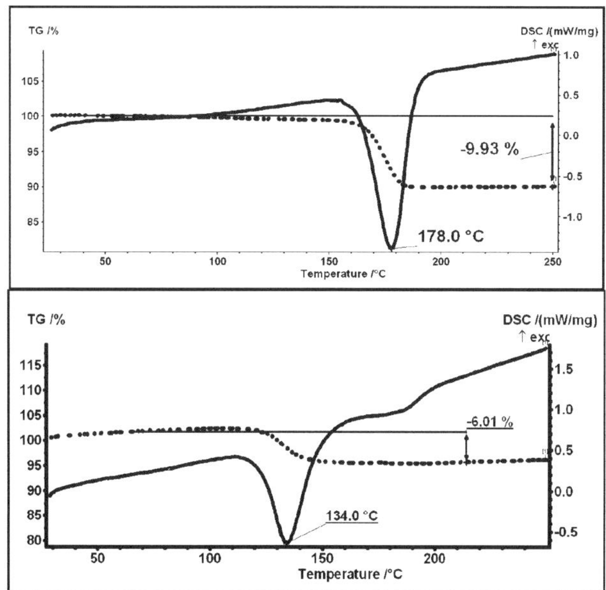

Figure 1. DSC-TG curves of AlH_3: above – the initial hydride, below – the ball milled hydride (30 min, 250 min^{-1}, 250 ml); the heating rate is 10 degrees/min, Ar flow

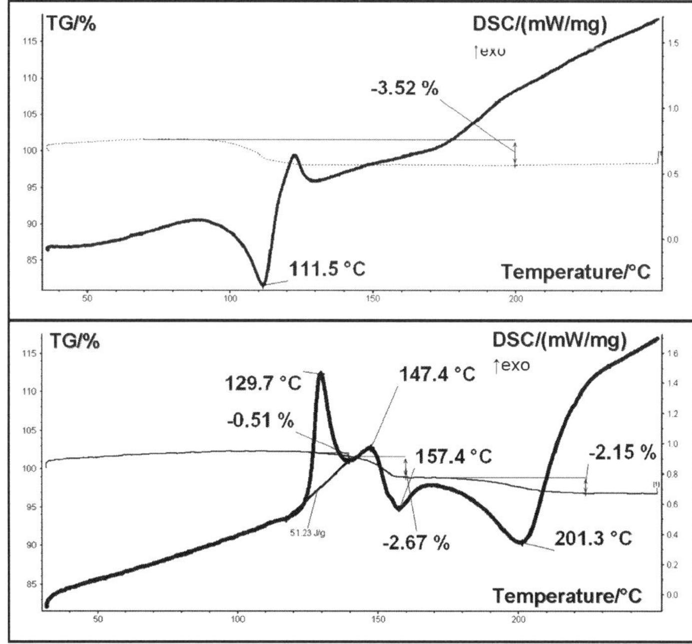

Figure 2. DSC-TG curves of the composites AlH_3 + xLiH after the ball milling (30 min, 250 min^{-1}, 250 ml, 10 degrees/min, Ar flow): x = 0.2 (top), x = 1 (bottom)

Addition of TiH_x (5–20 mol%) strongly decrease thermal stability of the AlH_3, catalyzing the process of its decomposition even at small energies of the mechanical activation (150–200 min^{-1}, Ar). The effect of the decomposition of the aluminum hydride is not appears and temperature of the dehydrogenation of the TiH_2 is less then at initial hydride.

Prepared by ball milling (150–200 min^{-1}, 5–30 min) composites of aluminum hydride with vanadium hydride (5–20 mol%) contain only a minor amount of metallic Al. The amount and the decomposition temperature of the aluminum hydride in the composite are the same as those in the ball milled pure AlH_3. Thus, the hydride VH_x doesn't change the thermal stability of AlH_3, while preserving a high weight content of hydrogen (~7 wt%).

After the ball milling of the mixtures of AlH_3 with $LiNH_2$ at 150 min^{-1} for 5–30 min under argon we observed AlH_3, Al, and Li_2NH. The DSC-TG curves of the composites show a "sharp" exo-effect at 150°C, and up to 5 mass% of H_2 releasing. This effect is probably correspond to a formation of a complex adduct $AlH_3{\cdot}xNH_3$. Degradation and decomposition of the composites with amide were observed.

4. Conclusions

Ball milling of AlH_3 at various energies under argon reduces the decomposition temperature of the hydride. Ball milling also causes a partial amorphization of the material and enhances its chemical reactivity.

Additions of LiH to AlH_3 at ball milling lead to a formation of the alanate phase and decrease the decomposition temperature down to 112°C.

Ball milling of composites of the hydrides MgH_2 and AlH_3 reduces the thermal stability both of AlH_3 and MgH_2.

Addition of TiH_x during ball milling catalyses the decomposition of AlH_3, while vanadium hydride having no effect on the thermal stability of AlH_3.

Lithium amide destabilizes AlH_3 with a release of up to 5 mass% of H_2 at 140–160°C.

Acknowledgement

The work is supported by INTAS (Grant № 05-1000005-7665).

References

1. Sandrock G. et al. Alkali metal hydride doping of α-AlH_3 for enhanced H_2 desorption kinetics, *J. Alloy. Compd.*, 2006, **421**: 185–189.
2. Kircher O., Fichtner M. Kinetic studies of the decomposition of $NaAlH_4$ doped with a Ti-based catalyst, *J. Alloy. Compd.*, 2005, **404–406**: 339–342.
3. Liang G., Huot J., et al. Hydrogen desorption kinetics of a mechanically milled MgH+5at.% V nanocomposite, *J. Alloy. Compd.*, 2000, **305**: 239–245.
4. Konovalov S.K., Bulychev B.M. The P,T-State diagram and solid phase synthesis of aluminum hydride, *Inorg. Chem.*, 1995, **34**: 172–175.

KINETICS OF THE ORIENTATIONAL LONG-RANGE ORDERING OF INTERSTITIAL HYDROGEN ATOMS IN METALS HAVING HEXAGONAL-CLOSE PACKED STRUCTURE

T.M. RADCHENKO, V.A. TATARENKO
Department of Solid State Theory, G. V. Kurdyumov Institute for Metal Physics, N.A.S. of Ukraine, 36 Academician Vernadsky Boulevard, UA-03680 Kyyiv-142, Ukraine
Tel. +380 44 424 12 21; Fax: +380 44 424 25 61;
E-mail: Taras.Radchenko@gmail.com

Abstract. Using the Onsager-type kinetics equation, the differential equation for the kinetics of orientational long-range order (LRO) parameter for the interstitial H-atoms' distribution within the hexagonal close-packed (h.c.p.) lattice of a metal (*Me*) is firstly derived. The numerical solutions of this equation enable to analyze the orientational ordering of H atoms. The decreasing of H concentration results in a decelerating of the orientational-LRO parameter change and in an increasing of its relaxation time. Temperature influence on the orientational-ordering process is more appreciable for lower hydrogen concentrations because of the open-chain (two by two) distribution of H atoms in h.c.p. metals.

Keywords: orientational long-range order, ordering kinetics, Onsager-type equation

1. Introduction

An ordering in the solid solutions is a topic of growing technological and scientific interest within the field of materials science. A long-range ordering as well as a short-range one exerts a strong influence on physical properties of the rare earth–hydrogen systems [1, 2]. The LRO kinetics is one of the diffusion processes occurring within the atomic-scale ranges. X-ray diffraction technique is the most convenient experimental instrument to provide us detailed information on this process. There are also different theoretical methods for this one. One of the models is based on the Onsager-type microscopic-diffusion equation [3]. Using such a model, kinetics of the structural transformations during ordering, clustering and coarsening has been studied in a whole series of publications of Khachaturyan and his followers [3–9]. Nevertheless, all obtained results relate to the substitutional alloys based on the cubic Bravais lattice. Therefore, theoretical investigation of ordering kinetics in the interstitial solutions based on the non-cubic lattice becomes relevant nowadays. In particular, it relates to the h.c.p. host-crystal–H structures. Moreover, experimental measurements have not been done for some rare earth–hydrogen h.c.p. solutions, and theoretical investigation is a better way.

A given work is concerned with the relaxation kinetics of the orientational long-range ordering of interstitial hydrogen atoms within the h.c.p. lattice of *Me*.

2. Kinetics model

Hydrogen atoms are located at the tetrahedral interstices of rare-earth h.c.p.-*Me* lattice, which is consisting of the two interpenetrating hexagonal Bravais lattices shifted relative to each other with fundamental periods of translations – a_0 and c_0 (Fig. 1). Each primitive unit cell having the volume of $(\sqrt{3}/2)a_0^2c_0$ and constructed on the basic vectors of translation, $\mathbf{a}_1 \equiv (a_0, 0, 0)$, $\mathbf{a}_2 \equiv (-a_0/2, a_0\sqrt{3}/2, 0)$, $\mathbf{a}_2 \equiv (0, 0, c_0)$, in a Cartesian co-ordinate system *Oxyz* contains two *Me* ions: one of which is located at the unit-cell 'origin' with the radius-vector $\vec{\rho}_1 = \mathbf{0}$, and another one is located in its site with the radius-vector [10, 11] $\vec{\rho}_2 = 2\mathbf{a}_1/3 + \mathbf{a}_2/3 + \mathbf{a}_3/2 = (a_0/2, a_0/(2\sqrt{3}), c_0/2)$. Tetrahedral interstices form the four interpenetrating sublattices within the h.c.p. lattice of *Me* (Fig. 1).

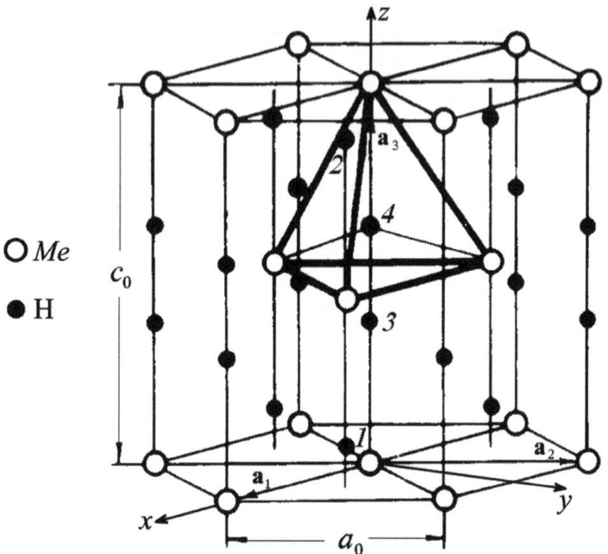

Figure 1. Location of tetrahedral interstices (solid circles) forming the four interpenetrating sublattices (p, $q = 1, \ldots, 4$) within the h.c.p. host-metal (*Me*) lattice with parameters, a_0 and c_0 (open circles are sites forming the two interpenetrating sublattices); $\mathbf{a}_1$, $\mathbf{a}_2$ and $\mathbf{a}_3$ are the basic translation vectors

Using the static concentration waves' approach [3], the single-site occupation-probability function for H atom in p-th tetrahedral-interstices' sublattice ($p = 1, \ldots, 4$) within the primitive unit cell at an 'origin' site $\mathbf{R}$ related to the basic vectors as $\mathbf{R} = H_1\mathbf{a}_1 + H_2\mathbf{a}_2 + H_3\mathbf{a}_3$, $P_p(\mathbf{R})$, can be written as follows [10, 11]:

$$\begin{pmatrix} P_1(\mathbf{R}) \\ P_2(\mathbf{R}) \\ P_3(\mathbf{R}) \\ P_4(\mathbf{R}) \end{pmatrix} = c + \frac{1}{2}\eta \begin{pmatrix} 1 \\ -1 \\ 1 \\ -1 \end{pmatrix}, \quad (1)$$

where c is an atomic fraction of the interstitial H atoms, η is an orientational LRO parameter, $\mathbf{k}=\mathbf{0}$ is wave vector of the reciprocal space, which 'generates' the orientationally ordered superstructure [3, 10, 11].

Long-range order relaxation kinetics is one of the microdiffusion-mediated processes occurring within the atomic-scale ranges. Therefore, in our study, we can use the kinetics model based on the Onsager-type microscopic-diffusion equation [3]. Within this model, the orientational-ordering kinetics of H atoms is described by a single-site occupation probability function, $P_p(\mathbf{R},t)$, where vector $\mathbf{R}$ refers to the 'origin' position of the primitive unit cell, in which a considered site is located, and index p denotes the interstitial sublattice. Then, inasmuch as the rate of $P_p(\mathbf{R},t)$ change with time t, $dP_p(\mathbf{R},t)/dt$, is proportional to the thermodynamic driving forces, $\delta F/\delta P_q(\mathbf{R}',t)$ (F – the configurational Helmholtz free energy [3]), and $P_p(\mathbf{R})+P_p^{\otimes}(\mathbf{R})\equiv 1$ ($\otimes$ symbolizes vacant tetrahedral interstice), one can write [3]:

$$\frac{dP_p(\mathbf{R},t)}{dt}=-\frac{1}{k_BT}\sum_{\mathbf{R}'}\left\{\left[c^2L_{p1}^{\mathrm{HH}}(\mathbf{R}-\mathbf{R}')-c(1-c)L_{p1}^{\mathrm{H}\otimes}(\mathbf{R}-\mathbf{R}')\right]\frac{\delta F}{\delta P_1(\mathbf{R}',t)}+\right.$$
$$+\left[c^2L_{p2}^{\mathrm{HH}}(\mathbf{R}-\mathbf{R}')-c(1-c)L_{p2}^{\mathrm{H}\otimes}(\mathbf{R}-\mathbf{R}')\right]\frac{\delta F}{\delta P_2(\mathbf{R}',t)}+$$
$$+\left[c^2L_{p3}^{\mathrm{HH}}(\mathbf{R}-\mathbf{R}')-c(1-c)L_{p3}^{\mathrm{H}\otimes}(\mathbf{R}-\mathbf{R}')\right]\frac{\delta F}{\delta P_3(\mathbf{R}',t)}+$$
$$\left.+\left[c^2L_{p4}^{\mathrm{HH}}(\mathbf{R}-\mathbf{R}')-c(1-c)L_{p4}^{\mathrm{H}\otimes}(\mathbf{R}-\mathbf{R}')\right]\frac{\delta F}{\delta P_4(\mathbf{R}',t)}\right\}\quad (p=1,\ldots,4). \tag{2}$$

Here, k_B is a Boltzmann constant, T is an absolute temperature, $L_{pq}^{\mathrm{HH}}(\mathbf{R}-\mathbf{R}')$ and $L_{pq}^{\mathrm{H}\otimes}(\mathbf{R}-\mathbf{R}')$, are the kinetic coefficients, which are proportional to the probability of an elementary microdiffusion-jump exchange of 'atoms' at issue between the site $(q,\mathbf{R}')$ and the site $(p,\mathbf{R})$ during the time unit.

Since the total H-atoms' number is fixed, there is a following restriction for the kinetic coefficients:

$$\sum_{p=1}^{4}\sum_{\mathbf{R}}\frac{dP_p(\mathbf{R})}{dt}=\frac{dN_{\mathrm{H}}}{dt}\equiv 0,\ \text{i.e.}\ \sum_{p=1}^{4}\sum_{\mathbf{R}}L_{pq}^{\mathrm{HH}}(\mathbf{R}-\mathbf{R}')\equiv 0$$
$$\text{and}\ \sum_{p=1}^{4}\sum_{\mathbf{R}}L_{pq}^{\mathrm{H}\otimes}(\mathbf{R}-\mathbf{R}')\equiv 0, \tag{3}$$

where N_{H} is the total number of H atoms in the system at issue.

The thermodynamic driving forces, $\delta F/\delta P_p(\mathbf{R}')$, can be written as follows:

$$\frac{\delta F}{\delta P_p(\mathbf{R}')}=\sum_{\mathbf{R}}\sum_{q=1}^{4}w_{pq}(\mathbf{R}-\mathbf{R}')P_q(\mathbf{R})+\ln\frac{P_p(\mathbf{R}')}{1-P_p(\mathbf{R}')}, \tag{4}$$

where $w_{pq}(\mathbf{R}-\mathbf{R}')$ are the 'mixing' energies defined by the pairwise interaction energies [10],

$$w_{pq}(\mathbf{R}-\mathbf{R}')=W_{pq}^{\mathrm{HH}}(\mathbf{R}-\mathbf{R}')+W_{pq}^{\otimes\otimes}(\mathbf{R}-\mathbf{R}')-2W_{pq}^{\mathrm{H}\otimes}(\mathbf{R}-\mathbf{R}'). \tag{5}$$

Fourier transforms of the 'mixing' energies (for $\mathbf{k}=\mathbf{0}$),

$$\tilde{w}_{pq}(\mathbf{0})=\sum_{\mathbf{R}}w(\mathbf{R}-\mathbf{R}')\exp(-i\mathbf{k}\cdot\mathbf{R})=\sum_{\mathbf{R}}w(\mathbf{R}-\mathbf{R}'), \tag{6}$$

are the elements of the interaction matrix [11],

$$\left\|\tilde{w}_{pq}(\mathbf{0})\right\|=\left\|\begin{matrix}\tilde{w}_{11}(\mathbf{0}) & \tilde{w}_{12}(\mathbf{0}) & \tilde{w}_{13}(\mathbf{0}) & \tilde{w}_{14}(\mathbf{0})\\ \tilde{w}_{12}(\mathbf{0}) & \tilde{w}_{11}(\mathbf{0}) & \tilde{w}_{14}(\mathbf{0}) & w_{13}(\mathbf{0})\\ \tilde{w}_{13}(\mathbf{0}) & \tilde{w}_{14}(\mathbf{0}) & \tilde{w}_{11}(\mathbf{0}) & \tilde{w}_{12}(\mathbf{0})\\ \tilde{w}_{14}(\mathbf{0}) & \tilde{w}_{13}(\mathbf{0}) & \tilde{w}_{12}(\mathbf{0}) & \tilde{w}_{11}(\mathbf{0})\end{matrix}\right\|, \tag{7}$$

with the following eigenvalues [11]

$$\begin{aligned}\lambda_1(\mathbf{0})&=\tilde{w}_{11}(\mathbf{0})+\tilde{w}_{12}(\mathbf{0})+\tilde{w}_{13}(\mathbf{0})+\tilde{w}_{14}(\mathbf{0}),\\ \lambda_3(\mathbf{0})&=\tilde{w}_{11}(\mathbf{0})-\tilde{w}_{12}(\mathbf{0})-\tilde{w}_{13}(\mathbf{0})+\tilde{w}_{14}(\mathbf{0}),\\ \lambda_2(\mathbf{0})&=\tilde{w}_{11}(\mathbf{0})+\tilde{w}_{12}(\mathbf{0})-\tilde{w}_{13}(\mathbf{0})-\tilde{w}_{14}(\mathbf{0}),\\ \lambda_4(\mathbf{0})&=\tilde{w}_{11}(\mathbf{0})-\tilde{w}_{12}(\mathbf{0})+\tilde{w}_{13}(\mathbf{0})-\tilde{w}_{14}(\mathbf{0}).\end{aligned} \tag{8}$$

The value $\delta F/\delta P_p(\mathbf{R}')$ possesses the same symmetry as the function $P_p(\mathbf{R}',t)$ [3]. Therefore (in case of the orientational LRO), $\delta F/\delta P_p(\mathbf{R}',t)$ as well as $P_p(\mathbf{R}',t)$ (1) can be represented as the superposition of the same concentration wave:

$$\begin{pmatrix}\delta F/\delta P_1(\mathbf{R})\\ \delta F/\delta P_2(\mathbf{R})\\ \delta F/\delta P_3(\mathbf{R})\\ \delta F/\delta P_4(\mathbf{R})\end{pmatrix}=\tilde{c}(\eta)+\frac{1}{2}\begin{pmatrix}1\\-1\\1\\-1\end{pmatrix}\tilde{\eta}(\eta). \tag{9}$$

Combination of expressions (4) and (9) gives:

$$\frac{\delta F}{\delta P_1(\mathbf{R})}=\frac{\delta F}{\delta P_3(\mathbf{R})}=c\lambda_1(\mathbf{0})+\frac{1}{2}\eta\lambda_4(\mathbf{0})+k_BT\ln\frac{c+\eta/2}{1-c-\eta/2}=\tilde{c}+\frac{1}{2}\tilde{\eta}, \tag{10a}$$

$$\frac{\delta F}{\delta P_2(\mathbf{R})}=\frac{\delta F}{\delta P_4(\mathbf{R})}=c\lambda_1(\mathbf{0})-\frac{1}{2}\eta\lambda_4(\mathbf{0})+k_BT\ln\frac{c-\eta/2}{1-c+\eta/2}=\tilde{c}-\frac{1}{2}\tilde{\eta}. \tag{10b}$$

Summation and difference of the last two equations give these functions, $\tilde{c}(\eta)$ and $\tilde{\eta}(\eta)$:

$$\tilde{c} = c\lambda_1(\mathbf{0}) + \frac{1}{2}k_BT\ln\frac{(c-\eta/2)(c+\eta/2)}{(1-c-\eta/2)(1-c+\eta/2)}, \tag{11a}$$

$$\tilde{\eta} = \eta\lambda_4(\mathbf{0}) + k_BT\ln\frac{(c+\eta/2)(1-c+\eta/2)}{(c-\eta/2)(1-c-\eta/2)}. \tag{11b}$$

Substituting expression (4) into Eq. (2), using equalities (3) and (11b), followed by the Fourier transformations of both (left and right) parts of Eq. (2), one can obtain the differential equation for the orientational-LRO parameter, η, of H atoms:

$$\frac{d\eta}{dt} = -c(1-c)\tilde{L}(\mathbf{0})\left[\eta\frac{\lambda_4(\mathbf{0})}{k_BT} + \ln\frac{(c+\eta/2)(1-c+\eta/2)}{(c-\eta/2)(1-c-\eta/2)}\right], \tag{12}$$

where

$$\begin{aligned}\tilde{L}(\mathbf{0}) = \frac{c}{1-c}&\left[\tilde{L}_{11}^{\mathrm{HH}}(\mathbf{0}) - \tilde{L}_{12}^{\mathrm{HH}}(\mathbf{0}) + \tilde{L}_{13}^{\mathrm{HH}}(\mathbf{0}) - \tilde{L}_{14}^{\mathrm{HH}}(\mathbf{0})\right] - \\ &-\left[\tilde{L}_{11}^{\mathrm{H}\otimes}(\mathbf{0}) - \tilde{L}_{12}^{\mathrm{H}\otimes}(\mathbf{0}) + \tilde{L}_{13}^{\mathrm{H}\otimes}(\mathbf{0}) - \tilde{L}_{14}^{\mathrm{H}\otimes}(\mathbf{0})\right]\end{aligned}. \tag{13}$$

Thus, at least if $c \ll 1$, an exchange of H atoms does not change η. As seen from Eq. (12), it is convenient to define both the reduced time $t^* = \tilde{L}(\mathbf{0})t$ and the reduced temperature $T^* = k_BT/|\lambda_4(\mathbf{0})|$ to solve this kinetics equation numerically in terms of t^* and T^*.

3. Results and conclusions

Numerical solution of Eq. (12) yields the curves for the time dependence of the orientational-LRO parameter for H atoms at the tetrahedral interstices of the h.c.p.-*Me* lattice. In Fig. 2a–d), curves represent the solutions for the following initial magnitudes of the orientational-LRO parameter: $\eta_0 \equiv \eta(t = 0) = 0.1$ and $\eta_0 = 0.2$.

Figure 2(a–d) demonstrate that both the H concentration and the temperature strongly affect the quantitative and qualitative changes of both the kinetics path and the equilibrium value. Four concentrations and four temperatures provide not only the different profiles of relaxation curves but also the equilibrium orientational-LRO parameter value. Decreasing of the H concentration results in both the decelerating of the orientational-LRO parameter change (at the initial evolution stage) and the increasing of its relaxation time.

The rate of an orientational-LRO parameter change is higher at the initial stage of the ordering process. The decreasing of an initial orientational-LRO parameter promotes the increasing of this rate and, as expected, does not influences on the equilibrium orientational-LRO parameter – its magnitude is the same for both η_0 magnitudes at a given T (Fig. 2).

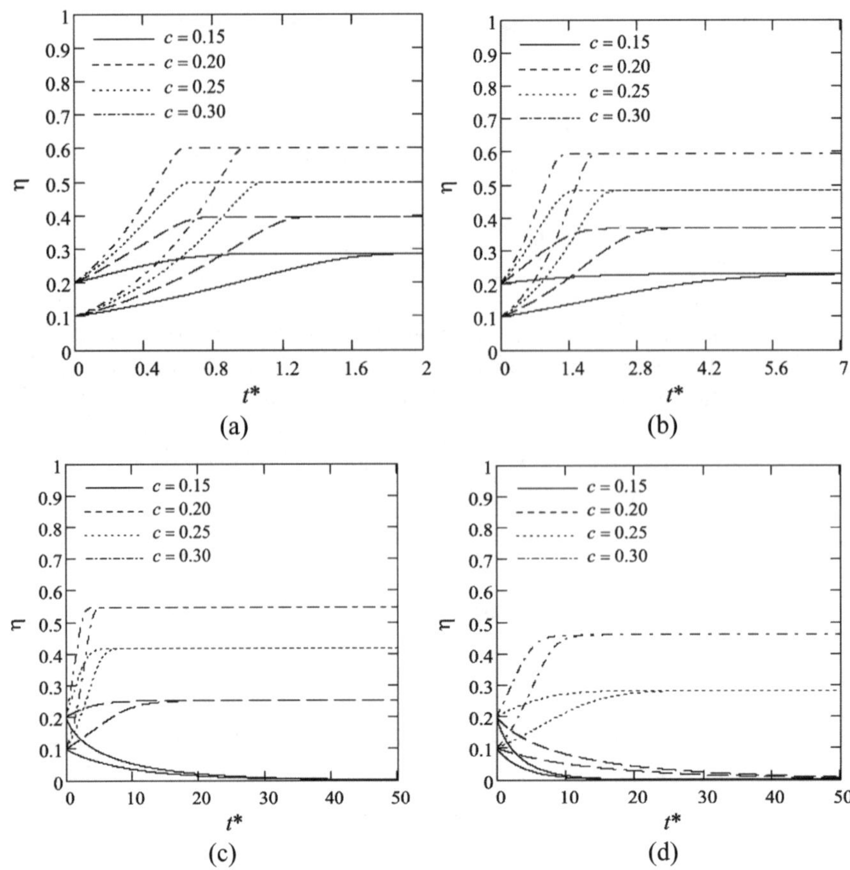

Figure 2. The reduced-time dependence of the orientational-LRO parameter, η, for different hydrogen concentrations, *c*, in h.c.p.-MeH_{2c} solid solution at some reduced temperatures: $T^* = 0.07$ (a), $T^* = 0.10$ (b), $T^* = 0.14$ (c), $T^* = 0.17$ (d)

Obviously, the increasing of *T* leads to the decreasing of instantaneous values of η and equilibrium ones. As shown in Fig. 2(c, d) (with additional calculations), at high temperatures and fixed concentrations, equilibrium η values may asymptotically verge towards zero, and the orientational disordering occurs. Besides, an influence of the temperature is more appreciable for lower H concentrations. In our opinion, it is caused by the open-chain (two by two) distribution of H atoms, for instance, in rare-earth h.c.p. metals.

Using derived equation of kinetics (12), one can predict the evolution of morphological imaging of the structural configurations described by the single-site occupation probabilities. Moreover, certainty of the Onsager-type kinetics coefficients enables to obtain results not only for *T*-dependent reduced time *t**, but also for the actual *t* scale. Vice versa, the relaxation times calculated from the independent experimental data [12, 13] allow an estimating of the Onsager-type kinetic coefficients, which determine the diffusivities [14].

Acknowledgements

The partial financial supports from NATO Reintegration Grant (RIG 981326) and Grant of N.A.S. of Ukraine for the young scientists (#4Г/30.07) are appreciated by the first author.

References

1. Vajda P. Hydrogen in rare earth metals, including RH_{2x}-phases. In: Gschneidner K.A., Jr., Eyring L., editors. Handbook on the Physics and Chemistry of Rare Earths. Amsterdam: Elsevier Science, 1995. Vol. **20**, 207–292.
2. Fukai Yuh. The metal–hydrogen system: Basic bulk properties. Berlin and Heidelberg: Springer-Verlag, 2003, 509 p.
3. Khachaturyan A.G. Theory of Structural Transformations in Solids. New York: J. Wiley & Sons, 1983, 574 p.
4. Chen L.Q., Khachaturyan A.G. Computer simulation of simultaneous ordering, decomposition and strain-induced coarsening. In: Chen H., Vasudevan V.K., editors. Kinetics of Ordering Transformations in Metals. Pittsburgh, PA: TMS, 1992, P. 197–206.
5. Chen L.Q, Khachaturyan AG. Kinetics of virtual phase formation during precipitation of ordered intermetallics. Phys. Rev. B, 1992, **46**: 5899–5905.
6. Poduri R., Chen L.Q. Computer simulation of the kinetics of order–disorder and phase separation during precipitation of δ' (Al_3Li) in Al–Li alloys. Acta Mater., 1997; **45**: 245–255.
7. Poduri R., Chen L.Q. Computer simulation of atomic ordering and compositional clustering in the pseudobinary Ni_3Al–Ni_3V system. Acta Mater., 1988; **46**: 1719–1729.
8. Wang Y., Banerjee D., Su C.C., Khachaturyan A.G. Field kinetic model and computer simulation of precipitation of $L1_2$ ordered intermetallics from f.c.c. solid solution. Acta Mater., 1998; **46**: 2983–3001.
9. Rubin G., Khachaturyan A.G. Three-dimensional model of precipitation of ordered intermetallics. Acta Mater., 1999; **47**: 1995–2002.
10. Bugaev V.N., Tatarenko V.A. Interaction and arrangement of atoms in interstitial solid solutions based on close-packed metals. Kiev: Naukova Dumka, 1989.
11. Tatarenko V.A., Tsinman K.L. Strain-induced effects on interactions of non-metal atoms introduced in metals having h.c.p. structure. 2. Interstitial atoms on tetrahedral interstices. Met. Phys. Adv. Tech., 1999; **18**(3): 273–311.
12. Tatarenko V.A., Radchenko T.M., Molodkin V.B. Microscopic characteristics of H diffusion and diffuse scattering of radiations in H.C.P.-*Ln*–H (from the Data on Electrical-Resistivity Relaxation). In: Veziroglu T.N. et al., editors. Hydrogen Materials Science and Chemistry of Carbon Nanomaterials: NATO Science Series, II: Mathematics, Physics and Chemistry. Dordrecht, The Netherlands: Kluwer. Vol. 172, 2004, P. 59–66.
13. Radchenko T.M., Tatarenko V.A. Comments Concerning Parameters of the Short-Range Order Evolution Determined from the Data on Kinetics of a Heat-Capacity Relaxation for Lu–H Alloy. In: Veziroglu T.N. et al. editors. Hydrogen Materials Science and Chemistry of Carbon Nanomaterials. Berlin: Springer, 2007, P. 229–234.
14. Radchenko T.M., Tatarenko V.A., Bokoch S.M. Diffusivities and kinetics of short-range and long-range orderings in Ni–Fe permalloys. Metallofiz Noveishie Tekhnol., 2006; **28**(12): 1699–1720 (in Russian).

EFFECT OF Mn AND Zr ON HYDROGEN ABSORPTION IN Mg-BASED NANOCOMPOSITES

X. YAO, Z.H. ZHU, G.Q. LU
ARC Centre for Functional Nanomaterials, University of Queensland, St Lucia, QLD 4072, Australia
Fax: 61 (07) 33656074, E-mail: x.yao@minmet.uq.edu.au.

X. YAO
School of Engineering, James Cook University, Townsville 4811, Australia

C.Z. WU, P. WANG, H.M. CHENG
Shenyang National Laboratory of Materials Science, Institute of Metal Research, Shenyang 110015, China

Abstract. This paper investigates the catalytic effect of MnZr metallic coupling on hydrogenation of magnesium. It is found that MnZr coupling greatly increased the hydrogenation kinetics, especially at low temperatures. The system can absorb 5wt% hydrogen within 30 minutes at 150°C. The effect of milling time on the microstructures and hydrogenation properties has also been investigated.

Keywords: hydrogen absorption, nanocomposite, capacity, kinetics, magnesium, manganese, zirconium, temperature

1. Introduction

Rapid growth in energy demand and depletion of energy resources, and the need for reducing greenhouse gas emissions all require an alternative and clean energy economy. An economy based on the clean energy carrier hydrogen is a promising one in this regard. A major challenge to realizing the hydrogen economy is the development of an efficient and safe system for practical hydrogen storage [1]. Magnesium and magnesium-based alloys are the most attractive materials with high hydrogen capacity, excellent reversibility, low cost and non-toxicity but limited for practical application so far due to their crucial limitations of high operational temperature and low kinetics for hydrogenation. To overcome these barriers, tremendous efforts have been devoted in the past decade to develop different strategies, such as nanostructuring [2], alloying [3], and the use of catalysts [4]. It has been proven that high-energy ball milling provides a good method to increase hydrogenation kinetics by reducing the grain size, refreshing the surface and introducing defects [5]. Our previous work demonstrated that the transition metals (TMs) play an important role to reduce remarkably the energy barrier of molecular hydrogen dissociation that enables it practical the hydrogenation of Mg-TMs system at low temperature [6–8]. In addition, it is proved that the coupling of TMs exhibits much more effective in catalysis on hydrogenation than the individual alone [9, 10]. However, the maximum

B. Baranowski et al. (eds.), *Carbon Nanomaterials in Clean Energy Hydrogen Systems.*

hydrogenation capacity is reduced by introduction of these TMs [9, 10]. The aim of this paper is to investigate a new couple of metallic catalyst of Mn and Zr on hydrogen absorption of Mg-based nanocomposites synthesized by high-energy ball milling, which proves a higher maximum hydrogenation capacity of the Mg-MnZr system with excellent kinetics.

2. Experimental procedure

All experiments and handling were conducted in a glovebox (MBraun) under Argon atmosphere with $O_2 < 1$ ppm and $H_2O < 1$ ppm, aiming to prevent the samples from oxidation.

Sample preparation: MgH_2 was obtained by heating Mg powders (purchased for Lancaster) to 400°C in a hydrogen storage rig and 90 wt% of the Mg powders was hydrided according to the volumetric calculations. Mn and Zr was a commercial product from Aldrich Co. MgH_2–MnZr composite sample was prepared using the composition of MgH_2-4.0 wt% (Mn + Zr), and mechanically milled by a SPEX8000 high-energy ball mill. A hardened steel crucible and six steel balls of 12 mm in diameter were used for milling. The ball to powder weight ratio was 12.5:1. One gram of the powder was removed after 30 hour milling. The rest of powder was then milled until to 60 hours.

Hydrogen absorption: The hydrogen storage properties (absorption capacity and kinetics) of the samples were evaluated by using an automated Sievert apparatus under a hydrogen pressure of 2.0 MPa at different temperatures. The microstructures of the samples were examined using JOEL 2010 TEM and XRD.

3. Results and discussion

3.1. HYDROGEN ABSORPTION

Figure 1 shows the hydrogen uptake of the composite samples milled for 30 and 60 hours, respectively, at different temperatures of 150°C, 200°C and 300°C. It is apparent that Mn and Zr catalysts milled with MgH_2 remarkably enhanced the hydrogenation kinetics. The MgH_2-MnZr sample shows very good kinetics at low temperatures, e.g. reaching 5.50 and 4.00 wt% H within 10 minutes at 200°C and

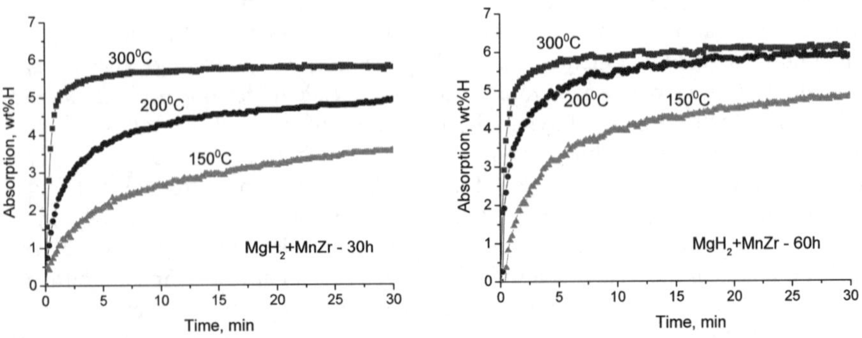

Figure 1. H-absorption of MgH_2-MnZr samples with milling times of 30 and 60 hours, respectively

150°C, respectively, is very attractive for on-board hydrogen storage in fuel cell cars because it is within the normal fueling time of gasoline-fueled commercial vehicles, e.g. in the order of a few minutes, and the operating temperature can be as low as 150°C that is easy to be realized for practical application.

From the comparison of Fig. 1(a, b), it is obtained that the milling time has a significant effect on the hydrogenation capacity at lower temperatures of 200°C and 150°C although no considerable difference can be observed at a high temperature of 300°C. This is mainly because of the reduction of particle (grain) size with milling time. At high temperature as 300°C, the hydrogenation rate is high enough to hydride the Mg particles even in large sizes, resulting in the no difference of the capacity between the two different milling times of 30 and 60 hours respectively. At low temperatures, however, the hydrogenation rates are limited to ensure that only the Mg particles in small sizes can be fully hydrided but the grains in the centre of large Mg particles will remain unchanged. This is the reason of the big difference of the capacity of the samples with different milling times, for example, the sample milled by 60 hours can absorb 4.83 wt% H but only 3.59 wt% H can be absorbed by the 30 hour-milled sample at 150°C.

Compared with our previous study on the catalytic effect of Fe and Ti [9], the couple of Mn and Zr shows much more effective for hydrogen absorption in Mg-based system as seen in Fig. 2. Figure 2 plots the hydrogenation kinetic curves of MgH_2-MnZr and MgH_2-FeTi systems at 300°C and 150°C, respectively (the data for MgH_2-FeTi is reproduced from Ref. [9]). It is obviously shown that there exists remarkable differences of the hydrogenation capacity of the two systems within a certain hydrogenation time for both high and low temperatures. At a high temperature as 300°C, the difference of the hydrogenation capacity of the two systems is about 1.0 wt% H. At a low temperature of 150°C, the difference becomes more significant, e.g. 2.2 and 1.7 wt% H for 30 and 60 hour-milled samples respectively. This suggests that MnZr is a much better couple of catalysts that enhanced the hydrogen storage properties for Mg, especially for those at low temperatures that are more important for practical applications.

In addition, it should be noted that the maximum capacity at high temperature (300°C) is about 6.0 wt% which is lower than the theoretical capacity 7.6 wt% for MgH_2. This suggests that the capacity of the Mg–MnZr nanocomposite system could be further improved, for example, adding carbon nanotubes is approved an effective way to increase the capacity of hydrogen storage [9–11].

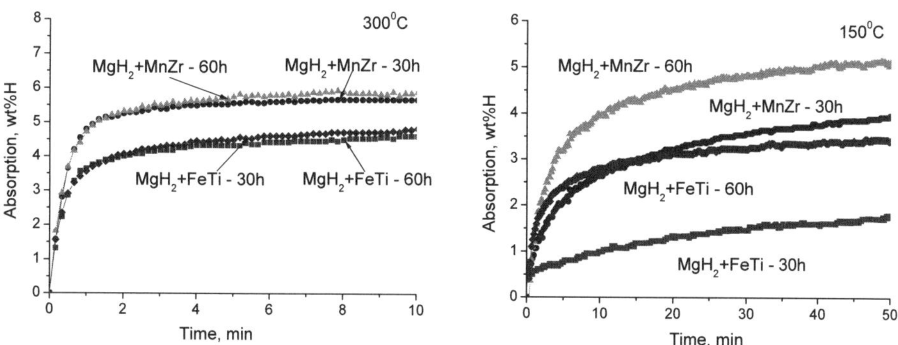

Figure 2. Comparison of H-absorption between MgH_2-MnZr and MgH_2-FeTi samples at 300°C and 150°C, respectively

To further explore this kinetics, the activation energy (E_A) for hydrogenation was determined. Figure 3 shows the plot of hydrogenation kinetics and the temperature of the two samples (R is the kinetics). It is given that E_A of the samples milled by 30 and 60 hours are 33.6 and 29.7 kJ/mol, respectively. These values are much less than that of pure Mg, e.g. 109 kJ/mol [11], indicated that adding MnZr additives is effective to reduce E_A and, in turn, significantly enhance the hydrogenation kinetics of Mg. Furthermore, it is also suggested that reducing the grain size of Mg alloys can decrease E_A and increase the kinetics.

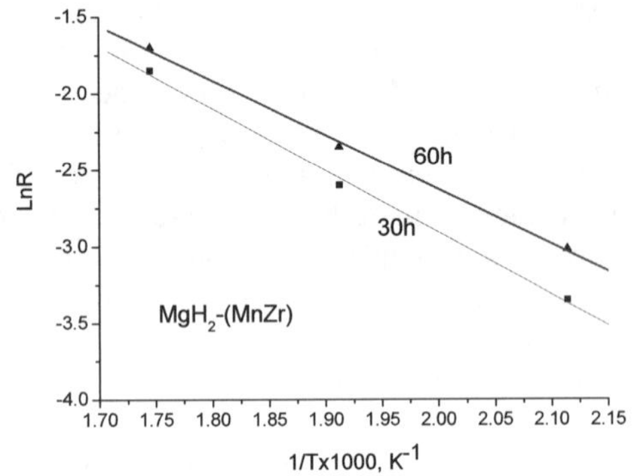

Figure 3. Relationship between hydrogenation rate and temperature of MgH_2-MnZr samples

3.2. MICROSTRUCTURE CHARACTERIZATION

As indicated in Fig. 4, XRD of the as-prepared MgH_2 sample shows that MgH_2 is present in the β-MgH_2 phase with a certain amount of Mg phase and a tiny amount of MgO phase. XRD patterns show the γ-MgH_2 metastable phase in milled samples, instead of the β-MgH_2 phase. This result agrees with the previous study [9] that suggested the γ-MgH_2 produced during ball milling. From the broadening of XRD patterns, it is easily calculated, by the method introduced in Ref. [12] with considering both the effect of crystalline size and the strains on the peak broadening, that the average grain sizes of the 30 and 60 hours milled sample are 19.5 and 16.2 nm, respectively.

This confirms that the grain size reduced with milling time and, in turn, enhanced the hydrogenation kinetics. Figure 5 shows the TEM images of 60 hour-milled sample and shows typical size of particles as several hundreds of nanometers and grains as ~16 nm.

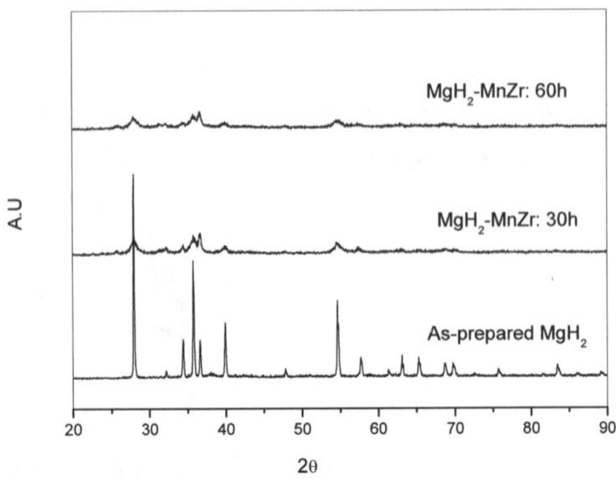

Figure 4. XRD patterns of the investigated system

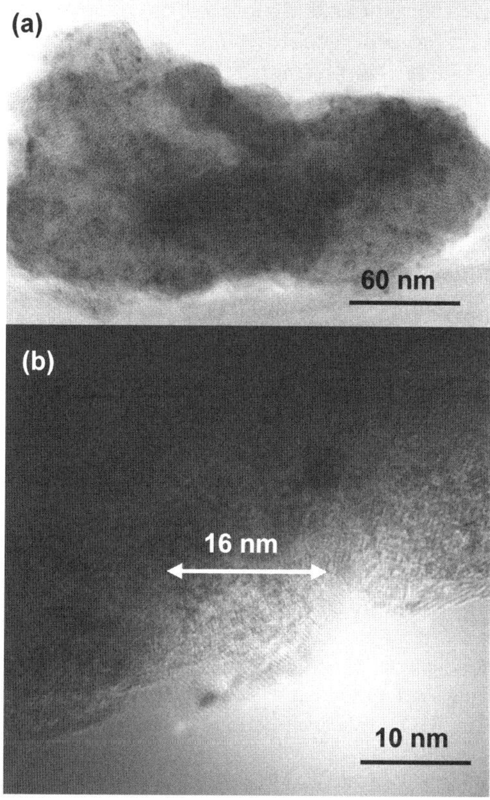

Figure 5. TEM images of a particle in Mg-MnZr system, (a) particle morphology (b) high resolution

3.3. HYDROGENATION MECHANISM

Our previous studies [6–8] demonstrated that the transition metals significantly reduce the energy barrier for molecular hydrogen dissociation and subsequently enable the hydrogenation of Mg-TMs at a low temperature possible. Mn and Zr play the similar role as previously studied other TMs such as Fe and Ti in this function, resulting in significant hydrogen absorption at a temperature as low as 150°C. However, Mg-MnZr has a remarkably higher hydrogenation capacity than that of Mg-FeTi under the same hydrogenation conditions. The possible reason is that Zr (~1.56 Å [13, 14]) and Mn (the radius of Mn is smaller, ~1.35Å, but has complex cubic structures) have a larger difference in size with Mg (~1.1 [13, 14]), and thus is very likely to be aggregated along the grain boundaries rather than milled into Mg structures. This is possible to cause a large deformation and vacancy along the grain boundaries. During the hydrogenation, the hydride-like structures that hinder atomic hydrogen diffusion will be not as easy as Mg-FeTi (~1.34 Å [13, 14]) to fully occupy the deformed grain boundaries. The will enable the atomic hydrogen diffusion along the boundaries, allowing more Mg grains at the central area of the Mg particles hydriding, e.g. increasing the hydrogenation capacity.

Further decrease of the grain size by increasing milling time is more significant in increasing the capacity than in improving the kinetics. This is because the further decrease of the particle size will enable more Mg be hydrided, but the hydrogenation kinetics is controlled by the hydrogen diffusion within the grains whose sizes are not reduced significantly (16.2 nm for 60 compared with 19.2 nm for 30 hour-milling).

4. Conclusions

The present study reveals that metallic catalyst of MnZr shows significant effect on enhancing the hydrogenation kinetics of Mg, due to reduce the activation energy for hydrogenation. Compared with MgH_2-FeTi system, the important feature of MgH_2-MnZr has remarkably higher hydrogenation capacity, especially at low temperatures, which is key to practical application of Mg-based materials. Decrease of grain size with milling time also improves the kinetics but not as significant as increasing the capacity.

Acknowledgements

This work is performed in ARC Centre of Excellence for Functional Nanomaterials (ARCOEFN). The authors greatly appreciate the financial support of Australian Research Council through ARCOEFN and Linkage Project.

References

1. Schlapbach L. and Zuttel A. Hydrogen storage materials for mobile applications. Nature, 2001; **414**: 353–358.
2. Seayad A. M. and Antonelli D. M. Recent advances in hydrogen storage in metal-containing inorganic nanostructures and related materials. Adv. Mater., 200; **16**: 765–777.
3. Hirscher M. and Becher M. J. Hydrogen storage in carbon nanotubes. J. Nanosci. Nanotechnol., 2003; **3**(1–2): 3–17.
4. Oelerich W., Klassen T., and Bormann R. Metal oxides as catalysts for improved hydrogen sorption in nanocrystalline Mg-based materials. J. Alloy. Compd., 2001; **315**(1–2): 237–242.
5. Zaluski L., Zaluska A., and Strom-Olsen J. O. Hydrogen absorption in nanocrystalline Mg2Ni formed by mechanical alloying. J. Alloy. Compd., 1995; **217**(2): 245–249.
6. Du A. J., Smith S. C., Yao X., and Lu G. Q. The Role of Ti as a catalyst for the dissociation of hydrogen on a Mg(0001) surface. J. Phys. Chem. B, 2005; **109**: 18037–18041.
7. Du A. J., Smith S. C., Yao X., and Lu G. Q. The catalytic role of sub-surface carbon in the chemisorption of hydrogen on a Mg(0001) surface: an ab-initio study. J. Phys. Chem. B, 2006; **110**: 1814–1819.
8. Du A. J., Smith S. C., Yao X., and Lu G. Q. Hydrogen spillover mechanism on Pd-doped Mg surface revealed by ab initio density functional calculation., J. Am. Chem. Soc., 2007; **129**: 10201–10204.
9. Yao X., Wu C. Z., Du A. J., Lu G. Q., Cheng H. M., Smith S. C., Zou J., and He Y. Mg-based nanocomposites with high capacity and fast kinetics for hydrogen storage. J. Phys. Chem. B, 2006; **110**: 11697–11703.
10. Yao X., Wu C. Z., Du A. J., Zou J., Zhu Z. H., Wang P., Smith S. C., Cheng H. M., and Lu G. Q. Metallic and carbon nanotubes-catalysed coupling of hydrogenation in magnesium. J. Am. Chem. Soc., 2007; submitted.
11. Wu C. Z., Wang P., Yao X., Liu C., Chen D. M., Lu G. Q., and Cheng H. M. Hydrogen storage properties of MgH2/SWNT composite prepared by ball milling. J. Alloy. Compd., 2006; **420**: 278–282.
12. Aymard L., Dehahaye-Vidal A., Portemer F., and Disma F. J. Alloy. Compd., 1996; **238**: 116–127.
13. Clementi E. and Raimondi D. L. Atomic screening constants from SCF functions. J. Chem. Phys., 1963; **38**: 2686–2689.
14. Slater J. C. Atomic radii in crystals. J. Chem. Phys., 1964; **41**: 3199–3204.

LEVELS OF ION-IMPLANTED DEUTERIUM RETENTION IN Zr AND IN THE TWO-LAYER Zr-Pd SYSTEM

I.M. NEKLYUDOV, A.N. MOROZOV, V.I. ZHURBA,
V.M. AZHAZHA, A.A. ANDREEV, E.V. PYATENKO
National Science Center "Kharkov Institute of Physics and Technology", 1 Akademicheskaya Str., Kharkov 61108, Ukraine
E-mail: morozov@kipt.kharkov.ua

V.G. KULISH
G.S. Skovoroda National Pedagogical University at Kharkov, 2 Blyukher Str., Kharkov, Ukraine

Abstract. Measurements on the maximum level of the deuterium retention in zirconium and in the two-layer Pd-Zr system during D_2^+ ion implantation at temperatures ~110 and ~300 K were carried out using the nuclear reaction method. The measurement data show that the presence of a Pd film on Zr decreases the level of maximum deuterium concentration in Zr.

Keywords: deuterium retention, zirconium, two-layer Pd-Zr system, nuclear reaction method, ion implantation

1. Introduction

Hydrogen accumulation in structural materials is an extra-dangerous phenomenon leading to the hydrogen degradation in them and unforeseen fractures. In this connection most of investigations on the hydrogen behavior in materials consist in searching the means to decrease the hydrogen concentration in them. Multilayer diffusion systems present a class of materials that can be considered as one of promising ways in solving this problem. In the present work the measurements on the on the maximum level of the deuterium retention in zirconium and in the two-layer Pd-Zr system during D_2^+ ion implantation at temperatures ~110 and ~300 K were carried out using the nuclear reaction method.

2. Experimental procedure

It is known, that the approach of deuterium nuclei at distances much less than in the deuterium molecule sharply increases the probability for the tunnel quantum transition through the Coulomb barrier and, consequently, the probability of fusion reactions. The most probable from these reactions are the following:

$$d + d \rightarrow T\ (1.01\ \text{MeV}) + p\ (3.02\ \text{MeV}) \tag{1}$$

$$d + d \rightarrow {}^3He\ (0.82\ \text{MeV}) + n\ (2.45\ \text{MeV}) \tag{2}$$

B. Baranowski et al. (eds.), *Carbon Nanomaterials in Clean Energy Hydrogen Systems.*

A condition, necessary to provide appreciable yields of these reactions is a high temperature. At room temperatures and pressures the reaction yield cross-sections are less by many orders of magnitude.

During irradiation of samples with deuterium ions the above-mentioned nuclear reactions, (1) and (2), take place. Experiments on deuterium saturation of samples and measurements of the nuclear radiation emission were carried out at the accelerating installation "SKIF" [1] with a necessary completion of the measuring chamber. The schematic presentation of the measuring chamber is given in Fig. 1. For recording of the of charged product emission from the nuclear fusion reactions we used a semiconductor surface-barrier detectors (122 mm^2 area) placed at a distance ~35 mm from the centre of the target to be irradiated. The detector axis was inclined at angle ~45° to the ion beam direction. The counting efficiency of each of detectors being used was calculated with taking into account the geometry and for different detectors it was from 1.5×10^{-3} to 3×10^{3} pulses/particle. Analysis of the pulse energy spectrum was performed using the multichannel amplitude analyzer AI-1024. In parallel with the recording by means of the multichannel analyzer, three single-channel analyzer **CAMAC** were used each of which recorded only the pulses being within the limits of the spectral line width of ^{3}He, T or p, respectively. The calibration of the energy spectrum of detectors with the use of radioactive isotopes ^{141}Am has been carried out by the energy lines of **α**-particles emitted by these isotopes [2].

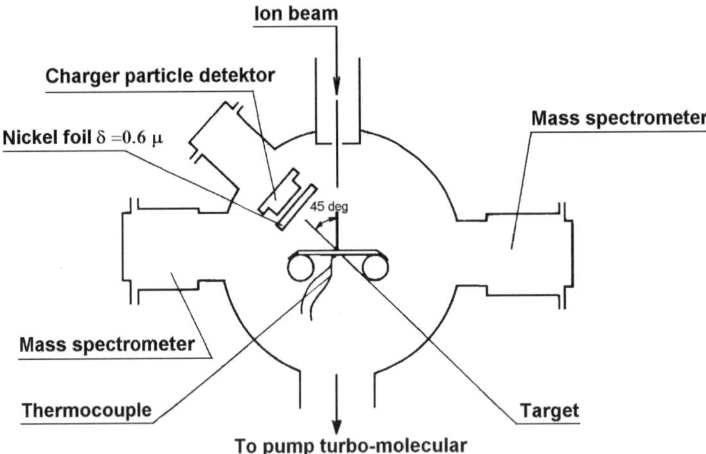

Figure 1. Schematic representation of the measuring chamber

To protect the detector from the low-energy phonon particle flux (reflected beam particles, diverse emission etc.) the working surface of the detector was shielded by the nickel film of 0.65–0.7 μm thickness. The presence of this film leaded to the shift of spectral lines and to the increase of their half-width as a consequence of energy loss during particle penetration through the film. As a result the proton peak was shifted from 3.02 MeV into the region of 2.95 MeV, tritium peak – from 1.01 to 0.86 MeV and helium peak – from 0.82 to 0.35 MeV. The value of these shifts corresponded to the design values calculated with taking into account the particle energy loss in the nickel film [3].

The energy spectra of charged particles obtained during irradiation of samples are shown in Fig. 2. As is seen from the figure, there are peaks in the spectrum corresponding to p, T and ^{3}He. Besides, there are two small peaks of α-particles with energies of 5,442 and 5,434 keV, corresponding to the radiation of the ^{241}Am source.

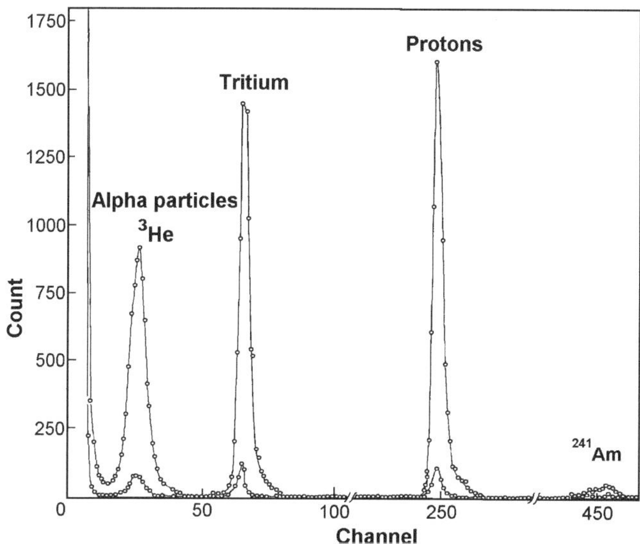

Figure 2. Energy spectra of charged particles of the zirconium sample irradiated with 24-keV D_2^+ ions for different doses

Control of deuterium concentration growth in the sample was carried out by plotting Y/ΔD as a function of dose, where Y is the number of particles of a given sort (fusion product) emitted from the target surface in the course of irradiation and recorded by the detector in time Δt, ΔD is the dose gain during the same time period, D is the irradiation dose. The relation between Y/ΔD and D can be obtained from the equation

$$\frac{Y}{\eta \cdot \cdot S \cdot \Delta t} = \sigma_i \cdot l_{e\phi} \cdot n_t \cdot \frac{\Delta N}{S \cdot \Delta t}, \tag{3}$$

where η – efficiency of particle counting by the detector

S – surface area being irradiated and, respectively, the area from which the emission of reaction products occurs

σ_i – cross-section for i-th fusion channel

L_{ef} – effective target thickness-implanted layer where the reactions occur

ΔN – number of deuterium particles impinging on the target surface in time Δt

n_t – average concentration of D atoms during the given Δt in the near-surface layer of l_{ef} thickness

As a result of transformation (3) taking into account, that ΔN/S = ΔD, we obtain

$$\frac{Y}{\Delta D} = \eta \cdot \sigma_i \cdot l_{e\phi} \cdot S \cdot n_t. \quad (4)$$

As under unchangeable experimental conditions, η, σ_i, $l_{e\phi}$, S are constants, formula (4) can be written in the

$$\frac{Y}{\Delta D} = k \cdot n_t. \quad (5)$$

From (5) it follows that Y/ΔD is proportional to the deuterium concentration in the near-surface metal layer.

For investigations we used the samples of 5 × 10 mm^2 from zirconium (foil of **δ** ~ 0.3 mm) and the two-layer Zr-Pd system (zirconium foil of **δ** ~ 0.3 mm with a palladium coating of ~0.9 µm thickness obtained by PVD (physical vapor deposition) method). The targets were irradiated with a D_2^+ ion beam of 24 keV energy and current density ~6–10 µA/cm^2 to dose ~2.5 × 10^{18} D/cm^2 at temperatures ~110 and ~300 K.

3. Results and discussion

Figure 3 shows the yield of tritium (product of d(d,T)p reaction) as a function of the dose of irradiation with D_2^+ ions for Zr samples. From the figure it is seen that the level deuterium retention at a temperature ~110 K increases by the approximately linear law and reaches the saturation within the dose range from 2.1 to 2.3 × 10^{18} D·cm^{-2}. Further increase of the irradiation dose does not lead to the growth of the level of deuterium retention in the implant layer.

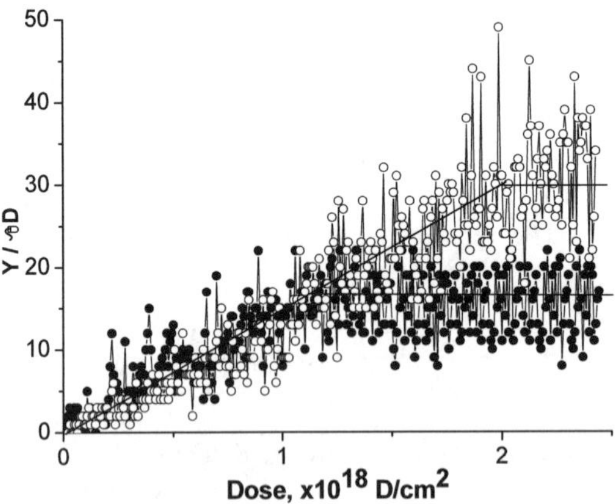

Figure 3. Yield of charged tritium particles in the d-d nuclear reaction as a function of the irradiation dose from Zr: $T_{irr.}$ ~ 110 K (–○–) and ~300 K (–●–)

The measurements of implanted deuterium retention level in zirconium samples at temperature ~300 K have shown that the saturation level is reached within the dose range from (1.1 to 1.2) × 10^{18} D·cm^{-2}. Taking into account that the Y/ΔD ratio in this case is two times less then the value of the same ratio for nitrogen temperatures it is evident that the deuterium concentration in the implant layer at room temperature is two times lower. The curves representing the tritium yield as a function of the irradiation dose for the two-layer Zr-Pd system at temperatures of samples ~100 and ~300 K are demonstrated in Fig. 4. The deuterium implantation took place from the zirconium side. It is seen that the deuterium saturation level in the implantation layer for the Zr-Pd system at temperature ~110 K is reached upon exceeding the irradiation dose $\geq 1 \times 10^{18}$ D/cm^2 that is comparable with the deuterium retention level in the implant layer for Zr at temperature ~300 K. At temperature ~300 K the deuterium saturation level for Zr-Pd samples is lower almost in four times as compared to the saturation level at temperature ~110 K.

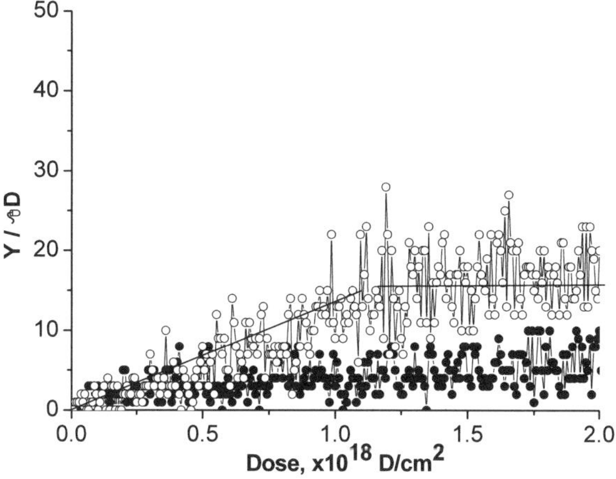

Figure 4. Yield of charged tritium particles in the d-d nuclear reaction as a function of the irradiation dose from the two-layer Zr–Pd system; $T_{irr.}$~ 110 K (–o–) and ~300 K (–•–)

4. Conclusions

The results obtained allow us to conclude that the presence of the Pd film on Zr (Zr-Pd two-layer system) decreases the level of maximum deuterium concentration in Zr.

Acknowledgements

This work is conducted with the support of the National Academy of Sciences of Ukraine (project #X-6-513).

References

1. V.V. Ruzhitsky, Yu.A. Gribanov, V.F. Rybalko, S.M. Khazan, A.N. Morozov, I.S. Martynov. Multipurpose experimental installation “SKIF”. Voprosy atomnoj nauki i tekhniki. Series: “Physics of radiation damages and radiation materials science”. 1989, **4**(51): 84–89 (in Russian).

2. V.F. Zelensky, V.F. Rybalko, A.N. Morozov, G.D. Tolstolutskaya, V.G. Kulish, S.V. Pistryak, I.S. Martynov. Experiments on the cold nuclear fusion in Pd and Ti saturated with deuterium by the ion implantation method. Voprosy atomnoj nauki i tekhniki. Series: "Physics of radiation damages and radiation materials science". 1990, **1**(53): 65–71 (in Russian).
3. N.N. Pucherov, S.V. Romanovsky, T.D. Chesnokova. Tables of the mass stopping power and paths of charged particles of 1-100MeV energy. Kyiv: Naukova Dumka, 1975, 397 p. (in Ukrainian).

SIMULATION OF PHASE DIAGRAMS IN $LaNi_5–H_2(D_2)$ SYSTEM AND INVERSION OF ISOTOPIC EFFECT

V.S. MARININ, YU.F. SHMALKO, K.R. UMERENKOVA
A.Podgorny Institute for Machine Building Problems of National Academy of Sciences of Ukraine, 61046 Kharkov, Ukraine

Abstract. Simulation of phase transitions in IMC–hydrogen systems on the basis of modified scheme of perturbation theory provides correct description of main characteristics of phase diagrams in the wide region of pressures of hydrogen isotopes (6 order and higher). The obtained data about thermodynamic parameters of α–β transition allows to describe inversion of isotopic effect in $LaNi_5–H_2(D_2)$ system. Data on solubility of hydrogen isotopes in $LaNi_5$ at pressures up to 500 atm are obtained; data on parameters of the critical point of α–β equilibriums (experimental data for which are absent) are also obtained. Proposed calculation scheme does not use fitting parameters or empirical correlations and leans upon atomic characteristics of the hydrogen subsystem and metallic matrix which are physically meaningful.

Keywords: hydrogen, metal hydrides, intermetallic compounds, phase diagrams, isotopic effect, lattice gas, thermodynamic perturbation theory

1. Introduction

Mathematical simulation of phase diagrams in hydrogen – metal hydride systems allows to reduce or exclude expensive and lasting investigations. However, until the present time methods for calculation thermal physical properties of metal hydrides and intermetallic compounds (IMC) were mostly empirical; this hindered development and creation of elements for metal hydride systems due to the fact that it was impossible to predict their parameters.

In the paper [1] of the series of works devoted to application of thermodynamic perturbation theory for description phase equilibriums in metal hydrides and IMC the following results are presented. Development of the model of non-ideal lattice gas of hydrogen ions based on the method of perturbation theory allows to reproduce the main peculiarities of phase diagrams for $IMC–H_2$ systems in the region of disordered phases. Obtained results for PCT diagrams (pressure – composition – temperature) for $LaNi_5–H_2$ systems agree with experimental data.

In the present work phase diagrams of $LaNi_5–D_2$ system are simulated in the frames of the proposed method with the aim of evaluating isotopic effect in IMC hydrides at the example of $LaNi_5$ intermetallic compound. Apart from rather practical applications (with isotope isolation, hydrogen enrichment etc.), the problem is of a large interest from physical point of view. In particular, are molecular statistic models applicable to description of fine peculiarities in phase equilibriums? Without doubt, one of them is inversion of isotopic effect in $LaNi_5$–$H_2(D_2)$ system at common environment temperatures.

2. Chemical potential of lattice gas in IMC–hydrogen system

The new approach to the problem of calculation of phase equilibriums in metal hydrides lies in determination of the properties of hydrogen sub-system of metal hydride and also of equilibrium H_2 molecular phase within the unified method – modified scheme of thermodynamic perturbation theory (MPT) [2]. This method represents properties of the systems being investigated in the form of series in parameters describing differences between the real interaction potential of particles and potential of zero approximation system.

Thermodynamic description of the hydrogen sub-system in the region of disordered phases (in particular, α and β phases) was carried out on the model of non-ideal (non-interacting) lattice gas of hydrogen atoms. For calculations both direct interaction between hydrogen atoms and indirect "deformation" contributions to potential energy due to the lattice expansion caused by hydrogen solution were taken into consideration. Equilibrium properties of the lattice gas were defined for the basic case of interstitial solution with a unique type of interstitials.

In the present wok we shall restrict our consideration to differences in phase equilibriums for $LaNi_5$–H_2 and $LaNi_5$–D_2 systems, that is, for two hydrogen isotopes-hydrogen-prothium H and deuterium D. Further, the term "hydrogen" will stand both for H and D.

The important fact should be noted: initial IMC crystalline structure in the majority of cases does not differ from the structure of the crystalline matrix in hydride (deuteride) phases of IMC–hydrogen systems in the region of disordered α and β phases.

In this case, chemical potential μ_X of hydrogen (X=H, D) component of IMC hydride/deuteride has the form (including the members of the second order):

$$\beta\mu_X^+(\theta,T) = \ln\frac{\theta}{1-\theta} + \frac{W_1\theta}{T(1+\alpha c_s\theta)} + \frac{W_2\theta^2}{T^2(1+\alpha c_s\theta)^2}, \tag{1}$$

where β=1/kT; $\mu_H^+ = \mu_H - \mu_H^{st}$; $\mu_H^{st}(T)$– chemical potential in the standard state [1]; $\theta = C/C_s$ – hydrogen concentration; $C = n_{IMC} \cdot c$ – hydrogen concentration in the form of X/IMC relation, that is, on a formula unit of IMC; n_{IMC} – number of atoms in the formula unit, c – hydrogen concentration in X/Me units, that is, per one atom of the matrix; $\alpha = c^{-1}(\Delta V(c)/V)$ – dilatation coefficient of IMC lattice with hydrogen solution. Values C_s [H/IMC] – sorption capacity of IMC, or the maximum number of interstitial positions of H(D) atoms in the phase region being investigated, and c_s [X/Me] – the maximum concentration c are bound by the relation $C_s = n_{IMC} \cdot c_s$.

W_1 and W_2 constants providing the relationship between macroscopic properties of IMC–hydrogen interstitial solutions and microscopic (atomic) characteristics of hydrogen subsystem and IMC metallic matrix are: $W_1 = 2I_1 n_M(\sigma_1^3/v_0)E_1 c_s$, $W_2 = (3I_2/4I_1^2)W_1^2$, where I_1 = –5.585, I_2= 1.262 – MPT parameters for H(D) gas; n_M – number of atoms in the matrix of the elementary cell; v_0 – volume of the cell at C=0; E_1 [K] and σ_1 [m] – parameters of X–X interaction potential $u_X(r) = kE_1\varphi(r/\sigma_1)$ [2].

3. Curves of homogenous phase decomposition (CT diagrams) of $LaNi_5$–$H_2(D_2)$ system

One can suppose that in the process of prothium and deuterium absorption in intermetallic compound $LaNi_5$ in the region of α–β equilibriums, mainly the sublattice of T-interstitials is filled; that is, thermodynamic functions of the lattice gas describe equilibrium isotherms with one "plateau" in double phase (α + β) regions.

In order to calculate decomposition curves for homogenous phases of $LaNi_5$–$H_2(D_2)$ systems it is necessary to find parameters in the equation (1). According to [3–6] the value of C_s is equal to 6.7 ($c_s = 1.12$) for both systems. Stable combinations of $E_1\sigma_1^3$ responsible for mutual atom–atom interaction between prothium and deuterium are related approximately as $(E_1\sigma_1^3)_D = 0{,}97(E_1\sigma_1^3)_H$ [4, 5, 7]. Dilatation coefficient of the $LaNi_5$ lattice is higher in the case of deuterium solution than in the case of prothium solution by about of 5% [4, 7]; with $\alpha_H = 0.20$ [6] the value of $\alpha_D = 0.21$. In this case W_1 and W_2 constants for the systems being investigated are equal: $W_1^{(H)} = -2.521\cdot10^3$ K, $W_2^{(H)} = 1.929\cdot10^5$ K^2 [1]; $W_1^{(D)} = -2.444\cdot10^3$ K, $W_2^{(D)} = 1.813\cdot10^5$ K^2.

According to the expression (1) for chemical potential of the H(D) lattice gas the parameters of the critical point of α–β equilibriums are: $T_c = -0.2163W_1/(1 + \alpha c_s)$, $C_c = \theta_c \cdot C_s$, where $\theta_c = 0.46/(1 + 0.54\alpha c_s)$. For $LaNi_5$–$H_2(D_2)$ systems this gives $T_c^{(H)} = 445$ K, $T_c^{(D)} = 428$ K and $C_c^{(H)} = 2.75$ H/$LaNi_5$, $C_c^{(D)} = 2.73$ D/$LaNi_5$; experimental data about positions of critical points are not available.

$C_i(T) = C_s\cdot\theta_i(T)$ branches of the decomposition curves of $LaNi_5$–$H_2(D_2)$ homogenous phases into disordered phases $i = \alpha,\beta$ are defined by equilibrium conditions:

$$\begin{cases} p_X(\theta_\alpha,T) = p_X(\theta_\beta,T); \\ \mu_X^+(\theta_\alpha,T) = \mu_X^+(\theta_\beta,T), \end{cases} \tag{2}$$

where p_X – pressure of lattice X-gas, X = H, D.

Calculations of CT diagrams for $LaNi_5$–$H_2(D_2)$ systems were performed starting from the consequence of conditions (2) – the rule of equal squares (confined by the $\mu_X^+(\tilde\theta)$ curve and straight line $\mu_X^{+(PL)}$), which at the specified temperature $T < T_c$ is fulfilled in $(\mu_X^+ - \tilde\theta)$ plane [2]:

$$p_X^{(\beta)} - p_X^{(\alpha)} = \text{const}\int_{\tilde\theta_\alpha}^{\tilde\theta_\beta}[\mu_X^{+(PL)} - \mu_X^+(\tilde\theta)]d\tilde\theta = 0\,\cdot \tag{3}$$

Here $\tilde\theta = \theta/(1+\alpha c_s\theta)$; dependences $A^{(PL)} = A^{(\alpha\alpha}(T) = A^{(\beta\beta}(T)$, where $A^{(i)}(T) \equiv A(\theta_i,T)$ – values of the function $A = p_X, \mu_X^+$ at the phase boundaries $i = \alpha,\beta$, that is, on the pressure "plateau" and "plateau" of lattice X-gas chemical potential. Having found coordinates $\tilde\theta_i(T)$ according to (3), it is possible to obtain the branch of decomposition curves $\theta i(T)$ according to the dependence inverse to $\tilde\theta(\theta)$, that is, values $\theta_i = \tilde\theta_i/(1-\alpha c_s\tilde\theta_i)$ satisfying equations (2).

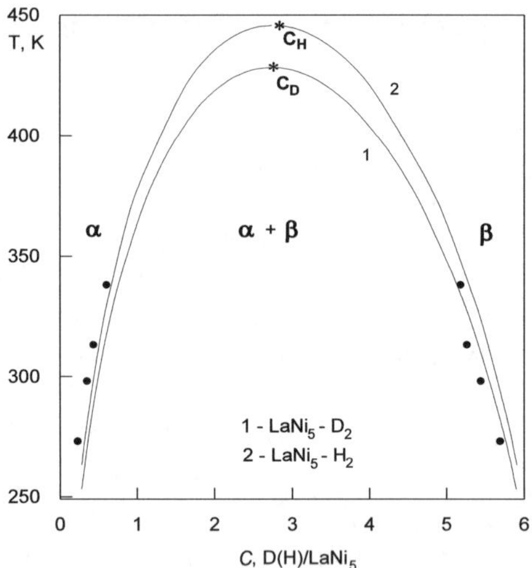

Figure 1. Decomposition curves of the homogenous phases (CT diagrams) for $LaNi_5$–$H_2(D_2)$ systems into disordered phases α and β phases (calculation). • – experimental data for $LaNi_5$–D_2 [3]. * C_H , * C_D – critical points of α–β equilibriums

In the Fig. 1 boundaries of the regions of disordered phases are shown (CT decomposition curves) for $LaNi_5$–H_2 and $LaNi_5$–D_2 systems calculated from equilibrium conditions of α and β phases according to the rule of equal squares (3). Values of critical parameters for α–β equilibriums that were found earlier correspond to coordinates of C_H and C_D points in the figure. Evaluation data for boundaries of (α + β) region in $LaNi_5$–D_2 system which were obtained from deuterium desorption isotherms are given for comparison according to figure given in [3].

4. $H_2(D_2)$ equilibrium pressure in the double-phase (α + β) region and inversion of isotopic effect

$p_{X_2}^{(PL)}(T)$ temperature dependences of decomposition pressure of hydride (deuteride) β-phases , that is, pressure of hydrogen ions on the "plateau" of the $p_{X_2}(C)$ isotherms in the heterogeneous (α + β) phase region may be presented in the form of conventional equitation

$$\ln p_{X_2}^{(PL)}(T) = -\frac{\Delta H_{\beta\to\alpha}^{(X)}}{RT} + \frac{\Delta S_{\beta\to\alpha}^{(X)}}{R}, \tag{4}$$

where $\Delta H_{\beta\to\alpha}^{(X)}$, $\Delta S_{\beta\to\alpha}^{(X)}$ correspond to enthalpy and entropy of β-phase decomposition. These parameters in MPT scheme have the following form:

$$\Delta H_{\beta\to\alpha}^{(X)} \cong H_{X_2}^0 + 2RT\Delta_{\beta\to\alpha}, \quad \Delta S_{\beta\to\alpha}^{(X)} \cong S_{X_2}^0 - 2R\Delta_{\beta\to\alpha}, \tag{5}$$

where $H_{X_2}^0$, $S_{X_2}^0$ – enthalpy and entropy of $H_2(D_2)$ in the standard state of ideal gas [8]; the value of $\Delta_{\beta\to\alpha}$ transition corresponds to relative difference in specific enthalpy h_X of the lattice X-gas $h_X^{(i)}(T) \equiv h_X^{st}(T) + h_X^{+}(\theta_i, T)$ on the boundaries of homogenous phases $\theta_\alpha(T), \theta_\beta(T)$.

As well as in the case of decomposition of hydride β-phase of $LaNi_5$–H_2 [1], we shall define parameters (5) for the range of operating parameters in the isolated (critical) point. we will receive the following expression for enthalpy of non-ideal lattice gas [2] and for critical concentration $\theta_c = 0.46/(1 + 0.54\alpha c_s)$ in the general form:

$$-\Delta^{(c)}_{\beta\to\alpha} = \left.\frac{\partial(\beta h^{+}_{X})}{\partial\theta}\right|_{\substack{T=T_c \\ \theta=\theta_c}} = \frac{1}{\theta_c(1-\theta_c)} + \frac{1}{\theta_c^2}\left[\ln(1-\theta_c) - \frac{0{,}763}{1+\alpha_X c_s}\right]. \tag{6}$$

According to (6) for $LaNi_5$–D_2 system at $\theta^{(D)}_c = 0.408$ the value of $\Delta^{(c)}_{\beta\to\alpha} = 2.72$. With account of parameters for $LaNi_5$–H_2 that were found earlier we will obtain the following set of thermodynamic parameters [1] of the $\beta \to \alpha$ phase transition in $LaNi_5$ hydrides and deuterides: enthalpy [kJ/mol] $\Delta H^{(H)}_{\beta\to\alpha} = 29.8$, $\Delta H^{(D)}_{\beta\to\alpha} = 31.7$; entropy [J/(K·mol)] $\Delta S^{(H)}_{\beta\to\alpha} = 104$, $\Delta S^{(D)}_{\beta\to\alpha} = 110$. $p^{(PL)}_{X_2}(T)$ dependences obtained for both systems according to (4) – (6), are compared in the Fig. 2 with experimental data [3] on the pressure on the "plateau" of H_2 and D_2 desorption isotherms. .

Pressure of hydrogen isotopes on the "plateau" of similar isotherms of solubility for metal–hydrogen and IMC–hydrogen systems may be lower for heavier isotope D_2 (normal isotopic effect) or higher (inverse isotopic effect) than for H_2. From Fig. 2 it is seen that depending on temperature in $LaNi_5$–H_2 system both normal and inverse effect can be observed. At the temperature T_i when these isotopic pressures are equal inversion of isotopic effect occurs. Intersection of calculated dependences $p^{(PL)}_{X_2}(T)$ takes place at $(\Delta H^{(D)}_{\beta\to\alpha} - \Delta H^{(H)}_{\beta\to\alpha}) \big/ (\Delta S^{(D)}_{\beta\to\alpha} - \Delta S^{(H)}_{\beta\to\alpha}) = 317$ K; experiment [3] gives $T_i = 308$ K in average for different samples

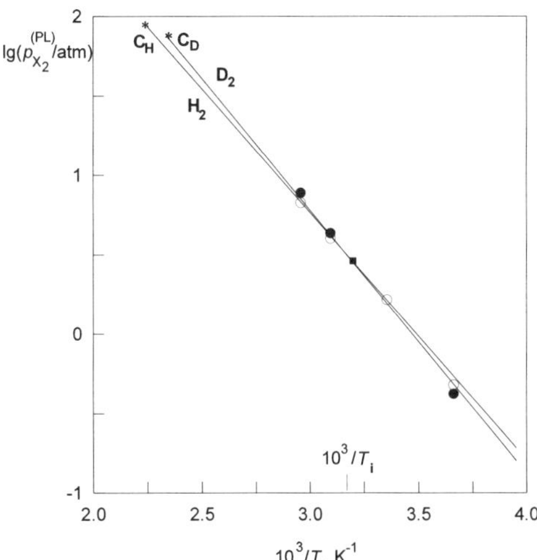

Figure 2. Logarithm of pressure of hydrogen isotopes on the "plateau" of $p_{X_2}(C)$ isotherms in double-phase region (α + β) as a function of inverse temperature, calculations according to (4). Experimental data on hydrogen desorption [3]:○ – H_2, ● – D_2. Critical points of α–β equilibriums – * C_H , * C_D (calculations); T_i – temperature of inversion of isotopic effect

From the equation (4) one can obtain the analytical expression for relationship between equilibrium pressures of hydrogen isotopes (at $T < T_c$, K):

$$\ln\left(p_{D_2}^{(PL)} / p_{H_2}^{(PL)}\right) = 0{,}72 - \frac{228{,}5}{T}, \qquad (7)$$

and pressure values at critical points of α–β-equilibriums: $p_{H_2}^{(c)} = 86$ atm, $p_{D_2}^{(c)} = 75$ atm. Experimental data on the parameters of critical points of these equilibriums in $LaNi_5$–$H_2(D_2)$ are absent yet.

Nature of inverse isotopic effect is mainly associated with lower energy of D–D interaction in the lattice in comparison with H–H interaction. Oscillation amplitude of D atoms in interstitials is less than the amplitude of H atoms [4]; their smaller volume leads eventually to the differences in $(E_1\sigma_1^3)$ combinations for interaction of isotopic atoms. From the other hand, differences in energies of zero oscillations of prothium and deuterium atoms in the metallic or IMC lattice are of a large importance. Competition of these two factors can explain the nature of isotopic effect (normal or inverse). As for effect inversion, in general, it should take place in any system, as straight lines in the Fig. 2 never run in parallel. However, such inversion can occur in the systems where significant temperature extension of double-phase region (α + β) exists, though this is not a sufficient condition. For example, for Pd–$H_2(D_2)$ system with inverse isotopic effect and very wide temperature range of α–β equilibriums (no less than 400 K) the dependences (2) may intersect at about 1,000 K which is much higher than the critical temperature of α–β transition (~570 K) [2].

5. Phase diagrams in the region of α–β equilibriums and in postcritical region

Phase PCT diagrams relating pressure p_{X_2} of the gaseous phase of hydrogen isotopes with C, T parameters of hydride and intermetallic compounds may be obtained from the equality condition of chemical potentials of the hydride X-subsystem $\mu_X(C,T)$ and molecular gaseous phase $\mu_{X_2}(p_{X_2},T)$ per an isotope atom:

$$\mu_{X_2}(p_{X_2},T) = 2\mu_X(C,T). \qquad (8)$$

After a number of rearrangements similar to [2], we can obtain (according to (8)) for $p_{X_2}(\theta,T)$ curves intersecting single- and double-phase regions of IMC–hydrogen systems:

$$\ln p_{X_2}(\theta,T) = \ln p_{X_2}^{(PL)}(T) + 2\beta[\mu_X^{+}(\theta,T) - \mu_X^{+(PL)}(T)], \qquad (9)$$

where $\mu_X^{+(PL)}$ – value of chemical potential of lattice X-gas on the "plateau" of $\mu_X^{+}(\theta)$ isotherms ; the value is defined by conditions (2) of the gas-fluid phase transition. Expression (9) at $\theta < \theta_\alpha$ and $\theta > \theta_\beta$ describes descending and ascending branches of $p_{X_2}(C)$ isotherms, respectively, and at $\theta_\alpha < \theta < \theta_\beta$ it gives the value $p_{X_2}^{(PL)}$ according to van't Hoff equation (4).

For isotherms $T \geq T_c$ and elevated pressures p_{X_2} we obtain

$$\ln p_{X_2}(\theta,T) + 0{,}18\frac{p_{X_2}(\theta,T)}{T} = Q_X + \frac{\Delta\Phi^0_{X_2}(T)}{R} + 2\beta[\mu^+_X(\theta,T) - \mu^{+(c)}_X], \quad (10)$$

where $\Delta\Phi^0_{X_2}(T) = \Phi^0_{X_2}(T) - \Phi^0_{X_2}(T^{(X)}_c)$; $\Phi^0_{X_2}(T) = -G^0_{X_2}(T)/T$, $G^0_{X_2}$ – Gibbs energy in the standard state of X_2 ideal gas [8]; $Q_X = \ln p^{(c)}_{X_2} + 0{,}18 p^{(c)}_{X_2}/T^{(X)}_c$.

Expression (10) provides PCT dependences above the critical point of α–β transition up to pressures of about 500 atm, that is, in the case when virial expansion of chemical potential of gaseous hydrogen phases [2] can be confined to second virial coefficient B_{X_2}.

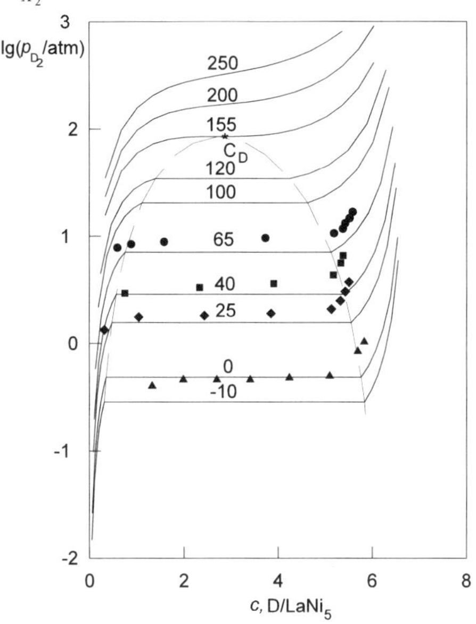

Figure 3. PCT dependences for $LaNi_5$–D_2 system, calculation according to (9) ($T < T_c$) and (10) ($T \geq T_c$). Temperatures on the isotherms are given in в °C. Dotted line represents the boundary of double-phase (α + β) region in the (lg p_{D_2} –C) plane. The critical point of α–β equilibriums: * C_D (calculation). Experimental data [3] on D_2 desorption at the temperatures: ▲ – 0°C, ♦ – 25°C, ■ – 40°C, ● – 65°C

Phase diagram of $LaNi_5$–H_2 system at temperatures below and above $T^{(H)}_c$ we obtained earlier, and it is presented in [1]. In the presented work additional phase equilibriums in $LaNi_5$–D_2 system are calculated. The set of calculated isotherms is shown in the Fig. 3 in comparison with available data on D_2 desorption [3]. In this figure predicted dependences at elevated temperatures and pressures are also shown both in double-phase and in overcritical regions. Experimental data on hydrogen solubility in $LaNi_5$ in this region of states are absent. At the critical temperature $\mu^{+(c)}_X = \mu^{+(PL)}_X(T^{(X)}_c)$ and $\Delta\Phi^0_{X_2} = 0$, and both expressions (9) and (10) describe the critical isotherm (155°C, Fig. 3) $p_{X_2}(\theta, T^{(X)}_c)$ with inflection point $p_{X_2}(\theta_c, T^{(X)}_c) \equiv p^{(c)}_{X_2} = p^{(PL)}_{X_2}(T^{(X)}_c)$.

6. Conclusions

Within the frames of thermodynamic perturbation theory it is possible to take into account hydrogen-hydrogen attraction adequately. This allows to reproduce not only qualitative parameters of phase equilibriums in IMC–hydrogen systems, but also to describe fine peculiarities, such as inversion of isotopic effect in $LaNi_5–H_2(D_2)$.

Nature of isotopic effect (normal or inverse) is defined mainly by competition of two factors – lower energies of D–D interaction in the lattice with comparison with H–H interaction, and from the other hand, by differences in energy of zero oscillations of prothium and deuterium atoms in the lattice.

Data on solubility of hydrogen isotopes at the pressure up to 500 atm and parameters of critical point α–β equilibriums were obtained.

References

1. V.S. Marinin, Yu.F. Shmalko, and K.R. Umerenkova, Development of the model of non-ideal lattice gas for description of phase equilibriums in intermetallic compounds, *Problemy mashinostroeniya*, 2006, **9**: 64–69 (in Russian).
2. V.S. Marinin, K.R. Umerenkova, and Yu.F. Shmalko, Simulation of the α–$\beta(\alpha')$ phase equilibrium of metal hydrides within the perturbation theory, *Functional Materials*, 2003, **10**: 607–614.
3. A. Biris and R.V. Bucur, The solubility of deuterium in $LaNi_5$, *Journal of the Less-Common Metals*, 1976, **49**: 477–482.
4. Hydrogen in Metals, ed. by G. Alefeld and J. Völkl, *Berlin, Springer*, 1978, **1, 2**.
5. A.F. Volkov, Investigation of interaction between hydrogen and $LaNi_5$ intermetallic compound, and optimization of treatment with hydrogen for obtaining activated powders, *Abstract of Ph.D. thesis, Donetsk, DPI*, 1989 (in Russian).
6. V.S. Marinin, Yu.F. Shmalko, and K.R. Umerenkova, Description of Phase Equilibriums in Intermetallic Compounds within the Perturbation Theory. In: *Hydrogen Materials Science and Chemistry of Carbon Nanomaterials, "NATO Security Through Science. Series A: Chemistry and Biology", New York, Springer*, 2007, P. 187–192.
7. V.S. Marinin, Yu.F. Shmalko, and K.R. Umerenkova, Phase equilibriums in transition metal hydrides and deuterides within the model of non-ideal lattice gas, *Atomic Science and Technology, Series "Physics of Radiation Damages and Radiation Material Engineering"*, 2006, **4**: 235–238 (in Russian).
8. Thermodynamic Properties of Individual Compounds, ed. by V.P. Glushko, *Moscow, Nauka*, 1978, **1** (in Russian).

FEATURES OF PHASE TRANSFORMATION IN FeCH SYSTEM AT HYDROGEN HIGH PRESSURE

L.M. POLTORATSKY
«The Siberian State Industrial University», 654007, Novokuznetsk, Russia

Abstract. With use of specially prepared (clear) materials – iron, hydrogen, carbon a series of experiments, concerned with heating and refrigerating of iron-carbon alloys, was accomplished. The experiments have shown, that at crystallization of a melt iron – carbon at sluggish refrigerating at pressure of hydrogen more 400 atm the carbon cannot crystallize by the way black lead. The high-carbon phase from a melt is secreted by the way diaphanous and adamantine crystals. In absence of hydrogen crystallization goes a routine way with formation of black lead.

Keywords: iron, hydrogen, carbon, pressure, diamond

1. Introduction

In the experiments on cooling of melted grey cast iron samples supersaturated with hydrogen at various rates (from 250°C/sec to 0.16°C/sec), the crystals well-seen through the light microscope had been found inside the matrix instead of the expected graphite inclusions.

Considering the characteristics of starting materials participating in formation of these crystals, they can be interpreted, with a definite confidence level, as a complex $Fe_nC_mH_k$ compound, i.e. iron carbohydride (Fig. 1(a)) or as diamond inclusions CA (Fig. 1(b)).

The question of existence of diamonds in iron-carbon alloys is under discussion from the beginning of the last century. At present, calculations confirm the possibility of cementite transformation into diamond in iron-carbon alloys at high temperatures and pressure [1]. It has been proved experimentally that dynamic synthesis of diamond in grey cast iron can be performed by two methods: plastic strain with explosion and explosion with temperature cycling [2]. It has been found that diamond crystal formation is possible on solidification of cast iron inoculated by graphite-spheroidizing elements – Mg, Ce and others [3].

There is information that hydrogen has influence on the diamond formation process in conditions of diamonds synthesis from a gas phase. Hydrogen influence on a diamond phase growth by the gas deposition method at low pressures has been studied in paper [4]. It has been found that atomic hydrogen introduction into a gas atmosphere provides a selectivity of a diamond phase crystallization. A substantial increase in the diamond growth velocity observed in practice as compared to the growth velocity in hydrocarbon pyrolysis, when atomic hydrogen concentration is below the equilibrium one, is caused by active participation of atomic hydrogen in the crystallization process itself. It is suggested that one of the main functions of atomic hydrogen lies in dehydrogenation and hydrogenation of the inoculating crystal surface to provide: formation of active carbonaceous areas for connection of acetylene (C_2H_2) or CH_3-radical to them in the first case, and stabilization of the formed diamond configurations in the second case [5, 6].

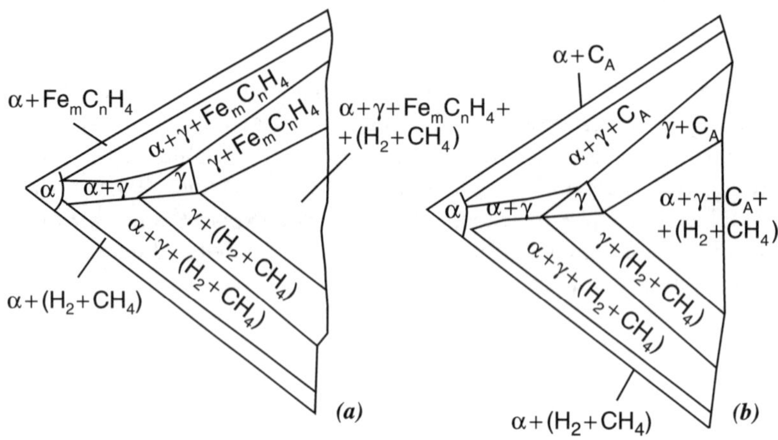

Figure 1. Isothermal section of Fe-C-H diagram on cooling to 700°C; interpretation of crystals as iron carbohidride *(a)* or diamond *(b)*

Study of the diamond synthesis in 'graphite-metal hydrides' systems allowed to assume that hydrogen presence in the reaction zone stimulates the process of their formation [7]. It can be presumed that hydrogen can have a catalyzing effect on diamond formation in iron-carbon alloys.

2. Results and discussion

In the presented studies the hypothetical possibility for diamond synthesis in iron-carbon alloys is considered, and the mechanism of hydrogen influence on the diamond phase formation is offered. For examination we used synthetic grey cast iron on ferritic matrix with the carbon content close to eutectic, and samples of pure iron pressed into fine clean graphite. Most of the samples had been saturated with hydrogen at high hydrogen pressure of 45–80 MPa in a wide temperature range. In test experiments the hydrogen had been cleaned through a palladium membrane. In individual cases electrochemical saturation with hydrogen had been used to produce concentrations up to 20 cm^3/100 g.

As evident from the results of the investigations [8], hydrogen prevents graphite dissolution in the matrix, and completely blocks graphite growth at high pressure. This is supported by the fact that heating of ferrite-based grey cast iron samples to near-melting temperatures, irrespective of heating and cooling rates, did not change the initial structure of the material under study at the specified pressures of the gas phase (Fig. 2). Since graphite did not dissolve in ferritic and austenitic matrix, the melting point of the samples coincided with the melting point of pure iron. The samples had been heated to their complete melting.

On cooling of melted samples at various rates (from 250°C/sec to 0.16°C/sec), the crystals well-seen through the light microscope had been found inside the matrix instead of the expected graphite inclusions. Attempts to elucidate the nature of these crystals by means of well-known methods of X-ray structure analysis have failed. The size of the inclusions allowed measuring of microhardness. However the diamond of the indenter practically made no mark on them at a load of 100 g (on Fig. 3 the dents of the indentor after microhardness measurement are seen), and internal loads, arising in crystals, destroyed them after removal of load in some cases.

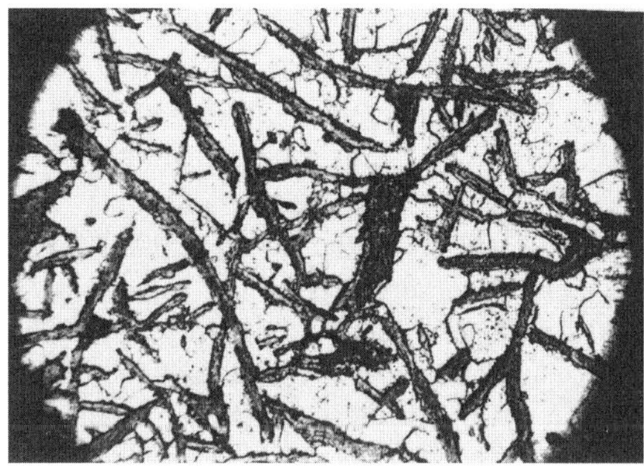

Figure 2. A microstructure of gray iron after heating to temperature 1,400^{0}C, hydrogen pressure 65 MPa

The possibility of the interpreting of the obtained crystals as diamond inclusions is stipulated by the following. It has been proved experimentally:

- Formation of diamond inclusions in iron-carbon alloys becomes possible at high temperatures and pressure [1].
- Dynamic synthesis of diamond in grey cast iron can be performed by two methods: plastic strain + explosion and explosion + temperature cycling [2].
- Formation of inclusions of diamond nature is possible on solidification of cast iron inoculated by graphite-spheroidizing elements – Mg, Ce and others [3].

There is information that hydrogen has influence on the diamond formation process in conditions of diamonds synthesis from a gas phase. Hydrogen influence on a diamond phase growth by the gas deposition method at low pressures has been studied in paper [4]. It has been found that atomic hydrogen introduction into a gas atmosphere provides selectivity of the diamond phase crystallization [9], it has found application in the diamonds synthesis technology. A substantial increase in the diamond growth velocity observed in practice as compared to the growth velocity in hydrocarbon pyrolysis, when atomic hydrogen concentration is below the equilibrium one, is caused by active participation of atomic hydrogen in the crystallization process. It is suggested that one of the main functions of atomic hydrogen lies in dehydrogenation and hydrogenation of the inoculating crystal surface to provide: formation of active carbonaceous areas for connection of acetylene (C_2H_2) or CH_3-radical to them in the first case, and stabilization of the formed diamond configurations in the second case.

Study of the diamond synthesis in 'graphite-metal hydrides' systems allowed to assume that hydrogen presence in the reaction zone stimulates the process of diamond formation [7]. Thus, the theoretical analysis makes it possible to come to a conclusion, that hydrogen has a catalyzing effect on the diamond formation process to inhibit the graphitization process. Hydrogen, the covalent bond energy of which with carbon is close to the C-C bond energy, can block available carbon covalent bonds on prismatic planes of the graphite crystal lattice. By blocking available carbon bonds, hydrogen can prevent the graphite dissolution in austenite as well as graphite precipitation from a liquid phase. Under these conditions, formation of diamond inclusions becomes potentially possible.

Under the action of hydrogen, formation of diamond inclusions is potentially possible; the nucleation centre of which can be the crystals observed in the pure iron saturated with hydrogen to the concentration of 20 cm^3/100 g (Fig. 4) [10].

The possibility of the diamond phase formation in conditions of the performed experiment is confirmed by:

- A phase metallographic analysis – with the help of the light radioscopy the optical transparence of the phase is observed very well, especially in places of chips, as well as by a specific faceting of crystals.
- Direct measurement of microhardness of crystals and matrix (for comparison in Fig. 3 the imprint of the diamond pyramid in the crystal is left at load 0,196 H, and the imprint in the matrix near the inclusion – at load 0,039 H).
- Rigid control of purity of initial materials and multiple repeatability of results.

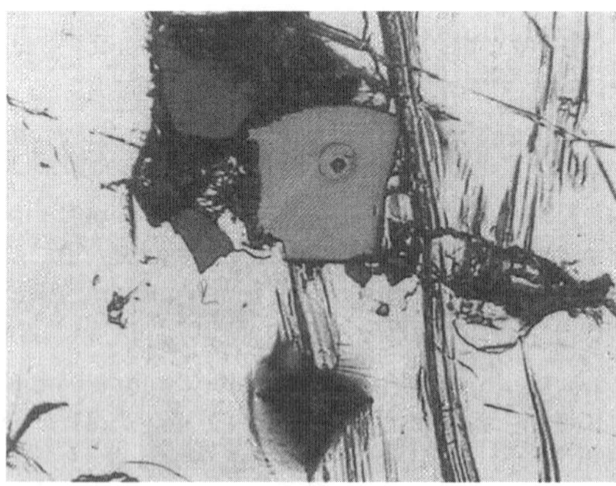

Figure 3. Microphotograph of crystals in ferritic matrix of Fe-C-H alloy with a close to eutectic carbon content (×500); an imprint of a diamond pyramid in the crystal – at a load 0,196H, in a matrix close to the inclusion – at a load 0,039H

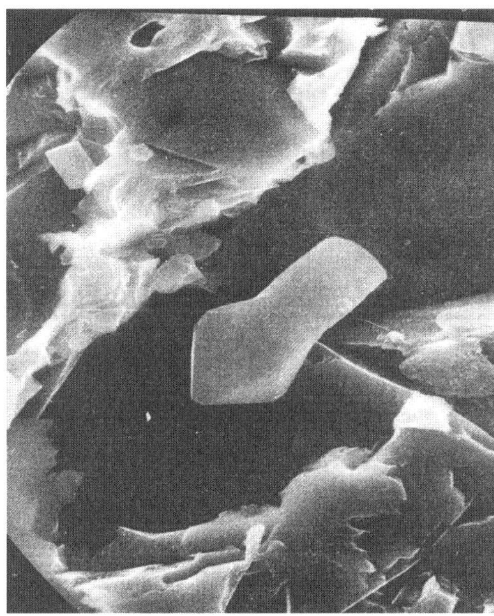

Figure 4. Microphotograph of crystals in a matrix of the hydrogen-enriched samples of pure iron (×5,000)

Potentially the hydrogen can have a catalyzing influence on the process of diamond formation, being a particular case of structure formation in the system iron–hydrogen and caused by carbon polymorphism.

3. Conclusions

For the first time the possibility of diamond-like structures generation during crystallization of iron-carbon alloys has been proved.

References

1. Gustafson P. The hydrogen in iron and alloys. Scandinavian J. Metallurgy. 1975 **(4)**: 259–267.
2. Sobolev V.V., Taran J.N., Gubenko S.I. Diamond synthesis in cast iron. M&HTM. 1993 **(1)**: 2–6.
3. Kimstach G.M. On the mechanism of diamond crystals formation in iron – carbon alloys. M&HTM. 1991 **(8)**: 6–7.
4. Spitsin B.V., Deregin B.V. Way of diamond faces build-up: A.S. 339134 USSR - № 3867534/23-02; declared 05.07.79. Discoveries. Inventions. Industrial Designs. Trade-marks. 1980 **(17)**: 323.
5. Frankliash M., Spear R. The diffusion of hydrogen in decarburised mild steel. Mater. Res. 1988 **3(1)**: 115–117.
6. Kitabatake M., Wasa K. The thermodynamics of the hydrogen – iron system. J. Vac. Sci. Technol. 1988. **A 6(3)**: 1793–1797.

7. Kulicova L.F., Shalimov M.D., Slesarev V.N., Jakovlev E.N. Synthesis of diamond in systems of graphite-hydrides of metals. Superhard Mater. 1988 **(1)**: 3–4.
8. Shapovalov V.I. Hydrogen influence on structure and characteristics of iron-carbon alloys. – Moscow: Metallurgy, 1982. 232 p.
9. Hibsman H.J. Diamond growth process: Pat. 3371996 US – № 56784; Pub. 06.09.85. B. **(36)**: 3.
10. Poltoratsky L.M. On structure formation in the Fe-C-H system. Transactions of SSIU «Perspective industrial technologies and materials». – Novosibirsk: Science, 2004. P. 329–334.

ELECTROCHEMICAL AND HYDROGEN SORPTION PROPERTIES OF AB_5–TYPE ALLOYS WHERE A – La, Ce; B – Ni, Co, Mn, Fe, Cu, Cr, Al

T.A. ZOTOV, I.A. ROMANOV, S.V. MITROKHIN,
V.N. VERBETSKY, O.A. PERTII
Moscow State University, Department of Chemistry,
Leninskie Gory 3, Moscow, 119992 Russia
Fax: (095) 932 88 46; E-mail: zotov@hydride.chem.msu.ru

Abstract. Like other accumulator types Ni-MH batteries are unable to keep high characteristics in the temperature range -30°C to +40°C. To solve this problem the negative electrode material – intermetallic compound (IMC) hydride – must have dissociation pressure of 0.1–1.5 atm in this temperature interval. The IMC composition was calculated with mathematic model. The model describes dependence of IMC hydrides properties on IMC composition. Using calculation results series of alloys were prepared and their hydrogen absorption and electrochemical properties were studied. The best alloy is $CeMn_{0.5}Co_{0.25}Al_{0.5}Ni_{3.75}$. The sample has discharge capacity 275 mAh/g, keeps 95% of this value at 40°C and 40% at -30°C.

Keywords: Ni-MH batteries, intermetallic compounds, IMC hydrides, mathematical model

1. Introduction

Today nickel metal hydride batteries (Ni-MH) are one of most power-capacious and reliable accumulators [1–6]. The discharge capacity of modern Ni-MH batteries is equal and in some cases greater than that of disposable non-rechargeable alkaline batteries. Like other accumulators, Ni-MH batteries can't preserve high capacity characteristics at temperatures from –30 to +40°C. Working temperature interval of accumulator depends on properties of negative MH-electrode. MH-electrode properties are determined by hydrogen sorption properties of intermetallic compound (IMC). The dissociation hydrogen pressure of alloy or IMC should be in the interval of 0.1–1.5 atm for its application in MH- electrode. Experimental search of alloy compositions satisfying these pressure conditions is very difficult. In our work we used mathematical model where hydride properties are function of compositions of alloy components. For this purpose the large database of properties for alloys of AB_5-type, where A: La, Ce, Mm, Gd, Nd, and B: Ni, Co, Al, Cu, Fe, Mn, Cr was analyzed. In a result of calculation the series of alloys were prepared. Their electrochemical and hydrogen sorption properties were measured at –30 to +40°C.

2. Alloys composition calculation

In this paper an attempt of IMC electrochemical properties prognosis was made by the calculation of thermodynamic parameters of their hydrides.

TABLE 1. Thermodynamic characteristics of the AB_5 IMC hydrides chosen for the data base

IMC	ΔH, kJ/mol H_2	ΔS, J/K mol H_2	Source
$Ce_{0.7}Nd_{0.3}Ni_{2.5}Cu_2$	15.7	72.2	[8]
$LaNi_{4.9}Cu_{0.1}$	28.5	97.4	[9]
$LaNi_{4.5}Cu_{0.25}Al_{0.25}$	39.0	115.0	[10]
$LaNi_4Cu$	48. 5	115.5	[9, 11]
$LaNi_3Cu_2$	31.0	96.0	[12]
$LaNi_2Cu_3$	34.8	109.0	[12]
$CeNi_{2.5}Cu_{2.5}$	24.1	97.1	[13]
$CeNi_{4.25}Cu_{0.75}$	25.2	86.6	[14]
$La_{0.3}Ce_{0.7}Ni_5$	25.3	109.0	[15]
$La_{0.2}Ce_{0.8}Ni_{4.7}Cu_{0,3}$	24.8	108.4	[14]
$La_{0.2}Ce_{0.8}Ni_5$	23.8	108.8	[14]
$CeNi_{2.5}Co_{2.5}$	20.7	81.1	[15]
$CeCo_5$	49.5	32.6	[15]
$CeNi_{2.5}Cu_{2.3}Al_{0.2}$	15.3	53.7	[8]
$CeCu_2Ni_{2.5}Al_{0.5}$	28.6	70.2	[8]
$CeNi_{4.4}Al_{0.6}$	18.9	77.7	[16]
$CeNi_{4.5}Al_{0.5}$	18.0	78.4	[16]
$CeNi_5$	22.2	111.0	[15, 14]
$CeNi_4Co$	30.5	133.0	[15]
$La_{0.1}Ce_{0.9}Ni_{4.95}Al_{0.05}$	27.6	97.9	[17]
$LaCo_5$	45.1	110.0	[15]
$LaNi_2Co_3$	43.9	128.0	[18]
$LaCo_2Ni_3$	31.0	16.0	[19]
$LaCu_{1.5}Ni_{3.5}$	35.3	116.6	[11]
$LaNi_4Al$	49.2	118.8	[19]
$LaNi_{4.6}Mn_{0.4}$	25.6	75.0	[9]
$LaCoNi_4$	32.0	112.0	[18]
$LaNi_{4.9}Al_{0.1}$	32.6	115.7	[8, 20]
$LaNi_2Al_{0.25}$	33.7	110.9	[21]
$LaNi_5$	30.9	107.7	[22, 23]
$NdNi_{4.5}Al_{0.5}$	47.0	154.9	[24]
$NdNi_4Al$	54.1	174.7	[24]
$NdNi_{3.5}Al_{1.5}$	54.3	173.6	[24]
$LaNi_{2.5}Fe_{2.5}$	37.2	97.5	[25]
$LaNi_4Fe$	34.5	92.7	[25, 9]

The mathematical model was used for calculation of ΔH and ΔS of hydrogen desorption reaction, because direct calculation hydride formation heat is very difficult task with inaccurate result.

And so we have used empirical and semiempirical model. The thermodynamic parameters change monotonously in the single phase region of solid solution according to the Wegard law. Therefore a large experimental thermodynamic data massive can be approximated with function of its composition. Such model was developed in our laboratory for calculation thermodynamic characteristics multicomponent stoichiometric alloys [7]. The lattice constants, ΔH and ΔS values of hydrogen desorption reaction can be described by polynomial:

$$F(x)=Y_0+\sum A_i\cdot x_i + \sum B_{ij}\cdot x_i\cdot x_j \qquad (j\geq i),$$

where x are IMC components concentration; Y_0, A, B are constants.

Experimental data of the parameters are used for determining the following coefficients of the polynomial (A_i,B_{ij}…). For the solving equation system the minimal number of initial data in data base is:

$$N=2i+(i^2-i)/2,$$

where i are independent metal concentrations.

The alloys constituting the data base in resent task are present in Table 1. The calculation conditions are: dissociation pressure 0.09–0.2 atm at –30°C and 1–3 atm at 40^0C. It was obtained near 250 results of calculations of the IMC compositions. Only three was chosen for the experimental study:

$CeMn_{0.25}Co_{0.75}Al_{0.5}Ni_{3.5}$
$CeMn_{0.5}Co_{0.5}Fe_{0.25}Al_{0.4}Ni_{3.35}$
$CeMn_{0.5}Co_{0.25}Al_{0.5}Ni_{3.75}$.

3. Experimental

Alloys were prepared by arc-melting the mixtures of pure initial metals under argon atmosphere. With the purpose of homogenization, the alloys were remelted four times. The structure and phase composition of alloys was examined by powder X-ray diffraction. The refinement of diffraction profiles was performed using the Rietveld method.

The hydrogen absorption properties were studied by measuring PCT isotherms using a Siewerts type apparatus at a hydrogen pressure below 100 atm. The thermodynamic characteristics (ΔH and ΔS) of IMC hydrides were calculated from Vant-Hoff equation by using desorption isotherms data at temperatures from –30°C to 40°C.

The electrochemical experiments were carried out in a three-electrode electrochemical glass cell with Hg/HgO electrode as the reference. The electrolyte was 6 M KOH solution. The MH electrodes were prepared by cold-pressing of mixture of IMC powder (20%) with copper powder (80%) in a pellet. Samples were charged with current density of 200 mA/g during 2–3 h. The resting time before discharging was 10 min. All samples underwent 2–3 charge-discharge cycles for the activation. The discharge capacity was checked at current densities of 100, 200 and 400 mA/g. Discharge curves were measured at temperatures 40°C, 23°C, 0°C and -30°C. Discharge curves at temperatures of 40^0C, 0^0C and -30^0C have only qualitative character because only working part of the cell was thermostated and the reference electrode was at room temperature. High rate discharge ability of alloys was calculated by following equation:

$$C_R = \frac{C_{400}}{C_{400} + C^*_{50}} \times 100\,\%,$$

where C^*_{50} is resting MH- electrode capacity that was measured after discharging of 400 mA/g with discharge current density of 50 mA/g.

The cycle life of electrode wasn't investigated but it was stated that after 15 charge-discharge cycles no loss in discharge capacity was noticed.

4. Results and discussion

X-ray diffraction analysis indicates that the main phase in all samples was $CaCu_5$-type phase. The cell parameters of IMC are presents in Table 2.

Hydrogen desorption isotherms of the samples are present in the Fig. 1. Hydrogen sorption properties of IMC hydrides are summarized in Table 3.

TABLE 2. Cell parameters of $CaCu_5$- type structure alloys

IMC	a (Å)	c (Å)	V (Å^3)
$CeMn_{0.5}Co_{0.25}Al_{0.5}Ni_{3.5}$	4.9394 ± 0.0004	4.0732 ± 0.0004	86.062 ± 0.013
$CeMn_{0.5}Co_{0.5}Fe_{0.25}Al_{0.4}Ni_{3.35}$	4.9481 ± 0.0006	4.0757 ± 0.0006	86.419 ± 0.020
$CeMn_{0.5}Co_{0.25}Al_{0.5}Ni_{3.75}$	4.9428 ± 0.0007	4.0727 ± 0.0007	86.171 ± 0.023

TABLE 3. Hydrogen sorption properties of IMC hydrides

IMC	ΔH_{des} (kJ/mol H_2)	ΔS_{des} (J/K mol H_2)	t (°C)	P_{dis} (atm)	H_2 content	
					mass %	H/IMC
$CeMn_{0.25}Co_{0.75}Al_{0.5}Ni_{3.5}$	28 ± 2	103 ± 6	40	4.84	1.1(0)	4.6
			25	2.25	1.1(3)	4.8
			0	1.03	1.1(9)	5.0
			-30	0.19	1.2(7)	5.3
$CeMn_{0.5}Co_{0.5}Fe_{0.25}Al_{0.4}Ni_{3.35}$	30 ± 1	99 ± 4	40	1.49	1.2(5)	5.3
			25	0.7	1.2(6)	5.3
			0	0.25	1.3(0)	5.4
			-30	0.05	1.3(3)	5.6
$CeMn_{0.5}Co_{0.25}Al_{0.5}Ni_{3.75}$	30 ± 2	97 ± 7	40	1.12	1.2(9)	5.4
			25	0.48	1.3(1)	5.5
			0	0.20	1.3(5)	5.6
			-30	0.04	1.3(8)	5.8

Analysing the obtained data we can see that increase of cobalt content by partial substitution of manganese and nickel leads to decreasing hydrogen capacity, increasing hydrogen dissociation pressure and decreasing hydrogen dissociation enthalpy; iron addition leads to same result but the change in the properties is less significant.

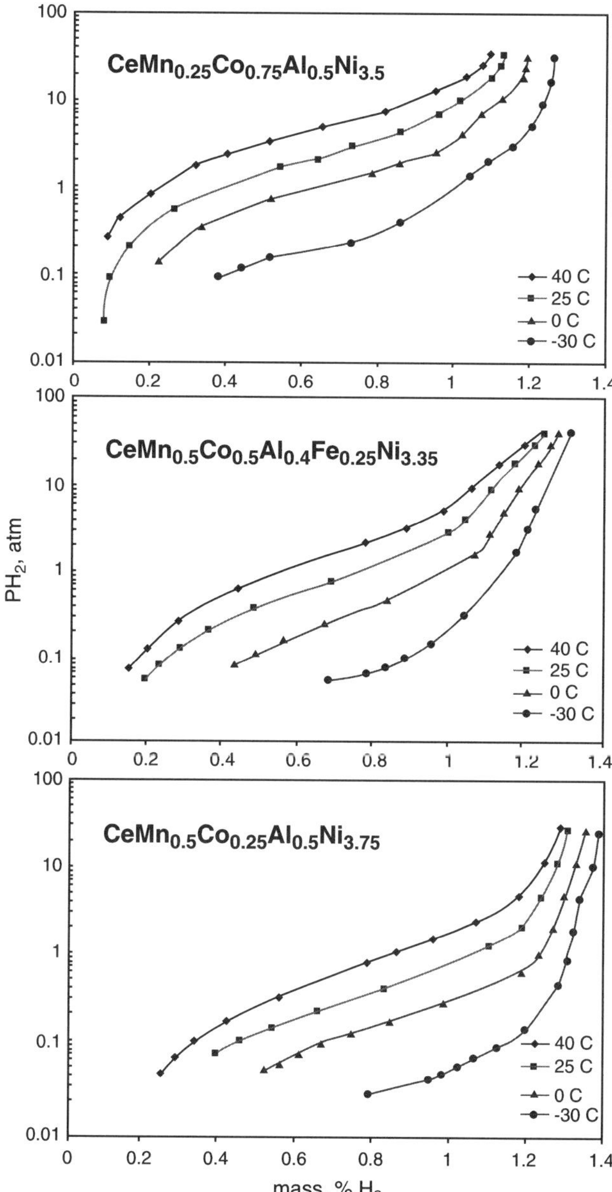

Figure 1. PCT isotherms of studied IMC hydrides

The hydrogen sorption properties of AB_5 alloys determined during experiment were slightly different from calculated values (Table 4). This divergence can be explained by two ways.

TABLE 4. The comparison of hydrogen desorption enthalpies (kJ/mol H_2) and hydrogen dissociation pressures of IMC hydrides (atm) at temperatures 40°C and –30°C that were obtained by the calculation (ΔH^C, P_{40}^C, P_{-30}^C) and by the experiment (ΔH^E, P_{40}^E, P_{-30}^E)

IMC	ΔH^C	ΔH^E	P_{40}^C	P_{40}^E	P_{-30}^C	P_{-30}^E
$CeMn_{0.25}Co_{0.75}Al_{0.5}Ni_{3.5}$	23	28	1.55	4.84	0.12	0.19
$CeMn_{0.5}Co_{0.5}Fe_{0.25}Al_{0.4}Ni_{3.35}$	24	30	1.2	1.49	0.13	0.05
$CeMn_{0.5}Co_{0.25}Al_{0.5}Ni_{3.75}$	24	30	2.6	1.12	0.1	0.04

At first it is imperfection of mathematic modeling. Using more initial data will increase the prognosis accuracy. And besides we understand that initial data were obtained in different laboratories with different accuracy. Secondly the divergence can happens due to the loss of Mn and Ce during the alloy melting.

However we suppose that hydrogen sorption properties of these IMC satisfied the MH-electrode requirements.

Maximum discharge capacity of samples 250–280 mAh/g is realized at room temperature at discharge current density 100 mA/g. Discharge curves of MH-electrodes at discharge current density 100 mA/g and different temperatures are present in Fig. 2, electrochemical properties of IMC are summarised in Table 5.

TABLE 5. Electrochemical properties of MH-electrodes with studied IMC

IMC	t (°C)	Discharge capacity		C_R (%)
		C_{100} (mAh/g)	C_{400} (mAh/g)	
$CeMn_{0.25}Co_{0.75}Al_{0.5}Ni_{3.5}$	40	227	225	97.0
	25	239	192	79.3
	0	211	117	52.0
	-30	88	31	33.7
$CeMn_{0.5}Co_{0.5}Fe_{0.25}Al_{0.4}Ni_{3.35}$	40	170	78	43.1
	25	218	86	39.6
	0	150	60	39.1
	-30	68	28	38.4
$CeMn_{0.5}Co_{0.25}Al_{0.5}Ni_{3.75}$	40	261	254	97.0
	25	275	241	86.1
	0	247	189	75.6
	-30	112	45	40.2

The dependence of discharge capacity at discharge current density of 100 mA/g on temperature is present in Fig. 3.

Analyzing electrochemical properties of samples we can make following conclusions:

- Discharge capacity of MH-electrode increase with nickel content in IMC
- Partial substitution of nickel and aluminum by iron leads to significant decrease of discharge characteristics

– Partial substitution of nickel and manganese by cobalt result in insignificant changes of the discharge ability but leads to its small decrease at low temperatures

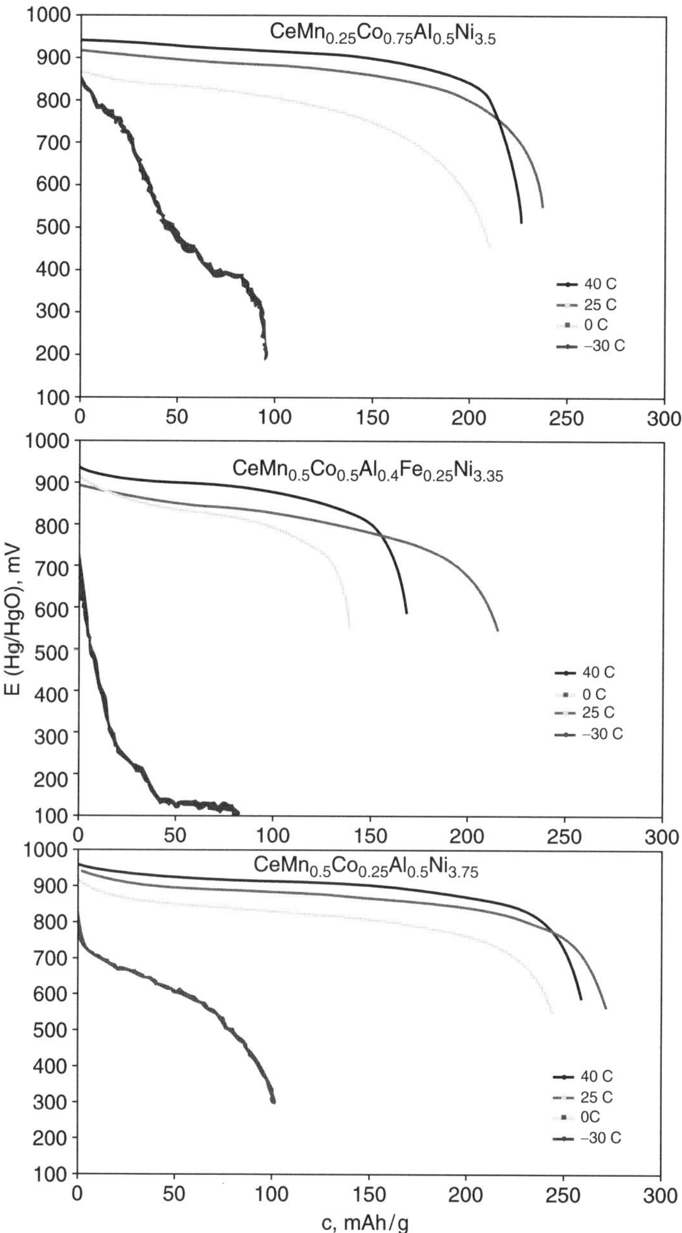

Figure 2. Discharge curves of MH-electrodes at discharge current density 100 mA/g

IMC $CeMn_{0.5}Co_{0.25}Al_{0.5}Ni_{3.75}$ shows best electrochemical properties among studied alloy series. Its hydrogen capacities at all temperatures are maximal among these alloys. And besides nickel concentration is highest in this series that yields a best electrocatalytic activity. Discharge capacity $CeMn_{0.5}Co_{0.25}Al_{0.5}Ni_{3.75}$ (275 mAh/g at 100 mA/g and room temperature) is comparable with commercial MH-electrode alloys (250–310 mAh/g). Discharge capacity $CeMn_{0.5}Co_{0.25}Al_{0.5}Ni_{3.75}$ depends on temperature nearly three times weaker than capacity of the commercial alloys [2].

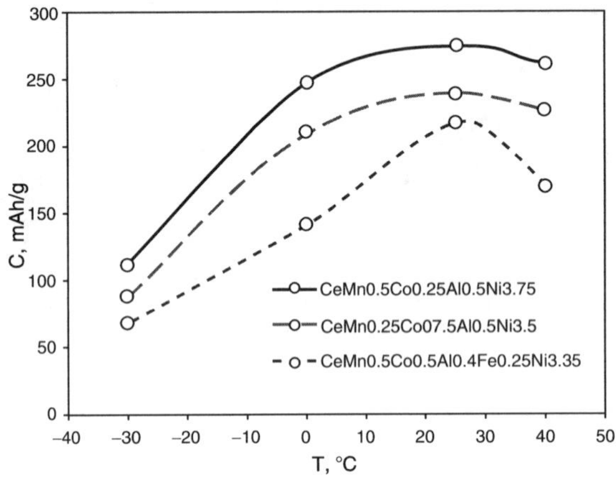

Figure 3. The dependence discharge capacity $C_{100\ mA/g}$ on temperature

5. Conclusions

The principal possibility of mathematic calculation the composition of IMC with required hydrogen properties is shown in this paper. Although the divergence between calculated and experimental hydrogen sorption characteristics all samples satisfied the working conditions of MH-electrodes. Hydrogen content in IMC hydrides is in the range of 1.1–1.3 mass%, discharge capacities at 100 mA/g and room temperatures are 218–275 mAh/g. The best sample $CeMn_{0.5}Co_{0.25}Al_{0.5}Ni_{3.75}$ has discharge capacity 275 mAh/g, keeps 95% of this value at 40°C and 40% at -30°C.

Acknowledgements

This work is supported in part by RFBR Grant 06-08-01598.

References

1. Brodd R.J., Bullock K.R., Leising R.A., Middaugh R.L., Miller J.R., Takeuchi E. Batteries, 1977 to 2002, *J. Electrochem. Soc.*, 2004, **151**(3): K1–K11.
2. Nickel-Metal Hydride Application Manual / http://data.energizer.com.
3. Petrii O.A., Vasina S.Ya., Korobov I.I. The electrochemistry of intermetallic hydrides, *Russ. Usp. Khim.*, 1996, **65**(3): 195–209.

4. Kleperis J., Wójcik G., Czerwinski A., Skowronski J., Kopczyk M., Beltowska-Brzezinska M. Electrochemical behavior of metal hydrides, *J. Solid State Electrochem.*, 2001, **5**: 229–249.
5. Rong L., Jianmin W., Shaoxiong Z., Jiuhong Q. Effects of cobalt content and preparation on electrochemical capacity of AB_5-type hydrogen storage alloys at different temperature, *J. Rare Earths.*, 2006, **24**: 341–345.
6. Sakai T., Miyamura H., Kuriyama N., Kato A., Oguro K., Ishikawa H. Metal hydride anodes for nickel-hydrogen secondary battery, *J. Electrochem. Soc.*, 1990, **137**(3): 795–799.
7. Mitrokhin S.V., Verbetsky V.N. Titanium based laves phase hydrides with high dissociation pressure, *Int. J. Hydrogen Energy*, 1996, **21**(11/12): 981–983.
8. Sinha V.K., Wallage W.E. Effect of neodymium, copper and aluminum on the hydriding characteristics of the $CeNi_5$, *J. Less-Common Met.*, 1984, **96**: 283–290.
9. Zhao Shuang, Lin Qin, Cher Ning. Calculation and prediction for the hydriding properties of $LaNi_{5-x}M_x$ alloys, *J. Alloy. Compd.*, 1999, **287**: 57–61.
10. Burnasheva V.V., Petrova L.A., Semenenko K.N. Interaction with hydrogen of IMC $LaNi_{5-x}(T_1,T_2)_x$, T_1, T_2 - Al, Cr, Fe, Cu, *Russ. J. Inorg. Chem.*, 1984, **29**: 869–873.
11. Semenenko K.N., Yakovleva N.A., Ganich E.A. The Tian-Calvet calorimetry as a method for investigation heterogeneous hydrogenation catalysts based on IMC disperse powders, *Thermodyn. Acta*, 2001, **169**: 161–166.
12. Ganich E.A., Yakovleva N.A., Semenenko K.N. The study of interaction with hydrogen of IMC $LaNi_{5-x}Cu_x$, x = 2-3, *Russ. J. Common Chem.*, 1999, **69**: 1288–1292.
13. Sinha V.K., Wallage W.E. Hydrogen sorption by the $Ce_{1+x}Ni_{2.5}Cu_{2.5}$ alloys, *J. Phys. Chem.*, 1984, **88**: 115–109.
14. Klyamkin S.N., Verbetsky V.N., Karin A.A. Thermodynamic particularities of some $CeNi_5$–based metal hydride systems with high dissociation pressure, *J. Alloy. Compd.*, 1995, **231**: 479 482.
15. Sarynin V.K. The study of interaction with hydrogen of IMC $CaCu_5$ and UNi_5. Ph.D. thesis, Moscow, 1980.
16. Sinha V.K., Wallage W.E. Hydrogen storage properties of the $CeNi_{5-x}Al_x$ alloys, *J. Phys. Chem.*, 1984, **88**: 102–105.
17. Burnasheva V.V., Petrova L.A., Semenenko K.N. Cerium and aluminium influence on phase equilibrium in system $Ce_xLa_{1-x}Ni_{5-y}Al_y$ – H_2, *Russ. Metalls*, 1993, **3**: 213–218.
18. Colinet C., Pasturel A. Enthalpies of formation and hydrogenation of $LaNi_{5-x}Co_x$ compounds, *J. Less-Common Met.*, 1987, **134**: 109–122.
19. Mendelsohn N.H., Gruen D.M. Hydrogen absorption in the ternary substituted AB_5 alloys with particular reference to $La_{1-x}Y_xNi_5$ and $LaNi_{5-x}Al_x$, *J. Rare Earths Met. Science Tech.*, 1978, **6**: 25–31.
20. Filatova E.A., Yakovleva N.A., Semenenko K.N. Calorimetric investigations of interactions in systems $LaNi_{4.9}Al_{0.1}$ и $LaNi_{4.9}Sn_{0.1}$, *Russ. J. Common Chem.*, 2001, **71**: 1084–1091.
21. Filatova E.A., Yakovleva N.A., Semenenko K.N. Calorimetric investigations of interactions in systems $LaNi_{4.75}Al_{0.25}$ – H_2 и $LaNi_{4.8}Sn_{0.2}$ – H_2, *Russ. News Sci. Acad., Chem. Series*, 1999, **10**: 1841–1848.
22. Van Vucht J.H.N., Kuijpers F.A., Bruning H.C.A.M. Reversible room-temperature absorbtion of large quantities of hydrogen by intermetallic compounds, *Phlips Res. Repts*, 1970, **25**: 133–140.
23. Kuijpers F.A., Van Mal H.H. Sorption hysteresis in the $LaNi_5$-H_2 and $SmCo_5$-H_2 systems, *J. Less–Common Met.*, 1971, **23**: 395–398.

24. Bobet J.-L., Pechev S., Chevalier B., Darriet B. Structural and hydrogen sorption properties of $NdNi_{5-x}Al_x$ and $GdNi_{5-x}Al_x$, *J. Alloy. Compd.*, 1998, **267**: 136–141.
25. Lamloumi J., Percheron–Guegan A., Thermodynamic, structural and magnetic properties of $LaNi_{5-x}Fe_x$ hydrides, *J. Less-Common Met.*, 1987, **130**: 111–122.

PECULIARITIES OF HYDROGEN INTERACTION WITH INTERMETALLIC COMPOUNDS AB_2

E.YU. ANIKINA
Chemistry Department, Lomonosov Moscow State University, 119992 Moscow, Russia
E-mail: anikina@hydride.chem.msu.ru

Abstract. In the present work we compare the thermodynamic functions of the hydrogen desorption from $ZrMn_2$-H, $ZrMn_{2.7}$-H and $Ti_{0.9}Zr_{0.1}Mn_{1..3}V_{0.5}$-H systems, which were calculated on the base of data obtained from the study of these systems by means of P-C measurements (C=H/AB_2) and the calorimetric method.

Keywords: intermetallic compounds, hydrides, thermodynamic, calorimetry, $ZrMn_2$, $ZrMn_{2.7}$, $Ti_{0.9}Zr_{0.1}Mn_{1.3}V_{0.5}$

1. Introduction

In the present work as subjects under study the alloys with hexagonal Laves phase structure C14 stoichiometric and nonstoichiometric compositions $ZrMn_2$, $ZrMn_{2.7}$ and $Ti_{0.9}Zr_{0.1}Mn_{1.3}V_{0.5}$ were chosen. These alloys are perspective materials for using them in the different stationary and mobile hydrogen accumulators, as they posses fast kinetics, a large hydrogen storage capacities and resistance to degradation during cycling. The thermodynamic properties were studied in terms of van't Hoff plots, which suggest temperature independence of partial molar enthalpy (ΔH) of hydrogen reaction with intermetallic compounds, and the calorimetric method.

2. Experimental

The intermetallic compounds (IMC) of AB_2-type were prepared from starting elements a purity grade higher than 99.9% by arc melting technique in an argon atmosphere. Because of the high volatility of manganese, 4% extra Mn was added compared to the desired final stoichiometry. The alloy samples were turned over and remelted some times and then annealed at about 950°C for 240 h in the sealed quartz ampoules under residual argon pressure 0.01 atm to ensure homogeneity. Powder X-ray diffraction analysis (Cu Kα radiation, Thermo Ariel diffractometer) indicated that all prepared intermetallic compounds had the hexagonal $MgZn_2$ Laves phase structure however small amounts of ZrO_2 (less then 5%) were found as minor phases for $ZrMn_2$ and $ZrMn_{2.7}$. The refined unit lattice parameters are presented in the Table 1 and they are in good agreement with the reference data [1–3].

TABLE 1. Unit cell parameters of the starting alloys

IMC	a (Å)	b (Å)
$ZrMn_2$	5.031	8.261
$ZrMn_{2.7}$	4.997	8.205
$Ti_{0.9}Zr_{0.1}Mn_{1.3}V_{0.5}$	4.920	8.070-

The chemical composition of alloys and their homogeneity were examined by electron microscopy and electron probe analysis. These results indicate that the starting alloys were well-crystallized, homogeneous, single phase samples with stoichiometry $ZrMn_2$, and $Ti_{0.9}Zr_{0.1}Mn_{1.3}V_{0.5}$ or normalized to AB_2 $Zr_{0.82}Mn_{0.18}Mn_2$ and $(Ti_{0.89}Zr_{0.11})(Mn_{1.39}V_{0.54}Ti_{0.07})$. It should be noted that all calculations in this work were carried out on AB_2 composition of IMC in order to have the possibility to compare the interaction of these compounds with hydrogen.

The study of the hydrogen interaction with the intermetallic compounds was carried out on the twin-cell differential heat-conducting calorimeter of Tean-Calvet type connected with a conventional Sieverts-type apparatus for gas dosed feeding. The apparatus scheme was described elsewhere [4]. The application of this installation permits to measure simultaneously the P-C-T and ΔH-C-T dependences (C=H/AB_2], P – equilibrium pressure, T – the reaction temperature, ΔH – enthalpy).

3. Result and discussion

3.1. CALORIMETRIC RESULTS

The calorimetric results are presented in Figs. 1–3 and the Table 2. As one can see from Figs. 1–3 and the Table 2 there is one region of constant values of the enthalpies in the $ZrMn_2$-H system at different temperatures of the reaction. In the |ΔHdes.|-C plot measured at 100°C there is one region where the enthalpy values are constant (~40 kJ/mol H_2) i.e. the $\alpha \leftrightarrow \beta$-transition (C < 2.0) occurs. It corresponds with the reference data good enough [5, 6].

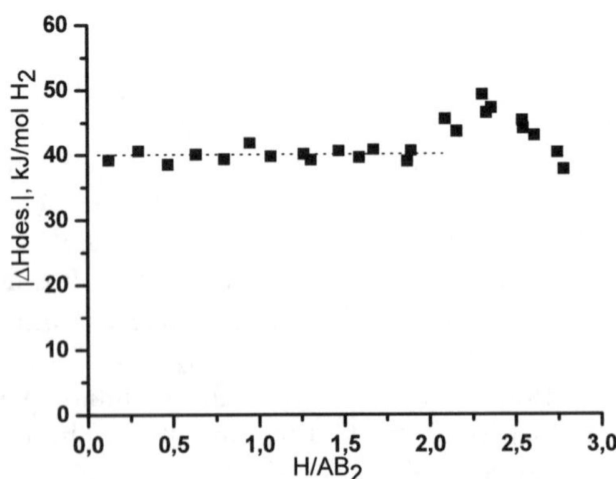

Figure 1. Desorption enthalpy vs. composition for $ZrMn_2$ at 100°C

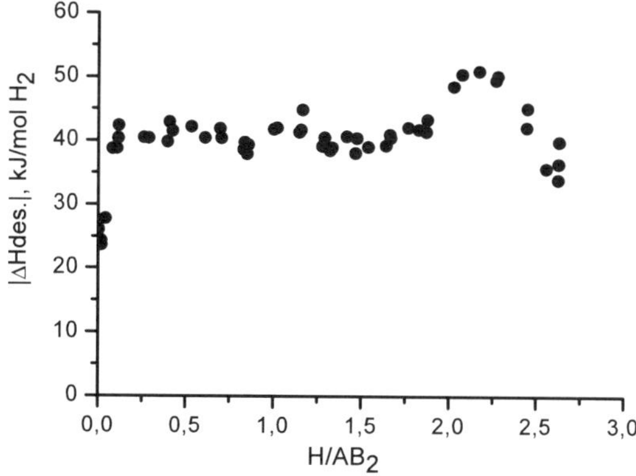

Figure 2. Desorption enthalpy vs. composition for $ZrMn_2$ at 170°

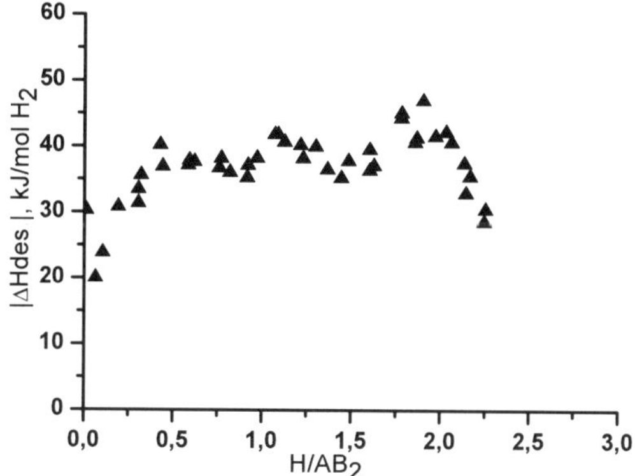

Figure 3. Desorption enthalpies vs. composition for $ZrMn_2$ at 245°

With rising of the experimental temperature from 100°C to 170°C the plot of the |ΔHdes.|-C dependence changes its shape (see Fig. 2). In comparison with the plot, obtained for 100°C (see Fig. 1), the values of the enthalpy in the plateau region $0.2 < C < 1.8$ does not remain constant, and the |ΔHdes.|=f(C) function acquires complex character.

One can see that, firstly, some fold appears on the plot of |ΔHdes.|-C dependences in the range of $C \sim 1.2$ that corresponds to the mid-point of the $\alpha \leftrightarrow \beta$ transition and, secondly, the enthalpy values within two half plateau do not remain constant, but some dispersion was observed. Further raise of the experimental temperature up to 245°C intensifies these phenomena, namely, the fold in the plateau middle becomes higher and the dispersion of the enthalpy values increases.

It is known from reference data [1, 3, 7] that hydrogen atoms in the $ZrMn_{2+X}$ compounds with hexagonal Laves phase structure C14 mainly occupy tetrahedral interstitial sites *24(l), 12(k)₁*, *6(h)₁* and *6(h)₂* with [Zr2Mn2] but for each individual AB_2 compound the tetrahedral holes exist which the hydrogen atoms prefer to occupy. For $ZrMn_2$ such holes are $12(k)_1$ and $6(h)_1$, besides the distribution of hydrogen atoms among these holes is approximately equal at 25–40°C ($12(k)_1$ -0.372, $6(h)_1$ – 0.369) [3]. Therefore it could be surmised on the base of the obtained calorimetric data and the reference data [5, 6] that at ambient temperatures the energy of hydrogen interaction with these tetrahedral holes are approximately close that correlates with wide flat plateau on the plot of the ΔH-C dependence at 100°C but when the reaction temperature raises it could be assumed that the difference among the energy of hydrogen interaction with tetrahedral holes $12(k)_1$ and $6(h)_1$ appears and as a result the dispersion of the enthalpy values in the plot of the ΔH-C dependence at 170°C and 245°C appear.

The $ZrMn_{2.7}$-H system was studied in the temperature range from 75°C to 180°C and it was determined that there are two regions with the constant values of enthalpy in the plot of ΔHdes.-C dependences at 75°C and 82°C so it is possible to assume the existence of two hydrides, β and γ, in the $ZrMn_{2.7}$-H system at this conditions. Further rising of the experimental temperature leads to disappear of one plateau that means one hydride forms at high temperatures and it could be drawn a conclusion that raise of the temperature results in leveling of energy of hydrogen interaction with occupied interstitial sites.

The $Ti_{0.9}Zr_{0.1}Mn_{1.3}V_{0.5}$-H system was studied from 63°C to 190°C. As one can see from Table 2 at 63°C there is one region of constant values of the enthalpy desorption in other words one plateau exists or one hydride forms. In the temperature range from 82°C to 140°C there are two regions of constant values of enthalpy desorption that is it could be supposed that in the $Ti_{0.9}Zr_{0.1}Mn_{1.3}V_{0.5}$-H system the formation of two hydrides takes place: β- and γ-hydrides. The study of this system at 150°C and 170°C revealed that at these temperatures one hydride phase forms again.

Compared the results of calorimetric measurements, obtained for three studied IMC-H systems with the same hexagonal $MgZn_2$ Laves phase structure, it may be marked that the transition from IMC with more simple chemical composition (from $ZrMn_2$) to IMC with the more complex chemical composition (to $Ti_{0.9}Zr_{0.1}Mn_{1.3}V_{0.5}$) results in the more complex interaction the intermetallic compounds with hydrogen, namely, depending on the temperature of hydrogen reaction with IMC it is possible the formation of one or two hydride phases moreover the complication of chemical composition of IMC results in that the formation of two hydrides takes place at higher temperature in other words the temperature interval, within the formation of two hydrides in the IMC-H system occurs, increases with a complication of chemical composition of the IMC. It should be noted some more peculiarity taking place in the studied systems IMC-H. For the $ZrMn_{2.7}$-H system at the conditions when there is a possibility of the formation of two hydride phases the values of |ΔHdes.| for $ZrMn_{2.7}H_{\sim 1}$ > |ΔHdes.| for $ZrMn_{2.7}H_{\sim 2}$ and vice verse for the $Ti_{0.9}Zr_{0.1}Mn_{1.3}V_{0.5}$-H system one can observe contrary picture, namely, |ΔHdes.| for $Ti_{0.9}Zr_{0.1}Mn_{1.3}V_{0.5}H_{\sim 1}$ < |ΔHdes.| for $Ti_{0.9}Zr_{0.1}Mn_{1.3}V_{0.5}H_{\sim 2}$.

TABLE 2. Temperature dependence of desorption enthalpy for the $ZrMn_{2.7}$-H and $Ti_{0.9}Zr_{0.1}Mn_{1.3}V_{0.5}$-H systems

Temperature, °C	$ZrMn_{2.7}$		$Ti_{0.9}Zr_{0.1}Mn_{1.3}V_{0.5}$	
	Range, C=H/AB_2	|ΔHdes.| (kJ/mol H_2)	Range, C=H/AB_2	|ΔHdes.| (kJ/mol H_2)
63	_0	–	0.20 < C < 2.20	30.96±
75	0.20 < C < 1.20 1.20 < C < 2.20	34.09 ± 0.80 30.55 ± 1.17	–	–
82	0.20 < C < 1.10 1.10 < C < 1.9	35.45 ± 0.95 32.15 ± 0.98	0.2 < C <1.34 1.60 < C < 2.20	30.12 ± 1.19 36.36 ± 0.87
100	–	–	0.30 < C < 1.60 1.63 < C < 2.10	29.71 ± 1.37 35.57 ± 0.74
120	0.30 < C < 1.60	31.34 ± 0.89	0.40 < C < 1.00 1.2 < C < 1.78	29.60 ± 0.82 34.37 ± 1.30
130	–	–	0.40 < C < 1.20 1.34 < C < 1.80	28.64 ± 0.75 34.39 ± 1.55
140	–	–	0.40 < C < 0.90 1.05 < C < 1.92	27.39 ± 0.50 33.52 ± 0.83
150	0.30 < C < 1.58	31.21 ± 1.49	0.35 < C < 1.25	28.92 ± 1.03
170	–	–	0.45 < C < 1.05	27.36 ± 0.68
180	0.30 < C < 1.30	29.43 ± 0.68	–	–

3.2. P-C-T relationships

Desorption pressure-composition-temperature (P-C-T) relations at different temperatures for $ZrMn_2$-H, $ZrMn_{2.7}$–H and $Ti_{0.9}Zr_{0.1}Mn_{1.3}V_{0.5}$-H systems are shown in Figs. 4–6, respectively. As it could be seen from these plots the $ZrMn_2$-H system has the largest hydrogen capacity and the most stable hydride forms in this system. The deviation from stoichiometric composition of IMC and increasing of a number of the chemical elements which form this IMC lead to decreasing of the hydrogen capacity of the studied alloys, the increase in plateau slopping and the formation of the less stable hydrides.

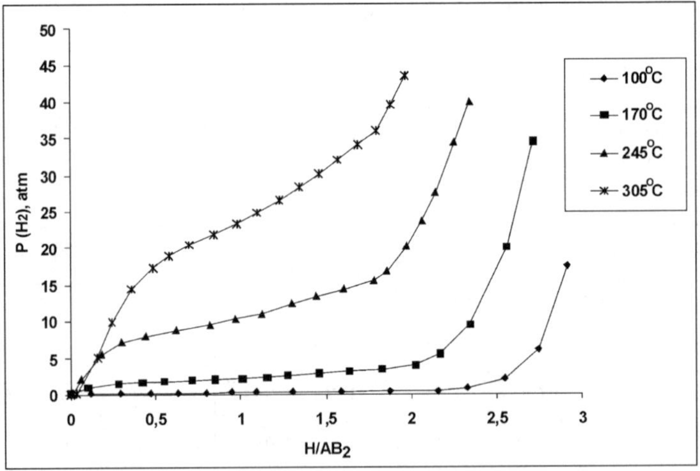

Figure 4. Desorption isotherms for the $ZrMn_2$–H system

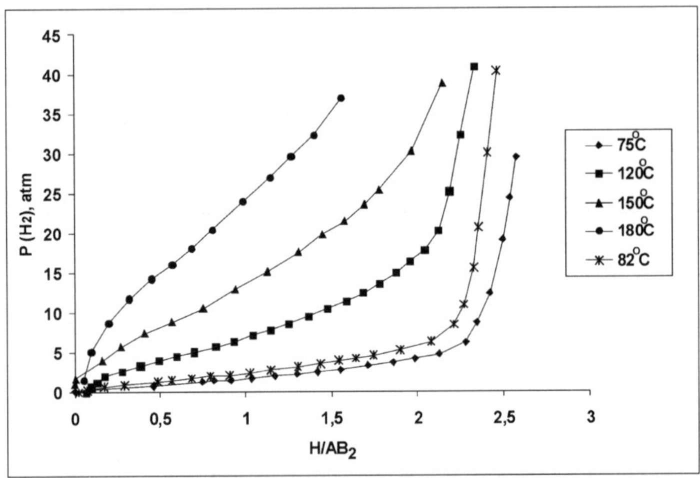

Figure 5. Desorption isotherms for the $ZrMn_{2.7}$–H system (AB_2-$Zr_{0.8}Mn_{0.18}Mn_2$)

The relative partial molar enthalpy ΔH_H and entropy ΔS_H changes for the hydrogen desorption were calculated from the van't Hoff plots based on the measured P-C-T relations. The change of $|\Delta H_H|$ and $|\Delta S_H|$ as a function of hydrogen concentration are presented in Figs. 7, 8, respectively. For the $ZrMn_2$-H system as one can see from Fig. 7 the partial molar enthalpy of hydrogen in the α-solid solution region ($0 < C < 0.5$) gradually increases up to the values |ΔHplat.| (40–41 kJ/mol H_2), then there is the wide

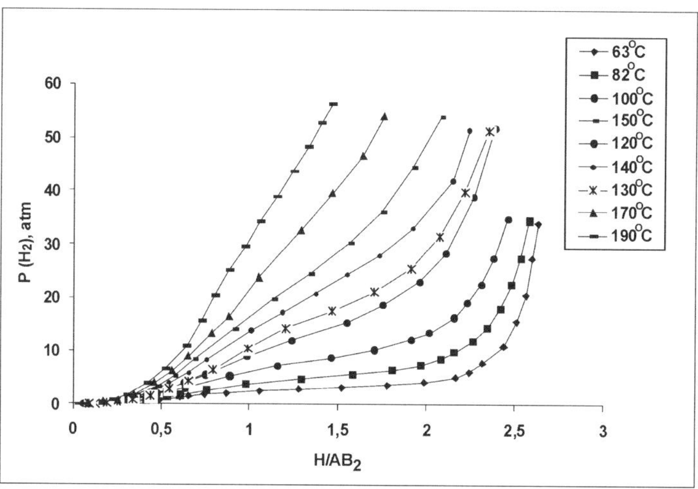

Figure 6. Desorption isotherms for the $Ti_{0.9}Zr_{0.1}Mn_{1.3}V_{0.5}$-H system ($AB_2$ – $(Ti_{0.89}Zr_{0.11})(Mn_{1.39}V_{0.54}Ti_{0.07})$)

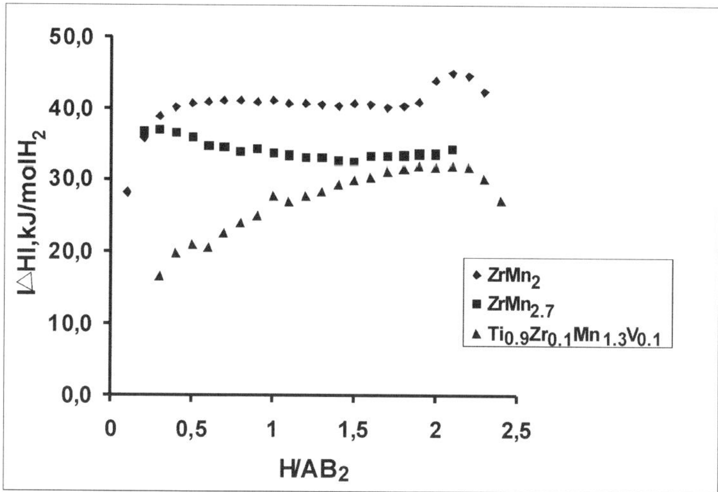

Figure 7. Partial molar desorption enthalpy calculated from van't Hoff plot for the $ZrMn_2$-, $ZrMn_{2.7}$- and $Ti_{0.9}Zr_{0.1}Mn_{1.3}V_{0.5}$-H systems

plateau region ($0.5 < C < 1.8$).The values of |ΔHplat.| slightly decrease with increase of C along the plateau that is in a good agreement with the reference data [5, 6] and our calorimetric data. The $ZrMn_2$-H system shows a maximum after $(\alpha + \beta)/\beta$ phase boundary. In the plot of the $|\Delta S_H|$-C dependence (see Fig. 8) one can see that in the dilute solution region the values of $|\Delta S_H|$ similarly increase until to reach the plateau region at C = 0.5 and then |ΔSplat.| slightly increases with the increase C along plateau (from ~96.0 J/Kmol H_2 to ~105.0 J/Kmol H_2).

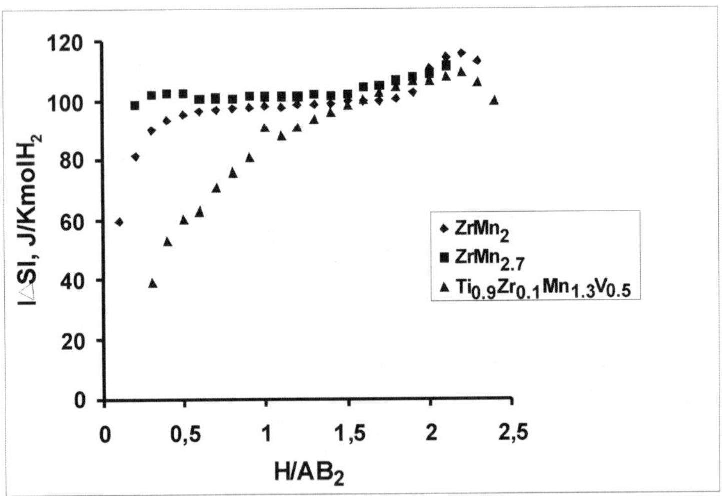

Figure 8. Partial molar desorption entropy calculated from van't Hoff plot. for the $ZrMn_2$-, $ZrMn_{2.7}$- and $Ti_{0.9}Zr_{0.1}Mn_{1.3}V_{0.5}$-H systems

For the $ZrMn_{2.7}$-H system the $|\Delta H_H|$-C dependence acquires the more complex character. It should be noted that $ZrMn_{2.7}$ is hyperstoichiometric intermetallic compound with the crystal structure C 14 Laves phase or normalazed to AB_2 $Zr_{0.82}Mn_{0.18}Mn_2$. It was found [8, 9] that excess manganese atoms substitute at the zirconium sites in the crystal structure C 14 that leads to a decrease of the cell parameters and the reduction of amount of [Zr2Mn2] tetrahedral sites; besides, a lowering of the zirconium concentration results in a decrease in the chemical activity of the available sites in the structure. During the hydrogenation of $ZrMn_{2.75}$ it was established [3] that hydrogen atoms occupied interstitial sites $6(h)_1$ (0.286), $12(k)_1$ (0.247), 24(l) (0.146) and $6(h)_2$ (0.045), that is a difference in the occupation of the tetrahedral holes $6(h)_1$ and $12(k)_1$ for $ZrMn_{2.7}$ is bigger than for $ZrMn_2$.

In the dilute phase of the $ZrMn_{2.7}$–H system some minor increase in the $|\Delta H_H|$ values is observed (at $C < 0.3$) and then the values of $|\Delta H_H|$ decrease with increasing the hydrogen concentration in the IMC. The change of the enthalpy values becomes discontinues at $C \sim 0.8$ and 1.5, where the small jumps in the dependence are observed. As regards to the changes of the entropy values taking place in this system they do not change such drastically. In the plot of $|\Delta S_H|$ as a function of hydrogen concentration there is some minor maximum in the solid solution region and then it can see two regions with practically constant entropy values, namely, $0.6 < C < 0.8$ (~100 J/Kmol H_2) and $0.9 < C < 1.5$ (~102 J/Kmol H_2) and then ($C > 1.5$) the $|\Delta S_H|$ values increase. It could be assumed that such changes of the thermodynamic functions are explained that in the $ZrMn_{2.7}$–H system two hydride phases form and the increase of hydrogen concentration in this system leads to formation of less stable hydride. The similar data were obtained from the calorimetric measurements (see Table 2).

In the Fig. 7 the plot of the $|\Delta H_H|$-C dependence is presented for the $Ti_{0.9}Zr_{0.1}Mn_{1.3}V_{0.5}$-H system. As one can see from this figure in the α-solution region ($0 < C < 0.5$) the values of $|\Delta H_H|$ increase smoothly, then there are two sharp changes in the plot of the $|\Delta H_H|$-C dependence at the hydrogen concentrations C = 0.6 and 1.0. The similar sharp changes in the $|\Delta S_H|$-C dependence are observed at the same hydrogen concentrations in $Ti_{0.9}Zr_{0.1}Mn_{1.3}V_{0.5}$ (see Fig. 8). Such behaviour, as in the case of the $ZrMn_{2.7}$–H system, could be associated with the beginning of an infill by hydrogen of a new type of the interstitial sites in the crystal structure of $Ti_{0.9}Zr_{0.1}Mn_{1.3}V_{0.5}$, which takes place at the hydrogen concentrations approximately 0.6 and 1.0 per formula unit of AB_2. Then there is the broad gentle maximum of the $|\Delta H_H|$-C and $|\Delta S_H|$-C dependences in the hydrogen concentration range $1.1 < C < 2.2$ (see Fig. 6). The increase of $|\Delta H_H|$ values observed over the hydrogen concentration interval is accounted for that the more stable hydrides form. The obtained result agrees with calorimetric data very well (see the Table 2). The complex character of the $|\Delta H_H|$-C and $|\Delta S_H|$-C dependences it could be explained by the complex chemical composition and the nonstoichiometric structure of this IMC. Earlier the $(Ti_{0.89}Zr_{0.11})(Mn_{1.47}V_{0.31}Ti_{0.22})D_{2.8}$–structure compositionally allied to IMC studied in this work was investigated [10] and it has been found, that in $(Ti_{0.89}Zr_{0.11})(Mn_{1.47}V_{0.31}Ti_{0.22})D_{2.8}$ deuterium atoms occupy three sites 24(l), $12(k)_1$ and $6(h)_1$ with $[A_2B_2]$. But since V and Ti, which have strong affinity to hydrogen, as well as Mn occupy some B sites there is a possibility to regard an interstitial site $[A_2B_2]$ as formed by active (Ti, Zr, V), which conditionally signed as A*, and inactive (Mn) to hydrogen components. On the base carried out calculations the authors [10] drew the conclusion that 24(l) transforms to $[A^*_4]$ and $[A^*_3B]$ in higher extent than $12(k)_1$ and $6(h)_1$ and therefore 24(l) positions are occupied by hydrogen at hydride formation in the first order. The authors have concluded that $(Ti_{0.89}Zr_{0.11})(Mn_{1.47}V_{0.31}Ti_{0.22})D_{2.8}$ with consideration of distribution deuterium atoms among interstitial sites may be written as $(Ti_{0.89}Zr_{0.11})(Mn_{1.47}V_{0.31}Ti_{0.22})D^1_{1.86}D^2_{0.69}D^3_{0.25}$, D^1, D^2, D^3 – deuterium atoms, arranged in 24(l), $12(k)_1$ and $6(h)_1$ respectively. It could be supposed that in our case that the similar complex structure exists for $Ti_{0.9}Zr_{0.1}Mn_{1.3}V_{0.5}$.

4. Conclusions

The $ZrMn_2$-H, $ZrMn_{2.7}$–H and $Ti_{0.9}Zr_{0.1}Mn_{1.3}V_{0.5}$-H systems were studied by means of the calorimetric and P-C isotherms methods. The |ΔHdes.|-C and |ΔSdes.|-C dependences were obtained for these alloys as well as the dependences of the enthalpy change |ΔHdes.|=f(C) at different temperatures.

The obtained data permit to draw a conclusion that the thermodynamic functions calculated from the P-C-T measurements and the calorimetric method for the studied AB_2-H systems are in the good agreement. For the $ZrMn_2$-H system for which the existence of one plateau in the plot of |ΔHdes.|-C was determined by means of both methods the values of the enthalpy changes with hydrogen concentration coincide. However when one deals with the more complex systems for which the formation of two hydride phases are observed within the definite temperature interval, the P-C-T measurement gives us the lowered values of |ΔHdes.|.

References

1. Shaltiel D., Jacob I., Davidov D. Hydrogen absorption and desorption properties of AB_2 Laves-phase pseudobinary compounds, *J. Less-Common Met.*, 1977. **53**: 117–131.
2. Fujitani S., Yonezu I., Saito T., Furukawa N., Akiba E. Relation between equilibrium hydrogen pressure and lattice parameters in pseudobinary Zr-Mn alloy systems, *J. Less-Common Met.*, 1991. **172–174**: 220–230.

3. Pontonnier L., Miraglia S., Fruchart D. Structural study of hyperstoichiometric alloys $ZrMn_{2+x}$ and their hydrides, *J. Alloy. Compd.*, 1992. **186**: 241–248.
4. Anikina E.Yu., Verbetsky V.N. Calorimetric investigation of the hydrogen interaction with $Ti_{0.9}Zr_{0.1}Mn_{1.1}V_{0.1}$, *J. Alloy. Compd.*, 2002. **330–332**: 45–47.
5. Luo W., Clewley J.D., Flanagan T.B., Oates W.A. Thermodynamic characterization of the Zr-Mn-H system: Part 1. Reaction of H_2 with single-phase $ZrMn_{2+x}$ C14 Laves phase alloys, *J. Alloy. Compd.*, 1992. **185**: 321–338.
6. Krupenchenko A.V., Magomedbekov E.P. Calorimetry of hysteresis transition in the IMC systems $AB_2 – H_2$ type, *Zh. Neorganicheskoi khimii*, 1994. **39**(1): 156–162 (in Russian).
7. Didisheim J.-J., Fischer P. Structural and magnetic phase transitions in $ZrMn_2D_3$, *J. Less-Common Met.*, 1984. **103**: 267–275.
8. Pourarian F., Sinha V.K., Wallace W.E., Kevin H. Kinetics and thermodynamics of $ZrMn_2$-based hydrides, *J. Less-Common Met.*, 1982. **88**: 451–458.
9. Sinha V.K., Yu G.Y., Wallace W.E. Hydrogen storage in some ternary and quaternary zirconium-based alloys with the C 14 structure, *J. Less-Common Met.*, 1985. **106**: 6–11.
10. Mitrokhin S.V., Smirnova T.N., Somenkov V.A., Glazkov V.P., Verbetsky V.N. Structure of (Ti,Zr) –Mn-V nonstoichiometric Laves phases and $(Ti_{0.9}Zr_{0.1}\ )(Mn0_{.75}V_{0.15}\ Ti_{0.11}\ )D_{2.8}$ deuteride, *J. Alloy. Compd.*, 2003. **356–357**: 80–83.

HYDROGEN SORPTION PROPERTIES OF METAL HYDRIDE AND GLASS PHASE

L. GRINBERGA, J. KLEPERIS
Institute of Solid State Physics of University of Latvia,
8 Kengaraga Str., Riga, LV-1063, Latvia
Fax: +371 67132778 ; E-mail: Liga.Grinberga@cfi.lu.lv

Abstract. This work is devoted to the development of a new composite material for hydrogen storage and to investigation of physical and chemical characteristics of material. The new composite material is formed by combination of AB_5 type lanthanum nickel alloy with a glass based on silica. Thermogravimetric and XRD measurements prove that the new material absorbs more hydrogen than both constituents separately. That is explained by hydrogen sorption in the composite material due to the hydrogen spillover phenomena, where the hydrogen molecules dissociated on the surface of catalyst in atoms and spills over to the inert surface. That causes the increase of catalytic activity of an alloy and the presence of hydrogen atoms at the interface metal-glass what enlarges an amount of adsorbed/absorbed hydrogen.

Keywords: hydrogen absorption, metal hydride alloys, composite, hydrogen spillover

1. Introduction

$LaNi_5$ has been one of the most investigated intermetallic compounds during the last decades. Despite of its satisfactory hydrogen capacity and easy activation, the binary compound is not suitable for applications due to its high plateau pressure and short lifecycle [1]. Modification of the physical and chemical properties of $LaNi_5$ is achieved by substituting lanthanum atom with a rare earth metal (e.g. Ce, Pr, Nd) or nickel with a transition metal (e.g. Al, Mn, Co, Cr). Specific sample preparation methods, such as melt–spinning, mechanical milling and others have been used to improve the hydrogenation kinetics of intermetallic compounds. The resulting alloys exhibit particular structural characteristics as nano-crystalline grain size with a high density of grain boundaries and lack a long-range order that provide fast hydrogenation kinetics and better lifecycle behaviour. The mechanical milling has been used for several hydrogen storage alloys; a good improvement in hydrogen activation and kinetics was observed [2, 3].

The diffusion of active species on the surface may play an important role in reactions on multifunctional catalyst transport phenomena. Especially, migration of hydrogen atoms from a metal, to an oxide or carbon surface that by itself has no activity for dissociate hydrogen adsorption, is important. It is in fact well known, that noble metals like Pt and Pd can adsorb and diffuse hydrogen in reactive forms over relatively large distances. This property named as spill-over effect, is widely exploited in catalysis [4].

Figure 1 represents a simplified hydrogen spill-over scheme. Oxide isn't absorbing hydrogen molecules, although if the molecule is adsorbed and dissociated on the catalyst, the formed hydrogen atoms are interacting with oxide phase.

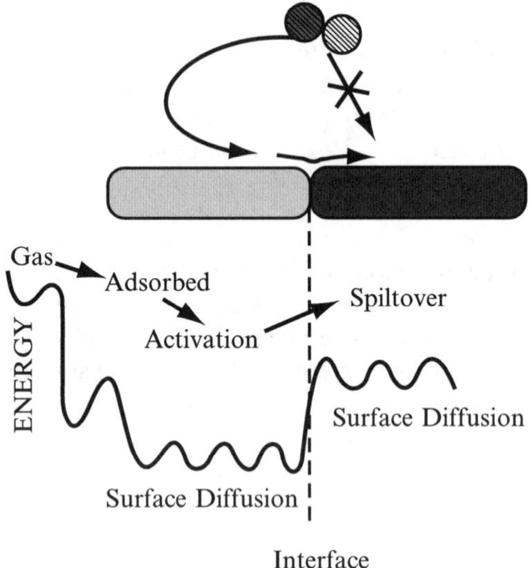

Figure 1. Schematic representation of spill-over process of a gaseous diatomic hydrogen from catalyst, and energy level diagram for those processes. Dark oval – non-adsorbing surface, light oval – catalyst, small circles – hydrogen molecule [5]

The spill-over of hydrogen involves a transfer of electrons to acceptors within the support; this process modifies the chemical nature of the support and can also activate a previously inactive material and/or induce subsequent hydrogen physisorption. Dissociation of hydrogen molecule on a metal and subsequent spill-over of atomic hydrogen to its support is highly dependent upon the chemical bridges formed at the interface. Hydrogen spill-over can be assessed in a number of ways, but perhaps the most common is simple calculation of the hydrogen to metal ratio, either the surface metal or total metal content. When spill-over occurs, the relation Hydrogen/Metal surface will typically exceed unity. In the case of materials that form hydrides, this relation will exceed the stoichiometric ratio of the hydride.

2. Experimental

The lanthanum rich mischmetal (LaMm) produced by the company Metal Rare Earth Limited of China that corresponds to formula $A_{0,96}B_{5,04}$ (where A=La, Ce, Nd, Pr and B=Ni, Co, Mn, Al, Cr) was used as catalyst and bulk material for hydrogen storage the composition of the sample.

Structural properties of the samples were studied by X-ray Diffractometer System X-STOE Theta/theta, using Kα Cu radiation $\lambda = 1.5406$ Å.

A thermogravimetric technique has been used to study the hydrogen sorption on prepared samples. The measurements were performed by equipment based on the Sartorius high pressure balance (HPB). The studied sample about 340 mg was initially degassed under vacuum down to 10^{-3} bars at the ambient temperature and flushed with helium gas. Afterwards, stepwise changes of the pressure inside the measuring device were applied. For each pressure at constant temperature the increase of the sample weight, was measured as a function of time due to hydrogen sorption.

3. Results and discussions

Analyzing treated data it was figured out that raw hydride material LaMm absorbing more hydrogen than classical $LaNi_5$ by itself. Comparison of both materials with the glass phase confirmed the same tendency – alloy LaMm and glass absorbed more hydrogen than $LaNi_5$ with glass (Fig. 2). The Fig. 2 also shows that pure $LaNi_5$ is absorbing and desorbing hydrogen when the pressure is changed. In contrast, the LaMm alloy is able to retain most of the hydrogen that were absorbed at high pressure. Only a slight decrease of the amount of absorbed hydrogen is observed when the pressure returns to one atmosphere.

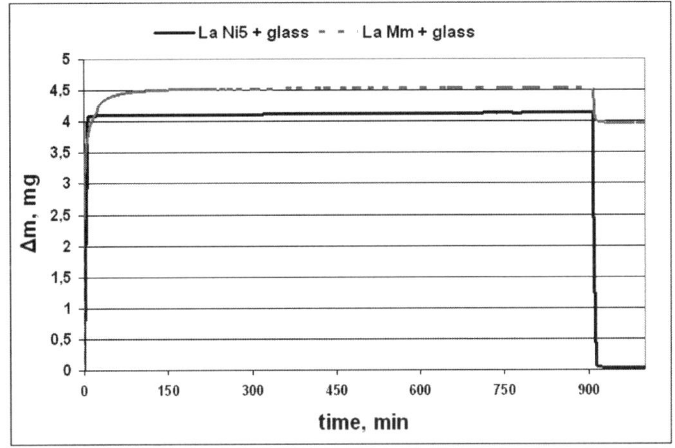

Figure 2. Change of weight of composite material at 23°C and 1 atm

That effect could be explained by the substitution of Ni with corresponding B-components: Al, Mn and Co. Also, a decrease of La constituent in the alloy could have the same effect. A substitution with Co also results in a higher cyclic stability of the materials. This substitution provides a higher resistance to decrepitation. In addition, substituted Mn facilitates the retention of high hydrogen sorption capacity [6].

The calculations of amount of absorbed hydrogen x and weight% for maximal change of weight (Δmtotal), for average change of weight after cycling (Δmav) and for change of weight at 1 atm (Δm 1 atm) of hydrogen in the raw material and composite, shows that both numbers are higher for composite than for pure alloy sample (Table 1). That could be explained with hydrogen spill-over and absorption in alloy-glass mixture.

TABLE 1. Calculations of HPB data for raw alloy (LaMm) and composite (LaMm + glass)

Δm	Δm (mg)	wt%	X
		LaMm	
Δmtotal	5,507E-03	1,582	6,966
Δmav	5,138E-03	1,477	6,499
Δm(1 atm)	4,883E-03	1,405	6,177
		LaMm + glass	
Δmtotal	5,048E-03	1,643	7,242
Δmav	4,616E-03	1,504	6,621
Δm(1 atm)	4,393E-03	1,433	6,302

The XRD results showed that the diffraction peaks of hydrogenated alloy LaMm were largely shifted to the smaller angles than that of the starting alloy, indicating that the α-phase of hydride is changed into the β-phase and the lattice parameters and cell volume of the hydride is larger than that of the starting alloy. For the composite the observed shift of XRD peaks after hydrogenation was even larger than that for pure LaMm alloy. Also corresponding lattice parameters and cell volume for hexagonal P6/mmm symmetry was larger of hydrogenated composite sample as for fully hydrogenated alloy LaMm (Fig. 3, Table 2) [7].

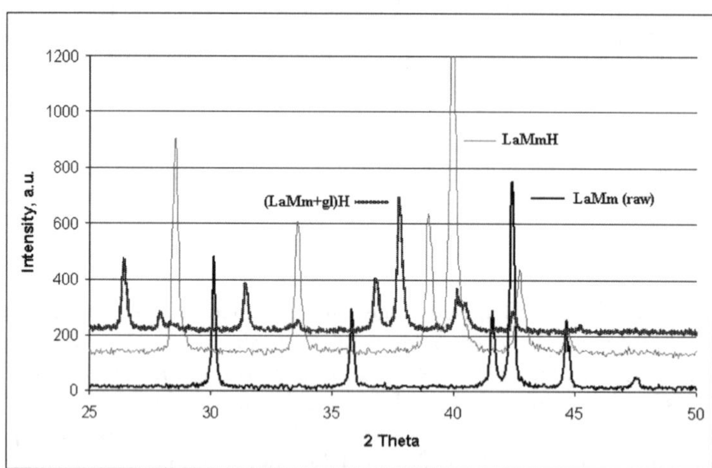

Figure 3. XRD plot of raw (LaMmraw), hydrogenated alloy LaMm (LaMmH) and hydrogenated alloy LaMm with glass phase (LaMm + glH), for better recognition intensity is increased by 100 units for LaMmH and by 200 units for (LaMm + gl)H

TABLE 2. Structural parameters of raw and hydrogenated samples

Sample	Cell size	a (Å)	c (Å)	V (Å3)
LaMm		5.0083	4.0567	88.12
LaMmhydrogenated		5.326	4.234	104.0
LaMm + glass hydrogenated		5.369	4.2754	106.78

4. Conclusions

Results of measurements with high pressure balance and calculations has showed that hydrogen uptake in the composite of an alloy LaMm with glass exceeds that of a pure alloy. It is explained with spill-over of hydrogen atoms from the AB_5 catalyst to glass phase. The following mechanism can be deduced: the hydrogen chemisorbs at the surface sites found on the alloy LaMm and dissociate into atoms. Bridges between the catalyst and glass particles allow to hydrogen atoms migrate onto the glass surface. Desorption occurs directly from the relatively lower energy glass sites without migration back to the catalyst. Hydrogen spill-over depends upon the glass-catalyst contact. The contact changes with the quality of the mixing and milling, as well as the position of alloy LaMm grains in the mixture.

It was observed from the X-ray diffraction patterns, that in the hydrogenation process the beta phase of the ball milled composite sample LaMm + glass occurred faster than in the pure alloy sample. Additionally, peaks shift to smaller angles are noticeably larger. It is possible to conclude, that the gamma hydride phase (γ) is formed only, when the alloy LaMm is mixed in composite with Pyrex glass. In the hydride γ phase the hydrogen atoms prefer the interstitial sites closer to Ni atoms in an elementary cell, producing inhomogeneous distribution and large lattice distortions [8].

Acknowledgements

Author (L.G.) acknowledge the European Union European Social Fund for support and L'Oreal for fellowship 'For Women in Science' in cooperation with UNESCO Latvia and Academy of Science of Latvia. Authors (L.G. and J.K.) acknowledge the Nordic Energy Research Project NORSTORE, European Regional Development Fund project Nr 066/033 and The Latvian State Research Program in Material Science VPP-05.

References

1. Bittner H.F., Badcock C.C. Electrochemical utilization of metal hydrides. *J. Electrochem. Soc.*, 1983, **130**: 193–198.
2. Ares J.R., Cuevas F., Percheron-Guégan A. Influence of thermal annealing on the hydrogenation properties of mechanically milled AB5-type alloys *J. Mat. Sci. Eng., B* 2004, **108**: 76–80.
3. Zaluski L., Zaluska A., Strom-Olsen J.O. Nanocrystalline metal hydrides *J. Alloy. Compd.*, 1997, **253–254**: 70–79.
4. Scarano D., Bordiga S., Lamberti C., Ricchiardi G., Bertarione S., Spoto G. Hydrogen adsorption and spill-over effects on H–Y and Pd-containing Y zeolites. An experimental and theoretical investigation. *J. Appl. Catal., A*, 2006, **307**: 3–12.

5. Conner Jr. W.C., Falconer J.L. Spillover in heterogeneous catalysis. *Chem. Rev.*, 1995, **95**: 759–778.
6. Zaluska A., Zaluski L., Ström-Olsen J.O. Structure, catalysis and atomic reactions on the nano-scale: a systematic approach to metal hydrides for hydrogen storage. *Appl. Phys. A-Mater.*, 2001, **72**(2): 157–165.
7. Kleperis J., Grinberga L., Vaivars G., Klavins J. Hydrogen absorption and adsorbtion in composite materials. *Latv. J. of Physics & Techn. Sciences*, 2006, **4**: 39–46. (in Latvian).
8. Nakamura Y., Nomiyama T., Akiba E. Phase transformation in La(CoxNi5-x)-H systems (x = 2, 3, 5) studied by in situ X-ray diffraction. *J. Alloy. Compd.*, 2006, **413**(1–2): 54–62.

FORMATION OF LINEAR STRUCTURES IN METAL-HYDROGEN INTERSTITIAL ALLOYS

N.Z. NAMORADZE,
Institute of Cybernetics, Georgian National Academy of Sciences, 5 Sandro Euli Str., Tbilisi, 0186, Georgia

I.G. RATISHVILI
E. Andronikashvili Institute of Physics, Georgian National Academy of Sciences,6 Tamarashvili Str., Tbilisi, 0177, Georgia

Abstract. Low-concentration α-phases of rare-earth metal hydrides are considered. A system of simple mathematical models aimed to describe equilibrium states of the subsystem of interstitial hydrogen atoms is presented. Description is based on the assumption that in the low-concentration interstitial metal-hydrogen systems atomic Me_aH_b complexes are formed and the interstitial alloy MeH_x can be considered as a dilute solution of these complexes in the metallic matrix. It is shown that in the suitable conditions the distribution of these complexes in lattices becomes spatially correlated that provides formation of some ordered structures. Consideration is applied particularly to the low-concentration rare-earth hydrides.

Keywords: low-concentration metal hydrides, metal-hydrogen atomic complexes, hydrogen ordering

1. Introduction

The aim of the present investigation can be formulated as an attempt to describe the hydrogen subsystem states in the low-concentration metal-hydrogen interstitial solid solutions. As an example we consider the low-concentration α-phases of rare-earth metal hydrides MH_x where N metal atoms (M-atoms) form a *hcp* lattice and hydrogen atoms (H-atoms) are distributed on the set of 2N tetrahedral interstitial positions (T-positions).

Rare-earth metals are known to form MH_x compounds with x in the range $0 \leq x \leq 3$. The concentration scale 0–3 contains three monophase regions associated with α, β and γ phases, and two biphase regions – $(\alpha + \beta)$ and $(\beta + \gamma)$. α-phase exists in the concentration range $0 \leq x \leq x_{max}(\alpha)$, β-phase – in the range $2 \leq x \leq x_{max}(\beta)$, and γ-phase – in the range $x_{min}(\gamma) \leq x \leq 3$. The biphase regions $(\alpha + \beta)$ and $(\beta + \gamma)$ exist in the ranges $x_{max}(\alpha) \leq x \leq 2$ and $x_{max}(\beta) \leq x \leq x_{min}(\gamma)$, respectively [1]. We shall consider below MH_x compounds with $x \leq x_{max}(\alpha)$.

Crystalline structure of compounds under consideration. Metals of the rare-earth row are known to form several types of crystalline structures and to undergo a number of polymorphic transformations [2]. In the high-temperature limit all members of the rare-earth row are characterized by cubic structures, but at room temperatures the high-temperature coordination of atoms remains equilibrium only for Eu and Yb, while the other metals transform into the hexagonal configurations differing in the sequences of basal planes. Particularly, the so-called "heavy rare-earth" metals (Gd, Tb, Dy, Ho, Er, Tm, Lu) have simple *hcp* lattices with the sequence of basal planes AB-AB-AB, while in the "light rare-earths" (La, Ce, Pr, Nd) the sequence of basal planes is ABAC-ABAC-ABAC (i.e. they have hexagonal lattices with unit cells doubled along the *c*-axis). Location of tetrahedral interstitial positions in both lattices is shown in Fig. 1(a, b). For comparison, in Fig. 1(c) are shown tetrahedral positions in the *fcc* lattices (presented as a sequence ABC-ABC-ABC of "hexagonal planes" perpendicular to [111] direction).

Low-temperature α-phases exist, particularly, in R-H compounds where R denotes the atoms of heavy rare-earth metals mentioned above, plus Sc-H and Y-H systems. It must be noted that light rare-earth metals do not form solid solution α-phases at room temperatures, and these systems will not be considered below. Thus, our consideration will include only the hydrides of heavy rare-earth metals.

In the lattices shown in Fig. 1 the metal atoms are surrounded with eight equally distanced T-positions, while each T-position is surrounded with four metal atoms. Thus, we have the totality of N metal atoms and totality of 2N T-positions. Both totalities can be represented as a system of MT_2-complexes, having e.g. configuration shown in Fig. 1(c); i.e. it can be assumed that each lattice site is related with two T-positions located in points (0 0 ±3/8) with respect to this site.

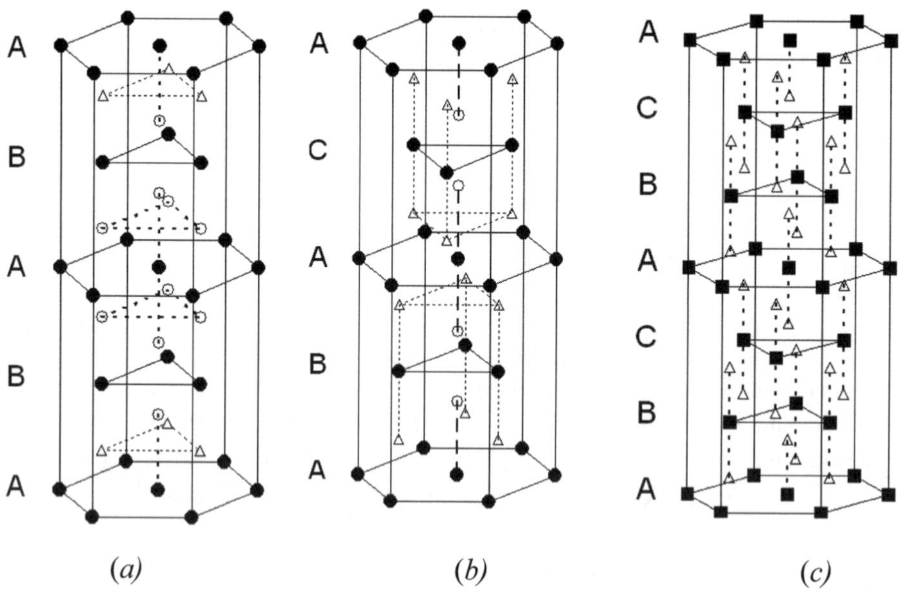

Figure 1. Location of tetrahedral interstitial positions in the simple hexagonal (gadolinium-type) lattice (*a*), in the doubled hexagonal (lanthanum-type) lattice (*b*), and in the *fcc*-lattice (*c*)

A number of experimental data. In the metal hydrides under consideration the concentration limits x_{max} (α) are sufficiently different for different compounds. At T = 500°C there are registered x_{max} (α) = 0.24 for Gd-H, and x_{max} (α) = 0.36 for Lu–H [1].

In a number of cases, at low temperatures the α-phase transforms into a partly ordered α∗-phase [1]. Particularly, α∗-phase is known to be formed in the hydrides of Er, Tm, Lu, Sc and Y. In Ho–H system are registered the traces of this phase [1].

It was suggested that in the α∗-phase there are created H-H pairs and that these pairs form a system of linear structures, chains along the *hcp* *c*-axis, consisting of linear MH_2-complexes oriented along the given *c*-axis [1].

It was established that the concentration limit of α-phase is temperature dependent and rises with increasing temperature – x_{max} (α) = x_{max} (α., T), while the upper limit of α∗-phase, which is lower than that of α-phase, in contrast, is temperature independent [1].

Both upper limits, x_{max} (α) and x_{max} (α∗), depend on the nature of metals forming the host lattice [1].

Some previous theoretical considerations. The mentioned experimental data concerning the state and structure of hydrogen subsystem in α and α∗ phases had been successfully considered basing on some analytical models [4–7]. Most of them discuss the conditions for formation of H-H pairs within the α∗-phase and established existence of linear structures formed by these pairs. We shall not analyze their conceptions and concentrate our attention on some problems which had been left open.

The problems under consideration. The assumed hydrogen subsystem structure in α*-phase based on the idea of existence of H-H pairs organized in linear structures implied, particularly, the existence of MH_2-complexes. But if we assume that in the low-concentration metal hydrides having a *hcp* lattice all solute H-atoms form MH_2 complexes, then a number of questions arise.

First of all, concerning the isolated MH_2 complexes we can ask: what are the temperature limits of stability of M-H bonds? What are the preferential configurations of these complexes in different conditions? What are the local distortion fields around an MH_2 complex in different metals?

At the next stage arise questions concerning MH_2-MH_2 interactions in different lattices. Particularly: What is the nature of MH_2-MH_2 interactions? What types of effects are caused by this interaction? In what conditions there are formed the superstructures of MH_2-complexes?

The aim of present investigation. In an ideal case all the mentioned questions have to be answered within the frames of a unique theoretical scheme based on the knowledge of electronic structure properties of the given interstitial alloys. This is difficult to do and, as a first step of recognizing the interrelations between the different sufficiently independent subsystems of M-H compounds, we prefer to use a set of phenomenological models, allowing us to establish qualitative relations between some physical characteristics of the M-H system in different conditions.

Particularly, the equilibrium configurations of isolated MH_2-complexes will be considered within the frames of the *configuration model*, while the role of MH_2-MH_2

interactions can be estimated basing on two models – *orientation model* and *cooperative model*, where the first one is devoted to describe influence of intercomplex interactions on the equilibrium orientations of MH_2 complexes, and the second – for estimation of possibilities of spatially ordered arrangement of MH_2-complexes in the given host lattices. The orientation model was described already in the article [7] and applied to Nb-H ordering system. We shall not reconsider it here.

In the present report we shall describe the general mathematical schemes of configuration and cooperative models and apply them mainly to the α-phases of rare-earth hydrides.

Basic assumptions. Throughout our consideration there will be postulated the fulfillment of the following restrictions.

1. All hydrogen atoms inserted in the metal lattice are located in tetrahedral interstitial positions and form MH_2-complexes.
2. MH_2-complexes can have a finite number of different spatial configurations. They can "rotate" around the central metal atom. Different configurations are energetically distinguishable.
3. Formation of superstructures of MH_2 complexes implies their spatially ordered distribution in the host lattice and their correlated orientations.
4. Interactions between the MH_2 complexes can be describe within the frames of the mean-field approximation.

2. Configuration model

2.1. POSSIBLE CONFIGURATIONS OF MH_2 COMPLEXES IN HCP LATTICES

Coordinate system. To describe the set of *hcp* lattice sites we shall use the orthogonal coordinate system where z-axis is oriented along the *c*-direction of the *hcp* unite cell and x and y axes are in the basal plane. If we choose x-axis to be oriented along the crystallographic direction along which the metal atoms are located regularly with an interval $u_0 = a$, then the orthogonal y-axis will be oriented along the direction where the nearest neighbors will be located at the distances $v_0 = a\sqrt{3}$. If along the z-direction we consider a simple [A-B-A-B] sequence of atomic planes parallel to the basal plane and assume an ideal *hcp* lattice, implying that shortest distances between a pair of atoms in the neighboring planes equals *a*, then the distance between the neighbor atoms along the z-direction will be $w_0 = a\sqrt{(8/3)} = 1.6333\ u_0$. If we consider the *hcp* lattice as consisting from *structural units* shown in Fig. 1 (i.e. consisting from doubled unite cells), then in the system with parameters $[u_0, v_0, w_0]$ the metal atoms forming this structural unite will be described by coordinates:

$$(0\ 0\ 0),\ (\pm1\ 0\ 0),\ (\pm1/2\ \pm1/2\ 0),\ (\pm1/2\ -1/6\ \pm1/2),$$

$$(0\ 1/3\ \pm1/2),\ (0\ 0\ \pm1),\ (\pm1\ 0\ \pm1),\ (\pm1/2\ \pm1/2\ \pm1). \qquad (2.1)$$

Representing the total *hcp* metal lattice as consisting of such structural unites we have to attribute to each of them 12 metal atoms.

Coordinates of tetrahedral positions. In the selected coordinate system with $[u_0, v_0, w_0]$ unites the set of eight nearest tetrahedral positions surrounding each metal atom will be described by coordinates (with respect to the given M-atom):

$$(0\ 0\ \pm 3/8),\ (\pm 1/2\ -1/6\ \pm 1/8),\ (0\ 1/3\ \pm 1/8). \tag{2.2}$$

Four metal atoms located around each tetrahedral position are separated from it by distances $[(3/8)\,w_0] = a\sqrt{(3/8)} = 0.6124\,a$.

We shall enumerate the H-atoms in different tetrahedral positions in the following way:

$$H^T_1 - (0\ 0\ 3/8),\quad H^T_2 - (0\ 0\ -3/8),$$
$$H^T_3 - (-1/2\ -1/6\ 1/8),\quad H^T_4 - (1/2\ -1/6\ 1/8),\quad H^T_5 - (0\ 1/3\ 1/8); \tag{2.3}$$
$$H^T_6 - (-1/2\ -1/6\ -1/8),\quad H^T_7 - (1/2\ -1/6\ -1/8),\quad H^T_8 - (0\ 1/3\ -1/8).$$

Possible configurations of MH_2-complexes. Taking into account the mentioned above coordinates of the possible positions of hydrogen atoms around a given metal atom located in a site of a *hcp* lattice, we can determine a number of different configurations of [H-M-H]-complexes. They are collected in Table 1. The corresponding figures describing different configurations are presented in Appendix I.

TABLE 1. Distinguishable configurations of MH_2 complexes

Configuration type	Tetrahedral positions	H-H distance d_{HH} (in u_0 unites)	Internal angle (in degrees)	Configuration degeneracy, $p(i)$
I	H^T_1-H^T_2	$\sqrt{(3/2)} = 1.22475$	180°	1
II	H^T_1-{H^T_6, H^T_7, H^T_8}	1	109°30'	6
III	H^T_1-{H^T_3, H^T_4, H^T_5}	$\sqrt{(1/2)} = 0.70711$	70°30'	6
IV	H^T_5-{H^T_6, H^T_7}	$\sqrt{(7/6)} = 1.08012$	123°40'	6
V	H^T_5-{H^T_3, H^T_4}	1	109°30'	6
VI	H^T_5-H^T_8	$\sqrt{(1/6)} = 0.40825$	39°	3

For comparison, we indicate that in the water molecule the H–H distance equals $d_{HH}(H_2O) = 1.515$ Å and the internal [H-O-H]-angle equals $\vartheta(H_2O) = 104°27'$.

Different configurations of MH_2-complexes deform in a different degree the neighbouring part of the host lattice.

In order to relate the hydrogen induced deformations with the set of configurations of MH_2-complexes, we have to consider projections of the corresponding H-H bond-vectors **L** on the z-direction, L_z, and on the basal plane, $L_\perp$. Different configurations have generally different values of the **L**-vector lengths and of the angles θ between the **L**-vector and z-axis. Particularly, following Table 1, for configuration I, H^T_1-H^T_2, we have $[\theta = 0,\ L = L_z = a\sqrt{(3/2)},\ L_\perp = 0]$, while for configuration V, H^T_3-H^T_5, we have $[\theta = \pi/2,\ L = L_\perp = a,\ L_z = 0]$. It is obvious that MH_2-complexes in these two configurations will deform the lattice in a different way. In the general case it can be assumed that lattice deformations are proportional to the averaged values of corresponding projections of **L**-vectors.

In Table 2 there are listed the **L**-vector lengths – L, and projections of **L**-vectors on the z-axis – L_Z, and on the basal plane – $L_\perp$. The values of L, L_Z, and $L_\perp$ are given in unites of the lattice parameter *a*.

In the simplest case all configurations of MH_2-complexes can be assumed to be energetically similar and equally probable. Then the mean values of L_Z, and $L_\perp$, denoted respectively, as $<L_Z>$ and $<L_\perp>$, can be determined as results of averaging over the total set of configurations:

$$< L_k > = \Sigma_j L_{ki}(j) p(j) / \Sigma_j p(j) ; \quad k = z, \perp ; \quad j = I - VI \tag{2.4}$$

TABLE 2. H-H bond lengths and bond components

Configuration type	L [a]	L_z [a]	$L_\perp$ [a]	Degeneracy $p(j)$
I	√(3/2)	√(3/2)	0	1
II	1	√(2/3)	1/√3	6
III	1/√2	1/√6	1/√3	6
IV	√(7/6)	1/√6	1	6
V	1	0	1	6
VI	1/√6	1/√6	0	3

In the considered case $\Sigma_\varphi p(j) = 28$, that provides following relations

$$<L_\perp> = [1 + (1/\sqrt{3})] (12/28) = 0.67600726\, a, \tag{2.5a}$$

$$<L_Z> = [\sqrt{(3/2)} + 6*\sqrt{(2/3)} + 15*(1/\sqrt{6})] * (1/28) = 0.43740888\, a, \tag{2.5b}$$

$$<L> = [\sqrt{(3/2)} + 12 + 6*\sqrt{(1/2)} + 6*\sqrt{(7/6)} + 3*(1/\sqrt{6})] * (1/28) = 0.8990311\, a \tag{2.5c}$$

These are the averaged characteristics of a MH_2-complexe in the high-temperature limit, when its interactions with the surrounding lattice and the H-H interactions inside the complexes both are neglected and different configurations are energetically indistinguishable. In these conditions from relations (2.5) it follows that the ratio of components $[<L_\perp>/<L_Z>] > 1$, implying that deformations of the lattice caused by MH_2-complexes have to be greater in the basal plane than along the z-axis. Such situation had been registered, particularly, in α-ScH_x hydrides which will be briefly discussed below.

In a number of lanthanides the host lattice deformations in the α-phase were found to be more significant along the z-axes than in the basal plane (see Appendix II). On supposing that deformations of the host lattices caused by interstitial hydrogen atoms are proportional to the projections of the **L**-vectors on crystallographic axes, we have to conclude that in these rare-earth hydrides configurations of MH_2-complexes with large L_Z-projections are more favourable energetically than those with large $L_\perp$-projections.

To illuminate this problem we have to reconsider the experimental data concerning the elasticity of the rare-earth metals and their α-hydrides.

2.2. CONFIGURATION-RELATED ENERGY OF MH_2-COMPLEXES IN METAL LATTICES

Deformability of the rare-earth metal lattices. To characterize quantitatively deformations of the host lattices induced by interstitial atoms there are used usually the following parameters:

$$[(\Delta a / a) / \Delta x \text{ at\%}] \equiv \delta_a (x) \text{ and } [(\Delta c / c) / \Delta x \text{ at\%}] \equiv \delta_c (x). \qquad (2.6)$$

The first one describes relative changes of the lattice parameters in the *basal plane* caused by the 1% changes of the interstitial atom concentration, while the second one – the relative changes of the lattice parameter *along direction perpendicular to the basal plane*.

To characterize the degree of the deformation anisotropy it is convenient to introduce an additional parameter:

$$(\Delta c / c) / (\Delta a / a) = [\delta_c(x) / \delta_a(x)] \equiv \delta_{ca}(x). \qquad (2.7)$$

Experimental investigations of the heavy rare-earth α-hydrides had shown that in the systems Y-H, Lu-H, Tm-H and Er-H the parameter of anisotropy $\delta_{ca}(x) > 1$, and it reduces along this range of hydrides. In the systems Dy-H, Tb-H, Sc-H this parameter, on the contrary, is less than 1, $\delta_{ca}(x) < 1$, and decreases along the range. In the system Ho-H, which is, in some sense, a boundary between these two groups, we have $\delta_{ca}(x) \approx 1$. These differences in the deformability of rare-earth hydrides are usually ascribed to the lattice contraction of the host lattices along the rare-earth row.

It was established as well, that parameters $\delta_c (x)$, $\delta_a (x)$ and $\delta_{ca} (x)$ are temperature dependent. Particularly, at increasing temperatures the ratio $\delta_{ca}(x)$ tends to 1.

In the case of heavy rare-earth metal α-hydrides the anisotropy of lattice deformations is characteristic for temperature-induced deformations as well. These deformations are described usually by parameters:

$$[(\Delta a / a) / \Delta T] \equiv \delta_a (T) , [(\Delta c / c) / \Delta T] \equiv \delta_c (T) \text{ and } \delta_{ca}(T) \equiv [\delta_c (T) / \delta_a (T)]. \qquad (2.8)$$

In the *Appendix II* is given the table (based on Table 3 in [1]) where are summarized the experimental data concerning the lattice deformability properties of heavy rare-earth metal α–hydrides. This table illuminates the interrelations between the lattice parameter values and the lattice elastic properties.

Comparing the rows describing Sc and Y hydrides, we can see that small differences in lattice parameters provide qualitative differences in the lattice deformability properties. Particularly, in both M-H compounds δ_c (T) is generally greater than δ_a(T) (except $ScH_{0.15}$ at 80 K, where $\delta_c(T) < \delta_a(T)$). It means that, at least at T ≥ 300 K, in both compounds the lattice is easily deformable preferentially along the *c*-axes than in the perpendicular direction. In these conditions it can be expected that for MH_2-complexes energetically favourable should be configurations with the large L_Z-projections, that results in the preferable deformation of the host lattice along the "easy" *c*-axes. In the case of YH_x-system this conclusion is true ($\delta_{ca}(x) > 1$), but in ScH_x-system there is established that $\delta_{ca}(x) < 1$. The latter relation can be explained if we additionally assume that in spite of the host lattice elastic properties, the preferable configurations of the MH_2-complexes are those with large $L_\perp$-projections. From this conclusion, together with the relations (2.5) which implies that in the absence of the lattice influence on the orientations of MH_2-complexes the mean "horizontal"-component is larger than the "vertical" one, and, consequently, should provide a larger distortion in the basal plane, it follows that in ScH_x-system the interaction of MH_2-complexes with the lattice are very weak. Alternatively, in the case of systems where the lattice deformations caused by the interstitial hydrogen atoms are stronger along the *c*-direction we have to assume that, this is caused by stronger interactions of MH_2-complexes with the surrounding lattice.

In the next item we discuss this problem from a phenomenological point of view.

The energy ascribed to the pairs of interstitial H-atoms in the hexagonal lattices. When in the *hcp* lattice is inserted a single H-atom, we can always point to some metal atom, with respect to which the inserted H-atom will occupy (0 0 3/8), or (0 0 –3/8) positions. But if we consider a couple of H-atoms located in the neighbourhood of a given metal atom, then even if one of them occupies (0 0 ±3/8)-type H^T_1 interstitial position, the second H-atom can be located in any of the remained seven neighbouring H^T_2-H^T_8 interstitial positions. The question is – what additional energy should be ascribed to the host metal lattice containing an MH_2-complex in one of the six mentioned above configurations?

We have pointed out already that different configurations of H-H pairs (characterized by **L**-vector components along the *c*-axis and in the basal plane), will cause different distortion fields in the host metal lattice. Correspondingly, different energy additions are required for realization of these configurations. As the rare-earth hexagonal lattices under consideration can be distorted more easily along the *c*-axis than in the basal plane, we shall assume that on introducing a pair of H-atoms in the metal lattice, the minimum additional elastic energy E_0 will be needed for location of these atoms in the H^T_1 and H^T_2 positions, respectively. Formation of an MH_2-complex in some other configuration requires of addition of the elastic energy greater than E_0.

Let us introduce two quantitative parameters describing the geometry of MH_2-complexes. These are: (1) the angle ϑ between the M-H bonds, and (2) the angle φ between the H-H bond and the z-direction (*c*-axis of the host lattice).

Configuration-associated energy of MH_2-complexes is a function of these two parameters: $E(MH_2) \equiv E(\vartheta, \varphi)$.

In the first approximation it can be assumed that

$$E(\vartheta, \varphi) = E_1(\vartheta) + E_2(\varphi), \tag{2.9a}$$

where

$$E_1(\vartheta) = m_1(R) f_1(\vartheta) \quad \text{and} \quad E_2(\varphi) = m_2(R) f_2(\varphi). \tag{2.9b}$$

$m_1(R)$ and $m_2(R)$ (R = La, Ce, …, Lu, Sc, Y) characterize the specific properties of the host lattices where α-phases are formed, while $f_1(\vartheta)$ and $f_2(\varphi)$ are supposed to be slightly dependent on R functions.

For illuminating the mathematical scheme we choose the simplest, physically meaningful expressions for $f_1(\vartheta)$ and $f_2(\varphi)$ functions, representing them as:

$$f_1(\vartheta) = k_1(\vartheta) f_1^0(\vartheta), \quad f_2(\varphi) = k_2(\varphi) f_2^0(\varphi), \tag{2.10a}$$

where

$$f_1^0(\vartheta) = 1/(1 - \cos\vartheta), \qquad f_2^0(\varphi) = \sin\varphi, \tag{2.10b}$$

while $k_1(\vartheta)$ and $k_2(\varphi)$ are supposed to be some polynomials in ϑ and φ, respectively, which have to be determined on comparing results of calculations with the experimental data.

These expressions allow us to estimate the relative values of the "additional energies" E(*j*) ascribed to MH_2-complexes in different configurations *j*. On supposing that $m_1(R) = m_2(R) = 1$ and $k_1(\vartheta) = k_2(\vartheta) = K = 0.01$ eV, it follows

$$E(j) = K f^0(\vartheta, \varphi), \tag{2.11a}$$

where

$$f^0(\vartheta, \varphi) = f_1^0(\vartheta) + f_2^0(\varphi). \tag{2.11b}$$

2.3. THERMODYNAMIC RELATIONS

Stat-sum of MH_{22}-complex states. The stat-sum Z(T) characterising the set of states of MH_2-complexes can be determined as

$$Z(T) = \Sigma_j\, p(j) \exp[-E(j) / k_B T], \tag{2.12}$$

where the sum Σ_j extends over the six geometrically different configurations, *j* = I - VI, and *p*(*j*) describes the degeneracy of states.

Numerical estimations. Basing on expressions (2.11a, b) it is easy to determine the configuration-associated energy E(*j*) and to calculate the corresponding Boltzmann factor, $\exp[-E(j)/k_BT]$, for different configurations of MH_2-complexes. Calculations have been performed for two temperatures – 80 and 300 K, and results are presented in Table 3.

TABLE 3. Calculated Boltzmann factor values at 80 and 300 K

Configu-rations	ϑ	$f_1^0(\vartheta)$	φ	$f_2^0(\varphi)$ (eV)	$f^0(\vartheta, \varphi)$ 80 (K)	E(*j*) 300 (K)	$\exp[-E(j)/k_BT]$	
I	180°	0.5	0°	0	0.5	0.005	0.484092	0.824102
II	110°	0.7452	35°15'	0.5771	1.3223	0.013223	0.146812	0.599519
III	70°	1.5198	54°45'	0.8166	2.3364	0.023364	0.037080	0.404945
IV	120°	0.6667	67°45'	0.9256	1.5923	0.015923	0.099225	0.540050
V	110°	0.7452	90°	1	1.7452	0.017452	0.079483	0.509026
VI	35°30'	5.3792	0°	0	5.3792	0.053792	0.000408	0.124764

In this table we have used the following approximations: $10^3 k_B \approx 0.08615$ (eV/K), ϑ(II), ϑ(V) ≈ 110°, ϑ(IV) ≈ 120°.

Basing on numerical results given in Tables 2 and 3 it becomes possible to estimate the temperature variations of the mentioned above stat-sum Z(T), which in the given conditions will take the following form:.

$$Z(T) = \exp[-E(I) / k_BT) + 6\,\{\exp[-E(II) / k_BT) + [-E(III) / k_BT] + \exp[-E(IV) / k_BT] + \\ + \exp[-E(V) / k_BT]\} + 3 \exp[-E(VI) / k_BT]. \tag{2.13}$$

Particularly, using Boltzmann factor values from Table 3, we obtain:

$$Z(80\text{ K}) = 2.6609181; \qquad Z(300\text{ K}) = 13.519634. \tag{2.14}$$

The probability of the given configuration j, denoted as $\omega_j(T)$, can be determined as

$$\omega_j(T) = [p(j) / Z(T)] \exp[-E(j) / k_B T]. \tag{2.15}$$

H-H bond components mean values. The latter expression makes it possible to estimate the mean values of the components of **L**-vector connecting the H-atoms in MH_2-complexes.

Modifying relation (2.4) on taking into account the probabilities of configurations (2.15), we obtain

$$< L_{ki}(T) > = \Sigma_j\, \omega_j(T)\, L_{ki}(j), \qquad k = z\,,\perp. \tag{2.16}$$

Expressions (2.16) and (2.15) containing stat-sum (2.13) and Boltzmann factors listed in Table 3 lead us to the following results:

$$< L\,(80\text{ K}) > = 0.369684\, a, \qquad < L_\perp(300\text{ K}) > = 0.412029\, a,$$

$$< L_z(80\text{ K}) > = 0.652989\, a\,, \qquad < L_z(300\text{ K}) > = 0.474415\, a. \tag{2.17}$$

To characterize the temperature-dependent anisotropy changes of orientations of MH_2-complexes in hexagonal lattices we introduce parameter $d_{z\perp}(T)$ determined as a ratio:

$$d_{z\perp}(T) = < L_z(T) > / < L_\perp(T) >. \tag{2.18}$$

From (14) it follows:

$$d_{z\perp}(80\text{ K}) = 1.766, \qquad d_{z\perp}(300\text{ K}) = 1.151. \tag{2.19}$$

As it can be expected, anisotropy significantly diminishes at increasing temperatures.

2.4. APPLICATION OF MATHEMATICAL SCHEME TO RARE-EARTH HYDRIDES

Relations between the mean values of H-H bond components and deformations of the host lattice. On implying that in the α-phases of rare-earth hydrides all interstitial hydrogen atoms form MH_2-complexes, it is natural to make the following step and assume that lattice deformations caused by interstitial H-atoms are proportional to the projections of H-H bonds on the crystallographic axes. It means, particularly, that the mentioned above parameters $\delta_c(x)$ and $\delta_a(x)$, describing the deformability properties of the lattice, should be proportional to the mean values, $<L_z(T)>$ and $<L_\perp(T)>$, respectively.

It is interesting to compare the ratio (2.18) to the corresponding experimental values of parameter $\delta_{ca}(x)$ shown in Appendix II.

First of all, as it was mentioned above already, there are two groups of hydrides. The first one contains hydrides Y-H, Er-H, Tm-H, Lu-H, and the second – Sc-H, Tb-H, Dy-H. In the first case the parameter $\delta_{ca}(x) > 1$, while for the second group $\delta_{ca}(x) < 1$. These relations are valid at all investigated temperatures (from T = 80 K to T ≈ 775–900 K). Thus, for the first group the proposed configuration-dependent energy expressions (2.9)–(2.11) are physically meaningful. For the second group, to obtain the desired relation $\delta_{ca}(x) < 1$, it seems to be necessary to modify functions $E_1(\vartheta) = m_1(R) f_1(\vartheta)$ and $E_2(\varphi) = m_2(R) f_2(\varphi)$, by varying the factors $m_1(R)$ and $m_2(R)$, and by inserting different polynomial functions $k_1(\vartheta)$ and $k_2(\varphi)$. But, on the other hand, the required result ($\delta_{ca}(x) < 1$) can be achieved if we reduce the configuration-dependent energy $E(\vartheta, \varphi)$ and make it negligeble with respect to thermal energy kT; in other words, if we reproduce the situation corresponding to relations (2.5).

An ambiguous situation is registered in Ho-H hydrides. In $HoH_{0.20}$ at 645 K we have $\delta_{ca}(x) > 1$, while at 900 K $\delta_{ca}(x) < 1$. It means that in this system the configuration-dependent energy $E(\vartheta, \varphi)$ at 600 K seems to be greater than the thermal kT energy and relations (2.9)–(2.11) are valid, while at 900 K $E(\vartheta, \varphi)$ becomes negligible with respect to kT, and relations (2.5) have to be used.

The next point, which has to be stressed, corresponds to the fact that in the case of both groups the degree of anisotropy diminishes at increasing temperatures. The calculated values (2.19) are of the same order that corresponding experimental data shown in Appendix II. This indicates that the magnitude of the phenomenological configuration-dependent energy $E(j)$ had been selected correctly.

Equilibrium configuration of MH_2-complexes. In the hydrides of Y, Er, Tm and Lu, where the "easy deformation direction" coincides with the *c*-axis of the metal lattice, it seems natural to suppose that Configuration I, where the pair of H-atoms occupy the tetra-positions (0 0 ±3/8), will be energetically favourable and will posses the probability maximum. But calculating the probabilities (2.15) for different configurations at temperatures 80 and 300 K (see the table below – Table 4), we find that at both temperatures the most probable is not Configuration I, but Configuration II.

TABLE 4. Probabilities of different configurations of MH_2-complexes

Configuration type	I	II	III	IV	V	VI
$\omega(j)$ 80 K	0.181927	0.331041	0.083610	0.223740	0.179223	0.000518
$\omega(j)$ 300 K	0.060956	0.266066	0.179714	0.239673	0.225905	0.027685

The numerical results, given in the last table, must be considered only as a qualitative information, because for applications to the real systems the configuration-dependent energy $E(\vartheta, \varphi)$ has to be defined more precisely than it was done in relations (2.9)–(2.11).

Limiting concentrations of α-hydrides. In Table 5 there are presented experimental values of high-temperature maximal equilibrium hydrogen concentrations in α-hydrides of *hcp* heavy rare-earth group metals [1], together with the crystallographic data of host lattices [2].

TABLE 5. Limiting hydrogen concentrations in solid-solution α-phases

Metal	Gd	Tb	Dy	Ho	Er	Tm	Lu	Sc	Y
$x_{max(\alpha)}(500^{o}C)$	0.24	0.26	0.28	0.30	0.32	0.34	0.36	≥0.35	≥0.32
a (A)	3.6360	3.6010	3.5903	3.5773	3.5588	3.5375	3.5031	3.3090	3.6474
c (A)	5.7826	5.6936	5.6475	5.6158	5.5874	5.5546	5.5509	5.7306	5.2733
c/a	1.550	1.581	1.573	1.570	1.570	1.570	1.585	1.571	1.594

We see that along the row the limiting equilibrium concentration in the α-phase at 500°C $m_{ix(\alpha)}(500^{o}C)$ in average is about $x_{max(\alpha)}(500^{o}C) \approx 0.3$ and increases from Gd to Lu, accompanying the reduction of the lattice parameters.

Concerning this limiting values of hydrogen concentration in the α-phases some comments have to be made from the point of view of the model discussed in the given section of the article.

One of our basic assumptions consists in the statement that hydrogen atoms within the considered *hcp* metal lattices can exist only in the form of MH_2-complexes. When we analyzed the set of different configurations of the given MH_2-complex and probabilities of these configurations, we were obliged to assume additionally that corresponding tetra-positions were free. It means that at least the six metal atoms nearest to the $[H^T_1\text{-}M\text{-}H^T_2]$-complex are not in connection with other H-atoms. Thus, each pair of H-atoms has to be related with seven "hydrogen-free" metal atoms. This condition can be fulfilled, generally, if hydrogen concentration does not exceed the value

$$x_{max}(\alpha) = 2/7 \approx 0.29\,, \tag{2.20}$$

that qualitatively coincides with the mean value of maximum hydrogen concentration in the α-phase at 500°C (see the last table).

Some words about deviations from the value (2.20) registered at the beginning and the end of the rare-earth row.

We have the following experimentally established relations for the metals and their hydrides under consideration:

(a) Increase of the nucleus positive charge along the row induces the decrease of ionic radii and correspondingly provide the decrease of lattice parameters. (We imply the row from lanthanum till lutetium.) The melting temperature of metals increases along the row [2].
In spite of the lattice contraction and increase of metal-metal interactions,

(b) The enthalpies of hydrogen solution in metal lattices diminishes from Gd to Lu. [1].

(c) The relative deformations caused by solute hydrogen atoms diminishes from Gd to Lu (see Appendix II).

(d) The upper concentration limits of α-phases increase from Gd to Lu (see Table 5).

These features of the M-H systems under consideration allow us to think about some different scenario of hydrogen behaviour in the metal lattices. In the "weakly-bounded" gadolinium the elastic interactions of hydrogen atoms (via the metal lattice) seems to be too weak to form the H-H pairs (or MH_2-complexes), while in the other side of the row, in the case of $ErH_x - LuH_x$ systems, where the H-H elastic interactions increase, the energy gain at hydrogen clustering becomes so strong, that a part of H-atoms may form the MH_3-complexes. (The higher order clusters seems to be not suitable coming from the valency of rare-earth metals).

In the first case hydrogen atoms are dissolved in the tetrahedral positions separately and each H-atom must be connected with four "hydrogen-free" metal atoms, forming a virtual HM_4-complex. Conditions for such type of solution can be fulfilled until the hydrogen concentration will be lower than

$$x_{max}(\alpha) = 1/4 = 0.25\,. \tag{2.21}$$

In the opposite case, in the lutetium hydrides, a number of MH_3-complexes are assumed to be formed. They may be imagined e.g. as a combination of our configurations II and V, including tetrahedral positions $H^T{}_1$, $H^T{}_6$ and $H^T{}_8$. This configuration will be 12-fold degenerate. All H-H distances in it will equal a. Coming from this model of MH_3-complexe it follows (by the analogy of solutions with MH_2-complexes) that the upper concentration limit will be equal to

$$x_{max}(\alpha) = 3/7 = 0.43. \tag{2.22}$$

This result suggests that in α-phases of hydrides Er-H, Tm-H, Lu-H, Sc-H and Y-H a part of hydrogen atoms are united in MH_3-complexes, while the remained H-atoms form, perhaps, the MH_2-complexes.

2.5 CONCLUDING REMARKS

1. The discussed above *configuration model* tends to describe the equilibrium state of hydrogen atoms inserted in rare-earth metal *hcp* lattices and distributed among tetrahedral interstitial positions. It was assumed that *all* H-atoms form MH_2-complexes which can exist in six energetically different configurations. Each j configuration is characterised by different projections of H-H bonds on the z-axis and basal plane of the host lattice, providing different local distortions of the metal lattice.
2. On assuming some simplest form of configuration-dependent energy $E(j)$, we become able to introduce temperature-dependent probabilities of different configurations, that allow us to calculate the equilibrium mean values of H-H bond projections. In other words, we obtain possibilities to estimate qualitatively the degree of lattice distortions caused by the interstitial H-atoms. On comparing with experimental data it becomes possible to conclude, that in the case of systems Y-H, Er-H, Tm-H and Lu-H configuration-dependent energies $E(j)$ (of the selected type) are sufficiently strong, providing H-H-bond projections on the z-axis to be greater than those in the basal plane. Correspondingly, in these systems the host lattice is distorted more significantly along the z-axis. In the case of systems Sc-H, Tb-H and Dy-H it seems that $E(j)$ energies are very week with respect to kT-energy and provide H-H bond projections along the z-axis to be smaller than those in the basal plane. As a result, distortions in the basal plane become larger than along the z-axis.
3. Analysis of the α-phase upper limits on the concentration scale had determined the boundaries of applicability of the MH_2-complex model. Indeed, from the point of view of this model the α-phase can exist until hydrogen concentration allows

MH_2-complexes to make "free-rotations" in the metal lattice, that requires hydrogen concentration to be lower than 2/7. This condition is approximately fulfilled in the systems Dy-H, Ho-H and Er-H. But to explain the concentration limits in the case of other hydrides we have to assume that besides the MH_2-complexes there are formed additionally: (1) a number of MH_3-complexes (in systems Tm-H, Lu-H, Sc-H and Y-H), or (2) HM_4-complexes (in systems Gd-H, Tb-H).

Thus, the general conclusion of the configuration model can be formulated as a statement that in rare-earth metal α-hydrides a small portion of interstitial hydrogen atoms is randomly dissolved in the host metal lattice, while the main part of them tend to form a number of MH_n-complexes characterized by a set of energetically different configurations.

3. Cooperative model

3.1. MATHEMATICAL DESCRIPTION

Model assumptions. The low-temperature ordered structures formed in the α*–phase of heavy rare-earth hydrides were intensively investigated during last two decades [1]. Basing on neutronographic investigations it was established that within the α*–phase there are formed some ordered linear structures of H-H pairs represented usually as "zigzag chains" along the z-axes. An element of this linear ordered configuration is shown in Fig. 2(a) [1].

We assume, first of all, that in α*–phase *all hydrogen atoms* form H-H pairs which are tightly bound to the corresponding metal atoms generating linear MH_2-comlexes. These H-M-H atomic groups can have different configurations, but prefer to be oriented along the hexagonal *c*-axis of the metal lattice (denoted as Configuration I in the previous section).

The next step of our assumptions consists in the statement that in α*-phase of heavy rare-earth metal hydrides the hexagonal metal lattice containing linear "zigzag" configurations of MH_2-complexes can be reproduced in a simplified form as a set of columns, oriented along *c*-axis and considered as a combination of structural units shown in Fig. 2(b). Each structural unit (SU) is a pair of *hcp* unit cells containing (or not) a pair of hydrogen atoms in the fixed linear configuration. All other tetrahedral positions, inside of the structural unit and on its edges, are assumed to be permanently empty.

This assumption has some arguments. Existence of linear "zigzag" chains of H-M-H complexes was confirmed by neutronographic experiments [1]. On the other hand, if we represent the hexagonal lattice consisting of N_M metal atoms (M-atoms) and N_H hydrogen atoms (H-atoms) as a set of "hydrogenated" SU, then taking into account that per each "hydrogenated" SU we have 2 H-atoms and 12 M-atoms, we obtain for the possible maximum hydrogen concentration in the α*-phase, $x(\alpha^*)_{maxim}$, the value:

$$x(\alpha^*)_{maxim} = N_H / N_M = 2/12 \approx 0.17\,, \tag{3.1}$$

which qualitatively agrees with some experimentally determined concentration limits [1]: $x(\alpha^*)_{maxim} \approx 0.2$ for YH_x and LuH_x, and $x(\alpha^*)_{maxim} \approx 0.11$ for TmH_x, but disagrees with those for systems ScH_x ($x(\alpha^*)_{maxim} \approx 0.35$), ErH_x ($x(\alpha^*)_{maxim} \approx 0.065$) and HoH_x ($x(\alpha^*)_{maxim} \approx 0.03$).

If hydrogen concentration x is lower than $x(\alpha^*)_{maxim}$, then some of SU will not contain MH_2-complexes and there arise a question concerning spatial distribution of two types of SU. For simplicity, we shall denote SU containing a pair of hydrogen atoms as SU_H, and SU without hydrogen atoms – as SU_0.

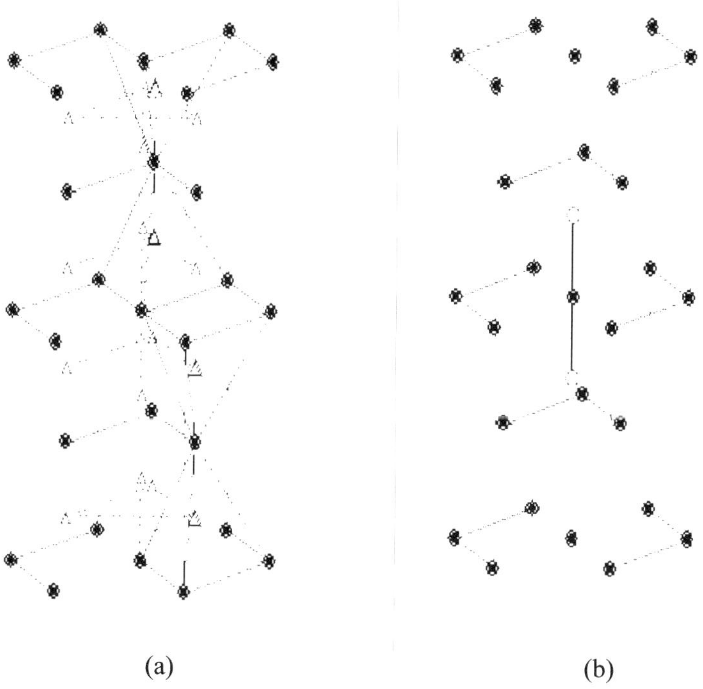

Figure 2. Doubled unit cell of a *hcp* lattice containing hydrogen interstitials (a) the element of experimentally established linear ordering in the α*-phase; (b) the Structural Unit (SU_H); a simplified representation of the experimentally established linear structure

Let us consider a column along the *c*-axis consisting of N structural units and containing N_M metal atoms and N_H H-atoms. It is assumed that $[N_H / N_M] \equiv x < x(\alpha^*)_{maxim}$. If we denote that the column contains N_0 empty structural units SU_0 and $(N - N_0)$ "hydrogenated" structural units SU_H, then the following relations can be written:

$$N_M = 12\,N \quad , \qquad N_H = 2\,(N - N_0)\ , \qquad N_0 = (N - 0.5\,N_H) \tag{3.2}$$

$$x(\alpha^*) = N_H / N_M = (1/6)\,[1 - (N_0 / N)] \tag{3.3}$$

The final step of our assumptions postulates the energy ascribed to each SU depending on its position in the column.

If SU_H follows SU_0 then its energy is $E_1 (SU_H) \equiv E$, but if it follows SU_H then its energy is $E_2 (SU_H) \equiv (E - \Delta E)$. The energy of SU_0 in all cases equals zero, $E (SU_0) \equiv 0$.

We do not consider interactions between the neighbouring columns.

Thermodynamic consideration. Equilibrium state of the column containing N SU and consisting of N_0 SU_0 and $(N - N_0)$ SU_H "particles", can be determined on analysing the corresponding free-energy expression

$$F(N) = E(N) - TS(N), \tag{3.4}$$

where $E(N)$ is the energy of a column consisting of N units, and $S(N)$ is its entropy.

Due to our last assumption it is obvious that energetically it is favourable to collect all SU_H in one group (then with respect to the state where each SU_H is separated from others by one or more SU_0, a gain in energy equal to $[(0.5\ N_H - 1)\ \Delta E]$ will be reached). Consequently, at T = 0 K, when the entropy contribution to the free-energy is zero, equilibrium state implies collection of all SU_H in a single group.

Let us now determine the entropy of distribution of the SU in columns. We shall consider a simplest case when it is assumed that all groups (chains) of SU_H contain a similar number of elements, e.g. n-elements. Neighbouring chains can be divided by a different number of SU_0. To prevent the confluence of neighbouring chains of "hydrogenated" SU we have to assume additionally that each chain of SU_H must start and terminate by the attached SU_0 elements. Thus, each "isolated" chain will consist of [n + 2] elements, containing n SU_H and 2 SU_0 at the ends. The number of such chains equals

$$\nu(n) = 0.5\ N_H / n\ . \tag{3.5}$$

Consequently, the number of hydrogen-free structural units equals

$$\nu_0 = [N - \nu(n)(n + 2)]. \tag{3.6}$$

It means that in the column consisting of N structural units we have to consider a statistical assembly containing in total $[\nu_0 + \nu(n)]$ elements. From (3.5) and (3.6) it follows that

$$[\nu_0 + \nu(n)] = [N - 0.5\ N_H (n + 1) / n]. \tag{3.7}$$

In this statistical assembly the number W_n of physically different states with the same energy will be equal to

$$W_n = (\nu_0 + \nu(n))! / (\nu(n)!\ \nu_0 !). \tag{3.8}$$

Thus, for entropy of given distribution of structural units in the column, determined as

$$S(N) = k_B \ln (W_n)\ , \tag{3.9}$$

a following expression is obtained

$$S(N) = k_B \{(\nu_0 + \nu(n)) \ln (\nu_0 + \nu(n)) - \nu_0 \ln \nu_0 - \nu(n) \ln \nu(n)\}. \tag{3.10}$$

For the energy of each "isolated" chain, we can write

$$E(n) = [n\,(E_0 - \Delta E) + \Delta E], \quad (3.11)$$

Total energy of the column consisting of N structural units and containing $\nu(n)$ chains can be presented as

$$E(N) = \nu(n)\,E(n) = \nu(n)\,[n\,(E_0 - \Delta E) + \Delta E] = 0.5\,N_H\,[(E_0 - \Delta E) + \Delta E / n]. \quad (3.12)$$

Inserting expressions (3.10) and (3.12) in (3.4) we obtain the free-energy of the column as a function of the lengths of chains n.

Equilibrium equation. The extreme value of the free-energy function is determined from the condition

$$\partial F(N) / \partial n = 0. \quad (3.13)$$

On calculating the derivatives $[\partial\nu(n)/\partial n]$, $[\partial\nu_0/\partial n]$ and $[\partial\,(\nu_0 + \nu(n))/\partial n]$ and inserting them in expressions $[\partial E(N)/\partial n]$ and $[\partial S(N)/\partial n]$ we obtain :

$$\partial E(N) / \partial n = [\partial\nu(n) / \partial n]\,\Delta E = -0.5\,N_H\,\Delta E / n^2, \quad (3.14a)$$

and

$$\partial S(N) / \partial n = k_B\,0.5\,N_H\,[1 / n^2]\,\ln\{[\nu(n)\,(\nu_0 + \nu(n))] / \nu_0^2\} \quad (3.14b)$$

Introducing (3.14a) and (3.14b) in (3.13) we obtain an equation

$$0.5\,N_H\,[1 / n^2]\,[-\Delta E - k_B\,T\,\ln\{[\nu(n)\,(N_0 + \nu(n))] / N_0^2\}] = 0, \quad (3.15)$$

that implies

$$\{[\nu(n)\,(\nu_0 + \nu(n))] / \nu_0^2\} = \exp[-(\Delta E / k_B\,T)]. \quad (3.16)$$

Solution of equilibrium equation. Let us introduce in equation (3.16) a following notation

$$\exp[-(\Delta E / k_B\,T)] = b. \quad (3.17)$$

Then relation (3.16) transforms into the equality

$$(\nu(n) / \nu_0)\,[1 + (\nu(n) / \nu_0)] = b. \quad (3.18)$$

On introducing the variable

$$[\nu(n) / \nu_0] \equiv \chi \quad (3.19)$$

equation (3.18) can be written as

$$\chi^2 + \chi - b = 0. \quad (3.20)$$

Its solution is

$$\chi_0 = -0.5(1 \pm f), \quad (3.21)$$

where

$$f = (1 + 4b)^{0.5}. \tag{3.22}$$

Expression (3.21) can be solved with respect to the physical variable n using relations (3.2), (3.3), (3.5) and (3.6). Finally we obtain

$$n_{01} = 2\ \{x(\alpha^*) / [x(\alpha^*)_{maxim} - x(\alpha^*)]\}\ \{f / [1 + f]\}, \tag{3.23a}$$

$$n_{02} = 2\ \{x(\alpha^*) / [\, x(\alpha^*)_{maxim} - x(\alpha^*)]\}\ \{f / [f - 1]\}\ . \tag{3.23b}$$

The solution (3.23a) gives an unphysical result and for estimation of concentration and temperature dependences of equilibrium chains lengths n_0 we have to use the second one. Thus we postulate that equilibrium chains of the structural units containing MH_2-complexes can be described basing on the expression:

$$n_0(x(\alpha^*), \mathrm{T}) = 2\ \{x(\alpha^*) / [\, x(\alpha^*)_{maxim} - x(\alpha^*)]\}\ \{f / [f - 1]\}, \tag{3. 24}$$

where the first braces represent the concentration, and the second – temperature dependences of equilibrium chain lengths.

In the following below item we shall see that the chain lengths depend, of cause, on the temperature T and the value of the energy parameter ΔE, but the dominant influence on equilibrium values of $n_0(x\ , \mathrm{T})$ is provided by the concentration of solid solution; more precisely, by the difference $[\, x(\alpha^*)_{maxim} - x\ ]$.

3.2. NUMERICAL RESULTS

Calculations for a low-concentration hydride with characteristic parameters corresponding to α-phase of YH_x-system were performed basing on expression (3.24). The total number of SU in the column was taken equal to $N = 10^5$ elements. The energy parameter stimulating chains formation ΔE was taken equal to 0.001, 0.005, 0.010 and 0.025 eV. The limiting concentration – $x(\alpha^*)_{maxim}$ – had been equal to 0.18.

Concentration dependence of equilibrium chain lengths n_0 for temperature T = 1,000 K and a fixed value of energy parameter ΔE = 0.001 eV is given in Fig. 3. *Temperature dependences* of chain lengths for a number of concentrations $x < x(\alpha^*)$ and the given values of energy parameter ΔE are given in the subsequent Figs. 4, 5. In all cases the chain lengths reduce at concentration lowering and temperature increasing.

3.3 DISCUSSION AND CONCLUSIONS

The attempts of description of interstitial hydrogen atoms behavior in the hexagonal metal lattices had been undertaken in previous years. Particularly, a number of investigations were based on the examination of the host lattice elastic and electric properties of hydrogen solid solutions in metals [3–9]. In contrast to these fundamental descriptions we propose a simple phenomenological model which accounts nevertheless for such an important characteristic of the metal-hydrogen system, as the limiting concentration of the α*-phase.

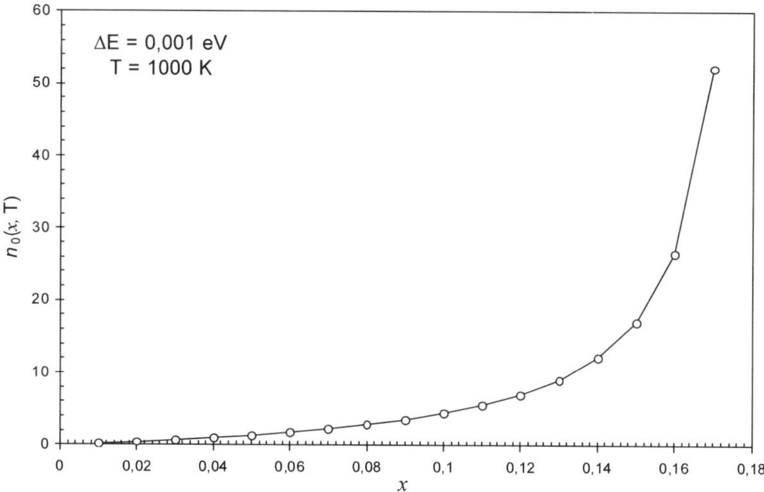

Figure 3. Concentration dependence of chain lengths at a given values of temperature T and of the energy parameter ΔE

Basing on our model some immediate conclusions can be done.

(a) In the high-temperature limit chain lengths are conditioned mainly by the alloy concentration and in the minor degree by temperature T and energy parameter ΔE.

(b) Temperature dependence of chain lengths can be seen at low temperatures and it is revealed more strongly at high values of the characteristic energy parameter ΔE.

Both these properties are easily explained by the structure of temperature-dependent factor in expression (3.24) and the exponential character of parameter b (expressions (3.17) and (3.22)).

The energy parameter values ΔE we have selected arbitrarily, just to estimate the role of this quantity on the equilibrium chain lengths $n_0(x, \mathrm{T})$. Its role is remarkable only at low temperatures. On comparing the calculated values of $n_0(\mathrm{T})$ with the experimentally estimated ones (collected in [1]) it can be deduced that the most reliable value of ΔE seems to be located in the range 0.001–0.005 eV

It has to be stressed that basing on Figs. 4, 5 we have to conclude that the strongest influence on the values of $n_0(\mathrm{T})$ is provided by the variations of hydrogen concentration, especially when we approach the upper limit of the α*-phase.

Some comments. The arrangement of structural units containing fixed MH_2-complexes in the chains within a given column was provided by a single empirical energy parameter ΔE. As no reorientations of MH_2-complexes within the structural units was taken into account and no interactions between the neighbouring columns were assumed we were failed to reproduce the order-disorder transition in the structural unit distribution, and we see only an exponential decreasing of chain lengths at temperature increasing.

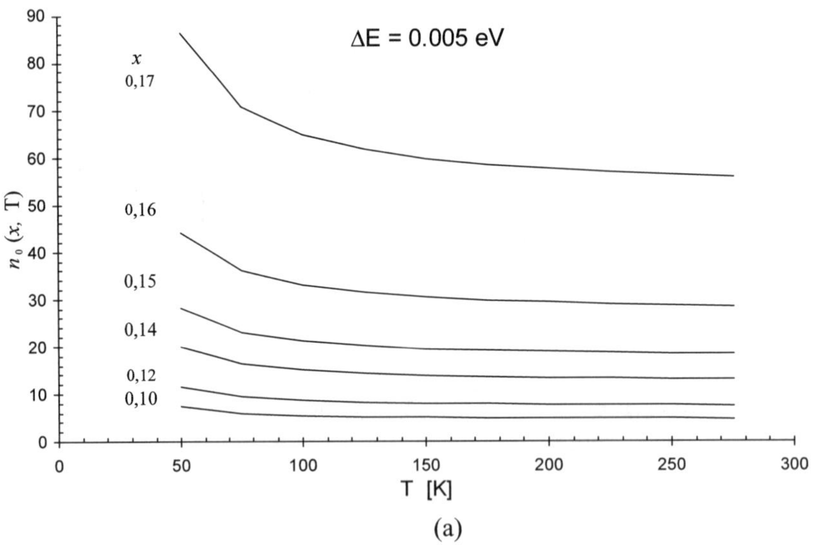

(a)

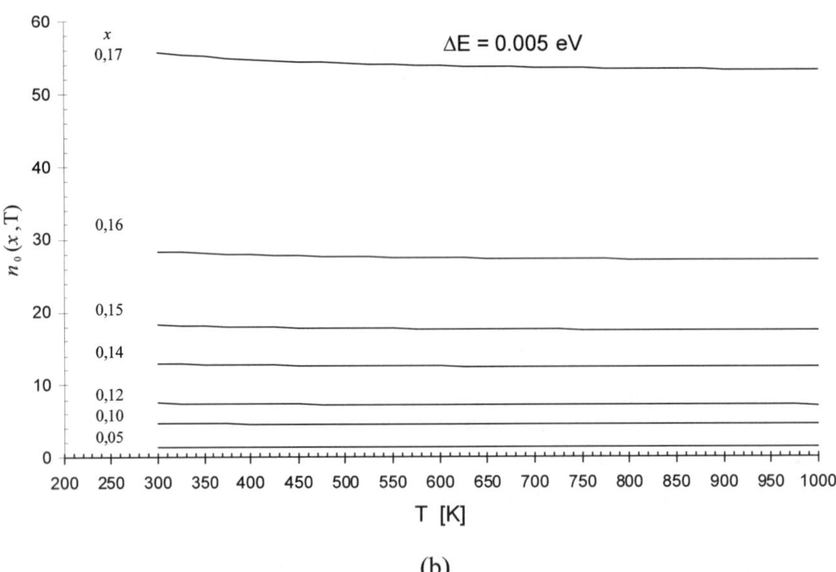

(b)

Figure 4. The low-temperature parts (a) and high-temperature parts (b) of n_0 (*x*, T) dependences at different concentrations for energy parameter ΔE = 0.005 eV

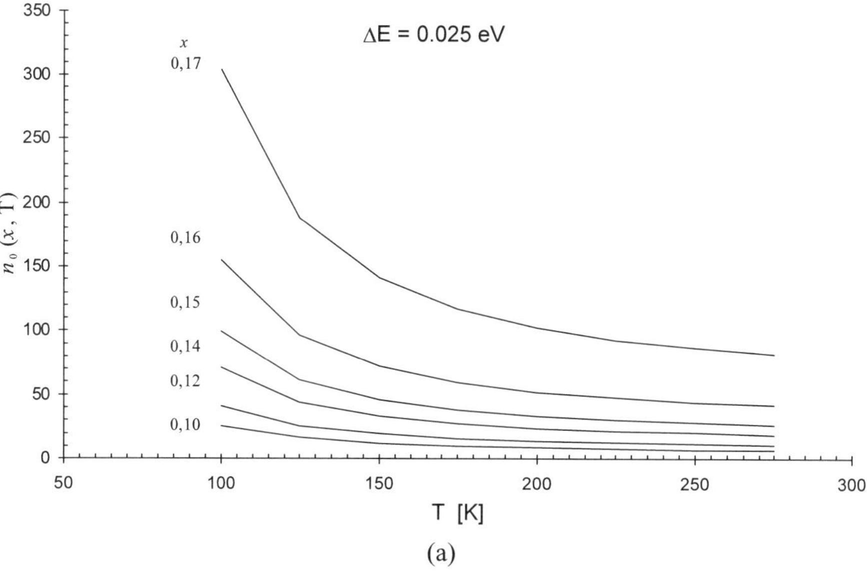

(a)

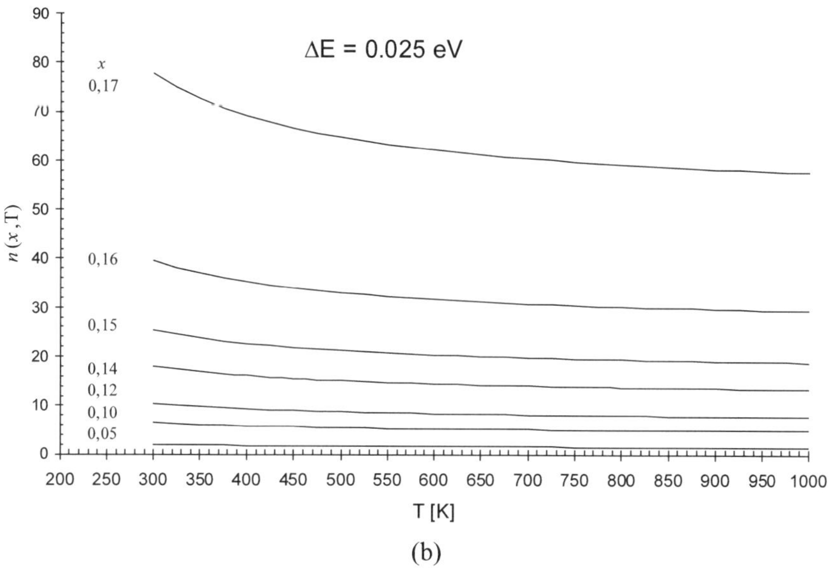

(b)

Figure 5. The low-temperature parts (a) and high-temperature parts (b) of n (x, T) dependence at different concentrations for energy parameter ΔE = 0.005 eV

4. General conclusions

In the present article we have analysed the consequences of an assumption that hydrogen atoms inserted in the rare-earth lattices form MH_2-complexes which can have different configurations and can be spatially arranged in linear structures. In order to quantitatively estimate the results of these assumptions we have discussed two models intended to describe the possible configurations of MH_2-complexes (*configuration model*) and the possibilities and conditions of the linear structure formations (*cooperative model*).

A general expression for the equilibrium chain lengths was obtained which visualize the role of concentration x, particularly in the vicinity of the limiting concentration $x(\alpha^*)_{maxim}$.

The influence of chain lengths on physical properties of metal hydrides will be analyzed separately in the future.

Acknowledgements

Present investigations of rare-earth metal-hydrides were performed within the frames of the STCU Project N 3867.

References

1. P. Vajda. "Hydrogen in rare-earth metals, including RH_{2+x} phases". In *Handbook on the Physics and Chemistry of Rare-Earths* (Eds. Gschneidner, Jr., L. Eyring), 1995, **20**, 207.
2. G.V. Samsonov. "Hand-book of Physical-Chemical Properties of Elements", (Kiev: Naukova Dumka: 1965) (in Russian).
3. G. Cannelli, R. Cantelli, F. Cordero. "Statistical model for the trapping of interstitials by substitutional (interstitial) atoms in solids". *Phys. Rev. B*, 1985, **32:**3573.
4. F. Cordero. "Anelastic (dielectric) relaxation of point defects at any concentration, with blocking effects and formation of complexes". *Phys. Rev. B*, 1995, **47:**7674.
5. V.A. Tatarenko, C.L. Tsynman. "Strain-induced and blocking effects in thermodynamics of the ordering and precipitation reactions within the off-stoichiometric close-packed-metal hydrides". *Sol. St. Ionics*, 1997, **101–103:** 1061.
6. V.A. Tatarenko. "Interaction of interstitial hydrogen atoms within h.c.p. metals and distribution of diffuse radiation-scattering intensity in reciprocal lattice". *Int. J. Hydrogen Energ.*, 2001, **26:** 469.
7. M.S.Blanter, E.B.Granovskiy. "Strain-induced interaction of hydrogen atoms in *hcp* rare earth metals". *J. Alloy. Compd.*, 2002, **335:** 1.
8. M.S.Blanter, E.B.Granovskiy, L.B.Magalas. "Ordering and strain-induced D-D interaction in Lu-D". *Phys. Stat. Sol. (b), 1-6 (2003)/DOI 10.1002/pssb.200301878.*
9. J. Garses, J.L. Gervasoni, P. Vajda. "Relaxation of internal stress field and hydrogen ordering in YH_x" *J. Alloy. Compd.*, 2005, **404–406:** 126.
10. I.G. Ratishvili. "Orientational model of hydrogen ordering in hydrides". *FMM*, 1991 (10): 74–86.

Appendix I. **Six different distinguishable configurations of MH_2-complexes**

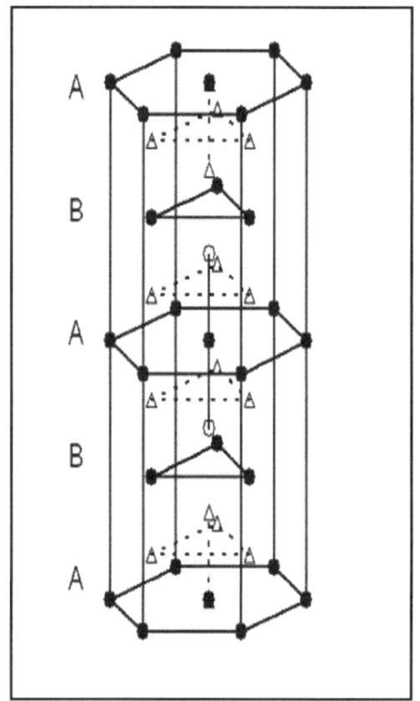

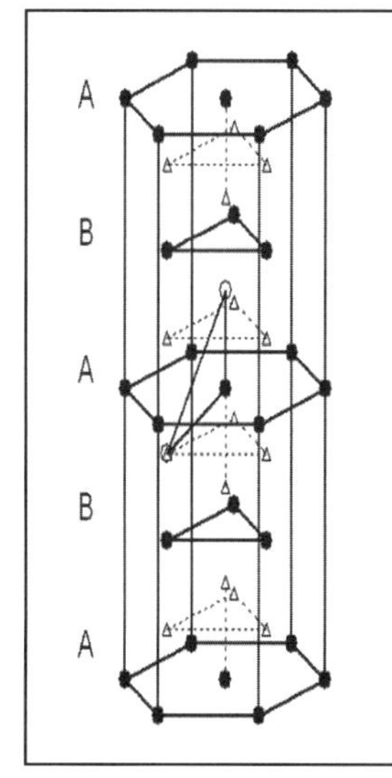

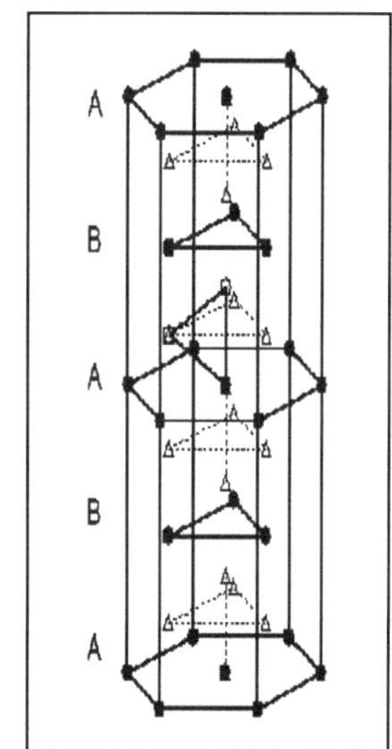

I
$H^T{}_1$–$H^T{}_2$

II
$H^T{}_1$–$H^T{}_6$, $H^T{}_7$, $H^T{}_8$
$H^T{}_2$–$H^T{}_3$, $H^T{}_4$, $H^T{}_5$

III
$H^T{}_1$–$H^T{}_3$, $H^T{}_4$, $H^T{}_5$
$H^T{}_2$–$H^T{}_6$, $H^T{}_7$, $H^T{}_8$

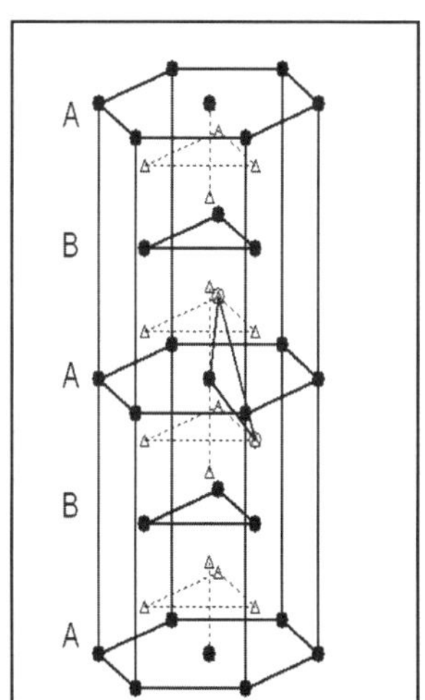

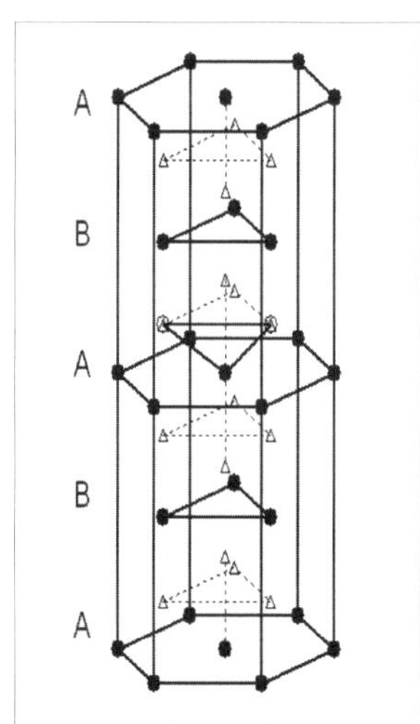

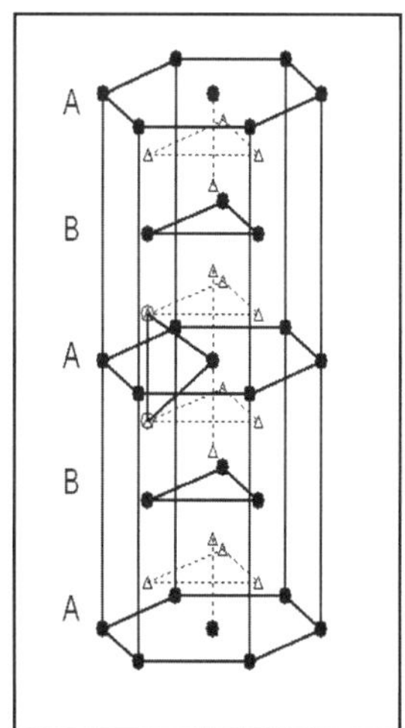

IV
$H^T{}_3$–$H^T{}_7$, $H^T{}_8$
$H^T{}_4$–$H^T{}_6$, $H^T{}_8$
$H^T{}_5$–$H^T{}_6$, $H^T{}_7$

V
$H^T{}_3$–$H^T{}_4$, $H^T{}_5$;
$H^T{}_6$–$H^T{}_7$, $H^T{}_8$;
$H^T{}_4$–$H^T{}_5$; $H^T{}_7$–$H^T{}_8$

VI
$H^T{}_3$–$H^T{}_6$
$H^T{}_4$–$H^T{}_7$;
$H^T{}_5$–$H^T{}_8$

Appendix II. **Deformability of lattices of the heavy rare-earth hydrides**

Elem-ents	H/M	Temper. T, K	Latt. param. a (A), Å	c (A), Å	Characteristics of the lattice deformability $\delta_a(x)$ (x10[4])	$\delta_c(x)$	$\delta_{ca}(x)$	$\delta_a(T)$ (x10[6])	$\delta_c(T)$	$\delta_{ca}(T)$
Sc	0.15	80	3.318	5.2775	3.47	2.85	0.8213	5.0	2.0	0.400
	0.15	300	3.3215	5.286	2.71	1.96	0.7232	5.0	8.5	1.700
	0.30	80	3.3295	5.2845	2.91	1.87	0.6426	3.6	6.7	1.861
	0.30	300	3.3340	5.2945	2.63	1.52	0.5779	6.6	7.75	1.174
Y	0.19	300	3.6636	5.7900	2.22	5.33	2.4009			
	0.18D[a]	300	3.6636	5.7780	2.15	4.57	2.1256			
	0.12	300	3.6542	5.7654	1.37	4.59	3.3504	5.3	20.2	3.811
	0.12	775	3.670	5.8175	2.28	3.81	1.6710	11.9	22.9	1.924
	0.22	300	3.6637	5.7940	1.93	4.76	2.4663	2.9	16.4	5.655
	0.22	775	3.678	5.838	2.23	3.68	1.6502	12.2	14.9	1.221
Tb	0.20	775	3.6515	5.7735	5.52	3.57	0.6467	7.1	16.3	2.296
	0.20	1000	3.6573	5.7947	5.33	3.22	0.6041	7.1	16.3	2.296
	0.2D	800	3.6511	5.7652	5.36	3.57	0.6660	7.8	16.7	2.141
	0.2D	1000	3.6568	5.7945	5.23	3.21	0.6138	7.8	16.7	2.141
Dy	0.20	675	3.6269	5.7255	4.15	3.21	0.7735	6.5	18.7	2.877
	0.20	1000	3.6334	5.7550	4.11	2.97	0.7226	6.5	18.7	2.877
	0.2D	700	3.6304	5.7318	4.57	2.49	0.5449	7.5	18.3	2.440
	0.2D	950	3.6372	5.7580	4.64	3.23	0.6961	7.5	18.3	2.440
Ho	0.20	645	3.6087	5.6985	3.38	3.47	1.0266	9.9	16.2	1.636
	0.20	900	3.6178	5.7220	3.94	3.16	0.8020	9.9	16.2	1.636
	0.2D	635	3.6103	5.7021	3.58	3.78	1.0559	10.8	15.8	1.463
	0.2D	875	3.6197	5.7237	4.19	3.31	0.7900	10.8	15.8	1.463
Er	0.035	300	3.5636	5.5960	3.43	5.47	1.5948			
	0.27	775	3.6035	5.7000	3.63	3.87	1.0661	13.8	16.7	1.210
Tm	0.08	300	3.5462	5.5795	3.20	5.95	1.8593			
	0.09D	300	3.5466	5.5868	2.73	4.36	1.5971			
	0.11	300	3.5478	5.5890	2.90	5.09	1.7552	6.2	25.0	4.032
	0.175	775	3.5720	5.6560	3.53	4.06	1.1501	8.7	14.8	1.701
Lu	0.205	300	3.5267	5.6071	2.98	5.05	1.6946			
	0.20	300	3.5250	5.6082	2.76	5.18	1.8768	9.2	13.2	1.435
	0.20	775	3.5412	5.6385	2.73	2.78	1.0183	10.9	10.9	1
	0.19D	300	3.5222	5.6040	2.68	5.18	1.9328	9.9	14.9	1.505
	0.19D	775	3.5400	5.6386	2.78	2.88	1.0360	12.3	12.2	0.992

[a]Index D denotes the metal-deuterium compounds.

EFFECT OF ALLOYING ELEMENTS ON THE FORMATION OF $FeTiH_4$: AN AB INITIO STUDY

F.G. SEN, A. KINACI, M.K. AYDINOL
Department of Metallurgical and Materials Engineering
Middle East Technical University, 06531, Ankara Turkey
Fax: 90 (312) 210 2518; E-mail: sfatih@metu.edu.tr

Abstract. FeTi can store 3.7 wt% hydrogen if it is synthesized in the form of $FeTiH_4$. In the present study, candidate alloying elements were investigated that would decrease the formation energy of this hydride, using first principles calculations within the generalized gradient approximation (GGA) to density functional theory (DFT). Selected alloying elements are Co, Mg, Mo, Ni, Nb, V and Zr, and they were substituted for Fe or Ti atoms separately. Their effects on the formation energy of $FeTiH_4$ were analyzed on the basis of electronic structure. It was found that Zr, Nb, Mo and Mg substitution for Fe decreases the formation energy of the hydride considerably. Stabilization is achieved by forming stronger bonds between hydrogen atoms and rest of the Fe atoms in the structure which is facilitated by the antibonding between the hydrogen atoms and the substituted high tendency hydride forming element.

Keywords: hydrogen storage, metal hydrides, FeTi, first principles calculations

1. Introduction

Hydrogen is regarded as the future energy source that would be used in vehicles, portable devices, housings and industry instead of carbon based fuels. Exhaustion of fossil fuels and the resulting environmental pollution caused by the use of fossil fuels, directed the scientists to search for alternative energy sources and hydrogen is the most promising candidate due to its high abundance and cleanliness. Hydrogen is considered to take the place of the combustion engines by fuel cells.

Hydrogen can be stored safely inside the metal itself and the amount of hydrogen stored per volume is greater than either gaseous or liquid storage. A metal-hydride is a metal-hydrogen alloy where the hydrogen is absorbed by a metal, an intermetallic compound or a substitutional alloy. There are various types of hydrogen absorbing intermetallic hydrides reviewed by Ivey and Northwood [1], such as AB, AB_5, AB_2, A_2B, A_2B_7 type alloys. In order to be used in hydrogen storage applications, a suitable metal hydride should have high storage capacity, low hydrogenation/ dehydrogenation temperature, good hydriding/dehydriding kinetics and low cost [1].

The primary difficulty of storing hydrogen in metallic hydrides, at the present time, lies in the thermodynamics and kinetics of absorption or desorption of hydrogen. Among AB intermetallics, FeTi–H system was investigated widely in the literature owing to its moderate kinetics with relatively high equilibrium hydrogen pressure (≈4.5 atm) at room temperature, but the low storage capacity of FeTi (up to 1.8 wt% in

B. Baranowski et al. (eds.), *Carbon Nanomaterials in Clean Energy Hydrogen Systems.*

the form of $FeTiH_2$) restricts its usage in practical applications. FeTi has a CsCl type crystal structure with a lattice parameter of 2.98 Å [2] and it was found experimentally that it can form the hydrides, FeTiH and $FeTiH_2$ [3].

In addition to experimental studies, computational methods, mainly ab initio methods were widely used in the investigation of metal hydride systems. In these studies [4–6] formation energies and electronic structures of hydrides were calculated and the relations with the hydride formation energy and the electronic structure of the crystal structure were reported. In a recent ab initio study [7], various crystal structures for FeTi hydrides were considered and relation between formation energies and electronic structure was investigated systematically. In this study the crystal structures of experimentally known hydrides, FeTiH and $FeTiH_2$ were predicted correctly. In addition, it was mentioned that FeTi can absorb 3.7 wt% hydrogen in the form of a less stable compound, $FeTiH_4$.

In the present study, candidate alloying elements on the bulk $FeTiH_4$ were investigated by using first principles calculations in order to increase the stability of this compound by decreasing its formation energy.

2. Computational methodology

In order to predict the ground-state crystal and electronic structures, we have used a projector augmented wave (PAW) method with the exchange correlation energy approximated in the generalized gradient approximation (GGA) to density functional theory (DFT). All of the calculations have been performed with the Vienna Ab Initio Simulation Package (VASP) [8–10]. Results were obtained using PAW-PBE [11] potentials provided with VASP.

$FeTiH_4$ was predicted [7] to have a tetragonal crystal structure with a space group of P4/*mmm*, where Fe is in *1a*, Ti in *1d* and H in *4i* Wyckoff positions, consisting of 6 atoms/unit cell as shown in Fig. 1. In order to investigate the effect of alloying elements on the formation energy of $FeTiH_4$, a 2 × 2 × 2 supercell consisting of 48 atoms were constructed. Alloying elements of Co, Mg, Mo, Nb, Ni, V and Zr were then substituted in place of either Fe or Ti separately in the supercell. The structure will have a formula unit $Fe_7MTi_8H_{32}$ for an M atom substituted for Fe and $Fe_8Ti_7MH_{32}$ for M atom substituted for Ti. During the calculations, the reciprocal space was sampled with 2 × 2 × 2 grid of k points. The obtained structure was then fully relaxed according to the atomic positions and cell parameters and the total energy of the system was calculated. We calculated the lattice parameters of $FeTiH_4$ in P4/*mmm* space group as a = 2.944 Å and c = 4.198 Å and the positional parameter of hydrogen at the *4i* site as z = 0.21. Effect of alloying elements on the formation energy was analyzed with the basis of crystal and electronic structure.

3. Results and discussion

We have also performed computations for the several crystal structures of monohydride and dihydride of FeTi. As a result of these, we have predicted the experimentally known structures as the ground state crystal structures. FeTiH has a crystal structure that belongs to the space group $P222_1$ and experimentally known $FeTiH_2$ has a crystal structure in the C*mmm* space group. All of the relaxed lattice parameters at the ground state structures of the hydrides were found to agree with the experimental values within 2% error margin [2, 12].

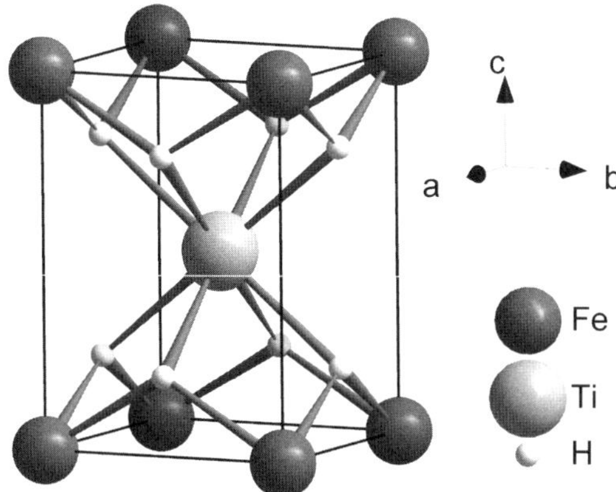

Figure 1. The crystal structure of $FeTiH_4$. P4/*mmm* tetragonal structure with Ti on the center, Fe on the corners and H on the cell surfaces

The formation energies of hydrides of FeTi were calculated by the energy of the reaction,

$$FeTi + \frac{n}{2}H_2 \rightarrow FeTiH_n.$$

We calculated the enthalpy of formation accurately for FeTiH and $FeTiH_2$ as -28.49 and -28.06 kJ/mol per H_2 respectively, compared with the experimentally found value for $FeTiH_2$, which is -28.1 [12]. The formation energy of the $FeTiH_4$ is calculated as -19.64 kJ/mol per H_2, which is higher than FeTiH and $FeTiH_2$.

The formation energies of the alloyed systems in which an element M substituted for an Fe atom, on the other hand, were calculated by the energy of the following reaction,

$$Fe_7MTi_8 + 16H_2 \rightarrow Fe_7MTi_8H_{32}.$$

The consequent formation energy per H_2 is then calculated by

$$\Delta E_f = \frac{1}{16}\left(E_{Fe_7MTi_8H_{32}} - E_{Fe_7MTi_8} - 16E_{H_2}\right).$$

The heat of formation equation for the Ti-site substitution will be similar to the above equation. The energy of H_2, to be used in formation energy calculations, is approximated by the energy of an H_2 molecule in a vacuum at zero Kelvin. Heats of formation of alloyed systems are given in Table 1 for each alloying element and for the respective position it was substituted for. Among the alloying elements some has a tendency to decrease the formation energy and some of them do not. Also some alloying elements decrease the heat of formation when substituted for Fe but does not have the same effect when it is substituted for Ti.

As it can be seen from Table 1, addition of Mg, Mo, Nb, V and Zr in place of Fe and Co, Mg, Ni and Zr in place of Ti, decrease the heat of formation of $FeTiH_4$. The decrease is more pronounced for the case that Fe is substituted rather than Ti. For instance Zr decreased the formation energy nearly two times, if Fe is substituted. Nb, V

TABLE 1. Heat of formation of $FeTiH_4$ systems that were alloyed by the corresponding element

Alloying element	ΔE_f Fe site	Ti site
Pure	-19.64	-19.64
Co	-18.59	-21.98
Mg	-29.10	-21.12
Mo	-27.35	-16.88
Ni	-18.01	-20.70
Nb	-35.03	-17.87
V	-32.81	-17.85
Zr	-40.11	-23.66

and Mo substitution for Fe decreased formation energy but these elements increased the formation energy if substituted for Ti. In the case of Co and Ni substitution for Ti, there is a slight decrease in the formation energy, but these alloying elements do not contribute a decrease if substituted for Fe atom. The addition of Mg and Mo would be an ideal selection in order to decrease the heat of formation, but not to stabilize the structure too much which may cause a decrease in the kinetics of hydriding/dehydriding.

In order to have a better understanding for the effect of the alloying elements on the formation energy, we have computed the electronic charge distributions in selected systems and take the difference between the hydrided and the metallic systems. The difference in electronic charge distribution therefore will show the charge transfer in these systems as hydrogen is inserted into the structure. In analyzing these charge differences, we have set the difference in pure FeTi system as the reference. Therefore the plots given in Fig. 2, reflects only the charge depleted or accumulated regions in the alloyed structure compared to the unalloyed case during hydriding. In Fig. 2, charge density difference plots of only Nb alloy are given but for other alloys similar conclusions can also be drawn.

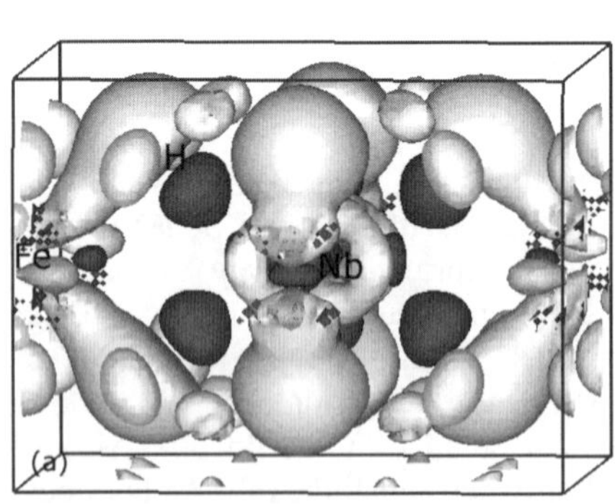

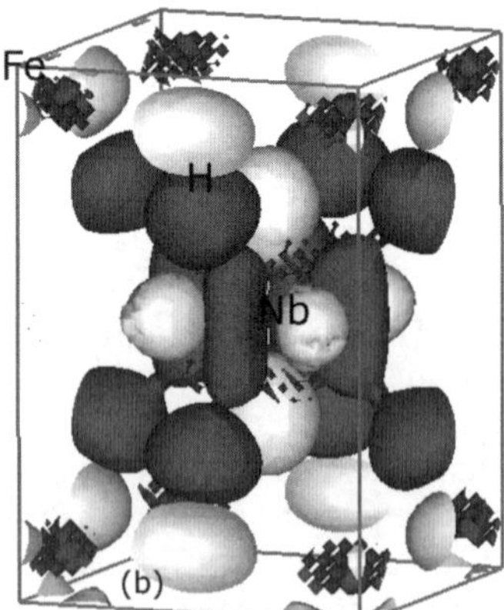

Figure 2. Charge density difference plots for the Nb Alloyed hydride with respect to the unalloyed charge difference taken as the reference. Nb is substituted for (a) Fe and (b) Ti. Dark and bright regions indicate charge depletion and accumulation respectively

It is seen from Fig. 2 that, as the alloy is hydrided the differences in the charge density distribution, therefore the state of bonding between the atoms, determines whether the formation energy increases or decreases. For the alloying elements put in place of Fe atom, causing a decrease in the formation energy have a common effect on the electronic structure. The orbitals of the alloying element pointing towards the hydrogen atoms are depleted, whereas there is some charge accumulation in the orbitals pointing away from hydrogen. This antibonding nature of the orbitals between the substituted atom and the hydrogen causes very strong bonding states to appear between the hydrogen atoms and the neighboring Fe atoms. This can be clearly seen in Fig. 2(a) for the Nb alloy and the very same was obtained for Zr alloy as well; both strongly reduce the formation energy of the hydride. In the alloy where, Zr was put in place of Ti, the same mechanism was found to be operative, but with very slightly stronger bonds between hydrogen and Fe, not very strong as in the previous case. Therefore Zr substitution in place of Ti did not decrease the formation energy as large as the Zr substitution in place of Fe. For the case that Nb was put in place of Ti, a similar antibonding Nb-H pair appears but this time no bonding is seen between hydrogen and the neighboring Fe and also Ti atoms. Therefore a weaker bond between Nb and H, and the absence of the strong interactions between the hydrogen atoms and the rest of the metals, Fe or Ti, caused a little increase in the formation energy. Another mechanism for the decrease of the formation energy was seen in Ni alloy which substitutes Ti. This time a stronger bonding state appears between Ni and the hydrogen without any change in the surrounding bonds. This effect therefore causes a slight decrease in the formation energy of the alloyed hydride.

It can be concluded therefore that, the effect of the alloying element on the formation energy of the hydride not only depends on the changes occurring in the electronic structure of the alloying element itself, but also on the changes occurring at the nearby metal atoms. In this respect atoms substituted for Fe has a more pronounced effect on the neighboring Fe atoms, thus causes a larger decrease in the formation energy. For the case that Ti is substituted, on the other hand, the electronic charge distribution was not changed considerably at the neighboring Ti and also at Fe sites. Therefore the effect of the alloying element remains limited.

It is also found that the alloying elements, substituted for Fe which causes a decrease in the formation energy of $FeTiH_4$ are all have a higher tendency [13] to form a hydride than Fe. When the alloying element of high tendency hydride former, substitutes the low hydride former Fe, it behaves completely oppositely and did not bond with hydrogen and furthermore forcing it to form strong bonds between the surrounding Fe atoms. Such a result is also reported [14] in pure XTi intermetallics (X = Fe, Co and Ni) that hydrogen interacts more strongly with X elements despite the fact that Ti is a hydride forming element whereas X's are non-forming.

4. Conclusions

In summary the influence of alloying elements on the stability of $FeTiH_4$ was investigated by means of total energy and electronic structure calculations. The heats of formation of alloyed systems were determined from the difference between the total energy of alloyed system and the total energies of the pure system, pure metals and hydrogen. It is found that, stabilizing the $FeTiH_4$ compound can be possible with alloying. Most effective alloying additions are Zr, Nb, Mo and Mg if they are substituted for Fe.

Acknowledgements

This study was supported by the Turkish Scientific and Technical Research Council (TUBITAK) Project no 106M174. The numerical calculations reported in this study were carried out at the ULAKBIM High Performance Computing Center at the Turkish Scientific and Technical Research Council (TUBITAK).

References

1. Ivey D.G., Northwood D.O., *J. Mater. Sci.*, 1983; **18**:321.
2. Thompson P., Reilly J.J., Hastings J.M., *J. Appl. Crystallogr.*, 1989; **22**:256.
3. Reilly J.J., Wiswall Jr. R.H., *Inorg. Chem.*, 1974; **13**(1):218.
4. Nagasako N., Fukumoto A., Miwa K., *Phys. Rev. B*, 2002; **66**:155106.
5. Miwa K., Fukumoto A., *Phys. Rev. B*, 2002; **65**:155114.
6. Smithson H., Marianetti C.A., Van der Ven A., Predith A., Ceder G., *Phys. Rev. B*, 2002; **66**:144107.
7. Kinaci A., Aydinol M.K., *Int. J. Hydrogen Energ.*, 2007; **32**:2466.
8. Kresse G., Hafner J., *Phys. Rev. B.*, 1993; **47**:558 ibid. 1994; **49**:14251.
9. Kresse G ., Furthmuller J., *Comput. Mat. Sci.*, 1996; **6**:15.
10. Kresse G., Furthmuller J., *Phys. Rev. B*, 1996; **54**(11):169.
11. Kresse G., Joubert D., *Phys. Rev. B*, 1999; **59**:1758.
12. Thompson P., Pick M.A., Reidinger F., Corliss L.M., Hastings J.M., Reilly J.J., *J. Phys. F*, 1978; **8**(4):L75–L80.
13. http://hydpark.ca.sandia.gov; Hydride information center, last checked May 2007.
14. Nambu T., Ezaki H., Yukawa H., Morinaga M., *J. Alloy. Compd.* 1999; **293–295**:213.

"HYSTERESIS" IN INTERACTION OF NANOCRYSTALLINE MAGNESIUM WITH HYDROGEN

K.B. GERASIMOV, I.G. KONSTANCHUCK, S.A. CHIZHIK
Institute of Solid State Chemistry and Mechanochemistry SB RAS, Kutateladze 18, Novosibirsk, 630128, Russian Federation
Tel.: 7-383-2-339-1842; Fax: 7-383-2-332-2847;
E-mail: gerasimov@solid.nsc.ru

J.-L. BOBET
Institut de Chimie de la Matière Condensée de Bordeaux ICMCB-CNRS, Université Bordeaux 1, 87 Av. Schweizer, F-33608 Pessac, France

Abstract. In recent years the kinetics of hydrogenation/dehydrogenation of magnesium have been significantly improved by the introduction of new methods of fabrication of nanocrystalline magnesium-catalyst composites. However, the equilibrium hydrogen pressure for decomposition of nanocrystalline MgH_2 was found to be somewhat lower than for conventional magnesium hydride. Moreover, the essential difference in equilibrium hydrogen pressure for absorption and for desorption of hydrogen by nanocrystalline magnesium was reported by many authors. This difference called "hysteresis" is a common phenomenon for hydrides, but it is not observed for magnesium powders with an ordinary particle size (more than 1 μm). The aim of the present work was to elucidate why the hysteresis arises in magnesium-hydrogen system with decreasing the particles size. It is shown, that the "hysteresis" observed in nanocrystalline magnesium is a seeming phenomenon, which is due to a hindered nucleation during hydriding at pressures closed to equilibrium. The pressures measured for desorption represent the real equilibrium. The plateau pressure of $Mg + H_2 \leftrightarrow MgH_2$ equilibrium is lower for nanocrystalline magnesium than that for conventional magnesium. This result is explained in terms of smaller surface energy of magnesium hydride in comparison with magnesium.

Keywords: hydrogen storage, nanocrystalline magnesium, gas-solid reactions, hysteresis

1. Introduction

Magnesium and magnesium-based alloys are regarded as very attractive materials for hydrogen storage due to high hydrogen capacity, relatively low cost and abundance of their raw materials. However, the practical application of magnesium hydride was restricted so far because of two main reasons: the reactions of hydrogenation/ dehydrogenation are slow and the pressure of hydrogen gas in equilibrium with magnesium hydride is low (1 bar at about 550 K), therefore a high temperature is required for the desorption of hydrogen.

In recent years, large success has occurred in kinetic properties of magnesium-based hydrogen storage materials from the introduction of new, non-conventional methods of fabrication of nanocrystalline magnesium-catalyst composites [1, 2]. These composites showed rapid absorption kinetics at temperature as low as 300 K and rapid desorption kinetics above 523 K at low hydrogen pressure (<0.15 bar) [3, 4]. However, the equilibrium pressure for the decomposition of nanocrystalline MgH_2 was observed to be somewhat lower than that for magnesium with ordinary particles size [5] and hence it required a temperature above 570 K for the desorption under a pressure of 1 bar. Moreover, an essential difference in equilibrium hydrogen pressures measured during the absorption and desorption of hydrogen was observed by many investigators for nanocrystalline magnesium (see [3, 5–10], for example).

This difference called "hysteresis" is a common phenomenon for hydrides [11], but it is not observed for magnesium powders with an ordinary particles size (more than 1 μm), if the duration of experiment is adequately long [12] and/or a catalyst is used to accelerate the reaction and so the reaching of the steady state [13]. Pure nanocrystalline magnesium has already a high rate of hydrogen sorption and desorption [5]. The addition of a catalyst results in further increase of a reaction rate. Therefore it is highly surprising, that a difference between equilibrium hydrogen pressure for formation and for decomposition of magnesium hydride was observed for nanocrystalline magnesium or magnesium-catalyst composites.

If experiments are carried out in conditions with good thermal transfer, exo- and endothermal effects can be excluded from consideration due to very slow rate of reaction near equilibrium. In this case, the hysteresis for hydrogen-metal systems is accounted for a partial inelasticity of a deformation caused by a movement of a metal-hydride interface during hydrogenation/dehydrogenation processes [14]. It is essential to emphasize, that it concerns the systems which are not complicated by other processes, except for transformation into hydride and back. Otherwise, the other cause of hysteresis may be (for example, decomposition, exfoliation, segregation, etc.) and this often takes place for solid solutions and intermetallics.

The negligible hysteresis for magnesium with an ordinary particles size means that the plastic accommodation of energy during the sorption and desorption of hydrogen is also negligible. There is no reason to expect a considerable increase in this energy for nanocrystalline magnesium.

The aim of the present work is to elucidate, why a difference in equilibrium hydrogen pressures for formation and decomposition arises in magnesium-hydrogen system with decreasing of the particles size.

2. Experimental details

The MgH_2 powder used in this work was prepared from magnesium (99.98% purity) at 653 K and 30 bar of hydrogen pressure (99.998% purity) during 5 days. According to X-ray diffraction (XRD) pattern, the powder consisted of approximately 95% MgH_2 and 5% of Mg. Nanocryctalline powders were prepared by milling in an AGO-2 planetary ball mill with hardened steel balls and vials. The vials were evacuated to a primary vacuum ($\sim 10^{-5}$ bar) and filled with hydrogen up to pressure of 10 bars before milling. The balls charge was approximately 200 g and the total mass of powders being treated was 3 g.

The crystalline structure and phase composition of powders were determined by XRD analysis (DRON-4 diffractometer) using Brag-Brentano geometry (CuKα radiation). The diffraction patterns were analyzed by a whole pattern fitting procedure using the program PowderCell 2.4 in order to determine the concentration of each phase, the crystallite size and the internal stress.

The hydrogen absorption/desorption kinetics were evaluated using Sievert's type system. The design of the system was similar to that used in [15]. The main features were: a reactor assembly with small internal volume due to the ceramic filler rod inside to minimize gas volume in the system (that is in an indeterminate temperature) and a large vessel immersed into Dewar vessel with ice-water mixture with volume approximately 20 times larger than the volume of other parts of the system, and more than 100 times larger than the volume of a sample place of the reactor. Due to these features the effective volume of the system change negligible with a sample temperature and a change in ambient temperature has only small effects on the pressure measured. The pressure was automatically monitored by an absolute pressure transducer with a measuring span of 25 bar and an accuracy better than 0.1%. The mass of the samples was approximately 100 mg in order to get good heat and mass transfer (hydrogen flow through the powder bed). The reactor was heated by a resistance furnace. The furnace temperature was controlled and measured with accuracy of 0.1 K.

3. Results and discussion

The crystallites size of MgH_2 quickly decreased with milling time and reached a steady state of 12–15 nm after milling for 30 min. During absorption-desorption cycling the crystallites size increased. It was demonstrated in [16] that milling of MgH_2 together with magnesium oxide led to a remarkable improvement of kinetics of hydrogen sorption/desorption due to a reduction of particles size. The effect was rationalized within the concept of "Process Control Agent", i.e. MgO reduced agglomeration processes during milling and the steady-state particle size of MgH_2 after prolonged milling was also reduced. In this work the efficiency of MgO for decreasing the growth rate of Mg crystallites during cycling was tested. Magnesium oxide of 99.99% purity was used for the experiments. According to our data, only a large amount of MgO distinctly reduced the growth of MgH_2 crystallites. Nevertheless, even for MgH_2 + 80 wt% MgO, the crystallites size increased from 12 to ~50 nm as a result of cycling at 523–613 K, in a comparison with growth from 15 to ~80 nm for MgH_2 milled without MgO after cycling in similar conditions. Since the crystallites size of Mg (MgH_2) increased during the experiments, the magnitude of the observed effects could not be related to a distinct crystallites size and the following discussion has only a qualitative character.

Figure 1 shows the experimental pressure-time curves for magnesium powder with an ordinary particle size (about 10 μm) and for the mixture MgH_2–80 wt% MgO milled for 30 min. While nanocrystalline magnesium reacted with hydrogen much faster than magnesium with an ordinary particle size under pressures exceeding the equilibrium pressure by ~5 (or more) bars, near the equilibrium pressure, the situation changed and the rate of hydrogen absorption by nanocrystalline magnesium turned out considerably lower, than the one for magnesium with an ordinary particles size. As a result, a negligible rate of hydrogen absorption indicating that the plateau pressure is reached, was not registered for nanocrystalline magnesium even when the experiments was continued over several days (curves 1 and 3 in Fig. 1). However, we showed that if during absorption process a higher pressure was applied for a short time (i.e. a few seconds) and then after reduced to the nominal pressure, the equilibrium pressure is reached even faster than for magnesium with an ordinary particles size. The difference between plateau pressures, measured for adsorption and desorption was weak in this case and the equilibrium pressure was noticeably lower, than for ordinary magnesium.

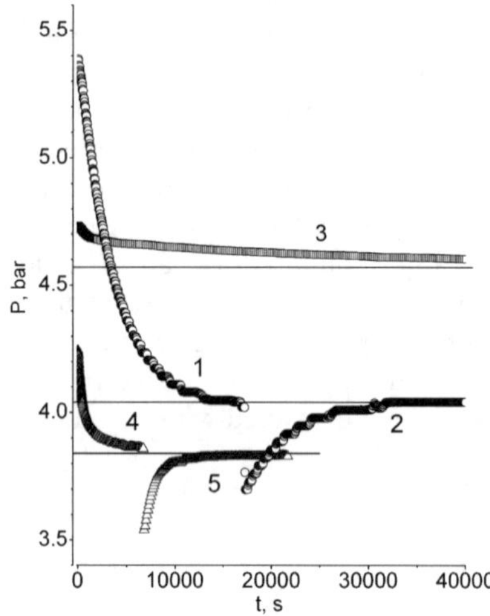

Figure 1. Pressure-time curves for reaching the equilibrium at 613 K. 1,2 – for conventional magnesium powder (1-hydriding, 2-dehydriding), 3 – hydriding of MgH_2 – 80 wt% MgO milled blend, 4 – hydriding of MgH_2 – 80 wt% MgO milled blend after short time hydriding at 18 bar, 5 – dehydriding of MgH_2 – 80 wt% MgO milled blend

The obtained data can be explained on the basis of the following assumptions. At a hydrogen pressure closed to the equilibrium one, the kinetic of hydriding of nanocrystalline magnesium is determined by the nucleation of MgH_2. The time of conversion of an individual magnesium particle into hydride after the nucleation is much less than the time of conversion of a whole sample. Partially hydrided at pressures closed to equilibrium one (during determination of P_{eq}), nanocrystalline magnesium powders mainly consist of two types of particles: some particles are completely converted into hydride and some other particles, which are still unreacted towards hydrogen. It is worth pointing out that particles containing both the hydride and the metallic form (coexistence of Mg and MgH_2 in a same particle) have never been observed. The last assumption was checked by the following experiments.

It is well-known since the work of Stander C.M. [17] that decomposition of MgH_2 is a nucleation and growth process. It is reasonable to expect that hydrogen desorption by hydrided magnesium powders with a large amount of particles, partially converted into hydride, will differ essentially from that of powders where such particles are absent. Figure 2 illustrates the dehydriding curves of nanocrystalline Mg – 80 wt% MgO mixture hydrided for different time (i.e. resulting in different degrees of transformation). The pressure of previous hydriding was being maintained in the range 2.75–2.65 bar for curves 1–4 and in the range 3.5–3.3 bar for curve 5. For curve 6, the pressure decreased during hydriding from 2.34 to 2.09 bar. This last case corresponds to the reaching of the equilibrium (i.e. the reaction rate at the end of hydriding was approximately the same as for curve 3 in Fig. 1). One can see that the maximum rate of decomposition or gas elimination are rather lower for curve 6 than for this of curve 1, while the amount of Mg transformed into MgH_2 for these two curves is almost the same. Moreover, the shape of curve 6 is similar to the one of curve 5 for completely hydrided powder and the curve 5 can be transformed to the curve 6 by simply changing the ordinate scale and vice versa. This could probably meant that partially converted particles (coexistence of Mg and MgH_2 in the same particle) are absent in the case of curve 6 (i.e. like in the case of completely hydrided powder).

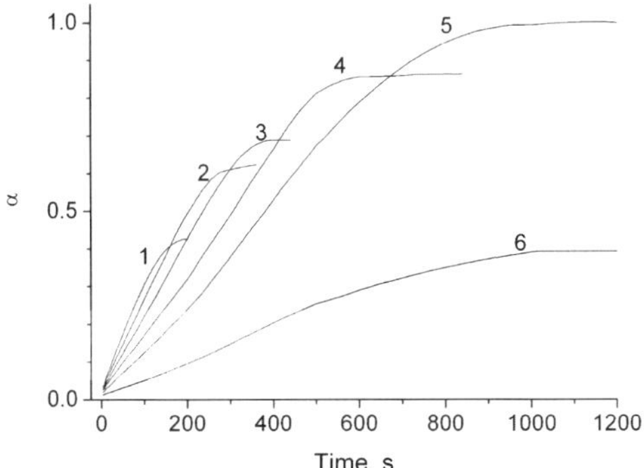

Figure 2. Dehydriding rates of MgH_2 – 80% MgO milled blend hydrided to different reaction fraction. T = 583 K. Pressure change was from 2.75 to 2.65 bar at previous hydriding in the case of (1–4) curves, from 3.5 to 3.3 bar in the case of curve (5) and from 2.34 to 2.09 bar in the case of curve (6). The last case reproduced reaching the equilibrium. α – relation of quantity of the hydrogen evolved at dehydriding to quantity of hydrogen absorbed at full hydriding

Considering the nucleation of MgH_2 as the rate determining step induces a powerful dependence of the reaction rate on hydrogen pressure applied for hydrogenation near equilibrium. Figure 3 shows the hydriding curve of Mg-80 wt% MgO at 583 K. The initial part of the curve (up to $\alpha = 0.8$) was measured in the pressure range

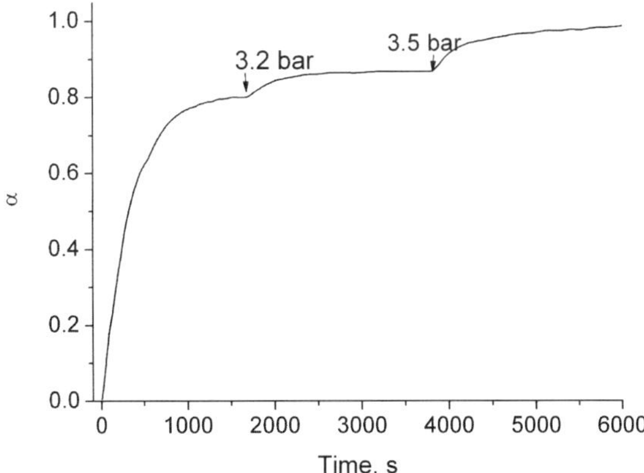

Figure 3. Hydriding curve for MgH_2 – 80% MgO milled blend at 583 K. Arrows mark the changes of pressure ranges. See text for details (α – degree of transformation of magnesium into hydride)

of 3.00–2.80 bar, after that, the pressure was increased up to 3.20 bar and the kinetic was recorded in the range of 3.20–3.06 bar and finally in the range of 3.50–3.27 bar. One can see that each increase in hydrogen pressure led to a large increase of absorption rate, while the overpressure (the difference between pressure in the reactor and the equilibrium pressure) changed only slightly. Indeed, the value of the equilibrium pressure measured in similar manner as for curve 4 in Fig. 1 was 1.52 bar. This means that the overpressure changed from 1.48–1.28 bar to 1.68–1.54 bar for the first increase of the pressure and to 1.98–1.75 bar for the second increase, whereas the absorption rate increased in several tens times after each increase of the pressure. This dependence cannot be explained if a chemisorption, diffusion or interface reaction is considered as the rate determining step. Only the hindered nucleation of MgH_2 can explain this strong dependence.

The hindered nucleation of MgH_2 during the reaction of nanocrystalline magnesium with hydrogen may be simply explained as being due to the constant rate of nucleation per unit area of magnesium surface. In this case the probability of nucleation for each individual particle varies as its squared linear dimension, i.e. quickly decreases with a reduction in sizes. Moreover, the nuclei of hydride are formed owing to a decomposition of supersaturated solution of hydrogen in magnesium. It is usually assumed for such nucleation (see [18], for example) that the supersaturated phase is in a state of quasi-equilibrium in which the distribution of nuclei of all sizes up to the critical one is essentially stationary with time. Such an assumption is reasonable, if only the number of hydrogen atoms in a particle of magnesium is sufficiently large. Hydrogen has a small solubility in magnesium. The estimation, carried out on the basis of data from the work of Nishimura et al. [19], shows that only about ten hydrogen atoms are dissolved in each particle of nanocrystalline magnesium of 15 nm of diameter at the equilibrium pressure. Under such condition, a joining of one hydrogen atom to a nucleus with critical size inevitably leads to an large decrease in the concentration of the solid solution being in equilibrium with the nucleus (disturbance of the quasi-equilibrium). As a result, the nucleus remains unstable after this joining instead of passing through the critical state and becoming free-growing.

Saita et al. [5] recently observed that the plateau pressure in hydrogen absorption and desorption for MgH_2 synthesized by hydriding of chemical vapor deposition (HCVD) was lower than that for the conventional product of MgH_2. The decrease was attributed to the great crystallinity of the HCVDed MgH_2 which was made of single crystals with nano-fibrous figure. This explanation of the effect is not in agreement with our results. The difference between equilibrium pressure of milled and conventional MgH_2 was a maximum (up to 40%) for initial absorption/desorption cycles and decreased with cycling. This means (from Saita's supposition point of view) that the as milled MgH_2 has the greater crystallinity than MgH_2 after cycling which is barely impossible. A noticeable contribution of a surface energy into the total energy of magnesium-hydrogen interaction for nanocrystalline magnesium seems to be a more reasonable explanation of the effect.

To explain the magnitude of effect, one need to suppose that the free surface energy of MgH_2 is less than that for Mg by 20–40%. Unfortunately, the surface energy of MgH_2 is unknown, to the best of our knowledge. Nevertheless, if a melting point of metal is not much lower than that of a binary compound of this metal, the surface energy of metal is typically larger than the surface energy of the compound [20]. For example, the values of 1,400, 600 and 800 mJ/m^2 for free surface energy of Zr, ZrO_2

and ZrC respectively [20] can illustrate our purpose. Therefore, the assumption that the surface energy of magnesium hydride is also lower than the surface energy of magnesium seems reliable.

The decrease of plateau pressure for nanocrystalline magnesium is of a great importance for hydrogen storage in magnesium based alloys. K.-F. Aguey-Zinsou et al. [16] supposed that one of the keys to enhancing the (de)hydriding properties of magnesium at temperatures below 523 K may be further decrease of magnesium particles size. The low decomposition pressure of MgH_2 is one of the main disadvantages which hamper its use for hydrogen storage. If the decrease of particles size inevitably leads to lowering of the decomposition pressure, to strive for further decrease seems not a promising direction.

4. Conclusions

"Hysteresis" in interaction of nanocrystalline magnesium with hydrogen is a seeming phenomenon which is due to a hindered nucleation during hydriding at hydrogen pressures closed to equilibrium. The pressures measured for desorption represent the real equilibrium.

The plateau pressure of $Mg + H_2 \leftrightarrow MgH_2$ equilibrium is less for nanocrystalline magnesium than that for conventional magnesium. The most probable reason of this is the smaller surface energy of magnesium hydride in comparison with magnesium.

These facts should be taken into consideration not only in scientific researches for correct determination of equilibrium hydrogen pressure for nanocrystalline magnesium hydride and, hence, for correct calculations of quantitative kinetic data of hydriding and dehydriding processes, but also in development of new nano-sized hydrogen storage materials.

Acknowledgements

The authors thank INTAS for financial support of the present study (the project INTAS 05 – 1000005 –7669).

References

1. Zaluski L, Zaluska A, Strom-Olsen JO. Nanocrystalline metal hydrides. *J Alloy Compd* 1997, **253–254**:70–79.
2. Zaluska A, Zaluski L, Strom-Olsen JO. Nanocrystalline magnesium for hydrogen storage. *J Alloy Compd* 1999, **288**:217–225.
3. Liang G, Huot J, Boily S, Van Neste A, Schulz R. Catalytic effect of transition metals on hydrogen sorption in nanocrystalline ball milled MgH_2-Tm (Tm = Ti, V, Mn, Fe and Ni) systems. *J Alloy Compnd* 1999, **292**:247–252.
4. Ma L-P, Wang P, Cheng H-M. Improving hydrogen sorption kinetics of MgH_2 by mechanical milling with TiF3. *J Alloy Compnd* 2007, **432**:L1–L4.
5. Saita I, Toshima T, Tanda S, Akiyama T. Hydrogen storage property of MgH_2 synthesized by hydriding chemical vapor deposition. *J Alloy Compd* 2007, **446–447**:80–83.
6. Gross KJ, Spatz P, Zuttel A, Schlapbach L. Mechanically milled Mg composites for hydrogen storage the transition to a steady state composition. *J Alloy Compd* 1996, **240**(1–2):206–223.

7. Gupta R., Agresti F., Lo Russo S., Maddolena A., Palade P., Principi G. Structure and hydrogen storage properties of MgH_2 catalyzed with La_2O_3. *J Alloy Compd* 2008, **450**:310–313.
8. Dehouche Z, Djaozandry R, Huot J, Boily S, Goyette J, Bose TR, Schulz R. Influence of cycling on the thermodynamic and structure properties of nanocrystalline magnesium based hydride. *J Alloy Compd* 2000, **305**:264–271.
9. Kandavel M, Ramaprabhu S. Correlation between hydrogen storage properties and amount of alloy particles in Mg-based composites. *J Alloy Compd* 2007, **438**:285–292.
10. Dehouche Z., Hamoudi S., Yoo Y., Belkacemi K. Effect of activated alloys on hydrogen discharge kinetics of MgH_2 nanocrystals. Proc. of 56th Canadian Chemical Engineering Conf., Sherbrooke, Quebec, Canada, October 15–18 2006, p.455.
11. Flanagan TB, Park C-N, Oates WA. Hysteresis in solid state reactions. *Prog Solid St Chem* 1995, **23**:291–363.
12. Kennelley JA, Varwing JW, Myers HW. Magnesium-hydrogen relationships. *J Phys Chem* 1960, **62**:703–704.
13. Akiba E, Nomura K, Ono S, Mizuno Y. Pressure-composition isotherms of Mg-Ni-H_2 alloys. *J Less-Common Met* 1982, **83**:L43–L46.
14. Balasubramaniam R. Hysteresis in metal-hydrogen systems. *J Alloy Compd* 1997, **253–254**:203–206.
15. Karty A, Grunzweig-Genossar J, Rudman PS. Hydriding and dehydriding kinetics of Mg in a Mg/Mg_2Cu eutectic alloy: pressure sweep method. *J Appl Phys* 1979, **50**:7200–7209.
16. Aguey-Zinsou K-F, Ares Fernandez JR, Klasse T, Bormann R. Using MgO to improve (de)hydriding properties of magnesium. *Mater Res Bull* 2006, **41**:1118–1126.
17. Stander CM. Kinetics of decomposition of magnesium hydride. *J Inorg Nucl Chem*, 1977, **39**:221–223.
18. Dunning WJ. Theory of crystal nucleation from vapor, liquid and solid systems. In: Garner WE, editor. Chemistry of the Solid State. London: Butterworth, 1955. pp. 159–183.
19. Nishimura C, Komaki M, Amano M. Hydrogen permeation through magnesium. *J Alloy Compd*, 1999, **293–295**:329–333.
20. Humenik M, Whalen TJ. Physicochemical aspects of cermets. In: Tinklepaugh JR and Crandall WB, editors. Cermets. New York: Reinhold, 1961, pp. 6–49.

HYDROGEN PERMEABILITY OF METALS AT PRESENCE OF INTERNAL STRESSES

N.M. VLASOV, I.I. FEDIK
Scientific Research Institute Scientific Industrial Association "Luch", Zheleznodorozhnaya 24, Podolsk Moscow region, 142100 Russia, Fax: +7(945) 5433363; E-mail: iifedik@luch.podolsk.ru

Abstract. Hydrogen permeability of metals at presence of internal stresses of different physical nature: temperature, concentration residual – has been investigated. Hollow cylinder from zirconium alloy is considered as illustration sample. The choice of such model system is conditioned by the following reason: zirconium alloys are used in nuclear power, cylindric claddings are integral elements of power plant constructions, internal stresses in hollow cylinder have logarithmic dependence on radial coordinate. The latter peculiarity significantly simplifies the solution of diffusion kinetics problems. Analytic dependences for hydrogen atoms concentration field with the account of internal stresses are received. Comparative analysis of hydrogen atoms diffusion flows through outer surface of hollow cylinder depending on the level and character of internal stresses distribution has been conducted. Hydrogen permeability of hollow cylinder is defined by two concurring processes: equilibrium concentration of hydrogen atoms on internal surface and the change of symmetry of diffusion equation, due to the field of stresses. Physical nature of the obtained results is discussed.

Keywords: internal stresses, hydrogen diffusion, permeability

1. Introduction

Investigation of interaction of hydrogen with metals presents great interest for perspective trends of modern technique. Successful solution of practical problems is impossible without realization of physical mechanisms of the current processes. One of them is hydrogen permeability of metals at presence of internal stresses. As the illustration, hollow cylinder made of zirconium alloy has been considered. The choice of such a model system is conditioned by the following reasons. First of all, zirconium alloys are used in nuclear power. It is conditioned by suitable combination of physical-mechanical properties and low cross-section of capture of thermal neutrons. Second, hollow cylinders (cylindric claddings) are integral elements of power plant constructions (e.g., fuel claddings of nuclear reactors). And at last, internal stresses in the similar geometry constructions quite often have logarithmic dependence on coordinates. Such dependence significantly simplifies the solution of diffusion kinetics problems with the account of internal stresses.

2. Internal stresses

The main types of internal stresses are temperature, concentration, residual. All of them take part in the processes of hydrogen reverse sorption and degradation of metal properties. Such influence is connected with acceleration of kinetics of diffusion. Really, internal stresses can accelerate (slow down) diffusion of hydrogen atoms in the volume of metal. Besides, these stresses change near-surface state of metal, which is manifested on the volume. It is known that influence of internal stresses on kinematics of diffusion goes through the first invariant of their tensor. Therefore, further we shall introduce analytic dependences of this characteristic concerning hollow cylinder. The occurrence of thermo-stresses is caused by non-uniform temperature deformation. For long hollow cylinder (the state of plane deformation) at free from radial stresses boundaries the first invariant of stresses tensor logarithmically depends on radial coordinates [1, 2]:

$$\sigma_{ll} = \frac{2\alpha(T_1 - T_2)\mu(1+\nu)}{(1-\nu)\ln\left(\frac{R}{r_0}\right)}\left\{1 - 2\ln\left(\frac{R}{r}\right) - \frac{2\ln\left(\frac{R}{r_0}\right)}{\left(\frac{R}{r_0}\right)^2 - 1}\right\}, \tag{1}$$

where α is coefficient of linear expansion, μ is shear modulus, ν – Poisson's ratio, $(T_1 - T_2)$ – temperature difference between inner and outer surfaces, r_0 and R are inside and outside radii of hollow cylinder. If $T_1 > T_2$, then $\sigma_{ll} < 0$ for $r = r_0$ and $\sigma_{ll} > 0$ for $r = R$. It means that material in the surrounding of inner surface is in the state of compression, and in the surrounding of outer surface it is in the state of spread. From the position of mathematical formalism it follows from evident inequality.

$$\frac{2\ln\left(\frac{R}{r_0}\right)}{\left(\frac{R}{r_0}\right)^2 - 1} < 1 \quad \text{at} \quad \frac{R}{r_0} > 1.$$

That is why definite contribution to the change of σ_{ll} sign is given by logarithmic member $\ln\left(\frac{R}{r}\right)$. If $T_1 < T_2$, then value σ_{ll} changes the sign. Concentration stresses are connected with non-uniform distribution of impurity atoms concentration. In continuous approximation concentration deformation is considered as continuous function of coordinates. Qualitative definition of this type of internal stresses is done analogously to temperature one. Coefficient of linear expansion corresponds to the change of parameter of crystal lattice due to single concentration of impurity atoms. The change in norms of constants in equations of thermo-elasticity gives possibility to write corresponding expression for the first invariant of concentration stresses tensor in hollow cylinder of identical geometry:

$$\sigma_{ll} = \frac{2\beta(C_1 - C_2)\mu(1+\nu)}{(1-\nu)\ln\left(\frac{R}{r_0}\right)} \left\{ 1 - 2\ln\left(\frac{R}{r}\right) - \frac{2\ln\left(\frac{R}{r_0}\right)}{\left(\frac{R}{r_0}\right)^2 - 1} \right\}, \tag{2}$$

where β is relative change of parameter of crystal lattice, $(C_1 - C_2)$ is concentration difference between inner and outer surfaces of hollow cylinder. The rest designations correspond to the earlier accepted. Note an interesting peculiarity of correlations (1) and (2). They differ from each other only in constants. That is why the influence of temperature and concentration stresses on the hydrogen atoms diffusion processes can be considered in single way with accuracy up to constants.

Concentration stresses exist in the material at non-uniform distribution of impurity atoms concentration. As the temperature increases, non-uniformity diffusely spreads and stresses disappear. However, at low temperatures (e.g., at room temperature) non-uniformity from substitution impurities stays unchanged due to the low diffusion value of coefficient. Diffusion coefficient of hydrogen atoms at this temperature for several orders exceeds the value of this characteristic for substitution impurities. Such peculiarity allows to consider diffusion migration of hydrogen atoms in the set up field of concentration stresses. The value $\beta(C_1 - C_2)$ presents non-uniform concentration deformation. The value of coefficient β of (2) is defined in the following way:

$$\beta = \frac{\Delta a}{a} \cong \frac{1}{3}\frac{\Delta \upsilon}{\upsilon}, \tag{3}$$

where υ is volume of an atom of the main metal, $\Delta\upsilon$ is change of the metal's volume at substitution of its atom by another impurity atom, *a* is parameter of crystal lattice. For system Zr-Sn at $r_{Zr} = 1.71 \cdot 10^{-10}$ m and $r_{Sn} = 1.86 \cdot 10^{-10}$ m [3] we get $\beta = 0.1$.

Residual stresses belong to quite important class of internal stresses. They occur in the material while conducting different technological operations. Let us consider the process of formation of residual stresses in the hollow cylinder. Banks of the cut of the hollow cylinder are moved to the angle ω and deficit material is placed there. At such an operation the surrounding region of the outer surface is in the state of compression and in the surrounding of inner surface it is in the spread state. The first invariant of tensor of residual stresses logarithmically depends on radial coordinate

$$\sigma_{ll} = \frac{\mu\omega(1+\nu)}{2\pi(1-\nu)} \left\{ 1 - 2\ln\left(\frac{R}{r}\right) - \frac{2\ln\left(\frac{R}{r_0}\right)}{\left(\frac{R}{r_0}\right)^2 - 1} \right\}, \tag{4}$$

where ω is turning angle of the banks of the hollow cylinder cut. The rest designations are identical to relations (1) and (2). The value $\omega/2\pi$ presents relative deformation while forming residual stresses in the hollow cylinder (angle ω is measured in radians). It is interesting to note that at the similar relations of the sizes R/r_0 coordinate dependence of internal stresses for relations (1), (2) and (4) is identical. Difference is characteristic only to constants which define nonuniform deformation:

temperature, concentration, while forming residual stresses. Similar coordinate dependences of expressions (1), (2), (4) significantly simplify the solution of the problems of diffusion kinetics.

3. Hydrogen segregation in the area of internal stresses

It is known that reverse hydrogen sorption depends on outside pressure and temperature. These parameters define curves of hysteresis, formation of hydrides in some metals, and also degradation of metal properties at hydrogenation and dehydrogenation. Moreover kinetics of diffusion processes significantly depends on the level and character of internal stresses distribution. This problem is paid less attention in different publications. It is somehow connected with the fact that in general case internal stresses have complex coordinate dependence that makes it difficult to receive analytical dependences for hydrogen atoms concentration field with the account of internal stresses. Introduced relations for σ_{ll} have logarithmic dependence on radial coordinate. Similar dependence enables getting analytical expressions for concentration fields of hydrogen atoms and, thus, investigate volume and surface effects taking into account internal stresses.

Binding energy (interaction potential) of hydrogen atom with the field of internal stresses is evaluated by the known relation [4]

$$V = -\frac{\sigma_{ll}}{3}\delta\upsilon, \tag{5}$$

where $\delta\upsilon$ is change of the metal's volume when placing hydrogen atom. This value for all the metals is approximately the same and makes $3\cdot 10^{-30}\,\mathrm{m}^3$. For $\sigma_{ll} > 0$ (positive dilatation) and $\delta\upsilon > 0$ (hydrogen atom increases parameter of crystal lattice) potential V takes negative value. This corresponds to attraction of hydrogen atom to the region of spread stresses and its release out from the region of compression stresses. After concentration with the account of the main types of internal stresses for hollow cylinder, from the solution of the problem it follows [5]:

$$\frac{1}{D}\frac{\partial C}{\partial t} = \frac{\partial^2 C}{\partial r^2} + \frac{1}{r}\frac{\partial C}{\partial r} + \frac{\nabla(C\nabla V)}{kT}, \quad r_0 < r < R, \tag{6}$$

$C = 0,\ t = 0,\ C = C_p$ at $r = r_0$, $C = 0$ at $r = R$, where D is coefficient of hydrogen atoms diffusion, k is Boltzmann's constant, T is absolute temperature, C_p is equilibrium concentration on the inner surface, V is total potential of hydrogen atom interaction with the field of temperature, concentration and residual stresses. Theses tresses can enter the final expression with different signs and thus strengthen (weaken) each other depending on the corresponding constants. Equilibrium concentration C_p on inner surface is defined by potential of interaction V:

$$C_p = C_0 \exp\left(-\frac{V}{kT}\right) \quad \text{at} \quad r = r_0, \tag{7}$$

where C_0 is concentration of hydrogen atoms with absence of internal stresses. Physical sense of initial and boundary conditions of the problem (6) is quite evident. In the initial moment of time there is no hydrogen in the material of hollow cylinder. Zero concentration is saved on the outer surface. It means that hydrogen atoms immediately leave outer surface as soon as they appear. Such statement of boundary conditions allows us better to set off the role of internal stresses of various physical nature. As the dependence σ_{ll} from radial coordinate is similar for all the types of internal stresses, we shall stop on the following variant: on the inner surface there are compression stresses, and on the outer – spread stresses. After not complicated transformations with the account of $\Delta V = 0$ and $\nabla V \sim 1/r$ we receive the simplified variant of the problem (6):

$$\frac{1}{D}\frac{\partial C}{\partial t} = \frac{\partial^2 C}{\partial r^2} + \frac{1-\alpha_1-\alpha_2-\alpha_3}{r}\frac{\partial C}{\partial r}, \quad r_0 < r < R, \tag{8}$$

$C = 0$, at $t = 0$, $C = C_p$ at $r = r_0$, $C = 0$ at $r = R$.

Constants α_1, α_2 and α_3 characterize the relation of hydrogen atoms binding energy with the field of internal stresses to the energy of thermal motion. For the adopted types of internal stresses these dimensionless parameters are as follows:

$$\alpha_1 = \frac{4\alpha(T_1-T_2)\mu(1+\nu)\delta\upsilon}{3kT\pi(1-\nu)\ln\left(\frac{R}{r_0}\right)};$$
$$\alpha_2 = \frac{4\beta(C_1-C_2)\mu(1+\nu)\delta\upsilon}{3kT\pi(1-\nu)\ln\left(\frac{R}{r_0}\right)}; \tag{9}$$
$$\alpha_3 = \frac{\omega\mu(1+\nu)\delta\upsilon}{3kT\pi(1-\nu)}.$$

All designations have been met earlier. As seen, logarithmic coordinate dependence quite strongly changes the diffusion equation with the account of internal stresses. In fact, their account leads to introduction of some dimensionless parameters which are easily calculated. And most of all: in the frames of linear theory with attraction of the superposition principle their algebraic addition is approved. All this gives the possibility to direct the diffusion kinetics in the volume at interaction of hydrogen with metals. If the values of dimensionless parameters are significantly less than one, then internal stresses are weak perturbations of the main diffusion flow. Opposite case corresponds to prevailing role of internal stresses. If the values of the parameters are close to one, then the influence of concentration gradients and internal stresses is juxtaposed. Let us evaluate the values of dimension less parameters for the systems Zr-Sn at the following values of the constants: $R/r_0 = 1.1$; $\alpha = 10^{-5} K^{-1}$; $T_1 - T_2 = 10^2 K$; $\mu = 4\cdot 10^{10}$ Pa; $\nu = 0.3$; $kT = 4.1\cdot 10^{-21}$ J $\delta\upsilon = 3\cdot 10^{-30} m^3$; $\beta = 0.1$; $\omega = 0.16 rad$; $C_1 - C_2 = 1.3\cdot 10^{-2}$ (at). Evaluations show that for adopted conditions numerical values of dimensionless parameters are close to the one with slight deviation to one or the other side. For better illustration we shall consider two cases of distribution of internal stresses. In the first case on the inner surface of hollow cylinder there are only compression stresses, which smoothly turn into spread stresses on the outer surface. The second case describes opposite situation: spread stresses on the inner surface are changed into compression stresses on the outer one. Mathematically it leads to the change of sign of dimensionless parameters of the equation (8). If $\alpha_1 + \alpha_2 + \alpha_3 = 1$, then from (8) we get:

$$\frac{1}{D}\frac{\partial C}{\partial t} = \frac{\partial^2 C}{\partial r^2},\ r_0 < r < R, \tag{10}$$

$C = 0,$ at $t = 0,$ $C = C_p$ at $r = r_0$, $C = 0$ at $r = R$.

Internal stresses change the law of formation of concentration field for hydrogen atoms: profile of concentration in the hollow cylinder is formed by the law for strip. Such regularity has simple physical explanation. For $\frac{\partial C}{\partial r} < 0$ the term $\frac{a_1 + a_2 + a_3}{r}\frac{\partial C}{\partial r}$ corresponds to the "source" of hydrogen atoms. The latters are displaced from the inner near-surface region and are attracted to the outer one. In this way inner stresses accelerate the process of hydrogen atoms concentration field formation. It follows from the equation (8).

Really, at $\frac{\partial C}{\partial r} < 0$ the speed of the concentration change is larger for $\alpha_1 + \alpha_2 + \alpha_3 = 1$, than for their zero value. If on the inner surface all the considered stresses are spreading and on the outer – compressed, then for $\alpha_1 + \alpha_2 + \alpha_3 = -1$ we get:

$$\frac{1}{D}\frac{\partial C}{\partial t} = \frac{\partial^2 C}{\partial r^2} + \frac{2}{r}\frac{\partial C}{\partial r},\ r_0 < r < R, \tag{11}$$

$C = 0,$ at $t = 0,$ $C = C_p$ at $r = r_0$, $C = 0$ at $r = R$.

And again internal stresses change the law of concentration field for hydrogen atoms: concentration profile in hollow cylinder is formed by the law for hollow sphere, $\frac{\partial C}{\partial r} < 0$ the term $\frac{a_1 + a_2 + a_3}{r}\frac{\partial C}{\partial r}$ corresponds to "discharge", for hydrogen atoms. Their diffusion migration is slowed down by the way of reaching the outer surface. The speed of formation of concentration profile decreases. This also follows from the equation (8). In fact at $\frac{\partial C}{\partial r} < 0$ the speed of concentration change is less for $\alpha_1 + \alpha_2 + \alpha_3 = -1$ compared to zero value.

Solution of the problems (10) and (11) are known. For convenience let us write them as follows:

$$C = C_p^1 f(r,t)\ (a), \qquad C = C_p^2 \frac{r_0}{r} f(r,t)\ (b), \tag{12}$$

where
$$f(r,t) = \frac{\frac{R}{r_0} - \frac{r}{r_0}}{\frac{R}{r_0} - 1} - \frac{2}{\pi}\sum_{n=1}^{\infty} \frac{\sin\left(\pi n\left(\frac{r}{r_0} - 1\right) \Big/ \left(\frac{R}{r_0} - 1\right)\right)}{n} \exp\left[-\frac{\pi^2 n^2 D t}{r_0^2\left(\frac{R}{r_0} - 1\right)^2}\right].$$

Formation process of concentration fields occurs according to (12a) and (12b) identically in time. It follows from the theory of equations of mathematical physics, as the equation (11) reduces to the equation (10) by simple change of variables. However at other equal conditions (for instance, at $C_p^1 = C_p^2$) the speed of concentration change according to (12b) slows down compared with (12a). At similar values $f(r,t)$ the similar value of concentration is received for the same time that of the larger value of concentration according to (12a).

The surface also has significant influence on the occurrence of diffusion processes. First of all it is conditioned by the influence of near-surface stresses on the change of the equilibrium concentrations according to the ratio (7). Saving through logic of statement we write this relation for two variants of summarized value of dimensionless parameters:

$$C_p^1 = C_0 \exp\left[\frac{1}{2} - \frac{\left(\frac{R}{r_0}\right)^2 \ln\left(\frac{R}{r_0}\right)}{\left(\frac{R}{r_0}\right)^2 - 1}\right]. \text{ for } \alpha_1 + \alpha_2 + \alpha_3 = 1,$$

$$C_p^2 = C_0 \exp\left[\frac{\left(\frac{R}{r_0}\right)^2 \ln\left(\frac{R}{r_0}\right)}{\left(\frac{R}{r_0}\right)^2 - 1}\right]. \text{ for } \alpha_1 + \alpha_2 + \alpha_3 = -1, \quad (13)$$

Internal stresses of different sign change equilibrium concentration of hydrogen atoms on the inner surface of hollow cylinder. At $\alpha_1 + \alpha_2 + \alpha_3 = 1$ dilatation on the inner surface is negative (compression stress) and that is why $C_p^1 < C_0$, that is equilibrium concentration is less than the corresponding value at the absence of stresses. Physically it means, that surface effect (decrease of equilibrium concentration) is concurring process, while forming concentration profile according to equation (10). At that, the role of surface effect is decisive. At $\alpha_1 + \alpha_2 + \alpha_3 = -1$ positive dilatation (spread stresses) increases equilibrium concentration of C_p^2 compared to C_0, that is $C_p^2 > C_0$. It means that the decrease of formation speed of concentration profile according to (11) is compensated by the increase of C_p^2. And again surface effect is decisive.

4. Hydrogen permeability of hollow cylinder

The influence of the considered volume and surface effects at interaction of hydrogen with metals is illustrated on the example of hydrogen permeability of hollow cylinder. Diffusion flows though outer surface in the condition $f(R,t) = 0$ for relation (12a) and (12b) are the following:

$$\left|\overline{j_1}\right| = DC_p^1 \frac{\partial f(r,t)}{\partial r}\Big|_{r=R}; \quad \left|\overline{j_2}\right| = DC_p^2 \frac{r_0}{R} \frac{\partial f(r,t)}{\partial r}\Big|_{r=R}; \quad (14)$$

where $\left|\overline{j_1}\right|$ and $\left|\overline{j_2}\right|$ – diffusion flows of hydrogen atoms through outer space for two cases correspondingly. Their relation characterizes hydrogen permeability depending on sign of dilatation field. For $\frac{R}{r_0} = 1.1$ we have $\frac{\left|\overline{j_1}\right|}{\left|\overline{j_2}\right|} = \exp(-0.098) \approx 1.$

Accepted relation of sizes R/r_0 saves hydrogen permeability at the change of the sign of dilatation field on the inner surface. Physically it means that equilibrium concentrations on the inner surface and stresses in the volume by their influence mutually compensate each other. If not to take into account surface effects, then at $C_p^1 = C_p^2$ we get for $\frac{R}{r_0} = 1.1$ $\frac{\left|\overline{j_1}\right|}{\left|\overline{j_2}\right|} = \frac{R}{r_0} = 1.1$.

In this case hydrogen permeability is defined only by relation of sizes of hollow cylinder. This relation shows that spread stresses increase diffusion flow from outer surface compared with compression stress. With the increase of the relation of sizes hydrogen permeability increases proportionally $\frac{R}{r_0}$. At similar surface concentrations hydrogen atoms are released from inner surface region and attracted to the outer one. This volume effect works stronger compared to the opposite character of internal stresses distribution.

5. Conclusions

There is considered hydrogen permeability of metals at presence of internal stresses of different physical nature: temperature, concentration, residual. Analytic dependences for hydrogen atoms concentration field in hollowed cylinder from zirconium alloy are received. Comparative analysis of hydrogen atoms diffusion flows through outer surface of hollow cylinder, depending on the level and character of distribution of internal stresses, has been conducted. It is shown that hydrogen permeability of hollow cylinder is defined by two concurring processes: equilibration concentration of hydrogen atoms on the inner surface of cylinder and changing the symmetry of diffusion equation due to the field of internal stresses.

References

1. Melan E., Parkus H., Wärmerspannungen infolge Stationärer Temperaturfelder, Springer, Wien, 1953.
2. Timoshenko S.P., Goodier J.N., Theory of Elasticity, McGraw-Hill, Tokyo, 1970.
3. Friedel J., Dislocations, Pergamon, Paris/Frankfurt, 1964.
4. Teodosiu C., Elastic Models of Crystal Defects, Springer, Heidelberg, 1982.
5. Vlasov N.M., Fedik I.I., Structural and impurity traps for hydrogen atoms, *Int. J. Hydrogen Energy*, 2006, **31**: 265.

COMPUTER MODELS OF NANOPARTICLES INCLUSIVE OF HYBRID ONES

G.S. YURJEV
Institute of Inorganic Chemistry SB RAS, 630090 Novosibirsk, Russia

Abstract. In our investigation we have examined the diffraction details, reasons for reflection broadening from nanoparticles, shift of reflections and appearance of additional reflections as compared with similar reflections of polycrystals. Understanding of these features is achieved via computer models of building units for nanoparticles. They can be spherical, hollow spherical, non-spherical, hybrid (core-shell) spherical particles and also icosahedrons constructed from tetrahedrons. Creation of the computer models is presented and their different projections and occurrence of interface (bonding between core and shell atoms) can be viewed on the PC display. The reality of interface is supported by interference in diffraction patterns which occurs at roughly equal number of atoms in the core and shell.

Experimental diffraction patterns from detonation nanodiamonds (DNDs) compared with theoretically computed accordingly to the proposed models are presented. There are studied nanodiamonds with respect to the reflection positions and their diffraction profile, and spherical hybrid nanodiamonds consisting of the diamond core and graphite shell. Model for a hypothetical spherical hybrid nanoparticle consisting of the diamond core and glycine shell and its theoretical diffraction pattern are also presented.

Original ideas in the present study are a complex of the programs for creation of computer models; a set of structural units for nanoparticles; observing of interface, compliance with the topology and complementarities of chemical bonds according to current notion of the pair potential of interatomic interaction.

Keywords: carbon, nanoparticles, diffraction, simulation

1. Introduction

The study of spatial structure of poly-nanocrystalline (NC) particles is linked with the complexity of interpretation of their diffraction patterns due to some shift of reflections and appearance of additional reflections in some cases, as compared with similar reflections of polycrystals, and their noticeable broadening (smoothness), as in the case of non-crystalline substances. The reflection broadening results from the comparability of the NC size with the diffracted radiation wavelength. The shift of reflections and appearance of extra reflections is linked with the diffraction on interatomic distances, which differ from the interplanar distances of polycrystals; for instance, on distances between NC atoms, including surface atoms, or between atoms of different structural elements the hybrid NCs consist of.

These problems can be solved with the help of 3D computer models of structural units of NCs, including hybrid crystals. The models are used to compute theoretical

B. Baranowski et al. (eds.), *Carbon Nanomaterials in Clean Energy Hydrogen Systems.*

diffraction patterns to compare with an experimental one both in the diffraction character (if there is an additional reflection) and in the position of reflections (at determination of crystal lattice parameters). Theoretical diffraction patterns are computed in accordance with the Debye analytical representation of 1915 [1], which describes the dependence of diffraction on interatomic distances in NC models:

$$I(s) = \sum_{i}^{N} \sum_{j}^{N} f_i(s) f_j(s) \sin sr_{ij}/sr_{ij} = \sum_{i=j}^{N} f_i^2(s) + 2\sum_{i\neq j}^{N} f_i(s) f_j(s) \sin sr_{ij}/sr_{ij};\; i(s) = \sum_{i\neq j}^{N} fi(s) fj(s) \sin sr_{ij}/sr_{ij}$$

here, $I(s)$ is the diffraction pattern consisting of two components of diffraction: on atoms in gaseous state, $\sum f_i^2(s)$, and on the atoms located at the r_{ij}, $i(s) = \{S(s)-1\}$, or interference contribution, where $S(s) = \{I(s)/\sum f_i^2(s)\}$ is the structural factor; $f(s)$ is the tabulated atomic scattering factor; $s = (4\pi/\lambda) \sin\theta$, $Å^{-1}$ is the recoil impulse; λ is the wavelength of diffracted radiation; θ^0 is half the Bragg diffraction angle; N is the number of atoms in a spatial model.

2. Scheme of diffraction experiment

Experimental diffraction patterns of NCs as well as of non-crystalline substances are weak in intensity. Besides, they can also show reflections from NCs that have noticeably different sizes. Therefore, the patterns have to be registered with a high-resolution θ–2θ diffractometer with the application of high-intensity synchrotron radiation from accelerators. Synchrotron radiation is the radiation of electrons moving with velocities close to relativistic in a magnetic field distorting their trajectory. The high resolution of diffraction space ($\Delta\lambda/\lambda = 4 \cdot 10^{-4}$) is provided by monochromatization of the incident and diffracted radiation. Figure 1 shows a schematic of diffractometer used in our experiments.

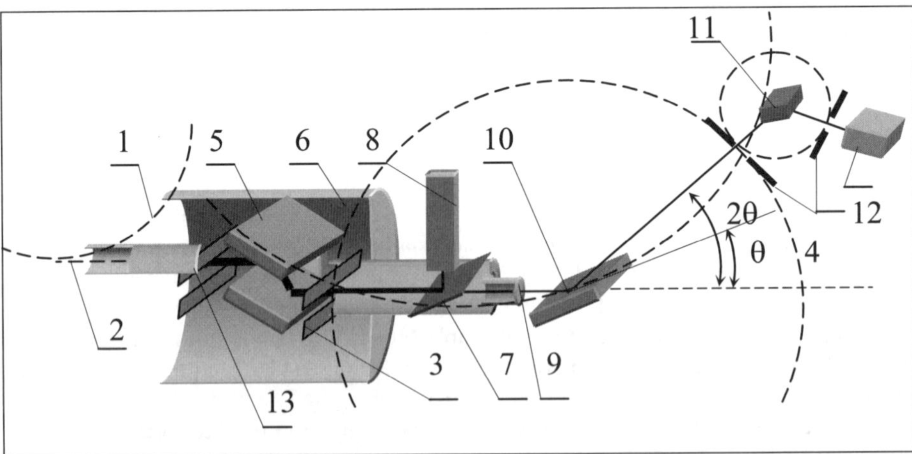

Figure 1. A schematic of diffractometer: 1 – storage ring orbit; 2 – SR beam; 3, 4 – diaphragms; 5, 11 – monochromators; 6 – vacuum volume of the "batterfly" monochromator (111)Si; 7 – thin lavsan film; 8 – entrance beam monitor detector; 9 – Be foil; 10 – sample; 12 – NaJ(Tl) scintillator; 13 – radiation gate. High degree of monochromatization $\Delta\lambda/\lambda \approx 3 \cdot 10^{-3}$–$10^{-4}$ at $\lambda = 1.5405$ Å is achieved by an additional spectral analyzer to discriminate scintillator impulses. Compton scattering and fluorescence are extracted by (111)Ge monochromator located in diffracted beam. Diffraction patterns are required to be recorded both with narrow and wide diaphragms and ratio between intensities of the patterns are being equivalent to ratio between the widths of diaphragms

3. Creation of the spatial computer model of nanoparticle (visualization of nanoparticles)

Computing theoretical diffraction patterns for the above-mentioned cases requires 3D modeling different structural units for NCs. Such units can be created on the base of unit cells (UCs) of crystalline substances, i.e. on the base of known parameters of crystal lattices, coordinates of atoms in the UCs, spatial groups of symmetry and atomic radii [2, 3]. Coordinates of atoms of such units are taken from an array of coordinates computed through translation of a unit cell along the x, y, and z axes of 3D space and restricted by the number of translations. Then there is computed the position of atoms in the space of the coordination spheres (CSs).

The CS parameters are the coordinates and number of atoms in the CS and also CS radii defined as the distance between the CS and the "centre" atom. For a spherical nanocrystal, its size is presented by the radius of the outer CS. The size of hollow nanosphere is given by the atoms of the inner to outer CSs [3]. For a non-spherical nanocrystal, its size is defined by the atoms of the 3D space limited by the number of UC translations along each of the axes (1D and 2D crystals can also be computed) [2]. In spherical hybrid nanocrystals consisting of a core and shell, the core is presented by an outer CS that confines it and the shell is presented by a hollow sphere. For an interface (chemical bond) to arise, the distance between the outer CS of the core and inner CS of the shell should be mean value relative to the radius of their atoms. Using known atomic coordinates and available program software we can present 3D spatial visualization of nanoparticles on the IBM-PC display.

4. Computer models supported by experimental diffraction pattern

For detonation nanodiamond (DND) powders, the NC size is linked with the degree of diffraction pattern broadening as well as with determination by the trial method at comparison of the areas under reflections in both theoretical and experimental patterns. The NC lattice parameters that define the reflection position are necessarily found via computing theoretical diffraction from a 3D model and comparison of reflection positions with experimental ones. It would be incorrect to define lattice parameters from reflections on an experimental diffraction pattern alone as it is done for polycrystalline substances.

This is confirmed by computation of a theoretical diffraction pattern the reflection positions in which depend on the NC size, the simulated atomic coordinates being unchanged (Fig. 2). If the character of an experimental pattern of, for instance, modified DNA [2] differs from a known pattern of natural polycrystalline diamond, it is necessary to construct a 3D model of hybrid NC from fragments of a specific size, shape (sphere, shell, and tetrahedron) and appropriate diamond lattice parameters. For a 3D model of spherical hybrid NC diamond consisting of a core and shell with different lattice parameters, an additional reflection between (220) and (311) (labeled with * in Fig. 3), missing in an experimental pattern of natural polycrystalline diamond, is reproduced in a theoretical diffraction pattern (Fig. 3).

The best result was obtained for a spatial spherical model consisting of the core made up of the 1st to 12th CSs with the lattice parameter *a* compressed by 5% (a = 3.3884 Å) and two shells surrounding a core. The first shell was made up of the13th to 18th CSs

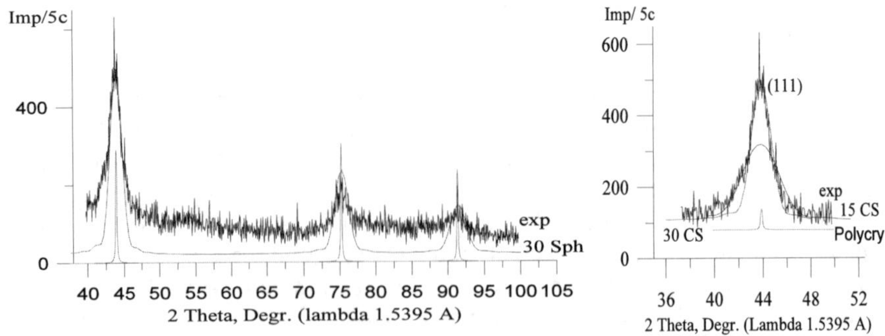

Figure 2. Left panel: experimental XRD pattern from synthesized powder DND sample [4] compared to the computed one for ND of size: $r = 27$ Å, made up of the 1st to 30th CSs, 14,811 atoms (red line). Right panel shows experimental diffraction (111) reflection compared with computed for ND of $r = 27$ Å (red line) and for ND of $r = 14.8$ Å, made up of the 1st to 15th CSs, 1,963 atoms (blue line). Calculated discrete reflections (rose line) belong to polycrystalline diamond. DND of $r = 27$ Å is formed by approx. equal (within 3.2%) number of atoms in the core $r_{core} = 21.5$ Å, 1st to 23rd CSs, 7,289 atoms and the shell $r_{sh} = 27$ Å, 24th to 30th CSs, $\Delta r = 5.63$ Å, 7,522 atoms

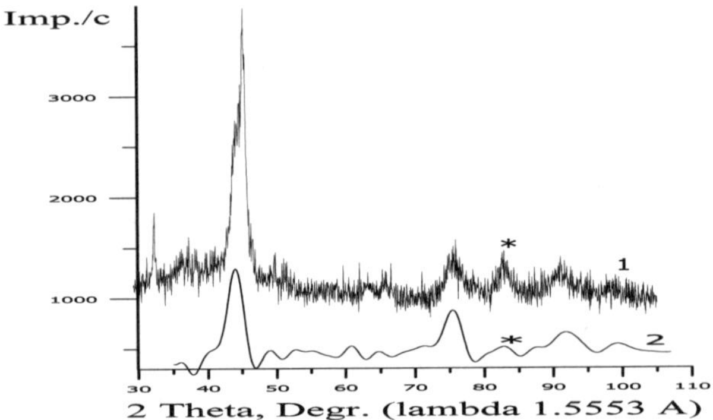

Figure 3. Experimental (1) and computer-simulated (2) XRD patterns from mechanically treated DND showing an additional diffraction reflection (labeled with *) missing in the natural diamond

with the parameter *a* compressed by 2.5% and the second shell was non-compressed. The first shell was of $\Delta R = 4.18$ Å ($R_{inner} = 13.11$ Å, $R_{outer} = 17.29$ Å), the second shell had R = 18.08 Å size and involved only the19th CS, in which ≈15% atoms were accidentally missing (the outer shell being loosely formed) relative to amount of atoms in an analogous CS in natural diamond. Introduction of vacancies into outer CS resulted in sharpening additional reflection (at $2\theta \approx 84°$) and better coincidence of its positions in the experimental and calculated patterns. Moreover, reproduction in the theoretical diffraction pattern of two weak reflections (at $2\theta \approx 63–66°$) indicates the best correlation between a proposed model of NP and its real structure.

For a hybrid NC of diamond/graphite, which consists according to computer models of a core (diamond) and shell (graphite), the simulated diffraction patterns from the core, shell and hybrid NC show the difference between them. The occurrence of interference and interface (Fig. 4, curve 1) proves that the computer model is realistic and that the hybrid NC system is stable.

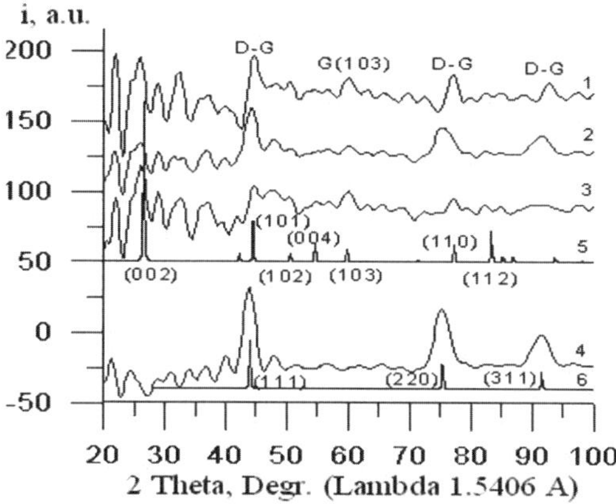

Figure 4. Computer-simulated interference patterns of hybrid diamond/graphite ($P6_3/mmc$) particle: (1) from two-phase structure; (2) additive superposition of diamond core and graphite shell patterns; (3) from a graphite shell (hollow); (4) from a diamond core; curves 5 and 6 showing discrete diffraction patterns from the polycrystalline graphite and polycrystalline diamond, respectively

Such hybrid NC assembly confirms that the topology and complementarities of chemical bonds (mutual consistency between chemical structures providing their interaction) have been observed, though they can be distorted within tolerable limits, i.e. the interatomic distances can be disproportionate, which is not forbidden at self-assembly and self-organization of bond geometry. Commensuration of bonds is checked by the value of interatomic distances, from minimal to maximal, in accordance with the atom radii.

This complies with the existing notion of the pair potential of interatomic interaction with a potential pit minimum at an interatomic distance of 2r (Fig. 5). A pit width is restricted by the smallest distance and from the side of big distances by 3r, the system being unstable at such a pair potential.

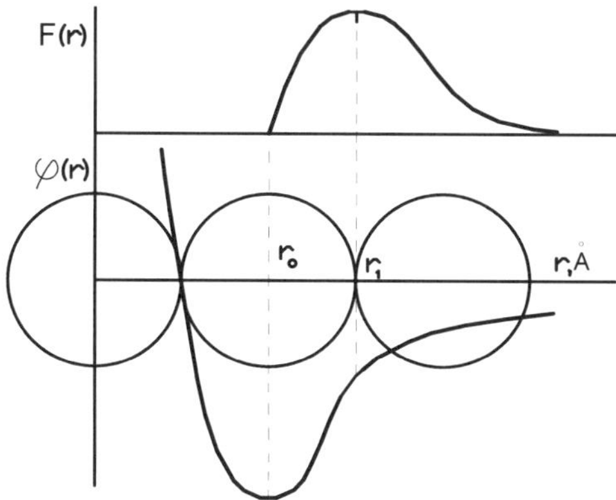

Figure 5. Bottom scheme showing a pair potential of the inter-particle interaction for non-metal atoms $\varphi(r)$; top curve describing interaction force F(r) between a pair of atoms

A pattern formed by additive superposition of computed independent diffractions of the core and shell differs from a computed diffraction pattern for the hybrid NC (non-additive superposition) (Fig. 4, curve 2). The difference is that there is interference in the hybrid NC case (diamond and graphite reflections). It causes diffraction on the distances between the core and shell atoms in the common continuum of the hybrid NC at roughly equal numbers of atoms in the core and shell [3].

Diffraction pattern (Fig. 4, curve 1) is compared to experimental (Fig. 6). Difference between them is possible to occur because of non-ideality of the real structure. The computer model is being realistic and the hybrid NC system is stable in the ground state because atomic coordinates in outer CS core and inner CS shell are complementary, though chemical bonds can be distorted within tolerable limits.

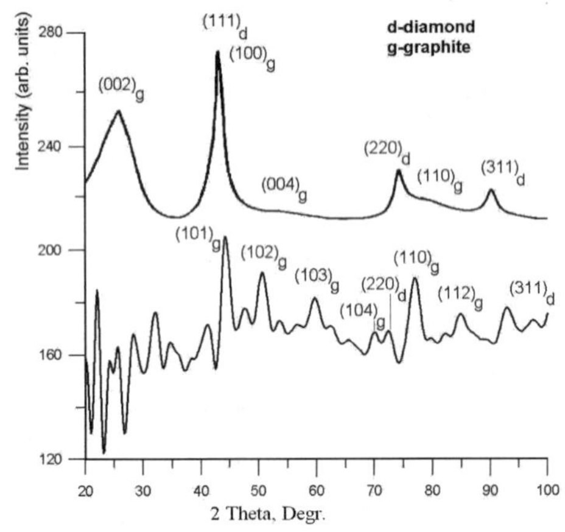

Figure 6. Computer-simulated *i(s)* interference part (bottom curve) of the diffraction pattern from a diamond/graphite nanoparticle with an approximately equal number of atoms in the core and shell as compared with the experimental diffraction pattern (top curve) presented in [5]

5. Computer models of hypothetical nanoparticles

Nowadays, synthesis of different core/shell hybrids as well as other nanoparticles has to be followed by XRD structural characterization based on computer 3D models. Theoretical calculation of the diffraction patterns provides controlling synthesized particles because XRD patterns from unified core/shell continuum and unbounded core and shell will differ distinctly (see Fig. 4). Calculated curves from complex particles as compared to the patterns from diamond core and shell diffracting independently of one other demonstrate changing positions and intensity of the reflections (diamond and graphite) due to interface.

We have tested not only spherical core/shell model for the modified DND, but also icosahedral model (Fig. 7) because tentative calculations showed a vague similarity between experimental and simulated patterns. To obtain more detailed information, models for icosahedrons of different size were treated on computer. Twenty tetrahedrons were combined via common planes (111) (by twinning) so that to form icosahedron. Tetrahedron was arranged into cube of size (8*a* x 8*a* x 8*a*). Atom in the centre of icosahedron is taken as common for all tetrahedrons and for this reason all interatomic distances in tetrahedrons were gradually deformed (a central part was compressed) from icosahedron vertex to the centre being an alternative for introducing dislocations. In the model, central atom of icosahedron should be absent to conform to an arrangement of the nearest four atoms.

Our calculations showed that experimental and simulated patterns (Fig. 7) corresponded in the diffraction positions together with an additional reflection (labeled with *). However, deformed profile of this reflection in the pattern from our model icosahedron (31,313 atoms, R = 38.38 Å, containing the 1st to 41st CSs) was inconsistent with the experimental finding. Experimental Laue picture [6] of icosahedral ND with an additional reflection is known. Figure 7 (left) displays an analogue (≈1,000 atoms) of model icosahedron.

Appearance of the additional reflection between (220) and (311) reflections in the simulated diffraction pattern of icosahedral nanodiamond is due to r_{ij} distances between atoms of nanoparticle including surface atoms diffracting accordingly Debye's formula (1). In this connection it should be noted that at definition of nanoparticle it is necessary not only to denote its size (less than 100 nm), but also to describe its characteristic property (for instance, structural factor) dependent on the structure, shape, extent of nanoparticle and interatomic (for icosahedron – in tetrhedrons) and interparticle (between tetrahedrons) interactions.

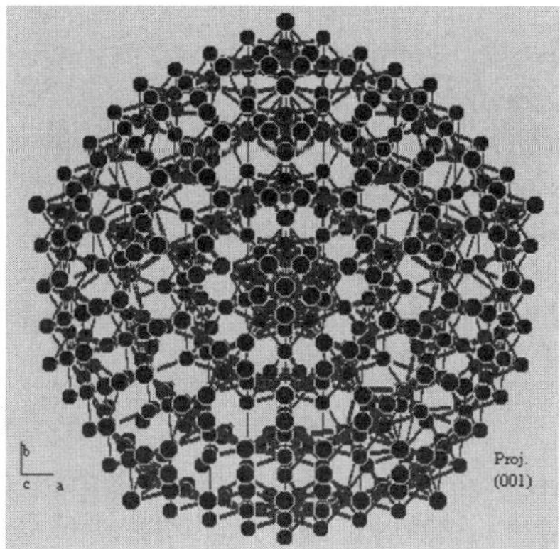

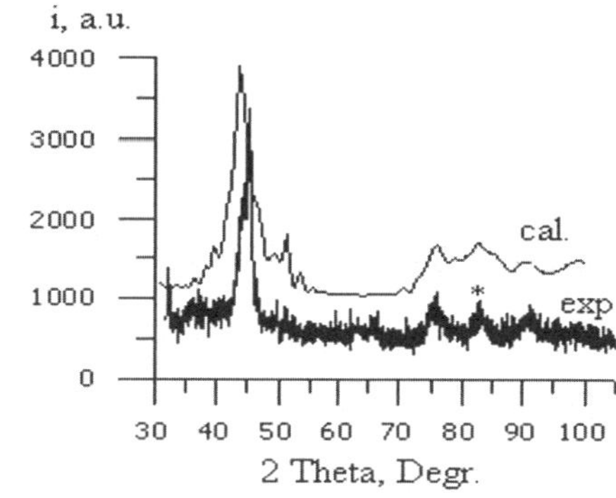

Figure 7. Left panel presenting 3D computer model icosahedral nanodiamond constructed from 20 nanocrystalline tetrahedral twins with the lattice parameters of the natural diamond; right panel showing a comparison between experimental (exp), the same as in Fig. 6, and calculated (cal) diffraction patterns

It is known that while atoms of carbon, nitrogen, oxygen, hydrogen etc. are components of the organic world, hybrid NCs of the synthetic world can theoretically be presented and later on synthesized as spherical or cylindrical particles, consisting of a diamond core (the inorganic world carbon) and an organic, e.g. glycine shell. Figure 8 demonstrates a constructed spherical model of hybrid nanoparticle consisting of the diamond core and glycine shell and calculated diffraction patterns.

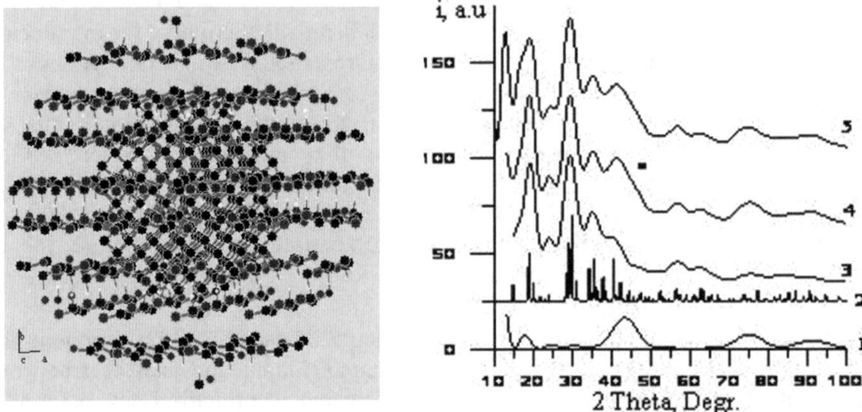

Figure 8. Left panel showing calculated XRD patterns from spherical model of diamond/glycine hybrid: 1 – diamond core of R = 6.65 Å, 275 atoms, consisting of six CSs; 2 – polycrystalline glycine; 3 – glycine shell made up of the 8th to 15th CSs (ΔR = 5.28 Å, R_{inner} = 7.15 Å, R_{outer}= 13.43 Å, 706 atoms); 4 – additive superposition of the core and shell patterns (* denotes manifestation of (111) diamond reflection); 5 – core/shell hybrid NP. Right panel shows a model of a spherical hybrid ND core/nanoglycine shell particle

6. Conclusions

At present, the state of chemical synthesis of hybrid nanoparticles cannot be provided by the knowledge of the ways and directions of chemical reactions or their rates. In this connection, the alone traditional physical research method to prove the reality of such particles is a method of direct determination of spatial structure of substances, i.e. the diffraction method. Using known atomic coordinates 3D spatial visualization of nanoparticles, individual structure details (for example, bordering CSs at assembling nanophases), measuring interatomic distances can be presented on the IBM-PC display.

Acknowledgements

I acknowledge the NOVA Publishers that concentrate our attention on the importance of research nanodiamond particles. I am very grateful to Budker Institute of Nuclear Physics of SB RAS (Novosibirsk) for making available for me the synchrotron radiation diffraction measurements. My thanks are also due to Acad. V.Ya. Shevchenko whose ideas of hybrid nanosized particles (centaurs) and particles with an equal amount of atoms inside and on the surface stimulate this work.

References

1. James R.W. The Optical Principles of the Diffraction of X-ray, Bell: London, 1950.
2. Yurjev G.S., Vereshagin A.L., Korchagin M.A. Structural study of detonation nanodiamond. *Diamond Relat. Mat.* 2005, **14**: 192–195.
3. Yurjev G.S., Vereshagin A.L., Kosov A.V. Computer-simulated diamond/graphite structure based on the diffraction pattern. *Diamond Relat. Mat.* 2007, **16**: 645–649.
4. Dolmatov V.Yu. *Superhard Mater.* 2005, **1**: 28–32.
5. Tomita S., Burian A., Dore J., Le Bolloch D.M., Fujii M., Hayashi M.S. *Carbon* 2002, **40**: 1469.
6. Son S.I., Chung S.J. *Z. Kristallogr.* 2004, **219**: 494–505 (in Russian).

AB-INITIO INVESTIGATION OF HYDROGEN ABSORPTION BY MAGNESIUM NANOPARTICLES

A.S. FEDOROV, G.N. CHURILOV
L.V. Kirensky Institute of Physics of Russian Academy of Sciences, Krasnoyarsk 660036, Russia
Fax: +7(3912)43-89-23; E-mail: alex99@iph.krasn.ru
A.A. KUZUBOV, M.V. SERJANTOVA
Siberian Federal University, Krasnoyarsk 660041, Russia

Abstract. Stability and binding energy of magnesium hydrides MgH_x (x = 0.5…2) is investigated by ab-initio calculations. Compare the binding energies of all known MgH_2 crystalline phases it is shown that these MgH_x partly saturated hydrides should split up to pure Mg and MgH_2 phases due to spinodal decomposition. Also the hydrogen atom hop rates between nearest minima in the pure Mg and basic MeH_2 α-phase are calculated. The finding explain the low rate of magnesium saturation by hydrogen. Moreover the hydrogen hop rate in Mg_7Sc and Mg_7Ti crystals are calculated. It is shown that the rate in Mg_7Sc between nearest minima in the vicinity of Sc atom is similar with that for pure MgH_2 α-phase whereas the analogous rate in Mg_7Ti is at ~2.5 * 10^6 times greater. It means that hydrogen diffusion inside Mg-Ti alloys would be essentially facilitated.

By calculation of the magnesium surface specific energy and using simple pressure estimation it is defined that for Mg nanoparticles of radius R > 5 nm the binding energy and geometry is practically coincide with these values for bulk material, so for such nanoparticle modeling it is possible to use results of the bulk material calculations what is far easier.

Keywords: ab-initio calculations, magnesium, hydrogen absoption, nanoparticles

1. Introduction

Discovery of reversible hydrogen absorption by palladium and some intermetallic alloys some decades ago has stimulated further investigations of hydrogen storage systems consist from metal hydrides. At that many researchers suppose that the solution of this problem should be sought in new metal hydride synthesis. Many researchers rely on that one of the best metal for hydrogen absorption is magnesium because of it is widespread metal and hydrogen content inside it reach 7.6 wt% what is enough for the practical using in motor vehicles according to the US Department of Energy demand (6.5 wt%). However, at present its using is limited due to slow kinetics of the hydrogenation and dehydrogenation reaction, even at elevated temperatures. The effective temperature of Mg hydrogenation/dehydrogenation is usually ~300°C. So to enhance the hydrogen sorption properties researchers try to reduce Mg particle size, to alloy magnesium with other elements, or to add catalysts to Mg particle surface [1].

The slow kinetics of hydrogen sorption is explained by two factors. The first one is explained by difficulty of the hydrogen molecules initial dissociation on the Mg surface due to low Mg catalytical activity. This trouble can be avoided by the Mg surface covering by some transition metals (TM) which can catalyze chemical dissociation of H_2.molecules. The second factor is connected with slow migration of H atoms through the Mg crystal. The both factors might be rate-limiting, and many reports are devoted to study one or another factor [3–7]. Unfortunately, despite of many experimental works devoted to that, the reason of slow hydrogen migration is still unclear. Though even the MgH_2 oscillation spectrum was experimentally defined by inelastic neutron scattering and calculated by density functional theory in [4], potential barriers for hydrogen atom movement inside Mg or its hydrides are unknown. Also the equilibrium of different MgH_x ($0 < x < 2$) phases is unknown.

The purpose of this work is to elucidate theoretically the reason of the slow hydrogen migration inside magnesium hydride and to propose a way to overcome it. The second aim is to define a geometry of most stable partly saturated magnesium hydride phases by comparison of their enthalpy. The third aim is to determine the metal particle size influence to the pressure inside it. For all of that we carry out density functional theory (DFT) calculations to determine binding energy E_{bind} of all structures and to find a transition state of hydrogen jumps in the hydrides.

2. Computational method

Assuming that enthalpy $H = E_{bind} + P\Omega$ under ambient conditions is close to it under temperature $T = 0$ K and pressure $P = 0$ conditions, zero temperature and pressure were used for all calculations. At that the enthalpy would be equal to the total binding energy E_{bind} which was defined during this energy minimum search. The calculations of MgH_x ($0 \le x \le 2$) binding energies were performed using density-functional theory (DFT) [10] within generalized-gradient approximation (GGA) using the functional of Perdew, Burke, and Ernzerhof (PBE) [11] for the exchange-correlation energy. At that periodical boundary conditions (PBC) were used. The calculations were implemented in the fully selfconsistent ab initio package VASP package [8 9]). There only valence electrons were considered using plane wave basis functions, at that their interactions with atom nuclei and core electrons were treated by ultrasoft scalar relativistic Vanderbilt-kind pseudopotential formalism [12]. The Vanderbilt pseudopotentials allow one to reduce significantly a maximal kinetic energy Ecutoff of basis functions without loss of accuracy. This lead to decreasing of the necessary number of plane waves and to essential reduction of the calculation time. The geometry optimization in all calculations has been carried out until the forces acting on all atoms become lower than 0.05 eV/Å. To improve the convergence of the binding energy the Gaussian broadening for occupation number of all electronic levels was used with a smearing width of 0.1 eV. For Brillouin-zone (BZ) sampling of hydrides the uniform Monkhorst-Pack (MP) scheme was used with 3 X 3 X 3 k-point mesh inside irreducible part of BZ, but for transition states of hydrogen atom in supercells the G-point was used only because of the supercell large size.

To find a transition state that is a potential barrier height $E_{barrier}$ for hydrogen atom movement from one potential minimum to nearest one the nudged elastic band method (NEB) [13] has been used. Then ratios k of hydrogen atom hopping between minima were calculated within the transition state theory by well-known formula [14–17] with taking into account zero-point motions ν_i of H atoms:

$$k = Ae^{-\frac{E_{barrier}}{kT}}, \tag{1}$$

where

$$A = \frac{kT}{\hbar}\frac{\prod_{i=1}^{3N-7}\left(1-e^{-\frac{\hbar\nu'_i}{kT}}\right)}{\prod_{i=1}^{3N-6}\left(1-e^{-\frac{\hbar\nu_i}{kT}}\right)} \tag{2}$$

Here T is the temperature, $E_{barrier}$ is the potential barrier height value for the atom hopping. The product $\prod_{i=1}^{3N-6}\left(1-e^{-\frac{\hbar\nu_i}{kT}}\right)$ and the total energy of zero vibrations $E_0 = \sum_{i=1}^{3N-6}\frac{\hbar\nu_i}{2}$ were calculated for local minimum position of the hydrogen atom, while the product $\prod_{i=1}^{3N-7}\left(1-e^{-\frac{\hbar\nu'_i}{kT}}\right)$ was obtained for hydrogen atom vibration in the saddle point of the transition state trajectory. To take into account zero-point motions of H atom, the potential barrier height in (1) was decreased to $E_{barrier} = E_{barrier}$-$E_0$.

3. Results and discussion

First of all the enthalpies of Mg_2H_x (x = 0, 1, 2, 2.5, 3, 3.5, 4) partly saturated magnesium hydrides have been calculated. Because of crystal structures of these phases are unknown, the structures of all known [2] MgH_2 phases were used as seed structures, see Table 1. There one can see that under different pressure Mg form five MgH_2 phases after full hydrogen saturation.

TABLE 1. Parameters of all known Mg crystal phases [2]

Phase	Stability conditions (GPa)	Symmetry group	Structure type	Cell parameters (Å)		
				a	B	c
α	<0.39	136: P42/mnm	TiO-r	4.501	4.501	3.010
β	>3.84	205: Pa3	Mod. CaF2	4.790	4.790	4.790
γ	>0.39	060: Pbcn	α-PbO2	4.501	5.420	4.917
δ	>6.73	029: Pbc21	Ortho.	4.860	4.635	4.751
ε	>10.26	062: Pnma	AlAu2	5.280	3.093	5.990

For the enthalpy calculations a different number of hydrogen vacancies were included in all these MgH_2 phases and the phase of minimal binding energy was chosen as most stable. Also the enthalpy of native hcp Mg structure has been calculated.

In the Table 2a, b results of the enthalpy calculations for all investigated partly saturated hydrides of the most stable α–phase with vacancies are shown. Also there the sum of the pure Mg enthalpy and the MgH_2 α-phase in the proper ratio are shown in the table odd lines. On can see that for all Mg_2H_x ($x = 1, 2, 2.5, 3, 3.5$) phases the enthalpy of the phase with vacancies is essentially higher than that for weighed mean of pure Mg and MgH_2 α-phase. It means that all partially saturated hydrides should be break up into pair of these phases, in other words spinodal decomposition is taken place.

TABLE 2A. Comparison of the binding energies for Mg_2H_x ($x = 1 - 3$) α-phase and for weighed total of hcp Mg_2 and Mg_2H_4 α-phases

Hydrogen concentration		Binding energy (eV)
Mg_2H_3	$Mg_2 + 3Mg_2H_4$	-56.282
	$4Mg_2H_3$	-52.589
Mg_2H_2	$Mg_2 + Mg_2H_4$	-20.798
	$2Mg_2H_2$	-19.919
Mg_2H	$3Mg_2 + Mg_2H_4$	-26.91
	$4Mg_2H$	-24.986

TABLE 2B. Comparison of the binding energies for (β, γ, δ, ε) phases of Mg_4H_x ($x = 1 - 3$) and for weighed total of hcp Mg_2 and Mg_4H_8 (β, γ, δ, ε) phases

Hydrogen concentration		Binding energy (eV)			
		$\beta - Mg_4H_8$	$\gamma - Mg_4H_8$	$\delta - Mg_4H_8$	$\varepsilon - Mg_4H_8$
Mg_4H_7	$2Mg_2 + 7Mg_4H_8$	-251.33	-254.32	-251.27	-246.24
	$8Mg_4H_7$	-244.73	-246.15	-244.64	-241.42
Mg_4H_6	$2Mg_2 + 3Mg_4H_8$	-111.21	-112.49	-111.18	-109.03
	$4Mg_4H_6$	-108.00	-108.02	-108.01	-108.36
Mg_4H_5	$6Mg_2 + 5Mg_4H_8$	-193.49	-195.63	-193.45	-189.86
	$8Mg_4H_5$	-185.59	-185.59	-185.61	-186.24
Mg_4H_4	$2Mg_2 + Mg_4H_8$	-41.14	-41.57	-41.13	-40.42
	$2Mg_4H_4$	-39.85	-39.84	-39.84	-40.07
Mg_4H_3	$10Mg_2 + 3Mg_4H_8$	-135.65	-136.94	-135.63	-133.47
	$8Mg_4H_3$	-129.58	-129.59	-129.57	-129.74
Mg_4H_2	$6Mg_2 + Mg_4H_8$	-53.37	-53.79	-53.36	-52.64
	$4Mg_4H_2$	-51.67	-51.67	-51.70	-51.52
Mg_4H	$14Mg_2 + Mg_4H_8$	-77.82	-78.24	-77.81	-77.09
	$8Mg_4H$	-75.09	-75.09	-74.95	-74.93

Then rates of H atom jumps between different neighbor minima for native Mg and MgH_2 α-phase have been calculated by NEB method. Because of inside the Mg hcp structure hydrogen atoms may fill both tetrahedral and octahedral holes, the hops and potential barriers for octahedral-octahedral, octahedral-tetrahedral and tetrahedral-tetrahedral jumps of two kinds have been calculated, see Figs. 1(a–c) and 2, lines 1–3.

In Figs. 1(d) and 2, line 4 the hop and the barrier value for hydrogen jump between filled and empty positions inside MgH_2 α-phase are shown. Also the hydrogen hope rates (see (1, 2)) at T = 300 K for all cases have been calculated, see caption of Fig. 2.

From defined hope rate values the reason of the hydrogen slow migration become obvious. Because of partially saturated hydride should be break up into native Mg and MgH_2 α-phase, the hydrogen diffusion rate would be defined by this rate inside MgH_2 α-phase where this rate is very small, see Fig. 2, line 4.

Be aspire to get large hydrogen diffusion rate inside Mg hydride, we calculated this rate inside of Mg_7Sc and Mg_7Ti alloys. A crystal structure with the unit cell of Mg_7TiH_x has been synthesized recently under high pressure [6]. The barrier value for hydrogen atom jump from filled to empty positions in the vicinity of Ti atoms which substitute some Mg atoms in α-phase have been calculated, see Fig. 2, line 5. This barrier was at ~0.4 eV smaller than that for the native MgH_2 α-phase. At the same time the barrier value for H jump in the vicinity of atom of Sc atom was almost the same as for MgH_2 case, so it is not shown. From this it follows that hydrogen diffusion inside Mg_7Ti alloy would be essentially larger in comparison with pure Mg. In addition, the presence of Ti atom having d-electron on Mg crystal surface should facilitate the hydrogen molecule dissociation. So the first factor (see above) of slow saturation would be overcome.

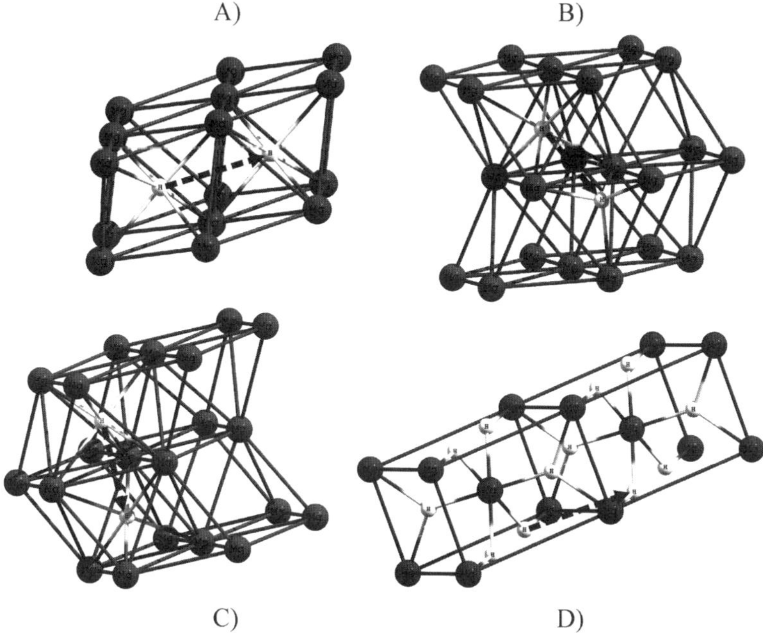

Figure 1. Hydrogen atom hops inside hcp Mg for octahedral-octahedral (A), octahedral-tetrahedral (B), another pair of octahedral-octahedral (C) holes. Hydrogen atom hop between nearest holes inside MgH_2 α-phase (D)

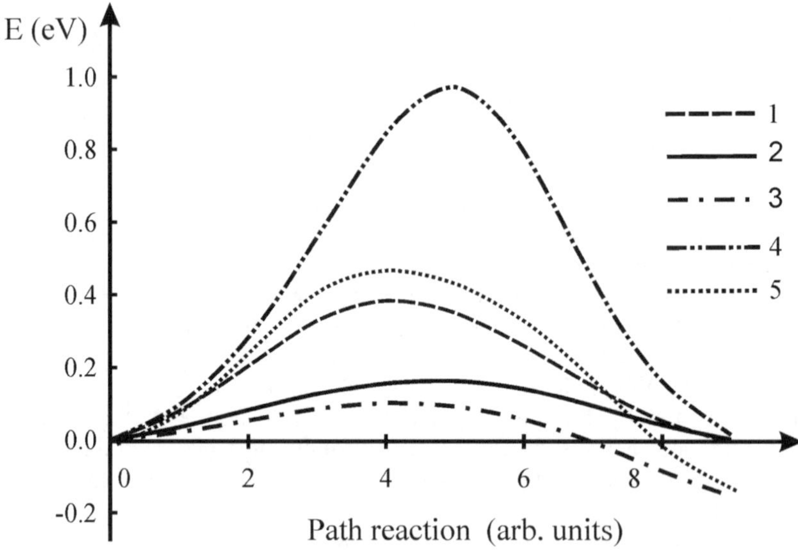

Figure 2. Barrier value for hydrogen atom jumps inside hcp Mg between octahedral-octahedral – 1, octahedral-tetrahedral – 2, another pair of octahedral-octahedral holes – 3. 4 – Barrier value for jump between nearest holes inside MgH_2 α-phase, see Fig. 1 also. 5 – Barrier value for hydrogen atom jump inside MgH_2 α-phase in the vicinity of Ti atom
The hydrogen hope rates at T = 300 K for (1), (2), (3), (4), (5) cases were equal to $9.21 * 10^8$, $1.41 * 10^9$, $1.63 * 10^{10}$, $1.1 * 10^{-2}$, $2.70 * 10^4$ (sec^{-1}) correspondingly

To investigate the Mg nanoparticle size influence to the stability of hydride the pressure P inside nanoparticles was calculated for different particle sizes. At first the surface specific energy E_{surf} of hcp Mg (0001) surface has been calculated. For that the slab having transverse sizes ~6.42 x 6.42 Å, thickness ~31.68 Å and vacuum space ~10 Å between adjacent slabs was used. After calculation of the slab binding energy E_{slab} the surface specific energy E_{surf} was defined: as $E_{surf} = \frac{E_{slab}\text{-}NE_0}{2S}$, were $E_0 = \frac{E_{bulk}}{N}$ is the binding energy per atom of bulk Mg super cell having N atoms inside and S is one-sided surface of the slab. The calculated surface energy E_{surf} for this surface is equal 0.27 eV/surface atom, what is very close to the value 0.27 eV/surface atom from [17]. In a similar manner surface energies for slabs with different surface orientations have been calculated. From these data the average surface energy $<E_{surf}>$ of nanoparticle was defined by averaging over different facets using equilibrium condition of the surface energy minimum, see [18]. According to this condition, surfaces of the nanoparticle different facets are inversely depended on their surface energies. Suppose that nanoparticles should have a shape having the minimal total surface energy $<E_{surf}>$, from equilibrium of the nanoparticle over its volume variation, it is easy to get the formula connected the pressure P inside the nanoparticle and its radius R: $P=\dfrac{2<E_{surf}>}{R}$.

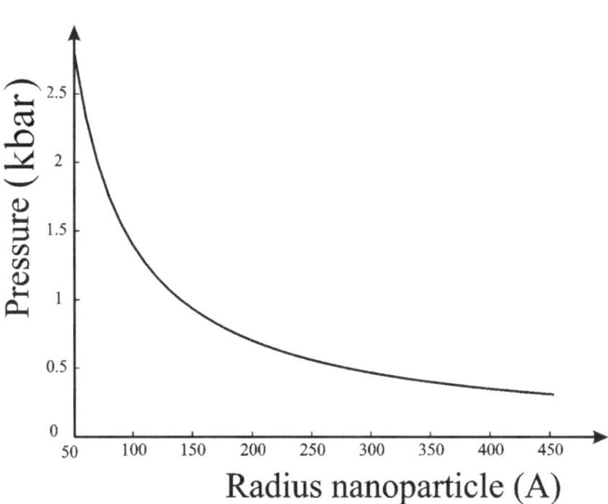

Figure 3. Dependence of the pressure inside Mg nanoparticle from its radius

Dependence of the pressure inside Mg nanoparticles from their radius is shown on Fig. 3. One can see that for nanoparticles of radius R > 5 nm the inner pressure is smaller than 5 Gpa, so for such nanoparticles one can use calculations of bulk material at pressure P < 5 Gpa what is far easier than calculations for nanoparticles.

4. Conclusions

The hydrogen absorption inside Mg crystal was theoretically investigated. It was shown that during hydrogen saturation of this subject the hydride structures should split up to the pure Mg and MgH_2 α-phases. Also the hydrogen hop rates ν inside Mg and MeH_2 phases were calculated. These data explain that the reason of the hydrogen slow diffusion was the high potential barrier value for hydrogen jumps inside MgH_2 α-phase. At T = 300 K the hop rate inside this phase was equal $\nu \approx 10^{-2}$ sec^{-1} only. To avoid this trouble the hop rates inside Mg_7Ti and Mg_7Sc alloys were calculated and was shown that the rate ν in vicinity of Ti atom is at ~$2.5 * 10^6$ times greater then that for initial MeH_2 α-phase. It means that hydrogen diffusion inside Mg-Ti dilute alloy would be essentially facilitated.

By calculation of the magnesium surface specific energy and using the simple estimation based on the nanoparticle equilibrium was shown that for Mg nanoparticles the inner pressure is negligible if the nanoparticle diameter larger than 10 nm, so it is possible to use bulk material calculations for stability and hope rate calculations.

References

1. J. Huot, G. Liang, R. Schulz, *Appl. Phys. A72*, **187** (2001).
2. P. Vajeeston, P. Ravindran, A. Kjekshus, H. Fjellvag, *Phys. Rev. Lett.*, **89**, 175506 (2002).
3. H.G. Schimmel, M.R. Johnson, G.J. Kearley, et al., *J. Alloy. Compd.*, **393**, 1–4 (2005).

4. H.G. Schimmel, M.R. Johnson, G.J. Kearley, et al., *Mat. Sci. Eng. B*, **108**, 38–41 (2004).
5. G. Liang, J. Huot, S. Boily, R. Schulz, *J. Alloy. Compd.*, **305**, 239 (2000).
6. A. Zaluska, L. Zaluski, J.O. Stro¨m-Olsen, *J. Alloy. Compd.*, **289**, 197 (1999).
7. J. Bloch, M.H. Mintz, *J. Alloy. Compd.*, **253–254**, 529 (1997).
8. G. Kresse, J. Hafner, *Phys. Rev. B47*, **558** (1993).
9. G. Kresse, J. Hafner, *Phys. Rev. B49*, **14251** (1994).
10. W. Kohn, L.J. Sham, *Phys. Rev.*, **140**, 1133 (1965).
11. J.P. Perdew, K. Burke, M. Ernzerhof, *Phys. Rev. Lett.*, **77**, 3865 (1996), ibid. **78**, 1396(E) (1997).
12. D. Vanderbilt, *Phys. Rev. B41*, **7892** (1990).
13. G. Henkelman, B.P. Uberuaga, H. Jónsson, *J. Chem. Phys.*, **113**, 9901 (2000).
14. N.B. Slater, *J. Chem. Phys.* **24**, 1256 (1956).
15. N.B. Slater, "*Theory of Unimolecular Reactions*" (Cornell University, Ithaca, NY) (1959).
16. G.H. Vineyard, *J. Phys. Chem. Solids, Pergamon Press*, **3**, 121–127 (1957).
17. D. Kyoi et al., *J. Alloy. Compd.*, **375**, 253 (2004).
18. S. Wei, M.Y. Chou, *Phys. Rev. B50*, **4859** (1994).

EFFECT OF PRESSURE ON THERMODYNAMIC PROPERTIES OF LIGHT ELEMENT HYDRIDES

YU.M. GORYACHEV, V.I. DEKHTERUK, M.I. SIMAN, L.I. FIYALKA
I. Frantsevych Institute for Problems of Materials Science of National Academy of Sciences of Ukraine03142 Kiev, Ukraine

Abstract. A model for calculation of thermal and elastic dynamic characteristics and of boron and aluminum hydrides has been developed. It was shown that they are related to the energy of atomization calculated by the method of group orbitals as linear combination of atomic orbitals (GOLKAO) and dependent on the hydride composition and pressure.

Keywords: boron hydride, aluminum hydride, thermodynamic function, elastic dynamic characteristics, enthalpy, atomization

1. Introduction

Light element hydrides, which have high ratios of hydrogen weight to the total hydride weight, are known to be of great interest for hydrogen power engineering and to attract the attention of many researchers thanks to their thermodynamic properties [1, 2].

The thermodynamic properties of compounds depend on their elastic-dynamic characteristics and factors of external action on them. We have calculated these characteristics for a number of hydrides

The following thermodynamic functions have been calculated by the authors: enthalpy ($\Delta H^o{}_{293\,K}$), heat capacity ($C^o{}_p$), entropy (S^o), and free energy (Gipps potential) $\Delta\Phi^o{}_{293\,K}$. [3–5].

2. Results and discussion

The procedure of calculation was as follows:

On the basis of the calculation results for the atomization energy (E_a, eV/at) of the investigated objects using a GO LCAO method, a quantitative analysis of their elastic-dynamic state was carried out in terms of approximation of the one-parameter model for the energy of atom-atom interaction [2]. The elastic compression modulus (β, Pa), melting point (T_m, K), Deby temperature (θ_D, K), coefficient of linear thermal expansion (α_T, K^{-1}), heat conductivity (λ,Wt/cm·K); and heat capacity coefficient (C_p, J/mol.K) were calculated. For standard conditions the following relations were derived:

$$\beta = (z \cdot E_a \cdot 1{,}602 \cdot 10^{-12}) / (27d^3);$$
$$T_m = (E_a \cdot 1{,}602 \cdot 10^{-12}) / (41 \cdot k);$$
$$\theta_D = 2{,}75 \cdot 10^{6} \cdot (E_a / A)^{0,5} \cdot \hbar / (d \cdot k);$$
$$\alpha_T = 3.\ k / (z \cdot E_a \cdot 1{,}602 \cdot 10^{-12});$$
$$\lambda_\phi = z^{0,5} \cdot E_a \cdot k \cdot 4.10^{-21} / (d \cdot \hbar);$$
$$C_p = (8.3 \cdot 10^{16} \cdot kT^2/\theta_D^2) \cdot (1.5 - T/\theta_D) \quad for\ T < \theta_D.$$

B. Baranowski et al. (eds.), *Carbon Nanomaterials in Clean Energy Hydrogen Systems.*

$$\Delta H_T = -E_a + \int C_p\, dT;\ \Delta S_T = \int (C_P/T)\, dT;$$
$$\Delta \Phi_T = \Delta H_T - T \cdot \Delta S_T.$$

Here E_a is the atomization energy; z is the coordination number of the nearest neighborhood; d is the average interatomic distance in this neighborhood (cm), k = $1.38 \cdot 10^{-16}$, erg.K^{-1}, $\hbar$ = $1.05 \cdot 10^{-27}$, erg.sec. (Boltzmann and Planck constants, respectively), A is the atomic weight in atomic units.

The results of calculation of the atomization energy, elastic-dynamic and thermodynamic properties of known boron hydrides and hypothetical aluminum hydrides in a solid state and under corresponding temperatures and normal pressure are presented in Table 1.

TABLE 1. Table Properties of hydrides at their melting points

M_AH_B	E_{aT} (eV/at)	-ΔH (eV/at)	-ΔΦ (eV/at)	T_m (K)	α_T, 10^{-6}	β, 10^{11}, (Pa)
B H	6.33	6,20	5.43	682	7.8	14.5
B_5H_7	5.29	5,21	4.79	400	9.4	12.1
B_5H_9	4.49	4.44	4.19	254	11	10.3
B_5H_{11}	3.87	3.84	3.67	177	12.8	8.87
B H_3	3.19	3.16	3.04	135	15.5	7.31
Al H	4.30	4.25	4.01	247	11.5	3.59
Al_5H_7	3.71	3.67	3.50	186	13.4	3.1
Al_5H_9	3.35	3.32	3.19	147	14.8	2.8
Al_5H_{11}	2.96	2.95	2.90	117	16.7	2.47
Al H_3	2.67	2.65	2.58	87	18.6	2.23

Under standard temperature and pressure, most of the investigated hydrides exist in a gas state. The stability of a solid state is only reached under high pressure.

In order to take into account the influence of pressure and temperature, we have developed the following relations:

$$\beta(P) = 1 \Big/ (1/\beta_0 - k \times P/\beta_0^2), \quad \beta(T) = 0.4\beta_0 \times (1.5 + \frac{Tm - T}{Tm}).$$

The analysis of the data obtained made it possible to establish a number of regularities for the systems studied, including the following.

The density of hydrides (d_1) depends on the pressure (P_1) and index (x) of hydrogen in them (MeH_x):

$$d_1 / d_0 = [1 - a\,(P_0/P_1)^{b} \cdot x],$$

where a = $4.25 \cdot 10^{-2}$; b = 0.22; d_0 is the density of metal at the pressure P_1; $P_0 = 10^{-4}$ GPa (standard pressure).

The dependence of the density of a solid on the pressure can be expressed as $d_p = d_0 \exp(P / \beta_p)$; where d_0 is the metal density under standard conditions, β_p is the Young modulus at the pressure P_0. The summarized results are listed in Table 2.

TABLE 2. Pressure and concentration dependences of hydride-to-metal densities ratios

P (GPa) / X	10^{-4}	10^{-3}	10^{-2}	10^{-1}	10^{2}
0	1.000	1.000	1.000	1.000	1.000
1	0.952	0.971	0.982	0.989	0.998
2	0.913	0,950	0.970	0.970	0.997
3	0.873	0.925	0.953	0,971	0.994
4	0.838	0.903	0.940	0.962	0.991

The dependence of the elastic compression modulus (Young modulus) on the pressure and the content of hydrogen can be described as:

$$\beta_p^{(x)} = ((\beta_0^0 + \beta_0^1 \cdot x) / (1 + x)) \cdot \exp [\, 2\Upsilon(1+x) \cdot p / /(\beta_0^0 + \beta_0^1 \cdot x)],$$

where Υ is the Gruneisen constant; β_0^0 is the elasticity modulus of metal under standard conditions (10^{-4} GPa = 1 atm); β_0^1 = 38 GPa is the reduced modulus of solid hydrogen.

The calculated equilibrium pressure of hydrogen over hydrides is presented in Table 3 versus temperature and composition.

TABLE 3. The dependence of the equilibrium hydrogen pressure over boron and aluminum hydrides on temperature and composition

T (K) / M_AH_B	400	600	800	1,000	1,200
BH	$6.7 \cdot 10^{-15}$	$5.6 \cdot 10^{-9}$	$5.2 \cdot 10^{-6}$	0.0003	0.005
B_5H_7	$8.5 \cdot 10^{-8}$	$2.9 \cdot 10^{-4}$	$1.7 \cdot 10^{-2}$	0.2	1
B_5H_9	$3 \cdot 10^{-4}$	$6.7 \cdot 10^{-2}$	1	4.9	14.5
B_5H_{11}	$3.3 \cdot 10^{-2}$	1.5	9.8	30.4	65
BH_3	2.6	25.7	81	161	255
AlH	$.4 \cdot 10^{-9}$	$3.6 \cdot 10^{-5}$	$3.7 \cdot 10^{-3}$	$5.9 \cdot 10^{-2}$	0.38
Al_5H_7	$1.3 \cdot 10^{-4}$	$3.8 \cdot 10^{-2}$	0.66	3.7	11.5
Al_5H_9	$1.9 \cdot 10^{-2}$	1	7.6	25.5	56.8
Al_5H_{11}	0.48	8.7	37	89	158
AlH_3	8	54	142	252	370

The calculation was performed according to the relation:

$$P_{H_2} = \{[(1+x) \cdot \exp(0.1 \cdot E_{AT} / (kT))]^{(2/x)}\} \cdot 10^3$$

The results obtained made it possible to determine the dominant trends of qualitative technological and quantitative effects on the processes of hydrogenation and dehydrogenation of compounds. For instance, in the simplest case of these processes, that is, in the case of the reaction

$$2\,BH_3 + 3\,H_2O = B_2O_3 + 6\,H_2,$$

an increase in pressure and temperature favors the reaction of hydrogenation whereas a decrease in pressure and temperature leads to dehydrogenation.

3. Conclusions

The relation between the equilibrium hydrogen pressure over boron and aluminum hydrides and their composition and temperature has been calculated. The dependence of thermal and elastic-dynamic properties of hydrides on their electron parameters has been established.

The principle characteristics of hydrogenation and dehydrogenation in the systems of boron and aluminum with hydrogen have been determined.

References

1. Cox J.D., Wagman D.D., and Medvedev V.A. CODATA Key Values for Thermodynamics, Hemisphere Publishing, New York, 1989.
2. Wagman D.D., Evans W.H., Parker V.B., Schumm R.H., Halow I., Bailey S.M., Churney K.L., and Nuttall R.L. The NBS Tables of Chemical of Chemical Thermodynamic Properties, J. Phys. Chem. Ref. Data, Vol. 11, Suppl. 2, 1982.
3. Goryachev Yu.M. Author's abstract of dissertation for the Ph.D. degree of ph.-m. sciences.- Kiev, IPMS NAS of Ukraine, 1983.- 33 p. (in Russian).
4. Goryachev Yu.M., Siman N.I., et al. Titanium hydrides and GOLKAO method. In: Electronic structure and properties of high-melting compounds and alloys.- Kiev, IPMS NAS of Ukraine, 2000.- P. 112–118 (in Russian).
5. Goryachev Yu.M., Dehtyaruk V.I., Siman N.I., and Fialka L.I. Hydrides of transition metals as accumulators of hydrogen for energetics.- Proceedings of NATO Advanced Research Workshop on Fuel cell.- Kiev, IPMS of Ukraine, June 6–10, 2004.

CONDUCTIVITY OF METALLIC HYDROGEN UNDER HIGH PRESSURE AND TEMPERATURE

V.T. SHVETS
Odessa State Academy of Refrigeration
1/3 Dvoryanska Str., 65082 Odessa, Ukraine

Abstract. The electrical resistivity of metallic hydrogen has been calculated. To do this, the perturbation theory in terms of electron–proton interaction for the reciprocal relaxation time characterizing the electrical conductivity has been employed. The second and third order terms as well as an approximate expression for the series sum have been calculated in detail. In doing so, the random-phase approximation with allowance for exchange interaction and correlations in the local-field approximation has been used for an electron subsystem and the exact solution of the Percus–Yevick equation for the hard sphere model has been employed for the proton subsystem. In this case, at given density and temperature, the only parameter of the theory is the hard-sphere diameter. To determine this parameter, the effective pairwise interproton interaction has been calculated. The hard sphere diameter has been determined from the dependence of the interproton interaction on the distance and the known temperature of the system. The dependence of the resistivity of metallic hydrogen on the density and temperature has been examined in a wide range of the latter quantities. In the entire range of densities and temperatures considered, the resistivity proves to be close to its limiting value for which the nearly free electron model is applicable.

Keywords: resistivity, pressure, temperature

1. Introduction

From the practical viewpoint, the interest in hydrogen is caused by the fact that it is the most widespread element in the Universe. The greater part of masses of planets of the Sun system falls on hydrogen in the metallic state. From the theoretical viewpoint, metallic hydrogen is a unique object, since it is the sole substance for which electron–ion interaction is known exactly, which makes it possible to most accurately calculate all the properties using a minimum number of simplifying assumptions. Yet, there is also a complicating circumstance consisting in that the Coulomb interaction is not weak. This fact is well known in the theory of electron gas. The random phase approximation is one of fundamental approximations of the theory of electron gas and corresponds to approximate summation of the entire perturbation-theory series in terms of Coulomb electron–electron interaction.

The possibility of the existence of metallic hydrogen was first predicted by Wigner and Huntington in 1935 [1]. It was supposed that hydrogen at high pressures should transform from a two-atom molecular state of an insulator to a single-atom state with

B. Baranowski et al. (eds.), *Carbon Nanomaterials in Clean Energy Hydrogen Systems.*

metallic conductivity. The real discovery of metallic hydrogen and a detailed study of the dependence of its resistivity on the density and temperature have been done only in 1996 [2]. In doing so, molecular hydrogen in a liquid state was subjected to shock compression to high pressures in a range of 0.93–1.80 Mbar at temperatures of 2,200–4,400 K. The metal–insulator transition with an electrical resistivity of the metallic phase of 500 μΩ cm was observed at a pressure of 1.4 Mbar and a temperature of 3,000 K. Further, in a pressure range of 1.4–1.80 Mbar and a temperature range of 3,000–4,400 K, the resistivity remained practically unaltered. In fact, the picture observed represented an insulator–semiconductor transition. In this case, the energy gap in molecular hydrogen did not vanish but decreased from 15 to 0.3 eV and practically became equal to the temperature of the sample under study. The metal–insulator transition was observed in nonideal hydrogen plasma under shock wave compression to pressures ~1.5 Mbar too [3, 4]. It is found that the conductivity increases sharply (by five orders of magnitude) at a density ~0.3–0.4 g/cm^3, reaching close to liquid-metal values ~1,000 μΩ cm.

It should be noted that experimental and theoretical investigations of tentative metallic state of hydrogen were also performed earlier. In [5], for example, the measurements of the electrical resistivity of molecular hydrogen at considerably lower pressures of 0.1–0.2 Mbar revealed an exponential temperature dependence characteristic of semiconductors with an energy-gap width of 12 eV. The first report on the discovery of metallic hydrogen was published in 1978 [6]. The authors of [6] reported the discovery of metallic hydrogen with an electrical resistivity of 1,000 μΩ cm at a pressure of 2 Mbar.

At present, the equilibrium properties of metallic hydrogen are investigated quite intensely [7–15]. Theoretically, electron transport phenomena in metallic hydrogen have been studied to a much lesser extent. In [16], for example, the resistivity of metallic hydrogen was calculated in a wide range of temperatures and densities but at a fixed value of the packing-fraction typical of liquid metals near the melting temperature and for the case where only the first term of expansion of the resistivity into a series in terms of the electron–proton interaction was taken into account. As will be shown below, it is the use of precisely the latter approximation that most greatly affects the results of numerical calculations. Of great interest is a computer simulation of different properties of metallic hydrogen including its resistivity [17]. Unfortunately, even upon computer simulation of the resistivity its values obtained experimentally and theoretically differ by more than an order of magnitude.

In studying equilibrium properties of metallic hydrogen an important role belongs to the application of the nearly free electron model. In this work we use this model for calculation of the electrical resistivity of metallic hydrogen. In doing so, we assume that hydrogen is in the metallic state with a zero energy gap rather than in the experimentally observed semiconductor state with an energy-gap width of 0.3 eV. Such a state is realized either at a higher pressure or at a higher temperature. Note that the core of the Jupiter with a radius equal to half the planet radius consists of hydrogen under pressures of 3–40 Mbar and at temperatures of 10,000–20,000 K.

2. Hamiltonian

Since hydrogen in the metallic state was obtained by shock compression of liquid molecular hydrogen and remained in the liquid state after the insulator–metal transition, the Hamiltonian of the electron subsystem of metallic hydrogen can be written in the form analogous to that of simple liquid metals

$$H = H_0 + H_{ie}. \tag{1}$$

To a great extent such a model is also applicable in the presence of molecular hydrogen along with atomic hydrogen. In a view of their electrical neutrality, the hydrogen molecules scatter conduction electrons far less than the protons do.

We assume that, much like liquid metals, electron gas is degenerate and the ion subsystem is disordered and static. The allowance for the dynamics of the proton subsystem brings about no great problems but its role is relatively small. In the nearly free electron approximation, the Hamiltonian of the ideal electron gas can be written as

$$H_0 = \sum_{\mathbf{k}} \varepsilon_k a_{\mathbf{k}}^{+} a_{\mathbf{k}}, \tag{2}$$

where $a_{\mathbf{k}}^{+}$ and $a_{\mathbf{k}}$ are the operators of creation and annihilation of electrons in the state with a wave vector $\mathbf{k}$,

$$\varepsilon_k = \hbar^2 k^2 / 2m \tag{3}$$

is the energy of free electron, and m is its mass.

The Hamiltonian of electron–proton interaction is taken in accordance with the diffraction model of metal in which the electron–electron interaction is taken into account via screening of electron–ion interaction

$$H_{ie} = V^{-1} \sum_{\mathbf{q}} W(q) \rho^{i}(\mathbf{q}) \rho^{e}(-\mathbf{q}). \tag{4}$$

Here, V is the volume of the system;

$$W(q) = -V(q) / \varepsilon(q) \tag{5}$$

is the screened potential energy of electron–proton interaction;

$$V(q) = 4\pi e^2 / q^2 \tag{6}$$

is the Fourier transform of the potential energy of Coulomb electron–proton interaction, e is the electron charge;

$$\varepsilon(q) = 1 + [V(q) + U(q)]\pi_0(q) \tag{7}$$

is the effective permittivity of electron gas in the random phase approximation;

$$U(q) = -2\pi e^2 / (q^2 + \lambda k_F^2) \tag{8}$$

is the potential energy of exchange interaction and correlations of electron gas; $\lambda \approx 2$ [18]; k_F is the Fermi wave vector;

$$\pi_0(q) = \frac{m k_F}{\pi^2 \hbar^2} \left(\frac{1}{2} + \frac{4k_F^2 - q^2}{8 k_F q} \ln \left| \frac{2k_F + q}{2k_F - q} \right| \right) \tag{9}$$

is the polarization function of free electron gas;

$$\rho^{e}(\mathbf{q}) = \sum_{\mathbf{k}} a_{\mathbf{k}}^{+} a_{\mathbf{k}+\mathbf{q}} \tag{10}$$

is the Fourier transform of the density operator of electron gas;

$$\rho^{i}(\mathbf{q})=\sum_{n}\exp(-i\mathbf{q}\mathbf{R}_{n}) \tag{11}$$

is the Fourier transform of the density operator of ions; and $\mathbf{R}_n$ is the radius vector of the *n*th ion.

Metallic hydrogen is the only system for which the unscreened potential of electron–proton interaction is known exactly. This potential is Coulomb. This circumstance fundamentally simplifies calculations of different properties of metallic hydrogen, since the problem of simulation of electron–proton interaction that gives rise to additional adjustable parameters in the theory and affects the reliability of the results obtained vanishes.

3. Electrical-resistivity

For simple disordered metals with relatively high conductivity, the electrical-resistivity coefficient R is determined by the known Drude formula that is a direct consequence of the nearly free electron model

$$R=\frac{m}{ne^{2}}\tau^{-1}. \tag{12}$$

Here, n is the density of electron gas and τ is the relaxation time of the process of electrical conductivity. The electrical resistivity coefficient of metallic hydrogen will be determined in the same manner.

Numerous works [19–28] are devoted to constructing the perturbation theory for electrical conductivity of liquid metals. Hereafter, the results of [28] will be used to perform concrete calculations.

In the framework of the theory of linear Kubo reaction and the method of two-time retarded Green's functions [28], the reciprocal relaxation time can be represented as the following series in terms of electron–proton interaction:

$$\tau^{-1}=\sum_{n=2}^{\infty}\tau_{n}^{-1}. \tag{13}$$

The general term of this expansion has the form

$$\tau_{n}^{-1}=\frac{N}{V^{n}}\sum_{\mathbf{q}_1,\ldots,\mathbf{q}_n}W(\mathbf{q}_1)\ldots W(\mathbf{q}_n)S(\mathbf{q}_1,\ldots,\mathbf{q}_n)\Gamma(\mathbf{q}_1,\ldots,\mathbf{q}_n). \tag{14}$$

The idea of such a representation and the corresponding designations were borrowed from the expansion of the ground-state energy of electron gas in simple crystalline metals into a series in terms of electron–ion interaction [29]. Here, $S(\mathbf{q}_1,\ldots,\mathbf{q}_n)$ is the n-particle structure factor of the ion subsystem, N is the number of ions in the system, and $\Gamma(\mathbf{q}_1,\ldots,\mathbf{q}_n)$ is the electron multipole for the process of electrical conductivity.

The second-order contribution to the reciprocal relaxation time for simple disordered metals, which was first obtained in the classical work of Ziman [30], has long been studied in much detail and has the form

$$\tau_2^{-1} = \frac{m}{12\pi^3\hbar^3}\int_0^{2k_F} W^2(x)S(x)x^3dx, \quad (15)$$

So, we directly turn to the consideration of the third-order contribution, which can be written as

$$\tau_3^{-1} = \frac{N}{V^3}\sum_{\mathbf{q}_1,\mathbf{q}_2,\mathbf{q}_3} W(\mathbf{q}_1)W(\mathbf{q}_2)W(\mathbf{q}_3)S(\mathbf{q}_1,\mathbf{q}_2,\mathbf{q}_3)\Gamma(\mathbf{q}_1,\mathbf{q}_2,\mathbf{q}_3). \quad (16)$$

For the noninteracting electron gas, the electron tripole obtained by the kinetic-equation method and characterizing the process of electrical conductivity has the form

$$\Gamma(\mathbf{k}_1-\mathbf{k}_2,\mathbf{k}_2-\mathbf{k}_3,\mathbf{k}_3-\mathbf{k}_1) = \frac{\pi\hbar}{3mNk_BT}(\mathbf{k}_1-\mathbf{k}_2)^2 n(\mathbf{k}_1)[1-n(\mathbf{k}_1)]\frac{\delta(\varepsilon_{\mathbf{k}_2}-\varepsilon_{\mathbf{k}_1})}{\varepsilon_{\mathbf{k}_2}-\varepsilon_{\mathbf{k}_3}}, \quad (17)$$

where T is the absolute temperature, k_B is the Boltzmann constant, and $n(k)$ is the Fermi–Dirac electron distribution function.

After some transformations [28], the expression for the third-order contribution can be reduced to the following integral in the sense of principal value:

$$\tau_3^{-1} = \frac{m^2}{24\pi^5\hbar^5k_F^2}\int_0^{\infty}\frac{f(k)}{k_F-k}dk. \quad (18)$$

Since the electron–proton interaction is known exactly, the principal approximation used further is the geometric approximation for the three-particle structure factor [28–32]

$$S(\mathbf{q}_1,\mathbf{q}_2,\mathbf{q}_3) = S(\mathbf{q}_1)S(\mathbf{q}_2)S(\mathbf{q}_3). \quad (19)$$

A consequence of this approximation is the following expression for the integrand function:

$$f(k) = \frac{1}{k_F+k}\sum_{n=0}^{\infty}(2n+1)A_nB_n^2(k), \quad (20)$$

where

$$A_n = \int_0^{2k_F} W(q)S(q)P_n\left(\frac{2k_F^2-q^2}{2k_F^2}\right)q^3dq, \quad (21)$$

$$B_n(k) = \int_{|k-k_F|}^{k+k_F} W(q)S(q)P_n\left(\frac{k^2+k_F^2-q^2}{2kk_F}\right)qdq, \quad (22)$$

where $P_n(x)$ is the Legendre polynomial of degree n.

4. Effective pairwise proton-proton interaction

The characteristic that is fundamentally important for numerical calculations of the electrical resistivity of metallic hydrogen is the effective pairwise ion–ion interaction. In the second order perturbation theory the expression for the effective pair potential in terms of electron–ion interaction is well known and has the form

$$V_{ef}(R)=\frac{e^2}{R}-\frac{v}{\pi^2}\int Q(q)\frac{\sin(qR)}{qR}q^2dq, \quad (23)$$

where

$$Q(q)=\frac{v}{2}V^2(q)\frac{\pi(q)}{\varepsilon(q)}. \quad (24)$$

The latter expression contains no adjustable parameters characterizing the ion subsystem and depends only on the Fermi wave vector, i.e., on the density of the system. The only universal approximation made upon its derivation is the random-phase approximation for an electron subsystem with allowance for exchange interaction and correlations of electrons in the local-field approximation.

The hard sphere diameter or the minimum distance to which protons can approach at a given temperature will be found from the condition of equality of the kinetic and potential energies of protons upon their closest approach [33]

$$V_{ef}(\sigma)=3k_BT/2\ . \quad (25)$$

Here, the kinetic energy is reckoned from the bottom of the potential well. The thus-found hard-sphere diameter is a function of the density and temperature. For this reason, all the quantities depending on the hard-sphere diameter likewise will be functions of the density and temperature.

5. Numerical calculations and discussion of results

If the exact solution of the Percus–Yevick equation for the hard-sphere model is employed for the pair structure factor of the ion subsystem, the expression for the electrical conductivity coefficient depends on three parameters: the hard sphere diameter σ, the packing fraction η, and the Fermi wave vector k_F. However, these three parameters are connected by two relations. For degenerate electron gas, $v=3\pi^2/k_F^3$. The expression for the volume occupied by hard-spheres can easily be obtained with the knowledge of the volume per ion $v=1/n$, $v\eta=\pi\sigma^3/6$. In other words, only one of these parameters is really independent. The hard sphere diameter is conveniently taken as such a parameter. If the temperature of the system is also known, the remaining parameter of the system can be determined as was shown in the preceding Section.

In the hard sphere model, the upper limit of the electron gas density corresponds to the value of packing density close to the greatest possible value of the latter. At high values of the packing fraction, there occurs a transition of metallic hydrogen from the liquid state to the crystalline state, which is not discussed in this work. At $r_s=0.9$, the density of electron gas is equal to the electron density in hydrogen atom. Such a density can be considered as extremely high. It is seen from the table that, from this viewpoint, the density corresponding to the transition of hydrogen into metallic state (0.32 mol/cm^3) [2, 12] is not high. The packing fraction is somewhat smaller than its values for the majority of liquid metals near their melting temperatures; the hard sphere diameter is approximately as large as two atomic diameters; and the Fermi wave vector is approximately equal to the Fermi wave vector of liquid Cr near the melting temperature of the latter.

Since metallic hydrogen in nature exists in a wide range of densities and temperatures, it is of interest to study the behavior of the resistance coefficient in a wide range of these parameters. It is such a dependence that is shown in the following figures.

It is seen in Fig. 1 that, with increasing density, the second and third order contributions into the electrical resistivity in terms of electron–proton interaction as well as the approximate sum of the series decrease rapidly. This is in good agreement with common physical concepts of the diminution of the role of interparticle interaction with

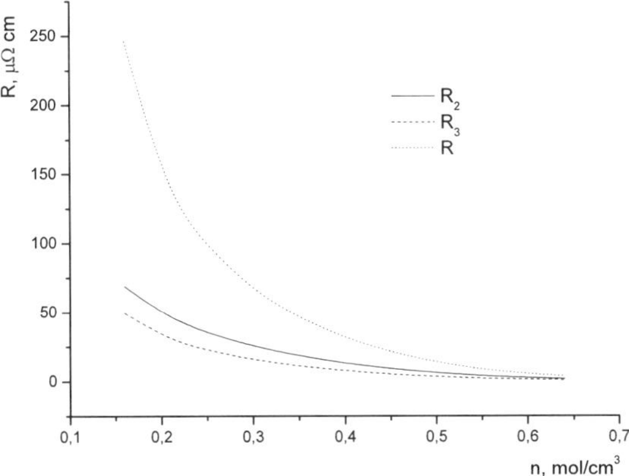

Figure 1. The electrical resistivity of metallic hydrogen at a temperature of 3,000 K as a function of density. R_2 is the second order contribution into the electrical resistivity in terms of electron–proton interaction, R_3 is the third order contribution, and $R = 1/(1 - R_3 / R_2)$ is the result of approximate summation of the perturbation theory series

increasing density of the system. For the electron gas density corresponding to the transition of hydrogen into metallic state, the electrical resistivity is $48\,\mu\Omega \cdot \mathrm{cm}$, which is considerably smaller than the experimental resistivity at this density. At the same time, the theoretical resistivity obtained is several times larger than the calculated results of other authors [16, 17]. The latter circumstance is connected with taking into account the higher-order perturbation-theory terms, which are signification in the entire range of densities and temperatures considered. When only the second order contribution is taken into account, the resistivity is equal to $20\,\mu\Omega \cdot \mathrm{cm}$. This value agrees well with the results of [16] obtained in the same approximation and with the results of computer simulation [17]. Note that at the point of hydrogen transition into the metallic state one should hardly expect good agreement with experimental electrical-resistivity data. This is connected with the presence, in the energy spectrum of electrons, of energy gap of width of about 3,000 K that is disregarded in the given calculations. At the same time, we expect that at densities higher than 0.32 mol/cm^3 or at temperatures above 3,000 K, the results of theoretical calculations would predict the real physical situation very accurately.

It is seen in Fig. 2 that, with increasing temperature, the second and third order contributions into the electrical resistivity in terms of electron–proton interaction as well as the approximate sum of the series slowly increase, as for typical metals.

A better agreement with the experimental electrical resistivity at the point of the metal–insulator transition likewise can be attained in the framework of the model of metal put forward in this study. To do this, it is necessary to assume that only part of hydrogen become transformed into the atomic state, i.e., to consider electron gas at lower densities. In this case, at the electron gas density corresponding to approximately 5% hydrogen in the atomic state, the resistivity in second order of perturbation theory approaches an experimental value of $500\,\mu\Omega\cdot\text{cm}$. Such a result is consistent with the assumption of the authors of [2] that upon the metal–insulator transition only 5% hydrogen is in atomic state. With taking into account the higher-order perturbation theory terms this percent well above and forms approximately 40%.

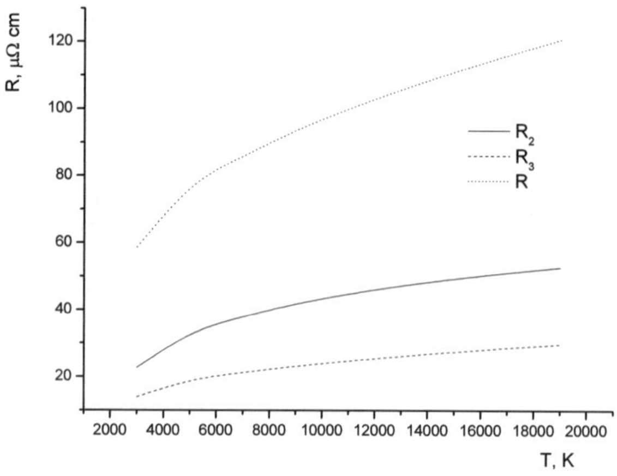

Figure 2. The electrical resistivity of metallic hydrogen at a density of 0.32 mol/cm^3 as a function of temperature. R_2 is the second order contribution into the electrical resistivity in terms of electron–proton interaction, R_3 is the third order contribution, and $R = 1/(1 - R_3 / R_2)$ is the result of approximate summation of the perturbation theory series

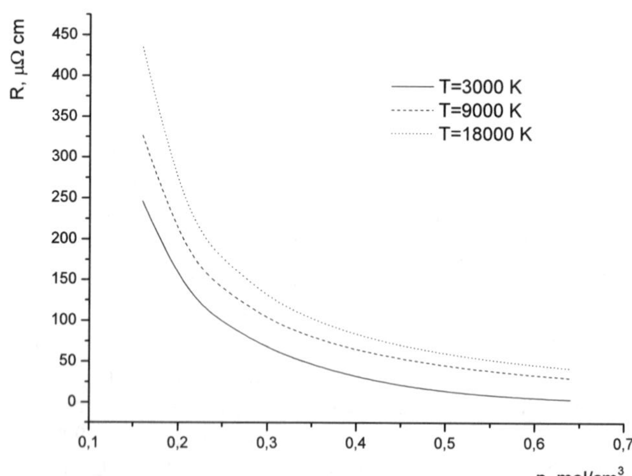

Figure 3. The electrical resistivity of metallic hydrogen as a function of the density at temperatures of 3,000, 9,000, and 18,000 K

The behavior of the density dependence of the resistivity at different temperatures is compared in Fig. 3. Figures 1 and 2 give also the results of approximate summation of the perturbation-theory series at different temperatures. Such a summation is necessary, since the third order term in electron–proton interaction is by no means small at all the temperatures and densities considered. It was assumed in this work that the perturbation theory series is a geometric progression with the denominator equal to the ratio of the third order term to the second order term in electron–proton interaction, R_3 / R_2. Then, the resistivity is $R = 1/(1 - R_3 / R_2)$. It is seen that the role of higher order terms in electron–proton interaction is more significant at low than at high densities.

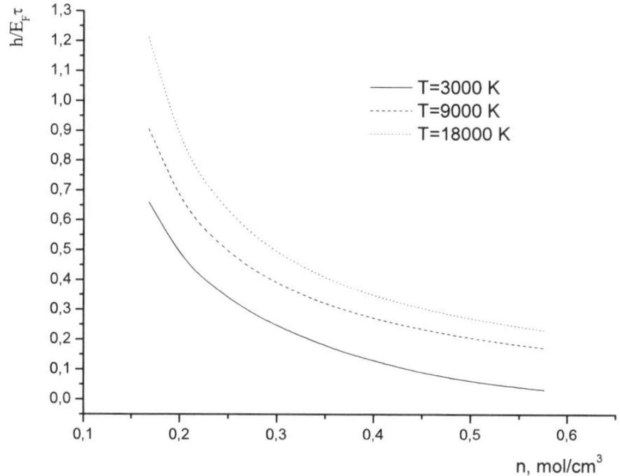

Figure 4. Dependence of the criterion of applicability of the nearly free electron model on the density at different temperatures

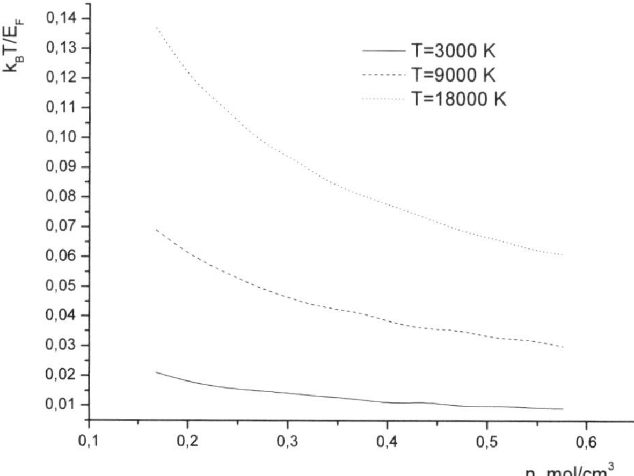

Figure 5. Dependence of the degeneracy parameter of the system on the density at different temperatures

The high resistivity values at low densities and high temperatures call for an analysis of the conditions of applicability of the nearly free electron model. The criterion of its applicability is the value of the dimensionless parameter $\hbar / \varepsilon_F \tau$. The model works well if this parameter is much smaller than unity. Figure 4 displays the dependence of this parameter on the electron gas density. It is seen from this figure that at low densities this parameter is approximately equal to unity, which indicates poor applicability of the nearly free electron model in this case. At high densities, including the density corresponding to the transition into the metallic state, its value is sufficiently small and corresponds to the values typical of the majority of simple metals. Another parameter that must be controlled is the parameter characterizing the degree of degeneracy of the electron gas, $k_B T / \varepsilon_F$. As is seen from Fig. 5, the electron gas is degenerate in the entire range of densities considered. At the same time, this parameter is already great enough for taking it into account led to substantial corrections to all calculated characteristics of metallic hydrogen.

6. Conclusions

Thereby, nearly free electron model allows to explain experimental value of the metallic hydrogen resistivity only with taking into account the higher order terms of perturbation theory. Herewith only approximately 40% of hydrogen must be in atomic state at the point of the metal–insulator transition.

References

1. E. Wigner and H. B. Huntington, *J. Chem. Phys.*, 1935, **3**: 764.
2. S. T. Weir, A. C. Mitchell, and W. J. Nellis, *Phys. Rev. Lett.*, 1996, **76**: 1860.
3. V. E. Fortov, V. Ya. Ternovoi, S. V. Kvitov, et al., *JETP Lett.*, 1999, **69**: 926.
4. V. Ya. Ternovoi, A. S. Filimonov, V. E. Fortov, et al., *Physica B*, 1999, **265**: 6.
5. W. J. Nellis, A. C. Mitchell, P. C. McCandless, et al., *Phys. Rev. Lett.*, 1992, **68**: 2937.
6. P. S. Hawke, T. J. Burgess, D. E. Duerre, et al., *Phys. Rev. Lett.*, 1978, **41**(14): 994–997.
7. E.G. Brovman, Yu. Kagan, and A. Kholas, *Sov. Phys. – JETP*, 1972, **34**: 1300.
8. S. A. Bonev and N. W. Ashcroft, *Phys. Rev. B Phys.*, 2001, **64**: 224112.
9. K. Nagao, S. A. Bonev, and N. W. Ashcroft, *Phys. Rev. B*, 2001, **64**: 224111.
10. C. F. Richardson and N. W. Ashcroft, *Phys. Rev. Lett.*, 1997, **78**: 118.
11. C. Pierleoni, D. M. Ceperley, and M. Holzmann, *Phys. Rev. Lett.*, 2004, **93**: 146402.
12. W. J. Nellis, *Rep. Prog. Phys.*, 2006, **69**: 1479.
13. V. S. Filinov, M. Bonitz, V. E. Fortov, et al., *Contrib. Plasma Phys.*, 2004, **44**: 388.
14. V. E. Fortov, *Phys.-Usp.*, 2007, **50**(4): 333–353.
15. V. T. Shvets, *JETP*, 2007, **104**: 655.
16. D. J. Stevenson and N. W. Ashcroft, *Phys. Rev. A: Gen. Phys.*, 1974, **9**(2): 782–791.
17. I. Kwon, L. Collins, J. Kress, and N. Troullier, *Phys. Rev. E*, 1996, **54**: 2844.
18. D. J. M. Geldart and S. H. Vosko, *Can. J. Phys.*, 1966, **44**: 2137.
19. B. Springer, *Phys. Rev.*, 1964, **136**: 115.
20. B. Springer, *J. Phys. C*, 1969, **2**: 288.
21. T. Neal, *Phys. Rev.*, 1968, **169**: 508.
22. T. Neal, *Phys. Fluids*, 1970, **13**: 249.
23. N. W. Ashcroft and W. Schaich, *Phys. Rev. B*, 1970, **1**: 1370.
24. N. W. Ashcroft and W. Schaich, *Phys. Rev. B*, 1971, **3**: 1511.
25. A. Bringer and D. Wagner, *Z. Phys.*, 1971, **241**: 295.
26. J. Popielawski, *Physica A*, 1974, **78**: 97.
27. J. Gorecki and J. Popielawski, *J. Phys. F*, 1983, **13**: 2107.
28. V. T. Shvets, *Phys. Met. Metallogr.*, 2000, **89**: 211.
29. E. G. Brovman and Yu. M. Kagan, *Usp. Fiz. Nauk*, 1974, **112**: 369.
30. J. M. Ziman, *Philos. Mag.*, 1961, **6**: 1013.
31. D. J. Stevenson, *Phys. Rev. B*, 1975, **12**: 3999.
32. W. H. Shih and D. Stroud, *Phys. Rev. B*, 1985, **31**: 3715.
33. V. T. Shvets, S. V. Savenko, Ye. K. Malynovski, *Condens. Matter Phys.*, 2006, **9**: 127.

SURFACE-MODIFIED AB_5 ALLOYS WITH ENHANCED HYDROGEN ABSORPTION KINETICS

M. WILLIAMS, M. LOTOTSKY, A. NECHAEV, V. LINKOV
South African Institute for Advanced Materials Chemistry, University of the Western Cape, Private Bag X17, Bellville, 7535, South Africa
V. YARTYS
Institute for Energy Technology, POB 40, Kjeller N-2027, Norway
Q. LI
Guangzhou Research Institute for Non-Ferrous Metals, Guangzhou 510651, China

Abstract. Surface engineering approaches were adopted in the preparation of advanced materials based on "low-temperature" AB_5 hydride-forming intermetallide. The approaches used included micro-encapsulation with palladium, as well as with palladium-containing mixed coatings. The influence of these modifications on surface morphology and hydrogenation characteristics of the material were investigated. It was found that palladium-nickel encapsulation improves the hydrogenation kinetics of the intermetallide in the absorption of hydrogen at low pressure and room temperature, and after long-term pre-exposure to air. The developed surface-modified materials may potentially find their application in highly-selective hydrogen extraction and purification processes.

Keywords: metal hydrides, AB_5, palladium, electroless deposition

1. Introduction

Large supply of pure hydrogen derived from gasification is a major stumbling block in its employment in energy-generating devices (e.g. fuel cells). This is due to surface deactivation and inefficiency of current hydrogen absorption media in gas feeds containing aggressive components and low hydrogen volume content. Breakthroughs in hydrogen separation and purification require novel materials which meet scientific and technological requirements (chemical and thermal stability; long process lifetimes; not prone to degradation). These requirements can potentially be met through hydride-forming metal alloys [1, 2].

Technologically-flexible "low-temperature" AB_5-type alloys based on rare-earth metals (A) and nickel (B) are potentially attractive materials for separation and purification of hydrogen due to good poisoning tolerance compared to other hydride-forming alloys, good reversibility in hydrogen absorption-desorption, express kinetics under mild conditions, tunable plateau pressures, reasonable cycle lifetimes, low hysteresis, reasonable hydrogen storage capacity and reasonable cost [1–4].

Hydrogenation/dehydrogenation performances in hydrogen storage alloys are surface state dependent [5]. AB_5 tends to exhibit susceptibility to large-to-trace volumes of these gases passivating the surface and, by such a way, inhibiting hydrogen

B. Baranowski et al. (eds.), *Carbon Nanomaterials in Clean Energy Hydrogen Systems.*

dissociation/recombination processes. As hydrogen purification from gasification mixtures would involve exposure to significant volumes of chemically-aggressive electrophilic gases (e.g. H_2S, CO, O_2, H_2O), so the rationale towards modification of AB_5 to improve their hydrogen sorption an poisoning tolerance comes into focus.

Several methods to eliminate the negative influence of impurities on H sorption properties of metal hydrides (MH) exist: creation of multi-phase materials [6], deoxidisers [7], surface modification [8, 11], including deposition of thin metal layers onto the surface of MH particles [9, 10]. The last method, when Platinum Group Metal (PGM) is used as a modifier, is especially efficient, since it allows increasing the resistance of MH materials even to such strong poisons as CO [11]. PGM surface coatings enable hydrogen to rapidly absorb through the surface of the alloy to the bulk material, while maintaining the hydrogenation activity even after extensive exposure to contaminated hydrogen. In particular, Pd exhibits high permeability towards hydrogen. It is impermeable to larger molecules, and as a result Pd also exhibits an infinitely high selectivity towards hydrogen which can be consequently absorbed in the bulk of the Pd-coated MH material bypassing a slow step of dissociative H_2 chemisorption. Pd encapsulation gives MH materials the ability to absorb hydrogen at low temperature without a special activation step, as well as enhanced absorption/desorption kinetics. Pd layers also compensate for losses in hydrogen absorption/desorption performances of the MH associated with formation of oxide layers and/or blocking of active centres on the surface of the material responsible for the transfer of molecular H_2 from the gas phase to H atoms in the bulk, and vice versa [10–13].

Nickel surface coatings are known to improve overall mechanical strength and corrosion resistance in hydrogen absorption-desorption of hydride-forming alloys. In addition, surface coatings of Ni on Ni-containing AB_5 alloys are known to increase the catalytic activity of the material towards hydrogen exchange reactions [14]. Ni also has the ability to lower the critical temperature and equilibrium pressure for β-phase formation, thus alleviating hydrogen embrittlement. Moreover, the addition of Ni to Pd results in the decrease of the difference in lattice parameters between α- and β-phases in Pd–H systems, so that any structural distortion that does occurs in the $\alpha \rightarrow \beta$ phase transition is not as severe as that observed in pure Pd [15]. It is expected that combined metal coatings containing both Pd and Ni will be especially efficient for improvement of hydrogen sorption performances of Ni-containing AB_5 alloys, due to higher activity of Pd towards dissociation-recombination of hydrogen molecules, its increased resistance to poisoning, and, from the other hand, because of peculiarities of the AB_5 alloys containing natural interface between their surface and bulk material, in the form of Ni clusters formed in the course of surface segregation processes.

Undesirable distortions in integrity of Pd surface coatings can be suppressed by alloying Pd with Group IB metals such as copper. Pd–Cu alloy coatings possess the high hydrogen selectivity and poisoning resistance needed to be considered as viable candidates for hydrogen separation from mixed gas feeds. Alloying Pd with Cu has been observed to reduce the deactivating effects of surface poisoning, particularly that of sulphurous compounds [16]. In addition, Pd-Cu alloys have been associated with enhanced resistance to thermal degradation and reduced material costs, due to a decrease in the amount of Pd used [15]. This alloy is associated with hydrogen diffusivity superior to that of pure Pd, albeit that the hydrogen solubility decreases somewhat. Some or all the enhanced properties observed upon alloying Pd with Cu may be attributed to a change in the electronic effects [17].

Although several coating techniques such as thermal deposition, chemical vapor deposition, and electrochemical plating can be successfully implemented in the surface modification of MH, electroless deposition is considered the easiest and most inexpensive way to deposit metallic coatings onto solid surfaces. Electroless deposition is a wet chemical reduction process in which aqueous metal ions, in alkaline or acidic baths, are autocatalytically reduced at a solid-liquid interface in the absence of an applied external electrical current to activate the process [18, 19].

The aim of this study is the development of approaches for the preparation of new materials, based on $MmNi_5$ alloys surface-modified using Pd and Pd-containing mixed coatings (Pd + Ni and Pd + Cu), for hydrogen extraction and purification.

2. Experimental

Commercial Mischmetal-Nickel AB_5-alloy powder (La,Ce,Pr,Nd)(Ni,Co,Al,Mn)$_5$, trademark DH4, produced by Guangzhou Research Institute of Non-Ferrous Metals, China, was employed as the core material. The alloy was allowed constant exposure to air throughout the experimental studies.

XRD studies of starting materials (core and surface-modified AB_5 alloy) and products of their hydrogenation were performed using a Bruker AXS D8 Advance, Cu-K_α, $\lambda_1 = 1.5406$ Å, $\lambda_2 = 1.5444$ Å, $\lambda_2/\lambda_1 = 0.5$, $2\theta = 10–90°$. XRD data was refined by Rietveld profile analysis with GSAS software [20].

Structural/morphological properties were characterized using scanning electron microscopy (Zeiss Ultra 55 HRSEM). Estimations of alloy composition and Pd loading were conducted using energy-dispersive x-ray emission spectroscopy (EDS).

Hydrogen sorption performances of core and surface-modified alloys were studied using a Sieverts'-type installation. The AB_5 core was loaded into reactor and activated by heating to 200°C in vacuum better than 10^{-5} mbar for 1.5 hours followed by hydrogenation at 20°C (final pressure 12 bar, calculated H sorption capacity corresponded to $H/AB_5 = 5.70$). The 1st hydrogenation cycle was very fast: more than 90% of the maximum H capacity was reached in ~1 minute, and reproduced in the course of the 2nd hydrogenation. Measurement of PCT characteristics were started from the 3rd cycle and included evacuation of the sample (T = 150°C, vacuum better 10^{-5} mbar, duration 5–10 hours) followed by consequent measurements of absorption and desorption isotherms at 30–150°C using standard volumetric procedures.

Surface modification was conducted using electroless deposition of Pd, Ni, and Cu either by single-film, mixed-bath, or sequential deposition, in N_2H_4- and NaH_2PO_2-based electroless plating baths [21–23]. Equivalent volume ratio was used when depositing Pd and Ni (Cu) using a mixed bath. A measured volume of reductant, N_2H_4 (1 M) or NaH_2PO_2 solution (10 g/l) was added separately. Plating time and agitation rate were fixed at 30 minutes and 300 rpm, respectively. Concentrations of components in Pd- Ni- and Cu-based baths, as well as volume ratio to be mixed is selected to provide 5.0×10^{-4} to 1.5×10^{-3} M of Pd^{2+}, 1×10^{-3} to 2.5×10^{-3} M of Ni^{2+} and 1.0×10^{-2} to 4.0×10^{-2} of Cu^{2+} in the bath. Examples of bath formulations for NaH_2PO_2-based electroless deposition baths are given in Table 1. The N_2H_4-based Pd plating solution had the same concentration of Pd^{2+} ions differing by only complexing agent (Na_2EDTA instead of NH_4Cl); the plating temperature was 60°C.

TABLE 1. Chemical constitution of hypophosphite-based electroless deposition baths

Pd-based plating solution		Ni-based plating solution		Cu-based plating solution	
$PdCl_2$	2 g/l	$NiCl_2$	0.76 g/l	$CuSO_4$	10 g/l
HCl	4 ml/l	$NiSO_4$	1 g/l	$NiSO_4$	1.6 g/l
NH_4OH	160 ml/l	$Na_2(EDTA)$	7.52 g/l	H_3BO_3	35 g/l
NH_4Cl	27 g/l	$Na_3[(OOCCH_2)_2C(OH)COO]$	1 g/l	$Na_3[(OOCCH_2)_2C(OH)COO]$	25 g/l
pH	9.0 ± 2	pH	9.0 ± 2	pH	10 ± 1
Plating temperature	50°C	Plating temperature	50°C	Plating temperature	70°C

The influence of surface modification of the AB_5 alloy on hydrogen sorption performances and poisoning resistance was studied by comparison of kinetics of hydrogen absorption at the same mild conditions (T ~ 20°C, P_{H2} ~ 5 bar), after exposure to air for ~2 weeks and evacuation before hydrogenation (10^{-3} mbar, 20°C). The sample weight was ~1.6 g, and the hydrogenation was conducted for 20–24 hours. The experimental results were processed by the application of formal kinetic analysis, using the Avraami-Erofeev equation [24]:

$$(H/AB_5) = (H/AB_5)_{\max} \cdot \left\{1 - \exp\left[-\left(\frac{t}{t_0}\right)^n\right]\right\}, \tag{1}$$

where (H/AB_5) is actual and $(H/AB_5)_{max}$ – maximum hydrogen concentration in the alloy (equal to 5.00, according to PCT studies); t is time; t_0 is the characteristic time of hydrogenation (reciprocal rate constant); and the index of power, n (should be integer or half-integer), can be interpreted as a value indirectly connected to reaction mechanism.

3. Results and discussion

SEM analysis of the starting AB_5 core alloy determined the particle size distribution as 10–70 μm with all particles possessing irregular shape, although some almost spherical particles were observed. In addition, EDS data suggested that the composition of the AB_5 core alloy corresponded to $La_{0.54}Ce_{0.31}Nd_{0.09}Pr_{0.05}Ni_{3.16}Co_{0.94}Al_{0.48}Mn_{0.43}$.

XRD studies of unmodified AB_5 alloy (Fig. 1) have shown that the material contains a major $CaCu_5$-type intermetallic phase with the periods of hexagonal unit cell a = 5.0030(1) Å, c = 4.0516(2) Å. The calculated occupancies of the components to yield best fit of the XRD data correspond to the formula $La_{0.40}Ce_{0.48}(Nd,Pr)_{0.16}Ni_{3.34}Co_{0.64}Al_{0.63}Mn_{0.58}$ that is rather close to the EDS results. The lattice periods are in good correspondence with literature data (a = 4.95…5.05 Å, c = 4.02…4.05 Å) for similar alloys $La_{0.8(1-x)}Ce_{0.8x}(PrNd)_{0.2}B_5$, where B_5 is $Ni_{3.55}Co_{0.75}Mn_{0.4}Al_{0.3}$, x = 0.1 [25]. The XRD pattern also shows the presence of trace amount of nickel that is typical for industrial-scale produced AB_5 alloys [26]. XRD studies of the hydrogenated sample show the presence of two main phases, viz. the major phase of β-hydride AB_5H_x characterized by the same symmetry of metal matrix as the starting AB_5 alloy and expanded unit cell ($\Delta a/a_0$ = 5.27%, $\Delta c/c_0$ = 2.82%, $\Delta V/V_0$ = 13.94%), also the α-solid solution of hydrogen in the AB_5 ($\Delta a/a_0$ = 1.04%,

$\Delta c/c_0 = 0.82\%$, $\Delta V/V_0 =$ 2.93%). The lower lattice expansion in the course of hydrogenation in comparison with the parent $LaNi_5$ alloy ($\Delta V/V_0 = 24\%$ for the β-hydride) is most probably caused by the high Co content in the studied alloy.

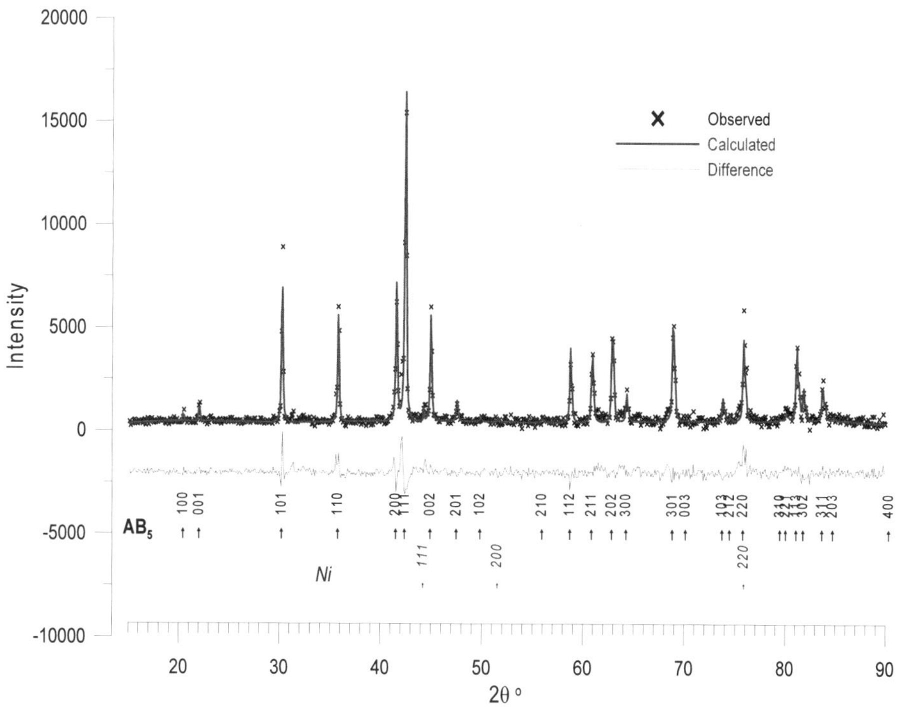

Figure 1. The indexed XRD pattern of the AB_5 core alloy

Figure 2 presents isotherms of H absorption and desorption for the activated AB_5 core alloy. Experimental data was fitted using the model of phase equilibria in metal-hydrogen systems [27]. PCT modelling assumed hydrogen in a metal matrix as a Van der Waals lattice gas, where parameters of H–H interaction have a random nature caused by inhomoheneities of the H storage material resulting in Gaussian distributions of the concentration-independent entropy and enthalpy terms of the equation of a pressure-concentration isotherm. Apparent PCT properties of the AB_5 alloy calculated from results of fitting the experimental PCT data are presented in Table 2. Typical AB_5 hydrogen sorption behaviour is demonstrated, characterized by the plateau slope, d(Ln P)/d(H/AB_5), less than 0.1, and hysteresis energy losses, $RT \ln\left(P_A / P_D \right)$, less than 1 kJ/mol H_2 ($T = 30\ldots100$°C).

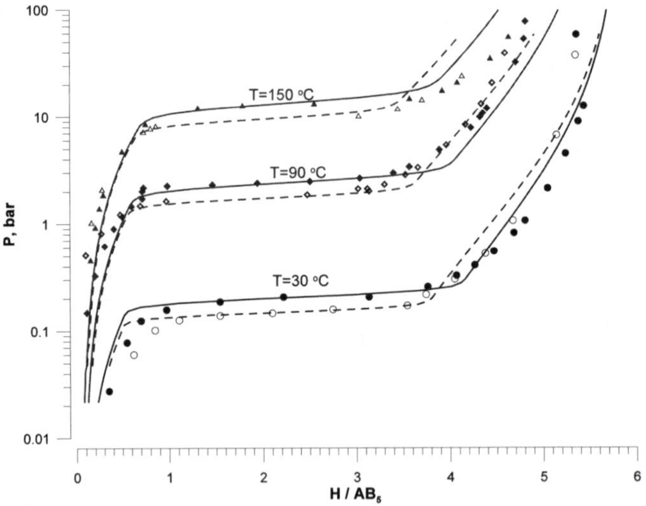

Figure 2. Hydrogen absorption/desorption isotherms for the AB_5 core alloy: experimental data (points) and calculated dependencies (curves). Filled symbols/solid lines correspond to absorption, and opened symbols/dashed lines – to desorption

TABLE 2. Summary of the calculated apparent PCT properties for the H storage alloy

Parameter	Absorption			
	$T = 30°C$	$T = 100°C$	$T = 150°C$	$T = 200°C$
Plateau midpoint (C_0), H/AB_5	2.316	2.263	2.205	2.126
ΔS^0, $J/moleH_2/K$ ($C = C_0$)	–104.13	–101.69	–100.03	–98.46
ΔH^0, $kJ/moleH_2$ ($C = C_0$)	–35.560	–34.181	–33.232	–32.322
d(Ln P)/d(H/AB_5)	0.082	0.115	0.140	0.172
P ($C = C_0$), bar	0.2051	3.366	13.28	37.56
C (P = 50 bar), H/AB_5	5.602	4.862	4.253	3.414
C (P = 1 bar), H/AB_5	4.601	0.431	0.272	0.202
Parameter	Desorption			
	$T = 30°C$	$T = 100°C$	$T = 150°C$	$T = 200°C$
Plateau midpoint (C_0), H/AB_5	2.139	2.053	1.985	1.911
ΔS^0, $J/moleH_2/K$ ($C = C_0$)	–101.40	–98.91	–97.23	–95.63
ΔH^0, $kJ/moleH_2$ ($C = C_0$)	–35.547	–34.175	–33.233	–32.335
d(Ln P)/d(H/AB_5)	0.071	0.095	0.119	0.152
P ($C = C_0$), bar	0.1484	2.414	9.469	26.65
C (P = 50 bar), H/AB_5	5.575	4.711	4.050	3.417
C (P = 1 bar), H/AB_5	4.431	0.447	0.280	0.211
Hysteresis energy losses, $RT \ln\left(P_A / P_D\right)$, $J/molH_2$ (C=C_0)	815	1,031	1,189	1,349

Figure 3 compares dynamics of hydrogen absorption by Pd-modified AB_5 alloys, prepared using sodium hypophosphite and hydrazine as reducing agents. Generally, modification by surface deposition of Pd resulted in significant improvements in hydrogen sorption dynamics compared to the unmodified AB_5 alloy. However, alloys modified by deposition of Pd from hypophosphite-based plating baths resulted in notably higher room temperature hydrogenation capacities ($H/AB_5 \sim 4.25$) and faster activation kinetics, after pre-exposure to air, compared to that prepared from hydrazine-based plating baths (H/AB_5 ~ 3.5). This observation may be a result of amorphous Pd deposits on AB_5 alloys, prepared from hypophosphite-based electroless plating baths (as a result of the impregnation of phosphorus atoms into the Pd layer), exhibiting higher catalytic activities towards the dissociation of hydrogen molecules and the subsequent absorption of hydrogen atoms into the core alloy compared to the crystalline Pd deposits on AB_5 alloys, prepared from hydrazine-based electroless plating baths.

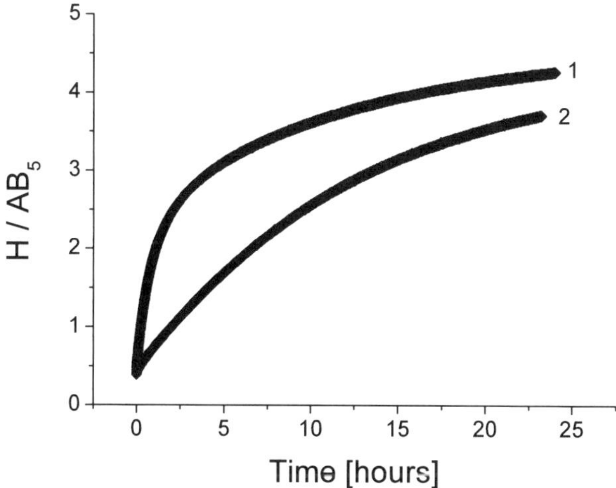

Figure 3. Dynamics of hydrogen absorption (T ~ 20°C, P_{H2} ~ 5 bar) by surface-modified samples of the AB_5 alloy, without pre-activation: 1 – Pd (30") hypophosphite derived; 2 – Pd (30"') hydrazine derived

Figure 4 shows SEM images of the unmodified AB_5 alloy and after its surface modification with Pd for 30 minutes. Pd particles were not visible on the surface of the unmodified AB_5 alloy, whereas nano-sized Pd particles (average size = 89.3 nm) were visible on the surface of the Pd-modified AB_5 alloy. Good Pd nanoparticle dispersion was observed over the surface of the AB_5 core, with minimal agglomeration.

EDS analyses (Table 3) show that very low Pd loadings <2 wt% were deposited onto the surface of the AB_5 intermetallide by electroless deposition. Trace levels of phosphorus formed as a result of the oxidation of sodium hypophosphite were also detected in the Pd-based coatings. It is known from [28] that phosphorus-containing coatings exhibit amorphous nature that is in correspondence with XRD data for the surface-modified alloys (see below). Cu levels were relatively high while the deposited Ni could not be resolved from the Ni surface matrix.

XRD studies of the surface-modified materials were difficult due to large interferences from the core alloy (AB_5-phase and traces of Ni). However, none of the measured XRD patterns showed a presence of the crystalline Pd that allows us to suppose its amorphous nature in the coatings (in agreement with that in [28]), which is due to the low temperature deposition using sodium hypophosphite as the reducing agent, and the subsequent inclusion of phosphorus into the metal coating. At the same time, the increased trace amounts of Ni (as compared to one detected in the unmodified alloy) in the alloy modified by Pd/Ni (both sequentially and by co-deposition) were observed. Similarly, in the alloys modified by Pd/Cu and Pd/Cu/Ni there were observed trace amounts of the Cu phase.

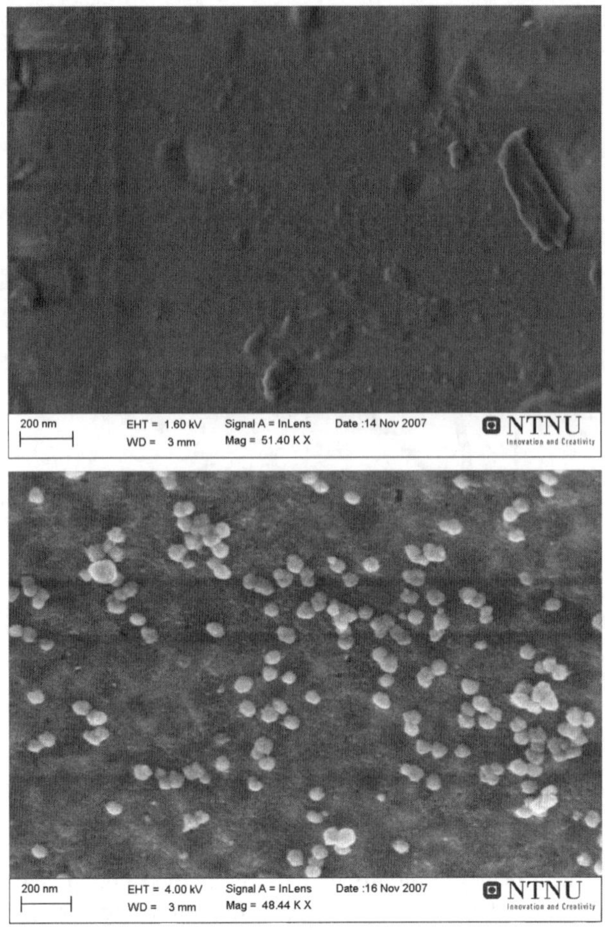

Figure 4. In-lens HRSEM images of (a) unmodified AB_5 alloy; (b) after 30 minute Pd deposition. Working distance – 3 mm; accelerating voltage – 4 kV

TABLE 3. Results of EDS analyses of the surface-modified AB_5 alloys (Figs. 5/2–6)

Modifier	Metal loading (weight%)			
	Pd	P	Cu	Ni
Pd	0.75	0.1	–	–
Pd/Cu (sequential deposition)	1.17	0.12	6.19	–
Pd/Cu (co-deposition)	1.93	0.21	5.69	–
Pd/Ni (sequential deposition)	1.32	0.17	–	n.a.
Pd/Ni (co-deposition)	1.27	0.22	–	n.a.
Pd/Cu/Ni (co-deposition)	1.01	0.28	2.09	n.a.

Figure 5 shows the dynamics of hydrogen absorption by surface-modified AB_5 alloys (curves 1, 2, 4–7). The data for the unmodified core alloy measured at the same conditions are presented as well (curve 3). As expected, the unmodified alloy absorbs hydrogen very slowly – the amount of absorbed hydrogen during 24 hours is less than $H/AB_5 = 1$, or 20% of maximum H sorption capacity of the activated alloy. Results of fitting the experimental data by equation (1) are presented in Table 4.

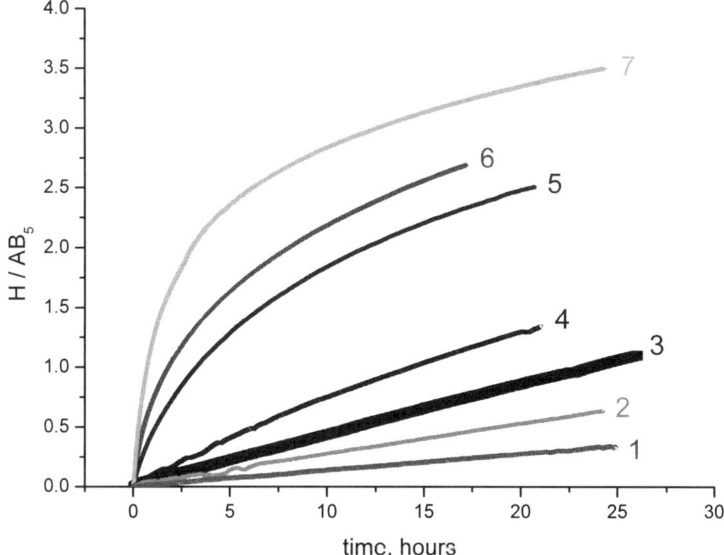

Figure 5. Dynamics of hydrogen absorption (T ~ 20°C, P_{H2} ~ 5 bar) by the core and surface-modified samples of the AB_5 alloy, without pre-activation: 1 – Pd/Cu, sequential deposition (SD); 2 – Pd/Ni (SD); 3 – unmodified alloy; 4 – Pd/Ni/Cu, co-deposition (CD); 5 – Pd/Cu (CD); 6 – Pd; 7 – Pd/Ni (CD)

TABLE 4. Results of fitting of the experimental data by the Avraami-Erofeev equation

Group	Sample # (Fig. 6)	t_0 (hours)	n	Pearson correlation coefficient, R^2
1	1	342(1)	1.013(1)	0.9993
	2	189.5(7)	0.974(2)	0.998
	3	103.0(1)	1.0173(6)	0.9997
	4	73.82(9)	0.9160(6)	0.9994
2	5	34.59(7)	0.645(1)	0.998
	6	26.75(4)	0.5612(5)	0.9992
	7	15.18(4)	0.450(1)	0.993

Two groups of materials can be divided in terms of their H absorption dynamic behaviour. The first includes unmodified alloy, alloy modified by sequential deposition of Pd and Ni (or Cu), as well as the alloy modified by co-deposition of Pd, Ni and Cu. These materials are characterised by slow hydrogen absorption reaching $H/AB_5 = 0.35\ldots1.35$ after 24 hours (estimated characteristic time of H absorption, $t_0 = 74\ldots340$

hours). Fitted n values belonging to this group are close to 1. Poor kinetic performances of the surface-modified alloys belonging to group 1 may have their origin in the formation of dense Ni (Cu) layer acting as an inhibitor to hydrogenation.

The second group includes samples modified by Pd and co-deposited Pd and Ni(Cu). These materials are characterised by significantly faster dynamics of hydrogenation, and absorb more than half of the maximum amount of hydrogen ($H/AB_5 = 2.5 \ldots 3.5$) in 24 hours, having characteristic time of hydrogenation, t_0, 3…7 times lower than for the unmodified alloy. Fitted n values are close to 0.5 that allows the assumption that hydrogenation of these materials has a mechanism which differs from that for materials of group 1 ($n \sim 1$).

The strongest effect of surface modification on H sorption properties was observed for the AB_5 alloy modified by co-deposition of Pd and Ni. Intensive formation of water vapour accompanied by a self-heating of the material has to be noted after post-analysis exposure to air, and was not observed for other samples. This illustrates that the material modified by co-deposition Pd and Ni has an activity higher than other materials in terms of processes in the transfer of hydrogen atoms between the bulk material and its surface.

It is probable that the influence of Ni in Pd/Ni co-deposition is due to the creation of an interfacial "bridge" between Ni surface clusters of the AB_5 alloy and deposited amorphous Pd layer. This interface may have increased the hydrogenation rate and quality of diffusional transport of H atoms, after H_2 dissociation on Pd particles, towards the bulk material. Similarly, the reverse process of passing the H atoms from the bulk to the modified surface followed by the associative H_2 desorption can be also facilitated.

4. Conclusions

AB_5 alloys surface-modified using hypophosphite-based Pd electroless deposition baths resulted in higher room temperature hydrogenation capacities and faster activation kinetics, compared to those prepared from hydrazine-based electroless plating baths. It may be a result of the amorphous Pd deposits on AB_5 alloys, prepared from the hypophosphite-based baths (due to the impregnation of phosphorus atoms into the Pd layer), exhibiting higher catalytic activities towards the dissociation of hydrogen molecules and the subsequent absorption of hydrogen atoms into the core alloy, compared to the crystalline Pd deposits, prepared from the hydrazine-based baths.

Pd-Ni mixed coatings deposited onto the surface of AB_5 promote faster hydrogenation performance on air-exposed non-activated alloys compared to that coated with Pd only.

The developed surface-modified materials are characterised by fast hydrogen absorption at the presence of gas impurities and may find their application in efficient hydrogen extraction/purification, storage and delivery systems.

Acknowledgements

This work is supported by South Africa – Norway Program of Research Co-operation (2007–2010), Project #180344.

References

1. Sandrock G. A Panoramic Overview of Hydrogen Storage Alloys from a Gas Reaction Point of View, *Journal of Alloys and Compounds*, 1999, **293–295**: 877–888.
2. Dantzer P. Properties of Intermetallic Compounds Suitable for Hydrogen Storage Application, *Material Science and Engineering*, 2002, **A329–331**, 313–320.
3. Mani N., Ramaprabhu S. Effect of Substitutional Elements on Hydrogen Absorption Properties in Mm-Based AB_5 Alloys, *Journal of Alloys and Compounds*, 2004, **363**: 275–291.
4. Mungole M.N., Balasubramaniam R. Effect of Hydrogen Cycling on the Hydrogen Storage Properties of $MmNi_{4.2}Al_{0.8}$, *International Journal of Hydrogen Energy*, 2000, **25**: 55–60.
5. Schlapbach L., Seiler A., Siegmann H.C., Waldkirch T.V., Zurcher P., Brundle C.R. Self Restoring of the Active Surface in $LaNi_5$, *International Journal of Hydrogen Energy*, 1979, **4**: 21–28.
6. Lototsky M.V. (1998) The effect of improvement hydrogen-sorption properties in multiphase polymetallic compositions, in: Hydrogen Power, Theoretical and Engineering Solutions. Proceedings of HYPOTHESIS-II Symposium, Grimstad, Norway, August 18–22, 1997.- Ed. by T.O. Saetre. Kluwer, Dordrecht/Boston, MA/London, pp. 343–352.
7. Da Silva E.P. Industrial Prototype of a Hydrogen Compressor Based on Metallic Hydride Technology, *International Journal of Hydrogen Energy*, 1993, **18**(4): 307–311.
8. Uchida H. Surface Processes of H_2 on Rare Earth Based Hydrogen Storage Alloys with Various Surface Modifications, *International Journal of Hydrogen Energy*, 1999, **24**: 861–869.
9. Chuang H.J., Chan S.L.I. Effect of Ni Encapsulation on the Properties of Ti–Zr-Based Hydrogen Storage Alloys, *Journal of Alloys and Compounds*, 2001, **314**: 224–231.
10. Willey D.B., Harris I.R., Pratt A.S. The Improvement of the Hydrogenation Properties of Nickel-Metal Hydride Battery Alloy by Surface Modification with Platinum Group Metals (PGMs), *Journal of Alloys and Compounds*, 1999, **293–295**: 613–620.
11. Bratanich T.I., Bulanov V.N., Skorokhod V.V., Klimenko V.P. Reversible Hydriding of $LaNi_{5-x}Al_x$ – Pd Composites in the Presence of Carbon Monoxide, *Powder Metallurgy and Metal Ceramics*, 2000, **39**(11–12): 575–583.
12. Geng M. Electrochemical Characterization of $MmNi_5$-Based Alloy Powder Coated with Palladium and Nickel-Palladium, *Journal of Alloys and Compounds*, 1994, **215**: 151–153.
13. Doyle M.L., Harris I.R, Pratt A.S, Willey D.B. (2000) Hydrogen Storage Materials, US Patent 6,165,643.
14. Slys I., Scherbakova L., Rogozinskaya A., Schur D., Rogozinskii A. Development of Methods of Deposition of Discontinuous Nickel Coatings on Powders of AB_5 type Alloys, Proceedings of the NATO Advanced Research Workshop on Hydrogen Materials Science and Chemistry of Carbon Nanomaterials, 2003, 131–136.
15. Lewis F.A. (1967) The Palladium/Hydrogen System, London, Academic, pp. 1–12, 48–49, 94–117.
16. Schulz R., Huot J., Liang G., Boily S., Lalande G., Denis M.C., Dodelet J.P. Recent Developments in the Applications of Nanocrystalline Materials to Hydrogen Technologies, *Materials Science and Engineering*, 1999, **A267**: 240–245
17. Piper J. Diffusion of Hydrogen in Copper-Palladium Alloys, *Journal of Applied Physics*, 1966, **37**(2): 715–721.
18. Volpe M., Inguanta R., Piazza S., Sunseri C. Optimized Bath for Electroless Deposition of Palladium on Amorphous Alumina Membranes, *Surface and Coatings Technology*, 2006, **200**: 5800–5806.
19. Mallory G.O., Hajdu J.B. (1990) Electroless Plating – Fundamentals and Applications, New York, William Andrew Publishing/Noyes, Vol. **16**, pp. 425–426.
20. Von Dreele R.B., Larson A.C. (2001) General Structure Analysis System, Los Alamos, NM, Regents of the University of California.

21. Changrong X., Xiaoxia G., Fanqing L., Dingkun P., Guangyao M. Preparation of Assymmetric Ni/Ceramic Composite Membranes by Electroless Plating, *Colloids and Surfaces A: Physicochemical and Engineering Aspects*, 2001, **179**: 229–235.
22. Yeung K.L., Christiansen S.C., Varma A.J. Palladium Composite Membranes by Electroless Plating Technique, Relationship Between Plating Kinetics, Film Microstructure and Membrane Performance, *Journal of Membrane Science*, 1999, **159**: 107–122.
23. Cheng D.H., Xu W.Y., Zhang Z.Y., Yiao Z.H. Electroless Copper Plating Using Hypophosphite as Reducing Agent, *Metal Finishing*, January, 1997: 34–37.
24. Forde T., Maehlen J.P., Yartys V.A., Lototsky M.V., Uchida H., (2006) Influence of Intrinsic Hydrogenation/Dehydrogenation Kinetics on the Dynamic Behaviour of Metal Hydrides: A Semi-Empirical Model and Its Verification, *International Journal of Hydrogen Energy*, doi: 10.1016/j.ijhydene.2006.07.015 (http://dx.doi.org//10.1016/j.ijhydene.2006.07.015).
25. Yuan X., Liu H.-S., Ma Z.-F., Xu N. Characteristics of $LaNi_5$-Based Hydrogen Storage Alloys Modified by Partial Substituting La for Ce, *Journal of Alloys and Compounds*, 2003, **359**(1–2): 300–306.
26. Baichtok Yu.K., Mordkovich V.Z., Dudakova N.V., Avetisov A.K., Kasimtsev A.V., Mordovin V.P. Technological Possibilities and State of the Art in the Developments of Hydride Thermal Sorption Hydrogen Compressors, *International Scientific Journal for Alternative Energy and Ecology*, 2004, **2**(10): 50–54.
27. Lototsky M.V., Yartys V.A., Marinin V.S., Lototsky N.M. Modelling of Phase Equilibria in Metal-Hydrogen Systems, *Journal of Alloys and Compounds*, 2003, **356–357**: 27–31.
28. Rao C.R.K, Trivedi D.C. Review: Chemical and Electrochemical Depositions of Platinum Group Metals and Their Applications, *Coordination Chemistry Reviews*, 2005, **249**: 613–631.

DEFORMATION OF ALUMINIUM MEMBRANE INDUCED BY HYDROGEN TRANSPORT

T.A. RYUMSHYNA
Donetsk Physical and Technical Institute of NAS of Ukraine
St. Luxemburg 72, 83114 Donetsk, 83114, Ukraine
E. LUNARSKA, O. CHERNYAEYWA
Institute of Physical Chemistry of PAS
ul. Kasprzaka 44/52, 01-224 Warsaw, Poland

Abstract. The formation of extremum on the permeation current curve under the application of the cathodic polarization during the electrochemical investigation of hydrogen transport through Al membranes was established. The analitical description of the effects on the base of hydrodinamical equation was proposed. Under the cathodic polarisation the system transfer in the instable state.The move to new external conditions realise by two interconnect processes with different velocities: hydrogen diffusion through Al plate and the deformation accomodation of material because of volume dilatation. Extremum forms at specific combination of parameters. The new model takes into account the influence of parameters and corresponds to experimental data.

Keywords: diffusion, hydrogen, stresses

1. Introduction

The investigation of hydrogen diffusion in material is one of the most important problems as material science, so electrochemistry. The hydrogen transport is the reason not only corrosion, but also deformation processes, causing to material degradation. Although there are the big number of articles, devoted to this problems [1–4] (though, certainly, not all material and varied conditions of their operation are explored), such problems, as the diffusions of hydrogen through the stressed material under external forces, the influence of hydrogen on the material characteristics are studied else not enough.

The interference effects of diffusion process and mechanical deforming have the special interest. On the one hand, external force fields influence on diffusion processes [3, 4], on the other hand, hydrogen atoms make the volume change in material and create the own stress fields, which can influence upon the further diffusion. Such a self influence leads to appearance of new effects. The study of hydrogen transport through Pd membrane [5] shows that the change of hydrogen pressure lead to the appearance of extremum on the pressure – time curves. This effect has got the name "uphill" diffusion or Lewis effect, and the nature of effect was given in the article [6]. The similar effects were discovered by B. Baranowsky for the PdPt alloys [7] and the description of effect was offered ibidem. However in Al similar effects did not observed. This work is a development of these studies in Al on base which is offered the new model of effect.

2. Experimental procedures

Authors [8–10] manage to realize the condition for penetrations of hydrogen into material under electrochemical conditions. The hydrogen permeability into Al foils (the purity 99.99% and thickness from 25 μm before 1 mm) was investigated by modified electrochemical method described in [11]. The membrane with active surface 0.5 cm^2 was placed in special camera, consisting of two electrochemical cells which contacted to each of sides a membrane.

After filling the cell by electrolyte the initial potential was enclosed this provided the gradient of concentrations between surfaces. As result the hydrogen penetration current was appeared written in output cell. After stabilization of the current density $\boldsymbol{J_0}$, the cathode polarization ***Jp*** was included in input cell that provided the arrival of new concentration of hydrogen in membrane. Created on output side of an Electrochemical condition on the plate surfaces provided the full eliminating of the anode dissolution and corrosions, so the measured current is a measure of flow of hydrogen getting through membrane. The typical current-time curve, measured in output cell is shown on Fig. 1. It was stated that the application of cathode polarization to input surface (on) tend to decrease of anode current before minimum magnitude $\boldsymbol{J_m}$, but then the current increase to saturation at $\boldsymbol{J_\infty}$. The position of minimum had firm renewed nature. The authors [8] were first one how had observed the appearance of extreme effect in aluminum and its alloys.

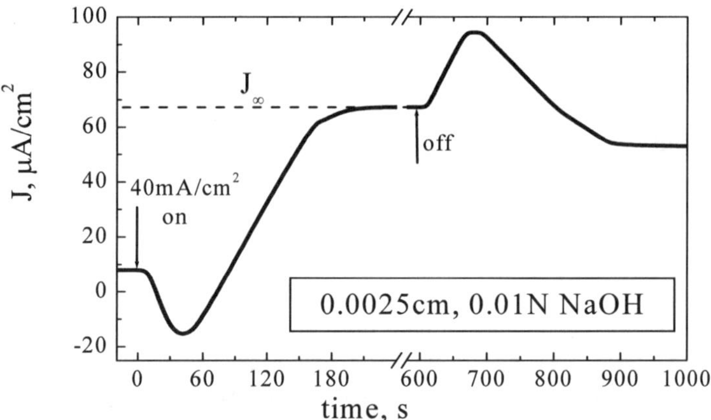

Figure 1. The appearance of minimum $\boldsymbol{J_m}$ on the permeation current curve at application of polarization on ingress side (on)

The influence of some parameters (the membrane thicknesses, magnitude and type polarizations) on position of extremes and parameter of permeation current – time curve are presented in the Tables 1 and 2.

TABLE 1. Influence of the membrane thickness on parameter of permeation current (J_p = 40 mA/cm^2)

h (mm)	0.025 mm	0.17 mm	0.5 mm	1.0 mm
T_{min} (sec)	40	58	80	120
ΔJ (mA/cm^2)	-23	-25	-30	-32
$(dJ/dt)_1$ (μA/cm^{2}sec)	-0.57	-0.43	-0.37	-0.26
$(dJ/dt)_2$ (μA/cm^{2}sec)	0.66	0.16	0.14	0.125

TABLE 2. Influence of the polarization on parameter of permeation current (h = 1 mm)

J_p (mA/cm^2)	20	40	80	100	120
T_{min} (sec)	140	80	30	20	0
ΔJ (mA/cm^2)	-25	-38	-26	-18	0
$(dJ/dt)_1$ (μA/cm^{2}sec)	-0.18	-0.48	-0.83	-0.9	–
$(dJ/dt)_2$ (μA/cm^{2}sec)	0.125	0.28	2.3	2.8	1.75

3. Discussion: Nature of effect

In initial state the considered system (the hydrogen flux passes through membrane) is open and stable. The application of initial voltage to membrane provides the constant initial gradient of concentration *grad n* and constant hydrogen flux *J* passes through plate

$$J = D\nabla n \quad , \tag{1}$$

where *D* is diffusion coefficient.

Because the flux is $J = n\left\langle v\right\rangle_d$, in initial state the certain constant concentration of hydrogen n_o passes through of section unit plate during unit time with an average "drift" velocity $\left\langle v\right\rangle_d$.

At application of cathode polarization, the big concentration of hydrogen enters to membrane, and system flux-membrane transforms to instable state and begins to reform to new external environment. Change to concentrations is described by Fick's equation:

$$\frac{\partial n}{\partial t} = -D\frac{\partial^2 n}{\partial t^2} \quad . \tag{2}$$

The time of transfer to stable state (the relaxation time τ_{dif}) according to principle of nonequilibrium thermodynamics [12] is $\tau_{dif} \approx h^2/D$, where *h* is linear size of considered region. For Al the diffusion coefficient is $D = 2–10·10^{-9}$ m^2/sec [10] and relaxation time along the length $h = 1$ mm equal to $\tau_{def} \approx 100–150$ sec.

However, when the hydrogen atoms include in plate, they bring to material the substantial volume change. So, if the volume of elementary cell of Al equals to $a^3 = (4.049\ A)^3 = 66.3809\ A^3$, the including hydrogen atom increase a cell volume on 2.9 A^3 [13]. The volume expansion coefficient equal to $\Delta V/V_o = 0.044$ is equal, and the corresponding linear expansion coefficient (or the coefficient of concentration deformation β) is equal to $\beta = 0.0146\ Al/H$.

When the matter have an ability to free expansion (dilatation) that is realized at soft fixing of the membrane edges which admiss the relative displacement of plate, the nonhomogeneous deformation state forms into material leafing to the appearance internal gradient stresses during the whole instable transition period.

Deformed correcting of material plate at the hydrogen penetration occurs very rapidly. The correction process is the propagation of deformation energy, which may be describe with help of the wave equation:

$$\frac{\partial^2 u}{\partial t^2} = \frac{1}{v_d^{\ 2}} \frac{\partial^2 u}{\partial x^2}, \tag{3}$$

where u is the elastic displacement; v_d is velocity of deformation wave equals to $v_d = \sqrt{E/\rho}$ (for Al v_d = 4 km/sec). Young's module for Al is $E = 70$ GPa, density of material is $\rho = 2.7$ g/cm^3. Along the thickness of plate 1 mm the new stable deformation state is fixed during the relaxation time $\tau_{def} = 10^{-6}$ sec. This time is much lesser then the relaxation time for the diffusion process.

The essential difference in velocities of two processes allows to consider the new state of system as quasistable (this possible when time $t < t_{min}$). This state is characterized by the presence of the deformation gradient in the inelastic medium, and, consequently, and stresses gradient. Because the observed minimum was revealed in time interval $t < \tau_{dif}$ this implies that the origin of the observed effect was determined by the influence of the gradient deformation field of surrounding medium (drift diffusion).

The change of hydrogen flux into membrane may be connected with change each of two parameters, defining its importance:

$$\frac{dJ}{dt} = \frac{d}{dt}(nv) = \frac{dn}{dt}v + \frac{dv}{dt}n. \tag{4}$$

The change of the flow is determined by the second term in the equation (4) at initial moments of time $t < t_{min}$, when concentration did not increase yet $n \sim const$. That is the drift velocity $\langle v \rangle_{dr}$ decreases because of the influence of the strained material. At the further evolution at $t > t_{min}$, the influence of increasing concentration given by first term in (4) begins to prevail. At $t > \tau_{dif}$ the new equilibrium state is established that corresponds to new boundary conditions (to a new concentration gradient) and the deformation gradient disappears.

Thus, there are some stages on the current-time curve (Fig. 1). Initial state, first stage started at polarization $t < t_{min}$, second stage $t > t_{min} < \tau_{dif}$, third stage under $t > \tau_{dif}$, corresponds to the stable state for new boundary conditions.

The initial state. The initial state of system is equilibrium one and is characterized by constant number of the current density ~200 μA/cm^2 which is established at the initial voltage creating an initial gradient of concentration ∇n_1 between plate surfaces. Thus a constant flux of hydrogen passes through a membrane. The magnitude of flux is determined by initial concentration of the particles n_0 moving with the some constant average drift velocity $\langle v_o \rangle_d$.

The estimations made on the basis of experimental data, resulted in [10] show, that at atomic concentration of hydrogen $c_o = 10^{-5}$ H/Al (corresponding volume concentration $n_o = 6 \cdot 10^{24}$ m^{-3}), the drift velocity of particles in the given conditions is $v_0 = 2 \cdot 10^{-5}$ m/sec.

1 Stage. The force – ***grad σ*** operates on the moved particles, because of presence of the stress gradient in front of the new diffusion flux. This force is caused by an elastic reaction of medium, and brings to braking of motion. The scheme of process on this stage is shown on the Fig. 2.

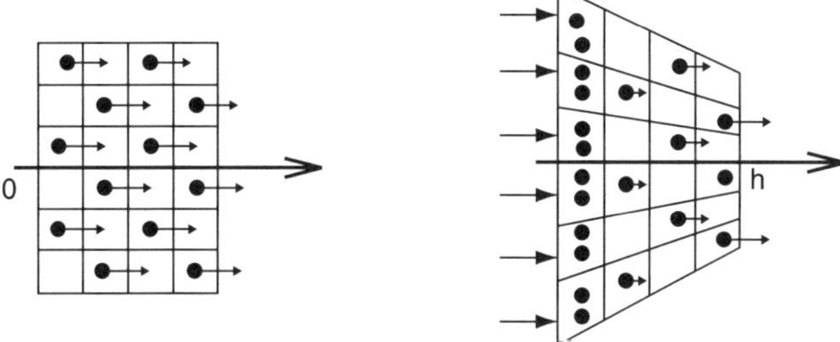

Figure 2. Scheme of deformation of plate and change of hydrogen velocity at polarization

The hydrodynamics equation (Euler's equation) applied to the flux motion gives:

$$\frac{\partial v_x}{\partial t} + v_x \frac{\partial v_x}{\partial x} = \frac{1}{\rho} \nabla \sigma , \tag{5}$$

where x is a coordinate in the direction of motion, and v_x is a coordinate of velocity toward.

The solution of equation allows to define the function $v(x, t)$, i.e. allows to find the velocity of the marked particle in any coordinate and at any time moment. Because the differential equation is nonlinear we simplify one and suppose the stationary flux in this time interval (i.e. the velocity of particles depend only on coordinate, but not time). In that case a deformation of a plate is described as two-dimensional straining in the membrane's plane decreasing along thickness on the linear law. Bernoulli's Equation for the stationary flux is:

$$\frac{v_o^2}{2} = \frac{v_s^2}{2} + \frac{\Delta\sigma}{\rho}, \tag{6}$$

where v_o is the initial velocity, v_s is the smallest velocity before a hydrogen front of in most strained part, $\Delta\sigma = E\,\varepsilon$ is the stress difference on edges of plate, $\varepsilon = \beta\,(c_p - c_o)$ is the most deformation of plate when the atomic concentration, $c_p(H/Al)$ is hydrogen concentration entered in material at polarization, c_o is the initial atomic concentration before polarizations, β is concentration factor of linear expansion of lattice. From (6) the minimum velocity in most strained area of plate is:

$$v_s = \sqrt{v^2{}_o - \frac{2E\varepsilon}{\rho}}\,. \tag{7}$$

The decrease of velocity is very significant. The slower particles reach to an output side later that will be shown as decrease in a flux. The minimal value of a current will be attained by that moment when the slowest particle will reach an exit. Note, that elastic resistance force of medium determined by a gradient of stresses Δσ/h acts opposite to the moved particle. Its value is very great and equals (at 10 %-s' share from possible number of penetrating particles at polarization) and equal to $4.2{\cdot}10^8$ N/m^3.

Because the gradient of elastic deformation of medium *grad σ* increase when the thickness of a plate decrease the force of resistance grows and the motion of particles is decelerated. Results of experiment for the time derivative of current density (which is proportional to the change of velocity) show the influence of membrane thickness and value of polarization on the process (Tables 1, 2).

2 Stage. The second stage of increase of current density after a minimum is determined by the contribution of the first term $\partial n/\partial t$ in equation (4), i.e. by the increase of hydrogen concentration under influence of cathodic polarization. The function $n(x, t)$ describing a increase of concentration in time was found by the solution of the diffusion equation (2) for an infinite plate thickness h when the hydrogen flux passes in the direction of x athwart the planes of membrane. The boundary and initial conditions corresponded experimental conditions of constant polarization on an entrance surface :
Boundary condition: $n = n_p$, at $t = 0$ and $n = n_0$, at $x = h$;
Initial condition; $n = n_0$, at $t = 0$ and $0 < x < h$
The solution is:

$$\frac{n(x,t)-n_0}{n_p-n_0} = \left\{1-\frac{4}{\pi}\sum_{n=0}^{\infty}\frac{(-1)^n}{2n+1}\exp\left[-(2n+1)^2\frac{\pi^2}{4}\frac{Dt}{h^2}\right]\cos\left(\frac{2n+1}{2}\right)\frac{\pi x}{h}\right\}. \tag{8}$$

Calculated concentration vs. time dependences on the output membrane surface of a different thickness are resulted on the Fig. 3.

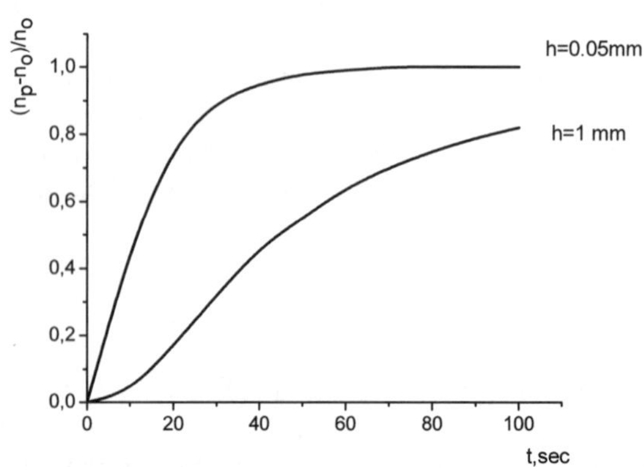

Figure 3. Kinetic dependences of relative concentration on the output side for different thicknesses of membranes (1 mm, 0.5 mm)

A significant acceleration of saturation of a membrane by new hydrogen concentration is observed when the thickness decrease or the polarization increase. The process of concentration accumulation starts to prevail of process of decrease the drift velocity much earlier when the membrane thickness decreases or polarization increase. Filling by new concentration occurs so quickly, that the decrease velocity has not time to be shown. Decrease of time of minimum t_{min} and time derivative of current density $(dJ/dt)_2$ shows the change of in the ratio of two processes (Tables 1, 2).

3 Stage. The third stage of current-time curve characterizes the established process for new conditions with the increased concentration gradient when the deformation gradient has already completed the action. The application of cathodic polarization which should lead to repeated growth of entering concentration, apparently, should result and in corresponding repeated increase of a saturation current. However, those it is not observed because of the following reasons. According to experiment [11], the value of effect depends on addition corrosion inhibitors. It testifies that is far from being all of 100% of concentration n_p have ability to come into membrane because of surface conditions. The growth of saturation current J_∞ doesn't proportional to the amount of polarization providing the concentration increase that indicates on the change of drift velocity. Dependence of a saturation current versus cathodic polarization for membranes of different thickness is resulted on Fig. 4. As identical polarization in identical conditions of experiment should provide equal amount of entering concentration, the increase J_∞ at decrease of thickness is connected with the increase of drift velocity that corresponds to general laws of hydrodynamics.

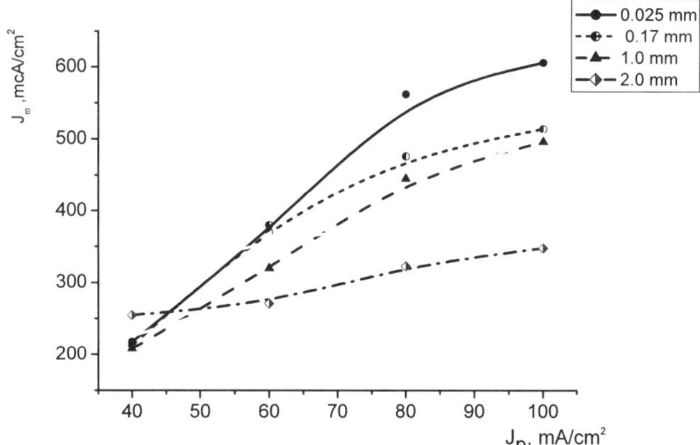

Figure 4. Influence of polarization on the satiation current J_∞ for 3rd stages for the membranes with a different thickness

4. Conclusions

1. The observed effects are determined by the instable state of the system at the application of cathode polarization. During transition of system with reference to new conditions there are two interconnected processes: process of diffusion of hydrogen

through a membrane and process of non-uniform deformation of the plate, occurring because of the volume changes brought by hydrogen. These processes occur to essentially various speeds. In result the extreme on a curve of a current of penetration is formed.
2. Change of parameters of external influences (amount of polarization) and parameters of system (thickness of a membrane) adjusts speeds of processes. At the certain combination of parameters conditions for formation of extreme are created.
3. The essential interconnection between diffusion processes of hydrogen in a material and mechanical deformation of the material should be taken into account at operation of the equipment in conditions of excited environments and mechanical loading.

Acknowledgement

Work is supported by Polish scientific Fund Kasa im. Mjanowskego.

References

1. Advances in Corrosion Science and Application. Proceeding of International Conference "MSCHS 2003", Zakopane, Poland, September 9–13, 2003.
2. Environmental Degradation of Engineering Materials. EDEM-2003, Bordeaux, France, 29 June, 2003.
3. Defect and Diffusion Forum. Proceeding of First International Workshop "Diffusion and Stresses". Hungary, 1995, V.129–130.
4. Defect and Diffusion Forum. Proceeding of First International Conference "Diffusion, Segregation and Stress in Material". Moscow, 2002, V.216–217.
5. Lewis F.A., Magennis J.P., et al. *Nature* 1983, **15**: 673–675.
6. Kandasamy K. *Int. J. Hydrogen Energy* 1995, **20**(6): 455–463.
7. Baranowski B. Diffusion in elastic media with stress field. Adv. Termodyn., Taylor & Franscis, New York, 1992.
8. Lunarska E., Chernyayeva O. Strain induced up-hill diffusion of hydrogen in Al. *Inzynieria Materiałowa* 2004, **25**(3): 144–150.
9. Lunarska E., Chernyaeva O., Ryumshyna T. *Izvestija RAN, «Physical» series*, 2003, **10**: 1417–1420 (in Russian).
10. Lunarska E., Chernyayeva O. *Int. J. Hydrogen Energy* 2006, **31**: 237–246.
11. Devanathan M., Stashurski Z. *J. Electrochem. Soc.* 1964, **111**: 619.
12. Landau L.D., Lifshits E.M. Theoretical Physics, V.**6**., Science, Moscow, 1988.
13. Alifeld G., Felkl I. Hydrogen in metals. "Mir", Moscow, 1981 (in Russian).

INFLUENCE OF METAL SUBSTITUTE ON ELECTROCHEMICAL AND SORPTION CHARACTERISTICS OF $LaNi_5$ BASED ALLOYS

L.G. SHCHERBAKOVA, YU.M. SOLONIN,
YE.N. SEVERYANINA
Frantsevich Institute for Problems of Material Sciences of NASU, 3 Krzhyzhanovsky Str., Kiev, 03142 Ukraine, Fax: 38 (044) 424 0381, E-mail: larisa_c@ukr.net

Abstract. A systematic investigation of the substituting components' influence on kinetic, sorption characteristics and the equilibrium hydrides formation potential in the $LaNi_5$ system alloys in alkaline electrolyte has been conducted. It was ascertained that increasing a substituting component radius and thus growth of an elementary cell volume results in increasing enthalpy of hydrides formation and anode shift of hydrides formation equilibrium potential. This shift increases with substituting component amount increasing. It was shown that the dependence between value of the hydrides formation enthalpy and hydrides formation (decomposition) equilibrium potential is linear. Inserting Co, Mn, Al into $LaNi_5$ leads to the growth of surface catalytic activity relative to hydrogen exchange reaction. In consequence of this the absorption of hydrogen by the alloy become easier that allows to receive on no activated electrodes high discharge capacity (280–320 mA.h/g) even at the initial charge/discharge cycles.

Keywords: photoelectrochemical cell, hydrides, kinetics, equilibrium potential

1. Introduction

Properties of hydrides forming $LaNi_5$ based materials when partial substituting Ni by Co, Al, Mn, Si is changing considerable. In particular, the forming hydrides stability becomes higher and their cyclic stability – better [1–3]. In the gas phase for all triple alloys $LaNi_{5-x}Me_x$ type (where M = Co, Al, Si, Mn, Cu) the hydride formation energy (ΔH) and hydrogen equilibrium pressure in the hydride ($P_{eq.}$) depend on the elementary cell volume [4]. Information on the hydride formation equilibrium potential ($E_{MH/H}$) and kinetic characteristics dependence on phase composition of $LaNi_5$ based doped alloys in alkaline electrolyte is limited and data are discrepant in many cases. Because of the importance of both these characteristics in the range of potentials near equilibrium potential for metal cathodes working in the pare with photoanodes obtaining such data is actual.

The object of this work was conducting systematic investigation of influence of substituting component (M = Co, Al, Mn) in $LaNi_{5-x}M_x$ alloys on running charge process, its kinetic, equilibrium potential, and discharge capacity of cathode material. The goal of such investigation – to explain the reasons for the choice of alloy optimal composition and a range of charge potentials when working in the pare with a photoanode in a photoelectrochemical cell.

B. Baranowski et al. (eds.), *Carbon Nanomaterials in Clean Energy Hydrogen Systems.*

2. Experimental details

$LaNi_5$, $LaNi_{4,5}Mn_{0,5}$; $LaNi_{4,5}Al_{0,5}$ и $LaNi_{2,5}Co_{2,5}$ alloys have been produced by the electro arc melting in argon protective atmosphere in a cupper water cooled crucible followed by remelting with a titanium getter no less than three times. After remelting the specimens were homogenized in a vacuum furnace during 60 hours. The gas chromatograph method revealed oxygen content in the specimens no more than 0.2%. Besides, $LaNi_{4,5}Al_{0,5}$ alloy was produced by sintering mixture of $LaNi_5$- $LaNi_{4,5}Al_{0,5}$ powders in the argon atmosphere (MG alloy).

The phase composition and structure of the specimens were investigated by the X-Ray phase analysis. Diffraction spectrum of the alloys' powders were obtained with use computerized X-Ray diffractometer ДРОН-3М in the monochromatic Cu K_α radiation at the $2\theta = 20^0–90^0$ angles range.

Electrodes were made by pressing intermetallic alloys compound and electrolytic cupper powders mixture in the form of tablets with diameter 9 mm. They were then investigated in 30% KOH solution with using potentiometer – galvanostat PGSTAT4-16, three-electrodes cell, Pt counter – electrode, and mercurial- oxide reference electrode. For surface activation and reduction of partially oxidized cupper powder the electrodes were electrochemically cycled (2–3 cycles) in the potential range -1.2–0.6 V ($\partial E/\partial \tau$ =2 mV/sec) in working electrolyte. The equilibrium potential value was determined by interrupting the charge current at different levels of saturation by hydrogen on the charge plateau. Discharge capacity ($C_{disch.}$) was calculated with using the discharge curves while reaching the -0.6 V potential value by the electrode. Quasi - stationary volt-ampere curves were obtained by stepped potential changing by 10 mV in each 2 minutes.

3. Results and discussion

The X-Ray phase analysis of the all $LaNi_{5-X}Me_X$ type alloys investigated specimens' diffractogram has revealed only the hexagonal $LaNi_5$ (the P6/mmm special group, the $CaCu_5$ structural type) diffraction reflexes that testifies to single phase structure of the specimens studied. Calculations of the radiographs have shown that the increasing of Co content in the alloy to 2.5 atoms results in the increasing of an elementary cell volume by 1.4% comparatively to the initial $LaNi_5$. Partial substituting nickel by 0.5 atoms of Al or Mn leads to even more visible increasing these parameters. The consequence of these results is that the changing of $LaNi_5$ elementary cell is increasing in dependence on substituting component in consecutive order Co < Mn < Al (Table 1).

Analysis of our previous [5] and taken from [6] hydrides decomposition enthalpy values ($\Delta H_{dec.}$) has shown that partial substitution of nickel leads to the increasing of $\Delta H_{dec.}$ value which depends on the substituting component kind (Table 1).

TABLE 1. Thermodynamic, electrochemical, and sorption characteristics of $LaNi_{5-X}Me_X$ alloys

№	Alloy	V_{cell} (A^3)	ΔV (%)	H_{dis} (kJ)/ Mol/H_2	$E_{M//HM}$ (V)	C_{disch} [a] (mA.h/g)	i_0 [b] (mA/cm^2)	β^c (V)
1	$LaNi_5$	86.92		31.6	-0.95	200	6.2	0.08 0.178
2	$LaNi_{2.5}Co_{2.5}$	88.13	1.4	37.6	-0.94	250	13.2	0.12
3	$LaNi_{4.5}Mn_{0.5}$	88.29	1.6	36.4	-0.92	315	22.0	0.12
4	$LaNi_{4.5}Al_{0.5}$	88.68	2.0	39.7	-0.91	320	22.5	0.13
5	$LaNi_{4.0}Al_{1.0}$	90.00	3.5	52.3	-0.85	260	–	–

[a]discharge capacity at the 3rd cycle; [b]i_0 exchange current; [c]β, Tafel curve slop

Working out the data in the -ΔH – $V_{яч}$ coordinates has testified that the hydride formation/decomposition enthalpy (Fig. 1) growth is near to linear while the increasing of the lattice volume under partial substituting nickel by different metals and the substituting component content in the $LaNi_5$ alloy.

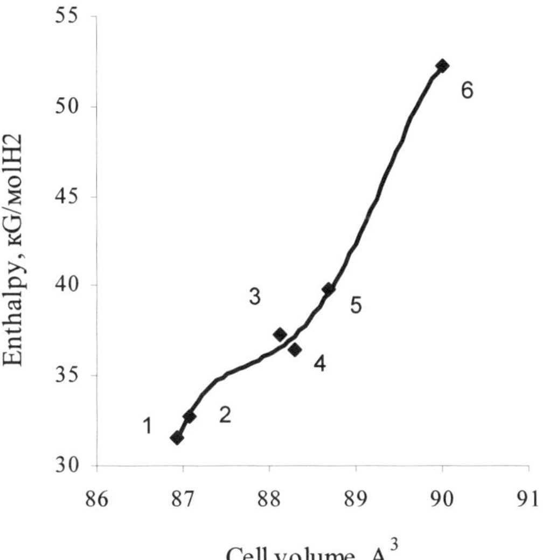

Figure 1. Change of the hydrides dissociation enthalpy of the alloys with the $LaNi_5$ lattice volume change when partial substituting nickel by other metals 1 – LaNi; 2 – $LaNi_{4.5}Co_{0.5}$; 3 – $LaNi_{2.5}Co_{2.5}$; 4 – $LaNi_{4.5}Mn_{0.5}$; 5 – $LaNi_{4.5}Al_{0.5}$; 6 – $LaNi_{4.0}Al_{1.0}$

Analysis of the obtained $LaNi_{5-X}Me_X$ alloys' charge curves at the first 3 cycles has shown that their behavior is different while charging process and depend on the substituting component (M) type. The maximum differences were observed at the first cycle of charge: on the cast $LaNi_5$ electrode the charge is running with low over tension $\Delta E = E_{charge} - E_{Mg/M\ equal}$ 0.05 V while the high values ΔE (0.1–0.15 V) are typical for the substituted alloys at the beginning and their essential lowering during the charging process. To the most extent this is typical for the alloys containing Mn and Al that can be connected with the electrode surface activation because of the dissolution of these components in the alkaline electrolyte during charging. At the subsequent cycles the following facilitation of the charging process for these alloys and decreasing the ΔE value was observed.

On the Fig. 2 the quasi stationary charge curves received on the electrodes activated by stepped cathode current (3 minutes) with the following interruption of polarization (3 minutes) are presented. The potentials on the charge curves correspond to the potential values that nave become stable after cutting off polarization i.e. to the equilibrium potentials of hydrides formation ($E_{M/MH}$). It seen that the anode shift of $E_{M/MH}$ value is more while inserting 0.5 atoms of Al, Mn than 2.5 Co atoms into $LaNi_5$. Thus, these metals lower the equilibrium pressure of the hydride phase dissociation more effectively and raise its stability.

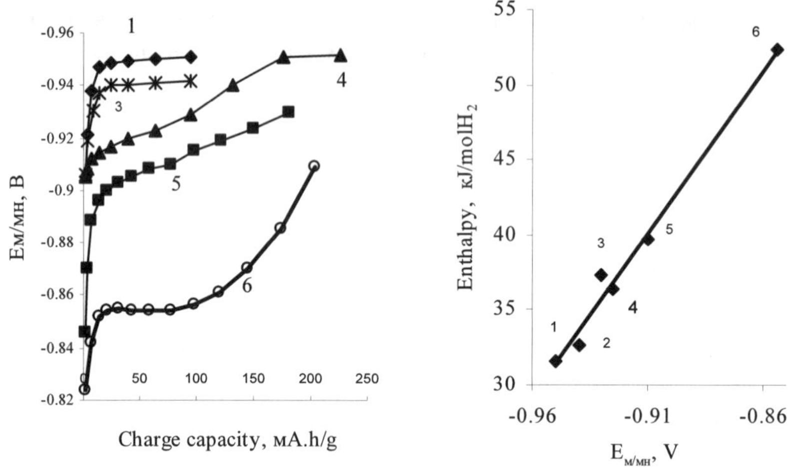

Figure 2. Quasi stationery curves of the $LaNi_{5-X}Me_X$ alloys type electrode charging it the 30% KOH solution 1 – $LaNi_5$, 2 – $LaNi_{4.5}Co_{0.5}$; 3 – $LaNi_{2.5}Co_{2.5}$; 4 – $LaNi_{4.5}Mn_{0.5}$; 5 – $LaNi_{4.5}Al_{0.5}$; 6 – $LaNi_{4.0}Al_{1.0}$

Figure 3. Dependence of the equilibrium hydride formation potential ($E_{M/MH}$) on enthalpy of the alloys hydrogenation/dehydrogenation (ΔH) (designation of the alloys are the same as on the Fig. 1)

The essential shifting of the hydrogenation ($E_{M/MH}$) and dehydrogenation ($E_{MH/M}$) equilibrium potentials in the anode direction (Table 1, Fig. 2) was observed when content of Al in the alloy was growing. The change of the $E_{MH/M}$ and $E_{MH/\ M}$ equilibrium potentials with increasing content of aluminum in the alloy were calculated from $LaNi_{4.5}Al_{0.5}$ and $LaNi_{4,0}Al_{1,0}$ alloys' hydrides formation and dissociation enthalpy values in accordance with the equation

$$Eeq. = - \Delta H/nF ,$$

where H – enthalpy, kG/mol; n – quantity of electrons participating in the reaction, F – Faraday constant, $9.65.10^4$ K.mol^{-1} .

It was determined that the calculated and experimental values are practically the same and equal 60 mV (Table 1). Such shift of the potential is equivalent to the decreasing of hydrogen equilibrium pressure in hydride by the order of magnitude (up to 0.001 MPa). Besides, the plateau potential values on the equilibrium charge and discharge curves of the $LaNi_{4.5}Al_{0.5}$ and $LaNi_{4,0}Al_{1,0}$ alloys have the equal values i.e. formation ($E_{M/MH}$) and dissociation ($E_{MH/M}$) equilibrium potentials for each alloy are the same.

Working out the data received in the $E_{M/MH} - \Delta H$ coordinates has shown that the dependence between these characteristics is linear i.e. the higher stability of the formed hydride the more anode the hydrogenation /dehydrogenation equilibrium potential (Fig. 3).

It is seen from the Table 1 that the discharge capacity ($C_{disch..}$) of the $LaNi_{5-X}Me_X$ alloys is dependant on the substituting component type : doping $LaNi_5$ by 0.5 atoms of Al and Mn make it easier to absorb hydrogen by the alloy that allows receiving the high discharge capacities 280–310 mA.h/g even at the first cycle. In the same time the $LaNi_5$ discharge capacity doesn't reach such values even after 3rd cycle.

Investigation of the kinetic characteristics of the cathode process at the charge potentials region has revealed that partial substituting nickel in $LaNi_5$ by Co, Al and Mn speeds up the cathode process running – the hydrogen exchange currents (i_{-0}) are increasing (Table 1). Working out the cathode curves in the lgi-E coordinates has shown (Fig. 4) that in contrast to $LaNi_5$ alloy in the case of $LaNi_{5-X}Me_X$ alloys the only linear section with tilt (β), close to 0.12 V is present at the wide potentials range. This is evidence of limitation of the cathode reaction in the $-1.0 \leq E \geq -0.85$ V potential range due to the single electron mechanism of charge carrying on this stage . Most probably that the water splitting stage will determine kinetics of hydrogen reduction in the alkaline electrolyte:

$$H_2O + e^- \rightarrow H_{abs.} + OH^- \text{ (Folmer reaction).}$$

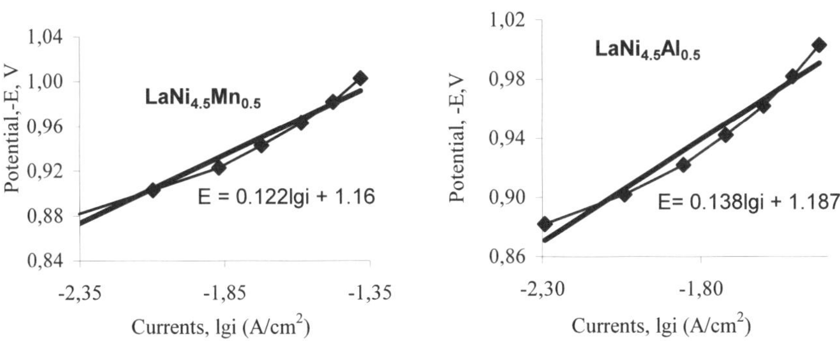

Figure 4. The cathode potentiodynamic curves received for different alloys in alkaline solution

At the range of potentials close to the hydride formation equilibrium potential (-0.89–0.96 V) the cathode currents of the $LaNi_{2,5}Co_{2,5}$ alloy electrode 2–2.5 times more and of the $LaNi_{4,5}Mn_{0,5}$ and $LaNi_{4,5}Al_{0,5}$ alloys electrode more than of the $LaNi_5$ electrode. Thus, using these alloys will allow carrying out the charging process in the electrochemical cell more effectively.

4. Conclusions

Since non-transition metals' radii: Al (r_{Al} = 0.143 nm) and in less degree Mn (r_{Mn} = 0.13 nm) are more than radius of Ni (r_{Ni} = 0.124 nm), substituting even less than 5% of Ni atoms in $LaNi_5$ alloy results in increasing crystal lattice cell of the alloy. In consequence of this, hydrogen atoms with effective radius 0.25–0.3 Å penetrate to alloys crystal lattice easily and form more thermodynamically stable hydrides. This leads to the increasing of hydrides decomposition enthalpy and the decreasing of the equilibrium hydrides formation potential in alkaline electrolyte. Influence of Co atoms that have the same as Ni atoms radius (r_{Co} = 0.124 nm) on these characteristics is considerably less. Change of hydrogenation process kinetics can be apparently connected with contribution of Al, Mn and Co atoms that have higher than Ni ionization potentials into energetic condition of the surface.

References

1. Sakai T., Oguro R., Miyambra H., *J. Less-Common Met.*, 1990; **161** (2): 193–202.
2. Willems J.J., Buschow K.H., *J. Less-Common Met.*, 1987; **129** (1): 13–30.
3. Meli F., Schlapbach L., *J. Less-Common Met.*, 1991; **172–174** (1–2): 1252–1259.
4. Sacai T., Hasama T., Miyamura H., *J. Less-Common Met.*, 1991; **171–174**: 1175–1184.
5. Shcherbakova L.G., Dan'ko D.B., V.B. Muratov V.B., Solonin Yu.M., Kolbasov G.Ya., Rusetskii I.A. Metal hydride use for solar energy accumulation. In: Hydrogen Materials Science and Chemistry of Metal Hydrides, NATO Science Series, Netherlands, Springer, 2007, p. 699–706.
6. Kolachev B.A., Shalin A.A., Ilyin R.E. Alloys. *Hydrogen Accumulators* (Moscow: Metallurgy, 1991), p. 352 (in Russian).

ELECTRONIC STRUCTURE OF THE FACE-CENTRED CUBIC $MoO_{1.9}$ PHASE OBTAINED DUE TO REDUCTION OF HYDROGEN BRONZE $H_{1.63}MoO_3$

O. YU. KHYZHUN*, V.L. BEKENEV, YU.M. SOLONIN
Frantsevych Institute for Problems of Materials Science, National Academy of Sciences of Ukraine, 3 Krzhyzhanivsky Str., Kyiv 03142, Ukraine
**Fax: +380 44 424 3364; E-mail: khyzhun@ipms.kiev.ua*

Abstract. The electronic structure of face-centred cubic (fcc) MoO_x (x = 1.9) oxide derived due to reduction of hydrogen bronze $H_{1.63}MoO_3$ has been studied using the X-ray emission spectroscopy (XES) and X-ray photoelectron spectroscopy (XPS) methods. For comparison, the electronic structure of molybdenum trioxide, MoO_3, and the $H_{1.63}MoO_3$ bronze was studied as well. The XES O $K\alpha$ and Mo $L\beta_{2,15}$ bands of fcc-$MoO_{1.9}$ and XPS valence-band spectra of the $MoO_{1.9}$, MoO_3 and $H_{1.63}MoO_3$ compounds were derived. Band structure calculations of fcc-MoO_2 have been fulfilled using the full-potential linearized augmented plane wave (FP-LAPW) method. The theoretical XES O $K\alpha$ and Mo $L\beta_{2,15}$ bands were calculated for fcc-MoO_2 employing the above method. A rather good agreement of shapes of experimental and theoretical XES O $K\alpha$ and Mo $L\beta_{2,15}$ bands for fcc molybdenum dioxide has been obtained.

Keywords: X-ray emission spectroscopy (XES), X-ray photoelectron spectroscopy (XPS), electronic structure, band-structure calculations, molybdenum oxide

1. Introduction

Molybdenum trioxide, MoO_3, is known to be a crystal with layered structure belonging to orthorhombic symmetry with lattice parameters a = 0.3963 nm, b = 1.3855 nm, c = 0.3696 nm [1]. MoO_3 is widely used as an important heterogeneous catalyst component [2–4], in the fuel industry in processes of desulfurization [5], in the solar cell technologies as a photochromic and electrochromic material [6]. Molybdenum dioxide, MoO_2, is crystallized in a distorted rutile-like monoclinic structure with lattice parameters a = 0.56109 nm, b = 0.48562 nm, c = 0.56285 nm, β = 120.95° [7]. MoO_3 can be reduced rather easily and the so-called Magnéli phases with crystal structures belonging to orthorhombic, monoclinic or triclinic symmetry are synthesized in the intermediate region between MoO_3 and MoO_2 compounds [8].

A new nonstoichiometric face-centred cubic (fcc) MoO_x oxide has been derived when heating molybdenum hydrogen bronze, H_xMoO_3, in hydrogen at temperatures 300–400°C [9]. The fcc-MoO_x oxide (with lattice parameter a = 0.408 nm) is crystallized in a nanocrystalline state (with particle sizes smaller than 100 nm) and possesses excellent gas sensor properties [9]. X-ray photoelectron spectroscopy (XPS) core-level studies

[10] have revealed that a surface of the fcc-MoO_x phase with the composition close to molybdenum dioxide (x = 1.9) consisted of molybdenum atoms in the +6, +5 and +4 valence states while those of the cubic hydrogen bronze $H_{1.63}MoO_3$ in the +6 and +5 valence states. In addition, XPS O 1s core-level binding energy measurements [10] has revealed that the charge states of oxygen atoms do not change in the sequence orthorhombic MoO_3 → cubic $H_{1.63}MoO_3$ → cubic $MoO_{1.9}$.

The purpose of the present paper was to employ possibilities of XPS and X-ray emission spectroscopy (XES) as well as of the full-potential linearized augmented plane wave (FP-LAPW) method [11] for investigation of the electronic structure of the fcc-MoO_2 phase.

2. Experimental

Thick cubic $MoO_{1.9}$ and $H_{1.63}MoO_3$ and orthortombic MoO_3 films were supported on an Al_2O_3 substrate. The film consisted of thin plate-like MoO_3 crystals was obtained after special activation used for densification and reorientation of the crystals in order to obtain a texture with some predominant crystal orientation [9]. The activation of MoO_3 crystals supported on an Al_2O_3 substrate was performed by their impregnation with palladium salt and following heating in air at 300°C. It has been found that, palladium did not take part in the reactions and it acted in fact as catalyst [9]. The content of palladium in the molybdenum trioxide thick film was minor (less than 0.1 wt%) and any presence of Pd in the film was detected using the XPS method [10]. The fcc-$MoO_{1.9}$ oxide was synthesized due to heating the cubic $H_{1.63}MoO_3$ bronze in hydrogen at 350°C [9].

XPS valence-band spectra were collected using an electron spectrometer equipped with a VG Clam2 hemispherical electron energy analyzer as described in Ref. [10]. The XPS spectra were measured at the constant pass energy of 20 eV. The X-ray emission O $K\alpha$ band, reflecting the energy distribution of the O 2p-like states, in the fcc-$MoO_{1.9}$ oxide under consideration was recorded using the technique employed previously in Ref. [12] for studying the band in hydroxyapatite crystals. In the present work, the XES O $K\alpha$ band was recorded with the spectrometer energy resolution better that 0.35 eV. The fluorescent X-ray emission Mo $L\beta_{2,15}$ band, reflecting the energy distribution of filled valence states of s,d-symmetry of molybdenum, in the molybdenum dioxide under consideration was derived using the technique described in details in Ref. [13]. The spectrometer energy resolution in the range corresponding to the energy of the Mo $L\beta_{2,15}$ band was found to be better than 0.4 eV.

3. Method of calculation

Band-structure calculations of cubic MoO_2 oxide were fulfilled employing the FP-LAPW method [11]. During the present calculations, positions of molybdenum and oxygen atoms have been chosen similar to those in the fcc structure of fluorite type. The generalized gradient approximation by Perdew et al. [14] was employed for calculation of the exchange-correlation potential. The *muffin-tin* sphere radii of 1.70 and 1.35 a.u. (1 a.u. = 0.0529177 nm) for Mo and O atoms, respectively, were used in the present calculations. The $Rk_{\max}$ parameter determining a number of linearized augmented plane waves has been chosen to be 7.0. The tetrahedron method by Blöchl et al. [15] was used for integration through the Brillouin zone, which sampling was done using 72 *k*-points within the irreducible part of the zone. The fitting of the calculated values of total energy using the procedure [16] have revealed the equilibrium lattice parameter a = 9.306418 a.u. for fcc-MoO_2. Using this lattice parameter, the total and partial densities of states (DOS) of the cubic molybdenum dioxide were calculated. For calculations of the XES Mo $L\beta_{2,15}$ and O $K\alpha$ bands, we have taken into account both spectrometer broadening and the broadening γ due to finite lifetime of the core-level and valence states. The broadening parameters were the following: S = 1 eV for spectrometer broadening; γ = 1.5 and 0.13 eV for the XES Mo $L\beta_{2,15}$ and O $K\alpha$ bands, respectively, as suggested in Ref. [17]. The W = 1.5 value for the parameter in Blokhin-Sachenko's formula [18] was used.

4. Results and discussion

Figure 1 presents experimental XPS valence-band spectra of the cubic $MoO_{1.9}$ oxide and molybdenum hydrogen bronze $H_{1.63}MoO_3$ under study. The spectra were excited by photons with energy of 140 eV. A typical evolution of the XPS valence-band spectra excited with photon energies ranging from 90 to 400 eV is shown in Fig. 2 for $H_{1.63}MoO_3$ bronze. From the above figure it is obvious that an increase of the relative intensity of the O 2*s*-like sub-band (with respect to the intensity of the maximum *A* of the XPS valence-band spectra) occurs with increasing the photon energies used for spectra excitation.

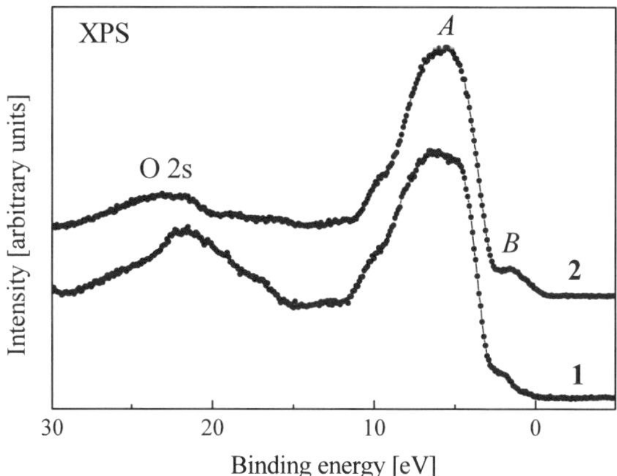

Figure 1. XPS valence-band spectra of (1) molybdenum hydrogen bronze $H_{1.63}MoO_3$ and (2) fcc-$MoO_{1.9}$ oxide (the spectra were excited using photon energy of 140 eV)

As one can see from Fig. 1, the relative intensity of the O 2*s*-like sub-band centred at about 22.0–22.5 eV below the Fermi level, E_F, is significantly smaller in cubic $MoO_{1.9}$ as compared with that in molybdenum hydrogen bronze $H_{1.63}MoO_3$. A decrease of the relative intensity of the XPS O 2*s*-like sub-band when going from $H_{1.63}MoO_3$ to $MoO_{1.9}$ can be explained by decreasing oxygen content in the above sequence of compounds. It is also necessary to mention that the XPS valence-band spectra of the $H_{1.63}MoO_3$ to $MoO_{1.9}$ compounds under consideration (Figs. 1 and 2) reveal the formation of an additional near-Fermi sub-band *B*, which is absent on the spectrum of orthorhombic MoO_3 (Fig. 3).

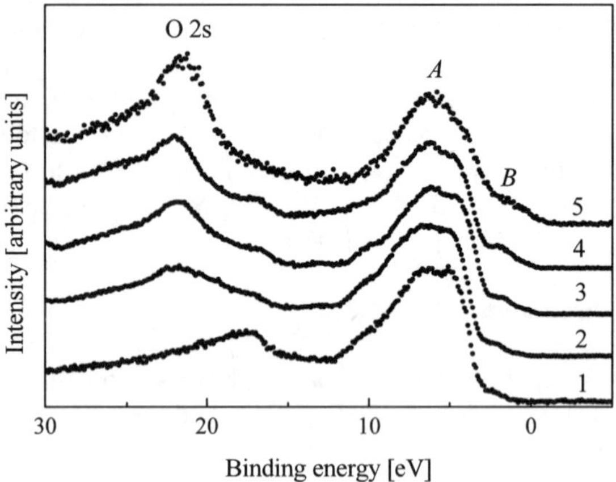

Figure 2. XPS valence-band spectra (excited by photons with energies of (1) 90 eV, (2) 140 eV, (3) 200 eV, (4) 280 eV and (5) 400 eV) of $H_{1.63}MoO_3$ bronze

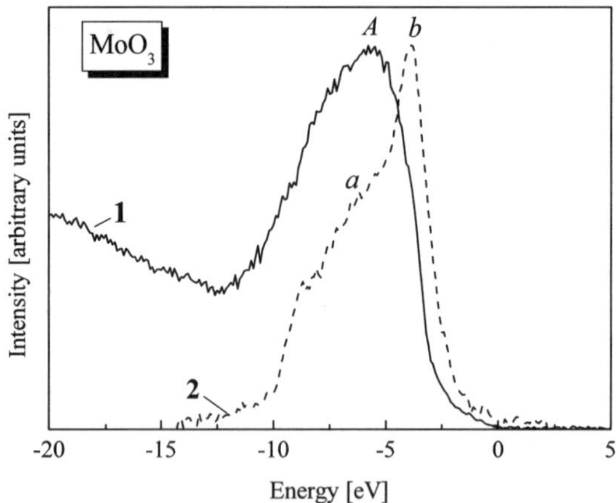

Figure 3. Comparison on a common energy scale of the XPS valence-band spectrum (solid curve 1) and X-ray emission O *K*α band (dashed curve 2) of orthorhombic MoO_3

The appearance of the near-Fermi sub-band on the XPS valence-band spectrum of cubic $MoO_{1.9}$ can be explained by contributions of the Mo 4*d*-like states (probably with some admixtures of the Mo 5*s*-like states) taking part in the formation of the new Mo–Mo bonds in this compound. It is worth to mention that, the formation of an additional fine-structure peculiarity was not observed on the high-energy slope of the XES O *K*α band of the cubic $MoO_{1.9}$ oxide investigated [10]. Therefore, the partial O 2*p*-like states do not take part in the formation of the near-Fermi sub-band *B* of the XPS valence-band spectrum of fcc-$MoO_{1.9}$. The similar effect is characteristic for substoichiometric monoclinic and hexagonal tungsten oxides $WO_{2.0\leq x<3.0}$ [19–21].

The origin of the formation of the near-Fermi sub-band *B* of the XPS valence-band spectra of $H_{1.63}MoO_3$ (Fig. 2) seems to be similar to that detected previously in XPS measurements of monoclinic and hexagonal tungsten hydrogen bronzes H_xWO_3 [22, 23]. In these works the formation of the near-Fermi sub-band on the XPS valence-band spectra of H_xWO_3 bronzes was explained by filling the W $d(t_{2g})$-like band, which is empty in WO_3, due to the extra hydrogen electrons. It is reasonably to assume that this effect is also true for the cubic $H_{1.63}MoO_3$ bronze investigated. The formation of the near-Fermi sub-band was also observed by Fleisch and Mains [24] on the XPS valence spectra of orthorhombic molybdenum trioxide partly reduced by ultraviolet irradiation in the presence of molecular hydrogen.

The present FP-LAPW calculations indicate that total DOS of fcc-MoO_2 is formed due to contributions of mainly Mo 4*d*- and O 2*p*-like states. As one can see from Fig. 4, four groups of bands can be identified on the curve of total DOS of fcc-MoO_2. The lowest group of bands (about 20 eV below E_F) originates mainly from the O 2*s*-like states, the second group (in the energy range from 9.1 to 2.5 eV below E_F) is formed mainly by the O 2*p*- and Mo 4*d*-like states (almost exclusively partial Mo $4d(t_{2g})$ states), the third group (positioned near the Fermi energy) is created exclusively by the Mo $4d(e_g)$ states and the fourth group (extended from 1.6 to 11.3 eV above E_F) is formed due to contributions mainly of the Mo $4d(t_{2g})$ and O 2*p* states.

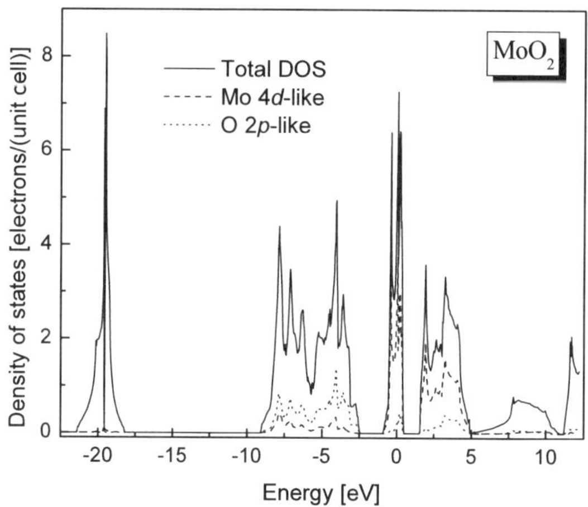

Figure 4. Total density of states (solid curve) and partial Mo 4*d*-like (dashed curve) and O 2*p*-like (dotted curve) densities of states of fcc-MoO_2

Due to the FP-LAPW results presented in Fig. 4, total density of states at the Fermi energy, $N(E_F)$, of 5.24 electrons/(unit cell) is characteristic of fcc-MoO_2. This value is about nine times bigger in comparison with the value of $N(E_F)$ obtained by Eyert et al. [25] for a usual monoclinic form of MoO_2. Our FP-LAPW results reveal that the cubic structure of MoO_2 is rather unstable and, probably, the fcc molybdenum dioxide can be preserved only in the form of a thick film as it was made in Ref. [9].

Theoretical X-ray emission O $K\alpha$ and Mo $L\beta_{2,15}$ bands, which were derived by the procedure mentioned in the Sect. 3 of the paper, are presented in Figs. 5 and 6, respectively. As one can see from the above figures, the shapes of the theoretical O $K\alpha$ and Mo $L\beta_{2,15}$ bands follow reasonably those of the corresponding experimental bands derived for the fcc $MoO_{1.9}$ thick film.

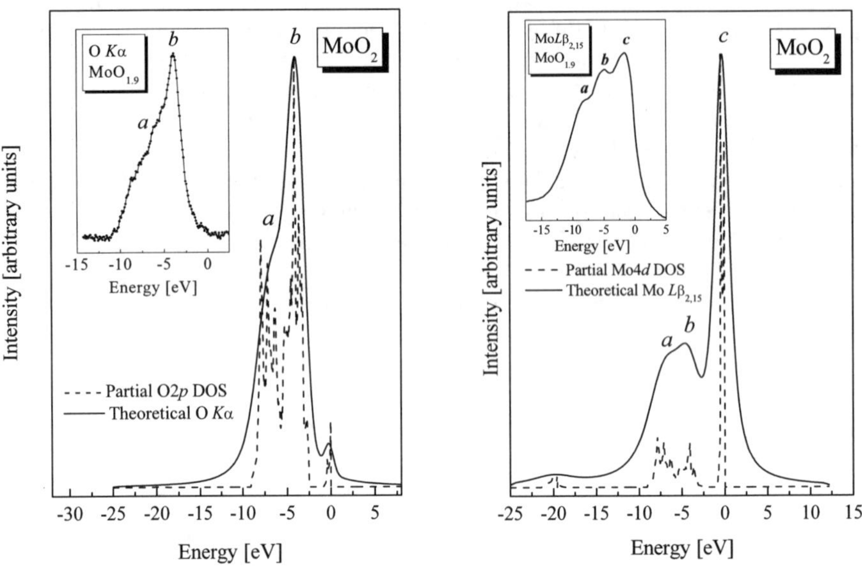

Figure 5. Theoretical O $K\alpha$ band (solid curve) and partial O 2*p*-like DOS (dashed curve) of fcc-MoO_2 (the experimental XES O $K\alpha$ band of fcc-$MoO_{1.9}$ is presented in the insert)

Figure 6. Theoretical Mo $L\beta_{2,15}$ (solid curve) and partial Mo 4*d*-like DOS (dashed curve) of fcc-MoO_2 (the experimental XES Mo $L\beta_{2,15}$ band of fcc-$MoO_{1.9}$ is presented in the insert)

5. Conclusions

Due to results of the present FP-LAPW calculations, main contributors to the valence band of the fcc-MoO_2 oxide are the O 2*p*- and Mo 4*d*-like states. The FP-LAPW method was found to be a very useful technique for theoretical calculations of the X-ray emission bands of fcc-MoO_2: a rather good agreement of shapes of experimental and theoretical O $K\alpha$ and Mo $L\beta_{2,15}$ bands for this oxide has been achieved. A significant part of density of states is detected at the Fermi level of fcc-MoO_2. The FP-LAPW results reveal that the cubic structure of MoO_2 is rather unstable. The formation of an additional near-Fermi sub-band, which is absent in MoO_3, was observed on the XPS valence-band spectrum of fcc molybdenum dioxide. It has been established that the sub-band is caused exclusively by the contributions of the Mo 4*d*(e_g) states.

References

1. Kihlborg L. Least squares refinement of the crystal structure of molybdenum trioxide // Arkiv Kemi, 1963, **21**: 357–364.
2. Machiels C.J., Sleight A.W. Kinetic isotope effect in the selective oxidation reaction of methanol to formaldehyde over some molybdate catalysts // J. Catal., 1982, **76**: 238–239.
3. Delmon B., Yates J.T. (editors) Transition-metal Oxides: Surface Chemistry and Catalysis, Studies in Surface Science and Catalysis. Amsterdam, Elsevier, 1989, **45**.
4. Yoshida S., Sakaki S., Kobayashi H. Electronic Processes in Catalysis: A Quantum Chemical Approach to Catalysis. New York, VCH, 1994.
5. Monk P.M.S., Martimer R.J., Rossiensky D.R. Electrochromism: Fundamentals and Applications. Weinheim, VCH Verlagsgesellschaft, 1995.
6. Granqvist C.G. Handbook of Inorganic Electrochromic Materials. Amsterdam, Elsevier, 1995.
7. Andersson G., Magnéli A. On the crystal structure of molybdenum trioxide, // Acta Chem. Scand., 1950, **4**: 793–797. a
8. Schlenker C. (editor) Low-Dimensional Electronic Properties of Molybdenum Bronzes and Oxides. Dordrecht, Kluwer, 1989.
9. Solonin Yu.M. The structural aspects of hydrogen interaction with activated textured molybdenum trioxide layer // Int. J. Altern. Energ. Ecol., 2002, (1): 73–76.
10. Khyzhun O.Yu., Strunskus T., Solonin Yu.M. XES, XPS and NEXAFS studies of the electronic structure of cubic $MoO_{1.9}$ and $H_{1.63}MoO_3$ thick films // J. Alloy. Compd., 2004, **366**: 54–60.
11. Blaha P., Schwarz K., Luitz J. WIEN97, A Full Potential Linearized Augmented Plane Wave Package for Calculating Crystal Properties. Vienna, Technical University, 1999.
12. Dubok V.A., Zhurakivsky E.A., Ivanchenko L.A., Podrushnyak E.P., Ulyanchich N.V., Khyzhun O.Yu. On some crystalline-structure defects in microcrystals of hydroxyapatite // Metallofiz. Noveishie Tekhnol., 1997, **13** (3): 32–38.
13. Khyzhun O.Yu., Zaulychny Ya.V., Zhurakovsky E.A. Electronic structure of tungsten and molybdenum germanides synthesized at high pressures // J. Alloy. Compd., 1996, **244**: 107–112.
14. Perdew J.P., Burke S., Ernzerhof M. Generalized gradient approximation made simple // Phys. Rev. Lett., 1996, **77**: 3865–3868.
15. Blöchl P.E., Jepsen O., Andersen O.K. Improved tetrahedron method for Brillouin-zone integrations // Phys. Rev. B, 1994, **49**: 16223–16233.
16. Murnaghan F.D. The compressibility of media under extreme pressures // Proc. Natl. Acad. Sci. USA, 1944, **30**: 244–247.
17. Khyzhun O.Yu., Bekenev V.L., Solonin Yu.M. Electronic structure of cubic MoO_x nanoparticles: experimental studies and band-structure calculations // Nanosistemy, Nanomaterialy, Nanotechnologii, 2007, **5**: 599–608. (in Russian).
18. Blokhin M.A., Shveitser I.G. X-Ray Spectroscopy Handbook. Moscow, Nauka, 1982 (in Russian).
19. Khyzhun O.Yu. XPS, XES and XAS studies of the electronic structure of tungsten oxides // J. Alloy. Compd., 2000, **305**: 1–6.
20. Khyzhun O.Yu., Solonin Yu.M. Electronic structure of the monoclinic and hexagonal trioxides of tungsten and hexagonal hydrogen tungsten bronze $H_{0.24}WO_3$ // Powder Metall. Met. Ceram., 2000, **39**: 287–294.
21. Solonin Yu.M., Khyzhun O.Yu., Graivoronskaya E.A. Nonstoichiometric tungsten oxide based on hexagonal WO_3 // Cryst. Growth Des., 2001, **1** (6): 473–477.
22. Colton R.J., Guzman A.M., Rabalais J.W. Electrochromism in some thin-film transition-metal oxides characterized by x-ray electron spectroscopy // J. Appl. Phys., 1978, **49**: 409–416.

23. Khyzhun O.Yu., Solonin Yu.M., Dobrovolsky V.D. Electronic structure of hexagonal tungsten trioxide: XPS, XES, and XAS studies // J. Alloy. Compd., 2001, **320**: 1–6.
24. Fleisch T.H., Mains G.J. An XPS study of the UV reduction and photochromism of MoO_3 and WO_3 // J. Chem. Phys., 1982, **76**: 780–786.
25. Eyert V., Horny R., Höck K.-H., Horn S. Embedded Peierls instability and the electronic structure of MoO_2 // J. Phys.-Condens. Mat., 2000, **12**: 4923–4946.

INTERACTION OF PSEUDO ALLOYS Mg–REM–Ni AND COMPOSITES ON THEIR BASE WITH HYDROGEN

D.N. BORISOV, P.V. FURSIKOV, B.P. TARASOV
Institute of Problems of Chemical Physics of RAS, Prospectus of the Academician Semenov, 1, Chernogolovka, Moscow Region, 142432 Russia
Tel./Fax: +7(496) 5221743; E-mail: btarasov@icp.ac.ru

Abstract. The hydrogen-sorbing properties of ternary eutectic pseudo alloys 72 wt% Mg – 8 wt% Mm(La) – 20 wt% Ni, obtained by sintering of hydrides MgH_2, $Mm(La)H_3$ and Mg_2NiH_4, taken in the proportion appropriate to quantity of components of mentioned hydrogenated alloys, and composites on their basis with intermetallic compounds $Mm(La)Ni_5$ are investigated.

Keywords: ternary eutectic pseudo alloy, hydrogen, intermetallic compound, composite, hydrogen, hydrogen sorption

1. Introduction

Speaking about hydrogen storage capacity of magnesium based alloys and alloys of ternary eutectic as Mg-REM-Ni it should be noted that they satisfy the up-to-date requirements with more than 5 wt% [1]. Highly refined structure of these alloys and the catalytic effect of REM and Ni [2] promote high rate of hydrogenation-dehydrogenation which can also be increased by different modifications of alloy. As the hydrogenation rate of magnesium alloys is limited by diffusion of hydrogen atoms through magnesium hydride the modification of alloy might decrease the grain size of magnesium and makes the distribution of the catalytic components more even. In this work we propose to form magnesium pseudo alloys (poly-metallic materials) by a sintering of hydrides of highly refined metals and composites on their base with addition of hydride forming intermetallic compounds of the AB_5 type such as La with Ni by means of mechanical treatment of the corresponding mixtures.

Thus the aim of this work was studying of processes of interaction between hydrogen and pseudo alloys 72 wt% Mg – 8 wt% Mm(La) – 20 wt% Ni (later – Mg-Mm(La)-Ni) and investigation of the influence of $Mm(La)Ni_5$ addition on hydrogenation kinetics of the obtained composites.

2. Experimental

The mixture of hydrides MgH_2, $Mm(La)H_3$, and Mg_2NiH_4, taken in proportion corresponding to the one of hydrogenated ternary eutectic alloy was used to prepare a pseudo alloy. Pseudo alloys were synthesized by agglomeration of hydrides in a pile at temperature 350°C. The composition of alloys was supervised by the chemical analysis and the X-ray testing.

The X-ray analysis of initial alloys and products of their interaction with hydrogen was spent on an automatic complex consisting from diffractometerADP-1 (CuK_{α}-emission) and the managing computer. The thermal analysis was carried out on installation STA 409 Luxx companies Netzsch. Registration as DSC curves, and curve of weight losses (TG) was carried out at speed of heating of 10 degrees/min in a argon stream.

3. Results and discussions

The products of interaction between hydrogen and ternary eutectic alloy Mg-Mm(La)-Ni are the mixture of three hydride phase: MgH_2, Mm(La)H_3 and Mg_2NiH_4 (intermetallic compound Mm(La)$_2$Mg$_{17}$ forms MgH_2, Mm(La)H_3 at hydrogenation) [1].

While agglomeration the gasification take place and pseudo alloy looks like porous material. After the first hydrogenation of this pseudo alloy at temperature of 300°C and pressure of 16 atm the presence of initial hydrides LaH_3, Mg_2NiH_4, and MgH_2 was observed by the X-Ray diffraction technique (Fig. 1).

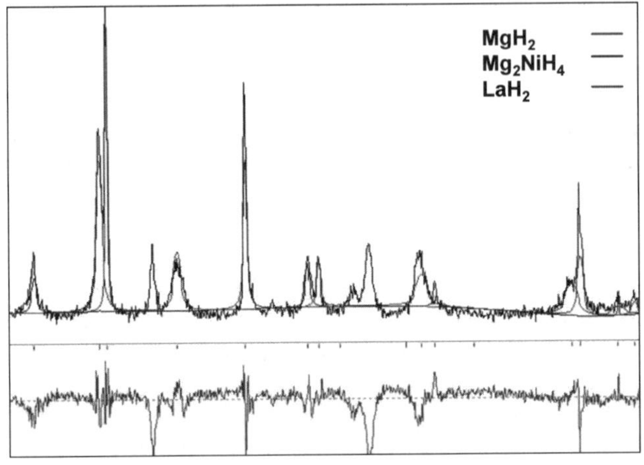

Figure 1. X-ray diffraction data of pseudo alloy

Thermal analysis of dehydrogenated pseudo alloys at temperature 250–300°C shows the presence of two endoeffects on the DSC curve (Fig. 2). The peak with the maximum at 460°C, shows evolving of residual hydrogen from MgH_2, and the peak with the maximum at 530°C shows the melting of the pseudo alloy. This temperature corresponds to the melting point of the ternary eutectic Mg-Mm(La)-Ni alloy with the same mass composition obtained by a common method [1].

From isotherms of the pseudo alloy–hydrogen system at the temperature of 300°C we found two plateaus corresponding to the phase changes in the Mg_2Ni–H_2 and Mg–H_2 systems (because of small quantities of components for La(Mm)Ni_5–H_2 and La(Mm)–H_2 systems plateau doesn't appear) (Fig. 3).

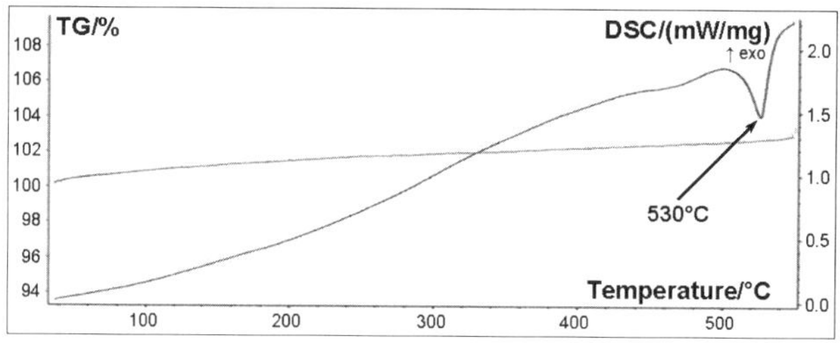

Figure 2. Thermogram of the pseudo alloy

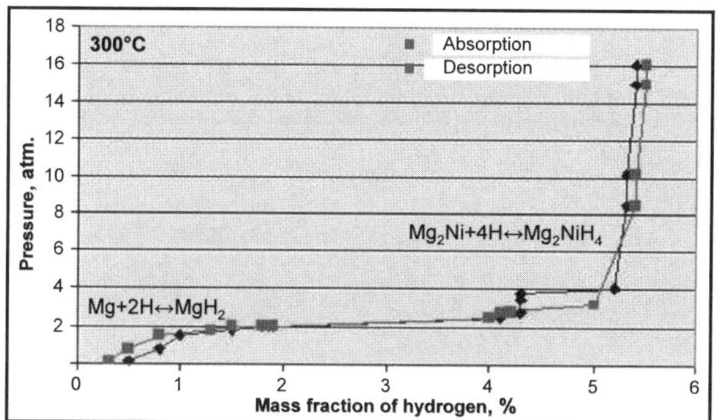

Figure 3. Sorption/desorption hydrogen isotherm for pseudo alloy–hydrogen system

The obtained pseudo alloys interact with hydrogen faster than the initial mixture of powders (Fig. 4), and the hydrogen capacity in the hydride of pseudo alloy being not less than that in the initial mixture of the hydrides. This fact shows that the method of forming poly-metallic hydrogen sorption composites on the Mg-REM-Ni base is perspective.

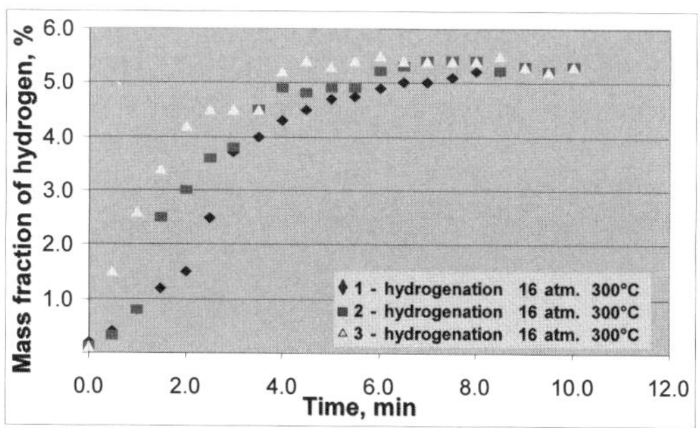

Figure 4. Hydrogen sorption curves of the pseudo alloys

Poly-metallic composite based on pseudo alloy with addition of 20 wt% low temperature intermetalic compound Mm(La)Ni_5, prepared by treatment in a planetary ball mill, shows higher hydrogenation rate in the first and following cycles as compared to the initial pseudo alloy (Fig. 5). In our previous works [1, 3] it was noted that an addition of intermetalic compound of the AB_5 type to poly-metallic composition with a ternary eutectic Mg-REM-Ni alloy had a positive effect on hydrogenation kinetics and decreased hydrogen evolution temperature.

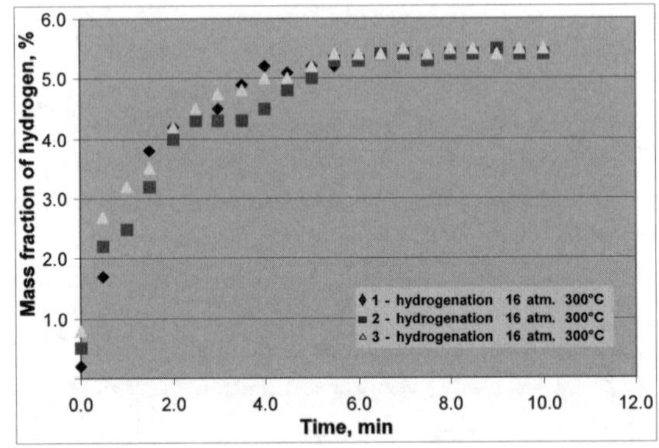

Figure 5. Hydrogen sorption curves of the pseudo alloy +20 wt% $Mm(La)Ni_5$

4. Conclusions

Forming of materials based on Mg-Mm(La)-Ni by preparation of a ternary eutectic pseudo alloy and an addition of the intermetalic compound $Mm(La)Ni_5$ into the poly-metallic composition based on Mg-REM-Ni result in increasing the hydrogen sorption rate and decreasing the dehydrogenation temperature.

References

1. Tarasov B.P., Fokin V.N., Borisov D.N., Gusachenko E.I., Klyamkin S.N., Yakovleva N.A., Shilkin S.P. Hydrogen storage by alloys of magnesium and rare-earth metals with nickel, *Alternativnaya Energetikay i Ekologiya*, 2004 **(1)**: 58–63 (in Russian).
2. Pezat M., Darriet B., Hagenmuller P. A comparative study of magnesium-rich rare-earth-based alloys for hydrogen storage, *J. of Less-Common Metals*, 1980 **74 (2)**: 427–434.
3. Klyamkin S.N., Lukashev R.V., Tarasov B.P., Borisov D.N., Fokin V.N., Yartys V.A. Hydrogen storage composites on based magnesium, *Materialovedenie*, 2005 **(9)**: 53–56 (in Russian).

NEUTRON DIFFRACTION INVESTIGATION OF HEXAGONAL AND CUBIC PHASES OF SYSTEM Ti-C-H

I. KHIDIROV, B.B. MIRZAEV, N.N. MUKHTAROVA, KH.M. KHOLMEDOV
Institute of Nuclear Physics, Uzbekistan Academy of Sciences, Tashkent, Uzbekistan
E-mail: khidirov@inp.uz
S.YU. ZAGINAICHENKO, D.V. SCHUR
Institute for Problems of Materials Science, National Academy of Sciences of Ukraine, Kiev, 03142, Ukraine
E-mail: shurzag@materials.kiev.ua

Abstract. It is established that at hardening from 1,000°C in samples TiC_xH_y at the concentration of carbon of $0.30 \leq x \leq 0.50$ and at a ratio $y \geq x$ the ordered hexagonal close-packed (HCP)-structure and at $y < x$ – face-centered cubic (FCC)-phase are stabilized. The crystal structure of the ordered HCP-phase corresponding to the formula $Ti_2C_{1-x'}H_{2-y'}$ ($1-x' = 2x$; $2-y' = 2y$) is described in the framework of sp. gr. $P\bar{3}m1$. Below 700°C the $Ti_2C_{1-x'}H_{2-y'}$ –phase with concentration of carbon $0.30 \leq x < 0.41$ decayed with precipitation of the hydride phase TiH_2. The ordered solid solution $Ti_2C_{1-x'}H_{2-y'}$ is stable at concentration $0.41 \leq C/Ti \leq 0.50$ (close to stoichiometry on carbon Ti_2C) but is metastable at concentrations $0.30 \leq C/Ti < 0.41$. Compounds TiC_xH_y at concentrations $0.45 \leq x \leq 0.50$ and $y < x$ have both disordered (sp. gr. Fm3m) and ordered FCC-structure (sp. gr.Fd3m) depending on temperature.

Cubic carbohydride $TiC_{0.47}H_{0.22}$ was prepared by means of quenching from 1,200°C followed by heat treatment using special regime for preventing the exit of hydrogen out of the lattice. It is shown that at the lower limit of homogeneity range of the cubic carbohydride hydrogen atoms occupy tetrahedral interstices 8(c) of the disordered cubic structure (sp. gr. Fm3m). It is found that carbon and hydrogen atoms are partially ordered when annealing at 900–800°C. In the ordered phase (sp. gr. Fd3m) carbon atoms occupy mainly (70%) octahedral interstices 16(c) and 30% of carbon and the most part of hydrogen atoms – another type of octahedral interstices 16(d).

Keywords: system Ti-C-H, hexagonal close-packed structure, face-centered cubic phase, ordered, disordered, neutron diffraction

1. Introduction

Carbon and hydrogen, each taken separately, have rather low limit solubility in the hexagonal lattice of α-Ti (0.5 mass% C, 0.2 mass% H) [1]. However simultaneous introduction of carbon and hydrogen into α-Ti lattice leads to sharp increasing their solubility (in tens times). So, study of solid solution of carbon and hydrogen in α-Ti is of interest from the standpoint of obtaining the solid solution with the greatest content of hydrogen. From this point of view cubic carbohydride Ti at the lower limit of carbon homogeneity region is of interest as here it has the greatest quantity of free interstices for hydrogen arrangement.

The aim of the present work is structural study of hexagonal and cubic Ti carbohydrides of a number of compositions by methods of neutron and X-ray diffraction over the temperature interval 1,200–450°C.

2. Experiment

Neutron diffraction patterns of powdery samples were obtained using the neutron diffractometer DN-500 of the INP AS RUz (λ = 0.1085 nm). Calculation and refinement of structure characteristics from neutron diffraction data was carried out by the Rietveld full-profile method [2]. X-ray diffraction patterns were obtained using the diffractometer DRON-3M (CuK_{α}- radiation).

The technique of samples preparation is similar to [3]. Composition of the samples was determined by chemical analysis; besides, composition of single-phase samples was controlled by means of minimizing the divergence factors of structure determination using neutron diffraction data.

3. Results

In the Table 1 the results of X-ray analysis of the samples after synthesis are present. The error in determining the lattice parameters does not exceed 0.15%. As the Table 1 shows, in the compounds $TiC_{0,05}H_{0,17}$ and $TiC_{0,2}$ $H_{0,16}$ there is the body-centered cubic (BCC) phase. In the Fig. 1 X-ray diagram of the sample $TiC_{0,2}H_{0,16}$ is submitted. The lattice parameter of the observed BCC-phase (3.345 Å) is greater by 1.2% than the lattice parameter of BCC pure β - Ti (3.282 Å) [1] or of β–phase of $TiH_{0.30}$ (3.305 Å) [6]. According to [7], the β–phase of titanium hydride is stable only at T ≥ 300°C, and at decreasing temperature it is not kept. So, the found BCC-phase is a solid solution of carbon and hydrogen in BCC-lattice of titanium. It is possible to explain the absence of BCC-phase at intermediate concentration (between $TiC_{0,05}H_{0,17}$ and $TiC_{0,2}$ $H_{0,16}$), apparently, by insufficient rate of hardening in air or in water.

As it may be seen from the Table 1, the single-phase solid solutions TiC_xH_y with hexagonal close-packed (HCP) structure can be received only at concentration satisfying to the condition $y \geq x \geq 0.30$, i.e. provided that the concentration of hydrogen is not less than concentration of carbon at values $x \geq 0.30$. In Fig. 2 the X-ray spectrum of the solid solution $TiC_{0,30}H_{0,34}$ is given. It is interesting to note that in X-ray diffraction pattern of this solid solution, as well as of other TiC_xH_y with HCP-structure (Table 1), there is superstructure reflection 001. Hence, the carbon atoms must be ordered in HCP – structure of a metal matrix. However the ordering of carbon atoms in the presence of titanium atoms should not "be seen" by X-ray analysis because of proportionality of intensity of X-ray diffraction maxima to the square of an element serial number [8]. The detection of the superstructural reflection in X-ray diagrams is explained by small rearrangement of the matrix lattice, i.e. by small displacements of metal atoms from positions, corresponding to HCP structure, because of the ordered arrangement of carbon atoms. Under condition x ≥ 0.45 > y the face-centered cubic (FCC) single-phase is observed.

TABLE 1. Phase composition and lattice parameters of the synthesized samples

Sample	Phase composition	Lattice parameters a (Å)	c (Å)	c/a
1	2		3	4
$TiC_{0,05}H_{0,17}$	BCC+	3,321		
	FCC+	4,266		
	HCP	2,954	4,689	1,587
$TiC_{0,10}H_{0,17}$	FCC+	4,301		
	HCP	2,952	4,693	1,590
$TiC_{0,10}H_{0,28}$	FCC+	4,266		
	HCP+	3,068	5,006	1,631
	α-Ti	2,951	4,684	1,587
$TiC_{0,16}H_{0,16}$	FCC+	4,269		
	HCP+	2,960	4,740	1,601
	α-Ti	2,951	4,684	1,587
$TiC_{0,20}H_{0,16}$	BCC+	3,345		
	FCC	4,2696		
$TiC_{0,20}H_{0,19}$	HCP+	2.985	4,815	
	FCC	3,780		
$TiC_{0,24}H_{0,25}$	HCP+	3,065	5,008	1,634
	FCC	4,275		
$TiC_{0,25}H_{0,18}$	FCC+	4,275		
	HCP+	3,065	5,008	1,634
	α-Ti	2,951	4,684	1,587
$TiC_{0,25}H_{0,33}$	FCC+	4,295		
	HCP	2,947	4,698	1,595
$TiC_{0,38}H_{0,30}$	HCP+	2,994	4,827	1,612
	HCP	3,018	4,892	1,621
$TiC_{0,40}H_{0,21}$	HCP+	2,998	4,838	1,614
	α-Ti	2,951	4,684	1,587
$TiC_{0,45}H_{0,22}$	FCC	4,271		
$TiC_{0,47}H_{0,22}$ [3]	FCC	4,275		
$TiC_{0,50}H_{0,21}$ [3]	FCC	4,301		
$TiC_{0,50}H_{0,28}$ [4]	FCC	4.310		
$TiC_{0,30}H_{0,34}$	HCP	3,065	4,970	1,621
$TiC_{0,30}H_{0,43}$	HCP	3.022	4.864	1.609
$TiC_{0,35}H_{0,43}$	HCP	3,024	4,894	1,619
$TiC_{0,38}H_{0,64}$	HCP	3.090	5.053	1.635
$TiC_{0,41}H_{0,41}$	HCP	3.026	4.905	1.621
$TiC_{0,43}H_{0,43}$	HCP	3.073	5.025	1,635
$TiC_{0,43}H_{0,55}$	HCP	3.078	5.026	1.635
$TiC0,_{45}H_{0,90}$	HCP	3.098	5.079	1,639
$TiC_{0,48}H_{0,70}$	HCP	3.093	5.064	1.637
$TiC_{0,45}H_{0,96}$ [5]	HCP	3.096	5.069	1.637

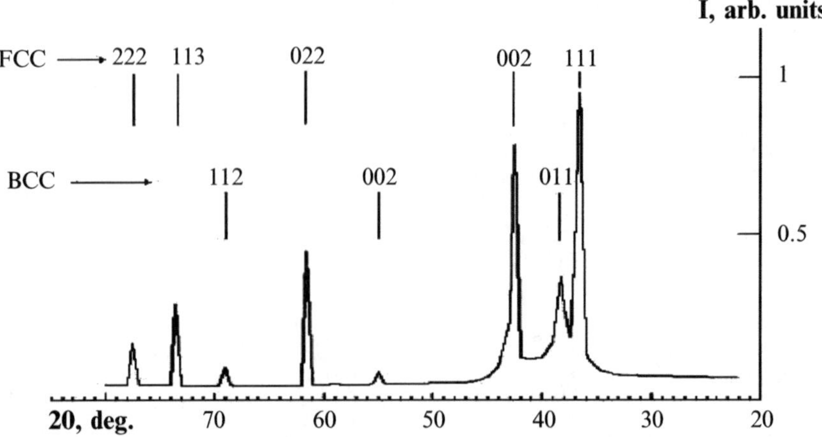

Figure 1. X-ray diffraction pattern of $TiC_{0,2}H_{0,16}$. Miller indices of reflecting plane of crystal lattice are indicated above reflections

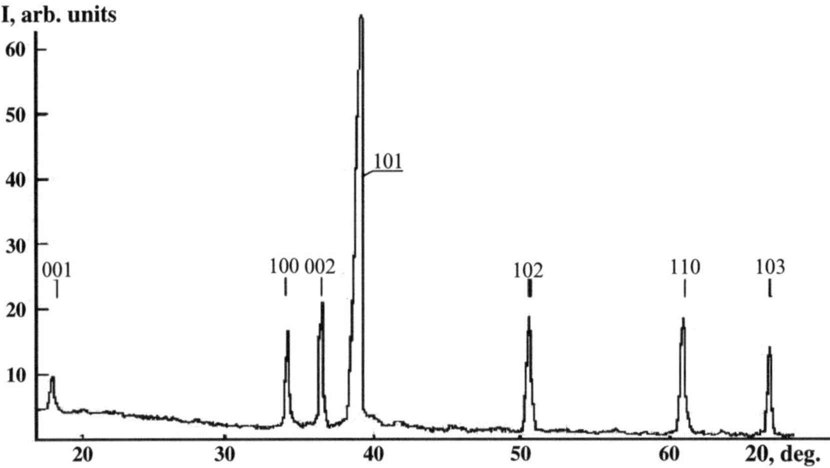

Figure 2. X-ray diffraction pattern of the solid solution $TiC_{0,30}H_{0,34}$

The crystal structures were studied for all single-phase hexagonal samples. In neutron diffraction patterns of all HCP solid solutions the superstructural reflections (in relation to the structure of α-Ti) are visible, and together with structural reflections they are indexed within the framework of space group (sp. gr.) $P\bar{3}m1$. In the Fig. 3(*a*) the neutron diffraction diagram of $TiC_{0,45}H_{0,64}$ is submitted. As it has been mentioned earlier, under conditions $x \geq 0.45 > y$ the FCC single-phase structure is observed (Fig. 3(b)). It should be noted that analogous result has been received in [5] for solid solutions TiC_xH_y of nearly stoichiometric composition on carbon. Within the framework of the assumed model the atoms might occupy the next positions:

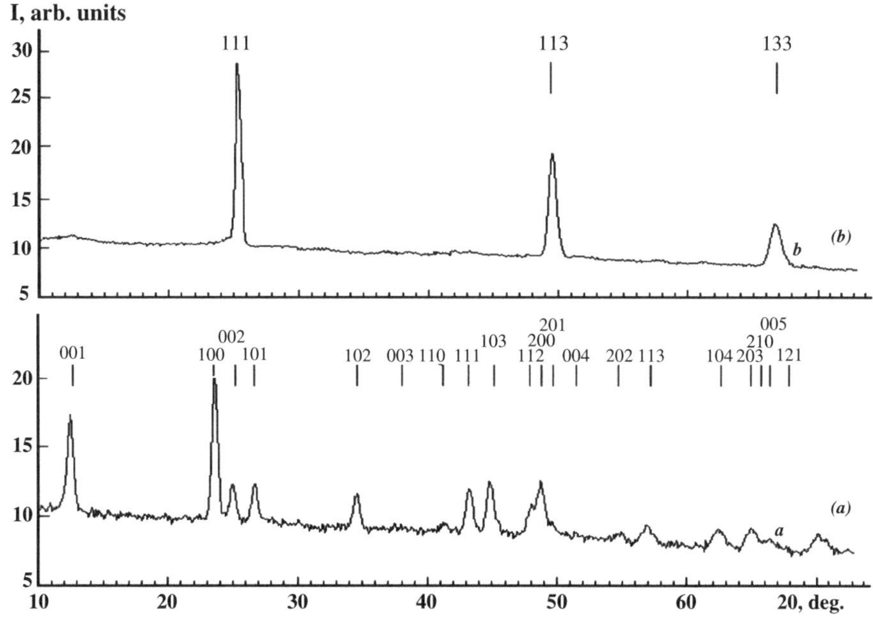

Figure 3. Neutron diffraction pattern of $TiC_{0,45}H_{0,64}$ (a) и $TiC_{0,45}H_{0,22}$ (b)

2 Ti atoms in position 2(d) with coordinates 1/3, 2/3, z_{Me}; 2/3, 1/3, $\bar{z}_{Me}$

m_1 C atoms in octahedral positions 1(a) with coordinates 0,0,0; m_2 C atoms in octahedral positions 1(b) with coordinates 0,0,1/2

n_1 H atoms in tetrahedral positions 2(d) with coordinates 1/3, 2/3, z_1; 2/3, 1/3, $\bar{z}_1$

n_2 H atoms in tetrahedral positions 2(d) with coordinates 1/3, 2/3, z_2; 2/3, 1/3, $\bar{z}_2$ where z_{Me}, z_1, z_2 are variable coordinates.

Besides, a part of hydrogen atoms can also be in octahedral positions 1(a) and 1(b). All mentioned variants of arrangement of carbon and hydrogen atoms are considered, occupancy of positions $n_{1,2}$ and $m_{1,2}$ being varied from zero up to the maximal values corresponding to the compound formula. At that, the values z_{Me}, z_1, z_2 were also changed till obtaining the good agreement between experimental and calculated intensities of diffraction maxima and minimal values of the discrepancy indices using Rietveld full-profile analysis. As results of calculations show, in single-phase hexagonal samples with concentrations $x \geq 0.43$ and $y \geq 0.55$ all carbon atoms are placed only in the octahedral positions 1(a) and hydrogen atoms – only in one type of tetrahedral positions 2(d). The structure of the above phase corresponds to the formula $Ti_2C_{1-x'}H_{2-y'}$, where $1 - x' = 2x$ and $2 - y' = 2y$. In Table 2 the structure characteristics of the solid solution $TiC_{0,45}H_{0,90}$ within the framework of sp. gr. $P\bar{3}m1$ are given. Parameters of long-range order [9] for carbon and hydrogen sublattices are $\eta_N = 0.90$ and $\eta_H = 0.90$ and correspond to the maximal values. At that, octahedral interstices 1(b) and tetrahedral interstices 2(d) with parameter $z_2 \approx 7/8$ remain vacant. The peculiarity of these HCP crystal structures is that the layers of empty and occupied octahedrons and tetrahedrons, which are placed perpendicularly to *c* axis, alternate. The ideal elementary cell of such solid solutions is shown in Fig. 4.

TABLE 2. Structure characteristics and discrepancy indices of solid solution $TiC_{0,45}H_{0,90}$

Atom	Position	Atom coordinates				B (Å^2)	ΔB (Å^2)	n	Δn
		x	y	z	Δz				
Ti	2(d)	1/3	2/3	0.234	0.001	1.79	0.75	2	0
C	1(a)	0	0	0		0.20	0.07	0.9	0.05
H	2(d)	1/3	2/3	0.632	0.004	2.90	0.30	1.81	0.10
$R_p = 1.34$; $R_{wp} = 1.6$; $R_{Бр} = 5.1\%$									

Notice: B – individual thermal factor; n – occupancy of a position

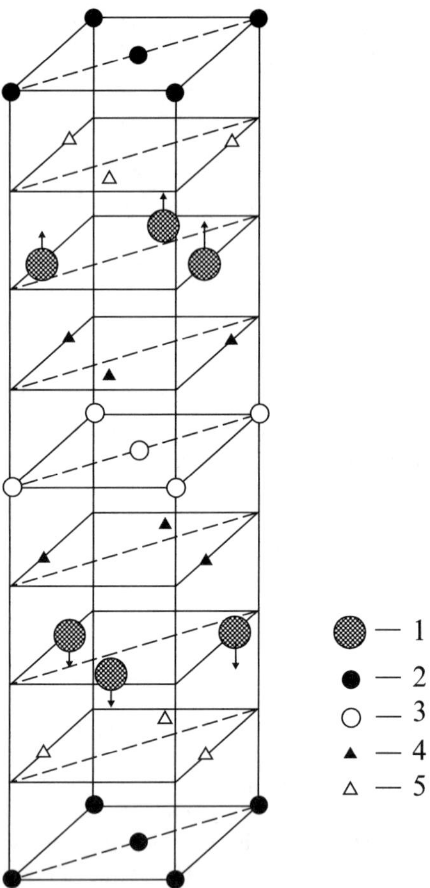

Figure 4. Schematic image of ideal unit cell corresponding to the composition Ti_2CH_2: 1 – Ti atoms; 2 – N atoms; 3 – vacancies of N atoms; 4 – H atoms; 5 – vacancies of H atoms. Arrows show the directions of shift of titanium atoms

In HCP solid solutions with carbon concentration $x \leq 0.43$ the hydrogen atoms, except for tetrahedral, partially occupy also octahedral interstices 1(b), free from carbon atoms (hydrogen atoms are partially disordered). In this case the good agreement between experimental and calculated intensities of diffraction reflections (Fig. 5) is observed.

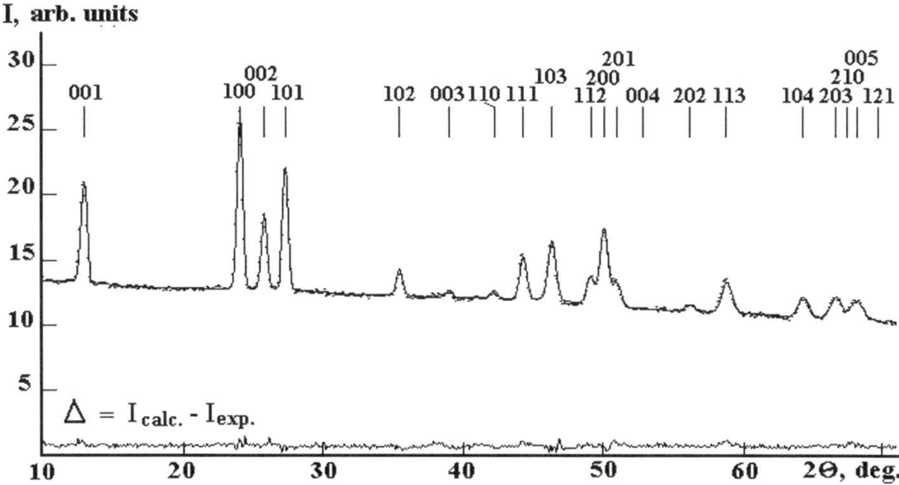

Figure 5. Neutron diffraction patterns of $TiC0_{43}H_{0,43}$.: solid line – calculated, dots – experimental, Δ – differential (experimental minus calculated)

As an example, the structural characteristics of the solid solution $TiC0_{43}H_{0,43}$ within the framework of sp. gr. $P\bar{3}m1$ are given in Table 3 .

TABLE 3. Structure characteristics and discrepancy indices of solid solution $TiC0_{43}H_{0,43}$

Atom	Position	Coordinates x	y	z	Δz	B (Å^2)	ΔB (Å^2)	n	Δn
Ti	2(d)	1/3	2/3	0.236	0.001	1.79	0.75	2	0
C	1(a)	0	0	0		0.20	0.07	0.86	0.04
H	2(d)	1/3	2/3	0.632	0.002	2.36	0.30	0.776	0.007
H	1(b)	0	0	1/2		2.90	0.30	0.084	0.007
$\eta_N = 0.86$, $\eta_H = 0.69$; $R_p = 0.93$; $R_{wp} = 1.2$; $R_{6p} = 3.6\%$									

In Table 4 occupation of tetrahedral and octahedral interstices with hydrogen atoms and the values of variable coordinate z are submitted. As Table 4 shows, occupancy of octahedral positions 1(b) by hydrogen atoms is insignificant. However they cannot be ignored. At any other variants of arrangement of nonmetal atoms, even at insignificant change of their position, the essential increase of the discrepancy indices on the Bragg reflections is observed. For example, in the assumption of arrangement of hydrogen atoms only in tetrahedrons, R – factor for $TiC0_{43}H_{0,43}$ grows up to 6.7% (from 3.6%) and for $TiC_{0,38}H_{0,64}$ – up to 11.3% (from 8.6%), and at that, the significant degradation of the agreement between experimental and calculated intensities is observed at the small angles where the influence of the thermal factor is insignificant. It is necessary to note small mistakes of definition of *z*-coordinate of titanium atom and also rather close their values (within the errors) for all structures: they insignificant fluctuate about average value $z_{aver.} = 0.234$. Significant disorder and relatively large errors are found for values of z for atoms of hydrogen in tetrahedrons. Apparently, small non-coherent background and thermal diffuse scattering from hydrogen atoms and also the technique of surveying powder patterns do not allow to determine coordinates of hydrogen atoms in the solutions more precisely.

TABLE 4. Occupancy of positions 2(d) and 1(b) with hydrogen atoms, z for H and Ti atoms, R_{Bragg} for solid solutions TiC_xH_y within the framework of sp. gr. $P\bar{3}m1$

№	Sample	n for H atoms		z in positions 2(d)		R_{Bragg} (%)
		2(d)	1(b)	z_H	z_{Ti}	
1	$TiC_{0,48}H_{0,70}$	1.40 ± 0.04	0	0.618 ± 0.002	0.232 ± 0.001	6.4
2	$TiC_{0,45}H_{0,90}$	1.80 ± 0.09	0	0.632 ± 0.005	0.234 ± 0.002	5.1
3	$TiC_{0,43}H_{0,55}$	1.10 ± 0.05	0	0.613 ± 0.005	0.238 ± 0.002	5.4
4	$TiC_{0,43}H_{0,43}$	0.776 ± 0.007	0.084 ± 0.007	0.631 ± 0.003	0.237 ± 0.001	3.6
5	$TiC_{0,41}H_{0,41}$	0.736 ± 0.008	0.084 ± 0.008	0.611 ± 0.003	0.232 ± 0.001	6.8
6	$TiC_{0,38}H_{0,64}$	1.224 ± 0.009	0.056 ± 0.009	0.613 ± 0.002	0.230 ± 0.001	8.6
7	$TiC_{0,35}H_{0,43}$	0.75 ± 0.01	0.11 ± 0.01	0.627 ± 0.004	0.237 ± 0.002	7.9
8	$TiC_{0,30}H_{0,43}$	0.75 ± 0.03	0.11 ± 0.03	0.627 ± 0.005	0.230 ± 0.002	8.8

As all investigated hexagonal solid solutions were obtained by means of hardening from 1,000°C, for obtaining the equilibrium state they were annealed in evacuated and sealed quartz ampoules at temperatures 700–600–500°C on 24 h. Such equilibrium annealing has not led to a change of phase composition and character of distribution of interstitial atoms in the samples with concentration of carbon $x \geq 0.41$. At the same time it leads to decay of hexagonal single-phase solid solutions with concentration of carbon $x < 0.41$ with precipitating hydride phase (Fig. 6). The neutron diffraction data of the decayed sample shows that it has broken up into HCP–phase of composition

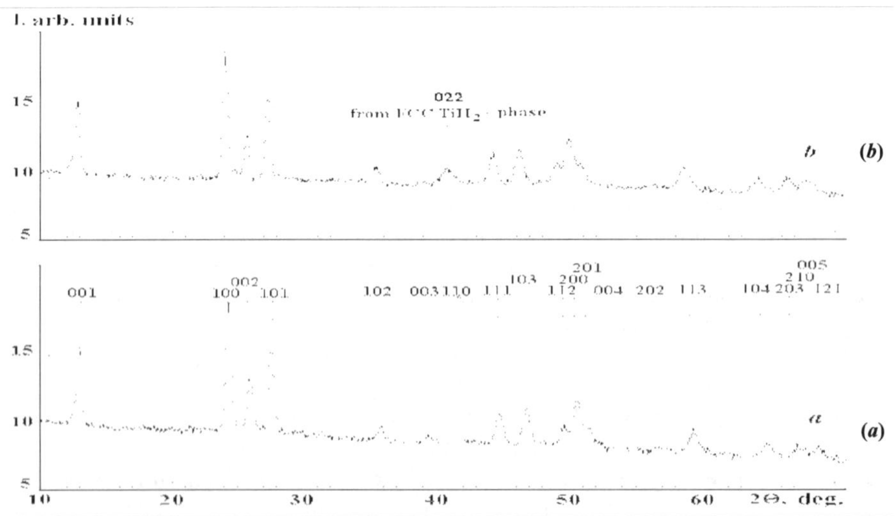

Figure 6. Neutron diffraction patterns of $TiC_{0,35}H_{0,43}$: a – after hardening from 1,000°C; b – after annealing at 700 ÷ 600°C during 48 h

$TiC_{0.41}H_{0.41}$ and FCC TiH_2–phase. Hence, within the temperature interval of 1,000–500°C the ordered solid solutions $Ti_2C_{1-x'}H_{2-y'}$ with carbon concentration $0.41 \leq x \leq 0.50$ (close to stoichiometry on carbon Ti_2C) is stable, at concentration $0.30 \leq x < 0.41$ it is metastable, and the lower limit of an equilibrium solid solution TiC_xH_y is near to composition $TiC_{0.41}H_{0.41}$.

Compounds TiC_xH_y with concentration $0.45 < x < 0.50$ and $y < x$ can have both disordered (sp. gr. Fm3m) and ordered (sp. gr. Fd3m) FCC-structure depending on temperature [3, 10]. In the neutron diffraction patterns of initial samples of titanium carbohydrides $TiC_{0.47}H_{0.22}$, $TiC_{0.50}H_{0.21}$ and $TiC_{0,50}H_{0,28}$ (received by quenching from temperature of 1,200°C) the same diffraction maxima were visible which were observed in X-ray diagrams. As an example, Fig. 7 shows the neutron diffraction pattern of titanium carbohydride $TiC_{0.50}H_{0.21}$. All these carbohydrides had face-centered cubic structure of NaCl-type (sp. gr. Fm3m) in which hydrogen atoms occupy only tetrahedral interstices 8(c) in the disordered cubic structure. According to neutron diffraction data, the samples $TiC_{0.47}H_{0.22}$ and $TiC_{0.50}H_{0.21}$ after quenching from 1,050–1,000°C are described in sp. gr. Fm3m, but the transition of a part of hydrogen atoms from tetrahedral to octahedral interstices and interchanges them by carbon atoms are observed. However, after the quenching of other samples the arrangement of hydrogen atoms in tetrahedral interstices is saved (Table 5). Hence, the temperature of hydrogen transition from tetrahedral to octahedral interstices depends on hydrogen and carbon concentrations.

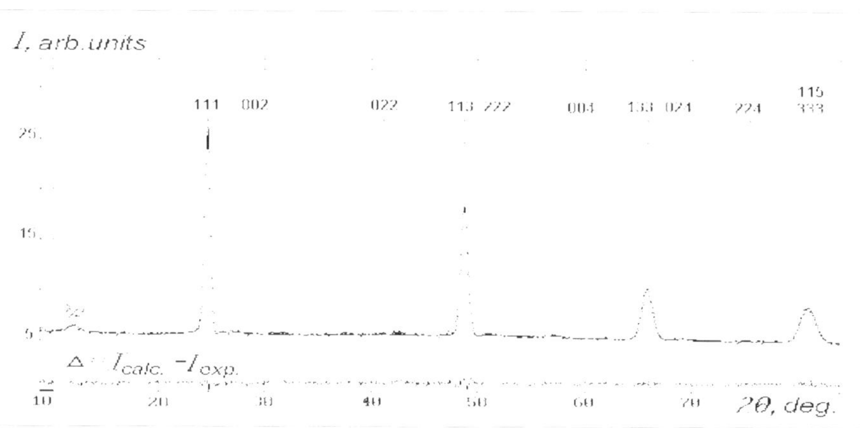

Figure 7. Neutron diffraction pattern of the initial sample $TiC_{0.50}H_{0.21}$. The notations are the same as in Fig. 5

TABLE 5. Structure characteristics of some carbohydrides in the framework of sp. gr. Fm3m

Atoms	Position	The number of atoms in the position (n) Quenching from 1,200°C $TiC_{0.47}H_{0.22}$	Quenching from 1,000°C $TiC_{0.47}H_{0.22}$	 $TiC_{0.47}H_{0.18}$
Ti	4(a)	4	4	4
C	4(b)	1.88	1.88	1.88
H	4(b)	0	0.42	0
H	8(c)	0.84	0.42	0.72
Discrepancy index, R (%)		$R_p = 3.7$; $R_{WR} = 1.9$; $R_{Br} = 2.5$	$R_p = 1.5$; $R_{WR} = 1.9$; $R_{Br} = 3.1$	$R_p = 2.3$; $R_{WR} = 3.1$; $R_{Br} = 2.1$

In all synthesized samples the ordering was observed at annealing temperatures of 900–600°C: it was pointed by arising of superstructure reflections (Fig. 8). The ordered structure of these carbohydrides has face-centered cubic unit cell (sp. gr. Fd3m) with parameter $a \approx 2\ a_0$ where a_0 is lattice parameter of the disordered structure. The crystal structures of the ordered phases of the carbohydrides TiC_xH_y both with $x \leq 0.50$, $y \leq 0.22$ and with $x \geq 0.55$, $y \leq 0.20$ are analogous [11]. They are described in the framework of the sr. gr. *Fd3m* in which carbon atoms mainly arrange over octahedral interstices 16(c), the rest small part of carbon and almost all hydrogen atoms are in another type of octahedral interstices, 16(d) (Table 6).

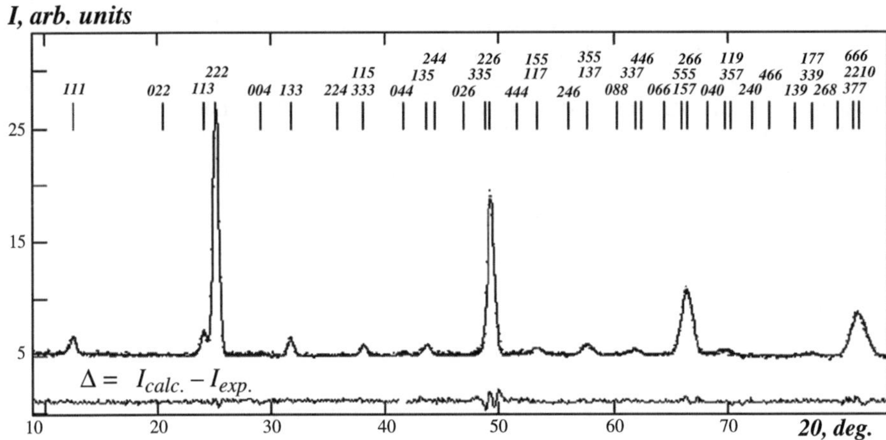

Figure 8. Neutron diffraction pattern of $TiC_{0.50}H_{0.21}$ after annealing at 800°C

TABLE 6. Structure characteristics of titanium carbohydride $TiC_{0.50}H_{0.21}$ in sp. gr. Fd3m after annealing at 800°C

Atoms	Position	The number of atoms in the position (n)	Δn
Ti	32	32	
C	16(c)	11.60	0.25
C	16(d)	4.40	0.25
H	16(d)	5.68	0.19
H	16(c)	1.04	0.19
R (%)		$R_p = 3.06$; $R_{wp} = 4.14$; $R_{Br} = 3.20$	

Using neutron diffraction data, the size of antiphase domains (APhD) in ordered carbohydrides was estimated after annealing at 700°C and 600°C. The size of APhD was determined using the half-width of superstructure reflection (111) and tool broadening, according to Sherrer formula [12]. The sizes of APhD in the titanium carbohydrides seems to agree closely and are equal (on average) $\overline{D}$ = 126 Å after the annealing at 700°C and $\overline{D}$ = 115 Å – at 600°C. Then the initial titanium carbohydrides were annealed in evacuated and sealed quartz ampoules at 450°C. Taking into account that at rather low temperature atoms diffusion mobility in titanium carbohydrides is not large, the annealing was carried out during 400 h. After this annealing the superstructure reflections corresponding to sp. gr. Fd3m arose in the neutron diagrams. It should be noted that there is splitting of the superstructure reflections. Figure 9 shows the neutron diagram of $TiC_{0.50}H_{0.21}$ after the annealing at 450°C during 400 h. Splitting of the superstructural reflections at the Bragg angles $2\theta > 30°$ into three symmetrically located reflections indicates that in the sample the long-period ordered APhD structure is formed [10]. As the resolution of neutron diffraction is poor because of small wavelength of monochromatizated neutrons ($\lambda = 1.085$ Å), neutron diagrams have been recorded repeatedly, and the pattern has been reproduced each time. It should be noted that at Bragg angles $2\theta > 30°$ almost all superstructure reflections are split. Hence, in the carbohydrides APhD ordering occurs in different directions. Using the distances between satellites of superstructure reflections, the period of the ordered APhD structures P has been determined according to [13]. The values of P determined by satellites of four superstructural reflections (133, 115, 135, 155) were ~10–12 .

Reflexes 133 and 024 of X-ray patterns (indices for initial Fm3m structure) indicate some distinctions in structure of the initial and annealed samples. After heat treatment of all samples these reflexes appreciably shift towards smaller angles (in comparison with initial positions). Figure 10 shows fragment of X-ray pattern corresponding to reflection 133 of $TiC_{0.50}H_{0.21}$ for initial sample (a) and annealed at 450°C during 400 h (b) when periodic APhD structures arose. The shift of reflections 133 and 024 is $\Delta 2\theta \approx 0.3°$ whereas the error in determining the Bragg angles does not exceed 0.03°. Therefore, the lattice parameter of carbohydrides having the ordered APhD structure increases noticeably ($\Delta a \approx 0.009$ Å).

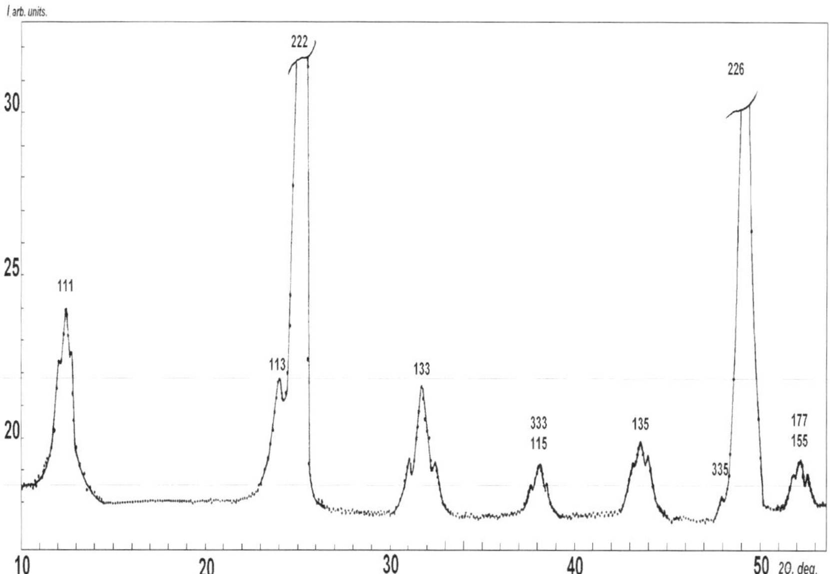

Figure 9. Neutron diffraction of carbohydride titanium $TiC_{0.50}H_{0.21}$ annealed at temperature 450°C during 400 h

As diffraction pattern recorded by X-rays is mainly defined by an arrangement of heavy atoms, that changes of the spectrum are related to changes of matrix. Besides increasing in lattice parameter after low-temperature annealing of $TiC_{0.47}H_{0.22}$ and $TiC_{0.50}H_{0.21}$, some fuzziness of $K_{\alpha1} - K_{\alpha2}$ – doublet is also observed. According to [13], a change of the cell dimension and arising of the microdistortions, resulting in degradation of doublet, is a result of formation of the long-period APhD structure.

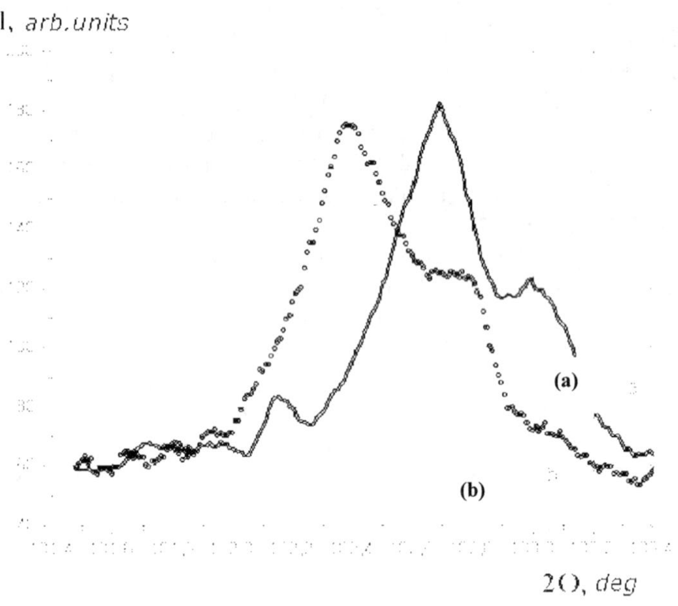

Figure 10. Fragment of the X-ray diffraction pattern (reflection 133): a -initial; b – annealed

4. Discussion of the results

Thus, it is shown that the single-phase interstitial solid solution TiC_xH_y can be obtained by hardening from 1,000° C only if concentration of C and H satisfies to conditions: $0.30 \le C/Ti \le 0.50$ and $H/Ti \ge C/Ti$. As distinct from hexagonal structure of α-Ti, described within the framework of space group $P6_3/mmc$, the solid solutions TiC_xH_y have the ordered hexagonal structure described within the framework of sp. gr. $P\bar{3}m1$. At that, carbon atoms occupy one of two types of octahedral interstitials, and in the solid solutions with high carbon concentration ($0.43 \le C/Ti \le 0.50$) hydrogen atoms occupy one of two types of tetrahedral interstitials. Such layered structure was observed for nitridehydrides of zirconium and titanium [14, 15], titanium carbohydride [5]. The analysis of results of all the works (including represented) allows to assume that in these compounds there is an effect of blocking of the neighboring interstices by interstitial atoms. This effect consists in that the carbon atom placed in octahedral interstice 1(b) does not permit hydrogen atom to occupy nearby tetrahedral interstices 2(d) with coordinate $z \approx 7/8$. According to the work [16], there is also effect of blocking of nearby interstitial positions by hydrogen atoms. At arrangement of hydrogen atom in tetrahedral interstice one or two nearest octahedral position 1(b) (in the framework of sp. gr. $P\bar{3}m1$) are blocked. Thus, it is possible to conclude that at least in the first coordination sphere of nonmetals the interaction carbon-hydrogen and hydrogen-hydrogen is the interaction of repulsion type. It is necessary to note that as a result of ordering interstitial atoms over the interstices of the certain types, metal atoms are essentially displaced ($z_{Me} \approx 0.234 \pm 0.002$) relatively of their ideal positions ($z_{id} =$ 0.250) in the direction of the layer filled with carbon atoms (Fig. 4). This displacement is found out even by X-Ray diffraction (Fig. 2). Hence, at least in the first coordination sphere the interaction titanium-carbon is of attraction character. In the solid solutions with low carbon concentration ($0.30 \leq C/Ti \leq 0.43$) hydrogen partly occupies, besides tetrahedral, also the octahedral interstices unfilled by carbon atoms (hydrogen atoms are partly disordered). Apparently, the found regularity can be explained by that solid solution $Ti_2C_{1-x}H_{2-y}$ is the system which can order and has wide homogeneity range. The temperature of order-disorder transition of the system increases with increasing carbon concentration, the ideal carbon compound being corresponded to the stoichiometry Ti_2C, and the structure of completely ordered phase corresponds to formula Ti_2CH_2. Solid solutions with lower carbon concentration have relatively low temperature of the disorder–order transition and, accordingly, have insufficient speed of hydrogen diffusion for recovery of full order during quenching. It seems interesting that in these solid solutions the process of ordering in carbon sublattice is completed earlier than in hydrogen sublattice. Apparently, it is an evidence of complex mechanisms of interaction between the atoms. At carbon concentrations $x < 0.30$ it is impossible to obtain a single-phase solid solution at any hydrogen concentrations. Hence, the ordered solid solutions TiC_xH_y are stable at concentrations $0.43 \leq x \leq 0.50$ (near carbon stoichiometry Ti_2C) but they are metastable at concentrations $0.30 \leq x \leq$ 0.41. The low limit of homogeneity range of the ordered solid solution TiC_xH_y is at the composition $TiC_{0.41}H_{0.41}$.

It should be noted that compounds TiC_xH_y at concentrations $0.45 \leq x \leq 0.50$ and $y < x$ have FCC-structure and are interstitial alloys (lattice of a matrix is differed from a lattice of initial metal) and only under condition of $y \geq x$ they turn into ordered solid solution of metalloids on the basis of α–Ti lattice. Also it is noticed that the composition of the lower limit of the solid solution does not side to pure titanium and lays at much more concentrations of carbon ($x \geq 0.30$). Besides, between titanium and carbon atoms in the first coordination sphere there is a rather large attraction force. All these facts allow to consider $Ti_2C_{1-x'}H_{2-y'}$-phase as hexagonal interstitial alloy.

It is interesting that at hardening from 1,000°C in the samples with low concentration of C and H ($x \leq 0.20$ and $y \leq 0.17$) the solid solution of C and H in BCC β- titanium (with impurity of FCC TiC_xH_y-phase) is found. It should be noted that in system Ti-N-H the formation of a BCC phase has not observed [15]. The basic known factor for stabilizing the β-phase is temperature, so this technological task was solved by heating (at punching) up to 900–1,000°C [17]. Other method consists in hydrogenating of titanium, doped by Al (up to 6 mass%) and V (up to 4 mass%), up to concentration $x_H \approx 0.25$ [18]. Our results show that for stabilization of β-phase it is enough to take concentrations both H and C in the certain parities. Though BCC phase is metastable, at the room temperature it has appeared enough stable. X-ray diffraction study of the BCC phase, being kept at the room temperature within a year, has shown that the process of aging is not observed.

Thus, the titanium carbohydrides at the lower limit of the homogeneity range, as well as carbohydrides with larger concentration of carbon, suffered the phase transition of "order-disorder" type. In the disordered state (in the framework of sp. gr. Fm3m), corresponding to the high temperature (1,200°C), the statistic arrangement of carbon atoms over octahedral interstices and of hydrogen atoms – over tetrahedral interstices is observed according to Hegg rule [13]. With decreasing temperature it is observed the ordered or partially ordered arrangement of carbon atoms over one of two types of octahedral interstices and hydrogen atoms – over the second type of octahedral interstices which are unfilled or partially filled by carbon atoms (in sp. gr. Fd3m). The change of a type of interstices filled by hydrogen atoms under ordering, may be explained by blocking effect: carbon atoms, when ordering over octahedral interstices, block all tetrahedral ones ($r_{octa\text{-}tetra}$ = 0.187 nm), probably, due to the repulsive force between interstitial atoms of various types. The distance between the two types of octahedrons is comparatively large ($r_{octa\text{-}octa}$ = 0.187 nm), thereby it is advantageous for hydrogen atoms to occupy octahedrons which are vacant or almost vacant from carbon atoms. It should be noted that filling up the tetrahedral interstices by hydrogen atoms in the disordered titanium carbohydride is found for the first time. It appears to see analogous situation also in titanium carbohydrides with a higher concentration of carbon atoms at sufficiently high temperature. As, in principle, hydrogen atoms can occupy both octahedral and tetrahedral interstices, titanium carbohydride with higher concentration of hydrogen may be prepared if certain conditions (high temperatures, high pressure of hydrogen, etc.) will be created.

For the first time in titanium carbohydrides we have found the formation of the long-period APhD structure analogous to that observed in substitutional solid solutions. This result enables us to conclude that both substitutional solid solutions and interstitial phases have identical mechanism of APhD ordering. Absence of APhD ordering in samples, annealed at higher temperatures ($T \leq 700°C$) may be explained by formation of ordered and randomly oriented APhD. At lower temperatures a slow formation of fine and oriented (ordered) APhD is observed. As the APhD size at temperature of 600°C is ~115 Å, one can suppose that sizes of the ordered domains should be considerably smaller than 100 Å.

5. Conclusions

1. It is established that at hardening from 1,000°C in the samples TiC_xH_y with carbon concentration $0.30 \leq x \leq 0.50$ and $y \geq x$ the ordered hexagonal close-packed (HCP) structure corresponding to the formula $Ti_2C_{1-x'}H_{2-y'}$ is stabilized, and at $y < x$ – the FCC-phase. Below temperature 700°C the $Ti_2C_{1-x'}H_{2-y}'$-phase with carbon concentration $0.30 \leq x < 0.43$ decays with separation of hydride phase TiH_2.
2. At temperatures $T < 700°C$ the lower limit of the $Ti_2C_{1-x'}H_{2-y'}$ homogeneity range is at concentrations $1\text{-}x' \approx 2\text{-}y' \approx 0.82$ ($x \approx y \approx 0.41$).
3. For the first time it is obtained the solid solution of carbon and hydrogen in the BCC – titanium lattice in metastable state which is stable enough at the room temperature.
4. At the lower limit of the homogeneity range of cubic Ti carbohydride five characteristic temperature ranges of structure changes are found:

- $T \geq 1{,}200°C$: titanium carbohydrides have disordered cubic structure described in sp. gr. Fm3m where carbon atoms statistically occupy octahedral interstices and hydrogen atoms – tetrahedral ones.
- $1{,}000°C \leq T \leq 1{,}100°C$: the disordered cubic structure is still kept in which the hydrogen atoms are arranged statistically over both octahedrons and tetrahedrons.
- $800°C \leq T < 1{,}000°C$: formation of structure with long-range order is observed; sp. gr. Fd3m; carbon and hydrogen atoms occupy octahedral interstices 16(c) and 16(d), respectively.
- $600°C \leq T < 800°C$: during the process of further ordering the partial decay of ordered cubic phase with segregation of α-*Ti* begins.
- $T \leq 475°C$: metastable ordered cubic phase forms in which periodic anti-phase domains is found, at these temperatures the process of decay is hindered.

Acknowledgements

The work was carried out at financial support of the Ukrainian Scientific – Technological Centre, project N Uzb-131 (j))

References

1. Kornilov I.I. Titanium. Moscow: Nauka. 1975. 308 p.
2. McCusker L.B., Von Dreele R.B., Cox D.E., et al. Rietveld refinement guidelines, *J. Appl. Crystallogr.*, 1999. **32**: 36–50.
3. Khidirov I., Sultanova S.Kh., Mirzaev B.B., et al. Phase transformation and ordering of anti-phase domains in the cubic titanium carbohydrides, *Altern. Energ. Ecol.*, 2004. **11**: 8–14 (in Russian).
4. Khidirov I., Sultanova S.Kh., Mirzaev B.B., Mukhtarova N.N. Neutron diffraction study of titanium carbohydride $TiC_{0.50}H_{0.28}$, Proceedings of International Conference for fundamental and applied problems of physics. Tashkent, Uzbekistan, 2004, 467–469.
5. Latergaus I.S., Karimov I.A., Em V.T., et al. Neutron diffraction study of hexagonal carbohydride of titanium, *Izv. AS UzSSR. Phys.-Math. Ser.*, 1985. **5**: 78–80.
6. Matysina Z.A., Schur D.V. Phase transformations $\alpha \rightarrow \beta \rightarrow \gamma \rightarrow \delta \rightarrow \varepsilon$ in hydrides TiH_x at increasing of hydrogen concentration, *Izv. VUZov. Fizika*, 2001. **11**: 89–94 (in Russian).
7. Eschenko R.H., Yelkina O.A., Syrenkov I.G., Taluts G.G. Structure transformations in alloys titanium-hydrogen and titanium-deuterium, *FMM*, 1995. **79**(2): 65–71 (in Russian).
8. Bacon G.E. Neutron Diffraction. Oxford: Clarendon. 1975. 636 p.
9. Khidirov I., Sultanova S.Kh., Mirzaev B.B., et al. Long-period structure in titanium carbohydrides. *Russ. J. Inorg. Chem.*, 2005. **8**(50): 1351–1355.
10. Gusev A.I. Order–disorder transformation and phase equilibrium in strongly non-stoichiometric compounds. *Uspekhi Fizicheskikh Nauk*, 2000. **140**(1): 3–40 (in Russian).
11. Latergaus I.S., Karimov I., Presman V.S., Em V.T. Phase transition order-disorder in titanium carbohydrides. *Neorganicheskie materiali*, 1987. **23**(9): 1471–1474 (in Russian).
12. Gorelik S.S., Rastorguev L.N., Skakov Yu.A. X-Ray and Electron-Optical Analysis. Moscow: Metallurgiya. 1970. 368 p.
13. Barrett C.S., Massalski T.B. Structure of Metals. Part 1. Oxford: Pergamon Press. 1980. 352 p.
14. Kalinin V.P., Andrievskiy R.A., Chervyakov A.Yu., et. al. Structure of $ZrND_{1.54}$. *Ukrain. Fizich. J.*, 1970. **15**(5): 842–844.
15. Khidirov I. Processes of ordering and decay in solid solutions of nitrogen and hydrogen in α-titanium, *J. Neorg. Khimii*, 2001. **46**(3): 499–505 (in Russian).

16. Rundqist S., Tellgren R., Andersson Y. Hydrogen and deuterium in transition metal-pelement compounds: crystal-chemical aspect of interstitial solid solubility and hydride phase formation. *J Less-Comm. Met*. 1984. **101**(1): 145–168.
17. Arkhipov V.E., Dubinin S.F., Eshchenko R.R., et al. Neutron diffraction study of hydrogenated titanium alloys VT6. *FMM*, 1993. **75**(6): 109–112 (in Russian).
18. Arkhipov V.E., Dubinin, S.F., Naish, B.E., et al. Neutron diffraction quantitative phase analysis of hydrogenated titanium alloys VT6. *FMM*, 1994. **78**(5): 147–155 (in Russian).

INFLUENCE OF VACUUM THERMAL TREATMENT ON STRUCTURE OF SOLID SOLUTIONS TiC_xH_y

I. KHIDIROV*, B.B. MIRZAEV, N.N. MUKHTAROVA
Institute of Nuclear Physics, Uzbekistan Academy of Sciences, Tashkent, Uzbekistan
**E-mail: khidirov@inp.uz*

V.A. BOGOLEPOV, A.F. SAVENKO, V. K.PISHUK
Institute for Problems of Materials Science, National Academy of Sciences of Ukraine, Kiev, 03142, Ukraine
E-mail: shurzag@materials.kiev.ua

Abstract. It is found complete thermoemission of hydrogen out of lattice of powder samples of the solid solution $TiC_{0.35}H_{0.43}$: complete evacuation of hydrogen atoms without change of the compound symmetry (which has been formed earlier by hydrogen) in vacuum at the temperature (275°C), lower than recrystallization temperature. After complete hydrogen thermoemission the change of direction of matrix atoms displacement relatively to their ideal position occurs; it may be explained by change of character of interaction between lattice atoms after "soft" removal of hydrogen. It is established that the complete H thermoemission, when the initial symmetry is kept, is observed only in strongly defective structures, i.e. at the lower limit of the homogeneity range of solid solutions TiC_xH_y.

Keywords: solid solutions, TiC_xH_y, vacuum evacuation, hydrogen-induced phases, crystal structure, neutron diffraction

1. Introduction

Since recent time interstitial solid solutions have been investigated as systems opened for interstitial atoms [1–3]. This direction is perspective as it opens an opportunity for artificial obtaining of new metastable hydrogen-induced phases which are absent in the phase diagram of binary interstitial alloys, and such phases can not be obtained by traditional methods of synthesis. Besides, the comparison of structural peculiarities of the hydrogen-induced and corresponding hydride phases opens new sides in understanding of crystal-chemistry of the compounds and of crystal-chemical behavior of hydrogen. However in the meanwhile there are insufficient researches in this direction.

The aim of the present work was neutron diffraction study of interstitial solid solutions $TiC_{0,45}H_{0,90}$, $TiC_{0,45}H_{0,64}$, $TiC_{0.35}H_{0.43}$ of the system Ti-C-H .

2.Experiment technique

Neutron diffraction patterns of powdery samples were obtained at the room temperature using the neutron diffractometer of INP AS RUz (λ = 1,085 Å). Calculations of the neutron diagrams and refinement of the sample structure were carried out by the Rietveld full-profile method [4].

B. Baranowski et al. (eds.), *Carbon Nanomaterials in Clean Energy Hydrogen Systems.*

The technique of samples preparation is analogous to [5]. Composition of the samples was controlled by minimizing the divergence factors of structure determination using neutron diffraction data. Dehydrogenation of the samples under persistent evacuation was carried out in the vacuum furnace of SShVL type. According to X-ray powder analysis, the investigated solid solutions TiC_xH_y were single-phase and had HCP-structure (a = 3.098 ÷ 3,024; c =5.079 ÷ 4,894 Å).

3. Results and discussion

In neutron diffraction patterns of all initial solid solutions (Fig. 1) superstructure reflections are visible; together with structural ones they are indexed within the framework of sp. gr. $P\bar{3}m1$; atoms Ti are located in positions 2(d) with z = 0.237 ± 0.002, atoms N – in octahedrons and atoms H – in tetrahedrons with z = 0.628 ± 0.004. For study of thermal stability the solid solutions were annealed in continuously pumped out vacuum not worse than $5.3 \cdot 10^{-3}$ Pa. Vacuum thermal treatment was started at the room temperature and then temperature was raised with a step of 25°; retention interval at each temperature was at first 24 h and then 48 h. Quantity of hydrogen in the samples before and after vacuum annealing was watched by slope of non-coherent background caused by scattering of neutrons on the nuclei of hydrogen atoms [6], and it was controlled by means of the analysis of experimental and calculated intensities of diffraction reflections. According to neutron structure analysis, up to temperature 250°C in the investigated solid solutions thermoemission of H is not found out practically. Since 275°C in solid solutions with large concentration of carbon it is found the separation of pure titanium and face-centered cubic phase, most probably, cubic carbohydride TiC_xH_y (Fig. 2). At the further increase of temperature the quantity of precipitating phases increases. At that, up to temperature 400°C the slope of non-coherent background is saved. So, solid solutions with high concentration of carbon (and hydrogen) $TiC_{0,45}H_{0,90}$, $TiC_{0,45}H_{0,64}$ in vacuum are thermostable only at temperatures T < 275°C, and the hydrogen in composition of HCD and FCC phases is kept up to temperature 400°C.

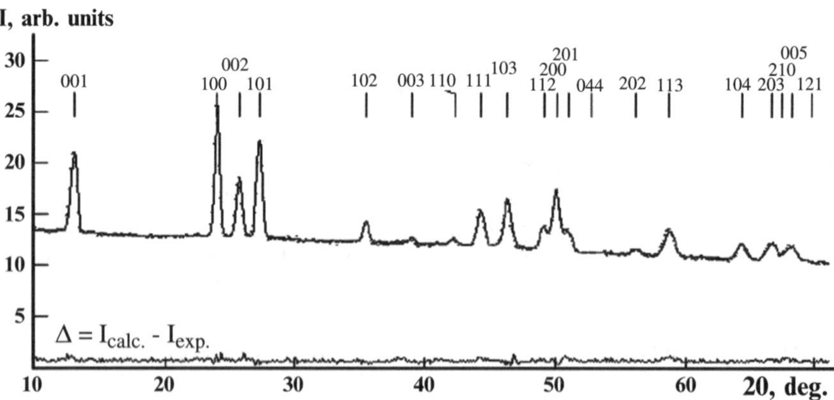

Figure 1. Neutron diffraction patterns of $TiC_{0,35}H_{0,43}$: solid line – calculated, dots – experimental, Δ – differential (experimental minus calculated). Miller indices of reflecting planes of crystal lattice are indicated above reflections

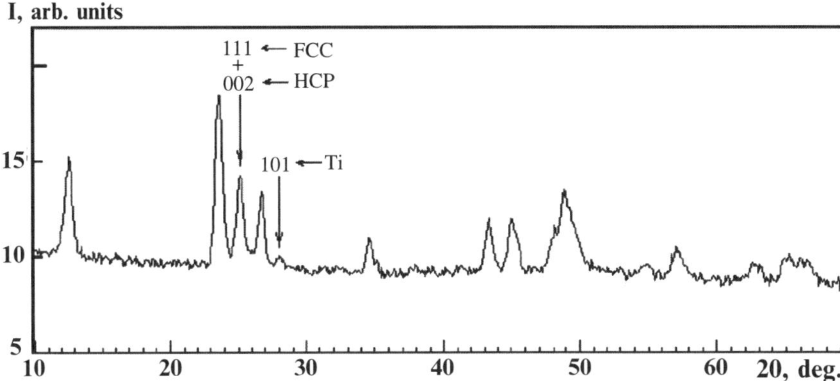

Figure 2. Neutron diffraction pattern of $TiC_{0.45}H_{0.64}$ after annealing during 48 h at temperature 275°C under persistent evacuation

In neutron diagram of the solid solution $TiC_{0,35}H_{0,43}$ after vacuum annealing at temperature 275°C it is observed the same diffraction reflections which were in the neutron diagram of initial hydrogen-containing sample (Fig. 1), but non-coherent background (caused by hydrogen) is completely absent (Fig. 3). Hence, at temperature 275°C in the solid solution $TiC_{0,35}H_{0,43}$ the complete removal of hydrogen out of crystal lattice with preservation of the initial hexagonal ordered $Ti_2C_{0.70}$-phase in metastable state is observed.

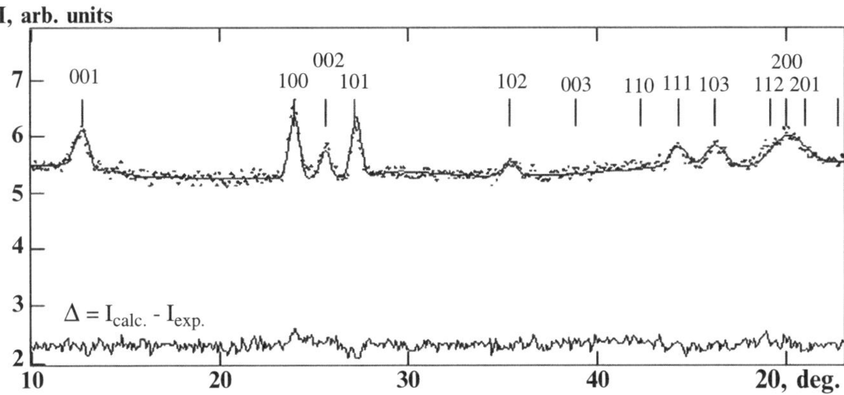

Figure 3. Neutron diffraction patterns of the ordered $Ti_2C_{0.70}$ – phase in metastable state; the sample was obtained from hydrogen-containing $Ti_2C_{0.70}H_{0.86}$ – phase (solid solution $TiC_{0.35}H_{0.43}$) by means of annealing at temperature 275°C under persistent evacuation. Notes are the same as in Fig. 1. Calculated diagram was obtained within the framework of sp. gr. $P\bar{3}m1$ where $z_{Ti} = 0.305$

It is necessary to note strong distortion of diffraction maxima, reduction of their intensity, wide scatter of points in the neutron diagram and gradual rise of background with increase of Wolf-Bragg angle. These effects testify to strong distortion of a crystal lattice (of static character). The best agreement between experimental and calculated intensities of neutron diffraction reflections (Fig. 3) and the minimal divergence factors can be obtained only under assumption that in the there is no hydrogen, so its composition corresponds to the formula $TiC_{0,35}$, but the initial ordered structure is saved. In Table 1 the results of processing of neutron diagram of $TiC_{0,35}$ are given. The Table 1 shows the large errors in determination of lattice parameters and high value of the Bragg discrepancy indices ($R_{Br} \approx 14\%$). Evidently, these results are caused by distortion of peak forms and wide scatter of points in the neutron diffraction pattern because of strong static distortions of the lattice. The large errors in determination of lattice parameters of the hydrogen-induced $Ti_2C_{0.70}$-phase do not allow to compare them with lattice parameters of the initial hydrogen-containing phase.

TABLE 1. Structure characteristics and discrepancy indices of structure determination of hydrogen-inducing ordered solid solution $TiC_{0,35}$ ($Ti_2C_{0.70}$-phase) in the framework of sp. gr. $P\bar{3}m1$

Atom	Position	Atom coordinates			
		x	y	z	Δz
Ti	2(d)	1/3	2/3	0.305	0.004
C	1(a)	0	0	0	
$a = 3.02 \pm 0.01$; $c = 4.89 \pm 0.01$ Å; $c/a = 1.62$; $B_{eff.} = 0.9 \pm 0.3$ Å^2; $R_p = 1.6$; $R_{wp} = 2.1$; $R_{Бр} = 13.6\%$					

Notice: $B_{eff.}$ – effective thermal factor

In structure of the hydrogen-induced $Ti_2C_{0.70}$–phase (as in the hydrogen-induced $Ti_2N_{0.52}$–phase [2]) it is observed a change of a sign of metal atoms displacement ($z_{Ti} = 0.305 \pm 0.004$) relatively to the ideal position ($z_{id} = 1/4$) in comparison with the hydrogen-containing $Ti_2C_{0.70}H_{0.86}$-phase where $z_{Ti} = 0.237 \pm 0.004$. If to assume that value of z_{Ti} in the hydrogen-induced $Ti_2C_{0.70}$–phase is the same as in hydrogen-containing phase $Ti_2C_{0.70}H_{0.86}$ than it is observed an essential growth of R_{Br} (up to 30% if $z_{Ti} = 0.250$ and to 33% if $z_{Ti} = 0.237$) and also rather bad agreement between experimental and calculated intensities of diffraction maxima in neutron diagrams (Fig. 4). The various directions of displacement of metal atoms in hydrogen-induced and hydrogen-containing phases, apparently, may be explained by various character of interatomic binding forces in the above phases.

So, it is possible to conclude that in solid solutions of Ti-C-H system complete thermoemission of hydrogen with preservation of initial symmetry, caused by presence of hydrogen, is observed in strongly defective structures. Apparently, in strongly defective solid solutions the interaction forces between atoms are weak and the temperature of hydrogen evacuation T_{ev} is below temperature of interstitial atoms redistribution or of recrystallization T_{recr}: $T_{ev} < T_{recr}$.

4. Conclusions

It is shown that phenomenon of complete thermoemission of hydrogen with preservation of the initial structure is observed only in strongly defective structures TiC_xH_y at temperature 275°C under persistent evacuation (not worse than 5.3 x 10^{-3} Pa); in these structures the interaction forces between atoms are weak and the temperature of hydrogen evacuation T_{evac} is lower than temperature of interstitial atoms redistribution or of recrystallization of the solid solution $T_{recryst}$: $T_{evac} < T_{recryst}$.

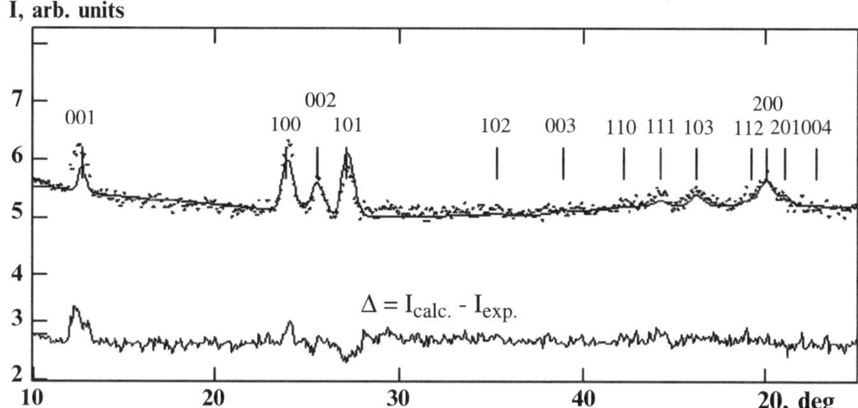

Figure 4. Neutron diffraction patterns of the ordered $Ti_2C_{0.70}$–phase in metastable state; the sample was obtained from hydrogen-containing $Ti_2C_{0.70}H_{0.86}$–phase (solid solution $TiC_{0.35}H_{0.43}$) by means of annealing at temperature 275°C under persistent evacuation. Notes are the same as in Fig. 1. Calculated diagram was obtained within the framework of sp. gr. $P\bar{3}m1$ where $z_{Ti} = 0.237$

In the solid solutions with the large concentrations of carbon and hydrogen at the same temperature the decay of the solid solution is observed, α-Ti and FCC titanium carbohydride TiC_xH_y being formed.

The crystal structure of metastable hydrogen-induced $Ti_2C_{0.70}$-phase is strongly distorted. It is found the effect of changing of direction of matrix atoms displacement relatively to their ideal position after complete thermo-emission of hydrogen out of lattice of the solid solution $TiC_{0.35}H_{0.43}$; it may be explained by change of character of interaction between the atoms after "soft" removal of hydrogen.

Acknowledgements

The work was carried out at financial support of the STCU, project № Uzb-131 (j).

References

1. Khidirov I., Rakhimov A.S., Loryan V.E., Kurbonov I.I. Receiving of metastable phases in the systems Ti–N and Zr-N by dehydrogenating hydrogen-containing solid solutions, *Neorganicheskie Materialy*, 1990. **26**(10): 2113–2116 (in Russian).
2. Khidirov I., Kurbonov I.I. Makhmudov A.Sh. Hydrogen-induced metastable phase in the systems Ti–N, *Metallofizika*, 1991. **13**(13): 43–47 (in Russian).
3. Khidirov I., Kurbonov I.I. Method of receiving of titanium nitride, Patent of Uzbek Republic, 1991. No. 2307.
4. McCusker L.B., Von Dreele R.B., Cox D.E., Louer D., Scardi P. Rietveld refinement guidelines, *Journal of Applied Crystallography*, 1999. **32**: 36–50.

5. Khidirov I., Sultanova S.Kh., Mirzaev B.B., Mukhtarova N.N. Phase transformations and ordering of anti-phase domains in cubic titanium carbohydrides at the lower limit of homogeneity range, *Alternative Energetics and Ecology*, 2004. **11**: 8-14.
6. Bacon G.E. Neutron Diffraction. Oxford: Clarendon Press. 1975. 636 p.

DETERMINATION OF ARRANGEMENT OF HYDROGEN ISOTOPES IN THE SOLID SOLUTION $TiN_{0.40}H_{0.19}D_{0.19}$ BY NEUTRON DIFFRACTION

I. KHIDIROV*, N.N. MUKHTAROVA
Institute of Nuclear Physics, Uzbekistan Academy of Sciences, Tashkent, Uzbekistan
**E-mail: khidirov@inp.uz*
L.N. PADURETS, A.L. SHILOV
Institute of General and Inorganic Chemistry of RAS, Moscow, 119991, Russia
E-mail: ashilov@rambler.ru
S.YU. ZAGINAICHENKO, D.V. SCHUR, A.F. SAVENKO
Institute for Problems of Materials Science, National Academy of Sciences of Ukraine, Kiev, 03142, Ukraine
E-mail: shurzag@materials.kiev.ua

Abstract. By neutron structure analysis the crystal structure of the solid solution with the combined isotope composition on the base of titanium – $TiN_{0.40}H_{0.19}D_{0.19}$ was studied (space group $P\bar{3}m1$). It was found isotopic effect in distribution of the hydrogen isotopes H and D: the isotopes arrange over the same type of tetrahedral interstices but have different coordinates z, i.e. the splitting of a set of tetrahedral positions 2d into hydrogen (H) and deuterium (D) is observed. It follows from this that in the ordered "layered" structure of the solid solution $TiN_{0,40}H_{0,19}D_{0,19}$ there can not be two different isotopes of the same element on one plane perpendicular to threefold axis. Crystal-chemistry analysis of environment of H and D is carried out and the essential differences in coordination polyhedrons surrounding the two isotopes are revealed. Analysis of the results of the present work and also literary data shows that in a number of metal-hydrogen systems hydrogen and deuterium should be considered as independent chemical components.

Keywords: solid solution, hydrogen isotope, $TiN_{0,40}H_{0,19}D_{0,19}$, crystal structure, neutron diffraction

1. Introduction

By neutron diffraction of cerium trihydride Ce_3D_2H the isotopes distribution over non-equivalent positions – in octa- and tetra-pores has been investigated, and the partially ordered state has been determined: the dominant arrangement of hydrogen in octahedrons and deuterium – in tetrahedrons of face-centered cubic lattice of a matrix [1]. The found effect is explained by reduction of the energy of local fluctuations when crossing a light isotope into octahedron and a heavy one – into tetrahedron. On the basis of thermodynamic reasons it was shown that octahedral interstices are more favorable to be occupied with easier isotope (hydrogen) [2]. In both works the interaction between interstitial atoms is considered as negligible.

In the work [3] by NMR-spectroscopy, thermal and X-ray phase analysis it was shown that the ordering of interstitial atoms in $HfV_2H_{4-x}D_x$ is regulated by short-range interaction, namely, by H-H blocking (with blocking radius near 2.1 Å) between H(D) atoms which are placed in different tetrahedral sub-lattices: **e** – (faceted on [Hf3V]) and **g** – (faceted on [2Hf2V]). In deuterohydrides the redistribution of hydrogen isotopes between coexisting phases occurs during ordering.

It is of interest to study solid solutions with combined isotopic composition if the third interstitial element presents. Firstly, if the atoms of the third element have the noticeably greater size than atoms of hydrogen, they inevitably will occupy octahedral voids. Secondly, in this case the interaction between atoms of the hydrogen isotopes and the third interstitial element is considerable and can noticeably influence on distribution of hydrogen isotopes. The aim of the present work is the study of isotopic effect in distribution of hydrogen isotopes in the solid solution $TiN_{0.40}H_{0.19}D_{0.19}$ by neutron diffraction.

2. Experiment

The sample $TiN_{0.40}H_{0.19}D_{0.19}$ was obtained by means of deuteration of the solid solution $TiN_{0.40}H_{0.19}$ [4] at the Sivert's type plant [5] in the Kurnakov Institute of general and inorganic chemistry of RAS. According to this technique, in the working volume of the plant a batch of powder $TiN_{0.40}H_{0.19}$ was placed and then it was pumped out up to pressure $1.33 * 10^{-2}$ Pa. Then pure deuterium was fed into standardized volume under the pressure necessary for obtaining prescribed composition. The absorption of deuterium was observed using a mercury manometer. The process was carried out at temperature 720 K. All deuterium was absorbed by initial nitridehydride during 10 min. The sample was soaked in the furnace till complete cooling of the reactionary tube up to the room temperature. The sample was unloaded in air. Content of deuterium in the sample was determined by absorbed volume of the gas. As a source of deuterium titanium deuteride was used.

Neutron diffraction patterns were obtained using the neutron diffractometer mounted at the thermal column of the atomic reactor WWR-SM of the INP AS RUz ($\lambda = 1.085$ Å) [6]. For phase analysis and estimation of homogeneity of solid solutions X-ray diffraction was used (CuK_α – radiation, $\lambda = 0.15418$ nm). Calculation of structure characteristics from neutron diffraction data was carried out by the Rietveld full-profile method [7].

According to X-ray diffraction patterns, the synthesized solid solution $TiN_{0.40}H_{0.19}D_{0.19}$ was single-phase and had the hexagonal (trigonal) unit cell with parameters $a = 3.009$, $c = 4.875$ Å (Fig. 1).

3. Results and discussion

The neutron diffraction pattern of the single-phase and homogeneous (on composition) solid solution $TiN_{0.40}H_{0.19}D_{0.19}$ was indexed in the framework of sp. gr. $P\bar{3}m1$ (Fig. 2). Neutron diffraction data can be explained well, only if to assume that all nitrogen atoms occupy octahedral positions 1(a) and atoms H and D – tetrahedral interstices 2(d) but with different coordinates z (Table 1). If to assume that isotopes H and D have identical coordinates z, then agreement between experimental and calculated interstices sharply decreases, particularly at the small Bragg angles (where the influence of thermal factor is still insignificant) and the divergence factor R_{Br} rises up to 11%.

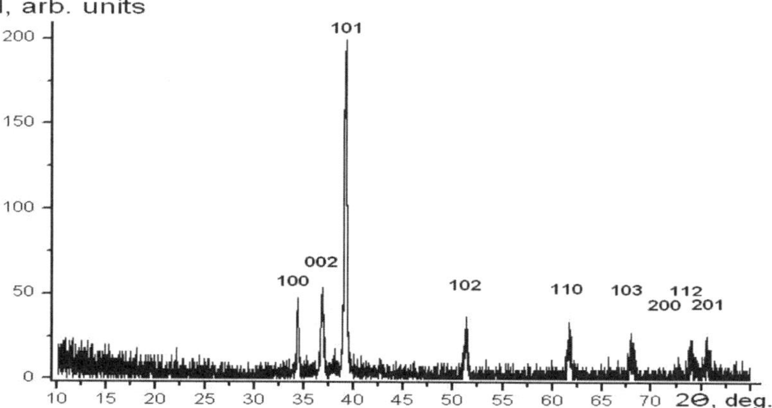

Figure 1. X-ray diffraction pattern of $TiN_{0.40}H_{0.19}D_{0.19}$. Above peaks the Miller indexes of the reflecting planes are indicated

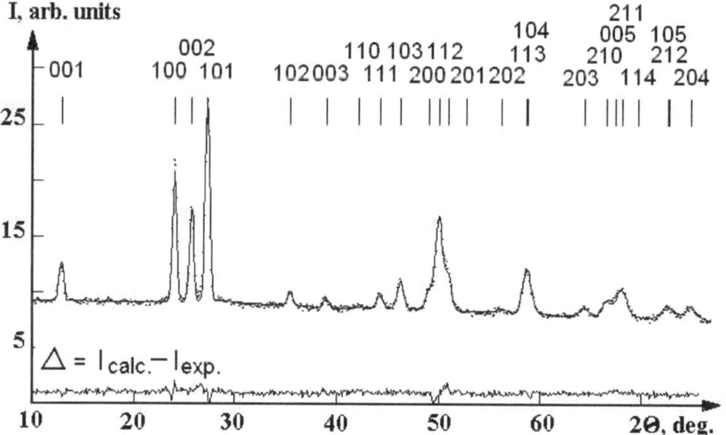

Figure 2. Neutron diffraction patterns of $TiN_{0.40}H_{0.19}D_{0.19}$: solid line – calculated, dots – experimental, Δ – differential

TABLE 1. Structure characteristics of solid solution $TiN_{0.40}H_{0.19}D_{0.19}$ in sp. gr. $P\overline{3}m1$

Atom	Position	x	y	z	Δz	N	Δ n
Ti	2d	0.3333	0.6667	0.2484	0.0021	2.000	0.000
N	1a	0.0000	0.0000	0.0000	0.0000	0.800	0.000
D	2d	0.3333	0.6667	0.5824	0.0035	0.316	0.015
H	2d	0.3333	0.6667	0.7945	0.0070	0.327	0.028
B_{ef} = 0.73 ± 0.04 Å; R_P = 2.07; R_{WR} = 2.69; R_{exp} = 1.51; R_{Br} = 6.69%							

n – occupancy of a position

So, as distinct from the ordered solid solutions TiN_xH_y [4, 8] and TiN_xD_y [9], in solid solution with combined isotopic composition $TiN_xH_{y/2}D_{y/2}$ the splitting up of tetrahedral position 2(d) into the hydrogen and deuterium is observed.

As it has been marked formerly, the ordering of hydrogen isotopes over octahedrons and tetrahedrons, caused by zero-order fluctuations of isotopes, was observed earlier in the hydride CeD_2H at low temperatures: $T < 300$ K [1]. Isotopic distribution (ordering) of hydrogen isotopes over interstices of the same type in a single-phase high-temperature state is found out for the first time.

The results of crystal structures decryption of some solid solutions of the system Ti-N-H(D) [4, 8, 9] have shown that all of them are isostructural. For more detailed description of the mentioned solid solutions it is convenient to use crystal-chemical approach: representation of a structure as a combination of Pauling polyhedrons on the basis of the closest packing of solid spheres. According to [10], metallic (atomic) radii of Ti, N and H are equal to 1.46; 0.71 and 0.46 ÷ 0.37 Å, respectively. The atomic radius of deuterium may be considered as close to 0.50 Å. Based on the sizes of atomic radii and concentration of elements, the large atoms of titanium are assumed to be a structure basis, i.e. two-layer spherical packing; in its octahedral and tetrahedral pores the rest atoms are arranged. According to [10], the radius of octahedral pore is ~41% and of tetrahedral one – ~22% of radius of the basis solid sphere. It gives $r_{oct} \approx 0.60$ and $r_{tetr} \approx 0.32$ Å, respectively, i.e. the size of octahedrons gives the fit for arrangement of N, and tetrahedrons – for H (D). It is convenient to describe the structure as a set of octahedrons and tetrahedrons similarly to the way used in [11]. Based on the atom coordinates determined by neutron diffraction for the studied solid solution, the regular system of points was constructed using multiplication formulas for sp. gr. $P\bar{3}m1$ [10].

For each polyhedron the distances from the central atom (N, H and D) up to the tops (atoms Ti) and also the polyhedron edges (distances Ti–Ti) were calculated. The distance $r_{1\text{-}2}$ between atoms 1 and 2 is defined by the formula for hexagonal system:

$$r_{1-2}^2 = (x_2 - x_1)^2 + (y_2 - y_1)^2 + (z_2 - z_1)^2 - (x_2 - x_{x1})(y_2 - y_1),$$

where x_n, y_n and z_n are coordinates of corresponding atoms. Calculations became simpler due to that atoms Ti, H and D are in the particular positions with multiplicity 2 - on the threefold axes and so have coordinates: (a/3, 2a/3, zc) and (2a/3, a/3, -zc), and N atom is in the symmetry center, i.e. it has multiplicity 1. The results are shown in the Table 2. For comparison the corresponding distances in the structure of α'-$TiN_{0.40}H_{0.21}$ [11] are also given.

N – octahedron. Table 2 shows that the sizes of N-octahedron corresponds to the crystal-chemical criteria: $r_{N\text{-}Ti} \approx 2.1$ Å, $r_{Ti\text{-}Ti} \approx 3$ Å. Since an threefold inversion axis passes through the center of octahedron, the pair of faces, which are the regular triangles ($r_{Ti\text{-}Ti} = a$), is placed in parallel with basic plane (001) (Fig. 3) and, thus, joining by the common faces, octahedrons form a layer of height $2z_{Ti}*c$ parallel to (001). This is the layer, in octahedral pores of which the nitrogen atoms are arranged. The second layer of octahedrons, adjacent on height, is empty, its height is equal to $(1\text{-}2z_{Ti})*c$; the difference is insignificant: height of the filled layer is 0.4968c (~2.42 Å) and of the empty one –0.5032c (~2.46 Å). In the interspaces between empty octahedrons of the unfilled layer the tetrahedrons are placed in which deuterium and hydrogen atoms are located. By virtue of the well-known effect of H-H blocking [12], D and H atoms can not array simultaneously in the same tetrahedron as it formally follows from Table 2. Therefore D and H tetrahedrons are considered as independent units.

TABLE 2. Interatomic distances r(Å) in the solid solutions

r	$TiN_{0.40}H_{0.19}D_{0.19}$	α′-$TiN_{0.40}H_{0.21}$ [11]
	N – octahedron	
N – Ti	2,120	2,098 (N_1)
		2,136 (N_2)
$Ti_1 - Ti_1'$	3,014	3,012
$Ti_1 - Ti_2$	2,983	2,921 (N_1)
		3,030 (N_2)
	D – tetrahedron	
$Ti_1 - D$	1,629	
$Ti_2 - D$	1,926	
$Ti_2 - Ti_2'$	3,014	
$Ti_1 - Ti_2$	3,008	
	H – tetrahedron	
$Ti_1' - H$	2,214	
$Ti_1 - H$	2,663	1,850
$Ti_2 - H$	1,753	1,850
$Ti_2 - Ti_2'$	3,014	3,012
$Ti_1 - Ti_2$	2,983	3,030

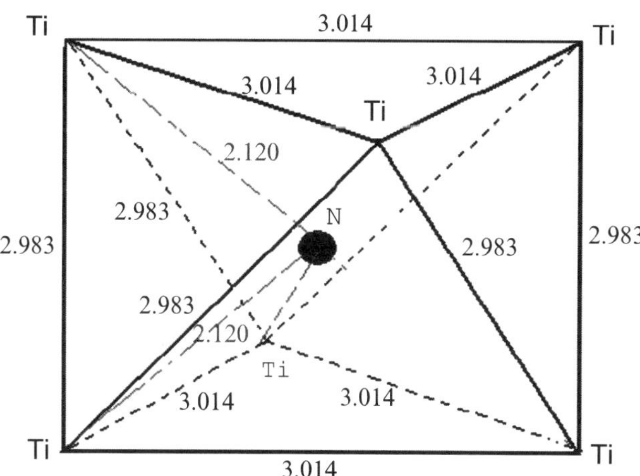

Figure 3. N – octahedron. The main interatomic distances are pointed out

<u>D – tetrahedron</u>. In accordance with radii, the distance r(D – Ti) ≈ 2 Å. So it should be noted the fact (Table 2) that one of the distances up to corner (Ti) is shortened to 1.629 Å (for comparison: the distance Ti–H(D) in titanium dihydride is 1.93 Å; in saturated hydrides of Ti intermetallic compounds – 1.87 ÷ 1.99 Å; in solid solution of hydrogen in titanium – 1.803 Å), thereby the deuterium atom is displaced from Ti – "bottom" of a tetrahedron towards a top (Fig. 4), i.e. towards the layer of N-octahedrons. Such displacement, perhaps, is caused by interaction forces between interstitial atoms D and N, at least in the first coordination sphere.

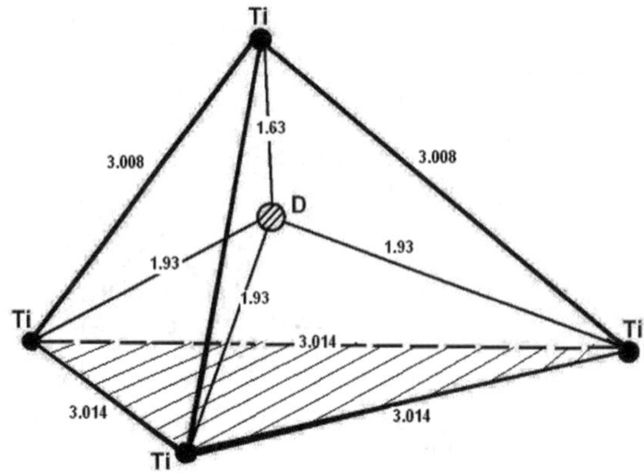

Figure 4. D – tetrahedron; the base parallel to (001) is hatched

H – polyhedron. With H – polyhedron (tetrahedron?) the situation is more intricate. First of all, if D-tetrahedrons are located in a layer of empty N-octahedrons, H-"tetrahedrons" are in the occupied layer (formally). At the theoretical value r(H-Ti) ≈ 1.9 Å, it should be noticed the very big difference in the sizes of distances H–Ti up to various Ti atoms (Table 2, Fig. 5). One can see that distances from H to atoms Ti_2 which form the base of a tetrahedron (i.e. parallel to (001)), are identical and are much shorter than distances up to apical atoms both of "own" and adjacent tetrahedrons (along axis c) having the same base. At that, the hydrogen atom is at a very small distance from this common base – 0.0429c (~0.21Å).

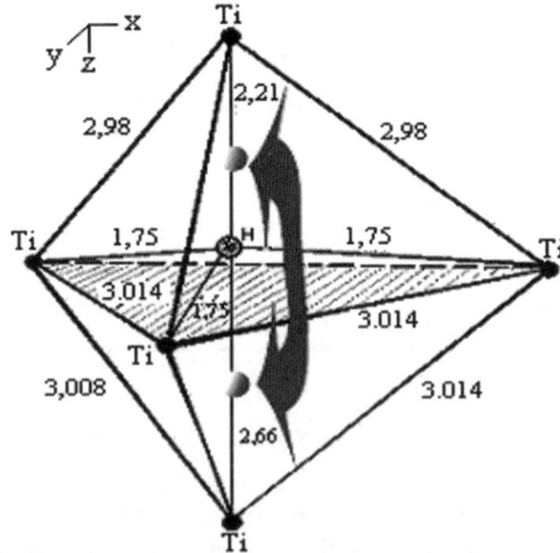

Figure 5. Coordination of H atom; the common base of two adjacent pyramid is hatched. Arrow and light balls symbolically indicate action of resonance mechanism

The values of H–Ti distances in $TiN_{0.40}H_{0.19}D_{0.19}$ allow one to conclude that coordination of hydrogen atom changes: in the first coordination sphere there are three Ti atoms at the distances about 1.7 Å; at the second sphere there are two Ti atoms (in apical position) at the distance 2.2 ÷ 2.7 Å. At that, it should be noted that the distance from H to Ti – top in the "own" tetrahedron is more than in adjacent one (2,663 against

2,214 Å). It leads to conclusion that coordination of hydrogen in this solid solution should be regarded as threefold (triangular). In crystal chemistry such coordination is frequently observed for atoms with small sizes; for example, it is usual for the boron atoms [13]. In hydride systems it is found for HoD_3, YD_3 and Th_4D_{15} [14].

If to include into the polyhedron the atoms from the second coordination sphere then one can see the united base and two tops laying at the opposite sides from it, so, it will be found unusual for such structures polyhedron-triangular bipyramid in the middle of which H-atom is placed (Fig. 5). Such coordination has been occurred only in solid solutions with combined isotope composition of hydrogen. But taking into account energy aspect, it seems of low-probability that at the large quantity of empty polyhedrons (amount of tetrahedrons is twice as large as amount of octahedrons) and at relatively little concentration of H atoms, they prefer to arrange not in empty tetrahedrons (and octahedrons) but in small triangular pores (if, certainly, to exclude occurrence of "chemism", i.e. chemical bonds of TiH_2 type). It seems more probably that the peculiarities of structure analysis methods have an important influence upon the results. Diffraction spectrum is some statistic information averaged over all irradiated volume. Therefore it may assume that actually the small atoms of hydrogen can both statistically arrange over usual tetrahedrons and also migrate from one pore to another, due to the well-known resonance (swing) mechanism of oscillation [13]. Note that a pair of tetrahedrons having common base in HCP lattice of $TiN_{0.40}H_{0.19}D_{0.19}$ can not be occupied by H atoms simultaneously since their similar arrangement leads to inadmissibly short distance H–H (from viewpoint of blocking effect). As a result of neutron diffraction experiment, it is given out the coordinates of some middle position of H-atoms, i.e. an approximate center of triangular bipyramid (or nearby center of triangular base), that we observe in practice. Unsymmetry of the mentioned middle position relative to the common base of tetrahedrons may be explained by non-equivalence of real hydrogen positions "upper" and "lower" tetrahedrons and also by various residence time of hydrogen atom in these tetrahedrons.

So, in the solid solution $TiN_{0.40}H_{0.19}D_{0.19}$ the third interstitial element (N) occupies octa-pores of one of two parallel layers but octa-pores of the second layer are empty. Atoms H and D (concentration of which is insignificant) are placed not in empty octahedrons but in tetrahedrons. At that, deuterium occupies tetrahedrons located in the layer of empty octahedrons being displaced from the geometrical centre of a tetrahedron toward the N-layer. Hydrogen, most likely, migrates over adjacent tetra-pores both of filled and empty layer of octahedrons.

Therefore, as distinct from the alloy CeD_2H [1] where isotopes arrange over interstices of different types: in octa- and tetra-pores with excess of light isotope in octa-pores, in the studied solid solution the isotopic splitting of the same type interstices takes place. Apparently, it is possible to explain this difference by that in the solid solution $TiN_{0.40}H_{0.19}D_{0.19}$ the interaction between interstitial atoms (N, H and D) is noticeable (as against from CeD_2H) and deformation interaction between interstitial atoms prevails rather than electronic, that leads to an arrangement of isotopes H and D at different distances from other atoms. At that, in the ordered "layered" structure of the solid solution $TiN_{0.40}H_{0.19}D_{0.19}$ there can not be two different isotopes of the same element on one plane parallel to (001).

4. Conclusions

The crystal structure of the solid solution with the combined isotope composition $TiN_{0.40}H_{0.19}D_{0.19}$ has been studied by neutron diffraction. This compound is isostructural with the ordered hexagonal α-phase of solid solutions of Ti-N-H (D) system [4, 8, 9]: the structure is described within framework of sp. gr. $P\bar{3}m1$ where the nitrogen atoms are located in octahedral positions 1 (a) and hydrogen (deuterium) atoms – in tetrahedral positions 2 (d). However in distribution of isotopes of hydrogen H and D isotopic effect is found: atoms H and D are arranged over tetrahedral interstices of the same type (2d) but with different coordinates z, i.e. the splitting of system of tetrahedral positions (2d) into hydrogen (H) and deuterium (D) is observed. Apparently, it is possible to explain this peculiarity by that in the studied solid solution the interaction between interstitial atoms plays the marked role, the interaction being of deformation character rather than electronic one. The presented data and results of the work [3] say that in a number of metal – hydrogen systems hydrogen and deuterium should be considered as independent chemical components.

Acknowledgements

The work was carried out at financial support of the Ukrainian Scientific - Technological Centre (project N Uzb-131 (j)) and Russian Fund of support of fundamental researches (project N 05-08-33589a).

References

1. Somenkov V.A., Entin I.P., Kost M.E., et al. Isotopic ordering in CeD_2H, *Fizika tverdogo tela*, 1975. **17**(8): 2368–2372 (in Russian).
2. Parshin P.P., Zemlyanov M.G. Isotopic effect for coordination of hydrogen atoms in metal hydrides, *Fizika metallov I metallovedenie*, 1981. **23**(7): 2136–2138 (in Russian).
3. Shilov A.L., Padurets L.N., Dobrokhotova Zh.V., et al. The mechanism of order-disorder transition in $HfV_2H_{4-x}D_x$, *Russian Journal of Inorganic Chemistry*, 1997. **42**(4): 459–463 (in Russian).
4. Khidirov I., Karimov I., Em V.T., et al. Neutron structure study of phase transition of order-disorder type in titanium nitridehydrides, *Izvestiya of AS of USSR. Inorganic Materials*, 1981. **17**(8): 1416–1420 (in Russian).
5. Kost M.E., Mikheeva V.I. Hydrides of transition metals, *Russian Journal of Inorganic Chemistry*, 1977. **22**(11): 2910–2924 (in Russian).
6. Shoyusupov, Sh., Getmanskiy, V.V., Khidirov, I., et al. Neutron diffractometer conjugated with computer IBM-PC, *Uzbek Journal of Problem of Energetics and Automatics*, 2002. **2**: 11–16 (in Russian).
7. McCusker L.B., Von Dreele R.B., Cox D.E., Louer D., Scardi P. Rietveld refinement guidelines, *Journal of Applied Crystallography*, 1999. **32**: 36–50.
8. Khidirov I. Processes of ordering and decay in the solid solution of nitrogen and hydrogen in α – titanium, *Russian Journal of Inorganic Chemistry*, 2001. **46**(3): 499–505 (in Russian).
9. Khidirov I., Padurets L.N. Ordering in nitrogen and deuterium solid solutions in α –Ti, *Russian Journal of Inorganic Chemistry*, 2001. **46**(9): 1561–1566 (in Russian).
10. Bokiy G.B. Kristallokhimiya. Moscow: Nauka. 1971. 400 p. (in Russian).
11. Belov N.B. Sketches on structural mineralogy. Moscow: Nedra. 1976. 344 p. (in Russian).
12. Khidirov I., Kurbonov I.I., Mukhtarova N.N. The crystal structure of new phase in the system Ti-N-H, *Kristallografia*, 1996. **41**(3): 450–456 (in Russian).
13. Shilov A.L., Padurets L.N. Volumetric effects and chemical bonding in hydrides of transition metals and their alloys, *Russian Journal of Inorganic Chemistry*, 2004. **49**(1): S45–S56 (in Russian).
14. Yartys' V.A., Burnasheva V.V., Semenenko K.N. Structure chemistry of hydrides of intermetallic compound, *Uspekhi khimii*, 1983. **52**(4): 529–562 (in Russian).

THE OBTAINING OF INTERMETALLIC COMPOUNDS FROM HYDRIDES OF TITANIUM AND ZIRCONIUM WITH NICKEL AND COBALT

A.G. ALEKSANYAN, H.G. HAKOBYAN, S.K. DOLUKHANYAN, N.L. MNATSAKANYAN, K.A. ABRAHAMYAN
A.B. Nalbandyan Institute of Chemical Physics of Armenian NAS, 5/2 P.Sevak Str., Yerevan 0014, Republic of Armenia

Abstract. The chemical and phase-structural transformations were researched upon obtaining intermetallic compounds of metals of IV and VIII groups using hydrides of titanium and zirconium. The possibility of alloys and IMC obtaining was demonstrated. The ability of obtained compact alloys and IMC to hydrogenation in combustion regime (SHS) forming the IMC hydrides was established.

Keywords: intermetallic compounds, alloys, hydrides

1. Introduction

Among functional materials of new generation, intermetallic compounds (IMC) are of valuable interest, because they are used as strengthening phases in industrial alloys of iron, nickel, titanium, etc., superconductors, materials with form memory, biologically and mechanically compatible implants, etc. A series of intermetallic compounds based on metals of fourth and eighth groups are known as basic for creation of hydrogen accumulators due to their high hydrogen-sorption properties. The quest and development of new ways for synthesis of these materials and their hydrides remains to be an essential technological task.

Earlier, the intermetallics and their hydrides had been synthesized in the Laboratory of high-temperature synthesis of IChPh of Armenian NAS, using the method of self-propagating high-temperature synthesis (SHS) [1–4, 7]. Recently, we elaborated a new method for obtaining the alloys from mixture of two or more hydrides, or of the hydride and metal, by compacting these mixtures and subsequent dehydrogenation [5].

Using this method, a series of alloys based on the metals of fourth group has been received. The method appeared also effective for producing intermetallics.

2. Experimental procedures

For conducting the experiments, the powder of fourth group metals were used synthesized by the SHS method: titanium hydride, TiH_2 (the hydrogen content 3.9–4.0 wt%), zirconium hydride, ZrH_2 (the hydrogen content 2.17 wt%), and powders of nickel (99.6 wt%), and cobalt (99.33 wt%). Dispersion of these powders was lower of 63 μm. Initial charges MeH_2 + Ni(Co) were prepared considering to obtain the alloys and intermetallics of different composition. Then, the charges were pressed and compacted to tablets of diameter 20 cm. Dehydrogenation/annealing was curried out in hermetic quartz reactor at temperatures 900–1,100°C.

The samples obtained after dehydrogenation were subjected to chemical, X-ray and DTA analyses. The chemical analysis of hydrogen content was carried out by pyrolysis method at temperatures 350–1,000°C. X-ray analysis of samples was performed on diffractometer DRON-0.5 using CuK_α radiation. The identification of crystal phases was performed using the ASTM and JCPDS card indexes. The thermal analysis was performed on Derivatograph Q-1500 (Hungary) at heating rate 20 degree/min.

3. Results and discussion

The total scheme of process for IMC and alloys receiving is:

$$MeH_2 + Ni(Co) \rightarrow MeNi(Co) + H_2 \uparrow.$$

The influences of pressing pressure, dispersion of initial reagents, dehydrogenating/ annealing temperature on the process of intermetallics formation (completeness and deepness of reaction, etc.) and on the mechanical properties of obtained IMC intermetallics (density of alloy, hardness, porosity, phase composition, etc.) were studied.

As X-ray analysis data shown, the compacting pressure did not essentially influence on the phase composition of final product.

Experiments, carried out for different compositions have shown that in the wide interval of ratio of initial mixture MeH_2/Ni(Co) the compacting and dehydrogenation/annealing procedures result in formation of phases corresponding to the phases of state diagram of the respective systems – Ti-Co, Ti-Ni, Zr-Co, Zr-Ni. For example, in system ZrH_2-Ni, intermetallics Ni_5Zr, NiZr, $NiZr_2$ were received, as well as the alloys including two and more phases.

TABLE 1.Characteristics of obtained IMC

Initial charge	IMC	Crystal structure and lattice parameters (Å)	Density (g/cm^3)	Hardness (HRA)
TiH_2 + Co	Ti_2Co	Cubic, a = 11.286	4.87	60.2 ± 1.4
TiH_2 + Ni	TiNi	Cubic, a = 2.996	4.79	41.8 ± 1.5
ZrH_2 + Ni	NiZr	Orthorhombic, a = 3.28; b = 10.042; c = 4.039	5.85	48.7 ± 1.5
ZrH_2 + Ni	$NiZr_2$	Tetragonal a = 6.425; c = 5.299; Cubic, a = 12.18	5.96	58.6 ± 1.5
ZrH_2 +7Ni	Ni_7Zr_2, Ni_7Zr_2, Ni	Orthorhombic, cubic, cubic	–	–
2ZrH_2 + Co	Zr_2Co	Tetragonal, a = 6.366; c = 5.477	4.87	58.2 ± 1.5

In Table 1, the characteristics of some alloys and intermetallics are presented. It is necessary to note that in system $2ZrH_2$–Ni, the formation of two-phase product is observed: a main (60–65%) phase of $NiZr_2$ with tetragonal crystal lattice and parameters a = 6.425Å and c = 5.299Å. The other reflections were identified as another modification of the same intermetallic compound $NiZr_2$ (Fig. 1). According to the card index JCPDS (41-0898), the second modification was cubic phase $NiZr_2$. The authors of work [6] prepared this cubic modification with parameter a = 12.27Å by rapid quenching from the melt spinning onto Cu-wheel at speed 60 m/sec. In our experiments, the second phase had a cubic structure with the parameter a = 12.18 Å.

For composition $2ZrH_2$-Co, the experiments were carried out with breaking the dehydrogenation/annealing process every 100 degrees. The changes in phase composition are presented in Fig. 2. It is necessary to note, that at compacting of initial samples already it is observed primary recrystallization of cobalt lattices. On Fig. 2 at 400°C the maximal peaks of two modifications of cobalt – hexagonal and cubic are marked.

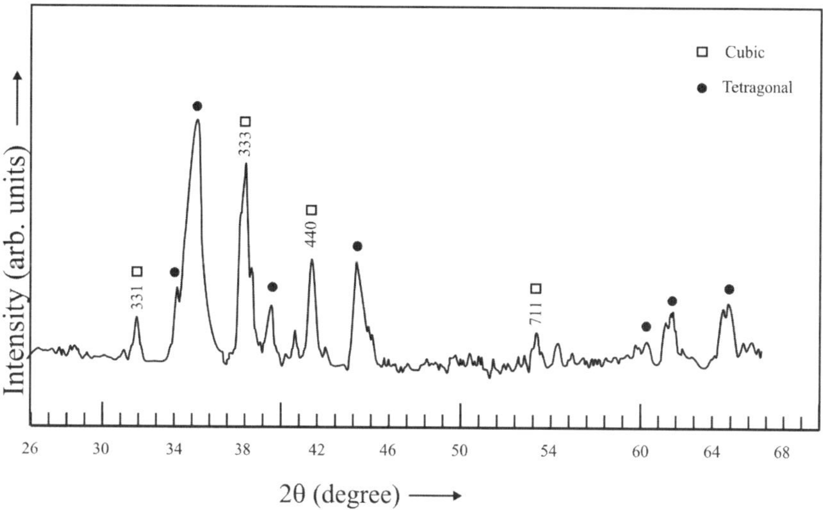

Figure 1. X-ray pattern of the products of system $2ZrH_2$–Ni

The obtained compact IMC samples kept their forms after dehydrogenation/ annealing. Some of titanium and zirconium based IMC had porous structure.

It was established, that the obtained samples of itermetallics without grinding interacted with hydrogen at pressure 5–10 atm and with deuterium at 30 atm in combustion regime (self-propagating high-temperature , SHS). This interaction resulted in formation of corresponding hydrides and deuterides of IMC (Table 2). Apparently, the process of dehydrogenation of initial mixture brings to formation of compact and strong intermetallic. At the same time, the obtained sample is enough porous promoting the hydrogen atoms penetration in the combustion front.

DTA data, the process of thermal decomposition of synthesized hydrides was single-stage, occurring in the interval 220–300°C.

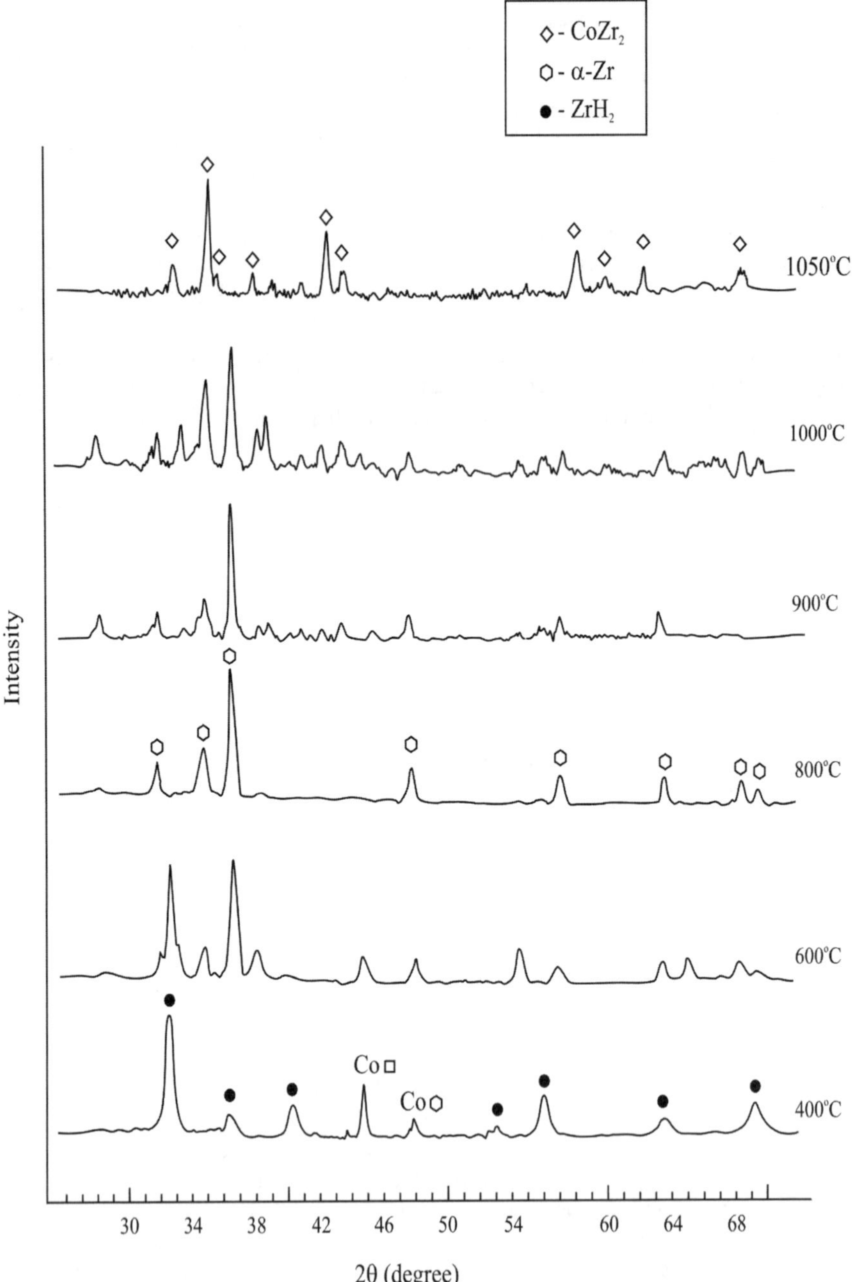

Figure 2. The changes in the phase composition in system $2ZrH_2$-Co at different temperatures in the process of intermetallic formation

TABLE 2. Characteristics of combustion process and of obtained IMC hydrides and deuterides

System	T_c(°C)	Hydrogen concent. (mass%)	Formula, crystal struture, parameters (Å)
$Ti_2Co + H_2$	425–440	1.96–2.03	Ti2CoH3, cubic, $a = 11.802$
$ZrNi + H_2$	340–350	1.83–1.93	$ZrNiH_3$-x, orthorhombic $a = 3.536;\ b = 10.43;\ c = 4.287$
$ZrNi + D_2$	340–350	3.32	$ZrNiD_{2.}$, orthorhombic $a = 3.536;\ b = 10.43;\ c = 4.287$
$Zr_2Ni + H_2$	430–440	1.83–1.91	$ZrNi_2H_{4.7}$, tetragonal $a = 6.850;\ c = 5657$
$Zr_2Ni + D_2$	430–440	3.84	$ZrNi_2D_{4.15}$, tetragonal $a = 6.850;\ c = 5657$
$ZrCo + H_2$	350–450	1.67–1.72	$ZrCoH_{3-x}$, orthorhombic $a = 3{,}37;\ b = 10.572;\ c = 4.318$
$Zr_2Co + H_2$	350–400	1.60–2.01	Zr_2CoH_5-x, tetragonal $a = 6.906;\ c = 5.550$

4. Conclusions

The possibility of the intermetallic compounds producing by compacting the mixture of binary hydrides of titanium and zirconium and metals of eighth group followed by dehydrogenation/annealing of initial charge was demonstrated.

Compared to the known methods of powder metallurgy for obtaining intermetallics (induction/arc smelting and combustion synthesis [7]), the elaborated method allows IMC receiving at temperatures not exceeding 1,050°C, with formation of strong samples of given shape and composition, corresponding to the phase state diagram.

It was demonstrated that the obtained compact IMC can interact with hydrogen in combustion regime forming the corresponding hydride.

Acknowledgements

The work is implemented at financial support of ISTC Grant A-1249 and of Ministry of Education and Science of Republic of Armenia, # 0567

References

1. Dolukhanyan S.K., Hakobian H.G. and Alexanian A.G. Combustion of metals in hydrogen and hydride production by self propagating high temperature synthesis. *Int. J. SHS*, 1992, **1**(4): 530–535.
2. Dolukhanyan S.K. Synthesis of novel compounds by hydrogen combustion. *J. Alloy. Comp.* 1997, **253–254**: 10–12.
3. Dolukhanyan S.K., Karimyan R.A., Akopyan A.G., Merzhanov A.G. Self-propagating high-temperature synthesis of intermetallides' hydrides on Ti-Co system base. *Zh. Neorg. Khim.*, 1983, **28**(5): 1101–1105 (in Russian).
4. Akopyan A.G., Dolukhanyan S.K., Karapetyan A.K., Merzhanov A.G. Zr-Ni system intermetallides hydrides synthesis in combustion regime. *Izv. Akad. Nauk SSSR., Neorg. Mater.*, 1983, **19**(6): 881–885 (in Russian).

5. Dolukhanyan S.K., Shekhtman V.SH., Aleksanyan A.G., Ter-Galstyan O.P., Abrosimova G.E., Sakharov M.K. Research of features of formation of structures of alloys in system Ti-Zr-H. *Khimicheskaya fizika*, 2007, **26**(11): 36–43 (in Russian).
6. Altounian Z. et al. *J. Appl. Phys*., 1987, **61**, 149.
7. Naiborodenko Yu.S., Itin V.I., Merzhanov A.G., Borovinskaya I.P., Ushakov V.P., Maslov V.P. Gasless combustion of metals mixture and intermetallides self-propagating high-temperature synthesis. *Izv. Vissh. Uchebn. Zaved.*, Fiz., 1973(6): 142–146 (in Russian).

HYDROGEN ABSORPTION PROPERTIES $ZrFe_2$ AND $ZrCo_2$ BASED ALLOYS

T.A. ZOTOV, R.B. SIVOV, S.V. MITROKHIN, E.A. MOVLAEV, V.N. VERBETSKY
Moscow State University, Department of Chemistry, Leninskie Gory 3, Moscow, 119992 Russia
Fax: (095) 932 88 46; E-mail: zotov@hydride.chem.msu.ru

Abstract. In this work $Zr_{1-x}Ti_x(Fe_{1-y}Al_y)_2$ system and the alloys $Zr_{0.8}Ti_{0.2}Co_{1.6}Al_{0.4}$ and $Zr_{0.9}Gd_{0.1}Co_2$ were studied with the purpose to investigate the influence of zirconium substitution by titanium or a rare earth metals on hydrogen sorption properties of IMC $ZrFe_2$ and $ZrCo_2$. The hydrogen absorption properties were studied by measuring PCT absorption and desorption isotherms using a high hydrogen pressure apparatus at a hydrogen pressure below 3,000 atm. Hydrogen desorption pressures and hydrogen storage capacity decrease with increasing aluminium content in $Zr_{1-x}Ti_x(Fe_{1-y}Al_y)_2$ alloys. The catalytic effect of gadolinium addition on acceleration of reaction of $ZrCo_2$ with hydrogen was established. The thermodynamic parameters of hydrogen desorption reactions were calculated for number alloys.

Keywords: hydrogen, intermetallic compounds, IMC hydrides, high pressure

1. Introduction

Recently high pressure hydrides were only of scientific interest. Now hydrogen storage and transportation at high pressure attracts attention of auto and energy companies. This makes the high pressure hydrides perspective materials for practical applications.

High pressure hydrides can also be used in metal hydride compressors, where hydrogen pressure raising is realized only by the temperature and alloy composition.

$ZrFe_2$ and $ZrCo_2$ with Laves phases structure were considered as hydride non-forming IMC. The reaction with hydrogen was realized for this compounds [1, 2] with the application of super high pressure of 10,000 atm. It was estimated that reaction with hydrogen at 100^0C starts at 1.1 GPa (10,856 atm). Equilibrium dissociation pressure is 0.35 GPa (3,454 atm) at 100^0C for $ZrFe_2$ hydride. Lattice expansion is 25.93% for $ZrFe_2$ hydride and 14.59% for $ZrCo_2$ hydride. Hydrogen content was calculated on base of lattice expansion rule to be 2.85 E^3/H. Hydride compositions are $ZrFe_2H_4$ and $ZrCo_2H_2$.

However, our investigation showed that IMC $ZrFe_2$ starts to react with hydrogen at room temperature and pressure near 800 atm. Hydride composition at hydrogen pressure of 1,900 atm is $ZrFe_2H_{3.5}$. Equilibrium absorption pressure is 690 atm and desorption pressure is 325 atm at room temperature [3].

The influence of aluminum substitution for cobalt in $Zr(Co_{1-x}Al_x)_2$ ($0.1 \leq x \leq 0.5$) alloys [4] and for iron in $Zr(Fe_{1-x}Al_x)_2$ alloys [5] shows that aluminium substitution results in reduction of equilibrium dissociation pressure, increase of desorption enthalpy and significant reduction of hydrogen content.

Influence of zirconium substitution by titanium or a rare earth metals on hydrogen sorption properties of IMC $ZrFe_2$ and $ZrCo_2$ was not described in literature. In this work $Zr_{1-x}Ti_x(Fe_{1-y}Al_y)_2$ system and the alloys $Zr_{0.8}Ti_{0.2}Co_{1.6}Al_{0.4}$ and $Zr_{0.9}Gd_{0.1}Co_2$ were studied with this purpose.

2. Experimental

Alloys were prepared by arc-melting the mixtures of pure initial metals under argon atmosphere. With the purpose of homogenization the alloys were remelted four times. The structure and composition of alloys was examined by a scanning electron microscope (SEM) with energy dispersive X-ray analyzer and powder X-ray diffraction. The refinement of diffraction profiles was performed using the Rietveld method.

The hydrogen absorption properties were studied by measuring PCT absorption and desorption isotherms using a high hydrogen pressure apparatus at a hydrogen pressure below 3,000 atm.

Calculation of amount of absorbed or desorbed hydrogen were done using a modified van-der-Waals equation (1),

$$[p+a(p)/V^{\alpha}][V-b(p)]=RT\,, \tag{1}$$

where a(p), b(p) and α(T) parameters were determined by analysis of large massive experimental data of hydrogen compressibility [6].
The thermodynamic characteristics (ΔH and ΔS) IMC hydrides were calculated from vant-Hoff equation for dependence of *lnf* on *1/T*:

$$RT\ln(f_p)=\Delta_r H - T\Delta_r S\,. \tag{2}$$

The values of fugacities were determined with equation (3), using the real molar hydrogen volumes, obtained from equation (1):

$$RT\ln(f_p)=RT\ln p-\int_0^p (V_{id}-V_{real})dp\,. \tag{3}$$

The hydriding reaction products were investigated by X-ray. At first the autoclave with hydrogen saturated sample was cooled to liquid nitrogen temperature. Then hydrogen pressure in autoclave was reduced to 1–2 atm. After that the sample holder was opened to air. In order to inhibit hydrogen recombination centers on hydride surface, cooled hydride was exposed to air for 5–10 min. IMC hydride was replaced from sample holder to metallic container cooled to liquid nitrogen temperature. All studied samples were inflammable at these conditions. For the additional passivation the container with sample was keep in liquid nitrogen for time more then 1 h. The sample was put out from liquid nitrogen directly before X-rays analysis. The analysis time was extremely shot, but sufficient for identification of lattice periods of hydride phases. The changes of sample volume during hydrogen absorption and desorption were taken into account in calculations of PCT isotherms.

3. Results and discussion

3.1. $(Zr\text{-}Ti)(Fe\text{-}Al)_2$ SYSTEM

Three alloys $Zr_{0.5}Ti_{0.5}Fe_{2-x}Al_x$ (x = 0.1, 0.4 and 0.8) and two alloys $Zr_{0.2}Ti_{0.8}Fe_{2-x}Al_x$ (x = 0.2 and 0.3) were prepared for investigate this system. The main phase in all alloys is C14 Laves phase, excepts alloy $Zr_{0.5}Ti_{0.5}Fe_{1.9}Al_{0.1}$ with C15 phase (Table 1).

TABLE 1. The structure of $(Zr\text{-}Ti)(Fe\text{-}Al)_2$ alloys

IMC	Phase	a, c (Å)	V (Å^3)
$Zr_{0.5}Ti_{0.5}Fe_{1.9}Al_{0.1}$	C15	7.052(6)	350.7(9)
$Zr_{0.5}Ti_{0.5}Fe_{1.6}Al_{0.4}$	C14	4.967(7), 8.085(8)	172.8(1)
$Zr_{0.5}Ti_{0.5}Fe_{1.2}Al_{0.8}$	C14	5.037(8), 8.193(6)	180.0(9)
$Zr_{0.2}Ti_{0.8}Fe_{1.8}Al_{0.2}$	C14	4.85(8), 7.91(5)	161.7(5)
$Zr_{0.2}Ti_{0.8}Fe_{1.7}Al_{0.3}$	C14	4.878(3), 7.93(8)	163.5(9)

The small (>2%) addition of ZrO phase (sp. gr. 225) was present in all alloys. The iron substitution of aluminium leads to evident increase of Laves phase cell parameters.

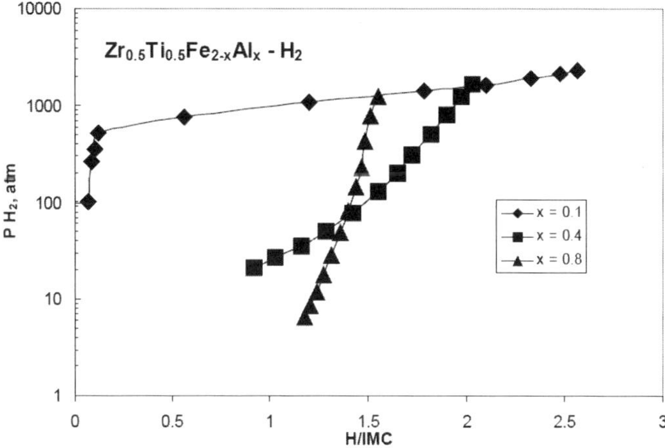

Figure 1. PCT desorption isotherms in the $Zr_{0.5}Ti_{0.5}Fe_{2-x}Al_x$–$H_2$ systems at room temperature

All $Zr_{0.5}Ti_{0.5}Fe_{2-x}Al_x$ alloys react with hydrogen. Active hydrogen absorption in first cycle starts at pressure of 1,940 atm for the sample with x = 1, but alloys with x = 0.4 and 0.8 reacted with hydrogen at pressure of 105–110 atm. The desorption isotherms at room temperature are presented at Fig. 1. Hydrogen desorption pressures and hydrogen storage capacity decrease with increasing aluminium content (Table 2). Although we couldn't detect the equilibrium desorption pressures of the $Zr_{0.5}Ti_{0.5}Fe_{1.2}Al_{0.8}$ hydride to calculate the thermodynamic parameters, we suppose that it's desorption enthalpy are maximal in the $Zr_{0.5}Ti_{0.5}Fe_{2-x}Al_x$ system. Also we can see that the hysteresis effect is disappearing with increasing aluminium concentration (Table 2). The hydrogen sorption activity also increases: the equilibrium absorption and desorption pressures in first cycle approach to pressures values at following cycles.

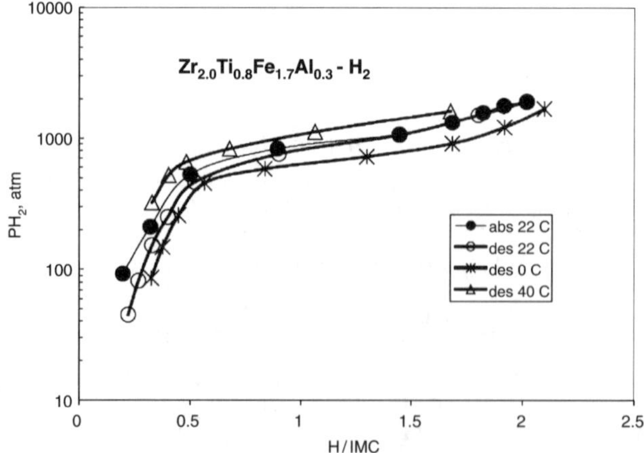

Figure 2. The interactions in $Zr_{0.2}Ti_{0.8}Fe_{1.7}Al_{0.3}$–$H_2$ system

The $Zr_{0.2}Ti_{0.8}Fe_{1.8}Al_{0.2}$ alloy doesn't react with hydrogen at pressure up to 2,300 atm. Whereas $Zr_{0.2}Ti_{0.8}Fe_{1.7}Al_{0.3}$ alloy begin to absorb hydrogen at pressure of 1,200 atm. This sample reach it's maximum hydrogen capacity only after second absorption–desorption cycle. The alloy $Zr_{0.2}Ti_{0.8}Fe_{1.7}Al_{0.3}$ is charactirised by negligible hysteresis and very small hydrogen desorption enthalpy (Fig. 2, Table 2).

TABLE 2. Hydrogen sorption properties of the $Zr_{0.5}Ti_{0.5}Fe_{2-x}Al_x$ alloys

IMC	P H2, atm 1abs/1des	P H2, atm 3abs/3des	H/IMC (mass%)	ΔH (kJ/mol H_2)/ΔS (J/K mol H_2)
$Zr_{0.5}Ti_{0.5}Fe_{1.9}Al_{0.1}$	~1,300/890	1,190/970	2.8_{2600}, 1.56	16.0 ± 1.7/117 ± 6
$Zr_{0.5}Ti_{0.5}Fe_{1.6}Al_{0.4}$	35/34	35/34	2.05_{1800}, 1.2	24.6 ± 1.5/112 ± 5
$Zr_{0.5}Ti_{0.5}Fe_{1.2}Al_{0.8}$	~1.5/~1	~1.5/~1	1.55_{1220}, 0.98	–
$Zr_{0.2}Ti_{0.8}Fe_{1.8}Al_{0.2}$	–	–	–	–
$Zr_{0.2}Ti_{0.8}Fe_{1.7}Al_{0.3}$	960/841	950/910	2.02_{1900}, 1.26	14.6 ± 1.9/111 ± 5

3.2. THE $ZrCo_2$-BASED ALLOYS

It was shown during preliminary investigations that IMC $ZrCo_2$ doesn't absorb hydrogen at pressures below 2,500 atm. It means that hydrogen absorption and desorption pressures of the $ZrCo_2$–H_2 system are the highest among describing in this paper IMC. Therefore the particle rare earth metal substitution of zirconium was applied for the expanding the lattice and increase hydrogen uptake ability. The gadolinium was taken for this purpose. This substitution is probable because the gadolinium solubility in zirconium is significant among other rare earth metals (~10% at 1,000^{0}C) and besides the gadolinium forms Laves phase with cobalt. Also the $Zr_{0.8}Ti_{0.2}Co_{1.6}Al_{0.4}$ alloy was prepared and studied.

The investigation SEM with EDX and XRD analyses shows that $Zr_{0.9}Gd_{0.1}Co_2$ consists from three phases: the main C15-type Laves phase, addition $Gd_{12}Co_{9.2}$ phase (sp. gr. 176) and small amount of zirconium oxide phase. By means of EDX the gadolinium concentration in Laves phase A-component is near 3–4 at.%.

The alloy $Zr_{0.8}Ti_{0.2}Co_{1.6}Al_{0.4}$ is crystalline in C14 and C15 Laves phase structures with small amount of zirconium oxide phase (Table 3).

TABLE 3. The structure of $ZrCo_2$-based alloys

IMC	Phases	a, c (Å)	V (Å^3)
$Zr_{0.9}Gd_{0.1}Co_2$	C15 97%	6.963(3)	337.6(3)
	$Gd_{12}Co_{9.2}$ 3%	11.2(3), 4.06(1)	44(4).0
$Zr_{0.8}Ti_{0.2}Co_{1.6}Al_{0.4}$	C14 54%	4.993(3), 8.043(5)	173.6(8)
	C15 46%	7.012(2)	344.7(9)

Both alloys interact with hydrogen at pressures below 100 atm. The $Zr_{0.9}Gd_{0.1}Co_2$–H_2 system has great hysteresis effect (Fig. 3): it's equilibrium absorption pressure at room temperature is 970 atm and desorption pressure is 230 atm, but in the $Zr_{0.8}Ti_{0.2}Co_{1.6}Al_{0.4}$-$H_2$ system hysteresis is insignificant.

TABLE 4. Hydrogen sorption properties of $ZrCo_2$-based alloys

IMC	P H2, atm 1abs/1des	P H2, atm 3abs/3des	H/IMC (mass%)	ΔH (kJ/mol H_2)/ΔS (J/K mol H_2)
$Zr_{0.9}Gd_{0.1}Co_2$	~1,050/205	970/230	2.22_{1990}, 1.03	23.4 ± 0.7/125 ± 2
$Zr_{0.8}Ti_{0.2}Co_{1.6}Al_{0.4}$	–	–/7.4	2.14_{2000}, 1.14	–

The appearance of wide alpha-region in PCT curves of the $Zr_{0.9}Gd_{0.1}Co_2$–H_2 (Fig. 3) can be explained by the presence of stable hydride forming addition phase. Furthermore hydrogen equilibrium absorption and desorption pressures changes small in following cycles at contrast of first cycle in both systems. We assume that it is catalytic effect of the rare earth metal or the aluminium presence.

There are two factors which can account for the catalytic effect of rare earth addition for the $Zr_{0.9}Gd_{0.1}Co_2$–H_2 system. At first the presence of the gadolinium atoms in A-component of Laves phase increases the hydride stability and decrease absorption and desorption hydride pressures. At second we believe that the hydrogen absorption starts at low pressure due to $Gd_{12}Co_{9.2}$ phase presence. This phase react with hydrogen in the first time with dispersion and enlarging active surface of the main phase at result.

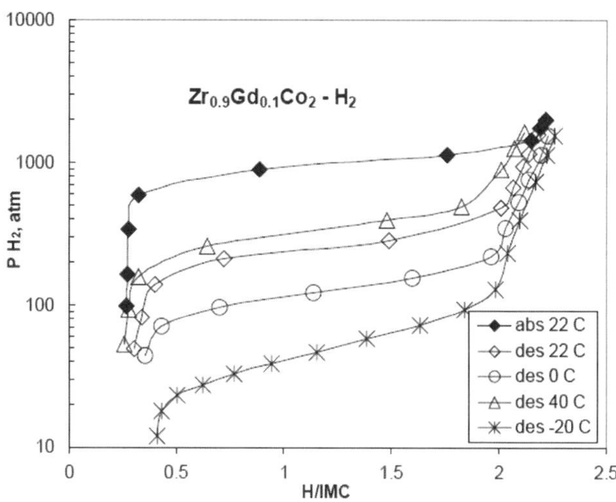

Figure 3. The interactions in $Zr_{0.9}Gd_{0.1}Co_2$–H_2 system

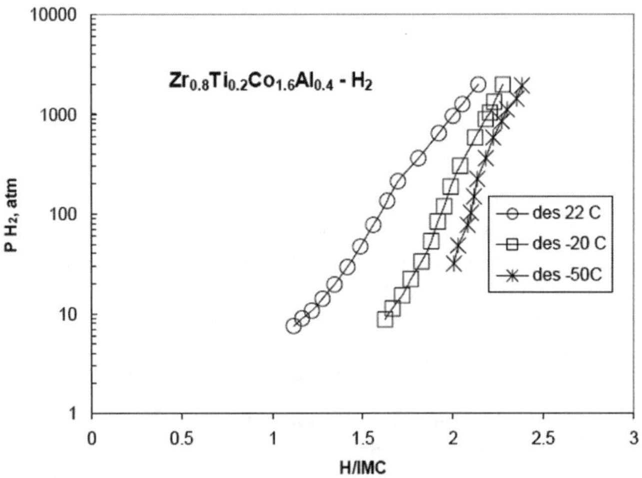

Figure 4. PCT desorption isotherms in $Zr_{0.8}Ti_{0.2}Co_{1.6}Al_{0.4}$-$H_2$ system

4. Conclusions

The influence titanium and aluminium substitution on structure and hydrogen sorption characteristics of $(Zr\text{-}Ti)(Fe\text{-}Al)_2$ system were investigated.

The catalytic effect of gadolinium addition on acceleration of reaction of $ZrCo_2$ with hydrogen was established.

The thermodynamic parameters of hydrogen desorption reactions were calculated for number alloys.

References

1. Filipek S.M., Jacob I., Paul-Boncour V., Percheron-Guegan A., Marchuk I., Mogilyanski D., Pielaszek J. *Polish J. Chem.*, 2001, **75**, **12**:1921–1926.
2. Filipek S.M., Paul-Boncour V., Percheron-Guegan A., Jacob I., Marchuk I., Dorogova M., Hirata T., Kaszkur Z. *J. Phys.-Condens. Mat.*, 2002, **14**:11261–11264.
3. Zotov T.A., Movlaev E.A., Mitrokhin S.V., Verbetsky V.N. *J. Alloy. Compd.* 2008, **459**: 200–224.
4. Bereznitsky M., Jacob I., Bloch J., Mintz M.H. *J. Alloy. Compd.*, 2002, **346**:217–221.
5. Bereznitsky M., Jacob I., Bloch J., Mintz M.H. *J. Alloy. Compd.*, 2003, **351**:180–183.
6. Hemmes H., Driessen A., Griessen R. *J. Phys. C Solid State*, 1986, **19**:3571–3585.

AB INITIO STUDY OF THE ELECTRONIC STRUCTURE AND PHONON DISPERSIONS FOR TiH_2 AND ZrH_2

V.I. IVASHCHENKO, L.A. IVASHCHENKO, P.L. SRYNSCKYY, L.A. GRISHNOVA, A.I. STEGNYY
Institute for Problems of Material Science, NAS of Ukraine, Krzhyzhanovsky Str. 3, 03142 Kyiv, Ukraine

Abstract. Pseudo-potential band-structure and phonon-structure calculations are carried out for the fluorite-type and tetragonal phases of TiH_2 and ZrH_2. Band structures, densities of states, total energy and phonon dispersion curves are computed. The total-energy curves as functions of the c/a ratio show two minima for $c/a < 1$ and $c/a > 1$. The results show that the Jahn-Teller splitting is responsible for the tetragonal distortion of the dihydrides in agreement with previous interpretations. Soft acoustic phonon modes in the long-wave range are observed in the calculated phonon dispersion curves of the cubic phases, pointing to the structural instability with respect to tetragonal distortion. Although the experimental findings show that the cubic-to-tetragonal phase transition in TiH_2 at 310 K is of the second order, the calculated characteristics clearly indicate the first-order transition that is close to the second-order one.

Keywords: ab-initio band structure, phonon spectrum, pseudo-potential method, titanium and zirconium dihydrides, phase transition

1. Introduction

TiH_2 undergoes a second-order phase transition from the face-centred cubic (FCC, Fm3m) structure to the body-centred tetragonal (BCT, I4/mmm) structure at approximately below 310 K [1–3]. There is no a single opinion on the phase stability of ZrH_2 [3]: whether the BCT ZrH_2 phase is stable up to the decomposition temperature, or ZrH_2 undergoes a first-order BCT-FCC phase transition higher 600^0C [3]. This is a common feature of all the dihydrides in the composition range $1.8 \leq x \leq 2$ for TiH_x and $1.62 \leq x \leq 2$ for ZrH_x. First the phase transition in TiH_2 was assigned to ordered filling of the tetrahedral holes [4]. Based on the results of band structure calculations, Switendick considered the FCC-BCT transition in TiH_2 as a result of the Janh-Teller splitting [5]. Given this finding, Ghatak, Ray and Tannous ascribed the second-order phase transition in TiH_2 to a splitting of the e_g density of states caused by the lattice distortion [6]. Since a phase transition for which the tetragonal strain ε_t is the only order parameter is first-order, ε_t must be coupled to an unknown order parameter [2]. In this connection, Kulikov [7] suggested that the spontaneous elastic strain ε_t is induced indirectly by the condensation of the Γ T_{2g} optic mode. Recent band structure calculations of both the FCC and BCT TiH_2 and ZrH_2 [8] confirmed previous interpretations [5] of the tetragonal distortion of the cubic unit cell. The review of the recent band structure calculations of TiH_2 and ZrH_2 was done in Ref. [6]. An analysis of the results of investigations of these phases allowed us to establish that there are no: (i) A common mechanism of the structural transformation in TiH_2 and ZrH_2 that could be able to account for most of experimental evidences; (ii) Comprehensive investigations of the phonon structure of the dihydrides; (iii) Bans structure calculations of the sub-stoichiometric cubic and tetragonal TiH_x and ZrH_x; (iv) A theoretical model of the inter-atomic interactions in the disordered dihydrides, where some octahedral holes are also occupied.

In the present work we aimed at filling these gaps in studying the structural and electronic properties of TiH_2 and ZrH_2. The electronic structures and phonon spectra of the cubic and tetragonal dihydrides were comprehensive investigated by using the first-principle plane-wave pseudo-potential method. Band structures, densities of states (DOS), total energy and, for the first time, phonon dispersion curves are presented and used to discuss the possible origins of the phase transitions and stability of TiH_x and ZrH_x.

2. Computational aspects

The electronic energy bands and phonon dispersions for the FCC and BCT TiH_2 and ZrH_2 phases were calculated using *ab initio* pseudo-potential approach [9]. The ultrasoft pseudo-potentials used in the calculations were generated according to the Vanderbilt scheme [10]. Plane waves up to a kinetic energy cut-off of 36 Ry were included in the basis set. Exchange and correlation was treated within the local density approximation (LDA) [11] and the generalized gradient approximation (GGA) [12]. For the FCC and BCT phases, Brillouin-zone (BZ) integrations have been performed using sets of special points corresponding to the (12 12 12) and (8 8 8) Monkhorst-Pack mesh, respectively. The same sets were used for calculating the phonon vibrational properties of the cubic phases using the density-functional perturbation theory described in Ref. [9]. The BZ integration was performed using the Gaussian broadening of Kohn-Sham energy levels with the parameters $\sigma = 0.01 - 0.05$ Ry. In the first approximation, the spearing function allows one to account for the effects of a finite temperature and small lattice distortions. To investigate substoichiometric structures we considered the cubic super cells of 12 atoms. This enabled us to model the substoichiometric (Ti_4H_7) and disordered ($Ti_4H_7{}^TH_1{}^O$, T-tetrahedral hole, O-octahedral hole) phases. The resulting density of states was computed using the linear tetrahedron method [13]. The bulk moduli and the equilibrium lattice parameters were obtained using the Murnaghan fit to the total energy as a function of the volume. The phonon dispersions were computed at the theoretical equilibrium geometry. The Parrinello-Rahman procedure was used for the relaxation of the atoms, while the symmetry was preserved [9]. Relaxation of the free atomic coordinates was considered accomplished, when atomic forces were less than 0.05 mRy/bohr. For the sake of comparison, we also carried out the total energy calculations based on the GGA using the (12 12 12) mesh for the BZ integration by the linear tetrahedron method.

3. Results and discussion

The calculated characteristics for both the FCC and BCT phases are summarized in Table 1. In Fig. 1 we show energy bands for TiH_2. The densities of states for TiH_2 and ZrH_2 are shown in Fig. 2. The agreement between our characteristics and the experimental ones is quite satisfactory. The previous theoretical results obtained by

Wolf and Herzig [8] and Ackland [14] also agree well with the present ones. It is seen that the LDA provides reduced lattice parameters and overestimated bulk modulii. Since in our case the GGA approach gives the more precise structural parameters, we concentrated on the GGA band structure calculations.

TABLE 1. Lattice parameters, a (Å), c/a ratios and bulk moduli, B (GPa) for the FCC and BCT phases of TiH_2 and ZrH_2. In the total-energy calculations the smearing σ = 0.01 Ry was used. The values in the parenthesis were calculated using the linear tetrahedron method for the Brillouin-zone integration

Dihydrides		GGA			LDA			Experiment[a]	
		a	c/a	B	a	c/a	B	a	c/a
	FCC	4.421	–	140	4.340	1.000	159	4.460	–
		(4.419)	–	(140)	4.399[b]	–	162[b]	$TiH_{1.95}$	
TiH_2		4.515	0.936	140	4.451	0.925	160	4.528	0.945
	BCT	4.367	1.036	140	4.283	1.040	159	$TiH_{1.99}$	
		(4.526)	(0.928)	(140)	4.513[b]	0.928[b]			
		(4.320)	(1.068)	(140)	4.303[b]	1.068[b]			
	FCC	4.807	–	131	4.804[b]	–	152[b]	4.780	–
		(4.806)	–	(132)				$ZrH_{1.55}$	
ZrH_2		4.997	0.884	132	5.008[b]	0.883[b]		4.975	0.894
		4.655	1.098	132	4.635[b]	1.113[b]		ZrH_2	
	BCT	(5.001)	(0.882)	(132)	4.860[c]	0.914[c]			
		(4.636)	(1.110)	(132)	4.610[c]	1.119[c]			
		5.000[c]	0.890[c]						
		4.640[c]	1.121[c]						

[a]The results were taken from Refs. 1–3, 8; [b]the LDA calculations [8]; [c]the LDA and GGA investigations [14].

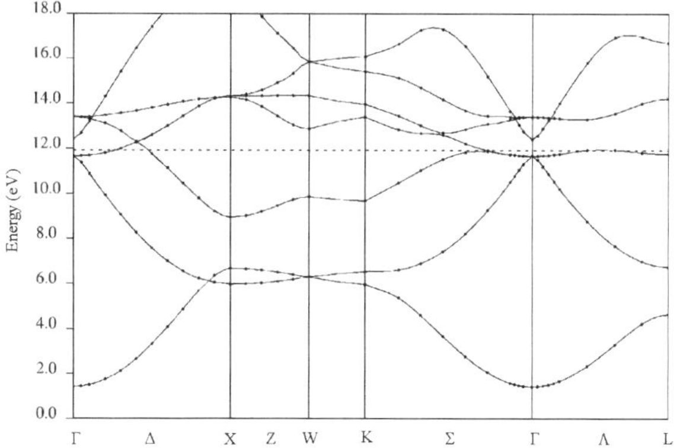

Figure 1. GGA energy bands for FCC TiH_2

Figure 1 shows that the Fermi energy crosses the two-fold degenerated flat band along the Γ-Λ-L direction. The band originates from the $\Gamma_{25'}$ state and changes along Γ-Λ-L in the range of 0.316 (TiH_2) and 0.570 eV (ZrH_2). The peak at the Fermi level (E_F) in TiH_2 is sharper than in ZrH_2 (cf. Fig. 2). For the tetragonal structures this peak is split into two peaks such that one of the peaks is below and one above E_F. Thus, E_F in the BCT phases is not situated in the peak of the DOS but in a local minimum, which leads to a reduced total energy. One can see that the structures in the DOS of the BCT phases around the Fermi energy are non-equivalent with respect to a tetragonal strain (c/a < 1 and c/a > 1). The calculated DOSs are in good agreement with the analogous spectra of Wolf and Herzig [8] and Ackland [15].

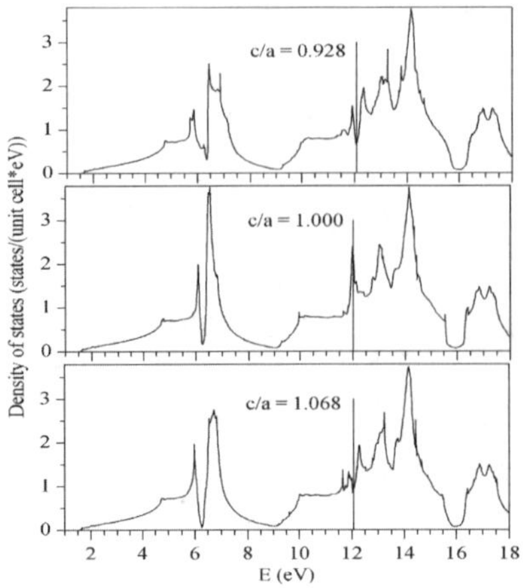

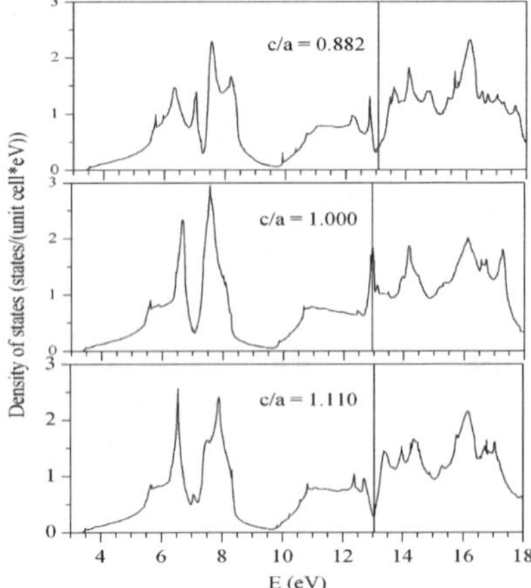

Figure 2. GGA density of states for FCC and BCT TiH_2 and ZrH_2. The vertical lines are the Fermi energies

A number of tetragonal structures with varying c/a ratios were examined using GGA, and the total energy curves for the dihydrides as function of the c/a ratios are shown in Fig. 3. They show the same double-minimum structure like that found previously [8, 14]. The total-energy curves show that the structures with c/a < 1 are most stable, which is consistent with experiment. For ZrH_2 the LDA and GGA calculations [14] and for TiH_2 the LDA investigations [8] show a deeper minimum for c/a > 1. Because the difference in the total energies is only a few tenths of a millirydberg, it is not clear whether this is physical relevant. We see that the minima in the total-energy curves decrease, and c/a ratios reach to unity, when the Gaussian broadening, σ, increases. This means that the high-temperature structures will be cubic.

In Fig. 3 we show the total energy as a function of the cell volume for both dihydrides. It is clearly seen that a reduction in the volume does not lead to a stabilization of the cubic structures. The tetragonal structures are stable in a wide range of compression. It should be noted that some trend towards the strengthening of the FCC phases at a large homogeneous expansion is observed.

The calculated GGA phonon dispersions along the high-symmetry directions in the BZ for the FCC TiH_2 and ZrH_2 are shown in Fig. 4. The soft modes in the transverse acoustic brunch around the Γ point of the BZ are clearly observed. The lower transverse acoustic bands are of the Δ_5, Σ_4 and Λ_3 modes for the Γ-Δ-X, Γ-Σ-X and Γ-Λ-L directions, respectively (in Fig. 4 these modes are not denoted). Using the well known relations between phonon frequencies, ω, and k-vectors in the long-wave limit for cubic crystals:

$$\omega^2\rho = ½\, C_{44}\, k^2 \ \{\Delta_5\};\ \omega^2\rho = ½\,(C_{11} - C_{12})k^2 \ \{\Sigma_4\};\ \omega^2\rho = ½\,(C_{11} - C_{12} + C_{44})\, k^2 \ \{\Lambda_3\},$$

(where ρ is a crystal density) we arrive to a conclusion that the C_{11}-C_{11} and C_{44} modulii will dramatic soften with decreased temperature and vice versa. Indeed, in the phonon dispersion curves calculated by using the large σ, the soft modes are not shown, which indicates an increase of the modulii with temperature (σ, in our case). We note also that the optical brunch of the phonon spectra does not contain any soft modes, and the optical modes are modified with changing σ insignificantly.

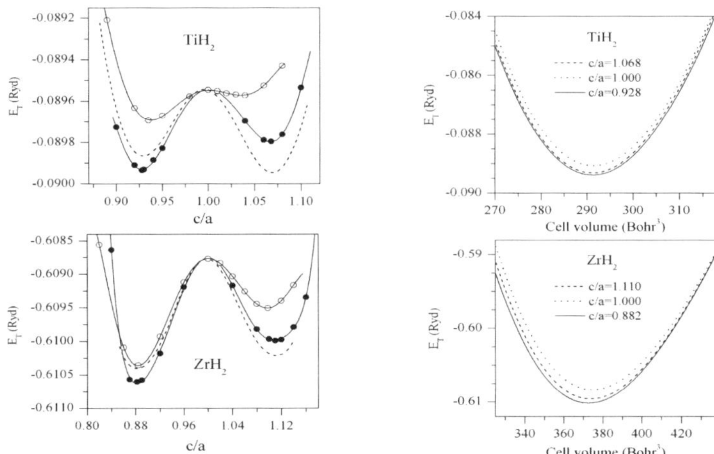

Figure 3. GGA total energy (E_T) as functions of the c/a ratio (left panel) and the volume (right panel). Total energies are relative to -119 Ryd (TiH_2) and -101 Ryd (ZrH_2)

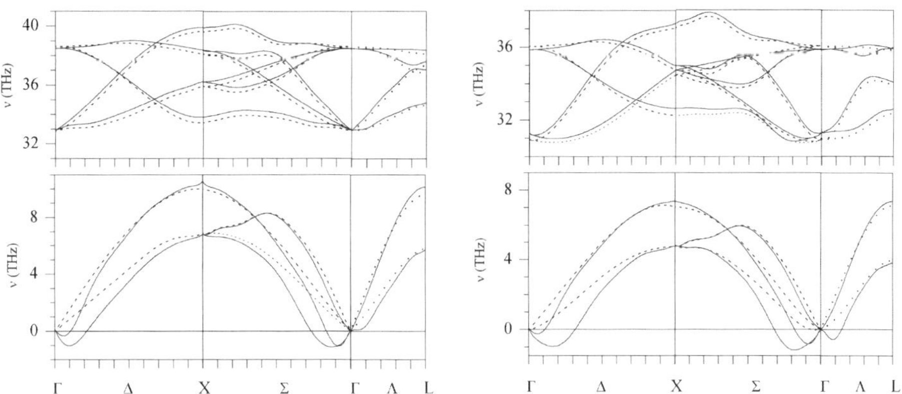

Figure 4. GGA calculated phonon dispersions for FCC TiH_2 and ZrH_2 at band smearing σ = 0.01 Ry (solid line) and σ = 0.04 Ry (dashed line). Negative frequencies ν mean $i|\nu|$

An origin of the soft acoustic transverse modes can be understood basing on an analysis of the atomic displacements and strains that could lead to splitting the double-degenerated Λ_3 band. For this purpose, for TiH_2 we have carried out structural optimizations, starting with the tetragonal simple cells with 12 atoms sited in various configurations. In each case this has the effect of increasing energy, showing that the BCT structure (I4/mmm) is most stable, and the H atoms shift only according to the tetragonal strain. Thus, a possible softening of the $\Gamma_{25'}$ optic mode [7] is not confirmed.

Returning to strains, easily to show that only tetragonal $\varepsilon_t = 2\varepsilon_{zz} - \varepsilon_{xx} - \varepsilon_{yy}$, $\varepsilon_e = \varepsilon_{xx} - \varepsilon_{yy}$ and monoclinic ε_{xy}, ε_{yz}, ε_{zx} strains are able to split the twofold degenerated Λ_3 band originated from the $\Gamma_{25'}$ state. The tetragonal and monoclinic strains are in fact the acoustic long-wave modes and determine the temperature behavior of the C_{11}–C_{12} and C_{44} modulii, respectively. Thus the interaction of the electrons in the Λ_3 band with the acoustic phonons gives rise to softening of the transverse acoustic modes and the C_{11}–C_{12} and C_{44} modulii.

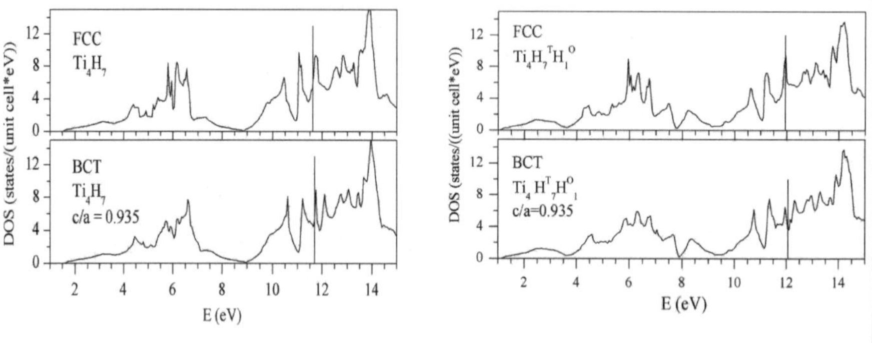

Figure 5. GGA density of states (DOS) for FCC and BCT TiH_x. Left panel – the substoichiometric phases; Right panel – the DOS for the structures, where one H atom is sited at the octahedral hole

In Fig. 5, we show the DOS of the substoichiomentric hydride $TiH_{1.75}$ with the cubic and tetragonal structures. We see that E_F shifts towards low energies as the hydrogen vacancy appears, and it does not fall at the peak of the DOS, as in the case of TiH_2 (cf. Fig. 2). As a consequence, in the tetragonally distorted $TiH_{1.75}$, E_F is just suited in one of the split peaks of the DOS. Correspondingly, the system will not gain in a total energy owing to the tetragonal distortion, and the crystal structure remains cubic.

Although H atoms occupy preferably tetrahedral holes (T), some of them can occupy octahedral holes (O) at high temperatures [1–3, 16]. Therefore one can suppose that it is the redistribution of H atoms between T and O holes that will results in the stabilization of the cubic structure. However we see from Fig. 5 that this is unlikely, since E_F in the cubic $Ti_4H^T{}_7H_1{}^O$ falls at the peak of the DOS.

The results presented above allow us to make some conclusions on the phase transition in TiH_2 and TiH_x.

(i) We see that the softening of the $\Gamma_{25'}$ optic phonon mode will not result in the cubic-to-tetragonal second-order phase transition, since this mode is not soft in the FCC phase. Moreover, the condensation of this mode gives rise to the structures that are not compatible with the BCT structure (I4/mmm).

(ii) The stabilization of the cubic phase due to the redistribution of H atoms between T and O holes was found to be unlikely.

(iii) The vacancies in the hydrogen sub-lattice stabilize the cubic phase. The DOS of $TiH_{1.75}$ presumes that in the range $1.8 \leq x < 2$, an order-disorder phenomenon can cause the second-order phase transition, which is consistent with the supposition [4].

(iv) The interaction of the electrons in the Λ_3 band with the acoustic phonons causes the softening of the transverse acoustic modes and the C_{11}–C_{12} and C_{44} modulii.

(v) Since the total-energy is not symmetric with respect to the tetragonal distortion ε_t (cf. Fig. 3), the phase transition in TiH_2 will be of first order close to second order. The second-order phase transition takes place provided there is no bands close to the two-fold degenerated Λ_3 band [6]. In our case, the Λ_3 band originates from the three-fold degenerated $\Gamma_{25'}$ state and, besides, other bands cross the Fermi level. As a consequence, the total energy is of an unsymmetric function of ε_t. It should be noted that the asymmetry is preserve in all total-energy curves calculated by using different values of σ (cf. Fig. 3).

We can examine the phase transition in TiH_2 based on the Landau theory of phase transition [17]. Due to the asymmetry of the total energy with respect to ε_t we have to take into account the third-order invariant in an expansion for the difference between the Gibbs free energy of the high-temperature configuration and that of the low-temperature configuration in a power series of order parameter [17, 18]:

$$\Delta G = \tfrac{1}{2}\, A(T)\varepsilon_t^2 + 1/3\, B(T)\varepsilon_t^3 + \tfrac{1}{4}\, C(T)\varepsilon_t^4 + \ldots, \qquad (1)$$

where $A(T) = a(T\text{-}T_0) = \tfrac{1}{2}\,(C_{11}\text{–}C_{12})$, and B,C – are combinations of effective nth-order elastic constants of the static crystal at temperature T [17, 18]. Thus, the order parameter, i.e., the physical quantity, which drives the phase transition Fm3M →I4/mm, is assumed to be a crystal deformation, namely the tetragonal strain. The elastic constant $\tfrac{1}{2}\,(C_{11}\text{–}C_{12})$, associated with the order parameter should be very small at a phase transition point ($T \approx T_0$) to stabilize a new low-temperature phase. This implies a softening of the transverse acoustic Σ_4 mode with lowering temperature in a high-temperature phase, which is consistent with our findings (cf. Fig. 4).

According our investigations, the phase transition in TiH_2 is of first order that is close to the second-order phase transition. This statement is not consistent with the experimental data [1-3]. However, since the asymmetry of the total energy is very small ~0.5 mRy, a discontinuity of the order parameter at the phase-transition point is suppose to be negligible. The further experimental investigations are needed to confirm or refute our supposition. In favor of our assumption points also to the fact that in ZrH_2 the phase transition is supposed to occur at high temperatures and has features of the first-order phase transition. The high transition temperature (~600–800^{0}C) in ZrH_2 [3] is due to the wide bandwidth of the Λ_3 band that is larger than in TiH_2. Here it should be stressed that Ghatak, Ray and Tannous [6] have reported that a rise in the bandwidth can favor a large value of transition temperature. Therefore we suppose that the tetragonal distortion in ZrH_2 will be stronger, and the difference between the total energy of the BCT phase and that of the FCC one will be higher than in TiH_2.

4. Conclusions

The present investigation provides a consistent description of cubic and tetragonally distorted titanium and zirconium dihydrides based on density functional theory. Lattice constants, Bulk moduli, c/a ratios are derived from total-energy calculations. The phonon dispersion curves for the cubic phases were calculated basing on the density-functional perturbation theory. The structural and electronic characteristics are in good agreement with the experimental and theoretical results. It was shown that deviation from stoichiometry leads to a stabilization of the cubic structure, and a redistribution of H atoms between tetrahedral and octahedral holes does not promote the stabilization of the cubic phase. Finally, it was deduced from the phonon dispersions and total energy curves that the phase transitions in TiH_2 and ZrH_2 are of first order that is close to the second-order phase transition. The order parameter is the tetragonal strain, and the elastic constant $\tfrac{1}{2}\,(C_{11}\text{–}C_{12})$, associated with order parameter, is supposed to soften smoothly with lowering temperature. These peculiarities indicate that the described structural transformations are proper ferroelastic transitions.

References

1. Yakel H.L. (1958) Thermocrystallography of higher hydrides of titanium and zirconium, *Acta Cryst.* **11**, 46–51.
2. Bale H.D., Peterson S.B. (1972) X-ray diffraction study of the structural transformation in TiH_2, *Solid State Commun.* **11**, 1143–1145.
3. Mueller W.M., Blackledge J.P., Libowitz G.G. (eds.) (1968) Metal Hydrides (Academic, New York) p. 278.
4. Azarkh Z.M., Gavrilov P.I. (1970) Structurnye izmenenia gidrida titana pri bolishih kontsetratsiah vodoroda, *Soviet Phys. Crystallogr.* **15**, 275–279.
5. Switendick A.C. (1970) Electronic band structures of metal hydrides, *Solid State Commun.* **8**, 1463–1467.
6. Ghatak S.K., Ray D.K., Tannous C. (1978) Studies of the structural stability of an electronic system with a two-fold degenerate band, *Phys. Rev. B.*, **18**, 5379–5384.
7. Kulikov N.I. (1985) Unusual behaviour of the band structure and electronic-induced phase transitions in metallic hydrides, *J. Less-Common Met.* **107**, 111–129.
8. Wolf W., Herzig P. (2000) First-principles investigations of transition metal dihydrides, TH_2: T = Sc,Ti,V,Y,Zr,Nb; energetics and chemical bonding, *J. Phys.-Condens. Mat.* **12**, 4535–4551.
9. Baroni S., Dal Corso, A., de Gironcoli S., Giannozzi P., Cavazzoni C., Ballabio G., Scandolo S., Chiarotti G., Focher P., Pasquarello A., Laasonen K., Trave A., Car R., Marzari N., Kokalj A., http://www.pwscf.org/
10. Vanderbilt D. (1990) Soft self-consistent pseudopotentials in generalized eigenvalue formalism, *Phys. Rev. B* **41**, 7892–7895.
11. Perdew J.P., Zunger A. (1981) Self-interaction correlation to density-functional approximations for many-electron systems, *Phys. Rev. B* **23**, 5048–5079.
12. Perdew J.P., Burke K., Ernzerhof M. (1996) Generalized gradient approximation made simple, *Phys. Rev. Lett.* **77**, 3865–3868.
13. Lehmann G., Taut, M. (1972) On the numerical calculation of the density of states and related properties, *Phys. Status Solidi B* **54**, 469–477.
14. Ackland G.J. (1998) Embrittlement and the bistable crystal structure of zirconium hydride, *Phys. Rev. Lett.* **80**, 2233–2236.
15. Ackland G.I. (1998) A reply to the comment by M. Gupta, *Phys. Rev. Lett.* **81,** 3301–3301.
16. Kudabaev Z.I., Torgeson D.R., Shevakin A.F. (1995) Nuclear magnetic resonance of the phase transition in TiH_2 and TiD_2, *J. Alloy. Compd.,* **231**, 233–237.
17. Landau L.D., Lifshitz E.M. (1958) Statistical Physics (Pergamon, London) Chap. XIV.
18. Hebbache M. (1994) Elastic phase transition in germanium and silicon, *Phys. Rev. B* **49**, 6522–6527.

INITIAL STAGE OF FORMATION OF ORDERED NANOCRYSTALLINE STRUCTURES IN FILMS OF ZIRCONIA-BASED SOLID ELECTROLYTE

V.B. MALKOV, V.N. STREKALOVSKIY
Institute of High-Temperature Electrochemistry, Ural Branch RAS
620219 Ekaterinburg, Sverdlovsk Region, Russia
A.V. MALKOV, O.V. MALKOV
"ROSNA" Scientific and Production Center, 620219, Ekaterinburg, Russia
V.G. PUCHIN
Institute of Metal Physics, Ural Branch RAS, 620219, Ekaterinburg, Russia
B.V. SHULGIN
Ural State Technical University, 620219, Ekaterinburg, Russia

Abstract. Methods of transmission electron microscopy were used for analyzing the initial stage of formation of ordered nanocrystalline solid-electrolyte structures ZrO_2-Y_2O_3 and ZrO_2-Sc_2O_3 produced by ion-plasma spluttering. At the initial stage of their formation the ordered nanocrystalline solid-electrolyte structures were found to be linear or bilinear ones. The electron diffraction analysis of the linear and bilinear nanocrystalline structures suggested a monocrystalline character of the nanocrystallites forming these structures. A model of a self-orientation mechanism responsible for formation of ordered nanocrystalline solid-electrolyte structures has been discussed.

Keywords: structure of films, films of zirconia-based solid electrolytes, ion plasma sputtering, transmission electron microscopy, linear ordered nanocrystalline structure, bilinear ordered nanocrystalline structure

1. Introduction

It is known that ordered nanocrystalline structures are formed during ion-plasma sputtering of ZrO_2-Y_2O_3 and ZrO_2-Sc_2O_3 solid-electrolyte films [1]. Investigations into formation of these nanocrystalline structures revealed linear and bilinear nanocrystalline entities representing the initial stage of formation of ordered nanocrystalline solid-electrolyte structures.

Transmission electron microscopy methods were used to study the initial stage of formation of ordered nanocrystalline structures in films of a zirconia-based solid electrolyte, which were prepared by ion plasma sputtering [1, 2].

2. Experimental and discussion

The transmission electron microscopy and electron diffraction techniques were used to study the initial stage of formation of ordered nanocrystalline structures in films of a zirconia-based solid electrolyte. The films were deposited by ion plasma sputtering on glass-ceramic substrates or fresh mica chips, which were covered first with a thin (~10 nm) layer of amorphous carbon.

It was found that at their initial stage of formation the ordered nanocrystalline structures in films of the solid electrolyte, which was based on zirconia stabilized to the cubic modification, were linear (Figs. 1 and 2) and bilinear (Fig. 3) entities made up of nanocrystallites. The nanocrystallites generally had an irregular round shape and were 20 ÷ 40 nm in size on the average. The nanocrystallites were characterized by a linear self-organization at the beginning and then, as the ordered nanostructure appeared, had a bilinear self-organization.

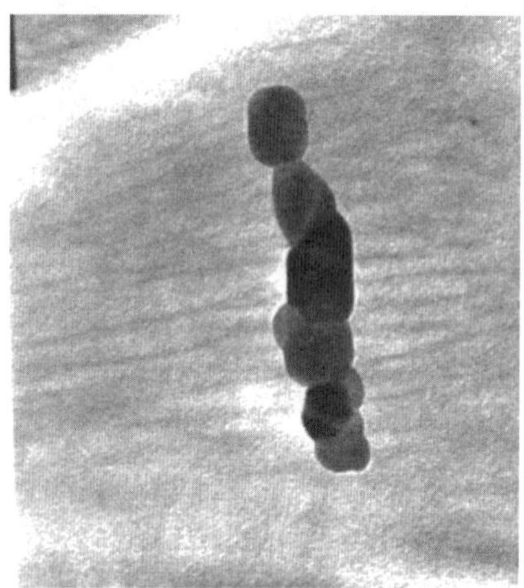

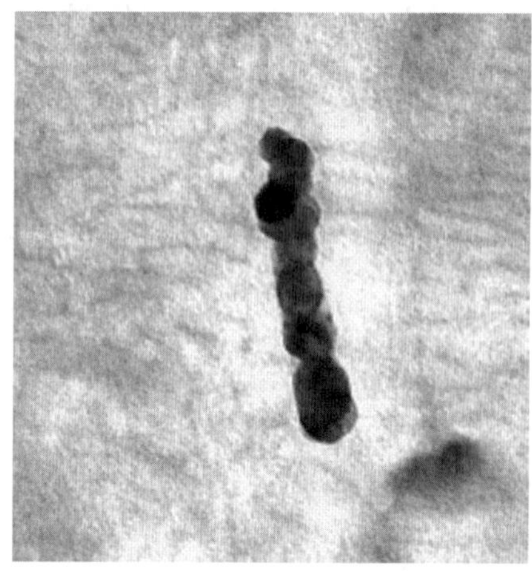

Fig.1 ×150,000 Fig. 2 ×150,000

Figure 1. Electron microscopic image of a linear ordered nanocrystalline structure in a film of the zirconia-based solid electrolyte

Figure 2. Electron microscopic photograph of a linear ordered nanocrystalline structure in a film of the zirconia-based solid electrolyte

Electron diffraction patterns of the linear and bilinear nanocrystalline structures were point patterns attesting to a monocrystalline character of the nanocrystallites forming these structures (Fig. 4). The continuity of the nanocrystallites was disturbed only in the bilinear nanostructures. The character of the electron diffraction patterns of the ordered nanocrystalline entities suggested a highly coordinated orientation of their constituent nanocrystallites.

The above experimental data can be useful for refinement of the model describing formation of the porous structure of films of zirconia-based solid electrolytes. The original model assumes formation of pores and submicropores as vacancies, which are trapped during sputtering of the solid-electrolyte films (sputtering at $T_s < 0.3\ T_{melt}$), are released to sinks in the form of nanocrystallite boundaries and are condensed [2– 5]. The refined model should be supplemented with formation of pores and submicropores caused by gapping of nanocrystallites at the very early stages of formation of the ordered nanostructures. Indeed, in Fig. 3, which shows a bilinear nanostructure, one can see voids between nanocrystallites and it is not improbable that these voids are retained at subsequent stages of formation of the solid-electrolyte films.

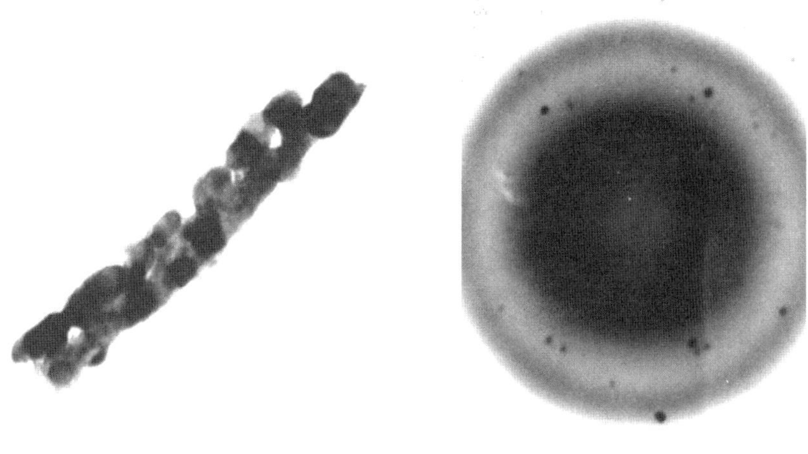

Fig. 3 ×150,000 Fig. 4

Figure 3. Electron microscopic image of a bilinear ordered nanocrystalline structure in a film of the zirconia-based solid electrolyte

Figure 4. Example of a point electron diffraction pattern of the linear and bilinear nanocrystalline structures

Since the nanocrystalline structures are separated from the substrate by a layer of amorphous carbon (~10 nm), one may state that these ordered nanocrystalline structures are formed due to the operation of a self-orientation mechanism, which is responsible for appearance of the linear and bilinear ordered nanocrystalline structures in the films of the zirconia-based solid electrolyte. Forces involved in self-orientation of the nanocrystallites during formation of the linear and bilinear nanostructures should be sufficiently high. Indeed, nanocrystallites comprising the ordered nanostructures of the solid electrolyte are coarse (20–40 nm in size) and their mutual ordering requires appropriate forces. This should be taken into account in construction of the interaction potential of the solid-electrolyte nanocrystallites.

3. Conclusions

Results of the study dealing with the initial stage of formation of a nanocrystalline structure in films of a zirconia-based solid electrolyte have led to the following conclusions.

1. Linear and bilinear ordered nanocrystalline structures are detected in films of the zirconia-based solid electrolyte. These structures represent the initial stage of formation of ordered nanocrystalline structures in the solid electrolyte films.

2. An ordered arrangement of the nanocrystallites is explained by a self-orientation mechanism, which operates at the initial stage of formation of ordered nanocrystalline structures in films of the zirconia-based solid electrolyte.
3. The initial stage of formation of films of the solid electrolyte, which is based on zirconia stabilized to the cubic modification, involves two types of the structure self-organization: (1) at the atomic level, leading to appearance of monocrystallites; (2) at the nanocrystalline level, leading to formation of ordered nanocrystalline entities.

References

1. Malkov V.B., Strekalovskiy V.N., Malkov A.V., Malkov O.V., Puchin V.G., Shulgin B.V. Hydrogen Materials Science and Chemistry of Carbon Nanomaterials // Proc. of NATO ARW on HMSCCN, Sevastopol, Ukraine, 2005.- P. 567–570 (Published by Springer, The Netherlands, 2007).
2. Georgy Hass. Physics of Thin Films. V 1.- Moscow: Mir, 1967.- 343 p.
3. Georgy Hass and Rudolf Tunn. Physics of Thin Films. V 4.- Moscow: Mir, 1970.- 440 p.
4. Palatnik L.S., Fuks M.Ya., Kosevich V.M. The Mechanism of Formation and the Substructure of Condensed Films.- Moscow: Nauka, 1972.- 320 p.
5. Chopra K.L. Electrical Phenomena in Thin Films.- Moscow: Mir, 1972.- 436 p.

THE INFLUENCE OF INTERSTITIAL IMPURITIES ON TEMPERATURE RANGES OF DEUTERIUM RETENTION IN AUSTENITIC STAINLESS STEEL 18Cr10NiTi

I.M. NEKLYUDOV, A.N. MOROZOV, V.I. ZHURBA
National Science Center "Kharkov Institute of Physics and Technology", 1 Akademicheskaya Str., Kharkov 61108, Ukraine
E-mail: morozov@kipt.kharkov.ua

V.G. KULISH, A.G. GALYTSKY
G.S. Skovoroda National Pedagogical University at Kharkov, 2 Blyukher Str., Kharkov, Ukraine

Abstract. The influence of nitrogen, oxygen and helium on the temperature range of deuterium retention in 18Cr10NiTi stainless steel (of AISI304L type) has been investigated. It is demonstrated that the introduction of oxygen, nitrogen or helium into 18Cr10NiTi steel extends the temperature range of deuterium retention up in temperature. It has been found that for 18Cr10NiTi stainless steel, pre-irradiated with helium ions, the increase in the temperature range of deuterium retention occurs in steps: on attainment of helium concentration of ~0.5 at.% He the temperature range increases by ~150 K, and on attainment of helium concentration of ~2.5 at.% He the temperature range increases by ~200 K. The introduction of oxygen into 18Cr10NiTi stainless steel results in the increase of the temperature range of deuterium desorption in the direction of rise in temperature. In the deuterium thermodesorption spectrum, this manifests itself by the occurrence of an additional low-amplitude peak with the maximum temperature $T_M \sim 560$ K. The introduction of nitric impurity into 18Cr10NiTi stainless steel results in the extension of the temperature range of deuterium retention towards higher temperatures.

Keywords: deuterium retention, 18Cr10NiTi stainless steel, interstitial impurities, oxygen, nitrogen, helium, ion implantation, thermodesorption

1. Introduction

As a constructional material for vessel internals of fission-type reactors WWER-1000, preference has been given to stainless steel 18Cr10NiTi, where the accumulation of hydrogen was neglected owing to its evolution at temperatures lower than the working temperature of the reactor operated at ordinary conditions. Garner et al. [1] have found the hydrogen retention in SS316 stainless steel exposed to reactor environment at temperatures exceeding the working temperatures. There is evidently no unambiguous explanation that can be offered now for this rather complicated and multilateral phenomenon. An enhanced bond of hydrogen dissolved in 18Cr10NiTi steel may be due to its interaction with transmutation elements (including helium), interstitial and substitutional impurities, vacancies and their complexes, dislocations and their pile-ups, grain and subgrain boundaries, phase constituents, etc. The present paper is concerned with the thermodesorption spectrometry studies into the influence of nitrogen, oxygen and helium on the temperature ranges of deuterium retention in 18Cr10NiTi austenitic stainless steel.

2. Experimental

The 18Cr10NiTi austenitic stainless steel samples, $10 \times 5 \times 0.15$ mm^3 in size, were fixed on the substrates (made of the same steel, ~0.3 mm in thickness), which simultaneously served as ohmic heaters. Prior to irradiation, the targets were subjected to a short-time (1 min) annealing at a temperature of ~1,350 K in vacuum ~2×10^{-5} Pa in order to degas the specimen and to remove impurities from its surface.

The interstitial impurities were introduced into the 18Cr10NiTi steel specimens by implanting 28 keV N_2^+ or O_2^+ ions at a temperature of ~300 K to doses ranging from 3.5×10^{16} to 1.8×10^{17} N/cm^2 and from 1.1×10^{17} to 2.3 x 10^{17} O/cm^2. Helium implantation into 18Cr10NiTi steel specimens was carried out by successive bombardment with ions of two different energies. This provided a uniform saturation with helium in a depth range of 110–220 nm, first by exposure to the ion beam with E ~ 70 keV to doses between 5×10^{15} and 1×10^{17} at.He/cm^2, and then by exposure to the beam with E ~ 30 keV to a dose being 0.7 of the original value. The helium concentrations in the layer for the mentioned dose range were calculated to vary from 0.5 to 10 at.%. The energy values were chosen with due regard for helium depth profiles computed by the use of the program TRIM [2] for nickel.

As the impurity ion concentration reached the assigned value, the specimen was cooled down to a temperature of ~140 K, and then deuterium was introduced by ion implantation, the exposure dose being the same for all the specimens, namely, 1.0 or 1.5×10^{16} D/cm^2. The given deuterium implantation dose corresponds to the phase of deuterium solid solution in the 18Cr10NiTi stainless steel and, accordingly, exerts no effect on the metal structure. Deuterium was implanted with the use of the D_2^+ ion beam having an energy of 14 keV and a current density of ~5 μA/cm^2. After irradiation, the specimens were heated up to a temperature of ~1,700 K at an average rate of ~5 K/s. In the course of heating, the desorption of deuterium and nitrogen was registered at dynamic conditions by two mass-spectrometers APDM-1 set for the masses 4 (D_2^+) and 28 (N_2^+). The temperature variations during heating were measured by means of a chromel-alumel thermocouple fixed to the specimen.

3. Results and discussion

It has been demonstrated in [3] that a preliminary implantation of helium into 18Cr10NiTi steel leads to the formation of traps capable of retaining hydrogen in the temperature range between 500 and 800 K. In this case, the implanted hydrogen first fills the highest-temperature trap of all available ones, and only after its occupation, it fills the trap having the temperature higher than the remaining traps. So, with an increasing dose of implanted hydrogen there occurs a successive occupation of progressively lower-temperature traps, this leading to the development of hydrogen thermodesorption spectrum in the direction of temperature drop. Obviously, having chosen a low deuterium implantation dose from the thermodesorption spectrum as a test dose, it is possible to estimate the upper boundary of the temperature range of deuterium retention in 18Cr10NiTi steel, and this was just done in the present work.

Figure 1 shows the thermodesorption spectra of deuterium implanted in 18Cr10NiTi steel to a dose of 1 x 10^{16} D/cm^2. It can be seen from the figure that the upper boundary of deuterium retention is dependent on the dose (concentration) of implanted helium. At the same time, as certain helium concentrations are attained with an increasing implantation dose, the rise in the temperature range of deuterium retention occurs in steps, by degrees of 150–200 K. In the helium concentration range under study two such steps are observed: one – in the 0.5 to 2.4 at.% He range (the temperature range of deuterium retention is extended from 400 to 600 K, and the maximum temperature of the highest peak is T_m ~ 560 K), and the other step is observed in the 2.5 to 6 at.% He range (the temperature range of deuterium retention is extended from 400 to 600 K, and the maximum temperature of the highest peak is T_m ~ 780 K).

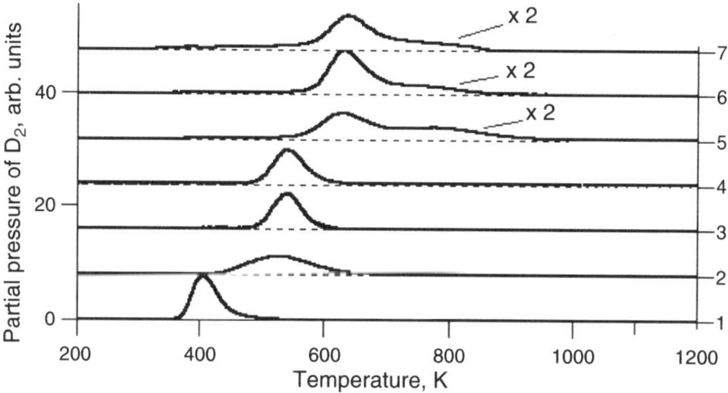

Figure 1. Thermodesorption spectra of deuterium implanted into 18Cr10NiTi steel to a dose of 1 × 10^{16} D/cm^2 versus preimplanted helium concentration. 1 – 0, 2 – 1, 3 – 1.7, 4 – 2.4, 5 – 2.53, 6 – 4.5, 7 – 6 at.% He.

Figure 2 gives the most typical curves of implanted deuterium desorption from 18Cr10NiTi stainless steel samples, which have been preimplanted with nitrogen ions. With an increase in the nitrogen interstitial impurity content the temperature range of deuterium retention in 18Cr10NiTi steel is extended up in temperature. Curve 1 corresponds to deuterium desorption from the sample not irradiated with N_2^+ ions, i.e., gives an idea of deuterium retention and emission from steel samples in the absence of the nitrogen interstitial impurity. In this case, the spectrum exhibits one peak of gas emission with the maximum temperature T_m ~ 410 K. Deuterium desorption begins at a temperature of ~330 K and is completed at ~600 K.

The introduction of nitrogen interstitial impurity at a level of 3.5 × 10^{16} N/cm^2 results in the peak spreading and the extension of gas emission range up in temperature (curves 2 and 3). A further increase in the nitrogen dose up to ~6.5 × 10^{16} N/cm^2 and ~1.1 × 10^{17} N/cm^2 has manifested itself in the change of the deuterium thermodesorption spectrum appearance, i.e., the spectrum now represents a complicated structure with poorly resolved, at least, two peaks. This appearance of the spectrum (curves 4 and 5) may be due to the formation of solid solution of nitrogen in 18Cr10NiTi stainless steel that precedes the nitride formation. The thermodesorption spectrum of implanted helium from the steel pre-irradiated with nitrogen ions to a dose of ~1.8 × 10^{17} N/cm^2 (curve 6) is qualitatively different from the previous spectra. It shows two well resolved peaks and an extended region of deuterium (substrate) desorption between them. It can be assumed that the peaks observed in the deuterium thermodesorption spectra may be attributed to: (i) deuterium emission from 18Cr10NiTi steel without any nitrogen interstitial impurity (T_m ~ 410 K); (ii) deuterium emission from the steel comprising nitrides (peak with the maximum temperature T_m ~ 640 K); (iii) deuterium emission from nitrogen solid solution in 18Cr10NiTi steel.

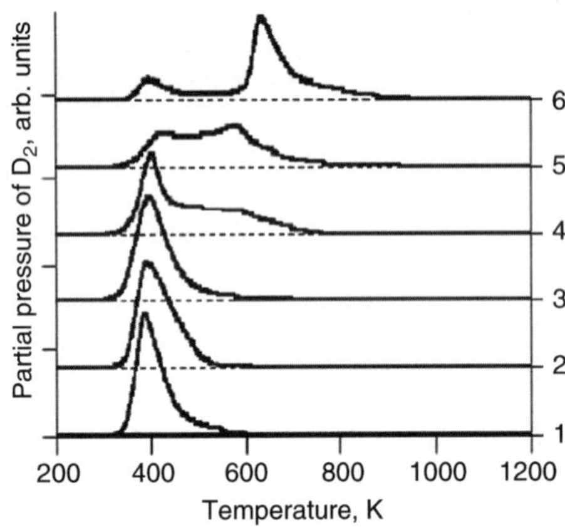

Figure 2. Thermodesorption spectra of deuterium implanted into 18Cr10NiTi steel to a dose of ~1.5 × 10^{16} D/cm^2 versus preimplanted nitrogen concentration. 1 – 0, 2 – 3.5 × 10^{16} N/cm^2, 3 – 5 × 10^{16} N/cm^2, 4 – 6.5 × 10^{16} N/cm^2, 5 – 1.1 × 10^{17} N/cm^2, 6 – 1.8 × 10^{17} N/cm^2

Figure 3 shows the spectra of deuterium emission from 18Cr10NiTi steel preimplanted with O_2^+ ions to two doses (curves 2 and 3). Curve 1 corresponds to the emission of implanted deuterium from the sample unirradiated with O_2^+ ions. In the absence of the oxygen interstitial impurity, the deuterium thermodesorption spectrum exhibits only one well-resolved peak with the maximum temperature T_m ~ 410 K. The gas emission begins at a temperature of ~ 330 K, and is substantially completed at ~ 650 K.

A preliminary O_2^+ ion irradiation of the samples to a dose of ~1.1 × 10^{17} O/cm^2 leads to a change of the deuterium thermodesorption spectrum that manifests itself in an insignificant intensity increase of the high-temperature part of the gas emission curve (curve 2). In this case, the maximum temperature of the peak and the temperature range of gas emission remain practically unchanged. With a rise in the exposure dose up to ~2.3 × 10^{17} O/cm^2, the deuterium emission spectrum has exhibited an additional inexplicitly resolved low-amplitude peak with the maximum temperature T_m ~ 560 K. It should be noted that a further increase in the O_2^+ ion irradiation dose has not led to an increase in the temperature range of gas emission.

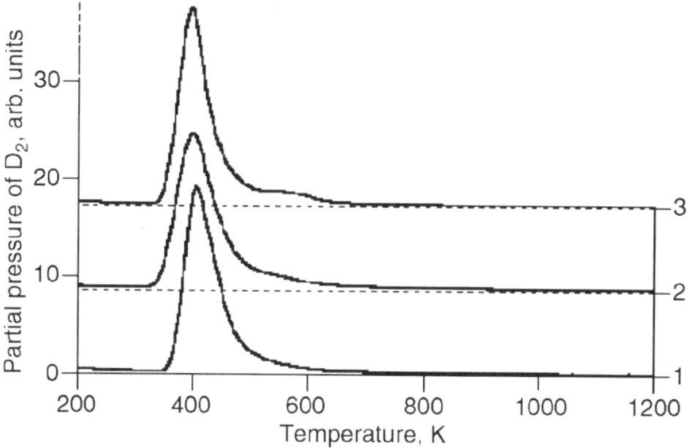

Figure 3. Thermodesorption spectra of deuterium implanted in 18Cr10NiTi steel at T_{irr} ~ 140 K to a dose of ~1.5×10^{16} D/cm^2 for the samples pre-irradiated to different doses of oxygen ions at T_{irr} ~ 300 K: 1 – 0, 2 – 1.1×10^{17} O/cm^2, 3 – 2.3×10^{17} O/cm^2

4. Conclusions

The undertaken studies have revealed a general regularity that shows up in the extension of the temperature range of implanted deuterium emission up in temperature, irrespective of the nature of interstitial impurity pre-introduced into the 18Cr10NiTi austenitic stainless steel. The degree of this extension depends on both the type and concentration of the interstitial impurity.

It has been found that for 18Cr10NiTi stainless steel pre-irradiated with helium ions the increase in the temperature range of deuterium retention occurs in steps, i.e., as the helium concentration attains ~0.5 at.% He, the temperature range increases by ~150 K, and as the helium concentration attains ~2.5 at.% He, the temperature range is extended by ~200 K more.

According to the literature data, the nitrogen interstitial impurity concentrations that may be accumulated in 18Cr10NiTi stainless steel components in the course of reactor service vary between ~0.01 and 0.02 wt%. The research results reported here encourage us to state that the nitrogen impurity of these values exerts practically no effect on the increase of temperature ranges of hydrogen retention in constructional materials of vessel internals of fission-type reactors WWER-1000 (18Cr10NiTi steel). In the general case, with an increasing nitrogen interstitial impurity concentration in 18Cr10NiTi steel there occurs an extension of the temperature range of deuterium release towards higher temperatures.

The introduction of oxygen into 18Cr10NiTi steel results in the extension of the temperature range of deuterium emission towards higher temperatures, this being reflected by the appearance of an additional peak with the maximum temperature T_m ~ 560 K.

The present results give evidence for the possibilities of using thermodesorption spectrometry not only to measure the quantity of hydrogen accumulated in the material, but also to estimate the structural condition of the latter.

References

1. F.A. Garner, B.M. Oliver, L.R. Greenwood, D.J. Edwards, S.M. Bruemmer, M.L. Grossbeck. Generation and Retention of Helium and Hydrogen in Austenitic Steels Irradiated in a Variety of LWR and Test Reactor Spectral Environments. 9th Meeting on Environmental Degradation of Materials, Tahoe, Utah, August, 2001. P. 54–72.
2. www.srim&trim.com.
3. A.N. Morozov, I.M. Neklyudov, V.I. Zhurba, V.G. Kulish. Hydrogen retention in austenitic corrosion-resistant steel preimplanted with helium ions. Materialovedenie, 2006, **3**: 47–50 (in Russian).

INTERACTION OF HYDROGEN WITH INTERSTITIAL IMPURITY IN TITANIUM POWDERS

O.M. SHAPOVALOVA, E.P. BABENKO*
Dnepropetrovsk National University,
Str. Nauchnaya, 13, Dnepropetrovsk, 49050, Ukraine
* *E-mail: v_babenko@dp.ukrtel.net*

Abstract. In this paper interaction of hydrogen with interstitial impurities and their mutual influence on the structure of titanium powders is considered.

Keywords: titanic powders, diffusion of impurity of introduction, titanium crystal lattice, not metal connections, hydrogen in titaniumm, electronic levels, valence electrons, factor of correlation

1. Introduction

Areas of applications of titanium hydride and the cycle of problems related to hydration of titanium are widening. In [1–3] the nature of hydrogen-caused brittleness of titanium was described quite completely, the influence of different factors on the kinetics of hydration of titanium and its alloys was established. This brings up to a necessity to thoroughly research interaction of hydrogen with interstitial impurities in titanium powders and their combined influence on their hardness. This problem is essential in connection with the wide implementation of domestically produced titanium hydrides and titanium powders, which are being exported.

2. Experimental

The purpose of this work is a study of influence of hydrogen on diffusion of interstitial atoms in titanium as well as their combined influence on the hardness of the powder as its main characteristic.

The object of the research was titanium powders of different ways of manufacturing.

Methods of research: chemical, spectral, correlation and regression analysis, measurement of hardness of titanium powders. For this purpose titanium powders produced domestically by different technologies were studied.

As one can see in Table 1, the smallest quantity of gaseous impurities are found in crystals of iodide titanium, the largest – in powders obtained by magnesium-thermal manufacturing, in the latter the impurities are seven times as high as in refined powders of improved purity and 4.8 times as high as in industrial electrolytic powders. However, hardness of these powders grows not in proportion the increase of the impurities, but only times 3.1 and 2 respectively.

TABLE 1. Contents of interstitial impurity in powders of different manufacture

Material	HB, MPa	Impurities (%)					∑Cl + H	∑O + N	∑O + N + C	∑ total gases
		Cl	N	C	H	O				
Crystals of iodided titanium	750	0.002	0.005	0.030	0.002	0.010	0.004	0.015	0.045	0.049
Crystals of electrolytic titanium of improved purity	867	0.054	0.012	0.005	0.010	0.029	0.064	0.041	0.046	0.110
Industrial electrolytic powders	1,270	0.041	0.017	0.014	0.007	0.083	0.048	0.100	0.114	0.162
Powders of sodium-thermal manufacturing	1,290	0.121	0.011	0.006	0.080	0.112	0.201	0.123	0.129	0.33
Powders of magnesium-thermal manufacturing	2,390	0.105	0.270	0.085	0.126	0.176	0.231	0.446	0.531	0.762

First, for establishing this disproportion we considered possible interactions of titanium with gaseous impurities with different speeds of diffusion. The sequence of lowering the speed of diffusion in α-titanium is as follows:

$$\mathrm{H} \rightarrow \mathrm{O} \rightarrow \mathrm{N} \rightarrow \mathrm{C} \rightarrow \mathrm{Cl}$$

$$1{,}5\mathrm{x}10^{-9} \rightarrow 1{,}6\mathrm{x}10^{-10} \rightarrow 1{,}2\mathrm{x}10^{-10} \rightarrow 0{,}5\mathrm{x}10^{-12} \rightarrow 0{,}35\times10^{-13}$$

As one can see, only hydrogen and oxygen have approximately comparable speeds of diffusion in α-titanium. They are first to fill in the octahedron pores in α-solid solution. Taking into account high activity of titanium to gases and their increased speeds of diffusion in titanium it can be presumed that, being a metal, titanium can form ion pair with each of the gases, distorting at that the crystal lattice of the metal the stronger, the more links are formed with this gas. Secondly, gases, while contacting with the surface of titanium powders, can form ion pairs between themselves and diffuse into the particle. Therefore three possible cases of forming of ion pairs were considered:

- Titanium with ions of one gas
- Titanium with ions of two gases
- Titanium with ions of a group of gases

Statistical analysis of the data o chemical composition of industrial titanium powders produced by different techniques has been done by correlation-regression method, linear, quadratic and power equations of dependence of the hardness on inclusion of interstitial impurities have been derived. Hardness according to Brinnel is the evaluation criterion for the quality of the powders.

Processing of the result of chemical analysis of the composition of the powder has shown that the most significant influence on the hardness of the titanium powders of all interstitial impurities must be caused by hydrogen with oxygen and, on a smaller scale, hydrogen with nitrogen. This has been corroborated by theoretical consideration of interaction of titanium with gases. If we range all interstitial impurities in titanium by growth of their atomic number, there will be a number of elements, whose last electronic levels are unfinished (Table 2).

H → C → N → O → Cl

1 → 6 → 7 → 8 → 17

TABLE 2. Electronic structure of titanium and interstitial impurities

Elements	Hydrogen	Carbon	Nitrogen	Oxygen	Chlorine	Titanium
Order number	1	6	7	8	17	22
Main unexcited state of the last electronic level of the atom	$1s^1$	$2s^2 2p^2$	$2s^2 2p^3$	$2s^2 2p^4$	$3s^2 3p^5$	$3d^2 4s^2$
Number of passed-over electrons	–	–	–	–	–	1; 2; 3; 4
The degree of oxidizing in the compound	–1	+4; -4; +2;	–3	–2	–1	–
Energy of ionization of the atom, eV	–	11.3	14.5	13.62	12.97	1e(4s)-6,83; 2e(4s)-13,57; 3e(3d)-27,52; 4e(3d)-44,12 5e(3p)-100,4
Stable state of the last level of the compound	$1s^2$	$2s^1 2p^3 \rightarrow 2s^2 2p^6$	$2s^2 2p^6$	$2s^2 2p^6$	$\rightarrow 3s^1 3p^4 3d^3 \rightarrow \rightarrow 3s^2 3p^6$	1-$3s^2 3p^6 3d^2 4s^1$; 2-$3s^2 3p^6 3d^2$; 3-$3s^2 3p^6 3d^1$; 4-$3s^2 3p^6$
Enthalpy ΔH, J/mol	43.6	4,617	5,493	519.14	517.5	–

It is known [4–6] that any system in thermodynamically unbalanced state tends to a more stable state. For this reason, for an atom of titanium it is easier to pass over electrons from the upper electronic levels as those having low energy of ionization of the atom, rather than to join and hold them (Table 2). Establishing a common electronic pair titanium-gas occurs by joining electrons of a titanium atom by atoms of non-metals, the latter become negatively charged ions, while positively charged ions of metal are formed due to losing electrons from the outer electronic levels.

H-1e to 1s-level; O-2e to 2p-level;

Cl-1e to 3p-level; N-3e to 2p-level; C-4e to 2s- and 2p-levels.

An atom of carbon for completion of its p-level firstly passes 1e from $2s^2$ to $2p^3$, and then joins electrons of an excited atom of titanium for completing s- and p-levels.

Being a metal, titanium can give up 2e from 4s-level, changing at that its valence, and, as chemically active element, can interact with one, two or a group of atoms of gases simultaneously. In all compounds under consideration atoms of interstitial impurities are oxidants, and titanium is the reducing agent.

As it is shown [4, 7, 8], atoms of gases in ion state have unfinished s- and p-levels, thus they form unpaired outer electrons. Separated valence electrons of an excited atom of titanium and excited atoms of interstitial impurities, having low ionization potential, under certain conditions can form common electron pair both with titanium and between themselves. This electron pair will circulate both around the nucleus of the atom of titanium and around the nucleus of gas.

The strongest links titanium can form with hydrogen and oxygen, since they form completed 1s- and 2p levels close to the nucleus. Hydrogen and oxygen for completing their levels join 1 and 2 electrons respectively from a titanium atom. This is confirmed by determination of the correlation coefficient between titanium and hydrogen and between titanium and oxygen, which are 0.40 and 0.67 respectively. Oxygen distorts the lattice of titanium stronger than hydrogen, therefore its correlation coefficient is higher than those of hydrogen, which is corroborated by higher values of their enthalpy: TiH – 43.6 J/mol. Nitrogen and carbon, like oxygen, are to finish completing outer electronic levels (2p-level) with electrons of titanium for a more stable state.

However nitrogen, for fully completing its 2p-level, needs to join three electrons and to spend by an order more energy on forming a pair than it is necessary for titanium and oxygen. This is confirmed by lower correlation coefficient (TiN – 0.58), than that for titanium and oxygen (TiO – 0.67), and the enthalpy of TiN is higher – 549.3 J/mol.

Carbon, for forming with titanium TiC compound, needs to join four electrons from a titanium atom for completing 1s-level and 2p-levels. Energy spent by electrons in this process is also high, 4,617 J/mol, while the correlation coefficient is insignificant.

Compound of titanium with chlorine and obtaining of chlorides under normal conditions is a complex process since titanium can give up from two to four electrons for forming electronic pairs, while chlorine can join only one electron on an energy level more remote from the nucleus. For this reason one atom of titanium may have two, three or four atoms of chlorine, with their 3p-level incomplete by one electron. These compounds are unstable and dissociate under any external influences. This explains low value of the correlation coefficient (0,14) though enthalpy of solid $TiCl_2$ is 517.5 J/mol.

Therefore all interstitial impurities in titanium powders can be ordered by descending of their correlation with titanium as follows [9]:

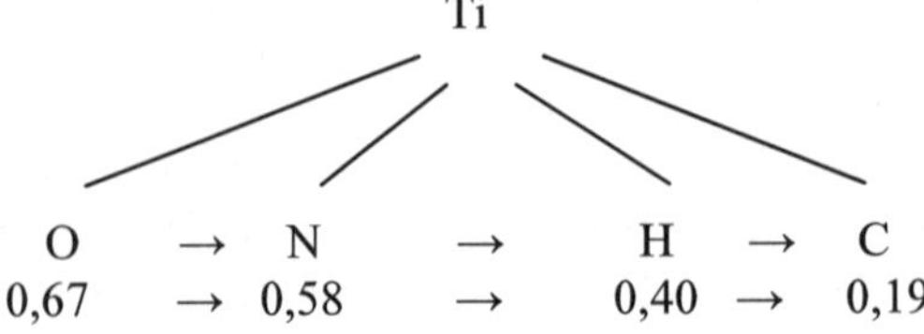

Atoms of gases in ion state have incomplete s- and p- levels, i.e. form unpaired outer electrons. Unpaired valence electrons of interstitial impurities on p-level, having lower ionization potential, under certain conditions can form ion links not only with atoms of titanium, but also between themselves. Common electronic pair of two unpaired electrons of different atoms of interstitial impurities is formed by co-valence link between two electrically negative ions. This electronic pair will circulate both around one atomic nucleus of gas and another nucleus of gas and will mutually increase distortion of the lattice of titanium.

One unpaired electron of s-level links of an atom of hydrogen links with one electron of p-level of an atom of oxygen by a co-valence link, thus forming an electronic pair circulating around both nuclei, and form a strong molecule of a hydrogen compound – hydroxide (OH). Around each nucleus of an element its own electronic cloud with certain electronic density is being formed. Thus atoms of gases, being positioned in octahedron pores of the crystal lattice of titanium, as interstitial impurities, cause its disturbance. Thus, created fields of distortions of the lattice impede motion of dislocations and increase strength and hardness properties of the powder. This same circumstance is corroborates both by sharp distortion of parameters of the lattice of the titanium powder and by appearing of different in their values correlations of hydrogen with interstitial impurities. While determining coefficients of correlations of hydrogen with interstitial impurities one can obtain the following sequence of values:

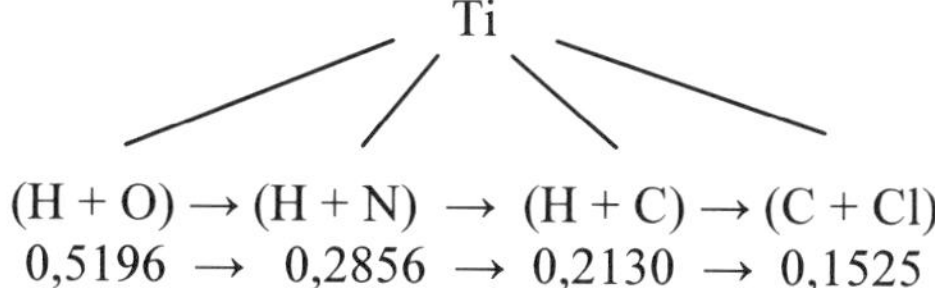

Such alignment groups of hydrogen with interstitial impurities and significant correlation coefficients are obviously related to possible forming of electronic pairs of hydrogen with gases. However, the strongest link corroborated by correlation-regression analysis is formed between oxygen and hydrogen.

By comparing characteristics of relations of presence of hydrogen with impurities in industrial powders different dependencies have been established (Fig. 1). The ratio of content of oxygen to hydrogen or vice versa was taken as a conventional unit, as well as contents of groups of atoms on condition that they are equal and for them actual difference between H/O and groups (Cl + H)/(O + N) is determined.

Comparing the ratios of contents of hydrogen to oxygen with the standard for powders of the same way of manufacturing we conclude that in powders of improved purity it exceeds unity (by 3.5 times), while in others it is below unity. This is caused by higher contents of hydrogen in refined powders. In hydrometallurgy processing, due to high activity of titanium to gases and high speed of diffusion of hydrogen in it, the metal has time to get saturated with hydrogen.

For studying of the influence of hydrogen on the hardness of the powder all gaseous impurities were split in different groups: (Cl + H), (O + N + C), (O + N), (Cl + N + O + C), (H/O), (O/H). If we are to consider ratios of sums between groups (Cl + H)/(O + N) or (Cl + H)/(O + N + C), we can obtain the following declining sequence of values:

$\sum$(Cl+H)/ $\sum$(O+N+C) s/t → i/p → m/t → e/l → iod.
1,56 → 1,39 → 0,43 → 0,42 → 0,08

$\sum$(Cl+H)/ $\sum$(O+N) s/t → i/p → m/t → e/l → iod.
1,63 → 1,56 → 0,51→ 0,48 → 0,26

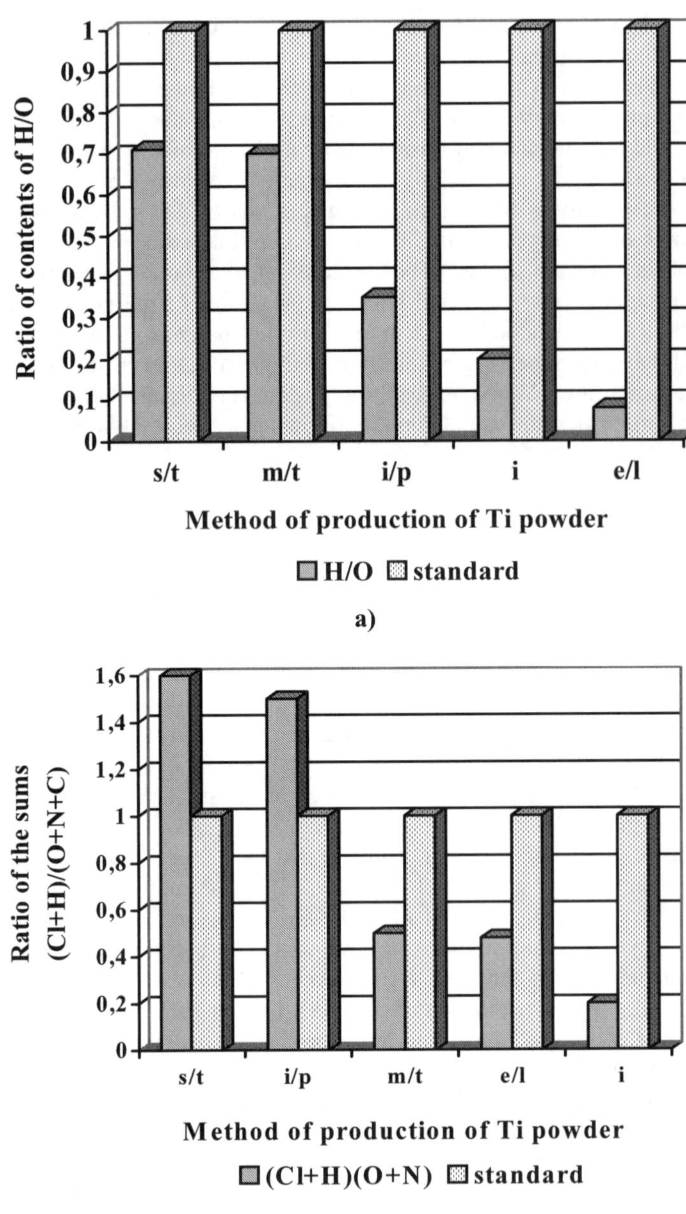

i – Iodide titanium; i/p – refined electrolytic powders of improved purity
e/l – Electrolytic titanium powders of industrial manufacturing
s/t – Titanium powders produced with sodium-thermal method
m/t – Titanium powders produced with magnesium-thermal method

Figure 1. Histograms of ratios:
(a) changes of contents of hydrogen and oxygen;
(b) changes of $\sum(Cl + H)/\sum(O + N + C)$ in titanium powders of different manufacturing

As it can be seen, the largest deviations of the values from the standard between groups of gaseous impurities is found in powders of sodium-thermal method of manufacture, and then, in powders of improved purity. This is seemingly related to the specific features of the way of manufacture, where powders contain increased quantity of chlorine and hydrogen, but also with the form, the degree of branching, size of the active surface of particles as well as with their granulometric composition.

It is known [6, 9] that interstitial impurities in titanium are positioned in interstices of octahedrons and tetrahedrons. In α-titanium octahedron- and tetrahedron-shaped pores have the following dimensions: 0.062 and 0.034 nm respectively, and in β-titanium 0.022 and 0.041 nm respectively. Authors believe, that interstitial atoms are differently positioned in pores of α- and β-titanium [9-11].

Positioning of interstitial atoms in titanium powders will also depend on the speed of diffusion of gases into the particle. The speed of diffusion of atoms of gases in titanium is different and, in turn, depends on the radiuses of the interstitial atoms and the sizes of the pores, which they get into (Table 3).

TABLE 3. Positioning of interstitial impurities in the lattice of titanium powders

Parameters	Titanium			Hydrogen	Oxygen	Nitrogen	Carbon	Chlorine
Atomic radius of an element (nm)	0.145			0.041	0.068	0.074	0.077	0.099
Radius of the pore (nm)	α-Ti	Octahedron	0.062	–	+	+	+	+
		Tetrahedron	0.034	–	–	–	–	–
	β-Ti	Octahedron	0.022	–	–	–	–	–
		Tetrahedron	0.041	+	–	–	–	–
Solubility of gases in titanium (%)	α-Ti	–		7.9	66.7	21.4	Under heating 1–1.9%	–
	β-Ti	–		44.0	5.5	6.5	882°C - 0.6%; 1,750°C - 3.1%	Under heating
Diffusion coefficient of gases in titanium (m^2/s)	α-Ti	–		$1.5 \cdot 10^{-9}$ [2]	$1.6 \cdot 10^{-10}$ [5]	$1.2 \cdot 10^{-10}$ [12]	$0.5 \cdot 10^{-12}$ [2] $7.9 \cdot 10^{-7}$ [12]	$0.5 \cdot 10^{-13}$ [2]
	β-Ti	–		$2.0 \cdot 10^{-9}$ [2]	$0.7 \cdot 10^{-10}$ [5]	$0.7 \cdot 10^{-10}$ [12]	$3.18 \cdot 10^{-7}$ [12]	$0.03 \cdot 10^{-13}$ [12]

\+ Pores, where interstitial impurities are positioned
– Pores, in which there are no interstitial impurities in lattices of titanium powders

It follows from Table 3 that the speed of diffusion of an atom of hydrogen is high both in α-Ti and in β-Ti in comparison with velocities of diffusion of other interstitial elements. Such an atom, getting into an octahedron-shaped pore of α-titanium, whose size is much larger than those of the atom of hydrogen itself, under the influence of its oscillation with large amplitude it easily leaves this pore. While in the tetrahedron-shaped pore of α-titanium an atom of hydrogen cannot penetrate since this pore is small for it. This is the cause of low solubility of hydrogen in α-titanium. In β-titanium this is reversed: octahedron-shaped pores (0.022 nm) are too small for hydrogen, while tetrahedron-shaped pores (0.041 nm) are comparable with the size of ionized atom. Besides, octahedron-shaped pores of α-Ti are elastic, they are easily stretched, and an interstitial atom, having got into it, cannot leave it easily.

Oxygen also has high speed of diffusion 1.6×10^{-10} m^2/s, however the radius of its atom is only 10% larger than the size of the octahedron-shaped pore and 54% larger than that of tetrahedron-shaped pore of α-Ti. Entering the lattice of β-Ti, an atom of oxygen is three times as large as an octahedron-shaped pore and 60.3% larger than tetrahedron-shaped pore. Therefore it is mainly found in α-titanium, and not in β-titanium. Besides, formation of oxides in titanium takes energy by an order higher than formation of hydrides of titanium. Nitrogen, carbon and chlorine have lower velocities of diffusion in the lattice of titanium, at that, enthalpy of their compounds with titanium are much higher than that of titanium hydrides. Therefore despite the high values of heat of formation of non-metal compounds, for oxygen, nitrogen and carbon it is thermodynamically more reasonable to be positioned in octahedron-shaped pores of α-Ti, distorting the crystal lattice the stronger, the larger is the difference in sizes between the interstitial atom and the pore.

While building analytical dependences with the use of "STATISTICA" program and determining linear and quadratic dependences of hardness on hydrogen and each interstitial impurity or of hardness on hydrogen and all possible groups, one can obtain the different outcomes (Fig. 2).

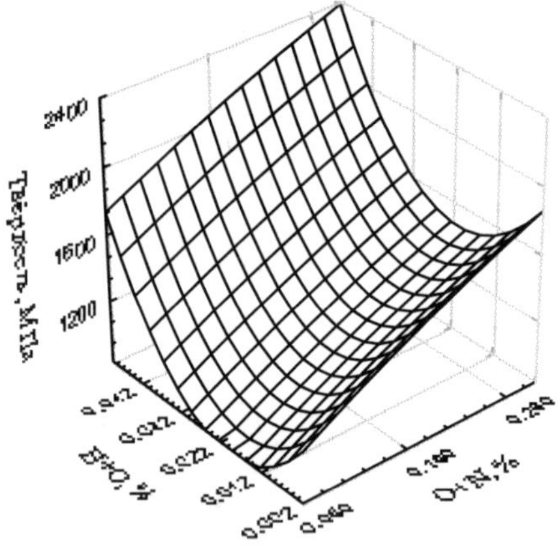

Figure 2. Combined influence of hydrogen

It has been established that quadratic equations reflect dependences of hardness of titanium powders on the contents of interstitial impurities more adequately than dependences obtained in correlation analysis. Building such dependences allows one to reveal simultaneous influence of hydrogen with interstitial impurities on the hardness of titanium powders, and also to establish, which of the groups have the strongest influence. Hydrogen with oxygen increases hardness of the powders sharper, than hydrogen plus oxygen and nitrogen. This confirms the assumption that octahedron-shaped pores are primarily filled in by hydrogen and oxygen. Chlorine with hydrogen does not distort the parameters of the lattice of titanium powders, and thus do not influence hardness of titanium.

3. Conclusion

Thus, the established analytical dependence of influence of interstitial impurities on the hardness of titanium powders has been corroborated by different methods of statistical analysis, the bond of titanium with interstitial impurities has been studied on electronic level, as well as the bond between interstitial atoms on the surface of titanium powder, and also the bond between titanium and various groups of impurities.

References

1. Rubcov A. N., Olesov Yu. G., Antonova M. M. Hydrogenation of the titanium materials. – Kiev: "Naukova dumka", 1971. – 126 p. (in Russian).
2. Luchinskiy G. P. Chemistry of titanium. – Moscow: "Chemistry", 1971. – 470 p. (in Russian).
3. Pul'tsin N. M. The titanium-gases interaction. – Moscow: "Metallurgy", 1968. – 374 p. (in Russian).
4. Samsonov G. V., Pryadkov I. F., Pryadkov L. F. Configurative model of a substance.– Kiev: "Naukova dumka", 1971.– 319 p. (in Russian).
5. Kolachev B. A. Physical metallurgy of titanium. – Moscow: "Metallurgy", 1968. – 184 p. (in Russian).
6. Homchenko G. P. Chemistry. – Moscow: "Vysshaya shkola", 1993. – 226 p. (in Russian).
7. Shapovalova O. M., Babenko E. P., Babenko Yu. V. Correlation between hardness and micro-hardness of titanium powders obtained from converted wastes // *Vistnyk DNU, Dnepropetrovsk*, 1999, (3): 3 (in Russian).
8. Shapovalova O. M., Babenko E. P. Influence of the oxygen content in titanium of the hydrogen absorption // *Visnyk Akademii mytnoi sluzhby Ukrainy, Dnepropetrovsk*, 2003 (2): 74–79 (in Ukrainian).
9. Shapovalova O. M., Babenko E. P. The production of hydrides in titanic powders under different ways of manufacturing // Hydrogen Materials Science and Chemistry of Carbon Nanomaterials. – Sevastopol, Ukraine, 2005. – P. 539–546.
10. Goroschenko Ya. G. Chemistry of titanium. – Kiev: "Naukova dumka", 1972. –287 p.
11. Shapovalova O. M., Babenko E. P. Consumption of gaseous impurities in manufacturing of titanium powders // *Vistnyk DNU*, Dnepropetrovsk, 2003 (7): 23–26 (in Russian).
12. Kiparisov S. S, Levinskiy Yu. V., Petrov A. P. Titanium carbide: production, properties, use. – Moscow: "Metallurgy", 1987. – 216 p. (in Russian).

THE PECULIARITIES OF FORMATION OF ALLOYS STRUCTURE IN THE SYSTEM Ti-Zr-Hf-H

S.K. DOLUKHANYAN, A.G. ALEKSANYAN,
O.P. TER-GALSTYAN, D.G. MAYILYAN
A.B. Nalbandyan Institute of Chemical Physics of National Academy of Sciences of Armenia, 0014 Yerevan, Republic of Armenia
V.SH. SHEKHTMAN, M.K. SAKHAROV, S.S. KHASANOV
Institute of Solid State Physics of Russian Academy of Sciences, 142432 Chernogolovka, Moscow District

Abstract. In Ti-Zr, Ti-Hf and Ti-Zr-Hf systems, the alloys were received by developed in the IChPh of Armenian NAS principally new, "hydride cyclization", method, and the alloys crystal structures were studied. The X-ray analyses proved that at hydride ratios in charge: TiH_2/ZrH_2; $TiH_2/HfH_2 \approx 2$; 1; 0.44; $Ti_{0.67}Zr_{0.33}$ (Ti_2Zr), $Ti_{0.67}Hf_{0.34}$ (Ti_2Hf), etc., the alloys of ω-phase formed, unusual for the state diagram of the systems Ti-Zr and Ti-Hf at ambient pressure. The stability of ω-phases of studied alloys under thermal action was defined. The interaction of α and ω-phase alloys with hydrogen was studied in SHS regime. Complex hydrogen-rich hydrides of FCC and BCT structures - $Ti_{0.9}Zr_{0.1}H_{1.87}$ ($Ti_9Zr_1H_{18.7}$), $Ti_{0.67}Zr_{0.33}H_{1.81}$ ($Ti_2ZrH_{5.42}$); $Ti_{0.5}Zr_{0.5}H_{1.37}$ ($TiZrH_{2.74}$) and $Ti_{0.3}Zr_{0.7}H_{1.96}$ ($TiZr_{2.3}H_{6.53}$) were received.

Keywords: crystal lattice of Ti alloys, α- and ω-phase, metal hydrides, annealing; dehydrogenation

1. Introduction

The metals of IVB group attract the permanent attention of researchers due to peculiarity of their characteristics, influencing a package of their physical properties. Phase transformations in these alloys were characterized by large variability and complexity. For binary and ternary systems inside this group, the continuous regions of solid solutions based on stable α (HCP,) and β (BCC,) modifications of pure metals are typical. In a series of experiments, the formation of hexagonal ω-phase as a specific structural form of Ti, Zr, Hf and their alloys was defined also. Many publications are devoted to this problem [1–7]. The appearance of this modification usually is considered caused by action of pressure (40–70 kbar) as a thermodynamic parameter, as well as by active role of local distortions, mechanical tensions, doping components.

The active role of hydrogen in structural transformation is worthy of special attention. For example, earlier it was shown [8, 9], that the hydrogen-saturation of Ti_2Zr α-solid solution stabilized the hydride, the structure of which was characterized with metal atoms ordered in $MgCu_2$ type cubic configuration of Laves' phase. In these experiments, the hydrogenation process consisted in a preliminary thermal treatment of α-alloy in vacuum followed by hydrogen giving at cooling stage. The achieved hydrogen concentration level was 50–57 at.%. The initial α-modification of solid solution was restored after hydrogen removing from this alloy by vacuum annealing.

In IChPh of Armenian NAS, a principally new technique based on a high efficiency "hydride cycle" is elaborated for receiving the alloys of refractory metals [10]. In the present work, the results of investigation of structures of alloys obtained in the system titanium-zirconium-hafnium-hydrogen by this technique are presented.

2. Experimental procedures

Synthesis of ω-alloys in hydride cycle. The essence of elaborated technique consists in usage of hydrides of maximal composition, MeH_2, as initial materials for the alloys preparation. Ready-made mixture of hydride powders in the defined ratio was compacted. Afterwards, the hydrogen is removed from the compacted sample by vacuum annealing. At temperature enough for active dissociation of hydrogen, the process results in formation of alloy of given composition consisting in choused metal components:

$$Me^{I}H_2 + Me^{II}H_2 = Me^{I}Me^{II} + 2H_2\uparrow$$

In the current experiments, the necessary hydrides of high hydrogen content were synthesized using SHS (self-propagating high-temperature synthesis) method; their certification manifested, that they belong to two known structural types:

TiH_2 (a = 4.454Å) – FCC lattice, structural type cF12 (CaF_2, fluorite family);

ZrH_2 (a = 4.943 Å; c =4.463 Å; c/a = 0.9) and HfH_2 (a = 4.892 Å; c = 4.203 Å; c/a = 0.86) – BCT-lattice, structural type tI6 (ThH_2 family).

In this scheme of experiment from the hydride powder compositions: $TiH_2 + ZrH_2$, or $TiH_2 + HfH_2$, or $TiH_2 + ZrH_2 + HfH_2$ the formation of two- and three-component hydrides of indicated metals is reliably reproduced. The practical efficiency of this technique is underlined by the significant decreased process temperature regard to the levels of standard melting and diffusion annealing regimes of refractory metals (lower by 500–800°C). Let us note that the total time of thermal treatment is also practically short, no more than 60–90 min. Structural characteristics of obtained alloys are presented in Table 1; certification of samples was performed on the diffractometers «DRON-0.5» and «SIMENS D-500».

The results of experiments evidenced the peculiar dependence of structure of the obtained alloys on the quantitative characteristics of the initial charge composition of hydrides mixture. Principally, the x-ray analysis data indicated the phase composition of the obtained alloys included the solid solutions based on known modifications of the metal components. At that, it is worthy of attention the active influence of composition of charge hydride mixture on the phase characteristics of synthesized alloy. So, at charge ratio (at.%): TiH_2/ZrH_2; $TiH_2/HfH_2 \approx 2$; 1; 0.44, the process results in formation of alloys: $Ti_{0.67}Zr_{0.33}$ (Ti_2Zr), and $Ti_{0.67}Hf_{0.34}$ (Ti_2Hf), etc., which do not correspond to the equilibrium diagrams of titanium-zirconium and titanium-hafnium at ambient pressure.

Instead of standard for these compositions solid solutions based on α-modification, structural certification allowed fixing the existence of essential amounts of ω-phase, unusual for equilibrium state diagrams for titanium-zirconium and titanium-hafnium at ambient pressure. Identification of obtained diffractograms shown that, the volume part of ω-phase is on the level of 70% for Ti_2Zr alloy, and 60% for Ti_2Hf (Fig. 1).

TABLE 1. Structural characteristics of obtained alloys

The formula of alloy	Main phase: crystal lattice, type of structure	Lattice parameter (Å), *c/a*
Ti	HCP, **α**-phase,	a = 2.95; c = 4.68
$Ti_{0.8}Zr_{0.2}$ (Ti_4Zr)	HCP, **α**-sol.sol.	a = 2.98; c = 4.743; ***c/a*** = 1.59
$Ti_{0.67}Zr_{0.33}$ (Ti_2Zr)	**ω**-phase, hexagon. AlB_2 type (70%),	a = 4.795; c = 3.020; ***c/a*** = 0.630
	HCP, **α**-sol.sol.	–
$Ti_{0.5}Zr_{0.5}$ (TiZr)	**ω**-phase, hexagon. AlB_2 type (70%),	a = 4.850; c = 3.017 ***c/a*** = 0.622
	HCP, **α**-sol.sol.	–
$Ti_{0.3}Zr_{0.7}$ ($TiZr_{2.3}$)	**ω**-phase, hexagon. AlB_2 type (70%),	a = 4.91; c = 3.13 ***c/a*** = 0.64
	HCP	–
$Ti_{0.2}Zr_{0.8}$ ($TiZr_4$)	HCP, **α**-sol.sol.	a = 3.196; c = 5.107; ***c/a*** = 1.59
$Ti_{0.8}\,Hf_{0.2}$	HCP, **α**-sol.sol.	a = 2.997; c = 4.782; ***c/a*** = 1.59
$Ti_{0.66}Hf_{0.3\,4}$	**ω**-phase, hexag., AlB_2 type (60%),	a = 4.8; c = 2.929 ***c/a*** = 0.61
	HCP, **α**-sol.sol.	–
$Ti_{0.34}Zr_{0.33}Hf_{0.33}$	**ω**-phase, hexagon., AlB_2 type (70%),	a = 4.82; c = 3.00 ***c/a*** = 0.622
	HCP, **α**-sol.sol.	–
$Ti_{0.66}Zr_{0.22}Hf_{0.12}$	**ω**- phase, hexagon., AlB_2 type (70%),	a = 4.76; c = 2.899 ***c/a*** = 0.609
	HCP, **α**-sol.sol.	–
Zr	HCP, **α**- phase	a = 3.25; c = 5.14

3. Discussion

On structural ω-modification in alloys. According to the literature [3], transformation of hexagonal α-modification of Ti and Zr to a new structure called ω-modification was observed in the experiments on pressure action of tens Kbar. This structure kept safe at room temperature after pressure remove. The results of x-ray structure analysis are known also indicating a metastable omega-phase appearance in some binary alloys of titanium or zirconium with transition metals (Cr, Mo, V, Mn, Nb, Ta). In this connection, the data are interesting showing that the structure of suboxide Ti_2ZrO in the metallic 'frame' of unit cell is identical to configuration of ω-phase [7].

According to the reference data [1], ω-phase structures of titanium and zirconium belong to the crystallographic family of hexagonal aluminum diboride, AlB_2: structural type **hP3,** space group **P 6/mmm,** coordinates of atoms:

1(a): 0, 0, 0
2(c): 1/3, 2/3, 1/2; 2/3, 1/3, 1/2;
lattice parameters: **ω-Ti a = 4.60Å; c = 2.82Å; c/a = 0.613**
ω-Zr a = 5.019Å; c = 3.089Å; c/a = 0.615

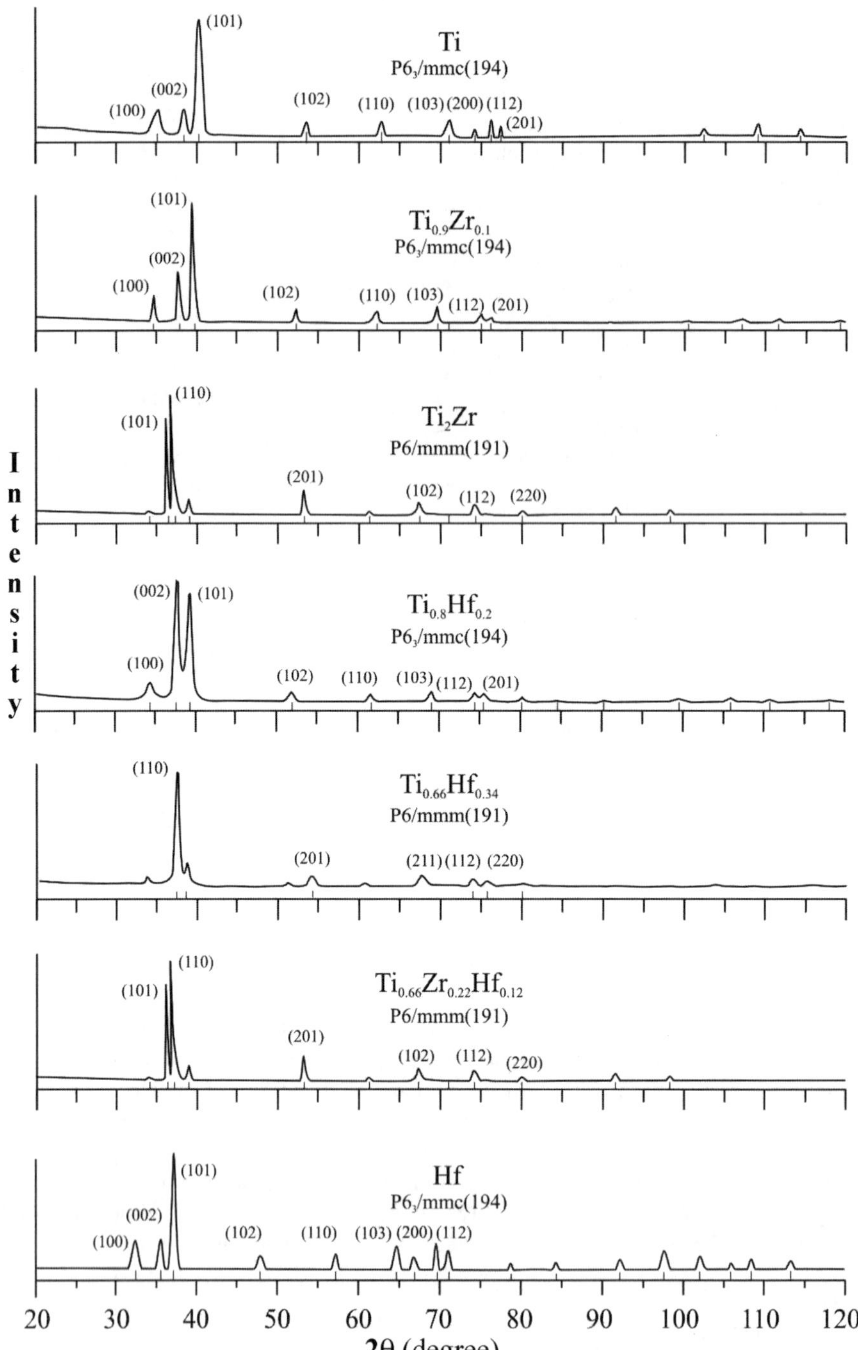

Figure 1. X-ray patterns of alloys

For this discussion important is that a main difference between **ω**-phase and **α**- and **β**-modifications of pure titanium (zirconium, hafnium) is a displacement of atoms of one element in two symmetrically nonequivalent positions, i.e. a specific ordering [2]. At that, inter-atomic distances in twofold position (on the half-height plane of hexagonal cell) correspond to value L = 0.577· **a** = **a**· 2\3 cos60°. The obtained numeral values are: L = 2.65 Å for ω-Ti; and L = 2.90 Å for ω-Zr.

For comparison: interatomic distances, L, on basal planes of low-temperature modifications correspond to the lattice period: in α-Ti a = 2.95 Å; in α-Zr a = 3.231 Å. Therefore, we can suggest that in pure metal, alike to binary alloy, this specific ordering allows to consider ω-phase as a peculiar intermetallic.

In works [4, 5], the results of investigations of the thermal stability of ω-phase in alloy TiZr are presented. The authors observed hexagonal ω-phase under pressure 40–70 Kbar. The annealing of this ω-phase under 400°C returned the alloy to the original crystal structure, i.e. the initial α-modification of solid solution restored.

We also studied the stability of ω-phase at thermal treatment. A special series of experiments was carried out on annealing the synthesized alloys having ω-phase structure: Ti_2Zr; Ti_2Hf; and of the other compositions. The regimes of annealing were chosen at several temperatures in the range 500–1,000°C and expositions between 30–120 min. The results of annealing have shown that the ω-phase transformation to the modification of α-sol. Sol., corresponding to the equilibrium state diagram, did not occur. In other words, the structure, formed in the 'hydride cycle', at high temperature appeared noticeably more stable, than the ω-phase obtained under action of high pressure.

The presented results show that the obtained ω-phases in alloys Ti_2Zr; TiZr, Ti_2Hf, and other compositions, might be considered as peculiar intermetallics having fixed positions of components.

Of a special interest is the analysis of dimension ratios and their transformations at phase conversions. In Fig. 2 a general view and dissection of a unit cell for alfa and omega phases are presented.

Let us consider the characteristics of interatomic distances and effective atomic radiuses according [2, 3, 6] and the obtained experimental data. Here $R_{ef.}$ was calculated as L/2, where L is interatomic distance in given plane.

Literary data:
α-Ti: a = 2.95 Å, $R_{ef.}$ = 1.475 Å. Atomic radius: R = 1.45 Å;
ω-Ti: a = 4.6 Å, z = 0; $R_{ef..}$ = 2.3 Å, z = ½; $R_{ef.}$ = 1.33 Å;
α-Zr: a = 3.231 Å; $R_{ef.}$ = 1.615 Å; Atomic radius: R = 1.59 Å;
ω-Zr: a = 5.019 Å; z = 0; Ref. = 2.51 Å; z = 1/2; $R_{ef.}$ = 1.45 Å.

Experimental data:
ω-Ti_2Hf: a = 4.8 Å; z = 0; $R_{ef.}$ = 2.4 Å, z = ½; $R_{ef.}$ = 1.38 Å;
ω-Ti_2Zr: a = 4.795 Å; z = 0; $R_{ef.}$ = 2.397 Å; z = ½; $R_{ef..}$ = 1.38 Å;
ω-$TiZr_2$: a = 3.99 Å, z = 0; $R_{ef.}$ = 1.995 Å; z = ½; $R_{ef.}$ = 1.15 Å.

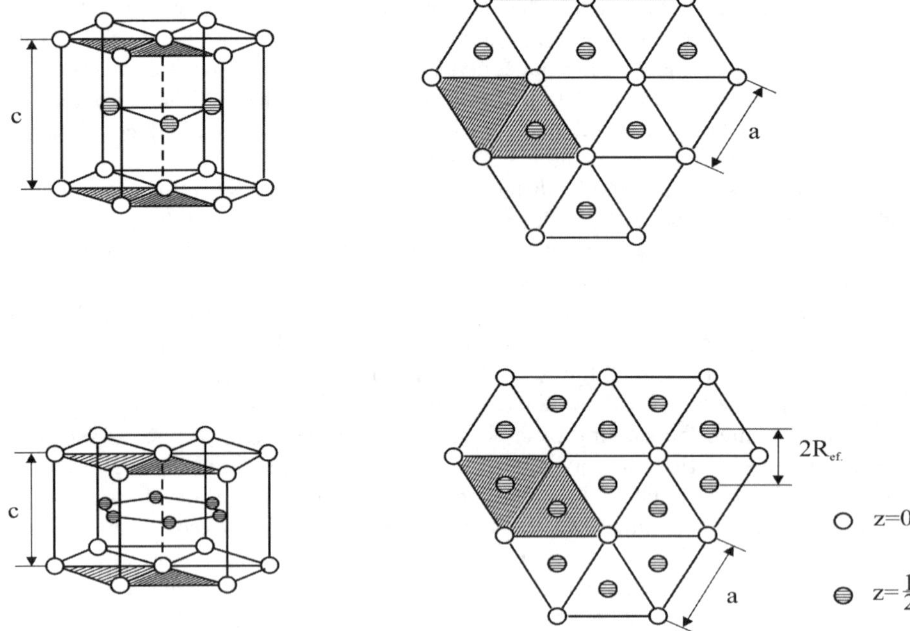

Figure 2. A general view and dissection of a unit cell for alfa and omega phases

The presented values for ω-modifications (both for pure elements Ti and Zr, and for the obtained alloys of transition metals of IVB group) show that in plane z = 0, the lattice parameter **a** is noticeably larger than the diameters of atoms in this plane. It means that in the basal plane of unit cell, the closest packing of atoms configuration absents. Let us note, that for the plane on the level z = 1/2, the closest packing exists (Fig. 2), moreover, the compression of atoms takes place, i.e. the radiuses are smaller than in the literature data.

Taking into account the specific effect of hydrogen on the alloys structure, the interaction of obtained α and ω-phase alloys with hydrogen has been studied in SHS regime. As a result, the next hydrogen-rich hydrides of FCC and BCT structures have been received: $Ti_{0.9}Zr_{0.1}H_{1.87}$ ($Ti_9Zr_1H_{18.7}$); $Ti_{0.67}Zr_{0.33}H_{1.81}$ ($Ti_2ZrH_{5.42}$); $Ti_{0.5}Zr_{0.5}H_{1.37}$ ($TiZrH_{2.74}$) and $Ti_{0.3}Zr_{0.7}H_{1.96}$ ($TiZr_{2.3}H_{6.53}$). In Fig.3, X-ray patterns of hydrides of alloys are presented. The hydrogen removing from these hydrides (under vacuum annealing at 700–1,000°C) brings to the again formation of initial α and ω-phases, i.e. the reversible reactions are realized:

$Ti_{0.9}Zr_{0.1}$ (α-phase) + H_2 ↔ $Ti_{0.9}Zr_{0.1}H_{1.87}$ (FCC)
Ti_2Zr (ω and α-phases) + H_2 ↔ $Ti_2ZrH_{5.42}$ (BCT)
$Ti_{0.66}Zr_{0.22}Hf_{0.12}$ (ω and α-phases) + H_2 ↔ TiZrHf $H_{6.53}$ (BCT)

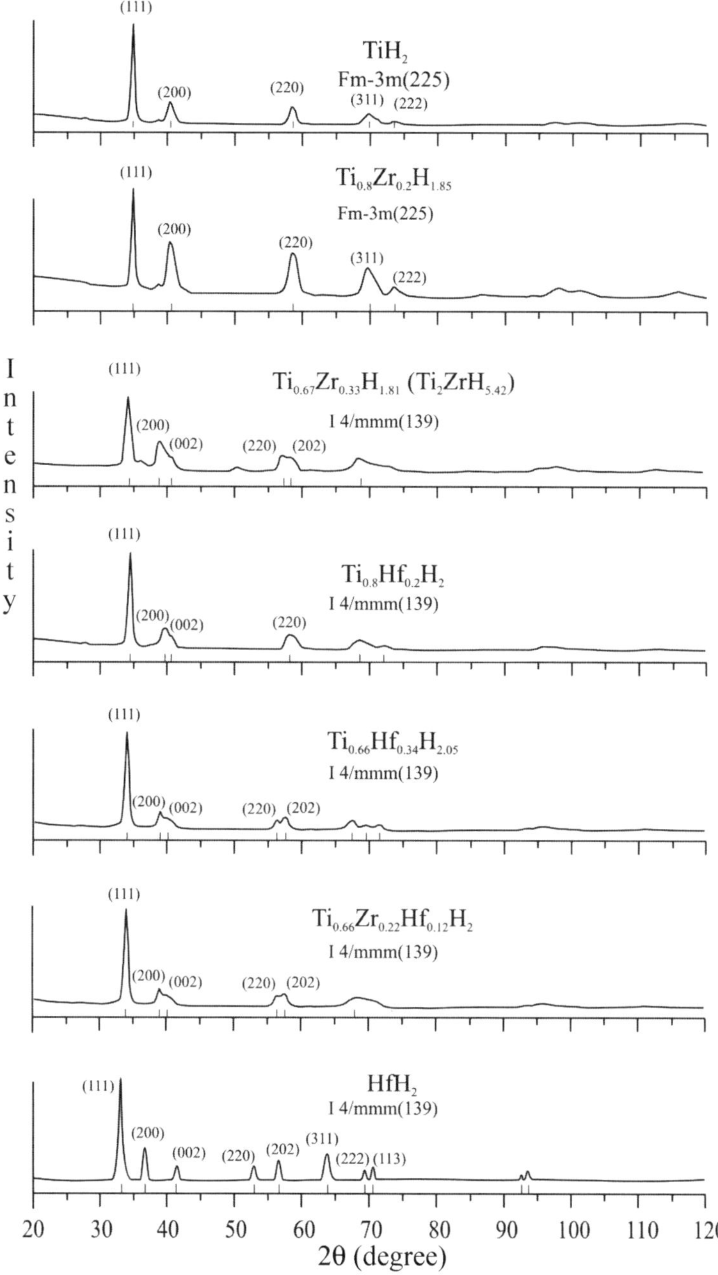

Figure 3. X-ray patterns of hydrides of alloys

The multiple repeating of cycle: the alloy hydride synthesis ↔ its complete dehydrogenation ended identically: the final products were the initial alloys of α and ω-structure.

A peculiar example was the titanium-hafnium composition, where the result of the reversible reaction depended on the form of initial sample, 'compacted' or "powder":

Ti_2Hf (ω and α-phase) + $H_2 \leftrightarrow Ti_2HfH_{2.74}$ (BCT) → Ti_2Hf (ω and α-phase) from compact sample; → Ti_2Hf (α-phase) from powder sample

A completely reversible reaction is realized at conditions of preliminary compacting of the hydride powder. Upon vacuum annealing of this hydride in the form of non-compacted powder, the end product of the reaction appeared α-modification of alloy.

TABLE 2. The parameters of crystal lattice of α and ω-phase alloys based on Ti_2Zr, TiZr and Ti_2Hf

Composition	Cryst. lattice	Calculation parameter	Experimental parameter
TiZr	a	4.81	4.85
ω-phase	c	2.95	3.017
+			
α- phase	a	3.08	3.009
	c	4.90	4.787
Ti_2 Zr	a	4.74	4.790
70% ω- phase	c	2.91	3.020
+			
30% α- phase	a	3.04	–
	c	4.83	–
Ti_2Hf	a	–	4.8
(from tablet of the hydride)	c	–	2.929
60% ω-phase			
+			
40% α-phase	a	3.03	3.052
	c	4.80	4.84
Ti_2Hf			
(from powder of the hydride) α-	a	3.03	3.052
phase	c	4.80	4.84

In Table 2, the parameters of crystal lattice of α and ω-phase alloys based on Ti_2Zr, TiZr and Ti_2Hf are presented. These data show that the parameters of α-phase alloy, obtained after dehydrogenation of $Ti_2HfH_{2.74}$ hydride powder and those of α-phase alloy obtained in the compacted sample (in mixture with ω-phase) coincide with the parameters calculated by the Vegard method. This result evidences the identical chemical composition of both alloys.

4. Conclusions

A series of experiments was conducted for receiving the alloys of IVB group elements using a special hydride cycle. It was established, that at definite ratio of initial mixture of hydrides, the obtained alloys have a specific ω-structure. It was shown, that this structure is noticeably more stable than ω-modifications of Ti and Zr, known as phases of high pressure.

Acknowledgements

The work is performed at financial support of ISTC Grant A-1249, and Ministry of Education and Science of Armenia, theme 0567.

References

1. Villars P. and Calvert L. *Pearson's Handbook of Crystallographic: Data for Intermetallic Phases* 1991, **3**: 2465, 2466, 2469 and 3238.
2. Gustav E.R., Schulze E.R. Metallophysik. Berlin: Akademie-Verlag, 1967.
3. Nosova G.I. Phase transformations in titanium alloys. Moscow: Metallurgia, 1968. pp. 120–162.
4. Tonkov E. High pressure phase transformations. Philadelphia, PA: Gordon & Breach, 1992, Vol. 2, pp. 682–691.
5. Bashkin I.O., Pagnuev A.Yu., Gurov A.F., Fedotov V.K., Abrosimova G.E., Ponjatovskij E.G. Phase transformations in the equal atomic TiZr alloy under pressure up to 70 kbar. *Fizika tverdogo tela* 2000, **42**: 163–169.
6. Gorelik S.S., Rastorguev L.N., Skakov Yu.A. X-ray and electro optical analyses. Moskow: Metallurgia, 1970 (in Russian).
7. Fykin L.E., Glazova V.V., Kornilov I.I., et al. Crystal structure of the Ti_2ZrO sub oxide. *Proceedings of AS USSR* 1968, **182**(3): 576–579.
8. Sokolova E.I., Padurez L.N., Shilov A.L. Abramov N.A. λ-phase of the $ZrTi_2$-D system. *Russian Journal of Inorganic Chemistry* 1990, **35**(12): 3200–3202.
9. Skripov A.V., Udovic T.J., Huang Q., et al. Neutron diffraction study of the location jf deuterium in the deuterium-stabilized $ZrTi_2D_{3.83}$ phase. *Journal of Alloys and Compounds* 2000, **311**: 234–237 .
10. Dolukhanyan S.K., Shekhtman V.Sh., Aleksanyan A.G., Ter-Galstyan O.P., Abrosimova G.E., Sakharov M.K. *Chemical Physics Reports*, 2007, **26**(11): 36–43.

RECEEIVING OF COMPACT CARBIDES AND CARBOHYDRIDES BASED ON TITANIUM AND VANADIUM

S.K. DOLUKHANYAN, N.N. AGHAJANYAN
A.B. Nalbandyan Institute of Chemical Physics of Armenian NAS, 5/2 P. Sevak Str., Yerevan 0014, Armenia

Abstract. Monophase carbides and carbohydrides based on titanium and vanadium have been received by the method of Self-propagating High-temperature Synthesis (SHS). It was shown, that vanadium insertion into system Ti-C-H influenced the structure-forming and thermal properties of final products. In the whole range of ratio of metals to carbon, carbohydrides of cubic FCC lattice are formed even at low concentrations of carbon. Solid, porousless, monophase samples of bimetallic carbide samples have been received by compacting and subsequent dehydrogenation the synthesized carbohydrides.

Keywords: combustion processes, self-propagating high-temperature synthesis (SHS), carbides, carbohydrides

1. Introduction

Intermediate phases of transition metals with carbon sub-lattice are of essential interest as hydride-forming compounds with special physical and chemical properties. The interest to carbohydrides is conditioned by widening of possibilities of new hydrogen containing materials, stable at higher operational temperatures than the existing binary hydrides of transition metals. These materials are perspective for using in nuclear power for biological protection.

Combustion processes in the system Me-C-H had been studied earlier by the method of Self-propagating High-temperature Synthesis (SHS). The main factors influencing the formation of carbohydrides of HCP and FCC crystal lattices had been stated by the authors. The differential-thermal analysis data proved the suggested by Samsonov G.V. increase of thermal stability of hydrides upon carbon insertion into crystal lattice of titanium hydride [1]. Really, the dissociation temperature of titanium carbohydrides and carbodeuterides appeared higher by 200°C of those for TiH_2 and TiD_2 [2–4]. Traditionally, carbohydrides are received by long-term homogenization of Ti + TiC mixture in vacuum at 1,000°C, followed by stepwise (100°C/min) cooling in hydrogen environment down to 500°C. It is a multi-stage laborious process, requiring complicate furnace equipment [1]. In this connection, the use of SHS in study of vanadium influence on the structure-formation and adsorptive properties of titanium carbohydrides was of particular interest. Multi-component interstitial phases are interesting because each element introduced into crystal lattice of binary compound can change its physical and chemical properties radically.

B. Baranowski et al. (eds.), *Carbon Nanomaterials in Clean Energy Hydrogen Systems.*

The aim of the present investigation was the obtaining of new hydrogen containing bimetallic compounds of transition metals. These materials are perspective for nuclear power as biological protection materials, and as composite solid alloys.

2. Results and discussion

The complicated carbohydrides with titanium and vanadium were synthesized in one technological stage: the mixture xTi + (1-x)V + yC, where $0.7 \leq x \leq 1$ and $0.35 \leq y \leq 0.5$, was pressed to cylindrical tablet (P = 10,000 kN) and ignited in the argon environment (P = 2 atm) then, hydrogen had been given (P = 10 atm). Based on the chemical and x-ray analyses, it was stated that the insertion of vanadium into system Ti-C-H essentially influenced the structure of the final products: in the whole researched range of metals and carbon ratio, only carbohydrides of cubic lattice were formed. At low concentration of carbon, a carbohydride of hexagonal structure did not form, as it was in the system Ti-C-H. At increasing of vanadium content, parameter of cubic crystal lattice decreased. With carbon content decreasing, the hydrogen content in FCC carbohydrides increased (Table 1). Undoubtedly, Ti/V ratio also influenced the hydrogen content.

Let us note that in the obtained bimetallic carbohydrides the sum of indexes on non-metals always is more than one. It is supposed that, the H atom occupies both tetra - and octapores in the crystal lattice, as it was stated for the system Ti-C-H (D) [5].

The process of carbohydrides annealing was studied on the Siverts' installation at 1,000°C, as well as the change of mass, volume and density at hydrogen removing from hydrogen-containing products (Table 1).

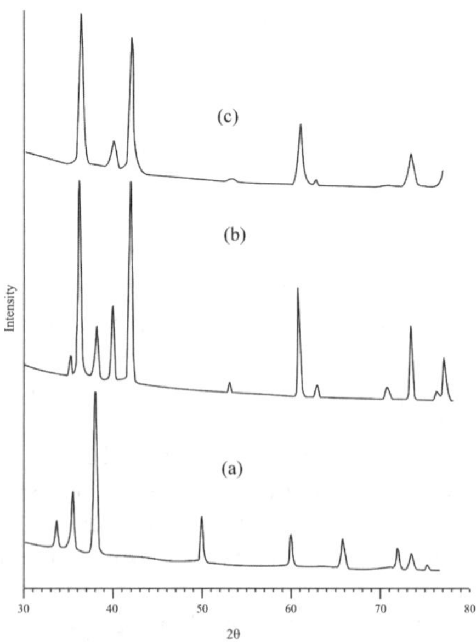

Figure 1. Difractograms of: (a) carbohydride $TiC_{0.4}H_{1.21}$ of HCP structure; (b) $TiC_{0.4}$ obtained after compacting and dehydrogenation; (c) $TiC_{0.4}$ after grinding of surface

TABLE 1. Characteristics of SHS carbohydrides and carbides after dehydrogenation

Initial charge	Empiric formula	Phase comp.	Non-metal content (wt.%)		Crystal lattice param. (Å)	Relative change of mass, volume, and density (%)		
			H_2	C		Δm	Δv	Δρ
Ti + H_2	TiH_2	FCC	4.06	–	4.458	–	–	–
Ti + C	TiC	FCC	–	20.0	4.327	–	–	–
Ti + 0.4C	$TiC_{0.4}H_{1.21}$	HCP	2.24	8.75	3.076 5.033	–	–	–
After annealing	$TiC_{0.4}$	FCC + HCP	–	8.75	4.295 2.937 4.709	1.95	25.97	32.45
0.9Ti + 0.1V + 0.35C	$Ti_{0.9}V_{0.1}C_{0.35}H_{1.02}$	FCC + FCC	1.91	7.85	4.286 4.45	–	–	–
After annealing	$Ti_{0.9}V_{0.1}C_{0.35}$	FCC	–	7.85	4.284	2.06	11.37	34.09
0.9Ti + 0.1V + 0.4C	$Ti_{0.9}V_{0.1}C_{0.4}H_{0.95}$	FCC	1.75	8.87	4.296	–	–	–
After annealing	$Ti_{0.9}V_{0.1}C_{0.4}$	FCC	–	8.87	4.281	1.74	23.71	28.64
0.9Ti + 0.1V + 0.5C	$Ti_{0.9}V_{0.1}C_{0.5}H_{0.67}$	FCC	1.22	10.8	4.296	–	–	–
After annealing	$Ti_{0.9}V_{0.1}C_{0.5}$	FCC	–	10.8	4.292	1.18	6.35	5.53
0.7Ti + 0.3V + 0.4C	$Ti_{0.7}V_{0.3}C_{0.4}H_{0.84}$	FCC	1.54	8.83	4.292	–	–	–
After annealing	$Ti_{0.7}V_{0.3}C_{0.4}$	FCC	–	8.83	4.278	1.47	5.38	4.09
0.7Ti + 0.3V + 0.5C	$Ti_{0.7}V_{0.3}C_{0.5}H_{0.55}$	FCC	0.98	10.9	4.281	–	–	–
After annealing	$Ti_{0.7}V_{0.3}C_{0.5}$	FCC	–	10.9	4.281	0.89	1.18	0.31
V + C	VC	FCC	–	17.0	4.118	–	–	–

The Fig. 1 presents the difractograms of: (a) carbohydride $TiC_{0.4}H_{1.21}$ of HCP structure; (b) sample $TiC_{0.4}$, obtained after compaction and dehydrogenation of carbohydride; and (c) the same sample after polishing the surface. The presented difractograms show, that at hydrogen removing from the crystal lattice, two-phase product was formed: non-stochiometric carbide $TiC_{0.4}$, and titanium of HCP structure (b). The polishing changed the ratio and intensity of phases (c).

The Figs. 2 and 3 present the difractograms of SHS carbohydrides $Ti_{0.7}V_{0.3}C_{0.4}H_{0.84}$ and $Ti_{0.9}V_{0.1}C_{0.5}H_{0.67}$ before and after dehydrogenation. As it is seen from the Figures, bimetallic carbohydrides have a mono-phase FCC structure. After annealing and complete removing of hydrogen, carbide holds its mono-phase, FCC structure.

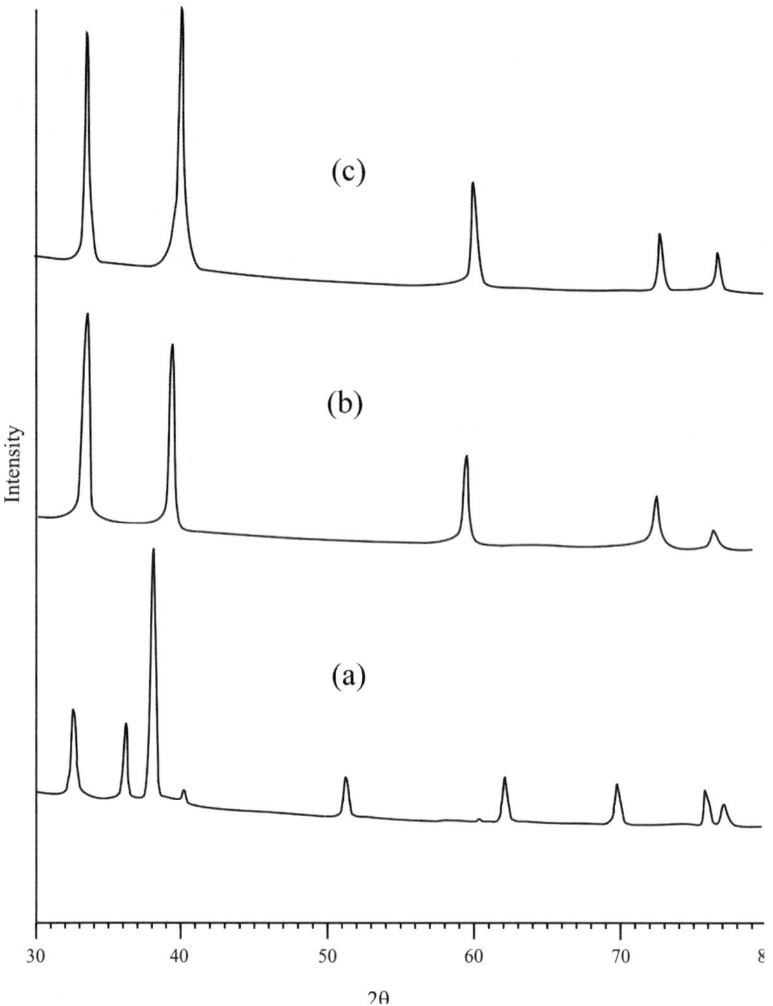

Figure 2. Difractograms of: (a) initial charge of metals and carbon; SHS carbohydride $Ti_{0.7}V_{0.3}C_{0.4}H_{0.84}$ before (b) and after (c) dehydrogenation

On the installation DERIVATOGRAPH Q-1500 D a process of differential-thermal decomposition of bimetallic carbides xTi-(1-x)V-yC-H was studied and the decomposition temperatures were determined. In Table 2, the Temperatures of DTA decomposition are shown.

In Fig. 4, DTA curves of decomposition of different bimetallic carbohydrides are presented. In Fig. 5 the DTA the curves of TiH_2 (a), $TiC_{0.4}H_{1.21}$ (b), $Ti_{0.7}V_{0.3}C_{0.4}H_{0.84}$ (c) decomposition are presented. It was demonstrated, that the vanadium insertion into crystal lattice of carbohydride brought to the decrease of the decomposition temperature. As it is seen in the Fig. 5, the main quantity of hydrogen is removed from TiH_2 at 610°C, from $TiC_{0.4}H_{1.21}$ at 835°C, from $Ti_{0.7}V_{0.3}C_{0.4}H_{0.84}$ at 380°C. It means that by the vanadium insertion into crystal lattice, a carbohydride can be synthesized with the low dissociation temperature, lower than that of binary titanium hydride.

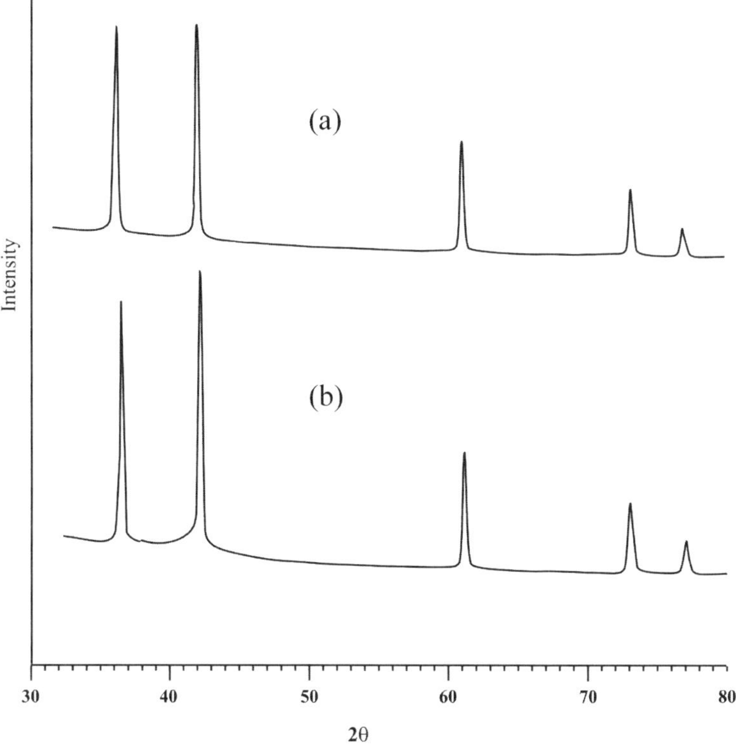

Figure 3. Difractograms of: SHS carbohydride $Ti_{0.9}V_{0.1}C_{0.5}H_{0.67}$ before (a) and after (b) dehydrogenation

TABLE 2. Temperature of decomposition of SHS produces

Empiric formula of material	Temperatures of DTA decomposition (°C)		
TiH_2	530	610	–
$TiC_{0,4}H_{1,21}$	510	570	835
$Ti_{0,9}V_{0,1}C_{0,35}H_{1,02}$	400	580	760
$Ti_{0,9}V_{0,1}C_{0,4}H_{0,95}$	410	470	–
$Ti_{0,9}V_{0,1}C_{0,5}H_{0,67}$	410	–	–
$Ti_{0,7}V_{0,3}C_{0,4}H_{0,84}$	380	580	–
$Ti_{0,7}V_{0,3}C_{0,5}H_{0,55}$	390	–	–

Simultaneously, the change of kinetic of distortion process was observed: $TiC_{0.4}H_{1.21}$ decomposes with three endo-effects (Fig. 5(b)), the bimetallic carbohydride – with two endo-effects (Fig. 4(b, d)), but upon increasing the content of carbon, decomposition proceeds with one endoeffect (Fig. 4 (c, e)).

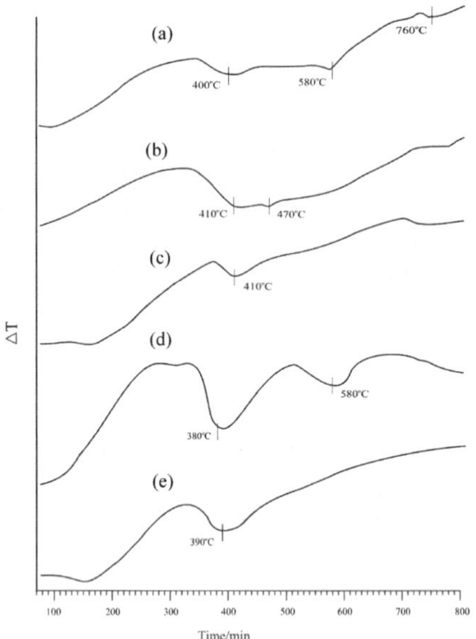

Figure 4. DTA curves of decomposition of different bimetallic carbohydrides: (a) $Ti_{0.9}V_{0.1}C_{0.35}H_{1.02}$; (b) $Ti_{0.9}V_{0.1}C_{0.4}H_{0.95}$; (c) $Ti_{0.9}V_{0.1}C_{0.5}H_{0.67}$; (d) $Ti_{0.7}V_{0.3}C_{0.4}H_{0.84}$; (e) $Ti_{0.7}V_{0.3}C_{0.5}H_{0.55}$

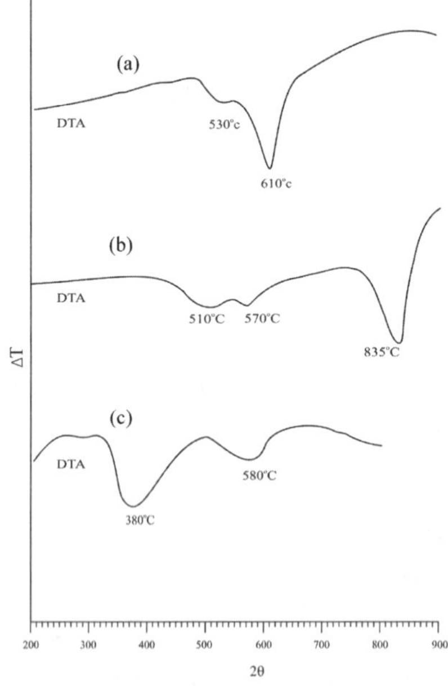

Figure 5. DTA curves of decomposition of: (a) TiH_2, (b) $TiC_{0.4}H_{1.21}$, (c) $Ti_{0.7}V_{0.3}C_{0.4}H_{0.84}$

The X-ray density of synthesized samples were calculated. The results are presented in Table 3. It can be noted that at dehydrogenation of carbohydrides, the X-ray density decreased. The density increased at increasing of the vanadium content in carbide. At identical ratio of metals, the carbon increasing in carbide also bring to the increase of X-ray density.

TABLE 3. X-ray density of carbides and carbohydrides

Material	Crystal lattice param. (Å)	Volume of unit cell (Å^3)	Molecular mass (g)	X-ray density (g/cm^3)
TiH_2	a = 4.458	88.597	49.88	3.738
TiC	a = 4.327	81.014	59.88	4.909
$TiC_{0.4}H_{1.21}$	a = 3.076 c = 5.033	41.242	53.894	4.339
$TiC_{0.4}$	for HCP a = 4.295 c = 2.937,	79.23	52.684	4.415
	for FCC a = 4.709	35.178	47.88	4.972
$Ti_{0.9}V_{0.1}C_{0.35}H_{1.02}$	a = 4.286	78.733	3.4196	4.505
	a = 4.454	88.359	49.88	3.748
$Ti_{0.9}V_{0.1}C_{0.35}$	a = 4.284	78.623	2.3895	4.424
$Ti_{0.9}V_{0.1}C_{0.4}H_{0.95}$	a = 4.296	79.285	53.94	4.517
$Ti_{0.9}V_{0.1}C_{0.4}$	a = 4.281	78.458	52.99	4.485
$Ti_{0.9}V_{0.1}C_{0.5}H_{0.67}$	a = 4.296	79.285	54.861	4.595
$Ti_{0.9}V_{0.1}C_{0.5}$	a = 4.292	79.064	54.191	4.551
$Ti_{0.7}V_{0.3}C_{0.4}H_{0.84}$	a = 4.292	79.064	54.442	4.572
$Ti_{0.7}V_{0.3}C_{0.4}$	a = 4.278	78.293	53.602	4.546
$Ti_{0.7}V_{0.3}C_{0.5}H_{0.55}$	a = 4.281	78.458	55.353	4.685
$Ti_{0.7}V_{0.3}C_{0.5}$	a = 4.281	78.458	54.803	4.638
VC	a = 4.118	69.833	62.951	5.986

3. Conclusions

By SHS method, monophase carbides and carbohydrides based on titanium and vanadium are obtained. It is shown, that insertion of vanadium into the system Ti-C-H influence structural and thermal properties of final products. By compaction and following dehydrogenation of synthesized carbohydrides, solid, porousless monophase samples of bimetallic carbides are obtained.

Influence of dispersion of initial powders of bimetallic carbohydrides, pressure of pressing, temperature of dehydrogenation on mechanical properties of final carbide samples is investigated.

References

1. Samsonov G.V., Antonova M.M., Morozov V.V. Triple compounds of systems Me-C-H and Me-N-H. *Poroshkovaya metallurgiya*, 1970 (4): 66–69.
2. Martirosyan N.A., Dolukhanyan S.K., Merzhanov A.G. Critical events at combustion of mixtures like Ати+Вв+Сг. *Fyzika goreniya I vzryva*, 1981 (4): 24–29 (in Russian).
3. Dolukhanyan S.K., Aghajanyan N.N., Hakobyan H.G., Shekhtman V.Sh., Ter-Galstyan O.P. "SHS-Carbohydride on the Base of Ti and Ti-V." International Symposium on Self-Propagating High-temperature Synthesis (SHS-99), Moscow, August 16–19, 1999, P. 58.

4. Agajanyan N.N., Dolukhanyan S.K., Shekhtman V.Sh., Arutjunyan Kh.S., Tamamyan N.G., Ter-Galstyan O.P., Thermal-radiation synthesis of titanium carbohydrides. *Khimicheskaya fizika*, 2001, **20**(3): 42–46 (in Russian).
5. Renaudin G., Yvon K., Dolukhanyan S.K., Aghajanyan N.N., Shekhtman V.Sh. Crystal structures and thermal properties of titanium carbo-deuterides as prepared by combustion synthesis. *Journal of Alloys and Compounds*, 2003, **356–357**: 120–127.

INTERACTION OF INTERMETALLIDE TiAl WITH AMMONIA

V.N. FOKIN, M.S. DULYA, E.E. FOKINA, D.N. BORISOV,
I.I. KOROBOV, A.A. POLETAEV, B.P. TARASOV
Institute of Problems of Chemical Physics of RAS
Prospectus of the Academician Semenov, 1, Chernogolovka,
Moscow Region, Russian Federation, 142432
Tel./Fax: +7(496) 5221743; E-mail: fvn@icp.ac.ru

Abstract. Processes of hydrogenation, hydronitridation, disproportionation, and dispergation which occur in system TiAl–NH_3 in an interval of the temperatures 100–500°C under pressure of 1.5 MPa were investigated in the presence of activator NH_4Cl. Phase transformations were researched and conditions of obtaining of fine-dispersed intermetallide powder with the average size of particles of 0.3 μm were determined.

Keywords: intermetallic compound, ammonia, hydronitridation, phase transformation, hydride, nitride, fine-dispersed powder

1. Introduction

The breaking point and other mechanical properties of the titanium widely used in the aerospace industry are considerably improved at alloying with aluminum [1]. In some cases for crushing cast titanium aluminides formed it is more expedient to use not so much external mechanical exposures, how many the internal stresses arising in a material at its interaction with hydrogen [2]. However such method of hydride dispergation concedes on a number of parameters to a method of ammoniac dispergation allowing, in particular, to receive powders with more advanced area of specific surface and, as consequence, with the smaller size of particles reaching of nanometer range in some cases [3].

In Ti–Al system the existence of several intermetallic compounds is revealed, from which TiAl having tetragonal structure (a = 3.988 Å, c = 4.076 Å) [4] interacts with hydrogen under pressure of 7,000 MPa at the temperature 300°C for 24 h with formation of the hydride appropriated to formula $TiAlH_2$. On the data of X-ray phase investigation, during hydrogenation TiAl disproportionates with formation of the titanium dihydride and the compound $TiAl_3$ not containing of hydrogen. The same reaction is marked and at realization of hydrogenation in conditions of lower pressure (150–180 MPa) at 400–600°C. After removal of pressure the product contains 0.7 wt% of hydrogen that corresponds to formation of the titanium dihydride [5, 6].

In the present work the experimental research of an opportunity of preparation of high-dispersed powders of TiAl was carried out in temperature interval of 100–500°C by treatment of intermetallic compound by ammonia generating hydrogen and nitrogen.

2. Experimental

The initial intermetallic compound TiAl was prepared by alloying the metal charge (purity: Ti – 99.98%, Al – 99.99%) in the furnace with a tungsten not spent electrode in an atmosphere of high purity argon with the subsequent annealing of alloy at 800°C for 2 weeks. The alloy was washed out by ethanol and ether for removal of oil traces and cleared from oxide film mechanically. For investigation the intermetallide samples were prepared as powders by mechanical crush of alloy regulus in a metal mortar with the subsequent screening of fractions with the size of particles of -200, -100 and -50 μm. The specific surface area of such powders makes 0.04–0.06 m^2/g. The composition of an initial alloy was established by X-ray phase analysis. The periods of an elementary cell of prepared intermetallide (a = 3.976 Å, c = 4.100 Å) were well coordinated to the literary data [4], as an impurity TiAl contained the trace quantities of the compound Ti_3Al. By results of the chemical analysis the alloy had composition $Ti_{1.02}Al$.

Ammonium chloride was dried by pumping out for 9 h at 150°C. The ammonia drained by metallic sodium had of 99.99% cleanliness.

The interaction of TiAl powders with ammonia (so-called process of hydronitridation) was investigated in the presence of 10 wt% (from quantity of intermetallide entered into reaction) of NH_4Cl as a reaction promoter at initial pressure of ammonia of 0.5–0.8 MPa.

The mixture of powders (0.8–1.0 g) was grinded in an agate mortar at room temperature and placed in a reactor-autoclave of a laboratory high-pressure setup (volume 60 ml). The reactor was evacuated up to residual pressure of ~1 Pa for 0.5 h at room temperature, filled by ammonia up to pressure of 0.5–0.8 MPa, and then the system was maintained at this temperature for 0.5 h. Further a reactor was heated up at various temperatures for 3 h, cooled up to room temperature, and again heated up. As a rule, at carrying out of process a pressure in system did not exceed 1.5 MPa. After realization of necessary number of heating-cooling cycles a reactor was cooled in an ammonia atmosphere up to room temperature, the gas phase was discharged into a buffer vessel. The solid products of reaction were unloaded in an inert atmosphere and analyzed.

The NH_4Cl removal from interaction products was carried out by two ways. On the first way the weighed sample (~0.1 g) was filled with 10 ml of absolute ethanol and subjected to mechanical hashing for 1 h at room temperature. Procedure repeated twice. By the second technique a mixture of products was vacuumized under pressure of ~1 Pa at 300°C for 3 h.

The X-ray graphical investigations were performed on a diffractometer ADP-1 (CuK_α-emission). The determination error of interplane distances did not exceed 0.005 Å.

The specific surface area of samples was determined on an amount of the low-temperature adsorption of krypton after a removal of the volatile products from a solid phase under a vacuum of 1.3×10^{-3} Pa at temperature of 300°C for 15 h and was calculated by the Brunauer-Emmet-Teller method. The determination error was ±10%.

The composition of the received phases was established by methods of volumetric and chemical analyses.

The content of hydrogen in the reaction products was defined by a standard method of burning the samples in a stream of oxygen. The content of nitrogen was determinated by the Kjeldal method. The content of chlorine was analyzed by a method of turbidimetry.

3. Results and discussion

It is established, that at interaction temperatures up to 200°C the intermetallide with the size of particles of 200 μm is rather stable in the ammonia atmosphere: in these conditions the metallic matrix of TiAl is kept, but there is a gradual accumulation of some of nitrogen and hydrogen in it that results in a formation of new introduction phases and gradual increase of a specific surface area of products. The sample which is found in an autoclave at 250°C under pressure of ~1.5 MPa contains maximal (for this series of experiences) quantity of hydrogen and trace quantities of nitrogen that corresponds to composition $TiAlH_{1.4}N_{<0.1}$. This sample has a specific surface area, exceeding the initial one on the order (0.3 against 0.04 m^2/g); the average size of particles is 5.0 μm. At that the parameters of a crystal lattice of a sample do not change practically. At the rise in reaction temperature up to 400°C there is a further increase of a specific surface area with preservation of a metallic matrix of intermetallide that testifies to increase of material dispersivity (the size of particles is 0.9–1.6 μm). At the same time in a metallic matrix there is an accumulation of nitrogen to simultaneous loss of hydrogen therefore the product of hydronitridation at 400°C has the composition $TiAlH_{0.4}N_{0.1}$ practically kept after pressure removal. On data of X-ray phase analysis, the ammoniac treatment of a alloy powder at ≥450°C results in the decomposition of hydridonitride phase with formation of cubic titanium nitride (a = 4.241 Å) and occurrence on the diffractogrames of the reflexes of new phase $TiAl_2$, i.e. alongside with process of nitridation in the mentioned conditions the reaction disproportionation takes place.

The data on investigation of intermetallide treatment with the sizes of particles of 100 and 50 μm with ammonia are given in Table 1.

The TiAl interaction with the size of particles of 100 μm with NH_3 begins already at 100°C (sample 1) and also is accompanied by formation of hydride phases with minor alteration of parameters of a TiAl crystal lattice (a = 3.992 Å, c = 4.066 Å). It is necessary to note, that products of the ammoniac treatment carried out at 100–150°C after pressure removal do not contain some hydrogen. During the further increase in temperature of ammoniac treatment up to 200°C the hydrogen accumulation in the intermetallide goes on proceeds (samples 2, 3) and under pressure the phase composition of a product (sample 3) corresponds to $TiAlH_{1.9}N_{<0.1}$, i.e. the sample contains the trace quantities of nitrogen. Thus the size of a specific surface area of the hydride phases obtained varies a little bit (0.4 m^2/g), and the size of particles makes 3.0–3.5 μm. At pressure removal there is a gas evolution, and a product of the hydronitridation which have been carried out at 250°C (sample 4) has the composition $TiAlH_{0.1}N_{<0.1}$.

At hydronitridation temperature of 300–350°C ammonia as the carrier of chemically connected hydrogen and nitrogen promotes the joint introduction of atoms of these elements in a crystal lattice of intermetallide, and the interaction product at 300°C (sample 5) has the maximal size of a specific surface area (5.7 m^2/g) and accordingly the smallest size of particles (0.3 μm) at formula composition $TiAlH_{0.4}N_{0.2}$

(after removal of pressure). At the same temperatures the decomposition of hydridonitride phase begins, and the further hydronitridation at 400–500°C (samples 7–9) proceeds similarly to interaction of TiAl with the size of particles of 200 μm with ammonia. The specific surface area of nitridation products obtained at 450°C makes 3.7 m^2/g (sample 8), at 500°C it is reduced up to 1.5 m^2/g (sample 9). The size of particles is 0.4 and 1.0 μm accordingly.

TABLE 1. Conditions and investigation results of intermetallide TiAl treatment by ammonia

Sample no.	Size of TiAl particles (mcm)	Conditions of treatment		Gross-composition of a product under pressure of 1.5 MPa	Specific surface area of sample (m^2/g)
		T (°C)	Time (h)		
1	100	100	25	$TiAlH_{0.7}$	0.5
2	100	150	30	$TiAlH_{1.5}$	0.4
3	100	200	30	$TiAlH_{1.9}N_{<0.1}$	0.4
4	100	250	30	$TiAlH_{0.9}N_{<0.1}$	0.5
5	100	300	30	$TiAlH_{0.6}N_{0.2}$	5.7
6	100	350	30	$TiAlH_{2.2}N_{0.2}$	3.1
7	100	400	30	$TiAlH_{0.5}N_{0.1}$	2.8
8	100	450	30	$TiAlH_{0.2}N_{0.1}$	3.7
9	100	500	30	$TiAlH_{0.1}N_{0.1}$	1.5
10	50	150	28	$TiAlH_{1.3}N_{<0.1}$	1.7
11	50	200	32	$TiAlH_{1.4}N_{<0.1}$	2.0
12	50	250	28	$TiAlH_{3.4}N_{0.2}$	2.5
13	50	300	28	$TiAlH_{0.6}N_{0.1}$	3.6
14	50	350	30	$TiAlH_{2.7}N_{0.1}$	2.0
15	50	400	30	$TiAlH_{0.6}N_{0.2}$	2.1

Ammonium chloride leaves easily at pumping out of samples at 300°C. Such operations reduce the nitrogen contents in some times (for sample 5 the nitrogen content decreases from 4.2 wt% up to 1.0 wt%) and practically completely delete from samples chlorine (the chlorine content in samples is <0.1 wt%). The NH_4Cl removal occurs effectively and at treatment of samples by ethanol. So, the nitrogen content in the sample 6 after treatment by ethanol has decreased with 3.6 up to 2.2 wt% also at practical absence in a sample of chlorine (<0.1 wt%). Efficiency of techniques of ammonium chloride removal is confirmed also with the data of X-ray phase analysis.

On an example of a 100-μm sample it is shown, that in absence of NH_4Cl the hydronitridation process practically does not go – after 30-h treatment at 300°C the reaction product had the gross-composition $TiAlH_{0.1}N_{<0.1}$ and the undeveloped specific surface area (<0.1 m^2/g).

From data of Table 1 it is visible that the ammonia treatment of intermetallide with the size of particles of 50 μm results in deeper hydrogenation of a metallic matrix in comparison with similar process for TiAl with larger size of particles, i. e. the products of the interaction which have been carried out at identical temperatures, as a rule, contain the bigger amount of hydrogen. The maximal composition on hydrogen is marked for a sample received at 250°C (under pressure sample 12 has the gross-composition $TiAlH_{3.4}N_{0.2}$). For an initial 50-μm alloy with rise in temperature of processing by ammonia the propensity of obtaining products to sintering is marked, that strongly complicates a course of all processes.

As a whole, the pattern of change of hydrogen amount in samples with the 50- and 100-μm size of particles of initial intermetallide is similar: the gradual increase of hydrogen quantity in samples passes through a maximum in the field of 200–250°C and then is reduced. The value change of specific surface area of the products, which were obtained from initial TiAl with the various sizes of particles, occurs also similarly: with rise in temperature the gradual increase of specific surface area reaches a maximum in the field of 300°C and is sharply reduced. Besides at temperatures of 250–300°C the crystal lattice of intermetallide contains a maximum quantity of the absorbed nitrogen which also promotes its destruction. Thus the fact of the high content of hydrogen in both series of the samples treated by ammonia at 350°C (samples 6 and 14) pays attention to itself, and thus of sharp reduction of quantity of a specific surface area. Probably, this temperature is optimum for decomposition of hydridonitride phase formed at lower temperatures.

Thus, the interaction of intermetallic compound TiAl with ammonia in the presence of NH_4Cl can be described by reactions the schemes of which (1–3) are submitted below:

$$TiAl + NH_3 \xrightarrow[NH_4Cl]{100\text{–}150^\circ C} TiAlH_x, \quad (1)$$

$$TiAl + NH_3 \xrightarrow[NH_4Cl]{200\text{–}350^\circ C} TiAlH_xN_y, \quad (2)$$

$$TiAl + NH_3 \xrightarrow[NH_4Cl]{400\text{–}500^\circ C} TiAl_2 + TiN_x. \quad (3)$$

The submitted schemes allow precisely enough to look after process of phase transformation from initial intermetallide to products of its full decomposition (TiN, $TiAl_2$) at increase of temperature and to note a special role of ammonia in these processes, consisting in decomposition of metal matrix TiAl.

Certainly, the mentioned schemes do not describe all processes occurring at intermetallide treatment by ammonia in the promoter presence. So, well known, that the formation of the hydride titanium anticipates the appearance of the titanium nitride. On diffractograms of interaction products the appropriate peaks are imposed against each other, and it is not obviously possible unequivocally to draw a conclusion on presence of hydride phase at products.

4. Conclusions

At investigation of interaction of the intermetallic compound TiAl with the size of particles of 200, 100 and 50 μm with ammonia in the presence NH_4Cl as process activator it is established, that

- Depending on interaction temperature in system TiAl–NH_3 there are the processes or of hydrogenation/hydronitridation with formation of hydride/hydronitride phase of initial intermetallide, or of disproportionation with isolation of a phase of new intermetallide.
- The use of an initial 100-μm powder is more preferable for chemical dispergation.
- At ammoniac treatment temperature of 300°C the hydridonitride of titanium aluminide is obtained which at pressure removal turns in fine-dispersed powder of intermetallide containing some of hydrogen and nitrogen and having size of particles of 0.3 μm.

Acknowledgement

The work is executed at financial support of the Russian Basic Research Foundation (grant No 07-03-00237).

References

1. Maeland A.J., Hauback B., Fjellvag H., and Sorby M. The structures of hydride phases in the Ti_3Al/H system, *Abstracts Book of 5th International Conference "ICHMS'97", Crimea, Ukraine*. 1997, P.318.
2. Fokin V.N., Tarasov B.P., Korobov I.I., and Shilkin S.P. The some generality of dispergation of intermetallic compounds and alloys on their basis, formed by the rare-earth and 3*d*-transitive metals, *Koordinatsionnaya Khim*., 1992, **18**: 526–534 (in Russian).
3. Fokin V.N., Shilkin S.P., Fokina E.E., Tarasov B.P., and Mozgina N.G. The synthesis and some properties of tetragonal nanocrystallic dihydride of the titanium, *Zh. Neorg. Khim.*, 1998, **43**: 13–16 (in Russian).
4. *Diagrams of state of double metal systems*: the Directory, Mashinostroenie, Moscow. 1996, **1**: 225 (in Russian).
5. Semenenko K.N., Verbetskii V.N., Zontov V.S., Ioffe M.I., and Tsutsuran S.V. The interaction of intermetallic compounds of the titanium with hydrogen, *Zh. Neorg. Khim.*, 1982, **27**: 1359–1362 (in Russian).
6. Hashi K., Ishikawa K., Suzuki K., and Aoki K. Hydrogen absorption and desorption in the binary Ti–Al system, *J. Alloy. Compd.*, 2002, **330–332**: 547–550.

HYDROGEN INDUCED MICROTWINS IN Cr ALLOYED FERRITE

K. LUBLINSKA, A. SZUMMER, K.J. KURZYDLOWSKI
Warsaw University of Technology, Faculty of Materials Science and Engineering, Woloska 141, 02-507 Warsaw, Poland
Fax: +4822 6281983; E-mail: klublin@meil.pw.edu.pl

Abstract. Hydrogen embrittles ferritic stainless steels at ambient temperature by a mechanism the details of which are still not fully explained. The present work was aimed at investigating more thoroughly the H-induced microstructural changes that take place in these steels. Polycrystalline and single crystals samples were examined after cathodic hydrogen charging using light, scanning and transmission electron microscopy. It was shown that hydrogen induces in the surface layer formation of needle-shaped microtwins in the ferrite grains. The orientation of the microtwins depends on the crystallographic orientation of a particular grain. The observed surface relief of these microtwins indicates that the hydrogen-induced transformation is connected with a volume increase. Inside the microtwins formation of microcracks can be noticed. A mechanism explaining formation of these microcracks was proposed.

Keywords: hydrides, microstructure, H-induced microtwins, hydrogen embrittlement, ferritic stainless steel

1. Introduction

Earlier studies showed that in the hydrogen charged ferritic and duplex stainless steels, formation of needle-shaped microtwins takes place [1–5] These microtwins, reveal a convex surface relief indicating a transformation associated with a volume change. Inside the microtwins, microcracks are formed indicating that hydrogen intake makes these materials embrittled. TEM observations have demonstrated substantial increase in the dislocation density inside and around the microtwins [1, 4]. The X-ray diffraction investigations have shown that hydrogen ingress expands the lattice parameter of the bcc structure of Cr alloyed ferrite by about 0.14% and is a source of high stresses exceeding the yield point [5].

The present study aimed at investigating more thoroughly formation of the H-induced microtwins taking place in Cr alloyed ferrite.

2. Experimental

The experiments were carried out on ferritic stainless steel grade 00H19T, on single crystals of ferritic stainless steel grade 0H16 and on duplex steel grade 0018N5M3S. Chemical composition of these steels are given in Table 1.

The specimens after initial grinding on abrasive papers were polished using diamond pastes and finally electropolished. Next, the specimens were cathodically charged with hydrogen at room temperature in 0.5M H_2SO_4 solution, with 1 mg/dm^3 As_2O_3, added as a promoter of hydrogen ingress, under current density 50 mA/cm^2,

B. Baranowski et al. (eds.), *Carbon Nanomaterials in Clean Energy Hydrogen System.*

TABLE 1. Chemical composition of the studied materials

Steel	C	S	P	Cr	Ni	Mn	Si	Mo	Ti
00H19T	0.025	0.011	0.021	19.30	0.15	0.36	0.41		0.41
0H16	0.060	0.014	0.006	16.00			0.10		
00H18N5M3S	0.042	0.019	0.005	17.90	4.90	0.28	1.72	2.65	

for a period of about 1 ÷ 2 hours using platinum anodes. Microstructures of samples were studied using light, scanning and transmission electron microscopy (LM, SEM, HRSEM, TEM). Microhardness measurements were done using diamond Vickers pyramid.

3. Results and discussion

Figure 1 shows the microstructure of the chromium ferritic stainless steel (grade 00H19T) before (Fig. 1(a)) and after (Fig. 1(b)) cathodic H-charging. It can be noted that hydrogen charging have produced at the surface the needle shaped microtwins with convex relief [2, 3]. Similar microtwins form in the ferrite grains in duplex stainless steel (grade 00H18N5M3S) (Fig. 3).

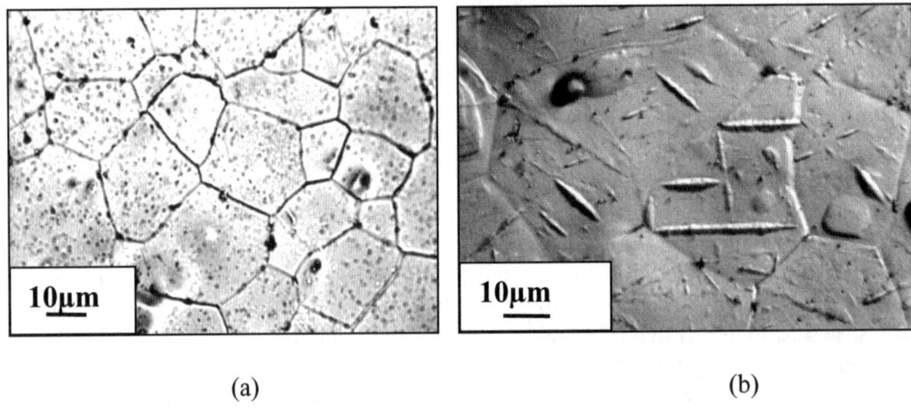

(a) (b)

Figure 1. Microstructure of ferritic 00H19T stainless steel. (a) Before and (b) after H-charging LM

The microtwins orientation depends on the crystallographic orientation of the sample surface. Figure 2(a) shows a single crystal sample with [100] orientation. In this case the long axes of microtwins are parallel to [002] or [020] directions. In the case of <111> orientation, the microtwins are parallel to <112> directions. Figure 2(b) shows the geometrical relationship between various directions of the microtwins (solid lines). As one can see these directions for the these crystal orientations [100], [110] and [111] lie on the same $(1\bar{1}0)$ plane.

It was found that the microhardness of the microtwin is higher than that of the parent ferrite grain (Fig. 3).

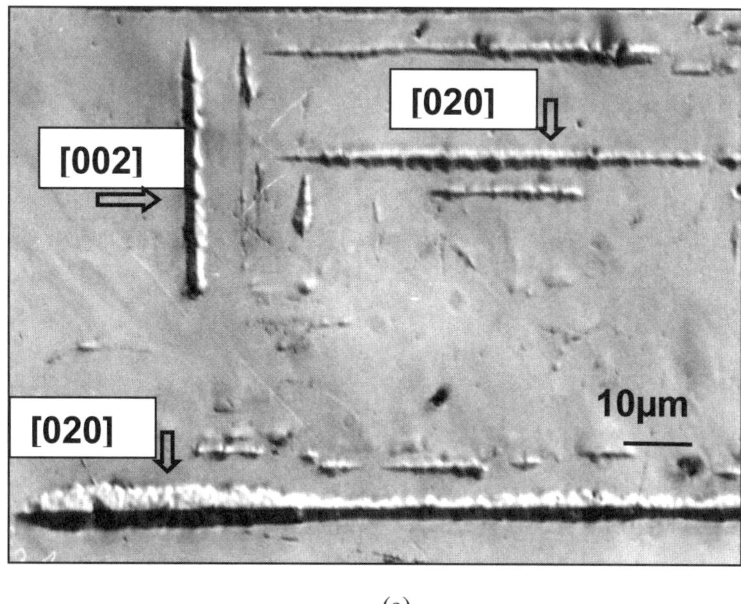

(a)

[$\bar{1}\bar{1}2$] [002] [002]
(110)
($1\bar{1}0$)
(100)
(111)

(b)

Figure 2. (a) Hydrogen induced "grain oriented" set of microtwins ⇩ in the form of needles forming the surface relief of ferritic steel single crystal (16%Cr). Matrix orientation [100]. (b) Geometrical relationship between the microtwins directions – solid lines on (110), ($1\bar{1}0$) and (111) planes

This is consistent with the TEM examinations of specimens after H-charging, which revealed heavily faulted regions with high dislocation density inside and around the microtwins (Fig. 4(a)). This can be explained in terms of the local volume increase of the H-induced microtwins, which cannot be accommodated only elastically but require also plastic deformation of the matrix leading to the dislocation density increase.

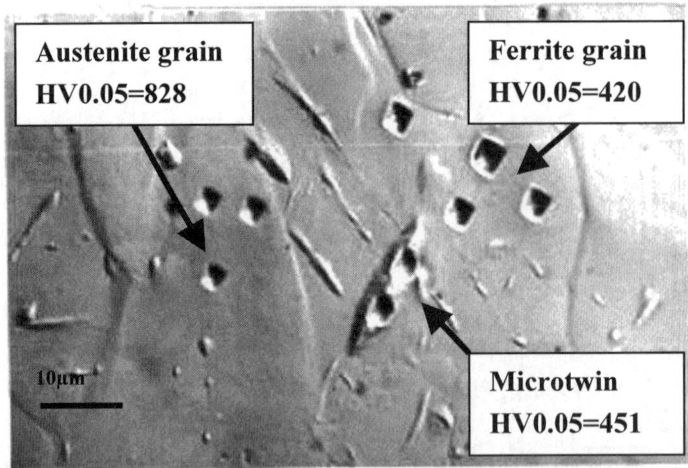

Figure 3. Microhardness indentations in H-charged duplex steel (grade 00H18N5M3S) and microhardness measurements results (HV0.05)

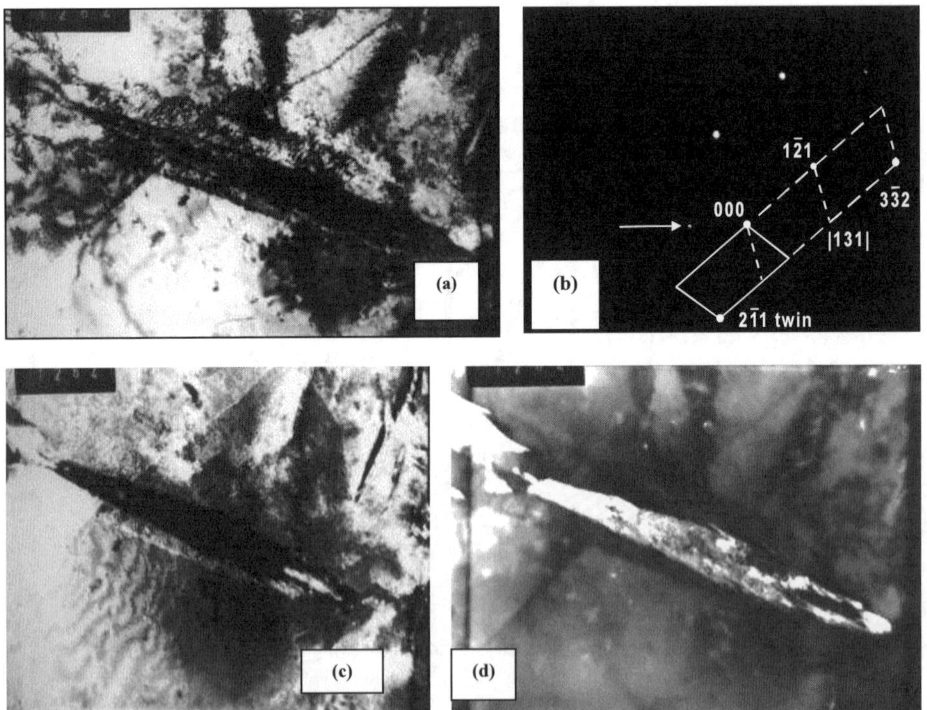

Figure 4. TEM image of H-induced microtwin in ferritic stainless steel. (a) Bright field, (b) diffraction pattern with matrix and twin (→) reflections, (c) dark field obtained from matrix reflection, (d) dark field obtained from twin reflection

The dark field image (Fig. 4(d)) obtained from one of the electron diffraction twin spot (Fig. 4(b)) demonstrates that the H-induced bright needle-shaped region has twin crystallographic orientation in relation to the "hosting" ferrite grain.

Figure 5(a, b) shows a high resolution SEM images of the "head" of the microtwin with several slip steps giving some insight into its growth mechanism.

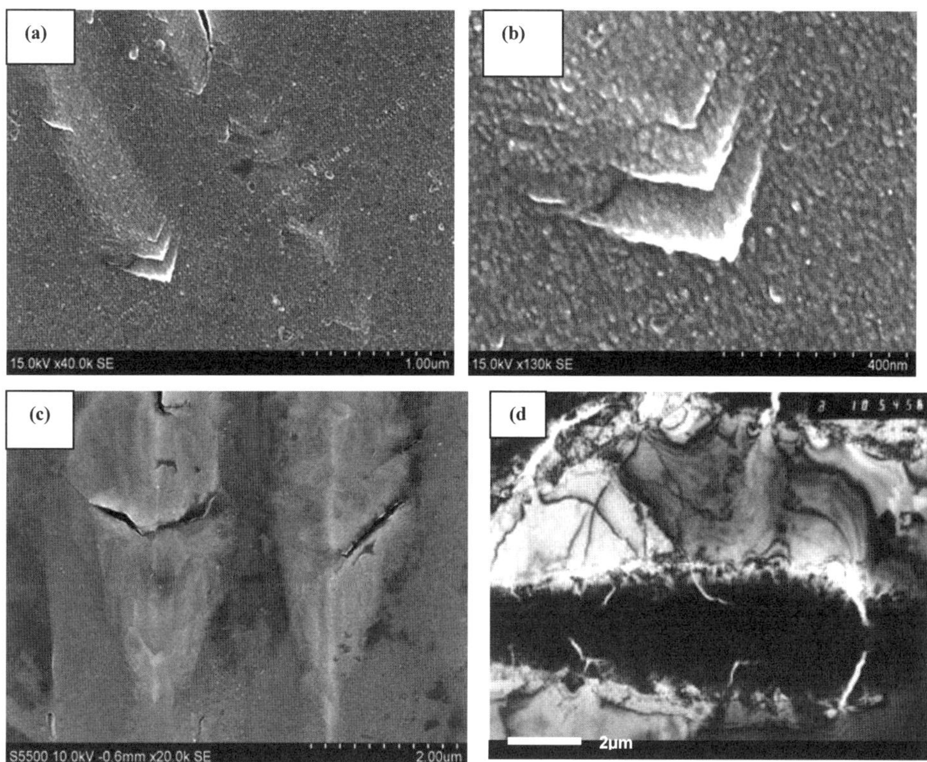

Figure 5. High resolution SEM image of the "head" of microtwin in ferritic stainless steel – (a, b) SEM (c) and TEM (d) image of microtwin with microcracks

Inside the microtwins one can frequently notice microcracks (Fig. 5) which are formed under the stresses related to a volume change due to formation of the H-induced microtwins. When H-charging is stopped, desorption of hydrogen causes the shrinkage of the microtwins which facilitates microcracks formation. This assumption is supported by the observation of disappearance of some microtwins soon after the H-charging (Figs. 6 and 7).

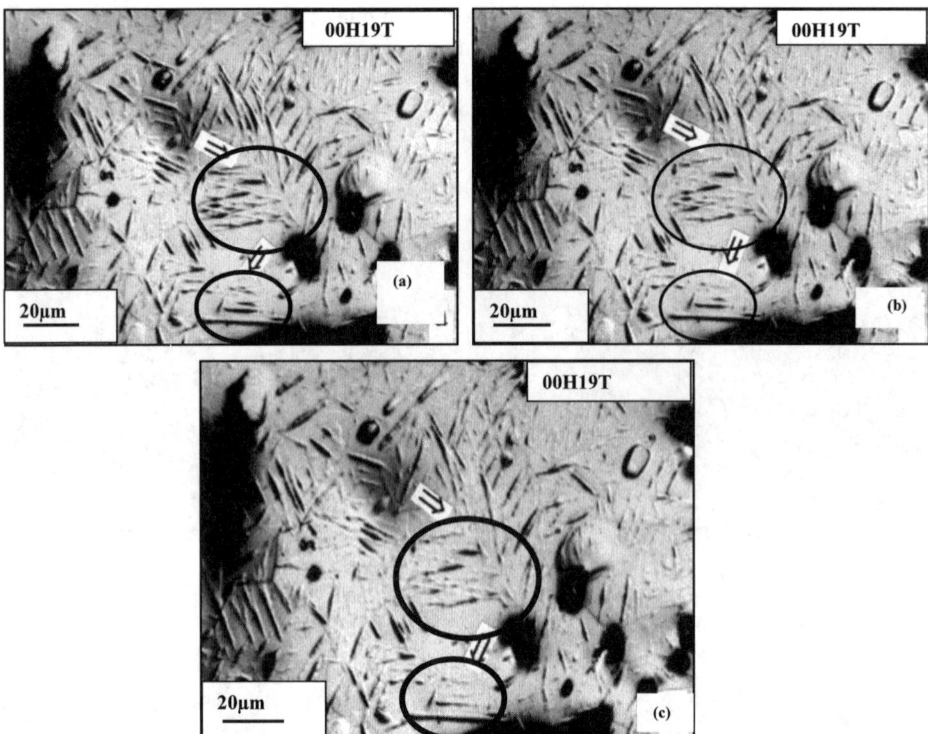

Figure 6. Microtwins in 00H19T ferritic stainless steel. (a) Immediately after H-charging was stopped, (b, c) several minutes later – some microtwins have disappeared

4. Conclusions

Cathodic hydrogen charging into ferritic stainless steels brings about the formation of grain-oriented needle-shaped microtwins with a pronounced surface relief revealing a transformation connected with volume change.

Inside the needles microcracks are formed, indicating that hydrogen entry makes these steels embrittled.

The hydrogen-induced microstructure changes are accompanied by a substantial increase in the dislocation density inside and around the needles.

A likely higher concentration of hydrogen inside the needles in the surrounding material expands their volume, thus bringing about a surface relief and plastic accommodation.

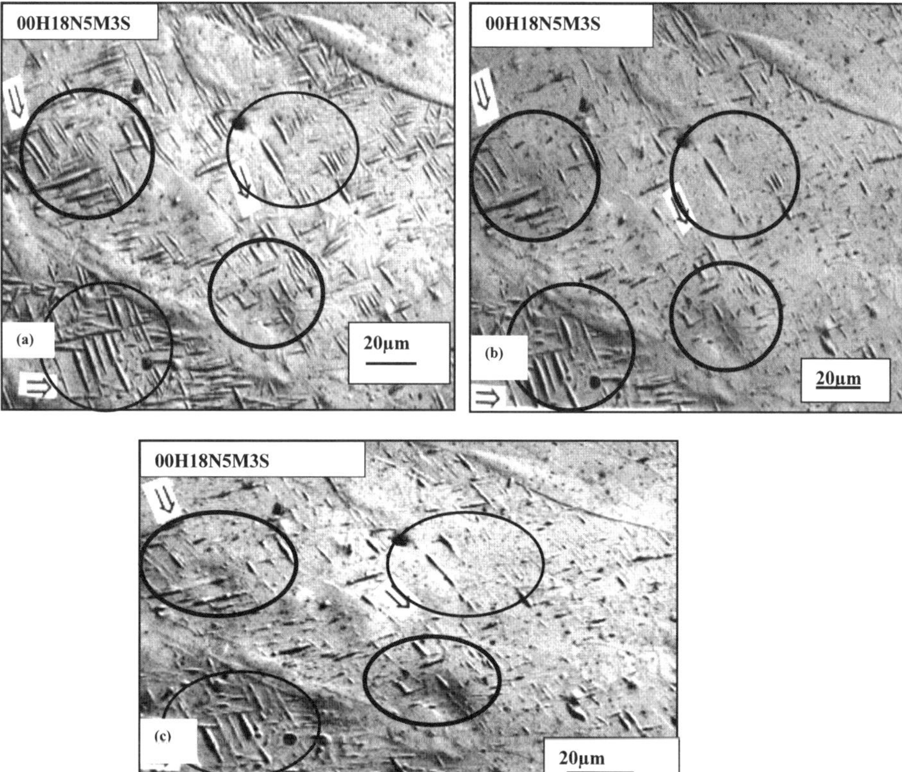

Figure 7. Microtwins in duplex steel (grade 00H18N5M3S). (a) Immediately after H-charging was stopped, (b, c) several minutes later – some microtwins have disappeared

Taken into account the properties of the hydrogen induced microtwins in the Cr alloyed ferrite, it is highly probable that these microtwins are hydrides [6]. The analogy exists in this respect with the Zr, in which hydride phase γ assumes similar form of the needle-shaped microtwins [7].

Acknowledgment

This work was supported by Polish Ministry of Science and Higher Education as grant PBZ KBN-117/T08/2005.

References

1. Tanino M., Komatsu H., Funaki S. Hydrogen induced martensitic transformation and twin formation in stainless steels, *Journal de Physique*, 1982. **43**: 4–503.
2. Szummer A., Lublinska K., Jezierska E., Hydrogen induced microstructure changes in single crystals of ferritic stainless steel, Fortschritte der Metallographie, **27** Eds. M. Kurz, M. Pohl, DGM Informationsgesellschaft, Verlag GmbH, 1995, P. 335–338.
3. Szummer A., Lublinska K., Jezierska E. Effect of cathodic hydrogen charging on structural changes in ferritic stainless steel, *Microstructural Science*, **23**, ASM Int. Metals Park, Ohio, USA, 1996, P. 141–147.
4. Szummer A., Jezierska E., Lublinska K. Hydrogen surface effects in ferritic stainless steel, *J. Alloys & Compounds*, 1999. **293-295**: 356–360.
5. Lublinska K., Szummer A., K., Kurzydlowski J. lattice parameter changes in hydrogen induced corrosion processes of ferritic stainless steel, *Surface Engineering*, 2005. **2A**: 167–174.
6. Lublinska K., Szummer A., K., Kurzydlowski J. In situ investigation of hydrogen effect on ferritic stainless steel, Hydrogen Economy and Hydrogen Treatment of Materials, Proc. V Inter. Conf. HTM-2007, Donetsk, Ukraine, 2007, P. 507–511.
7. Cotterill P., The hydrogen embrittlement of metals, *Progress in Materials Science*, 1961. **9**(4): 205–250.

HYDROGEN SORPTION IN LAYERED NANOPOROUS GaSe CRYSTALS

Z.D. KOVALYUK, A.P. BAKHTINOV, V.N. VODOP'YANOV, A.V. ZASLONKIN, V.V. NETYAGA
Chernivtsy Department of the Frantsevich Institute for Problems of Materials Science of the National Academy of Sciences of Ukraine, Iryna Vilde Str. 5, Chernivtsi, 58001, Ukraine

Abstract. By using the methods of atomic force microscopy (AFM) as well as infrared (IR) absorption and impedance spectroscopy the surface morphology, optical and electrical properties of layered GaSe crystals with nano-scale defects filled with hydrogen from gaseous phase are investigated. Hydrogen sorption is under investigations for layered GaSe crystals with surface defects different in morphology, density, and size. Changes in the impedance and IR spectra at hydrogen desorption from nano-porous H_xGaSe crystals are investigated. It is found that there are resonance variations in frequency dependences of the impedance, a giant change of the conductivity along the C axis (even to $10^7 \div 10^8$ times) typical for "*semiconductor – metal*" transitions, and an intensive light scattering in the IR spectral range. These phenomena are related to polytype phase transitions occurring in "soft" nano-porous H_xGaSe crystals at the transition of the hydrogen from liquid state of aggregation to gaseous one.

Keywords: hydrogen sorption, layered crystals, nanoporous crystals, GaSe, phase transition, optical absorption, impedance spectroscopy

1. Introduction

Investigation of the mechanisms of interaction between hydrogen and layered or porous nano-structured materials opens a possibility for their application as materials for hydrogen energetics. Materials with nano-scale cavities (nano-pores) having high specific surface and providing high room temperature sorption of hydrogen are prospective for accumulation and transportation of molecular hydrogen [1]. On the basis of a computer simulation by the method of molecular dynamics it was concluded in [2] that at sorption of hydrogen by a system of deformed carbon nanotubes it was possible the existence of molecular hydrogen in liquid state of aggregation at a temperature ≈80 K and a pressure P ~ 10 MPa.

Processes of electrochemical intercalation with hydrogen of layered InSe and GaSe single crystals and physical properties of the obtained intercalates have been investigated [3–5]. The carried out comprehensive studies of surface morphology, nuclear magnetic resonance (NMR) spectra, and X-ray diffraction analysis showed that at electrochemical intercalation of hydrogen it entered into the interlayer spaces and into the layers of GaSe crystals in molecular form [6]. At increasing the hydrogen density in H_xGaSe to $x = 2$ (x is the amount of inserted hydrogen atoms per the formula unit of the crystal) the H_2 molecules demonstrate a tendency to the transition to an "ordered" state in the gap and the interlayer pressure increases to a value $P \approx 9.8$ MPa at $T = 80$ K [5]. From the measured NMR spectra of H_xGaSe crystals it was concluded that in the van der Waals (vdW) gaps the inserted hydrogen existed as a "quasi-liquid" phase that at $T < 130$ K and concentrations $x > 2$ transforms into a "bound" phase [3]. In the latter phase the H_2 molecules occupy translation ordered sites in the crystal structure of H_xGaSe.

Cleaved vdW surfaces of the crystal are characterized with a low density of surface defects [7]. Electrochemical intercalation of such crystals was carried out at optimum conditions when the density of surface defects during hydrogen insertion is practically the same [6]. Nano-scale linear defects and cavities (nano-pores) at vdW surfaces of GaSe can be created due to flowing plastic deformation in the basal (0001) and pyramidal crystallographic planes of the crystal [8]. Processes of hydrogen sorption in layered $A^{III}B^{VI}$ crystals with nano-dimensional defects and physical properties of such crystals are not investigated. In this work we investigate processes of hydrogen sorption in nano-porous GaSe crystals, which have been filled with molecular hydrogen from gaseous phase, morphology of the vdW surfaces of these crystals and their optical and electrical properties.

2. Experimental procedures

The morphology of a surface of layered GaSe crystals was investigated by means of an atomic force microscope (AFM) Nanoscope IIIa Dimension 3000 (Digital Instrument) in tapping mode. The probe tip radius did not exceed 10 nm.

The spectra of IR absorption have been measured by using an IR dispersion spectrometer in the spectral range 650–6,000 cm^{-1} with a resolution not more than 3 cm^{-1}. The impedance measurements were carried out by means of a device Solartron FRA 1255 in the frequency range 10^{-1} to 10^6 Hz. The complex measurements of optical transmission (T) and conductivity across the layers ($\sigma_{||C}$) of GaSe have been done during the same experiment for the same sample in the temperature range 78–300 K at continuous pumping-out from the cryostat's camera containing investigated samples at a residual pressure $\sim 1 \cdot 10^{-3}$ Pa. The samples under investigations had in-plane size of 5×10 mm^2 and 0.5–1.0 mm thick. They were placed on the sapphire substrate and made fixed to the cryostat's diaphragm. The part of the sample destined to admittance measurements between the upper and lower (0001) planes of the crystal was shielded of light. Silver paste was applied as a contact material to the sample. Layered ε-GaSe crystals grown by the Bridgman method have been saturated with molecular hydrogen at a temperature $T = 79$ K and a pressure $P \sim 0.1$ MPa. Nano-scale defects at vdW surfaces were created because of thermal annealing of these crystals in a H_2 atmosphere at temperatures $T < 900$ K and a hydrogen pressure $P < 1$ MPa. Processes of hydrogen desorption from H_xGaSe crystals were investigated using measurements of the pressure at heating H_xGaSe crystals in a vacuum.

3. Results and discussion

The morphology of the surfaces of GaSe crystals exposed to thermal treatment at the same thermodynamic conditions in a H_2 atmosphere and in vacuum is different (Figs. 1, 2). In the first case we have found a created aggregation of a high amount of small-sized (<1 nm) defect formations (Fig. 1) over the whole sample bulk whereas the vacuum annealing leads to the formation of more large-scale cavities at the upper surface of the crystal.

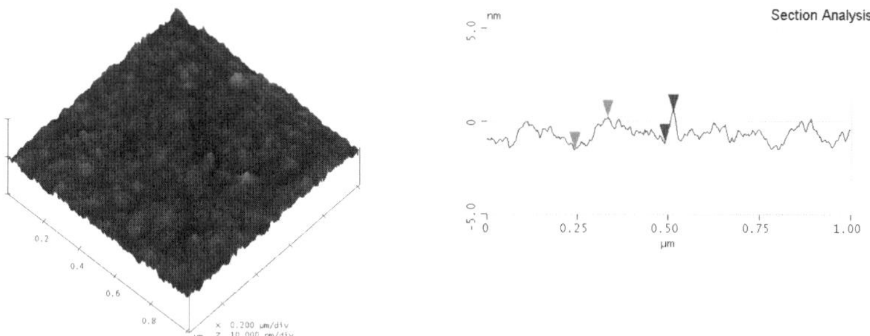

Figure 1. AFM patterns of GaSe cleaved surface after annealing at a temperature T = 720 K in molecular hydrogen atmosphere

With increasing temperature the interlayer interaction in GaSe decreases ($C_{11}/C_{44} \approx 11.4$) and the anisotropy of the elastic properties of the crystal becomes higher. A strain appeared due to vacuum annealing relaxes in upper layers of the crystal. This process causes the morphology of the crystal surface (Fig. 2). During such annealing Se atoms fall out into the interlayer space and remain in the crystal [9].

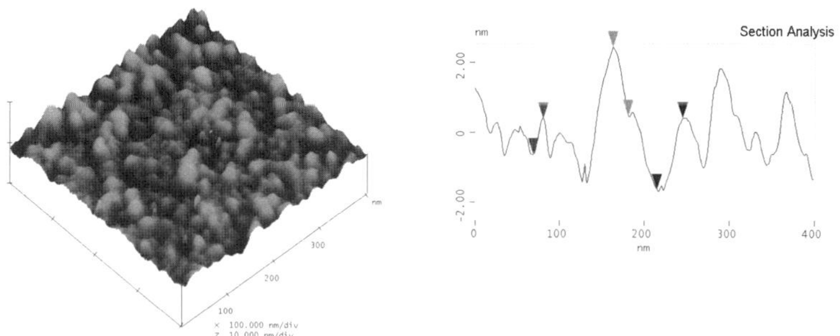

Figure 2. AFM patterns of GaSe cleaved surface after annealing at a temperature T = 720 K in vacuum

It is known that molecular hydrogen is weakly adsorbed at surfaces of semiconductors with Ga dangling bonds. Therefore in technologies of semiconductor compounds, which contain gallium, atomic hydrogen is usually used for elimination of oxides and formation of Ga–H bonds at molecular beam epitaxy [10]. Atomic hydrogen usually was generated by cracking of molecular H_2 at hot tungsten filament in cracking cell at high temperatures (~1,800 K) [11], which are much higher than those applied by us at stepped annealing. At electrochemical H^+ ion intercalation of GaSe bonds of Ga–H are also not created [4, 6]. In Ga-containing semiconductor compounds hydrogen inserted in considerable quantity does not form chemical compounds. For instance, after annealing GaN epitaxial layers with high density of implanted hydrogen it was found the existence of extended defects and nano-pores filled with hydrogen [12]. In our case molecular hydrogen was inserted into GaSe crystals at the liquid nitrogen temperature and a hydrogen pressure ≈0.1 MPa. Upon thermal annealing the formation of defects at vdW surfaces of GaSe occurs at deformation influence of hydrogen with participation of chalcogen atoms according to the chemical reaction $H_2 + Se \leftrightarrow H_2Se$. The annealing temperature was selected taking into account the temperature dependence of the equilibrium constant of this reaction. During annealing hydrogen participates in a mass transfer of the chalcogen. If a part of the chalcogen to move off from the temperature annealing zone, an excess of Se vacancies appears in the GaSe crystal. At decreasing the annealing temperature these vacancies interact each other and with structure defects and create pores. In solids pores are a result of coalescence of mobile vacancies that can be created at irradiation of metals with high-energy particles or due to different mechanical and thermal treatments of materials. The mechanism of pores nucleation is analogous to that of vapor condensation. Nucleation of vacancy pores and their coalescence in layered crystals occurs under conditions of high mobility and anisotropic diffusion of Se vacancies. Because of this process the pores condense at the basal (0001) planes of GaSe.

An increase of the density of defects in the anion planes can lead to flowing plastic deformation in the pyramidal crystallographic planes of GaSe [8] like that as it takes place for crystals having hexagonal lattice. These planes are oriented at an angle relative to the (0001) basal plane. On the vdW surface, the slippage planes pass through Se atoms, are oriented along the directions <1120>, and form triangles at the intersections (Fig. 3). We have observed a nano-scale plastic flowing in GaSe [8]. Such the kind of plastic deformation begins to rival with the basal flowing deformation at low temperatures and when force acting across a (0001) plane is available. Such a deformation was observed at measurements of the conductivity $\sigma_{||C}$ at $T \approx 80$ K for H_2-filled GaSe crystals with a high amount of basal defects. Thermally induced strains in these crystals were appeared at their cooling because of a deformation influence from the direction of the electrical wire contact. An analysis of the morphology of the surface, which contains nano-cavities formed by the intersection of the pyramidal planes of sliding plastic deformation, has shown that plastic flowing along these planes is accompanied by a displacement of separate GaSe layers in a direction perpendicular to the C axis [8].

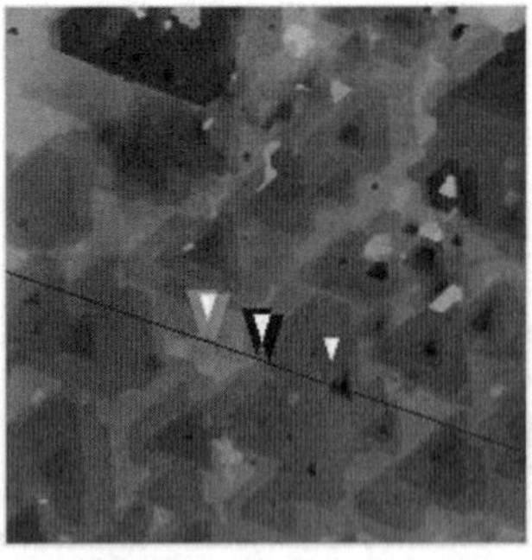

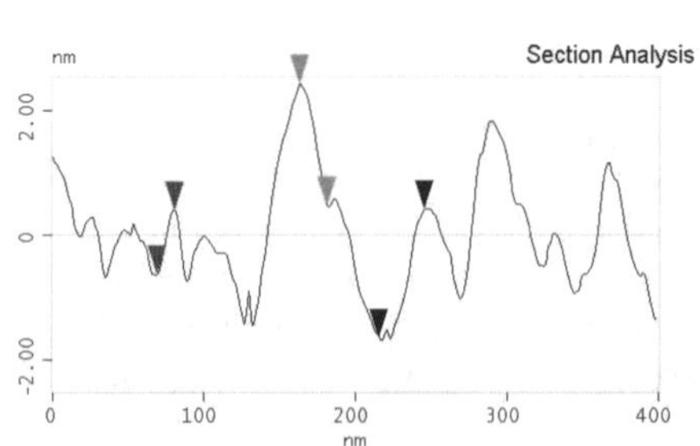

Figure 3. AFM patterns of GaSe cleaved surface at flowing plastic deformation along the pyramidal crystallographic planes

Infrared absorption spectra at T = 80 K of hydrogen-filled nano-porous H_xGaSe crystals demonstrate the presence of active optical vibrational modes (Fig. 4, curve 2) which are absent in the spectra for free of defects GaSe crystals (Fig. 4, curve 1).

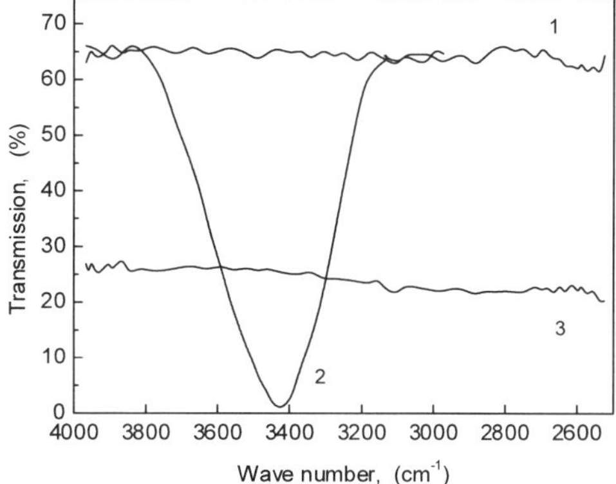

Figure 4. Optical transmission spectra of hydrogen-filled GaSe crystals measured at T = 80 K (1, 2) and T ~ 200 ÷ 220 K (3): 1 – free of defects GaSe crystal, 2 – nano-porous GaSe crystal, 3 – layered nano-porous GaSe crystal after desorption of hydrogen due to heating with a rate β ~ 10 K·min^{-1}

Evolution of the manifestation of these modes on reaching a temperature T = 80 K is shown in Fig. 5.

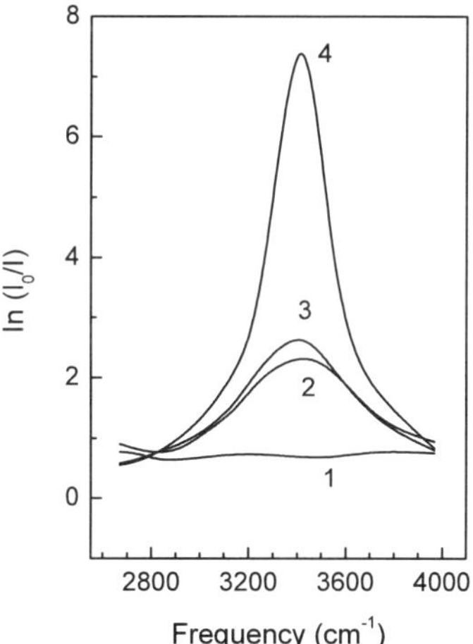

Figure 5. Optical absorption spectral dependences for hydrogen-filled GaSe crystals measured after cooling to T = 80 K in: 1 – 900 s, 2 – 1,200 s, 3 – 1,500 s, and 4 – 2,000 s. Specra's scanning time ≈100 s.

The intensity of the optical absorption band increases during some relaxation time τ after the cooling. The display of the active mode in the optical absorption spectra is related to the structure of the samples being investigated. They were observed only for H_2-filled nano-porous (pores dimension <1 nm) GaSe crystals, which flowing plastic deformation along the pyramidal crystallographic planes was absent for. The relaxation time τ value is defined by processes of structure relaxation of H_xGaSe crystals after cooling. For crystals with different density of basal defects it varies in the range 10^3–10^4 s. As the time passes a steady wide intensive absorption band (3,100–3,800 cm^{-1}) becomes formed (Fig. 4, curve 2). Similar absorption bands we had also observed in electrochemically intercalated with hydrogen samples H_xGaSe ($x > 1.5$).

In the case when gaseous hydrogen was inserted into free of defects GaSe crystals, not subjected to thermal annealing in a H_2 atmosphere according to the described above technique, it resulted in an increase of the interlayer spacing and, therefore, of the lattice parameter *c*. For example, after a hydrogen insertion at a temperature T = 873 K and a pressure P = 0.3 MPa the lattice parameter *c* was increased from 15.94 to 15.97 Å. It indicates on its insertion into the vdW gaps where it produces a pressure above 9.8 MPa at T = 80 K [5]. However, in the IR spectra of these crystals the optical active modes are not observed (Fig. 4, curve 1). At such the way of the insertion into GaSe crystals nonpolar molecules H_2 have no dipole moment and, therefore, are "inactive" in the optical absorption spectra [13]. It is known that optical absorption spectra of free of defects GaSe crystals do not contain peculiarities in the spectral range 2,000–5,000 cm^{-1} [14]. At T = 80 K a wide absorption band in the range 3,200–3,400 cm^{-1} has been observed in spectra of defect GaSe crystals previously annealed in vacuum at T = 1,000–1,100 K. It was supposed to be due to dislocations and stacking faults in these crystals [9].

It is reasonable to assume that the observed by us peculiarities of the IR absorption spectra are related to the presence of molecular hydrogen in the nano-porous layered crystal. H_2 molecules fill the interlayer spaces and pores in GaSe and become polarized in the presence of dangling bonds of Ga atoms. In this case the H_2 molecules do not dissociate and may be in a state of "weak" bonding to the GaSe lattice [15]. As a result, a long-range dipole-dipole interaction, acting with the aid of the GaSe lattice, appears between them. According to the estimations in [5], at T = 80 K and a pressure P = 9.8 MPa the amount of liquid hydrogen between the layers can achieve of about 25%. According to the Kelvin's equation, capillary condensation of vapours of wetting liquids takes place at more low pressures than the pressure of saturated vapour of these liquids over a flat. The pressure of vapour condensation in nano-dimensional pores is immediately related to the tension of *liquid–vapour* surface, which is defined by form and size of the pores [16]. The H_2 molecules very quickly (for some ps) enter into the nano-pores owing to the nano-capillary nature of adsorption [1] and can be in a liquid state in nano-porous GaSe crystals at T > 80 K. A long-range dipole-dipole interaction between the hydrogen-filled pores in H_xGaSe favours to the formation of a three-dimensional superlattice of the pores as it was observed in metals [17]. It is known that the symmetry and the crystallographic axes of such the superlattice and the initial crystal must coincide. The superlattice will be stable if the configuration energy of the system at its formation of pores and vacancies will be minimum [18]. The stability of pores superlattice is supposed to be due to elastic interaction between the pores and can be realized in crystals with anisotropic elastic properties. At T < 4 K the molecular hydrogen creates in the gap of free of defects H_xGaSe crystals a structure consisting of the lattice of the anisotropic layered crystal and the lattice of the molecular H_2 [5]. In this case the hydrogen in solid phase forms a hexagonal lattice (c = 6.12 Å and a = 3.75 Å) which, along with β-GaSe, belongs to the space group D^4_{6h}. A superlattice of the pores in H_xGaSe ($x > 2$) can explain the results of X-ray investigations [4] where intensive reflection lines from the hexagonal lattice created by textured intercalant inclusions in the crystal.

Long times of the formation of the optical active modes (Fig. 5) can be related to the formation of a superlattice of H_2-filled pores in the crystal. In this case a structure rearrangement of the ε-polytype of GaSe into the β-polytype takes place. It is put into effect by means of a slide of the layers in the basal planes of the crystal under conditions of high interlayer pressure. As a result of such rearrangement, a stable system consisting of the porous GaSe ("host") filled with H_2 molecules ("guest") becomes created. A "host" influence on "guest" properties consists in polarization of the H_2 molecules. A "guest" influence on "host" properties leads to a change of the electrical potential of vdW surfaces [19] what we observed experimentally. Additionally, this influence is apparent in the spectral shift of the optical absorption maximum (Fig. 5) to higher values of energy in comparison to results in [9].

It is known that there is a possibility to observe of optical absorption of OH hydroxyl groups in the investigated spectral range [14]. Such groups can be inserted into porous crystals during hydrogen insertion and they are very sensitive to the structure of porous materials. We eliminate any correlation between OH groups and the spectral peculiarities (Fig. 5) as vibrational spectra of hydroxyl groups manifest themselves at room temperature whereas the peculiarities in the spectra of nano-porous H_xGaSe crystals were not found at T = 300 K. In addition, for vibrational spectra of hydroxyl OH and Se–H groups there is a spectral shift of the absorption band maximum with changing temperature. In the spectra of the hydrogen-filled H_xGaSe crystals the shaped band and its position remain constant at heating these crystals from 80 to 175–190 K.

Desorption of molecular hydrogen from nano-porous H_xGaSe crystals depends on the rate of heating the crystals and deformations appeared at interaction of their defect sub-system with hydrogen. H_2 molecules in these crystals are located in the interlayer spaces and pores. In the interlayer space they are connected to the layers by vdW forces. For H_xGaSe crystals with basal type defects, in which there is no plastic flowing deformation along the pyramidal crystallographic planes, at a low rate of heating ($\beta \approx 0.5$–$1.0\ K{\cdot}min^{-1}$) from T = 80 K a desorption of H_2 molecules takes place in the temperature range 175–230 K. It results in a decrease of the interlayer pressure and disappearance of the optical active modes in the IR spectrum (Fig. 6).

A part of the adsorbed H_2 molecules remains in H_xGaSe crystals at their further heating to room temperature with the same rate. After room temperature keeping the crystals in vacuum for several days the samples were repeatedly cooled to T = 80 K and the effect of the formation of the active vibrational modes in the IR absorption spectrum has been reproduced. At heating these crystals with a rate $\beta \approx 0.5\ K{\cdot}min^{-1}$ the second maximum of hydrogen desorption is observed in the temperature range 360–390 K. From the results of a simulation of desorption processes of molecular hydrogen

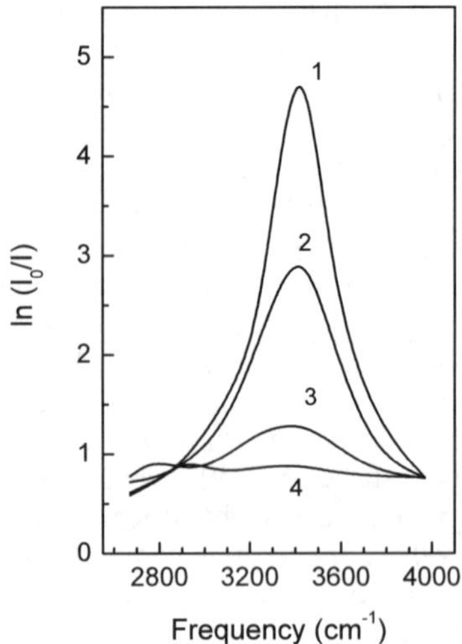

Figure 6. Optical absorption spectral dependences for hydrogen-filled GaSe crystals measured at heating the sample with a rate $\beta = 0.5$ K·min^{-1} at temperatures: 1 – 182 K, 2 – 190 K, 3 – 195 K, and 4 – 205 K. Specra's scanning time ≈100 s

from layered and carbon materials with nano-scale pores it follows that the presence of such a maximum of thermally stimulated desorption at T > 300 K is caused by desorption of H_2 from nano-pores of a size below 1 nm [1]. The obtained by us results are in agreement to the NMR data for H_xGaSe [4] where an abrupt change of the half-width of the NMR spectral line had been observed at T ≈ 370 K. This peculiarity was supposed to be due to getting out of hydrogen from the layers (a "bound" state) into the interlayer space and intensive deintercalation of the hydrogen from the crystal.

The transition of hydrogen from the liquid state of aggregation into the gaseous one (phase transition I) takes place in a "soft" (with respect to elastic properties) layered GaSe crystal sensitive to dynamic deformation effects. Investigation of electrical characteristics of these crystals in a wide frequency range by the method of impedance spectroscopy makes it possible to have analyzed the contribution of sorption processes to a change of the charge state of vdW surfaces of layered crystals. Continuous measurements of impedance spectra (time of a measurement less than 2 min) enabled to establish variations of electrical properties of H_xGaSe crystals in the range of the phase transition. At T = 300 K the dark conductivity $\sigma_{||C}$ of GaSe crystals is caused by basal defects and related to them potential barriers. The presence of such barriers leads to a frequency dependence of $\sigma_{||C}$ in the low-frequency range (10–10^4 Hz) (Fig. 7, curve 1). Insertion of H_2 molecules into the vdW gaps decreases the overlapping between the wave functions of carriers in different layers and the conductivities of the H_xGaSe crystals (Fig. 7, curve 2) one or two order of magnitude lower than those of the GaSe (Fig. 7, curve 1). At cooling from 300 to 80 K the conductivity of the GaSe samples decreases (Fig.8, curve 1) whereas that of the nano-porous H_xGaSe crystals increases (Fig. 8, curve 2). It is caused by a change of the charge state of vdW surfaces at formation of a pores superlattice under conditions of dipole–dipole interactions between the pores.

If at a low temperature (T = 80 K) in the crystals with a high amount of the pores a plastic flowing deformation along the pyramidal crystallographic planes takes place, it results in hydrogen desorption through the created extended defects and a fast decrease of interlayer pressure. Because of this process the liquid hydrogen is found to be in an overheated state. Its overheat is also possible at high rates of crystal heating ($\beta \approx 10$ $K{\cdot}min^{-1}$). In this case the interlayer pressure quickly decreases because of fast desorption of H_2 from the vdW gaps of the crystal being accompanied by a vigorous boil of the liquid H_2 and quick desorption of the gas from the crystal. The arisen elastic wave causes a displacement of weakly bounded GaSe layers with respect to each other along the (0001) plane and in the direction along the C axis. The variation of the conductivity frequency dependence $\sigma_{||C}(\omega)$ at heating the sample with a rate $\beta \approx 10$ $K{\cdot}min^{-1}$ is shown in Fig. 9.

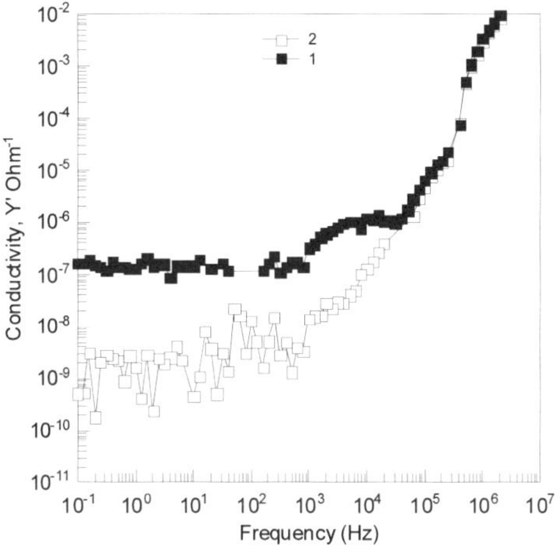

Figure 7. Frequency dependences of dark conductivity across the layers for GaSe crystals measured at T = 300 K: 1 – initial GaSe sample, 2 – hydrogen-filled nano-porous H_xGaSe sample

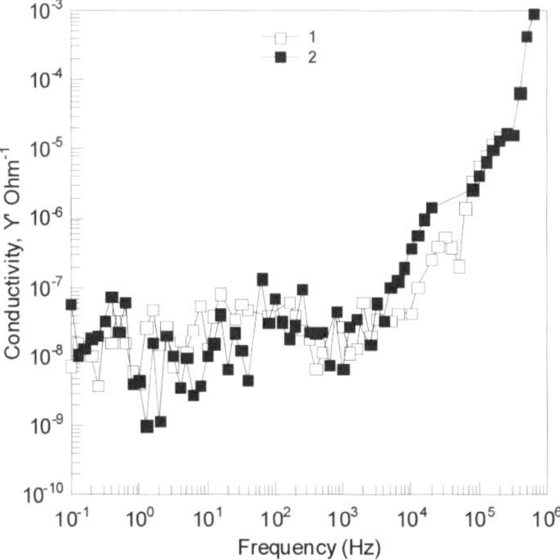

Figure 8. Frequency dependences of dark conductivity across the layers for GaSe crystals measured at T = 80 K: 1 – initial GaSe sample, 2 – hydrogen-filled nano-porous H_xGaSe sample

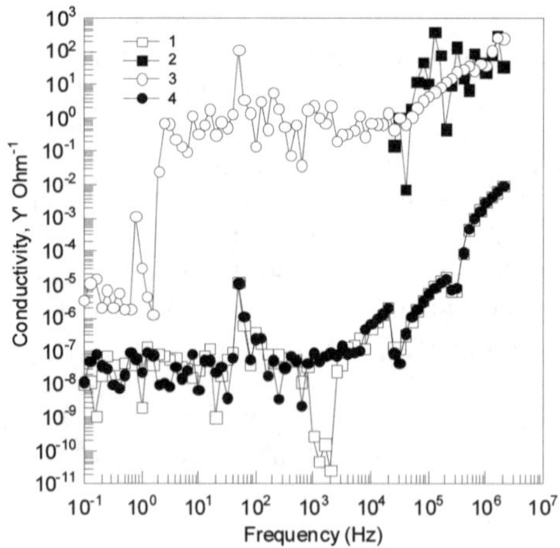

Figure 9. Frequency dependences of dark conductivity across the layers for H_xGaSe crystals measured at heating the sample with a rate $\beta = 10$ K·min^{-1} at temperatures: 1 – 193 K, 2 – 200 K, 3 – 227 K, and 4 – 250 K. Scanning time of impedance spectra ≈100 s

After an abrupt decrease of the interlayer pressure at T ≈ 170 K the pores superlattice destroys what is accompanied by a change of the charge at the vdW surfaces and increasing the potential to a value >10 V at the contacts to the H_xGaSe sample. Additionally the active optical modes in the IR spectrum disappear. The low-frequency impedance spectrum at T ≈ 193 K shows the resonance decrease of the conductivity at f = 1.26 kHz (Fig. 9, curve 1). It is known that at passing a shock wave through a GaSe sample the material particles obtain a velocity $U_p \sim 10^4$–10^5 cm·s^{-1} [20]. This magnitude depends on dynamic pressure, is determined by the elastic properties of the material and can be changed in materials with nano-pores filled with a liquid [21]. The observed low-frequency peculiarities of the impedance can be related to the Peierls' phase transitions between the polytypes, which occur at displacement of the layers along the (0001) plane. The ε- and β-polytypes of GaSe have the different spatial symmetry and, therefore, the band structure. The corresponding energy gaps differs by some tens millielectron volt. It determines differences in their electrical properties. The resonance behaviour of the low-frequency impedance spectra (Fig. 9, curve 1) is related to the peculiarities of propagation of electromagnetic waves in deformed layered media through the boundaries of layers with different potential. After desorption of the H_2 molecules from the pores of the H_xGaSe sample (T = 200–225 K, $\beta \approx 10$ K·min^{-1}) a porous vdW surface contains dangling Ga bonds. The chalcogen vacancies in GaSe act as donors [9]. Conductivity along the C axis of defect GaSe crystals can be considered in the framework of the Anderson model [22]. A decrease of the interlayer distance in H_xGaSe leads to the fact that the carrier quasi-Fermi level crosses the mobility edge. Such a "*semiconductor - metal*" transition is accompanied by an abrupt increase of the conductivity (up to 10^7–10^8 times) (Fig. 9, curves 2 and 3) and is similar to that observed in GaSe when a pressure above 250 kbar was applied along the C axis [23].

At the polytype phase transformations in GaSe after the hydrogen desorption fluctuations of the structure inhomogeneities with dimensions comparable to the light wave-length appear. Due to intensive light scattering with these inhomogeneities the optical transmission of the nano-porous H_xGaSe crystals (Fig. 4, curve 3) decreases. After heating the crystal to T = 255 K the transmission increases to a value typical for ε-GaSe (Fig. 4, curve 1).

4. Conclusions

1. Nano-porous gallium selenide layered crystals can be obtained by means of thermal annealing in molecular hydrogen atmosphere. At this technique the formation of pores takes place due to chalcogen vacancies, appeared in the anion planes, their condensation and coalescence at different temperatures. Time and temperature of the annealing determine the dimension and morphology of the pores.
2. At cooling to T = 80 K hydrogen-filled nano-porous H_xGaSe crystals (pores dimension below 1 nm) and at the absence of plastic flowing deformation along the pyramidal crystallographic planes of the crystal a superlattice of the pores filled with liquid hydrogen becomes created. The formation of the superlattice is promoted by the anisotropy of the crystal elastic properties, high mobility of the defects in the basal (0001) planes of H_xGaSe, and dipole–dipole interaction between H_2 molecules. We suppose that the appearance of active optical vibrational modes in the IR optical absorption spectra is caused by the superlattice.
3. At heating of the H_xGaSe crystals with a rate $\beta \approx 0.5$–$1.0\ K{\cdot}min^{-1}$ the hydrogen desorption takes place in the temperature ranges 175–230 K and 360–390 K. In the temperature range 175–230 K it is due to getting out molecular hydrogen from the vdW spaces where it was bound to the layers with molecular interaction. At temperatures 360–390 K the hydrogen desorption takes place from the nano-dimensional pores of H_xGaSe.
4. Sorption processes lead to changes of the charge state of the layer surfaces and manifest themselves in the impedance frequency dependences. At flowing plastic deformation along the pyramidal crystallographic plane of the H_xGaSe crystal and at its fast heating there is a dynamic deformation influence on the crystal. It is caused by the transition of the hydrogen from liquid state of aggregation to gaseous one. In this case the observed changes of the electrical and optical properties are because of the polytype phase transformations in nano-porous H_xGaSe crystals.

References

1. Han S.S., Kim H.S., Han K.S., et al. Nanopores of carbon nanotubes as practical hydrogen storage media. *Appl. Phys. Lett.*, 2005, **87**: 213113.
2. Han S.S., Kang J.K., Lee H.M., et al. Liquefaction of H_2 molecules upon exterior surfaces of carbon nanotube bundles. *Appl. Phys. Lett.*, 2005, **86**: 203108.
3. Kovalyuk Z.D., Prokipchuk T.P., Seredjuk A.I., Tovstyuk K.D. Investigation of impurity state in hydrogen intercalates of gallium selenium by NMR method. *Fiz. Tverd. Tela*, 1987, **29**: 2191–2193 (in Russian).
4. Kovalyuk Z.D., Prokipchuk T.P., Seredjuk A.I., Tovstyuk K.D., et al., NMR in hydrogen subsystem of H_xGaSe introduction junctions. *Fiz. Tverd. Tela*, 1988, **30**: 2510–2511 (in Russian).
5. Zhirko Yu.J., Kovalyuk Z.D., Pyrlja M.M., Boledzyuk V.B. (2007) Application of layered InSe and GaSe crystals and powders for solid state hydrogen storage. In: Hydrogen Materials Science and Chemistry of Carbon Nanomaterials, Eds. Veziroglu T.N et al., Springer, 325–340 p.
6. Zhirko Yu.J., Kovalyuk Z.D., Klad'ko V.N., et al. Investigations of the processes of intercalation with hydrogen in InSe and GaSe layered crystals. In: Proceedings of the Fifth International Conference "Hydrogen Economy and Hydrogen Treatment of Materials" Donetsk, Don. STU, Ukraine, May 21–25, 2007, **2**: 606–610 (in Russian).
7. Terhell I.C.I., Polytypism in the III-VI layer compounds. *Prog. Cryst. Growth Charact.*, 1983, **7**: 55–110.
8. Bakhtinov A.P., Vodop'yanov V.N., Slyn'ko E.I., et al. Self-organization of PbTe and SnTe nanostructures on the van der Walls GaSe(0001) surface. *Tech. Phys. Lett.*, 2007, **33**: 86–90.
9. Shigetomi S., Ikari T., Nishimura N. Annealing behavior of layered semiconductor p-GaSe single crystal. *Phys. Stat. Sol.* (a), 2001, **185**: 341–348.
10. Tejedor P., Crespillo M.L., Joyce B.A. Influence of atomic hydrogen on step stability during homoepitaxial growth on vicinal GaAs surfaces. *Appl. Phys. Lett.* 2006, **88**: 063101.
11. Suzuki Y., Shimoda M., Okada Y., et al. Formation of quantum dot structures by atomic hydrogen assisted selective area molecular beam epitaxy. *Jpn. J. Appl. Phys.*, 1997, **36**: L1538–L1540.
12. Radu I., Singh R., Scholz R., et al. Formation of nanovoids in high-dose hydrogen implanted GaN. *Appl. Phys. Lett.*, 2006, **89**: 031912.
13. Huber K.P., Herzberg G. (1979) Molecular Spectra and Molecular Structure. IV, Van Nostrand Reinhold, New York, 716 p.
14. Allakhverdiev K., Fernelius N., Gashimzade F., et al. Anisotropy of optical absorption in GaSe studied by midinfrared spectroscopy. *J. Appl. Phys.*, 2003, **93**: 3336–3339.
15. Vol'kenstein F. (1987) Electronic Processes on the Surface of Semiconductors at Chemisorptions. Nauka, Moscow, 432 p. (in Russian).
16. Dourdain S., Gibau A. On the capillary condensation of water in mesoporous silica films measured by x-ray reflectivity. *Appl. Phys. Lett.*, 2005, **87**: 223105.
17. Evans J.H. Observation of regular void array in high purity molybdenum irradiated with 2 MeV nitrogen ions. *Nature*, 1971, **229**: 403.
18. Brown L.M. A simple explanation for the stability of voids in materials under irradiation. *Ser. Met.*, 1972, **6**: 387–394.
19. Raissi F., Farivar R. Room-temperature hydrogen gas sensor. *Appl. Phys. Lett.*, 2005, **87**: 164101.

20. Lu X.Z., Lee S., Garuthara R., Alfano R.R. Shock-wave-induced collision broadening of the photoluminescence spectra in GaSe. *Appl. Phys. Lett.*, 1987, **51**: 1789–1791.
21. De Ryck L., Groby J.-P., Leclaire P., et al. Acoustic wave propagation in a macroscopically inhomogeneous porous medium saturated by a fluid. *Appl. Phys. Lett.*, 2007, **90**: 181901.
22. Kyazym-zade A.G., Abasova A.Z., Salmanov V.M., et al. Self-organization of packing defects and Anderson localization in A3B6—type layered crystals. *Mat. Sci. Eng. B*, 2002, **88**: 282–285.
23. Dunn K.J., Bundy F.P. Pressure-induced metallic and superconducting state of GaSe. *Appl. Phys. Lett.*, 1980, **36**: 709–710.

INVESTIGATION OF ALLOYS FORMATION FROM BINARY HYDRIDES IN Zr-Hf SYSTEM

D.G. MAYILYAN
A.B. Nalbandyan Institute of Chemical Physics of National Academy of Sciences of Armenia, 0014 Yerevan, Republic of Armenia

Abstract. Using a principally new method based on a special "hydride" cycle, Zr-Hf alloys were received. Obtained alloys are of HCP structure in accordance with the binary alloys phase diagram of Zr-Hf system. It was shown that all alloys in this system trend to formation of high hydrogen content hydrides.

Keywords: Zr-Hf system, alloys of transition metals, "hydride" cycle, Self-propagating High-temperature Synthesis (SHS), metal hydrides

1. Introduction

Formation of continuous solid solutions at interaction of hafnium and zirconium is conditioned by isomorphic crystal lattices of their two modifications, favor size factor, and the fact, that they are analogous elements. Dependences of principal properties of Zr-Hf alloys on the composition obey the common regularities existing in continuous solid solutions. The temperatures of solidus and liquidus, as well as of polymorphic $\alpha \leftrightarrow \beta$ transformations change in accordance with continuous curves [1, 2].

The traditional methods for alloys formation are the induction, electrical arc, electrical beam fusion, and powder metallurgy [1, 3]. All these methods are laborious and technologically complicated. They demand deep vacuum, inert environment at high temperatures, long-time and recurrence processes, etc.

In the Laboratory of Refractory Metals and High-temperature Synthesis of IChPh of Armenian NAS, a principally new method was developed for obtaining the alloys of transition metals which consists in synthesis of their binary hydrides followed by compaction and dehydrogenation the hydrides [4, 5]. In the present work, the results of investigation of alloys formation in Zr-Hf system by this method are presented. The results of investigation of interaction of the obtained alloys with hydrogen are presented also.

2. Results and discussion

The experiments were carried out through the next, earlier developed, scheme:

1. Preliminarily, in constant pressure reactor, the stochiometric hydrides of zirconium (ZrH_2) and hafnium (HfH_2) had been synthesized by the method of self-propagating high-temperature synthesis (SHS) [6, 7].
2. The powders of binary hydrides were mixed and compacted at different pressures.
3. In a hermetic quarts reactor, the binary hydrides mixture was dehydrogenated at temperature of active dissociation of hydrogen (vacuum annealing).

The dehydrogenation process can be presented by the next equation:

$$xZrH_2 + (1\text{-}x)HfH_2 \rightarrow Zr_xHf_{1\text{-}x} + H_2\uparrow \quad (1)$$

In the elaborated experimental cycle, a number of Zr_xHf_{1-x} (x = 0.2–0.8) alloys were received (Table 1).

The structure of obtained alloys was studied by X-ray analysis on diffractometer DRON-0.5 (Fig. 1). In the Table 1 the crystallographic parameters of obtained alloys taken from the diffraction pictures are presented. The obtained alloys are solid solutions based on α-zirconium or hafnium.

TABLE 1. Crystallographic parameters of obtained alloys

The formula of alloy	**Parameter of cryst. lattice (Å), c/a**		
	a	**c**	**c/a**
Zr	3.22	5.130	1.59
$Zr_{0.8}Hf_{0.2}$	3.23	5.139	1.56
$Zr_{0.7}Hf_{0.3}$	3.224	5.110	1.58
$Zr_{0.66}Hf_{0.34}$	3.219	5.116	1.59
$Zr_{0.5}Hf_{0.5}$	3.210	5.098	1.59
$Zr_{0.3}Hf_{0.7}$	3.196	5.081	1.59
$Zr_{0.2}Hf_{0.8}$	3.195	5.068	1.58
Hf	3.18	5.040	1.58

Upon heating the compacted hydrides mixture, a strong activation of metal occurs due to cleave of Me-H links. The free links promote the diffusion and mutual dissolution of metal components. It is important also that under mechanical pressure at compaction of hydrides mixture, the fine (micron and submicron sizes) particles of powders densely contact because of high fragileness of initial hydrides' powders. Nanoscale (20–80 nm) crystallites of used hydride powders [8] in their turn actively support the formation of alloys at relatively low temperatures owing to solid-phase diffusion. All these conditions are rather favorable for alloys formation. The temperature of alloys formation by the described technique is at the least, by 600–800°C lower than in the traditional methods [1, 3].

In the earlier studied systems of IV group transition metals, Ti-Zr, Ti-Hf, etc., along with α-modification alloys, unusual for the binary alloys phase diagram ω-phase had been received. Unlike to these systems, ω-phase was not obtained in our experiments in Zr-Hf system. The binary alloys phase diagram of this system differs from that of the previous systems by the higher temperature of α↔β transformation than the temperature of dehydrogenation. If in the above-described "hydride" technique, ω-phase in the systems Ti-Zr, Ti-Hf was formed in the way of high-temperature β-modification transformation into the low-temperature modification, the increase of the dehydrogenation temperature must bring to ω-phase formation in the system Zr-Hf also. However, in the conducted investigations, it was demonstrated that the increase of temperature of the initial hydrides dehydrogenation did not result in ω-phase formation. Therefore we can suggest that more probably, the ω-phase absence is conditioned by the sizes of zirconium and hafnium atomic radiuses (so called size factor).

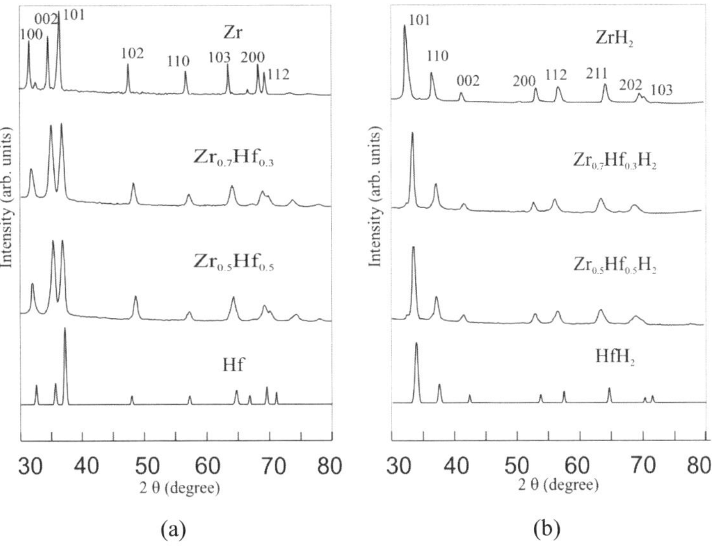

Figure 1. Diffraction pictures of obtained alloys (a) and their hydrides (b)

In works [9, 10], the information was presented concerning ω-phase formation at high-pressure treatment (up to 70 kbar) of Ti-Zr and Zr-Hf α-alloys. In work [10], the existence of ω-phase in the system Zr-Hf was indicated based on stepwise temperature change upon superconductivity transitions, but it had not been validated by X-ray analysis.

Thereafter, we studied the interaction of obtained alloys with hydrogen in SHS regime. It was shown, that all alloys in Zr-Hf system trend to formation of high-hydrogen content hydrides, H/Me = 1.8–2.08:

$$Zr_xHf_{1-x} + H_2 \leftrightarrow Zr_xHf_{1-x}H_y \quad (y = 1.8\text{–}2.08) \qquad (2)$$

All of the obtained alloys' hydrides (Table 2) were of BCT crystallographic structure (structural type ThH_2), isomorphous to dihydrides of zirconium and hafnium. Hydrogen removing from the alloys hydrides resulted in return to α-alloys with HCP structure (equation 2).

TABLE 2. Some characteristics of hydrides

The formula of hydride	Content H_2 (wt%)	Parameter of cryst. lattice (Å), c/a		
		a	c	c/a
ZrH_2	2.14	3.504	4.450	1.27
$Zr_{0.8}Hf_{0.2}H_{1.8}$	1.63	3.480	4.440	1.28
$Zr_{0.7}Hf_{0.3}H_{1.87}$	1.57	3.490	4.427	1.27
$Zr_{0.66}Hf_{0.34}H_2$	1.56	3.481	4.418	1.27
$Zr_{0.5}Hf_{0.5}H_2$	1.49	3.478	4.364	1.25
$Zr_{0.3}Hf_{0.7}H_2$	1.31	3.465	4.370	1.26
$Zr_{0.2}Hf_{0.8}H_{2.08}$	1.28	3.451	4.357	1.26
HfH_2	1.11	3.456	4.327	1.25

In Tables 1 and 2, the lattice parameters of obtained α-alloys and their hydrides are shown. It is seen, that, as it was expected because of a little difference in atomic radiuses, the change of alloys' and their hydrides' composition resulted in negligible changes in lattice parameters **a** and **c**, as well as in the axes ratio, **c/a**.

3. Conclusions

Using a new method [4, 5] with a special «hydride» cycle: hydrogenation of pure metals, dehydrogenation of compact mixture of binary hydrides, Zr-Hf alloys were received. It was established, that the obtained alloys have HCP structure of α-zirconium (hafnium), in accordance with the binary alloys phase diagram of Zr-Hf system [1, 2].

It was shown that in combustion regime, all alloys in the system Zr-Hf trend to formation of high hydrogen content hydrides (H/Me = 1.8–2.08).

The above presented results indicate that at using a special, "hydride" cycle, an earlier unknown, effective manner of alloys formation in the system Zr-Hf was realized.

Acknowledgments

Author expresses its thanks to Prof. S.K. Dolukhanyan, Dr. A.G.Aleksanyan and Prof. V.Sh. Shekhtman for support and valuable consultations during implementation of the work.

The work was implemented in the frame of theme 0567, approved by Armenian Ministry of Education and Science, and at financial support of ISTC Grant A-1249.

References

1. Savickij E.M., Burkhanov G.S. Metal science of refractory metals and alloys. Moscow: *Nauka*, 1967.
2. Elliot R.P. Structures of binary alloys, v. 2. Moscow: *Metalurgia*, 1970.
3. Kislij P.S., Kuzenkova M.A. Sintering of refractory compounds. Kiev: *Naukova dumka*, 1980.
4. Dolukhanyan S.K., Shekhtman V.Sh., Aleksanyan A.G., Ter-Galstyan O.P., Abrosimova G.E., Sakharov M.K. Investigation of peculiarities of formation of alloys' structure in system Ti-Zr-H. *Chemical Physics Reports*, 2007; **26**(11): 36–43.
5. Dolukhanyan S.K., Aleksanyan A.G., Shekhtman V.Sh., Mantashyan A.A., Mayilyan D.G., Ter-Galstyan O.P. New technique for producing the alloys based on transition metals. *Chemical Journal of Armenia*, 2007; **60**(4): 25–32.
6. Dolukhanyan. S.K., Nersesyan M.D. Nalbandyan A.B., Borovinskaya I.P., Merzhanov A.G. Combustion of transition metals in hydrogen. *Dokl. AS USSR* 1976; **231**(3): 675–678.
7. Dolukhanyan S.K., Aleksanyan A.G., Hakobyan H.G. Interaction of hafnium with hydrogen and nitrogen in the combustion regime. *International Journal of Hydrogen Energy*, 1995; **20**(5): 391–395.
8. Shekhtman V.Sh., Dolukhanyan S.K., Abrosimova G.E., Abrahamyan K.A., Aleksanyan A.G., Aghajanyan N.N., Ter-Galstyan O.P. The nanocrystalline forming by combustion synthesis of Ti (Zr) hydrides. *International Journal of Hydrogen Energy*, 2001; **26**: 435–440
9. Bashkin I.O., Pagnuev A.Yu., Gurov A.F., Fedotov V.K., Abrosimova G.E. Ponjatovskij E.G. *Fizika tverdogo tela*, 2000; **42**(1): 163–169.
10. Bashkin I.O., Nefedova M.V., Ponjatovskij E.G., Tissen V.G. *JETPh Letters,* 2003; **78**(2): 91–94.

NEW TECHNIQUE FOR PRODUCING THE ALLOYS BASED ON TRANSITION METALS

A.G. ALEKSANYAN, S.K. DOLUKHANYAN, A.A. MANTASHYAN, D.G. MAYILJAN, O.P. TER-GALSTYAN
A.B. Nalbandyan Institute of Chemical Physics of National Academy of Sciences of Armenia, 0014 Yerevan, Republic of Armenia
V.S. SHEKHTMAN
Institute of Solid State Physics of Russian Academy of Sciences, 142432 Chernogolovka, Moscow District

Abstract. In the present work, an essentially new approach to production of alloys of the refractory metals is presented. It consists in co-compacting and dehydrogenation-sintering of mixture of powders of two and more hydrides of transition metals of III, IV and V groups, or of mixture of powder of any of these hydrides with a transition metal of III–VIII groups. It was shown, that the heating of these mixtures after compacting in vacuum up to temperature higher than the temperature of a hydride dissociation, results in formation of strong, porousless, compact binary and triple alloys. The dependence of an alloy's structure formation on the components ratio in initial charge is established. The thermodynamic and kinetic estimations of reaction of alloy formation from the hydrides are given. The interaction of some of received alloys with hydrogen in combustion mode is studied, and the hydrides of these alloys are synthesized.

Keywords: hydride, alloys, dehydrogenation

1. Introduction

The existing traditional methods for producing binary and multi-component alloys are based on the technologies of fusion (induction, electric arc, electron beam, etc.), or of powder metallurgy. The complexity of obtaining various high quality alloys is connected with the certain difficulties such as necessity of deep vacuum and inert environment at high temperature, long-lasting and manifold repetition of processes, etc. In powder metallurgy, also there is a problem of fine powders preparation. The quest of new effective methods for producing binary and multi-component alloys of prescribed physical and technical characteristics remains an actual task of modern Material Science.

Recently in the Laboratory of High-temperature Synthesis and Technology of Inorganic Compounds of the Institute of Chemical Physics, Armenian NAS, a new method was developed for receiving the alloys using the hydrides of transition metals as raw materials [1]. These hydrides were obtained by method of self-propagating high-temperature synthesis (SHS) in systems Me-H [2–4], and this circumstance makes this new technology highly beneficial towards the above mentioned traditional technologies.

This essentially new approach appeared a productive and effective method for preparing the alloys of refractory metals not only of IV group, but also for preparation of intermetallics and of other types of alloys with different metallic additives

The developed cycle of operations: SHS-synthesis of fine (consisting in nano-scale, 20–80 nm, crystallites) hydrides of metals of maximal hydrogen content, co-compacting of a mixture of powders of hydrides, followed by dehydrogenation, allowed preparing a number of alloys. In systems Ti-Zr, Ti-Hf, Ti-Zr-Hf, the alloys of α- and ω-phases had been received [2, 3].

Each of these operations provides optimum conditions for formation of an alloy:

- SHS-synthesis of hydrides of transition metals provides wide assortment of fine dispersion powders (consisting of nano-scale crystallites, 20–80 nm) of hydrides of transition metals.
- The ability of obtained fine powders of hydrides to compacting provides consolidation of reagents$Me'H_2$+$Me''H_2$; $Me'H_2$+ Me″. with good contact between crystallites.
- The nano-scale dimensions of crystallites of hydrides provide more active interaction between crystallites and more active course of diffusion-chemical processes.
- The removing of hydrogen and the other processes proceeding at dehydrogenation-sintering (phase transitions, formation of defects which accelerate diffusion, re-crystallization processes, etc.) promote the formation of alloys of any crystal structure.

The purpose of the present work is to describe (a) the experiments proving the fact of formation of alloys in different systems: Ti-Zr, Ti-Hf, Ti-V, Ti-Nb, Ti-Al, Ti-Fe, Ti-Ni, Ti-Co, Zr-Ni, Zr-Co, Zr-Y; (b) the basic characteristics of process of formation of alloys based on the transition metals of IV group from their hydrides; (c) the interaction of the received alloys with hydrogen in combustion regime (SHS) resulting in formation of hydrides of these alloys. The determination of factors which influence on the process of alloy formation in the elaborated cycle is an essential problem. As such factors, the following parameters were considered: the temperature of heating, the relative density of samples, the pressure of co-compacting, the quality of initial hydride and metal powders, which are important in the directed synthesis of the given (certain) alloy.

2. Experimental procedures

To realize the experiments on preparation of alloys, the special installation has been designed and prepared. It consisted of a quartz reactor, a furnace, devices for controlling the vacuum and temperature in the reactor during dehydrogenation and the alloy formation (Fig. 1).

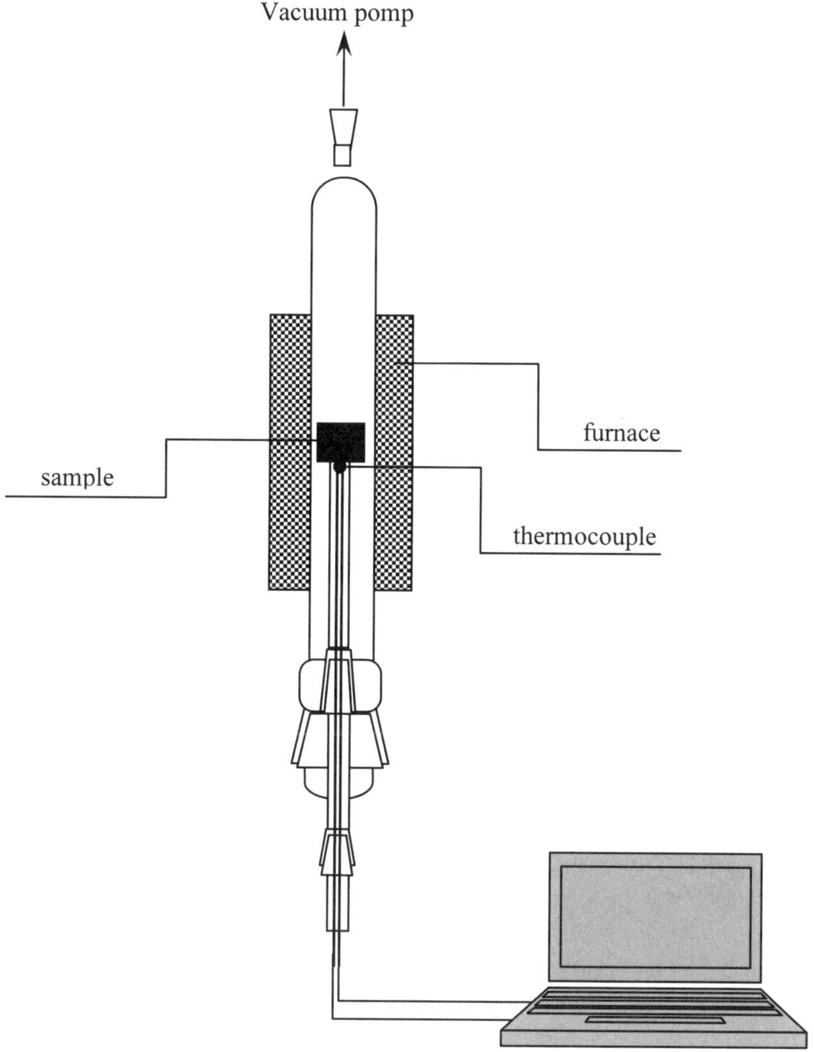

Figure 1. Scheme of dehydrogenation reactor

The initial hydrides were synthesized in combustion regime by SHS method in reactors of constant pressure.

The full cycle of experiment included:

- Preparation of powders of transition metal hydrides by SHS method [2–4]
- Grinding and mixing the powders of hydrides
- Compacting this mixture at different pressures
- Removing the hydrogen (dehydrogenation and vacuum annealing)
- Synthesis of hydrides of the obtained alloys by SHS method

The conditions of formation of solid solutions and intermetallics have been studied in detail using the variety of hydrides. The influence of process parameters: chemical peculiarity and ratio of initial components, phase transformations, etc. on the processes of dehydrogenation and the alloy formation have been established. The processes of obtained alloys combustion in hydrogen have been investigated. The regularities of interaction of these alloys with hydrogen and formation of corresponding hydrides of high hydrogen-content in the SHS-MODE are established.

The certification of samples at all stages of full cycle has been performed using the following analyses: chemical (pyrolysis method), differential-thermal (Derivatograph Q-1500), X-ray (DRON-0.5 and "Siemens D-500"). The microstructure of alloys has been studied on the scanning electron microscope CamScanMV230 (VEGA TS 5130MM). The density and microhardness of some alloys have been measured.

3. Results

3.1. INTERACTION OF TWO METAL HYDRIDES

A number of porousless, compact, binary alloys of metals of fourth group in α-modification (in accordance with the state diagram) and in ω-modification (unusual for the state diagram) had been received. The alloys of titanium and the metals of fifth group, and the zirconium-yttrium, zirconium-niobium alloys had been received (Table 1).

TABLE 1. Characteristics of binary alloys

Initial charge	Binary alloys	Phase composition
$xTiH_2 + (1\text{-}x)ZrH_2$	$Ti_xZr_{(1-x)}$: $Ti_{0.8}Zr_{0.2}$; $Ti_{0.66}Zr_{0.34}$; $Ti_{0.4}Zr_{0.6}$; $Ti_{0.2}$ $Zr_{0.8}$; etc.	α and ω
$xTiH_2 + (1\text{-}x)HfH_2$	$Ti_xHf_{(1-x)}$: $Ti_{0.8}$ $Hf_{0.2}$; $Ti_{0.6}Hf_{0.4}$; etc.	α and ω
$xZrH_2 + (1\text{-}x)HfH_2$	$Zr_xHf_{(1-x)}$: $Zr_{0.8}$ $Hf_{0.2}$; $Zr_{0.66}$ $Hf_{0.34}$; etc.	α
$xTiH_2 + (1\text{-}x)VH_{0.8}$	$Ti_xV_{(1-x)}$: $Ti_{0.8}V_{0.2}$; $Ti_{0.7}V_{0.3}$ $Ti_{0.5}V_{0.5}$	α and β
$xTiH_2 + (1\text{-}x)NbH_{1.7}$	Tix $Nb_{(1-x)}$: $Ti_{0.8}$ $Nb_{0.2}$; $Ti_{0.7}$ $Nb_{0.3}$; $Ti_{0.5}$ $Nb_{0.5}$	α and β
X $ZrH_2 + (1\text{-}x)YH_2$	$Zr_{0.92}Y_{0.08}$	α
$ZrH_2 + 2.5NbH_{1.7}$	$Zr_{0.3}$ $Nb_{0.7}$	β

3.2. INTERACTION OF THREE HYDRIDES

In Table 2 the characteristics of obtained triple alloys based on IV group metals are presented.

TABLE 2. Characteristics of obtained triple alloys

Initial charge: $xTiH_2 + y\ ZrH_2 + z\ HfH_2$; $x + y + z = 1$	Triple alloys	Phase composition	Lattice parameters of main phase
$0.34TiH_2 + 0.33ZrH_2 + 0.33HfH_2$	$Ti_{0.34}Zr_{0.33}Hf_{0.33}$	α and ω	a = 3.205; c = 5.095 *c/a* = 1.59
$0.66TiH_2 + 0.22ZrH_2 + 0.12HfH_2$	$Ti_{0.66}Zr_{0.22}Hf_{0.12}$	ω and α	a = 4.76; c = 2.899 *c/a* = 0.609

3.3. INTERACTION OF HYDRIDE OF IV GROUP METAL WITH THE METALS OF III–VIII GROUPS

In Table 3 some characteristics of alloys of Ti and Zr with Al, Cr, Re etc. are presented.

TABLE 3. Characteristics of alloys based on Ti and Zr with the III–VIII group metals

System	Formula of alloy	Phase composition
$xTiH_2 + (1\text{-}x)V$	Sol. Solutions: $Ti_{0.8}V_{0.2}$; $Ti_{0.7}V_{0.3}$; $Ti_{0.4}V_{0.6}$; etc.	α and β
$xTiH_2 + (1\text{-}x)Nb$	Sol. Solutions: $Ti_{0.8}Nb_{0.2}$; $Ti_{0.7}Nb_{0.3}$; $Ti_{0.4}Nb_{0.6}$; etc.	α and β
$xTiH_2 + (1\text{-}x)Re$	Sol. Solutions: $Ti_{0.95}Re_{0.05}$; $Ti_{0.85}Re_{0.15}$	α and β
$TiH_2 + Fe$	TiFe	Intermetallic
$xZrH_2 + (1\text{-}x)Y$	$Zr_{0.92}Y_{0.08}$	α
$TiH_2 + Al$	TiAl	Intermetallic
$xTiH_2 + Ni$	TiNi	Intermetallic
$xTiH_2 + Co$		Intermetallic
$xZrH_2 + (1\text{-}x)Ni$		Intermetallic
$xZrH_2 + (1\text{-}x)Ni$	Zr_2Ni, ZrNi	Intermetallic
$xZrH_2 + (1\text{-}x)Co$	Zr_2Co, ZrCo	Intermetallic

Figures 2 and 3, the X-ray patterns of some obtained alloys are presented.

3.4. INTERACTION OF COMPACT ALLOYS WITH HYDROGEN AND DEUTERIUM

Some of obtained compact alloys (presented in Tables 1–3) interact with hydrogen in combustion (SHS) regime under hydrogen pressure 10–30 atm. and the combustion temperatures 480–600°C. This interaction results in formation of hydrides of alloys with rather high hydrogen content. The data concerning composition, structure and distortion temperature of formed hydrides of alloys are shown in Table 4.

In Fig. 3, the X-ray patterns of obtained hydrides of alloys are presented.

4. Discussion

Concerning the fact of observed in our experiments formation of binary and triple alloys based on the elements of IVA group from their hydrides, we can to assume a simultaneous realization of several different processes: phase transitions with creation of: point defects, dislocations, internal and external surfaces; chemical diffusion; re-crystallization processes with formation of structure corresponding to the given system.

Process of formation of alloys proceeds according the reactions:

$$xMe'H_2 + (1\text{-}x)Me''H_2 \rightarrow Me'_xMe''_{(1\text{-}x)} + H_2\uparrow \text{ or}$$

$$xMe'H_2 + (1\text{-}x)Me'' \rightarrow Me'_xMe''_{(1\text{-}x)} + H_2\uparrow$$

Formation of alloys in the given scheme includes the processes:

Phase transitions and re-crystallization. The general scheme of phase transitions during formation of an alloy, for example, from hydrides of titanium and zirconium, is shown below.

$xTiH_2$ (FCC)+(1-x)ZrH_2 (BCT)→xTi (HCP)+(1-x)Zr (HCP)→$Ti_xZr_{(1-x)}$ (HCP α, or ω-phases).

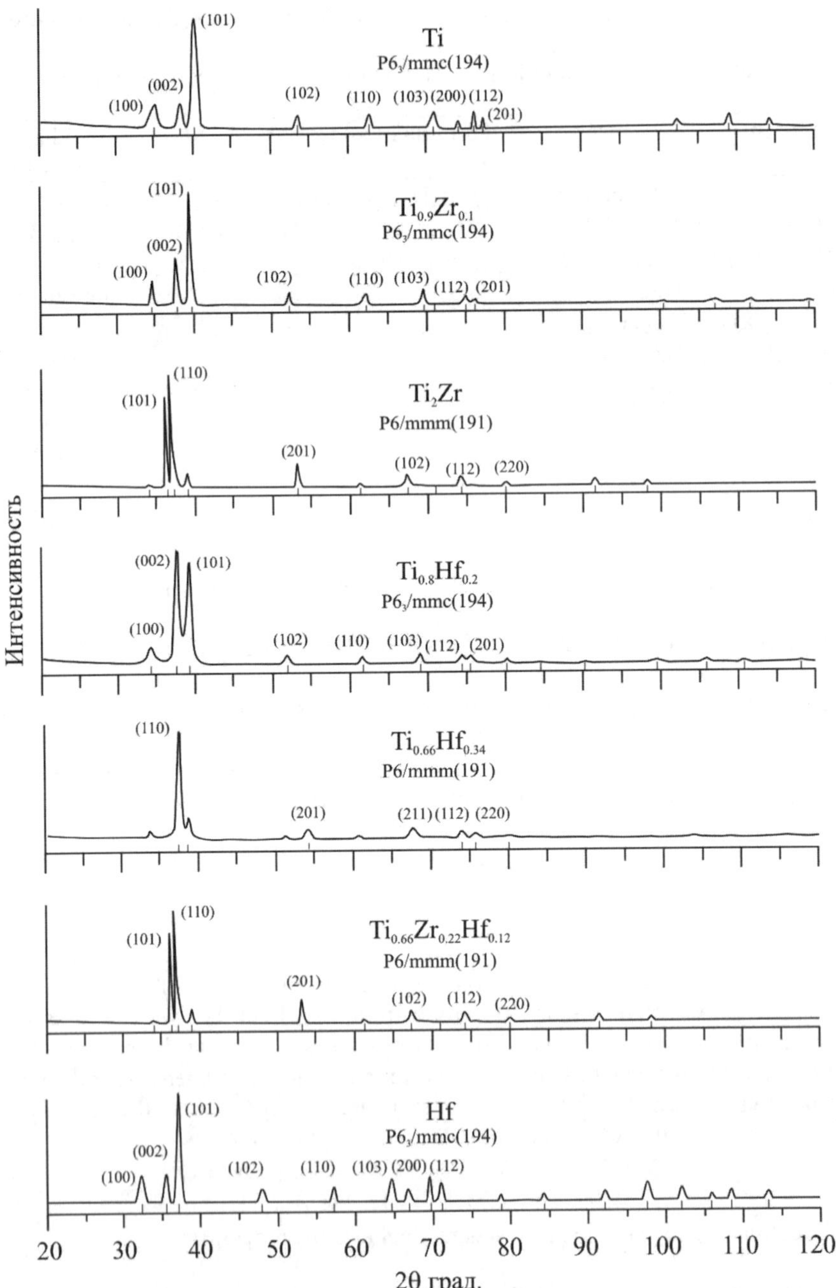

Figure 2. X-ray pattern in the system Ti-Zr-Hf, obtained by the technique of compacting-dehydrogenation.

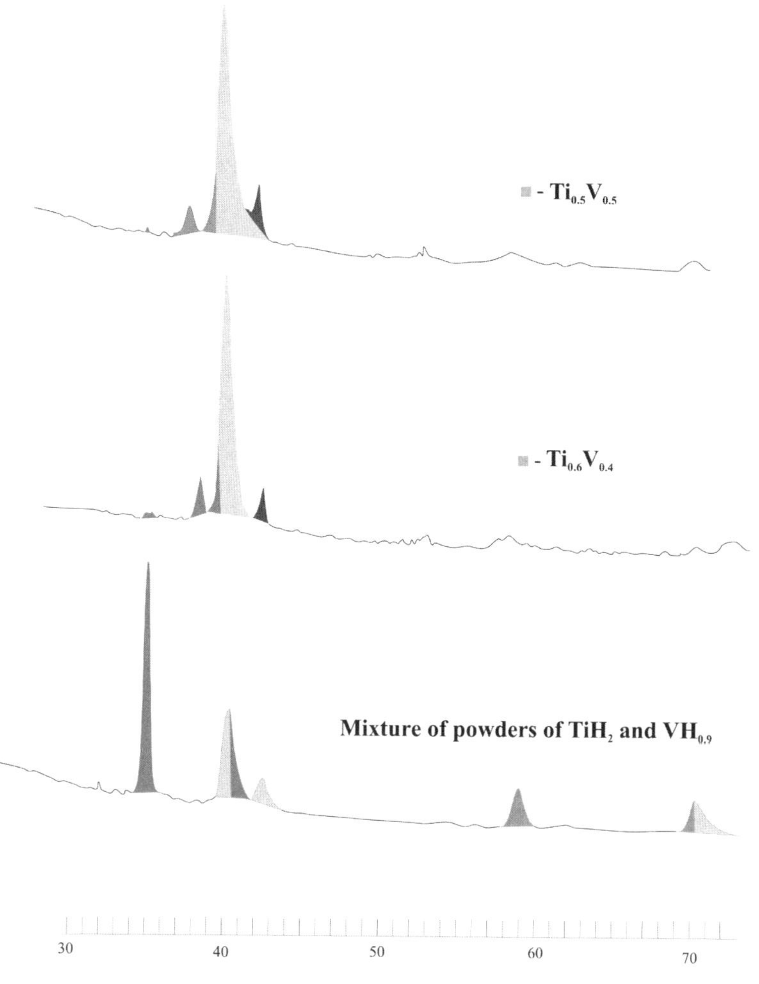

Figure 3. X-ray pattern in the system Ti-V obtained by the technique of compacting-dehydrogenation

TABLE 4. Characteristics of hydrides and deuterides of alloys

Calculated formula	H(D) Cont (wt%)	H/Me	Main phase: crystal lattice, type of structure, parameters	Temperature of endoeffects (°C)		
				I	II	III
$Ti_{0.8}Zr_{0.2}H_{1.85}$	3.32	1.85	FCC, type CaF_2, a = 4.509	460	650	790
$Ti_{0,88}Zr_{0,12}D_{1,93}$	6,8	1.93	FCC, type CaF_2, a = 4,497			
$Ti_{0.67}Zr_{0.33}H_{1.92}$	3.04	1.92	BCT, type ThH_2, a = 3.289; c = 4.470; ***c/a*** = 1.39			
$Ti_{0.67}Zr_{0.33}D_{1.68}$	5.28	1.68	BCT, type ThH_2, a = 3.279; c = 4.421; ***c/a*** = 1.35			
$Ti_{0.5}Zr_{0.5}D_{1.93}$	5.24	1.93	BCT, type ThH_2, a = 3.282; c = 4.84; ***c/a*** = 1.366			

Calculated formula	H(D) Cont (wt%)	H/Me	Main phase: crystal lattice, type of structure, parameters	Temperature of endoeffects (°C)		
				I	II	III
$Ti_{0.2}Zr_{0.8}H_2$	2.4	2.0	BCT, type ThH_2, a = 3.442; c = 4.452; ***c/a*** = 1.29	530	770	790
$Ti_{0.2}Zr_{0.8}D_{1.92}$	4,45	1,92	BCT,typ ThH_2, a = 3.466; c = 4.491; ***c/a*** = 1.3			
$Ti_{0.8}Hf_{0.2}H_2$	2.63	2.0	FCC, CaF_2, a = 4.497	430	660	
$Ti_{0.8}Hf_{0.2}D_{1.86}$	4.79	1.86	FCC; type CaF_2			
$Ti_{0.66}Hf_{0.34}H_{2.19}$ (from ω-phase)	2.32	2.1	BCT, type ThH_2, a = 3.244; c = 4.445; ***c/a*** = 1.36	470	670	730
$Ti_{0..66}Hf_{0.34}D_{2.0}$	4.13	2.0	BCT, type ThH_2, a = 3.226; c = 4.448; ***c/a*** = 1.379			
$Ti_{0.34}Zr_{0.33}Hf_{0.33}H_{2.13}$ (from ω-phase)	1.98	2.13	BCT, typeThH_2 a = 3.271; c = 4.421; ***c/a*** = 1.35	420	760	–
$Ti_{0..66}Zr_{0.22}Hf_{0.12}D_{1..88}$	4.89	1.88	BCT, type ThH_2, a = 3.271; c = 4.421; ***c/a*** = 1.35			
$Ti_{0..34}Zr_{0.33}Hf_{0.33}D_{2..2}$	4.01	2.2	BCT, type ThH_2, a = 3.376; c = 4.322; ***c/a*** = 1.28			
$Zr_2NiH_{4.7}$	1.91	1.56	Tetragonal; a = 6.809; c = 5.698 ***c/a*** = 0.84	270		
$Zr_2NiD_{4.15}$	3.32	1.38	Tetragonal; a = 6.919; c = 5.549; ***c/a*** = 0.80			
$ZrNiH_3$	1.8	1.5	Orthorhombic; a = 3.509; b = 10.454; c = 4.287			
$ZrNiD_{2.58}$	3.32	1.29	Orthorhombic; a = 3.517; b = 10.47; c = 4.276			

Liberation of hydrogen from compacted mixtures of initial hydrides results in phase transformations: FCC crystal lattice of titanium hydride, TiH_2, at temperature of dissociation (600–620°C) transforms to hexagonal compact packed lattice (HCP) of titanium; dehydrogenation of zirconium hydride ZrH_2, of body-centered tetragonal lattice (BCT) at temperature of dissociation (800–820°C) transforms to hexagonal compact packed lattice (HCP) of zirconium [5].

Liberation of hydrogen from crystal lattice increases defects of structure in powder particles. Increase of free energy due to the accumulated dot defects, surface dislocations and volumetric defects accelerates diffusion processes.

The heating of a sample up to the temperature higher than the dissociation temperature results in phase transitions α→β (polymorphic transformation). For example, in titanium the α→β transition occurs at 882°C, and in zirconium – at 865°C. The latent heat of allotropic transformations in titanium is 3,330 ± 295 J/mol (795 ± 70 cal/mol) [9].

It is necessary to note, that the phase transitions are accompanied with the re-crystallization processes. Besides the primary re-crystallization, which probably, accompanies the plastic deformation, a collective re-crystallization also occurs during compacting and heating of a sample.

In this connection, extremely high frequency of exchange of places is achieved dependent on the energy of activation At the moment of phase transition, the crystal lattice, due to reorganization of structural units, is in non-stable state: as a result of occurrence of straining, or of liberation of active atoms at decomposition of the main phase. At that, the process of formation of an alloy may be activated. The energy of defects in crystal lattice of initial particles and of the structural imperfections, which have arisen both during SHS-synthesis of hydrides and during cold pressing at compacting, also can influence on the process of formation of an alloy at dehydrogenation-sintering of compact mixtures of hydrides. The kinetics of the hydride decomposition also may influence on this process.

In the described experiments, re-crystallization at dehydrogenation-sintering begins after achievement of temperatures 700–1,000°C, and the alloys of refractory metals are formed in short time (no more than in one hour). The temperature of alloys formation at sintering of hydrides is lower, at least, by 600–800°C than those demanded in the traditional methods: induction or arc fusion, diffusion sintering of the metals, etc. Thus, we may assume that the alloys are formed at essentially low temperatures owing to a set of all the processes which accompany dehydrogenation and the sintering.

The developed cycle of experiments – synthesis of metal hydrides of maximal hydrogen content by SHS method, co-compacting and dehydrogenation of their mixture – allowed us to receive a number of the alloys corresponding to the state diagrams, but in systems Ti-Zr, Ti-Hf, Ti-Zr-Hf, the alloys were received having structure of ω-phase which is unusual for the state diagrams of these systems. It is known, that ω-phase is a phase of a high pressure: in titanium and zirconium, it usually appears at pressure exceeding 0.6 T/m^2 [9]. In alloys Ti-Zr, Ti-Hf, according to [6], also, ω-phase is fixed after processing under pressure 50–70 kbar.

We established [7] that there is an obvious dependence of structure of the obtained ω-alloys from the composition of initial mixtures (Tables 1–3). It was shown, that the ω-phase alloys, unusual for the state diagram for the systems Ti-Zr, Ti-Hf [8, 10], etc., are formed at ratio of concentrations of initial binary hydrides of titanium, zirconium and hafnium 1:1 and 2:1. The ω-phase alloys received by this technique are stable at higher annealing temperatures, than the ω-alloys received at high pressures.

Reactionary solid phase diffusion. Upon heating the mixture of compact hydrides, for example, $xTiH_2 + (1\text{-}x) ZrH_2$, a strong activation of metals takes place due to disruption of Me-H link at temperatures a little bit higher than the dissociation temperature of hydrides (800–1,000°C). Upon heating in hydrogen environment, a clearing from oxide film which usually is present on the surface of fine powders takes place. «Open links» and the cleared surface of powders promote proceeding of active diffusion processes. Therefore, all the accompanying processes become more active also: mutual dissolution of directly contacting reaction materials: Me' and hydride of Me", formation of germs of new phases, re-crystallization processes including collective re-crystallization. From the scientific publications, it is known, that the hydrides of transition metals possess high activity in the temperature interval of their intensive dissociation [11–14].

The opportunity of influence of hydrogen diffusion through particle surfaces cannot be excluded also: free, active atoms of hydrogen, going out from lattices of adjoined dissimilar particles, promote more active interactions on the adjoined surfaces, and the germs of a new phase is formed with lattice differing from those of both interacting phases.

It is important also that at compacting of mixtures of two and more hydrides under pressure, the grains of different hydride powders insert one into another owing to their high fragility, forming a strong skeleton. The nano-scale sizes of grains of hydrides' powders (20–80 nm) also actively promote formation of alloys at rather low temperatures due to diffusion. All the above mentioned creates rather favorable conditions for realization of solid-phase diffusion mechanism of formation of alloys in conditions of our experiments.

As is known, chemical reactions in solid phases proceed at direct interaction of particles of crystal substances. Among these reactions are reactions of oxidation, reduction, and thermal decomposition of solid phases. According to the traditional conceptions at reactionary solid-phase diffusions, two basic processes occur: formation of solid solution and formation of new compounds.

In our case we can suppose that a reaction of MeMe′ alloy formation proceeds by the steps:

1. The initial materials are in compact mixture of particles of hydrides MeH and Me'H or of a hydide and metal, MeH and Me. The heating of this mixture up to the dissociation temperature results in decomposition of a hydride and "denudation" of active metallic centers. Migrating hydrogen stimulates the processes of diffusion and diffusion mass transfer. Unstable reaction layer is formed, the life-time of which dependences on the temperature. Further temperature increasing brings to the appearance of crystallic MeMe' and formation of structures characteristic for a given alloy.
2. Formation of a solid solution in a lattice of solvent metal occurs due to atomic diffusion, i.e. moving of atoms of one substance in a crystal lattice of another.
3. The activated atomic diffusion transforms into the reactionary diffusion resulting in formation of saturated solid solutions or intermetallics. Some alloys and intermetallics can be formed by super-saturation of solid solutions up to new phases with structure differing from the structure of solvent metal.

Liquid-phase diffusion. At first glance in our case, liquid-phase diffusion is improbable because the temperatures of heating of hydrides mixtures (up to 1,000°C) are much lower than the temperatures of melting of the used metals (except TiH_2-Al system). Though we can suggest that at hydrogen dissociation, the phase transition begins on a surface of powder and, probably, some energy is liberated due to specific volume decreasing more than by 20%. Due to the accumulated dot defects, surface dispositions and volumetric defects, some additional energy can be liberated also. These effects can locally increase the temperature of a grain surface. The melting of thin layers of metal, activated during dissociation, can provide the beginning of metals fusion.

Thermodynamic and kinetic estimations of reaction of alloys formation from hydrides. As it has been shown above, at interaction (sintering) of hydrides of two metals (for example, TiH_2 + ZrH_2) at temperature of heating about 1,000°C, the titanium-zirconium alloys are formed. At the indicated temperatures, the decomposition of hydrides takes place (T_{dis} = 620–820°C), which is accompanied by liberation of hydrogen. Actually, the disruption of Ti-H and Zr-H links occurs and Ti-Zr link is formed, characteristic for the alloy. It is obvious, that in this case activation barriers of alloys formation should be essentially lower, than in traditional methods of formation of alloys directly from two metals (Ti and Zr).

For thermodynamic and kinetic estimations and comparison of two probable ways of alloys formation from powders of hydrides (TiH_2 + ZrH_2) and of metals (Ti + Zr), the interactions can be presented by the chemical equations, i.e. by chemical reactions of alloys formation:

$$TiH_2 + ZrH_2 \rightarrow TiZr\ (alloy) + 2H_2 + \Delta H_1 \quad (1)$$

$$Ti + Zr \rightarrow TiZr\ (alloy) + \Delta H_2 \quad (2)$$

We shall note, that difference of activation energies of these two reactions of the same alloy formation is defined by a difference of their thermal effects: according to Poliani [13] rule, at transition from one reaction to another in the series of similar reactions, a change of activation energy (the kinetic factor) is proportional to the difference of thermal effects of these reactions: $\Delta E = \alpha\Delta H$; where ΔH is a difference between thermal effects, and ΔE is a difference between activation energies of these reactions, E_2-E_1.

For quantitative estimation of difference in the activation energies of reactions (1) and (2), naturally, coefficient α has to be known. We do not know its value in our reactions, but obviously, it has to be positive (>0).

The difference in thermal effects of reactions (1) and (2) may be calculated by subtraction of reaction (2) from (1). The result will be presented by equation (3):

$$TiH_2 + ZrH_2 \rightarrow Ti + Zr + 2H_2\uparrow + \Delta H_3 \quad (3)$$

The thermal effect, ΔH_3, corresponds to the difference of thermal effects of reactions (1) and (2): $\Delta H_3 = \Delta H_1 - \Delta H_2$. This approach allows excluding the necessity of knowing the heats of alloys formation in system Ti-Zr. Obviously, the correctness of evaluation is defined by formation of the same alloy with the identical crystal lattice in both reactions (1) and (2). ΔH_3 calculated from the standard heats of formation of components of the equation (3) (T = 298 K), is equal:

ΔH_3 = -1,219.73 kJ/mol (-291.33 kcal/mol). In these calculations the data on standard heats of formation are used from works [5, 14]: ΔH_{TiH2} = -130.3; ΔH_{ZrH2} = -166; ΔH_{Ti} = -471.43; ΔH_{Zr} = -608.34; ΔH_{H2} = -218.13 kJ/mol. Let us note that the presented data concerning the heats of formation of metallic states Ti and Zr and of their hydrides clearly demonstrate that the links in metals are more strong than the Me-H links in hydrides ($\Delta H_{Me} << \Delta H_{Me\text{-}H}$).

Hence, the above presented evaluations show that the difference of the thermal effects of reactions (1) and (2) is high and is ΔH (ΔH_3) = -291.33 kcal/mol. Consequently, we can state, that in accordance with the above shown Polany equation, activation energy of reaction (1) has to be essentially lower of the activation energy of reaction (2). That's why the interaction of hydrides TiH_2 and ZrH_2 resulting in alloy formation can proceed in more "soft" conditions (at lower temperatures) than if the alloy is formed from the same (Ti and Zr) metals.

5. Conclusions

The stated above results show that at use of a special "hydride" cycle, an earlier unknown but highly effective way of formation of alloys of transition metals and their hydrides is founded.

A more general conclusion from the results on metal alloys formation, described in the present work is the following: the metal alloys formation from hydrides is a more easy way, than the formation of alloys directly from the same metals, because the chemical links in hydrides are less strong, than links in the metals.

The described method results in obtaining not only of the traditional alloys in accordance with the state diagram, but also in formation of alloys of structure, unusual for the state diagram of different metal systems.

Hence, the results of implemented researches state that on the kinetics of condensation and sintering, on the mechanisms of the mass transfer at formation of alloys of refractory metals from their hydrides, the following characteristics of hydrides essentially influence:

- The nature of chemical link Me-H and crystal structure
- Phase transitions at dehydrogenation
- Structure defects in initial hydrides and the defects appeared at compacting
- Physical properties of hydrides (plasticity at compacting even at not very high pressures, high dispersion ability (till micron and submicron sizes due to nano-scale grains)
- Chemical properties (high hydrogen content, rather low temperatures of decomposition)
- Thermodynamic and kinetic properties of initial hydrides

Acknowledgments

Work is implemented at financial support of the International Science and Technology Center (ISTC, Grant A-1249), and in the frame of scientific theme 0567 of the Ministry of Science and Education of Republic of Armenia.

References

1. Dolukhanyan S.K., Shekhtman V.Sh., Aleksanyan A.G., Ter-Galstyan O.P., Abrosimova G.E., Sakharov M.K. Investigation of peculiarities of formation of alloys' structure in system Ti-Zr-H. *Chem. Phys. Rep.*, 2007; **26**(11): 36–43.
2. Dolukhanyan S.K., Nersesyan M.D., Nalbandyan A.B., Borovinskaya I.P., Merzhanov A.G. Combustion of transition metals in hydrogen. *Dokl. AS USSR*, 1976; **231**(3): 675–678.
3. Dolukhanyan S.K., Hakobyan H.G., Alexanyan A.G. *Int. J. SHS*, 1992; **1**(4): 530–535.
4. Dolukhanyan S.K., Alexanian A.G., Hakobian H.G. Interaction of hafnium with hydrogen and nitrogen in the combustion regime. *Int. J. Hydrogen Energ.*, 1995; **20**(5): 391–395.
5. Antonova M.M. Properties of metal hydrides. Handbook. Kiev: Naukova Dumka, 1975.
6. Korzhov V.P., Karpov M.I., Shekhtman V.Sh., Aleksanyan A.G., Dolukhanyan S.K. Proceedings of the World European Powder Metallurgy Association, the Exhibition "EuroPM 2005",12-th International Exhibition and Conference of Powder Metallurgy, 2–5 Oct., 2005. Prague, Czech Republic, pp. 180–188.
7. Shekhtman V.Sh., Dolukhanyan S.K., Abrosimova G.E., Abrahamyan K.A., Aleksanyan A.G., Aghajanyan N.N., Ter-Galstyan O.P. *Int. J. Hydrogen Energ.*, 2001; **26**: 435–440.
8. Kornilov I.I. Titanium, Moscow: Nauka, 1975.
9. Elliot R.P. Structures of binary alloys, v. 2, Moscow: Metalurgia, 1970.
10. Villars P., Calvert L.D. Pearson's handbook of crystallographic data for intermetallic phases, 2nd Edition, ASM International, Ohio, 1991, **3**: 2465, 2466, 2469 and 3238.
11. Bashkin I.O., Pagnuev A.Yu., Gurov A.F., Fedotov V.K., Abrosimova G.E., Ponjatovskij E.G. *Fizika tverdogo tela*, 2000, **42**(1): 163–169.
12. Evans V.G., Polanyi V. *Nrans. Far. Soc.* 1938, **34**: 11.
13. Verjatin U.D. et al. Thermodynamic properties of non-organic materials. Moscow: Atomizdat, 1965.
14. Lunin V.V., Askhabova Kh.N., Romanowski B.V. *Dokl. AS USSR*, 1980, **250**(4): 896–898.

FORMATION OF POROUSLESS COMPACT PRODUCES FROM HYDRIDES OF REFRACTORY METALS

S.K. DOLUKHANYAN, A.G. ALEKSANYAN,
K.A. ABRAHAMYAN, S.S. MARDANYAN
A.B. Nalbandyan Institute of Chemical Physics of National Academy of Sciences of Armenia, 0014 Yerevan, Republic of Armenia
V.SH. SHEKHTMAN, M.I. KARPOV, V.P. KORZHOV
Institute of Solid State Physics of Russian Academy of Sciences, 142432 Chernogolovka, Moscow District

Abstract. The compacting ability of SHS-hydrides of titanium, zirconium, hafnium, etc. was studied in dependence of the process conditions. The dependences of physical and mechanical properties of obtained compact hydrides from the dispersion and chemical composition of hydride powders, pressure of pressing, temperature of dehydrogenation, etc. were determined. The compacted hydrides are characterized by the density and hardness comparable with the corresponding theoretical parameters of the initial metal. The compacted metal hydrides are of practical interest because they combine the advantages of metal, for instance, the hardness and stability, with the unique moderating ability of hydrogen. By dehydrogenation of the compacted hydrides, strong samples of IVa group metals were prepared. Nano-scale sizes of hydride germs provided the formation of dense samples due to mutual diffusion during the pressing and dehydrogenation processes. The influence of pressing (compacting) pressure, temperature at dehydrogenation, dispersion of initial powder, on the mechanical properties of final metal samples (solidity, micro-hardness, porosity, structure, etc.) was established.

Keywords: self-propagating high-temperature synthesis, hydrides of transition metals, compact hydrides, dehydrogenation, protective materials against neutron radiation

1. Introduction

The idea on compacting the powders of refractory metals and alloys' hydrides and their using as raw materials for obtaining the high-strength produces from refractory metals has arisen in the Laboratory of High-temperature Synthesis and Technology of Inorganic Compounds of the Institute of Chemical Physics, Armenian NAS, during investigation of processes of Self-propagating High-temperature Synthesis (SHS) in Me-H systems and development of essentially new high-efficiency method for synthesis of transition metals' hydrides [1–3].

The spectrum of application of transition metal hydrides is extremely wide. One of the major is the nuclear technique where they are used as biological protection from ionizing radiation, in particular, against neutrons. It is known that the best moderators of neutrons are hydrogen-containing materials as: water, paraffin, polyethylene, and hydrides of some metals – Ti, Zr, V, etc. Water and organic compounds are not stable enough, especially at high temperatures. At interaction with neutrons, hydrogen shows unique properties. The mass of H atom is closer to the neutron mass than the mass of any other element. H atom can absorb the energy of neutron and has a large scattering cross-section. The effectiveness of a protective material in moderation of neutrons is determined by rather high concentration of H atoms, N_H (Table 1). Mass consumption of hydrides is limited by the economic reasons (for instance, high cost of hydrides receiving), and some technological peculiarities of hydrides exploitation. To overcome the economic limitations, the development of highly productive and cheap methods for hydrides preparation is required. SHS technique represents such a method.

Usually, the hydrogen content in a volume unit of SHS hydride is rather high, close to the theoretical, evaluated from the X-ray and densimeter data. However, in bulk hydride powder, it is much lower. Therefore, a priori, the usage of powders of SHS hydrides as shielding materials cannot be effective because the bulk density of a hydride cannot ensure the high atomic concentration of hydrogen. Hence, a necessity had arisen to compact the hydrides and an idea had appeared to investigate the plastic deformation of SHS hydrides at pressing, to obtain compact samples with high atomic concentration of hydrogen.

TABLE 1. Atomic concentrations of hydrogen in different hydrogen-containing materials

Compound	At. concentration of hydrogen, $N_H \times 10^{22}$ at. H/cm^3	Hydrogen content (wt%)	Temperature of dissociation (^{0}C)
ZrH_2	7.34	2.01	800–1,000
TiH_2	9.5	4.02	600–800
MgH_2	6.7	7.6	
VH_2	11.4	3.78	
$ZrV_2H_{4.3}$	7.06	1.85	550–600
$ZrNiH_3$	7.73	1.96	250–350
$ZrCoH_3$	7.64	1.95	250–350
$TiFeH_2$	5.5	1.5	80
$LaNi_5H_7$	7.6	1.5	30–50
Mg_2NiH_4	5.9	3.8	200–250
Gas.H_2 at 100 atm	0.65		
Liquid H_2	4.2		
H_2O	6.7		
Paraffin C_nH_{2n+2}	7,8–8.0	0.87–0.91	–
$LiBH_4$	7.6	0.67	–
Arom. hydrocarbons	4.7–6.7	0.85–1.05	–
Polystyrene $(C_8H_9)_n$	5.4	1.05	–

The hydrides under study were synthesized in “High pressure bomb” in SHS regime. The obtained hydride powders were subjected to dry grinding in agate mortar or ball mill. The grinding was performed during 10–120 min, until the dispersion of the initial powders was lower than 5–10 μm. SHS hydrides of metals of IV group are characterized by high hydrogen content and due to this, by high dispersancy to micron and submicron sizes in a short grinding time. It should be noted that SHS hydride powder particles are formed by germs of nano-scale sizes, 20–80 nm.

For the sample certification, we used the chemical, differential-thermal (Derivatograph Q-1500), and X-ray (Diffractometers DRON-0.5 and Simens D-500) analyses. The compacting of samples was performed on the hydraulic press at loading up to 450 kN. In collet press-forms, the tablets with diameter 22–25 mm and thickness 5–8 mm have been pressed from the powders of hydrides of titanium, zirconium, etc.

The metallographic certification of samples was performed in the Institute of Solid State Physics of RAS. The HRA hardness was measured on the Rockwell instrument (diamond pyramid, loading 60 kg). Ten measurements on both sides of cylindrical sample have been performed. The density was determined by hydrostatic weighing based on the geometrical parameters and weight of the sample.

The microstructure was investigated on digital electronic scanning raster electron microscope CamScan MV2300, VEGA TS 5130MM, equipped with YAG detectors of secondary and reflected electrons, and with energy disperse x-ray microanalyzer with semi-conductor Si(Li) detector Link INCA Energy. Calculations of the x-ray microanalysis results were carried out with the INCA Energy 200 program and the software package developed in IEM of the RAN.

2. Experimental procedures

Compacting of hydrides. The dependence of compacting ability of powders of Ti, Zr, and other metal hydrides on the process conditions has been studied. The dependence of physical and mechanical properties of compact hydrides on the dispersion of hydrides powders and the pressing pressure has been stated. Figure 1 shows the microstructure of TiH_2 powder.

In Table 2, the characteristics of compact hydrides are presented. At pressing under 300–400 kN, solid, porousless samples of hydrides were obtained with density, close to theoretical [4].

In Figs. 2–4, the microstructures of relief surfaces of TiH_2, ZrH_2, and $Ti_{0.8}Hf_{0.2}H_2$ compact samples, respectively, are presented. The structures of tablets are very dense at considering both the section and the break surfaces. The powder particles densely adjoin each other. The sizes of pores both on the surface and inside the tablets do not exceed 5 μm. As the data in the table show, the hydrides are characterized by high hardness, close to that of metal. Such compact metal hydrides are of practical interest, because they combine the advantages of metals (hardness and stability) with hydrogen’s unique neutron-moderating ability.

Figure 1. Microstructure of TiH_2 powder

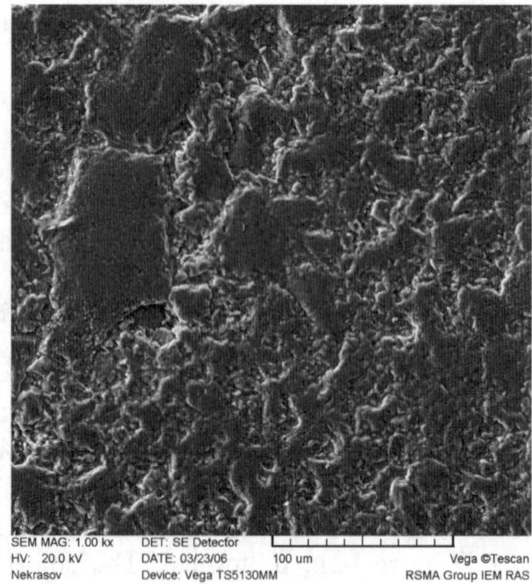

Figure 2. Microstructure of surface of pressed cylindrical sample compact of TiH_2

TABLE 2. The characteristics of compact hydrides

MeH_2	H content (wt%)	Lattice, parameters (Å)	Temper. range of dissociation (°C)	Pressing pressure (kN)	Density (g/cm^3)	Hardness (HRA)
TiH_2	4.01	fcc, type CaF_2	480–610	300	3.71	49.8 ± 0.8
ZrH_2	2.16	bct, type ThH_2, a = 3.495; c = 4.463	450–810	350	5.39	41.6 ± 1.0
HfH_2	1.09	bct, type ThH_2 a = 4.911; c = 4.505	420–830	400	11.35	45.7 ± 0.7

Preparation of compact samples of transition metals and alloys. The above presented results served a basis for the further development of this direction. As the next step of investigations, we studied the possibility of preparation of compact produces from refractory metals and alloys by compacting and subsequent dehydrogenation of their hydrides. This task appeared no less interesting and important. In the scientific

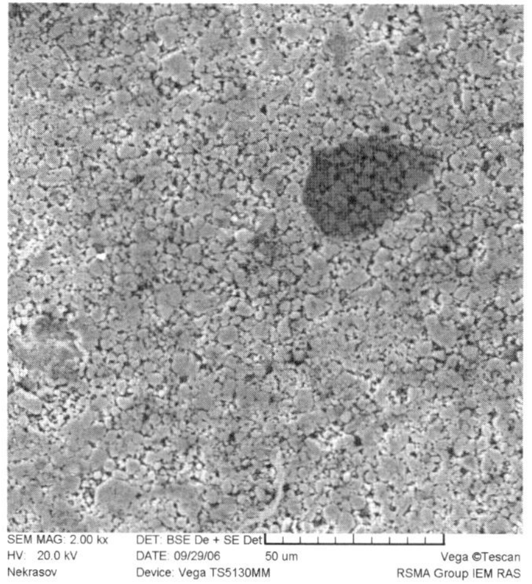

Figure 3. Microstructure of pressed cylindrical sample compact of ZrH_2 powder

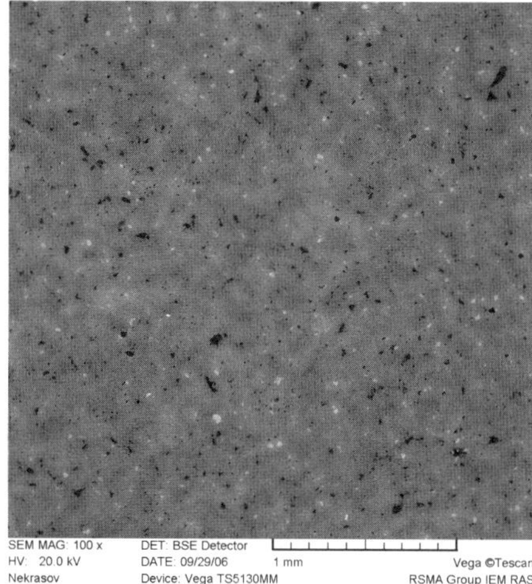

Figure 4. Microstructure of pressed cylindrical sample of $Ti_{0.8}Hf_{0.2}H_2$

publications, practically, there was no information concerning the usage of metal hydrides as raw materials for preparation of compact produces from refractory metals and alloys. Such investigations have been performed for the first time. The method is based on the reaction $MeH_2 \leftrightarrow Me + H_2\uparrow$, for example, $TiH_2 \leftrightarrow Ti + H_2\uparrow$ or $ZrH_2 \leftrightarrow Zr + H_2\uparrow$.

From the compacted sample, hydrogen was removed at rather low temperature (lower than 1,000°C). At this, the hydrogen outcome did not result in loosening the sample, quite the contrary: at temperature of hydrogen dissociation, sintering of grains took place. At that, the sintering preferentially occurred on the surfaces of unit grains, which became highly activated due to hydrogen dissociation. Actually, a reaction-activated sintering happened at essentially low temperature. Samples of titanium, zirconium, hafnium, etc. have been received.

To investigate the process of dehydrogenation-sintering of compact hydrides, a hermetic quarts device was made, allowing evacuation down to 10^{-3} Hg mm and heating up to 1,050°C. The device was provided with a system for gas inlet-outlet (Fig. 5). The temperature was measured by platinum/platinum-rhodium thermo-couple. The device was supplied with a computer registration of temperature and kinetics of the process. Dehydrogenation was performed at 800–1,050°C. The completeness of the process was controlled by the vacuum level, chemical and x-ray analyses.

The dependence of physica-chemical and mechanical characteristics of the obtained compact metal samples on the chemical composition of the hydride, the pressing pressure, the powder dispersion (micro- and nano-dimensions), the sample density, the temperature and rate of dehydrogenation-sintering was studied.

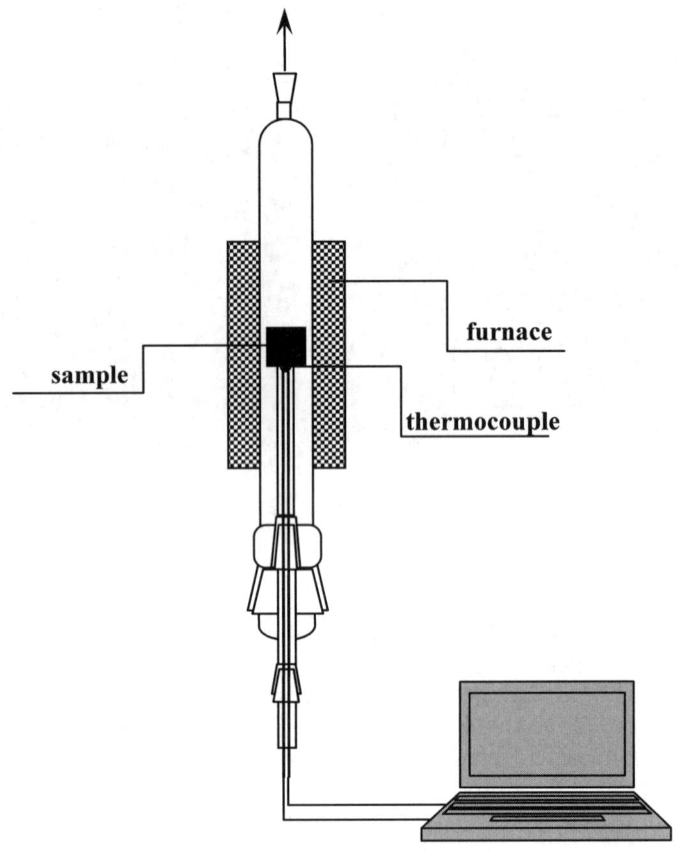

Figure 5. Scheme of dehydrogenation reactor

Titanium. At titanium hydride compacting, the properties of the obtained samples did not depend on the pressing pressure above 250 kN. At more than doubling the pressing pressure, the density and hardness of the sintered samples remained within the limits 4.45–4.50 g/cm^2 and 74.0–75.5 HRA, respectively. This density is close to the theoretical density of titanium.

The duration of titanium hydride grinding practically did not influence the characteristics of the sintered product, either. Whether ground by hand for 10–20 min, or in a ball mill for 40 min, the titanium hydride powder grains did not exceed 10 μm, since the high hydrogen content provides quick dispersion down to these sizes. The density and hardness of the sintered samples, obtained from titanium hydride powder after 10 min grinding and dehydrogenation, and were 4.25 g/cm^3 and 61.5 HRA, respectively. After 1 h grinding, at other identical parameters of the process, the density and hardness of the sintered samples were 4.54–4.55 g/cm^3 and 73.0–75.5 HRA, respectively. The microstructure of the obtained titanium samples is shown in Fig. 6.

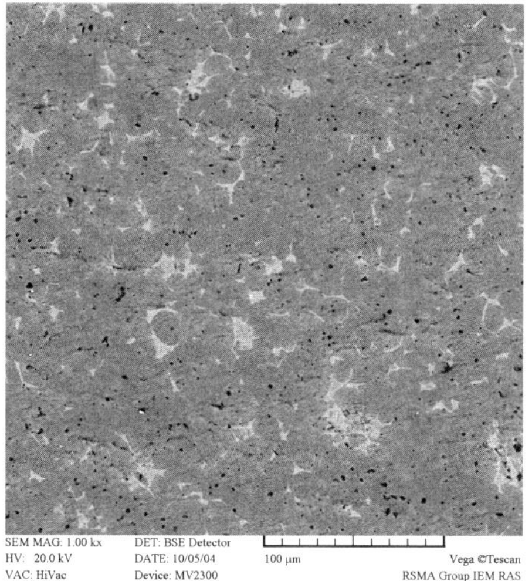

Figure 6. Microstructure of compact (sintered) titanium

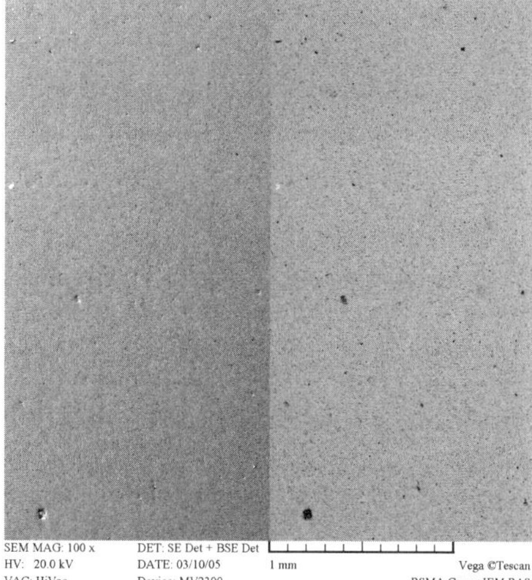

Figure 7. Microstructure of compact zirconium

Zirconium. Identical results have been obtained after the dehydriding of compact ZrH_2 samples. Zirconium samples of high density were received upon sintering at 950°C (and above) of its hydride powder grinded down to <10 µm. The density of the zirconium samples achieved 6.02–6.91 g/cm^3, and hardness – 63.2–74.2 HRA, depending on initial powder dispersion. Figure 7 shows the microstructure of compact zirconium after the dehydriding; the pictures on the left were made in secondary electrons (the relief regime); the pictures on the right represent the surface in the mode of phase contrast (shooting in reflected electrons).

3. Discussion

The powders of refractory metals and their compounds, as well as of the composite materials, based on them, are classified as hard-sintering materials. Construction produces of them can be made mainly by power-consuming hot-pressing. The technique of cold pressing of powders of refractory metals and their further sintering at temperature above 1,800—2,200°C is a power-consuming process, too: to obtain solid produces by methods of powder metallurgy, fine metal powders with particle size under 5 µm are needed. The grinding of a metal down to these sizes is a rather laborious and expensive process.

At present, SHS is the most productive among all the existing methods for receiving hydrides, and it can provide production of a wide assortment of hydrides with high hydrogen content. It is known, that hydrogen, inserted into crystal lattice of metal, embrittles it. At this, the higher the hydrogen content in the hydride, the easier and faster its grinding (30–60 min. vs. 100 h for the metal or alloy). The usage of hydride powder as raw material for receiving solid produce made of refractory metal or alloy is more preferential because the hydride grinding down to as low as 10 mm is not a problem, and does not contaminate the powder.

In the described experiments, the hydride compacting followed by hydrogen removing at temperature 700–1,050°C and short exposition (no more than 1 h) resulted in formation of porousless compact metal sample. These temperatures are much lower than those for melting metals, and at least by 800–1,000°C lower than those required in the traditional methods of metal produce production by diffusion sintering. To explain this effect, let us consider some processes, which may take place at dehydrogenation.

At first, the heating of the samples results in the structural (phase) transformations with volume change. For example, the face-centered cubic (FCC) lattice of titanium hydride, on reaching the dissociation temperature (600–650°C), converts into hexagonal close-packed (HCP) lattice of titanium: TiH_2 (FCC)→Ti(HCP) + H_2; the body-centered tetragonal BCT lattice of zirconium hydride, on reaching the dissociation temperature (800–850°C), converts into close-packed hexagonal (HCP) lattice of zirconium ZrH_2 (BCT)→Zr(HCP)+H_2. The process goes on continuously until the complete disappearing of the hydride. Germs of new phases are formed (here– of metals). At the increase of temperature, a lot of new germs form, combine and amalgamate (solute in each other), forming a strong frame.

As it was indicated above, the samples of compact hydrides under study are practically porousless and of high density (close to the theoretical), due to the mechanical contact of small particles of powder. One may suggest, that at the initial stage of destruction of hydrides' crystal lattice, the substance is in a chaotic (amorphous is not excluded) state, after which re-crystallization processes take place. Some of the germs grow due to their neighbors, resulting in formation of a dense metal sample.

During heating of the compacted hydride powder, for example, TiH_2 or ZrH_2, at temperature a little higher than that for hydride dissociation (800–1,000°C), an essential activation of metal occurs due to Me-H link disruption. Simultaneously, at heating in hydrogen environment, purification from the oxide film, which usually covers the fine powder surface, happens. The "open links" and the purified surface of the powders facilitate diffusion and intersolution of metal particles at rather low temperatures. Actually, a solid-phase diffusion of titanium or other metal powders takes place resulting in formation of a compact, practically porousless refractory metal sample.

It is essential, that at compacting, due to high brittleness, hydride particles diffuse in each other, forming a solid frame. At this, the existence of H in the crystal lattice increases the capability of germs to plastic deformation, promoting formation of deformation-strengthened produces. The integrity of tablet pressed from hydride powder is provided by dense mechanical contacts of fine particles. At sintering metals by the methods of powder metallurgy for activation of compacting and sintering kinetics, little additions, plasticizers, are often used. In our technique, when metal hydride powder is used as a raw material, plasticizers are not needed, because the high hydrogen content in crystal lattice embrittles the mould and adds plasticity at pressing. Probably, at pressing, additional grinding of hydride germs and their mutual intrusion takes place, too. Essential is the large specific surface of micron and sub-micron particles of the hydride powder. All these ensures the maximal compressing of the samples. Nano-scale dimensions of germs (20–80 nm) of the used hydride powders also actively promote formation of dense produces owing to mutual diffusion at dehydrogenation-sintering.

All these create rather favorable conditions for realization of solid-phase diffusion mechanism of formation of produces at relatively low temperatures in the conditions of our experiments.

The energy of defects and irregularities, formed during the synthesis of hydrides in SHS regime or cold pressing processes in the crystal lattice of initial particles, also can influence the alloy formation process during dehydrogenation-sintering of compacted hydride powders. The free energy increase at the expense of accumulated point defects, surface dislocations and volume defects can be accompanied by the subsequent energy liberation, also promoting the sintering of metal particles at relatively low temperature.

4. Conclusions

1. Compact hydrides were received with the density close to the theoretical. From the compacted hydrides, the strong samples of IVa group metals were prepared. Nano-scale sizes of powders of the used hydrides promote formation of dense produces due to interdiffusion at pressing and dehydrogenation of hydrides.
2. The influence of pressing (compacting) pressure, dehydrogenation temperature, and dispersion of initial hydride powder on the mechanical properties of final produces, such as solidity, micro-hardness, porosity, structure, etc., have been shown.
3. In principle, the elaborated method for obtaining the produce of a given shape from the hydrides of refractory metals is rather promising. However, the introduction of this technique into industry, the outcome to customers is related with a wide range of engineering problems, appearing at practical making of a specific produce. Integration of efforts of scientists and engineers of different specialties is necessary for the successful use and advancement of this technique.

Acknowledgments

The work is performed at financial support of ISTC Grant A-1249, and Ministry of Education and Science of Armenia, theme 0567.

References

1. Dollukhanyan S.K., Nersesian M.D., Nalbandian A.B., Borovinskaya I.P. and Merzhanov A.G. Transition Metals Combustion in Hydrogen., *Dokl. AN SSSR* 1976; **231**(6): 675–678.
2. Dolukhanyan S.K., Hakobyan H.G. and Alexanyan A.G. Combustion of Metals in Hydrogen and Hydride Production by Self Propagating High Temperature Synthesis. *Int. J. SHS* 1992; **1**(4): 530–535.
3. Dolukhanyan S.K., Alexanian A.G. and Hakobian H.G. Interaction of Hafnium with Hydrogen and Nitrogen in the Combustion Regime. *Int. J. Hydrogen Energ.* 1995; **20**(5): 391–395.
4. Mueller W.M., Blackledge J.P. and Libowitz, G.G., Eds. Metal Hydrides. Academic, New York, 1968, P.429.

XRD STUDY OF HYDROGENATED FULLERITES AND METALS FROM IV–V GROUPS

A.A. ROGOZINSKAJA, I.I. TIMOFEEVA,
L.A. KLOCHKOV, A.A. ROGOZINSKIY,
A.D. ZOLOTARENKO, A.F. SAVENKO
Frantsevich Institute for Problems of Materials Science of NASU,
3, Krzhyzhanovsky Str., Kyiv, 03142 Ukraine

Abstract. The present work shows the results of XRD studies of different materials after their interaction with hydrogen.

Keywords: hydrogenated fullerite, XRD analysis, transformation of fullerite lattice

1. Introduction

Hydrogenation of different elements is a complex method related both to electron and crystalline structure of element and peculiarities of hydrogen atoms behavior in interaction.

As is shown in [1, 2], at fullerite hydrogenation the hydrogen atoms can be located at different interstices of the fullerite lattice depending on thermodynamical conditions. One can expect that under conditions used for fullerite hydrogenation in the present work (temperature, hydrogen pressure) hydrogen would be located at tetrahedral interstices of the fcc fullerite lattice analogously to fcc-metals. The present work shows the results of XRD studies of different materials after their interaction with hydrogen.

2. Experimental

Fullerite hydrogenation was performed in purified hydrogen under constant hydrogen pressure 130 at in the temperature range 300–450°C.

Beginning from 400–420°C, interaction of hydrogen with fullerite proceeds intensively. Hydrogenation was carried out under different exposure, from 50 to 5,000 min. The hydrogen content was determined by the volumetric method using the Sivert's apparatus. With increasing exposure the hydrogen content in fullerite is increased. After 50 min hydrogenation the hydrogen content made up 1.5 mass% ($C_{60}H_{12}$), after 3,000 min hydrogenation the hydrogen content was 5–6 mass% ($C_{60}H_{36}$-$C_{60}H_{48}$).

Increasing exposure up to 5,000 min allowed the production of fullerite hydrides containing approximately 7 mass% hydrogen ($C_{60}H_{60}$).

The produced fullerite C_{60} hydrides were studied by the XRD method.

X-ray survey of the samples was performed on the diffractometer DRON-3 in filtrated CuK_{α}-radiation under a voltage of 35 kV on the tube and a current of 20 mA. To refine the results of measurements, the reflection lines from each sample were recorded three times.

The XRD analysis has shown that all produced samples have the fcc-structure of starting fullerite. Different values of fullerite lattice parameters have been obtained depending on the hydrogen content in fullerite. With increasing hydrogen content, the lattice parameter and volume are increased (Fig. 1, Table 1).

The lattice parameter calculated by the reflection lines (311) and (420) is increased from 14.031 Å for starting fullerite to 15.183 Å at the hydrogen content of 5.3 mass%. Consequently, the volume of the lattice cell, V, is increased.

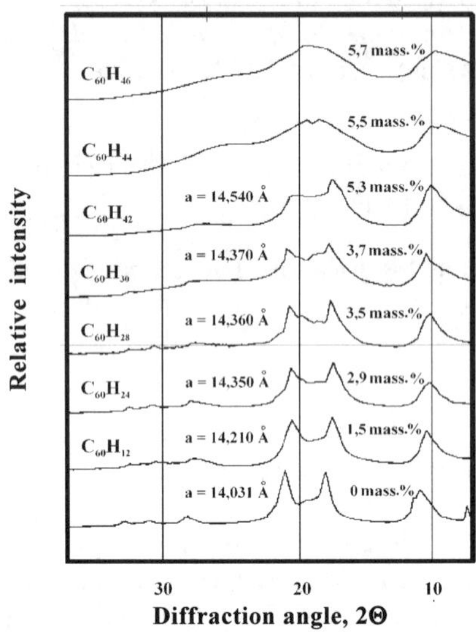

Figure 1. Transformation of fullerite lattice in its saturation with hydrogen

TABLE 1. Lattice parameters and volume of the lattice cell of hydrogenated fullerite

Sample	***a* (A°)**	**V (A°)³**	**Mass content % H₂**
start	14,031	2,762	–
$C_{60}H_{12}$	14,210	2,869	1.5
$C_{60}H_{24}$	14,350	2,955	2.9
$C_{60}H_{28}$	14,360	2,961	3.5
$C_{60}H_{30}$	14,370	2,967	3.7
$C_{60}H_{42}$	14,540	3,080	5.3
$C_{60}H_{44}$	Amorphous phase		5.5

The samples containing over 5.5 mass% H_2 become X-ray amorphous, the lines (111), (220) and (311) degenerate into a diffuse halo characteristic of amorphous materials.

The X-ray diffraction pattern of the samples containing 6 and 7 mass% H_2 shows a weak but clear line with d = 2.28 Å. It is conceivable that small domains of fullerite hydride with the hcp-lattice appear in the interstitial cavities of X-ray amorphous substance due to hydrogen supersaturation.

Hence hydrogenated fullerite retains its lattice at the hydrogen content up to 5.3 mass%.

The further increase in the hydrogen content renders fullerite amorphous. The results obtained after fullerites hydrogenation have been compared to the peculiarities of fcc-metals hydrogenation. The process of fcc-metals hydrogenation has been studied repeatedly [3].

The present work considers peculiarities of hydrogenation of fcc-metals produced under different conditions, namely, by melting, rolling, drawing, chemical method.

The hydrides of transition metals, Ti, Zr, V, Nb, Ta have been studied.

The metals subjected to rolling and drawing exhibit texture and, besides the main metal, contain little amounts of its oxides.

The study of the metals exposed to hydrogen flow (at corresponding temperatures) has revealed the following.

Hydrogenation of melted titanium gives rise to the surface titanium hydride with fcc-structure of CaF_2 type and lattice parameter a = 4.458 Å corresponding to the composition $TiH_{1.95}$ [3, 4]. Titanium oxides are also kept. However texture in the titanium hydride is not observed (Table 2, Fig. 2).

The similar results have been also obtained for other rolled metals, Zr, Nb, V, Ta. In all cases their hydrides do not reach the limit composition and correspond to those falling within the region of homogeneity. For example, the hydride produced from melted zirconium has lattice parameter a = 4.781 Å corresponding to the composition ZrH_2. Hydrogenation of textured rolled zirconium gives rise to the hydride with lattice parameter a = 4.779 Å corresponding to the composition $ZrH_{1.97}$.

Hydrogenation of vanadium foil at 1,400°C is responsible for the solid hydrogen solution in vanadium that shows up as an increase in its lattice parameter. Before and after hydrogenation the lattice parameter was a = 3.027 and 3.054 Å, respectively. Besides that, the formation of vanadium hydride with tetragonal lattice and parameters a = 3.031 Å, c = 3.410 Å which correspond to the composition $VH_{1.82}$ occurs.

TABLE 2. Results obtained from the X-ray diffraction pattern of hydrogenated titanium wire

Experiment			$TiH_{1.95}$[a]			TiO_2(rutile)[a]			Ti_4O_7[a]		
I	2θ	d	I	d	hkl	I	d	hkl	I	d	hkl
62	26.9	3.33							100	3.35	020
50	27.6	3.23				100	3.25	110			
7	31.0	2.88							60	2.91	122
50	34.5	2.60							90	2.63	200
25	35.3	2.54	90	2.54	111						
20	37.0	2.43				50	2.48	101	90	2.48	120
37	37.8	2.38							50	2.38	222
100	39.5	2.28				8	2.29	200			
20	40.5	2.227	100	2.227	200						
25	42.5	2.13							90	2.13	024
20	53.8	1.71				60	1.69	211			
10	56.0	1.64				20	1.62	200			
12	59.0	1.56	80	1.57	220						
12	62.0	1.50				10	1.48	002			
10	68.5	1.37				20	1.36	301			
15	70.0	1.34	40	1.34	311						
7	74.1	1.28	30	1.29	222						

[a]Literature data

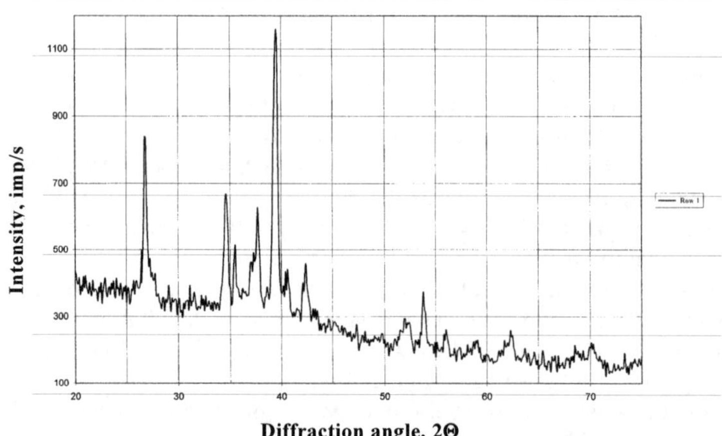

Figure 2. X-ray diffraction pattern of hydrogenated titanium wire

Hydrogenation of niobium provides niobium hydride with cubic lattice (lattice parameter $a = 3.445$ corresponding to $NbH_{0.9}$).

Hydrogenation of rolled tantalum gives rise to tantalum hydride with orthorhombic structure and parameters a = 4.727 Å, b = 4.779 Å, c = 3.430 Å corresponding to $TaH_{0.8}$ [3, 4].

In this case, like in all the previous ones, texture in hydrogenated metals is not observed. Most likely this is due to the rearrangement of the metal crystalline lattice caused by hydrogenation.

The fact that hydrogen saturation of the rolled metal is difficult compared, for example, to the melted metal has arrested our attention. This may be due to both the texture of rolled metals and the presence of an oxide layer.

3. Conclusions

- Fullerites saturation with hydrogen gives rise to solid solutions that manifests itself as an increase in fullerites lattice parameters.
- Fullerite containing over 5.5 mass% H_2 becomes amorphous.
- Metals rolling accompanied by partial oxidation hinders metals saturation with hydrogen in hydrogenation.

References

1. Schur D.V., Zaginaichenko S.Yu., Matysina Z.A., Shaposhnikova T.I., Kopylova L.I., Kamenetskaya E.A. The structural features of C-H bonds formation on the surface of fullerene C_{60} frame, XI International Conference, ICHMS'2007, "Hydrogen Materials Science and Chemistry of Carbon Nanomaterials", 22–28 September, Sudak, Crimea, 2007, P. 730–731.
2. Schur D.V., Matysina Z.A., Zaginaichenko S.Yu. Carbon nanomaterials and phase transformations in these materials. Nauka i obrazovanie, Dnepropetrovsk, 2007, 680 (in Russian).
3. Myuller V., Blekledzh D., Litovits Dzh. Hydrides of metals. Atomizdat, Moscow, 1975, 430.
4. International Tables for X-ray Crystallography "Alphabetical Index of Inorganic Compounds", International Centre for Diffraction Date, Swarthmore, PA, 1998, P. 427.

HYDROGEN SOLUBILITY IN BOROHYDRIDES AND METALS HYDRIDES

Z.A. MATYSINA
Dnepropetrovsk National University, Str. Nauchnaya, 13, Dnepropetrovsk, 49050 Ukraine
S.YU. ZAGINAICHENKO, D.V. SCHUR, T.A. TRIFONOVA
Frantsevich Institute for Problems of Materials Science of NASU, 3, Krzhyzhanovsky Str., Kyiv, 03142 Ukraine
Fax: 38 (044) 424-0381; E-mail: shurzag@ipms.kiev.ua

Abstract. The theoretical investigation of the hydrogen solubility in borohydrides and hydrides of different alkali metals has been carried out. The possibility of interstitial component ordering is taken into account. The calculation of free energies of phases, determination of the thermodynamic equilibrium conditions of phases permits us to ascertain the dependence of hydrogen solubility on temperature and pressure, to study the P-T-c diagrams of hydrogen solubility in the phases and, in particular, to elucidate the functional peculiarities of isopleths and isotherms of hydrogen solubility in studied phases. The plots of these dependences have been constructed. Comparison of the theoretical results with experimental data on the isopleths and isotherms of hydrogen solubility in $LiBH_4$, $NaBH_4$, KBH_4, Na-NaH-H, LiH, $2LiNH_2$ + MgH and $LiNH_2$ + LiH systems has been carried out.

Keywords: borohydride, alkali metal hydride, phase transformation, hydrogen solubility, P-T-c diagram, isopleth, isotherm

1. Introduction

The borohydrides and hydrides of metals as interstitial phases (solid solutions and interstitial compounds) with their unique properties attract attention of scientists and technical workers in connection with the possibility of controlling their properties to a considerable extent through interstitial additions. The knowledge of the interstitial hydrogen atoms solubility limit and of the arrangement character of interstitial atoms in the crystal lattice are of paramount importance to create materials with the required properties. The limiting concentration of the gas atoms depends heavily on the pressure and temperature. The study of P-T-c diagrams is the fundamental method of investigation of introduction phases. The view of P-T-c plots depends considerably on the character of distribution of gas atoms over the positions of different type, on the crystalline structure, on the composition and degree of atomic ordering in the case of alloys. The redistribution of gas atoms in crystal and phase transitions can modify these plots.

The experimental investigation of chemical reactions with increase in reactions temperature

$$\mathrm{LiBH}_{x_1} \rightarrow \mathrm{LiBH}_{x_2} + \frac{x_1 - x_2}{2}\mathrm{H}_2 \rightarrow \mathrm{LiH}_{x_3} + \mathrm{B} + \frac{x_1 - x_3}{2}\mathrm{H}_2, \quad x_1 > x_2 > x_3, \qquad (1)$$

where

$$0 \le x_1 \le 4, \quad 0 \le x_2 \le 2, \quad 0 \le x_3 \le 1, \qquad (2)$$

provides a possibility to use borohydrides as materials for the accumulation and storage of hydrogen. A great number of scientific papers [1–11] are devoted to investigation of these phase transformations. The $LiBH_4$, $LiBH_2$ phases and the mixed phase LiH + B are realized respectively in the neighbourhood of temperatures 108°C, 200°C and 450°C. Over somewhat different temperature interval the similar chemical reactions can proceed in the case of borohydrides of another alkali and nonalkali metals [12–25], as $NaBH_4$, KBH_4, $Be(BH_4)_2$, $Ca(BH_4)_2$, $Cu(BH_4)_2$, $Mg(BH_4)_2$, $Mn(BH_4)_2$, $Zn(BH_4)_2$, $Al(BH_4)_3$, $Sc(BH_4)_3$, $Ti(BH_4)_3$, $Hf(BH_4)_4$, $Zr(BH_4)_4$. In these compounds the hydrogen atoms form the tetrahedral configuration around the boron atoms. The weight content of hydrogen in the named phases and also in the hydrides of alkali metals LiH, NaH is sufficiently high (see Table 1) [11, 23–25].

TABLE 1. The weight content of hydrogen in different borohydrides

Phase	Hydrogen (wt%)
$LiBH_4$	18.3
$NaBH_4$	10.6
KBH_4	7.4
$Be(BH_4)_2$	20.7
$Ca(BH_4)_2$	9.7
$Cu(BH_4)_2$	9.2
$Mg(BH_4)_2$	14.8
$Mn(BH_4)_2$	9.5
$Zn(BH_4)_2$	8.5
$Al(BH_4)_3$	16.8
$Sc(BH_4)_3$	13.5
$Ti(BH_4)_3$	13.1
$Zr(BH_4)_4$	10.7
$Hf(BH_4)_4$	11.3
LiH	12.6
NaH	4.1

The calculation of hydrogen solubility in borohydrides and hydrides of different alkali metals, the theoretical investigation of conditions of the thermodynamic equilibrium of the phases of chemical reaction (1), the elucidation of the conditions of phase transformations realization with increasing temperature, the study of P-T-c diagrams of hydrogen solubility in the phases and, in particular, the determination of special features of functional dependences of P-T isopleths and P-c isotherms of hydrogen solubility in these phases are of interest to chemical-process engineers and technical workers.

2. Free energies of the phases α, β, γ

The calculation of the phases free energies and the investigation of thermodynamic equilibrium conditions have been performed to solve the raised problem. The calculation of free energies has been carried out using the molecular-kinetic theory by the method of average energies in a model of pair interatomic interactions.

The calculation of free energies of the phases

$$\alpha = MBH_{X_1}, \quad \beta = MBH_{X_2}, \quad \gamma = MH_{X_3} + B, \quad M = Li,\ Na,\ K \tag{3}$$

on the one site of crystal lattice gives the following formulae

$$\begin{aligned} f_\alpha &= U_\alpha^o + 2U_\alpha c + 2kT\Delta_\alpha - kTc_\alpha \ln\lambda_\alpha, \\ f_\beta &= U_\beta^o + U_\beta c + kT\Delta_\beta - kTc_\beta \ln\lambda_\beta, \\ f_\gamma &= U_\gamma^o + U_\gamma c + \frac{1}{2}kT\Delta_\gamma - kTc_\gamma \ln\lambda_\gamma, \end{aligned} \tag{4}$$

where U_α^0, U_β^0, U_γ^0 are energetic parameters, determined by interaction of M and B atoms, which independent of hydrogen concentration,

$$U_\alpha = u_{BH}^{(\alpha)} + u_{MH}^{(\alpha)}, \quad U_\beta = u_{BH}^{(\beta)} + u_{MH}^{(\beta)}, \quad U_\gamma = 3u_{MH}^{(\gamma)}, \tag{5}$$

$u_{BH}^{(i)}$, $u_{MH}^{(i)}$ are the interaction energies of nearest atomic pairs BH and MH in the phases i = α, β, γ; $c_i = c_\alpha, c_\beta, c_\gamma$ are concentrations of hydrogen atoms in the relation to the number of their positions,

$$0 \le c_i \le 1, \tag{6}$$

$$\Delta_i = c_i \ln c_i + (1-c_i)\ln(1-c_i), \quad i = \alpha,\ \beta,\ \gamma, \tag{7}$$

k is the Boltzmann's constant, T is absolute temperature, λ is the activity of hydrogen atoms in the phases.

3. Hydrogen solubility in the phases α, β, γ

The concentrations c_i of hydrogen atoms in the relation to the number of their positions are expressed by the concentrations x_i of hydrogen atoms in the relation to the number of boron atoms in the phases α, β and to number of M atoms in the phase γ by the formulae

$$x_1 = 4c_\alpha, \quad x_2 = 2c_\beta, \quad x_3 = c_\gamma. \tag{8}$$

The energies $U = U_\alpha, U_\beta, U_\gamma$ and the activities $\lambda = \lambda_\alpha, \lambda_\beta, \lambda_\gamma$ depend on external pressure P of the medium, where the phases are located

$$U = U_o(1-\alpha P), \quad \alpha \le 1, \tag{9}$$

$$\lambda = DP^{1/2}, \tag{10}$$

where α and D are the pressure-independent coefficients [26–31].

The minimization of free energies (4) gives the formula

$$\ln(DP^{1/2}) = \ln\frac{c}{1-c} + \frac{U_o(1-\alpha P)}{kT}, \tag{11}$$

or

$$c = \left[1 + \frac{1}{DP^{1/2}} \exp\frac{U_o(1-\alpha P)}{kT}\right]^{-1}, \qquad (12)$$

which define the P-T-c diagram of hydrogen solubility of phases being investigated.

The formula (11) is convenient to use for construction of isopleths, the formula (12) is suitable for construction of isotherms. In the first case when the value of pressure P is prescribed for different concentrations *c* the temperature T is calculated and the dependence $\ln P\left(\frac{1}{kT}\right)$ is found. In the second case we also preassign the value of pressure P for different temperatures T, calculate the concentration *c* and as a result of which define the dependence $\ln P(c)$.

The borohydrides and hydrides of various metals have been studied experimentally [1–20, 32–51]. The experimental plots of isopleths and isotherms of hydrogen solubility for the systems $LiBH_4$, $NaBH_4$, KBH_4, Na-NaH-H and Li-H and also of the similar isotherms for recently studied systems $2LiNH_2 + MgH_2$, $LiNH_2 + LiH$ [11, 38, 48—51] are shown in Fig. 1.

It is evident from Fig. 1(a, b) that isopleths present the linear dependence $\ln P(1/T)$ and the slope of lines of this dependence is determined by the chemical nature of metal and its concentration. The decrease of metal M concentration leads to the convergence of isopleths.

As can be seen from Fig. 1(c–e), the plateau (horizontal section of the curve) is observed on the plots of pressure P as a function of hydrogen concentration *c* and the plateau length decreases with a rise in temperature. Furthermore, the pressure decreases sharply when $c \rightarrow 0$, and conversely, the pressure increases drastically when $c \rightarrow 1$ (100 at.% H).

The formulae (11) and (12) allow the explanation and substantiation of established experimentally functional dependences of isopleths and isotherms of borohydrides and hydrides of alkali metals.

The plots that demonstrate the isopleths of hydrogen solubility in phases, constructed by the formula (11), for the following values:

(a) $U = 1$ eV, $D = 0.1$ $Pa^{-1/2}$, $\alpha = 0$ and concentrations $c = 0.5; 0.6; 0.7; 0.8; 0.9$
(b) $D = 0.1$ $Pa^{-1/2}$, $\alpha = 0$, $c = 0.8$, $U = 0.8; 1; 1.2$ eV

are shown in Fig. 2(a, b). It is obvious from Fig. 2(a) that the increase of concentration *c* leads to the higher passage of isopleths (their level increases), what is in agreement with experimental data (Fig. 1(b)), the distance between isopleths tended to increase with increase in concentration *c*, what is also in accordance with experimental results (Fig. 1(b)), the slope of isopleths, constructed for the energy U = const, does not vary. The different values of the energy U produce the change in the slope of isopleths, as in experimental plots of Fig. 1(a). This is confirmed by the design plots of Fig. 2(b) and it is evident from these plots that the isopleths pass below, their level reduces and it this case their slope increases (an increase of tgδ) as the energy U increases. It can be assumed that in MBH_4 phases at the transition from $LiBH_4$ to $NaBH_4$ and further to KBH_4 the energy U increases at the expense of which the slope of isopleths undergoes a rise and its level is raised, what is observed on the experimental plots of Fig.1(a).

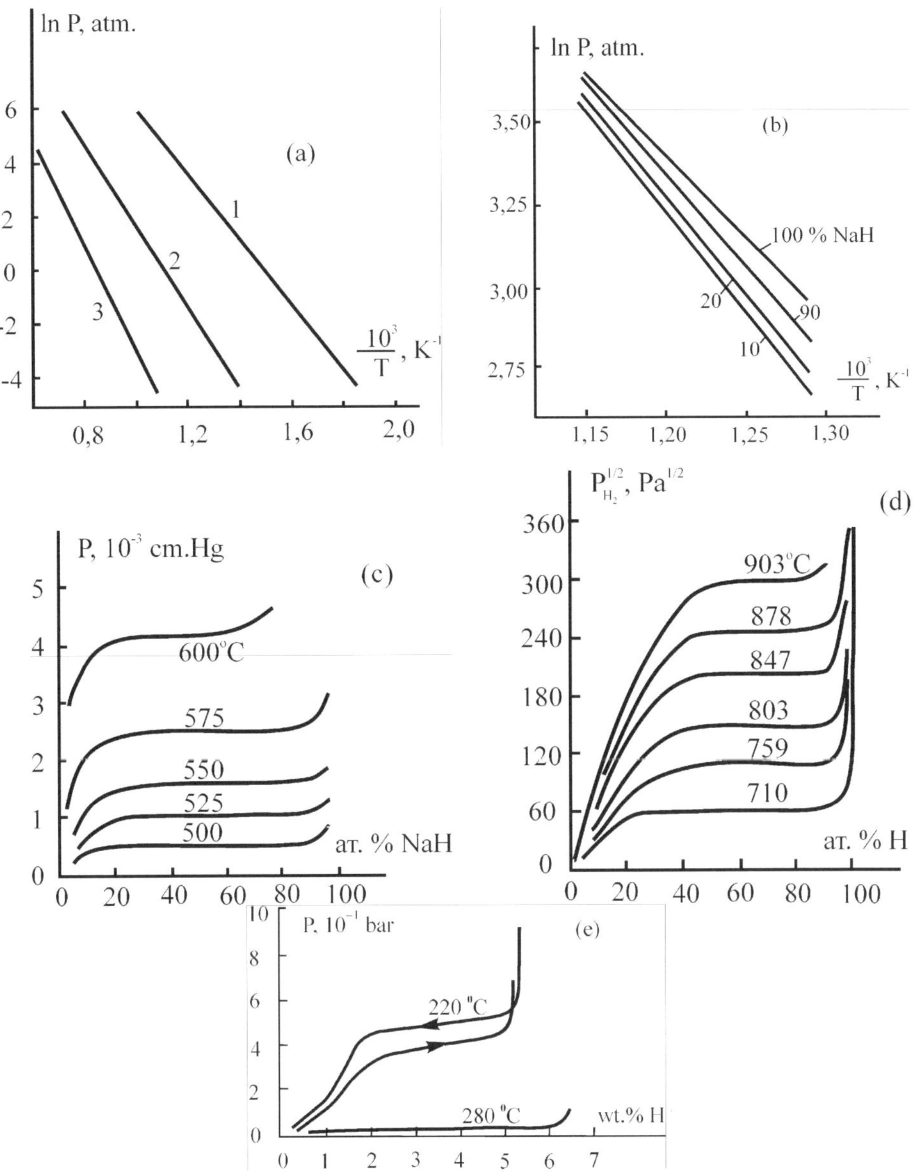

Figure 1. Experimental plots:
(a) Isopleths for borohydrides of alkali metals $LiBH_4$, $NaBH_4$, KBH_4 (straight lines 1, 2, 3)
(b) Isopleths for Na-NaH-H system
(c) Isotherms for this system
(d) Isotherms of hydrogen solubility in lithium
(e) Isotherms for nitrohydrides of
$2LiNH_2 + MgH_2$ system (upper curves) and
$LiNH_2 + LiH$ system (lower curve)

The plots of design isotherms of hydrogen solubility in the phases α, β, γ, constructed by the formula (12), for the following values:

U = -1 eV, α = 0, D = 1 $Pa^{-1/2}$ (curves 1 and 2), D = 10 $Pa^{-1/2}$ (curves 1′ and 2′) and temperatures kT = 1 eV (curves 1 and 1′), kT = 100 eV (curves 2 and 2′) are shown in Fig. 3. It is obvious from this Fig. 3 that the isotherms run higher with increase in temperature, as in experimental plots of Fig. 1(c, d), and also the value lnP decreases sharply when $c\rightarrow0$, and conversely if $c\rightarrow1$, the value lnP increases steeply, what also corresponds to the experimental data. Moreover, over the wide range of concentrations *c* the slope of isotherm is steady, but not horizontal, as in experimental plots of Fig. 1(c, d).

The isotherms have been constructed also for the positive energies U and for the coefficients α different from zero. These plots are shown in Fig. 4(a–c).

It follows from Fig. 4 that the plateau (horizontal section of the curve) appears and is lengthened with a reducing temperature, the plots are flat over the wide temperature range of hydrogen concentrations, and also if $c\rightarrow0$ the value lnP decreases sharply, but if $c\rightarrow1$ the value lnP increases. All these data are in agreement with experiments (Fig. 1(c, d)), but experimental curves pass higher with a rise in temperature. This can be caused by the slight manifestation of the pressure dependence of the energy U, i.e. by the reduced effect of pressure on the energies of the interatomic interactions at the sufficiently high temperatures (by this is meant that the coefficient α decreases with increasing temperature). Actually the plots of Fig. 4(b) show that the decrease of coefficient α from 10^{-2} to 10^{-3} and 10^{-4} Pa^{-1} increases the level of isotherms, in this case this rise increases with the temperature drop (Fig. 4(d)). Besides that, it is obvious from Fig. 4(d) that the simultaneous decrease of the coefficient α and temperature T is favourable to the existence of plateau on the isotherms.

4. Conclusions

Hence, the developed statistical theory of the temperature phase transitions at the performance of the chemical reaction (1) permits to substantiate and explain the temperature course of this reaction, allows to study the P-T-c diagram of hydrogen solubility in phases of reaction (1) in more detail and makes possible the investigation of functional dependence of isopleths and isotherms of hydrogen solubility.

It is significant that the regularities of isotherms course, manifested experimentally at different temperatures, as evident from calculations, depend on the coefficient α to a large extent, because this factor determines the change of energies of interaction between hydrogen atoms and crystalic matrix atoms under changes of external all-round pressure. This dependence decays at high temperatures.

Notice that experimental investigation of isopleths and isotherms of hydrogen solubility in the phases of chemical reaction (1) allows one with the derived formulae (11), (12) to evaluate by the plots run the phase parameters as the energies of interaction between hydrogen and matrix atoms, the character of their dependences on external all-round pressure, the activity of hydrogen atoms in the phases. And vice versa the knowledge from independent experiments of physical properties for each specific phase makes it possible to predetermine in detail the P-T-c diagram on hydrogen solubility in each phase and in particular to define the common view of isopleths and isotherms of hydrogen solubility.

It should be pointed out that the borohydrides of alkali metals are considered as an irreversible systems of hydrogen storage what must be taken into account with their use, for example, in transport [51, 52]. It this case as the result of the chemical reaction (1) the hydrogen is not completely evolved: the part of hydrogen atoms remains linked to the lithium atoms. Besides that, the thermal dehydrogenation is a very power-intensive process because the high temperatures are required for its realization. It should be also mentioned that the processing method for the consumed borates and the production method of metastable solutions from them (such liquid phase can be pump out at the service station) have been proposed recently [52] with the aim of finding of technical moments design for the problem of investigation of alkali metal borohydrides as hydrogen storage system.

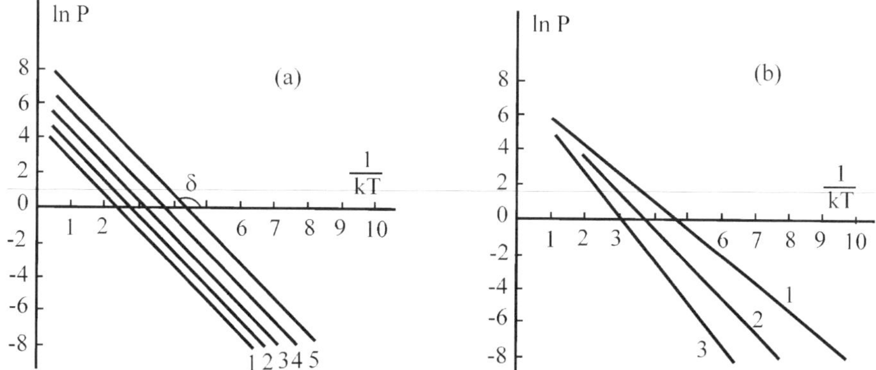

Figure 2. The design isopleths of hydrogen solubility in α, β, γ phases plotted by formula (11) for different coefficients:

(a) U = 1 eV, D = 0.1 $Pa^{-1/2}$, α = 0 and c = 0.5; 0.6; 0.7; 0.8; 0.9 (straight lines 1, 2, 3, 4, 5, correspondingly)

(b) D = 0.1 $Pa^{-1/2}$, α = 0, c = 0.8 and U = 0.8; 1; 1.2 eV (straight lines 1, 2, 3, correspondingly)

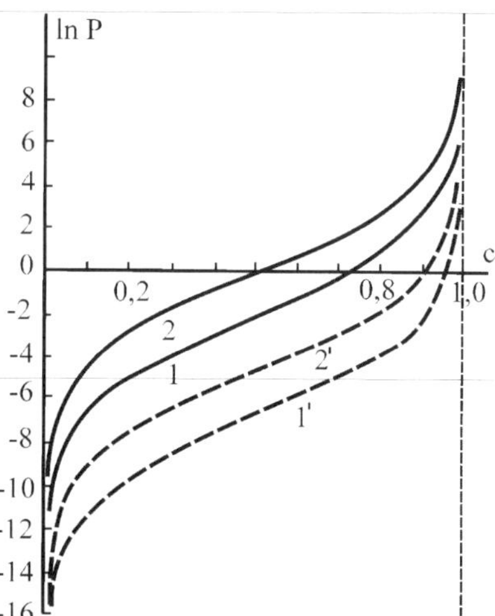

Figure 3. The design isotherms of hydrogen solubility in α, β, γ phases plotted by formula (12) for parameters equal to

U = -1 eV, α = 0, D = 1 $Pa^{-1/2}$ (continuous curves 1 and 2)

D = 10 $Pa^{-1/2}$ (dotted lines 1' and 2') and for temperatures kT = 1 eV (curves 1 and 1'), kT = 100 eV (curves 2 and 2')

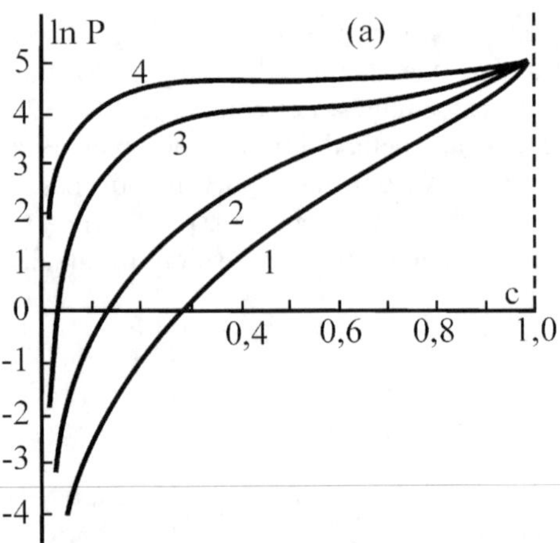

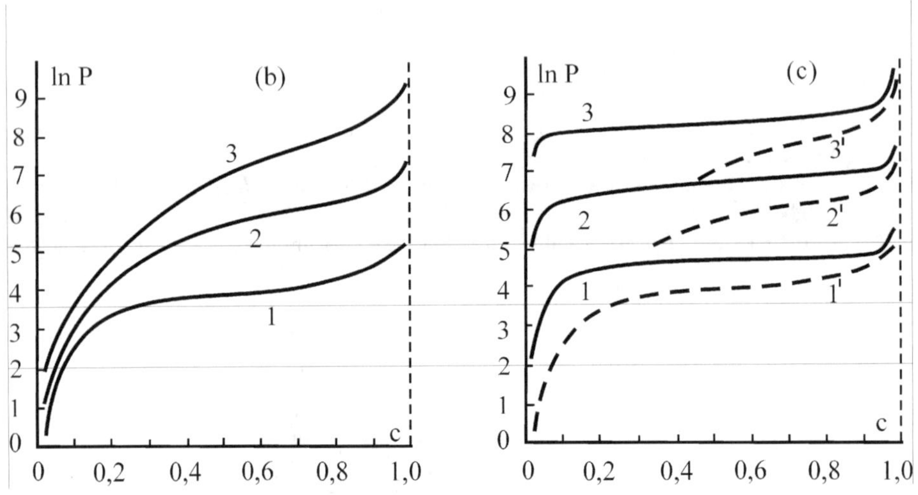

Figure 4. The design isotherms of hydrogen solubility in α, β, γ phases plotted by formula (12) for parameters U, D, α, T equal to

(a) U = 1 eV, D = 1 $Pa^{-1/2}$, $\alpha = 10^{-2}$ Pa^{-1} and kT = 1; 0.5; 0.25; 0.1 eV (curves 1, 2, 3, 4, correspondingly)
(b) U = 1 eV, D = 1 $Pa^{-1/2}$, kT = 0.25 eV and $\alpha = 10^{-2}$, 10^{-3}, 10^{-4} Pa^{-1} (curves 1, 2, 3, correspondingly)
(c) U = 1 eV, D = 1 $Pa^{-1/2}$, kT = 0.1 eV and $\alpha = 10^{-2}$, 10^{-3}, 10^{-4} Pa^{-1} (curves 1, 2, 3, correspondingly)

Curves 1', 2', 3' corresponds to the plots (b)

References

1. Amendola S.C., Sharp-Goldman S.L., Januja M.S., Spencer N.C., Kelly M.T., Petillo P.J., Binder M. Hydrogen storage by $NaBH_4$ // *Int. J. Hydrogen Energ.*, 2004, **29**: 263.
2. Lodziana Z., Vegge T. Structural stability of complex hydrides: $LiBH_4$ revisited // *Phys. Rev. Lett.*, 2004, **93** (14): 145501.
3. Vajo J.J., Skeith S.L., Mertens F. Reversible storage of hydrogen in destabilized $LiBH_4$ // *J. Phys. Chem. B Condens. Matter Mater. Surf. Interfaces Biophys.*, 2005, **109** (9): 3719–3722.
4. Chandhuri S., Muckerman J.T. A critical first step for reversible hydrogen storage in $NaAlH_4$ // *J. Phys. Chem. B Condens. Matter Mater. Surf. Interfaces Biophys.*, 2005, **109** (15): 6952–6957.
5. Laversenne L., Lŭtoffŭ J.M., Percheron-Guegan A., Auroux A., Bonnetot B. Borohydrides for hydrogen storage. Thermolysis of hydrolysis advantages and disadvantages // Abstr. IPHE Conf. - Italy. - 2005.
6. Meisner G.P., Scullin M.L., Balogh M.P., Pinkerton F.E., Meyer M.S. Hydrogen release from mixtures of lithium borohydride and lithium amide: a phase diagram study // *J. Phys. Chem. B Condens. Matter Mater. Surf. Interfaces Biophys.*, 2006, **110** (9): 4186–4192.
7. Au M. Jurgensen A., Zeigler K. Modified lithium borohydrides for reversible hydrogen storage // *J. Phys. Chem. B Condens. Matter Mater. Surf. Interfaces Biophys.*, 2006, **110** (13): 7062–7067.
8. Au M. Jurgensen A., Zeigler K. Modified lithium borohydrides for reversible hydrogen storage // *J. Phys. Chem. B Condens. Matter Mater. Surf. Interfaces Biophys.*, 2006, **110** (51): 26482–26487.
9. Züttel A. Properties of boranates. First Workshop of the Helmholtz Initiative Functional Materials for Mobile Hydrogen Storage. - 2006.
10. Talyzin A.V., Anderson O., Sundqvist B., Kurnosov A., Dubrovinsky L. High-pressure phase transition in $LiBH_4$ // *J. Solid State Chem.*, 2007, **180** (2): 510–517.
11. Laversenne L., Bonnetot B. Hydrogen storage using borohydrides // *Ann. Chim. Sci. Mat.*, 2005, **30** (5): 495–503.
12. Stock A., Sfitterlin W., Kurzen F. Boron hydrides. XX. Potassium diborane $K_2(B_2H_6)$ // *Z. Anorg. Allgem. Chem.*, 1935, **225**: 225–242.
13. Schlesinger H.I., Sanderson R.T., Burg A.B. Metallo borohydrides I. Aluminium borohydride // *J. Am. Chem. Soc.*, 1940, **62**: 3421–3425.
14. Burg A.B., Schlesinger H.I. Metallo borohydrides II. Beryllium borohydride // *J. Am. Chem. Soc.*, 1940, **62**: 3425–3429.
15. Schlesinger H.I., Brown H.C. Metallo borohydrides III. Lithium borohydride // *J. Am. Chem. Soc.*, 1940, **62**: 3429–3435.
16. Wiberg E. Magnesium borohydride // *Z. Naturforschung*, 1950, **5b**: - 397–398.
17. Schlesinger H.I., Herbert C., Brown H.C., Abraham B., Bond A.C., Davidson N., Finholt A.E., Gilbreath J.R., et. al. New developments in the chemistry of diborane and the borohydrides // *J. Am. Chem. Soc.*, 1953, **75** (1): 186–190.
18. Brown H.C. Hydroboration. - New York: Benjamin, 1962.
19. James B.D., Wallbridge M.G.H. Metal tetrahydroborates // *Prog. Inorg. Chem.*, 1970, **11**: 97–258.
20. Greenwood N.N. The chemistry of boron // *Compr. Inorg. Chem.*, 1973, **8**: 745–762.
21. Maybury P.C., Mitchell R.W., Hawthorne M.F. Hydrogen adducts of cobalt and nickel boride // *J. Chem. Soc. Chem. Commun.*, 1974, 534–535.
22. Liu J., Fang Z.Z., Sohn H.Y. A new Li-Al-N-H system for reversible hydrogen storage // *J. Phys. Chem. B Condens. Matter Mater. Surf. Interfaces Biophys.*, 2006, **110** (29): 14236–14239.
23. Gilbert M., Brown P.J., Joachim H., Schaeibel J.H. Complex hydrides for hydrogen storage studies of the $Al(BH_4)$ system. - Project ID # STP24. - Metal hydride Center of Excellence. - 2007.
24. Soloveichik G.l. Metal borohydrides as hydrogen storage materials // *Material Mat.*, 2007, **2** (2): 11–15.
25. Balema V.P. Mechanical processing in hydrogen storage research and development // *Material Mat.*, 2007, **2** (2): 16–18.

26. Smirnov A.A. Molekulyarno-kineticheskaya teoriya metallov. Moscow: Nauka. 1966. 488 p. (in Russian).
27. Matysina Z.A. Atomniy poryadok i svoystva splavov. Dnepropetrovsk: DGU. 1981. 112 p. (in Russian).
28. Matysina Z.A., Milyan M.I. Teoriya rastvorimosti primesey v uporyadochennyh fazah. - Dnepropetrovsk: DGU, 1991. -180 p. (in Russian).
29. Matysina Z.A., Schur D.V. Vodorod i tverdofaznie prevrascheniya v metallakh, splavakh, fulleritakh. Dnepropetrovsk: Nauka I obrazovanie. 2002. 420 p.
30. Matysina Z.A., Zaginaichenko S.Yu. Defecty structury kristallov.- Dnepropetrovsk: Nauka i obrazovanie. 2003.- 284 p. (in Russian).
31. Matysina Z.A., Zaginaichenko S.Yu., Schur D.V. Rastvorimost primesey v metallah, splavah, intermetallidah, fulleritah.- Dnepropetrovsk: Nauka i obrazovanie. 2006. - 514 p. (in Russian).
32. Finholt A.E., Bond A.C., Jr., Schlesinger H.I. Lithium aluminium hydride, aluminium hydride and lithium gallium hydride and some of their applications in organic and inorganic chemistry // *J. Am. Chem. Soc.*, 1947, **69**: 1199–1203.
33. Finholt A.E., Bond A.C., Jr., Wilzbach K.E., Schlesinger H.I. The preparation and some properties of hydrides of elements of the fourth group of the periodic system and of their organic derivatives // *J. Am. Chem. Soc.*, 1947, **69**: 2692–2696.
34. Shull C.G., Wollan E.O., Morton G.A., Davidson W.L. Neutron diffraction studies of NaH and NaD // *Phys. Rev.*, 1948, **73**: (8): 842–847.
35. Ahmed M.S. C. The state of ionization in lithium hydride // *Phil. Mag.*, 1951, **42**: 997–1004.
36. Hurd D.T. An introduction to the chemistry of hydrides. - New York, London: Wiley, Chapman & Hall, 1952.
37. Herd D.T. Vvedenie v himiyu gidridov. - Moscow: IL, 1955 (in Russian).
38. Banus M.D., McSharry J.J., Sullivan E.A. The sodium–sodium hydride hydrogen system at 500-600°C // *J. Am. Chem. Soc.*,1955, **77**: 2007–2010.
39. Libowitz G.G., Gibb T.R.P. The ionic character of transition metal hydrides // *J. Phys. Chem.*, 1956, **60** (4): 510–511.
40. Messer C.E., Damon E.B., Maybury P.C., Mellor J., Seales R.A. Solid-liquid equilibrium in the lithium - lithium hydride system // *J. Phys. Chem. Itnaka.*, 1958, **62**: 220–222.
41. Cochran W. Some electron distribution determinations by X-ray diffraction // *Rev. Mod. Phys.*, 1958, **30**: 47–50.
42. Roos A. Li, its hydrides and their applications // *Chim. Ind.*, 1961, **85**: 86–91.
43. Pretzel F.E., Lewis W.B., Szklarz E.G., Vier D.T. Properties of lithium hydride – vacancy formation, cavitation and lithium precipitation in irradiated lithium // *J. Appl. Phys.*, 1962, **33** (1): 510–518.
44. Messer C.E., Eastman J.C., Mers R.G., Maeland A.J. Ternary perovskite phases in systems of lithium hydride with barium, strontium and calcium hydrides // *Inorg. Chem.*, 1964, **3**: 776–778.
45. Bell N.A., Coates G.E. Lithium and sodium beryllium hydride // *J. Chem. Soc. A.*, 1968, 628–631.
46. Mueller W.M., Blackledge J.P., Libowitz G.G. (Eds.) Metal hydrides. - New York, London: Academic, 1968.
47. Aronson S., Salzano F.J. The solid state reaction of lithium hydride and aluminium // *Inorg. Chem.*, 1969, **8**: 1541–1542.
48. Tallefsene M., Kadlis E., Jiles E. // *J. Less-Common Met.*, 1985, **110**: 107–117.
49. Kolpakov B.A., Il'in A.A., Lavrenko V.A., Levinsky Yu.V. Gidridnie sistemy. Spravochnik. - Moscow: Metallurgiya, 1992. - 352 p. (in Russian).
50. Chase M.W. (Ed.) Thermochemical tables // J. Phys. Chem. Ref. Data Monograph, N9 (Am. Chem. Soc., Am. Inst. Phys., Nat. Inst. Stand. and Tech., 1998).
51. Luo W., Ronnebro E. Towards a viable hydrogen storage system for transportation application // *J. Alloy. Compd.*, 2005, **404–406**: 392–395.
52. Ranjard J.F., Glipa X. Procédé pour maintenir liquide a la température ambiante une solition aqueuse de borate de sodium. - French Pat. 2 859648-A1. (2003/9/03).

ACOUSTOMICROSCOPY ANALYSIS OF STRUCTURE PARAMETERS AND PHYSICAL CHARACTERISTICS OF HYDROGEN CONTAINING MATERIALS IN CONDENSE STATE TO DETECTION AND STUDY THEIR'S LIMIT STATES

A.I. KUSTOV
Voronezh State Pedagogical University, 86, Lenin Str., 394043, Voronezh, Russia
I.A. MIGEL
Voronezh Military Air Engineering Institute, 27 Starikh Bolshevikov Str., 394064, Voronezh, Russia
E-mail: akvor@yandex.ru

Abstract. The problem of reveals the limit state of solid materials including the hydrogenous ones has become urgent recently. The state of the material when it is close to the loss of stability is called limit. Different limit criteria are applied depending on the operation conditions (level of the temperature, stress, etc.). They are often different characteristics of physical – mechanical properties. Therefore, the state is limit on the reaching which the material inevitably changes its properties on engaging some structural parameter. It is necessary to reveal such states at the possibly initial stages. First the local structure changes inner stress, physical-mechanical parameters (e.g. the velocity of acoustic waves (AW), ratio $\Delta V / V\%$, modulus E and G, etc.) can be used as the limit criteria.

Keywords: method of visualization, acoustic microscopy, subsurface layers, nanopore density, elastic-mechanical steel parameters, nondestructive evaluation

1. Introduction

At present the problem of nondestructive express control of physic-mechanical properties of materials in condenses state is one of the most important scientific problems. Objects of investigations of the most interest were the different metal materials, used in processing of fuel elements too. That is, dependences of such properties from structural parameters of material play very important role.

The paper deals with the perspectives of the application of acoustic microscopy methods for studying the changes of physical-mechanical properties of materials in a condensed state under hydrogen influence. The basic principles of the methods as well as the results of the experiments of studying the structure of materials in a condensed state and its transformation upon changing the composition and types of thermomechanical treatment are given in the article. The high sensitivity to non-heterogeneity and defects upon acoustic visualization and in the regime of determining physic-mechanical properties are demonstrated. The paper deals with the study of process modifications of the elastic-mechanical characteristics of materials, the damage accumulation, the output of the limit state criteria for them.

In this paper the results of experimental work of investigation of physic-mechanical properties materials with the help of scanning acoustic microscope (SAM) [1, 2] are given. The essence of the subjected methods is first inlayer visualization of subsurface structures of the objects under investigation and second in the definition of velocity values of acoustic waves and calculate the elasticity constant of solid material. The nondestructive methods of investigation of structure and characteristics used have no limits by the nature of materials, including nanostructural materials. The state of the material when it is close to the loss of stability is called limit. Different limit criteria are applied depending on the operation conditions (force of the temperature, stress, etc.). They are often different characteristics of physical–mechanical properties. Therefore, the state is limit on the reaching which the material inevitably changes its properties on engaging some structural parameter. It is necessary to reveal such states at the possibly initial stages. First the local structure changes inner stress, physical-mechanical parameters (e.g. the velocity of acoustic waves (AW)) can be used as the limit criteria. The problem of reveals the limit state of solid materials including the hydrogenous ones has become urgent recently. The paper deals with the study of process modifications of the elastic-mechanical characteristics of materials, the output of the limit state criteria for them.

2. Results and discussion

Base methods of acoustomicroscopy defectoscopy are the visualization and V(Z)-curve methods. The image analyses assignment us possibility to calculate grain dimensions, to observe their transformation from time or with external influences. V(Z)-method allow to determine some elastic-mechanical parameters (υ_R, E, G, $\Delta V/V\%$) [3, 4].

Example of the obtained by SAM image of microstructure of BHC-2M steel is represented on Fig. 1.

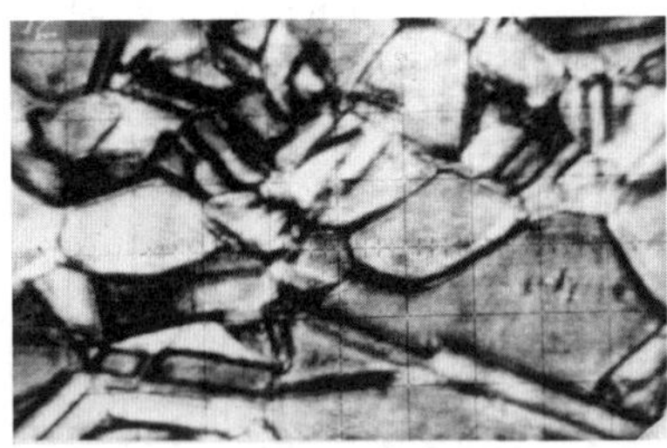

Figure 1. Acoustical image of subsurface layers *BHC-2M* steel (H_2O, f = 407 MHz, scale: 28 μm/div., Z = -12 μm)

The comparative optical photo with the same magnification gives the image of a polished surface without revealing structure elements. Structure transformation has been observed after deformational and thermal influences.

In Fig. 2 present V(Z)-curve for 60Г steel. In according to principles as account earlier [5, 6], obtained curve permit to determine the significance of velocities (υ_R) of surface acoustic waves (SAW). For this calculation we use characteristic distance (ΔZ_N) between maximums, situated on the right of the main one.

In order to calculate the value of the elastic modules in the local area of the studied material one should use the tabular values ρ_S of density and of Poisson coefficient (ν), or their quantity, determined by one of the certain standard methods. Then, with taking into account the definition υ_R in material:

$$E = \upsilon_R^2 \cdot \frac{2 \cdot \rho_s \cdot (1+\nu)^3}{(0,87 + 1,12 \cdot \nu)^2} \quad (1) \quad \text{and} \quad G = \upsilon_R^2 \cdot \rho_s \cdot \left(\frac{1+\nu}{0,87+1,12\nu}\right)^2 \quad (2)$$

V(Z)-curve method allow to distinguish quantity properties of steel with different extent of deformation (Fig. 3) or to conduct texture characterization [6, 7]. In Fig. 4 present the example of exposed the texture (anisotropy of elasticity modulus in 60Г steel).

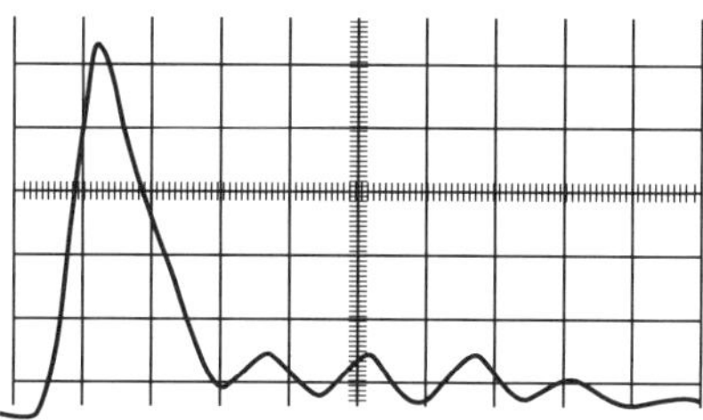

Figure 2. V(Z)-curve for investigation region in 60Г steel (H_2O, f = 407 MHz, scale: horizontal – 6.2 μm/div., vertical – 0.5 V/div., value of maximum – 6).

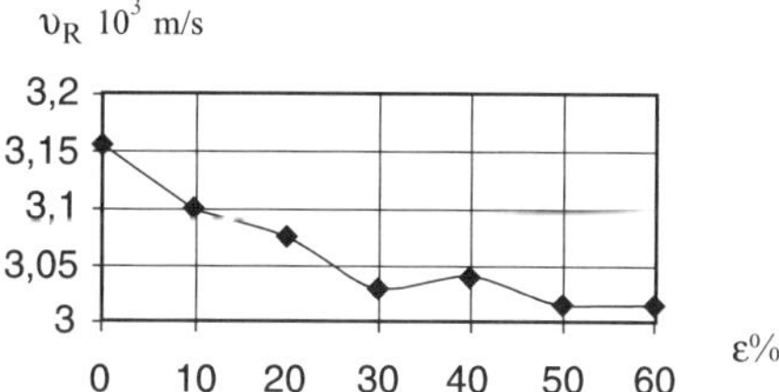

Figure 3. Dependence of SAW velocity from deformation degree in 18ХГТ steel

One of the important parameter of solid state sample is the grain dimension (d_3). The dependences of conventional fluid point ($\sigma_{0,2}$) from dimension d_3 in one of low carbon steel type are present in Fig. 5. The results were obtained with standard definition. As follow from that diagram, linear dependence $\sigma_{0,2}$ from grain dimension observes in 10–150 μm interval.

Received dependences $\sigma_{0,2}$ from d_3 are in good accordance with low Hall-Peach:

$$\sigma_{0,2} = \sigma_0 + k\, d_3^{-1/2}, \qquad (3)$$

where k and σ_0 – the constants for that material.

However, we due to turn the attention at the fact, that the experimental curve has break by d_3 = 40 μm (with linear correlation coefficient ~0.9). The nature of this break demand further investigation. Analogical dependence was obtained for 15Х2НМФА steel too (Fig. 6).

If we estimated grain dimension from acoustic image, then we calculate critical significance $\sigma_{0,2}$ for this material. That significance may be to determine with method V(Z)-curve. Since change of grain dimension transforms value of SAW velocity. Dependence υ_R from d_3 are demonstrated in Fig. 7.

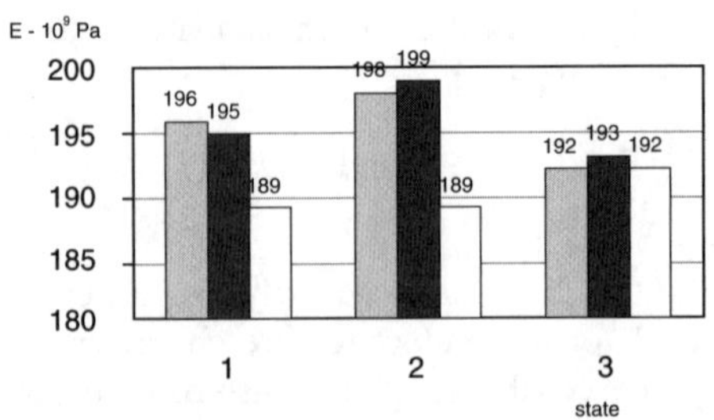

Figure 4. Anisotropy of elasticity modulus E_{eff} after different type TMO in specimens of 60Г steel: 1 – trail deformation (57%); 2 – patenting; 3 – nothing deformation (in left and middle columns measurements were carry out in direction, which is along to direction of rolling; in right columns – across to it)

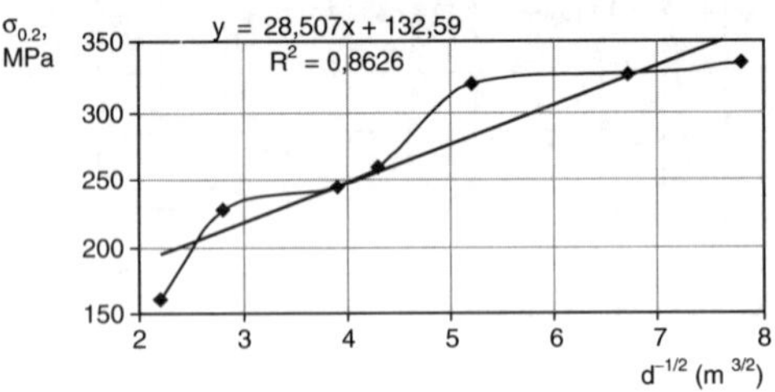

Figure 5. The strength properties dependence ($\sigma_{0,2}$) from grain dimension (d_3) by 14XГC steel

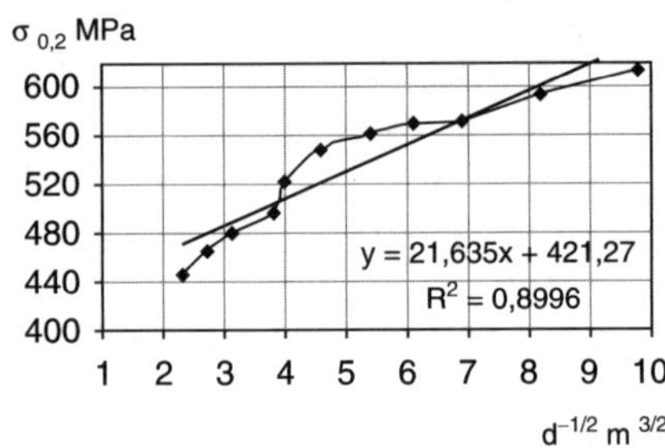

Figure 6. The conventional fluid point ($\sigma_{0,2}$) dependence from grain dimension (d_3) by 15X2HMФA steel

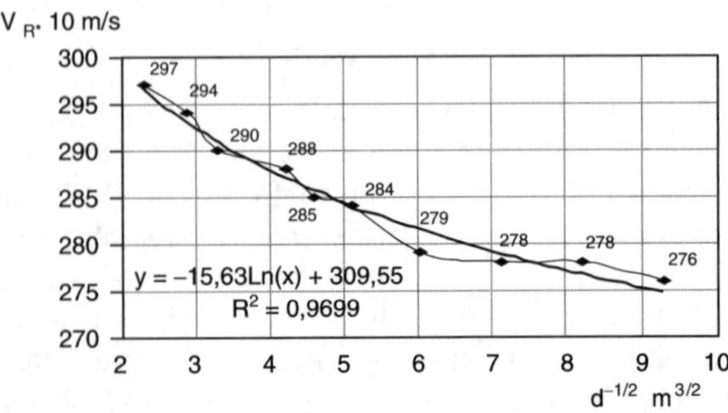

Figure 7. The velocity (υ_R) dependence from grain dimension (d_3) by 15X2HMФA steel

That allow define $\sigma_{0,2}$ and grain dimensions from υ_R measuring. Obtained dependence is similar to straight line, but the approximately coefficient for logarithmic curve is some above, then it (0.97).

The limit state reveal at the initial stages can be carried out by sensitive to micro- and nanoinhomogeneous methods. The limit criteria are possible to be evaluated according to the quantity, form, and distribution in volume of such inhomogeneities. Acoustic microscope defectoscopy methods actively developing in the in the latest 10–15 years can be referred to such methods [3, 7]. They allow both to get subsurface acoustic images of object microstructures and to define their physical–mechanical characteristics. Therefore, it was suggested to apply a scanning acoustic microscope (SAM) of the reflective type for the nondestructive control of the hydrogenous materials. In first stage of our work we studied model objects – steel samples with changed properties of subsurface layers. Alterations were realized with creation of pitting corrosion. If the mechanical stress were apposition to these samples, the critical stress which connected with form of cracks became lower (on 1.5–2.8 time). In Fig. 8 the acoustic image of subsurface layers of the steel sample (06Х14Н6МД2Т), subjected to corrosion is presented. The pitting with the special dimensions of 3–15 μm are visualized with high resolution. Besides, the depth of demonstration of the defects, their dimensions and quantity in the raster allow to judge by the limit state. It was not possible to reveal such inhomogeneities both at the optical images and with the help of some other scientific methods. The other example of the output of limit state criterion is connected with the appearance in hydrogenous metals of the cracks of flocken type.

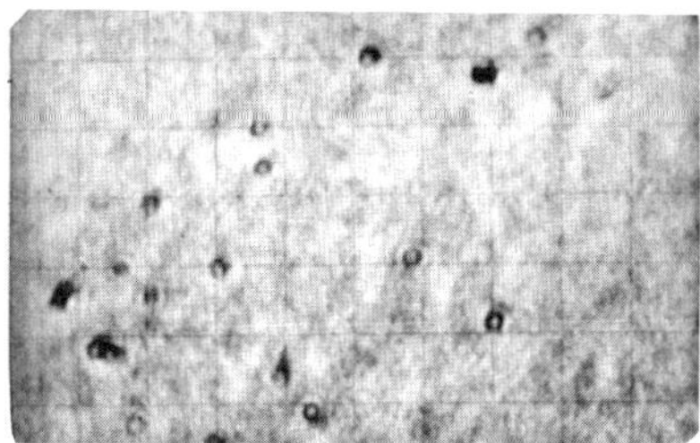

Figure 8. The pitting corrosion visualization by steel surface with acoustomicroscopy method (scale 24 μm/div., Z = -7 μm)

Visualization method allow to calculate the depth of penetration microdefects in specimen experimentally and carry out modeling of their form (Fig. 9). For example, form of pitting surface in 06Х14Н6МД2Т steel is defined by logarithmic curve (with correlation coefficient 0.99). If approximate was conducted for the obtained series of curves, we may to estimate penetration depth of microdefects (thickness of disturb layer). This regime have allowed to measure micro-defects density in raster. That is characteristic dependence from thickness of taking down layer (Fig. 10).

It is also possible to apply acoustic microscopy methods based on the use of V(Z) – curves for the study of the limit state of the materials in a solid state. It makes possible to calculate the values of velocities SAW and to define the quantity of the coefficient changes of AW fading in the material at the relative modification ΔV/V% of height of the main maximum of the V(Z)–curves.

The control technology of limit state with appliance of V(Z)–curves consists of the analysis of their form transformation. The values of velocities of surface acoustic waves (SAW) and the change of the level of their fading were calculated due to the special dependencies. The limit state of material conclusion was made according to the dimension of local fluctuations of physical–mechanical parameters.

In Fig. 11 the results of definition of penetration depth are presented. This meanings connect with velocity of SAW. If pitting are in our objects, that conduct to change of υ_R (~7–8%).

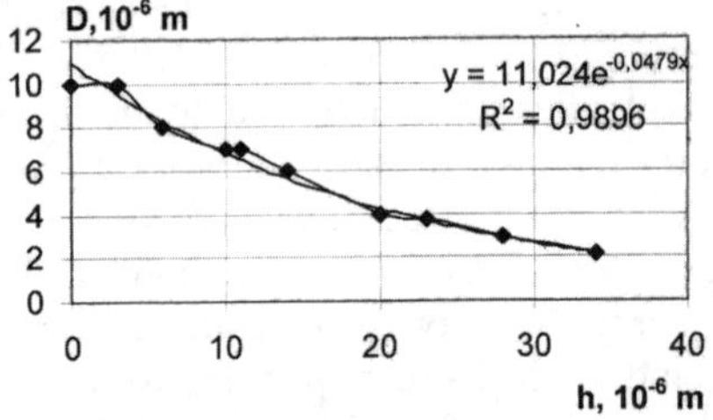

Figure 9. The change of pitting diameter dimension in steel depend from thickness of taking down layer

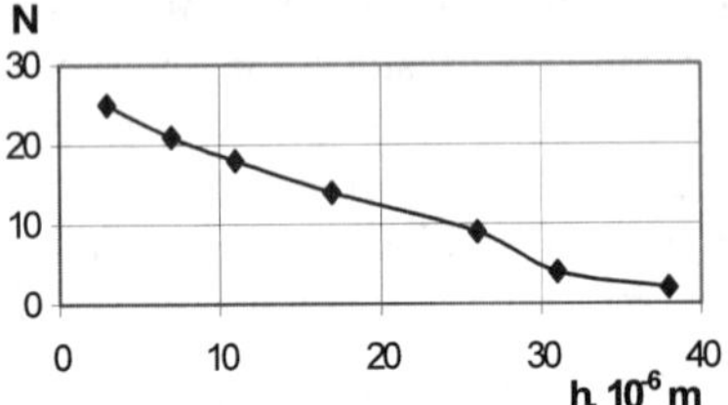

Figure 10. The dependence of quantity N pittings from visualization depth (acoustical method)

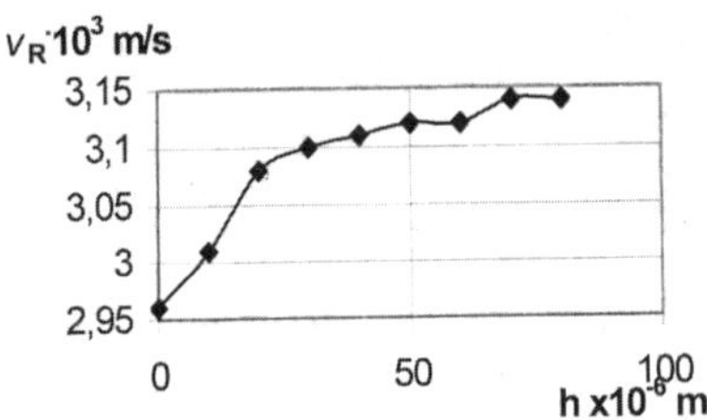

Figure 11. Dependence of SAW velocity in steel with pittings from depth penetration (acoustical method)

The results of the experiments illustrating the possibility to define the thickness of the modified layer caused by hydrogen diffusion in the metal materials according to the value of velocity SAW or the ratio ΔV/V% were presented before. The difference between the initial and modified material according to these characteristics can reach 5–20%. The results of modeling experimental for nitrogen steel are presented in Fig. 12 (steel 08X18H10T).

Obtained experimental curve υ_R are approximated by parabolic dependence with coefficient 0.986. We can to ensure definition depth of transformed layer, if shell analyses that dependence.

It is possible to estimate the value and some other sample characteristics, for example the hydrogen concentration, according to the changes of the velocity SAW and the relative height ΔV/V% of the maximum V(Z)–curve.

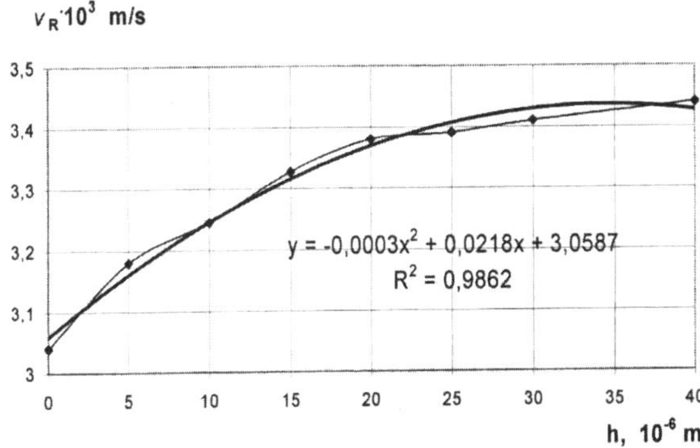

Figure 12. The dependence of SAW velocity (υ_R) in steel (08X18H10T) on the depth penetration

The dependencies both between the hydrogen concentration and the number of the microcracks emerged, and the values of velocity SAW on depth of the hydrogen penetration into the sample, period of diffusion, etc. are received at that. The limit hydrogen concentration in this material can be evaluated concerning the value of velocity SAW from the received dependence for steel. In Fig. 13 presented curve of υ_R – dependence from time exposure of steel (40XH) specimens in hydrogen. The optimum time dimensions are in interval ~55–67 h.

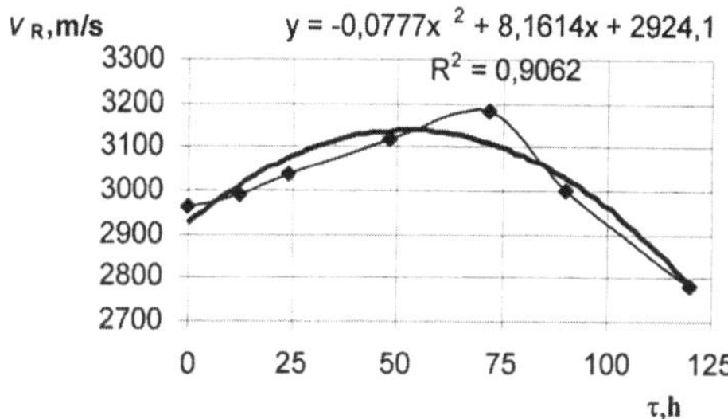

Figure 13. Dependence of SAW velocity from time exposure (τ) of steel specimens in hydrogen

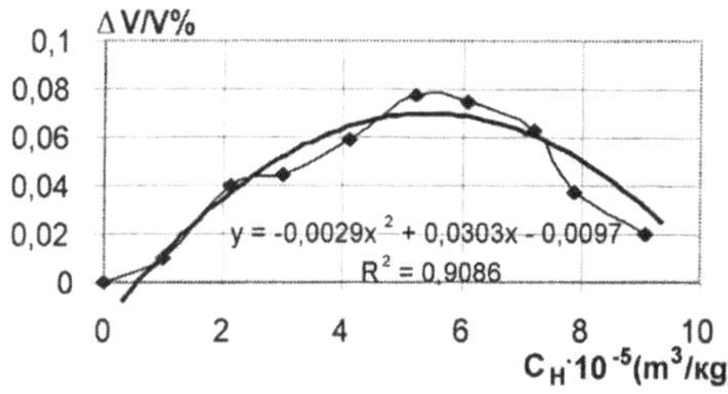

Figure 14. The dependence of relative height ($\Delta V/V\%$) on steel (55XH) from content of hydrogen

As examples for a number of steels it is possible to consider dependencies of the number of originating defects of the flocken type [7], change of the velocity values of surface acoustic waves (SAW), characteristics of the absorption process ($\Delta V/V$), sizes of appearing heterogeneities upon concentration of the diffusion hydrogen. In Fig. 14 the dependence of relative height (ΔV/V%) V(Z)-curves from hydrogen concentration in steel (55XH) are presented. This experimental curve was approximated by trend of polynomial of 2 degrees. The trend permit to estimate a content of hydrogen at $(5.2–5.8)\cdot10^{-5}$ m^3/kg meaning as dangerous.

3. Conclusions

Acoustic microscopy methods have prospects for studying the materials on condensed state, their properties and dependences from structural parameters. The results obtained have proved the possibility of output of the limit state criteria of the hydrogenous materials by applying acoustic microscope defectoscopy methods.

The main criteria of the limit is the size of the grains of the structure, their orientation, the presence of the texture. The methods AMD used in this work give the opportunity to determine the solidity of steel ($\sigma_{0,2}$) safely and evaluate their changes because of the effects of deformation or concentration.

The essential advantage in AM experiments is the simultaneous use of different work time table SAM. It allows increasing the reliance of the results to a high degree.

The methods of revealing the defects with the help of SAM systems have a great independent impotence.

Their efficacy is confirmed with the results of the pitting corrosion in steel.

The particular attention should be paid to further working out devoted to studying diffusive layers in steel including ones connected with hydrogen as well.

Acknowledgements

This work has been supported by Russian Academy of Sciences and administration of VSPU.

References

1. Wilson R.G., Weglein R.D. (1994) Acoustic microscopy of materials and surface layers, *Appl. Phys.*, 1994, **55**(9): 3261–3275.
2. Kustov A.I. (1994) Study of structure and physical-mechanical properties of solid materials by the acoustomicroscope methods, *Materials of VII Rus. conf. "Damped materials"*, pp. 89–97.
3. Budanov A.V., Kustov A.I., Migel I.A. The study of changes of physical-mechanical properties of materials and condensed state under hydrogen influence using fault detection acoustic microscopy methods. In: Hydrogen Materials Science and Chemistry of Metal Hydrides // Proc. of NATO ARW. Netherlands: Kluwer, 2004, **172**: 203–215.
4. Kustov A.I., Migel I.A., Sukhodolov B.G. Study of effect of different types of heat mechanical treatment on structure and properties of steels and alloys, *Metallovedenie i termoobrabotka metallov*, 1998, **4**: 128–137 (in Russian).
5. Budanov A.V., Kustov A.I., Migel I.A. The study of the effect of hydrogen on physical-mechanical properties of steel by acoustic microscopy methods. In: Hydrogen Materials Science and Chemistry of Metal Hydrides // Proc. of NATO ARW. Netherlands: Springer, 2002, **71**: 131–140.
6. Kustov A.I. (1998) Detection of inhomogenetics by acoustomicroscope methods, *Fizika. i khimiya stekla*, **24**(6): 809–816. (in Russian).
7. Kustov A.I., Migel I.A. Investigation of the physical properties of materials for fuel elements and work up of limit state criteria for hydrogen containing solid materials with acoustic microscope defectoscopy methods. In: Hydrogen Materials Science and Chemistry of Metal Hydrides // Proc. of NATO ARW. Netherlands: Springer, 2007: 451–458.

INVESTIGATION OF THE GRANULOMETRIC COMPOSITION AND STRUCTURE OF AlN AND Si_3N_4 POWDERS AFTER HYDROGEN–THERMAL TREATMENT

V.V. SKOROKHOD, I.A. MOROZOV, R.A. MOROZOVA, A.I. ITSENKO, G.S. OLEYNIK, T.V. DUBOVIK, A.A. ROGOZINSKIJ, A.A. ROGOZINSKAYA, V.V. LYCHKO
I. Frantsevich Institute for Problems Materials Science of Ukrainian National Academy of Sciences, 3 Krzhyzhanovsky Str., .03142, Ukraine
E-mail: imorozov@i.ua

Abstract. A principle new approach to the non-metal nitride powders activation is proposed in this paper. This approach consists in hydrogen-thermal treatment (HTT) of precursor non-metal nitrides powders at 500–1,150°C in presence of hydrogen and catalysts. The goal of hydrogen-thermal treatment of powders is to extract the irrelevant impurities, to grind the powder particles and to improve the powder structure. Evolution of AlN and Si_3N_4 powders structure in HTT at different temperatures (500–1,150°C) has been studied. HTT of non-metal nitrides powders was shown to favour the removal of sorbed impurities from the particle surface and to decrease in starting temperature of powder sintering.

Keywords: non-metal nitride, hydrogen, thermal treatment, sorbed impurity, sintering

1. Introduction

Non-metallic nitrides are characterized by high mechanical strength, thermal resistance, chemical stability as well as by specific dielectric properties and high heat conductivity. This makes the base for using them in developing composite materials of wide purposes [1].

Low concentration and low mobility of defects as well as low rate of diffusion processes are peculiar for crystal lattice of non-metal nitrides. Moreover it should be kept in mind that in sintering not all powder surface takes part. The considerable powder surface is passivated with atoms of chemisorbed impurities which deteriorate the sintering. Therefore to obtain porousless materials the sintering processes must be activated. Both physical processes (which are directed to forming different defects and to their relaxation during sintering), and chemical reactions (related to the new phases formation during sintering) create the foundation for the activation methods [2].

2. Experimental details

A principle new approach to the sintering activation is proposed of the authors. This approach consists in hydrogen-thermal treatment (HTT) of precursor non-metal nitrides powders at 500–1,150°C in presence of hydrogen and catalysts. The goal of hydrogen-thermal treatment of powders is to extract the irrelevant impurities, to grind the powder particles and to improve the powder structure. The use of such powders will allow to improve the technology of materials production and to develop new materials with assigned structure and necessary service characteristics.

In connection with this it is interesting to study the granulumetric composition and structure of nitride powders of AlN and Si_3N_4 after Hydrogen–Thermal Treatment. HTT of powders was carried out in the unit of Siverts type [3]. The time intervals for HTT at temperature 500–1,150°C (with step 50°) are 30, 60 and 90 min at every temperature point. Powders of non-metallic nitrides were studied by electron microscopy and selected area diffraction as well as X-ray analysis.

3. The results and discussion

The results obtained revealed the AlN, both initial and after HTT, to be only as wurtzite modification. In monocrystalline particles the dislocation density does not exceed 10^6 cm^2. Electron-probe microanalysis showed that both kinds of AlN powder contain only oxygen impurity. It should be noted that the oxygen content in AlN after HTT considerably less (about two times) as compared with initial state.

Carrying out investigation revealed the granulometric compositions of AlN powder in initial state and after HTT to be considerably different. Initial powder contains only 7% (by volume) of AlN as monocrystalline particles and their size range is 0.5–2.0 μ. Particles have polyhedral shape. Another AlN particles are as aggregates of two types. The first are the loose flat aggregates of 5.0–7.0 μ size. The particles are linked by film bridges. The percentage of such aggregates is about 20% (by volume). The other type are the porousless bulky aggregates of 10.0–20.0 μ (and more) size. In these aggregates the particles are linked by complete film connections.

The HTT radically changed the granulometric composition of AlN powder. The main powder component after HTT at 1,150°C is monocrystalline AlN particles (about 90%). As in initial powder these have polyhedral shape. Their size range is 0.5–2.0 μ and most of them are of 0.5–1.0 size. Porousless bulky aggregates of particles in treated powder are completely absent and only about 10% of aggregates of 3.0–5.0 μ (and less) size look as loose ones.

Thus the results of study show that HTT of AlN powder favour to:

- Purification from oxygen
- Change in granulometric composition due to breakdown of film connections between particles in aggregates
- Formation of well-regulated structure

Typical structure components of initial AlN powder and of that after HTT at 1,150°C are shown in Figs. 1 and 2 [4].

The electron microscopy and selected area diffraction study showed that silicon nitride both in initial powder and after HTT is a mix of α- and β-modifications. The powder of initial Si_3N_4 consists mostly of flat aggregates of particles. Aggregate size range is 5.0–30.0 μ. Aggregates look as chains of particles linked each other with bridge connections. Among such chains are porousless bulky aggregates of 1.0–3.0 μ size. Particles have rounded shape, sometimes with elements of faceting. Their size is 0.1–0.3 μ. Among such particles are anisometric crystals of needle-like shape. Their form-factor is of 3–10 and their volume part in powder does not exceed 5–7%.

After HTT of Si_3N_4 powder at 800°C the size range of aggregates is the same as in initial powder. However HTT results in some loosening of aggregates and fragments with more dense packing of particles are formed. These fragments are separated each

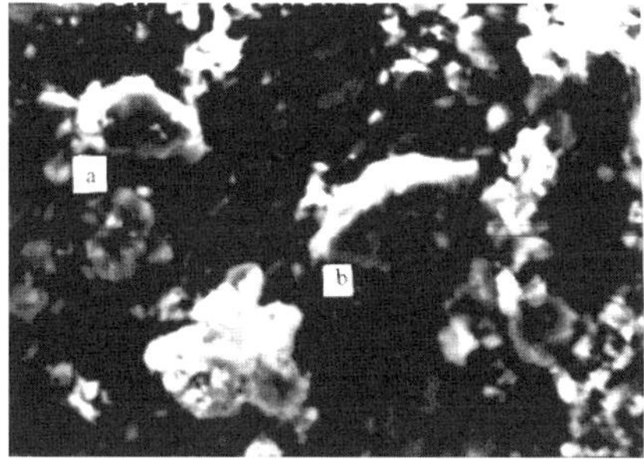

Figure 1. Typical structure components of precursor AlN powder: (a) monocrystalline particles; (b) dense aggregates

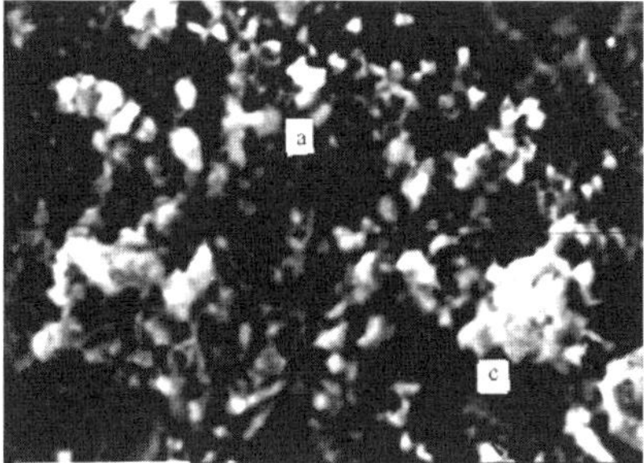

Figure 2. Typical structure components of AlN powder after HTT at 1,150°C: (a) monocrystalline particles; (c) loose aggregates

other by loose layers formed by particles with point contacts between them. Size of fragments is 1.0–5.0 μ. After HTT of powders at 900°C further decay of aggregates into smaller high-dense fragments with loose interlayers takes place. There are bridges between particles in the fragments. Silicon nitride powder after HTT at 1,150°C for 60 min consists of aggregates of 3.0–10.0 μ size particles. The aggregates are formed by bulky porousless granules of 0.5–1.0 μ size. Typical are the particles located in surface layer of granules to be regularly faceted. There are physical contacts between particles in granules. Similar contacts are between anisometric grains and small aggregates of particles.

The study showed that in HTT the initial stage heating of Si_3N_4 powder sintering proceeds, i.e. the sintering in the state of free poured powder takes place. With this the Si_3N_4 powder grains three-level microstructure: aggregate (I) – granule (II) – monocrystalline particle (III) (Fig. 3). Structure transformations proceed mostly by sintering of particles in granules. The particles sintering in HTT proceeds through mass transfer in gas phase. To this indicates the regular faceting of particles located at the surface of aggregates.

The results obtained allow to conclude the structure-morphological changings in Si_3N_4 powder with increase in temperature of HTT to proceed in the following sequence:

- Loosening of particles aggregates having mechanical contacts and formation of fragments of more packed particles (800°C)
- Further fragmentation of aggregates and initial stage of bridge connection formation between particles in fragments (900°C)
- Full breakdown of initial aggregates and formation the monolithic granules with bridge (physical) connections between granules (1,150°C)

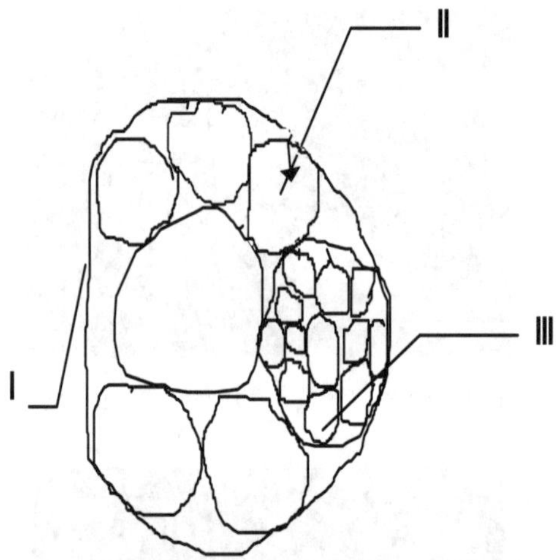

Figure 3. Schematic view of structure components of Si_3N_4 powder after HTT at 1,150°C

It should be pointed out that optimal time interval of HTT for all powders examined is 60 min. This conclusion results from that the increase in time interval does not change markedly the powder structure. The effect of HTT on the process of powder loosening and on the sintering temperature depends in general on the following factors: (a) dispersity and phase composition of powder; (b) type of chemical binding in powder. The influence of the first factor is displayed in the case of Si_3N_4. This powder contains the metastable phase (α-modification). The sintering of such powder already at 900°C is characterized both by high powder dispersity and by developing the initial stage of phase transformation $\alpha \rightarrow \beta$. Phase transformation always causes the activation of mass transfer processes. The effect of the second factor is displayed when comparing the features of behavior of nitrides under study by the following. The sintering of Si_3N_4 begins at 900°C and for AlN the sintering does not proceed even at 1,150°C. This can be explained first of all by type of chemical binding: Si_3N_4 has only covalent bond type and AlN has ion-covalent bond type.

4. Conclusions

Evolution of AlN and Si_3N_4 powders structure in HTT at different temperatures (500–1,150°C) has been studied. HTT of non-metal nitrides powders was shown to favour the removal of sorbed impurities from the particle surface. This results in loosening of powder particles aggregates containing sorbed impurities. HTT of non-metal nitrides powders was shown to favour the removal of sorbed impurities from the particle surface and to decrease in starting temperature of powder sintering.

References

1. Samsonov G.V. Non-metal nitrides. Moscow: Metallurgy. 1969. 265 p. (in Russian)
2. Uscocovitch D.P., Samsonov G.V., Ristitch M.M. Activated sintering. Beograd: International Institute for science of sintering. 1974. 347 p. (in Russian).
3. Smith D.P. Hydrogen in metals. Chicago, IL: University of Chicago Press. 1948. 346 p.
4. Patent 54168A Ukraine, MCI C 04 B 35/38. Method of treatment of AlN powders. Morozov I.A., Morozova R.O., Dubovik T.V. and other (Ukraine); Frantsevich Institute of Problem Materials Science NAS Ukraine. No. 2002054377; Declar. 28.05.2002; Publ. 17.02.2003. Bull. No. 2 (in Russian).

THERMAL CONDUCTIVITY OF HYDRIDE POWDERS LAYERS INTERACTED WITH HYDROGEN

A.I. SOLOVEY, YU.I. SHANIN
FSUE Scientific Research Institute of Science and Industrial Association "Luch", Podolsk, Moscow area, Russia, 142100
Tel./Fax: 7+10+(4967)634582; E-mail: syi@luch.podolsk.ru

Abstract. The results of experiments on measurements of reacting with hydrogen hydride beds thermal conductivity coefficient are presented in the article. Installation, designs of measuring cell, and experimental technique for measurements of thermal conductivity of various hydride layers (including various metal fillers as well) are briefly described. Experiments were carried out with use of different hydrides, including $LaNi_5$. Powders with particle size 50 ÷ 200 μm had effective thermal conductivity (1.25 ± 0.05) W/(m·K), values which are consistent with data of other authors. Increase of powder thermal conductivity by introduction of aluminum corrugation can improve a layer effective thermal conductivity by a factor of 3 ÷ 5.

Keywords: hydride, hydrogen, hydride layer, coefficient of thermal conductivity, experiment

1. Introduction

Powder materials thermal conductivity is well studied [1]. The coefficient of thermal conductivity of a "powder-gas" system is determined by porosity, particles size, pressure and gas composition, and decreases with powder bed porosity increase, decrease of powder particles size and reduction of gas pressure.

The thermal conductivity of $LaNi_5$ 100 μm powder with porosity $\varepsilon = 0.5 \div 0.57$ changes from 0.9 W/(m·K) at hydrogen pressure 0.1 MPa to 1.5 W/(m·K) at hydrogen pressure 1 MPa [2].

There are some features in methods of investigation of powders reacting with gas thermal conductivity [2].

To increase hydride beds thermal conductivity, various structured fillers (disks, finned high thermal conductivity materials, foamed metals, metal powders sintered as composites etc.) are used. By changing filler fraction, it is possible to adjust smoothly the resulted (effective) thermal conductivity of a hydride bed [3, 4].

2. Installation, experimental technique

2.1. INSTALLATION

Installation for thermal conductivity measurement of intermetallic compound (IMC) powders at various hydrogen pressure was designed and made (its scheme is illustrated in Fig. 1). The installation includes a unit for preparation of pure hydrogen (14) from $LaNi_5$ powder, system of calibrated volumes (9, 12) with a measuring cell (16) and

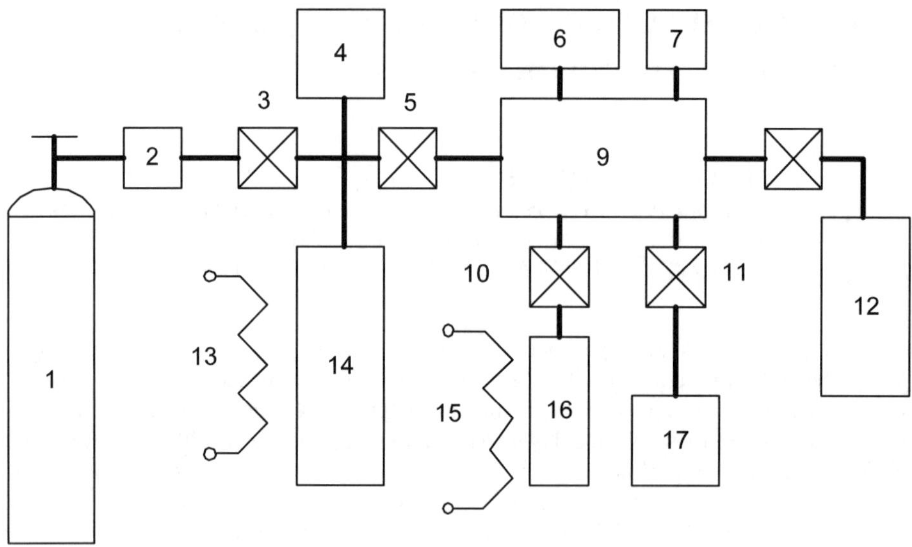

Figure 1. The diagram of installation for measurement of IMC powders thermal conductivity during reaction with hydrogen: 1 – bottle with hydrogen; 2 – gas pressure regulator; 3, 5, 8, 10, 11 – gas valves; 4, 6 – standard pressure gages; 7 – pressure gauge; 9, 12 – calibrated volumes; 13, 15 – heaters; 14 – source of pure hydrogen; 16 – measuring cell; 17 – roughing-down pump

system of evacuation by roughing-down pump (17). Hydrogen pressure in the system is controlled by standard pressure gages (4, 6) and rheostat pressure gauge (7). Furnaces (13, 15) are used to fill with hydrogen of calibrated volumes and for preparation of powders for experiments during evacuation by a roughing-down pump.

The technique of investigation is similar to presented in [2], but was simplified. Hydrogen volume was dosed from measuring capacity; absorption of hydrogen was calculated by change of pressure in it. When the technique was developed, well studied $LaNi_5$ powder was checked as a reference material.

2.2. DESIGNS OF MEASURING AND MODELLING CELLS

Investigation of thermal conductivity of IMC powders was carried out in a cylindrical measuring cells 8 mm ID. The measuring cell design is presented in Fig. 2. The cylindrical cell is a copper tube 16 mm OD. The internal volume of a cell 8 mm in diameter (4) is filled with IMC powder. The chromel-copel thermocouple (3) is placed in a cell the center in a stainless steel cover 1.5 mm in diameter. The second thermocouple (2) takes temperature inside a cell case (1). The furnace (5) covered by layer of thermal insulation (6) heats a cell.

Effective thermal conductivity of a hydride bed was investigated with use of modelling cell from a corrosion-proof pipe 22 mm OD and 20 mm ID, and aluminum corrugation inserted into it. Two chromel-copel thermocouples were placed on a cell axis and on an external pipe surface. The height of measuring and modelling cells was 100 mm.

Potentiometer KSP-4 registered hydrogen pressure from rheostat pressure gauge EDD-10K and temperature from thermocouples.

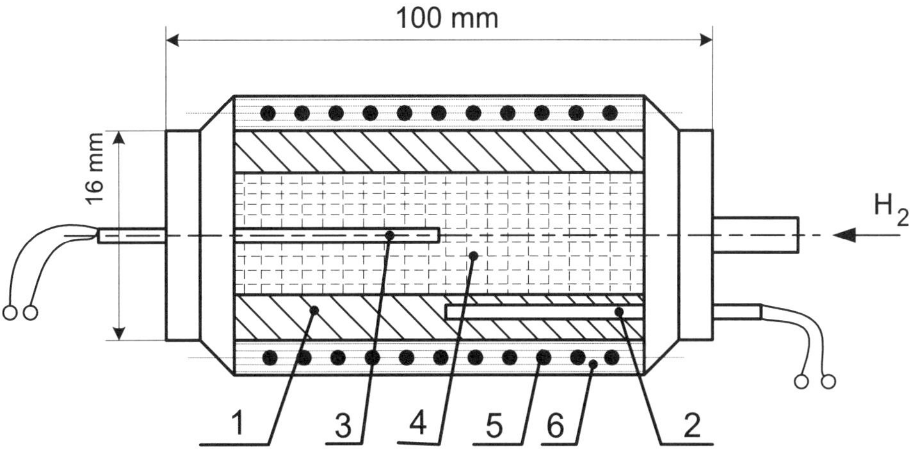

Figure 2. Measuring cell design: 1 – cell case; 2 – thermocouple inside a cell case; 3 – thermocouple taking powder temperature; 4 – IMC powder; 5- heater; 6 – thermal insulation

Experimental technique. Methods of regular heating rate were used in experiments. Measurements were realized in two versions. In the first version, the cell was evacuated from the beginning, then filled by calibrated volume of hydrogen to the specified pressure, and calibrated volume of hydrogen was absorbed in a cell. In the second version, a cell with hydride powder was heated for a certain time, then heating was disconnected. At that time process of some of hydrogen sorption and desorption takes place sequentially in a powder.

During hydrogen sorption non-stationary change of temperatures was measured in two points through a hydride layer thickness. Only regular part of non-stationary process of temperature change was used for data processing.

Thermal conductivity coefficients were calculated from the equation of thermal balance for a cell with maximum random error ±20%.

3. Results and discussion

3.1. THERMAL CONDUCTIVITY OF $LANI_5$ AND $ZRCRFE_{1.2}$ POWDERS

$LaNi_5$ powder with initial particle size 60 ÷ 200 μm and bulk density of 4,300 kg/sm^3 and $ZrCrFe_{1.2}$ powder with particles size less than 200 μm were used in experiments. The level of initial pressure during hydrogen sorption in cells changed in a range of 0.4 ÷ 1.1 MPa for $LaNi_5$ and 0.2 ÷ 0.6 MPa for $ZrCrFe_{1.2}$. Maximum temperature of a cell was reached for 15 ÷ 30 s. Effective coefficient of thermal conductivity was calculated from the empirical formula:

$$\lambda_{eff} = 1.72(k\overline{P} + 5\overline{t}) / \Delta t\,, \tag{1}$$

where $\overline{P}, \overline{t}$ – rates of pressure and temperatures change; Δt – temperatures difference between the cell center and a wall, k – empirical factor describing features of a cell (it was determined by calibration).

λ_{eff} of $LaNi_5$ and $ZrCrFe_{1.2}$ were calculated from the equation (1) as λ_{eff} = 1.25 ± 0.05 W/(m·K) and λ_{eff} = 0.5 ± 0.05 W/(m·K), respectively, taking into account linear relationship between temperature and pressure change rates. Obtained data are in good agreement with data of other authors [2].

3.2. THERMAL CONDUCTIVITY OF $LANI_5$ POWDERS WITH ALUMINUM FOIL FILLER

Comparative experiments on a modelling cell with and without aluminum corrugation were carried out on the same $LaNi_5$ powder with particle size less than 200 μm and bulk density 3,400 kg/cm^3. Calculation of coefficient of thermal conductivity was carried out only during hydrogen desorption when cell was heated.

Effective coefficient of thermal conductivity was determined from the empirical formula:

$$\lambda_{eff} = 1.2(1.7 \cdot 10^{-3}\overline{P} + 35\bar{t}) / \Delta t\,, \qquad (2)$$

that gives λ_{eff} = 5.0 ± 0.5 W/(m·K). The effective coefficient of powder thermal conductivity without corrugation is 1 ÷ 1.5 W/(m·K), i.e. aluminum corrugation increases λ_{eff} by a factor of 3 ÷ 5.

3.3. APPLICATION OF MICROENCAPSULATING AND FOAMED METAL

Powder of $MmNi_{4.15}Fe_{0.85}$ alloy with particle size less than 100 μm was encapsulated by chemical deposition of copper coating with thickness less than 5 μm. High porosity permeable cellular copper with open porosity ~95% was used as foamed metal.

Efficiency of powders microencapsulating and foam metal was checked experimentally on special installation. Installation was made of two reactors – high-temperature and low-temperature connected by the pipeline through a valve. The high-temperature reactor was filled by hydride of $LaNi_{4.6}Al_{0.4}$, and investigated materials was loaded into a low-temperature reactor. The low-temperature reactor alloy was charged by hydrogen during heating of high-temperature reactor up to 200°C for 10 min, and a low-temperature reactor was cooled by water. After charging, the valve between reactors was closed, and a high-temperature reactor was cooled up to a room temperature. In experiment the valve between reactors was opened, the high-temperature reactor was cooled by flowing water, and the temperature from thermocouple in a low-temperature reactor was registered by KSP-4.

Indirect measurements of thermal conductivity on modelling installation have been carried out by this technique under identical conditions, the results were following:

- 1st experiment (powder $MmNi_{4.15}Fe_{0.85}$ with particle size 100 μm, an aluminum corrugation in a reactor) – cooling of a reactor up to -16°C for 4 min, average power of cooling of 30 W
- 2nd experiment (microencapsulated $MmNi_{4.15}Fe_{0.85}$ powder) – cooling of a reactor up to -12°C for 4 min, average power of cooling of 22 W
- 3rd experiment (filling of foam metal in a reactor by $MmNi_{4.15}Fe_{0.85}$ powder on vibration table) – cooling of a reactor up to -25°C for 3 min., average power of cooling of 40 W

The quantity of metal hydride in all experiments was the same – 110 g.

3.4. COMPACTS FROM IMC POWDER

Thermal conductivity measurements were carried out in hydride layers by a stationary method at various hydrogen pressures. Cylindrical specimens 13.5 mm in diameter and 6 ÷ 9 mm in height were prepared by IMC powders compaction at specific pressure 4.6 t/cm^2. $LaNi_{4.6}Al_{0.4}$ and $MmNi_{4.15}Fe_{0.85}$ powders of two fractions were used: 70 and 70 ÷ 200 µm. 20 vol% of copper fibers 2 ÷ 4 µm in length and 50 µm in diameter were added to increase their thermal conductivity. To exclude heat effects of hydrogen interaction with specimens during experiments in hydrogen environment, measurements were made after not less than 3 h had elapsed after hydrogen input.

Results of thermal conductivity measurements depending on hydrogen pressure are shown in Tables 1, 2.

TABLE 1. Thermal conductivity of compacted powder $LaNi_{4.6}Al_{0.4}$, W/(m·K)

Powder particle size	Hydrogen pressure (MPa)		
	0	0.11	0.4
70 µm	3.6	9.6	18.7
70 ÷ 200 µm	4.4	17.6	18.7
200 µm + 20% Cu	8.4	14.9	17.7

TABLE 2. Thermal conductivity of compacted powder $MmNi_{4.15}Fe_{0.85}$, W/(m·K)

Powder particle size	Hydrogen pressure (MPa)	
	0	0.11
70 µm	2.3	9.0
70 ÷ 200 µm	1.9	11.8
200 µm + 20% Cu	4.7	9.1

The analysis of thermal conductivity dependence on initial powder particle size showed increase of thermal conductivity with particle size increase. This fact can be explained by increase of contact resistance during compaction of more fine powders. Insignificant deviations from such dependence concern to compacts made of oxidized powders.

4. Conclusions

Installation for measurement of IMC powders thermal conductivity during sorption and desorption of hydrogen was designed and made. The measured values of $LaNi_5$ powder effective thermal conductivity, 1.25 ± 0.05 W/(m·K), is in a good agreement with published data.

On a modelling cell of modular sorber 20 mm in diameter it was shown that addition in $LaNi_5$ powder of aluminum corrugation with high thermal conductivity increased effective thermal conductivity by a factor of 3 ÷ 5.

The possibility to improve essentially hydride beds thermal conductivity by addition of foamed metal, by microencapsulating and by compaction was demonstrated.

References

1. Dulnev G.N., Zarichnyak Yu.P. Thermal conductivity of mixtures and composite materials. Leningrad: Energy. 1974. 264 p. (in Russian).
2. Goodell P.D. Thermal conductivity of hydriding alloy powders and comparisons of reactor systems, *J. Less-Common Met.*, 1980. **74**: 175–184.
3. Suda S., Komazaki Y., Kobajashi N. Effective thermal conductivity of metal hydride beds, *J. Less-Common Met.*, 1983. **89**: 317–324.
4. Shanin Yu.I. Improvement of Hydride Heat Devices Overall Performance. In: Hydrogen Materials Science and Chemistry of Carbon Nanomaterials // Proc. of NATO ARW, Netherlands: Springer, 2007, P. 383–397.

ANALYSIS OF RESTRICTIONS PREVENTING IMPLEMENTATION OF HYDRIDE HEAT PUMPS

YU.I. SHANIN, O.I. SHANIN
FSUE Scientific Research Institute of Science and Industrial Association "Luch", Podolsk, Moscow area, Russia, 142100
Tel./Fax: 7+10+(4967)634582; E-mail: syi@luch.podolsk.ru

Abstract. In article the analysis of the restrictions preventing effective operation the HHP is lead. For perfection the HHP it is necessary to improve properties of hydrides (especially capacity on hydrogen), processes of heat and mass transfer in hydride layers and the devices operating in consisting of the HHP. The various constructive measures directed on improvement of operation and competitiveness the HHP are considered.

Keywords: hydride, hydrogen, hydride bed, hydride sorber, hydride heat pump

1. Introduction

Now, despite of intensive searches hydric materials, metal hydrides are the most comprehensible, as from the point of view of capacity, and technological commercial production. Use of hydrides in heat devices and machines has a number of features and restrictions. Hydride heat pumps (HHP) allow to receive heat and cold in the temperature conditions close to ambient temperatures and to use heat waste products and low-potential sources of heat. The HHP will consist of the generators-sorbers representing grouped hydride layers. Advantage the HHP is an opportunity of management processes in hydride layers using of heat streams.

In article the various restrictions caused by specificity of hydrides are considered at their use in HHP. From all circle of known hydride storage devices of hydrogen we shall consider only what on temperature of formation are close to an ambient temperature, and equilibrium pressure do not exceed 5 MPa. The intermetallides used in hydride technical devices are considered only and is narrower in the HHP.

2. Results and discussion

2.1. SCIENCE OF MATERIALS RESTRICTIONS

Use of alloys-storage of hydrogen is based on their reversible interaction with hydrogen. These reactions are chemical reactions with all imperfections inherent in them. Chemical reaction is characterized by cleanliness of initial components (and also a degree of influence of the impurity which are being them), speed of reactions of interaction, completeness of reaction and heat of formation. Interaction of metal with hydrogen includes some stages [1], beginning from a mass transfer of hydrogen in a gas phase and finishing diffusion of atoms of hydrogen in volume of metal and transition of a solid solution in hydride. A limiting stage of process of hydrogenation is

B. Baranowski et al. (eds.), *Carbon Nanomaterials in Clean Energy Hydrogen Systems.*

diffusion of hydrogen in metal, influence of diffusion amplifies with temperature pulldown.

The basic requirement to alloys forming hydride at their use as actuating fluids in the HHP – convertibility of reaction with hydrogen. Reaction proceeds with allocation of heat at formation of hydride and absorption of heat at a dissociation of hydrogen. It is possible to direct process in any direction and to operate its speed change of temperature and pressure. The amount of hydrogen absorbed by an alloy is the major (but not unique) the characteristic of its technical application. Requirements to alloys-storage of hydrogen are formulated in numerous articles and generalized in the handbook [1].

Limiting on hydrogen capacity of an opportunity of formation of convertible hydrides do not exceed opportunities of hydride generative metals and their binary compounds [2]. The elements-catalysts added to hydride generative elements, only soften conditions of formation of hydrides. By search of new more technological materials it is necessary to head for the most absorbing hydrogen a material which with the help of alloying additives is resulted in corresponding conditions of its technical application.

For the HHP the basic question determining productivity and, finally, competitive ability is the question on amount of useful hydrogen transferable between reactors the HHP. At a modern level for substantial improvement of characteristics the HHP it is necessary capacity of hydride on hydrogen and the amount useful transferable hydrogen to increase a minimum in two times (with 1.5 up to 3.03.5 wt% and with 0.8 ÷ 1 up to 1.6 ÷ 2.0% accordingly for a range of operating temperatures close to room temperature).

The curve behavior of hydrogenation-dehydrogenation (*P-C-T*) determines the basic characteristics of hydride. Values of the top and bottom pressure in working area of temperatures should be technically comprehensible.

The hysteresis of processes of interaction of metals and intermetallides with hydrogen is characteristic for all systems. Distinguish three types of the hysteresis phenomena [1]: (a) hysteresis of hydrogen pressures; (b) hysteresis of solubility; (c) thermal hysteresis. The value of a hysteresis essentially depends on structure of an intermetallide and temperature of its interaction with hydrogen. Activation of an initial material by an annealing is connected through its temperature with value of a hysteresis [1]. Full clearness about mechanisms of an result hysteresis is not present. Presence of a hysteresis between direct and return reactions of interaction with hydrogen, and also absence of a plateau of pressure in the field of (α + β)-phases, affect the amount of hydrogen participating in a cycle the HHP.

The kinetics of hydrogenation-dehydrogenation is determined by the interconnected processes [1]: kinetic factors of the phenomenon (surface processes, diffusion and phase transformations), a heat flow and a mass flow. Speeds of absorption/desorption of hydrogen depend on a kind of supervising processes ((1) surface reaction; (2) diffusion of hydrogen; (3) chemical reaction; (4) rise and growth of a new phase). Chemism reactions of hydrogen sorption makes a demands to kinetics of these reactions: speed of responses should be not lower than speeds heat-and mass transfer of processes in hydride layers. Especially it concerns to area of temperatures below indoor temperature.

Activation of initial alloys is necessary, if on their surface are formed oxide a film. Process of activation of an initial alloy imposes restrictions on conditions of its carrying out: values of temperatures and pressure, quantity of temperature cyclings. Conditions of activation should allow to lead it within the limits of the technical device containing forming hydride an alloy.

Propensity IMS to degradation due to "poisoning" with impurity in hydrogen can be reduced a rational alloying. "Poisoning" is fraught with reduction of sorption capacity of hydrogen. The long cycling of hydrides can lead to a disproportionation of initial alloys and change of initial structure, that also leads to reduction of sorption capacity of hydrides.

The problem of strong change of volume of some hydrides concerns to number of technical problems of use of hydrides at absorption/desorption (up to 20 ÷ 30%), that makes demands to a design of hydride containers.

Article [3] is devoted to the analysis of the presenting requirements to hydride at its use in thermal devices, and questions of a choice of effective hydrides.

2.2. THERMODYNAMIC RESTRICTIONS

Enthalpy ΔH and entropy ΔS of hydride formation are the basic values determining position by a curve of equilibrium pressure in coordinates LnP-$1/T$. The enthalpy value of formation can play a positive role in case of generation of heat and a cold (where favorably to have big ΔH) and to have an adverse effect in the hydrogen compressor (where favorably to have low ΔH). These two characteristics define ideal efficiency of HHP.

Theoretically in a cycle the HHP it is possible to receive factor of transformation of energy above unit ($COP_{th.} \approx \Delta H_B/\Delta H_A$, where A – high-temperature hydride, B – low-temperature hydride, at $\Delta H_B > \Delta H_A$ $COP > 1$). In practice to select pair hydrides with such properties difficultly, therefore for realized the HHP $COP < 1$.

The COP of a cycle the HHP is connected with thermodynamic and thermalphysic properties of hydrides and with temperature cycle indexes. Ways of increase of factor of transformation of energy COP are:

- Increase in an enthalpy of formation of low-temperature alloy ΔH_B
- Organization of a cycle for decrease in deviations of operating temperatures from an average level of temperature T_m that promotes reduction of losses of the heat going on preheating (precooling) of hydrides
- Weight reduction of a design HHP, participating in processes of a sorption of hydrogen, at maintenance of sufficient durability of hydride containers (reactors-sorbers)
- Application of regeneration for useful use of heat (cold) of preparatory stages of a cycle the HHP

The value of equilibrium pressure above hydride in the field of temperatures of operation of HHP should be not below atmospheric pressure for maintenance of necessary intensity and duration of reactions.

2.3. RESTRICTIONS OF THERMALPHYSIC PROPERTIES OF HYDRIDE LAYERS

In spite of the fact that initial hydride generative alloys have sufficient for technical applications factor of heat conductivity (~15 ÷ 30 W/(m·K)), necessity of their use as a powder results in values of effective heat conductivity $\lambda_{eff.} = 1 \pm 0.5$ W/(m·K), that is insufficient for carrying out in HHP of heat exchanging processes of the necessary intensity.

Generally heat conductivity of metal hydride powders in the complex way depends on heat conductivity of an initial alloy, porosity ε of a covering, pressure P of hydrogen in system and degrees of hydrogenation of a powder.

Heat conductivity of hydride layers should be up to date $\lambda_{eff.} = 5 \div 15$ W/(m·K). It can be reached entering additives of height heat-conducting metals, or formation (with possible sintering) powders of initial alloys under pressure. At addition of metals is put the task of creation high-porous heat-conducting open structure with its optimization on weight and anisotropy of properties in a direction of streams of heat and mass [4]. At formation of hydride generative materials under pressure (including incapsulation and addition of high-temperature powders) alongside with achievement of necessary heat conductivity is put a task to provide sufficient gas permeability of the formed structure for hydrogen. Gas output speed should exceed speed of heat-conducting processes.

It is important to notice to necessity of reduction of contact resistance, both at connection of powders, and at their contact to heat-conducting parts of designs of HHP.

The detailed analysis of models of calculation of effective heat conductivity of systems «hydride + metal additives» with reference to the HHP is given in paper [4].

2.4. CONSTRUCTIVE RESTRICTIONS

Designing of HHP is first of all correct designing of hydride layers. In practice:

(1) It is necessary to provide a good supply (removal) of hydrogen and heat in (from) a zone reactions, i.e. it is necessary to provide thin enough layers with the advanced surface of heat exchange and their uniform warming-up. Processes of absorption and allocation of hydrogen by hydride layers have non-stationary character. Duration and intensity of non-stationary processes heat and mass transfer in hydride layers is characterized by dimensionless criteria: Fourier number – **Fo** and Biot number – **Bi**. Their connection with physical parameters **Fo** $= a\tau/\delta^2_{ch.}$, **Bi** $= \alpha\delta_{ch.}/\lambda_{eff.}$ allows to formulate requirements to hydride layers. Here a denotes factor of thermal conductivity of hydride layer, τ designates time, α is factor of heat emission and under $\delta_{ch.}$ understand the reduced characteristic thickness of a hydride layer in a direction of effect of a heat flow. Thus, it is impossible to forget that the hydride reactor-sorber is the vessel working both under pressure, and in conditions of cyclically extending and narrowed actuating fluid (hydride).

(2) Factors of the reduced heat emission from the heat-carrier and a hydride layer should be comparable on value.

Also it is necessary to agree processes of heat exchange between the reactors-sorbers the HHP operating in pair.

At designing reactors-sorbers it is necessary to aim to that moving of front of sorption reaction did not result in essential change of a way of flow of hydrogen in a hydride bed. The similar problem exists in radial hydride beds where streams of heat and hydrogen can be directed to the different sides. Especially it affects operation of a low-temperature sorber during generation of a cold when in the beginning of process it is necessary for heat to pass through all hydride bed for maintenance of hydrogen desorption, and at the end of process at displacement of reaction front to a wall is at a loss the output of hydrogen since it is necessary for it to overcome resistance of all bed.

Therefore in refrigerator HHP the refrigerating part of a cycle should be designed better from a point of view of increase in intensity of heat exchanging processes.

(3) Hydraulic resistance to flow of hydrogen should be minimized. Analyzing the equation laminar flow of hydrogen in pores of hydride powder it is possible to note a number of constructive measures which promote reduction of resistance of hydrogen flow: reduction of thickness δ of a bed of hydride (characteristic size $\delta_{ch.}$), passing hydrogen; reduction of pores tortuosity (a_{sin}) and increase in their diameter (d_p); increase in porosity of covering ε. Presentation of these requirements speaks about necessity of use for porous beds of structures with the open, organized porosity of small relative size ($l/d_p < 100$, at $\delta_{ch} = 5$ mm $d_p < 50$ μm; where l – the resulted length of pore).

(4) For the purposes of uniform distribution of temperature on volume of a hydride layer in conditions given heat (-cold) productivity it is necessary to minimize heating of the heat-carrier on length of a heat exchanging surface.

2.5. RESTRICTIONS OF HEAT ENGINES

In HHP takes place the problem of carry of not maximal to possible amount of hydrogen at operation of hydrides in pair. On the one hand it is necessary to increase its amount, on the other hand it is necessary to reach necessary levels of temperature and capacity. That is the optimization problem takes place.

The cycle of HHP has stages of preliminary heating and cooling. It is necessary to spend heat which cannot be useful on these stages is used. In HHP is appropriately application of partial regeneration of heat. For use of heat (cold) transferable hydrogen also it is possible to apply partial regeneration.

It is necessary to remember, that efficiency of HHP (*COP*) falls at expansion of a range of operating temperatures and deviations from average temperature. It is necessary to achieve the compromise between a special-purpose designation of the engine and its efficiency.

There is a problem of reception of the big output powers from a unit of weight of hydride. In opinion of authors [5] HHPs become competitive with other kinds of heat engines at achievement of "output of an alloy" at a level of 800 ÷ 1,000 W/kg. Today this number changes in limits from 40 up to 250 W/kg.

3. Conclusions

The analysis of the restrictions arising by development of HHP shows, that at a modern level they are not competitive because of small amount it is useful transferable hydrogen that is connected to insufficient hydrogen capacity of hydride generative alloys.

References

1. Kolachev B.A., Shalin R.E., Il'in A.A. Hydrogen Storage Alloys. Hand-book. Moscow: Metallurgy. 1995. 384 p. (in Russian).
2. Vlasov N.M., Solovej A.I., Fedik I.I., Tchernikov A.S. Hydrogen Storage Alloys and Their Hydrogen Reversible Sorption Capability. – In abstract book of VIII International Conference of ICHMS'2003. - IHSE. Kiev - 2003. P. 154–155.

3. Shanin Yu.I. Improvement of Hydride Heat Devices Overall Performance. In: Hydrogen Materials Science and Chemistry of Carbon Nanomaterials // Proc. of NATO ARW, Netherlands: Springer, 2007, P. 383–397.
4. Shanin Yu.I. Optimization of Hydride Heat Pumps Operation. – In the same place as [3], P. 851–859.
5. Ron M., Josephy Y. Optimization of a Hydrogen Heat Pump. - Z. für Physikalische Chemie Neue Folge. 1989, Bd. 164, P. 1475–1484.

HYDRIDE HEAT PUMP APPLICATION FOR ADDITIONAL COTTAGE HEATING

YU.I. SHANIN, A.I. SOLOVEY, O.I. SHANIN
FSUE Scientific Research Institute of Science and Industrial Association "Luch", Podolsk, Moscow area, Russia, 142100
Tel./Fax: 7+10+(4967)634582; E-mail: syi@luch.podolsk.ru

Abstract. The problems of a metal hydride heat pump (MHHP) practical application for an additional heat supply and air-conditioning of a small country house are considered in the article. The plan of application and description of MHHP operation in various thermodynamic cycles are presented. Hydrides couples provided required temperature conditions were selected. Calculated characteristics of ~10 kW MHHP are presented. The MHHP design is illustrated schematically. Experimental unit for MHHP modules tests and test results are reported.

Keywords: hydride, hydrogen, metal hydride heat pump, module, experiment

Nomenclature

d	Diameter (m)
G	Coolant flow (kg/s)
P	Pressure (Pa)
Q	Quantity of heat (J)
T, t	Temperature (°C, K)
COP	MHHP efficiency in a refrigerator cycle (useful cold to used heat ratio)
HTH, LTH	High-temperature and low-temperature hydrides
MHHP	Metal hydride heat pump

Indexes

h, m, l	Temperature levels: high, middle, low

1. Introduction

Metal hydride heat pump (MHHP) operation with the same hydrides couple both for additional heat and cold generation is known [1]. The MHHP can give additional heat in a heat transformation mode (i.e. when two levels of temperature – low T_l and middle T_m are specified), and coolant temperature is increased from middle T_m to high temperature level T_h. If the MHHP is working in a refrigerator cycle (i.e. when two levels of temperature – high T_h and average T_m are specified), additional heat is taken away for heating from an environment at temperature T_l and used at the middle level of pump temperature T_m.

B. Baranowski et al. (eds.), *Carbon Nanomaterials in Clean Energy Hydrogen Systems.*

2. Results and discussion

2.1. THE COMBINED MHHP-BASED HEATING UNIT

The designed MHHP is used as an additional element of a gas water-heating boiler. The diagram of installation includes a water-heating gas apparatus, heat pump and heat exchanger.

Figure 1 illustrates the schematic diagram of one of the developed versions of MHHP-based heating unit (the main house heating system is not shown in the diagram) which includes:

- Natural gas burner
- High-temperature (HT) side with HTH, temperature gauge and heat exchanger TO1 (is not shown in the diagram)
- Low-temperature (LT) side with LTH, temperature gauge and two heat exchangers TO2 and TO3 (are not shown in the diagram)
- Hydrogen line connecting LT and HT sides
- Additional source of low-temperature heat, for example, heat of soil with heat exchanger TO4 (sign 10 in Fig. 1)
- Heating system (sign 11 in Fig. 1) with working coolant (water) at temperature level 45–70°C
- Water supply system consisting of a hot water tank and heat exchanger for water heating (is not shown in the diagram)

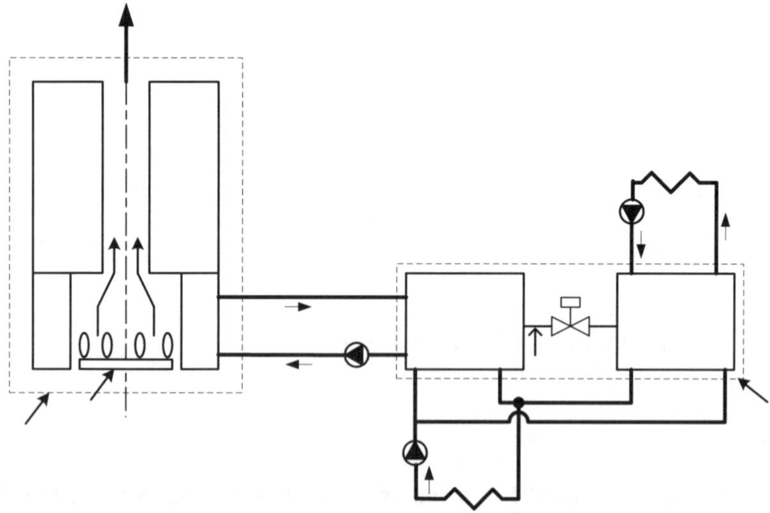

Figure 1. Schematic diagram of house additional heating with use of MHHP: 1 – water-heating apparatus; 2 – torch; 3 – combustion materials; 4 – MHHP; 5 – MHHP high-temperature side; 6 – MHHP low-temperature side; 7 – hydrogen lines; 8 – valve (version without valves is also possible); 9 – pumps; 10 – heat exchanger with soil; 11 – heating appliance (Elements of automatics are not shown in the diagram)

The MHHP works cyclically and consists of two parts – high temperature (HT) and low-temperature (LT) sides. In one of possible heating modes, water in the water-heating apparatus is overheated to 100–120°C (Fig. 2(a)). It is used for heating of a

MHHP HT side. A MHHP LT side uses heat of a soil with temperature 0–15°C. The middle temperature level of water in the heating system is 45–65°C. A radiator of the house heating system is heated from a heat exchanger and from MHHP.

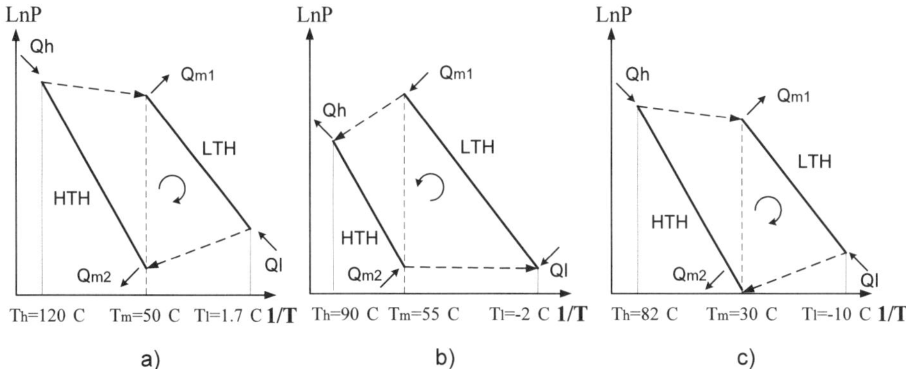

Figure 2. Schematic illustration of the designed MHHP cycles: (a) refrigerating cycle with heat generation at a level of $T_m = 45 \div 65$°C; (b) cycle of heat transformer with rising of temperature to $T_h = 90$°C; (c) refrigerating cycle (air-conditioning) with working temperature $T_l = -10$°C. The cycle direction is shown by a circular arrow

The refrigerator cycle of installation operation, when additional heat is taken away on a middle level, had been developed. The MHHP works cyclically and has two couples of reactor-sorbers.

For specified temperature levels of cycles, suitable hydrides couples have been chosen: $LaNi_{4.6}Al_{0.4}$ or $LaNi_{4.5}Al_{0.5}$ as high-temperature alloy, and $MmNi_{4.5}Mn_{0.4}$ or $TiF_{e0.8}Mn_{0.2}$ as low-temperature one. Hydrides were selected with use of the computer code [2] (only hysteresis effect was accounted for) with the further experimental verification. Results are presented in the Table 1.

TABLE 1. Temperature levels of MHHP operation with hydride couple consisting of $LaNi_{4.6}Al_{0.4}$ and $TiFe_{0.8}Mn_{0.2}$ at various cycles

MHHP operating mode	T_{hmax} (°C)	T_{lmin} (°C)	T_m (°C)	P_{max} (atm)	P_{min} (atm)	*COP*
Refrigerating cycle – 1 (heat generation), Fig. 2(a)	106.5	1.7	50	4.9	0.75	0.42
Heat transformer, Fig. 2(b)	90	-2.1	50	5.3	0.51	0.41
Refrigerating cycle – 2 (air conditioning), Fig. 2(c)	82.2	-9.8	30	2.2	0.31	0.445

1. Cycles were plotted accounting for hysteresis of hydrides *P-C-T* characteristics
2. The weight of a design is assumed to be equal to hydride weight in a reactor-sorber

Properties of hydrides couple ($LaNi_{4.6}Al_{0.4}$-$TiFe_{0.8}Mn_{0.2}$, see Table) provide all thermodynamic cycles (Fig. 2) for the specified temperature ranges. The cycle of heat transformation, when gas boiler heats water to 50–60°C, and water passed through the heating system, can be heated again to the necessary level by MHHP (Fig. 2(b)). Besides, this hydrides couple during a hot season can be used as a system for air conditioning (Fig. 2(c)). At this mode, the water-heating boiler heats water for hot water supply to 90°C, which is partially used in a MHHP HT side. The level of middle temperature corresponds to a level of an ambient temperature, and on a MHHP LT side the cycle temperature can reach negative temperatures.

2.2. MODULE AND MHHP DESIGNS

The MHHP has a modular-type design. Two modules have been designed with different tube sizes used as a case. In one case ∅51x1 mm tubes were used for a reactor-module, in the other – ∅32 x 1 mm (see Fig. 3). The central aluminium drain-pipe wrapped outside by the filtering fine grid is inserted inside reactors. To increase thermal conductivity, copper foil plates 0.2 mm in thickness were inserted in a radial direction with 2 mm space. The reactor free volume had been filled up with milled initial alloy (particle size <1 mm) and compacted by vibration (filling porosity 35–40%). Reactor-modules were arranged in a hexagonal lattice with step 1.05*d* in units and inserted in a shell-and-tube heat exchangers (19 modules at d = 51 mm and 37 modules in d = 32 mm, respectively). Heat exchangers with pass partition plates provide coolant flow inside it. 10 kW MHHP has 100 kg of hydride forming alloys, its sizes are ~Ø350 x 1,100 mm and weight ~220 kg.

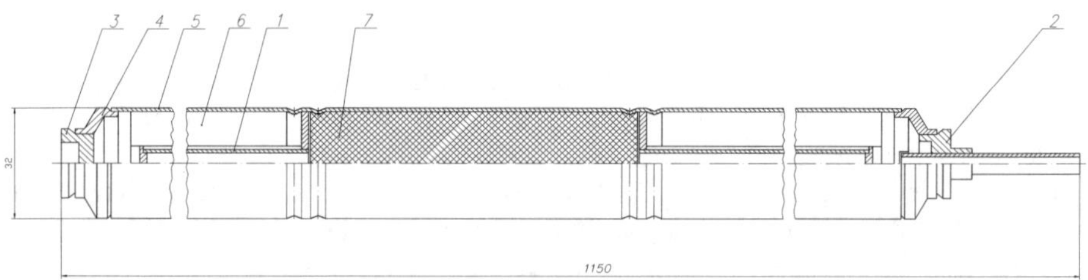

Figure 3. Schematic drawing of the module with external tube ∅32 mm in diameter: 1 – collector; 2 – inlet; 3 - plug fitting; 4 – bottom; 5 – sleeve; 6 – insert; 7 – silicate cotton. Hydrides are not shown

The version of a MHHP (presented schematically in Fig. 4) with 37 reactors-modules has been made. Calculated characteristics of the designed MHHP are the following:

- Metal hydrides couple: HTH – $LaNi_{4.6}Al_{0.4}$, LTH – $TiFe_{0.8}Mn_{0.2}$
- Specific thermal power (designed) 100 W/(kg MeH)
- Number of modules – 37
- Loading:
  - Total – 100 kg
  - LTH – 47.7 kg
  - HTH – 52.3 kg
- Designed cycle time (stationary operation) – 25.2 min
- Average MHHP power in a cycle – 11.7 kW
- Water flow:
  - Through a high-temperature side during heating – G ~ 1 kg/s
  - Through a high-temperature side during regeneration – G ~ 0.65 kg/s
  - Through a low-temperature side during heating – G ~ 0.37 kg/s
  - Through a low-temperature side during regeneration – G ~ 0.6 kg/s;
- Total MHHP weight ~ 220 kg.

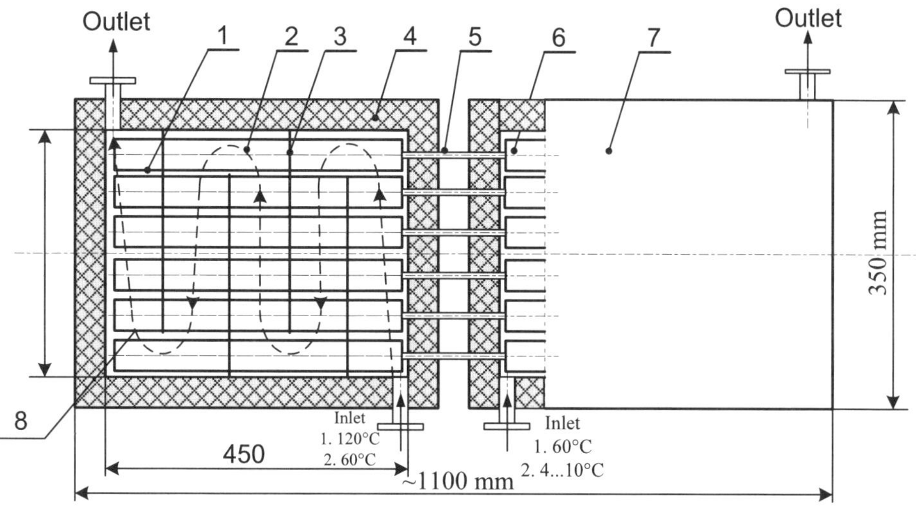

Figure 4. MHHP constructive diagram: 1 – MHHP high-temperature side; 2 – high-temperature reactor (module); 3 – partition; 4 – heat insulation; 5 – tubes for hydrogen travel; 6 – low-temperature reactor (module); 7 – MHHP low-temperature side; 8 – line of coolant flow

2.3. THE MHHP EXPERIMENTAL MODULE

The module consists of two reactor-sorbers (low-temperature and high-temperature) which structurally repeat a filling of the regular module (with a tube case Ø32 x 1 mm and 400 mm in length, made of stainless steel). 0.8 kg of $LaNi_{4.6}Al_{0.4}$ powder (high-temperature) and 0.63 kg of $TiFe_{0.8}Mn_{0.2}$ (low-temperature) are charged into reactors.

For the technological purposes and convenience of work, vacuum valves and a compound vacuum and pressure gauge are included in the hydrogen line connecting both reactors. They provide measurement of each alloy absorptive ability.

The reactor has an external coaxial tube-type cooling jacket attached to water collectors of the test unit.

2.4. THE TEST UNIT

In experiments, the refrigerator cycle of the MHHP module working at a level of high temperature ~130°C was realized. The test unit (Fig. 5) includes: experimental module, electric system, vacuum system, systems of water-conditioning and control. Water is conditioned in three separate lines: cold water line – low temperature level (5 ÷ 20°C), hot water line – middle temperature level (40 ÷ 60°C) and overheated water line – high temperature level (100 ÷ 130°C).

The control system including valves, electric valves and one-coil transformer provides changing of temperature and water flow and control of water flow direction in a certain sequence.

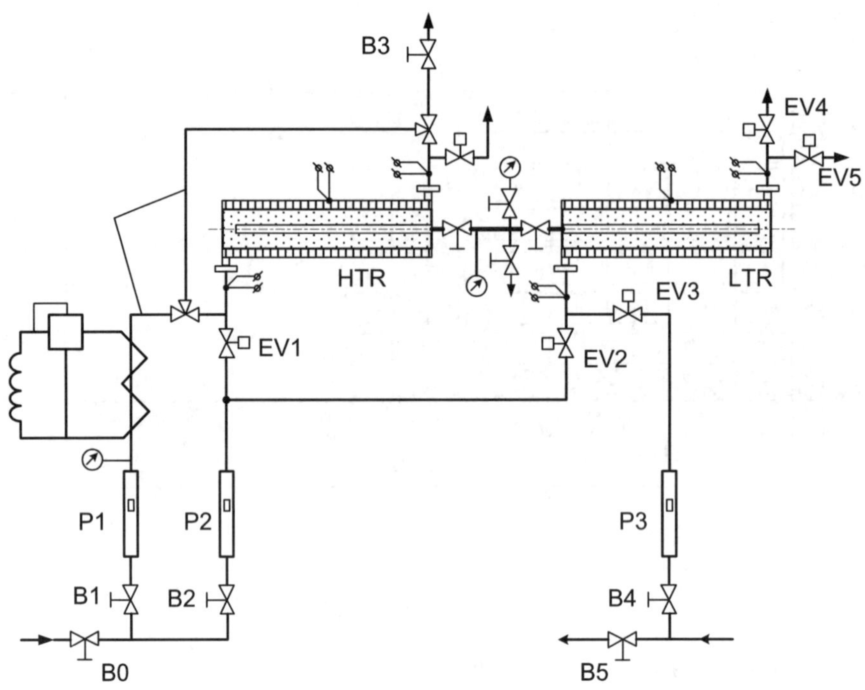

Figure 5. The diagram of the test unit: B0–B5 – valve; BB1–BB4 – vacuum valve; BT1–BT2 – three-way valve; M1–M2 – manometer; MB – pressure-and-vacuum gage; HTR – high-temperature reactor-sorber; LTR – low-temperature reactor-sorber; P1–P3 – flowmeter; t1–t6 – thermocouple; EV1–EV6 – electric valve; TEH – tubular electric heater; ATF – autotransformer

2.5. TECHNIQUE AND TEST RESULTS

Heat effects were determined by calorimetry in two stages. At the first stage, the high-temperature reactor-sorber with closed hydrogen valve was heated to an isothermal condition of overheated water at 120–130°C. The low-temperature reactor-sorber has temperature of hot water $T_m = 45 \div 55$°C. After opening a valve hydrogen flows in a low-temperature reactor, is absorbed by alloy heating the coolant. Useful instant and time-averaged heating power were determined as a difference of temperatures of water in a reactor cooling jacket at known flow. Power of heating reached the maximum value of 140 W for 1.5 min and then exponentially decreased. Average power at this stage is 110 W. After hydrogen transfer, valve is closed, reactors reaches stable conditions at the other temperature levels. Hot water in the high-temperature reactor is maintained at the temperature of $T_m = 45 \div 55$°C, cold water in the low-temperature reactor is maintained at the temperature of $T_l = 15 \div 20$°C. Opening of a valve causes flow of hydrogen and its absorption in a high-temperature reactor. The maximum power reaches 40 W for 1.5 mines, average power is 28 W. The coefficient of performance of energy with accuracy ±20% is 1.25 (i.e. *COP* = 0.25).

3. Conclusions

The work showed that successful choice of hydride couples provides operation of the multipurpose MHHP working as: (1) an additional heating device; (2) a conditioner. Besides, heat generation is possible both in direct (more rigid temperature operating conditions of water-heating boiler), and in the reverse direction (heat transformation at lower coolant temperature).

References

1. Yu.I. Shanin. Simulation of operation heat or cold-making unit with the hydride heat pump. – In: Theses of 8-th Int. Conf. "The Hydrogen Material Science and Carbon Nanomaterial Chemistry". Sudak, Crimea, Ukraine, 14–20 Sept., 2003, P.304–307.
2. Yu.I. Shanin. Preselection of hydrides for hydride heat pumps. - In: Theses of 6-th Int. Conf. "The Hydrogen Material Science and Chemistry of Metal Hydrides". Katsiveli, Crimea, Ukraine, 02–08 Sept., 1999, P.304–305.

THE INFLUENCE OF NANOSTRUCTURAL OXIDE FILMS ON WEAR-RESISTANCE OF TITANIUM MATERIALS

A.M. PETROVA, M.B. SHTERN
Institute for Problems of Material Sciences of Ukrainian National Academy of Sciences, Krzhyzhanovskogo Str., 3, Kyiv 03142, Ukraine
Fax: 424 2131; E-mail: tkach@ipms.kiev.ua

Abstract. The wearing mechanism of titanium material of the Ti–Cr–TiC composition obtained by different ways of leading chromium and carbon into titanium as separate elements Cr and TiC or as a compound Cr_3C_2 was investigated. The formation of nanostructural oxide films acting like a solid lubrication at friction on the surface of titanium materials was shown. Defensive properties of the oxide films on the friction surface of Ti–Cr_3C_2 material are stronger. The structure of such a material consists of small titanium carbide engagings that have a strong connection to titanium-chromium base and work till the abrasion.

Keywords: titanium, nanostructure, oxide film, wearing, friction, carbide

1. Introduction

The study of wearing out and friction of titanium materials containing chromium and carbide of titanium is depicted in papers [1–3]. It is shown the positive influence of alloying material of chromium and solid inclusions of titanium carbide on its wear-resistance.

The paper is aimed at examining the mechanism of wearing for Ti–Cr–TiC materials obtained by different ways of leading chromium and carbon into titanium as separate elements Cr and TiC or as a compound Cr_3C_2. In the last case chromium carbide interacts with titanium in the process of sintering forming at that the its carbide and alloying titanium base with the chromium.

2. Results and discussion

Examination was carried out on the compositions Ti–TiC, Ti–Cr, Ti–Cr–TiC, Ti– Cr_3C_2 containing an equal amount of input Ti, Cr and C and obtained by pressing and subsequent sintering in vacuum 0.13 Pa from the mixture of powders of electrolytic titanium of −180 + 40 μm fraction and TiC, Cr, Cr_3C_2 of 10 μm fraction. Density of the materials amounted to 4.3–4.4 g/cm^3.

Tests were realized on the friction machine M-22M working by the scheme shaft-bush, in the air, sliding race of 1 m/s, graduate increase of loading from 2.3 to 15.0 MPa, and for the Ti–Cr_3C_2 material – up to the loading at which a catastrophic wearing out is observed. Hardened steel 45 (HRC 45-50) was used as a rider. Wear-resistance of friction couple was estimated by the wearing intensity of the sample and ride, measurable to 25 μm/km. The temperature of friction surface was calculated with the help of chromel-alumel thermocouple at the edge of contact zone for sample and rider. The results of the test are given in the Table 1.

TABLE 1. Wearing out (*I*), friction coefficient (μ), the temperature in the friction zone (*t*) of the friction couples, titanium material – hardened steel 45 at *V* = 1.0 m/sec, at graduated loading

P (Mpa)	Ti–TiC			Ti–Cr			Ti–Cr–TiC			Ti–Cr $_3$C$_2$		
	I(μm/km)	μ	*t* (°C)	*I* (μm/km)	μ	*t* (°C)	*I* (μm/km)	μ	*t* (°C)	*I*(μm/km)	μ	*t* (°C)
2.5	Cat.	–	–	330	0.15	75	70	0.27	80	<25	0.32	80
5.0				90	0.17	100	<25	0.33	190	<25	0.35	110
7.5				25	0.3	150	<25	0.35	230	<25	0.32	120
10.0				80	0.39	300	<25	0.37	250	<25	0.34	220
12.5				2,000	0.42	380	<25	0.37	300	<25	0.32	300
15.0							<25	0.38	330	<25	0.38	330

Wearing tests (the table) showed an extreme wearing out of titanium material Ti–TiC while the first loading of $P = 25$ MPa as well as unsteady character of titanium-chromium alloy wearing and high wear-resistance at all the following loadings for the material, containing TiC and Cr conjointly.

The structure, microhardness of friction surface and materials was investigated by cut transversely to friction surface. The X-ray analysis of wearing products of friction couples as well as from the surface of friction was carried out.

The X-ray analysis of wearing products of friction couples (Ti–Cr_3C_2)–hardened steel 45 shows the presence of oxides, hydroxides, nitrides, ferric, titanium, chromium hydrides of the following compounds (in keeping with the decrease of line number at rontgenogram): Fe_3O_4, Fe_2O_3, $FeCr_2O_4$:, TiO_2; TiN; TiH_2, TiC, β .Fe_2O_3 .H_2O, Cr_2O_3, CrO_3, Cr_2N. Higher contents of iron compounds in the wearing products is the evidence of prevailing wearing of steel rider. X-ray analysis of the friction surface of Ti–Cr alloy discloses the oxides Fe_3O_4, Fe_2TiO_4, Cr_2O_3, TiO_2 and the compound Fe_2Ti. The presence of titanium material of the compound containing iron on the friction surface may be explained by the transition of rider material on conjugate surface with its next interaction with titanium and formation of Fe_2Ti compound and of compound oxide Fe_2TiO_4 of spinel type. Nitrides, oxides, titanium and chromium hydrides are the products of the interaction between titanium-chromium basis of material and air; titanium carbide is present in the wearing products as a result of wearing out of titanium material. Chromium, ferric and titanium oxides formed in the process of friction as well as compound oxide of spinel type participate on the formation of surface wear-resistance nanostructural films; nitrides, hydroxides and carbides bring the elements of abrasive wearing.

The structures of the surface layers and friction surfaces of the investigated materials are presented at the Figs. 1 and 2. The friction surface of Ti–TIC (Fig. 1(a)) material is uneven and has the areas of deep pits at those places where the titanium carbide inclusions come out on the friction surface, which apparently are painted out. The size of carbide inclusions exceeds significantly the one to be input and comes to 100 μm. That is a result of coagulation in the process of sintering. There is no strain. Microhardness from the friction surface deep into the material is equal and amounts to

2.4–3.5 MPa. The friction surface is loosened (Fig. 2). Light parts as a de facto contact between sample and rider are narrow and interrupted.

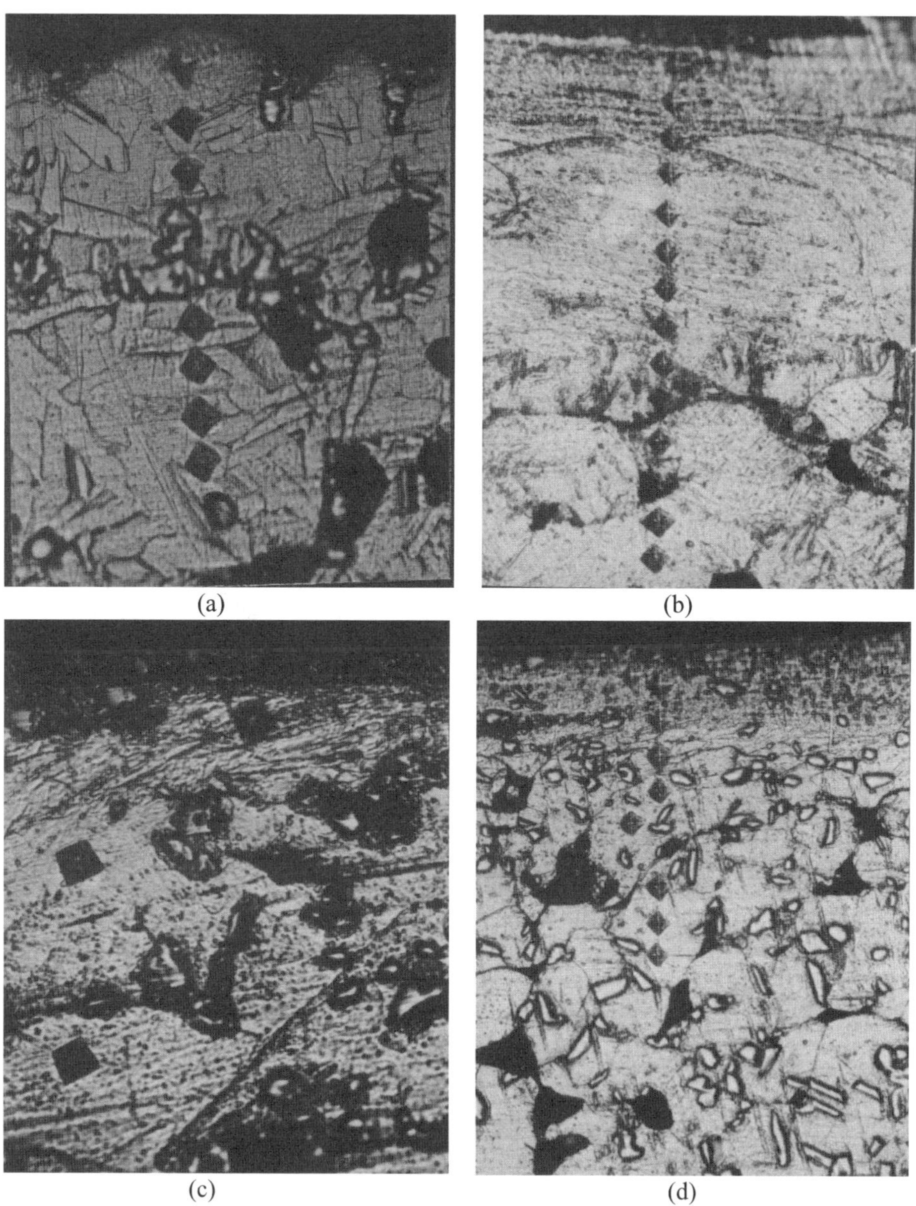

Figure 1. Microstructure of the surface layer after the friction of Ti–TiC (a). Ti – Cr (b), Ti–Cr–TiC (c), and Ti–Cr_3C_2 (d) materials

×250

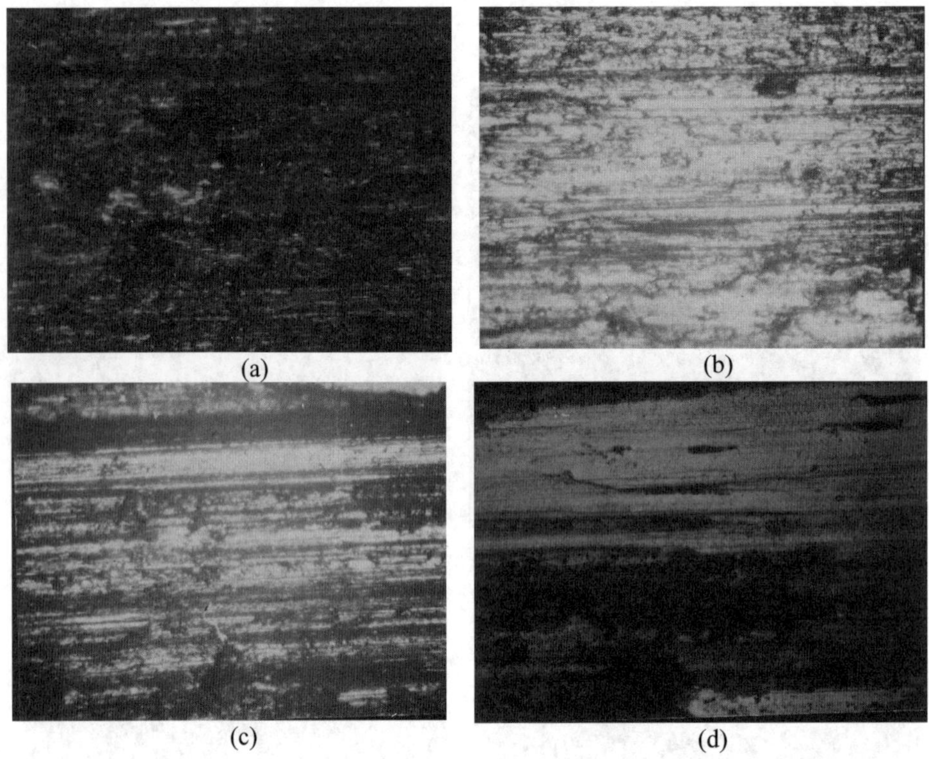

Figure 2. The friction surface of Ti–TiC (a), Ti–Cr (b), Ti–Cr–TiC (c), and Ti–Cr_3C_2 (d) materials ×63

Plastic strain shows itself within the surface layer of titanium-chromium alloy Ti–Cr. It is expressed by the crush of base grains which are situated directly within the surface layer as well as by its oblongness along the friction way and by the typical strain lines curve deep into the material. A significant shear strain causes the separation of the thin surface layer of the material which is fixed on the picture as thin light band under which a dark area of the destroyed material may be observed. Microhardness of the strained layer is the same as of the sample's core, amounts 3.2–4.2 MPa that 1.0 MPa higher than the titanium in Ti–TiC material. The increase of microhardness can be explained by the influence of alloying element of chromium on the microhardness of titanium. The structure of titanium-chromium alloy is single-phase and large-grained. Plastic strain causes the flow of the material and results in enclose of the pores of the sintered sample with the compaction of the material under the friction surface. The friction surface of alloy is loosened (Fig. 2(b)) with the sign of material's pits and separate small areas of white oxide films.

At the input of chromium and titanium carbide conjointly into titanium the plastic strain of the surface layer loses significantly its tenseness, the lines of strain are directed to the friction surface at angle of 45° and are observed in thin surface layer of the deep micrometer (Fig. 1(c)). There are some small inclusions of chromium and titanium compounds in the surface layer. The friction surface is uneven (Fig. 2(c)), although the parts of titanium pits are not as evident as in case of Ti–TiC material. Oxide film on the friction surface (Fig. 2(c)) contains a number of parallel dark tracks appeared as a result of abrasive carbide inclusions.

Some insignificant parts of the material are subjected to plastic strain within the Ti–Cr_3C_2 (Fig. 1(d)). These parts are situated under the surface layer strengthened by dispersed particles which differ in looks from the bigger inclusions of titanium carbide and are the secondary compounds appeared in the process of friction. The depth of the strengthened layer is 5–7 μm and microhardness is 15–20 MPa higher than the core of the sample. The inclusions of TiC in size of 5–10 μm at the outlet on the friction surface preserve the strong connection with the metal base of the material and work till the abrasion. The friction surface (Fig. 2(d)) is even, without any pit-parts of structural components. Therefore, some small inclusions of titanium carbide that are formed within the material have a strong connection to the base and prevent significantly the plastic strain at friction rather that while putting the titanium carbide into the charge. The decrease of plastic strain as well as the wear of the material creates the conditions for the oxygen, hydrogen and nitrogen penetration deep into the material which results in formation of the layer strengthened by titanium oxides, hydrides and nitrides which in their turn make the plain strain of the surface layer go down and increase the wear-resistance of the material.

The friction surface of the material (Fig. 2(d)) contains wide light bands of the oxide films that are destroyed by the abrasive particles in some places. Microhardness of the friction track grows up to 130–250 MPa . It is an evidence of significant strengthening of the titanium material in zones of direct contact between the material and rider.

With the rise of loading to the extreme wearing out that takes place at $P = 30.0$ MPa wide light bands disappear from the friction surface of Ti–Cr_3C_2 material and the presence of interchangeable permanent thin light and dark bands which may be an evidence of active destruction of the oxide film that gets milder under the action of upcoming high temperature in the friction zone as well as gets more mobile and speeds away from the friction surface (Fig. 3).

3. Conclusions

Therefore, Ti–Cr , Ti–Cr–TiC, Ti–Cr_3C_2 nanostructural oxide films formed at friction of titanium materials perform the function of solid lubricant. Its defensive properties depend on applies loadings, sheer strain of surface layers, size and durability of the connection between carbide titanium inclusions with titanium-chromium basis.

Defensive properties of the oxide films formed at the friction of Ti–Cr_3C_2 material are stronger. The titanium carbide is not put directly into the initial charge but is formed in the process of sintering while putting the chromium into the titanium carbide. The structure of such a material contains some small inclusions of titanium carbide that are formed during sintering and that have a strong connection to the titanium-chromium base.

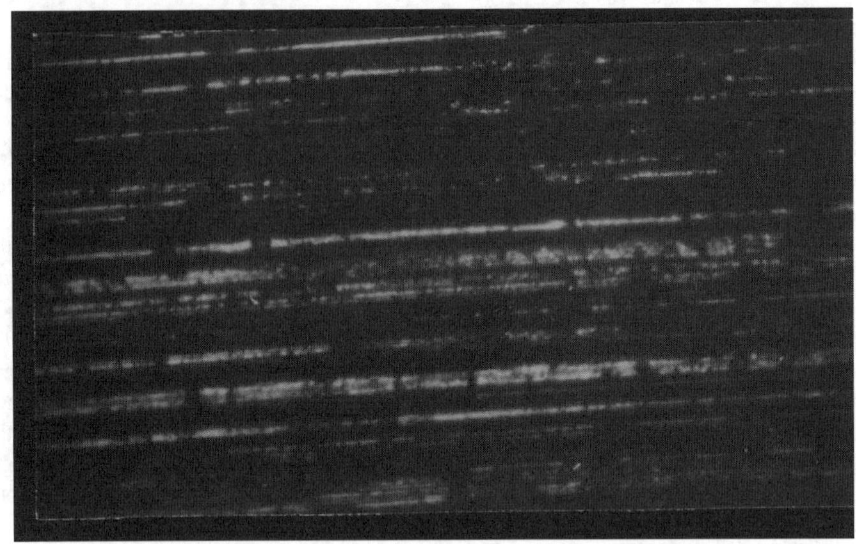

Figure 3. The friction surface of the Ti–Cr_3C_2 material after the wearing out to the loading up to the 30.0 MPa

×63

References

1. Radomysel'skii I.D., Titarenko S.V., Petrova A.M., Polotai V.V. The studying of friction and wearing for the sintered titanium materials, *Powder Metallurgy*, 1977, **6**: 61–66.
2. Petrova A.M., Polotai V.V. To the wearing out mechanism of titanium-chromium alloys in the air, Abstract of the report on All-Union Conference "Technical control over the tribotechnical characteristics of the friction units", Kishinev, 1985, pp. 116–117.
3. Petrova A.M., Polotai V.V. Effect of the chromium contents on the tribotechnical characteristics of titanium-chromium alloys, *Powder Metallurgy*, 1987, **5**: 51–56.
4. Kolesnichenko L.F., Polotai V.V., Zabolotnyi L.V. Methodology of the complete studying of the friction and wearing of the ceramic materials, *Powder Metallurgy*, 1970, **3**: 61–66.

AB INITIO SIMULATIONS OF LIQUID AND AMORPHOUS SiC AND SiCN

V.I. IVASHCHENKO, L.A. IVASHCHENKO, P.L. SRYNSCKYY, L.A. GRISHNOVA
Institute for Problems of Material Science, NAS of Ukraine, Krzhyzhanovsky Str. 3, 03142 Kyiv, Ukraine

Abstract. We carried out *ab initio* molecular dynamics simulations of amorphous and liquid silicon carbide (l-SiC, a-SiC) and silicon carbon nitride (l-SiCN, a-SiCN). In the liquid phases, the tendency of carbon towards low coordination dominates the structural characteristics. Silicon coordination in both liquids is comparable with that in pure l-Si. It is found that the network of the amorphous phases deviates from an ideal tetrahedral geometry because a considerable number of C atoms are threefold coordinated. Silicon atoms are slightly over-coordinated in both the samples. In a-SiCN the Si network are strongly distorted and no preference for tetrahedral coordination. In contrast to a-SiC, in the amorphous Si-C-N system, the Si-Si nearest neighbour correlations are weakly shown. Our simulation is in good agreement with most measurements.

Keywords: *ab-initio* molecular dynamics, liquid and amorphous SiC and SiCN, coordination number, pair correlations

1. Introduction

The synthesis of materials containing silicon, carbon and nitrogen has grown in interest in recent years. Several procedures have been developed to produce SiCN ($Si_xC_yN_z$) films that have excellent properties such as high oxidation resistance, wide band gaps, and promising mechanical characteristics [1–4]. The properties of the films are very sensitive to deposition conditions. Despite numerous experimental investigations of these films, so far there is no a common opinion about the bonding picture of such films. In particular, many researchers came to a conclusion that Si-C, Si-N and C-C bonds are predominant [1, 2], whereas other group of researchers suggested that Si-C bonds are almost missing [3, 4].

Using tight-binding molecular dynamics (MD) simulations [3], different models of the amorphous network of a-$Si_xC_yN_{1-x-y}$ were generated to study the elastic and mechanical properties. *Ab initio* MD simulations of a-SiC were performed by Finocchi et al. [5]. Chemical ordering in a-SiC was comprehensive studied using Monte Carlo simulations based on the Tersoff potential [6]. Despite the useful information presented in these investigations there is still no the comprehensive comparatively analysis of the structural changes in amorphous silicon carbide when incorporating nitrogen. Besides, so far, the SiC and SiCN liquids were not investigated with a first principles method. Therefore we have used *ab initio* MD simulations to carry out the comparative analysis of the geometry and the bonding picture of the Si-C and Si-C-N structures in the liquid and amorphous states.

2. Computational aspects

Simulations of the liquid and amorphous Si-C ($Si_{0.5}C_{0.5}$) and Si-C-N ($Si_{0..375}C_{0..375}N_{0.25}$) phases were done with computational cubic cells of 64 atoms. For the SiC structures, the macroscopic densities were fixed at the calculated density of 3C-SiC (3.212 g/cm^3). In the case of SiCN materials, the average density of the $Si_3C_3N_2$, $Si_4C_2N_2$ and $Si_2C_4N_2$ phases with the zinc-blende structure (3.033 g/cm^3) was used. Our calculations were performed in the framework of the plane-wave pseudo-potential method using the EXPRESSO code [7]. For all the components we used the ultra-soft pseudo-potentials generated according to the scheme [8]. The electronic wave functions and the charge density were expended in plane waves with energy cut-offs of 24 Ry and 96 Ry, respectively. The exchange-correlation energy and potential used to implement LDA (the local density approximation) are those determined by Ceperley and Alder, as interpolated by Perdew and Zunger [9]. For the crystalline and non-crystalline structures, Brillouin-zone integration was performed using sets of special points corresponding to the (888) and (111) Monkhorst-Pack mesh, respectively. In the case of the non-crystalline materials, the initial rock-salt structures were considered. For SiCN structures, eight silicon and eight carbon atoms were randomly substituted by nitrogen. The initial SiC and SiCN structures were heated to 4,000 and 3,500 K, respectively, and equilibrated for about 0.7 ps. The systems showed a diffusive behaviour and were therefore characterized as liquids (melts). The liquids were then cooled to 500 K for 1.0 ps and again equilibrated for 0.8 ps at 500 K. The time step was 5 a.u.

3. Results and discussion

The partial pair correlation functions (PCF) of the liquid and amorphous silicon carbide and silicon carbon nitride are shown in Fig. 1. The structural characteristics of the generated samples are presented in Table 1. The bond-angle distributions of silicon and carbon atoms (i.e., having a carbon or silicon atom as the vertex atom) are shown in Fig. 2. The partial coordination numbers and the numbers of chemical bonds are presented in Fig. 3.

Although the peaks of the PCF of the liquids are considerable widened and flattened, compared to the PCF of the amorphous phases, we can identify several distinct features. In both systems each kind of correlation is shown in the view of the separate peaks. For l-SiC, the over-coordination of Si atoms gives rise to the large average Si-Si bond-length. It is seen from Table 1 that lowering of the coordination at cooling leads to reducing this characteristic from 2.45 to 2.31 Å. We note also that, in l-SiC, the coordination of C-atoms is lower than in the amorphous sample. An increase in $n_{C\text{-}C}$ means a rise of the part of C sp^3 atoms, for which reason $l_{C\text{-}C}$ in a-SiC is higher than in the respective liquid. The fact that the C sp^2 configuration is dominant over other configurations in l-SiC is also confirmed by the bond-angle distribution (cf. Fig. 2), where the peak around the graphite bond angle of 120^0 is clearly seen. The partial g(θ) formed by Si triplets is quite broad, much broader than the respective distribution in the amorphous alloys and no preference for tetrahedral coordination is shown.

For the amorphous sample, the $g_{C\text{-}C\text{-}C}(\theta)$ function clearly indicates the presence of C sp^3 and C sp^2 atoms. Regarding Si atoms, we find that 12.5% are threefold, 25% are fivefold coordinated. Increasing coordination leads to lower values for the bond angles and vice versa. The Si-Si-Si bond angle distribution has tree main features around 80^0,

103^0 and 120^0. Given this information and allowing for the total $g_{A\text{-}Si\text{-}B}(\theta)$ function, we come to a conclusion that Si atoms form a distorted tetrahedral network.

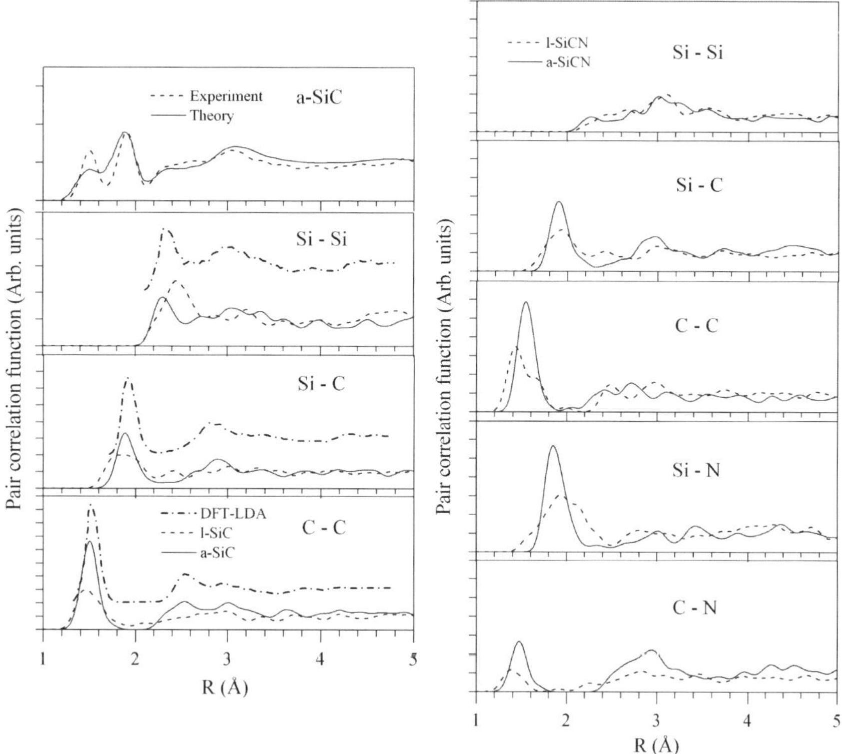

Figure 1. Partial pair correlation functions (PCF) for the amorphous and liquid Si-C (left panel) and Si-C-N (right panel) systems. The experimental total PCF for a-SiC was taken from Ref. 10. The DFT-LDA PCFs [5] are also shown. The PCFs are normalized to unity at $R \to \infty$

Without going to a detailed analysis of the PCF of a-SiC, we note that our curves are in good agreement with the analogous distributions obtained by Finochhi et al. [5] (cf. Fig. 1). For a-SiC, both simulations predict the partially chemically ordered structures.

In experiment, a-SiC are obtained in the view of thin films produced with the help of various deposition techniques [1–4], or in the view of a thin layer formed by an amorphization of the top layers of the bulk SiC samples [10]. The network of such a-SiC structures is characterized by different extents of chemical ordering depending on a preparation technique. In Fig. 1 we compare our PCF for a-SiC with the analogous function of an amorphous layer obtained by amorphization [10]. We see that the theoretical sample has a larger number of homo-nuclear bonds, compared to the experimental sample. The positions of the PCF peaks in both the samples are in good agreement.

Nitrogen incorporation changes the picture of atomic correlations. For l-SiCN, the prominent maximum around 110^0 in the bond angle distribution formed by C triplets are observed, showing that the C sp^3 network in l-SiCN is less distorted than in l-SiC. One can see that Si atoms do not have a dominant tetrahedral coordination, as in the case of l-SiC. The first peak in the PCF related to C atoms (at 1.43 Å) has shoulder around 1.6 Å. We suppose that the later structure is due to the C atoms sited in strongly distorted tetrahedral surrounding, whereas the peak at 1.43 Å can be assigned to C sp^1 and C sp^2 atoms. The average coordination numbers of Si atoms in l-SiC and l-SiCN are 5.69 and 5.21, respectively (cf. Table 1), which is slightly less than in l-Si (~6.0).

TABLE 1. Positions of the first peaks ($l_{A\text{-}B}$, A, B = Si, C, N) in the partial PCF (in Å) and partial coordination numbers ($n_{A\text{-}B}$) of the liquid and amorphous Si-C and Si-C-N structures. The partial coordinations are indicated by the areas under the first peaks (up to the minimum) in the respective PCF. The bondlengths of the respective crystalline materials are: $l_{Si\text{-}Si}$ = 2.35 Å (c-Si), $l_{Si\text{-}C}$ = 1.89 Å (3C-SiC), $l_{C\text{-}C}$ = 1.46 Å (graphite), $l_{C\text{-}C}$ = 1.46 Å (diamond), $l_{Si\text{-}N}$ = 1.71–1.79 Å (β-Si_3N_4) and $l_{C\text{-}N}$ = 1.43–1.49 Å (β-C_3N_4)

	l-SiC		a-SiC		l-SiCN		a-SiCN	
	$n_{A\text{-}B}$	$l_{A\text{-}B}$	$n_{A\text{-}B}$	$l_{A\text{-}B}$	$n_{A\text{-}B}$	$l_{A\text{-}B}$	$n_{A\text{-}B}$	$l_{A\text{-}B}$
Si-Si	3.63	2.45	2.00	2.31	1.58	2.48	0.66	2.27
			2.00[a]	2.32[a]				
Si-C	2.06	1.88	2.12	1.89	1.71	1.94	1.71	1.90
			2.05[a]	1.89[a]				
Si-N	–	–	–	–	1.92	1.93	1.75	1.84
Si	5.69	–	4.12	–	5.21	–	4.12	–
			4.05[a]					
C-Si	2.06	1.88	2.12	1.89	1.71	1.94	1.71	1.90
C-C	1.44	1.46	1.69	1.50	1.33	1.43	1.66	1.54
			1.80[a]	1.50[a]				
C-N	–	–	–	–	0.25	1.39	0.42	1.47
C	3.50	–	3.81	–	3.29	–	3.79	–
			3.85[a]					
N-Si	–	–	–	–	2.88	1.93	2.63	1.84
N-C	–	–	–	–	0.37	1.39	0.62	1.47
N	–	–	–	–	3.25	–	3.25	–

[a]DFT-LDA MD simulations [5]

The distinct Si-C, C-C, Si-N and C-N peaks in the PCF of a-SiCN are clearly detected at distances 1.90, 1.54, 1.84 and 1.47 Å, respectively (cf. Fig. 1). The first coordination peak of the Si-Si pair correlation function is remarkably broad, indicating that Si atoms accommodate at rather different relative positions. Figure 3 shows that Si atoms are mainly substituted by nitrogen in the nearest neighbour surrounding of silicon. The number of C atoms sited around silicon atoms also decreases, but to a lesser extent. Regarding C atoms, we see that the formation of some few C-N bonds occurs mainly owing to breaking C-Si bonds. The C-C coordination remains practically unchanged despite a reduction in the carbon content (cf. Table 1). These findings enable us to suppose that when incorporating nitrogen, N preferentially bonds to Si. As a results, the number of Si and, to laser extent, C atoms around silicon decreases. Thus, the main bonds in a-SiCN are Si-N, Si-C, C-C, and C-N.

These results are consistent with the bonding pictures suggested for amorphous SiCN films [1–4]. In particular, it was established that Si-N bonds are dominant over other bonds in the films deposited by magnetron sputtering [2]. For the CVD films the main bonds are Si-C, Si-N and C-C [1]. For the a-SiCN films deposited by ion-beam sputtering [3] the major part of the bonding is between Si and N and between C and C or N. However the small proportion of Si-C bonds in the films [3, 4] is in contradiction with our results. From the tight-binding MD model of a-$Si_{0.33}C_{0.33}N_{0.33}$ [3] it follows that C-N correlations are dominant over C-C correlations, which is not consistent with our conclusions. As for the rest, both the simulations give more or less similar results.

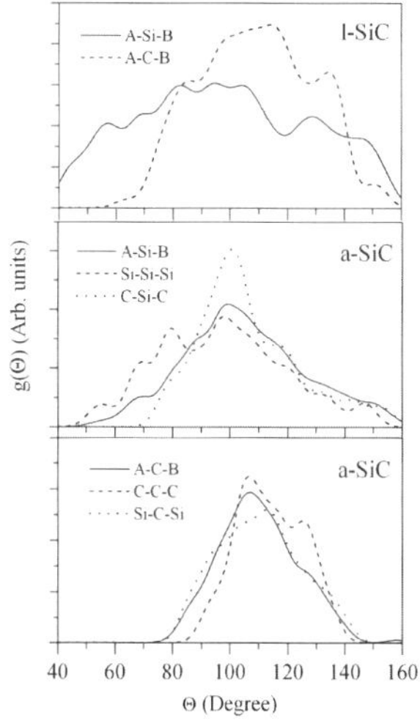

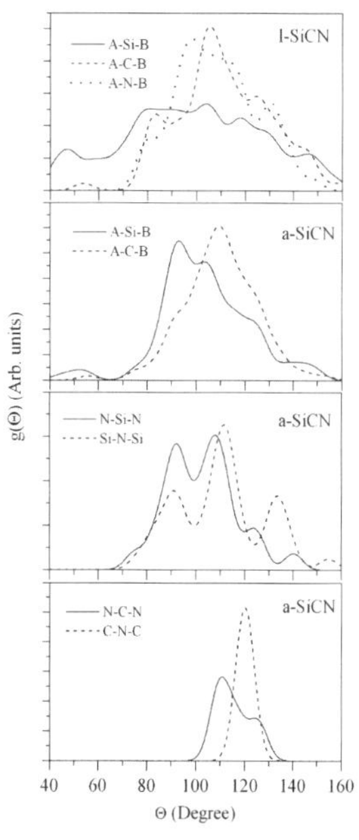

Figure 2. Partial bond-angle distribution functions for the A-Si-B and A-C-B (A, B = Si, C, N) triplets

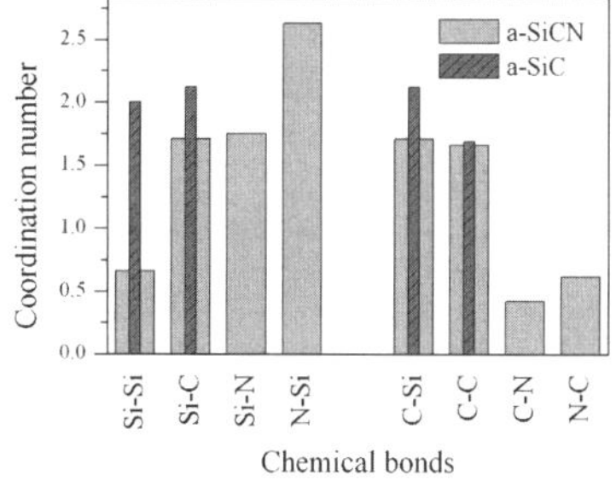

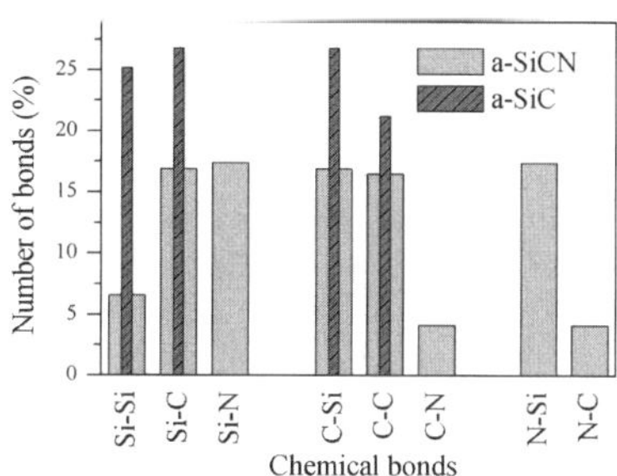

Figure 3. Coordination numbers (left panel) and number of bonds (right panel) of a-SiC and a-SiCN

The tetrahedral character of the a-SiCN network formed by C atoms is confirmed by the bond angle distribution functions g(θ), shown in Fig. 2. There, the main structure falls at $\sim 110^0$. The $g_{A\text{-}Si\text{-}B}(\theta)$ function is characterized by three features around 90^0, 106^0 and 124^0, indicating the presence of threefold, fourfold and fivefold coordinated Si atoms. The three-peak structure of the $g_{Si\text{-}N\text{-}Si}(\theta)$ points to a strongly distorted Si-N network formed by the abnormally coordinated Si atoms. Nevertheless we assign the peaks at $\sim 113^0$ in the $g_{Si\text{-}N\text{-}Si}(\theta)$ and at $\sim 108^0$ in the $g_{N\text{-}Si\text{-}N}(\theta)$ to the distorted fragments like β-Si_3N_4, since these bond angles are close to those in the β-Si_3N_4 phase (112.6–123.7^0 for N and 107.4–112.6^0 for Si) [11]. The peak at 120^0 in the $g_{C\text{-}N\text{-}C}(\theta)$ function (cf. Fig. 2) shows that N atoms in the C-N network are threefold coordinated. A comparison of the N-C-N and C-N-C bond angle distributions shows that N bonds to

C sp^3 and C sp^2 atoms. We revealed that the bond angles and bond lengths for the N sp^2–C sp^3 network are similar to those for the β-C_3N_4 phase ($l_{C\text{-}N}$ ~ 1.43–1.49 Å; θ ~ 112.2–123.9^0 for N and 106.9–112.2^0 for C) [11]. This prompts us to infer that the N sp^2–C sp^3 correlations in a-SiCN are very close to those inherent to the β-C_3N_4 structures.

4. Conclusions

We have performed *ab initio* MD simulations of SiC and SiCN structures in the liquid and amorphous states. Unlike the SiC structures, in the SiCN phases, the Si-Si first neighbour correlations are practically lacking. The liquids are characterized by the strongly distorted over-coordinated Si-Si network without a preference for tetrahedral coordination. Carbon atoms form the network with the predomination of the C sp^2 - (l-SiC) and C sp^3 -, C sp^1 - (l-SiCN) configurations. In the amorphous samples, C atoms are mostly sited in the tetrahedral surrounding. In a-SiC, silicon retains a coordination number close to four, indicating persistence of tetrahedral order. In a-SiCN, Si atoms do not show preference for tetrahedral coordination. Nitrogen incorporation leads to a formation of Si-N and C-N bonds that are inherent to the β-Si_3N_4 and β-C_3N_4 structures. Nitrogen mostly bonds to silicon. The main chemical bonds in a-SiCN are Si-N, Si-C, C-C and C-N.

References

1. Bendeddoche A., Berjoan R., Beche E., Merle-Mejean T., Schamm S., Serin V., Taillades G., Pradel A., Hillel R. (1997) Structural characterization of amorphous SiC_xN_y chemical vapor deposited coatings, *J. Appl. Phys.* **81**, 6147–6154.
2. Vlcek J., Kormunda M., Cizek J., Perina V., Zemek J. (2002) Influence of nitrogen-argon gas mixture on reactive magnetron sputtering of hard Si-C-N films, *Surf. Coat. Technol.* **160**, 74–81.
3. Lehmann G., Hess P., Wu J.-J., Wu C.T., Wong T.S., Chen K.H., Chen L.C., Lee H.-Y. (2001) Structure and elastic properties of amorphous silicon carbon nitride films, *Phys. Rev. B* **64**, 165305–165310.
4. Zhang D.H., Gao Y., Wei J., Mo Z.Q. (2000) Influence of silane partial pressure on the properties of amorphous SiCN films prepared by ECR-CVD, *Thin Solid Films* **377–378**, 607–610.
5. Finocchi F., Galli G., Parrinello M., Bertoni C.M. (1992) Microscopic structure of amorphous covalent alloys probed by *ab initio* molecular dynamics: SiC, *Phys. Rev. Lett.* **68**, 3044–3047.
6. Kelires P.C. (1992) Short-range order and energetics of disordered silicon-carbon alloys, *Phys. Rev. B.* **46**, 10048–10061.
7. Baroni S., Dal Corso A., de Gironcoli S., Giannozzi P., Cavazzoni C., Ballabio G., Scandolo S., Chiarotti G., Focher P., Pasquarello A., Laasonen K., Trave A., Car R., Marzari N., Kokalj A., http://www.pwscf.org/.
8. Scandolo S., Giannozzi P., Cavazzoni C., Gironcoli S., Pasquarello A., Baromi S. (2005) First-principles codes for computational crystallography in the Quatum-ESPRESSO package, *Z. Kristalogr.* **220**, 574–579.
9. Vanderbilt D. (1990) Soft self-consistent pseudopotentials in generalized eigenvalue formalism, *Phys. Rev. B* **41**, 7892–7895.
10. Perdew J.P., Zunger A. (1981) Self-interaction correlation to density-functional approximations for many-electron systems, *Phys. Rev. B* **23**, 5048–5079.
11. Ishumaru M., Bae I.-T., Hirotsu Y., Matsumura S., Sickafus E. (2002) Structural Relaxation of Amorphous Silicon Carbide, *Phys. Rev. Lett.* **89**, 055502–055504.
12. Wang C.-Z., Wang E.-G., Dai Q. (1998) First principles calculations of structural properties of β-$Si_{3-n}C_nN_4$ (n=0,1,2,3), *J. Appl. Phys.* **83**, 1975–197

MODELLING OF DIFFUSION TDS-SPECTRUM PEAK OF DEHYDRIDING WITH SIZE REDUCTION AND HEAT ABSORPTION EFFECTS

YU.V. ZAIKA, N.I. RODCHENKOVA
Institute of Applied Mathematical Research of Karelian Research Centre RAS, Pushkinskaya Str. 11, Petrozavodsk, 185910 Russia

Abstract. In the paper we consider the problem with moving bounds that models the kinetics of hydrogen desorption from hydrides of metals. Phase change, desorption processes on the surface, size reduction, and heat absorption on the phase bound are taken into consideration. Corresponding boundary-value problem with moving bounds is presented for the experimental method of thermal desorption spectrometry (TDS). Computational algorithm based on difference approximations is developed. The results of numerical modeling are presented.

Keywords: boundary-value problem with moving bounds, mathematical modelling, kinetics of hydrogen desorption

1. Introduction

In the context of hydrogen energy problems, materials suitable for hydrogen storage and transporting are being intensively searched for. Experimental research demands that adequate mathematical models be developed. This would help to specify physical-chemical ideas about hydrogen interaction with solids, discover the limiting factors and considerably reduce the expenses on experimental research by means of numerical simulation for different sets of parameters and experimental conditions (including those extremal). We do not make an extensive literature review, but only mention the papers [1–6], which deal with the subject of the paper. The paper is continuation of [5] (but presentation is independent). Mathematical models in the form of systems of ordinary differential equations, when diffusion can be considered relatively fast (TDS-spectrum may have few peaks), are presented in [5]. This essentially simplify the parameter identification.

The aim of this paper is to present the mathematical model of the TDS-dehydriding, and the numerical method for boundary-value problem solution, when metal hydride decomposition, size reduction effect, heat absorption effect, diffusion and desorption are main factors. Physical-chemical details may be precised in reference book [7]. Note that the "abstract" hydride does not exist. There are metallic and salt-like hydrides. For example, instead of the size reduction effect we may observe the volume increase [7]. The hydride heating law is a functional parameter (control). This law can be varied for the purpose of hydrogen flux stabilization (hydride accumulators for hydrogen engines). The paper is written in such a way that it is simply to make necessary changes in model depend upon specific situation. The problem has an applied aspect. Difficulties of solving the inverse problem of parameter identification by using experimental data are taken into account. Hence, the authors use minimal mathematical tools.

B. Baranowski et al. (eds.), *Carbon Nanomaterials in Clean Energy Hydrogen Systems.*

Let us briefly describe the TDS-experiment. A sample of powder metal hydride is placed into the chamber and relatively slowly heated under continuous evacuation. The desorption flux of hydrogen from the sample (measured by a mass spectrometer) gives information about hydrogen interaction with the metal. Equilibrium regularities are sufficiently well studied (PTC-diagrams). Much attention is paid to kinetics of hydriding/dehydriding of metals. We consider the material from [2] as the modelled object. Under heating ErH_3 changes as follows: $ErH_3 \to ErH_2 \to Er$ with corresponding peaks on the TDS-spectrum (diagram of measured hydrogen desorption flux with respect to current temperature). In the further considerations, the material is not important. A specialist decides how to interpret the word "phase" (α and β) in each specific case. For example, this may mean the pairs $ErH_2 \to ErH_3$, $Er \to ErH_2$. The effects of heat transport by diffusing particles, thermal diffusion, mechanical tensions and deformations are considered as secondary effects, i.e. comparable with the experimental errors. Sorption is exothermal while hydride decomposition is endothermal. We suppose that heating of the powder is almost uniform (as, for instance, in [2]). These specifications determine the range of model adequacy.

2. Basic assumptions and notation

Density increases (due to the volume decrease) while model hydride decomposes owing to lattice rearrangement. For some hydrides this effect amounts to several tens of percents. Let the sample volume in the β phase (hydride) be V, while in the α phase (solid solution of hydrogen atoms in the metal) the same sample occupies the volume γV. Here, the size reduction coefficient is denoted by γ, $\gamma < 1$. We suppose that $\gamma = \mathrm{const}$ within the peak of TDS-spectrum $T \in [T^-, T^+]$. We assume that the temperature range $T^+ - T^-$ is rather narrow. The heating rate is relatively small. Denote the radius of a shrinking spherical particle of powder metal hydride by $L(t)$, and the radius of the shrinking hydride core by $\rho(t)$. When some amount of hydride decomposes, its volume decreases. Hence, the decrease of the whole particle volume can be expressed in terms of the amount of decomposed hydride: $(1-\gamma)\left[\rho_0^3 - \rho^3\right] = L_0^3 - L^3$. Here, the common factor $4\pi/3$ has been cancelled, $L_0 = L(0)$, $\rho_0 = \rho(0)$, initial time instant $t = 0$ corresponds to the beginning of the desorption peak. The mechanism of inside reduction results in the tension growth. Powder of metal hydrides becomes finer after a few cycles of hydriding/dehydriding. We consider only fine enough powders, where tensions are not critical. Let $c(t,r)$ be the concentration of corpuscular hydrogen dissolved in the outside metal skin (in α phase) with width $L-\rho$; Q – the concentration of hydrogen in hydride (β phase). Traps are the defects of metal structure, in particular, micropores. Let $z(t,r)$ be the concentration of hydrogen captured by traps; a_1, a_2 – the coefficients of reversible capture; μ – the kinetic constant; p – the pressure of molecular hydrogen in the chamber; s – the adhesion coefficient of hydrogen to the surface. The outside heating of the particle is usually linear, i.e. $T_L(t) \equiv T(t,L) = T_0 + vt$. We consider that hydrogen dissolved in hydride has left the sample before the beginning (remember that $t=0$ corresponds to the beginning of the TDS-spectrum peak). So $Q = c_\beta^{crit} = \text{const}$ is the critical concentration. During further heating, β phase decomposes and desorption becomes faster.

Atoms combine in molecules H_2 and leave the particle. So, we model the desorption flux density with a square dependence on the atoms concentration in the near-to-surface layer (bulk desorption model): $J(t) = b(T)c^2(t,L)$. We also assume that all parameters depend on temperature in the Arrhenius way, in particular, desorption coefficient is $b(T) = b_0 \exp\{-E_b/[RT]\}$, where E_b is the activation energy, $b_0 = const$, $R = const$. We assume that a_i are constant within the peak ($T \in [T^-, T^+]$). The pressure of the molecular hydrogen in the chamber initiates flux density $\mu s(T)p(t)$ of atoms H on the surface (in the near-to-surface bulk). Dehydriding allows to suppose that $\mu sp < bc_L^2$, $c_L \equiv c(t,L)$. Only some of desorbed hydrogen returns if pumping is strong. Pressure $p(t)$ and the desorption density $J(t)$ are dependent: $p(t) = \theta_1 \int_0^t J(\tau)\exp\{(\tau - t)/\theta_0\}\, d\tau$. This is the standard expression in the theory of measuring. If a δ-impulse of hydrogen is made in the chamber, the pressure will trigger off and then will decrease exponentially. The constants θ_i are the characteristics of the vacuum system. If vacuum system is powerful enough then below $\mu sp \approx 0$. Formally, $J = (\dot{p} + p/\theta_0)/\theta_1$, but the measurements of $p(t)$ are noisy. Hence, $J(t)$ is a solution of integral equation of the first kind.

We model a single particle. Taking the particle radii distribution into account, we obtain rather smooth model peaks. The distribution compensates the error of the spherical approximation. Numerically, it is sufficient to consider 5–10 values of L_0 to obtain rather smooth mean curve. The size distribution is considered to be Gaussian.

3. Volume size reduction effect

Both bounds are moving: $\rho = \rho(t)$, $L = L(t)$, $(1-\gamma)\left[\rho_0^3 - \rho^3\right] = L_0^3 - L^3$. We do not make the model complicated, so the outside skin "decreases as a whole". We suppose that the metal structure defects are small ($a_i = 0$). We model the flux density of hydrogen atoms (in the line of $\mathbf{r}_0$ along radius, $|\mathbf{r}_0| = 1$) using expression

$$j(t,r) = -D(T)\frac{\partial c}{\partial r} + c(t,r)v(t,r),$$

where $D(T) = D_0 \exp\{-E_D/[RT]\}$ is diffusion coefficient, $v(t,r)$ is the medium movement velocity coming from the opposite direction ($v < 0$, we define it more exactly later). Taking spherical symmetry into account, the diffusion equation becomes

$$\frac{\partial c}{\partial t}=\frac{1}{r^2}\frac{\partial}{\partial r}\left(r^2\left[D(T)\frac{\partial c}{\partial r}-c(t,r)v(t,r)\right]\right).$$

We specify the expression for function $v(t,r)$. In time instant $t=t_0$ we choose the sphere of radius $r>\rho$ and determine the rate of sphere volume change $V(r)$ during the size reduction. Let $R(t)$ be the radius of the sphere ($R(t_0)=r$), then we get

$$\dot V(r)\equiv\left.\frac{d}{dt}\right|_{t_0} V(R(t))=4\pi r^2\dot R(t_0).$$

We suppose that mechanical tensions are not critical and there is no cracking. Then the following equalities take place $\dot V(r)=\dot V(L)$, i.e. $r^2\dot R=L^2\dot L$. A vanishing volume is balanced by outside layers. Hence, the medium movement velocity along the radius is presented in the form $v(t,r)=L^2\dot L/r^2=(1-\gamma)\rho^2\dot\rho/r^2$, $r\in[\rho,L]$. At that $v(t,\rho)=(1-\gamma)\dot\rho$ is the part of phase bound movement velocity $\dot\rho$.

Let $S(r)$ be the sphere area of radius r, $T_L(t)\equiv T(t,L(t))$. Differentiating the material balance expression on time, we get:

$$QV(\rho)+4\pi\int_{\rho(t)}^{L(t)} r^2c(t,r)dr+\int_0^t S(L)b(T_L)c^2(\tau,L)d\tau=\text{const},$$

$$Q\rho^2\dot\rho+\int_\rho^L r^2c_t(t,r)dr+c(t,L)L^2\dot L-c(t,\rho)\rho^2\dot\rho+b(T_L)c^2(t,L)L^2=0,$$

$$[Q-\gamma c(t,\rho)]\rho^2\dot\rho-\rho^2D(T[t,\rho])\left.\frac{\partial c}{\partial r}\right|_\rho+b(T_L)c^2(t,L)L^2+L^2D(T_L)\left.\frac{\partial c}{\partial r}\right|_L=0.$$

First two items describe the processes on the phase bound $r=\rho$, the third item describes the processes on the surface $r=L$ (we have two groups of items which correspond to different bounds $r=\rho$ and $r=L$). There is no accumulation H on the surface $b(T_L)c^2(t,L)=-D(T_L)c_r(t,L)$. As a result of fluxes balance on the outside bound we get the Stefan condition

$$[Q-\gamma c(t,\rho)]\dot\rho(t)=D(T[t,\rho])\left.\frac{\partial c}{\partial r}\right|_{\rho(t)}.$$

The phase bound movement equation may be independently derived from the balance $Q\delta V-c(t,\rho)\gamma\delta V=-S(\rho)D(T[t,\rho])c_r(t,\rho)\Delta t+o(\Delta t),\quad \delta V=-\Delta V>0.$ Then, the nonlinear boundary condition of the third kind will be a corollary.

4. Simple model of heat absorption

Heating of the metal layer of the particles is faster than diffusion in the context of TDS-method. Hence, we find temperature distribution in the growing metal skin with solution of H in the "quasi-stationary" form $T(t,r) = A^0(t) + B^0(t)/r$, $r \in [\rho, L]$, $B^0(t) < 0$. The heating rate is small: temperature distribution slightly changes in comparison with the uniform distribution. There is a small decrease of T at $r = \rho$ as compared with $r = L$. The bracket in the heat conduction equation $\nu T_t = \lambda(T_{rr} + 2T_r / r)$ is supposed to be equal to zero. Then we formally get "quasi-stationary".

The decomposition is endothermal. We consider the initial variant of hydride decomposition. Suppose that the hydride temperature in powder particle remains the same while hydride decomposes. But the surrounding metal skin changes its temperature. On small time segment $[t, t + \Delta t]$, the hydride spherical layer decomposes. Its volume is equal to $\delta V = [\rho^3 - (\rho + \Delta\rho)^3] 4\pi / 3$, $\Delta\rho < 0$, $\Delta\rho = \rho(t + \Delta t) - \rho(t) \approx \dot{\rho}(t)\Delta t$. The quantity of absorbed heat $\vartheta \delta V$, which is proportional to this volume, is equal to magnitude $P 4\pi\rho^2 \Delta t$ to within $o(\Delta t)$. Here, $P = \lambda T_r = -\lambda B^0 / \rho^2$ is for the heat flow density. Comparing these two expressions and dividing by $\Delta t \to 0$, determine coefficient B^0: $B^0(t) = \kappa \dot{\rho}(t) \rho^2(t)$, $\kappa = \vartheta / \lambda$. Magnitude κ describes heat conductivity properties of metal and decomposition energy. We suppose that κ is positive constant within the peak of TDS-spectrum. Coefficient $A^0(t)$ is determined by controllable temperature $T_L(t)$ on the surface. For temperature in hydride core T_* symmetry condition takes place $T_{*,r}(t,0) = 0$. Model $T_* = A_*^0(t) + B_*^0(t)/r$ $(r > 0)$ coincide with this condition at $B_*^0 = 0$. For variant considered there is $T_* = A_*^0(t) = const$. Thus, hyperbolic distribution $T(t,r)$ with respect to r slightly decreases at $r \to \rho + 0$ as compared with $T(t,L)$, as decomposition is endothermal. The core is heated practically uniformly: $T_* = T_0$. There is a temperature discontinuity on the phase bound owing to large difference in heat conductivity ability (for instance, hydride is salt-like). We have $B^0(t) \to 0$ at $\rho(t) \to 0$. Degassing stage begins and proceeds at uniform heating of the particle.

When necessary one can make the following approach: consider the heating of a hydride core (using simple mathematical tools where possible). The outside heating is supposed to be linear ($T_L(t) \equiv T(t,L) = T_0 + vt$, $T \in [T^-, T^+]$). We will find the solution of the heat conduction equation for the hydride core

$$\nu_* \frac{\partial T_*}{\partial t} = \lambda_* \left(\frac{\partial^2 T_*}{\partial r^2} + \frac{2}{r} \cdot \frac{\partial T_*}{\partial r} \right), \quad \left. \frac{\partial T_*}{\partial r} \right|_{r=+0} = 0, \ \lambda_* \neq \lambda_*(T_*),$$

when $T_*(t,r) = \varphi(r)t + \psi(r)$, $r \in (0, \rho)$. Each spherical layer is heated with a constant rate, as with the surface. We do not use such a model in the skin of α phase because of heat absorption "distorting" impact under the hydride decomposition. The metal has relatively large heat conductivity coefficient λ. So we consider "quasi-stationary" in metal $T(t,r) = A^0(t) + B^0(t)/r$, as above. For functions $\varphi(r)$, $\psi(r)$ we have:

$$\varphi'' + \frac{2}{r}\cdot\varphi' = 0,\ \frac{v_*}{\lambda_*}\cdot\varphi = \psi'' + \frac{2}{r}\cdot\psi',\ \varphi = C_1 + \frac{C_2}{r},$$

$$\psi = \frac{C_1 v_*}{6\lambda_*}\cdot r^2 + \frac{C_2 v_*}{2\lambda_*}\cdot r + C_3 + \frac{C_4}{r}.$$

Taking into account $\varphi'(+0) = \psi'(+0) = 0$, we get moving parabola with velocity $C_1 = v_*$. Its domain is converged on time:

$$T_*(t,r) = v_* t + C_3 + \frac{v_* v_*}{6\lambda_*}\cdot r^2,\quad v_* = C_1,\quad r \in [0, \rho(t)].$$

This approximation agrees with physical ideas. Without the hydride core heating ($v_* = 0$), we have simple variant examined earlier $T_* = \text{const}$.

Let ϑ be the absorbed heat under the volume unit hydride decomposition. Then, to within $o(\Delta t)$: $\Delta\rho \approx \dot{\rho}\Delta t < 0$,

$$\vartheta\left[\rho^3 - (\rho + \Delta\rho)^3\right]\frac{4\pi}{3} = \lambda\left.\frac{\partial T}{\partial r}\right|_{\rho} 4\pi\rho^2\Delta t - \lambda_*\left.\frac{\partial T_*}{\partial r}\right|_{\rho+\Delta\rho} 4\pi(\rho + \Delta\rho)^2\,\Delta t.$$

Dividing on $\Delta t \to 0$, we find equation for $B^0(t)$: $\vartheta\dot{\zeta}(t) = 3\lambda B^0(t) + v_* v_* \zeta(t)$, $\zeta \equiv \rho^3$. Coefficient $A^0(t)$ is determined using $T_L(t)$: $A^0(t) = vt + T_0 - B^0(t)/L(t)$. Note that the number of the necessary model parameters has grown. To make the values v_*, C_3 more precise, one can use conjunction conditions at $t = 0$, which is linear with respect to v_*, C_3: $T_*(0,\rho_0) = T(0,\rho_0)$, $\lambda T_r(0,\rho_0) = \lambda_* T_{*r}(0,\rho_0)$.

5. Local equilibrium and initial data

In the equilibrium situation, concentration of atoms H in the metal is maintained at $\bar{c} = \eta Q < Q$. At high temperature, hydride decomposes sufficiently fast, and on the phase bound the local equilibrium $c(t,\rho) = \bar{c}$ takes place under the condition of diffusion barrier. Function $\varphi(r) = A + B/r$, which is independent of time, satisfies the diffusion equation $c_t = D(t)\left(c_{rr} + 2c_r/r\right)$, where $D(t) \equiv D(T(t))$. Take it as initial distribution $c(0,r)$ in the layer $r \in [\rho_0, L_0]$: as a result of bulk desorption, the initial metal skin grows slowly, and heating is uniform before the intensive hydride decomposition begins ($t = 0$). Thus, the kind of dissolved hydrogen distribution (hyperbolic quasi-stationary) changes slightly with respect to time when the bound moves very slowly at the preliminary stage. Denote $c(0, L_0) = c_0$, when:

$$\varphi(\rho_0)=\bar{c},\quad \varphi(L_0)=c_0<\bar{c}\Rightarrow\varphi(r)=A+\frac{B}{r}=\frac{L_0c_0-\bar{c}\rho_0}{L_0-\rho_0}+L_0\rho_0\cdot\frac{\bar{c}-c_0}{L_0-\rho_0}\cdot\frac{1}{r}.$$

Parameter c_0 is determined from the boundary condition on the surface at $t=0$:

$$b(T_0)c_0^2=-D(T_0)\frac{\partial\varphi}{\partial r}\bigg|_{L_0}=D(T_0)\cdot\frac{\bar{c}-c_0}{L_0-\rho_0}\cdot\frac{\rho_0}{L_0}.$$

This is a square equation with respect to c_0. The roots are positive and have the different signs. Hence, the near-to-surface concentration is determined unambiguously.

Using $\varphi(r)$, from Stefan condition at $t=0$ we find $\dot{\rho}_0$ and initial temperature distribution $T(0,r)$, $r\in[\rho_0,L_0]$ (see item 2 of algorithm in Sect. 9).

Note that initial distribution $\varphi(r)$ in a sufficiently thin metal cover does not practically influence on the model TDS-peak, since almost all hydrogen is located in the hydride core. Initial data is allowed to be arbitrary within reasonable limits.

6. Boundary-value problem of TDS-dehydriding

The defects are small ($z(t,r)=0$, $a_i=0$). We compactly present the model:

$$\frac{\partial c}{\partial t}=\frac{1}{r^2}\frac{\partial}{\partial r}\left(r^2\left[D(T[t,r])\frac{\partial c}{\partial r}-c(t,r)v(t,r)\right]\right),\quad t\in(0,t_*),\quad r\in(\rho(t),L(t)),$$

$$c(0,r)=A+\frac{B}{r}=\frac{L_0c_0-\bar{c}\rho_0}{L_0-\rho_0}+L_0\rho_0\cdot\frac{\bar{c}-c_0}{L_0-\rho_0}\cdot\frac{1}{r},\quad r\in[\rho_0,L_0],\quad c_0\equiv c(0,L_0),$$

$$v(t,r)=(1-\gamma)\frac{\rho^2\dot{\rho}}{r^2}=\frac{L^2\dot{L}}{r^2},\quad [Q-\gamma c(t,\rho)]\dot{\rho}(t)=D(T[t,\rho])\frac{\partial c}{\partial r}\bigg|_{\rho(t)},\quad \rho(0)=\rho_0,$$

$$D(T_L)\frac{\partial c}{\partial r}\bigg|_{L(t)}=-b(T_L)c^2(t,L),\quad c(t,\rho(t))=\bar{c}<Q,\quad T_L(t)\equiv T(t,L),$$

$$T(t,r)=A^0(t)+B^0(t)r^{-1},\quad r\in[\rho(t),L(t)],\quad (1-\gamma)\left[\rho_0^3-\rho^3\right]=L_0^3-L^3,$$

$$A^0(t)+B^0(t)L^{-1}(t)=T_L(t),\quad B^0(t)=\kappa\rho^2(t)\dot{\rho}(t),\quad T_L(t)=T_0+vt.$$

Magnitudes Q, $\bar{c}$, κ, γ are considered as constant within the TDS-peak ($T\in[T^-,T^+]$). Diffusion and desorption coefficients depend on temperature in the Arrhenius way: $D(T)=D_0\exp\{-E_D/[RT]\}$, $b(T)=b_0\exp\{-E_b/[RT]\}$.

After the hydride core has vanished ($\rho(t_*)=0$), we switch to the degassing problem without moving bounds ($r\in[0,L_*]$, $L_*=L(t_*)$), with symmetry condition $c_r(t,0)=0$ and uniform heating of metal particle $T(t,r)=T_L(t)\equiv T(t)$:

$$\frac{\partial c}{\partial t} = D(t)\left(\frac{\partial^2 c}{\partial r^2} + \frac{2}{r}\cdot\frac{\partial c}{\partial r}\right), \quad t \in (t_*, t^*), \; r \in (0, L_*), \quad \varphi(r) = c(t_*, r),$$

$$D(t)\frac{\partial c}{\partial r}\bigg|_{L_*} = -b(t)c^2(t, L_*), \quad \frac{\partial c}{\partial r}\bigg|_{r=+0} = 0, \quad D(t) \equiv D(T(t)), \quad b(t) \equiv b(T(t)).$$

Time instant t^* is defined using degassing condition $J(t^*) = b(t^*)c^2(t^*, L_*) \approx 0$.

Note that in spherical coordinates we have singularity at $r \to +0$. Concerning the hydride decomposition stage, this singularity is insignificant: the volume of the hydride core decreases as ρ^3. We can numerically switch to the metal degassing stage when $\rho < L/10$. This switch does not influence on desorption flux diagram $J(t)$. The experimental errors of determination $J(t)$ keeps within 10–20% at best. We have $c_r / r = (c_r(t,r) - c_r(t,0))/r \approx c_{rr}(t,r)$ at $r << 1$ on degassing stage $t > t_*$.

7. Some generalizations

We have presented the boundary-value problem for α phase above. One can add the diffusion equation and boundary conditions for the hydride core. The parameters, concerning hydride, will be marked with an asterisk. At the same time * is moved up if we also have an index. We model the following dehydriding process. Let Q be the "base" concentration of hydrogen atoms in hydride MeH_σ. This concentration is caused by strong enough chemical bonds (as in salt-like hydrides). The parameter σ is not necessarily a natural number. The concentration may be smaller only during decomposition: the lattice is rearranged and the hydrogen solution is formed in the metal. Hydride with hydrogen concentration Q is an "independent" material and may absorb hydrogen under the corresponding outside pressure. Let $u(t,r)$ be the concentration of the diffusing hydrogen, which is not bonded chemically, $c_*(t,r) = u(t,r) + Q$. We consider that u is sufficiently small, and so diffusion coefficient D_* in MeH_σ (depends on temperature in the Arrhenius way) does not depend on u:

$$\frac{\partial u}{\partial t} = \frac{1}{r^2}\cdot\frac{\partial}{\partial r}\left(r^2 D_*(T_*[t,r])\frac{\partial u}{\partial r}\right) - a_1^* u + a_2^* w,$$

$$\frac{\partial w}{\partial t} = a_1^* u - a_2^* w, \quad t \in (0, t_*), \; r \in (0, \rho(t)).$$

Here $w(t,r)$ is the concentration of atoms H in defects (traps). If necessary, one can consider the limits of their capacity: add the factor $(1 - w / w_{\max})$ to a_1^*. The defects are considered as secondary effects, therefore, suppose that $a_i^* = \text{const}$ within the TDS-peak. More detailed modeling of defect influence is an independent problem. Decomposition and desorption splash begins when temperature reaches a certain value under evacuation conditions. The moving phase bound (different phases correspond to different lattices) is determined by critical concentration $u^{\rm crit} \approx 0$, ($c_*^{\rm crit} \approx Q$). Close to the phase bound concentrations c, u ($D_* < D$) decrease owing to desorption and diffusion in α phase. The lattice MeH_σ becomes unstable, and its rearrangement takes place. The heating of the particle surface is made monotonous and sufficiently slow in order to the range $[T^-, T^+]$ is small and the critical temperature point should not be reached: it is necessary to provide a "smooth" hydrogen consumption. As for the singularity of spherical system at $r = 0$, then during computational modeling stage the differentiation operator should be presented as

$$\frac{\partial}{\partial r}\left(D_*[t,r]\frac{\partial u}{\partial r} \right) + \frac{2}{r} \cdot D_*[t,r]\frac{\partial u}{\partial r}.$$

The second item is approximated by expression $2(D_* u_r)_r$ at $r \approx 0$ using symmetry $u_r(t,0) = 0$. The boundary conditions for the core are: $u(t,\rho) = 0$, $u_r(t,0) = 0$.

Similarly (without *), we write down the diffusion equations with reversible capture in metal at $r \in (\rho(t), L(t))$, $(1-\gamma)[\rho_0^3 - \rho^3] = L_0^3 - L^3$:

$$\frac{\partial c}{\partial t} = \frac{1}{r^2} \cdot \frac{\partial}{\partial r}\left(r^2 [D(T)\frac{\partial c}{\partial r} - c(t,r)v(t,r)] \right) - a_1 c + a_2 z \,,$$

$$\frac{\partial z}{\partial t} = a_1 c - a_2 z \,, \quad z = z(t,r), \; T = T(t,r), \; v(t,r) = (1-\gamma)\dot{\rho}\rho^2 / r^2.$$

Phase bound movement equation (the Stefan condition):

$$[Q - \gamma(c(t,\rho) + z(t,\rho))]\dot{\rho}(t) = D(T_\rho)\frac{\partial c}{\partial r}\bigg|_\rho - D_*(T_\rho^*)\frac{\partial u}{\partial r}\bigg|_\rho ,$$

$$T_\rho \equiv T(t,\rho(t)), \; T_\rho^* \equiv T_*(t,\rho(t)).$$

The diffusion inflow of hydrogen atoms from the hydride core reduces the bound movement and maintains concentration $c_*(t,\rho) = Q$ ($u(t,\rho) = 0$).

The boundary condition on the surface remains the same. The diffusion inflow of H atoms is taken away by molecular desorption. The condition on the phase bound is:

$$k(T_\rho^*)Q[1 - c(t,\rho)/\bar{c}] = -D(T_\rho)\frac{\partial c}{\partial r}\bigg|_{\rho(t)} + D_*(T_\rho^*)\frac{\partial u}{\partial r}\bigg|_{\rho(t)} .$$

Another possible variant is $k = k(T_\rho)$. The left part is interpreted as the model of the hydride decomposition self-control. The highest possible flux density is bounded by kQ. Factor $1 - c/\bar{c}$ shows that the number of unbound sites decreases and the hydride decomposes slowly when concentration $c(t,\rho) < \bar{c}$ grows. When potential hydride decomposition velocity (kQ) is significantly higher than the diffusion abilities, then we have local equilibrium $c(t,\rho) \approx \bar{c}$ and parameter k is eliminated from the model.

The concentrations $\bar{c}$, Q were supposed independent on $T \in [T^-, T^+]$. For instance, the left and right walls of PTC-diagrams are almost vertical for erbium hydride. When necessary, one can find dependence $\bar{c}$, Q on temperature in reference book [7]. We should only consider that additional items are appeared under derivation of equations (when we differentiate the material balance). Further complications of the model are: diffusion coefficient depends on concentration (when diffusion coefficients are relatively large), possible accumulation of atoms H on the surface in compliance with fluxes balance of penetration on the surface and solution to the bulk (we get dynamic boundary conditions in the form of differential equations for surface concentration). If we maintain essential pressure $p(t)$ $\left(\mu s p - b c_L^2 = D c_r(t,L)\right)$, then the temperature is controlled exactly. And instead of setting $T_L(t)$, one can write a heat exchange model (the Newton's model) between gas and metal. But authors does not know real-world problems, which would require such complication of models, and any experimental material, which would allow to hope that multiparametric identification problems may be correctly solved.

8. Transformation of boundary-value problem

As mentioned above, it is possible to neglect by small core of β phase and numerically switch to the degassing stage at $\rho << L$. Calculation accuracy requirement is reduced at the desorption decrease stage. So, the authors choose the following variant: moving and dilative segment $[\rho(t), L(t)]$ is transformed in $[0,1]$. Then the analytical grid is constructed. Consider the change of variables:
$(t,r) \leftrightarrow (t,x)$, $r = \rho(t) + x[L(t) - \rho(t)]$, $u(t,x) = c(t, r(t,x))$.

The boundary-value problem for TDS-dehydriding after change of variables:

$$\frac{\partial u}{\partial t} = \frac{1}{(L-\rho)^2} \cdot \frac{\partial}{\partial x}\left(D(T) \frac{\partial u}{\partial x} \right) + f(t,x) \cdot \frac{\partial u}{\partial x}, \quad t \in (0, t_*), \; x \in (0,1),$$

$$f(t,x) \equiv \left[\dot{\rho}(1-x) + \dot{L}x + \frac{2}{\rho + x(L-\rho)} \cdot D(T) - w(t,x) \right] \cdot \frac{1}{L-\rho},$$

$$w(t,x) \equiv v(t, r(t,x)) = (1-\gamma) \cdot \frac{\rho^2 \dot{\rho}}{(\rho + x(L-\rho))^2} = \frac{L^2 \dot{L}}{(\rho + x(L-\rho))^2},$$

$$u(0,x) = \frac{L_0 u_0 - \bar{u}\rho_0}{L_0 - \rho_0} + L_0 \rho_0 \cdot \frac{\bar{u} - u_0}{L_0 - \rho_0} \cdot \frac{1}{\rho_0 + x(L_0 - \rho_0)}, \quad u_0 \equiv u(0,1), \quad \bar{u} = \bar{c},$$

$$[Q - \gamma u(t,0)]\dot{\rho}(t) = \left(\frac{D(T)}{L-\rho} \cdot \frac{\partial u}{\partial x} \right)\Bigg|_{x=0}, \quad \rho(0) = \rho_0, \quad (1-\gamma)\left[\rho_0^3 - \rho^3\right] = L_0^3 - L^3,$$

$$\left(\frac{D(T)}{L-\rho} \cdot \frac{\partial u}{\partial x} \right)\Bigg|_{x=1} = -b(T_L)u^2(t,1), \quad u(t,0) = \bar{u} < Q, \quad T_L(t) \equiv T(t, L(t)) = T_0 + vt,$$

$$T(t,x)=A^0(t)+B^0(t)/(\rho+x(L-\rho)),\quad A^0+B^0/L=T_L,\quad B^0=\kappa\rho^2\dot{\rho}.$$

Formally, one can leave behind a physical meaning of $\rho(t)$ as the phase bound and $L(t)$ as the particle radius. We consider them as functional parameters.

We generally write down the boundary-value problem in a standard form:

$$\frac{\partial u}{\partial t}=\frac{\partial}{\partial x}\left(\Phi_1\frac{\partial u}{\partial x}\right)+\Phi_2\frac{\partial u}{\partial x},\quad u(0,x)=\psi(x),\quad u(t,0)=\bar{u},$$

$$\left.\left(\Phi_3\frac{\partial u}{\partial x}\right)\right|_{x=1}=-b(T_L)u^2(t,1).$$

But here "coefficients" $\Phi_i=\Phi_i(t,x;u(\cdot))$ are nonlinear functionals with respect to solution u. If function $u(t,x)$ is formally known, we get nonlinear ordinary differential equation for $\rho(t)$: $\dot{\rho}=G(t,\rho,\dot{\rho})$ from the Stefan condition (temperature $T(t,0)$ depends on $\rho(t)$, $\dot{\rho}(t)$). Value ρ_0 is given, and $\dot{\rho}_0$ is computed (see items 1, 2 in section 9). Functions $L(t)$, $f(t,x)$, $T(t,x)$ are unambiguously determined using $\rho(t)$. Hence, algorithm will be based on the implicit scheme and will be iterative.

9. Algorithm of numerical solution

We consider the TDS-dehydriding boundary-value problem in variables (t,x).

1. We determine the initial data $\psi(x)\equiv u(0,x)$, $x\in[0,1]$. For that, we find concentration value $u_0=u(0,1)$, using boundary condition at $t=0$

$$\left.\frac{D(T_L)}{L-\rho}\cdot\frac{\partial u}{\partial x}\right|_{x=1}=-b(T_L)u^2(t,1),\quad T_L(t)=T_0+vt,$$

and initial hyperbolic distribution $\left(c(0,r)=A+B/r\right)$

$$\psi(x)\equiv u(0,x)=\frac{L_0u_0-\bar{u}\rho_0}{L_0-\rho_0}+\frac{\bar{u}-u_0}{L_0-\rho_0}\cdot\frac{L_0\rho_0}{\rho_0+x(L_0-\rho_0)},\quad \bar{u}=\bar{c}<Q.$$

We get square equation with one positive root u_0:

$$u_0^2+\frac{D(T_0)\rho_0}{b(T_0)L_0(L_0-\rho_0)}\cdot u_0-\frac{D(T_0)\rho_0\bar{u}}{b(T_0)L_0(L_0-\rho_0)}=0.$$

2. Calculate initial bound movement velocity $\dot{\rho}_0=\dot{\rho}(0)$ from the Stefan condition

$$[Q-\gamma\bar{u}]\dot{\rho}_0=\frac{D(T(0,0))}{L_0-\rho_0}\cdot\psi'(0),\quad \psi'(0)=-\frac{L_0(\bar{u}-u_0)}{\rho_0}.$$

Here the diffusion coefficient is nonlinear function with respect to $\dot{\rho}_0$. We remember that temperature $T(t,r) = A^0(t) + B^0(t)/r$ in variables (t,x) is

$$T(t,x) = T_L(t) + \kappa\rho^2\dot{\rho}(L-\rho)(1-x)[L(\rho + x(L-\rho))]^{-1}, \quad T_L(t) = T_0 + vt.$$

So, $T(0,0) = T_0 + \kappa\dot{\rho}_0\rho_0(L_0-\rho_0)/L_0 \equiv f_0(\dot{\rho}_0)$. We get equation for $\dot{\rho}_0$, taking into account that diffusion coefficient $D(T)$ depends on temperature in Arrhenius way

$$\dot{\rho}_0 = -\frac{L_0(\bar{u}-u_0)}{(Q-\gamma\bar{u})(L_0-\rho_0)\rho_0} \cdot D_0 \exp\{-E_D/[Rf_0(\dot{\rho}_0)]\}.$$

This is a problem about fixed point $(\dot{\rho}_0 = F_0(\dot{\rho}_0))$. It is efficiently solved by simple iteration method in model parameters range (Sect. 11). Value $\dot{\rho}_0$ determine $T(0,x)$.

3. Calculate $\rho_1 = \rho_0 + \tau\dot{\rho}_0$ and value $L_1 \approx L(\tau)$ from $(1-\gamma)\left[\rho_0^3 - \rho_1^3\right] = L_0^3 - L_1^3$. On segment $r \in [\rho_1, L_1]$ we define temperature $T(\tau,r)$, using in expression $B^0(\tau) = \kappa\dot{\rho}_1\rho_1^2$ initial value $\dot{\rho}_0$ instead of unknown yet $\dot{\rho}_1$. Suppose in variables (t,x)

$$T(\tau,x) = T_L(\tau) + \kappa\rho_1^2\dot{\rho}_0(L_1-\rho_1)(1-x)[L_1(\rho_1 + x(L_1-\rho_1))]^{-1}, \quad x \in [0,1].$$

4. We solve the diffusion problem on layer $t = \tau$ using implicit method and substitute only known $\dot{\rho}_0$ and $\dot{L}_0$ for magnitudes $\dot{\rho}_1$, $\dot{L}_1$ from expression $(1-\gamma)\rho_0^2\dot{\rho}_0 = L_0^2\dot{L}_0$. Approximation order is $O(h^2+\tau)$.

Consider a grid $x_i = ih$, $i = 0,1,\ldots,N$, $t_n = n\tau$, where $h = 1/N$ is the step with respect to x, τ is the step with respect to time. Consider grid approximations $U_i^n \approx u(t_n,x_i)$, $w_i^n \approx w(t_n,x_i)$, $L_n \approx L(t_n)$, $\rho_n \approx \rho(t_n)$, $\dot{\rho}_n \approx \dot{\rho}(t_n)$, $a_i^n = \left(D(T[t_n,x_i]) + D(T[t_n,x_{i-1}])\right)/2$. We have $U_i^0 \approx u(0,x_i)$ from initial data. In order to approximate the diffusion equation with order $O(h^2+\tau)$ in node (x_i,t_n) we use standard four-point pattern $(x_{i\pm1},t_{n+1})$, (x_i,t_n), (x_i,t_{n+1}). Then we get the system, which is solved by the sweep method: $U_i^{n+1} = \alpha_{i+1}^{n+1}U_{i+1}^{n+1} + \beta_{i+1}^{n+1}$, $i = 1,\ldots,N-1$, $n \ge 1$, where α_{i+1}^{n+1}, β_{i+1}^{n+1} are the sweep coefficients. Stability conditions holds true. First, we calculate α_2^{n+1} and β_2^{n+1} using known formulas, and then calculate the remaining coefficients α_i^{n+1} and β_i^{n+1}, $i = 3,\ldots,N$. We use the bound condition at $x = 1$ in order to find U_N^{n+1}. Then calculate

$$U_i^{n+1},\ i = N-1,\ldots,1,\ U_0^{n+1} = \bar{u} = \eta Q < Q.$$

5. Now we know the approximation value of concentration gradient $\widetilde{c}_r \approx c_r(\tau, \rho_1)$. We substitute in the bound movement equation (the Stefan condition) $[Q - \gamma \bar{c}]\dot{\rho}_1 = D(T(\tau, \rho_1))\widetilde{c}_r$ temperature expression $T(\tau, \rho_1)$ as a function of variable $\dot{\rho}_1$ owing to expressions

$$A^0(\tau) + B^0(\tau)/L(\tau) = T_L(\tau), \quad B^0(\tau) = \kappa\dot{\rho}_1\rho_1^2.$$

In variables (t, x) we get the equation for $\dot{\rho}_1$:

$$f_1(\dot{\rho}_1) \equiv T_L(\tau) + \kappa\rho_1\dot{\rho}_1(L_1 - \rho_1)L_1^{-1},$$

$$[Q - \gamma\bar{u}]\dot{\rho}_1 = \frac{D(f_1(\dot{\rho}_1))}{L_1 - \rho_1} \cdot \widetilde{u}_x, \quad \left.\frac{\partial u}{\partial x}\right|_{x=0} \approx \widetilde{u}_x = \frac{-3U_0^1 + 4U_1^1 - U_2^1}{2h}.$$

Problem $\dot{\rho}_1 = F_1(\dot{\rho}_1)$ is solved by iteration method, where $\dot{\rho}_0$ is initial value $(\tau << 1)$.

6. Return to item 3. There we had to use value $\dot{\rho}_0$ instead of $\dot{\rho}_1$. Now we substitute current approximation of $\dot{\rho}_1$ in $B^0(\tau) = \kappa\dot{\rho}_1\rho_1^2$ and go to item 4. It is expedient to make a few iterations of the kind. For rougher calculations one can cancel this item.

7. We go to the next time layer $t = 2\tau$ and work analogously to items 3–6: $\rho_2 = \rho_1 + \tau\dot{\rho}_1$, and so on. Stop criterion is $\rho(t_*) \approx 0 \;\; (\rho < L/10)$.

8. We go to metal degassing stage with dissolved H.

10. Model with surface desorption

There are different arguments in favour of the surface. Among them are the presence of dirt and oxidation films, which can significantly change properties of material. The fact, that experimental data depend on hydride pounding, is also an argument for the significant role of the surface. Let $q(t)$ be the concentration of corpuscular hydrogen on the particle surface. Then we model the desorption flux density with a square dependence on the surface concentration: $J(t) = b(T)q^2(t)$. One more equation is needed, which would connect the surface concentration with the near-to-surface bulk concentration. We assume that a condition of quick solution take place in the range of TDS-peak $T \in [T^-, T^+]$: $c(t, L) = gq(t)$, $g = const$. It can be obtained from the condition that the difference of the solution into the bulk and the return to the surface fluxes is taken away by diffusion: $D(T_L)c_r(t, L) = k^-(T_L)q(t) - k^+(T_L)c(t, L)$, provided that both terms at the right part are significantly greater than the left part and the activation energies of $k^+(T_L)$, $k^-(T_L)$ are almost equal, i.e. the surface is isotropic. Then $g = k^- / k^+ = const$, $T \in [T^-, T^+]$.

To diffusion equation and the Stefan condition on the phase bound we add $c(0,r)=\bar{c}$ and $\dot{\Theta}L^{-2} = \mu s(T_L)p - b(T_L)q^2 - D(T_L)\frac{\partial c}{\partial r}\Big|_L$, $\Theta \equiv qL^2$, $c(t,L) = gq(t)$.

We also can use hyperbolic distribution $c(0,r)=\varphi(r)$.

11. Numerical experiments

We assume that the evacuating system is powerful enough and hence a very small amount of desorbed hydrogen returns to the surface, i.e. $\mu sp \approx 0$. Desorption flux density curves $J(t)=b(T_L(t))c^2(t,L)$ are represented in the figures. The heating is linear. The figures illustrate the impact of different kinetic parameters. Following the TDS-experiment, we use coordinates $\{T,J\}$: $T=T_L(t)\leftrightarrow t$. Parameter values correspond to the curves in the following way: the first value is for the curve with the highest maximum, the last is for that with the lowest maximum. We define $T_0=400\text{K}$, $Q=4\cdot10^{22}\,\text{cm}^{-3}$. A parameters, which is common for all curves, are in the table:

$b_0 = 7\cdot10^{-10}$ c	$E_b = 145\text{KJ}$	$\dot{T} = 0.3$	$L_0 = 5\cdot10^{-3}\,\text{cm}$	$\gamma = 0.75$
$D_0 = 8\cdot10^{-3}$ c	$E_D = 50\text{KJ}$	$\eta = 0.15$	$\rho_0 = 0.96L_0$	$\kappa = 10^5$

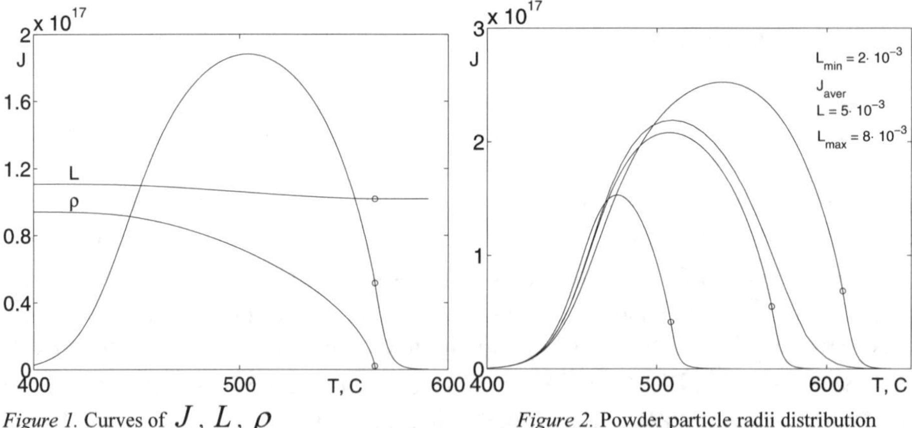

Figure 1. Curves of J, L, ρ

Figure 2. Powder particle radii distribution

The tiny circle shows the finishing time of the hydride decomposition. The particle radii size distribution is considered to be Gaussian: the mean radius of the particle is $\bar{L}=5\cdot10^{-3}\,\text{cm}$, $\sigma=10^{-3}\,\text{cm}$. On numerical modeling stage we use a discrete set of 10 radii values. In Fig. 1 one can see a desorption flux density curve J with respect to current temperature $T=T_L(t)$ together with the scaled curves of the particle radius

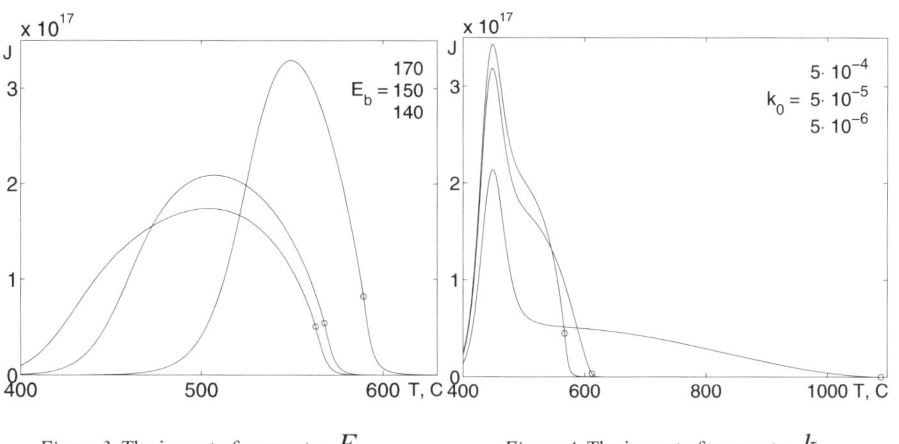

Figure 3. The impact of parameter E_b

Figure 4. The impact of parameter k_0

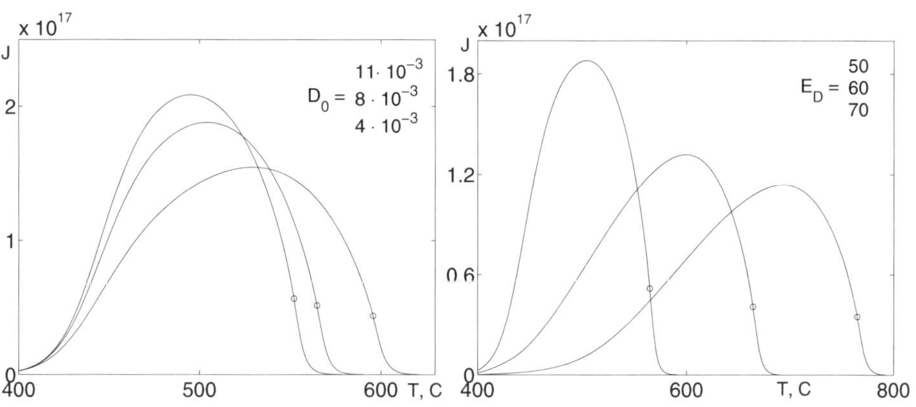

Figure 5. The impact of parameter D_0

Figure 6. The impact of parameter E_D

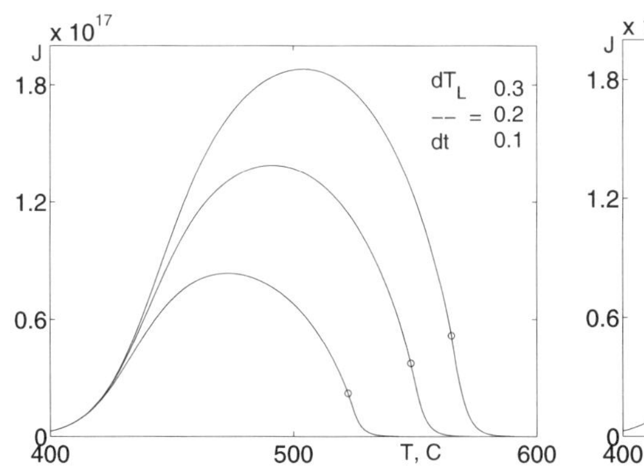

Figure 7. The impact of heating rate

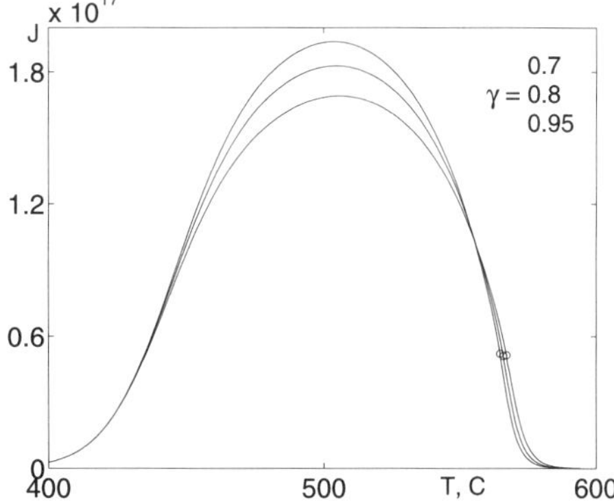

Figure 8. The impact of coefficient γ

$L \cdot \max J / [1.5 \max L]$, hydride core radius $\rho \cdot \max J / [2 \max \rho]$. Curves $J(T)$ can be seen in Fig. 2 when $E_b = 150$. The curves differ from each other: $L = L_{\min} = 2 \cdot 10^{-3}$ (the highest maximum); with powder particle radii size distribution (J_{aver}); $L = 5 \cdot 10^{-3}$ (mean radius); $L = L_{\max} = 8 \cdot 10^{-3}$ (the lowest maximum). Figure 3 shows that parameter E_b influence on flux dynamic, curve shape is changed. Changes are: $E_b = 170;\ 150;\ 140$ ($L = 5 \cdot 10^{-3}$). Figure 4 illustrates the influence of parameter k_0. The smaller the hydride decomposition density, the lower the desorption density curve. At the same time, this curve moves to the right and become wider. The slower diffusion (Figs. 5, 6, $\kappa = 10^6$), the lower the curve and it moves to the right. The influence of heating rate is illustrated in Fig. 7. The influence of size reduction coefficient on the desorption flux density curve is illustrated in Fig. 8.

Thus, parameter variations agree with physical ideas, and the model may be used for kinetic parameters identification of dehydriding of a certain material.

Acknowledgements

This work was supported by grants from the Basic Research program, Branch of Mathematical Sciences, Russian Academy of Sciences (Program 3), and by Russian Science Support Foundation.

References

1. Castro F.J., Meyer G. (2002) Thermal desorption spectroscopy (TDS) method for hydrogen desorption characterization (I): theoretical aspects, *J. Alloy. Compd.*, V. **330–332**, P. 59–63.
2. Gabis I., Evard E., Voit A., Chernov I., Zaika Yu. (2003) Kinetics of decomposition of erbium hydride, *J. Alloy. Compd.*, V. **356–357**, 353–357.
3. Zaika Yu.V., Chernov I.A. Modelling of TDS-spectra of dehydrating. In: Hydrogen Materials Science and Chemistry of Carbon Nanomaterials, NATO Science Series, Netherlands: Kluwer, 2004, **172**, 415–426.
4. Gabis I., Voit A., Evard E., Zaika Yu., Chernov I., Yartys V. (2005) Kinetics of hydrogen desorption from powders of hydrides of metals, *J. Alloy. Compd.* **404–406**, 312–316.
5. Zaika Yu.V., Rodchenkova N.I. (2006) Modelling high-temperature peak of TDS-spectra of dehydrogenation, *Mathematical modelling*, V. **18**, No. 4, P. 100–112 (in Russian).
6. Zaika Yu.V., Rodchenkova N.I. TDS-spectra of hydride powder decomposition: modelling with size reduction effect. In: T.N. Veziroglu et al., editors. Hydrogen Materials Science and Chemistry of Carbon Nanomaterials, NATO Science Series, Netherlands: Springer, 2007, P. 619–631.
7. Kolachev B.A., Iljin A.A., Lavrenko V.A., Levinskij Yu.V. (1992) Hydride systems: book of reference – Moscow: Metallurgiya, 1992.

PHYSICOCHEMICAL BASICS OF BIOSYSTEMS ORGANIZATION AND DEVELOPMENT

L.V. BOBUKH, I.E. DZHEBYAN, T.N. SIVOLAP, K.A. BOBUKH
National Metallurgical Academy of Ukraine
4 Gagarin Avenue, Dnepropetrovsk, 49600 Ukraine
E-mail: luce@i.ua

Abstarct. Energy-elemental principles of organization and functioning of living H, p-elemental systems out of a biotic environment are scientifically based. Basics of an influence of changing factors of environment (C elements, E energy, S configuration of a way, V speed) on processes of organization and development of different forms of living organisms during several geological epochs are determined and graphically expressed. Creation of artificial biosystems as conjugation of elements and energies being task-oriented (on nanolevel) by intelligent natural biosystem (human being) is determined.

Keywords: biosystem, energy, element, state diagram, kinetics, self-organization, functioning, environment, nanotechnology

1. Introduction

The main purpose of modern science is cognition of a matter of life organization and developing. Elaborating main theses of 'Theory of Energy-Elemental Systems' authors marked common (for physical and biological systems) constituents, which aggregate defines structure and possessing properties of systems:

- Qualitative as well as quantitative assortments of the elements (C)
- Qualitative as well as quantitative assortments of the energies (E)
- Energy-elemental way configuration (S)
- Speed of change of energy-elemental state (V)

2. Results and discussion

2.1. ENERGY-ELEMENTAL SYSTEMS (PHYSICAL, BIOLOGICAL): THE FUNCTION OF ENERGY-ELEMENTAL STATE OF SYSTEMS AND ITS CHANGE

The objects of research (physical, biological) are considered to be energy-elemental systems $\left(\begin{smallmatrix} e_1, & e_2, & \ldots & e_n \\ c_1, & c_2, & \ldots & c_n \end{smallmatrix}\right)$ and the coupling of qualitative as well as quantitative assortments of the elements $c_1, c_2 \ldots c_n = C$ and the energies $e_1, e_2 \ldots e_n = E$. By way of the characteristics of energy-elemental state the function of the system's state f(E,C) and its change Δf(E,C) are elaborated. That sort of energy-elemental approach is being based on three fundamental laws, namely, mass conservation law, energy conservation law, matter motion law. At the same time it allows to present these laws not only as unified mathematical equation of speed transition from one energy-elemental state into another $V=\frac{\Delta f(E,C)}{\tau}$ (1) but also as graphical expressions (Figs. 1–9).

A scheme of the arrangement of atoms of hydrogen elements H, elements p, s, d, f as against each other according to the quantity of energy-elemental function f (E, C) inherent in the concrete element is represented in the Fig. 1 [1, 2]. The complex examination of the quantities of atomic mass, atomic radii, electro-negativity of elements and the type of the chemical link formed by them in systems, the temperature of melting one-element systems is the fundamental basis of the proposed order of elements (taking into consideration elements' position in periods and groups). Based on systematization of elements behaviour (by Mendeleyev) that complex analysis allows to develop a systematization of the energy-elemental systems, e.g. to express the order of formation built upon H, p, s, d, f elements, systems that have various aggregative states under natural energy conditions.

Besides, that complex analysis allows pointing to the different energy-elemental inertance of systems. The lowest values of the quantity Δf (E, C) reside to the systems of H, p elements. Thus, energy-elemental system (gas mixture H_2, Cl_2) holds low values of energy-elemental inertance. The change of a state function Δf (E, C) (chemical compound of HCl formation) occurs under 'soft' energy constituent (light quantum) against a background of relatively low temperatures ~25°C:

$$H_2 \quad + \quad Cl_2 \xrightarrow{h\nu} 2HCl$$

(H-element) (p-element)

As for the systems on *d*-elements basis, these ones hold higher values of energy-elemental inertance. The changing of a state function Δf (E, C) (chemical compound Fe_3C formation) occurs under 'hard' energy constituent (temperature) against a background of relatively high temperatures ~1,650°C:

$$3Fe \quad + \quad C \xrightarrow{t^0} Fe_3C$$

(d-element) (p-element)

By way of changing order of complex behaviour of separate H, p, s, d, f elements against each other, the order of behaviour change (in particular, aggregative state) of systems against each other is represented (Fig. 1). The analysis of given regularities enables one mark the elements that form systems with the lowest values, such as f (E, C) and Δf (E, C): the element H as well as p elements of periods 2, 3, groups IV–VI (H, C, O, N, P, S, Si – biogenic elements). Systems based on biogenic elements are capable of changing their energy-elemental state under the natural changing of characteristics of the environment and taken in as biosystems.

Physical systems on the base of s, d, f elements that hold high values, such as f (E, C) and Δf (E, C) conserve their energy-elemental state relatively steady under the natural change of environmental characteristics. As a rule, the change of energy-elemental state of physical systems is attained by artificial way and realized under technological processes of manufacture.

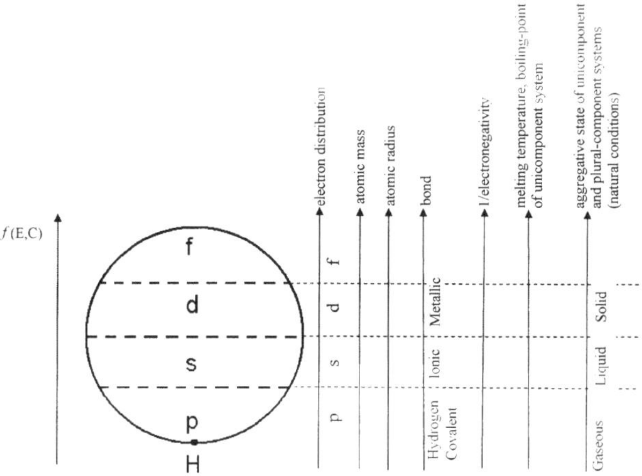

Figure 1. A scheme of the qualitative values of energy-elemental function f(E,C) of systems formed by different elements

2.2. DIAGRAMS OF ENERGY-ELEMENTAL STATE OF SYSTEMS

The diagrams that are practiced in physical metallurgy (Fig. 2) are based on the principle of formation and use that have been illustrated in work [3] for the state description in which the physical system is being found at the moment.

For open systems: $f(E, C) \neq \text{const}$; $\Delta f(E, C) \neq 0$; τ and V have concrete values. It is convenient to carry out the description of continuous transition of the open system (biosystem) from one state into another using diagram which represents 'a block' out of many diagrams and being guided by $\Delta\tau$ duration.

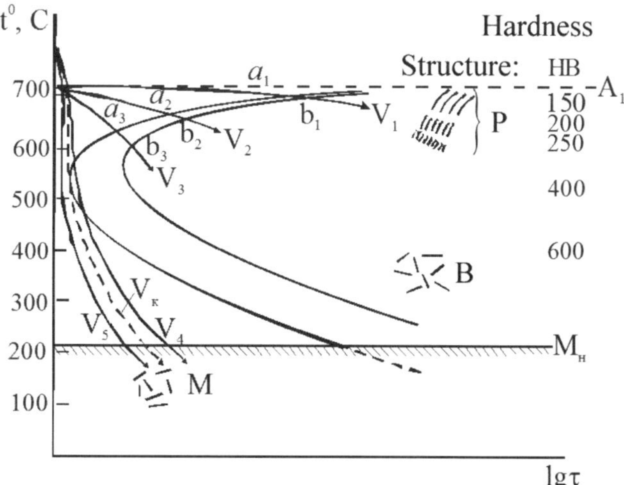

Figure 2. A superposition to a diagram of isothermal decay of austenite of cooling curves, schemes of structures and approximate solidity of those [3]. P – pearlite, B – bainite, M – martensite

A solid model represented in the Fig. 3 is worked up by the authors [4]; within its bounds (limited capacity) the state of biosystem's functioning is provided, its life as *f*(E; C).

The highest degree of stability of the biosystem (a degree of proximity to the equilibrium state) resides to those points, which are positioned on the line of equilibrium A-A (Fig. 3). The least degree of stability of the biosystem (the greatest remoteness from the line of equilibrium A-A) corresponds to the points, represented within the capacity which lays between the dotted and the firm surfaces of the diagram.

Changes in the position of some point that describes the equilibrium of the system during a certain period of time $\Delta\tau = \tau_3 \div \tau_c$ is represented (Fig. 3) on the cross-section plane that is perpendicular to the coordinate plane E-C of the kinetic diagram. If the biosystem is not effected by 'exiting' factors of external environment (critical quantities $C_{кр}$, $E_{кр}$), then the period of the stable biosystem in equilibrium (lifetime) may be rather long $\Delta\tau = \tau_3 \div \tau_c$. If diversion (line 1) from equilibrium has been observed in the biosystem under the effect of external factors (C and/or E changes), it means that the biosystem has been twice (b points) in the non-equilibrium state (illness), but quantity $\Delta\tau$ remained the same: $\Delta\tau = \tau_3 \div \tau_c$. If the biosystem, under the effect of changes in external factors $C_{кр}$ and/or $E_{кр}$ has revealed processes of diversion (line 2) from equilibrium (the system state is described by the point *n*, that is located outside the diagram), the biosystem has stopped its functioning for rather a short period of time $\Delta\tau' = \tau_3 \div \tau'_C < \Delta\tau$.

The introduction and the use of the energy-elemental function concept permitted to represent a graphical change of a state of the separate biosystem (Fig. 4).

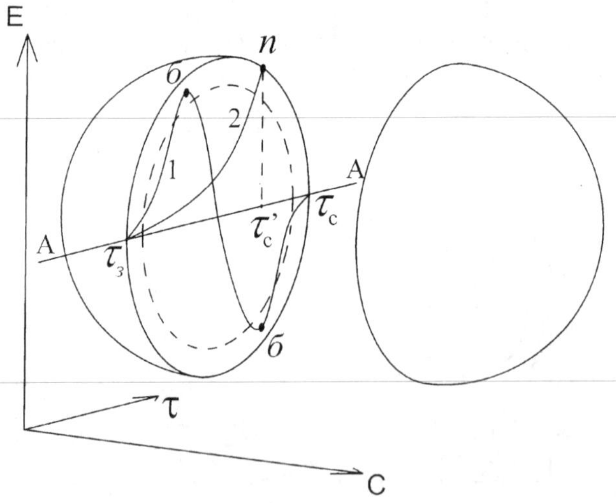

Figure 3. Energy-elemental model of biosystem; *f* (E, C) – function of energy-elemental state; τ – time of biosystem's existence from its origin (з) to death (c); A-A – line of energy-elemental equilibrium of biosystem

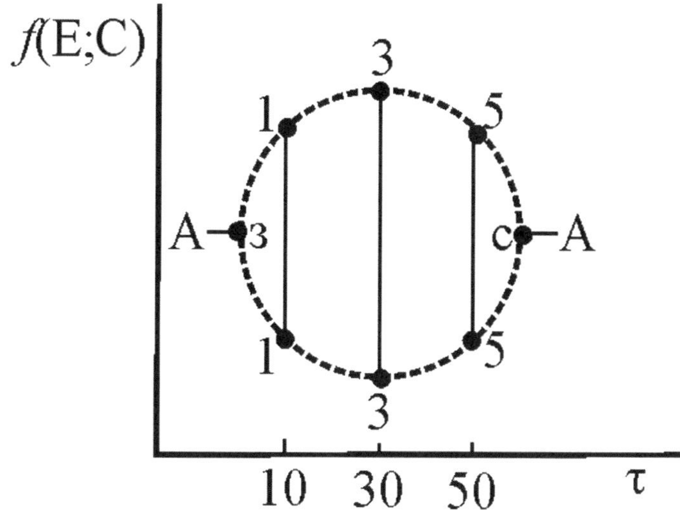

Figure 4. A nature of the threshold valuation change of the chemical elements C and the quantity E provides with the functioning of biosystem using the following duration $\Delta\tau$: $\tau 3$ – conception, $\tau 10$ – tenth anniversary, $\tau 30$ – thirtieth anniversary, $\tau 50$ – fiftieth anniversary; τc – death

2.3. KINETICS OF ENERGY-ELEMENTAL SYSTEMS. A GRAPH OF LE CHATELIER-BROWN PRINCIPLE

The state function $f(E, C)$ enables one to characterize an effect on behaviour of a system, separate elements, energies as well as their unified change. Namely, the unified change of elemental and energy state $f(E, C)$ occurs under materially existing systems. The introduction of the concept of a state function $f(E, C)$ and its change $\Delta f(E, C)$ for real energy-elemental system's description enables one to represent Le Chatelier-Brown principle in a graphical expression (Fig. 5) [1, 5]. The outer line A (the driving one) and the inner B (the driven one) are real, open, communicating, exchanging energies and elements systems that are being examined. If upon the system B being in equilibrium state the external energy-elemental influence is exerted, a 'mirror-like' change will occur in it against the external influence. If the case is that configuration of the energy-elemental way of the driving system A happens to be complicated (spiral within a spiral, for example), the driven system B according to Le Chatelier-Brown principle will pass a 'mirror-like' energy-elemental way of a similar configuration (Fig. 6).

The driving system A, holding some speed of energy-elemental function change, assigns proportional speed of energy-elemental function to the driven system B:

$$\frac{\Delta f_A(E,C)}{\tau} = \frac{\Delta f_B(E,C)}{\tau}; \qquad V_A = V_B$$

When the principle of speed proportionality of communicated systems is disrupted, the systems become segregated. The segregation can be caused by change which is accumulated within internal system B and leads to its energy-elemental change of inertance. Otherwise, such sudden overfalls of the energy-elemental state's speed change of the external system A which the internal system B is not capable of reiterating (not enough speed to plane energy-elemental state).

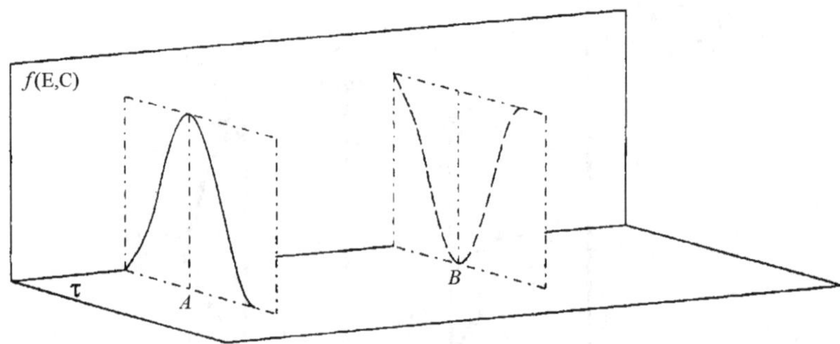

Figure 5. A graph of influence of the energy-elemental function change *f*(E, C) of the driving system *A* (environment) on the energy-elemental function change of driven system *B*. Le Chatelier-Brown principle graph (a simple one)

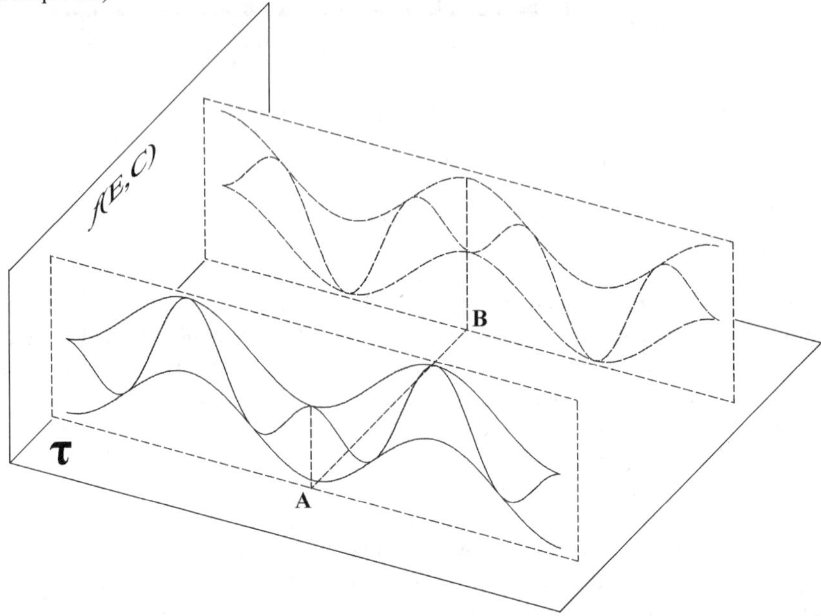

Figure 6. Chatelier-Brown principle graph (a complicated one). A is a driving energy-elemental system (environment); B is a driven energy-elemental system (biosystem)

2.4. PRINCIPLES OF ORGANIZATION AND DEVELOPMENT OF BIOSYSTEMS (NANO, MICRO, MACRO LEVEL)

Within the framework of A-C sections fundamental constituents are designated, the aggregate of which defines the system's behaviour change:

- Qualitative and quantitative elements assortment (C)
- Qualitative and quantitative energy assortment (E)
- Configuration of the energy-elemental way (S)
- Speed of the energy-elemental state change (V)

We are going to examine the importance of each of these constituents in organization and development of biosystems.

Qualitative and quantitative elements assortment of biosystems is established from nano to macro level (Fig. 7). Fundamental chemical elements that organize DNA are: H, C, O, N and P. Taking into consideration only these elements (100%), the authors have made a fresh account and expressed approximate content of elements for biosystems (such as land animals, macro level), atomic mass%: 54.66 H; 30.26 C; 9.07 O; 5.57 N; 0.42 P. Liberal share of atoms of elements that organize the system is due to hydrogen and carbon. H, C, O, N, P (H, p-elements of 2–3 periods of IV–VI groups) are among elements that constitute systems with the low values of energy-elemental function $f(E,C)$ and its change $\Delta f(E,C)$ (Fig. 1) [1, 2].

Qualitative and quantitative energies assortment under the influence of which the organization of H, p elements of biosystem is going on, assigned by energies assortment of natural land conditions (gravitation; magnetic, electric fields of Earth; thermal, light and other electromagnetic radiations; several other influences that refer to the energy ones). Spiral changing of the energy constituent is the determinative factor that initiate spiral form of H, p-elements arrangement (nano level) of biosystem (DNA) and also spiral form of energy-elemental state change which refer to the developing (micro, macro level) biosystem (cell, organism) (Fig. 7).

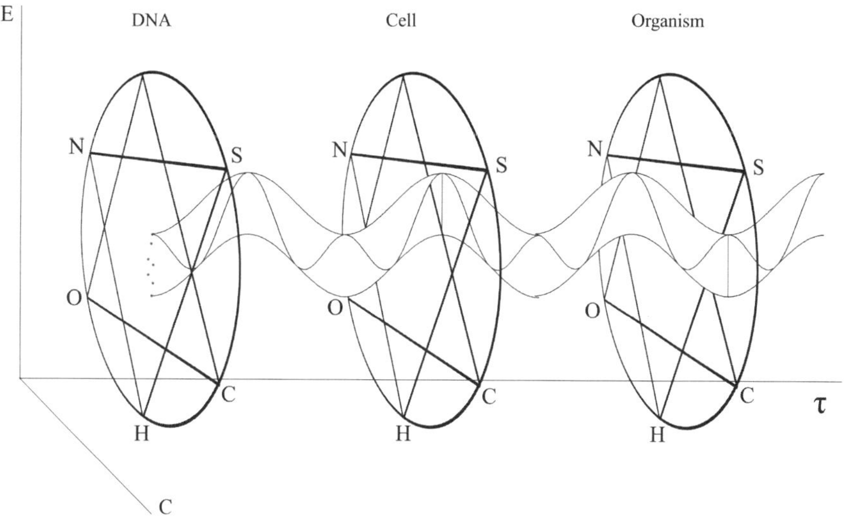

Figure 7. A mechanism of organization and change (nano, micro, macro level) of the state of the matter (biosystem)

Configuration of the way (spiral within a spiral) along which the energy-elemental state of the biosystem is changing, defined by Earth revolving on its axis as well as round the Sun [1].

Speed of the energy-elemental state change of biosystems (as well as physical systems (Fig. 2)) is due to the factor which influence on biosystems' behaviour (equation 1, Figs. 5, 6). The speed of energy-elemental state of a system change must be proportionate to the speed of energy-elemental state of environment change. Only if that is the case, both environment and biosystem are taken as communicable (elements and energies exchanging) systems and the principle of continuous influence of environment upon organizational and developing processes of biosystems are provided.

If the change of each constituent (C, E, S, V) or their complex in environment occurs in spurts (not in helical fashion), and biosystem (by virtue of its inertance) is not capable of equalizing its energy-elemental equilibrium, the connection (environment ↔ biosystem) will be broken and biosystem will cease its functioning.

General provisions of energy-elemental systems theory [4–6] are used for graphical expression (Figs. 8, 9) and analysis of such processes as organization, developing and biosystems evolution.

Initial presuppositions:

(a) Qualitative elements assortment that Earth atmosphere was consisted of: H, C, O, N. Nowadays most of scientist agree in views that nearly 4 billion years ago in Earth atmosphere methane (CH_4) was prevailing against a background of probable presence of hydrogen (H_2), ammonia (NH_3), water vapor (H_2O) and carbonic acid (CO_2).

Virtually, as soon as the earth has begun to exist the geological record of the planet was started. It is possible to learn about it owing to archaic sediments. The results of vital functions methanotrophic bacteria are already displayed within these strata.

(b) Qualitative and quantitative energies assortment of environment is being limited by examination of thermal and ultra-violet constituent.

Based on longstanding data, a most important conclusion that can come to nowadays is that evolution of life was formed of irreversible processes of the Earth planet getting colder: That process that began 4 billion years ago.

Ultra-violet radiation from the Sun had been far more intense at that time than at the present day as there was no protective ozone layer in its way.

(c) Configuration of the energy-elemental way.

Energy-elemental state of environment is measured in some ideal theoretical way ‘spiral within a spiral’.

The influence of the Moon on the configuration of external spiral as well as the influence of planets of the Solar system, other star systems on the configuration of external spiral is not being examined.

(d) Speed of the energy-elemental state change.

When the Earth was formed (nearly 4 billion years ago), a time required for one turn of revolving on its axis averaged 6 hours. 1 billion years ago the earth year averaged 540 astronomical days rather than 365 astronomical days usual to us. Astronomic day averaged 15 hours. Five hundred million years ago the duration of astronomical day was 20.8 hours contrary to our current 24. Speed of our planet revolving on its axis is being reduced.

Using kinetic regularities of the energy-elemental system theory, taking into consideration (a)–(d) presuppositions, graphs (Figs. 8, 9) of energy-elemental function of the environment change influence f (E,C) on organization and development of biosystems processes (nano, micro, macro level) are worked out here.

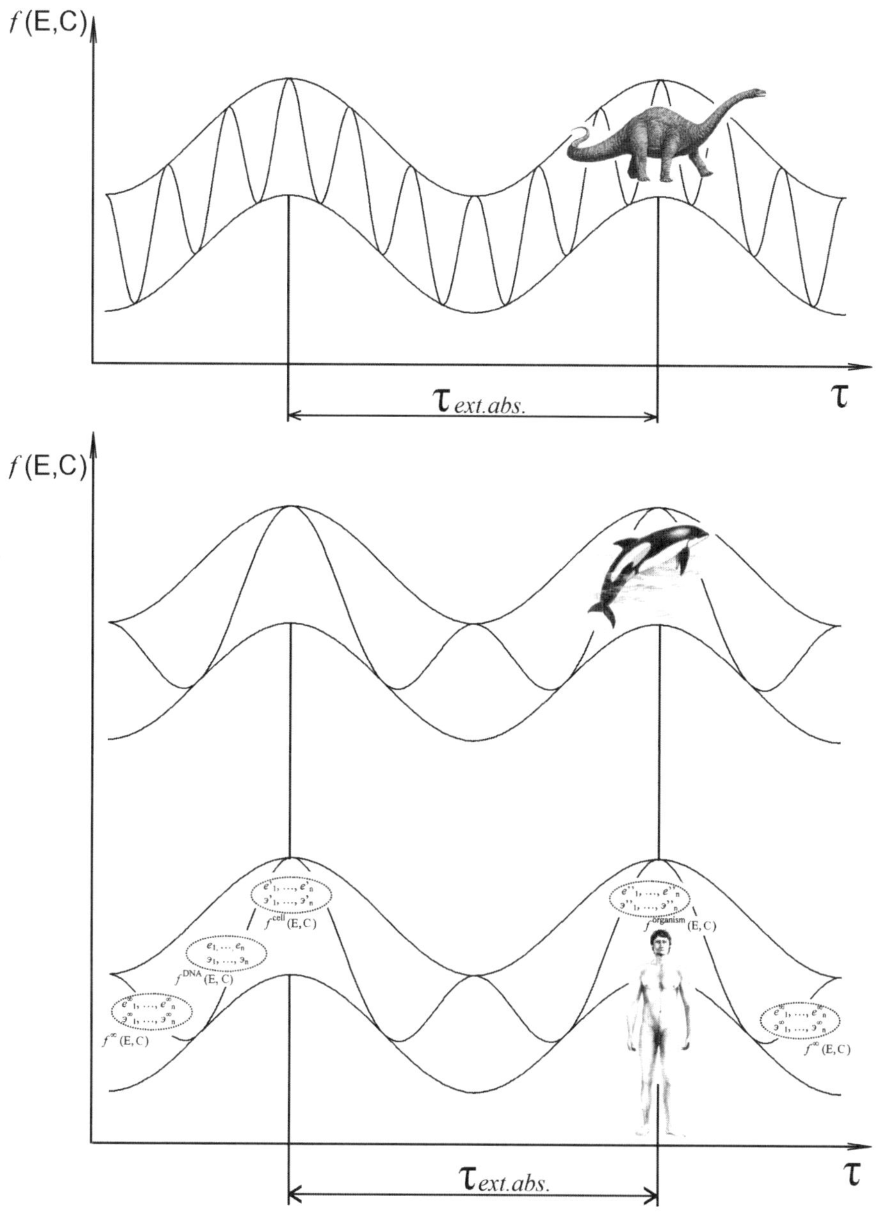

Figure 8. Organization and development of biosystems (DNA→ cell→ organism) conditioned by energy-elemental state of environmental function change Δf(E,C). $\tau_{ext.abs.}$ is a duration being examined by an exterior observer

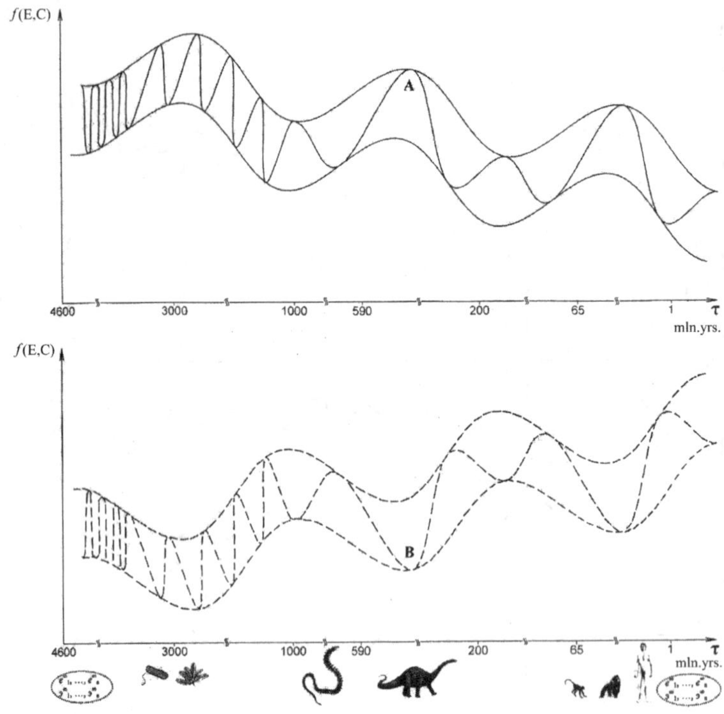

Figure 9. Energy-elemental state *f*(E,C) function change of planet Earth (A) and biosystems (B)

The basis of qualitative elements assortment which is taking part in organization and development of biosystems both 4 billion years ago and at present time put together such elements (C) as: H, C, O, N, P. Quantitative assortment of elements in environment and its aggregative state can be different. Basin accounting: the base is H_2O_{liq} ; quantity of dissolved N_2 is up to 23.3 g, O_2 is up to 70 g in 1 m^3 H_2O under 0°C and 1 atom. Environment accounting (% mixture acc. to capacity): 78 N_2 ; 21 O_2 ; $2 \cdot 10^{-5}$ - 3 H_2O_{gas}. Biosystems that have different values of energy-elemental function *f*(E,C) and thus, different behaviours are developing in environments that have different quantitative elements and energies assortment (Fig. 8). Aquatic environment (relatively high values of *f*(E,C)) is forming and providing the development of biosystem-of-a-dolphin whereas air provides us habitat-of-a-man. The environment (air) with similar levels of values *f*(E,C) but with different speed change Δ *f*(E,C) is forming and providing the development of biosystems with different values as well (dinosaur, human).

The energy level (E is thermal, ultra-violet constituents) which has decreased for 4 billion years which in a graph (Fig. 9) is expressed by decrease of a general level *f* (E,C) in time. Graphically was also taken into account the decrease of speed of the Earth revolving on its axis (internal spiral). Complex influence of factors change (C, E, S, V) on processes of organization and development of biosystems using the duration from 4 billion years ago till now is represented here. Under general decrease *f* (E,C) of environment A, the value of *f* (E,C) level of biosystems B is being increased.

3. Conclusions

1. By way of characteristics of energy-elemental state, a function of a state of the system f(E,C) and its change Δf(E,C) is worked out.
2. A scheme of comparison of qualitative values of energy-elemental function f(E, C) for systems formed by different assortments of elements C (p, s, d, f) and energies E is represented (Fig. 1).
3. Mathematical equation (1) of speed change of energy-elemental state of a system is worked out.
4. Kinetic diagrams of a change of energy-elemental state of physical and biological (Figs. 3, 4) systems are worked out with application f(E, C).
5. A qualitative graphical expression (with application f(E, C)) of Le Chatelier principle is proposed (Figs. 5, 6).
6. A possible influence of changing factors of environment (C – elements, E – energy, S – configuration of a way, V – speed) on processes of organization and development of different forms of living organisms during several geological epochs is graphically expressed (Figs. 7–9). That permitted us to base on science energy-elemental principles of DNA organization of environment and to describe basic mechanisms of development living system (DNA → cell → organism) out of a biotic environment.
7. Life origin and life on its own is being explained not as spontaneous, or accidental occurrence, but as a process of energy-elemental system transition from one state into another, which is regularly organized and being developed by the Universe in time.
8. Regulations developed can become fundamental (chemistry, physics, mathematics) theoretical base:
   - Cognition and description of nanoprocesses in physical and biological systems
   - Learning of wildlife objects through properties and rules peculiar to nanolevel of energy-elemental systems on H, p-elements basis under influence of changing energy-elemental state of environment
   - Ecology as a science due to definition of influence of energy-elemental states of environment (physical system) on biosystem internal energy-elemental state
   - Proving (mathematically and graphically) the integrity of the theory of biosystems' evolution, which includes all levels connected to planetary processes of evolution of the Earth – from energy-elemental (problems of life origin, i.e. nanolevel) up to ecosystems and biosphere
   - Determining creation of artificial biosystems as conjugation of elements and energies being task-oriented (on nanolevel) by intelligent natural biosystem (human being)

References

1. Bobukh L.V., Bobukh K.A., Belskaya E.N., Kikot V.S. Physical and chemical fundamentals of energy-element system state changing. IX International Conference "Hydrogen Materials Science. Chemistry of Carbon Nanomatetials". 5–11 September 2005, Sevastopol, Crimea, Ukraine, P. 748–751.
2. Bobukh L.V., Bobukh K.A. Dzhebyan I.E. Fiziko-chimicheskiye zakonomernosti organizacii i razvitiya biosistem. Herald of Dnepropetrovsk National University. 2006. N9/2. P. 12–17.
3. Gulyaev A.P. Materialovedenie. Moscow, Metallurgy. 1969. P. 541.
4. Bobukh L.V., Bobukh K.A. The common regularities and kinetic diagram of the change of states of living and lifeless matter, *Dopovidi NAN Ukrainy*, 2001. **9**: 127–131 (in Russian).

5. Bobukh L.V., Bobukh K.A., Bobukh T.A. The development of theoretical basis of processes of change of energetic state of physical and biological systems, *Inzhenernaya ekologiya*, M. 2004. **3**: 56–60 (in Russian).
6. Bobukh L., Bobukh K. Development of the physicochemical basis for materials science. Hydrogen materials science and chemistry of metals hydrides. *NATO Science Series. II Mathematics, Physics and Chemistry*, 2002. **82**: 171–177.

CLADDING OF IRON POWDER WITH THE AIM OF PREPARING OF SOFT MAGNETIC MATERIALS WITH IMPROVED PROPERTIES

O.A. PANASYUK, G.A. BAGLYUK, V.A. MASLYUK,
L.M. APININSKAYA, N.M. VERGELES, O.V. VLASOVA
Frantsevich Institute for Problems of Materials Science of NASU,
3, Krzhizhanovsky Str., Kyiv 03142, Ukraine
E-mail: panasyuk@ipms.kiev.ua

Abstract. The technological processes of the fixing nickel, copper and tinny covering on particles of iron powder, applicable for fabrication magnetic-soft material by methods to powdered metallurgy are designed and explored. The magnetic characteristic magnetic-soft material, made from iron powder, particles which beforehand cover nickel or copper hang by nature covering, its composition, mode annealing temperature and so on are installed. Varying all these forming possible get magnetic-soft powdered material with necessary given characteristic.

Keywords: iron powder, nickel coating, copper coating, composition, temperature, coercive force, thickness

1. Introduction

It is known that cladding of disperse particles provides the formation of both adhesion-active and barrier layers in the preparation of composite materials and imparts special properties to powders. In the present work, the deposition of copper, nickel, and tin coatings on iron powder particles was performed to improve the magnetic characteristics of powder soft magnetic materials.

2. Experimental results

The work was performed using PZHR 3.315.24 iron powder (fraction $\geq 250\ \mu m$). Coatings were obtained on powder particles by the methods of chemical reduction and electrochemical deposition.

Nickel, copper, and tin coatings were obtained on iron powder particles by the methods of direct chemical reduction of the indicated metals from solution of their salts using different reducers and complexing agents.

Nickel coatings were obtained from solutions of nickel chloride and nickel sulfate using sodium hypophosphite as a reducer. Investigations were carried out with both acid (pH = 2–4) and alkali (pH = 8–10) solutions. Agents controlling pH and complexing agents were added to each solution.

Nickel coatings obtained in the process of reduction from both acid and alkali solutions were amorphous, half-lustrous, light, and contained phosphorous, the amount of which changed depending on the conditions of the process.

B. Baranowski et al. (eds.), *Carbon Nanomaterials in Clean Energy Hydrogen Systems.*

Iron particles clad with nickel were near-spherical in shape (Fig. 1).

Iron particles were poured in acryl resin and, after its curing, a metallographic sample was prepared by the metallographic method for photographing with a microscope. It is seen from Fig. 1 that nickel forms a rather dense and uniform coating on the surface of an iron particle. The powder clad with nickel was heat treated in hydrogen at a temperature of 460°C for 1 h. The phosphorous contents in the coatings deposited from acid and alkali solutions range from 9% to 10% and from 2% to 4%, respectively (Table 1).

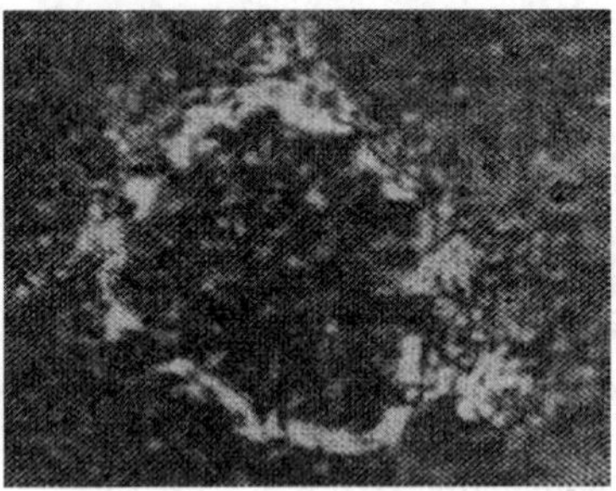

Figure 1. Iron particle with a nickel coating. X 800

TABLE 1. Chemical composition powders clad with nickel, copper, tin

	Powder	Chemical (mass%)		composition		Hc (A/м)
		P	Ni	Cu	Sn	
1	PZhRV clad with nickel from acid solution	0.39	3.6	–	–	6.05
2	PZhRV clad with nickel from alkali solution	0.25	2.5	–	–	6.1
3	PZhRV clad with copper	–	–	4.3	–	6.9
4	PZhRV clad with tin	–	–	–	1.4	7.0

The presence of phosphorous in a nickel coating influences the magnetic properties of the material. Specifically, at a phosphorous content of ~3%, the material is less magnetic than electrolytic nickel, whereas at a phosphorous content of 10–11%, coatings appear to be nonmagnetic. These data are known from the literature [1]. The values of the coercive force of obtained iron powder materials with nickel coating and without it are presented in Table 2.

TABLE 2. Magnetic properties (coercive force) of iron powder materials with a nickel coating and without it

Magnetic properties	Chemically reduced nickel: From acid solutions (content of P is 9–10%)	Chemically reduced nickel: From alkali solutions (content of P is 2–4%)	Uncoated iron powder material in the as-delivered state
Coercive force, Oe	Nonferromagnetic	7.2	6.7

These data indicate that materials from iron powder the particles of which were coated with chemically reduced nickel containing to 4% of phosphorous have improved magnetic properties and, due to phosphorous, are more compact as a result of the formation of the liquid phase during sintering.

It was established that after heat treatment (at 400–460°C for 1 h in hydrogen), in the nickel coating, phosphorous is present in the form of a solid solution of phosphorous in nickel and phosphides, i.e. the coating is not monophase, and this determines to a large degree properties of the powder.

The magnetic properties of the coatings depend on their composition and structural state. The largest coercive force of the powders clad with nickel was observed after annealing in hydrogen in the temperature range 360–400°C for 1 h. In the same temperature range, structural transformations proceed in nickel coatings containing phosphorous, i.e., the formation of a two-phase system, which consists of the solid solution of phosphorous in nickel and nickel phosphide Ni_3P is completed.

The magnetic properties also depend on the degree of distortion of the lattice of β-nickel, which is caused by phosphorous inclusions, the content of which depends on the phosphorous content in the solution. The values of magnetization of particles of the iron powder clad with nickel and products from it are also related to the change of sizes of crystals in the coatings.

2.1. DEPOSITION OF COPPER COATING

Copper coatings on iron powders were obtained by the following two methods:

(1) chemical reduction; (2) electrochemical deposition.

Among the chemical methods of deposition of coatings, chemical deposition of copper coatings occupies the second place (after deposition of nickel coatings). The advantages of chemically deposited copper coatings are as follows:

– The process is performed at room temperature.

– Chemically deposited copper is more elastic than other chemical coatings and can serve both as an independent coating and as a sublayer for nickel, tin, chromium, etc. [2].

Solutions for chemical deposition of copper coatings contain the following components:

– complex copper salts; – agents controlling pH; – a reducer; – stabilizing additives.

Copper sulfate $CuSO_4{\cdot}5H_2O$ is used as a copper salt, glycerol, oxalates, carbonates, ammonia, trilon B, and salts of tartaric acid serve as complexing agents, and formaldehyde is used as a reducer.

The process is based on the reduction of copper by formaldehyde from a complex copper salt in an alkali medium.

The reduction of copper proceeds by the reaction:

$$Cu^{2+} + 2HCHO + 4OH^- \rightarrow Cu + H_2 + 2HCOO^- + 2H_2O$$

The reduction rate of copper depends on the concentration of copper in the solution, content of a complexing agent, concentration of the reducer, and the alkali content.

The reduction of copper begins at pH > 2 and is intensified with increasing pH.

The deposition rate of copper coating depends on the concentration of formaldehyde and rises substantially as its content increases.

Formaldehyde is the only reducer which catalyzes the reduction reaction of copper at room temperature, makes it a self-catalyzed reaction, i.e., a reaction proceeding independently of the formed (reduced) copper.

The data presented in Table 3 indicate the influence of the nature of complexing agents (tartrate and glycerol) and the deposition time of copper on the thickness of the copper layers.

A copper layer with a thickness 2 mg/cm^2 was obtained using tartrate (Seignette salt) for 1.5 h and using glycerole for 3.2 h (Table 3).

TABLE 3. Compositions of solutions deposition for copper coatings of iron powder

Components	Concentration of components in solutions (g/l)	
	Solution I	Solution II
Copper sulfate	30	50
Sodium hydroxide	50	50
Glycerole	–	50
Potassium–sodium	–	–
Tartrate (Seignette salt)	100	–
Formalin 40% (ml/g)	30	15
pH	11.5	11.5
T°C	20–25	20–25
Deposition rate of the copper coating, mg/cm for 2 h	2.0–2.5	0.8–1

In further work, solution II (Table 3) with glycerol, as a complexing agent, was used. The choice of this solution, in which the copper deposition rate in smaller than that in solution I, is explained by the fact that it is much more stable in operation, which is very important in deposition of coatings on powders since the exact ratio of the volume of solution to the area of the coated surface (density of charging) is not always retained. For deposition of copper coatings, the optimal density of charging was ~2.5 dm^2/l. Since the surface of the powder to be coated cannot be exactly determined, the stability of solutions for coating with copper is of great importance. The solution with glycerol can be repeatedly corrected. The prepared solution can be hold for a long time. Formalin is poured before the deposition of copper is begun. A stabilizing additive, intended for preventing the reduction of copper in the volume of the solution and favoring its maximum deposition of powder particles, was additionally introduced in the solution for coating with copper.

As a stabilizer, thiourea in amounts of 0.1 ml/l was used. It was introduced in a bath every 1–1.5 h of operation. The deposition rate of copper in such solutions ranged from 1.5 to 2 μm/h.

Such a thickness of copper coatings makes it possible to use them for the formation of an initial contact layer on iron powder particles, on which consequently copper can be deposited by the electrochemical method without using cyanic electrolytes. In order to obtain thicker copper layers on powder particles, the process time was increased to 2–3 h. and temperature was also raised to 38–40°C. However, as temperature increases, the danger of self-decomposition of the solution also increases [2]. That is why temperature was not raised above 40°C. As a result of chemical decomposition, a copper layer ~8μm thick was obtained on iron powder. The density of charging was no larger than

0.25–0.3 dm^2/l. The iron powder (250 g) with a deposited layer of chemically reduced copper was heat treated (annealed) at temperatures 600°C, 800°C, and 900°C. An optimal temperature was 800°C (Fig. 2).

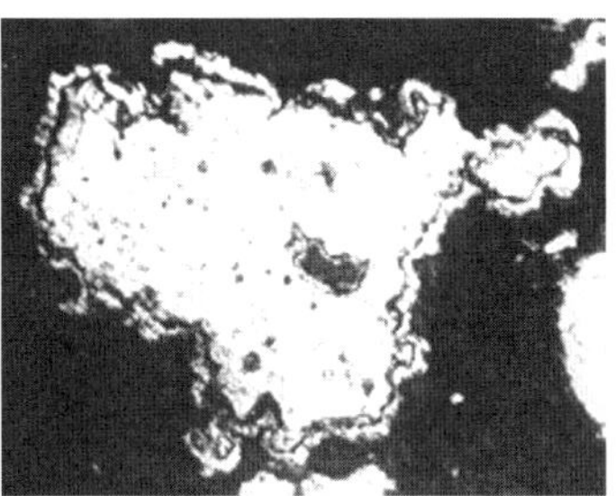

Figure 2. Iron powder particle coated with copper after heat treatment at 800°C. X 800

The flowability, apparent density, and green density of the obtained powder with a copper coating were measured and compared with those of powders coated with nickel and tin (Table 5).

Thus, it was established that solutions with glycerol operating at room temperature are stable in operation and, therefore, are most suitable for the coating of iron powders with copper.

For the iron powders coated with copper, the optimal temperature of annealing n hydrogen was found to be equal to 800°C. By varying the thickness and quality of copper coatings obtained by different methods, and selecting the annealing temperature for such powdcr, wc managed to obtain the powder with a copper coating, the use of which in composite materials allows to control the anisotropy of the magnetic properties of powder soft magnetic materials.

2.2. DEPOSITION OF TIN COATINGS ON IRON POWDER (TINNING)

The tinning process is a contact process and is based on the deposition of tin from solutions of its complex salts. Most often, tin chloride with ammonium alum or thiourea, as complexing agents, are used [3].

The reduction of tin compounds to the metal in aqueous solutions is possible using a number of strong reducers, since tin has no catalytic properties with respect to oxidation of water-containing compounds.

In acid or neutral media, the process proceeds only in the initial period of deposition as long as the surface of iron particles or as-deposited copper serve as a catalyst. While tin is reduced from solution, the surface of iron powder particles is coated with tin and ceases to act as a catalyst of the process. That is why the precipitation of tin is slowed down or stops. Only in a strong alkali medium containing to 40 g/l of NaOH, at a temperature of no lower than 90°C, a rather dense tin deposit can be obtained.

The simplest and most efficient method of deposition of tin coatings consists in performing a disproportionation reaction of tin in strong alkali hot (to 90°C) solutions. The first reaction is as follows: $2\ Sn^{2+} \rightarrow Sn^0 + Sn^{4+}$, i.e., tin from $SnCl_2$ is reduced to metallic tin and is oxidized to Sn^{4+}.

The formation of metallic tin (Sn°) by this reaction suggests the possibility of its using for depositing tin coatings. The second reaction is the following: $Sn^{4+} + 2e^{-} \rightarrow Sn^{2+}$.

The deposition of tin begins in the temperature range 46–70°C depending on the concentration of the initial salt of tin and pH of the medium. Thus, tin can be deposited by the disproportionation reaction.

If, as an additional reducer, sodium hypophosphite (10–15 g/l) is added in an alkali solution, then, at a temperature of 46°C and above, sodium hypophosphite is decomposed by water with the formation of atomic hydrogen by the reaction

$$NaH_2PO_2 + H_2O \rightarrow NaH_2PO_3 + 2H.$$

The formed atomic hydrogen interacts with tin ions

$$Sn^{2+} + 2H \rightarrow Sn + 2H^{+};$$

$$Sn^{4+} + 4H \rightarrow Sn + 4H^{+}.$$

Along with these primary reactions, a secondary reaction of the action of atomic hydrogen on HPO_3^{-} ions, which leads to the reduction of phosphorous containing in the HPO_3^{2-} ion to the elementary state, proceeds

$$H_3PO_3 + 3H \rightarrow P + 3H_2O.$$

Phosphorous precipitating on the surface of iron powder particles forms tin phosphides Sn_2P, Sn_3P, and Sn_5P_2 with the reduced tin. The overall reaction of the reduction of tin with atomic hydrogen with the formation of tin phosphides can be represented by the following equation:

$$H_2PO_2^{-} + Sn^{2+} + 7H \rightarrow Sn_3P + 5H^{+} + 2H_2O$$

By varying the composition of the solution and its operating regime, metallic coatings can be deposited on iron powder particles from strong alkali solutions stable at room temperature. The operation of alkali solutions was checked during deposition of tin on pure iron powder and iron powder with an as-deposited copper coating. It was established that the quality of the tin coating deposited on the iron powder preliminarily coated with copper was higher. This is explained by the fact that, due to a potential difference which arises between copper and tin, on the surface coated with copper, a denser tin coating is deposited. Deposition occurs only on the surface of the iron powder coated with copper which is catalytically active at the initial moment at a rate of ~8 μm/h. The coating is a dense crystalline tin deposit (Fig. 3(*a*, *b*)).

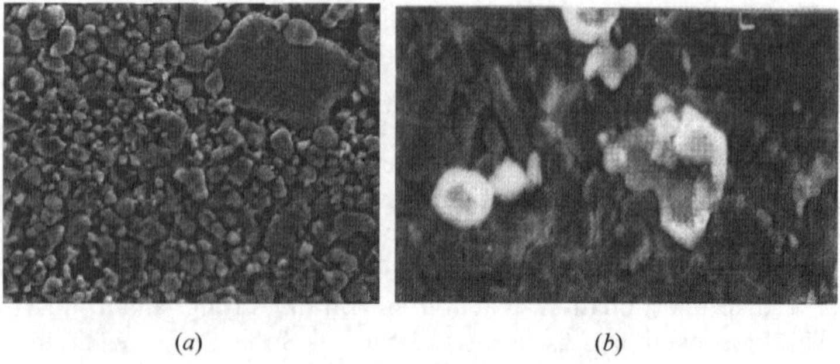

(*a*) (*b*)

Figure 3. Iron powder particle coated with tin. X 800

The deposition rate of tin depends on the preliminary preparation of powder (deposition of a copper coating) and a hydroxide cation (in the presence of KOH, the process proceeds much faster than in the presence of NaOH).

The deposition of tin by the disproportionation method was performed in a solution containing $SnCl_2$ (40 g/l) and NaOH (160 g/l) at a temperature of 90°C. The deposition rate attained 8 μm/h.

In Table 4, characteristics of iron powder clad with tin obtained from different solutions for deposition of tin coating are presented.

TABLE 4. Characteristics of iron powder clad with tin from different solutions

Solution composition	Coating thickness (μm)	Tin content (%) (chemical analysis)	Appearance of coated powder
$SnCl_2$ – 10–20 g/l Ammonium alum – 300 g/l T_{room} τ – 1 h	1.3–2	0.65	Light grey with a yellow undertone
$SnCl_2$ – 8–20 g/l H_2SO_4 – 30–40 g/l Thiourea – 35–45 g/l T_{room} τ – 1 h	3–4	–	Grey through which yellow is visible
$SnCl_2$ – 37–40 g/l NaOH – -160 g/l T – to 90°C τ – 1 h	7–8	3.6	Homogeneous light grey powder
$SnCl_2$ – 37–40 g/l NaOH – 160 g/l sodium hypophosphite – 100 g/l T – to 90°C τ – 1 h	10	3.8	Homogeneous light grey powder

From Table 4 it follows that the disproportionation method showed a good performance (in alkali solution) in the preparation of tin coatings on iron powders. The coating obtained by deposition for 1 h is light grey, uniform and its thickness is 8 μm.

The addition of hypophosphite (100 g/l) increases the deposition rate to 10–12 μm for 1.0–1.5 h, and the obtained coatings are more compact, even, and smooth.

The disadvantages of such solutions are high alkalinity and high temperature, which is inconvenient in operation. Nevertheless, this method is rather efficient for depositing high-quality tin coatings on iron powder particles.

High-quality tin coatings are obtained on iron powder particles by deposition of tin on a copper sublayer deposited by the chemical method. The deposition of tin occurs on the catalytically active copper surface of iron powder at a rate of 8 μm/h. The coating is a dense crystalline deposit containing a mixture of pure β-tin and tin phosphides.

Characteristics of iron powder with a nickel, copper, and tin coatings are presented in Table 5.

TABLE 5. Characteristics of iron powder with a nickel, copper and tin coatings

	Coating thickness (μm)	Contents of elements (%)	Flowability (sec)	Apparent density (g/cm^3)	Density (g/cm^3) (P = 800
Uncoated iron powder in the as-delivered condition		–	55.52	2.28	6.8
Fe + Ni acid	4.5	P – 0.39	46.76	2.72	
Fe + Ni acid + 400°C in hydrogen for 1 h		Ni – 3.6	41.89	2.86	
Fe + Ni alkali	4.5	P – 0.09	49.93	2.63	
Fe + Ni alkali + 400°C in hydrogen for 1 h		Ni – 2.7	50.03	2.63	
Fe + Cu (chem.)	8	Cu – 4.3	49.83	2.51	7.25
Fe + Cu (chem.) + 800°C in hydrogen for 1 h	8	Cu – 4.3	53.95	2.74	–
Fe + Sn (standard l ti)		Sn – 0.65	57.68	2.57	
Fe + Sn (disproportionation)		Sn – 3.6	49.64	2.78	
Fe + Sn (disproportionation +		Sn – 3.8	50.44	2.8	

3. Conclusions

Thus, methods and technological deposition conditions of nickel, copper and tin coatings in iron powder particles by chemical deposition were investigated.

A number of solutions the for deposition of high-quality coatings on iron powder particles are proposed. Some technological parameters of iron powder with particles clad with nickel, copper and tin were studied and compared with those of the initial iron powder. It was established that soft magnetic materials prepared from iron powder preliminarily clad with one of the indicated metals have better magnetic properties, which can be varied depending on the nature and composition of coatings, deposition method and heat treatment conditions.

References

1. Gorbunova K. M., Nikiforova A. A. Physicochemical Fundamentals of the Process of Chemical Nickel Deposition (*Moscow: Izdatelstvo AN SSSR: 1960*) (in Russian).
2. Vishenkov S. A. Chemical and Electrochemical Methods of Deposition of Metallic Coatings (*Moscow: Mashinostroenie: 1975*) (in Russian).
3. Sviridova V. V. Chemical Deposition of Metals from Aqueous Solutions (*Minsk: Izdatelstvo Universitetskoe: 1987*) (in Russian).

AUTHOR INDEX

SUBJECT INDEX